Engineering
Electronics

Engineering Electronics
A Practical Approach

Robert Mauro
Department of Electrical Engineering
Manhattan College

PRENTICE HALL
Englewood Cliffs
New Jersey 07632

Library of Congress Cataloging-in-Publication Data

MAURO, ROBERT

Engineering electronics : a practical approach / Robert Mauro.
    p.   cm.

  Includes index.
  ISBN 0-13-278029-1
  1. Electronics. I. Title.
TK7816.M35   1989   87-28091
621.381—dc 19   CIP

Editorial/production supervision: Joan McCulley/Colleen Brosnan
Interior design: Lorraine Mullaney with assistance from Jayne Conte
Cover design: Linda Conway
Manufacturing buyer: Mary Noonan
Page layout: Lorraine Mullaney
Cover photo: Courtesy of NEC Electronics, Inc.

 © 1989 by Prentice-Hall, Inc.
A Division of Simon & Schuster
Englewood Cliffs, New Jersey 07632

Printed in the United States of America

10 9 8 7 6 5 4 3 2 1

ISBN 0-13-278029-1

Prentice-Hall International (UK) Limited, *London*
Prentice-Hall of Australia Pty. Limited, *Sydney*
Prentice-Hall Canada Inc., *Toronto*
Prentice-Hall Hispanoamericana, S.A., *Mexico*
Prentice-Hall of India Private Limited, *New Delhi*
Prentice-Hall of Japan, Inc., *Tokyo*
Simon & Schuster Asia Pte. Ltd., *Singapore*
Editora Prentice-Hall do Brasil, Ltda., *Rio de Janeiro*

Contents

Contents

Contents

CHAPTER 8

Amplifier Stability and Oscillators 527

CHAPTER 9

Power Electronics 625

CHAPTER 11

**Digital Timing
and Memory
Circuits 879**

APPENDIX I

**A Review
of Basic Network
Concepts 954**

APPENDIX II

**Transmission-Line
Effects 972**

**Answers to
Selected Odd
Problems 989**

Index 997

Preface

Engineering Electronics: A Practical Approach is a text designed to provide the reader with a strong understanding of analog and digital electronics using both discrete and integrated components. It is a direct outgrowth of the author's teaching and practical engineering experience, and it has been written to provide the proper blend of theory and practice needed for the reader to become a skilled electronics engineer.

The earlier chapters are meant to be used for an introductory core course in an electrical engineering program. However, because of the many new analysis techniques introduced in these chapters and also because of the wealth of up-to-date information given in the later chapters, the book should also provide excellent self-study material for practicing engineers. Furthermore, with careful selection of the material to be presented, this text can also be useful in two-year engineering technology programs.

The material contained in the book is rigorous and provides the reader with a carefully chosen mathematical justification for all information presented so that he or she should never feel that any of the material or the formulas have been "pulled out of the air." Although the text delves quite heavily into the aspects of electronics that are pertinent to a pedagogically sound understanding of electronic principles, it is sparse in the inclusion of material that is not strictly relevant to the study of electronics. As such, for example, in the area of digital electronics little time is spent covering the introductory subjects of Karnaugh maps, combinational circuit design, sequential circuit design, and so on. These matters are fully explained in other courses in nearly all EE programs. On the other hand, in this area a good deal of the text is devoted to a detailed investigation of the electronic circuit behavior of logic gates, flip-flops, semiconductor memories, and other digital logic devices.

It was the author's intention to provide the reader with a complete basic understanding of electronics. The text contains a minimum of extraneous material. It employs a practical engineering approach to the design and analysis of electronic circuits. Great care is taken in developing the operational theory of the devices examined to ensure that the reader has a good physical and mathematical understanding of their behavior. Additional attention is given to the development of the device models to be sure that the reader has an equally good understanding of where the models come from and what their limitations are.

The use of simplified models and simplified equivalent circuits is stressed for the practical solution of real electronics problems. The text is structured to teach the reader how to "look at" complex electronic circuits and reduce them to single-loop analysis problems where feasible, instead of having to rely solely on complex mesh equations, nodal analysis equations, or computer solutions. It emphasizes the use of intuitive methods for the analysis and design of electronic circuits. This book is light on formulas and strong on the use of techniques that provide for a physical understanding of the operation of electronic circuits. Along these lines, the simplified equivalent circuits developed in the early chapters are used throughout the text for multitransistor circuit analysis, frequency response analysis, transient response analysis, feedback circuit analysis, and dc bias analysis.

In examining the performance of circuits in this book, the author frequently combines the dc and ac responses of the circuit to give the reader a better idea of the actual waveshapes that will be obtained in the laboratory when probing the circuit point by point with an oscilloscope. This is important, as it helps the reader to understand more fully the relationship between small-signal circuit performance, dc biasing, and device saturation and cutoff.

In Chapters 1 and 2 the diode is introduced. In particular, in the first chapter the characteristics of the ideal diode and the *pn* junction diode are covered, as are the applications of diodes in clipping and clamping circuits. Both graphical and modeling techniques are presented for the analysis of circuits containing diodes. The large- and small-signal analysis of diodes are also dealt with in this chapter, as is the subject of zener diodes. The chapter concludes with a detailed investigation of the device physics of the *pn* junction diode.

In Chapter 2 we examine the rectification properties of diodes. The role of the transformer in electronic power supply design is reviewed and the subjects of half-wave and full-wave rectification are covered in some detail. The capacitive-input filter is discussed, and the design of zener-diode-regulated power supplies is examined.

Chapter 3 begins with a discussion of the properties of electronic amplifiers. Next, the principles of operation of the bipolar junction transistor (BJT) are introduced and the device physics are then carefully investigated. Techniques are presented in this chapter for determining the operating point of BJT circuits, and the question of dc bias point stability is covered. Both graphical and modeling techniques are employed in the analysis of these problems. The chapter concludes with a discussion of graphical techniques for the analysis of BJT circuits containing ac sources.

In Chapter 4 the investigation of the ac performance of BJT circuits is continued but this time from a small-signal approach. The hybrid parameter model is developed and related to the graphical characteristics of the transistor, and the small-signal equivalent circuits for the BJT are derived. The application of these equivalent circuits to the analysis of multitransistor circuit problems is also considered.

The theory of operation of the junction field-effect transistor (JFET) and the metal-oxide-semiconductor field-effect transistor (MOSFET) are introduced in Chapter 5, and graphical analysis techniques for assessing the operating point stability of various FET circuits is considered. A detailed small-signal model for the field-effect transistor is developed from its volt-ampere characteristics, and this model is simplified and used to derive the equivalent circuits for the FET. The chapter concludes with an investigation of the factors governing the design of multistage BJT and FET amplifier circuits.

Chapter 6 examines the frequency response of electronic circuits by employing conventional network analysis methods and then by using circuit impedance techniques. Simple approximations are used to analyze the frequency response of multistage transistor circuits. The relationship between the transient response and the frequency response of circuits is carefully developed.

Chapter 7 deals with the important subjects of feedback and the operational amplifier (op amp). The chapter begins with a detailed discussion of the effects of feedback on the performance of electronic amplifiers and presents techniques for determining the input impedance, output impedance, and gain of feedback circuits. The ideal operational amplifier is introduced and elementary linear applications of the op amp are discussed. Use of the op amp in conventional active filter and in switched-capacitor active filter designs is also covered in some detail, as are the nonlinear applications of the operational amplifier. Also presented in this chapter is an in-depth look into the internal structure of modern operational amplifiers. Topics discussed include current sources, current mirrors, active loads, and techniques for analyzing the ac and dc internal circuit performance of op amps. The chapter concludes with a discussion of second-order effects in op amps, including input and output impedance levels, dc offset effects, bandwidth, and slew rate.

The question of amplifier stability and also that of oscillator design is addressed in Chapter 8. Several different amplifiers are examined and the effect of feedback on their stability is discussed. Amplifier compensation techniques are introduced and the trade-offs of using different compensation schemes are compared. Techniques for the generation of sinusoidal oscillations in discrete transistor circuits and in op amps are presented, and the questions of amplitude and frequency stabilization in oscillator circuits are carefully examined. Methods for producing nonsinusoidal oscillations are also considered.

Chapter 9 deals with the subject of power electronics. Single-ended and push-pull linear power amplifiers are investigated, and techniques are presented for estimating the distortion levels in both the BJT and FET versions of these circuits. Linear and switching regulated power supplies are covered, and power supply overload protection schemes are discussed. The control of ac power using SCRs and triacs is investigated, and a detailed examination of thermal problems in power semiconductor circuits is included.

Chapter 10 is concerned with digital electronics and the chapter begins with a discussion of the switching characteristics of diodes, BJTs, and FET devices. The transfer characteristics and switching performance of TTL, ECL, MOS, and the CMOS logic families are carefully examined. Interfacing techniques between logic families are discussed, and an in-depth analysis of transmission-line effects in digital circuits is presented.

Chapter 11 deals with the subjects of digital timing and memory circuits. The use of logic gates and Schmitt triggers as digital oscillators is examined, and their use as monostable multivibrators (one-shots) is also investigated. Techniques for the electrical debouncing of mechanical switches are presented. The architecture of semiconductor memories at the system level is discussed, and the design of both read/write memories (RAMs) and read-only memories (ROMs) is covered. The cell designs of MOS flip-flops are presented, and the expressions for the switching performance of these cells is derived. The design of RAM and ROM memory arrays at the chip level is also covered, as are the operation of programmable read-only memories (PROMs), erasable programmable read-only memories (EPROMs), and electrically alterable read-only memories (EAROMs). A detailed discussion of dynamic memories is included. Applications of ROMs and programmable logic arrays (PLAs) to the design of sequential circuits are also presented. The chapter and the text conclude with a discussion of gate arrays.

Two detailed appendices are also included which provide the reader with important background information. Appendix I contains a review of specific network theory points that are crucial to a full understanding of electronics, and covers such topics as voltage and current dividers, superposition, Thévenin and Norton equivalents, frequency response, transient response, and resonance. The second appendix reviews techniques for the analysis of transmission-line problems. It contains a derivation of the transmission-line equations, a development of the concept of the reflection diagram and the reflection coefficient, and a discussion of the use of impedance diagrams for the analysis of transmission-line circuits containing nonlinear sources and loads.

Chapters 1 through 5 taken together provide a basic introductory course in electronics and cover the low-frequency operation of diodes, BJTs, and FETs. The only prerequisites required for this material are a first-year college mathematics course and a course in basic network theory.

Starting with Chapter 6, the text begins to work with transfer functions, Laplace transforms, and the sinusoidal steady state. Because of this, it is a good idea for the reader intending to go beyond Chapter 5 to have had a second course in network analysis. Since an appendix covering these subjects is included at the end of the text, it may be possible to use this material to replace a formal networks course covering these subjects. Chapters 7 through 9 contain advanced material in analog electronics. Depending on the speed of the instructor, and on his or her willingness to choose judiciously the topics to be covered, it may be possible to cover Chapters 6 through 9 in a second course in electronics.

Chapters 10 and 11 can be used to form a course in digital electronics. Because this material is virtually independent of that covered in Chapters 7 through 9, it may, if desired, be covered earlier.

The textbook features a large number of solved text examples, as well as numerous exercises at the end of each section. A wide range of practical homework problems are also provided at the end of each chapter. Answers are given for selected homework problems at the end of the text, and a solutions manual for instructors is available from the publisher.

This text was written at Manhattan College, where the author is a member of the Electrical Engineering Department. He would like to thank the many people who assisted in the development of this text. Professor Borrmann provided many valuable discussions and was kind enough to read and comment on most of the manuscript. This book was used in note form by many members of the faculty and by numerous students. Their criticisms and suggestions are gratefully acknowledged. A special note of thanks is also due to two graduate students; Greg Nardozza, for testing many of the theoretical analysis methods presented in the later chapters, and to Lisa Governali, for assembling the solutions manual.

The author is also indebted to the staff at Prentice Hall for their invaluable assistance in the preparation of this manuscript, especially Bernard Goodwin, Tim Bozik, Elizabeth Kaster, Joan McCulley, and Colleen Brosnan.

In closing, I wish especially to thank my wife, Jean, and my children, Luke and Kate, for their tolerance, support, and understanding during the many years involved in the preparation of this manuscript.

Robert Mauro

Engineering
Electronics

CHAPTER 1

Introduction to Electronics: The Diode

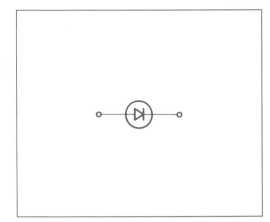

1.1   INTRODUCTION

In the twentieth century no other area of engineering has experienced the phenomenal growth and rapid changes seen by the electronics industry. The evolution from the vacuum tube to the transistor, and then from the transistor to the integrated circuit, has occurred with amazing speed. Today we accept as commonplace the miracles of radio, television, satellite communication, calculators, and computers, scarcely considering that none of these advances would have been possible without the incredible developments that have taken place in the field of electronics.

To the engineering student contemplating the study of electronics, mastery of this subject might at first appear to be a formidable, if not impossible task. At least it seemed so to this author when he first began his career in electrical engineering some years ago. Fortunately, things are not quite that bad. Two factors are in the student's favor. First, most electronic systems, regardless of their apparent complexity, can be reduced to an interconnection of basic building blocks. Once this fact is recognized, it becomes clear that expertise in this subject lies not in exhaustively studying all possible electronic circuits but in developing an understanding of their basic building blocks.

A second point to remember is that while new electronic devices are constantly being introduced, their application areas remain basically the same. As a result, if you thoroughly learn the fundamental concepts of electronic circuit theory using currently available components, then, later in your career when these components are replaced by new ones that are yet to be developed, you will easily be able to incorporate them into your designs.

Before beginning the formal text material on electronics, you should be reminded that the author is assuming that you are familiar with basic network theory. Consult Appendix I at the end of this text to review the following topics: voltage and current dividers, superposition, Thévenin and Norton equivalents, the transient and frequency response of electrical networks, Laplace transforms, resonance, and transformers.

1.2 THE IDEAL DIODE

The subject of electronics is concerned with the analysis and design of circuits containing diodes, transistors, integrated circuits, energy sources, and transformers as well as the more familiar R, L, and C components. We begin the study of this subject by examining the most fundamental electronic device, the diode.

As shown in Figure 1.2-1a, a diode is a two-terminal device; i_D defines the current flow through the diode and v_D the voltage drop across it. The side of the diode where i_D enters is called the anode, and that where it leaves is known as the cathode. Figure 1.2-1b illustrates the volt-ampere or V–I characteristics of a typical electronic diode. As the graph indicates, the diode is a device that permits current to flow easily in one direction while almost completely preventing its flow in the other. In this way the operation of the diode is quite similar to that of the mechanical check valve shown in Figure 1.2-2.

Here, when the pressure on the left-hand side of the check valve (P_1) is greater than that on the right (P_2), the door flips open and water flows through the valve. On the other hand, when the pressure P_2 exceeds P_1, the door remains closed and no water flows. It should be noted that when the check valve is open and water is flowing, P_1 and P_2 are approximately the same, so that P_v is nearly zero. Conversely, when the valve is closed and no water flows, P_v is negative.

Figure 1.2-1

(a) Electronic Symbol for the Diode   (b) Typical Volt-Ampere Characteristics of a Diode

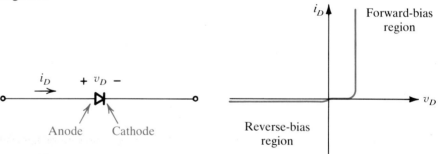

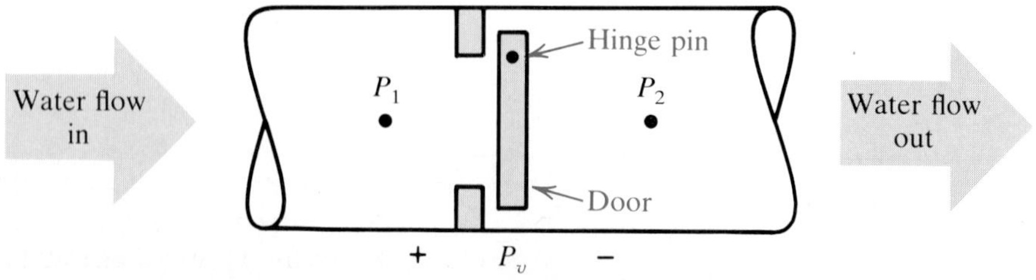

Figure 1.2-2
Water flow in a pipe containing a check value.

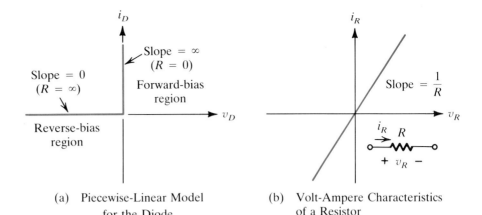

(a) Piecewise-Linear Model
for the Diode

(b) Volt-Ampere Characteristics
of a Resistor

Figure 1.2-3

The volt-ampere characteristics of the electronic diode may be described in a similar fashion. Following Figure 1.2-1b, when v_D is greater than zero, the diode is operating in the forward-bias region and large currents can flow through it. On the other hand, when v_D is less than zero, the diode is reverse biased (or operating in the reverse-bias region) and the current flow through it is nearly zero.

To better understand the behavior of this device, consider the idealized diode volt-ampere characteristics given in Figure 1.2-3a. Here we are approximating the diode's *V–I* performance by two straight-line segments. As a result, it is possible to represent the diode by two very simple models: one when it is operating in the forward-bias region and a second when it operates in the reverse-bias region.

To develop these models, let us recall that the volt-ampere characteristics for a resistor have the form given in Figure 1.2-3b. This follows directly from Ohm's law,

$$i_R = \frac{v_R}{R} \qquad (1.2\text{-}1)$$

where R = resistance, ohms
$\quad i_R$ = current flow through the resistor
$\quad v_R$ = voltage drop across the resistor

As is apparent from this equation, the slope of the straight line in Figure 1.2-3b is equal to $1/R$. Thus, a device whose volt-ampere characteristic is a straight line passing through the origin may be modeled by a resistor whose resistance is simply equal to 1 over the slope of the line. Applying this result to each of the line segments in Figure 1.2-3a, we can see that the ideal diode may be represented by a short circuit (or zero resistance) when it is operating in the forward-bias region, and by an open circuit (or infinite resistance) when it operates in the reverse-bias region.

1.3 THE SEMICONDUCTOR DIODE

For many applications the ideal diode characteristics discussed in Section 1.2 are all that is necessary to carry out an accurate analysis of a specific diode network. There are, however, instances where the nonideal characteristics of actual diodes can have important effects on the behavior of the circuit. In this section we examine the volt-ampere characteristics of several different commercially available diodes to see how their performance deviates from the ideal and to develop an understanding of when these differences are significant and when they can be neglected.

Semiconductor diodes are a very important class of commercially available diodes. Their behavior from a semiconductor physics viewpoint is presented in some detail in Sections 1.10 and 1.11, while their basic device characteristics are discussed in this section.

The volt-ampere characteristics of a typical semiconductor diode are described quite well by the equation

$$i_D = I_s(e^{v_D/V_T} - 1) \qquad (1.3\text{-}1)$$

where I_s = reverse saturation current
$\qquad V_T = kT/q$
$\qquad k$ = Boltzmann's constant $(1.38 \times 10^{-23} \text{ J/K})$
$\qquad T$ = absolute temperature, Kelvin
and $\quad q$ = electronic charge $(1.6 \times 10^{-19} \text{ C})$

At room temperature (300 K) the constant V_T can readily be shown to be equal to about 0.026 V, or 26mV. Equation (1.3-1) is sketched in Figure 1.3-1 for several different forward-bias current levels.

The shape of these curves may be understood as follows. In the forward-bias region, that is, when i_D is greater than zero, as soon as v_D exceeds a few V_T, then $(e^{v_D/V_T} - 1) \simeq e^{v_D/V_T}$ and the current grows exponentially with further increases in applied voltage. When the polarity of v_D is reversed, the diode is said to be operating in the reverse-bias region. Here the exponent of the exponential term is negative and for v_D more negative than a few V_T, the -1 term dominates in equation (1.3-1), and the resulting current, i_D, is approximately equal to $-I_s$.

Although the i_D versus v_D curve has an exponential character in the forward-bias region, the specific shape of this curve depends on the relative size of the current scale used. When the current scale for i_D is comparable to I_s, as shown in Figure 1.3-1a, the curve has the classic exponential shape. However, when the current scale used is very large in comparison to the reverse leakage current, the curve stays relatively flat along the v_D axis until the voltage across the diode exceeds a particular value, and then i_D appears to grow nearly vertically with further increases in v_D. This situation is illustrated in Figure 1.3-1b and c for the case where the forward diode current levels shown are on the order of $10^6 I_s$ and $10^{11} I_s$, respectively.

Using these results, let's examine the volt-ampere characteristics of a small-signal silicon diode. For most applications, the forward current levels in such a diode are typically on the order of 1 to 100 mA, and the reverse leakage current for this size diode is theoretically about 0.5 pA (see Section 1.11). As a result, because

$$i_D \simeq I_s e^{v_D/V_T} \qquad (1.3\text{-}2)$$

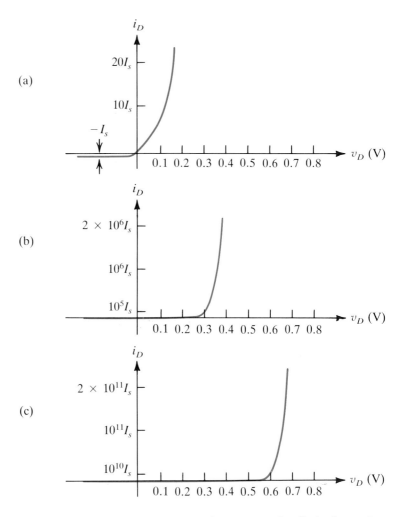

Figure 1.3-1
Volt-ampere characteristics of a diode drawn using different current scales.

in the forward-bias region, the voltage drop across the diode for a given current i_D flowing through it is

$$v_D = V_T \ln \frac{i_D}{I_s} \qquad (1.3\text{-}3)$$

Using this equation, the diode voltage drop at a current level of 10 mA is about 620 mV. At 100 mA, because of the logarithmic nature of equation (1.3-3), the voltage drop does not increase that much over its value at 10 mA. In particular, at 100 mA, v_D is only 675 mV. Thus, when a silicon diode is forward biased, regardless of the current flow through it, the voltage drop across it will nearly always turn out to be somewhere in between 0.6 and 0.7 V.

The V–I characteristics of a typical small-signal silicon diode are illustrated in Figure 1.3-2. Also shown for comparison are the volt-ampere characteristics of a small-signal germanium diode. As illustrated in this figure, the reverse leakage currents of germanium diodes are much larger than those of comparable silicon devices, typically being on the order of 1 μA. As a result, following equation (1.3-3), for forward-bias currents in the range 1 to 100 mA the drop across a germanium diode is about 0.2 to 0.3 V, significantly smaller

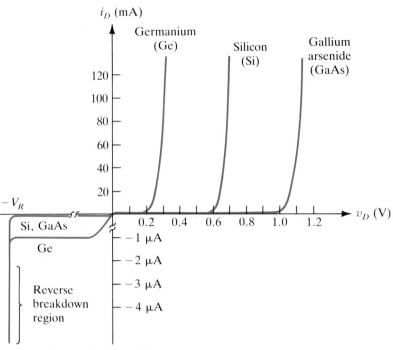

Figure 1.3-2

Volt-ampere characteristics of several different semiconductor diodes.

Note: Upper and lower current scales are not the same

than the 0.6- or 0.7-V drop found across a silicon diode carrying similar currents.

Because of the high leakage currents of germanium diodes and transistors, silicon is much more popular than germanium for the manufacture of semiconductor devices. Therefore, in this book (for the most part) only silicon devices will be studied.

For completeness the *V–I* characteristics of a gallium arsenide (GaAs) diode have also been included in Figure 1.3-2. While the use of gallium arsenide materials in electronic devices is far below that of silicon, GaAs does at least bear mentioning in this introductory chapter since it is used extensively in the manufacture of light-emitting diodes (LEDs) and is finding increasing application in the fabrication of high-speed electronic circuits.

The reverse breakdown region shown in Figure 1.3-2 is not explicitly described by equation (1.3-1) but results from second-order effects explained more fully in Section 1.11. It occurs when the reverse voltage across the diode exceeds the breakdown voltage $(-V_R)$ specified for the device. For most applications, operation of the diode in the breakdown region is undesirable, and using our valve analogy represents a failure of the valve to prevent the backflow of water in the presence of high reverse pressures. However, as we will see in Section 2.3, there are some instances where operation in the reverse breakdown region can be utilized to great advantage. In fact, there are diodes designed specifically to operate in this portion of the volt-ampere characteristic. These are known as zener diodes, and much more will be said about them in Chapter 2.

1.3-1 Following equation (1.3-1), find the effective reverse leakage current in the GaAs diode shown in Figure 1.3-2. *Answer* 4×10^{-20} A

1.3-2 For a semiconductor diode, what is the change in diode voltage corresponding to a tenfold increase in diode current flow? *Answer* 60 mV

1.4 ANALYSIS OF DC CIRCUITS CONTAINING DIODES

For the circuit illustrated in Figure. 1.4-1, let us suppose that we wish to determine the diode current flow, i_D, and the diode voltage drop, v_D. Applying Kirchhoff's voltage law to this circuit, we obtain

$$V_{BB} = i_D R + v_D \qquad (1.4\text{-}1)$$

This equation has two unknowns, i_D and v_D, so that it cannot be solved directly. A second equation relating these two quantities is required. This second relationship is, of course, the diode equation (1.3-1), which is repeated below for completeness.

$$i_D = I_s(e^{v_D/V_T} - 1) \qquad (1.4\text{-}2)$$

Substituting equation (1.4-2) into (1.4-1) yields

$$V_{BB} = I_s R (e^{v_D/V_T} - 1) + v_D \qquad (1.4\text{-}3)$$

In principle, this equation can be solved since it contains only one unknown. However, as a practical matter the solution is extremely difficult to obtain since this expression is a nonlinear transcendental equation. This problem is amenable to numerical solution on a computer, and if you are interested, you are invited to pursue this approach.

As an alternative, this problem may be solved graphically by plotting equation (1.4-1), called the dc load-line equation, on the same set of axes as the diode volt-ampere relationship. The resulting simultaneous graphical solution of these two equations is shown in Figure 1.4-2.

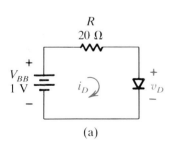

(a)

Figure 1.4-1
Simple Diode Circuit

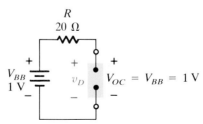

(b) Circuit for Computing V_{OC}; Replace Diode by an Open Circuit; the Polarity of V_{OC} Is the Same As That for v_D

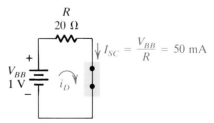

(c) Circuit for Computing I_{SC}; Replace Diode by a Short Circuit; the Direction of I_{SC} Is the Same As That for i_D

To plot the load-line equation, we first observe that this equation is a straight line, and that as a result, only two points on the line are needed to be able to sketch it. The two simplest points to use are those where the line intersects the axes, the so-called i_D and v_D intercepts. To find the v_D intercept, simply set i_D equal to zero in equation (1.4-1). This gives the v_D intercept as $v_D = V_{BB} = 1$ V. Using the same approach for the i_D intercept and setting v_D equal to zero yields i_D as V_{BB}/R or 50 mA for this problem. By drawing a straight line between these two points, we may complete our sketch of the load-line equation. Of course, the intersection of the two curves gives the simultaneous solution of both equations. This point, specifically $i_D = 17$ mA and $v_D = 0.67$ V, is called the operating point or Q (quiescent) point of the circuit.

In carrying out the graphical analysis of circuit problems containing diodes (and other circuit elements, too, for that matter), it is extremely useful to observe that the load line may be sketched without ever explicitly writing out the load-line equation. This is because the v_D and i_D intercepts of the load line are simply equal to the open-circuit voltage (V_{OC}) and the short-circuit current (I_{SC}) for the network connected to the diode. In calculating both of these quantities, the directions assumed for V_{OC} and I_{SC} need to be the same as those defined originally for v_D and i_D, respectively. The technique for computing V_{OC} and I_{SC} for this particular example is illustrated in Figure 1.4-1b and c.

The solution of diode circuit problems by graphical means is a time-consuming process. Fortunately, there are alternative solution methods available. The technique of modeling the diode by a simplified equivalent circuit is probably one of the best approaches. In developing a model for the diode, what

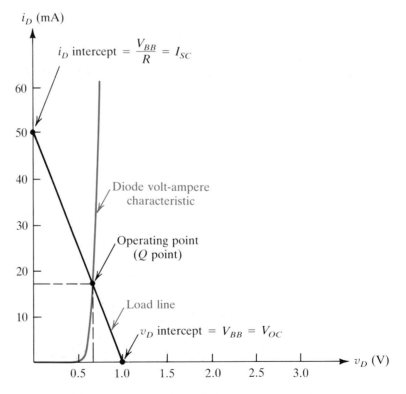

Figure 1.4-2
Graphical solution for the diode circuit problem given in Figure 1.4-1a.

Chapter 1 / Introduction to Electronics: The Diode

we are attempting to do is to replace its nonlinear characteristics by a simpler linear (or at least piecewise linear) equivalent circuit that will give us reasonably accurate results. In the remainder of this section we investigate several different diode models and try to get an understanding of both where the models come from and when the use of specific models is appropriate.

In general, when the diode is forward biased, it will be represented by one model, and when it is reverse biased, by another. The simplest model for the diode is obtained by replacing the volt-ampere characteristic given in Figure 1.2-1b by the ideal one given in Figure 1.2-3a. It should be clear to you by studying Figure 1.2-3a that in this instance when the diode is forward biased, it can be modeled by a short circuit and when it is reverse biased, by an open circuit. These results are summarized in Figure 1.4-3a.

Because of the similarity between the operation of the diode and that of an electronically controlled switch, it is at times useful to use the term ON to describe the state of a forward-biased diode and OFF to describe the diode when it is reverse biased. This terminology will be used extensively in describing the operation of diode circuits throughout the remainder of this book.

To develop a "better" model for the semiconductor diode, we want to obtain one whose composite volt-ampere characteristics more closely approximate those of an actual diode. Two more sophisticated models are shown in Figure 1.4-3b and c. Clearly, the model in Figure 1.4-3c should give the best results since its piecewise linear V–I behavior most nearly follows that of the original diode. However, this model is also the most complex of the three shown in the figure.

Before proceeding with a comparison of the performance of these models, it will be useful to make sure that we understand the origin of each. Since we have already taken a fairly detailed look at the simple diode model illustrated in Figure 1.4-3a, let's turn our attention to the slightly more involved model presented in part (b) of the figure. Here, when the voltage across the diode is less than V_o, the diode is considered to be OFF and may be represented by an open circuit. When the diode is ON, according to the graph in Figure 1.4-3b, the voltage across the diode remains constant at V_o volts regardless of the current flow through it; this type of volt-ampere behavior is best modeled by a fixed battery of voltage V_o. For silicon diodes, in the absence of specific graphical data, the value of V_o will be assumed to be 0.7 V. For germanium (Ge) and gallium arsenide (GaAs) diodes, V_o values of 0.3 V and 1.2 V are reasonable values to use.

For the more accurate diode model shown in Figure 1.4-3c, the reverse-bias or OFF condition of the diode is represented by a current source equal to the diode reverse leakage current, I_s. The equivalent circuit for the case where the diode is forward biased or ON is obtained by observing that in this portion of the volt-ampere curve, an increase in the current through the diode causes a corresponding increase in diode voltage. This indicates the presence of a resistor in the model. In addition, the fact that the diode has a voltage drop V_o across it when there is very little current flowing through it necessitates the inclusion of a battery in series with the resistor in the equivalent circuit.

To determine the values of r_D and V_o to use in the model, note that the voltage v_D across the diode equivalent circuit when it is ON in Figure 1.4-3c can be written as

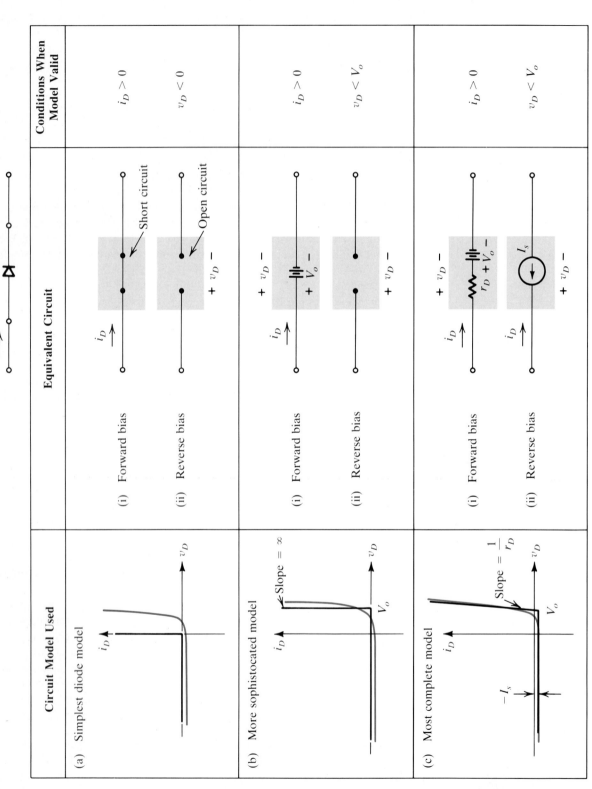

Figure 1.4-3 Piecewise linear models for representing the semiconductor diode.

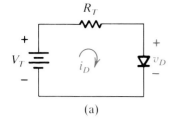

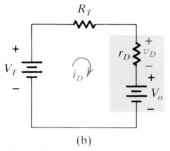

Figure 1.4-4

Replacing a diode by its equivalent-circuit model.

$$v_D = i_D r_D + V_o \tag{1.4-4a}$$

Rewriting this expression as

$$i_D = \frac{1}{r_D} v_D - \frac{V_o}{r_D} \tag{1.4-4b}$$

we can see that V_o is obtained from the v_D intercept of the straight line sketched in Figure 1.4-3c, while the resistor value, r_D, may be found by measuring its slope. For a given diode both r_D and V_o vary with the position of the Q point, and typical values for r_D and V_o are on the order of 0.1 to 10 Ω and 0.6 to 0.8 V, respectively, for silicon small-signal diodes.

Now that we have a basic understanding of where the models come from, let's try to develop a feeling for when the use of a particular model is appropriate. Consider, for example, the analysis of the simple diode circuit shown in Figure 1.4-4a using the complete diode model given in Figure 1.4-3c. If we assume that the diode is ON, the equivalent circuit shown in Figure 1.4-4b applies. Writing the expression for i_D in this circuit, we obtain

$$i_D = \frac{V_T - V_o}{R_T + r_D} \tag{1.4-5}$$

From an examination of this result we see that r_D can be neglected when it is much smaller than R_T, and similarly, V_o can be neglected when it is much smaller than V_T. Thus, in general the diode can be replaced by a short (when it is forward biased) as long as the diode voltage, V_o, and diode resistance, r_D, are much smaller than the external voltage and external resistance of the Thévenin equivalent of the circuit connected to the diode.

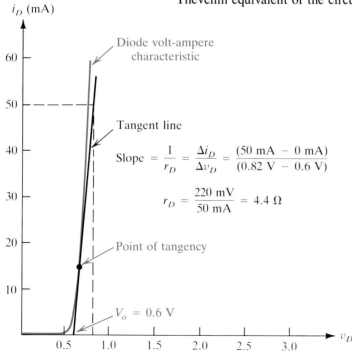

(a) Use of the Tangent Line to Develop the Model for the Diode in Figure 1.4-1a

Circuit Model Used	i_D (mA)	v_D (V)
Actual diode (graphical)	17	0.66
Short circuit	50	0.0
$V_o = 0.7$ V, $r_D = 0$ (silicon diode)	15	0.7
$V_o = 0.6$ V, $r_D = 4.4$ Ω (data from tangent line to curve)	16	0.67

(b) Comparison of the Operating Points Obtained Using Different Circuit Models for the Diode in Figure 1.4-1a

Figure 1.4-5

Use of the tangent line to develop the model for the diode in Figure 1.4-1a.

To illustrate this point let us reexamine the silicon diode problem presented in Figure 1.4-1a, which we originally solved graphically in Figure 1.4-2. To develop the best complete model for this diode, draw a tangent line to the curve in the neighborhood of the expected operating point (in this case around 10 to 20 mA). The values of V_o and of r_D are determined from this straight-line approximation to the curve to be 0.6 V and 4.4 Ω, respectively (Figure 1.4-5a). In solving this circuit problem using the battery model for the diode, it will be assumed that V_o is about 0.7 V.

The results of the solution to this problem using each of the three different diode models in Figure 1.4-3 is summarized in Figure 1.4-5b. As expected, for this circuit the simplest short-circuit model for the diode is inadequate since V_o cannot be neglected in comparison to $V_{BB} = 1$ V. Furthermore, and again as expected, the inclusion of r_D in the diode model improves the overall accuracy of the results.

EXAMPLE 1.4-1

For the circuit illustrated in Figure 1.4-6a, determine the operating point of the diode.

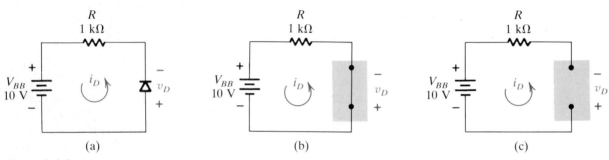

(a)   (b)   (c)

Figure 1.4-6
Example 1.4-1.

SOLUTION

To begin the analysis of this circuit, it is first necessary to guess at the state of the diode. By examining this circuit it should be relatively apparent that the diode is OFF, but to illustrate the analysis technique, we will "guess" initially that the diode is ON.

Since we are assuming the diode to be ON, we will replace it by the forward-bias model (in this case a short circuit) and will check to see if the diode current is greater than zero (Figure 1.4-6b). For this circuit the current i_D is -10 mA, and because the current is less than zero, the original assumption that the diode was ON is incorrect. Therefore, the diode must be OFF. To confirm this fact and to actually calculate the voltage across the diode, we now need to replace the diode by the reverse-bias model (an open circuit) and calculate the voltage v_D in Figure 1.4-6c. Carrying out this procedure, we find that v_D is -10 V, and because this voltage is less than zero, as predicted, the diode is reverse biased, or OFF.

1.4-1 In Figure 1.4-1a, find i_D and v_D if V_{BB} is changed to 2 V. *Answer* $i_D = 40$ mA, $v_D = 0.75$ V

1.4-2 Find i_D for the circuit in Figure 1.4-1a considering the diode to be germanium with $r_D = 0$. *Answer* 35 mA

1.5 ANALYSIS OF DIODE CIRCUITS CONTAINING LARGE-SIGNAL AC SOURCES

In this section we investigate the behavior of diode circuits containing large-amplitude time-varying signals. For our purposes, the definition of large signals will be those with amplitudes greater than 1 or 2 V. As in Section 1.4, we begin the analysis of these circuits by using graphical techniques, and after that progress to the solution of these same problems by making use of the diode models developed in Section 1.4.

The method for handling diode circuits containing ac sources is essentially the same as that employed in Section 1.4 for dc sources. For example, to analyze the current flow in the circuit illustrated in Figure 1.5-1 by graphical techniques, we first write down the load-line equation as

$$v_{in} = i_D R + v_D \tag{1.5-1}$$

The major difference between this equation and the dc load-line equation (1.4-1), is that the i_D and v_D intercepts of equation (1.5-1) vary with time. As a result, the instantaneous operating point of the circuit also changes with time. The position of the load line on the diode characteristic at five important instants of time is shown in Figure 1.5-1b, and these specific points in time are defined on the v_{in} waveshape in part (c) of the figure. Note that although the intercepts of the load line vary, the slope of the load line remains fixed at $-1/R$, so that all the load lines shown in Figure 1.5-1b are parallel to one another.

At each point in time the intersection of the load line with the diode characteristic curve gives the corresponding values of i_D and v_D for the circuit being examined. By sweeping the load line back and forth between its extreme limits and noting the values of i_D and v_D, the sketches shown in Figure 1.5-1c can readily be obtained.

This graphical method, while useful for educational purposes, is extremely awkward for attacking practical diode problems containing ac sources. Fortunately, the diode models developed in Section 1.4 can also be applied to analyze diode circuits containing time-varying input signals. However, when using them in this application, the analysis technique needs to be modified slightly. In this instance, rather than solving for the state of the diode, we are interested in determining the values of the input signal for which the diode is ON and those values of the input signal for which it is OFF. Once these two sets of values are known, the diode can be replaced by the appropriate model in each of

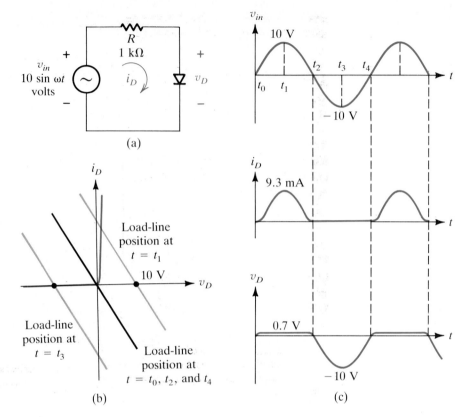

Figure 1.5-1
Graphical analysis of a diode circuit containing a large-signal ac source.

Step 1. To find out when the diode is ON or forward biased:

a. Consider the diode to be ON and replace it by its equivalent circuit (a short for the simple model).

b. Find the range of input signal values for which i_D is greater than zero. This is the range of input signal values for which the diode is ON.

c. For this range of input signal values, and with the diode still replaced by its appropriate forward-bias model, determine the required voltages and currents in the circuit.

Step 2. To find out when the diode is OFF or reverse biased:

a. Consider the diode to be OFF and replace it by its equivalent circuit (an open for the simple model).

b. Find the range of input signal values for which v_D is less than zero (or $v_D < V_o$ for the fancier models). This is the range of input signal values for which the diode is OFF.

c. For this range of input signal values, and with the diode still replaced by its appropriate reverse-bias model, determine the required voltages and currents in the circuit.

Note: Once the range of input signal values for which the diode is ON have been determined in step 1, it can be understood that the range of values for which the diode is OFF are simply those values not included in step 1.

Figure 1.5-2
Use of models for the analysis of diode circuits containing large-signal ac sources.

these regions and the required voltages and currents determined. This procedure is outlined step by step in Figure 1.5-2 and is illustrated in the example that follows.

EXAMPLE 1.5-1

Sketch the diode voltage and current as functions of time for the circuit given in Figure 1.5-1a, replacing the diode by the ideal model. Compare these results with those obtained graphically and shown in Figure 1.5-1c.

SOLUTION

Closely following Figure 1.5-2 and starting at step 1, we initially assume the diode to be ON and replace it by a short circuit (Figure 1.5-3a). We next calculate the values of v_{in} for which i_D is greater than zero. From this circuit the current i_D can be written as

$$i_D = \frac{v_{in}}{R} \tag{1.5-2}$$

and thus the diode is ON, or i_D is greater than zero, whenever v_{in} is greater than zero. As a result, whenever v_{in} is above the zero-volt level in the figure, the diode voltage v_D equals zero and following equation (1.5-2), the diode current has the same shape as the applied input voltage, v_{in} (see Figure 1.5-3c).

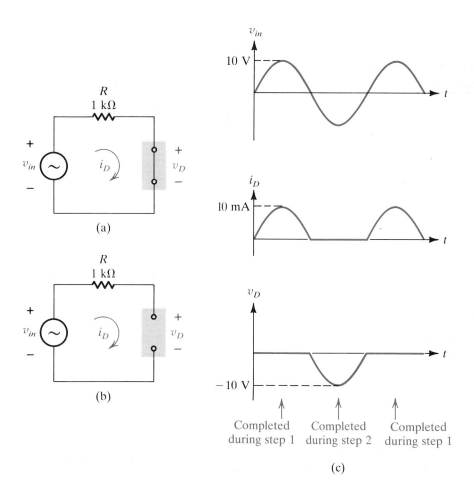

Figure 1.5-3
Example 1.5-1. Analysis of diode circuits containing large-signal ac sources.

Proceeding to step 2 in Figure 1.5-2, we replace the diode by its reverse-bias model, an open circuit, and compute the range of values of v_{in} for which the diode voltage, v_D, is less than zero. The equivalent circuit for this case is shown in Figure 1.5-3b, and for this circuit

$$v_D = v_{in} \qquad (1.5\text{-}3)$$

As expected, v_D is less than zero and the diode is OFF whenever v_{in} goes negative. It is also apparent for this circuit that i_D equals zero for these values of v_{in}. Using these results we can now complete the sketches for i_D and v_D in Figure 1.5-3c.

Exercises

1.5-1 In Example 1.5-1 (Figure 1.5-1a) the diode is shunted with a 2-kΩ resistor R_1. Find the peak current flow i_{R1} through R_1 when v_{in} is positive and also when it is negative. *Answer* Peak $i_{R1} = 0$ when $v_{in} > 0$ and 3.3 mA when $v_{in} < 0$.

1.5-2 Repeat Exercise 1.5-1 if the diode in Figure 1.5-1a is connected in series with R_1 instead of in parallel. *Answer* Peak $i_{R1} = 3.3$ mA when $v_{in} > 0$ and 0 when $v_{in} < 0$.

1.6 CLIPPING (OR LIMITING) CIRCUITS

In the design of signal-processing systems it is sometimes necessary to prevent the output signal from exceeding or falling below a predetermined voltage level. This function is often performed by a diode network known as a clipping or limiting circuit.

In a previous section of this chapter we have already seen a circuit that can perform clipping at the zero-volt level (Figure 1.5-1). With the addition of one or more batteries to this type of diode circuit, it is also possible to vary the level at which the clipping occurs or to shift the dc value of the input signal before it is applied to the clipping circuit.

Several different clipping circuits are illustrated in Figure 1.6-1, and for all of these circuits the battery V_{BB} determines the clipping level. For the battery orientation shown in parts (a) and (b), the clipping level is positive. Reversing the battery polarity as shown in parts (c) and (d) causes the clipping to occur at the $-V_{BB}$ voltage level. The orientation of the diode in these circuits determines whether the clipping occurs above or below the V_{BB} voltage level. When the diode's anode is located at the top of the circuit as in parts (a) and (c), the diode is forward biased and clips off all voltages above V_{BB}, while for the circuits in parts (b) and (d), the diode direction is reversed and clipping occurs when v_o attempts to drop below the clipping level.

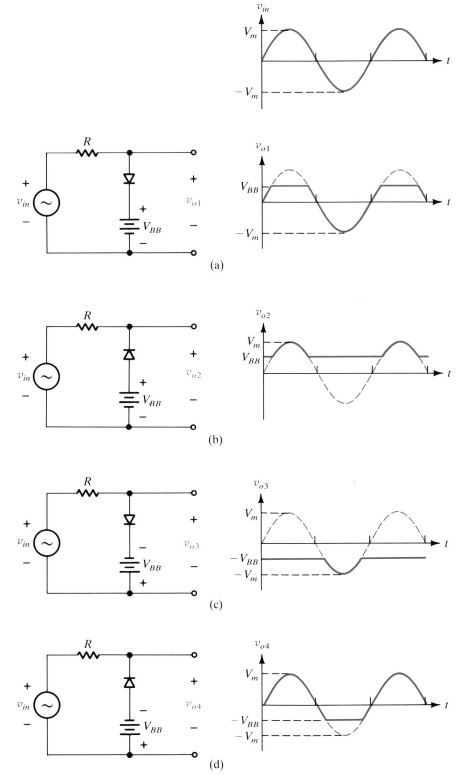

Figure 1.6-1
Clipping at the zero volt level.

EXAMPLE 1.6-1

Describe the operation of the clipping circuit in Figure 1.6-1a for the case where v_{in} is a 10-V peak amplitude sine wave and $V_{BB} = 5$ V.

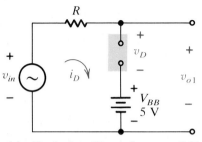
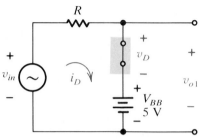

Figure 1.6-2
Example 1.6-1.

(a) Equivalent Circuit for $v_{in} < 5$ V   (b) Equivalent Circuit for $v_{in} > 5$ V

SOLUTION

Assuming the diode to be ideal, when the diode is OFF, the circuit in Figure 1.6-2a applies, from which it follows that

$$v_D = v_{in} - V_{BB} - i_D R = v_{in} - 5 \qquad (1.6\text{-}1)$$

Therefore, v_D is less than zero and the diode is OFF for all v_{in} less than 5 V. Furthermore, following Figure 1.6-2a it is also apparent that

$$v_{o1} = v_{in} - i_D R = v_{in} \qquad (1.6\text{-}2)$$

since $i_D = 0$ when the diode is OFF. As a result, as long as the diode remains OFF, v_{o1} follows the applied input v_{in}.

When v_{in} exceeds 5 V the diode conducts and the equilvalent circuit given in Figure 1.6-2b is valid, so that the output voltage may be written as

$$v_{o1} = v_D + V_{BB} = 5 \text{ V} \qquad (1.6\text{-}3)$$

Therefore, once v_{o1} attempts to exceed $V_{BB} = 5$ V, the diode turns ON and clips off the output at the V_{BB} voltage level. The dashed portion of the v_{o1} waveshape indicates what the output would have looked like had the diode not turned ON.

Exercises

1.6-1 Consider that v_{in} is a 30 V(p-p) triangle wave and that $V_{BB} = 10$ V in Figure 1.6-1c. If $R = 1$ kΩ, what is the peak current flow in the resistor and the maximum reverse voltage across the diode? *Answer* 25 mA, 5 V

1.6-2 In Example 1.6-1, what is the maximum voltage drop across R? *Answer* 5 V

1.7 CLAMPING (OR DC RESTORATION) CIRCUITS

When electrical signals are transmitted from one point to another in space, or when they are passed through electronic circuits containing capacitors, the original dc levels associated with these signals are often lost. The clamping circuits discussed in this section provide a simple way to reinsert specific dc levels into the signal without distorting the time-varying aspects of the waveform.

    Chapter 1 / Introduction to Electronics: The Diode

A relatively simple clamping circuit is illustrated in Figure 1.7-1a. In general, for this circuit the output voltage v_o (which is also the same as the diode voltage) may be written as

$$v_o = v_D = v_{in} - v_C \qquad (1.7\text{-}1)$$

If we assume that the capacitor is initially uncharged, then at $t = 0+$, because the capacitor voltage cannot change instantaneously, $v_C(0+) = 0$, so that

$$v_o = v_{in} \qquad (1.7\text{-}2)$$

As a result, as v_{in} initially increases, the diode turns ON and the capacitor charges with a time constant $\tau = RC$, where R is determined by the resistance of the diode. If we assume that the diode is ideal, that is, that $R = 0\ \Omega$, the capacitor charges instantaneously with its voltage following v_{in} during the first quarter of the cycle (Figure 1.7-1b). Once v_{in} passes the maximum at $t = T/4$ and begins to decrease, from equation (1.7-1), v_D becomes less than zero and the diode turns OFF.

As v_{in} decreases further, v_D remains less than zero, the diode stays OFF, and the capacitor remains charged at V_m volts. In fact, after the initial charging of the capacitor, the diode never again turns ON and the equivalent circuit of Figure 1.7-1c applies. From this circuit it is clear that in the steady state the output v_o is given by the expression

$$v_o = v_{in} - v_C = v_{in} - V_m \qquad (1.7\text{-}3)$$

Using this result, we can see that the output is an undistorted replica of the input except that its dc level is at $-V_m$ volts and not zero volts (Figure 1.7-1d). Furthermore, as the amplitude of the input signal changes, the dc level of the output will automatically adjust itself so that the resulting output waveform will

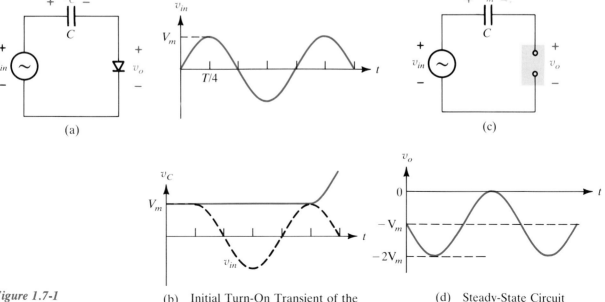

Figure 1.7-1
Diode clamping circuit.

(a)

(b) Initial Turn-On Transient of the Circuit in (a)

(c)

(d) Steady-State Circuit Output Voltage

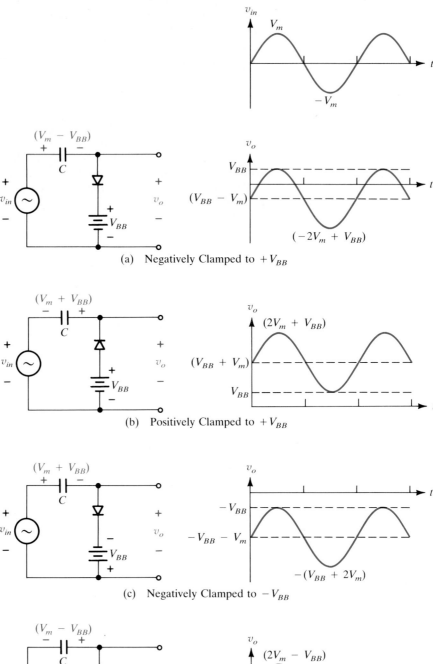

(a) Negatively Clamped to $+V_{BB}$

(b) Positively Clamped to $+V_{BB}$

(c) Negatively Clamped to $-V_{BB}$

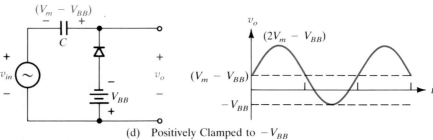

(d) Positively Clamped to $-V_{BB}$

Figure 1.7-2
Basic types of clamping circuits.

always have its maximum value at zero volts. Therefore, this circuit is said to clamp the output negatively to zero volts.

Not shown for simplicity in the circuit in Figure 1.7-1a is a resistor (usually of high ohmic value) that is often connected in parallel with the diode. This resistor permits the capacitor to discharge should the input signal be removed or should the amplitude of the input be reduced. Without this resistor if, for example, the input signal were reduced to zero, the diode would remain reverse biased forever at $-V_m$ volts, and the capacitor could never discharge.

Figure 1.7-2 illustrates several different types of clamping circuits. In these circuits the battery V_{BB} determines the clamping level and the diode orientation the clamping direction (i.e., whether the input voltage is clamped above or below V_{BB}). In particular, when the diode points down, as in parts (a) and (c), the output is clamped negatively (or below) V_{BB}, while when it points up, as in parts (b) and (d), the output is clamped positively (or above) V_{BB}.

EXAMPLE 1.7-1

For the clamping circuit illustrated in Figure 1.7-3a, the input signal v_{in} shown in part (b) of the figure is a 14-V(p-p) amplitude triangle wave sitting on top of a -5-V dc level. Determine the steady-state output voltage.

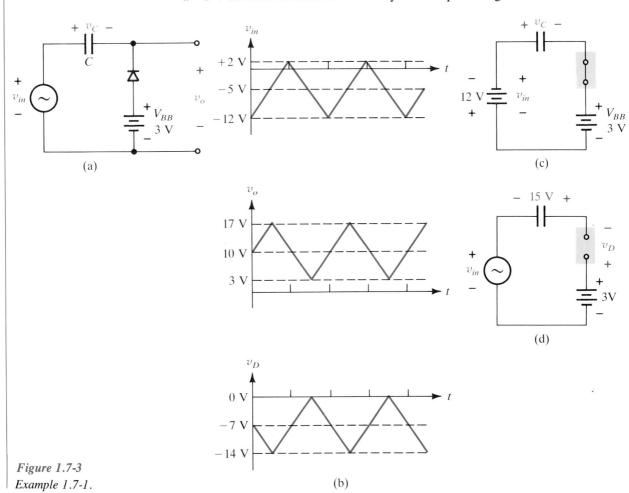

Figure 1.7-3
Example 1.7-1.

Basically, this circuit is identical to that in Figure 1.7-2b and therefore we should expect that the output will be a triangle wave that is clamped positively to $V_{BB} = +3$ V. To see why this is the case, we need to observe that initially, that is, when the circuit is first turned on, the diode will conduct on the negative half of the v_{in} cycle and the capacitor will charge up to a peak negative voltage

$$v_C = -12 - 3 = -15 \text{ V} \tag{1.7-4}$$

in accordance with the equivalent circuit shown in Figure 1.7-3c when v_{in} reaches its maximum negative value. Once this initial charging is completed, thereafter, the diode voltage v_D will always be less than zero, the diode will remain OFF, and the equivalent circuit given in Figure 1.7-3d will be valid. Following this figure the steady-state output voltage is seen to be

$$v_o = 15 + v_{in} \tag{1.7-5}$$

This waveshape is sketched in Figure 1.7-3b along with that for the voltage across the diode, noting that

$$v_D = 3 - v_o \tag{1.7-6}$$

Exercise

1.7-1 For the circuit in Example 1.7-1 (Figure 1.7-3) find the positive and negative peaks in v_o if a 2-V battery is connected in series with the capacitor with its polarity oriented in the same direction as v_C. Also, find v_C. *Answer* Maximum $v_o = 17$ V, minimum $v_o = 3$ V, $v_C = -17$ V

1.8 SMALL-SIGNAL ANALYSIS OF DIODE CIRCUITS

Most electronic circuits can be characterized as two-port networks where an input signal is applied to one port, or one side of the circuit, and an output is generated at the other. The ac analysis of these types of circuits can be greatly simplified when the amplitude of the input signal is small. It is the purpose of this section to develop the mathematical theory that will permit us to carry out the small-signal analysis of electronic circuits. Initially, these techniques will be applied to the analysis of diode circuits. The main motivation for introducing this material at this time is to lay the groundwork for the later extension of these methods to the small-signal analysis of bipolar and field-effect transistor circuits.

The circuit shown in Figure 1.8-1a illustrates the type of diode problem that we are interested in learning how to analyze. This specific circuit is biased at a particular dc Q point by the battery V_{BB} and is driven by a small signal ac input v_{in}. It can be analyzed graphically by plotting the load-line equation

$$v_{in} + V_{BB} = i_D R + v_D \tag{1.8-1}$$

on the diode volt-ampere characteristics. Of course, the position of the load line on the graph varies with v_{in}. If we set the ac signal source $v_{in} = 0$ in equation (1.8-1), the resulting load-line equation is known as the dc load line. The intersection of this load line with the diode characteristic determines the dc

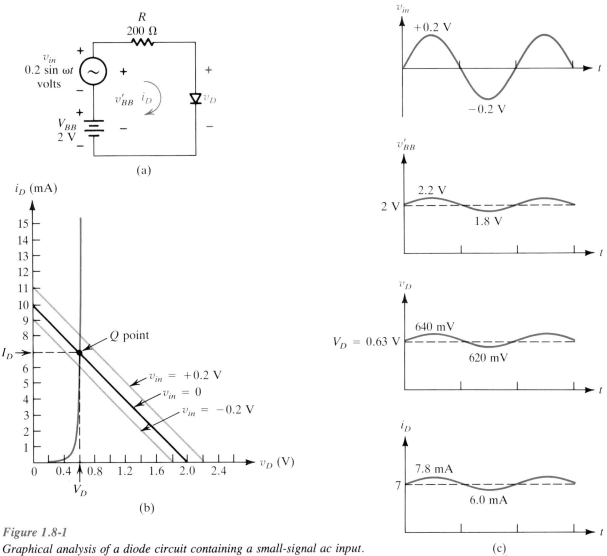

Figure 1.8-1
Graphical analysis of a diode circuit containing a small-signal ac input.

quiescent point for the circuit, which for this example is at $I_D = 7.0$ mA and $V_D = 0.63$ V.

In the previous paragraph the expressions for the current flow through the diode and the voltage drop across the diode were written with capital letters and capital subscripts to indicate that these quantities represent the dc or the quiescent operating values for this circuit. This notation, as well as that used to describe ac and composite (or ac + dc) signals, is summarized in Figure 1.8-2. It would be advisable for you to take a few minutes to study each of the terms in this figure before proceeding with the remainder of this section since this notation will be used extensively in the remainder of the text.

If we return now to the original problem illustrated in Figure 1.8-1a, we can see that as v_{in} varies between plus and minus 0.2 V, the load line will move back and forth about the $v_{in} = 0$ line as shown. As v_{in} varies, the corre-

1. Capital letter, capital subscript = I_D
   Pure dc quantity

2. Lowercase letter, lowercase subscript = i_d
   Pure ac quantity

3. Capital letter, lowercase subscript = I_d
   ac amplitude

Example: $i_d(t) = I_d \sin \omega t$

4. Lowercase letter, capital subscript
   i_D = dc + ac = total signal

Example: $i_D = I_D + i_d = I_D + I_d \sin \omega t$

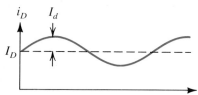

5. dc power supply voltages (batteries) have
   double subscripts

Examples: V_{CC}, V_{DD}, V_{BB}

Figure 1.8-2
Notation used in this text (illustrated for the diode current i_D).

sponding values of v_D and i_D can be read off from the load-line intersection points to sketch the diode voltage and current as functions of time (Figure 1.8-1c). It is very important for you to notice that the time-varying portions of both v_D and i_D will be very nearly sinusoidal when the amplitude of v_{in} is small, while for larger values of v_{in} these waveshapes will become distorted. We will see shortly that simple diode models, at least when v_{in} is small, can be developed which will permit the analysis of the circuits, such as that in Figure 1.8-1a, by inspection.

To begin the development of these small-signal models, note in Figure 1.8-1b that the movement of the instantaneous operating point about the Q point is small when the amplitude of v_{in} is small. As a result, it is possible to approximate the diode curve shown in the figure by the first few terms of a Taylor series expansion about the Q point. The general form of this power series expansion for i_D as a function of v_D may be written as

$$i_D = i_D\big|_{Q\,pt} + \frac{di_D}{dv_D}\bigg|_{Q\,pt} (v_D - V_D) + \frac{d^2 i_D}{dv_D^2}\bigg|_{Q\,pt} \frac{(v_D - V_D)^2}{2!} + \cdots \quad (1.8\text{-}2)$$

At first this expression does not appear to be any simpler to work with than the original diode equation, but if the deviations from the Q point are small enough, all terms after the linear one will be small and can, as a first-order approximation, be dropped. The expression for i_D in this instance reduces to

$$i_D \simeq i_D\big|_{Q\,\text{pt}} + \frac{di_D}{dv_D}\bigg|_{Q\,\text{pt}} (v_D - V_D) \tag{1.8-3}$$

where $i_D\big|_{Q\,\text{pt}} = I_D$ and where from calculus $di_D/dv_D\big|_{Q\,\text{pt}}$, the first derivative of i_D evaluated at the Q point is also equal to the slope of the line tangent to the diode curve at that point. Because the latter term is a constant and has units of $1/\text{resistance}$, we will represent it by a constant $1/r_D$. Substituting these results into equation (1.8-3) yields

$$i_D = I_D + \frac{1}{r_D}(v_D - V_D) = \underbrace{\frac{1}{r_D}v_D}_{\text{slope}} - \underbrace{\left(\frac{V_D}{r_D} - I_D\right)}_{i_D \text{ intercept}} \tag{1.8-4}$$

This last equation, which approximately represents the original diode characteristic in the neighborhood of the Q point, is of course a straight line and even more specifically is actually the equation of the line drawn tangent to the diode characteristic at the Q point (Figure 1.8-3).

It is important to realize after all of this math that the result obtained here is essentially the same as the one we got earlier in Section 1.4 when we developed our large-signal model for the diode when it was ON. In fact, by comparing the sketch given in Figure 1.8-3 with that presented in Figure 1.4-3c, it is apparent that the electrical model represented by this straight-line approximation to the diode V–I curve is once again simply a battery in series with a resis-

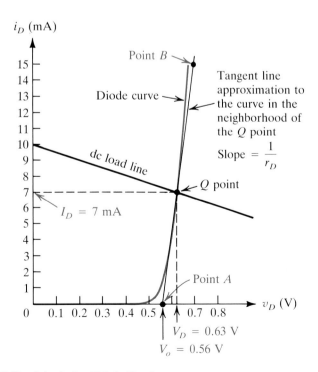

Figure 1.8-3
Developing the small-signal model for the analysis of the example given in Figure 1.8-1.

tor. As before, the battery value is obtained directly from the v_D intercept of the straight line ($V_o = 0.56$ V for this example), while the resistor value is equal to the reciprocal of the slope of the tangent line. From the graph the value of r_D may be computed directly by taking any two points on the line and forming the ratio of the change in v_D to the change in i_D. In particular, for points A and B on the line we obtain

$$r_D = \frac{\Delta v_D}{\Delta i_D} = \frac{(0.76 - 0.56)\ V}{(15 - 0)\ mA} = \frac{200\ mV}{15\ mA} = 13.3\ \Omega \qquad (1.8\text{-}5)$$

If the diode characteristic truly follows equation (1.4-2), the resistance r_D can also be computed by differentiating this expression and evaluating it at the Q point as follows:

$$r_D = 1 / \left(\frac{di_D}{dv_D} \Big|_{Q\ pt} \right) = \frac{V_T}{I_s e^{v_D/V_T}} \qquad (1.8\text{-}6)$$

Because the diode is forward biased, v_D/V_T is much greater than 1, so that

$$i_D = I_s(e^{v_D/V_T} - 1) \simeq I_s e^{v_D/V_T} \qquad (1.8\text{-}7)$$

and substituting this result into equation (1.8-6) we obtain

$$r_D = \frac{V_T}{I_D} \qquad (1.8\text{-}8)$$

For the specific Q point of this circuit, r_D turns out to be (26 mV)/(7 mA), or about 3.7 Ω. The discrepancy between this result and that given in equation (1.8-5) is due to the large semiconductor bulk resistance associated with this particular diode. The origin of this resistance and techniques for its minimization are discussed further in Section 1.11. For now it is enough to say that equation (1.8-5) provides the more accurate value of r_D for use in the model because it includes the resistance of the semiconductor external to the *pn* junction.

To continue the small-signal analysis of the original circuit given in Figure 1.8-1a, we replace the diode by its electrical equivalent circuit in Figure 1.8-4a. In all probability you could now analyze this circuit by inspection, but for reasons that will become more apparent later, the following systematic analysis procedure is suggested. Solve the problem by using superposition, evaluating the dc solution first. Use this result to develop the model for the ac portion of the analysis. Once both the dc and ac solutions have been found, the total solution (i.e., the dc + ac solution) is obtained by adding each of the individual results.

To begin the dc analysis, set the ac source in Figure 1.8-4a equal to zero. This results in the equivalent circuit given in Figure 1.8-4b. Using this circuit, the diode's quiescent current and quiescent voltage are found to be

$$I_D = \frac{(2 - 0.56)\ V}{213.3\ \Omega} = 6.75\ mA \qquad (1.8\text{-}9a)$$

$$V_D = (6.75\ mA)(13.3\ \Omega) + 0.56 = 0.65\ V \qquad (1.8\text{-}9b)$$

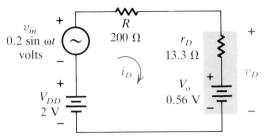

(a) Overall Model for the Small-Signal
      Analysis of the Circuit in Figure 1.8-1

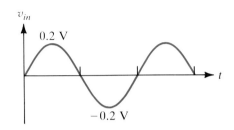

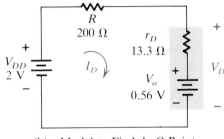

(b) Model to Find dc Q Point

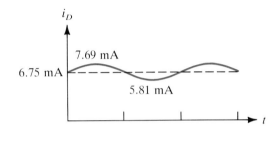

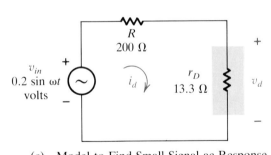

(c) Model to Find Small-Signal ac Response

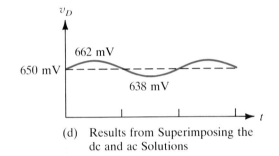

(d) Results from Superimposing the
      dc and ac Solutions

Figure 1.8-4
*Analysis of the circuit in Fig-
ure 1.8-1b using small-signal
ac model concepts.*

Although this dc analysis technique is certainly valid, it is essential to realize that this level of detail is generally not required in finding the circuit Q point. Most of the time graphical data on the diode will be unavailable, and in this instance it will be necessary to consider the diode as a 0.7-V battery (if it is silicon) and to employ this approximation to calculate I_D. Substituting this result into equation (1.8-8b), we can then calculate r_D for use in the ac analysis.

To compute the ac portion of the diode current and the diode voltage, we set all of the dc sources in Figure 1.8-4a equal to zero. Using the resulting circuit given in Figure 1.8-4c, we can write down i_d and v_d as

$$i_d = \frac{v_{in}}{r_D + R} = \frac{200 \text{ mV}}{213 \text{ }\Omega} \sin \omega t = 0.94 \sin \omega t \text{ mA} \qquad (1.8\text{-}10\text{ a})$$

$$v_d = \frac{r_D}{r_D + R} v_{in} = 12 \sin \omega t \text{ mV} \qquad (1.8\text{-}10b)$$

The "total" sketches for i_D and v_D in Figure 1.8-4d were obtained by applying the superposition principle and adding the dc and ac results given in equations (1.8-9) and (1.8-10), respectively. The easiest way to make these sketches is to place dashed lines at the dc levels of i_D and v_D and then to sketch the appropriate amplitude sine waves about each of these dc levels. The maximum and minimum amplitudes indicated on these sketches were obtained by taking each signal's dc level and adding or subtracting the corresponding peak ac amplitudes from it. Notice that the results presented in Figure 1.8-4d compare favorably with those obtained graphically and shown in Figure 1.8-1c.

Before leaving the subject of ac small-signal diode circuit analysis, we need to consider the question of just how small a small signal has to be to be able to approximate equation (1.8-2) by the linear equation in (1.8-3). If all the terms in (1.8-2) are retained, and if the derivatives of i_D in equation (1.8-7) are substituted into (1.8-2), we obtain

$$i_D = i_D\big|_{Q\text{ pt}} + \left.\frac{I_s e^{\,v_D/V_T}}{V_T}\right|_{Q\text{ pt}} (v_D - V_D)$$
$$+ \left.\frac{I_s e^{\,v_D/V_T}}{V_T^2}\right|_{Q\text{ pt}} \frac{(v_D - V_D)^2}{2!} + \cdots \qquad (1.8\text{-}13a)$$

which on substitution of $v_d = v_D - V_D$ yields

$$i_D = I_D + I_D \frac{v_d}{V_T} + \frac{I_D}{2!}\left(\frac{v_d}{V_T}\right)^2 + \frac{I_D}{3!}\left(\frac{v_d}{V_T}\right)^3 + \cdots \qquad (1.8\text{-}13b)$$

But if we recall from equation (1.8-8) that $r_D = V_T/I_D$, we may put the equation for i_D into a form where the effect of the nonlinear terms is more readily seen,

$$i_D = \underbrace{I_D}_{\substack{\text{dc} \\ \text{term}}} + \underbrace{\frac{1}{r_D} v_d}_{\substack{\text{linear} \\ \text{ac term}}} + \frac{1}{r_D}\underbrace{\left[\frac{v_d\,v_d}{2!\,V_T} + \frac{v_d}{3!}\left(\frac{v_d}{V_T}\right)^2 + \cdots\right]}_{\text{nonlinear ac terms}} \qquad (1.8\text{-}14)$$

From this equation we can see that the nonlinear ac terms can be neglected when (v_d/V_T) is much less than 1. As a result, small-signal analysis methods may be applied accurately whenever the ac diode voltage is less than about 5 or 10 mV. For example, the actual error made by using a straight-line model for a silicon diode when the ac diode voltage is 10 mV is less than 10%.

Exercises

1.8-1 Following Figure 1.8-1, estimate the amplitude of the ac current i_d if the amplitude of v_{in} is increased to 0.3 V and R is decreased to 150 Ω. *Answer* 1.3 mA

1.8-2 Using modeling techniques, estimate the amplitude of the ac voltage drop across the diode in Figure 1.8-1a if R is reduced to 100 Ω. *Answer* 4 mV

1.9 THE ZENER DIODE

Diodes that are manufactured specifically for operation in the reverse break-down region are usually referred to as zener diodes. The most common application for these devices is as voltage references, although they are also often used in waveshaping circuits, where batteries might otherwise have to be employed. In this section we illustrate several of these applications and also develop models for the zener to assist in the analysis of circuits containing this type of diode.

The symbol for the zener diode is illustrated in the circuit of Figure 1.9-1a, and its volt-ampere characteristic is given in part (b) of the figure to assist in finding the operating point for this circuit. The characteristics of the zener diode are exactly the same as those of the regular diode in the forward and also in the reverse region as long as v_D remains greater than the zener breakdown voltage, $-V_{ZZ}$. When the diode is operating in the zener region, the voltage across it remains relatively constant as long as the diode current is less than $-I_{ZK}$, and it is precisely this characteristic that makes the zener valuable as a voltage reference.

The circuit given in Figure 1.9-1a may be analyzed by writing Kirchhoff's voltage law for the loop as follows:

$$V_{BB} = -i_D R - v_D \qquad (1.9\text{-}1)$$

If we plot this load-line equation on the device characteristics in Figure 1.9-1b, we can see that the circuit Q point is $V_D = -5$ V and $I_D = -5.0$ mA. For this circuit it is apparent that the diode is operating in the reverse breakdown or zener region. By noting that the v_D intercept is at $-V_{BB}$, we can see that the diode Q point will remain in the zener region as long as the battery V_{BB} is larger than about 5 V. If V_{BB} is exactly 5 V, the operating point of the circuit

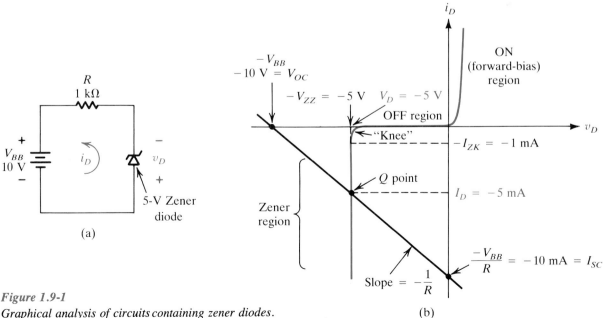

Figure 1.9-1
Graphical analysis of circuits containing zener diodes.

will be down in the knee of the curve, and as a result, the voltage across the diode will be somewhat less than -5 V. To guarantee that the diode is "solidly zenering," or that the operating point is well down into the zener portion of the characteristic, the diode specifications usually include a minimum reverse zener current, I_{ZK}, which must be flowing in the diode if the voltage across it is going to really be close to the zener voltage, $-V_{ZZ}$, specified for the diode. The requirement to have a minimum zener current in the diode is especially important for the voltage reference applications of the zener, to be discussed later in this section.

The results obtained from this simple example may suggest possible zener diode models to you. In the forward-bias region when the diode is ON, and in the reverse-bias region where it is OFF, the diode models are the same as those for the regular semiconductor diode (Figure 1.9-2). But the range of

Figure 1.9-2
Models for the zener diode.

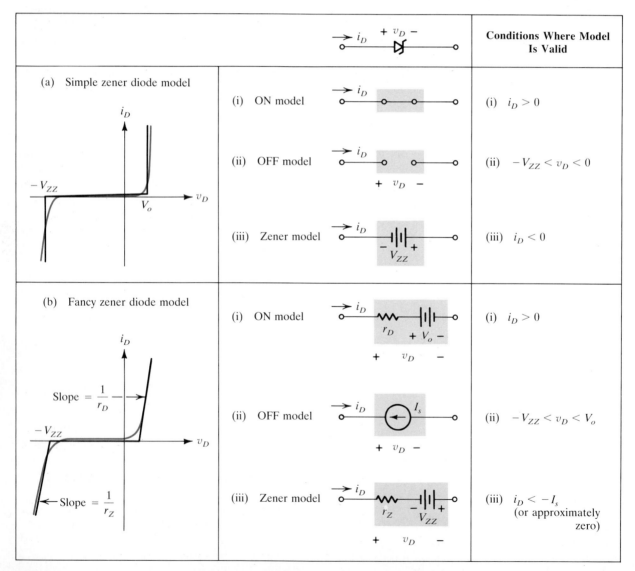

Chapter 1 / Introduction to Electronics: The Diode

voltages for which the diode is considered to be OFF needs to be modified. Previously, the diode was OFF as long as v_D was less than zero (or possibly V_o) volts, while for the zener diode the range of voltages for which the diode is OFF is given by the inequality

$$-V_{ZZ} < v_D < V_o \tag{1.9-2}$$

where the right-hand part of the inequality guarantees that the diode is not forward biased, while the left-hand side of the inequality ensures that it is not zenering.

To model the diode when it is in the zener region, if the volt-ampere characteristic is assumed to be vertical, the diode can be represented by a battery because the voltage across it is constant regardless of the current flow through it. If greater accuracy is required, we can take the slight tilt of this line into account by including a resistor, r_Z, in series with the battery. The mathematical justification for this model is based on the same type of power series expansion that was used to develop the small-signal model for the conventional diode in Section 1.8. The actual value of r_Z to use in the model is obtained by noting that $1/r_Z$ equals the slope of the line drawn tangent to the curve at the Q point. You should also note that since this portion of the diode characteristic is itself very nearly straight, the model (which of course is a straight-line approximation to the curve) will be valid for large deviations from the Q point. In fact, the parameters V_{ZZ} and r_Z will be relatively independent of the actual Q-point location in the zener region. For this reason, manufacturers often specify a particular value for r_Z along with the other zener diode parameters. Two possible zener diode models are given in Figure 1.9-2. In general, the simpler model for the diode in the zener region (i.e., a battery) is OK to use as long as the external resistances are much larger than r_Z.

To apply these models to the analysis of circuits containing zeners, we must employ techniques paralleling those used previously for circuits containing conventional diodes. To carry out a dc Q-point analysis, and more specifically to determine whether or not the diode is zenering, we will assume it to be in the zener region, replace it by its appropriate model (a battery), and then check if i_D is less than zero in order to confirm our original assumption. To see how this is actually carried out, let's repeat the analysis of the problem given in Figure 1.9-1a, but this time with the aid of the diode model in Figure 1.9-2a. If we assume that the diode is zenering, we may redraw this circuit as shown in Figure 1.9-3. From this picture the diode current is given by

$$i_D = -\frac{V_{BB} - V_{ZZ}}{R} = -5.0 \text{ mA} \tag{1.9-3}$$

which is the same result as that obtained previously by graphical analysis. The fact that the diode current is negative substantiates our original assumption that the diode was operating in the zener region.

To analyze zener diode circuits containing large-signal ac sources, we again use an approach paralleling that employed in the case of ordinary diodes. The first task is to determine the range of input signals for which the diode is ON, OFF, and zenering. Once these are established, the diode is replaced by the appropriate model in each case in order to calculate the required voltages and currents. This technique is illustrated in the example that follows.

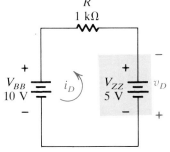

Figure 1.9-3

Solution of zener diode Q-point problems with the use of models.

EXAMPLE 1.9-1

The circuit given in Figure 1.9-4a contains a 5-V zener diode and is driven by a 20-V(p-p) sine wave. Sketch the resulting zener current and voltage. You may assume the zener diode to be ideal.

SOLUTION

To determine when the diode is ON, we replace it by a short circuit (Figure 1.9-4b) and determine those values of v_{in} for which i_D is greater than zero. For this circuit $i_D = v_{in}/R$, and the diode is ON as long as v_{in} is greater than zero. In this region $v_D = 0$ V. To see when the diode is OFF, we replace it by an open circuit (Figure 1.9-4c) and observe that $v_D = v_{in}$ for this case. As a result, the diode will be OFF as long as v_{in} is less than zero but greater than -5 V. When v_{in} drops below -5 V, the diode Q point will move into the zener region, and for this case the circuit model given in Figure 1.9-4d applies. To confirm this we note that the diode current for this circuit may be written as

$$i_D = \frac{v_{in} + V_{ZZ}}{R} \tag{1.9-4}$$

and because i_D must be negative for the diode to be in the zener region, the input signal must be less than $-V_{ZZ}$ or -5 V. The actual diode voltage and current produced by the 20-V(p-p) sinusoidal input to this circuit are given in Figure 1.9-4e.

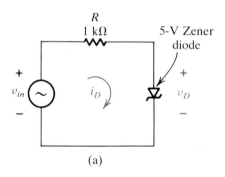

(a)

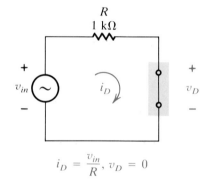

$$i_D = \frac{v_{in}}{R}, \, v_D = 0$$

(b) Equivalent Circuit When Diode Is ON

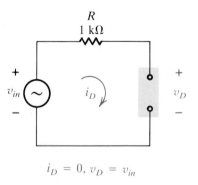

$$i_D = 0, \, v_D = v_{in}$$

(c) Equivalent Circuit When Diode Is OFF

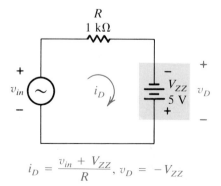

$$i_D = \frac{v_{in} + V_{ZZ}}{R}, \, v_D = -V_{ZZ}$$

(d) Equivalent Circuit When Diode Is Zenering

Figure 1.9-4
Example 1.9-1.

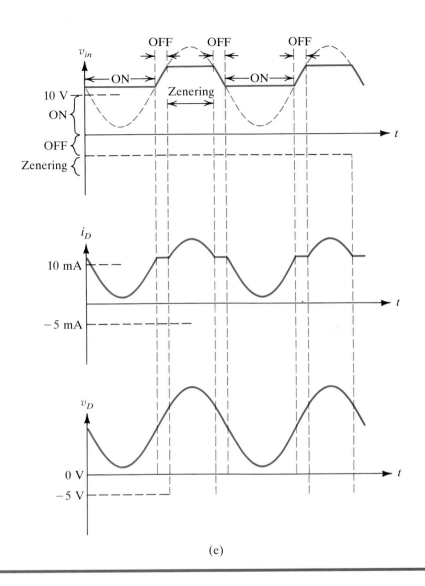

Figure 1.9-4
(Continued)

(e)

At the beginning of this section we noted that one of the major applications of the zener diode was as a voltage reference or as a regulator in a power supply. When employed as a voltage regulator (see Figure 1.9-5a), the diode is biased in its zener region and the circuit is designed for it to stay there despite substantial variation in either the unregulated supply voltage, V_{BB}, or the load, R_L. This circuit is called a shunt regulator since the zener is shunted across the output terminals, and as long as the diode stays in the zener region, the output voltage will remain fixed at the zener voltage, V_{ZZ}. When analyzing this type of circuit, since i_D and v_D are always negative, it is convenient to define the zener current, i_Z, and the zener voltage, v_Z, as the negatives of these quantities (see Figure 1.9-5b).

For a regulator to work properly, the zener current must be kept within well-defined limits. This current must be greater than I_{ZK} for the diode to re-

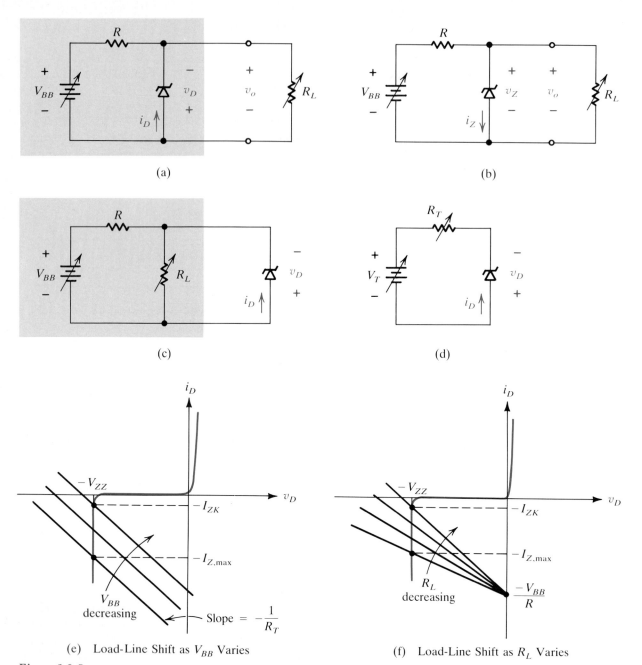

Figure 1.9-5

Zener diode in a voltage-regulated power supply.

(a)

(b)

(c)

(d)

(e) Load-Line Shift as V_{BB} Varies

(f) Load-Line Shift as R_L Varies

main solidly in the zener region, but in addition, it is necessary that i_Z not become too large. The zener diode, like all other electronic devices, can dissipate only a specific amount of power. Should the average power dissipated in the diode exceed this limit, irreparable damage can be done to the device. If we for the moment consider that the zener voltage and current are approximately constant, the average power dissipated in the diode is given by the product $v_Z \times i_Z$. Therefore i_Z must be kept small enough so that the power dissipation in the zener does not exceed its maximum rating. For convenience we will call the maximum allowed zener current $I_{Z,\,max}$.

When examining the operation of zener voltage regulator circuits, it is convenient to divide the analysis into two distinct problem classes: those for which V_{BB} varies with R_L fixed, and those for which V_{BB} is fixed but R_L varies. Of course, in a real power supply both of these quantities may actually vary simultaneously, but for now we assume that only one changes at a time.

To see the effect of the variation of each of these quantities, we will redraw the regulator in Figure 1.9-5a as shown in Figure 1.9-5c and then convert this circuit to its Thévenin equivalent (Figure 1.9-5d). For this circuit

$$R_T = \frac{R \cdot R_L}{R + R_L} \tag{1.9-5a}$$

$$V_T = V_{BB}\frac{R_L}{R + R_L} \tag{1.9-5b}$$

The load line for this case is the same as that given by equation (1.9-1) and sketched in Figure 1.9-1.

As V_{BB} varies with R_L fixed, the slope of the load line does not change, but the v_D and i_D intercepts both vary as shown in Figure 1.9-5e. When V_{BB} becomes too small, the diode does not have enough current to stay in the zener region, while too large a value for V_{BB} will cause the zener current to exceed $I_{Z,max}$.

As R_L varies with V_{BB} fixed (see Figure 1.9-5f), both the slope and v_D intercept of the load-line change, but the i_D intercept remains constant. For this case when R_L is too small, it takes away too much current from the diode and causes it to drop out of the zener region, while with large values of load resistance, care must be taken not to exceed the zener's power-handling capability.

The graphical analyses just carried out are not the best way to attack zener diode voltage regulator problems. They were presented here only to help you develop a good physical picture of what happens as V_{BB} and R_L vary. As we shall see in the example that follows, zener diode shunt regulator problems are best analyzed by making use of the simple zener diode models already developed.

EXAMPLE 1.9-2

In this example we consider the case of a 5-V zener-regulated power supply circuit which is driving a fixed 25-Ω load (see Figure 1.9-6a). If the diode remains in the zener region, we may replace it by a 5-V battery to obtain the overall equivalent circuit given in part (b) of the figure. The diode has the following characteristics: $V_{ZZ} = 5$ V, $I_{ZK} = 10$ mA, and a maximum allowed power dissipation $P_D = 1$ W. Find the allowed range of input voltages for which this circuit will work properly.

SOLUTION

To determine the values of V_{BB} for which this circuit will work, note that as V_{BB} is reduced, the zener current I_Z will get smaller and eventually the zener will shut off. At the other extreme, when V_{BB} gets to be too large, the zener current will become so big that the power dissipation in the diode will exceed its maximum rating and the diode will be destroyed.

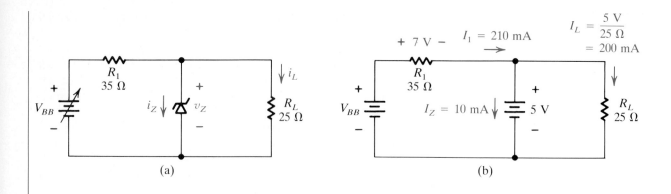

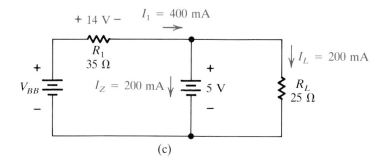

Figure 1.9-6
Example 1.9-2.

(c)

To find the lower limit for V_{BB}, note from Figure 1.9-6b that when the diode is just on the verge of turning OFF, I_Z is at about 10 mA and therefore I_1 is 210 mA, since I_L is fixed at 200 mA as long as the diode stays in the zener region. If I_1 is 210 mA, the voltage drop across R_1 is 0.21 × 35 = 7.35 V and V_{BB} is 7.35 + 5 = 12.35 V.

The maximum allowed value of V_{BB} occurs when the diode has the largest possible value of current in it (see Figure 1.9-6c). Because this diode is a 1-W device, $I_{Z,\,max}$ is 1 W/5 V or 200 mA. For this case I_1 is 400 mA and the corresponding voltage drop across R_1 is 14 V. The value of V_{BB} required to produce these conditions is V_{BB} = 14 + 5 = 19 V.

In summary, the circuit illustrated in Figure 1.9-6a will produce a constant 5-V output even though the input voltage V_{BB} is varied all the way from 12.3 to 19 V.

Exercises

1.9-1 For the circuit in Figure 1.9-3, find the power dissipated in the resistor and the zener. *Answer* 25 mW, 25 mW

1.9-2 In Example 1.9-2 (Figure 1.9-6) R_L is made adjustable. Find the value of R_L for which I_Z = 100 mA, given that V_{BB} = 12 V. *Answer* 50 Ω

1.9-3 In Figure 1.9-6c, find the value of V_{BB} for which I_Z = 100 mA. *Answer* 15.5V

1.10 INTRODUCTION TO SEMICONDUCTOR PHYSICS

1.10-1   Basic Concepts of Insulators, Conductors, and Semiconductors

In isolated atoms the electrons that surround the nucleus reside in specific orbits or specific energy levels associated with their distance from the nucleus. When a number of such atoms are brought together to form a solid, these discrete energy levels split into large numbers of closely spaced levels known as energy bands.

In equilibrium the electrons occupy the lower orbital positions or energy levels, and for an electron to move to a higher orbit, it must gain additional energy from the external (thermal, electrical, and optical) energy sources that act on the material.

If an electron gains enough energy, it may be able to escape totally from the influence of the nucleus and become a "free electron" within the solid. The ease with which this process occurs determines whether we classify that material as an insulator, a semiconductor, or a conductor.

In a solid, the uppermost energy band containing electrons at $T = 0$ K is called the valence band, and the next available band above it is termed the conduction band. In general, electrons that are able to enter the conduction band are free to travel throughout the solid and are no longer bound to or associated with a specific atom. The energy required to elevate an electron from the valence band into the conduction band depends on the size of the energy gap, E_G, between these two bands.

In insulators this gap size is quite large and it is very difficult to remove electrons from their parent atoms. Diamond is an excellent example of a material in this class. Its energy gap is 5.5 eV and its resistivity at room temperature is about 10^{12} Ω-cm.

Semiconductors have band structures that are quite similar to those of insulators, the major difference being that their energy gaps are much smaller. As a result, while semiconductors are insulators at 0 K, at room temperature the thermal energy they absorb allows a considerable number of electrons to enter the conduction band. If we define n_i as the free electron concentration in the semiconductor, the relationship between the gap energy and n_i may be expressed as

$$n_i(T) = AT^{3/2}e^{-E_G/2KT} \tag{1.10-1}$$

where A = constant that depends on the material
T = temperature, K
k = Boltzmann's constant = 8.6×10^{-5} eV/K
E_G = energy gap between the conduction band and the valence band

Table 1.10-1 illustrates the band gap energy, carrier concentration, and resistivity of several selected materials.

The band structure of a conductor is quite different from that of an insulator or a semiconductor. Rather than being separated by an energy gap, the valence and conduction bands overlap. As a result, large numbers of free electrons are available at all temperatures, causing the resistivity of conductors to be quite low. Copper, for example (Table 1.10-1), has a resistivity at 300 K

1.10-1 Basic Electrical Properties of Several Selected Insulators, Semiconductors, and Conductors (at $T = 300$ K)

Material	E_G (eV)	n_i (cm^{-3})	ρ (Ω-cm)	Permittivity ϵ (F/m)
Diamond	5.5	10	10^{12}	4.86×10^{-11}
GaAs	1.43	9×10^6	4×10^8	1.16×10^{-10}
Si	1.11	1.4×10^{10}	2.5×10^5	1.05×10^{-10}
Ge	0.67	2.5×10^{13}	43	1.41×10^{-10}
Cu	—	8.5×10^{22}	1.67×10^{-6}	8.85×10^{-12}

that is about 10^{11} times less than that of the intrinsic semiconductor silicon and more than 10^{17} times smaller than that of an insulator such as diamond.

1.10-2 Intrinsic Semiconductors

Most semiconductor materials that are used for electronic applications are fabricated as single crystals because the electrical properties of these structures can be accurately controlled. A crystal is a material containing a periodic array of atoms and having a typical interatomic spacing of about 5 Å or 5×10^{-10} m. Crystals that are free from all impurities are known as intrinsic or pure semiconductors.

Silicon is the most popular semiconductor material, although germanium and gallium arsenide also find application in the power and high-speed circuit areas, respectively. For the purpose of this discussion, however, and in fact throughout most of this book, we focus our attention on silicon devices.

Silicon is a tetravalent material; that is, it has four valence electrons in its outermost orbit. In its crystalline state silicon has a tetrahedral structure in which each atom of the crystal is situated physically at the center of a tetrahedron formed by its four nearest-neighbor atoms. Each neighboring atom is bound to the central atom by a covalent bond in which one pair of the valence electrons is shared with each neighbor.

At room temperature there is a small but finite probability that some of the valence electrons will absorb enough thermal energy to break their covalent bonds. When this occurs it results in the creation of both a free electron and a free hole. The free hole, which is actually a broken silicon bond, has a net charge of $+1e$, where $e = 1.6 \times 10^{-19}$ C. This, of course, is to be expected since the hole is essentially an ionized atom.

The hole is termed "free" because the position of this ionized atom can move about in the crystal. It does this by taking an electron from an adjacent silicon atom. When this occurs the original hole is neutralized so that it disappears, and a new hole appears at the site from which the electron was taken. This process can be viewed as that of an electron jumping from one atom to the next, or more simply as that of a hole moving in the opposite direction.

The properties of an intrinsic semiconductor may be summarized as follows:

1. For an intrinsic semiconductor the free holes can move around just like the free electrons, although their speed may be a little slower because their motion is associated with the process of electrons jumping from one atom position to another.

2. In an applied field the holes move in a direction opposite to that of the electrons so that they appear to have a net positive charge.

3. For an intrinsic semiconductor the number of free holes equals the number of free electrons or

$$n = n_i = p \qquad (1.10\text{-}2)$$

where n_i is given by equation (1.10-1) and n equals the number of free electrons per cubic centimeter and p the number of free holes per cubic centimeter.

1.10-3 Doped or Impurity Semiconductors

Intrinsic semiconductors, while of great interest from a physics viewpoint, have limited utility as electronic devices because their characteristics vary so greatly with temperature. By adding small amounts of impurity to an intrinsic semiconductor, it is possible to create extremely versatile materials whose electronic properties exhibit a much smaller variation with temperature. These materials are known as doped or impurity semiconductors.

Consider, for example, what happens when we add a small amount of a pentavalent atom such as phosphorus to a silicon crystal. Due to their similar chemical properties, the phosphorus atoms are able to fit right into the crystal positions normally occupied by some of the silicon atoms. The phosphorus doping levels used are typically on the order of 10^{17} cm^{-3} and since the concentration of silicon atoms is about 5×10^{22} cm^{-3} there is only about one phosphorus atom for every 500,000 atoms of silicon. Thus, each phosphorus atom is surrounded by many, many silicon atoms.

At each point in the crystal where a phosphorus atom is located, four of its five valence electrons form covalent bonds with the four nearest silicon atoms. The fifth valence electron of the phosphorus atom orbits far away from its nucleus, and as a result of the shielding produced by the nearby silicon atoms, this electron is only loosely attracted to the nucleus. In fact, at room temperature nearly all of these outermost valence electrons will have escaped from the influence of their respective phosphorus atoms.

As a result, at room temperature the following situation exists. To begin with, nearly all the phosphorus atoms are ionized. This creates approximately 10^{17} free electrons per cubic centimeter. In addition, once ionized these phosphorus atoms tend to remain ionized permanently, creating a fixed or bound hole at each phosphorus site.

Besides the free electrons donated to the crystal by the phosphorus atoms, there are also additional free electrons contributed by the broken bonds of the silicon atoms themselves (approximately 1.4×10^{10} cm^{-3}), so that the overall concentration of free electrons in this material at room temperature may be written as

$$n = N_D + n_i = 5 \times 10^{17}/\text{cm}^{-3} + 1.4 \times 10^{10} \text{ cm}^{-3} \simeq 5 \times 10^{17} \text{ cm}^{-3}$$
$$(1.10\text{-}3)$$

where N_D is the phosphorus (or donor) doping concentration and n_i is the intrinsic carrier concentration.

Because no free holes are contributed by the phosphorus atoms, it might at first appear that the concentration of free holes would be equal to its intrinsic value of $n_i = 1.4 \times 10^{10}$ cm^{-3}. Actually, the number is considerably less than this. When the number of free electrons increases above the equilibrium value, there will be many more of these electrons wandering about in the crystal. This increases the probability that one of them will combine with a hole to form a neutral atom, and this decreases the number of free holes present in the material. Quantitatively, this relationship is expressed by the law of mass action, which states that the product of the number of free electrons times the number of free holes is a constant, n_i^2, or that

$$n \cdot p = n_i^2 \qquad (1.10\text{-}4)$$

This result is independent of the doping levels in the semiconductor, and substituting equation (1.10-3) into (1.10-4), we may therefore express the density of free holes as

$$p = \frac{n_i^2}{N_D} = \frac{(1.4 \times 10^{10})^2}{5 \times 10^{17}} = (3.9 \times 10^2) \text{ cm}^{-3} \qquad (1.10\text{-}5)$$

This type of pentavalent doping is known as donor doping since the phosphorus atom donates its outermost valence electron to the silicon crystal, and because this material has many more free electrons than free holes, it is called a negative or n-type semiconductor. When an electrical current flows in this material, it will mostly be due to the movement of the electrons and hence the free electrons in an n-type semiconductor are known as the majority carriers, and the free holes are called the minority carriers.

By doping the silicon with a trivalent material such as boron, a semiconductor having properties opposite to those of the n-type silicon can be produced. This type of doping is called acceptor doping, and the description of its effect on the silicon crystal directly parallels the discussion just presented for the case of donor doping. Boron has three valence electrons, and when it is used to dope silicon, one of the covalent bonds that it forms with its four nearest silicon neighbors has only one electron. A strong tendency exists for this bond to acquire a second electron, and as a result, at room temperature, nearly all of these bonds will be completed by taking an extra electron from elsewhere in the crystal. When this has been done, two things occur. First, the additional electron in the vicinity of the boron atom gives rise to a net charge of $-1e$, forming a bound electron at this position in the crystal. Second, because each of these electrons is taken from a silicon atom in the crystal lattice, this process also creates an ionized silicon atom or a free hole.

Because there are a large number of free holes and a very small number of free electrons in this material, it is called a positive type or p-type semiconductor. The specific doping technique is called acceptor doping because the boron, in this case, accepts an extra electron from the silicon crystal lattice. For p-type material the holes are the majority carriers and the electrons are the minority carriers.

Before proceeding to Section 1.11, let's summarize the characteristics of doped semiconductors:

1. The basic definitions are

n = number of free electrons per cubic centimeter

p = number of free holes per cubic centimeter

n_i = number of intrinsic carriers per cubic centimeter (holes or electrons for undoped materials)

N_D = number of donor atoms per cubic centimeter

N_A = number of acceptor atoms per cubic centimeter

2. For an intrinsic semiconductor

$$n_o = p_o = n_i \qquad (1.10\text{-}6)$$

where $n_i = 1.4 \times 10^{10}$ cm^{-3} for silicon at room temperature. The subscript o is used here to indicate that the semiconductor is in thermal equilibrium with its surroundings.

3. For an n-type semiconductor

$$n_{no} = N_D \qquad (1.10\text{-}7a)$$

and

$$p_{no} = \frac{n_i^2}{N_D} \qquad (1.10\text{-}7b)$$

In these equations the first subscript, n, is used to identify the semiconductor as n-type material, and as above, the second subscript, o, indicates that the semiconductor is in equilibrium.

4. For a p-type semiconductor

$$p_{po} = N_A \qquad (1.10\text{-}8a)$$

and

$$n_{po} = \frac{n_i^2}{N_A} \qquad (1.10\text{-}8b)$$

where the subscript p identifies the semiconductor as p-type material.

Exercises

1.10-1 Find n_i at 100°C for an intrinsic silicon semiconductor. **Answer** 1.38×10^{12} cm^{-3}

1.10-2 A silicon semiconductor is doped with 5×10^{10} boron atoms/cm³. Find n and p at 300 K. **Answer** $n = 6.4 \times 10^{10}$ cm^{-3}, $p = 3.06 \times 10^9$ cm^{-3}

1.11 THE *PN* JUNCTION DIODE

The volt-ampere characteristics of the *pn* junction semiconductor diode were introduced in Section 1.3, but little theoretical justification for its performance was offered at that time. In this section we attempt to give the reader a quantitative understanding of the behavior of this device.

1.11-1 Current Flow Mechanisms in Semiconductors

The two major components of current flow in a semiconductor are those associated with the drift and the diffusion of charge carriers. The drift portion of this current is caused by the movement of free holes and electrons in the presence of an electric field, while the diffusive portion of the current flow is re-

1.11-1 Electrical Properties of Several Important Semiconductor Materials at 300 K

Material	n_i (cm^{-3})	Mobility (cm^2/V·s)		ρ (Ω-cm)	σ (S·cm^{-1})	Diffusion Constant (cm^2/s)	
		μ_n	μ_p			D_n	D_p
Si	1.4×10^{10}	1300	500	2.5×10^5	4.4×10^{-6}	34	13
Ge	2.5×10^{13}	3800	1800	43	2.2×10^{-2}	98	46
GaAs	9×10^6	8500	400	4×10^8	1.3×10^{-8}	220	10

lated to the movement of these charge carriers from more crowded to less crowded regions in the material.

When an electric field is applied to a solid, both electrons and holes move at a constant velocity in response to the field. The holes move in the same direction as the applied field, while the electrons move in the opposite direction. To a first-order approximation the velocity of both of these charge carriers is proportional to the field strength, so that we may write

$$v_n = \text{electron drift velocity} = -\mu_n \mathscr{E} \qquad (1.11\text{-}1a)$$

and
$$v_p = \text{hole drift velocity} = \mu_p \mathscr{E} \qquad (1.11\text{-}1b)$$

where $\mathscr{E}$ is the electric field strength and μ is a constant known as the charge carrier's mobility. As mentioned earlier in Section 1.10, v_p is smaller than v_n because the movement of free holes is, by its very nature, a much slower process than the movement of free electrons. Table 1.11-1 gives the electron and hole mobilities of several of the more popular semiconductors.

Because the movement of charge constitutes a current, the overall drift current associated with the application of an electric field to a semiconductor may be written as

$$J_{\text{drift}} = -qnv_n + qpv_p = q(n\mu_n + p\mu_p)\mathscr{E} = \sigma\mathscr{E} \qquad (1.11\text{-}2)$$

where J_{drift} is the current density or current flow per unit area, and σ is a constant known as the material's conductivity.

Besides the hole and electron motion associated with drift of charge carriers in an electric field, a significant portion of the current flow in semiconductors is caused by the diffusion of these carriers from more crowded to less crowded regions in the material. This component of the current flow is known as the diffusion current.

Diffusive forces are probably less well known than the electrostatic drift forces, but their effect on the behavior of the free charges in a semiconductor can be understood just as easily. To begin with, the movement of particles as a result of diffusion has nothing to do with the electrical charge on them but is related only to their distribution in space. Particles, like people, tend to diffuse from more crowded to less crowded regions. Furthermore, the amount of diffusion is proportional to the concentration gradient or rate of change in concentration with distance. The constant of proportionality is called the diffusion constant.

Following Figure 1.11-1 the diffusive current flows associated with the

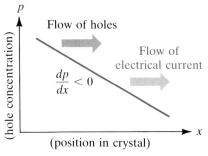

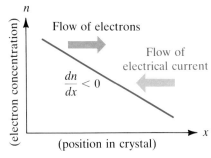

(a) Current Flow Associated with the Diffusion of Holes

(b) Current Flow Associated with the Diffusion of Electrons

Figure 1.11-1

nonuniform concentration of holes and electrons in a semiconductor may be written as

$$J_{\text{diffusion, holes}} = \text{diffusion current density for holes} = -qD_p\frac{dp}{dx} \qquad (1.11\text{-}3a)$$

$$J_{\text{diffusion, electrons}} = \text{diffusion current density for electrons} = qD_n\frac{dn}{dx} \qquad (1.11\text{-}3b)$$

where D_p and D_n are the diffusion constants for holes and electrons, respectively. (See Table 1.11-1 for data on these constants for selected semiconductors.) The minus sign in equation (1.11-3a) is needed because when dp/dx is negative, as shown in Figure 1.11-1a, the holes and hence the electrical current associated with their motion will flow toward the right. The minus sign makes this current come out positive when dp/dx is negative.

1.11-2 Recombination Time and the Diffusion Equation

In a semiconductor the free holes and the electrons are continuously disappearing by recombining with one another. At the same time new electron–hole pairs are also being created by thermal generation, and in equilibrium the recombination rate (or death rate) must equal the generation rate (or birth rate) if a steady-state population is to be achieved.

To illustrate the relationship between the carrier recombination and the carrier concentration profile, consider the elemental semiconductor volume shown in Figure 1.11-2. Here holes are assumed to be entering the volume from the left and leaving from the right at the x and $x + \Delta x$ surfaces, respectively. At any point in time in the steady state the net flow of holes into this volume, ΔV, must be equal to the recombination rate inside ΔV. This statement may be expressed mathematically as

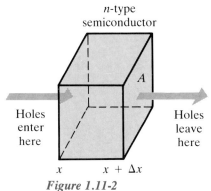

Figure 1.11-2
Hole flow in an elemental volume $\Delta V = A\,\Delta x$.

$$\underbrace{-D_pA\frac{dp}{dx}\bigg|_{x=x}}_{\substack{\text{rate of flow}\\\text{into }\Delta V}} + \underbrace{D_pA\frac{dp}{dx}\bigg|_{x=x+\Delta x}}_{\substack{\text{rate of flow}\\\text{out of }\Delta V}} = \underbrace{\frac{p - p_o}{\tau_p}\Delta x}_{\substack{\text{recombination}\\\text{within }\Delta V}} \qquad (1.11\text{-}4a)$$

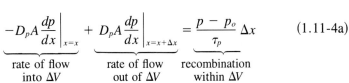

where we have neglected the effect of electric fields. In this equation τ_p is known as the recombination time for holes, and p_o represents the equilibrium hole concentration. Similar definitions exist for τ_n and n_o. Dividing both sides of equation (1.11-4a) by $(D_p A\,\Delta x)$ and taking the limit as Δx goes to zero, we obtain

$$\frac{d^2(p-p_o)}{dx^2} = \frac{p-p_o}{D_p\,\tau_p} = \frac{p-p_o}{L_p} \qquad (1.11\text{-}4b)$$

where L_p equals the diffusion length for holes. Equation (1.11-4b) is known as the diffusion equation for holes, and from a similar derivation, the corresponding equation for electrons is seen to be

$$\frac{d^2(n-n_o)}{dx^2} = \frac{(n-n_o)}{D_n\,\tau_n} = \frac{n-n_o}{L_n} \qquad (1.11\text{-}4c)$$

1.11-3 *PN Junction in Equilibrium*

A semiconductor diode can be constructed by joining a piece of *n*-type material and a piece of *p*-type material together. The *p* side of the diode is called the anode and the *n* side the cathode. Figure 1.11-3a illustrates the internal charge distribution in each of these semiconductor materials prior to connecting the two pieces together. The charges with the circles around them represent the bound holes and electrons in the semiconductor, while those without the circles are the free charges.

When these *p*- and *n*-type semiconductors are joined as shown in Figure 1.11-3b, several things happen simultaneously. Because of the large concentration gradients at the junction, free holes immediately begin to diffuse from the *p* material into the *n,* and free electrons diffuse from the *n* into the *p*. These free charges meet in the vicinity of the junction and combine to form neutral atoms. As a result, as this diffusion process continues, the free electrons and free holes begin to disappear near the junction, uncovering the bound positive and negative charges as shown in the figure.

Because the region in the vicinity of the junction is now devoid of free carriers, it is often referred to as the depletion region or space-charge region. These bound charges create an electric field (or voltage) which acts as a barrier to limit the further diffusion of charges across the junction. Initially, this voltage is zero, but as more free electrons and holes recombine near the junction, this potential gets larger, and eventually it reduces the majority carrier diffusion to a trickle.

The variation in charge density with position in the crystal is sketched in Figure 1.11-3c. Here $-w_p$ and w_n represent the boundaries of the depletion region in the *p* and *n* materials, respectively. From Poisson's equation,

$$\nabla \cdot \mathscr{E} = \frac{d\mathscr{E}}{dx} = \frac{\rho}{\epsilon} \qquad (1.11\text{-}5a)$$

so that the electric field intensity is simply $1/\epsilon$ times the integral of the charge density. In a similar fashion, because

$$\mathscr{E} = -\nabla V = -\frac{dV}{dx} \qquad (1.11\text{-}5b)$$

the voltage across the junction is the negative integral of the field intensity. Both of these quantities are also sketched in Figure 1.11-3c.

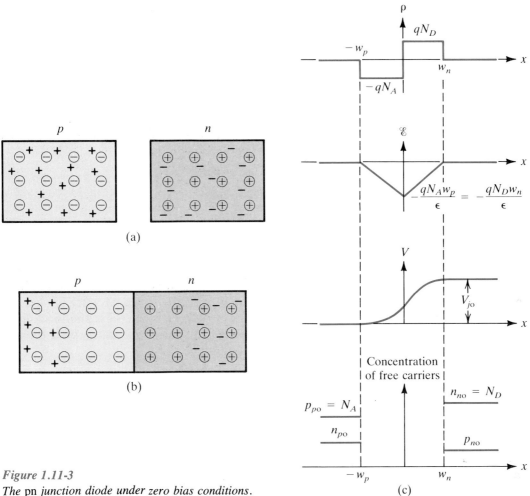

Figure 1.11-3
The pn *junction diode under zero bias conditions.*

(a)

(b)

(c)

The voltage across the junction in equilibrium may be directly related to the carrier concentrations on both sides of the junction by noting that the net flow of holes and the net flow of electrons across the junction must each be zero separately. As a result, using the expressions for the flow of holes, we may write

$$J_{\text{diffusion, holes}} + J_{\text{drift, holes}} = -qD_p\frac{dp}{dx} - qp\mu_p\frac{dV}{dx} = 0 \qquad (1.11\text{-}6)$$

This equation is valid at any point x in the diode. Integrating both sides of x along the length of the depletion region, we find after some manipulation that

$$V_{jo} = V_T \ln\frac{p_p(-w_p)}{p_n(w_n)} \qquad (1.11\text{-}7a)$$

where $V_T = kT/q = D_n/\mu_n = D_p/\mu_p$. This expression for V_T is known as the Einstein relation, and for $T = 300$ K, V_T is approximately equal to 26 mV. In a similar fashion, V_{jo} can also be related to the electron concentrations by the expression

$$V_{jo} = V_T \ln \frac{n_n(w_n)}{n_p(-w_p)} \qquad (1.11\text{-}7b)$$

Because, for example, $p_p(-w_p)$ and $p_n(-w_n)$ are equal to N_A and n_i^2/N_D, respectively, equation (1.11-7a) may also be expressed in terms of the doping concentrations as

$$V_{jo} = V_T \ln \frac{N_A N_D}{n_i^2} \qquad (1.11\text{-}8)$$

To illustrate the size of this built-in voltage, V_{jo}, consider the specific example of a silicon diode with $N_A = N_D = 10^{17}$ cm^{-3}. Substituting this data into equation (1.11-8) yields $V_{jo} = V_T \ln [10^{34}/(2 \times 10^{20})] = 820$ mV. Generally, for all reasonable doping levels this voltage is always on the order of several hundred millivolts.

Interestingly enough, V_{jo} cannot be measured directly because when this is attempted, the voltage drops across the metal–semiconductor contacts at both ends of the diode exactly cancel V_{jo}, so that the terminal voltage measured across the leads is always zero. This result is not all that surprising since if the terminal voltage were nonzero, power could be drawn from the diode forever, and this would violate conservation of energy.

EXAMPLE 1.11-1

a. Derive an expression for the width of the depletion region of the *pn* junction diode in Figure 1.11-3 in terms of the voltage V_{jo} across the junction in equilibrium.

b. Evaluate this width for a silicon diode in which $N_A = 10^{17}$ cm^{-3} and $N_D = 10^{16}$ cm^{-3}. The permittivity of silicon is 1.062×10 F/cm.

SOLUTION

Because the uncovered bound charges on either side of the depletion region are each associated with the recombination of a free electron from the *n* side and a free hole from the *p* side, the total number of uncovered bound positive charges to the right of the junction is equal to the total number of uncovered bound negative charges on the left. Therefore, w_n and w_p in Figure 1.11-3c are related to one another by the expression

$$w_p N_A = w_n N_D \qquad (1.11\text{-}9)$$

The sketches for the electric field and voltage across the junction in this figure follow directly from the fact that $\mathscr{E}$ is $1/\epsilon$ times the integral of the charge density, while V is the negative integral of the electric field. As a result, the field sketch represents the growth of area under the charge density curve with distance. Hence it has a triangular shape with a maximum height equal to the area under the left-hand side of the curve or $(-qN_A w_p/\epsilon)$. In similar fashion, the equilibrium voltage V_{jo} across the junction is equal to the negative of the area under the curve for the electric field, so that

$$V_{jo} = \tfrac{1}{2}(w_p + w_n)\frac{qN_D w_n}{\epsilon} \qquad (1.11\text{-}10a)$$

or

$$w_n^2 + w_n w_p = \frac{2\epsilon V_o}{qN_D} \tag{1.11-10b}$$

Combining equations (1.11-9) and (1.11-10b) yields

$$w_n = \sqrt{\frac{2\epsilon V_{jo}}{qN_D^2(1/N_D + 1/N_A)}} \tag{1.11-11a}$$

and

$$w_p = \sqrt{\frac{2\epsilon V_{jo}}{qN_A^2(1/N_D + 1/N_A)}} \tag{1.11-11b}$$

and adding these two expressions, we find that the overall width of the depletion region is

$$w = w_n + w_p = \sqrt{\frac{2\epsilon V_{jo}}{q}\left(\frac{1}{N_D} + \frac{1}{N_A}\right)} \tag{1.11-12}$$

For the particular silicon diode in question, following equation (1.11-8), the built-in junction potential V_{jo} is equal to

$$V_{jo} = V_T \ln \frac{N_A N_D}{n_i^2} = 26 \text{ mV} \ln \frac{(10^{17})(10^{16})}{(1.4 \times 10^{10})^2} = 760 \text{ mV} \tag{1.11-13}$$

and substituting this result into equation (1.11-12), we find that

$$w = \sqrt{\frac{(2)(10^{-12} \text{ F/cm})(0.76 \text{ V})}{1.6 \times 10^{-19} \text{ C}}\left(\frac{1}{10^{17}} + \frac{1}{10^{16}}\right) \text{cm}^{-3}} = 3.2 \times 10^{-5} \text{ cm} \tag{1.11-14}$$

1.11-4   Derivation of the Diode Equation

When an external voltage is applied to the diode, equations (1.11-7a) and (1.11-7b) are still valid if V_{jo} is replaced by $(V_{jo} - V)$, where V is the externally applied diode voltage. In using these equations, the charge concentrations at the edges of the depletion region will no longer be equal to their equilibrium values. However, because po and no are so large, the changes in these quantities may be neglected, and using this fact, $p_n(w_n)$ and $n_p(-w_p)$ may therefore be approximated as

$$p_n(w_n) = p_{po}e^{-(V_{jo}-V)/V_T} = (p_{po}e^{-V_{jo}/V_T})e^{V/V_T} = p_{no}e^{V/V_T} \tag{1.11-15a}$$

and

$$n_p(-w_p) = n_{po}e^{V/V_T} \tag{1.11-15b}$$

These expressions determine the heights of the minority carrier densities at the edges of the space-charge layer. Using these results and solving equations (1.11-4b) and (1.11-4c), the minority carrier concentrations on both sides of the depletion region are found to be

$$p_n(x) = p_{no} + [p_n(w_n) - p_{no}]e^{-(x-w_n)/L_p}$$
$$= p_{no} + p_{no}(e^{V/V_T} - 1)e^{-(x-w_n)/L_p} \tag{1.11-16a}$$

$$n_p(x) = n_{po} + [n_p(-w_p) - n_{po}]e^{(x+w_p)/L_n}$$
$$= n_{po} + n_{po}(e^{V/V_T} - 1)e^{(x+w_p)/L_n} \tag{1.11-16b}$$

The relationship between the current flow through a diode and the voltage applied across it may be developed by recognizing that the terminal current is equal to the current flow through any cross section of the diode. Because of this fact the diode current may be computed at the point where it is easiest to calculate, which in this case is at the edge of the depletion region. In general, the current flow at any point within the diode consists of drift and diffusion components for both the holes and the electrons. As a result, the total current flow at the $x = w_n$ boundary of the depletion region, for example, may be written as

$$i_D = A[J_{\text{diffusion, holes}}(w_n) + J_{\text{diffusion, electrons}}(w_n)$$
$$+ J_{\text{drift, holes}}(w_n) + J_{\text{drift, electrons}}(w_n)] \qquad (1.11\text{-}17)$$

If we neglect the generation and recombination of carriers in the space-charge layer, the current components entering on one side of the depletion region will be equal to those leaving from the other side. This permits us to rewrite the expression for the diode current solely in terms of the minority carrier currents as

$$i_D = A[J_{\text{diffusion, holes}}(w_n) + J_{\text{diffusion, electrons}}(-w_p)$$
$$+ J_{\text{drift, holes}}(w_n) + J_{\text{drift, electrons}}(-w_p)] \qquad (1.11\text{-}18)$$

The drift current terms for both the holes and the electrons (which is proportional to the product of the number of carriers and the electric field) may be neglected because both the number of carriers and the size of the $\mathscr{E}$ field are small at the edges of the depletion region. As a result, the expression for the current through the diode may be approximated as

$$i_D \simeq A[J_{\text{diffusion, holes}}(w_n) + J_{\text{diffusion, electrons}}(-w_p)] \qquad (1.11\text{-}19)$$

and substituting equations (1.11-16a) and (1.11-16b) into this equation, we have

$$i_D = A\left(-qD_p \frac{dp}{dx}\bigg|_{x=w_n} + qD_n \frac{dn}{dx}\bigg|_{x=-w_p}\right)$$

$$= qАn_i^2\left(\frac{D_p}{L_p N_D} + \frac{D_n}{L_n N_A}\right)(e^{v_D/V_T} - 1) = I_s(e^{v_D/V_T} - 1) \qquad (1.11\text{-}20)$$

where

$$I_s = qАn_i^2\left(\frac{D_p}{L_p N_D} + \frac{D_n}{L_n N_A}\right) \qquad (1.11\text{-}21)$$

The term I_s in these equations is called the reverse leakage current or the reverse saturation current. Notice that it is a strong function of temperature since n_i^2 increases rapidly with T [equation (1.10-1)]. To compute a typical value for I_s, consider a silicon diode in which $N_A = N_D = 10^{16}$ cm^{-3}, $A = 0.01$ cm^2, and $L_n = L_p = 0.002$ cm. Substituting these data, we find that the theoretical reverse leakage current for this diode is about 0.75 pA at room temperature. Actual small-signal silicon diodes typically have reverse leakage currents that are on the order of 1 pA to 1 nA.

Chapter 1 / Introduction to Electronics: The Diode

1.11-5 *PN* Junction under Reverse-Bias Conditions

If an external potential is applied as shown in Figure 1.11-4a, the diode is said to be reversed biased. When the battery voltage is initially connected, some of the holes from the *p*-type material are attracted by the battery to the left-hand side of the diode, where they meet up with, and recombine with, electrons from the battery. Similarly, electrons are drawn out of the *n*-type material on the right-hand side of the diode, where they flow into the battery. As a result, the depletion region gets wider as more charge is uncovered, and the voltage across the junction increases (by V_{BB} volts).

The large electric field across the junction under reverse-bias conditions has two major effects. First, it reduces the majority-carrier diffusion across the junction to nearly zero, and second, it sweeps out any minority carriers located near the edge of the depletion region. Both of these facts combine to reduce the minority-carrier concentrations at the edges of the space-charge layer to nearly zero, as predicted by equations (1.11-15a) and (1.11-15b). Of course, far from the depletion region, these concentration levels return to their equilibrium levels as illustrated in Figure 1.11-4c.

The current flow under reverse-bias conditions is obtained by direct substitution of $V_D = -V_{BB}$ into equation (1.11-20). For this situation the exponential term in this expression is nearly zero, so that

$$i_D = -I_s \tag{1.11-22}$$

This current flow is associated with the diffusion of the minority carriers toward the edges of the depletion region (due to the shape of these charge distributions) followed by the drift of these carriers across the depletion region in the $\mathscr{E}$ field. The net current flow contributed by the majority carriers under reverse-bias conditions is essentially zero.

Because of the definition of I_s in equation (1.11-21), it is apparent that the current which flows in the diode when it is reverse biased is independent of the applied reverse voltage. However, as already noted, it is a strong function of temperature. In fact, for silicon diodes, I_s approximately doubles for every $10°$ C increase in temperature.

1.11-6 *PN* Junction under Forward-Bias Conditions

In Figure 1.11-5a the battery V_{CC} is connected to forward bias the diode. The battery pulls electrons out of the *p*-type material on the left (creating new holes at the left-hand boundary of the diode) and forces additional electrons into the *n*-type semiconductor on the right. These new holes in the *p*-type material and electrons in the *n*-type material are pushed toward the junction by the battery and tend to cover up the exposed charge in the depletion region, thereby reducing both the junction potential and the width of the depletion region (Figure 1.11-5b). This lowering of the voltage across the junction (by V_{CC} volts) greatly increases the majority-carrier diffusion and raises the minority-carrier densities at the edges of the depletion region, as illustrated in Figure 1.11-5c.

The current flow in the diode under forward-bias conditions may be obtained by a direct examination of equation (1.11-20). When V_{CC} is more than a few V_T, the exponent of the exponential term in this equation is large and posi-

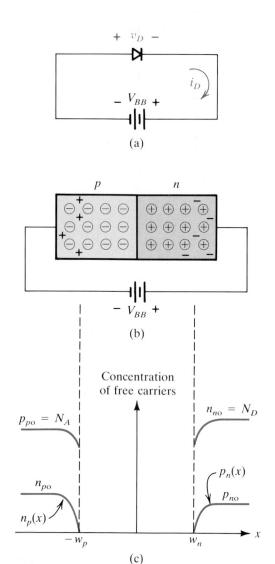

Figure 1.11-4
The pn *junction under reverse-bias conditions.*

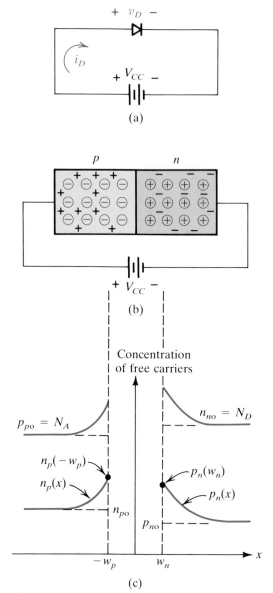

Figure 1.11-5
The pn *junction under forward-bias conditions.*

tive. As a result, this exponential term is much greater than unity, and the overall current flow in the diode may be approximated as

$$i_D \simeq I_s e^{v_D/V_T} = I_s e^{V_{CC}/V_T} \qquad (1.11\text{-}23)$$

1.11-7   Volt-Ampere Characteristics of an Actual *PN* Junction Diode

Before leaving this discussion on the theory of operation of the *pn* junction diode, it may be useful to compare the volt-ampere characteristics of the ideal semiconductor diode described by equation (1.11-20) with those of an actual semiconductor diode in order to point out several second-order effects that were not included in our derivation of the diode equation. The ideal diode volt-ampere curve associated with equation (1.11-20) is sketched in Figure 1.11-6 along with that of an actual *pn* junction diode to facilitate a direct comparison between them.

When the diode is forward biased and i_D is not too large, the behavior of the typical commercially available semiconductor diode is quite similar to that of the ideal *pn* junction device. However, as the forward current increases, the curve of the commercial diode begins to deviate considerably from that of the ideal diode's exponential characteristic. In fact, for large currents the commercial diode's volt-ampere characteristic is nearly a straight line. To model this diode properly and to understand the origin of this second-order effect, we need to observe that the voltage difference between the actual and the ideal semiconductor curves increases almost linearly with increasing current flow through the diode. As a result, an actual semiconductor diode can be modeled as an ideal *pn* junction diode connected in series with a resistor.

The physical origins of this resistance can best be understood by examining Figure 1.11-7. In a *pn* junction diode the depletion region is only a few micrometers thick, and the remainder of the semiconductor outside the depletion region electronically behaves like a resistance. It is this resistance that gives rise to the additional voltage drop across the diode as the current flow through it is increased. Obviously, to minimize this resistance, or diode bulk resistance as it is often called, we must make the *p* and *n* regions as thin as possible. This is not as easy as it sounds because it is very difficult to fabricate a device that is only 10^4 or even 10^5 Å thick. As a result, commercially available small-signal semiconductor diodes typically have bulk resistances that vary from about $0.1\,\Omega$ to $10\,\Omega$.

In addition to its forward-bias problems, commercially available diodes also suffer from reverse breakdown effects of the type illustrated in Figure

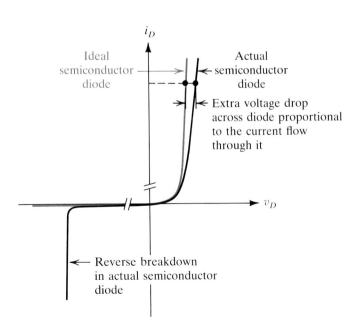

Figure 1.11-6

A comparison of the volt-ampere characteristics of an ideal and an actual semicon-ductor diode.

1.11 / The *PN* Junction Diode

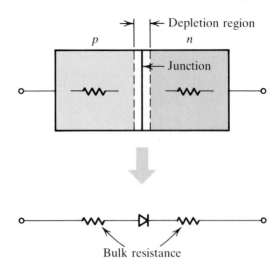

Figure 1.11-7

The bulk resistance of a semi-conductor diode.

Bulk resistance

1.11-6. At some critical voltage the diode, which has been preventing the flow of nearly all current in the reverse direction, suddenly collapses and begins to conduct. Depending on the voltage at which this occurs, the source of this breakdown can be traced to either zener or avalanche effects.

Zener breakdown generally occurs at less than 5 or 10 V and is found in diodes that have heavily doped *pn* junctions. For such diodes the depletion region will be very narrow because it is possible to uncover large numbers of bound charges in very short distances. As a result, very intense electric fields can be produced at the junction. In fact, even at low reverse voltages, these fields can become so large that they can actually tear electrons loose from atoms within the depletion region, forming free electron–hole pairs. These free electrons and holes will be swept across the junction and will contribute to the reverse leakage current of the diode.

In conventionally doped diodes the depletion region is much wider and field strengths are insufficient to cause zener breakdown. Instead, for this type of diode, when high reverse voltages are applied, the minority carriers passing through the depletion region are accelerated to very high velocities. These velocities can become so great that when they collide with neutral atoms in the depletion region, they can actually knock electrons out of these atoms, producing additional free carriers in the process. These free carriers can then be accelerated to create additional free carriers, and the number of such carriers can rapidly multiply, producing large increases in reverse diode current. This type of chain reaction is very familiar to most skiers and is called the avalanche effect.

Most of the time diode reverse breakdown is undesirable because it represents a failure of the diode to prevent the flow of current in the reverse direction. But as we saw in Section 1.9, there are situations where this breakdown can be used to advantage. By carefully controlling the doping levels it is possible to produce diodes that will break down at predetermined voltages. These diodes can be used as voltage references because of the nearly vertical character of this portion of the *V–I* characteristic. Regardless of whether the breakdown is due to zener or avalanche effects, these types of reference diodes are usually referred to as zener diodes.

1.11-8 Metal–Semiconductor Contacts

To connect a semiconductor device to other parts of the circuit, some type of metallic contact to the semiconductor is usually required, and unless special care is exercised, this connection point will often exhibit rectifying characteristics similar to those of a conventional *pn* junction diode. This type of metal–semiconductor contact is known as a Schottky diode or Schottky barrier and is named after William Schottky, an early investigator in the field of metal–semiconductor interfaces. The internal structure of a typical Schottky diode is shown in Figure 1.11-8a, and its circuit symbol is given in part (b) of the figure.

A principal difference between a Schottky diode and an ordinary *pn* junction diode is that the Schottky device operates much faster than the semiconductor diode. In a *pn* junction diode the turn-off time depends on the time required to change the minority-carrier charge distribution from that shown in Figure 1.11-5c to that presented in Figure 1.11-4c. A significant part of the time needed to accomplish this changeover depends on the lifetimes or the recombination times of the minority carriers in the *p* and *n* materials. Typical recombination times are on the order of several hundred nanoseconds to several microseconds.

While the volt-ampere characteristics of a Schottky diode are similar to those of an ordinary *pn* junction diode, its basic operating principle is somewhat different. In a Schottky diode, rather than being dependent on the minority carriers, nearly all of the current flow across the junction in Figure 1.11-8a is associated with the movement of the majority carriers (or the electrons for the diode shown) in both the metal and the semiconductor. Because the redis-

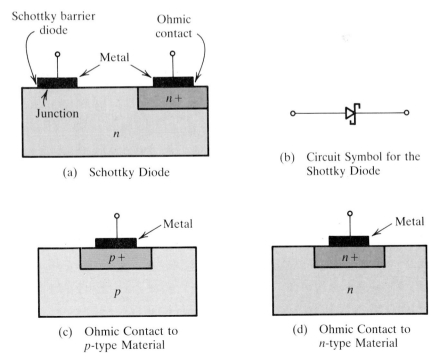

(a)   Schottky Diode

(b)   Circuit Symbol for the Shottky Diode

(c)   Ohmic Contact to *p*-type Material

(d)   Ohmic Contact to *n*-type Material

Figure 1.11-8
Metal semiconductor contacts.

tribution of the majority carriers in metals and semiconductors occurs with a time constant $\tau = \sigma/\epsilon$ that is typically only a few picoseconds, Schottky diodes have switching times that are virtually instantaneous compared to those of ordinary semiconductor diodes.

Besides their extremely rapid switching times, Schottky diodes also have larger reverse leakage currents and correspondingly smaller forward voltage drops than comparable silicon pn junction devices. For example, I_s in a small silicon Schottky diode is typically about 10 μA. As a result, for a forward current of 100 mA through this type of diode, the voltage drop across it will only be $V_T \ln (100/0.01)$, or 240 mV, compared with a nearly 700-mV drop across a similar silicon pn junction rectifier. This characteristic of the Schottky diode makes it ideally suited for applications in high-efficiency power supply designs, where its low-voltage drop translates into reduced power losses. Schottky diodes are also used together with bipolar junction transistors to make devices known as Schottky transistors. Their main application area is in high-speed digital logic circuits (Section 10.4).

Not all metal–semiconductor contacts have rectifying properties. When a heavily doped semiconductor (having N_A or N_D greater than about 10^{17} cm^{-3}) is placed in contact with a metal, the contact exhibits an ohmic behavior in which the voltage drop across it is small regardless of the direction of the current flow through it. Figure 1.11-8c and d illustrate the techniques for producing ohmic metallic contacts to p- and n-type semiconductors, respectively. Basically this is accomplished by interposing a region of highly doped semiconductor material (designated by $p+$ and $n+$ in the figure) in between the metal and the semiconductor to be contacted.

The reason this type of contact exhibits an ohmic rather than a rectifying behavior has to do with the width of the depletion region formed at the junction. Because the doping level in the semiconductor is so high, this depletion region is extremely narrow. As a result, charge carriers are able to "tunnel" through the potential barrier in the reverse-bias direction almost as easily as when the diode is forward biased.

Exercises

1.11-1 Find the average velocities of holes and electrons in a 1-cm-long piece of intrinsic silicon that has a 2-V battery connected across its ends. *Answer* $v_p = 10$ m/s, $v_n = 26$ m/s

1.11-2 The electron concentration in gold is 5.9×10^{22} cm^{-3} and its resistivity is 2.35×10^6 Ω-cm. Determine the mobility of electrons in gold. *Answer* 45 cm^2/V-s

1.11-3 Find the reverse leakage current in a silicon diode operating at room temperature if $A = 0.05$ cm^2, $N_A = N_D = 10^{16}$ cm^{-3}, and the recombination times for holes and electrons are both 1 μs. *Answer* 1.48 pA

1.11-4 A semiconductor diode has a reverse leakage current of 10 pA.
a. Calculate i_D when the voltage across the diode is 0.4 V.
b. Repeat part (a) if the polarity of the voltage is reversed.
Answer (a) 0.048 mA, (b) 10^{-11} A

1.3-1 A silicon diode operating at room temperature has a reverse leakage current of 1 nA. Find the applied forward voltage necessary to produce current flows of **(a)** 1 mA; **(b)** 1 μA; **(c)** 1 nA.

1.3-2 Repeat Problem 1.3-1 for the case of a germanium diode with a reverse leakage current of 100 nA.

1.3-3 When a silicon diode is forward biased at a particular operating current, how much must the applied voltage be increased in order to increase the current flow by a factor of 10?

1.3-4 In terms of I_s, the reverse leakage current, what are the approximate values of current flow corresponding to applied diode voltages of ±200 mV? You may consider that the diode is silicon.

1.3-5 For the circuit shown in Figure 1.4-1, V_{BB} is adjustable, $R = 1$ kΩ, the diode is silicon, and the circuit is operating at room temperature. If I_s for the diode is 2 nA, estimate the required value of V_{BB} needed to produce a current flow through R of **(a)** 0.1 mA; **(b)** 10 mA; **(c)** −1 nA.

1.3-6 Repeat Problem 1.3-5 for the case where an identical diode is connected in parallel with the first.

1.4-1 The diode in the circuit shown in Figure 1.4-1a has the volt-ampere characteristics given in Figure 1.4-2. Select R to produce current flows of **(a)** 10 mA; **(b)** 5 mA; **(c)** 0 mA.

1.4-2 For the circuit of Problem 1.4-1, find the current i_D and the voltage v_D if R is **(a)** 100 Ω; **(b)** infinite **(c)** 0 Ω.

1.4-3 **(a)** Sketch the V–I characteristics of the series connection of a 1-kΩ resistor and a 2-V battery.
(b) Using the volt-ampere characteristics developed in part (a) together with the load-line equation for the external portion of the circuit, solve graphically for V and I in this circuit.

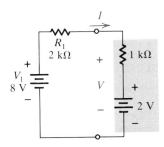

Figure P1.4-3

(c) Repeat the solution for V and I using conventional current analysis techniques.

1.4-4 The diode in Figure 1.4-1a is replaced by a nonlinear device whose current flow is $v/2$ kΩ for $i < 1$ mA and $(v - 1.75$ V$)/0.25$ kΩ for $i > 1$ mA.
(a) Sketch the V–I characteristics of this device.
(b) Find the device small-signal resistance at an operating voltage of 3 V.
(c) If V_{BB} is adjustable, find the value of V_{BB} needed to place the operating point at $V = 3$ V and $I = 5$ mA.
(d) Develop a model for this device that is valid in this region, and using this model, recompute V and I.

1.4-5 **(a)** The volt-ampere characteristic shown in Figure 1.4-2 is that of a single diode. If two such diodes are connected in series, sketch their composite volt-ampere characteristics on the same graph.
(b) If a 200-Ω resistor in series with a 2-V battery is connected to the composite diode circuit to forward bias them, find the power dissipated in the resistor.

1.4-6 Repeat Problem 1.4-5 for the case where the two diodes are connected in parallel.

1.4-7 For the circuit in Figure 1.4-1a, $V_{BB} = 50$ V and $R = 1$ kΩ.
(a) Find i_D using an ideal diode model.
(b) Find i_D considering that the diode is silicon (but still using $r_D = 0$). Is this result significantly different from that obtained in part (a)? If not, why not?
(c) As a first-order approximation using the result in part (b), compute the diode resistance r_D. Use this resistance together with the fact that the diode-bulk resistance is 10 Ω to recompute i_D. Is the new value of i_D found significantly different from that in part (a)? If not, why not?

1.4-8 Repeat Problem 1.4-7 for $R = 50$ Ω. Are V_o, r_D, and/or both important now? Why?

1.4-9 Repeat Problem 1.4-7 for $R = 50$ Ω and $V_B = 2$ V. Are V_o, r_D, and/or both important now? Why?

1.4-10 For the circuit shown in Figure 1.4-4a, $V_T = 2$ V, $R_T = 100$ Ω, and the diode is shunted with a 100-Ω resistor. Determine i_D and v_D given that $V_o = 0.5$ V and $r_D = 50$ Ω for the diode.

1.4-11 In Figure 1.4-4a, $V_T = 1.5$ V, $R_T = 70$ Ω, and the single diode is replaced by a two-diode series connection D_1 and D_2. The forward-bias equivalent circuits for these diodes are $V_{o1} = 0.7$ V and $r_{D1} = 10$ Ω for D_1, and $V_{o2} = 0.3$ V and $r_{D2} = 20$ Ω for D_2.

(a) Sketch the linearized $V-I$ characteristics for each of these diodes.

(b) Find the I_{D1}, V_{D1}, I_{D2}, V_{D2}, and the voltage across R_T using graphical techniques.

(c) Repeat part (b) using equivalent-circuit techniques.

1.4-12 For the circuit in Figure 1.4-6a, $V_{BB} = 5$ V, $R = 100$ kΩ, and $I_s = 10$ μA.

(a) Sketch the approximate diode $V-I$ characteristic in the reverse-bias region and find I_D and V_D graphically.

(b) Repeat part (a) using modeling techniques.

1.4-13 For the circuit in Figure 1.4-4a, $V_T = 50$ V, $R_T =$ kΩ, and the diode is replaced by two diodes (D_1 and D_2) connected in parallel. Given that D_1 is a germanium diode ($V_o = 0.3$ V, $r_D = 5$ Ω) and D_2 a silicon diode ($V_o = 0.6$ V, $r_D = 10$ Ω), determine the voltage across and the current through each of the diodes D_1 and D_2.

1.5-1 For the circuit in Figure 1.5-1a, v_{in} is replaced by a 30-V (p-p) sawtooth waveshape and the diode direction is reversed.

(a) Accurately sketch v_D if the diode is ideal.

(b) Repeat part (a) if the diode is assumed to be silicon with $r_D = 0$.

(c) Repeat part (a) if the diode is again assumed to be silicon but with $r_D = 100$ Ω.

1.5-2 Repeat Problem 1.5-1 if the diode reverse breakdown voltage is assumed to be 10 V.

1.5-3 For the circuit shown in Figure 1.5-1a, a 2-kΩ resistor R_2 is shunted across the diode and v_{in} is replaced by a 20-V (p-p) triangle wave. Sketch the current flow through R and the voltage across each resistor.

1.5-4 Repeat Problem 1.5-3 if v_{in} is replaced by a 40-V (p-p) sine wave and the direction of the diode reversed.

1.5-5 In Figure 1.5-1a, v_{in} is changed to a 30-V (p-p) sawtooth and a resistor $R_2 = 2$ kΩ is added in series with the diode. Sketch the voltage across the diode–R_2 combination.

1.5-6 Repeat Problem 1.5-5 if the diode direction is reversed and the signal source is replaced by a 20-V (p-p) triangle wave.

1.5-7 A diode having the $V-I$ characteristics given in Figure 1.4-2 is forward biased by connecting it in series with a 0.7-V battery and a voltage source $v_{in} = 0.2 \sin \omega t$ V.

(a) What is the maximum value of the diode current?

(b) What is the minimum value of the diode current?

(c) Is the waveshape of the current sinusoidal?

1.5-8 In Figure 1.5-1a the diode is replaced by a two-diode series combination D_1–D_2 and v_{in} is changed to a 6-V (p-p) triangle wave. $V_o = 0.75$ V and $r_D = 100$ Ω for D_1 and D_2.

(a) Sketch and label the current i that flows.

(b) Sketch and label v_o, the voltage across both diodes.

(c) Calculate the average power dissipated in D_1.

1.5-9 For the circuit shown in Figure 1.5-1a, $v_{in} = 2 \sin \omega t$ V, $R = 50$ Ω, and the diode is silicon with $r_D = 25$ Ω. Carefully sketch and label v_D.

1.5-10 In Figure 1.5-1a, $v_{in} = 3 \sin \omega t$ V and the diode is shunted with a 2-mA current source (arrow up). Carefully sketch and label i_D and v_D.

1.6-1 In Figure 1.6-1c, v_{in} is a 20-V (p-p) triangle wave, $V_{BB} = 10$ V, and $R = 1$ kΩ. Carefully sketch and label v_{o3} and i_D if the position of R and the diode are switched.

1.6-2 For the circuit shown in Figure 1.6-1b, v_{in} is a 30-V (p-p) sawtooth waveshape and $V_{BB} = 7$ V. Carefully sketch and label i_D, the voltage across the resistor, and v_D.

1.6-3 For the circuit given in Figure 1.6-1a, $V_{BB} = 3$ V, v_{in} is a 20-V (p-p) triangle wave, and the diode is labeled D_1. A second diode–battery branch (D_2, $V_{BB2} = 6$ V) is connected in parallel with the first so that the anode of D_2 and the positive terminal of V_{BB2} are toward the bottom.

(a) For what values of v_{in} is D_1 ON?

(b) For what values of v_{in} is D_2 ON?

(c) Where are D_1 and D_2 both OFF?

(d) Sketch the output voltage v_{o1}.

1.6-4 For the circuit in Figure 1.6-1a, $v_{in} = 20 \sin \omega t$ V, $R = 1$ kΩ, and $V_{BB} = 10$ V. Sketch v_{in} and directly under it sketch i_D.

1.6-5 The diode in the circuit shown in Figure 1.6-2b can be represented by an open circuit when it is off and a 0.7 V battery when it is on. Carefully sketch and label $v_{o2}(t)$ given that v_{in} is a 10 V (p-p) amplitude triangle wave and $V_{BB} = 2$ V.

1.6-6 For the circuit given in Figure 1.5-1a, the series connection of a diode and a 5-V battery is shunted across the resistor R. The components are oriented so that the diode anode and the battery negative terminal are on the right. Sketch the voltage across R if $v_{in} = 20 \sin \omega t$ V.

1.7-1 **(a)** In Figure 1.7-1a, v_{in} is a 30-V (p-p) sawtooth waveshape. Sketch v_C and v_o.

(b) Repeat part (a) if the diode direction is reversed.

1.7-2 Switch the position of the capacitor and the diode in Figure 1.7-2a and let $V_{BB} = 10$ V and $v_{in} = 20 \sin \omega t$ V. Sketch the steady-state voltages v_D and v_o.

1.7-3 For the circuit shown in Figure 1.7-2a, a resistor $R_1 = 2$ kΩ is connected in series with v_{in} and a second resistor $R_2 = 1$ kΩ is shunted across the v_{in}–R_1 combination. Sketch the steady-state voltage across the diode if $V_{BB} = 5$ V and $v_{in} = 20 \sin \omega t$ V.

1.7-4 For the circuit in Figure 1.7-2b, $V_{BB} = 5$ V and v_{in} is a 20-V (p-p) square wave. Sketch v_o in the steady state.

1.7-5 In Figure 1.7-2b, let $v_{in} = V_m \sin \omega t$ and $V_{BB} = V_m/2$. Starting at $t = 0$ with the capacitor initially uncharged, explain the operation of the clamper circuit and indicate what the voltage v_o looks like in the steady state.

1.8-1 A silicon diode has a quiescent current of 2 mA flowing in it.

(a) Determine its small-signal ac resistance.

(b) What is the overall small-signal resistance of this diode if its bulk resistance is equal to 10 Ω.

1.8-2 For the circuit shown in Figure 1.8-1a, R is changed to 1 kΩ and the diode is shunted with a 1-kΩ resistor. Determine the approximate ac voltage across the diode if $v_{in} = 100 \sin \omega t$ mV and $V_{BB} = 2$ V.

1.8-3 Repeat Problem 1.8-2 if the direction of the diode is reversed.

1.8-4 **(a)** In Problem 1.8-2, determine the dc voltage across the diode needed to make the ac part of the output voltage equal to 33.3 $\sin \omega t$ mV. The diode is silicon and has a reverse leakage current of 1 nA.

(b) What is the value of V_{BB} needed to produce the conditions described in part (a)?

1.8-5 For the diode circuit in Figure 1.4-1a, $V_{BB} = 3$ V and $R = 100\ \Omega$. An additional circuit consisting of the series connection of a voltage source $v_{in} = 2 \sin \omega t$ V, a capacitor C, and 300-Ω resistor is connected across the diode terminals. Using the diode characteristics in Figure 1.4-2, sketch i_D and v_D if C is an ac short.

1.8-6 Repeat Problem 1.8-5 by finding the Thévenin equivalent of the circuit within the shaded region and then solving the problem graphically.

1.9-1 The diode in Figure 1.9-1 has a $V_{ZZ} = 5$ V and an $I_{ZK} = 2$ mA.

(a) Find i_D by inspection.

(b) Sketch the V–I characteristics for this diode and confirm the answer in part (a) by using graphical techniques.

(c) Find i_D and v_D if R is changed to 10 kΩ.

1.9-2 The circuit given in Figure 1.9-5a represents a regulated power supply. The load draws 1 A. The zener is a 10-V 5-W device. The minimum zener current (I_{ZK}) is 200 mA.

(a) What is the value of the load resistance R_L?

(b) Select R so that the supply operates properly at the minimum value of V_{BB}, which is to be 13 V.

(c) Find the maximum value that V_{BB} can have without damaging the zener.

1.9-3 For the circuit given in Figure 1.6-1a, $R = 1$ kΩ, $V_{BB} = 10$ V and the diode is replaced by a 5-V zener. In addition, a resistor $R_2 = 1$ kΩ is shunted across the diode–battery branch.

(a) Find the v_{in} value for which the diode just turns ON.

(b) Find the v_{in} value for which the diode just zeners.

(c) Find i_R for the condition in part (b).

1.9-4 The circuit shown in Figure 1.9-5a is a zener-regulated power supply in which $V_{BB} = 20$ V, $R_L = 50\ \Omega$. The zener characteristics are $V_{ZZ} = 10$ V, $I_{ZK} = 20$ mA,

and $P_D = 10$ W. Find the range of values of R for which the circuit will perform satisfactorily.

1.9-5 In Figure 1.9-5a, V_{BB} and R, together with the zener diode, form a regulated power supply. Let $V_{BB} = 30$ V and $R = 10\ \Omega$ and determine the allowed range of values of R_L for which the circuit will function as 10-V regulated power supply. The zener characteristics are $V_{ZZ} = 10$ V, $P_D = 15$ W, and $I_{ZK} = 0.2$ A.

1.9-6 For the circuit in Figure 1.9-4a, $v_{in} = 20 \sin \omega t$ V and the diode is an ideal 10-V zener. Sketch v_D.

1.9-7 **(a)** Repeat Problem 1.9-6 for the case where v_{in} is a 20-V (p-p) square wave. Also sketch i_D.

(b) Estimate the average power delivered by the source and also the average power dissipated in R and in the zener.

(c) Is energy conserved in this circuit? Explain.

1.9-8 In Figure 1.9-4a, the diode is replaced by the series connection of two zeners, D_1 and D_2 (cathodes touching, with D_1 on top). If $V_{ZZ1} = 7$ V and $V_{ZZ2} = 4$ V, carefully sketch the voltage across the D_1–D_2 combination given that $v_{in} = 20 \sin \omega t$ V.

1.9-9 Reverse the direction of the zener diode in Figure 1.9-4a and shunt a 1-kΩ resistor across the diode. Given that v_{in} is a sawtooth varying between 30 and 40 V and that $V_{ZZ} = 10$ V, sketch i_Z.

1.9-10 In Figure 1.4-6a the diode has a reverse leakage current of 1 μA.

(a) Find i_D and v_D if R is increased to 1 MΩ.

(b) Repeat part (a) if the diode is shunted with an ideal 5-V zener diode D_2 (cathode up). Also find i_{D2}.

1.10-1 Find the number of free electrons per cubic centimeter in diamond at 75° C.

1.10-2 A silicon semiconductor is doped with $N_A = 10^{15}$ cm^{-3} acceptor atoms. Find n_{po} and p_{po} at 75° C.

1.10-3 Determine the constant A in equation (1.10-1) for silicon at room temperature.

1.10-4 Discuss the validity of each of the statements given below. In a p-type material:

(a) Most of the current flow is carried by electrons.

(b) There are a very large number of bound holes.

(c) The number of free electrons is smaller than that for an intrinsic material of the same size.

For Problems 1.10-5 through 1.10-9, in each case select the best answer and explain why that answer was chosen.

1.10-5 When the temperature of an intrinsic semiconductor is raised, its resistance:

(a) Increases.

(b) Decreases.

(c) Remains the same.

1.10-6 When the number of donor atoms is an n-type material is increased:

(a) The number of electrons increases, decreases, remains the same.

(b) The number of holes increases, decreases, remains the same.

(c) The resistivity increases, decreases, remains the same.

1.10-7 If a dc voltage is applied to an n-type semiconductor, the free holes and electrons:

(a) Move in opposite directions.

(b) On the average have no net motion.

(c) The question has no meaning because there are no free holes in an n-type semiconductor.

1.10-8 In an intrinsic semiconductor nearly all the current is carried by the holes.

(a) True.

(b) False.

(c) Not enough information to judge.

1.10-9 In an intrinsic semiconductor:

(a) The number of free holes is > the number of free electrons.

(b) The number of free holes is < the number of free electrons.

(c) The number of free holes is = to the number of free electrons.

(d) You cannot tell—it depends on the doping.

1.11-1 A bar of n-type silicon having a cross-sectional area of 0.1 cm² is doped with 10^{16} phosphorus atoms/cm³. When a potential of 1 V is applied to the bar, a current of 500 mA flows. Determine the length of the bar.

1.11-2 A sample of intrinsic silicon 0.1 cm long with a cross-sectional area of 0.05 cm² is to be used as a thermistor. If the mobility is assumed to vary as $T^{-1.5}$ with temperature (where T is in degrees Kelvin), sketch the resistance of the sample as a function of temperature from 0 to 100° C.

1.11-3 Considering that the mobility of electrons and holes in silicon varies inversely with the square of the temperature (in K), find the resistance of a bar of silicon at 350 K whose resistance at room temperature is 1 Ω.

1.11-4 Using the fact that I_s for a diode doubles for every 10° C increase, show that $\Delta V/\Delta T$ is about -2 mV/° C if the current flow i_D through the diode is held constant.

1.11-5 Starting with equation (1.11-6), prove equation (1.11-7a).

1.11-6 A silicon pn junction diode is made from p and n materials having conductivities of 200 $(\Omega\text{-cm})^{-1}$ and 25 $(\Omega\text{-cm})^{-1}$, respectively.

(a) Find the reverse leakage current of the diode if its cross-sectional area is 2×10^{-2} cm², and if $\tau_n = \tau_p = 0.5 \ \mu S$.

(b) What is the width of the depletion region in this diode under zero-bias conditions?

1.11-7 A silicon pn junction diode is doped such that $N_A = N_D = 10^{17}$ cm^{-3}.

(a) Find V_{jo} at room temperature.

(b) Find V_{jo} and w at 100° C.

1.11-8 Calculate w_n and w_p in Example 1.11-1 and show that their sum equals the total width found in the example.

1.11-9 When a silicon diode is carrying a current of 100 mA, the voltage drop across it is 1.0 V. In addition, its reverse leakage current is 1 nA.

(a) Find the effective small-signal resistance of the diode at a current flow of 100 mA.

(b) Estimate the diode's bulk resistance.

CHAPTER 2

Power Supplies

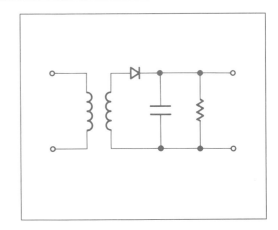

2.1 INTRODUCTION

In most of the world today, electrical power is transmitted as an ac voltage that varies sinusoidally in time. However, electronic circuits generally require dc rather than ac power to bias the transistors and ICs that make up these circuits. As a result, when designing electronic systems, some type of circuitry is needed to convert from ac to dc power.

The block diagram shown in Figure 2.1-1 is that of a typical electronic power supply. Depending on the quality of the dc required at the output, one or more of the blocks illustrated in this figure may be eliminated from specific designs.

In this chapter we attempt to develop an understanding of each of the blocks given in Figure 2.1-1. Let's start with the transformer. The ac voltage that the utility companies provide has a fixed amplitude and a fixed frequency. In the United States, the amplitude of this voltage is generally 120 V rms (or about 170 V peak), and the frequency is 60 Hz. For most power supply designs, the desired dc output voltage is such that we will need to change this 120-V input signal to some other value. This is usually accomplished with an electrical device known as a transformer, which not only provides us with a simple method for changing from one voltage level to another, but affords electrical ground isolation to minimize shock hazards.

The circuit shown in Figure 2.1-2 illustrates a simple two-winding power transformer. We will refer to the winding on the left, where the power enters, as the primary, and that on the right as the secondary. To the extent that the transformer is ideal, the voltage and current in the primary are uniquely related to the voltage and current in the secondary by the equations

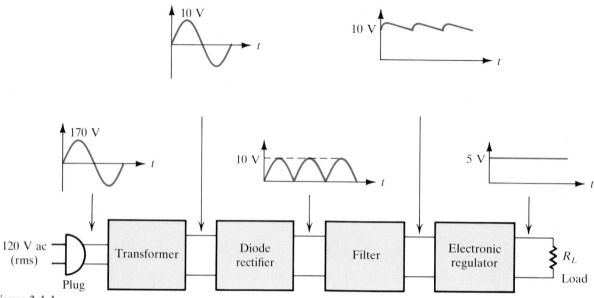

Figure 2.1-1
The block diagram of a typical electronic power supply.

$$v_2 = \frac{1}{N} v_1 \qquad (2.1\text{-}1a)$$

and

$$i_2 = N i_1 \qquad (2.1\text{-}1b)$$

where N is the turns ratio. The turns ratio of a transformer is a constant and is equal to the ratio of the number of turns on the primary to the number of turns on the secondary.

The phase relationship between the voltage on the primary and that on the secondary is indicated by the dots on the windings. If the dots line up with the same voltage polarity signs on each side of the transformer, the primary and secondary voltages will be in phase. Similarly, if the current enters the dot on the primary and leaves the dot on the secondary, both of these currents will also be in phase with the primary voltage. If because of the direction assumed for i_2, for example, the current i_2 flowed into the dot instead of out, its waveshape shape would be 180° out of phase from v_1 and would be inverted in Figure 2.1-2b.

If the transformer has more turns on the primary than on the secondary, the turns ratio, N, is greater than 1. This type of transformer is called a step-down transformer since the voltage on the secondary is less than that on the primary. Note, however, that although the voltage on the secondary is stepped down by N, the current in the secondary is stepped up by the same factor. Hence the term "step-down transformer" describes what happens to the voltage, not the current, in going from the primary to the secondary.

In specifying the type of transformer that is needed for a particular design, some measure of the maximum power that the transformer can deliver is needed. Usually, this is specified in terms of the volt-ampere ratings of the windings. The VA or volt-ampere rating of a winding is equal to the product of the rms voltage on that winding times the maximum-allowed rms current that the winding can safely carry. Rms quantities are used in these definitions because we are interested in the average power that the transformer can deliver. A complete review of rms concepts is contained in Section 4 of Appendix I.

The conversion of ac electrical power into dc power involves an alteration of the basic sinusoidal waveshape of the input signal in order to produce an output that has a nonzero dc average. Mathematically, the average value or dc average of periodic function $f(t)$ is defined as

$$f_{\text{avg}} = \frac{1}{T} \int_0^T f(t) \, dt = \frac{\text{net area under 1 cycle of } f(t)}{\text{time for 1 cycle}} \qquad (2.1\text{-}2)$$

where T is the period of $f(t)$.

Purely sinusoidal signals, since they have equal areas above and below the time axis, have zero average values. Linearly amplifying or transforming these ac signals will not alter this symmetry because the resulting outputs will still be sine waves. To obtain dc electrical power from an ac voltage source, we must therefore pass the voltage through some type of nonlinear circuit to alter the symmetry of the waveshape and produce a nonzero dc component.

Diodes are the nonlinear circuit elements that are most commonly used to alter the symmetry of the input ac voltage. One example of a diode rectifier circuit is illustrated in Figure 2.1-3. It is known as a half-wave rectifier since it allows only one-half of the applied input waveform to pass through to the output. To understand how this circuit operates, consider the diode to be ideal and notice that it conducts whenever v_s is greater than zero. During this time, since the diode can be replaced by a short circuit when it is ON, v_L equals v_s. On the negative half of the cycle, that is, when v_s is less than zero, the diode turns OFF and the output voltage, v_L, is zero. Mathematically, these results may be summarized as

$$v_L = \begin{cases} V_m \sin \omega t & 0 < t < T/2 \\ 0 & T/2 < t < T \end{cases} \qquad (2.1\text{-}3)$$

The pertinent waveshapes for this half-wave rectifier circuit are shown in Figure 2.1-3b. Notice that the peak inverse voltage (PIV) across the diode is equal to V_m, the maximum secondary voltage.

By examining the output voltage, v_L, sketched in Figure 2.1-3, it should be apparent that this waveform has a nonzero dc average. The actual value of this dc component may be computed by substituting equation (2.1-3) into equation (2.1-2) and carrying out the integration as follows:

$$V_{\text{dc}} = \frac{1}{T} \left[\int_0^{T/2} V_m \sin \omega t \, dt + \int_{T/2}^T 0 \, dt \right] = \frac{V_m}{\pi} \qquad (2.1\text{-}4)$$

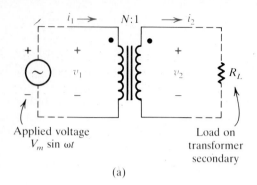

Applied voltage
$V_m \sin \omega t$

Load on
transformer
secondary

(a)

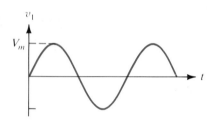

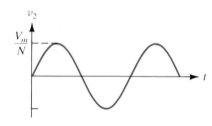

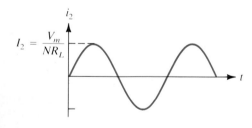

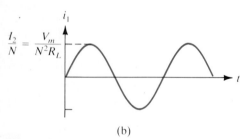

(b)

Figure 2.1-2
The ideal transformer.

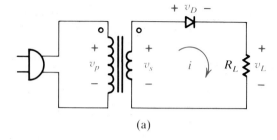

(a)

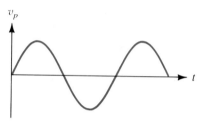

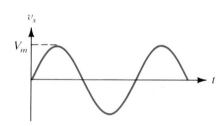

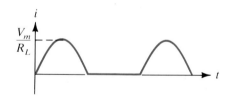

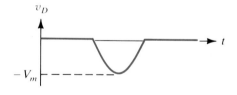

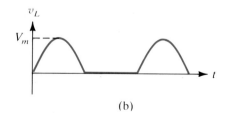

(b)

Figure 2.1-3
The half-wave rectifier.

EXAMPLE 2.1-1

The circuit shown in Figure 2.1-4a is an automobile battery charger. The battery V_{BB} will be charged if, on the average, more current flows into it than out of it. The average charging current (or dc current component of i_D) is indicated by the ammeter. In analyzing this circuit assume that the transformer has a 0.1-Ω secondary winding resistance, r_w, and that the diode can be represented by a resistor–battery series equivalent circuit with $r_D = 0.1\ \Omega$ and $V_o = 0.7$ V.

a. Determine the reading on the ammeter, assuming that the battery voltage does not change as it is charged.
b. Find the total charge delivered to the battery after 2 hours of operation.

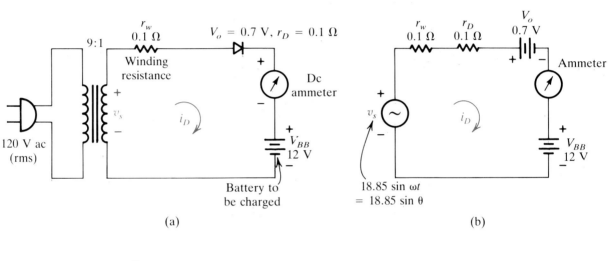

(a)

(b)

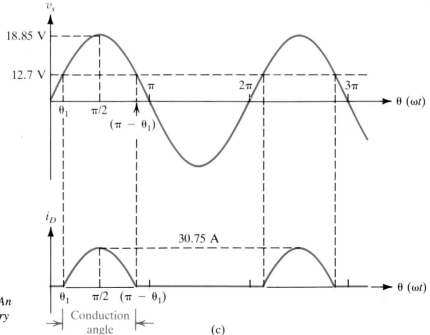

Figure 2.1-4
Example 2.1-1. An automobile battery charger system.

(c)

SOLUTION

The equivalent circuit for the battery charger during the time that the diode is ON is given in Figure 2.1-4b. Notice that the transformer has been replaced by an equivalent sinusoidal voltage source having a peak amplitude of $120 \times 1.414/9 = 18.85$ V. Because the input signal is periodic, we will need to examine only one representative cycle of this sine wave to understand the steady-state behavior of this circuit. In doing this, it will be convenient to re-place ωt in the sine wave by θ so that we can describe points on the sine wave in angular terms rather than at points in time. Using this notation, the first peak in the sine wave occurs at $\pi/2$ radians and the entire cycle is completed in 2π radians.

In Figure 2.1-4b the diode is assumed to be ON. For this model to be valid, the charging current, i_D, must be greater than zero. From the circuit, this current may be written as

$$i_D = \frac{v_s - V_o - V_{BB}}{r_w + r_D} \tag{2.1-5}$$

so that the diode is ON, or i_D is greater than zero, whenever

$$v_s > V_o + V_{BB} = 12.7 \text{ V} \tag{2.1-6}$$

Using equations (2.1-5) and (2.1-6), we may sketch the diode current as shown in Figure 2.1-4c. The peak value of the diode current is obtained by direct substitution into equation (2.1-5) when v_s takes on its maximum value of 18.85 V. The resulting current is $(18.85 - 12.7)/0.2$, or 30.75 A. Despite this rather large peak diode current, we will see that the average dc current charging the battery is rather small. This is because the diode is only ON over a small portion of the entire sine-wave cycle. To compute the point in the cycle where the diode first turns ON, we note that the diode is forward biased whenever the secondary voltage is greater than 12.7 V, so that it first conducts when

$$v_s = 18.85 \sin \theta = 12.7 \tag{2.1-7a}$$

or

$$\theta_1 = \sin^{-1} \frac{12.7}{18.85} = 42° = 0.736 \text{ rad} \tag{2.1-7b}$$

Using this result, we may now determine the average charging current by directly substituting equation (2.1-5) into equation (2.1-2) to obtain

$$I_{dc} = \frac{1}{T} \int_0^T i_D(t) \, dt = \frac{1}{2\pi} \int_0^{2\pi} i_D(\theta) \, d\theta = \frac{1}{2\pi} \int_{\theta_1}^{\pi-\theta_1} \frac{18.85 \sin \theta - 12.7}{0.2} \, d\theta$$

$$= \frac{1}{2\pi} (-94.35 \cos \theta - 63.5\theta) \Big|_{\theta_1}^{\pi-\theta_1} = 5.43 \text{ A} \tag{2.1-8}$$

You should note that this is the current which the dc ammeter would indicate.

The total charge delivered to the battery in 2 hours is obtained by recalling that current flow is the rate of flow of charge, and more specifically, that 1 A of current represents a charge flow of 1 C/s. The total charge delivered to the battery in 2 hours or in 7200 seconds would be 5.43 C/s × 7200 s, or 30,096 C.

The half-wave rectifier that we have examined is perhaps the simplest circuit available for converting ac to dc, but it is relatively inefficient since it produces very little dc voltage in comparison to the peak ac signal amplitude present at the input. To improve the efficiency of this conversion process, it is necessary to make the output waveform versus time look more like the straight line that would be obtained from a pure dc signal. One way to accomplish this is with a full-wave rectifier. This circuit consists of two half-wave rectifiers operating on alternating halves of the sine-wave cycle. Figure 2.1-5 illustrates one specific type of full-wave rectifier circuit that makes use of a center-tapped transformer. Because of the position of the dots on the transformer winding, v_{s1} and v_{s2} (the secondary voltages) have equal amplitudes but are 180° out of phase; that is, $v_{s1} = -v_{s2}$. Consequently, when diode D_1 is ON, D_2 is OFF, and vice versa, and the net current in the load is a full-wave rectified sine wave. The important waveshapes for this circuit are given in Figure 2.1-5d.

To understand the operation of this circuit, we will replace the transformer voltage v_{s1} by a voltage source v_A, and v_{s2} by $-v_A$, since it is 180° out of phase from v_{s1}. When v_A is positive, D_1 turns ON, and we may replace the circuit in Figure 2.1-5a by that given in Figure 2.1-5b. Applying Kirchhoff's voltage law to the outside loop in this equivalent circuit, we can see that $v_{D2} = -2v_A$, so that D_2 is OFF as long as v_{s1} is positive. With D_2 OFF, $i_2 = 0$, and

$$i_L = i_1 = \frac{v_{s1}}{R_L} \tag{2.1-9}$$

When v_{s1} goes below zero, v_{s2} becomes positive, and the picture is simply reversed, with D_2 turning ON and D_1 going OFF. The equivalent circuit for this case is illustrated in Figure 2.1-5c. For this circuit, $i_1 = 0$, and the load current is now given by

$$i_L = i_2 = \frac{v_{s2}}{R_L} \tag{2.1-10}$$

and is again positive.

The waveforms for i_1 and i_2 are sketched in Figure 2.1-5d. By applying Kirchhoff's current law to the node at the right in Figure 2.1-5a, we see that $i_L = (i_1 + i_2)$, and we may sketch the load current by graphically adding up the individual sketches for i_1 and i_2. As shown in the figure, the load current that results from this addition process is positive during both half-cycles of the input sine wave. This waveshape is called a full-wave rectified sine wave and is obtained by flipping the lower half of a sine waveform up about the time axis. The output voltage v_L has the same waveshape as i_L because $v_L = i_L \times R_L$.

One other important set of waveshapes still to be determined are those for the voltages v_{D1} and v_{D2} across the diodes. To get the voltage v_{D1} across D_1, for example, we can apply Kirchhoff's law to the upper loop in Figure 2.1-5a to obtain

$$v_{D1} = v_{s1} - v_L \tag{2.1-11}$$

so that v_{D1} can be sketched by subtracting the waveshape for v_L from that for v_{s1} (Figure 2.1-5d). As expected, when v_{s1} is positive and the diode D_1 is conducting, v_{D1} is zero, while when the diode is OFF the reverse voltage across it

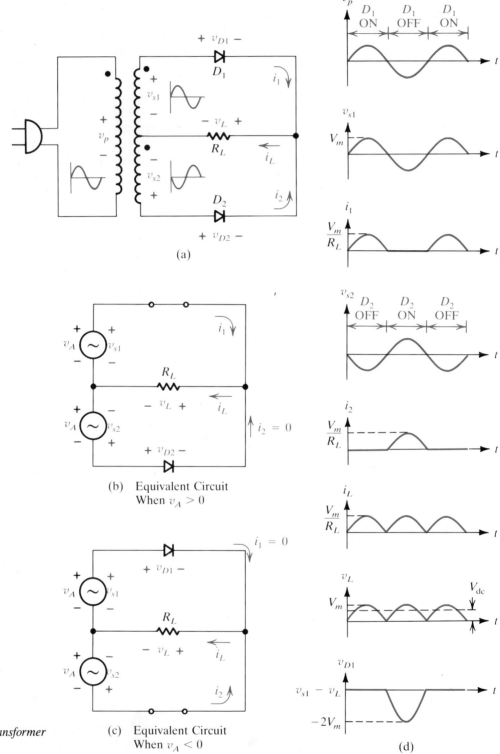

Figure 2.1-5
The center-tapped transformer full-wave rectifier.

(a)

(b) Equivalent Circuit
When $v_A > 0$

(c) Equivalent Circuit
When $v_A < 0$

(d)

is twice that of the input. Consequently, the peak inverse voltage or PIV across the diode is $2V_m$.

The dc average of the voltage across the load can be computed directly by substituting a mathematical expression for v_L into equation (2.1-2), but there is a much easier way to get the answer. By comparing the half-wave and full-wave waveshapes for v_L in Figures 2.1-3b and 2.1-5b, respectively, it is apparent that the area under the full-wave curve, and also the dc average of this waveform, is just twice that for the half-wave case. This makes the dc average voltage at the output of a full-wave rectifier simply equal to $2 \times V_m/\pi$.

The center-tapped transformer full-wave rectifier circuit just discussed enjoyed great popularity in the early days of radio and television when vacuum-tube diodes were used extensively. The advent of inexpensive semiconductor diodes has made the bridge-type full-wave rectifier circuit much more popular than that employing a center-tapped transformer. A typical bridge-style full-wave rectifier is illustrated in Figure 2.1-6a. It employs four rather than two diodes to produce the full-wave rectification of the input voltage, but is able to do this with a much smaller transformer than that needed for a comparable center-tapped design. This is because the bridge design uses only a single secondary winding, and also because the current flow in this winding is a sine wave rather than a half-wave rectified sine wave so that the required secondary VA rating can also be smaller. Since the transformer is a much more expensive component than the semiconductor diodes in the rectifier circuit, most modern power supplies use a full-wave bridge design.

The operation of this circuit may be understood by examining Figure 2.1-6 carefully. When v_s is positive, diodes D_1 and D_2 conduct. If we assume that the diodes are ideal, D_1 and D_2 can be replaced by short circuits, and the overall equivalent circuit reduced to that given in Figure 2.1-6b. From this picture it should be clear that i_1 and i_2 will be greater than zero, and that D_1 and D_2 will remain ON as long as the secondary voltage, v_s, is positive. By redrawing the circuit as illustrated in Figure 2.1-6c, we can also see that D_3 and D_4 will remain OFF as long as the secondary voltage is positive. In addition, when D_3 and D_4 are OFF, the voltage across them is simply given by $v_{D3} = v_{D4} = -v_s$, so that their peak inverse voltage is equal to the maximum value of the secondary voltage and not twice this maximum value as was the case for the center-tapped full-wave design examined previously.

The analysis of this circuit's operation when the secondary voltage becomes negative is essentially the same if the roles of the two diode pairs are interchanged. When v_s is negative, diodes D_3 and D_4 turn ON and D_1 and $D2$ go OFF, so that the resulting current flow through the load is still positive. As a result, i_L is a full-wave rectified sine wave. The pertinent current and voltage waveshapes for this circuit are given in Figure 2.1-6d.

If you are having difficulty understanding how the bridge diode circuit produces a full-wave rectified voltage at the output, perhaps the two additional equivalent circuits given in Figure 2.1-6 may help. When v_s is greater than zero, using Figure 2.1-6e, i_L is simply v_s/R_L and looks like a positive-going sine wave. When v_s is negative, the circuit in Figure 2.1-6f applies and $i_L = -v_s/R_L$; but since v_s is negative, and since a minus times a minus is a plus, i_L is again a positive-going sine wave.

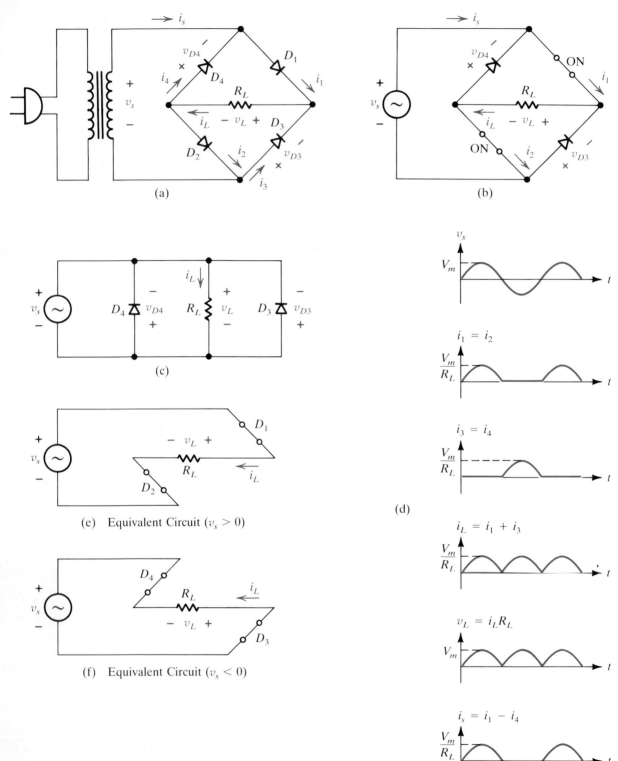

Figure 2.1-6
A full-wave bridge rectifier.

2.1-1 a. Find the average value of a half-wave rectified square wave with a peak amplitude of V_m volts.
b. Repeat part (a) for a full-wave rectified square wave.
Answer $V_m/2$, V_m

2.1-2 In Example 2.1-1 (Figure 2.1-4), what will the ammeter read if a 170-V peak amplitude square wave is used instead of a sine wave at the input to the transformer? **Answer** 15.375 A

2.2 CAPACITIVE-INPUT FILTERING FOR POWER SUPPLIES

The addition of filtering circuits to the rectifiers discussed in Section 2.1 makes the output voltage more like that from an actual dc source. One of the most popular filters, the so-called capacitive-input filter, simply consists of a capacitor connected across the output terminals of the power supply. Interestingly enough, this type of filter, used in conjunction with a voltage-regulator IC, is often all that is necessary to construct a high-performance dc power supply.

The circuit given in Figure 2.2-1a illustrates the design of a simple half-wave rectifier capacitive-input power supply. It has been redrawn in part (b) of the figure with R_L temporarily removed and with the transformer replaced by a voltage source equal to the transformer secondary voltage. If we assume that the capacitor is initially uncharged and that v_s is given by the expression

$$v_s = V_m \sin \omega t \qquad \text{for } t > 0 \qquad (2.2\text{-}1)$$

we can trace the operation of this circuit as follows.

For $t > 0$, as v_s begins to increase, the diode will turn ON and the capacitor will charge, with v_C following v_s. This will continue until v_s reaches its first peak at $t = T/4$. At this point, as the input voltage begins to drop, v_D, which equals $(v_s - v_C)$, will become negative and the diode will turn OFF. With the diode OFF, the capacitor is essentially connected to an open circuit so that it will remain charged to the peak value, V_m, of the input voltage. Because the input voltage, v_s, will never again exceed V_m, the diode can no longer turn ON and the capacitor will simply stay charged at V_m volts forever. Note that the output for this case is a constant dc voltage once the initial charging of the capacitor is completed. It should also be mentioned that the operation of this circuit is virtually identical to the clamping circuit discussed in Section 1.7 (see Figure 1.7-1) except that the output is taken across the capacitor instead of across the diode.

The current, i_D, in the circuit in Figure 2.2-1b can be computed by recalling that

$$i = C\frac{dv}{dt} \qquad (2.2\text{-}2)$$

for a capacitor, so that the diode current (which is the same as the capacitor current for this circuit) may be obtained simply by differentiating the capacitor voltage waveform in Figure 2.2-1c. For all $t > T/4$, the diode current is zero since v_C is constant. This fact correlates well with our understanding of this circuit since we had previously seen that the diode was OFF for all $t > T/4$.

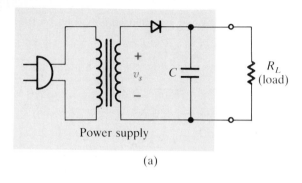

Power supply

(a)

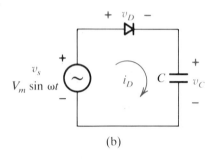

(b)

Figure 2.2-1
A capacitive-input power supply.

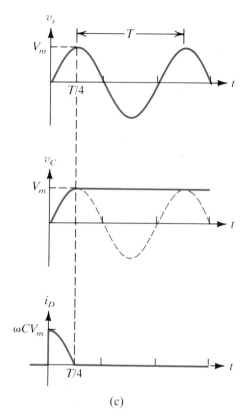

(c)

To compute the initial charging current through the diode for $t < T/4$, we observe that

$$v_C = V_m \sin \omega t \qquad (2.2\text{-}3)$$

in this interval because the capacitor and the input voltages are equal in this region. By substituting equation (2.2-3) into (2.2-2), we see that the diode current initially has the shape of a cosine wave with

$$i_D = C \frac{d}{dt} (V_m \sin \omega t) = \omega C V_m \cos \omega t \qquad (2.2\text{-}4)$$

and that it has a maximum amplitude given by $\omega C V_m$.

When the load resistor, R_L, is reconnected to the circuit as shown in Figure 2.2-2a, the situation changes. The load, being in parallel with the capacitor, tends to discharge the capacitor exponentially toward zero volts during the time that the diode is OFF. This takes four or five time constants, τ, with τ given by the product $R_L C$. For good power supply design, this time constant must be chosen so that it is large compared with the period, T, of the applied sine wave. If this is the case, the input voltage, v_s, will rise above v_C, and turn the diode back ON before the capacitor can discharge appreciably. This recharges the capacitor back to the peak input voltage V_m, and at this point the diode again turns OFF and the cycle repeats. The dashed curve in Figure 2.2-2b represents the waveshape of v_s and the solid curve that of v_o in the steady state.

In the time interval from T_1 to T_2, the diode conducts and charges the ca-

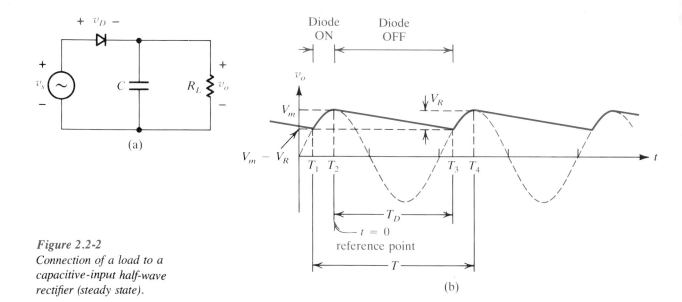

Figure 2.2-2

Connection of a load to a capacitive-input half-wave rectifier (steady state).

pacitor back up to the peak amplitude of the input voltage. As v_s decreases thereafter, v_D becomes negative and the diode shuts OFF. With the diode OFF, the capacitor now begins to discharge exponentially toward zero with a voltage

$$v_o(t) = V_m e^{-t/\tau} \tag{2.2-5}$$

using the point $t = T_2$ as the zero-time reference point for the start of the exponential decay. As the voltage on the capacitor falls, the input voltage, v_s, eventually comes back up, and once it reaches v_o the diode turns back ON and the recharging process is repeated. This occurs at $t = T_3$ in Figure 2.2-2b. To explicitly determine the discharge time, T_D in the figure, we note that at $t = T_D$,

$$v_s = v_o \tag{2.2-6a}$$

or

$$V_m \cos \omega t \,|_{t=T_D} = V_m e^{-t/\tau} \,|_{t=T_D} \tag{2.2-6b}$$

This is a transcendental equation and cannot be solved by algebraic means.

To solve equation (2.2-6b) in a simple way, we will need to make one or two reasonable approximations. In analyzing this circuit we are, for the most part, interested in obtaining an estimate of the magnitude of the ripple voltage, V_R, in order to see how good our capacitive filter is in producing a quality dc output (see Figure 2.2-2b). In a good power supply design, this ripple voltage is usually small. As a result, the diode conducts for only a limited portion of the entire cycle, and as a first-order approximation, it is reasonable to assume that the diode discharge time is nearly equal to T, the period of the sine wave. To the extent that this approximation is valid, the final capacitor voltage after discharging for T seconds is obtained by evaluating equation (2.2-5) at $t = T$, where we have

$$V_m - V_R = V_m e^{-T/\tau} \tag{2.2-7}$$

This equation, of course, may now be directly solved for the ripple voltage, V_R. Before actually carrying out this calculation, one other extremely useful approximation will be made. In the exponential term, τ is almost always much greater than the sine wave period T, and thus the exponent in equation (2.2-7) is usually very small.

Any exponential of the form $\exp(x)$ can be written in a power series as

$$e^x = 1 + x + \frac{x^2}{2!} + \frac{x^3}{3!} + \cdots \tag{2.2-8}$$

in which for small x (e.g., for $x < 0.1$) the higher-order terms in the series can be neglected and $\exp(x)$ approximated as

$$e^x \simeq 1 + x \tag{2.2-9}$$

You should verify this approximation for yourself by using your calculator to compute, for example, $\exp(0.01)$, $\exp(0.02)$, $\exp(-0.01)$, and so on.

Once you are convinced of the validity of this approximation, we can continue the ripple analysis by rewriting equation (2.2-7) as

$$V_m - V_R \simeq V_m\left(1 - \frac{T}{\tau}\right) \tag{2.2-10a}$$

or

$$V_R = V_m\frac{T}{\tau} = \frac{V_m T}{R_L C} \tag{2.2-10b}$$

This result is most useful for calculating the ripple produced at the output of half-wave capacitive-input filters connected to resistive loads; for full-wave rectifiers, T is replaced by $T/2$.

In electronic circuits the load often consists of transistors, ICs, zener diodes, and so on, and cannot therefore be represented by a simple resistor R_L. For such nonresistive loads, the result given in equation (2.2-10b) needs to be put into a slightly more general form. To carry out this modification, consider a load of the type illustrated in Figure 2.2-3a. Here the circuit connected to the power supply is specified only in terms of the load current, i_o, that it draws from the supply.

For this circuit, when the diode turns ON, the capacitor charges to the peak amplitude of the input V_m. Between peaks the diode is OFF, and the capacitor discharges through the load. The equivalent circuit during this time interval is given in Figure 2.2-3b. Using equation (2.2-2), we see that

$$i_o = -C\frac{dv_o}{dt} \simeq -C\frac{\Delta v_o}{\Delta t} \tag{2.2-11}$$

If the load current, i_o, is constant, dv_o/dt will also be a constant and v_o will drop linearly with time during the discharge (Figure 2.2-3c). Solving equation (2.2-11) for Δv_o, the ripple voltage for this case may be written as

$$V_R = |\Delta v_o| = \frac{i_o \Delta t}{C} = \frac{i_o T}{C} \tag{2.2-12}$$

If the constant-current discharge waveshape given for v_o in Figure 2.2-3d is compared with the resistive discharge curve in Figure 2.2-2c, we can see that the two results are virtually identical, at least to the extent that the exponential

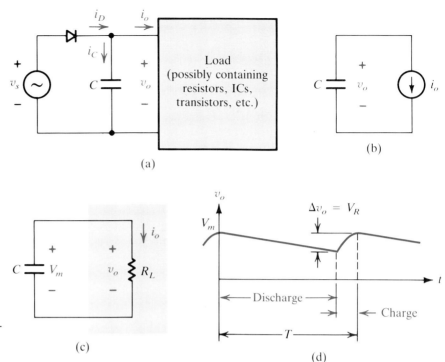

Figure 2.2-3
Ripple analysis for a capacitive-input filter connected to a nonresistive load.

(a)

(b)

(c)

(d)

decay in Figure 2.2-2c may be approximated by a straight line. This, of course, is valid [see equations (2.2-8) and (2.2-9)] as long as the exponent in equation 2.2-5 is small or as long as τ is much greater than T.

The ripple equation for the resistive loading case in equation (2.2-10b) can also be derived by making use of the constant-current discharge analysis just developed. If we consider that the capacitor equivalent circuit during resistive discharge looks like that given in Figure 2.2-3c, we may calculate the power supply ripple by making one simplifying approximation: namely, that the ripple voltage is so small that the voltage across the capacitor stays close to V_m. To the extent that this approximation is valid, the discharge current produced by the resistive load is

$$i_o = \frac{v_o}{R_L} \simeq \frac{V_m}{R_L} \tag{2.2-13}$$

By substituting this result into equation (2.2-12), we obtain the same result for the case of resistive loading as we did originally in equation (2.2-10b). In summary, the best way to carry out the ripple analysis in capacitive-input power supplies is to calculate the discharge current from the capacitor (if it is not explicitly given) and to substitute this result into the basic equation for the capacitor [equation (2.2-11)] in order to calculate the ripple voltage.

The diode current for the capacitive-input half-wave rectifier of Figure 2.2-3 is shown in Figure 2.2-4. The details of this waveshape may be understood by applying Kirchhoff's current law to the node at the right of the diode in Figure 2.2-3a and noting that

$$i_D = i_o + i_C \tag{2.2-14}$$

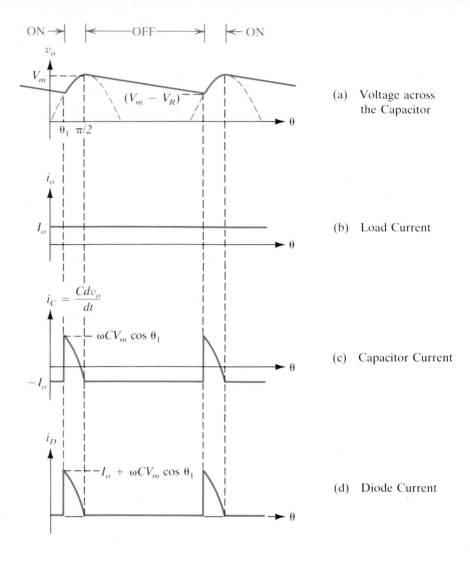

(a) Voltage across the Capacitor

(b) Load Current

(c) Capacitor Current

(d) Diode Current

Figure 2.2-4
Diode current in a capacitive-input power supply.

where i_o is nearly a constant and i_C is equal to $C(dv_C/dt)$. During the capacitive discharge time, the capacitor current is equal to $-i_o$ because the diode is OFF, while when the diode is ON, the capacitor current is given by equation (2.2-4) and has a peak amplitude of

$$\omega C V_m \cos \theta_1 \qquad (2.2\text{-}15)$$

where θ_1, the angle at which the diode turns on, is

$$\theta_1 = \sin^{-1} \frac{V_m - V_R}{V_m} \qquad (2.2\text{-}16)$$

[If you have forgotten where this expression comes from, review Example 2.1-1, especially equation (2.1-7).]

Following equation (2.2-14), the diode current is obtained by adding the curves for i_o and i_C. Notice that the peak diode current

$$i_o + \omega C V_m \cos \theta_1 \qquad (2.2\text{-}17)$$

can be much larger than the average diode current, so that considerable care must be taken to ensure that this quantity does not exceed the diode's maximum current-handling capability.

The average current through the diode can be computed by several different techniques, the most obvious being by a direct application of equation (2.1-2) to the diode current waveshape in Figure 2.2-4d. A second, easier approach makes use of the fact that in the steady state the average current flow into the capacitor from the diode must equal the average flow out into the resistor. Were this not the case, the average charge on the capacitor would grow continuously, and the voltage on the capacitor would go toward plus or minus infinity with time. Using this fact, we can see that the average current flow through the diode is approximately equal to V_m/R_L.

EXAMPLE 2.2-1

The circuit given in Figure 2.2-1a is to be designed to deliver 10 V dc to a 20-Ω resistive load.

a. Assuming the diode to be ideal, and the input line voltage to be 120 V ac (rms), determine the required transformer turns ratio.
b. Select a suitable capacitor value if the ripple at the output is to be less than 5% of the output voltage.

SOLUTION

Because the dc voltage at the load is to be 10 V, the peak secondary voltage will also need to be 10 V, assuming that the diode is ideal. The applied input voltage is 120 V rms or 170 V peak, so that the required transformer turns ratio is 170 V/10 V, or 17:1.

To determine the size of the capacitor needed, note that the current drawn by the load resistor will be 10 V/20 Ω or 500 mA. For this design, we are to select the capacitor so that with 500 mA flowing out of it during the discharge portion of the cycle, the voltage across the capacitor drops by only 5% of 10 V, or by about 0.5 V. Using equation (2.2-12), we see that the required value of C is

$$C = \frac{i_o \, \Delta t}{\Delta v_o} = \frac{(500 \text{ mA})(16.6 \text{ ms})}{0.5 \text{ V}} = 16.6 \times 10^{-3} \text{ F} \qquad (2.2\text{-}18)$$

or about 16,000 μF.

EXAMPLE 2.2-2

For the diode circuit in Example 2.2-1, calculate the required diode ratings. Specifically, you are to determine:

a. The average diode current.
b. The peak diode current.
c. The minimum allowed reverse breakdown voltage.

SOLUTION

To a first-order approximation, the average diode current is equal to the average current delivered to the load, which for this particular circuit is about 500 mA. The peak diode current is determined by direct substitution into equation (2.2-17), but first we need to calculate the diode cut-in angle from equation (2.2-16). Using V_R equal to 0.5 V and $V_m = 10$ V, we find that it is

1.253 rad. Substituting this result into equation (2.2-17), we obtain a peak diode current of

$$i_D = 0.5 \text{ A} + (377 \text{ rad/s})(16 \times 10^{-3}\text{F})(10 \text{ V})(0.312) = 19.3\text{A} \qquad (2.2\text{-}19)$$

To determine the peak inverse voltage across the diode, examine Figure 2.2-1a again and note that

$$v_D = v_s - v_C = 10 \sin \omega t - 10 \qquad (2.2\text{-}20)$$

when v_D is less than zero. From this equation the maximum reverse voltage across the diode is 20 V for this circuit.

Exercises

2.2-1 For the half-wave rectifier in Figure 2.2-3, estimate the ripple voltage across the capacitor if $C = 2000 \ \mu\text{F}$, $i_o = 1$ A, and v_s is a 500-Hz 20-V(p-p) amplitude triangle wave. ***Answer*** 1 V

2.2-2 Repeat Exercise 2.2-1 if v_s is changed to a square wave having the same frequency and amplitude. ***Answer*** 0.5 V

2.2-3 For the circuit illustrated in Figure 2.2-1a, $C = 2000 \ \mu\text{F}$, the transformer turns ratio is 6 : 1, the load current is 500 mA, and the input is a 170-V 60-Hz triangle wave. Find the approximate power dissipated in the diode, given that it is a silicon device. ***Answer*** 0.35 W

2.3 REGULATED POWER SUPPLY DESIGN USING ZENER DIODES

In a conventional power supply, as the current drawn by the load increases, the ripple across the capacitor will also increase, and the dc average voltage at the output terminals will fall from the V_m to about $(V_m - V_R/2)$. Some power supplies incorporate additional electronic circuitry called regulators to prevent the output voltage (or sometimes the output current) from varying when the load is changed.

The ability of a power supply to resist variation in its output voltage as a result of load changes is known as the load regulation of the supply. Expressed as a percent, it is defined as

$$\text{load regulation} = \frac{V_{\text{no load}} - V_{\text{full load}}}{V_{\text{no load}}} \times 100\% \qquad (2.3\text{-}1)$$

From this equation we can see that the regulation of an ideal power supply is 0%, since $V_{\text{no load}}$ and $V_{\text{full load}}$ will be the same. By way of contrast, the regulation of the simple capacitive-input power supply discussed in Example 2.2-1 is

$$\text{load regulation} = \frac{10 - 9.75}{10} \times 100 = 2.5\% \qquad (2.3\text{-}2)$$

The addition of electronic voltage regulation to a capacitive-input style power supply can improve its load regulation by at least one or two orders of magnitude.

In Section 2.2 we saw that it was possible to reduce the ripple voltage at the power supply output to any desired value by making the filter capacitor larger and larger [equation (2.2-12)]. However, there are several problems associated with this method of ripple reduction. First, as the value of the capacitance is increased, the physical volume and the cost of the capacitor may both become prohibitively large. Second, as the value of the capacitance is increased, the diode's peak current also increases sharply.

A better method for improving the regulation of a power supply is to keep the capacitor reasonably small and add an electronic voltage regulator to the power supply in order to improve the load regulation and reduce the output ripple. Simple zener diode regulators are discussed in this section, and the more elaborate transistor and integrated-circuit linear and switching regulator designs are presented in Chapter 9.

The regulation effect of the zener diode can best be understood by reexamining the circuit given in Figure 1.9-1a, paying special attention to the location of the load line on the diode characteristics in part (b) of the figure. As long as the load line intersects the diode curve in the zener region, the voltage appearing across the diode will be approximately constant. This requires that V_{BB} be large enough to keep the Q-point current above I_{ZK}. In addition, V_{BB} must not get too large, or else power dissipation in the diode may become excessive. The technique for determining the allowed range of V_{BB} values was illustrated in Example 1.9-2. For that specific circuit it was shown that any input voltage between 12.3 and 19 V would keep the output voltage constant at 5 V without damaging the zener.

To illustrate the technique for designing a zener-regulated power supply, the circuit discussed above has been redrawn in Figure 2.3-1 with V_{BB} replaced by a half-wave capacitive-input power supply. The turns ratio of 8.95 : 1 was purposely chosen so that the peak voltage on the capacitor would be 19 V (the same as the maximum allowed value for V_{BB} in Example 1.9-2). What we want to illustrate here is the technique for selecting the capacitor in this circuit.

The load connected to the capacitor will cause the voltage v_C to ripple approximately as shown in Figure 2.3-1b, and as in Example 1.9-2, the zener circuit will continue to work properly as long as v_C is above 12.3 V. The main question to be answered is: How big does the capacitor need to be to guarantee that the ripple does not cause v_C to go below 12.3 V?

If the capacitive discharge current were constant, the capacitor value could be calculated directly from equation (2.2-12). But i_1 is given by

$$i_1 = \frac{v_C - v_o}{R_1} = \frac{v_C - 5}{35} \tag{2.3-3}$$

as long as the diode stays in the zener region. This equation is sketched in Figure 2.3-1c, from which it is apparent that i_1 is not constant. Therefore, equation (2.2-12) is not really applicable to this problem because the discharge rate varies considerably over each cycle. However, to obtain a reasonable engineering solution to this problem, we will use the worst-case discharge rate of 400 mA for i_1. If we then substitute this value into equation (2.2-12), the resulting answer for C will be larger than is really needed. But if this capacitor size is used, it will guarantee that the resulting ripple voltage will be less than

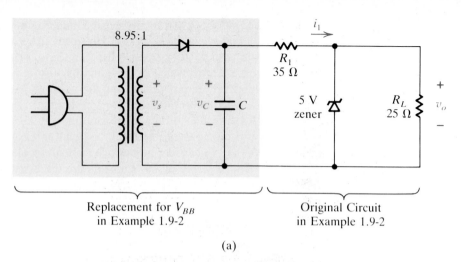

Figure 2.3-1
A zener diode regulated power supply.

(b)

(c)

$(19 - 12.3)$, or 6.7 V. Following this procedure, the required value of C is

$$C = \frac{i \Delta t}{\Delta v} = \frac{(0.4 \text{ A})(16.6 \times 10^{-3} \text{ s})}{6.7 \text{ V}} \approx 1000 \ \mu\text{F} \qquad (2.3\text{-}4)$$

EXAMPLE 2.3-1

Design a 5-V 1-A zener regulated power supply using the circuit given in Figure 2.3-2 by selecting reasonable values for R_1 and C. The zener diode can dissipate up to 10 W and requires a minimum zener current of 100 mA.

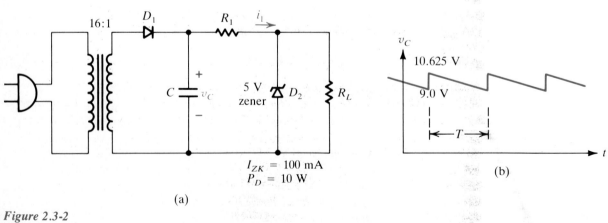

$I_{ZK} = 100 \text{ mA}$
$P_D = 10 \text{ W}$

(a)

(b)

Figure 2.3-2
Example 2.3-1.

SOLUTION

By specifying that the power supply maximum output current is 1 A, we are really saying that the circuit must work with values or R_L having resistances from 5 V/1 A or 5 Ω all the way up to $R_L = \infty$ (an open circuit). This variation in the allowed values of R_L will be important in determining the required value for R_1.

Because the transformer has a 16 : 1 turns ratio, the peak voltage on the capacitor will be 170/16, or 10.625 V. As with previous circuits that we have examined, the actual capacitor voltage will ripple below this point (Figure 2.3-2b). In selecting R_1, consider that when it gets too small, excessive current will flow into the zener, while when it is too large, there will not be enough current available to keep the zener operating point out of the knee in the characteristics. The worst-case load values corresponding to each of these conditions will be R_L equal to infinity and R_L equal to 5 Ω, respectively.

Under no-load conditions, all the current through R_1 flows into D_2, and if the average value of this current is too large, excessive power will be dissipated in the zener. For the specific diode in this problem, the average zener current must be kept below 10 W/5 V, or 2 A. To relate this to the minimum allowed value for R_1, we must realize that i_1 is given by equation (2.3-3) and its waveshape will be similar to that given in Figure 2.3-1c. A precise solution of this equation will be difficult, but here again we can perform a worst-case solution of the problem by using the maximum value of the current i_1 in place of its average value. The solution for the minimum allowed value for R_1 using this approximation is (10.625 − 5.0)/2, or 2.8 Ω.

The maximum allowed value for R_1 will depend on how far the voltage on the capacitor drops. As this value has not yet been specified, we will arbitrarily assume it to be 9 V (we will see later that this value is actually determined by how much ripple can be tolerated in the output). With the minimum capacitor voltage set at 9 V and with the maximum load of 5 Ω connected, the zener is on the verge of going OFF and has a current of I_{ZK} or 100 mA in it. The load current is 1 A, so that i_1 must be 1.1 A. Using equation (2.3-3) again, we see that R_1 must be (9 − 5)/1.1, or 3.6 Ω. Summarizing these results, for this circuit to work properly, the allowed range of values for R_1 is

$$2.8\ \Omega < R_1 < 3.6\ \Omega \qquad (2.3\text{-}5)$$

We will select a value of R_1 equal to 3 Ω for this design.

Once R_1 has been chosen, the required capacitor size may be determined by using the maximum value for the current i_1 in equation (2.2-12) with $i_o = i_1 = (10.625 - 5)/3 = 1.875$ A and $V_R = 10.625 - 9 = 1.625$ V. The resulting value for C is (1.875 A) × (16.6 ms)/1.625 V, or about 20,000 μF.

The previous zener diode power supply analysis assumed that the zener portion of the diode characteristic was vertical (see Figure 1.9-1), so that variation in the diode Q point produced no changes in the output voltage as long as the diode remained in the zener region. Actual zener diodes do not have perfectly vertical zener characteristics, and as a result, variations in the Q-point current do cause changes in the zener voltage. Example 2.3-2 illustrates how to determine the resulting output ripple by graphical means, and Example 2.3-

3 suggests how the diode model introduced in Section 1.9 (see especially Figure 1.9-2b) can be used to carry out this same analysis.

EXAMPLE 2.3-2

The zener diode in the regulated power supply shown in Figure 2.3-3 has the volt-ampere characteristics given in part (d) of the figure. From the graph we can see that this diode is nominally a 5-V device, and further, because the line in the zener region is severely tilted, that it has a rather large zener resistance, r_Z. Assuming that the capacitor and transformer have been selected so that the voltage across the capacitor has the waveshape given in part (e), sketch the corresponding zener current, i_Z, and zener voltage, v_Z.

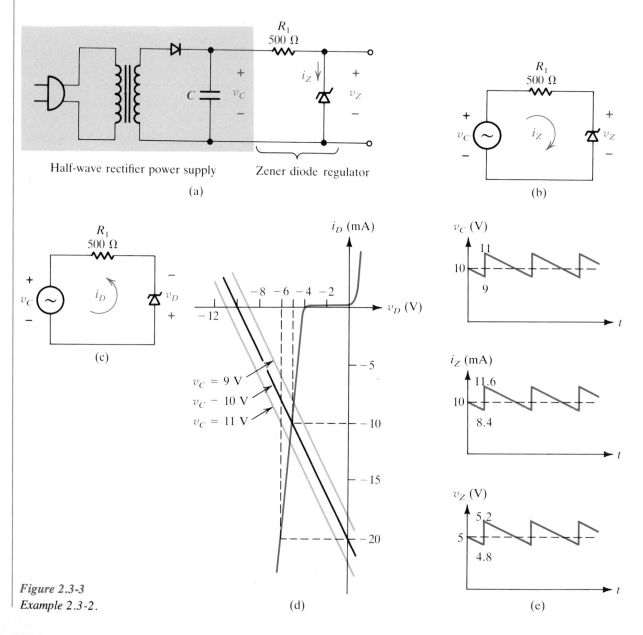

Figure 2.3-3
Example 2.3-2.

The circuit given in Figure 2.3-3a may be redrawn as shown in part (b), with the half-wave rectifier power supply replaced by an equivalent voltage source v_C. The load line equation for this circuit may be written as

$$v_C = -i_D R_1 - v_D \qquad (2.3\text{-}6a)$$

or by making the substitution $i_Z = i_D$ and $v_Z = -v_D$ as

$$v_C = i_Z R_1 + v_Z \qquad (2.3\text{-}6b)$$

(Figure 2.3-3c). The load line given in equation (2.3-6a) is sketched in Figure 2.3-3d. As v_C varies, the intercepts change as shown, but the slope, $-1/R_1$, remains constant. The corresponding values of i_D and v_D (and hence i_Z and v_Z) can be read directly from the Q-point variation with changes in v_C. The required sketches for i_Z and v_Z are given in Figure 2.3-3e. By examining the sketches for the v_C and v_Z, we can see that the input and output ripple voltages are 2 V and 0.4 V, respectively, so that a fivefold reduction in the ripple has been achieved with the use of the zener regulator in this circuit. You should also note that the tilt of the characteristic in the zener region was exaggerated for this example, so that the ripple reduction factors with real zener diodes would actually be much greater than this.

Power supply circuits of the type described in Figure 2.3-3a can also be analyzed by means of the small-signal models introduced for the zener diode in Section 1.9. The basic approach is similar to the small-signal analysis technique that we discussed in Section 1.8 for conventional diodes. To carry out this type of analysis, we begin by replacing the input signal, in this case v_C, by the series connection of two sources, one representing the dc average of the input voltage and the second the ac part of the signal. In addition, the diode is replaced by its equivalent circuit, and then superposition is applied to calculate the ac and dc parts of the solution. The total solution is obtained by adding the individual solutions together.

EXAMPLE 2.3-3

Repeat Example 2.3-2 using the appropriate model for the zener diode.

SOLUTION

The dc average of the input signal is $(11 + 9)/2$, or 10 V. The remaining part of v_C constitutes the ac part of the input signal and is a 2-V(p-p)-amplitude sawtooth waveshape.

The model for the zener is obtained by drawing a tangent line to the zener curve in the neighborhood of the Q point, which for this case is near the point $I_Z = 10$ mA and $V_Z = 5$ V. The zener battery voltage, V_{ZZ}, to be used with the model is read directly from the v_D intercept as 4 V, and the zener resistance, r_Z, is determined from the slope of the tangent line as 100 Ω.

For this particular problem the total solutions for i_Z and v_Z could probably be written down directly by inspection of Figure 2.3-4a; but to reemphasize the basic approach to the solution of these types of problems, we will again use superposition. Setting the ac sources equal to zero in Figure 2.3-4a, we obtain

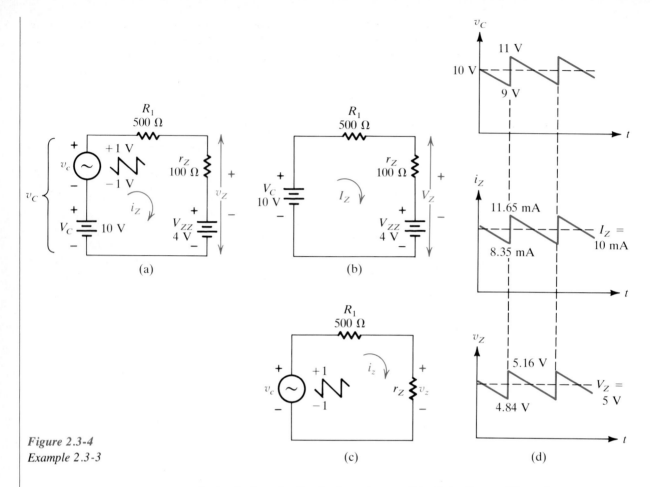

Figure 2.3-4
Example 2.3-3

the equivalent dc circuit given in part (b) of the figure. The quiescent current for this circuit is

$$I_Z = \frac{V_C - V_{ZZ}}{R_1 + r_z} = \frac{6 \text{ V}}{600 \text{ }\Omega} = 10 \text{ mA} \tag{2.3-7a}$$

and the corresponding value of V_Z is given by

$$V_Z = V_{ZZ} + I_Z r_z = 4 \text{ V} + (10 \text{ mA})(100 \text{ }\Omega) = 5 \text{ V} \tag{2.3-7b}$$

The ac portion of the solution is obtained with the aid of Figure 2.3-4c as

$$i_z = \frac{v_c}{R_1 + R_Z} \tag{2.3-8a}$$

and

$$v_z = v_c \frac{r_z}{R_1 + r_z} \tag{2.3-8b}$$

Substituting the fact that v_c is 2 V(p-p), we find that the output ac current and output ac voltage are 3.3 mA(p-p) and 0.33 V(p-p), respectively. The total solutions for i_z and v_z are obtained by adding the dc and ac solutions given in equations (2.3-7) and (2.3-8). The resulting waveshapes are sketched in Figure 2.3-4d; compare them with the graphical results given in Figure 2.3-3e.

The solution for v_z given in equation (2.3-8b) helps us to see that the output ripple voltage is produced by a voltage division between R_1 and r_z, and thus the output ripple is minimized by making $R_1 \gg r_z$.

Exercises

2.3-1 Consider Figure 2.3-1.
a. Determine the average power dissipated in the zener diode assuming that the current i_1 flowing through the resistor R_1 has the form given in Figure 2.3-1c.
b. Repeat part (a) if R_L is removed.
Answer (a) 0.51 W, (b) 1.52 W

2.3-2 For the zener-regulated power supply in Figure 2.3-2, $R_1 = 3\ \Omega$, $C = 20{,}000\ \mu F$, and $R_L = 10\ \Omega$. Estimate the output ripple amplitude if D_2 is a 7-V zener with an $r_z = 0.1\ \Omega$. *Answer* 32 mV(p-p)

PROBLEMS

Unless otherwise noted, all diodes and other circuit components may be assumed to be ideal. If a diode is specified as being silicon, $V_o = 0.7$ V and $r_D = 0$ for this diode.

2.1-1 Derive expressions for:
(a) The rms value of the triangular current i_1 having an amplitude I_m.
(b) The average value of i_1.
(c) The rms value of the half-wave rectified triangular current i_2 having an amplitude I_m.
(d) The average value of i_2.

2.1-2 For the circuit shown in Figure 2.1-5, $R_L = 10\ \Omega$, the transformer turns ratio is $2 : 1 : 1$, and v_p is a 120-V (rms) sine wave.
(a) Sketch i_1, i_2, and i_p (the current flowing into the dot on the primary).
(b) Determine the required volt-ampere rating of the primary and secondary windings.
(c) Find the average power delivered to the load.

2.1-3 For the waveshape shown in Figure P2.1-3:
(a) Calculate the dc average.
(b) Calculate the rms value.
(c) Assume that this is the voltage connected across a 1-kΩ resistor and calculate the power dissipated in it.

2.1-4 Sketch and carefully label the current in the transformer primary for each circuit shown in Figure P2.1-4. All inputs are 120 V ac (rms).

Figure P2.1-3

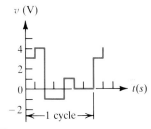

(a)

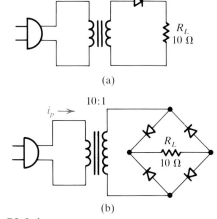

(b)

Figure P2.1-4

2.1-5 For the circuit shown in Figure P2.1-3a, v_s is a 10-V (p-p) square wave, and the diode is silicon with $r_D = 20\ \Omega$.
(a) Find the dc load voltage under no-load (R_L infinite) and full-load ($R_L = 1\ k\Omega$) conditions.
(b) Calculate the load regulation of this circuit [equation (2.3-1)].

2.1-6 A full-wave center-tapped power supply is to be built using rectifiers having peak current ratings of 2 A and PIVs of 50 V. The input is 120 V ac (rms).
(a) Determine the smallest turns ratio that can be used in this circuit.
(b) What is the smallest value of R_L permitted using the turns ratio selected in part (a)?
(c) Find the average dc power delivered to the load for the N and R_L selected in parts (a) and (b).

2.1-7 The circuit shown in Figure 2.1-4a is to be used to charge a 12-V automobile battery. The transformer is changed and how has a 10 : 1 turns ratio and a secondary winding resistance of 1 Ω; the diode is silicon and r_D is zero.
(a) Find the dc current delivered to the 12-V battery.
(b) Find the dc power delivered to the 12-V battery.
(c) Sketch v_D and i_D.
(d) A current-limiting resistor R_1 is connected in series with the diode. How big should it be to limit the peak charging current to 3 A?

2.1-8 A full-wave battery charger is constructed by removing the load R_L in Figure 2.1-6 and replacing it by an ammeter in series with a 10-V battery that is to be charged (plus terminal on the right). The diodes are silicon with $r_D = 1\ \Omega$, and the transformer turns ratio and secondary winding resistance are 10 : 1 and 1 Ω, respectively. The open-circuit voltage at v_s is a 30-V (p-p) triangle wave.
(a) Sketch and label the charging current i_L.
(b) Determine the steady-state dc ammeter reading.

2.2-1 In Figure 2.1-5, v_p is a 120-V (rms) 60-Hz sine wave and the transformer turns ratio is 10 : 1 : 1. R_L is shunted by a 5000-μF capacitor, and the average current through the load R_L is 2 A. Sketch and carefully label the voltage across the capacitor.

2.2-2 Design a full-wave (two-diode) capacitively filtered power supply to deliver 500 mA to a 100-Ω load. Consider that the diodes are silicon with $r_D = 1\ \Omega$.
(a) What must C be to keep the ripple below 5% of the output voltage?
(b) What is the required transformer secondary rms voltage?
(c) What are the maximum steady-state forward currents and reverse voltages that the diodes must withstand?

2.2-3 In Figure 2.2-2a, v_s is a 100-Hz 0- to 20-V sawtooth voltage source. In addition, R_L is 100 Ω and $C = 1000\ \mu$F. Sketch v_s, v_o, and i_D assuming the diode to be ideal.

2.2-4 For the circuit shown in Figure P2.2-4, v_{in} is a 100-Hz 20-V(p-p) triangle wave. Carefully sketch and label v_{in}, v_o, v_s, and i_{D1}. Consider the diodes to be silicon.

2.2-5 For the circuit shown in Figure P2.2-4:
(a) Find the rms value of the current in the secondary of the transformer.
(b) Determine the required VA ratings of the transformer primary and secondary.

2.2-6 Design a power supply with a capacitive input filter and full-wave rectifier bridge to supply 24 V dc at 400 mA from a 120-V ac (rms) 60-Hz input. The output ripple is to be less than 1% [0.24 V(p-p)].

2.2-7 For the circuit given in Figure 2.2-2a, $R_L = 500\ \Omega$, $C = 10\ \mu$F, and v_s is a 20-V (p-p) 1 kHz square wave. Carefully sketch and label the steady-state voltages v_s, v_D, and v_o.

2.2-8 In the power supply shown in Figure P2.2-4, consider that the value of C is unknown and that v_{in} is a 120-V (rms) 60-Hz sine wave.
(a) Select C to keep the ripple voltage at less than 10% of the peak dc output voltage.
(b) What is the approximate dc current in the load?
(c) Using the results of part (a), sketch and carefully label the transformer primary voltage, v_o, and i_{D1}.

2.2-9 For the circuit shown in Figure P2.2-4, v_s is a 17-V 60-Hz sine wave. Find the rms value of the diode current i_{D1} (see the rms table in Figure I.4-1 of Appendix I).

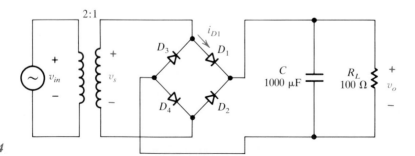

Figure P2.2-4

2.2-10 Estimate the dc average and the percent ripple in the output voltage for the circuit shown in Figure P2.2-10. The diode is silicon.

2.2-11 For the circuit shown in Figure P2.2-10, replace R_1 by a short circuit and recompute the dc average and the percent ripple in v_o.

2.2-12 For the circuit shown in Figure P2.2-10, replace R_1 by a 2-H inductor and recompute the dc average and the percent ripple in v_o.

2.2-13 The circuit shown in Figure P2.2-13 is a half-wave-style voltage doubler that consists of a clamper circuit followed by a peak detector. Explain how this circuit works, and in doing so sketch the steady-state voltages v_{in}, v_{C1}, v_{C2}, v_{D1}, v_{D2}, and v_o. You may assume that R_L is very large.

2.3-1 Design a 10-V 1-A regulated power supply having the general form shown in Figure 2.3-2a. Consider that the transformer turns ratio is 7 : 1, and that D_2 is a 10-V 15-W zener with $I_{ZK} = 0$.
(a) Select a reasonable value for R_1. What is the maximum power dissipation that it must handle?
(b) Choose C to make the ripple in v_C less than 0.5 V.
(c) What is the minimum allowed value of R_L before the supply drops out of regulation?

2.3-2 The circuit shown in Figure 2.3-2a consists of a half-wave peak rectifier driving a zener regulator in which $R_L = 50 \ \Omega$.

(a) Choose the value of R_1 so as to make the power delivered to the zener diode equal to 2W. (For this computation, the voltage across the capacitor can be assumed to be constant in time.)
(b) Using the value of R_1 selected in part (a), compute the smallest allowed value for C that will keep the zener current above 100 mA over the entire sine-wave input cycle.
(c) What is the value of the peak inverse voltage across D_1?

2.3-3 You are to design an 8-V regulated power supply having the general form shown in Figure P2.3-3. You may assume that C is large enough to neglect ripple.
(a) What is v_C?
(b) Select the zener voltage required.
(c) Select the zener power-handling capability (dissipation) so that the zener will be OK under both full-load and no-load conditions.
(d) If $I_{ZK} = 0.5$ A, for what range of load resistors (R_L) will the supply remain regulated?

2.3-4 The full-wave bridge portion of the power supply in Figure P2.3-3 is replaced by a center-tapped transformer-style full-wave rectifier circuit. The transformer turns ratio is 9 : 1 : 1, C is very large, $V_{ZZ} = 10$ V, and $R_L = 5 \ \Omega$.
(a) Find the power delivered to the load.
(b) Determine v_C, I_{R1}, and I_Z neglecting ripple.
(c) Find the maximum power dissipation in the zener for values of R_L between 5 Ω and infinity.

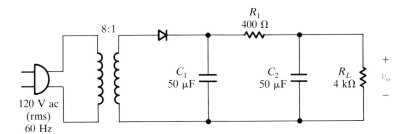

Figure P2.2-10

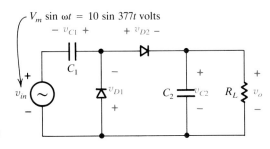

Figure P2.2-13

2.3-5 For the circuit shown in Figure P2.3-3, the diodes are silicon with $r_D = 0.1\ \Omega$. If $V_{ZZ} = 8$ V, $I_{ZK} = 0.1$ A, and $r_Z = 0.5\ \Omega$, determine the load regulation for currents between 0 and 2 A (R_L between ∞ and 4 Ω).

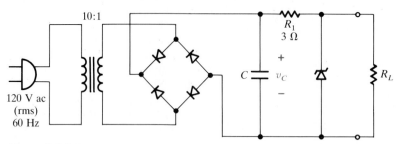

Figure P2.3-3

CHAPTER 3

Introduction to the Bipolar Junction Transistor

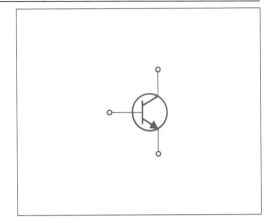

3.1   FUNDAMENTALS OF ELECTRONIC AMPLIFIER DESIGN

Electronic amplifiers are used to provide power gain between the signal applied to the input of an amplifier and the signal delivered to the load. At first one might think that this type of amplification could be achieved more easily and less expensively with a passive device such as a transformer. However, while a transformer can be used to step up the voltage or current level of a signal, its power gain is always less than or equal to 1.

In an electronic amplifier, on the other hand, the power gain can be greater than 1. Initially, this statement may appear to violate the law of conservation of energy, since more power seems to leave the amplifier than originally went in. However, when the dc power entering from the power supply is included, the signal power delivered to the load is always found to be less than the sum of the powers contributed by the dc supply and the signal source. Of course, the difference between the net power entering the amplifier and that delivered to the load is equal to the power that is dissipated in the amplifier as heat.

The simplest type of electronic amplifier has a single set of input terminals and a single set of output terminals, and as such it may be considered to be a two-port network. Generalized two-port networks are rather involved, and for simplicity in analyzing the amplifiers in this section, we will assume that their amplification properties are unilateral, that is, that signals travel only from the input to the output with no feedback occurring from the output back to the input. In addition, we will assume that one of the input terminals of the amplifier and one of its output terminals are each connected to a common reference point called ground.

To the extent that these assumptions are valid, we may represent our amplifier by the equivalent circuits illustrated in Figure 3.1-1, where R_i is called the amplifier's input impedance, R_o is known as its output impedance, and R_s and R_L are the source and load impedances, respectively. In general, all these impedances can be complex quantities. However, for simplicity, in the remainder of this section we will, for the most part, assume that they are purely resistive.

Notice that the output portion of both of the equivalent circuits in Figure 3.1-1 contains controlled or dependent sources. In this book all controlled sources are represented by diamond-shaped symbols to distinguish them from independent sources, which are represented by circles.

3.1-1 Effect of the Source and the Load on the Design of an Amplifier

The characteristics of the source and the load have a dramatic impact on the amplifier that needs to be constructed. Taking this fact into account, the design of a multistage electronic amplifier of the type illustrated in Figure 3.1-2a can generally be broken down into three distinct parts:

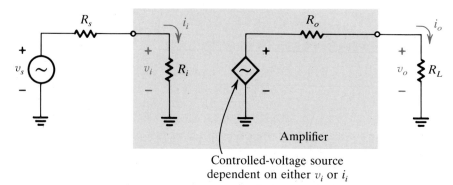

Controlled-voltage source
dependent on either v_i or i_i

(a)   Thévenin Equivalent of an Electronic Amplifier
with All Impedances Considered Resistive

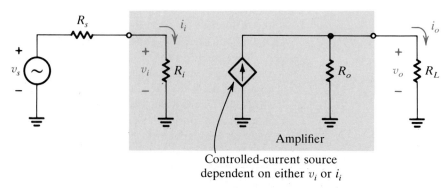

Controlled-current source
dependent on either v_i or i_i

(b)   Norton Equivalent of an Electronic Amplifier
with All Impedances Considered Resistive

Figure 3.1-1

1. The design of the input stage to best match the characteristics of the source to those of the amplifier.
2. The design of the output stage to match the characteristics of the amplifier to the load that it must drive.
3. The design of the intermediate amplifier stages to provide any extra power gain that is needed above and beyond that supplied by the input and output stages.

The requirements for the input stage of this amplifier depend primarily on the type of source that is to be connected to it. Consider, for example, the situation illustrated in Figure 3.1-2b, in which the input signal to the amplifier comes from some type of electrical transducer that converts the quantity being measured, be it of mechanical, thermal, chemical, or electromagnetic origin, into a corresponding electrical signal.*

Depending on whether the quantity being measured is converted into a

* Typical examples of commonly employed transducers include accelerometers, thermistors, microphones, antennas, phonograph cartridges, tape heads, and various types of chemical and physiological measurement electrodes.

Figure 3.1-2

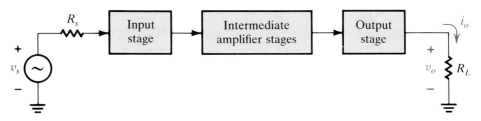

(a)   Multi-stage Electronic Amplifier

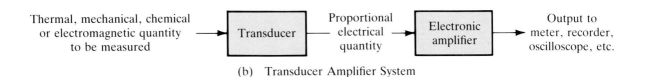

(b)   Transducer Amplifier System

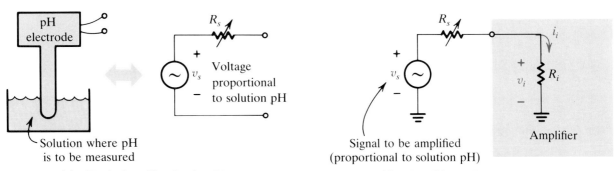

(c)   Equivalent Circuit of a pH
Electrode Chemical Transducer

(d)   Amplifier Driven by a Voltage
Source Input Signal

voltage or a current, it may be better to represent the transducer by either a Thévenin or a Norton equivalent circuit. At first it might not seem to make much difference which model is used since one can easily be transformed to the other, but in fact this is a very important decision. For example, consider the problem of amplifying the signal coming from a pH electrode or pH transducer. This type of chemical transducer is used to measure the alkalinity or acidity of solutions, and its equivalent circuit has the form illustrated in Figure 3.1-2c. In this equivalent circuit there is a well-defined relationship between the open-circuit voltage, v_s, and the pH of the solution. Thus, the pH of a particular solution may be determined by accurately measuring v_s. In addition, this equivalent circuit contains a source resistance, R_s, that is very large, and which varies greatly with temperature and from one electrode to the next. Therefore, to measure the pH of a solution accurately with such an electrode, it will be necessary to connect it to an amplifier whose output voltage or current will be proportional to v_s and will be independent of the source resistance, R_s (Figure 3.1-2d).

Based on our previous discussion, the output signal from this amplifier can be chosen to depend on either v_i or i_i (Figure 3.1-1). Let's first consider the case where the amplifier output circuit contains a voltage-controlled voltage source that is proportional to v_i. Following Figure 3.1-2d, we then have

$$v_i = v_s \frac{R_i}{R_i + R_s} \tag{3.1-1}$$

If
$$R_i \gg R_s \tag{3.1-2a}$$

then
$$v_i \simeq v_s \tag{3.1-2b}$$

and v_i, and hence the amplifier output signal, will be independent of R_s.

When equation (3.1-2b) is satisfied, we say that the amplifier does not load the source. In general, this result is valid as long as

$$R_i \geq 10R_s \tag{3.1-3}$$

since v_i will then be within 10% of v_s.

For situations where equation (3.1-3) is not satisfied, v_i will be significantly smaller than v_s and will be a function of R_s. Thus, to amplify accurately the source voltage v_s independent of its source impedance, we require an amplifier with a high input impedance whose output voltage or current is proportional to the amplifier input voltage.

Let's consider whether it is possible to use the current i_i in this pH transducer example as the input quantity to be amplified. Following Figure 3.1-2d, the input current may be written as

$$i_i = \frac{v_i}{R_i + R_s} \tag{3.1-4}$$

and by examining this result, it should be clear that this type of amplifier will not work satisfactorily since i_i depends strongly on R_s. The only way to eliminate this dependence is to make R_i much greater than R_s, and since this also makes i_i very small, this is not the best solution.

Consider next the situation where the signal to be amplified is a current

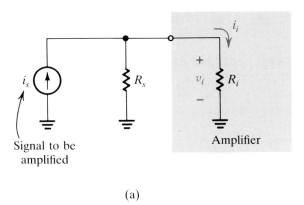

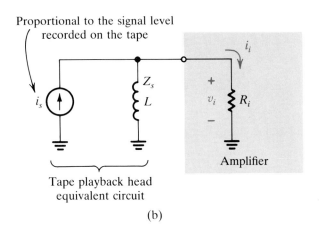

(a)

(b)

Figure 3.1-3
An amplifier driven by a current source input signal.

source rather than a voltage source (Figure 3.1-3a). Here again, let us assume that signal amplification independent of R_s is desired, and let's determine whether a voltage or a current amplifier is most appropriate in this situation. Following Figure 3.1-3a, we may write

$$i_i = \frac{R_s}{R_s + R_i} i_s \qquad (3.1\text{-}5)$$

Clearly, for

$$R_i \ll R_s \qquad (3.1\text{-}6a)$$

$$i_i \simeq i_s \qquad (3.1\text{-}6b)$$

Therefore, the signal current i_s can be amplified independent of the source resistance R_s if a low-input-impedance amplifier is used whose output-controlled source is proportional to i_i (see Figure 3.1-1).

If v_i is used in place of i_i as the amplifier input variable, then because

$$v_i = \frac{R_s R_i}{R_s + R_i} i_s \qquad (3.1\text{-}7)$$

the amplifier output signal will depend on R_s. The only way to eliminate this dependence is to make R_i much smaller than R_s, and this is not the best solution to the problem because it also makes v_i very small. As a result, when amplifying a signal whose equivalent circuit can best be represented by a current source, a current amplifier with a low input impedance is needed.

Thus far we have indicated that signal amplification independent of the source impedance is important because of the large variability or even the nonlinearity that may exist in Z_s. However, even when the source impedance is linear and constant, it may still be important for the signal amplification to be independent of Z_s because of the variation of this impedance with frequency. Consider, for example, the case of a magnetic tape recorder whose playback head has an equivalent circuit of the form illustrated in Figure 3.1-3b. In this equivalent circuit the current source i_s is directly proportional to the original signal level recorded on the tape, and the source impedance of the tape head, Z_s, is simply equal to $j\omega L$, where L is the inductance of the coil in the playback head. From this figure it is apparent that faithful reproduction of the signal originally recorded on the tape requires that the tape head be connected to a low-input-impedance amplifier where $R_i \ll Z_s$, so that i_i will be equal to i_s.

If R_i is not much smaller than ωL over all signal frequencies of interest, signal distortion will result. For example, if a high-input-impedance (voltage-style) amplifier is connected to the tape head instead of using a current amplifier, the effective input voltage to this amplifier will be $v_i = L(di_s/dt)$. As a result, in this case the amplifier output will be proportional to the derivative of the original signal recorded on the tape, not to the signal itself.

In certain applications it is not critically important that the signal amplification be independent of the source impedance. A good example is the phonograph amplifier shown in Figure 3.1-4a, whose overall equivalent circuit is given in part (b) of the figure. This circuit will work fine even if R_i is not much greater than R_s, since in this case the amplifier will just load the source, making the sound from the loudspeaker somewhat lower than might otherwise be expected. To compensate for this, all that needs to be done is to raise the volume control a little since this amplifier is not part of a calibrated system. Were this, on the other hand, the vertical amplifier of an oscilloscope, this loading would be intolerable because it would destroy the measurement accuracy of the instrument.

3.1-2 Impedance Matching

For some sources (e.g., antennas) the source impedance is very well defined, and in these instances, rather than trying to make the amplifier output signal independent of R_s and R_L, maximum power transfer may be a more important design consideration. Let us again examine Figure 3.1-4b, but this time with an eye toward maximizing the power transferred from the source into the am-

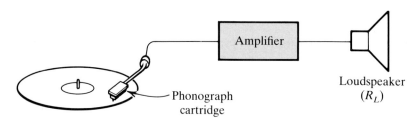

(a)   Phonograph System to Be Examined

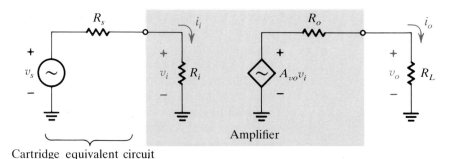

Cartridge equivalent circuit

Figure 3.1-4
An example of a system where loading may not create a serious problem.

(b)   Overall Equivalent Circuit of the Phonograph System in (a) Considering the Cartridge Impedance to Be Resistive

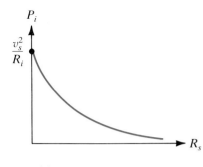

(a) R_i Fixed, R_s Variable

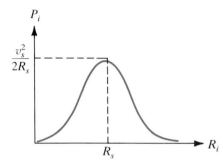

(b) R_s Fixed, R_i Variable

Figure 3.1-5
The power delivered to an amplifier as a function of the source and input resistance.

plifier. For this circuit the instantaneous power delivered to the amplifier can be written as

$$P_i = \frac{v_i^2}{R_i} = v_s^2\left(\frac{R_i}{R_i + R_s}\right)^2 \frac{1}{R_i} = v_s^2 \frac{R_i}{(R_i + R_s)^2} \qquad (3.1\text{-}8)$$

This equation is sketched in Figure 3.1-5 with R_i fixed and R_s variable in part (a), and with R_s fixed and R_i the variable parameter in part (b). The location of the peak in the curve in Figure 3.1-5b can readily be found by differentiating equation (3.1-8) with respect to R_i, setting the result equal to zero. When this is done, the maximum of the curve is found to occur at $R_i = R_s$. From these two sketches, the following facts are clear:

1. If R_s is adjustable while R_i is fixed, make R_s equal to zero in order to deliver maximum power to the amplifier.
2. If R_i is adjustable while R_s is fixed, make R_i equal to R_s in order to get maximum power into the amplifier.

These two statements, especially the second one, while seemingly simple, create a great deal of confusion for electronic circuit designers. Statement 2 is really valid only when a transformer is used to match the source and input impedances, *not* when a resistor is connected in series with or shunted across the original input impedance of the amplifier in order to make R_i equal to R_s. Although this second approach does result in maximum power being delivered to the overall resistor network, it *does not* deliver maximum power to the amplifier.

Thus far we have examined carefully the relationship between the characteristics of the source and the amplifier's input stage. Next we look at the factors affecting the output side of the amplifier.

Frequently, one wishes to maintain a constant output voltage for a given input signal level in spite of large variations in the value of the load, and depending on the character of the amplifier output stage, this may or may not be possible. If, for example, the output circuit has the form illustrated in Figure 3.1-6a, where A_{vo} is a constant, then

$$v_o = A_{vo} v_i \frac{R_L}{R_L + R_o} \qquad (3.1\text{-}9)$$

Figure 3.1-6

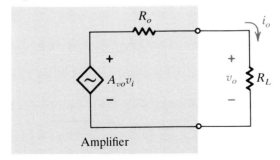

(a) Thévenin Equivalent Circuit at the Output of an Electronic Amplifier Where v_T is a Voltage-Controlled Voltage Source

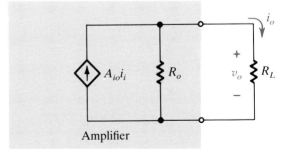

(b) Norton Equivalent Circuit at the Output of an Electronic Amplifier Where i_n is a Current-Controlled Current Source

Clearly, for
$$R_o \ll R_L \qquad (3.1\text{-}10\text{a})$$
$$v_o \simeq A_{vo} v_i \qquad (3.1\text{-}10\text{b})$$

and, as required, is independent of the load resistance. If the amplifier has an output equivalent circuit of the form shown in Figure 3.1-6b, where A_{io} is a constant,

$$v_o = A_{io} i_i \frac{R_L R_o}{R_L + R_o} \qquad (3.1\text{-}11)$$

Here again, to make v_o independent of R_L, R_o (the amplifier output impedance) needs to be much smaller than R_L.

Consider next the opposite situation—where we require that the current delivered to the load by the amplifier be independent of the load resistance. Again using Figure 3.1-6a, but this time solving for i_o, we have

$$i_o = A_{vo} v_i \frac{1}{R_L + R_o} \qquad (3.1\text{-}12)$$

from which it is apparent that i_o will be independent of R_L as long as the amplifier output impedance R_o is much greater than the load resistance R_L. A similar relationship exists for the case where the amplifier output equivalent circuit has the form given in Figure 3.1-6b.

Sometimes maximum power gain is more important than making the signal power delivered to the load independent of R_L. When this is the case, a transformer can be used in the output stage to match the transistor output impedance to that of the load.

EXAMPLE 3.1-1

The power gain, G, of an amplifier such as that shown in Figure 3.1-7a may be defined as the ratio of the power delivered to the load to that supplied to the circuit by the source, v_s. Using this definition, the input power may be written as

$$P_{in} = \frac{v_s^2}{R_s + R_i} \qquad (3.1\text{-}13\text{a})$$

while that delivered to the load is

$$P_L = \frac{v_o^2}{R_L} \qquad (3.1\text{-}13\text{b})$$

so that
$$G = \frac{P_L}{P_{in}} = \left(\frac{v_o}{v_s}\right)^2 \frac{R_s + R_i}{R_L} \qquad (3.1\text{-}13\text{c})$$

a. Calculate P_{in} and P_L in terms of v_s, and find the power gain G for the circuit shown.
b. Repeat part (a) if a 99-kΩ resistor is connected in series with R_s to "match" the source impedance to R_i.
c. Repeat part (a) if impedance-matching transformers are inserted at the input and output terminals of the amplifier.

SOLUTION

For the circuit in Figure 3.1-7a, the instantaneous input power supplied by v_s is

$$P_{in} = \frac{v_s^2}{101 \text{ k}\Omega} \tag{3.1-14a}$$

The corresponding power delivered to R_L may be written as

$$P_L = \frac{v_o^2}{10 \ \Omega} = \left[(99v_s)\left(\frac{10}{1010}\right) \right]^2 \frac{1}{10 \ \Omega} = \frac{v_s^2}{10.4 \ \Omega} \tag{3.1-14b}$$

so that

$$G = \frac{v_s^2/10.4}{v_s^2/101 \text{ k}\Omega} = 9.7 \times 10^3 \tag{3.1-14c}$$

Connecting a 99-kΩ resistor in series with R_s in no way improves the per-

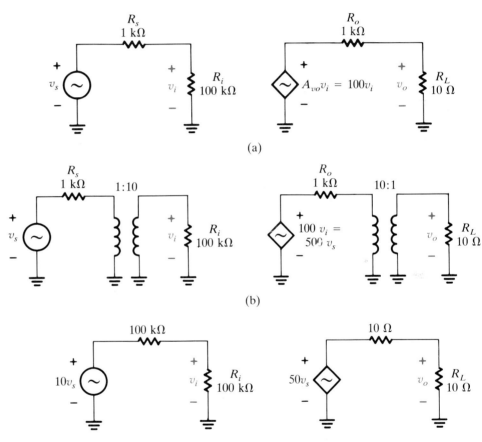

(a)

(b)

(c) Equivalent Circuit at the Amplifier Input

(d) Equivalent Circuit at the Load

Circuit Configuration	P_{in}	P_L	G
Original Circuit	$v_s^2/101 \text{ k}\Omega$	$v_s^2/10.4 \ \Omega$	9.7×10^3
99 kΩ in series with R_s	$v_s^2/200 \text{ k}\Omega$	$v_s^2/40.8 \ \Omega$	4.9×10^3
Matching transformers added	$v_s^2/2 \text{ k}\Omega$	$625v_s^2/10 \ \Omega$	1.25×10^5

(e)

Figure 3.1-7
Example 3.1-1.

3.1 / Fundamentals of Electronic Amplifier Design

formance of the circuit. This is not the proper way to match impedances. At any rate, the effective resistance in series with R_i is now 100 kΩ, so that the power gain may now be written as

$$G = \frac{P_L}{P_{in}} = \frac{v_s^2/40.8 \ \Omega}{v_s^2/200 \ \text{k}\Omega} = 4.9 \times 10^3 \tag{3.1-15}$$

Thus, by adding a 99-kΩ resistor in series with R_s, the power gain of the amplifier has been reduced by a factor of about 2.

To match the amplifier to its source and the load impedances properly, transformers having the turns ratios indicated in Figure 3.1-7b need to be connected at the amplifier input and output terminals. With these transformers in place, the effective impedance seen by v_s is 2 kΩ, so that

$$P_{in} = \frac{v_s^2}{2 \ \text{k}\Omega} \tag{3.1-16}$$

Using the input equivalent circuit given in Figure 3.1-7c, it is apparent that $v_i = 5v_s$. Similarly, the equivalent circuit at the load has the form shown in Figure 3.1-7d, so that

$$P_L = \frac{v_L^2}{10} = \frac{(25v_s)^2}{10 \ \Omega} = \frac{625v_s^2}{10 \ \Omega} \tag{3.1-17a}$$

and

$$G = \frac{625v_s^2/10 \ \Omega}{v_s^2/200 \ \text{k}\Omega} = 1.25 \times 10^5 \tag{3.1-17b}$$

The results from this example are summarized in Figure 3.1-7e, and as indicated by these data, the use of impedance-matching transformers has improved the amplifier power gain by a factor of nearly 13 times. In addition, the actual power delivered to the load for a given value of v_s has increased by about 625 times.

3.1-3 Basic Categories and Models of Ideal Amplifiers

Table 3.1-1 summarizes the amplifier impedance characteristics needed to make the overall circuit gain independent of R_s and R_L. Depending on whether the input signal is a voltage or current source and also on whether the output voltage or the output current is to be held constant, the type of amplifier that is needed falls into one of four categories: voltage, current, transconductance, and transresistance amplifiers.

When R_s and R_L are completely absent from the gain expressions, the amplifiers are said to be ideal, and when the inequalities listed in the second and fifth columns of the table are satisfied, an actual electronic amplifier can have performance characteristics closely approaching those of the ideal circuit.

Based on the definitions in the right-hand column of this table, one can readily develop the ideal amplifier models illustrated in Figure 3.1-8. For example, for an ideal voltage amplifier the output resistance is zero and the Thévenin output voltage is $A_{vo}v_i$. Thus, the output equivalent circuit for this type of amplifier is simply an ideal voltage source $A_{vo}v_i$. Furthermore, the am-

Table 3.1-1 Classification of Amplifiers into Different Types Depending on the Input Signal to Be Amplified and the Signals to Be Held Constant at the Load

Source Parameter to Be Amplified	Amplifier Input Impedance Needed to Make Amplifier Output Independent of R_s		Output Parameter of Interest	Amplifier Output Impedance Needed to Make Output Independent of R_L		Circuit Gain Specification
	Approximate	Ideal		Approximate	Ideal	
Source voltage, v_s	$R_i \gg R_s$	$R_i = \infty$	v_o	$R_o \ll R_L$	$R_o = 0$	$\frac{v_o}{v_s}$ = voltage gain (A_{vo}) [dimensionless]
			i_o	$R_o \gg R_L$	$R_o = \infty$	$\frac{i_o}{v_s}$ = transconductance gain (G_m) [siemens]
Source current, i_s	$R_i \ll R_s$	$R_i = 0$	v_o	$R_o \ll R_L$	$R_o = 0$	$\frac{v_o}{i_s}$ = transresistance gain (R_m) [ohms]
			i_o	$R_o \gg R_i$	$R_o = \infty$	$\frac{i_o}{i_s}$ = current gain (A_{io}) [dimensionless]

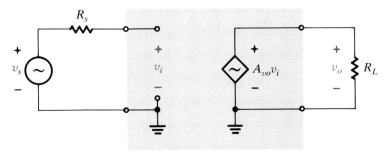

(a) Ideal Voltage Amplifier (v_o/v_s = Voltage Gain = A_{vo})

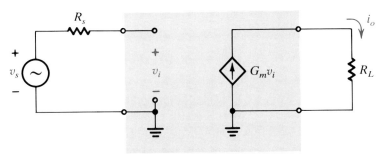

Figure 3.1-8

(b) Ideal Transconductance Amplifier (i_o/v_s = Transconductance Gain = G_m)

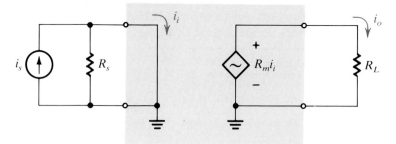

(c) Ideal Transresistance Amplifier (v_o/i_s = Transresistance Gain = R_m)

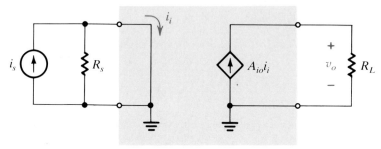

Figure 3.1-8 (*Continued*)

(d) Ideal Current Amplifier (i_o/i_s = Current Gain = A_{io})

plifier input impedance is infinite, so that it can be represented by an open circuit. The resulting overall equivalent circuit for an ideal voltage amplifier is
given in Figure 3.1-8a. The equivalent circuits for the ideal transconductance,
transresistance, and current amplifiers are given in Figure 3.1-8b through d
and can be developed similarly.

Exercises

3.1-1 For the amplifier shown in Figure 3.1-1a, R_s = 1 kΩ, R_i = 500 Ω,
R_o = 0, R_L = 10 Ω, v_s = 1 sin ωt V, and the controlled voltage source at the
output equals $20v_i$.
(a) Find P_L.
(b) Add a resistor R_x = 500 Ω in series with R_i to "match impedances" and recompute P_L.
(c) Repeat part (a) if a transformer with a 2:1 turns ratio is connected between
R_s and R_i.
Answer (a) 2.22 W, (b) 1.25 W, (c) 2.5 W

3.1-2 A two-stage transconductance amplifier is to drive a 10-kΩ load. The
source impedance is 200 kΩ, and R_i = 1 mΩ, R_o = 10 kΩ, and G_m = 40 mS
for each amplifier stage. Find the overall voltage gain of the amplifier. *Answer*
6.67×10^4

3.1-3 For the amplifier described in Exercise 3.1-2, find the current gain. *Answer* 8×10^6

3.1-4 For the amplifier described in Exercise 3.1-2, find the overall amplifier
power gain and show that it is equal to the product of A_v and A_i from Exercises
3.1-2 and 3.1-3. *Answer* 5.33×10^{11}

Chapter 3 / Introduction to the Bipolar Junction Transistor

3.2 THEORY OF OPERATION OF THE BIPOLAR JUNCTION TRANSISTOR

The bipolar junction transistor (BJT) was invented in the late 1940s by scientists and engineers at the Bell Telephone Laboratories in the United States. Like its predecessor, the vacuum tube, the BJT has the rather unique ability to amplify the power level of an electrical signal, and as such it can deliver more output power to a load than is present at its input.

The BJT is a three-terminal device that may be constructed by taking two diodes and sticking them together "in a special way." When the diodes are connected at their cathodes we wind up with a *pnp* transistor (Figure 3.2-1a), and by connecting the diodes at their anodes an *npn* transistor is produced. The symbols for the *pnp* and *npn* transistors are given in parts (b) and (c) of Figure 3.2-1. There are basically no differences between *pnp* and *npn* transistors except that all the voltage and current directions are opposite from one another. In particular, *npn* transistors usually operate with positive power supply voltages while *pnp* circuits use negative power supplies. Because students generally find it easier to work with positive voltages and currents, we will for the most part use *npn* transistors in this book except when there is a special advantage in having a *pnp* device in the circuit.

The three leads on the bipolar junction transistor are known as the emitter, the base, and the collector; the relationship of these names to the physical operation of the BJT will be seen shortly. Notice that the arrow on the emitter lead of the transistor symbol indicates the orientation of the base–emitter diode and also shows the direction in which the current flows in the emitter.

If, as suggested previously, a transistor were actually constructed by joining two diodes together, it would not work very well. This is because a transistor is more than just a series of *pn* junctions. To make a good-quality transistor it is necessary for the doping levels and device dimensions to be carefully controlled. In these categories the two most important requirements are that:

1. The emitter is heavily doped in comparison to the base.
2. The base is made very thin compared to the minority carrier diffusion length in the base.

(a) Making a *pnp* Transistor

Figure 3.2-1
"Making" pnp *and* npn *transistors from semiconductor diodes.*

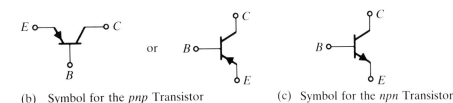

(b) Symbol for the *pnp* Transistor

(c) Symbol for the *npn* Transistor

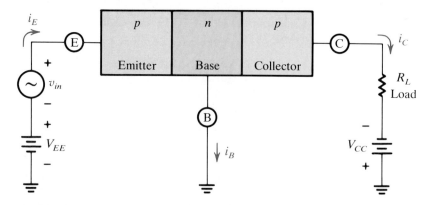

Figure 3.2-2

Biasing arrangement for a pnp transistor. (V_EE and V_cc are the biasing voltages and v_in is the input signal to be amplified. The signal v_in is considered to be much smaller than V_EE so that the base-emitter junction is always forward-biased.)

Besides designing the transistor properly, it is also necessary to bias the transistor correctly if it is going to be able to function as a power amplifier. In particular, the external biasing circuitry must be connected so that:

3. The base–emitter junction is forward biased.
4. The base–collector junction is reverse biased.

The rationale for each of these four requirements will be seen shortly.

A simplified diagram of one possible biasing arrangement for a *pnp* transistor satisfying requirements 3 and 4 is given in Figure 3.2-2, and a detailed diagram of the current components flowing in this transistor is shown in Figure 3.2-3. The basic operating principle of the transistor may be explained as follows. The forward-biased emitter–base junction causes a rather large current

Figure 3.2-3

The important current components in the pnp transistor of Figure 3.2-2.

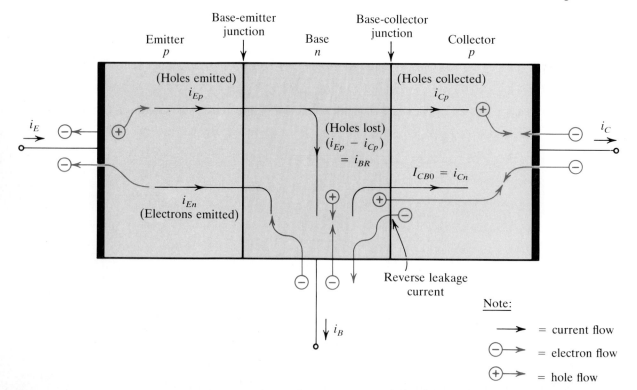

to be "emitted" from the emitter into the base region. This current diffuses across the base and is "collected" by the collector of the transistor. For a properly constructed and properly biased transistor, nearly all the current "emitted" by the emitter is "collected" by the collector, so that i_C is almost equal to i_E, and i_B is nearly zero. For the circuit given in Figure 3.2-2, notice that the battery V_{EE} forward biases the base–emitter diode, while the battery V_{CC} is used to reverse bias the base–collector junction. Also note that the terms i_E, i_B, and i_C are used to represent the emitter, base, and collector currents, respectively.

3.2-1 Detailed Description of the Operating Principles of the Bipolar Junction Transistor

The detailed operation of the bipolar junction transistor may be understood with the aid of Figure 3.2-4, which shows the minority-carrier charge distributions in the emitter, base, and collector regions when the base–emitter junction is forward biased and the base–collector junction is reverse biased. For simplicity, the width of the base–emitter and base–collector depletion regions have been assumed to be zero. In this figure $n_e(x)$, $p_b(x)$, and $n_c(x)$ are defined as

$n_e(x)$ = minority-carrier concentration of electrons in the emitter

$p_b(x)$ = minority-carrier concentration of holes in the base

$n_c(x)$ = minority-carrier concentration of electrons in the collector

Because of the biasing arrangement used in Figure 3.2-2, the base–emitter junction is forward biased and the base–collector junction is reverse biased. As a result, following equations (1.11-15a) and (1.11-15b),

$$n_e(0) = n_{eo}e^{v_{EB}/V_T} \tag{3.2-1a}$$

$$p_b(0) = p_{bo}e^{v_{EB}/V_T} \tag{3.2-1b}$$

$$p_b(w) \simeq 0 \tag{3.2-1c}$$

and $$n_c(w) \simeq 0 \tag{3.2-1d}$$

where n_{eo} and p_{bo} are the equilibrium minority-carrier concentrations of electrons and holes in the emitter and base, respectively.

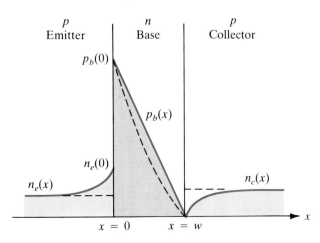

Figure 3.2-4
Minority carrier charge distributions in the emitter, base, and collectors of a BJT biased in the forward active region.

Solving the continuity equations (1.11-9b) and (1.11-9c) subject to the boundary conditions given in equations (3.2-1a) through (3.2-1d) leads to the following expressions for the minority-carrier distributions in the transistor:

$$p_b(x) \simeq p_b(0)\left(1 - \frac{x}{w}\right) = p_{bo}e^{v_{EB}/V_T}\left(1 - \frac{x}{w}\right) \qquad \text{for } 0 \le x \le w \qquad (3.2\text{-}2a)$$

$$n_e(x) = n_{eo}(e^{v_{EB}/V_T} - 1)e^{x/L_n} + n_{eo} \qquad \text{for } x \le 0 \qquad (3.2\text{-}2b)$$

and
$$n_c(x) = n_{co}(1 - e^{-(x-w)/L_n}) \qquad \text{for } x \ge w \qquad (3.2\text{-}2c)$$

These results are sketched in Figure 3.2-4. Notice that the distribution of holes in the base is nearly a straight line. Actually, the rigorous solution of the continuity equation in the base contains two exponential terms, but because the base is assumed to be very thin compared to L_p, the exponentials may be approximated by the first two terms in their power series [equation (2.2-8)], and this leads to the expression given in equation (3.2-2a). The details of this calculation are left as a homework exercise for the student (Problem 3.2-9).

Because the base–emitter junction is forward biased, most of the current flow across this junction is associated with the diffusion of holes from the emitter into the base, and electrons from the base into the emitter (Figure 3.2-3). As a result, the overall emitter current may be written as

$$i_E = i_{Ep} + i_{En} \qquad (3.2\text{-}3)$$

In designing a high-quality transistor, we want most of the current "emitted" by the emitter to be "collected" by the collector, and for this to be the case, at a minimum, it will be necessary that $i_{Ep} \gg i_{En}$. Mathematically, following equations (1.11-3a) and (1.11-3b), i_{Ep} and i_{En} may be expressed as

$$i_{Ep} = -qAD_p\frac{dp_b(x)}{dx}\bigg|_{x=0} = \frac{qAD_p}{w}p_{bo}e^{v_{EB}/V_T} \qquad (3.2\text{-}4a)$$

and
$$i_{En} = qAD_n\frac{dn_e(x)}{dx}\bigg|_{x=0} \approx \frac{qAD_n}{L_n}n_{eo}e^{v_{EB}/V_T} \qquad (3.2\text{-}4b)$$

By examining these results, it is apparent that i_{Ep} will be much greater than i_{En} if the emitter is doped much more heavily than the base. This will make $n_{eo} \ll p_{bo}$ and will cause most of the emitter current to be associated with the injection of holes from the emitter into the base.

For later use, it is convenient to define the emitter injection efficiency γ as the fraction of the emitter current in a *pnp* transistor that is due to the flow of holes. Following this definition and substituting equations (3.2-4a) and (3.2-4b) into this expression, the emitter injection efficiency may be written in terms of the transistor parameters as

$$\gamma = \frac{i_{Ep}}{i_E} = \frac{i_{Ep}}{i_{Ep} + i_{En}} \simeq 1 - \frac{i_{En}}{i_{Ep}} = 1 - \frac{D_n n_{eo} w}{D_p p_{bo} L_n} \qquad (3.2\text{-}5)$$

using the approximation $1/(1 + x) = 1 - x$ for small x.

Paralleling the development used to describe the emitter current, the collector current can also be written as

$$i_C = i_{Cp} + i_{Cn} \qquad (3.2\text{-}6)$$

where i_{Cp} is the component of the collector current associated with the diffusion of holes from the base into the collector, and i_{Cn} is the component due to the diffusion of electrons from the collector into the base. Ideally, to the extent that $p_b(x)$ is a straight line, $i_{Cp} = i_{Ep}$. Actually, however, due to recombination in the base, some of the holes are lost and the shape of the distribution for $p_b(x)$ is more like the dashed curve in Figure 3.2-4, and as a result, $i_{Cp} < i_{Ep}$.

This loss in current in traveling from the emitter to the collector may be estimated by using the straight-line approximation to $p_b(x)$ and noting that

$$i_{Ep} \quad - \quad i_{Cp} \quad = \sum_{\text{base}} \frac{p_b(x) - p_{bo}}{\tau_p} \Delta V \qquad (3.2\text{-}7)$$

$$\begin{array}{ccc} \text{hole current} & \text{hole current} & \text{hole current lost in base} \\ \text{into base} & \text{out of base} & \text{due to recombination} \end{array}$$

in accordance with equation (1.11-4a) with the d/dt term set equal to zero. The term on the right-hand side of equation (3.2-7) must be summed over the entire base volume in order to get the net loss in emitter current associated with the recombination of holes in the base.

Because this loss of holes is due to their recombination with electrons flowing in from the base lead, the right-hand side of equation (3.2-7) is also equal to a part of the base current, i_{BR}, that is known as the base recombination current. Following Figure 3.2-4, i_{BR} may therefore be written as

$$i_{BR} = \sum_{\text{base}} \frac{p_b(x) - p_{bo}}{\tau_p} A \, \Delta x \simeq \frac{A}{\tau_p} \int_0^w p_b(x) \, dx = \frac{Aw}{2\tau_p} p_{bo} e^{v_{EB}/V_T} \qquad (3.2\text{-}8)$$

where A is the transistor cross-sectional area.

If we define the base transport factor, T, as the fraction of the emitted hole current reaching the collector, then on substituting equations (3.2-4b) and (3.2-8), the base transport factor may be written in terms of the transistor design parameters as

$$T = \frac{i_{Cp}}{i_{Ep}} = \frac{i_{Ep} - i_{BR}}{i_{Ep}} = 1 - \frac{w^2}{\tau_p D_p} = 1 - \left(\frac{w}{L_p}\right)^2 \qquad (3.2\text{-}9)$$

Combining equations (3.2-9) and (3.2-5), the hole portion of the collector current may be related to the total emitter current by the expression

$$\alpha_F = \frac{i_{Cp}}{i_E} = \gamma T$$

$$= \left(1 - \frac{D_n n_{eo} w}{D_p p_{bo} L_n}\right)\left[1 - \left(\frac{w}{L_p}\right)^2\right] \simeq 1 - \frac{D_n n_{eo} w}{D_p p_{bo} L_n} - \left(\frac{w}{L_p}\right)^2 \qquad (3.2\text{-}10)$$

since each of the right-hand terms in the parentheses is small. The quantity α_F in this expression is called the transistor current gain, and because the collector and emitter currents are nearly equal in a good-quality transistor, the α value is almost 1. The subscript F on α is used to indicate that this is the current gain when the transistor is operated in the normal forward direction. By interchanging the roles of the emitter and the collector, the transistor can also be made to operate as a power amplifier. However, because the doping levels are no longer optimal, the transistor alpha, now known as α_R or reverse alpha, will be quite small.

As we have just mentioned, it is desirable that the transistor alpha be as close as possible to 1. As a result, each of the terms next to the 1 in equation (3.2-12) represents transistor losses and should be made as small as possible. The first loss term, $(D_n n_{eo} w / D_p, p_{bo} L_n)$, represents the fraction of the emitter current associated with the flow of electrons across the base–emitter junction. This part of the emitter current can never become part of the collector current because the collector only "collects" holes that are "emitted" by the emitter. The size of this term may be minimized by heavily doping the emitter and lightly doping the base.

The second loss term in equation (3.2-10), $(w/L_p)^2$, is associated with the recombination of the holes in the base. As is apparent from this expression, its effect may be minimized by making the base thin in comparison to the diffusion length for holes.

Thus far we have said nothing about the term i_{Cn} in equation (3.2-6). This term represents the leakage current that flows across the reverse-biased base–collector junction, and substituting equation (3.2-2c) into equation (1.11-3b), i_{Cn} may be written as

$$i_{Cn} = qAD_n \frac{dn_C(x)}{dx}\bigg|_{x=w} = \frac{qAD_n n_{Co}}{L_n} \tag{3.2-11}$$

This expression is quite similar to that for I_s in a *pn* junction diode [equation (1.11-21)].

Because i_{Cn} is the current that flows in the transistor when the emitter is open circuited, or when $i_E = 0$, the symbol I_{CBO} is frequently used in place of i_{Cn}. Here the *CB* portion of the subscript on I_{CBO} indicates that the current flow is between the base and the collector, while the O in the subscript tells us that this current is measured with the emitter open circuited. Using this symbol for i_{Cn} and $\alpha_F i_E$ for i_{Cp}, the overall expression for the collector current in equation (3.2-3) may be rewritten as

$$i_C = \alpha_F i_E + I_{CBO} \tag{3.2-12}$$

As with I_s in a diode, I_{CBO} for a transistor is a strong function of temperature, and whereas it may usually be neglected at room temperature for silicon transistors, it can become important at high temperatures.

Rather than expressing i_C in terms of i_E, it is frequently useful to write the collector current as a function of the base current instead. To do this, we need only recognize in Figure 3.2-3 that

$$i_E = i_B + i_C \tag{3.2-13}$$

and substituting this result into equation (3.2-12), we obtain

$$i_C = \alpha_F(i_B + i_C) + I_{CBO} \tag{3.2-14a}$$

or

$$i_C = \frac{\alpha_F}{1 - \alpha_F} i_B + \frac{1}{1 - \alpha_F} I_{CBO} \tag{3.2-14b}$$

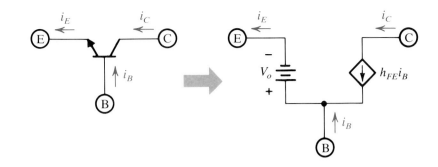

Figure 3.2-5
Model for the npn *bipolar junction transistor operating in the active region (base-emitter junction forward-biased, base collector junction reverse-biased).*

In this expression it is convenient to define the transistor h_{FE} as

$$h_{FE} = \frac{\alpha_F}{1 - \alpha_F} \qquad (3.2\text{-}15)$$

Here the subscript F again denotes that this is the forward current gain. The additional subscript E is used to indicate that this is the current gain between the base and the collector of the transistor, with the emitter considered to be the common terminal. This quantity (h_{FE}) is sometimes referred to as the beta (β_F) of the transistor.

Using the definition of h_{FE} given in equation (3.2-15) we may rewrite equation (3.2-14b) in terms of h_{FE} as

$$i_C = h_{FE}i_B + (1 + h_{FE})I_{CBO} \qquad (3.2\text{-}16)$$

In a well-designed transistor the alpha is typically on the order of 0.99, so that the h_{FE} will nominally be $0.99/(1 - 0.99)$, or about 100. Equation (3.2-16) illustrates why the BJT is often called a current amplifier, since any current i_B entering the base is amplified nearly 100-fold as it travels to the collector.

For silicon transistors operating at normal temperatures, the second term in equation (3.2-16) is usually quite small. Therefore, except in Section 3.4, where we discuss Q-point stabilization techniques, we will neglect the term $(1 + h_{FE})I_{CBO}$ and write the transistor collector current as

$$i_C = h_{FE}i_B \qquad (3.2\text{-}17)$$

This expression, together with equation (3.2-13), is all that is needed to carry out the analysis of nearly any BJT circuit. Using these equations, the simplified *npn* transistor model given in Figure 3.2-5 follows directly. (A similar model may also be drawn for the *pnp* transistor by reversing the polarity of the battery V_o, flipping the direction of the controlled current source, and changing the directions of i_B, i_C, and i_E.)

EXAMPLE 3.2-1

The circuit shown in Figure 3.2-6 is a transistor amplifier. The transistor is silicon and has an h_{FE} of 100. V_{EE} and V_{CC} are used to bias the transistor, and v_{in} is the signal to be amplified.

a. Determine the dc operating point. Specifically, find I_E, I_C, I_B, V_{EB}, and V_{CB}. (Note that V_{EB}, for example, indicates the voltage from emitter to base, or $V_E - V_B$.)

b. Find the ac quantities i_e, i_c, i_b, v_{eb}, and v_{cb}.

c. Sketch the total (dc + ac) waveforms for these voltages and currents.

SOLUTION

The circuit in Figure 3.2-6a is best analyzed by substituting in the model given in Figure 3.2-5 for the transistor to yield the equivalent circuit given in Figure 3.2-6b. (Note, however, that because this is a *pnp* transistor, the current directions as well as the directions of the controlled source and the battery must be reversed.) As was the case for the diode circuits that we examined in Chapters 1 and 2, we will analyze transistor circuits by applying the superposition principle. To carry out the dc analysis for the Q point, we set $v_{in} = 0$ and solve for the required dc voltages and currents with the aid of Figure 3.2-6c.

Analyzing the loop on the left in Figure 3.2-6c, we see that

$$I_E = \frac{V_{EE} - V_o}{R_E} = \frac{2\text{ V}}{1\text{ k}\Omega} = 2.0\text{ mA} \tag{3.2-18}$$

Notice for this loop that $V_{EB} = 0.7$ V. Combining equations (3.2-13) and (3.2-17), we also have

$$I_B = \frac{I_E}{1 + h_{FE}} \tag{3.2-19a}$$

and

$$I_C = h_{FE}I_B = \frac{h_{FE}}{1 + h_{FE}}I_E \tag{3.2-19b}$$

so that we obtain $I_B = 0.0198$ mA and $I_C = 1.98$ mA. As discussed previously, notice that I_E and I_C are nearly equal and that I_B is much smaller than either of them.

To calculate the transistor collector to base voltage, V_{CB}, we observe from the right-hand side of Figure 3.2-6c that

$$V_{CB} = I_C R_C - V_{CC} \tag{3.2-20}$$

and on substituting 1.98 mA for I_C and 20 V for V_{CC}, we find that V_{CB} is $(9.9 - 20)$, or -10.1 V. These dc quantities are placed as dashed lines on the sketches given in Figure 3.2-7 and will serve as the dc levels onto which we will add the ac results calculated in the paragraphs that follow.

For the ac portion of the analysis, we again go back to Figure 3.2-6a, and this time we set all the dc sources equal to zero to obtain the ac equivalent circuit given in part (d) of the figure. From the input loop, we can readily see that $v_{eb} = 0$ and that

$$i_e = \frac{v_{in}}{R_E} = \frac{50\sin\omega t\text{ mV}}{1\text{ k}\Omega} = 50\sin\omega t\ \mu\text{A} \tag{3.2-21}$$

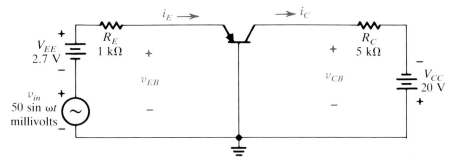

(a) Original Circuit for Example 3.2-1

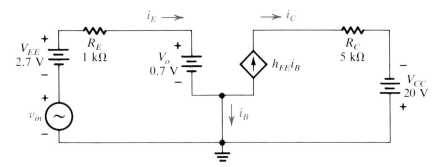

(b) Equivalent Circuit with Transistor Model in Place

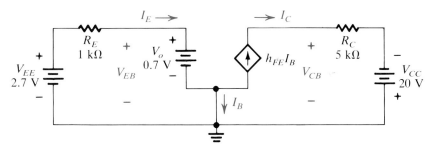

(c) Equivalent Circuit for the dc Analysis

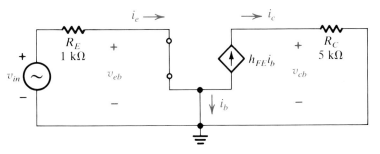

Figure 3.2-6
Example 3.2-1

(d) Equivalent Circuit for the ac Analysis

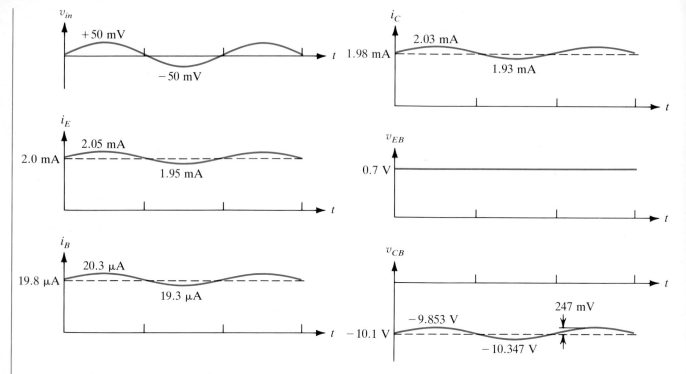

Figure 3.2-7
Pertinent input and output wave shapes for the circuit in Example 3.2-1.

Using the ac equivalents of the equations in (3.2-19), the base and collector currents are calculated as*

$$i_b = \frac{1}{1 + h_{FE}} i_e = 0.495 \sin \omega t \ \mu A \qquad (3.2\text{-}22a)$$

and

$$i_c = \frac{h_{FE}}{1 + h_{FE}} i_e = 49.5 \sin \omega t \ \mu A \qquad (3.2\text{-}22b)$$

The ac output voltage, v_{cb}, is found by applying Ohm's law in the output loop to yield

$$v_{cb} = i_c R_c = (49.5 \sin \omega t \ \mu A)(5 \ k\Omega) = 247.5 \sin \omega t \ mV \qquad (3.2\text{-}23)$$

The sketches of the pertinent waveshapes in this circuit, given in Figure 3.2-7, were obtained by adding the dc and ac portions of the results. Notice that because the resistive portion of the base–emitter diode model was assumed to be zero (it was represented by a battery), the sketch for v_{EB} shows no ac variation.

*The use of h_{FE} instead of h_{fe} in the ac models for the BJT (bipolar junction transistor) is not strictly valid, but the errors that this approximation introduces are not large enough to warrant a detailed discussion at this time. See Section 4.3 for the definition of h_{fe}, the small-signal current gain.

Exercises

3.2-1 Repeat Example 3.2-1a if $R_E = 2 \ k\Omega$. *Answer* $I_E = 1$ mA, $I_C = 1$ mA, $I_B = 0.01$ mA, $V_{EB} = 0.7$ V, $V_{CB} = -15$ V

3.2-2 Find v_{cb} in Example 3.2-1 if $R_E = 2$ kΩ and $v_{in} = 1 \sin \omega t$ volts. *Answer* 2.5 sin ωt volts

---

3.3 DC Q-POINT ANALYSIS TECHNIQUES

In this section we investigate the volt-ampere characteristics of the bipolar junction transistor, relating these curves where possible to the semiconductor physics of the device. Next, we use these characteristic curves to develop an understanding of the possible states of the transistor (paralleling the ON–OFF states of the ordinary diode), and we learn to evaluate the transistor's operating point using a graphical approach. Several examples of this technique will be shown, and as with the diode, this approach will eventually be replaced by one which makes use of simple models that are valid for each of the transistor's states.

3.3-1 Graphical Analysis of BJT Circuits

The fundamental transistor Q-point analysis problem is illustrated in Figure 3.3-1. This type of circuit is known as a common-emitter configuration because the emitter of the transistor is at ac ground potential. It is probably the most popular of all bipolar transistor circuit configurations since, as we shall see later, it has a large voltage gain and can be biased with a single power supply voltage. The circuit shown in the figure has six unknowns: I_B, I_C, I_E, V_{BE}, V_{CE}, and V_{CB}. Fortunately, two of these can easily be eliminated by applying Kirchhoff's laws to the transistor to obtain

$$I_E = I_B + I_C \tag{3.3-1a}$$

and

$$V_{CB} = V_{CE} - V_{BE} \tag{3.3-1b}$$

As a result, only four unknowns, I_B, I_C, V_{BE}, and V_{CE}, actually need to be evaluated, with the other two following directly from the use of equations (3.3-1a) and (3.3-1b). To determine these four unknown quantities, four equations will be required. Two of these equations can be obtained by applying Kirchhoff's

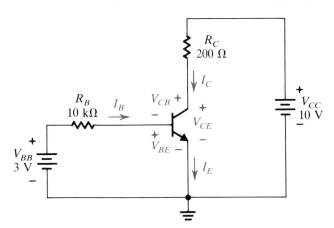

Figure 3.3-1
The common-emitter bipolar junction transistor circuit configuration.

voltage law to the input and output loops of the transistor circuit in Figure 3.3-1 to yield

$$V_{BB} = I_B R_B + V_{BE} \tag{3.3-2a}$$

$$V_{CC} = I_C R_C + V_{CE} \tag{3.3-2b}$$

Equation (3.3-2a) has two unknowns, and to solve it, another relationship between I_B and V_{BE} will need to be found. A similar requirement exists for I_C and V_{CE} in equation (3.3-2b). As with the case of the diode, these additional equations will be provided in graphical form by obtaining the device's input and output volt-ampere characteristics.

To understand the transistor's volt-ampere behavior, let's initially examine what happens to the input characteristic, i_B versus v_{BE}, as the collector-to-emitter voltage is varied in Figure 3.3-2a. Because the base–emitter junction of the transistor is basically a diode, we should not be too surprised to find that the i_B versus v_{BE} characteristic looks a lot like that of an ordinary diode. The variation of these curves with changes in v_{CE} needs a little more explaining.

When v_{CE} equals zero, a large portion of the current injected into the base by the emitter never makes it to the collector since there is no voltage on the collector to attract these charge carriers. Therefore, for a given value of v_{BE}, V_{BE1} in the figure, the resulting base current, I_{B3}, is rather large. As v_{CE} increases, so does the collection efficiency of the collector, and as a result, the base current for a given value of v_{BE} drops. It is useful to note that once the collector-to-emitter voltage exceeds a few tenths of a volt, the collection efficiency is at a maximum and nearly all of the emitter current is being swept

Figure 3.3-2

The volt-ampere characteristics of the bipolar junction transistor.

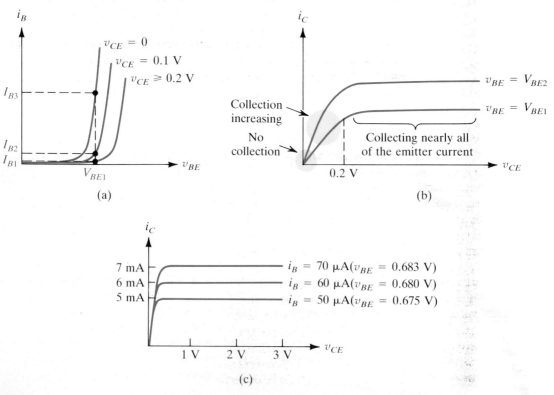

(a)

(b)

(c)

up by the collector. You should recall that the base current, I_{B1}, that still flows at this point is due to recombination losses in the base and to the diffusion of holes, for this *npn* transistor, from the base into the emitter.

The output characteristics of the transistor can best be understood by examining Figure 3.3-2b. The curves shown are obtained by applying a fixed value of base–emitter voltage to the transistor, $v_{BE} = V_{BE1}$, for example, then plotting the collector current i_C, as v_{CE}, the collector-to-emitter voltage, is varied. When v_{CE} is close to zero, only a small fraction of the current injected by the emitter into the base is "collected" by the collector. As the voltage on the collector increases, so does its collection ability, and as shown in the figure, i_C increases accordingly. Once v_{CE} exceeds a few tenths of a volt, nearly all the current being emitted by the emitter is being collected by the collector, so that for all practical purposes, the collector current no longer increases with further increases in v_{CE}.

The slight upward tilt of the curves in this region is caused by a phenomenon known as base-width modulation and can be explained as follows. Because the collector junction is reverse biased, as v_{CE} increases, so does the width of its depletion region. This reduces the effective width of the base and also reduces base recombination, so that more current is collected by the collector—hence the slight increase in collector current with increasing v_{CE}.

If v_{BE} is increased to V_{BE2} in Figure 3.3-2b, the current injected by the emitter into the base is correspondingly increased, so that the collector current plateaus at a higher level. Therefore, for a given set of v_{BE} values, there is a corresponding set of i_C versus v_{CE} characteristic curves. However, because the collector current is exponentially related to the applied base–emitter voltage, very small changes in v_{BE} create large changes in i_C. As a result, labeling the output family of curves with different v_{BE} values is not very useful, and as a practical matter, these curves are usually drawn with i_B and not v_{BE} as the input variable. A typical set of transistor output curves is illustrated in Figure 3.3-2c for several different values of i_B. The approximate corresponding values of v_{BE} are included in parentheses for comparison.

EXAMPLE 3.3-1

Find the Q point of the circuit shown in Figure 3.3-1 using the transistor volt-ampere characteristics given in Figure 3.3-3.

SOLUTION

Following the same techniques used previously for graphically analyzing diode circuits, the load-line equations given in (3.3-2a) and (3.3-2b) have been sketched in parts (a) and (b), respectively, on the transistor characteristics given in Figure 3.3-3. The output load-line equation may be sketched as usual by determining the V_{CE} and I_C intercepts of the equation and drawing a straight line through these points. Using equation (3.3-2b), these can be seen to be at $V_{CE} = V_{CC} = 10$ V and $I_C = V_{CC}/R_C = 50$ mA, respectively.

The input load line is a bit more difficult to sketch because the V_{BE} intercept at 3 V is not on the graph. Some other point on the graph which also satisfies equation (3.3-2a) will be required. Consider, for example, the point where $V_{BE} = 1.0$ V. From equation (3.3-2a) the corresponding base current satisfying this equation is $I_B = (3.0 - 1.0)/10 \text{ k}\Omega = 0.20$ mA. Using this point ($I_B = 0.20$ mA, $V_{BE} = 1.0$ V) together with the I_B intercept at $V_{BB}/R_B = 0.3$ mA, we may sketch the input load-line equation as shown in Fig. 3.3-3a.

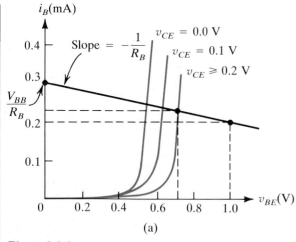

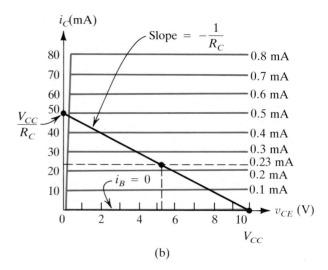

Figure 3.3-3
Volt-ampere characteristics for the silicon transistor in Figure 3.3-1.

Despite the fact that both load lines are now in place, it is still not immediately clear where the transistor's Q point lies because each of the load lines intersects several different volt-ampere curves. To determine which curves to use in order to evaluate the actual circuit Q point, let us begin by examining Figure 3.3-3a, and initially assume that we are on the $V_{CE} = 0$ V curve, that is, that the quiescent collector-to-emitter voltage is equal to zero for this circuit. If $V_{CE} = 0$, the circuit's Q point is at the intersection of the load line with the $V_{CE} = 0$ curve or at $V_{BE} = 0.53$ V and $I_B = 0.25$ mA.

To determine the corresponding output portion of the Q point, we must find out where the $I_B = 0.25$ mA output characteristic curve intersects the output load line. If we examine Figure 3.3-3b, we will see that no specific curve corresponding to $I_B = 0.25$ mA is shown; but by interpolation, we could, if necessary, sketch it in approximately. It would be located slightly above the dashed line at $I_B = 0.23$ mA in the figure. Using this approximation, the output load line intersects this curve at about $I_C = 25$ mA and $V_{CE} = 4.8$ V. This result is inconsistent with the original assumption that V_{CE} was equal to zero, and as a result, this guess as to the value of V_{CE} was incorrect. A similar sequence of events would follow were we to assume a value of 0.1 V for V_{CE}, but the assumption that V_{CE} is > 0.2 V will lead to a set of consistent results.

For this case, using Figure 3.3-3a, we can see that the input portion of the Q point is at about $I_B = 0.23$ mA and $V_{BE} = 0.72$ V for this silicon transistor. To determine the output values for the transistor's Q point, we must locate the intersection of the output load line and the $I_B = 0.23$ mA curve. Following Figure 3.3-3b, we find that this occurs at $I_C = 23$ mA and $V_{CE} = 5.2$ V. Following our original assumption, the resulting value of V_{CE} is greater than 0.2 V, and therefore these answers do represent the actual Q point of the transistor. These results may be summarized as $I_B = 0.23$ mA, $V_{BE} = 0.72$ V, $I_C = 23$ mA, and $V_{CE} = 5.2$ V.

The analysis of Example 3.3-1 by graphical techniques is, to say the least, cumbersome, and it is fortunate that the simple model described in Figure 3.2-5 works very well for analyzing this particular problem. In general, however, depending on the particular location of the Q point, this model may not always correctly describe the behavior of the transistor. Additional models will be needed. To develop an understanding of what these models look like, and when their use is appropriate, we need to take a closer look at the transistor's output volt-ampere characteristics.

The output, or collector characteristics, have three major regions of interest that we need to define. These are called the active, cutoff, and saturation regions. When the transistor is operating in the active region, the base–emitter junction is forward biased and the base–collector junction is reverse biased. This is the normal operating mode of the transistor when it is being used as a linear amplifier, and hence the terms "active region" and "linear region" are sometimes used interchangeably to describe this part of the transistor's volt-ampere characteristics.

The portion of the output characteristics where the transistor is in the active region may be determined by using the simple two-diode picture given in Figure 3.3-4 for the case of an *npn* transistor. Here, if the base–emitter junction is forward biased, v_{BE} is equal to V_o, and v_{BC} must be less than zero to keep the base–collector junction reverse biased. Therefore, v_{CE}, which equals $v_{BE} - v_{BC}$, should be greater than a few tenths of a volt in order to guarantee that the transistor is in the active region. As a rough approximation, we will say that the base–collector junction is reverse biased whenever v_{CE} is greater than zero. In addition, since the base–emitter junction is forward biased, i_B will be greater than zero. These two requirements, that is, that both v_{CE} and i_B be greater than zero, are sufficient to guarantee that the transistor is in the active region. You should recall that a transistor operating in the active region has a collector current given by equation (3.2-17). The active-region portion of the output volt-ampere characteristics is illustrated in Figure 3.3-5, and the

Figure 3.3-5

The operating regions of the bipolar transistor:

Figure 3.3-4

A simple two-diode model for understanding the state of the bipolar transistor.

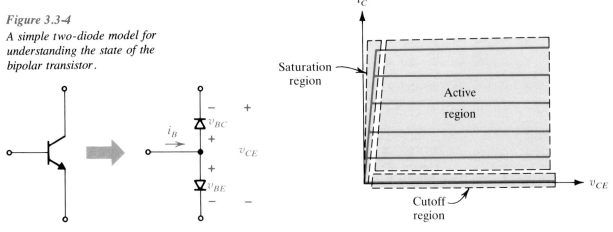

State of the Transistor	Status of the Junctions	Reasonable *npn* Approximations	Conditions to Test for (to Ensure Are in This State)	Approximate *npn* Model in That State
Active	Base-emitter is forward biased, base-collector is reverse-biased	$V_{BE} \approx V_o = \begin{cases} 0.7 \text{ V for Si} \\ 0.3 \text{ V for Ge} \end{cases}$ $I_C = h_{FE} I_B$	$V_{CE} > 0$ V (approx.) I_B or $I_C > 0$	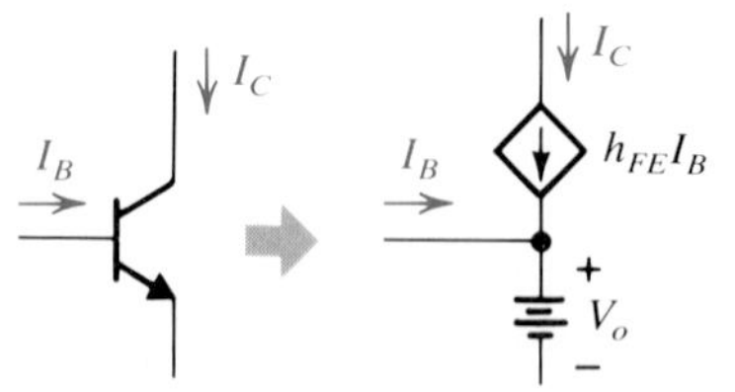
Cutoff	Base-emitter is reverse biased, base-collector is reverse-biased	$I_B = 0,\ I_C = 0,\ I_E = 0$	$V_{BE} < 0$ V, $V_{CE} > 0$ V	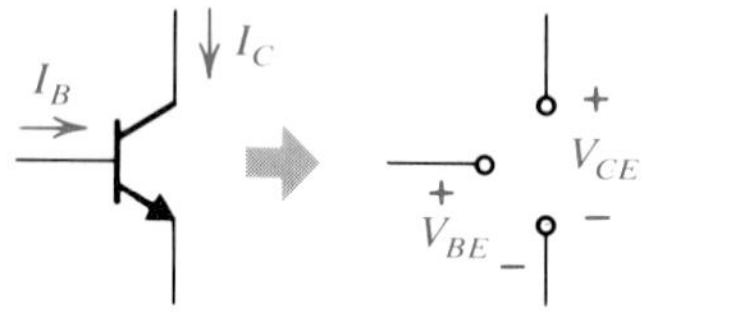
Saturation	Base-emitter is forward biased, base-collector is forward biased	$V_{BE} \approx V_o = \begin{cases} 0.7 \text{ V for Si} \\ 0.3 \text{ V for Ge} \end{cases}$ $I_C = I_{C,\text{sat}}$ $V_{CE} \approx \begin{cases} 0.2 \text{ V for Si} \\ 0.1 \text{ V for Ge} \end{cases}$	$I_B > I_{B,\text{min}}$ where $I_{B,\text{min}} = \dfrac{I_{C,\text{sat}}^{*}}{h_{FE}}$	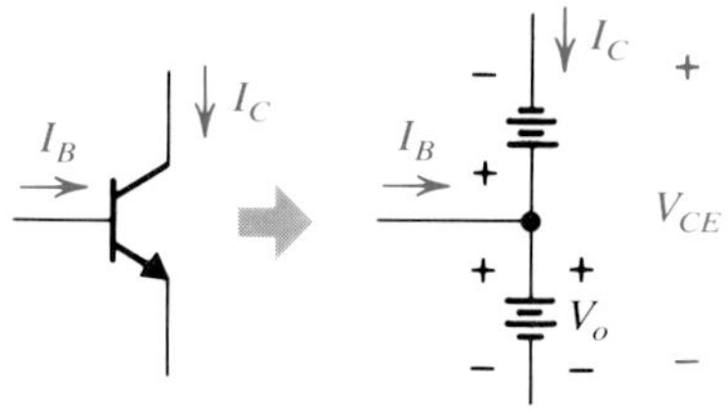

*$I_{C,\text{sat}}$ is the collector current that flows in the transistor when it is replaced by its saturation model.

Figure 3.3-6

The three important states of the bipolar transistor.

transistor's performance characteristics in the active region are summarized in Figure 3.3-6.

In the cutoff and saturation regions, the transistor is not operating as a linear device, and as a result, there is no longer a one-to-one correlation between the shape of the input and output signals. When operating in these modes, as we shall see shortly, the transistor's behavior is similar to that of a switch. This analogy is illustrated in Figure 3.3-7.

To cut off the transistor, both the base–emitter junction and the base–collector junction are reverse biased. As long as v_{BE} is less than V_o volts, the base–emitter diode will be off, and for all v_{CE} greater than zero, the base–collector diode will also be off. With $v_{BE} < 0$, i_B will be zero, and from equation (3.2-24) this will guarantee that i_C will also be zero. The cutoff region of the collector characteristics is found in the lower portion of Figure 3.3-5. In cutoff, no matter what voltage v_{CE} is across the transistor, the collector current through it will be zero. This behavior is the same as that of a switch operating in the open position (Figure 3.3-7).

To saturate a transistor, both the base–emitter and base–collector junctions must be forward biased. Again using our simple diode picture of the transistor, we can see that if both diodes are forward biased, then, to the extent that the voltage drops across them are equal (Figure 3.3-7c), the collector-to-emitter voltage, v_{CE}, will be approximately zero. Thus, the saturation region is located on the extreme left-hand side of the collector characteristics in Figure 3.3-5. Continuing the transistor-switch analogy, we can see that because a saturated

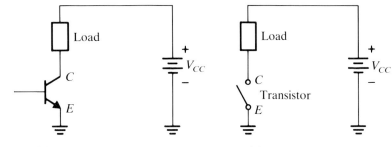

(a) Using the Transistor to   (b) Transistor in Cutoff
     Switch Power to a Load

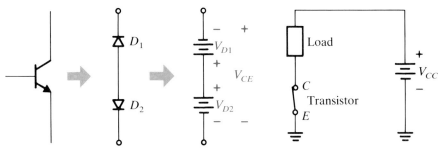

Figure 3.3-7

The operation of the BJT as a switch in the cutoff and saturation regions.

(c) Saturation Model for a BJT   (d) Transistor in Saturation

transistor has no voltage drop across it, although substantial current may flow through it, its behavior closely resembles that of a switch operating in the closed position (Figure 3.3-7d).

For commercially produced transistors, the base–collector voltage drop is usually a little smaller than the base–emitter drop because the doping level in the collector is lighter than that in the emitter. As a result, when a transistor saturates, v_{CE} is actually slightly greater than zero, being about 0.2 V for silicon transistors and 0.1 V for germanium devices.

The conditions necessary to place a transistor in saturation can best be understood by reexamining the circuit given in Figure 3.3-1, except that for this analysis we will assume that the battery V_{BB} is variable and can be adjusted to produce any desired value of base current in the transistor (Figure 3.3-8a). The transistor collector characteristics have been redrawn in Figure 3.3-8b along with the output load line for this problem. The Q point for this circuit depends on the base current flowing in the transistor. The Q-point locations corresponding to several different values of base current are indicated on the load line as Q_0 through Q_8, and the detailed data associated with each of these Q points are summarized in Table 3.3-1.

When I_B is equal to zero, the transistor Q point is at the intersection of the load line and the $i_B = 0$ curve. This point is denoted as Q_0 in Figure 3.3-8b, and here $I_C = 0$ and $V_{CE} = 10$ V, which corresponds to the cutoff state for the transistor. If I_B is increased to 0.1 mA, the Q point moves up on the load line to Q_1. Here $I_C = 10$ mA and $V_{CE} = 8$ V, and as shown by its position on the load line, the transistor enters the active region. Even in the absence of any graphical data, we could confirm that the transistor is in the active region by noting that I_B is greater than zero, as is V_{CE}. These are the conditions stipulated in Figure 3.3-5 to guarantee that the transistor is in the active region.

As I_B is increased from 0.1 through 0.4 mA, the transistor remains in the active region and its Q point moves along the load line from positions Q_1 through Q_4 (see Figure 3.3-8b and Table 3.3-1). When I_B reaches 0.5 mA, the collector current is 50 mA and V_{CE} is reduced to approximately zero volts. The transistor is now in the saturation region of the collector characteristics at the operating point Q_5.

If I_B is increased further, to 0.6 mA, for example, something very interesting happens. The Q point for this case is determined by the intersection of the $i_B = 0.6$ mA curve with the load line. This specific curve has been drawn darker than the others in Figure 3.3-8b to emphasize that the vertical part of the transistor characteristic is also part of the $i_B = 0.6$ mA curve, and therefore that the actual Q point for this case, Q_6, is at nearly the same location as Q_5. As a result, the collector current is still 50 mA, and V_{CE} remains at about zero volts. In fact, the location of all Q points for which I_B is greater than 0.5 mA will coalesce at the same Q point as the $I_B = 0.5$ mA operating point,

Figure 3.3-8

Variation of the transistor Q point with V_{BB}.

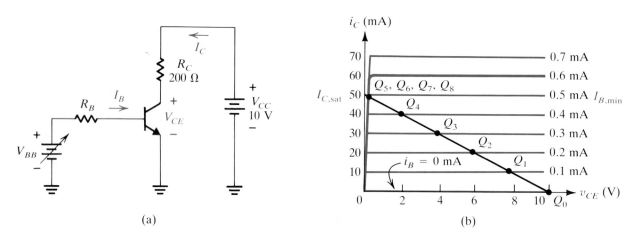

(a)   (b)

3.3-1 Tabulation of Transistor Q-Point Data for the Circuit of Figure 3.3-8

I_B Applied to Base (mA)	Resulting I_C (mA)	Resulting V_{CE} (V)	Q-Point Location	Comments
0	0	10	Q_0	Transistor is cut off
0.1	10	8	Q_1	Transistor is active
0.2	20	6	Q_2	" " "
0.3	30	4	Q_3	" " "
0.4	40	2	Q_4	" " "
0.5	50	0	Q_5	Transistor just saturates
0.6	50	0	Q_6	Transistor is saturated
0.7	50	0	Q_7	" " "
0.8	50	0	Q_8	" " "

Q_5. Therefore, the transistor is saturated as long as the base current into it exceeds this minimum value of 0.5 mA. This current is designated $I_{B,\text{min}}$.

To determine the minimum value of I_B required to saturate the transistor without resorting to a complete graphical analysis, we notice that $I_{B,\text{min}}$ can be obtained from the saturation collector current by

$$I_{B,\text{min}} = \frac{I_{C,\text{sat}}}{h_{FE}} \tag{3.3-3}$$

where

$$I_{C,\text{sat}} = \frac{V_{CC}}{R_C} \tag{3.3-4}$$

for this example. In general, $I_{C,\text{sat}}$ is the collector current that flows in the transistor when it is replaced by its saturation model (Figure 3.3-5). The h_{FE} of the transistor needed to solve equation (3.3-3) can be determined by evaluating equation (3.2-17) in the neighborhood of the Q point as

$$h_{FE} = \left. \frac{i_C}{i_B} \right|_{Q\,\text{pt}} = \frac{30 \text{ mA}}{0.3 \text{ mA}} = 100 \tag{3.3-5}$$

For the circuit being examined, using equations (3.3-5) and (3.3-4), $I_{C,\text{sat}}$ and $I_{B,\text{min}}$ are $10/0.2 = 50$ mA and $50/100 = 0.5$ mA, respectively. Therefore, the transistor circuit shown in Figure 3.3-8 will be saturated for all values of base current greater than 0.5 mA.

The technique required to determine the state of a specific transistor circuit is actually quite similar to that employed for finding the operating points of diode circuits. Based on a visual examination of the transistor circuit whose Q point is to be determined, an initial guess is made as to the state of the transistor. The transistor is next replaced by the model corresponding to the state selected, and the pertinent currents and voltages are calculated for the circuit with this model in place. These results are then compared against the entries in column 4 of the table given in Figure 3.3-6 to determine if the original guess was correct. Do not worry if this method is still a bit unclear to you; the examples that follow should help to clarify this technique.

EXAMPLE 3.3-2

Repeat the analysis of the circuit given in Figure 3.3-1 by using the transistor equivalent-circuit method just discussed. You may consider that the transistor is silicon and that it has an $h_{FE} = 100$.

SOLUTION

From an examination of the circuit in Figure 3.3-1, it should be clear that the base–emitter junction is forward biased, and therefore the transistor is either in the active or saturated state. Let us assume the transistor to be active and replace it by its equivalent model (Figure 3.3-9). From the input loop,

$$I_B = \frac{V_{BB} - V_o}{R_B} = \frac{2.3}{10} = 0.23 \text{ mA} \tag{3.3-6a}$$

and

$$V_{BE} = V_o = 0.7 \text{ V} \tag{3.3-6b}$$

The collector current is obtained directly as

$$I_C = h_{FE}I_B = 23 \text{ mA} \tag{3.3-7}$$

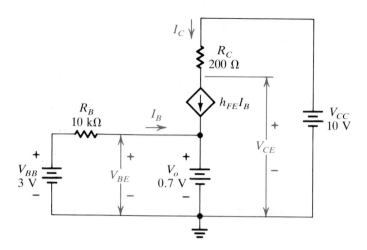

Figure 3.3-9
Equivalent circuit for Example 3.3-2.

Applying Kirchhoff's law to the output loop, we may calculate the collector-to-emitter voltage from

$$V_{CE} = V_{CC} - I_C R_C = 10 - 4.6 = 5.4 \text{ V} \qquad (3.3\text{-}8)$$

From equation (3.3-1b) the base–collector voltage, V_{BC}, is $(0.7 - 5.4)$ or -4.7 V, so that the base–collector junction is reverse biased. From Figure 3.3-6, using the fact that the base–emitter junction is forward biased and that the base–collector junction is reverse biased, we can see that our original assumption that the transistor was in the active region was correct. You should observe that these results, specifically that $I_B = 0.23$ mA, $V_{BE} = 0.7$ V, $I_C = 23$ mA, and $V_{CE} = 5.4$ V, compare quite favorably with those obtained earlier by graphical techniques in Example 3.3-1.

EXAMPLE 3.3-3

The transistor shown in Figure 3.3-10 is silicon and has an $h_{FE} = 100$. Prove that it is saturated.

SOLUTION

In the circuit shown in Figure 3.3-10a, the node labeled $+10$ V illustrates a notation that is commonly used in electronic circuit schematics. It indicates that both R_B and R_C are to be connected to the $+10$-V power supply, and implies that the other end of this supply is connected to ground. If it helps you to visualize the circuit better, you can, in effect, consider that there is a $+10$-V battery connected between this node and ground (Figure 3.3-10b). To many people, Figure 3.3-10b is still confusing because both R_B and R_C are connected together at the top. This circuit need not present a problem, because it is simply a way of indicating that the potential at the top of each of these resistors is $+10$ V. In fact, if it is easier for you to understand, you can split the V_{CC} source into two separate batteries, as shown in Figure 3.3-10c, and analyze this circuit instead. All these circuit schematics represent the same physical circuit. Use the one that is the easiest for you to work with.

In finding the Q point for this circuit, we are asked to show that this transistor is saturated, so we will begin the analysis of this problem by replacing the transistor by its saturated model (Figure 3.3-10d). The actual base current

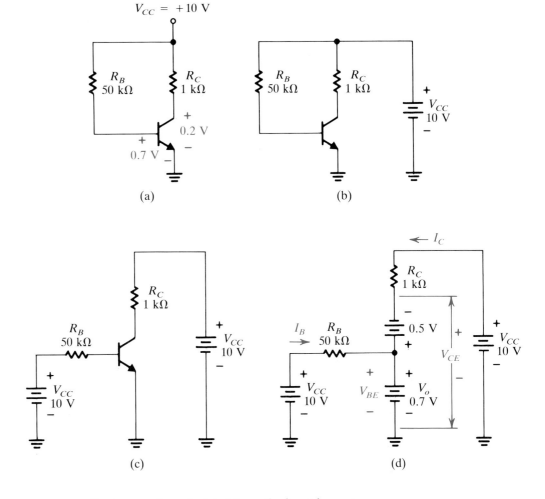

Figure 3.3-10
Example 3.3-3.

flowing can be calculated from the input loop as

$$I_B = \frac{V_{CC} - V_o}{R_B} = \frac{9.3}{50} = 0.186 \text{ mA} \qquad (3.3\text{-}9)$$

If the transistor is saturated, as the model shown in the figure assumes, the saturation collector current can be obtained from the collector loop as follows:

$$I_{C,\text{sat}} = \frac{V_{CC} - V_{CE}}{R_C} = \frac{10 - 0.2}{1 \text{ k}\Omega} = 9.8 \text{ mA} \qquad (3.3\text{-}10)$$

Substituting this result into equation (3.3-3), we can see that the minimum base current, $I_{B,\text{min}}$, needed to saturate this transistor is 9.8 mA/100 or 0.098 mA. From equation (3.3-9), it is apparent that the actual base current flowing in this transistor is well in excess of $I_{B,\text{min}}$, so that the transistor is saturated.

Notice that for this example, as for many others that we will examine, it is not really necessary to go through all the trouble of drawing the equivalent circuit in Figure 3.3-10d. The entire analysis can be carried out directly from the original circuit given. If we are assuming the transistor to be saturated, we simply label V_{BE} and V_{CE} as 0.7 and 0.2 V, respectively, as shown in Figure

3.3-10a. The base current is obtained by adding up the voltage drops from the top to the bottom on the left-hand side of the figure, setting their sum equal to V_{CC}:

$$V_{CC} = I_B R_B + V_{BE} = I_B R_B + 0.7 \qquad (3.3\text{-}11)$$

You should observe that this equation is the same as (3.3-9) and is just another way of applying Kirchhoff's law to the input loop of the circuit. The collector current is obtained in the same way by adding up all the voltages on the right-hand side of the circuit, setting their sum equal to V_{CC} to yield

$$V_{CC} = I_C R_C + V_{CE} = I_{C,\text{sat}} R_C + 0.2 \qquad (3.3\text{-}12)$$

Once these two equations are obtained, the remainder of the analysis follows exactly as before.

3.3-3   Q-Point Analysis of BJT Circuits Using DC Equivalent Circuits

In the remainder of this section, we discuss an equivalent-circuit technique that will permit us to carry out the dc analysis of active multitransistor bipolar transistor circuits by inspection. In addition, and of even greater significance, as we will see later in the book, is the fact that this method can also be extended to cover the dc analysis of field-effect transistors as well as the ac analysis of all types of transistor circuits.

To begin the development of this technique, consider the generalized BJT circuit given in Figure 3.3-11a, in which R_B and V_{BB} represent the Thévenin equivalent of the base input circuit. The transistor is replaced by its active region model in part (b) of the figure, and in part (c) the original current source, $h_{FE} I_B$, has been split into two equivalent current sources, each having the same value of $h_{FE} I_B$. At first the two circuits in Figure 3.3-11b and c may not appear to be equivalent, but on closer study we will see that they are in fact the same because both circuits are described by identical node equations. For example, in the original circuit there is a current $h_{FE} I_B$ flowing into the base node, and in the new circuit, $h_{FE} I_B$ still flows into this node. Similarly, in the original circuit, a current $h_{FE} I_B$ flows out of the collector node; this is still the case for the new circuit as well. Therefore, for all practical purposes, the two circuits are electronically the same. To see why Figure 3.3-11c is useful, let's develop the equivalent circuits that we "see" looking into the base, the emitter, and the collector of the transistor. Because it is the easiest, we begin at the collector.

The collector portion of Figure 3.3-11c has been redrawn in Figure 3.3-12, and we will find its Thévenin equivalent. For this circuit the Thévenin or open-circuit voltage is

$$V_{OC} = V_{CC} - h_{FE} I_B R_C \qquad (3.3\text{-}13)$$

and the short-circuit current is

$$I_{SC} = \frac{V_{CC}}{R_C} - h_{FE} I_B \qquad (3.3\text{-}14)$$

By dividing V_{OC} by I_{SC}, we find that the Thévenin resistance, or circuit output impedance, is R_C. This impedance could also have been found by setting the

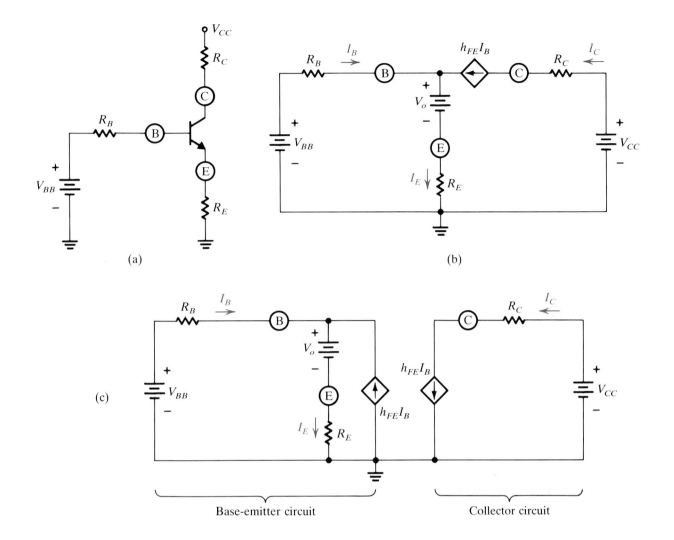

(a)

(b)

(c)

Base-emitter circuit   Collector circuit

Figure 3.3-11
General bipolar junction transistor circuit.

Figure 3.3-12
The equivalent circuit looking back into the collector.

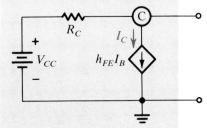

sources equal to zero, that is, by shorting V_{BB} and V_{CC}, opening $h_{FE}I_B$, and looking back into the terminals of the network. In general, you should be very careful when setting controlled sources equal to zero. The controlled current source $h_{FE}I_B$ is open in this case because when V_{BB} is set equal to zero, I_B and hence $h_{FE}I_B$ are also made zero.

To develop an equivalent circuit for what we "see" looking into the base terminal of the transistor, we need to reexamine the base portion of Figure 3.3-11c. The goal in analyzing this circuit is to replace the portion of the circuit to the right of the base terminal by some simpler circuit that will tell us more about how the transistor loads the external circuit containing V_{BB} and R_B. For this circuit, V_B, the voltage from base to ground, can be written as

$$V_B = \underbrace{V_o}_{V_1} + \underbrace{(1 + h_{FE})I_B R_E}_{V_2} \qquad (3.3\text{-}15)$$

In Figure 3.3-13a, the original equivalent circuit in Figure 3.3-11c has been replaced by the series connection of two elements whose values are yet to be

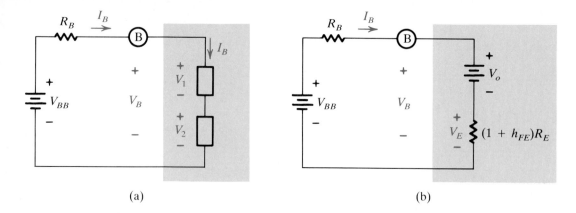

(a)   (b)

Figure 3.3-13

The equivalent circuit looking into the base.

determined. This is a simple series circuit, and the current through each of the elements is the same and is equal to I_B. The voltages across these elements, V_1 and V_2, represent the two voltage terms in equation (3.3-15). Let's try to figure out what each of these elements is. The box labeled V_1 has a constant voltage V_o across it regardless of the current that flows through it. This element, of course, is a battery V_o. The second box has a voltage drop across it equal to

$$V_2 = [(1 + h_{FE})I_B]R_E = [(1 + h_{FE})R_E]I_B \qquad (3.3\text{-}16)$$

Because this element has a voltage drop across it proportional to the current I_B flowing through it, it can be represented by a resistor of value $V_2/I_B = (1 + h_{FE})R_E$. As a result, when a resistor in the emitter is "reflected" into the base, its impedance is magnified by the factor $(1 + h_{FE})$. The overall equivalent circuit seen looking into the base is shown in Figure 3.3-13b.

To obtain the equivalent circuit seen looking back into the emitter, we have redrawn the base–emitter circuit of Figure 3.3-11c in Figure 3.3-13a. To find the Thévenin equivalent of the circuit within the shaded region in Figure 3.3-14a, we need to find the open-circuit voltage and the short-circuit current. Under open-circuit conditions (i.e., with R_E removed), I_E, and therefore I_B, are both zero, so that the current source, $h_{FE}I_B$, is zero as well. For this case the

Figure 3.3-14

The equivalent circuit looking back into the emitter.

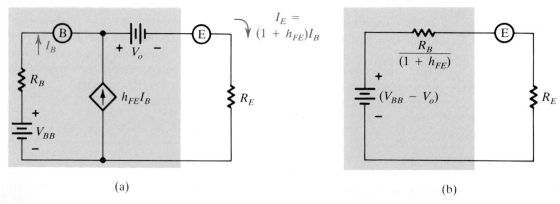

(a)   (b)

open-circuit voltage may be written as

$$V_{OC} = V_{BB} - V_o \tag{3.3-17}$$

To find the Thévenin resistance, or the circuit's output impedance, we will first need to determine the short-circuit current. With R_E replaced by a short, the base current is now equal to

$$I_B = \frac{V_{BB} - V_o}{R_B} \tag{3.3-18a}$$

so that the short-circuit current at the emitter terminals is

$$I_{SC} = \frac{(1 + h_{FE})(V_{BB} - V_o)}{R_B} \tag{3.3-18b}$$

Dividing equation (3.3-17) by (3.3-18b), we obtain the Thévenin resistance as $R_B/(1 + h_{FE})$. The overall equivalent circuit is given in Figure 3.3-14b.

Sometimes when working on an engineering problem, the mathematics becomes so involved that we lose sight of the fundamental nature of what we have proven. Therefore, before moving on to application examples of these equivalent circuits, it might first be useful to summarize our observations.

1. The equivalent circuit seen looking into the base of the transistor is obtained by taking the entire circuit in the emitter and reflecting it into the base. Similarly, the equivalent circuit looking back into the emitter is found by reflecting the circuit from the base into the emitter.
2. Impedances reflected from the emitter into the base, such as R_E in Figure 3.3-13b, are magnified by the factor $(1 + h_{FE})$.
3. Impedances reflected from the base into the emitter, such as R_B in Figure 3.3-14b, are divided by $(1 + h_{FE})$.
4. Voltages reflected back and forth across the base–emitter junction, V_{BB} and V_o, for example, are unchanged (see Figures 3.3-13b and 3.3-14b).
5. The collector circuit has no effect on what happens in the base or the emitter as long as the transistor stays in the active region. This result is readily apparent from the models in Figures 3.3-13 and 3.3-14, and also from Figure 3.3-11c, since the collector circuit in this figure is completely split off by itself.
6. The output impedance in the collector circuit is just equal to the collector resistor R_C. The collector current is $h_{FE}I_B$, where I_B is determined by examining the equivalent base circuit. The collector current may also be found by working with the emitter equivalent circuit and using the fact that I_C and I_E are approximately equal.

EXAMPLE 3.3-4

The circuit illustrated in Figure 3.3-15a contains an emitter resistor. Find I_B, I_C, V_B, V_E, V_C, V_{BE}, and V_{CE} given that $h_{FE} = 100$ for the transistor.

SOLUTION

In Figure 3.3-15b we have drawn the input equivalent circuit and have replaced the R_1–R_2 resistive divider by its Thevenin equivalent. By inspection, the current flow in this circuit is

$$I_B = \frac{V_{BB} - V_o}{R_B + (1 + h_{FE})R_E} = \frac{9.3 \text{ V}}{755 \text{ k}\Omega} = 0.0123 \text{ mA} \tag{3.3-19}$$

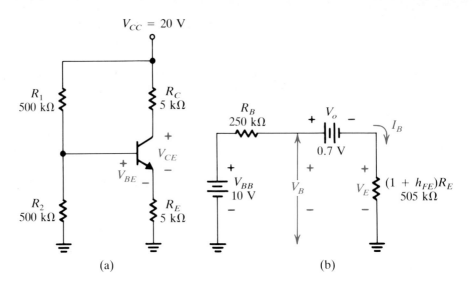

Figure 3.3-15
Example 3.3-4.

(a)

(b)

The voltage V_E in this circuit is seen to be $I_B(1 + h_{FE})R_E = (0.0123 \text{ mA})$ $(505 \text{ k}\Omega) = 6.22$ V, and V_B is simply $(V_o + V_E)$ or 6.92 V.

Assuming the transistor to be in the active region, the collector current is $h_{FE}I_B = 100(0.0123 \text{ mA}) = 1.23$ mA. From Figure 3.3-15b the collector voltage is given by

$$V_C = V_{CC} - I_C R_C = 20 - (1.23)5 = 13.5 \text{ V} \qquad (3.3\text{-}20)$$

Using this result, the collector-to-emitter voltage is

$$V_{CE} = V_C - V_E = 13.5 - 6.22 = 7.28 \text{ V} \qquad (3.3\text{-}21)$$

Exercises

3.3-1 If R_B in Figure 3.3-1 is adjustable, and if $h_{FE} = 100$, find the value of R_B needed to just saturate the transistor. *Answer* 4.6 kΩ

3.3-2 For the circuit given in Figure 3.3-10a, find the value of V_{CE} if R_B is removed from V_{CC} and connected to the collector of the transistor. *Answer* 3.78 V

3.3-3 Find I_C for the circuit in Figure 3.3-15 if $h_{FE} = 20$. *Answer* 0.524 mA

3.4 Q-POINT STABILITY

In the design of BJT circuits, we are usually interested in placing the Q point at a specific location. If maximum signal swing at the output is desired, the Q point needs to be positioned at the center of the load line, while if minimum power consumption from the dc supply is more important, as is often the case with portable electronic equipment, it may be more desirable to locate the Q point somewhere near cutoff at the bottom of the load line (Figure 3.4-1). Regardless of the specific position that is finally selected, it is usually important that the Q point remain fixed at this location despite variations that might occur in the transistor's parameters.

The Q point of a BJT is a function of the external circuit in which the transistor is connected, as well as being a function of the transistor h_{FE}, the base–emitter voltage drop V_o, and the base–collector reverse leakage current I_{CBO}. All of these device parameters vary widely with temperature. In addition, the h_{FE} also changes significantly from one device to the next, with production spreads of 3 or 4:1 being common. Thus, in designing BJT circuits having stable Q points, biasing techniques need to be developed that make the Q-point position relatively independent of the transistor's h_{FE}, V_o, and I_{CBO} values.

The circuit given in Figure 3.4-2, known as a fixed-bias design, is an ex-

Figure 3.4-1
Location of the Q point on the load line.

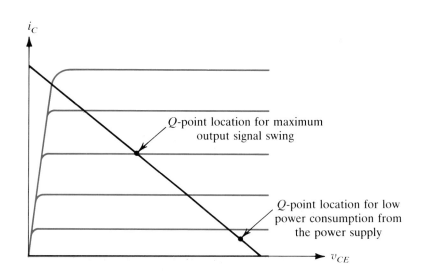

Q-point location for maximum output signal swing

Q-point location for low power consumption from the power supply

Figure 3.4-2
A fixed bias transistor circuit.

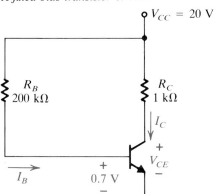

(a)

Q-point Location	I_B (μA)	h_{FE}	I_C (mA)	V_{CE} (V)	Comment
QP1	100	100	10	10	Q point at center of load line
QP2	100	150	15	5	Transistor still active
QP3	100	200	20	~0	Transistor saturated
QP4	100	300	20	~0	Transistor saturated

(b)

ample of a poor way to achieve Q-point stability. In analyzing this circuit, we will consider that the transistor is silicon and that it has a nominal h_{FE} of 100. Using the input loop equation and solving for I_B, the quiescent base current is

$$I_B = \frac{V_{BB} - V_{BE}}{R_B} = \frac{20 - 0.7}{200} \simeq 100 \ \mu A \qquad (3.4\text{-}1)$$

and with $h_{FE} = 100$, the corresponding collector current and collector-to-emitter voltage are $I_C = 10$ mA and $V_{CE} = 10$ V. These data are listed in Figure 3.4-2b, and the position of the load line and the Q point for this particular h_{FE} value are also shown on the volt-ampere characteristics in Figure 3.4-3a. Notice that the Q point is located at the approximate center of the load line at its intersection with $i_B = 100 \ \mu A$ transistor curve.

If another transistor of the same type, but with a different h_{FE}, is substituted into this circuit, the Q point shifts markedly. As shown in Figures 3.4-3b and 3.4-2b, an h_{FE} increase to 150 causes a significant Q-point shift to the left on the load line, while a further increase to 200 or 300 will actually cause the transistor to saturate, rendering it useless as a linear amplifier. It is interesting to observe that the base current for all these h_{FE} values remains fixed at 100 μA. In fact, this is why this biasing arrangement is called a fixed-bias circuit. Unfortunately, however, to keep the output portion of the Q point stable

Figure 3.4-3

Graphical evaluation of the Q point for the fixed bias circuit in Figure 3.4-2.

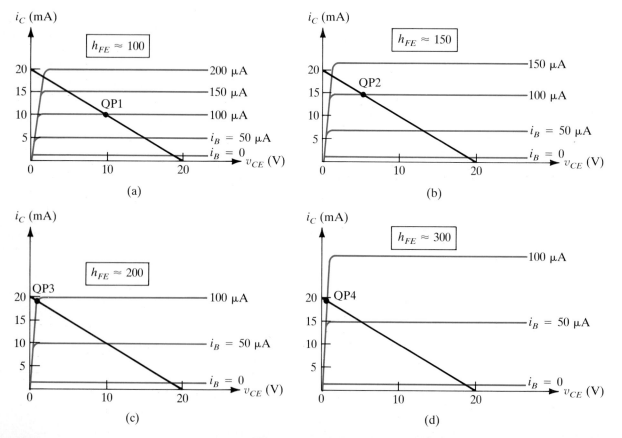

(a)   (b)   (c)   (d)

we need to maintain the collector current, not the base current, at a fixed value. The self-bias circuit illustrated in Figure 3.4-4 can help us to achieve this type of Q-point stability through the negative feedback associated with the addition of the emitter resistor R_E.

In Figure 3.4-4b, we have drawn the input equivalent circuit and have replaced the R_1–R_2 resistive divider by its Thévenin equivalent. Using this equivalent circuit, the base current is

$$I_B = \frac{V_{BB} - V_o}{R_B + (1 + h_{FE})R_E} \tag{3.4-2}$$

and neglecting I_{CBO}, we may write the collector current as

$$I_C = h_{FE}I_B = \frac{h_{FE}(V_{BB} - V_o)}{R_B + (1 + h_{FE})R_E} \tag{3.4-3}$$

Notice that if
$$(1 + h_{FE})R_E \gg R_B \tag{3.4-4}$$

the R_B term may be neglected in the denominator, and because h_{FE} is usually much greater than 1, the h_{FE} in the numerator of equation (3.4-3) nearly cancels the $(1 + h_{FE})$ term in the denominator, so that the collector current can be approximated as

$$I_C \simeq \frac{V_{BB} - V_o}{R_E} \tag{3.4-5}$$

At this point, several rather important observations should be made. To begin with, to the extent that equation (3.4-5) is valid, the position of the Q point is now independent of the transistor h_{FE}. This occurs because although increases in h_{FE} tend to increase I_C, from equation (3.4-2), these same h_{FE} increases also cause I_B to decrease, so that the product $I_C = h_{FE}I_B$ remains relatively constant.

A second point to observe is that when the inequality in equation (3.4-4) is

Figure 3.4-4
A self-biased bipolar junction transistor circuit.

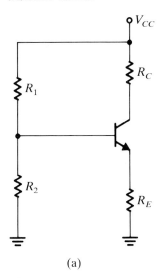

(a)

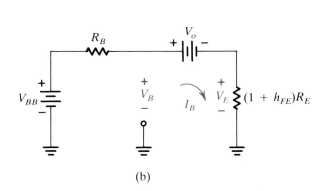

(b)

satisfied, the transistor does not load the $R_B - V_{BB}$ portion of the circuit since $I_B \ll I_{R1}$ and I_{R2} (or $I_B R_B \ll V_{BB}$). As a result, in this instance the voltage at the base of the transistor, V_B, is approximately equal to V_{BB} (see Figure 3.4-4b). Using this fact, V_E and I_E can be written as

$$V_E = V_{BB} - V_o \qquad (3.4\text{-}6a)$$

and
$$I_E = \frac{V_{BB} - V_o}{R_E} \qquad (3.4\text{-}6b)$$

Notice that because $I_C \simeq I_E$, this approach yields the same result for I_C as obtained earlier [equation (3.4-3)].

Thus far we have seen that selecting R_E and R_B to satisfy the inequality in equation (3.4-4) results in a circuit whose Q point is relatively independent of the transistor's h_{FE}. In addition, equation (3.4-5) also illustrates that the effect of the base–emitter voltage on the position of the Q point can be minimized by choosing V_{BB} to be much greater than V_o.

The analysis for the transistor collector current in Figure 3.4-4 has thus far neglected the effect of I_{CBO}. To include this term and to quantify the effects of changes in h_{FE} and V_o, we must use the full expression for I_C given in equation (3.2-16) and must work directly from Figure 3.4-4a since the model in part (b) of the figure does not include I_{CBO}. Applying Kirchhoff's voltage law to the input loop (after the R_1–R_2 network is replaced by its Thévenin equivalent), we obtain

$$V_{BB} = I_B R_B + V_o + (I_C + I_B)R_E \qquad (3.4\text{-}7a)$$

or on substituting equation (3.2-16),

$$V_{BB} = I_B R_B + V_o + [(1 + h_{FE})I_B + (1 + h_{FE})I_{CBO}]R_E \qquad (3.4\text{-}7b)$$

Solving this equation for I_B yields

$$I_B = \frac{(V_{BB} - V_o) - (1 + h_{FE})R_E I_{CBO}}{R_B + (1 + h_{FE})R_E} \qquad (3.4\text{-}8)$$

If we substitute this result back into equation (3.2-16), after some mathematical manipulation we find that

$$I_C = \frac{h_{FE}(V_{BB} - V_o)}{R_B + (1 + h_{FE})R_E} + \frac{(1 + h_{FE})(R_B + R_E)I_{CBO}}{R_B + (1 + h_{FE})R_E} \qquad (3.4\text{-}9)$$

In this expression it is important to recognize the first term is the major contributor to the collector current; the second term just gives a correction factor that accounts for the effect of the reverse leakage current I_{CBO}.

In determining the effects that changes in h_{FE}, V_o, and I_{CBO} have on I_C, we will consider that each of these parameters varies individually with the other two held constant, and that the overall change in I_C produced by all three of them acting simultaneously is just equal to the sum of the variations associated with each of them acting alone. Basically, this approximation is fairly accurate as long as the changes produced by each of these parameter variations is small. This approach, besides simplifying the mathematics, also vividly demonstrates the design techniques that will be needed to minimize the effect that these transistor parameters have on the collector current.

Let's begin by examining the effect that changes in h_{FE} have on I_C since this is usually the major factor tending to alter the collector current. If we define h_{FE1} and h_{FE2} as the smaller and larger values of h_{FE}, respectively, and I_{C1} and I_{C2} as the corresponding collector currents, then using equation (3.4-9) and neglecting the second term in this expression, we may write the change in collector current associated with a particular change in h_{FE} as

$$\Delta I_C = I_{C2} - I_{C1} = (V_{BB} - V_o)\left[\frac{h_{FE2}}{R_B + (1 + h_{FE2})R_E} - \frac{h_{FE1}}{R_B + (1 + h_{FE1})R_E}\right]$$

$$= \frac{h_{FE}(V_{BB} - V_o)}{R_B + (1 + h_{FE1})R_E}\frac{(h_{FE2} - h_{FE1})(R_B + R_E)}{h_{FE1}[R_B + (1 + h_{FE2})R_E]} \qquad (3.4\text{-}10)$$

Noting that the first term on the right-hand side of this expression is equal to I_{C1}, the relative change in the collector current, $\Delta I_C/I_{C1}$, may be written as

$$\frac{\Delta I_C}{I_{C1}} = \frac{h_{FE2} - h_{FE1}}{h_{FE1}}\frac{R_B + R_E}{R_B + (1 + h_{FE2})R_E} \qquad (3.4\text{-}11)$$

By considering that equation (3.4-4) must generally be valid for a properly designated BJT circuit, equation (3.4-11) simplifies to

$$\frac{\Delta I_C}{I_{C1}} = \frac{\Delta h_{FE}/h_{FE1}}{h_{FE2}}\left(1 + \frac{R_B}{R_E}\right) \qquad (3.4\text{-}12)$$

when $(1 + h_{FE})R_E \gg R_B$. In this expression the term $\Delta h_{FE}/h_{FE1}$ represents the relative change in h_{FE}, and following equation (3.4-12), it is apparent that the relative change in I_C can be considerably smaller than the corresponding change in h_{FE} if R_E and h_{FE} are large.

In a similar fashion, the effect of changes in V_o on I_C can be determined by evaluating equation (3.4-9) for two different values of $V_o(V_{o1}$ and $V_{o2})$ and then subtracting these two equations to yield

$$\Delta I_C = I_{C2} - I_{C1} = \frac{-h_{FE1}}{R_B + (1 + h_{FE1})R_E}(V_{o2} - V_{o1})$$

$$= \frac{-h_{FE1}(V_{BB} - V_o)}{R_B + (1 + h_{FE1})R_E}\frac{V_{o2} - V_{o1}}{V_{BB} - V_{o1}} \qquad (3.4\text{-}13a)$$

or

$$\frac{\Delta I_C}{I_{C1}} = \frac{\Delta V_o}{V_{BB} - V_{o1}} \qquad (3.4\text{-}13b)$$

But

$$V_{BB} - V_{o1} = V_{E1} = I_{E1}R_E \simeq I_{C1}R_E \qquad (3.4\text{-}14)$$

so that the relative change in I_C associated with changes in V_o may be written as

$$\frac{\Delta I_C}{I_{C1}} = \frac{-\Delta V_o}{V_{E1}} = \frac{-\Delta V_o}{I_{C1} R_E} \qquad (3.4\text{-}15)$$

In other words, to minimize the changes in I_C when V_o varies, the dc emitter voltage, V_{E1}, must be much greater than the change in V_o.

The effect of changes in I_{CBO} on the position of the Q point may be computed by evaluating equation (3.4-9) at the two extreme values of I_{CBO} (I_{CBO1} and I_{CBO2}), with h_{FE} and V_o considered to be constants. Carrying out this procedure, we obtain

$$\Delta I_C = \frac{(1 + h_{FE1})(R_B + R_E)\Delta I_{CBO}}{R_B + (1 + h_{FE1})R_E}$$

$$= \underbrace{\frac{(1 + h_{FE1})(V_{BB} - V_o)}{R_B + (1 + h_{FE1})R_E}}_{\approx I_{C1}} \frac{(R_B + R_E)\Delta I_{CBO}}{V_{BB} - V_o} \qquad (3.4\text{-}16a)$$

Thus

$$\frac{\Delta I_C}{I_{C1}} = \frac{(R_B + R_E)\,\Delta I_{CBO}}{I_{E1} R_E} \simeq \frac{I_{CBO2}}{I_{C1}}\left(1 + \frac{R_B}{R_E}\right) \qquad (3.4\text{-}16b)$$

considering that $\Delta I_{CBO} \simeq I_{CBO2}$ because of the exponential growth of this leakage current with temperature. In examining equation (3.4-16b) we can see that the effect of changes in I_{CBO} on I_C can be minimized by making R_E large, and also by making the quiescent current I_{C1} large compared with I_{CBO2}.

By combining equations (3.4-12), (3.4-15), and (3.4-16b), the overall relative change in collector current due to variations in h_{FE}, V_o, and I_{CBO} may be approximated as

---

$$\frac{\Delta I_C}{I_{C1}} = \left(\frac{\Delta h_{FE}}{h_{FE1} h_{FE2}} + \frac{I_{CBO2}}{I_{C1}}\right)\left(1 + \frac{R_B}{R_E}\right) - \frac{\Delta V_o}{I_{C1} R_E} \qquad (3.4\text{-}17)$$

This expression is valid only when equation (3.4-4) is satisfied. For situations where $(1 + h_{FE})R_E$ is not much greater than R_B, the complete expression for I_C in equation (3.4-9) needs to be used in order to evaluate the Q-point shift. As a practical matter, the latter approach is not all that complicated when analyzing given BJT circuits, but it can be extremely difficult to use for new circuit designs.

All the results obtained thus far indicate that R_E should be made as large as possible for optimum Q-point stability. But making R_E too large creates problems, too. To begin with, as R_E increases in Figure 3.4-4a, the voltage gain of the amplifier decreases (Section 3.6). This problem can be circumvented to a degree by connecting a bypass capacitor in parallel with R_E that is large enough to look like a short to the ac signal. Although this solves the gain problem, an additional problem remains. Even though R_E is bypassed, there is still a large dc voltage drop across it, and this voltage subtracts from V_{CC} to reduce the allowed signal swing at the output. As a result, R_E should not be made any larger than is needed to ensure the required Q-point stability.

EXAMPLE 3.4-1

The circuit shown in Figure 3.4-5 is to be designed so that the Q point is at the center of the load line, and the overall dc power consumption is as small as possible. The circuit is to operate over the temperature range -50 to $+125°C$, with less than a 20% overall Q-point shift. Select the resistors R_1, R_2, R_C, and R_E considering that the ac circuit gain (v_c/v_{in}) is to be as large as possible. For the transistor, I_{CBO} is 5 nA at room temperature, and the h_{FE} takes on values of 50 and 200 at -50 and $125°C$, respectively.

SOLUTION

In designing this circuit, let us arbitrarily assume that the percent changes that occur in I_C due to the variations in h_{FE}, I_{CBO}, and V_o are 5%, 5%, and 10%, respectively. The change in temperature over the operating range of this transistor circuit is $175°C$, and using the fact that the temperature coefficient for V_{BE} is approximately -2 mV/°C (Section 1.11), the maximum change in V_o that occurs is $(-2$ mV/°C$)(175°C)$, or -0.35 V. This change in V_o is to produce at most only a 10% shift in I_C, and following equation (3.4-17), this requires that

$$\frac{-\Delta V_o}{I_{C1} R_E} = 0.1 \tag{3.4-18a}$$

or

$$I_{C1} R_E = \frac{-0.35 \text{ V}}{-0.1} = 3.5 \text{ V} \tag{3.4-18b}$$

This is the minimum dc emitter voltage that is needed to ensure that $\Delta I_C/I_{C1}$ due to ΔV_o is less than 10%.

Because the relative change in I_C associated with the changes in h_{FE} must be less than 5%, again following equation (3.4-17), we also have

$$\frac{\Delta h_{FE}/h_{FE1}}{h_{FE2}}\left(1 + \frac{R_B}{R_E}\right) = \frac{150 - 50}{(50)(150)}\left(1 + \frac{R_B}{R_E}\right) = 0.05 \tag{3.4-19a}$$

Figure 3.4-5
Example 3.4-1.

or

$$\frac{R_B}{R_E} = 2.33 \tag{3.4-19b}$$

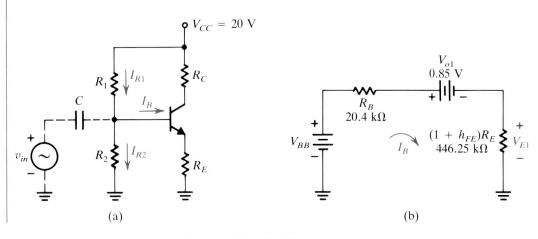

(a)

$V_{CC} = 20$ V

R_1 I_{R1}

C I_B R_C

v_{in} R_2 I_{R2} R_E

(b)

V_{o1}
0.85 V

R_B
20.4 kΩ

V_{BB}

I_B $(1 + h_{FE})R_E$
446.25 kΩ V_{E1}

3.4 / Q-Point Stability

131

Since I_{CBO} doubles for every 10°C increase (Section 1.11), the expression for I_{CBO} at 125°C is

$$I_{CBO2} = I_{CBO}(125) = I_{CBO}(25) \; 2^{(125-25)/10} = 5.12 \; \mu A \qquad (3.4\text{-}20)$$

For the change in I_C associated with variations in I_{CBO} to be less than 5%, following equation (3.4-17) we require that

$$\frac{I_{CBO2}}{I_{C1}}\left(1 + \frac{R_B}{R_E}\right) = 0.05 \qquad (3.4\text{-}21a)$$

so that
$$I_{C1} \geq \frac{I_{CBO2}(1 + R_B/R_E)}{0.05} = 0.34 \; mA \qquad (3.4\text{-}21b)$$

For safety we will design for $I_{C1} = 0.4$ mA, and for this value of I_C it follows immediately from equation (3.4-18b) that

$$R_E = \frac{3.5 \; V}{0.4 \; mA} = 8.75 \; k\Omega \qquad (3.4\text{-}22)$$

Substituting this result into equation (3.4-19b), we also find that $R_B = 20.4$ kΩ.

For the Q point to be located at the center of the load line, we require that V_{CE} be 10 V, so that

$$V_{RC} = 20 - 10 - 3.5 = 6.5 \; V \qquad (3.4\text{-}23a)$$

and
$$R_C = \frac{6.5 \; V}{0.4 \; mA} = 16.25 \; k\Omega \qquad (3.4\text{-}23b)$$

The Thévenin input voltage, V_{BB}, applied to the base may be determined with the aid of Figure 3.4-5b, using the lowest temperature to obtain the largest (or worst-case) value of V_{BB} that is needed. In particular, following this figure and noting that $I_{B1} = 0.4$ mA/50 $= 0.008$ mA we obtain

$$V_{BB} = I_{B1}R_B + V_o + V_E = (0.008)(20.4 \; k\Omega) + 0.85 \; V + 3.5 \; V = 4.51 \; V \qquad (3.4\text{-}24)$$

Using this result, the resistors R_1 and R_2 may now be determined as follows. The open-circuit voltage V_{BB} is given by the expression

$$V_{BB} = \frac{R_2}{R_1 + R_2} V_{CC} \qquad (3.4\text{-}25)$$

and by multiplying both sides of this equation by R_1, we obtain

$$V_{BB}R_1 = \frac{R_1 R_2}{R_1 + R_2} V_{CC} = R_B V_{CC} \qquad (3.4\text{-}26a)$$

so that
$$R_1 = \frac{(20.4 \; k\Omega)(20 \; V)}{4.51 \; V} = 90.4 \; k\Omega \qquad (3.4\text{-}26b)$$

Furthermore, because $R_B = R_1 \| R_2$, it also follows that

$$\frac{1}{R_2} = \frac{1}{R_B} - \frac{1}{R_1} = \frac{1}{20.4 \text{ k}\Omega} - \frac{1}{90.4 \text{ k}\Omega} \qquad (3.4\text{-}27a)$$

or $\qquad\qquad R_2 = 26.3 \text{ k}\Omega \qquad\qquad\qquad\qquad\qquad (3.4\text{-}27b)$

This completes the circuit design. As a check on the accuracy of this design, the collector currents I_{C1} and I_{C2} are found by direct substitution into equation (3.4-17) to be 0.3920 mA and 0.4676 mA, respectively, so that

$$\frac{\Delta I_C}{I_{C1}} = \frac{0.4676 - 0.392}{0.392} \times 100 = 19.3\% \qquad (3.4\text{-}28)$$

This result is just within the 20% maximum allowed Q-point shift.

Exercises

3.4-1 For the circuit in Figure 3.4-2, consider that $R_B = 10$ MΩ and $R_C = 50$ kΩ. If $h_{FE} = 50$ and $I_{CBO} = 10$ nA at 25°C, and if h_{FE} increases to 75 at 75°C, find the percent shift in I_C associated with this change in temperature. *Answer* 75%

3.4-2 For the circuit in Figure 3.4-5, consider that $R_B = 20$ kΩ, h_{FE} is constant at 50, I_{CBO} is zero, and I_C is 1 mA at 25°C. Determine the required values of V_{BB} and R_E if I_C is to change by less than 4% when the temperature increases to 100°C. *Answer* 4.85 V, 3.75 kΩ

3.5 ANALYSIS OF BJT CIRCUITS CONTAINING AC SOURCES

In Chapters 1 and 2 we demonstrated that the ac performance of diode circuits could be analyzed by both graphical and modeling techniques. We saw that the graphical method provided an excellent insight into the operation of these circuits, and was most useful when we were first learning about the diode's characteristics. But because these graphs tended to be tedious to work with, we eventually abandoned them in favor of a modeling approach. This presentation order will also be followed in studying the ac performance of BJT circuits. In the remaining sections of this chapter, we concentrate on graphical techniques for understanding the response of transistor circuits to ac sources, and in Chapter 4 we develop detailed small-signal models for the BJT. These models along with the ac equivalent circuits for the transistor will then be applied to the analysis and design of relatively sophisticated multitransistor circuits.

The common-emitter amplifier illustrated in Figure 3.5-1 is to be analyzed graphically. Let's begin with the dc analysis of this circuit. By setting the ac source equal to zero, the input and output dc load-line equations for this circuit may be written as

$$V_{BB} = I_B R_B + V_{BE} \qquad\qquad (3.5\text{-}1a)$$

and $\qquad\qquad V_{CC} = I_C R_C + V_{CE} \qquad\qquad (3.5\text{-}1b)$

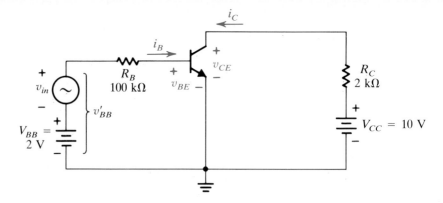

Figure 3.5-1

A common-emitter bipolar junction transistor amplifier.

Plotting these equations on the graphical characteristics in Figure 3.5-2, the transistor Q-point data is seen to be $I_B = 15$ μA, $V_{BE} = 0.65$ V, $I_C = 1.7$ mA, and $V_{CE} = 6.7$ V. Notice that to determine I_C and V_{CE}, the $i_B = 15$ μA line had to be sketched in on the transistor output characteristics, between the 10-μA and 20-μA curves.

To determine the total (or dc + ac) response of the circuit, we need to write down the dynamic load-line equations governing the operation of this circuit. For this simple circuit, they may be written down, by inspection of Figure 3.5-1, as

$$v'_{BB} = V_{BB} + v_{in} = i_B R_B + v_{BE} \tag{3.5-2a}$$

and
$$V_{CC} = i_C R_C + v_{CE} \tag{3.5-2b}$$

By plotting these equations on their respective transistor characteristics in Figure 3.5-2, we obtain the overall circuit responses given in Figure 3.5-3. Note that for the output circuit, both the dc and dynamic load lines are identical. The load lines for the input circuit have the same slope but different i_B and v_{BE} intercepts. When $v_{in} = 0$, the dc and dynamic input load lines overlap, and as v_{in} changes, the dynamic load line moves as shown in Figure 3.5-2b. For $v_{in} = +1$ V, its i_B intercept is at 3 V/100 kΩ or 30 μA, and when v_{in} is -1 V, it is at 10 μA. As v_{in} varies between $+1$ and -1, the corresponding values of i_B and v_{BE} are read off from the intersection of the dynamic load line with the

Figure 3.5-2

Graphical characteristics for the transistor in Figure 3.5-1.

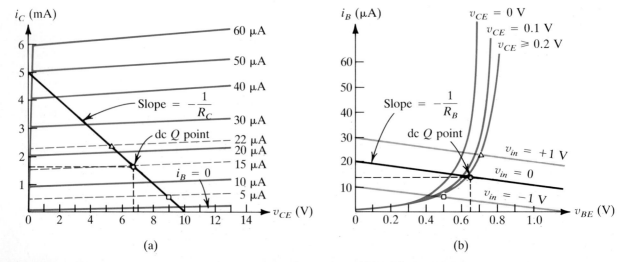

(a)   (b)

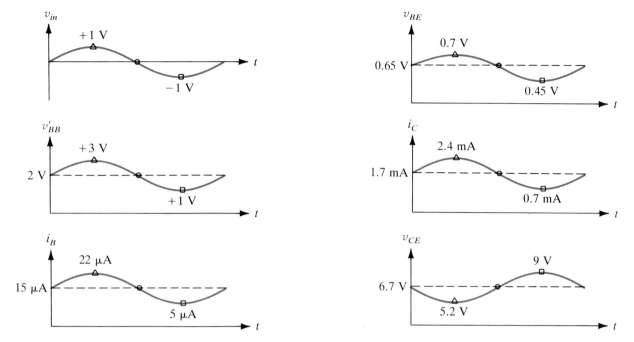

Figure 3.5-3
Pertinent waveshapes for the circuit in Figure 3.5-1.

$v_{CE} \geq 0.2$ V curve in Figure 3.5-2b. This curve is used because the transistor is in the active region. Note that the points corresponding to $v_{in} = 0, +1$, and -1 are labeled with a circle, a triangle, and a square, respectively, on both the transistor characteristics and the waveform sketches given in Figure 3.5-3.

The waveshapes for i_C and v_{CE} are found by observing that as v_{in} varies, the transistor base current shifts accordingly, and this causes the operating point to move on the output load line as shown in Figure 3.5-2a. Specifically, when $v_{in} = +1$ V, $v'_{BB} = +3$ V, $i_B = 22$ μA, and $v_{BE} = 0.7$ V. In addition, the point $i_C = 2.4$ mA and $v_{CE} = 5.2$ V is read off directly from the intersection of the $i_B = 22$ μA curve (shown in dashed lines in Figure 3.5-2a) with the dynamic load line. A similar set of points may be found for the case when $v_{in} = -1$ V; these are $v'_{BB} = 1$ V, $i_B = \mu$A, $v_{BE} = 0.45$ V, $i_C = 0.7$ mA, and $v_{CE} = 9$ V. Using these points together with the dc Q point results, we may approximately sketch the important circuit waveshapes, as shown in Figure 3.5-3.

In examining these waveshapes, we make the following observations. First, although the input v_{in} is a sine wave, the output waveshapes are only approximately sinusoidal since the transistor volt-ampere characteristics are somewhat nonlinear. Second, the ac voltage gain of the common-emitter circuit is negative; that is, as v_{in} increases, the ac part of v_{CE} decreases. If we define the voltage gain of this circuit as

$$A_v = \frac{v_{ce}(p-p)}{v_{in}(p-p)} \qquad (3.5\text{-}3)$$

the specific gain of this circuit is $(5.2\text{-}9)/[1-(-1)]$, or about -1.9.

EXAMPLE 3.5-1

Repeat the analysis of the circuit shown in Figure 3.5-1 by using the simple linear transistor model given in Figure 3.2-5.

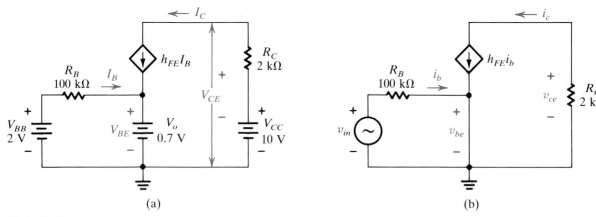

(a)   (b)

Figure 3.5-4
Example 3.5-1.

Using the definition of h_{FE} given in equation (3.2-17), we see from the V–I characteristics that the transistor's h_{FE} is approximately 100. Substituting the model for the transistor, the dc analysis of this circuit may be carried out with the aid of Figure 3.5-4a. The Q-point values for this circuit are

$$V_{BE} = V_o = 0.7 \text{ V} \tag{3.5-4a}$$

$$I_B = \frac{V_{BB} - V_{BE}}{R_B} = 13 \ \mu\text{A} \tag{3.5-4b}$$

$$I_C = 100 I_B = 1.3 \text{ mA} \tag{3.5-4c}$$

and $\qquad V_{CE} = V_{CC} - I_C R_C = 7.4 \text{ V} \tag{3.5-4d}$

On comparing these results with those obtained graphically, we find that there are several reasonably substantial differences. Principally, this is because this circuit has very poor Q-point stability, and as a result, the small difference between the graphical value of V_{BE} and the model's assumed V_{BE} of 0.7 V produces large discrepancies in I_B, I_C, and V_{CE}.

To carry out the ac part of the analysis, we reconnect v_{in} and replace all the batteries by short circuits, as shown in Figure 3.5-4b.* From this circuit the relevant ac parameters are

$$v_{be} = 0.0 \text{ V} \tag{3.5-5a}$$

$$i_b = \frac{v_{in}}{R_B} = 10 \sin \omega t \ \mu\text{A} \tag{3.5-5b}$$

$$i_c = 100 i_b = 1 \sin \omega t \ \text{mA} \tag{3.5-5c}$$

and $\qquad v_{ce} = -i_c R_C = -2 \sin \omega t \text{ V} \tag{3.5-5d}$

Combining equations (3.5-4) and (3.5-5), we may write the expressions for the

---

* The use of h_{FE} in the ac transistor model is not strictly valid, but the errors this approximation introduces are not large enough to warrant a detailed discussion here. See Section 4.3 for the definition of h_{fe}, the ac small-signal transistor current gain.

total (i.e., the dc + ac) circuit responses as

$$v_{BE} = 0.7 \text{ V} \tag{3.5-6a}$$

$$i_B = (13 + 10 \sin \omega t) \ \mu\text{A} \tag{3.5-6b}$$

$$i_C = (1.3 + 1 \sin \omega t) \text{ mA} \tag{3.5-6c}$$

and $$v_{CE} = (7.4 - 2 \sin \omega t) \text{ V} \tag{3.5-6d}$$

You should sketch these results for yourself to demonstrate that they compare quite favorably with the sketches given in Figure 3.5-3 that were obtained using graphical analysis techniques.

Transistor circuits of the type shown in Figure 3.5-1 are used only rarely in actual engineering designs because they have such poor Q-point stability. The addition of a properly chosen emitter resistor substantially improves the dc performance of the circuit. Unfortunately, the inclusion of an emitter resistor makes the circuit very difficult to analyze graphically. To appreciate this difficulty, consider the circuit shown in Figure 3.5-5a. The dc input and output load-line equations for this circuit are obtained directly from part (b) of this figure as

$$V_{BB} = I_B R_B + V_{BE} + I_E R_E \tag{3.5-7a}$$

and $$V_{CC} = I_C R_C + V_{CE} + I_E R_E \tag{3.5-7b}$$

Ordinarily, equation (3.5-7a) would be sketched onto the transistor input characteristics to obtain the quiescent base current and base–emitter voltage, and then equation (3.5-7b) would be drawn on the output characteristics to determine the quiescent collector current and collector-to-emitter voltage. How-

Figure 3.5-5
A BJT connected to an unby-passed emitter resistor.

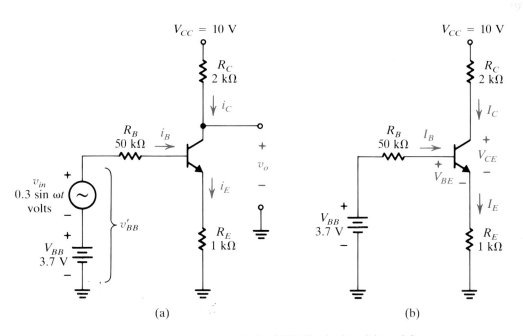

(a) (b)

ever, these equations cannot be sketched directly because the presence of the term $I_E R_E$ introduces an additional unknown I_E. Even if we attempt to eliminate this unknown by substituting $(I_B + I_C)$ in place of I_E, the resulting equations

$$V_{BB} = I_B R_B + V_{BE} + (I_B + I_C) R_E \qquad (3.5\text{-}8a)$$

and

$$V_{CC} = I_C R_C + V_{CE} + (I_B + I_C)R_E \qquad (3.5\text{-}8b)$$

are still impossible to sketch directly.

To carry out the graphical analysis of this problem, we will need to eliminate I_C from equation (3.5-8a) and I_B from equation (3.5-8b). To accomplish this, we will go back to the original equations in (3.5-7) and approximate I_E by $(1 + h_{FE})I_B$ in equation (3.5-7a) and by I_C in (3.5-7b). This results in the following load-line equations:

$$V_{BB} = I_B[R_B + (1 + h_{FE})R_E] + V_{BE} \qquad (3.5\text{-}9a)$$

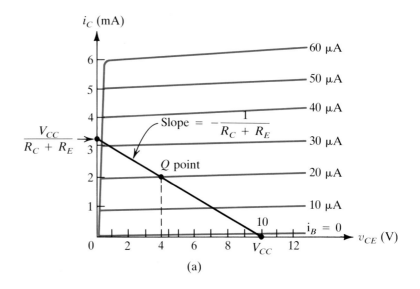

(a)

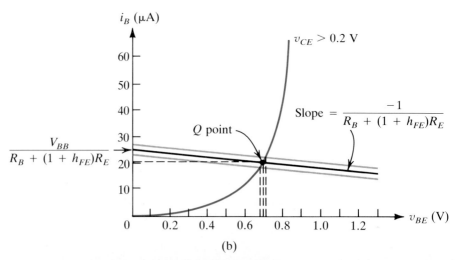

(b)

Figure 3.5-6

Transistor characteristics and load lines for a BJT amplifier with an unbypassed emitter resistor.

138   Chapter 3 / Introduction to the Bipolar Junction Transistor

$$\text{and} \qquad V_{CC} = I_C(R_C + R_E) + V_{CE} \qquad (3.5\text{-}9\text{b})$$

The mathematical purist will, of course, realize that we are no longer carrying out a strictly graphical analysis of this problem since we have mixed the modeling and graphical concepts together, but this slight bending of the rules will permit us to develop a good graphical picture of how this circuit operates. Notice that equation (3.5-9a) now has only two unknowns, I_B and V_{BE}, so that a graphical solution is possible by plotting it on the transistor's input characteristics. A similar approach can be used for the output load-line equation (3.5-9b). Both of these equations are sketched on the transistor characteristics in Figure 3.5-6, from which we find that the circuit Q-point values are $I_B = 20\ \mu A$, $V_{BE} = 0.7\ V$, $I_C = 2.0\ mA$, and $V_{CE} = 4.0\ V$.

In a similar fashion, the dynamic load-line equations are obtained by inspection of Figure 3.5-5a as

$$v'_{BB} = i_B R_B + v_{BE} + i_E R_E \qquad (3.5\text{-}10\text{a})$$

$$\text{and} \qquad V_{CC} = i_C R_C + v_{CE} + i_E R_E \qquad (3.5\text{-}10\text{b})$$

$$\text{or} \qquad v'_{BB} = i_B[R_B + (1 + h_{FE})R_E] + v_{BE} \qquad (3.5\text{-}11\text{a})$$

$$\text{and} \qquad V_{CC} = i_C(R_C + R_E) + v_{CE} \qquad (3.5\text{-}11\text{b})$$

Figure 3.5-7
Waveforms for the circuit in Figure 3.5-5.

The dynamic load line for the output circuit is the same as for the dc analysis, whereas the input dynamic load line is parallel to the dc load line but moves back and forth as v_{in} varies. The i_B intercepts for the input load line are

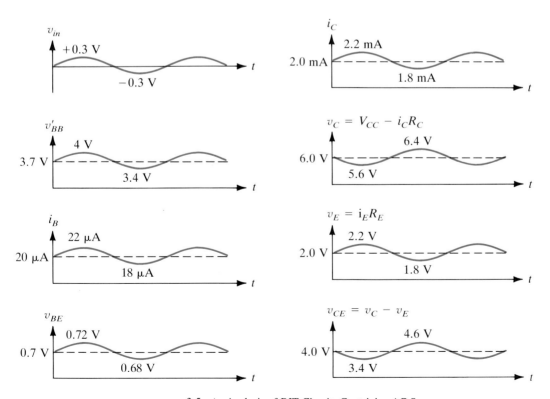

26.4 μA when $v_{in} = +0.30$ V and 22.5 μA when $v_{in} = -0.3$ V. The position of these load lines is indicated in Figure 3.5-6, and the resulting circuit waveforms are sketched in Figure 3.5-10. We can see that because the ac signal v_{in} is so small, the deviations about the Q point are very nearly sinusoidal. For this circuit, we define the voltage gain as $A_v = v_c/v_{in} = v_o/v_{in}$, so that $A_v = (5.8 - 6.6)/[0.3.5 - (-0.3)] = -1.33$. Notice that this gain is considerably smaller than it was without R_E present in Example 3.5-1.

EXAMPLE 3.5-2

Repeat the analysis of the circuit given in Figure 3.5-5 using the simple transistor model in Figure 3.2-5.

SOLUTION

Inserting the *npn* transistor model into the original circuit in Figure 3.5-5, we obtain the circuit shown in Figure 3.5-8a, and applying the superposition principle to this problem, we may redraw the dc and ac equivalent circuits as illustrated in Figure 3.5-8b and c. The dc circuit shown in Figure 3.5-8b was simplified by using the equivalent circuit given in Figure 3.3-13c. Using this dc circuit, we may calculate the Q-point values as

$$I_B = \frac{3.7 - 0.7}{151 \text{ k}\Omega} = 19.8 \text{ } \mu\text{A} \tag{3.5-12a}$$

$$V_{BE} = V_o = 0.7 \text{ V} \tag{3.5-12b}$$

$$I_C = h_{FE} I_B = 1.98 \text{ mA} \tag{3.5-12c}$$

$$V_C = V_{CC} - I_C R_C = 6.04 \text{ V} \tag{3.5-12d}$$

$$V_E = I_E R_E = 2.0 \text{ V} \tag{3.5-12e}$$

and

$$V_{CE} = V_C - V_E = 4.04 \text{ V} \tag{3.5-12f}$$

The ac response of the circuit to the source v_{in} is found by examining Figure 3.5-8c. The base current is obtained by applying Kirchhoff's law to the input loop, which yields

$$v_{in} = i_b R_B + (1 + h_{FE}) i_b R_E \tag{3.5-13a}$$

or

$$i_b = \frac{v_{in}}{R_B + (1 + h_{FE}) R_E} = 2 \sin \omega t \text{ } \mu\text{A} \tag{3.5-13b}$$

Using this result, and following Figure 3.5-8c, the other important ac quantities for this circuit are seen to be

$$v_{be} = 0 \tag{3.5-13c}$$

$$i_c = 100 i_b = 0.2 \sin \omega t \text{ mA} \tag{3.5-13d}$$

$$v_c = i_c R_C = -0.4 \sin \omega t \text{ V} \tag{3.5-13e}$$

$$v_e = (1 + h_{FE}) i_b R_E = 0.2 \sin \omega t \text{ V} \tag{3.5-13f}$$

and

$$v_{ce} = v_c - v_e = -0.6 \sin \omega t \text{ V} \tag{3.5-13g}$$

By adding the dc and ac results together you should be able to sketch the total circuit responses for i_B, v_{BE}, i_C, v_C, v_E, and v_{CE}. Furthermore, by comparing

Chapter 3 / Introduction to the Bipolar Junction Transistor

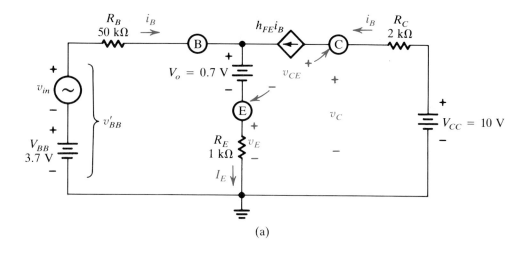

(a)

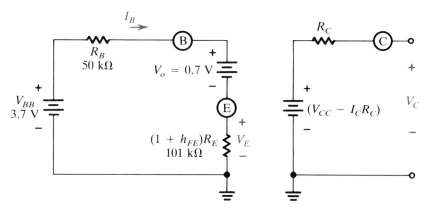

(b)   Equivalent Circuit for Carrying Out the DC Analysis

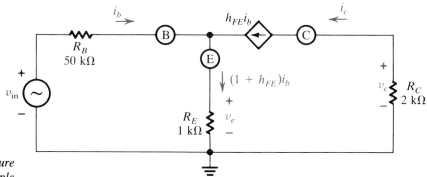

Figure 3.5-8

Equivalent circuit of the transistor amplifier in Figure 3.5-5. For use with Example 3.5-2.

(c)   Equivalent Circuit for Carrying Out the AC Analysis

these waveforms with the results given in Figure 3.5-7, which were generated by graphical analysis techniques, you should be convinced that even the use of these relatively simple transistor models provides a reasonably accurate way to analyze electronic circuits containing bipolar junction transistors.

3.5-1 Using graphical analysis, find the Q point of the circuit in Figure 3.5-5 if a 1-V battery is connected in series with R_E with its negative terminal connected to ground. The transistor volt-ampere characteristics are given in Figure 3.5-6. *Answer* $I_B = 14 \ \mu A$, $V_{BE} = 0.62$ V, $I_C = 1.4$ mA, and $V_{CE} = 4.8$ V

3.5-2 What are the total voltages v_o and v_{CE} if R_C is increased to 3 kΩ in Figure 3.5-5? *Answer* $v_o(t) = (4 - 0.6 \sin \omega t)$ V, $v_{CE}(t) = (2 - 0.8 \sin \omega t)$ V

3.6 ANALYSIS OF BJT CIRCUITS CONTAINING BYPASS AND COUPLING CAPACITORS

Bypass capacitors, as their name implies, are used to bypass the ac portion of a circuit's signal current to ground. Capacitors of this type are frequently used in parallel with emitter resistors in order to maintain a high ac gain, and are also connected in collector feedback paths where dc feedback is desired for Q-point stability, but where ac signal feedback is unwanted (Figure 3.6-1a). Bypass capacitors are also employed in dc power supplies to shunt undesired ripple signals to ground.

Coupling capacitors are used to provide ac coupling between transistor stages while affording dc isolation. Because the capacitor looks like an open to dc, this type of circuit coupling prevents the dc output from one stage from upsetting the Q point of the stage that follows. Coupling capacitors are also used at the input and output stages of an amplifier, where they are often connected in series with the signal source and the load to prevent these components from upsetting the Q point of the amplifier, and also to block any dc from entering the load or the signal source. Several examples of BJT circuits containing coupling capacitors are illustrated in Figure 3.6-1b.

The addition of coupling and bypass capacitors to a BJT amplifier makes the circuit rather difficult to analyze graphically. Consider, for example, the transistor amplifier illustrated in Figure 3.6-2a. Except for the addition of the bypass capacitor, C_E, this circuit is the same as that presented in Figure 3.5-5. Because the capacitor looks like an open to dc, both of these circuits have identical dc load-line equations and dc voltages and currents [see equations (3.5-10a) and (3.5-10b)]. However, the dynamic performance of these two circuits are quite different from one another.

To develop the dynamic load-line equations for the amplifier in Figure 3.6-2a, we will need to replace the capacitor by an equivalent-circuit element that will allow us to establish the relationship between i_E and v_E. If we assume that the impedance of the capacitor is small so that it looks like a short circuit to the ac portion of the emitter current, the voltage across the capacitor will be constant and equal to the the dc emitter voltage V_E. In fact, as far as the circuit connected to C_E is concerned, there is no way to distinguish this capacitor from a battery of voltage V_E. Furthermore, the Thévenin equivalent of this battery and the resistor R_E in parallel with it is just equal to a Thévenin voltage source $V_T = V_E$ because the circuit's Thévenin impedance is zero (Figure 3.6-2b).

The advantage of replacing the capacitor by a battery is that the dynamic

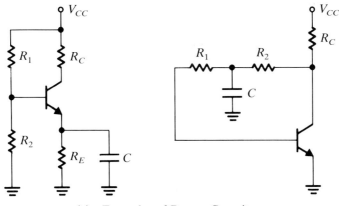

(a)   Examples of Bypass Capacitors

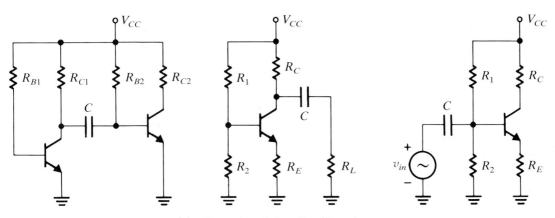

(b)   Examples of Coupling Capacitors

Figure 3.6-1

load-line equations can now be written directly from the circuit in Figure 3.6-2b as

$$v'_{BB} = i_B R_B + v_{BE} + V_E \qquad (3.6\text{-}1a)$$

and
$$V_{CC} = i_C R_C + v_{CE} + V_E \qquad (3.6\text{-}1b)$$

These load lines are sketched on the transistor characteristics in Figure 3.6-3 along with the circuit's dc load-line equations. Notice that the slope of the dynamic and dc load lines are different, and also that these two load lines intersect at the Q point.

To prove this second point formally, consider, for example, the output dynamic load-line equation (3.6-1b). The value of v_{CE} corresponding to a collector current of $i_C = I_C$ is found by direct substitution as

$$V_{CC} = I_C R_C + v_{CE} + V_E \qquad (3.6\text{-}2a)$$

or
$$v_{CE} = V_{CC} - I_C R_C - I_E R_E \simeq V_{CC} - I_C(R_C + R_E) \qquad (3.6\text{-}2b)$$

Comparing this result with equation (3.5-10b), we can see that the value of v_{CE} in equation (3.6-2b) is just equal to the quiescent collector-to-emitter voltage,

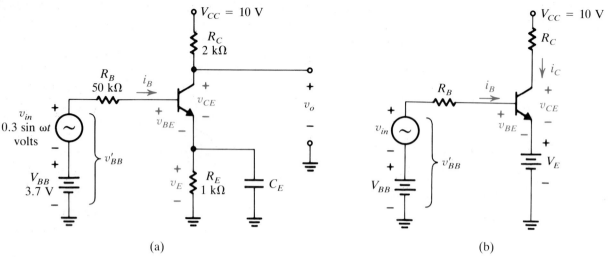

(a)   (b)

Figure 3.6-2
A BJT common-emitter amplifier with a bypassed emitter resistor.

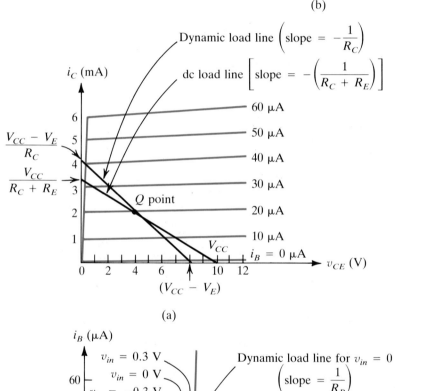

(a)

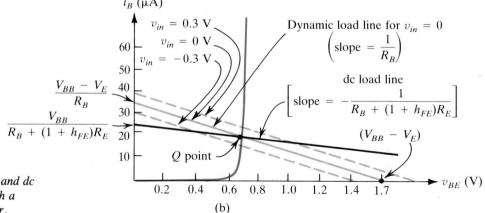

Figure 3.6-3
Position of the dynamic and dc load lines for a BJT with a bypassed emitter resistor.

(b)

V_{CE}, so that the Q point ($i_C = I_C$, $v_{CE} = V_{CE}$) is also a point on the dynamic load-line equation. Hence both the dc and dynamic load lines pass through, and intersect at, the Q point.

As v_{in} varies, the input dynamic load line moves up and down on the transistor input characteristics in Figure 3.6-3b. The values of i_B corresponding to the peak amplitudes of v_{in} at +0.3 V and −0.3 V are 25 μA and 15 μA, respectively, and because v_{in} is sinusoidal and has a fairly small amplitude, the base current is nearly sinusoidal, too. This waveshape, together with the other important voltages and currents for this circuit, are sketched in Figure 3.6-4. Notice that the addition of the emitter bypass capacitor has increased the circuit voltage gain to −3.33 compared with a gain of only −1.33 without the bypass capacitor (Figure 3.5-7).

EXAMPLE 3.6-1

Using the bipolar transistor model from Figure 3.2-5, calculate the ac voltage gain of the transistor amplifier in Figure 3.6-2a, both with and without the emitter bypass capacitor, C_E, in place. Assume a transistor h_{FE} of 100.

SOLUTION

The ac equivalent circuits for the unbypassed and bypassed cases are found in Figure 3.6-5a and b, respectively. Analyzing both circuits simultaneously, we may calculate the base current of each as

Figure 3.6-4
Waveforms for the circuit in Figure 3.6-2 containing a BJT amplifier with a bypassed emitter resistor.

Unbypassed: $i_b = \dfrac{v_{in}}{R_B + (1 + h_{FE})R_E}$ (3.6-3a)

Bypassed: $i_b = \dfrac{v_{in}}{R_B}$ (3.6-3b)

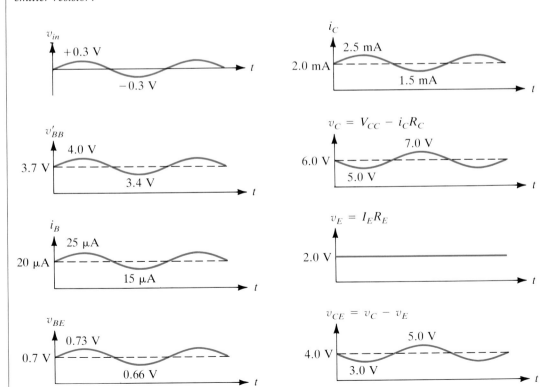

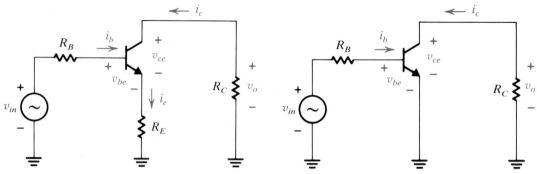

(a)   ac Equivalent Circuit with R_E Unbypassed   (b)   ac Equivalent Circuit with R_E Bypassed

Figure 3.6-5
Example 3.6-1.

The collector current for each circuit is simply

$$i_c = h_{FE}i_b \qquad (3.6\text{-}4)$$

where the appropriate value for i_b is determined from equation (3.6-3). The output voltage v_o is equal to

$$v_o = -i_c R_C \qquad (3.6\text{-}5)$$

and substituting equations (3.6-3) and (3.6-4) into (3.6-5), we may evaluate the ac gain for each case as

Unbypassed: $$A_v = \frac{v_o}{v_{in}} = \frac{-h_{FE}R_C}{R_B + (1 + h_{FE})R_E} \qquad (3.6\text{-}6a)$$

Bypassed: $$A_v = \frac{v_o}{v_{in}} = -\frac{h_{FE}R_C}{R_B} \qquad (3.6\text{-}6b)$$

By substituting in the component values, the amplifier gains for the unbypassed and bypassed cases are -1.32 and -3.92, respectively. Once again, these results compare favorably with those obtained graphically.

The analysis technique for circuits containing coupling capacitors is quite similar to that already developed for the case of bypass capacitors. Figure 3.6-6a illustrates a typical transistor amplifier circuit containing a capacitively coupled load. Using the dc equivalent circuit given in Figure 3.6-6b and applying Kirchhoff's voltage law, the dc load-line equations may be written as

$$V_{BB} = I_B R_B + V_{BE} \qquad (3.6\text{-}7a)$$

$$V_{CC} = I_C R_C + V_{CE} \qquad (3.6\text{-}7b)$$

By plotting these equations on the graphical characteristics given in Figure 3.6-7, we can rapidly evaluate the quiescent voltages and currents for this circuit as $I_B = 5.7$ mA, $V_{BE} = 0.66$ V, $I_C = 570$ mA, and $V_{CE} = 8.6$ V. In addition, because the capacitor is an open circuit to dc, the dc current in R_L is zero, so that $V_L = 0$, and V_1, the voltage across the capacitor C, is equal to $(V_{CE} - V_L)$ or V_{CE}.

To obtain the dynamic load-line equations for this circuit, we again recall

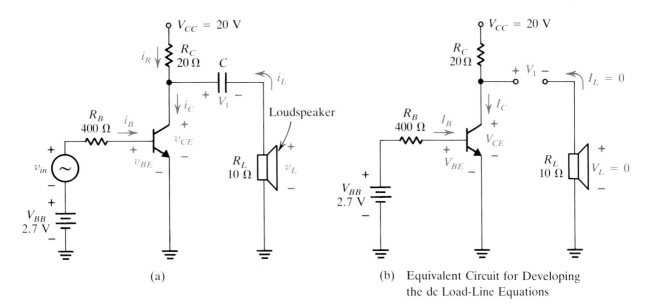

(a)

(b) Equivalent Circuit for Developing
the dc Load-Line Equations

Figure 3.6-6

*A BJT amplifier with a capac-
itively coupled load.*

that because the voltage across the capacitor is essentially constant at $V_1 = V_{CE}$ volts, we can replace it by an equivalent voltage source equal to this value (see Figure 3.6-8a).

To obtain the Thévenin resistance for the circuit within the shaded region, note that when V_{CC} and V_{CE} are set equal to zero in Figure 3.6-8a, the resistance between point c and ground is equal to the parallel combination of R_C and R_L. The Thévenin or open-circuit voltage between those same two points is found by applying the superposition principle to the circuit within the shaded region. When this is done, we see that V_T is given by

$$V_T = V_{CC}\frac{R_L}{R_C + R_L} + V_{CE}\frac{R_C}{R_C + R_L} \tag{3.6-8}$$

The overall Thévenin equivalent of the circuit in Figure 3.6-8a is presented in part (b) of the figure, and the dynamic load-line equation for this circuit can

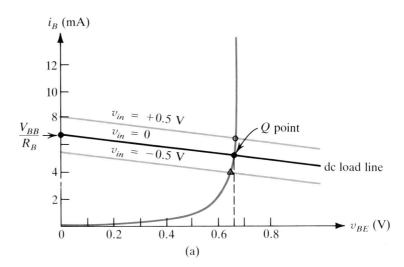

Figure 3.6-7

*Position of the dc and dynamic
load lines for the transistor
circuit in Figure 3.6-6 with a
capacitively coupled load.*

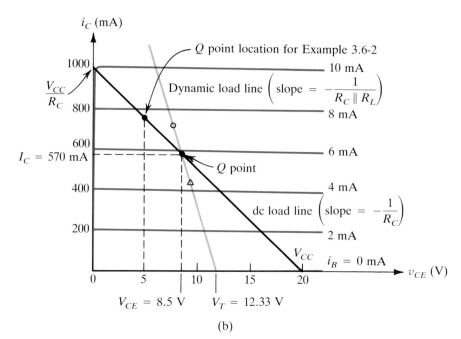

i_C (mA)

Q point location for Example 3.6-2

Dynamic load line $\left(\text{slope} = -\dfrac{1}{R_C \| R_L}\right)$

$\dfrac{V_{CC}}{R_C}$

$I_C = 570$ mA

Q point

dc load line $\left(\text{slope} = -\dfrac{1}{R_C}\right)$

$i_B = 0$ mA

V_{CC}

v_{CE} (V)

Figure 3.6-7 (Continued)

$V_{CE} = 8.5$ V $V_T = 12.33$ V

(b)

Figure 3.6-8
Equivalent collector circuit for the BJT amplifier of Figure 3.6-6.

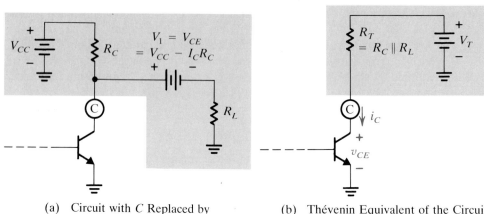

(a)   Circuit with C Replaced by
       an Equivalent Battery

(b)   Thévenin Equivalent of the Circuit
       within the Shaded Region in (a)

be written by inspection as

$$V_T = i_C R_T + v_{CE} \qquad (3.6\text{-}9)$$

EXAMPLE 3.6-2

For the circuit illustrated in Figure 3.6-6a the transistor has the characteristics given in Figure 3.6-7. Determine the value of R_B that will allow for maximum undistorted signal swing in the output (i.e., that will place the Q point in the middle of the dynamic load line).

SOLUTION

In selecting the Q-point location that will permit maximum signal swing at the output before transistor saturation or clipping occurs, at first it might appear plausible to place the Q point at the center of the dc load line. However, a close examination of Figure 3.6-9a reveals that placement of the Q point at this

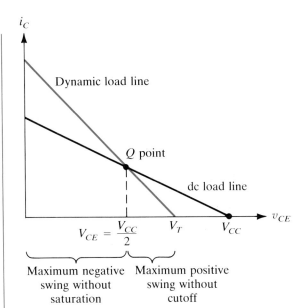

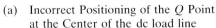

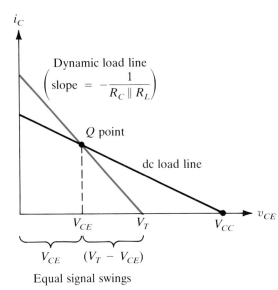

(a) Incorrect Positioning of the Q Point at the Center of the dc load line

(b) Correct Positioning of the Q Point at the Center of the Dynamic Load Line

Figure 3.6-9

Positioning of the Q point for maximum signal swing in the output.

location for a capacitively coupled amplifier will cause the amplifier to cut off prematurely since the allowed swing in the positive direction will be less than that permitted in the negative direction. To achieve maximum undistorted signal swing in the output, the Q point needs to be located at the center of the dynamic load line rather than at the center of the dc load line. Following Figure 3.6-9b, this occurs when

$$V_{CE} = V_T - V_{CE} \tag{3.6-10}$$

Substituting V_T from equation (3.6-8), we find that

$$V_{CE} = V_{CC}\frac{R_L}{R_C + R_L} + V_{CE}\frac{R_C}{R_C + R_L} - V_{CE} \tag{3.6-11a}$$

or after a little algebra that

$$V_{CE} = V_{CC}\frac{R_L}{R_C + 2R_L} \tag{3.6-11b}$$

Using this result, the required dc collector-to-emitter voltage needed for maximum output signal swing is $V_{CE} = 20(10/40) = 5.0$ V. The corresponding quiescent collector and base currents are found from the dc load lines in Figure 3.6-7b to be 750 mA and 7.5 mA, respectively. Using the fact that V_{BE} is about 0.66 V when $I_B = 7.5$ mA (Figure 3.6-7a), it therefore follows from the base loop in Figure 3.6-6 that the required value of R_B is $(2.7 - 0.66)$ V/7.5 mA or 272 Ω. Ideally, the maximum undistorted output for this Q-point location is 10 V(p-p).

Exercises

3.6-1 In Figure 3.6-2, connect a 2-kΩ resistor in between the emitter of the transistor and R_E so that only the 1-kΩ resistor is bypassed to ground.

a. Using graphical procedures, determine the Q point. The transistor graphical characteristics are given in Figure 3.6-3.

b. What are the i_C and v_{CE} intercepts of the dynamic load line on the transistor output characteristics?
Answer (a) $I_B = 8.6~\mu A$, $V_{BE} = 0.67$ V, $I_C = 0.86$ mA, $V_{CE} = 5.7$ V; (b) v_{CE} intercept is at 9.14 V, i_C intercept is at 2.3 mA

3.6-2 For the transistor amplifier in Figure 3.6-6a, a 5-Ω resistor is connected between the emitter and ground. What are the dynamic load lines for this circuit? *Answer* $v_{in} + 2.7$ V $= (0.9~k\Omega)~i_B + v_{BE}$, 16.4 V $= (11.66~\Omega)i_C + v_{CE}$

3.6-3 Repeat Exercise 3.6-2 for the case where the 5-Ω resistor is bypassed to ground. *Answer* $v_{in} + 1.35$ V $= (0.4~k\Omega)i_B + v_{BE}$, 15 V $= (6.66~\Omega)i_C + v_{CE}$

3.6-4 a. For the circuit discussed in Exercise 3.6-2, find the ac again v_l/v_{in}.
b. Repeat part (a) when the 5-Ω resistor is bypassed to ground.
Answer (a) -0.73, (b) -1.65

PROBLEMS

Unless otherwise noted, all transistors in this chapter are silicon with $h_{FE} = 100$ and $I_{CBO} = 0$.

3.1-1 Prove that the power gain of an ideal transformer is 1.

3.1-2 The circuit shown in Figure 3.1-4b has $R_s = 1~M\Omega$, $R_i = 100~\Omega$, $A_{vo} = 100$, $R_o = 0~\Omega$, and $R_L = \infty$.
(a) Compute the voltage gain v_o/v_s.
(b) An ideal voltage amplifier is available with $R_i = \infty$, $A_{vo} = 1$, and $R_o = 0$. Repeat part (a) if this buffer is inserted between the source and the amplifier.
(c) Repeat part (a) if a transformer having the turns ratio required to maximize the voltage gain is inserted between the source and the amplifier.

3.1-3 For the circuit in Figure 3.1-4, $R_i = 10~k\Omega$, $A_{vo} = 100$, and $R_o = 5~k\Omega$. Because the output controlled source in this amplifier is a voltage-controlled voltage source, this circuit is often called a voltage amplifier (especially when $R_i \gg R_s$ and $R_o \ll R_L$). Draw the other representations of this circuit as a transresistance amplifier, a current amplifier, and a transconductance amplifier.

3.1-4 Given a three-stage transconductance amplifier where each stage has an $R_o = 50~k\Omega$, $G_m = 5$ mS, and $R_i = 1~M\Omega$:
(a) Calculate the overall voltage gain, v_o/v_s, if $R_s = R_L = 10~k\Omega$.
(b) Repeat part (a) if R_o and R_i are both infinite.
(c) Determine the overall power gain for the circuit conditions in part (a).

3.1-5 Two voltage amplifier circuits (1 and 2) are available for the design of an amplifier whose load and source impedances are 10 Ω and 50 $k\Omega$, respectively. The characteristics of circuit 1 are $R_i = 100~k\Omega$, $A_{vo} = 50$, and

$R_o = 10~k\Omega$; the characteristics of circuit 2 are $R_i = 100~k\Omega$, $A_{vo} = 2$, and $R_o = 10~\Omega$.
(a) If the overall amplifier design can contain only a single stage, determine the power gain when each of these circuits is used as the amplifier.
(b) Explain why the power gain of circuit 2 is so much better than that of circuit 1 in this particular amplifier design.

3.1-6 The amplifier discussed in Problem 3.1-5 is to be designed to deliver an average power of 3 W to R_L when $v_s = 4 \sin \omega t$ mV.
(a) Using any combination of circuit 1 and circuit 2 amplifiers, determine the minimum number of stages required.
(b) Sketch your final circuit design and indicate the actual power delivered to the load for your circuit.

3.1-7 A voltage amplifier circuit is available having an $R_i = 100~k\Omega$, $A_{vo} = 150$, and $R_o = 100~\Omega$. Show that the design in Problem 3.1-6 can easily be accomplished with two of these stages if transformer coupling to the load is permitted. In particular:
(a) What turns ratio will result in maximum power being delivered to the load?
(b) What is the average power delivered to the load using the turns ratio selected in part (a)?

3.1-8 A current amplifier circuit is available in which $R_i = 10~k\Omega$, $R_o = 50~k\Omega$, and $A_{io} = 100$. An amplifier is to be constructed by cascading as many of these stages as needed. Find the minimum number of stages N required to achieve an overall voltage gain of at least 500,000 if R_s and R_L are 10 $k\Omega$ and 50 $k\Omega$, respectively.

3.1-9 Find the power gain, G, of the circuit described in Problem 3.1-8 if three stages of amplification are used.

3.1-10 What is the overall current gain of the amplifier in Problem 3.1-8 if four stages of amplification are used?

For Problems 3.2-1 through 3.2-4, select the best answer, and in each case explain the reason for your choice.

3.2-1 If a transistor is made by joining two diodes:
(a) It will work as long as you keep the leads short.
(b) It will work, but the collector breakdown voltage will be low.
(c) It will not work because the doping will be incorrect.
(d) It will not work because the effective base width will be too large.

3.2-2 In designing a BJT, we dope the emitter heavily in order to:
(a) Reduce reverse leakage current in the collector.
(b) Maximize the base–emitter breakdown voltage.
(c) Insure that most of the emitter current is due to carriers emitted from the emitter into the base.
(d) Minimize temperature effects in the transistor.

3.2-3 In a BJT the base is made thin in order to:
(a) Ensure that carriers injected from the emitter reach the collector.
(b) Minimize the collector reverse leakage current.
(c) Increase the base–emitter breakdown voltage.
(d) Reduce the base resistance.

3.2-4 In a BJT the collector junction reverse leakage current:
(a) Is more important at low temperatures than at high temperatures.
(b) Is due to electrons diffusing over from the emitter.
(c) May be minimized by making the base thin.
(d) Is due to electron–hole pair generation at the collector junction.

3.2-5 For an ideal BJT, $\alpha_F = 1$. Explain why $\alpha_F < 1$ for an actual transistor.

3.2-6 An ideal *pnp* silicon transistor has emitter, base, and collector doping levels of 5×10^{18}, 10^{17}, and 10^{15} cm^{-3}, respectively. In addition, the width of the base is 1 μm and the minority carrier diffusion lengths in the emitter and the collector are each 2 μm, while that in the base is 15 μm. Find α_F, h_{FE}, and I_{CBO} assuming that the transistor cross-sectional area is 0.01 cm^2.

3.2-7 When the collector and emitter leads are reversed, a BJT still works, but not as well. Explain this statement.

3.2-8 Find the emitter injection efficiency, base transport factor, and h_{FE} of a *pnp* transistor whose emitter, base, and collector doping levels are 10^{19}, 10^{17}, and 10^{15} cm^{-3}. Con-

sider that the diffusion lengths in the n and p materials are 10 μm and 8 μm, respectively, and also that the width of the base is 0.5 μm.

3.2-9 (a) Solve the continuity equation (1.11-9b) rigorously for $p_b(x)$ in the base of an *npn* transistor using the boundary conditions in equation (3.2-1) and show that

$$p_b(x) = [p_b(0) - p_{bo}] \frac{\sinh [(w - x)/L_p]}{\sinh (w/L_p)} + p_{bo}\left[1 - \frac{\sinh (x/L_p)}{\sinh (w/L_p)}\right]$$

(b) Using the fact that $w \ll L_p$, show that this expression for $p_b(x)$ reduces to that given in equation (3.2-2a).

3.2-10 (a) Using the solution for $p_b(x)$ given in Problem 3.2-9 but dropping the second term on the right-hand side of the equation (because it is small), derive expressions for I_{Ep} and I_{Cp}.
(b) Using the results from part (a), show that the base transport factor may be written as

$$T = \frac{I_{Cp}}{I_{Ep}} = \text{sech}\frac{w}{L_p}$$

(c) Demonstrate that the expression for T in part (b) reduces to that given in equation (3.2-9) for $w/L_p \ll 1$.

3.2-11 Repeat Example 3.2-1 if $V_{EE} = 1.7$ V, $R_E = 500 \ \Omega$, and $v_{in} = 1 \sin \omega t$ volts.

3.2-12 For the transistor amplifier examined in Example 3.2-1, determine the current gain, voltage gain, and power gain of the circuit.

3.3-1 (a) Find the Q point (I_{BE}, V_{BE}, I_C, and V_{CE}) for the transistor circuit given in Figure 3.3-1 if $V_{CC} = 15$ V, $R_C = 7$ kΩ, $R_B = 100$ kΩ, and $V_{BB} = -2$ V.
(b) Repeat part (a) if $V_{BB} = 2$ V.
(c) Repeat part (a) is $V_{BB} = 2$ V and $R_B = 10$ kΩ.

3.3-2 In a BJT circuit the transistor will be saturated if you "dump in enough base current." Using the graphical characteristics in Figure 3.3-3 and the circuit in Figure 3.3-1 with $R_C = 1$ kΩ and R_B and V_{BB} adjustable, discuss why the transistor is saturated for $I_B > 0.1$ mA.

3.3-3 For the circuit shown in Figure P3.3-3, select R_1, R_2, and R_E to place the Q point at $I_C = 3$ mA and $V_{CE} = 10$ V. You may assume that $h_{FE} = 20$.

3.3-4 For the circuit shown in Figure P3.3-4, select R_B and R_C to place the Q point at $I_C = 10$ mA and $V_{CE} = 5$ V.

3.3-5 For the circuit shown in Figure P3.3-4, $R_C = 1$ kΩ, $R_E = 0$, and $R_B = 24$ kΩ. In addition, a resistor $R_2 = 5$ kΩ is connected in series with a 10-V battery (negative terminal up) between the base and ground. Find the Q point of the transistor.

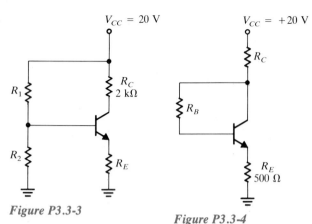

Figure P3.3-3

Figure P3.3-4

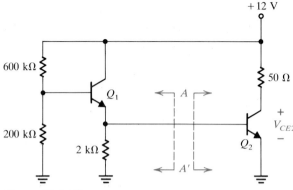

Figure P3.3-13

3.3-6 For the circuit shown in Figure P3.3-4, $V_{CC} = 24$ V, $R_C = 10$ kΩ, and $R_E = 270$ Ω. If a silicon transistor is used with $h_{FE} = 45$ and $V_{CE} = 5$ V, find R_B.

3.3-7 For the circuit in Figure P3.3-3, $R_1 = R_2 = 9$ kΩ, $R_C = 2.8$ kΩ, and $R_E = 0.7$ kΩ. Prove that the transistor is saturated.

3.3-8 In Figure P3.3-3, $R_C = 10$ kΩ, $R_E = 0$, $R_1 = 1$ MΩ, and $R_2 = 500$ kΩ. In addition, a -20-V battery (negative terminal up) is connected between the bottom of R_2 and ground. Find the transistor Q point.

3.3-9 For the circuit in Figure 3.3-4, $R_C = 2$ kΩ, $R_B = 20$ kΩ, and $R_E = 1$ kΩ. In addition, a resistor $R_2 = 5$ kΩ is connected from base to ground. Find V_{CE}.

3.3-10 In Figure P3.3-3, $R_C = 0$, $R_E = 20$ Ω, $R_1 = 1$ kΩ, and R_2 is replaced by a 10-V zener diode (cathode up). Find V_B, V_E, I_E, I_C, I_B, I_{R1}, I_Z, and V_{CE}. (*Hint:* Find each of the unknowns in the order given.)

3.3-11 For the circuit given in Figure P3.3-3, $R_C = 2$ kΩ, $R_1 = 386$ kΩ, $R_2 = \infty$, and $R_E = 0$. In addition, a resistor $R_L = 2$ kΩ is connected from the collector to ground. Find the circuit Q point.

3.3-12 The battery V_{BB} in Figure 3.5-1 is zero, v_{in} is a 10-V (p-p) square wave, $R_B = 10$ kΩ, $R_C = 5$ kΩ, and $V_{CC} = 20$ V. Carefully sketch and label i_B, i_C, and v_{CE}. (*Hint:* Repeat the Q-point calculation twice.)

3.3-13 For the circuit shown in Figure P3.3-13, find V_{CE2}. (*Hint:* As one possible approach, draw the equivalent circuit seen to the left and right of the points A–A' and proceed from there.)

3.3-14 For the circuit shown in Figure P3.3-14, both transistors are silicon. Q_1 is an *npn* transistor and Q_2 is a *pnp*. The h_{FE} values of both transistors are 100. Find the following quantities in the order given: V_1, V_2, I_{R3}, I_{E1}, V_{E1}, V_{RB1}, I_{B1}, I_{C1}, I_{B2}, I_{C2}, and V_{C2}.

3.3-15 For the circuit in Figure P3.3-3, $R_1 = R_2 = 50$ kΩ, $R_C = 5$ kΩ, and $R_E = 2$ kΩ. Find the transistor Q point.

3.4-1 Q-point stability can be achieved either by using an emitter resistor or by connecting a collector feedback resistor to the base. The circuit in Figure P3.3-3 has previously been shown to have an I_C given by equation (3.4-9), where $R_B = R_1 \| R_2$.

(a) Demonstrate that I_C for the transistor in Figure P3.3-4 (with $R_E = 0$) is identical to the result given in equation (3.4-9) if the term R_C is replaced by R_E.

(b) Based on the result in part (a), explain why the expression for the percent change in I_C for circuit (2) is identical to that given in equation (3.4-17) if R_C is replaced by R_E.

3.4-2 For the circuit in Figure P3.3-3, $R_C = 5$ kΩ. Select the remaining components to make $I_C = 1$ mA and $V_{CE} = 10$ V when $h_{FE} = 100$, and so that there is less than a 10% shift in I_C when h_{FE} changes to 200.

3.4-3 In Figure P3.3-3, $V_{CC} = 10$ V, $R_C = R_E = 3$ kΩ, $R_1 = 16.5$ kΩ, and $R_2 = 10$ kΩ. Find the worst-case Q-point shift in the temperature range 25° to 75°C. Assume that $V_{BE} = 0.7$ V and $I_{CBO} = 100$ nA at 25°C. Also assume that $h_{FE,min} = 100$ and $h_{FE,max} = 200$.

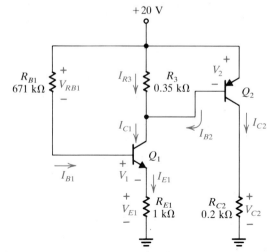

Figure P3.3-14

3.4-4 For the circuit given in Figure P3.3-4, $V_{CC} = 20$ V, $R_C = 5$ kΩ, and R_B and R_E are to be determined. Select these resistors for $I_C = 2$ mA when $h_{FE} = 100$, and so that less than a 10% shift in I_C occurs when h_{FE} increases to 200.

3.4-5 For the circuit shown in Figure P3.3-3, $R_C = 5$ kΩ, $R_E = 2$ kΩ, $R_1 = 200$ kΩ, and $R_2 = 50$ kΩ. If β_F (or h_{FE}) can vary from 20 to 100, and V_{BE} can vary from 0.60 to 0.70 V, find the minimum and maximum values of I_C.

3.4-6 In Figure P3.3-3, $R_C = 3$ kΩ and $R_1 = 60$ kΩ. Select R_2 and R_E so that $I_C = 3$ mA when $h_{FE} = 100$, and the Q-point shift due to the reverse leakage current is less than 5%. The maximum value of I_{CBO} is 5 μA.

3.4-7 Determine the value of V_{CE} at 100°C for the circuit in Figure 3.4-2a given that $V_{CC} = 3$ V, $R_C = 500$ Ω, and $R_B = 100$ kΩ. You may assume that $I_{CBO} = 0$.

3.4-8 (a) Complete the design of the circuit in Figure P3.3-3 (including the selection of R_C) given that $V_{CC} = 15$ V and $R_E = 1$ kΩ. The Q point is to be at $I_C = 2$ mA and $V_{CE} = 5$ V with $h_{FE} = 100$. In addition, the percent change in I_C is to be less than 3% when I_{CBO} increases to its maximum value of 10 μA.
(b) Find the percent change in V_{CE} if h_{FE} doubles.

3.4-9 For the circuit shown in Figure P3.3-3, $R_E = 500$ Ω, $R_1 = 330$ kΩ, $R_2 = \infty$, and I_{CBO} and V_{BE} are 100 nA and 0.65 V at room temperature. In addition, h_{FE} takes on values of 100 and 40 at 25 and -25°C, respectively.
(a) Calculate the Q point $(I_B, V_{BE}, I_C,$ and $V_{CE})$ at 25°C.
(b) Calculate the percent shift in I_C if the temperature decreases to 25°C below zero.

3.4-10 In Figure P3.3-3, $V_{CC} = 10$ V, and a -10-V battery (negative terminal up) is inserted between the lower terminal of R_2 and ground. Neglecting the effect of I_{CBO}, select the remaining components to place the nominal Q point at $V_{CE} = 5$ V and $I_C = 2$ mA when $h_{FE} = 100$. In addition, the percent shift in I_C is to be less than 10% when h_{FE} decreases to 50.

3.4-11 The circuit shown in Figure 3.4-2a is a fixed bias amplifier in which $R_C = 50$ kΩ, $R_B = 19.3$ mΩ, and $I_{CBO} = 10$ nA and $V_{BE} = 0.7$ V at 25°C.
(a) Find the transistor Q point at 25 and 100°C.
(b) Repeat part (a) if h_{FE} doubles to 200 at 100°C.

3.5-1 For the transistor shown in Figure P3.5-1, carefully sketch and label v_s, v'_{BB}, i_B, v_{BE}, i_C, and v_{CE}.

3.5-2 Repeat Problem 3.5-1 if a 2-kΩ resistor is connected in series with the emitter lead of the transistor.

3.5-3 The transistor in Figure P3.3-3 has the V–I characteristics given in Figure P3.5-1, and for the circuit $R_1 = 200$ kΩ, $R_2 = 100$ kΩ, $R_C = 1$ kΩ, $R_E = 0$, and $V_{CC} = 6$ V. In addition, a voltage source $v_{in} = 1 \sin \omega t$ V

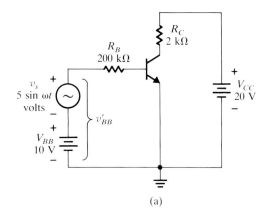

(a)

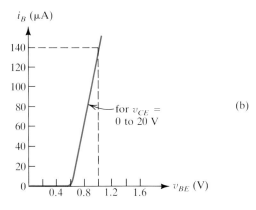

(b)

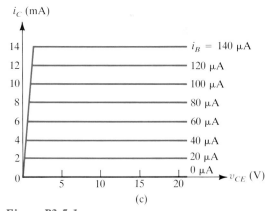

(c)

Figure P3.5-1

(plus on the right) is inserted between the R_1–R_2 junction and the base.
(a) Use the values of R_1, R_2, and V_{CC} (with $v_{in} = 0$) to get the Thévenin equivalent V_{BB} and R_B and write the dc Kirchhoff law equation around the base-to-emitter loop.
(b) Draw the dc load line on the input diode characteristics and find the values of the Q point (V_{BE}, I_B).

(c) Graphically determine the maximum and minimum values of i_B.

3.5-4 (a) Find the value of v_{in} that will just saturate and just cut off the transistor in Figure 3.5-8 given that $V_{CC} = 20$ V, $R_C = 3$ kΩ, $R_B = 100$ kΩ, $V_{BB} = 0$ and that a resistor $R_2 = 100$ kΩ is shunted from the base to ground..

(b) Carefully sketch and label v_o if v_{in} is a square wave whose amplitude varies between -5 and $+10$ V.

3.5-5 (a) Repeat part (a) of Problem 3.5-4.

(b) Carefully sketch and label v_o if v_{in} is a square wave whose amplitude varies between $+5$ and $+10$ V.

3.5-6 The transistor in Figure P3.3-3 has the V–I characteristics given in Figure P3.5-1, and for the circuit $R_1 = 166$ kΩ, $R_2 = 10.64$ kΩ, $R_C = 2$ kΩ, and $R_E = 0$. In addition, a voltage source $v_{in} = 0.35 \sin \omega t$ V (plus on the right) is inserted between the R_1–R_2 junction and the base. Carefully sketch and label i_B, i_C, and v_{CE}.

3.5-7 (a) R_C is increased to 5 Ω for the circuit given in Figure 3.5-1. Sketch i_C and v_{CE} given that $v_{in} = 1 \sin \omega t$ V.

(b) Repeat part (a) with V_{BB} selected to place the Q point at the center of the output load line.

(c) Determine the amplifier gain v_{ce}/v_{in} for the conditions in part (b).

3.6-1 The circuit shown in Figure P3.6-1 has the volt-ampere characteristics given in Figure P3.5-1.

(a) Write an expression for the input dc load-line equation.

(b) Sketch this equation on the input characteristics given.

3.6-2 For the circuit in Figure P3.6-1:

(a) Write an expression for the input dynamic load-line equation.

(b) Sketch the load line for $v_{in} = 0$, $+0.1$ V, and -0.1 V.

(c) Sketch v_{in}, v_{BE}, and i_B versus time.

3.6-3 The output volt-ampere characteristic for the transistor in Figure P3.6-1 is that given in Figure P3.5-1c.

(a) Write an expression for the output dc load line equation. Sketch the load line assuming $I_B = 75$ μA.

(b) Repeat part (a) for the dynamic load-line equation.

(c) Given that $i_B = (75 + 50 \sin \omega t)$ μA, carefully sketch i_B, i_C, and v_{CE} versus time.

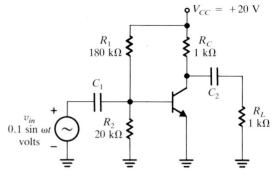

Figure P3.6-1

3.6-4 For the circuit shown in Figure 3.5-5, $V_{CC} = 20$ V, $V_{BB} = 3$ V, $v_{in} = 1 \sin \omega t$ V, $R_B = 20$ kΩ, $R_E = 0.2$ kΩ, and $R_C = 1.75$ kΩ. The transistor V–I characteristics are given in Figure P3.5-1. Carefully sketch and label v_{in}, $v_{BB'}$, i_B, v_{BE}, i_C, v_{CE}, v_E, and v_C.

3.6-5 Repeat Problem 3.6-4 if a very large bypass capacitor is connected across R_E.

3.6-6 For the circuit illustrated in Figure 3.6-6a, the transistor has the characteristics given in Figure 3.6-7.

(a) Sketch the waveforms for v_{CE} and v_L given that $v_{in} = 0.5 \sin \omega t$ V.

(b) Determine the average power delivered to the loudspeaker.

3.6-7 For the circuit shown in Figure P3.5-5, $V_{CC} = 20$ V, R_E is bypassed to ground with a large capacitor, and v_{in} and V_{BB} are chosen so that the total (dc + ac) base current in the transistor is $i_B = (40 + 20 \sin \omega t)$ μA. The transistor volt-ampere characteristics are given in Figure P3.5-1.

(a) Draw the dc load line on the output characteristics that relates i_C and v_{CE}. Be sure to indicate the Q-point location.

(b) Draw the dynamic (or ac) load line on the output characteristics that relates i_C and v_{CE}.

(c) Sketch i_B and v_{CE}.

3.6-8 Using the value of R_B selected in Example 3.6-2 determine the amplitude of v_{in} that will produce the maximum undistorted output voltage at v_L for the circuit in Figure 3.6-6.

CHAPTER 4

Small-Signal Analysis of Bipolar Transistor Circuits

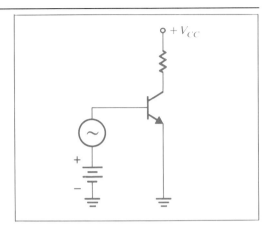

4.1 INTRODUCTION

In Chapter 3 we saw that the large-signal behavior of BJT circuits was best handled by graphical analysis because this was the only way to account properly for the nonlinear nature of the transistor's volt-ampere characteristics. However, the graphical approach is cumbersome, and as the signal amplitude is reduced, the use of small-signal linear models proves to be a much more effective technique. In fact, these models even work reasonably well when the signal amplitudes are quite large (see, for example, Example 3.6-2). Furthermore, although the graphical method appears to give you an excellent description of the actual circuit behavior, it is important to remember that the graphical data supplied by the manufacturer represents only the "typical," not the actual, characteristics of the device with which you are working. Therefore, even when the signal levels are not that small, the use of linear modeling techniques instead of graphical analysis methods usually makes more sense.

In Section 3.1 we introduced a very simple model for the bipolar junction transistor and used it for the analysis of both the dc and ac parts of transistor circuit problems. As we will see in Section 4.2, this model, while sufficiently accurate for initially developing an understanding of transistors, needs to be extended and modified to make it applicable to the general small-signal analysis of BJT circuits.

In the solution of actual engineering problems, the one-for-one replacement of each transistor in the circuit by its equivalent model can result in a network that is very difficult to analyze. To simplify the ac analysis of these types of problems, in Section 4.4 we develop equivalent circuits for the bipolar transistor similar to those already presented in Chapter 3 for the dc analysis of BJT circuits. These equivalent circuits effectively represent the Thévenin or Norton equivalents seen "looking into" the base, emitter, and collector terminals of the transistor, and they prove to be extremely useful for rapidly analyzing single- and multitransistor circuit problems.

The design of transistor amplifiers is often a very arduous task, and in Section 4.5 we present engineering guidelines for the construction of multitransistor amplifier circuits. As we will see, this type of design problem can usually be broken down into the design of the input, the intermediate, and the output stages of the amplifier. The design of the input and output stages is concerned with matching the amplifier characteristics to those of the source and the load, while the design of the intermediate amplifier stages is usually a function of the overall amplifier power gain that is required.

The circuits that are analyzed in this chapter are assumed to be operating in the midband frequency range. These frequencies are defined as those which are low enough so that the transistor models are valid, but high enough so that all coupling and bypass capacitors are shorts to the ac signals. The behavior of transistor circuits operating outside the midband frequency range is deferred until Chapter 6.

4.2 SMALL-SIGNAL AC MODEL FOR THE BIPOLAR JUNCTION TRANSISTOR

In Chapter 3 we developed a relatively simple model for the bipolar transistor that we used for carrying out both the dc and ac analysis of BJT circuits (Figure 3.2-5). As we saw, at least for the class of problems examined, this model provided rather good results. However, in choosing the examples used in Chapter 3, the author was careful to select problems that were appropriate for use with that model. Only a slight extension of this model is required to make it applicable to a wide range of transistor circuit analysis problems.

To illustrate how these changes may be incorporated into the original model, consider the *npn* transistor illustrated in Figure 4.2-1a along with the small-signal model developed previously in Chapter 3 in which the base–emitter junction was represented by an ac short circuit (Figure 3.6-2). From the i_E versus v_{BE} characteristic curve for this transistor given in Figure 4.2-2a, it is apparent that the use of a base–emitter short in the ac model is not strictly correct, because as i_E varies, v_{BE} also changes. To improve the accuracy of the model, let's use the approach we employed in developing the small-signal model for the *pn* junction diode in Section 1.8.

Following equations (3.2-4a) and (3.2-4b), the overall expression for i_E for this *npn* transistor can be written as

$$i_E = i_{Ep} + i_{En} = qA\left(\frac{D_p}{w} + \frac{D_n}{L_n}\right)e^{v_{BE}/V_T} = I_{EBO}e^{v_{BE}/V_T} \qquad (4.2\text{-}1)$$

where $v_{BE} = V_{BE} + v_{be}$ and I_{EBO} is the base–emitter junction reverse leakage current with the collector open circuited. Following Figure 1.8-1 and paralleling the development in equation (1.8-12), we may expand the exponential in

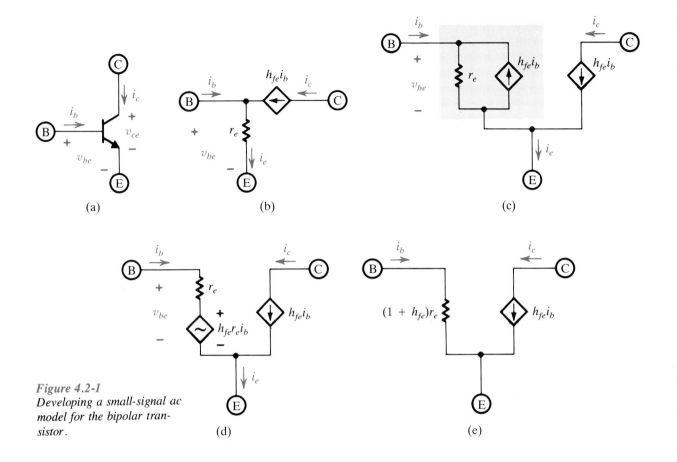

Figure 4.2-1
Developing a small-signal ac model for the bipolar transistor.

(a) (b) (c) (d) (e)

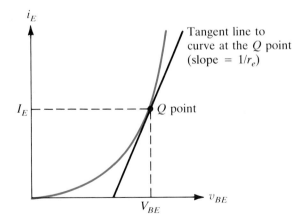

$\dfrac{V_{be}}{V_T}$	Relative 2nd Harmonic Distortion, I_2/I_1 (%)	Relative 3rd Harmonic Distortion, I_3/I_1 (%)
0.0	0.0	0.0
0.1	2.4	0.0
0.25	6.2	0.2
0.5	12.4	10.0
1.0	24.0	3.9

(a) Relationship between the Emitter Current and the Base-Emitter Voltage for the Bipolar Transistor

(b) Distortion in the Base Current of a BJT as a Function of the ac Base-Emitter Voltage

Figure 4.2-2

equation (4.2-1) in a power series and write the overall emitter current as

$$i_E = I_{EBO}(e^{v_{BE}/V_T} - 1) = I_{EBO}(e^{V_{BE}/V_T}e^{v_{be}/V_T} - 1)$$

$$= I_{EBO}\left[e^{V_{BE}/V_T}\left(1 + \frac{v_{be}}{V_T} + \left(\frac{v_{be}}{V_T}\right)^2\frac{1}{2!} + \left(\frac{v_{be}}{V_T}\right)^3\frac{1}{3!} + \cdots\right) - 1\right]$$

$$= I_{EBO}(e^{V_{BE}/V_T} - 1) + I_{EBO}e^{V_{BE}/V_T}\left[\frac{v_{be}}{V_T} + \left(\frac{v_{be}}{V_T}\right)^2\frac{1}{2!} + \left(\frac{v_{be}}{V_T}\right)^3\frac{1}{3!} + \cdots\right]$$

$$= \underbrace{I_E}_{\substack{\text{dc} \\ \text{term}}} + \underbrace{\frac{1}{r_e}v_{be}}_{\substack{\text{linear ac} \\ \text{term}}} + \underbrace{\frac{1}{r_e}\left(\frac{v_{be}^2}{2!V_T} + \frac{v_{be}^3}{3!V_T^2} + \frac{v_{be}^4}{4!V_T^3} + \cdots\right)}_{\text{nonlinear ac terms}} \tag{4.2-2}$$

where $I_E = I_{EBO}e^{V_{BE}/V_T}$, $r_e = V_T/I_E$, and $1/r_e$ is the slope of the line tangent to the i_E versus v_{BE} curve at the Q point in Figure 4.2-2a. For small signals, in particular for those of v_{be} much less than V_T, the expression for i_E can be approximated as

$$i_E = I_E + i_e \simeq I_E + \frac{v_{be}}{r_e} \tag{4.2-3a}$$

or
$$i_e = \frac{v_{be}}{r_e} \tag{4.2-3b}$$

From this result, we can see that an additional element, a resistor of value r_e, needs to be included in the emitter branch of the transistor ac small-signal model to account for the fact that the base–emitter voltage, v_{be}, increases in response to an increase in the emitter current i_e (Figure 4.2-1b).

To understand when equation (4.2-3) is a good approximation to (4.2-1), and to obtain a quantitative measure of the distortion produced when the ac part of v_{BE} gets too large, consider what happens when v_{be} is a sine wave $V_{be} \sin \omega t$. If we substitute this expression for v_{be} into equation (4.2-2) and apply the appropriate trigonometric identities, after considerable manipulation we may write i_E as

$$i_E = \underbrace{I_E}_{\substack{\text{dc} \\ \text{term}}} + \underbrace{I_1 \sin \omega t}_{\substack{\text{linear term} \\ \text{(fundamental)}}} + \underbrace{\underbrace{I_2 \sin (2\omega t + \phi_2)}_{\substack{\text{second harmonic} \\ \text{distortion}}} + \underbrace{I_3 \sin (3\omega t + \phi_3)}_{\substack{\text{third harmonic} \\ \text{distortion}}} + \cdots}_{\text{nonlinear ac terms}} \tag{4.2-4}$$

The first term in this expression represents the dc value of the emitter current and the next one the linear or usual small-signal current term. This is called the linear term because it has the same frequency as that of the applied signal v_{be}. Notice that because equation (4.2-2) is a nonlinear function of v_{be}, additional terms appear in equation (4.2-4) at harmonics of the applied signal frequency. It is the presence of these nonlinear distortion terms that makes the ac part of the emitter current nonsinusoidal even though the applied base–emitter voltage is a sine wave. The amplitude of these nonlinear distortion terms varies strongly with that of the applied ac base–emitter voltage. A detailed quantitative analysis of these amplitudes is beyond the scope of this book, but the results of this analysis are summarized in Figure 4.2-2b, in which the relative amounts of second and third harmonic distortion are listed as a function of the signal amplitude V_{be}. Using these data, we can see that if the nonlinear distortion is to be less than 5 or 10% of the fundamental, V_{be}/V_T must be smaller than 0.25, or V_{be} itself must be less than about 6 mV.

If the ac base–emitter voltage is within these bounds, equation (4.2-3) is a good approximation to (4.2-2), and the transistor model given in Figure 4.2-1c should work reasonably well. The relative importance of the resistor r_e in this model will depend on the external circuit connected to the transistor.

Notice in this equivalent circuit that the controlled current source is written as $h_{fe}i_b$. In this chapter and for the remainder of this book, following equation (4.3-11), h_{fe} rather than h_{FE} is used to represent the small-signal common-emitter current gain of the bipolar junction transistor.

To emphasize the isolation that exists between the collector and the rest of the transistor, the equivalent circuit of Figure 4.2-1b has been redrawn in part (c) of the figure with the original current source $h_{fe}i_b$ split into two equivalent current sources, each of value $h_{fe}i_b$. You should recall that the same approach was used in Section 3.3 when we developed the equivalent dc circuits for the BJT (see Figure 3.3-11, for example).

A further simplification of this circuit is possible by replacing the circuit within the shaded region by its Thévenin equivalent, as illustrated in Figure 4.2-1d. Here, because the voltage drop across the controlled voltage source is proportional to the current flow through it, it can be replaced by an equivalent resistor

$$R = \frac{v}{i} = \frac{h_{fe}r_e i_b}{i_b} = h_{fe}r_e \qquad (4.2\text{-}5)$$

This version of the BJT model is presented in Figure 4.2-1e.

For completeness, there is one additional second-order effect that needs to be included in our model for the BJT, and to understand its origin, consider the structural picture of the bipolar transistor illustrated in Figure 4.2-3a. This particular device is known as a planar transistor because of the flat surface on which the base and emitter are located. Notice that the actual area where the transistor activity occurs (shown in dashed lines in the figure) is only a small part of the overall device. Most of the transistor's bulk is needed to provide

Figure 4.2-3

Inclusion of $r_{bb'}$ in the bipolar transistor model.

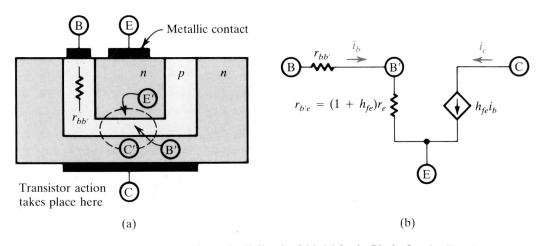

(a)

(b)

room for electrical connection to the external leads. In fabricating the transistor, it is rather easy to make low-resistance contact with the collector (C') and emitter (E') because of their large surface areas. The base connection is considerably more difficult. Because of the long path between B (the external base lead connection point) and B' (the real base of the transistor), and also because of the relatively low doping levels used in the base, the semiconductor material between B and B' exhibits a rather large resistance.

This ohmic resistance, designated by $r_{bb'}$, is known as the base-spreading resistance and has been included in the small-signal model given in Figure 4.2-3b. For simplicity, in this model the base–emitter small-signal resistance $(1 + h_{fe})r_e$ will also be referred to as $r_{b'e}$. Typical values of $r_{bb'}$ vary from about 10 to 200 Ω, while $r_{b'e}$ is on the order of 1 to 10 kΩ. Over the years many different names have been used to identify these two resistors. In particular, $r_{bb'}$ is sometimes called r_b or r_x, while the symbol r_π is often used in place of $r_{b'e}$. However, because of their clear relationship to the nodes where they are connected, we stay with the terms $r_{bb'}$ and $r_{b'e}$ in this book.

EXAMPLE 4.2-1

Figure 4.2-4
Circuit for Example 4.2-1.

The transistor in Figure 4.2-4 is silicon and has an $r_{bb'} = 50$ Ω and $h_{FE} = h_{fe} = 100$. Assume that C_1 and C_2 are large enough to be considered as ac shorts at the frequency of the input signal v_{in}. Sketch v_{BE}, i_B, i_C, v_C, v_E, and v_{CE} and find the voltage gain of the amplifier.

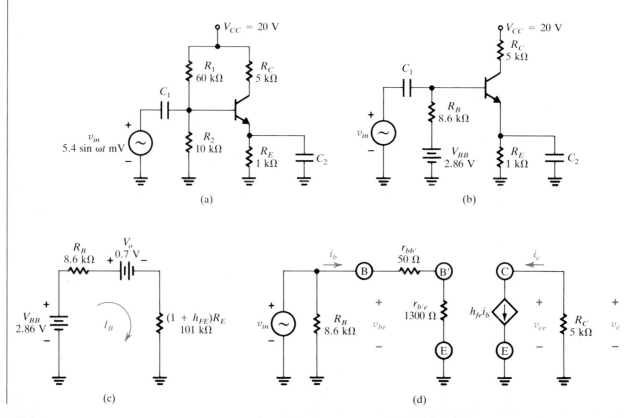

SOLUTION

The dc analysis of this circuit may be carried out with the aid of Figure 4.2-4b and c. Using these circuits, the results are $I_B = 19.7$ μA, $I_C = 1.97$ mA, $I_E = 1.99$ mA, $V_E = 1.99$ V, $V_C = 10.15$ V, and $V_{CE} = 8.16$ V.

The ac analysis of this problem follows directly from Figure 4.2-4d. Notice that because $I_E = 2$ mA, r_e is 26 mV/2 mA, or 13 Ω, so that $r_{b'e}$ is about 1.3 kΩ. Furthermore, since C_1 is a short circuit, the applied input voltage, v_{in}, appears directly across the base–emitter junction, so that

$$v_{be} = v_{in} = 5.4 \sin \omega t \text{ mV} \tag{4.2-6a}$$

$$i_b = \frac{v_{in}}{r_{bb'} + (1 + h_{fe})r_e} = 4 \sin \omega t \text{ } \mu\text{A} \tag{4.2-6b}$$

$$i_c = h_{fe}i_b = 0.4 \sin \omega t \text{ mA} \tag{4.2-6c}$$

$$v_c = v_{ce} = -i_c R_c = -2 \sin \omega t \text{ V} \tag{4.2-6d}$$

and $$v_e = 0 \tag{4.2-6e}$$

The important waveforms for this circuit are sketched in Figure 4.2-5. If we define the voltage gain of this circuit as $A_v = v_c(\text{p-p})/v_{in}(\text{p-p})$, then $A_v = -4$ V/10.8 mV, or about -370.

Notice that had we used the transistor model developed in Chapter 3, the base current would have been infinite because the base–emitter input impedance for that model was assumed to be zero. In the examples in Chapter 3, this assumption did not create serious errors because the problems were all chosen so that a large resistance was always connected in series with the base of the transistor. In the design of actual transistor amplifiers, however, this resistance is usually made as small as possible to achieve a large power gain.

Figure 4.2-5
Waveforms for Example 4.2-1.

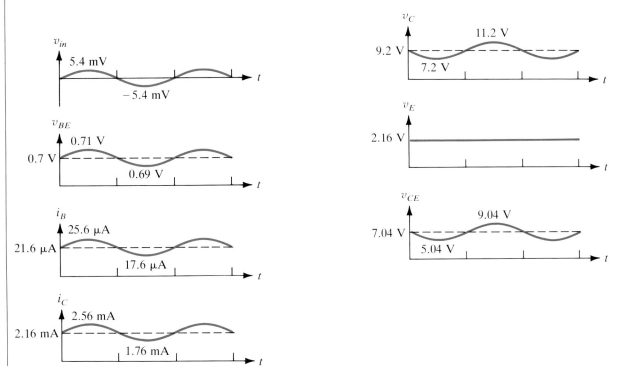

Hence for these more practical circuits, the transistor's input impedance becomes very important.

Before leaving this example, it is useful to point out that $r_{bb'}$ did not have much of an effect on the operation of this circuit, because $r_{b'e}$ is so much larger than $r_{bb'}$. In fact, $r_{bb'}$ can generally be neglected for BJT circuits whose signal frequencies are in the middle- and low-frequency ranges. However, when the transistor is operating at high frequencies, as we shall see in Chapter 6, the effect of $r_{bb'}$ can be quite pronounced.

Exercises

4.2-1 Using the first three terms in the expansion of equation (4.2-2), estimate the percent second harmonic distortion in i_E if $v_{be} = 4 \sin \omega t$ mV. **Answer** 3.8%

4.2-2 A transistor with $\alpha_F = 0.95$ is biased so that $I_E = 5$ mA. What are $r_{b'e}$ and h_{fe} for the transistor model in Figure 4.2-3? **Answer** $r_{b'e} = 104$ Ω, $h_{fe} = 19$

4.2-3 For the BJT amplifier discussed in Example 4.2-1, what is the maximum allowed amplitude of v_{in} for which no distortion in the output occurs? **Answer** 19 mV

4.2-4 a. Find the dc collector current in the transistor in Figure 4.2-4a if R_2 is changed to 5 kΩ.
b. What is $r_{b'e}$ for this transistor?
c. What is the amplifier voltage gain v_c/v_{in}?
Answer (a) 0.8 mA, (b) 3.28 kΩ, (c) -150

4.3 HYBRID PARAMETER SMALL-SIGNAL MODEL FOR THE BIPOLAR JUNCTION TRANSISTOR

The bipolar junction transistor is basically a nonlinear device. However, when it is biased in the active region and when the applied ac signals are small, the BJT can be shown to be approximately linear and may be analyzed by conventional two-port network theory. A review of two-port networks is presented below, and the application of these concepts to the analysis of small-signal BJT circuits is developed thereafter.

4.3-1 Review of Linear Two-Port Network Theory

A linear two-port network is a circuit that contains only linear elements and no independent sources. As shown in Figure 4.3-1, a two-port network has two sets of terminals, and generally, for electronics applications, the terminal pair on the left is designated as the network input or input port, while that on the right is called the output port.

A two-port network is described by two equations relating the four network variables I_1, V_1, I_2, and V_2. In general, these equations may be written in the

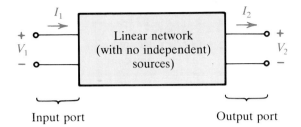

Figure 4.3-1
Linear two-port network.

Input port   Output port

form

$$X_1 = k_{11} Y_1 + k_{12} Y_2 \tag{4.3-1a}$$

and

$$X_2 = k_{21} Y_1 + k_{22} Y_2 \tag{4.3-1b}$$

where X_1, X_2, Y_1, and Y_2 represent any combination of these network variables and k_{11}, k_{12}, and so on, are constants known as the system parameters. These system parameters or system constants, as they are also called, depend only on the internal character of the network, and as such they are independent of any external circuitry that is connected onto the input or output terminals.

In the two-port equations (4.3-1a) and (4.3-1b), the variables on the left-hand side of the equations are considered to be the dependent variables and those on the right-hand side the independent variables. To write down a specific set of these two-port equations, two dependent variables need to be chosen from the four network variables (I_1, I_2, V_1, and V_2). From probability theory, the process of selecting two items from a set of four can be accomplished in any one of six different ways, and as such, there are six different possible sets of two-port network equations having the form given in (4.3-1a) and (4.3-1b). However, as we shall see shortly, some of these are more useful than others, especially in characterizing or modeling electronic devices.

In particular, because the BJT is a low-input-impedance, high-output-impedance device, for parameter measurement reasons it turns out to be most convenient to model this transistor using one particular set of parameters known as the system's hybrid or h-parameters. In this representation V_1 and I_2 are considered to be the dependent variables and are written in terms of I_1 and V_2 as

$$V_1 = \underbrace{h_{11} I_1}_{V_{11}} + \underbrace{h_{12} V_2}_{V_{12}} \tag{4.3-2a}$$

and

$$I_2 = \underbrace{h_{21} I_1}_{I_{21}} + \underbrace{h_{22} V_2}_{I_{22}} \tag{4.3-2b}$$

Following these equations, the h-parameters may be evaluated mathematically as follows. To compute h_{11}, for example, we set V_2 equal to zero in equation (4.3-2a) and solve for h_{11}, which yields

$$h_{11} = \left. \frac{V_1}{I_1} \right|_{V_2=0} \tag{4.3-3a}$$

Thus h_{11} is the network input impedance that is measured with the output short circuited. For simplicity, it is also called the network's short-circuit input impedance. The other definitions of the h-parameters follow similarly with

$$h_{12} = \left.\frac{V_1}{V_2}\right|_{I_1=0} = \text{open-circuit reverse voltage gain} \qquad (4.3\text{-}3b)$$

$$h_{21} = \left.\frac{I_2}{I_1}\right|_{V_2=0} = \text{short-circuit forward current gain} \qquad (4.3\text{-}3c)$$

and $\qquad h_{22} = \left.\frac{I_2}{V_2}\right|_{I_1=0} = \text{open-circuit output impedance} \qquad (4.3\text{-}3d)$

Notice that the dimensions of h_{11} and h_{22} are in ohms and in siemens, respectively, while h_{12} and h_{21} are dimensionless. In fact, the name "hybrid" was chosen for these parameters specifically to indicate the mixed nature of the units associated with the coefficients.

The hybrid representation for the two-port network illustrated in Figure 4.3-2 follows readily from the other equivalent circuits already developed for the z- and y-parameter representations. Let's try to understand why this model is so well suited to the representation of the BJT. Consider, for example, the measurement of h_{11}. Following equation (4.3-3a), this parameter is just equal to the ratio of V_1/I_1, measured with V_2 set equal to zero, or with the output short circuited. As mentioned in regard to the y-parameters, experimentally V_2 is set equal to zero by placing a capacitor across the output terminals. Because the output impedance of a BJT is large, it will be relatively easy to simulate this short circuit at the output. Similarly, because of the particular relationship

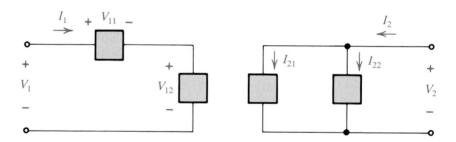

(a) Block Diagram Model for the h-Parameter Representation of a Two-Port Network

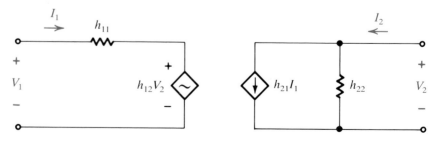

(b) Equivalent Circuit h-Parameter Model of a Two-Port Network

Figure 4.3-2

Chapter 4 / Small-Signal Analysis of Bipolar Transistor Circuits

between the BJT's impedance levels and the experimental conditions needed to determine the other h-parameters, it will also be equally easy to measure h_{12}, h_{21}, and h_{22} (see Problem 4.3-3). Besides being easy to measure experimentally, the use of h-parameters to represent the BJT also leads to a relatively simple circuit model since only two of the four parameters turn out to be important in most instances.

4.3-2 Development of the h-Parameter Small-Signal Model for the BJT

The physical models presented for the bipolar junction transistor in Chapter 3 and in Section 4.2 were all based on a careful examination of the device physics. The model described in this section is developed from a purely mathematical viewpoint and will permit us to take into account several of the second-order effects found in the transistor that, until now, we have neglected.

By examining the graphical characteristics of the bipolar junction transistor (e.g., Figure 3.3-3), we can see that it is possible to treat the BJT as a two-port network in which the base–emitter leads are taken to be the input terminals, and the collector–emitter leads the output terminals (Figure 4.3-3a). Furthermore, if we consider that i_B and v_{CE} are the independent variables, and that v_{BE} and i_C are the dependent variables, we can express v_{BE} and i_C as functions of the independent variables, where

$$v_{BE} = v_{BE}(i_B, v_{CE}) \tag{4.3-4a}$$

$$i_C = i_C(i_B, v_{CE}) \tag{4.3-4b}$$

These equations, of course, are just the mathematical expressions of the transistor's volt-ampere characteristics.

In developing a small-signal model for the transistor, we are interested in being able to predict its behavior when the input signal causes only small deviations of the operating point about the Q point. If we assume that the input signal changes the independent variables i_B and v_{CE} by $(i_B - I_B)$ and $(v_{CE} - V_{CE})$, respectively, the new values for the dependent variables v_{BE} and i_C can be expressed as two-dimensional Taylor series expansions of the functions in equations (4.3-4) as follows:

$$v_{BE} = v_{BE}\bigg|_{Q\,\text{pt}} + \frac{\partial v_{BE}}{\partial i_B}\bigg|_{\substack{Q\,\text{pt} \\ v_{CE}=V_{CE}}} (i_B - I_B) + \frac{\partial^2 v_{BE}}{\partial i_B^2}\bigg|_{\substack{Q\,\text{pt} \\ v_{CE}=V_{CE}}} \frac{(i_B - I_B)^2}{2!} + \cdots$$

$$+ \frac{\partial v_{BE}}{\partial v_{CE}}\bigg|_{\substack{Q\,\text{pt} \\ i_B=I_B}} (v_{CE} - V_{CE}) + \frac{\partial^2 v_{BE}}{\partial v_{CE}^2}\bigg|_{\substack{Q\,\text{pt} \\ i_B=I_B}} \frac{(v_{CE} - V_{CE})^2}{2!} + \cdots \tag{4.3-5a}$$

and

$$i_C = i_C\bigg|_{Q\,\text{pt}} + \frac{\partial i_C}{\partial i_B}\bigg|_{\substack{Q\,\text{pt} \\ v_{CE}=V_{CE}}} (i_B - I_B) + \frac{\partial^2 i_C}{\partial i_B^2}\bigg|_{\substack{Q\,\text{pt} \\ v_{CE}=V_{CE}}} \frac{(i_B - I_B)^2}{2!} + \cdots$$

$$+ \frac{\partial i_C}{\partial v_{CE}}\bigg|_{\substack{Q\,\text{pt} \\ i_B=I_B}} (v_{CE} - V_{CE}) + \frac{\partial^2 i_C}{\partial v_{CE}^2}\bigg|_{\substack{Q\,\text{pt} \\ i_B=I_B}} \frac{(v_{CE} - V_{CE})^2}{2!} + \cdots \tag{4.3-5b}$$

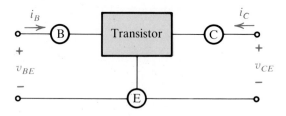

(a)   Transistor Being Modeled

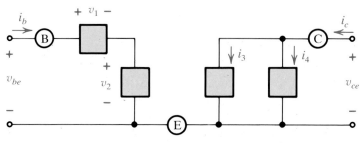

(b)   Building Blocks of the Hybrid Model

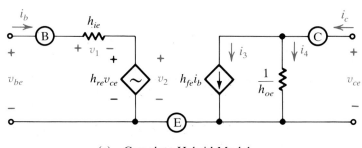

(c)   Complete Hybrid Model

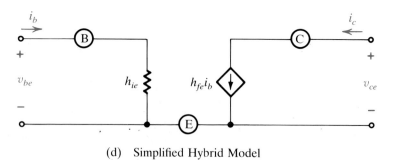

Figure 4.3-3
Common-emitter hybrid small-
signal model for the bipolar
transistor.

(d)   Simplified Hybrid Model

Although at first glance apparently complex, this development exactly parallels that done previously for the small-signal analysis of the diode in Section 1.8 [see especially equation (1.8-2)]. In particular, if the deviations about the Q point are small, we can, as a first-order approximation, neglect the quadratic and higher-order terms so that

$$v_{BE} \simeq V_{BE} + \left.\frac{\partial v_{BE}}{\partial i_B}\right|_{\substack{Q\text{ pt} \\ v_{CE}=V_{CE}}} (i_B - I_B) + \left.\frac{\partial v_{BE}}{\partial v_{CE}}\right|_{\substack{Q\text{ pt} \\ i_B=I_B}} (v_{CE} - V_{CE}) \quad (4.3\text{-}6a)$$

and

$$i_C \simeq I_C + \left.\frac{\partial i_C}{\partial i_B}\right|_{\substack{Q\text{ pt} \\ v_{CE}=V_{CE}}} (i_B - I_B) + \left.\frac{\partial i_C}{\partial v_{CE}}\right|_{\substack{Q\text{ pt} \\ i_B=I_B}} (v_{CE} - V_{CE}) \quad (4.3\text{-}6b)$$

Bringing V_{BE} onto the left-hand side of equation (4.3-6a) and I_C onto the left-hand side of (4.3-6b) and further defining $\Delta v_{BE} = v_{BE} - V_{BE}$ as the change in the base–emitter voltage, with similar definitions for the other circuit variables, we can rewrite equations (4.3-6) as

$$\Delta v_{BE} = \left.\frac{\partial v_{BE}}{\partial i_B}\right|_{\substack{Q\text{ pt} \\ v_{CE}=V_{CE}}} \Delta i_B + \left.\frac{\partial v_{BE}}{\partial v_{CE}}\right|_{\substack{Q\text{ pt} \\ i_B=I_B}} \Delta v_{CE} \quad (4.3\text{-}7a)$$

and

$$\Delta i_C = \left.\frac{\partial i_C}{\partial i_B}\right|_{\substack{Q\text{ pt} \\ v_{CE}=V_{CE}}} \Delta i_B + \left.\frac{\partial i_C}{\partial v_{CE}}\right|_{\substack{Q\text{ pt} \\ i_B=I_B}} \Delta v_{CE} \quad (4.3\text{-}7b)$$

The partial derivative terms in these equations are constants, and paralleling the two-port discussion earlier in this section, these constants are known as the transistor's hybrid or h-parameters. As mentioned earlier, the term "hybrid" is used to indicate the mixed nature of the units associated with these coefficients.

For the common-emitter transistor configuration under consideration, these equations can be rewritten in terms of the h-parameters as

$$\Delta v_{BE} = h_{ie}\,\Delta i_B + h_{re}\,\Delta v_{CE} \quad (4.3\text{-}8a)$$

$$\Delta i_C = h_{fe}\,\Delta i_B + h_{oe}\,\Delta v_{CE} \quad (4.3\text{-}8b)$$

These equations are identical to those developed in equations (4.3-2a) and (4.3-2b) if the substitutions $V_1 = \Delta v_{BE}$, $I_1 = \Delta i_B$, $I_2 = \Delta i_C$, and $V_2 = \Delta v_{CE}$ are made.

In the general two-port equations given in (4.3-2a) and (4.3-2b), numerical subscripts are used to refer to the different h-parameter coefficients. However, for electronic circuit applications, literal subscripts are employed with these parameters because they convey more information. In particular, the subscripts i, r, f, and o indicate that the coefficient is an input, reverse, forward, or output parameter, respectively. Furthermore, the second subscript e on all the coefficients is used to indicate that the parameters are being measured in a common-emitter circuit configuration with the base as the input and the collector as the output (Figure 4.3-3a). Thus, in comparing these two sets of equations, h_{11} has the same meaning as h_{ie}, h_{12} as h_{re}, h_{21} as h_{fe}, and h_{22} as h_{oe}.

The h-parameters of a BJT may be evaluated with the aid of equations (4.2-8) as follows. To find h_{ie}, for example, we hold v_{CE} constant (or alternatively, make $\Delta v_{CE} = 0$) and solve equation (4.3-8a) for h_{ie} to obtain

$$h_{ie} = \left.\frac{\Delta v_{BE}}{\Delta i_B}\right|_{\substack{Q \text{ pt} \\ v_{CE}=V_{CE}}} = \text{input impedance with output shorted} \quad (4.3\text{-}9)$$

The units of h_{ie} are volts/amperes or ohms, and this parameter is measured by taking the ratio of the change in v_{BE} to the change in i_B at the Q point along the constant v_{CE} line. Assuming a transistor Q point at $I_B = 200\ \mu\text{A}$, $V_{BE} = 0.7\ \text{V}$, $I_C = 22\ \text{mA}$, and $V_{CE} = 8\ \text{V}$, we may evaluate h_{ie} with the aid of Figure 4.3-4a. For this case h_{ie} is a measure of the slope of the line tangent to the ($v_{CE} > 0.2\ \text{V}$) curve at the Q point. Using points A and B on the tangent line, h_{ie} is $(0.76 - 0.62)\ \text{V}/(250 - 150)\ \mu\text{A}$, or $1400\ \Omega$.

To find h_{re}, we set Δi_B equal to zero in equation (4.3-8a) so that

$$h_{re} = \left.\frac{\Delta v_{BE}}{\Delta v_{CE}}\right|_{\substack{Q \text{ pt} \\ i_B=I_B}} = \text{reverse voltage gain with input open} \quad (4.3\text{-}10)$$

At first glance, on examination of Figure 4.3-4a, h_{re} appears to be zero, because the changes in collector-to-emitter voltage about the point $V_{CE} = 8\ \text{V}$ do not seem to produce any changes in v_{BE}. However, if we expand these characteristics along the v_{BE} direction in the neighborhood of the Q point (Figure 4.3-4b), although rather small, we may now numerically evaluate h_{re} as $(0.7005 - 0.6995)/(9 - 7)$, or 5×10^{-4}. Because it is the ratio of two voltage changes, h_{re} is dimensionless.

The parameter h_{fe} is obtained from equation (4.3-8b) as

$$h_{fe} = \left.\frac{\Delta i_C}{\Delta i_B}\right|_{\substack{Q \text{ pt} \\ v_{CE}=V_{CE}}} = \text{forward current gain with output shorted} \quad (4.3\text{-}11)$$

and the h_{fe} for the specific transistor being examined can be found by forming the ratio of the change in i_C to the change in i_B with these quantities evaluated along the constant v_{CE} line in Figure 4.3-4c. Using points C and D on this line,

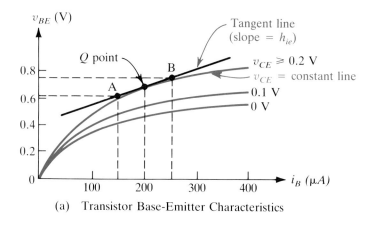

(a)   Transistor Base-Emitter Characteristics

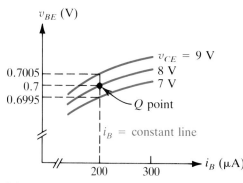

(b)   Transistor Base-Emitter Characteristic–
Expanded in the Vicinity of $v_{BE} = 0.7$ V

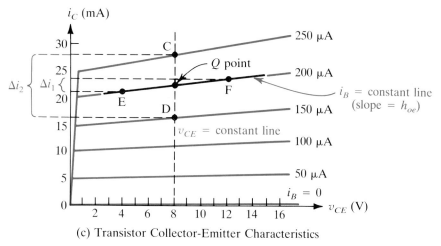

(c) Transistor Collector-Emitter Characteristics

Figure 4.3-4
Graphical measurement of the hybrid parameters.

h_{fe} is found to be $(26 - 17)$ mA/$(250 - 150)$ μA, or 90. As with h_{re}, h_{fe} is dimensionless since the quantity h_{fe} is the ratio of two currents.

Notice that for this transistor circuit, the dc h_{FE}, defined by equation (3.3-5), is equal to 22 mA/0.2 mA or 112. The dc and ac current gains, h_{FE} and h_{fe}, are identical only when the i_B curves on the transistor output characteristics are uniformly spaced and parallel to the v_{CE} axis.

The last hybrid parameter, h_{oe}, is obtained by setting i_B equal to zero in equation (4.3-8b) to yield

$$h_{oe} = \left. \frac{\Delta i_C}{\Delta v_{CE}} \right|_{\substack{Q \text{ pt} \\ i_B = I_B}} = \text{output admittance with input open} \qquad (4.3\text{-}12)$$

Because h_{oe} is the ratio of a current to a voltage, its units are in siemens. To evaluate the parameter h_{oe} for this transistor, we use Figure 4.3-4c and form the ratio of the change in collector current to the change in collector–emitter voltage along the i_B equals a constant line about the Q point. Using points E and F on this line, h_{oe} is found to be $(23.5 - 21.5)$ mA/$(12 - 4)$ V, or 0.25 mS. The reciprocal of h_{oe}, the resistance $1/h_{oe}$, is 4 kΩ. This value of $1/h_{oe}$ is smaller than for actual BJT devices and was selected this way to allow for easy graphical calculation of this parameter. The characteristics of the common-emitter h-parameters are summarized in Table 4.3-1.

Referring again to equations (4.3-8a) and (4.3-8b), when the changes i_B, v_{BE}, and so on, are sinusoidal, we may rewrite these equations as

$$v_{be} = \underbrace{h_{ie}i_b}_{v_1} + \underbrace{h_{re}v_{ce}}_{v_2} \qquad (4.3\text{-}13a)$$

and

$$i_c = \underbrace{h_{fe}i_b}_{i_3} + \underbrace{h_{oe}v_{ce}}_{i_4} \qquad (4.3\text{-}13b)$$

Using these two equations, the hybrid small-signal model can be immediately drawn in block diagram form as shown in Figure 4.3-3b, with v_{be} the sum of the two voltages v_1 and v_2, and i_c the sum of the two currents i_3 and i_4. To evaluate the circuit elements represented by each of the boxes in the figure, let's begin with the block labeled v_1. The current flow through this block is i_b, and the voltage drop across it is $h_{ie}i_b$; because the voltage drop across this block is proportional to the current flow through it, it can be represented by a resistor h_{ie}.

The box labeled v_2 has a voltage drop across it equal to $h_{re}v_{ce}$. This voltage drop is independent of the current i_b flowing through the block but depends on a voltage that exists elsewhere in the circuit; it is therefore represented by a voltage-controlled voltage source, $h_{re}v_{ce}$.

At the output side of the model, the current $i_3 = h_{fe}i_b$ depends on a current flowing elsewhere in the circuit. As a result, it is represented by a current-controlled current source, $h_{fe}i_b$. The last element in the model, represented by the box with the current i_4 flowing through it, has a current flow $i_4 = h_{oe}v_{ce}$,

4.3-1 Summary of the Common-Emitter Hybrid Parameter Characteristics

Parameter Symbol	Parameter Name	Typical Value (in Low-Power Transistors)	Units	Relationship to Model in Figure 4.2-3b
h_{ie}	Input resistance	2×10^2 to 10^4	Ohms	$r_{bb'} + (1 + h_{fe})r_e$
h_{fe}	Forward current gain	20 to 500	Dimensionless	Same
h_{oe}	Output admittance	5×10^{-6} to 50×10^{-6}	Siemens	Not included in this model
h_{re}	Reverse voltage gain	10^{-6} to 5×10^{-4}	Dimensionless	Not included in this model

which is proportional to the voltage, v_{ce}, across it. This box can therefore be represented by a resistance equal to $1/h_{oe}$.

The complete hybrid model is given in Figure 4.3-3c. For most applications, the elements containing the terms h_{re} and h_{oe} can be neglected, so that this model reduces to that developed in Section 4.2, where $h_{ie} = r_{bb'} + r_{b'e}$. This simplified model is presented in Figure 4.3-3d, but before we accept these approximations, it will be instructive for us to examine the complete model in order to develop an understanding of when these terms may be properly neglected. Consider, for example, the circuit illustrated in Figure 4.3-5, and let us calculate the voltage gain of this circuit given that $h_{ie} = 2$ kΩ, $h_{re} = 10^{-4}$, $1/h_{oe} = 100$ kΩ, and $h_{fe} = 100$.

The small-signal equivalent circuit is shown in Figure 4.3-5b, and by examining this figure, it should be clear that $1/h_{oe}$ can be neglected whenever it is much greater than R_C. For engineering purposes, a factor of 10 larger is usually sufficient to permit it to be neglected. Therefore, for this circuit we can approximate R'_C as $R_C = 5$ kΩ.

To understand the effect of the term $h_{re}v_{ce}$, consider that

$$v_{ce} = i_c R'_c = -h_{fe}i_b R'_c \qquad (4.3\text{-}14a)$$

so that
$$h_{re}v_{ce} = -(h_{re}h_{fe}R'_C)i_b \qquad (4.3\text{-}14b)$$

Figure 4.3-5

Application of the hybrid model to the analysis of a bipolar transistor amplifier.

Here, because this voltage is proportional to i_b, and also because of the negative sign in equation (4.3-14b), the source $h_{re}v_{ce}$ can be replaced by an equivalent negative resistance

$$R = -(h_{re}h_{fe}R'_C) \qquad (4.3\text{-}15)$$

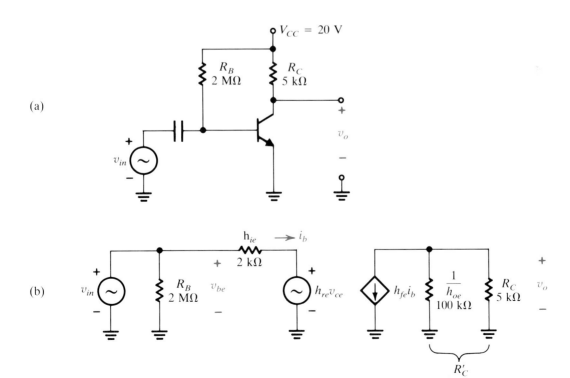

Clearly, this resistance is important when it is comparable to h_{ie}. For this example, $R = (10^{-4})(100)(5 \text{ k}\Omega) = -50 \ \Omega$ and can therefore be neglected in comparison to h_{ie}. In general, following equation (4.3-15), the controlled source $h_{re}v_{ce}$ may be replaced by a short whenever

$$h_{re} \ll \frac{h_{ie}}{h_{fe}R'_C} \tag{4.3-16}$$

For this circuit, we can neglect the controlled source at the input as long as h_{re} is much less than 2×10^{-3}.

EXAMPLE 4.3-1

For the circuit shown in Figure 4.3-6a, assume the transistor to have the volt-ampere characteristics given in Figure 4.3-4a and c.

a. By using the hybrid model for the transistor, sketch v_{BE}, i_B, i_C, and v_{CE}, given that v_{in} is a 20-mV(p-p) triangle wave.
b. Find the voltage gain of the amplifier.

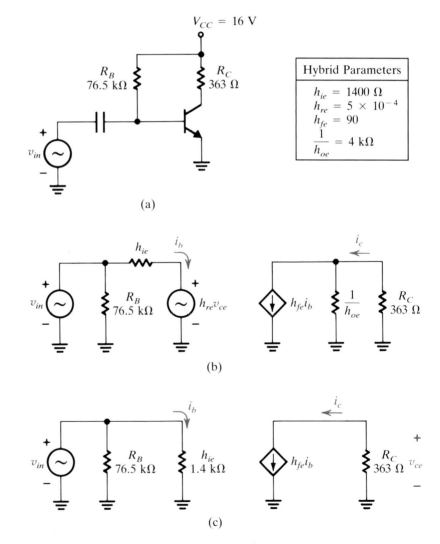

(a)

(b)

(c)

Figure 4.3-6
Circuit for Example 4.3-1.

Chapter 4 / Small-Signal Analysis of Bipolar Transistor Circuits

SOLUTION

The transistor volt-ampere characteristics are repeated in Figure 4.3-7, and the dc load lines for this circuit are sketched on these graphs. Using these load lines, it is seen that the transistor quiescent point is located at $V_{BE} = 0.7$ V, $I_B = 200$ μA, $I_C = 22$ mA, and $V_{CE} = 8.0$ V. This is the same Q point as shown on the transistor curves in Figure 4.3-4, and therefore the h-parameters are the same as those previously calculated. These values are listed in Figure 4.3-6a, and the ac small-signal model for this circuit is given in Figure 4.3-6b. For this circuit the voltage source $h_{re}v_{ce}$ is equivalent to a resistance of

Figure 4.3-7
Transistor volt-ampere characteristics for Example 4.3-1.

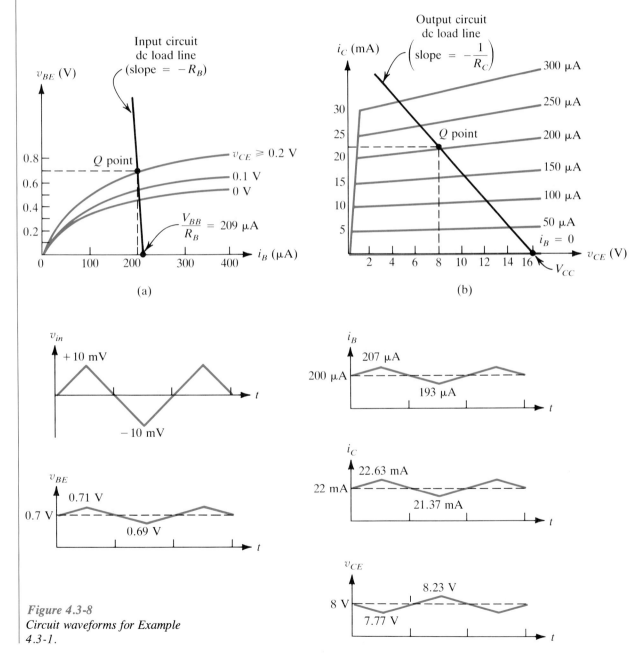

Figure 4.3-8
Circuit waveforms for Example 4.3-1.

−16.3 Ω and is therefore negligible. If we also neglect $1/h_{oe}$, we may redraw the model for this circuit as shown in Figure 4.3-6c. From this figure the pertinent currents and voltages in this circuit are

$$v_{be} = v_{in} \tag{4.3-17a}$$

$$i_b = \frac{v_{in}}{h_{ie}} \tag{4.3-17b}$$

$$i_c = h_{fe} i_b = \frac{h_{fe} v_{in}}{h_{ie}} \tag{4.3-17c}$$

and $$v_{ce} = -i_c R_C = -\frac{h_{fe} R_C}{h_{ie}} v_{in} \tag{4.3-17d}$$

Using these results and the fact that v_{in} is a 20-mV(p-p) triangle wave, we may sketch the required waveshapes as shown in Figure 4.3-8. The peak ac amplitude of the base current is 10 mV/1.4 kΩ = 7 μA, and that of the collector current is 90 × 7, or 0.54 mA. In addition, the ac collector-to-emitter voltage is $-i_c R_C$, or −0.23 V peak, and from equation (4.3-17d), the amplifier voltage gain is −(100)(363)/1400, or about −23.

Exercises

4.3-1 Find the h-parameters for the circuit given in Figure E4.3-1. *Answer* $h_{11} = 5$ kΩ, $h_{12} = 0.5$, $h_{21} = 0.5$, $h_{22} = 0.1$ mS

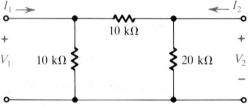

Figure E4.3-1

4.3-2 A bipolar junction transistor has the following h-parameters: $h_{ie} = 2$ kΩ, $h_{re} = 10^{-4}$, $h_{fe} = 150$, and $h_{oe} = 0.05$ mS. If $i_b = 5 \sin \omega t$ μA and $v_{ce} = -10 \sin \omega t$ volts, find v_{be} and i_c. *Answer* 9 sin ωt mV, 0.25 sin ωt mA

4.3-3 For the BJT amplifier in Figure 4.3-6, find the voltage gain v_{ce}/v_{in} if h_{re} is 10^{-2} instead of 10^{-4}. *Answer* −35

4.3-4 Evaluate the transistor h-parameters for the circuit illustrated in Figure 4.3-6 if R_B is increased to 153 kΩ. The volt-ampere characteristics for this transistor are given in Figure 4.3-7. *Answer* $h_{ie} = 2.9$ kΩ, $h_{re} = 0$, $1/h_{oe} = 8$ kΩ, $h_{fe} = 110$

4.4 SMALL-SIGNAL EQUIVALENT CIRCUITS FOR THE BJT

The simplified hybrid parameter model discussed in Section 4.3 and illustrated in Figure 4.3-3d provides the starting point for the development of a very powerful analysis tool for the study of bipolar transistor small-signal amplifiers. The basic method is very similar to that used in Section 3.3 for the dc analysis of BJT circuits. Paralleling this approach, we begin by deriving three separate equivalent circuits: specifically, those seen looking into the base, emitter, and collector of the transistor. These circuits are then used to solve several different single and multitransistor problems in order to illustrate the widespread applicability of these techniques.

To begin the development of these equivalent circuits, consider the generalized transistor circuit shown in Figure 4.4-1a. In part (b) of this figure, the transistor has been replaced by its simplified hybrid model, and the current source $h_{fe}i_b$ between the collector and emitter has been split into two equivalent current sources. You should recall that this is the approach we used in Section 3.3 when we developed the dc equivalent circuits for the BJT (Figure 3.3-11).

The base–emitter portion of this circuit has been redrawn in Figure 4.4-2a with the parallel combination of R_E and the current source $h_{fe}i_b$ replaced by their Thévenin equivalent circuit. Because the voltage across the controlled source is proportional to the current i_b flowing through it, the controlled source can be replaced by an equivalent resistor,

$$R = \frac{v}{i_b} = h_{fe}R_E \qquad (4.4\text{-}1)$$

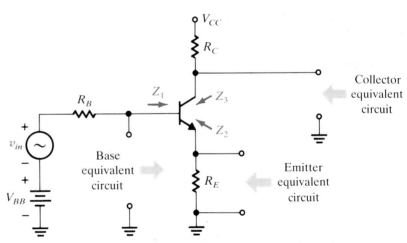

(a)   Generalized Bipolar Transistor Circuit to Be Analyzed

Figure 4.4-1

Developing the small-signal bipolar transistor equivalent circuits seen looking into the base, emitter, and collector.

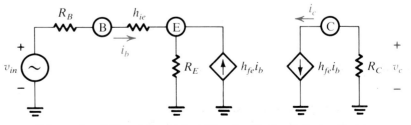

(b)   ac Small-Signal Equivalent Circuit with the Collector-Emitter Current Source Split into Two Equivalent Current Sources

The combined base equivalent circuit with this resistor and R_E is shown in Figure 4.4-2b. The utility of this equivalent circuit is immediately obvious since we now have a single-loop network in which the unknown voltages and currents can be found by inspection. Specifically, using this circuit, we can readily see that

$$i_b = \frac{v_{in}}{R_B + h_{ie} + (1 + h_{fe})R_E} \tag{4.4-2a}$$

$$v_b = \frac{h_{ie} + (1 + h_{fe})R_E}{R_B + h_{ie} + (1 + h_{fe})R_E} v_{in} \tag{4.4-2b}$$

$$v_{be} = \frac{h_{ie}}{R_B + h_{ie} + (1 + h_{fe})R_E} v_{in} \tag{4.4-2c}$$

and $$v_e = \frac{(1 + h_{fe})R_E}{R_B + h_{ie} + (1 + h_{fe})R_E} v_{in} \tag{4.4-2d}$$

In addition, the impedance Z_1 seen looking into the base terminal of the transistor is

$$Z_1 = h_{ie} + (1 + h_{fe})R_E \tag{4.4-3}$$

To develop the emitter equivalent circuit, we have redrawn the base–emitter circuit given in Figure 4.4-1b as shown in Figure 4.4-3a. Notice that interchanging the position of the resistor R_E and the current source $h_{fe}i_b$ does not alter the circuit because these components are in parallel. This circuit is best analyzed by obtaining the Thévenin equivalent of the portion of this circuit shown within the shaded region.

To find the Thévenin or open-circuit voltage notice that

$$i_e = (1 + h_{fe})i_b = 0 \tag{4.4-4a}$$

or $$i_b = 0 \tag{4.4-4b}$$

Figure 4.4-2

Equivalent circuit seen looking into the base (to the right of terminals B and G).

with R_E removed. Furthermore, with $i_b = 0$, there is no voltage drop across the resistors R_B and h_{ie}, and therefore

$$v_{oc} = v_e|_{i_e=0} = v_{in} \tag{4.4-5}$$

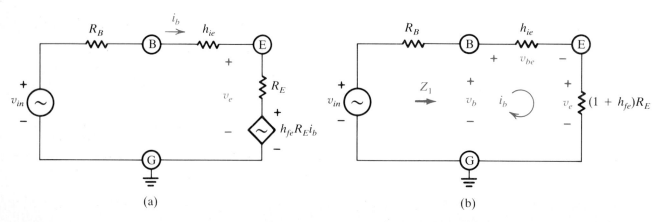

(a)   (b)

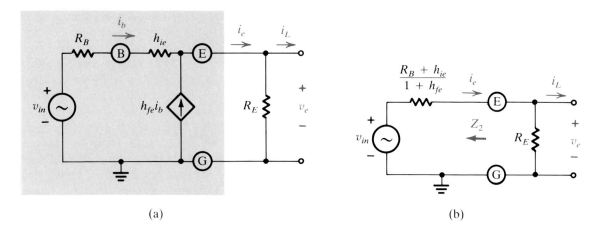

(a)   (b)

Figure 4.4-3

Equivalent circuit seen looking back into the emitter (to the left of terminals E and G).

To find the Thévenin or output impedance, we begin by calculating the short-circuit current. For this case, since the voltage across the output terminals is zero, the base current is given by

$$i_b = \frac{v_{in}}{R_B + h_{ie}} \tag{4.4-6}$$

so that the short-circuit current may be written as

$$i_{sc} = i_e|_{v_e=0} = 1 + h_{fe})i_b = (1 + h_{fe})\frac{v_{in}}{R_B + h_{ie}} \tag{4.4-7}$$

Recalling that the Thévenin impedance is equal to v_{oc}/i_{sc}, we find that

$$R_T = \frac{v_{oc}}{i_{sc}} = \frac{v_{in}}{(1 + h_{fe})[v_{in}/(R_B + h_{ie})]} = \frac{R_B + h_{ie}}{1 + h_{fe}} \tag{4.4-8}$$

for this circuit. The overall Thévenin equivalent for the circuit within the shaded region in Figure 4.4-3a is given in part (b) of the figure. This is the most useful emitter equivalent circuit. From this figure it is clear that the impedance, Z_2, seen looking back into the emitter (with the source v_{in} set equal to zero) is $(R_B + h_{ie})/(1 + h_{fe})$.

Using this circuit, the emitter current and emitter voltage may be written down immediately as

$$i_e = \frac{v_{in}}{[(R_B + h_{ie})/(1 + h_{fe})] + R_E} = \frac{(1 + h_{fe})v_{in}}{R_B + h_{ie} + (1 + h_{fe})R_E} \tag{4.4-9a}$$

and

$$v_e = v_{in}\frac{R_E}{R_E + [(R_B + h_{ie})/(1 + h_{fe})]} = v_{in}\frac{(1 + h_{fe})R_E}{R_B + h_{ie} + (1 + h_{fe})R_E} \tag{4.4-9b}$$

Notice that the solution for v_e is the same whether the base equivalent circuit or emitter equivalent circuit is used [see equations (4.4-9b) and (4.4-2d), respectively], and also that the emitter current in equation (4.4-9a) is simply $(1 + h_{fe})$ times the base current calculated using the base equivalent circuit [equation (4.4-2a)].

When students first derive these emitter equivalent circuit results, they are frequently puzzled as to why the base equivalent circuit given in Figure 4.4-2a cannot be used for both the base and the emitter equivalent circuits. The answer is very subtle and has to do with the fact that while both circuits (i.e., Figures 4.4-2b and 4.4-3b) have the same open-circuit voltage, the circuit in Figure 4.4-2b is invalid if any additional loading is placed across the emitter terminals. This is just another way of saying that the equivalent circuit given in Figure 4.4-2b does not correctly represent the output impedance seen looking back into the emitter terminals of the transistor. To see why this is so, we need only go back to Figure 4.4-2a, in which we replaced the controlled source $h_{fe}R_E i_b$ by an equivalent resistor $h_{fe}R_E$. The fundamental assumption here was that the current flowing through this controlled source was i_b; this is true, and hence the circuit in Figure 4.4-2b is valid only when there is no additional loading connected in parallel with R_E. In deriving the equivalent emitter circuit in Figure 4.4-3b, no such restrictions were imposed, and therefore this circuit truly represents the emitter equivalent circuit for the transistor under any arbitrary loading conditions, as long as the transistor remains in the active region.

To obtain the equivalent circuit seen looking back into the collector, we have redrawn this portion of Figure 4.4-1b in Figure 4.4-4b. This circuit as it stands is the Norton equivalent of the collector circuit. The circuit output impedance, labeled Z_3 in Figures 4.4-4a and 4.4-1a, is obtained by looking back into the collector terminal with the applied voltage v_{in} set equal to zero. If $v_{in} = 0$, the base current i_b is zero [see equation (4.4-2a)], and hence the collector current i_c or the current source $h_{fe}i_b$ is also zero. If this current source is replaced by an open circuit, it is then apparent that Z_3 is infinite, and thus the overall collector circuit output impedance, Z_4 in Figure 4.4-4b, is just equal to R_C.

Before applying these equivalent circuits to the analysis of single- and multitransistor BJT circuits, it may be a good idea first to summarize what we have learned so far in this section. By carefully reviewing the characteristics of

Figure 4.4-4

Equivalent circuit seen looking back into the collector (to the left of terminals C and G).

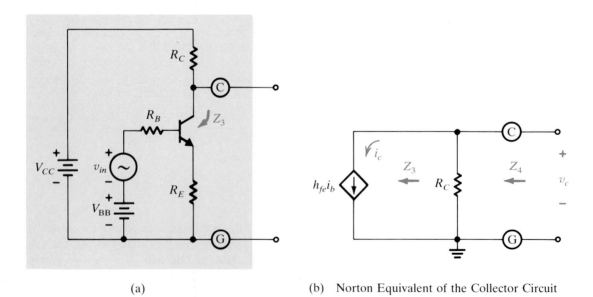

(a)　　　　　　　　　　(b) Norton Equivalent of the Collector Circuit

each of the equivalent circuits in Figures 4.4-2b, 4.4-3b, and 4.4-4b, the following results should be apparent:

1. The equivalent circuit in the base of the transistor is obtained by taking the circuit found in the emitter and reflecting it into the base. Similarly, the equivalent circuit in the emitter is found by reflecting the base circuit into the emitter.

2. Impedances reflected from the emitter into the base are multiplied by the factor $(1 + h_{fe})$.

3. Impedances reflected from the base into the emitter (including h_{ie}) are divided by the factor $(1 + h_{fe})$.

4. (a) Voltages reflected from the base into the emitter (see, for example, v_{in} in Figure 4.4-3b) or from the emitter into the base are unchanged.

   (b) Current sources reflected from the base into the emitter are multiplied by $(1 + h_{fe})$. Current sources reflected from the emitter into the base are divided by $(1 + h_{fe})$. To understand why this is so, consider the relationship between the Norton and Thévenin equivalent circuits, and recall that $i_N = v_T/R_T$.

5. (a) Impedances in the base and the emitter do not appear directly in the collector equivalent circuit.

   (b) The impedance Z_3 seen looking back into the collector terminal is infinite, so that the overall collector circuit output impedance (Z_4 in Figure 4.4-4b) is just equal to R_C.

6. In the collector circuit the collector current i_c is found either from $h_{fe}i_b$ (where i_b is obtained from the base equivalent circuit in Figure 4.4-2b) or from $i_c \simeq i_e$ (where i_e is found from the emitter equivalent circuit in Figure 4.4-3b).

EXAMPLE 4.4-1

The transistor amplifier shown in Figure 4.4-5 is silicon and has an h_{fe} of 100 and an h_{ie} equal to 1 kΩ.

a. Is the circuit operating as a small-signal amplifier if $v_{in} = 0.5 \sin \omega t$ volts?
b. Determine the voltage gain v_L/v_{in}.
c. Find the current i_L in the load.

SOLUTION

If we assume that all the coupling and bypass capacitors are shorts at the frequency of the applied input signal, the small-signal equivalent circuit may be drawn as shown in Figure 4.4-5b. Notice that only part of the emitter resistor is bypassed to ground so that the ac impedance in the emitter circuit is essentially R_{E1}, and it is this impedance that is magnified by $(1 + h_{fe})$ when it is reflected into the base circuit. By using the voltage-divider equation, we may calculate the ac base emitter voltage as

$$v_{be} = \frac{1}{51.5}(0.5 \sin \omega t \text{ V}) = 9.7 \sin \omega t \text{ mV} \qquad (4.4\text{-}10)$$

and referring back to Figure 4.2-2b, it should be clear that the amplitude of v_{be} is low enough for the small-signal model to work fairly well for attacking this problem.

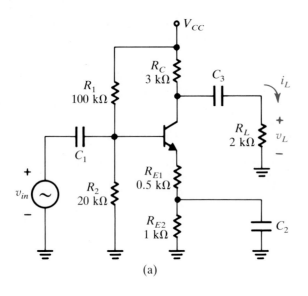

(a)

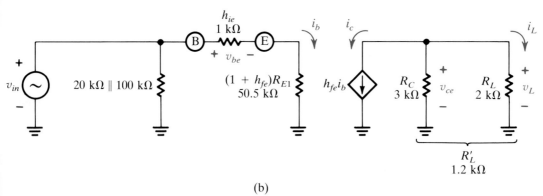

(b)

Figure 4.4-5
Circuit for Example 4.4-1.

Continuing the analysis of the circuit in Figure 4.4-5b, we see that

$$i_b = \frac{v_{in}}{51.5 \text{ k}\Omega} \tag{4.4-11a}$$

$$i_c = h_{fe}i_b = \frac{100v_{in}}{51.5 \text{ k}\Omega} \tag{4.4-11b}$$

and

$$v_L = -i_c R'_L = -\frac{120 \text{ k}\Omega v_{in}}{51.5 \text{ k}\Omega} \tag{4.4-11c}$$

so that

$$A_v = \frac{v_L}{v_{in}} = -2.33 \tag{4.4-11d}$$

Notice that because C_3 is a short to ac, $v_{ce} = v_L$, and therefore you will see the same ac signal at the collector and across the load R_L.

To calculate the current i_L flowing through the load, we can apply the current-divider rule to obtain

$$i_L = -\tfrac{3}{5}i_c = -1.17 \sin \omega t \text{ mA} \tag{4.4-12}$$

Of course, this is the same answer you would have gotten if you had taken the voltage v_L and divided it by the load resistance R_L.

EXAMPLE 4.4-2

The circuit shown in Figure 4.4-6 is known as a common-collector amplifier. More commonly, it is called an emitter follower because, as we shall see shortly, the voltage at the emitter closely follows the applied input voltage, both in amplitude and in phase. Assume that C_1 and C_2 are ac shorts and that $h_{ie} = 1$ kΩ and $h_{fe} = 100$ for the transistor.

a. Find the circuit voltage gain, v_e/v_{in}, if the load R_L is initially unconnected.
b. Recompute the voltage gain if a 1-kΩ load is connected as shown in the figure.
c. Determine the value of R_L that will cause the voltage gain to drop to one-half that found in part (a).
d. Find the input impedance, Z_{in}, of this emitter follower.

Figure 4.4-6
Circuit for Example 4.4-2.

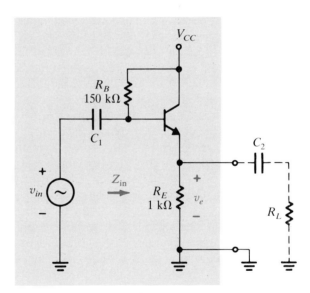

(a)   Emitter-Follower Transistor Amplifier

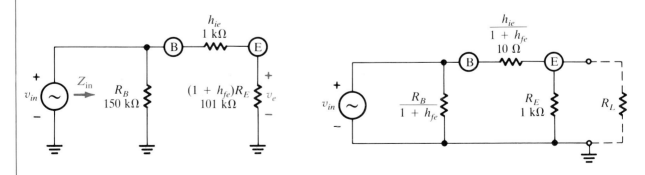

(b)   Base Equivalent Circuit with R_L Disconnected

(c)   Thévenin Equivalent of the Circuit within the Shaded Region

The base equivalent circuit for this amplifier, with R_L disconnected, is given in Figure 4.4-6b. By application of the voltage-divider rule, the circuit voltage gain is $A_v = v_e/v_{in} = 101/102 = 0.99$. Note that the resistor R_B has no effect on the circuit voltage gain because it is in parallel with the applied voltage source v_{in}; however, as we shall see in part (d) of this example, it does alter the circuit input impedance.

The effect of adding a 1-kΩ load resistor at the output terminals of the amplifier causes an additional resistor $(1 + h_{fe})R_L$ to be reflected into the base equivalent circuit in parallel with the reflected emitter resistor $(1 + h_{fe})R_E$. Because the parallel combination of these two resistors (50.5 kΩ) is still so much larger than h_{ie}, the addition of this load has relatively little effect on the circuit gain. Specifically, the voltage gain for this loaded emitter follower is $A_v = 50.5/51.5$, or 0.98. Thus, for both of these circuits the gain is almost 1 and the voltage at the emitter of the transistor very nearly follows the applied input voltage.

At first glance this type of circuit might seem to be rather useless since it appears that the transistor amplifier could simply be replaced by a piece of wire, and the source just directly connected to the load R_L. However, the two circuits are not exactly equivalent. Although both circuits have approximately the same voltage gain, so that the same power would be delivered to the load with each of these circuits, the loading on the source is much less for the emitter-follower circuit. Therefore, the emitter-follower circuit provides a much larger power gain. For example, if $v_{in} = 1 \sin \omega t$ volts, considering that the circuit input impedance Z_{in} is equal to $(150\ \text{k}\Omega)\|(51.5\ \text{k}\Omega)$, or 38.3 k$\Omega$, the average power delivered to the amplifier is $(0.707)^2/(38.3\ \text{k}\Omega) = 13.1\ \mu\text{W}$. Using $A_v = 0.98$ for this circuit, the corresponding power delivered to the load is 480 μW. Therefore, the power gain of this emitter-follower amplifier is 480/13.1, or 36.6. The power gain of a wire is 1.

To determine the value of R_L for which the voltage gain drops to one-half its open-circuit value, recall that this occurs when a load equal to the Thévenin output impedance is connected across the output terminals. Thus, for this circuit (see Figure 4.4-6c), the required output load is about 10 Ω.

In summary, an emitter-follower amplifier has the following characteristics: a voltage gain of nearly unity, a rather high input impedance, and a very low output impedance. A high input impedance is useful when we want to avoid loading a signal source, while a circuit having a low output impedance is desirable when we need to drive a heavy load. As a result, emitter followers are often used to match high-impedance sources to low-impedance loads.

EXAMPLE 4.4-3

Sometimes, sufficient amplifier gain cannot be achieved with a single transistor. The circuit given in Figure 4.4-7 is a two-transistor amplifier in which the collector output from the first stage is capacitively coupled into the base of the next. The capacitor C_2 prevents the dc collector voltage of Q_1 from upsetting the Q point of Q_2, and vice versa. Assuming that both transistors have $h_{ie} = 1$ kΩ, and $h_{fe} = 100$, calculate the overall circuit gain v_{c2}/v_{in}.

SOLUTION

Starting at the left-hand side of the amplifier, we may draw the equivalent circuit for the base of Q_1 as shown in Figure 4.4-7b. Because R_{B1} is so much bigger than h_{ie}, it can be neglected and i_b can be written, to a good approximation, as

$$i_{b1} = \frac{v_{in}}{3 \text{ k}\Omega} \tag{4.4-13a}$$

from which the collector current in Q_1 is

$$i_{c1} = h_{fe}i_{b1} = \frac{100v_{in}}{3 \text{ k}\Omega} \tag{4.4-13b}$$

Using the collector equivalent circuit in Figure 4.4-7c, and neglecting R_{B2} in this figure, we may immediately write down the base current into the second transistor as

$$i_{b2} = -\frac{2}{3}i_{c1} = -\frac{200v_{in}}{9 \text{ k}\Omega} \tag{4.4-13c}$$

One point that often causes confusion is the meaning of the minus sign in this equation. Students sometimes think that the transistor Q_2 will be cut off since i_{b2} is negative. However, i_{b2} is only the ac part of the overall base current in Q_2. For this transistor to be cut off, the total base current, i_{B2}, not just its ac

Figure 4.4-7
Circuit for Example 4.4-3.

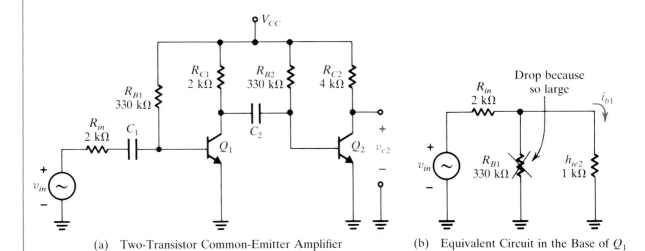

(a) Two-Transistor Common-Emitter Amplifier   (b) Equivalent Circuit in the Base of Q_1

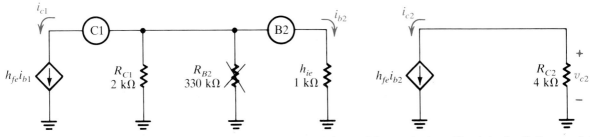

(c) Equivalent Circuit in the Collector of Q_1 and the Base of Q_2 (d) Equivalent Circuit in the Collector of Q_2

component, must be negative. Thus the minus sign in equation (4.4-13c) in no way indicates that Q_2 is cut off. It just means that the ac part of this base current is $180°$ out of phase from the applied input voltage v_{in}.

The collector current in Q_2 is calculated with the aid of Figure 4.4-7d as

$$i_{c2} = h_{fe}i_{b2} = -\frac{20,000v_{in}}{9 \text{ k}\Omega} \qquad (4.4\text{-}14a)$$

from which the collector voltage, v_{c2}, is

$$v_{c2} = -i_{c2}R_{C2} = \frac{(80,000 \text{ k}\Omega)v_{in}}{9 \text{ k}\Omega} = +8,888v_{in} \qquad (4.4\text{-}14)$$

Following this expression, the overall voltage gain of the amplifier is $A_v = v_{c2}/v_{in} = 8888$. Notice that in addition to being very large, the amplifier voltage gain is also positive because there is a $180°$ phase reversal in going through each of the common-emitter amplifiers Q_1 and Q_2 so that an overall phase shift of $360°$ occurs in v_{in} in passing through this amplifier. Of course, a phase shift of $360°$ is actually equivalent to a phase shift of zero since $\sin(\omega t + 360°) = \sin \omega t$.

EXAMPLE 4.4-4

Figure 4.4-8 illustrates a very interesting electronic circuit. It is called a cascode amplifier, and as we shall see later when we investigate the frequency response of transistor circuits, its main virtue is its excellent high-frequency performance. Calculate the voltage gain at each of the collectors (i.e., find v_{o1}/v_{in} and v_{o2}/v_{in}).

SOLUTION

To analyze this circuit, notice that with respect to ac, it may be redrawn as shown in Figure 4.4-8b, since the base of Q_2 is at ac ground potential. By inspection of this circuit, the base current into Q_1 is

$$i_{b1} = \frac{v_{in}}{h_{ie1}} \qquad (4.4\text{-}15a)$$

and $$i_{c1} = h_{fe1}i_{b1} \qquad (4.4\text{-}15b)$$

To compute the output voltage at the collector of Q_1, we observe that as far as Q_1 is concerned, the transistor Q_2 looks like a collector load resistor equal to the impedance seen looking back into the emitter of Q_2 (Figure 4.4-8c); of course, this is just the equivalent resistance in the base divided by $(1 + h_{fe})$, or

$$Z_{e2} = \frac{h_{ie2}}{1 + h_{fe2}} \qquad (4.4\text{-}16)$$

so that Q_1's collector voltage is

$$v_{o1} = -i_{c1}Z_{e2} = -\frac{h_{fe1}v_{in}}{h_{ie1}} \frac{h_{ie2}}{1 + h_{fe2}} \qquad (4.4\text{-}17)$$

If $h_{ie1} = h_{ie2} = 1 \text{ k}\Omega$, and if $h_{fe1} = h_{fe2} = 100$, the output voltage v_{o1} is simply an inverted version of the applied input voltage v_{in}.

To calculate the output voltage at the collector of Q_2, we use the fact that

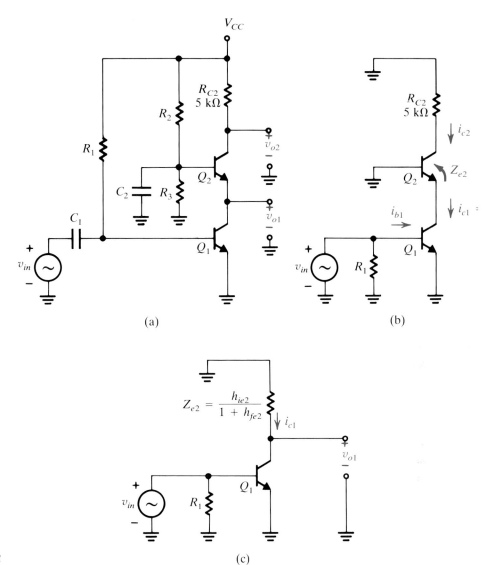

Figure 4.4-8
Circuit for Example 4.4-4

(a)

(b)

(c)

the emitter current of Q_2 equals the collector current of Q_1 so that both transistors have approximately the same collector current. As a result, v_{o2} may be written as

$$v_{o2} = -i_{c2}R_{C2} = -\frac{100v_{in}}{1\text{ k}\Omega}(5\text{ k}\Omega) = -500v_{in} \qquad (4.4\text{-}18)$$

Exercises

4.4-1 For the circuit in Figure 4.4-6a, find i_b and v_e if a 100-Ω resistor is capacitively coupled in parallel with R_E. The input signal v_{in} for this problem is to be considered to be 20 sin ωt mV. *Answer* i_b = 1.96 sin ωt μA, v_e = 18 sin ωt mV

4.4-2 For the circuit given in Figure 4.4-7 that was analyzed in Example 4.4-3, find the ac voltage at the collector of Q_1 in terms of v_{in}. **Answer** $-22v_{in}$

4.4-3 Find the overall circuit voltage gain in Figure 4.4-7 if $h_{ie2} = 2$ kΩ instead of 1 kΩ as in Example 4.4-3. **Answer** 6666

4.4-4 Find v_{o1} in Figure 4.4-7 if a 10-Ω resistor is connected from the emitter of Q_1 to ground. **Answer** -16.7

4.4-5 For the cascode amplifier shown in Figure 4.4-8a, find v_{o1}/v_{in} and v_{o2}/v_{in} if a 20-Ω capacitively coupled load resistor is shunted from the collector of Q_1 to ground. **Answer** $-0.666, -333.3$

4.5 MULTISTAGE BJT AMPLIFIER DESIGN CONSIDERATIONS

In Section 3.1 we discussed the basic principles involved in the design of electronic amplifiers, and in this section we apply these principles to the design of multistage BJT circuits. Regardless of the specifics of the amplifier to be constructed, conceptually the overall design process can be broken down into three distinct parts:

1. Design of the input stage
2. Design of the output stage
3. Design of the intermediate amplifier stages

For the most part, the construction of the input and output stages is determined by the characteristics of the source and the load and the required input and output impedance levels. The design of the intermediate amplifier stages generally depends on the overall power gain that is needed.

The amount of power transferred from the source to the amplifier, from the amplifier to the load, or between one stage and another depends on the relative impedance levels. As we demonstrated in Section 3.1, maximum power is transferred between two circuits when the output impedance of first circuit matches the input impedance of the second circuit. In matching the impedance levels of two circuits whose input and output impedances are not initially equal, a transformer must be used. Maximum power transfer cannot be achieved by connecting additional resistors in series with or in parallel with these circuits in order to make the two impedance levels the same.

EXAMPLE 4.5-1

For the circuit shown in Figure 4.5-1, assume that $h_{ie} = 1$ kΩ and $h_{fe} = 100$.

a. Determine the value of R_E that will deliver maximum signal power to the amplifier and which will, as a result, deliver maximum ac power to the load in the collector of this circuit.
b. Find the power gain of the amplifier.

SOLUTION

At first, to get maximum power into the amplifier, one might be tempted to adjust the value of R_E to make the overall amplifier input impedance equal to R_s. Because this is equivalent to adding a resistor of value $(1 + h_{fe})R_E$ in series

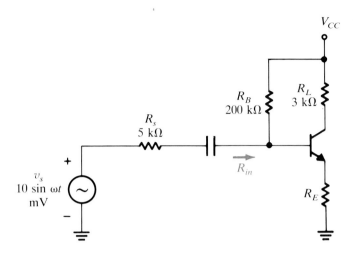

Figure 4.5-1
Circuit for Example 4.5-1.
Selection of R_E to maximize
the ac power delivered to R_L.

with the base, this is exactly the wrong thing to do. Although this will allow the source to deliver maximum power to the circuit, most of this power will be dissipated in the emitter resistor and will never reach the load. The optimum value of R_E that will maximize this circuit's power gain is zero. For $R_E = 0$, the base and collector currents will both be at a maximum, and this will deliver maximum signal power to the collector resistor. Realizing that some people may still be unconvinced, we will solve this problem using both values of R_E [i.e., with $R_E = 0 \ \Omega$ and with $R_E = 40 \ \Omega$ (the value that matches R_s to R_{in})].

For $R_E = 40 \ \Omega$, the transistor base current, neglecting the current shunted into R_B, is

$$i_b = \frac{v_s}{R_s + h_{ie} + (1 + h_{fe})R_E} = \frac{v_s}{10 \text{ k}\Omega} \tag{4.5-1}$$

from which the collector current, collector voltage, and average power delivered to the load are

$$i_c = h_{fe}i_b = \frac{100v_s}{10 \text{ k}\Omega} \tag{4.5-2a}$$

$$v_c = -i_c R_C = -\left(\frac{100v_s}{10 \text{ k}\Omega}\right)(3 \text{ k}\Omega) = -30v_s \tag{4.5-2b}$$

and $$P_L = \frac{V_c^2}{2 R_L} = \frac{(0.3)^2}{6 \text{ k}\Omega} = 15 \ \mu\text{W} \tag{4.5-2c}$$

The power delivered by the source is

$$P_{in} = \frac{V_s^2}{2 \times 10 \text{ k}\Omega} = 0.005 \ \mu\text{W} \tag{4.5-3}$$

so that the circuit power gain with $R_E = 40 \ \Omega$ is $15/0.005 = 3000$.
If R_E is set equal to zero, then

$$i_b = \frac{v_s}{R_s + h_{ie}} = \frac{v_s}{6 \text{ k}\Omega} \tag{4.5-4a}$$

$$i_c = h_{fe}i_b = \frac{100v_s}{6 \text{ k}\Omega} \tag{4.5-4b}$$

and $$v_c = -i_c R_C = -\frac{(300 \text{ k}\Omega)v_s}{6 \text{ k}\Omega} = -50v_s \tag{4.5-4c}$$

so that the power delivered to the load increases to

$$P_L = \frac{V_c^2}{2R_L} = \frac{(0.5)^2}{6 \text{ k}\Omega} = 41.6 \ \mu\text{W} \tag{4.5-4d}$$

Since the power in from the source is $V_s^2/12$ kΩ, or 0.0083 μW, the circuit power gain for this case is 5000.

Thus far we have examined the design trade-offs involved in the selection of the input stage of the amplifier. Next we take a look at the factors governing the design of the output stage.

In the design of the output circuit, one often wishes to maintain the output voltage constant in spite of large variations in Z_L. In this situation it is desirable that the amplifier's output impedance be as small as possible, and for BJT designs where a low output impedance is needed, an emitter-follower output stage is often used.

Sometimes maximum power gain is more important than a low output impedance, and for this case, when warranted, a transformer can be used in the output stage to match the transistor output impedance to that of the load. The circuit shown in Figure 4.5-2a illustrates this type of design, and the equivalent circuit given in part (b) of the figure shows that the turns ratio in this case must be selected to make $1/h_{oe}$ equal to N^2R_L. The improvement in power gain that is obtained by using a transformer depends on the degree of mismatch that exists between the circuit output impedance and the load (see Figure 4.5-2c and Problem 4.5-7). Unless this mismatch is substantial, the addition of the transformer to the circuit is usually unwarranted.

At this point in our discussion of BJT amplifier design techniques, we have examined the criteria for selecting the amplifier input and output stages and

Figure 4.5-2
Effectiveness of transformer coupling.

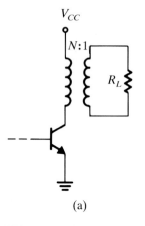

(a)

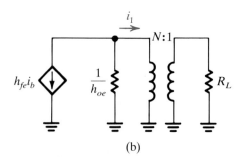

(b)

$\dfrac{1}{h_{oe}}$	Transformer Improvement Factor in Power to the Load
100 R_L	25.5
10 R_L	3.0
R_L	1.0
0.1 R_L	3.0
0.01 R_L	25.5

(c)

have also seen how a transformer can be used to advantage when maximum power gain in the circuit is important. Thus far, however, little has been said about which specific circuits should be used for the design of the intermediate stages of the amplifier.

In general, the intermediate stages are a cascade of similar transistor circuits whose design should be chosen to achieve the maximum power gain possible with the minimum number of components. When using bipolar transistors, three cascaded circuit configurations are possible: the common emitter, the common collector, and the common base. A three-transistor intermediate-stage design for each of these circuit types is illustrated in Figure 4.5-3. It should be noted for these amplifiers that some of the biasing circuitry has purposely been left out in order to simplify the gain calculations. Furthermore, for all the transistors in these amplifiers, we assume that $1/h_{oe}$ is infinite, $h_{ie} = 1\text{ k}\Omega$, and $h_{fe} = 100$.

We will begin the comparative analyses of these three circuits by starting with the common-emitter configuration given in Figure 4.5-3a. For this circuit

$$i_{b1} = \frac{v_s}{R_s + h_{ie}} \simeq \frac{v_s}{R_s} \tag{4.5-5a}$$

and
$$i_{c1} = \frac{h_{fe}v_s}{R_s} \tag{4.5-5b}$$

If we assume that R_C is much greater than h_{ie}, nearly all the collector current from the first stage will flow into the base of the next stage, so that

$$i_{b2} \simeq -i_{c1} = -\frac{h_{fe}v_s}{R_s} \tag{4.5-5c}$$

$$i_{c2} = -\frac{h_{fe}^2 v_s}{R_s} \tag{4.5-5d}$$

$$i_{b3} \simeq \frac{h_{fe}^2 v_s}{R_s} \tag{4.5-5e}$$

$$i_{c3} = \frac{h_{fe}^3 v_s}{R_s} \tag{4.5-5f}$$

and
$$v_{c3} = \frac{-h_{fe}^3 R_L}{R_s} v_s \tag{4.5-5g}$$

Using these results, the average power delivered to the load R_L is

$$P_L = \frac{(h_{fe})^6 V_s^2}{2} \frac{R_L}{R_s^2} \tag{4.5-6a}$$

while the average input power is

$$P_{in} \simeq \frac{V_s^2}{2R_s} \tag{4.5-6b}$$

so that the circuit power gain may be written as

$$G = \frac{P_L}{P_{in}} = (h_{fe})^6 \tag{4.5-6c}$$

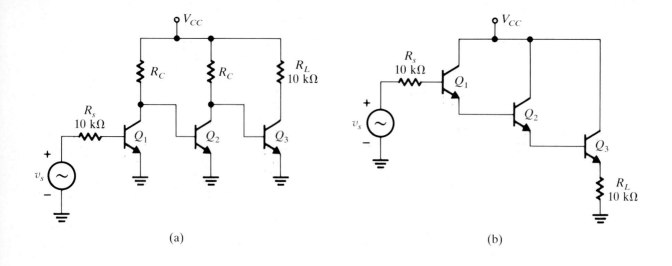

(a)   (b)

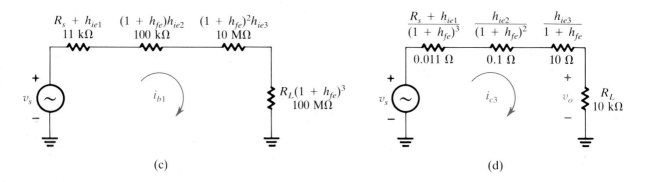

(c)   (d)

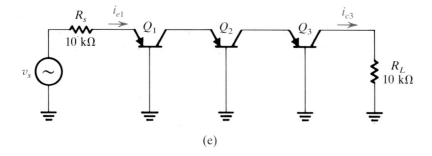

(e)

Figure 4.5-3
Comparison of three cascaded
three-stage amplifiers.

The power gain for the triple Darlington (or common-collector) circuit in Figure 4.5-3b is found quite easily by using the input and output equivalent circuits given in Figure 4.5-3c and d, respectively. The circuit input impedance, Z_{in} in Figure 4.5-3c, is approximately given by the expression

$$Z_{in} \simeq (h_{fe})^3 R_L \qquad (4.5\text{-}7a)$$

so that

$$i_{b1} \simeq \frac{v_s}{(h_{fe})^3 R_L} \qquad (4.5\text{-}7b)$$

Chapter 4 / Small-Signal Analysis of Bipolar Transistor Circuits

and the input power to the circuit is

$$P_{in} = \frac{V_s^2}{2(h_{fe})^3 R_L} \qquad (4.5\text{-}7c)$$

To a first-order approximation, the output voltage, v_o (see Figure 4.5-3d), is nearly equal to v_s, so that the power delivered to the load may be written as

$$P_L = \frac{V_s^2}{2R_L} \qquad (4.5\text{-}8)$$

from which the circuit power gain is seen to be

$$G = \frac{P_L}{P_{in}} = (h_{fe})^3 \qquad (4.5\text{-}9)$$

In addition to the fact that the common-collector power gain is considerably less than that for the cascaded common-emitter configuration, this circuit has the added disadvantage that it has no voltage gain.

The analysis of the cascaded common-base circuit shown in Figure 4.5-3c is very simple. Because the base of Q_1 is grounded, Z_{in} for this circuit is $h_{ie}/(1 + h_{fe})$. Since this impedance is much smaller than R_s, the input current from the source is approximately

$$i_{e1} = \frac{v_s}{R_s} \qquad (4.5\text{-}10a)$$

and the input power to this circuit is

$$P_{in} = \frac{V_s I_{e1}}{2} = \frac{V_s^2}{2R_s} \qquad (4.5\text{-}10b)$$

Furthermore, the emitter and collector currents of all the transistors in the cascade are nearly equal, so that

$$i_{c3} \simeq i_{e1} = \frac{v_s}{R_s} \qquad (4.5\text{-}11a)$$

As a result,

$$P_L = \frac{I_{c3}^2 R_L}{2} = \frac{V_s^2 R_L}{2R_s^2} \qquad (4.5\text{-}11b)$$

and the circuit power gain is

$$G = \frac{P_L}{P_{in}} = \frac{R_L}{R_s} \simeq 1 \qquad (4.5\text{-}11c)$$

Notice that a single-stage common-base design would have had the same power gain as this three-transistor design, so that the cascaded common-base configuration is unsuited to the generation of substantial voltage, current, or power gain.

In summary, the design of multistage electronic amplifiers is best carried out by splitting the design problem into three distinct parts: the design of the input, the intermediate, and the output stages. With regard to each of these amplifier stages:

1. The design needed for the input stage depends on the character of the source. If the input signal is best characterized as a voltage source, and if

the amplification is to be independent of R_s, then to avoid loading, the amplifier input impedance must be high. This suggests the use of a common-collector input stage or perhaps a common-emitter stage with an unbypassed emitter resistor. On the other hand, when the input signal is a current source, amplification independence from R_s is achieved by using a circuit that has a low input impedance. In this instance the input stage should probably be a common-base circuit if a very low input impedance is needed or a common-emitter circuit with a bypassed emitter resistor if intermediate input impedance levels are satisfactory.

2. The intermediate amplifier stage is used to produce any additional signal amplification that is needed to achieve the required overall amplifier power gain. Based on the comparison of the three amplifiers examined in Figure 4.5-3, it is apparent, at least for the specific values of R_s and R_L used with these examples, that the cascaded common-emitter amplifier offers significantly higher power gain than either the common-collector or common-base designs. Furthermore, an exhaustive analysis of these circuits for different values of R_s and R_L, as well as for circuits containing combinations of the common-emitter, common-collector, and common-base designs, demonstrates that the cascaded common-emitter circuit is always at least as good as any of these other circuit combinations.

3. In the design of the output stage, if the output voltage is to remain constant as R_L varies, the amplifier will need to have a low output impedance, and in this instance an emitter-follower output stage will probably work best. If, on the other hand, it is desired that the output current rather than the output voltage remain constant as R_L varies, a common-emitter or common-base output stage is preferred since their equivalent circuits look more nearly like current sources.

EXAMPLE 4.5-2

Design a BJT amplifier having the following characteristics:

$$\text{power gain} > 10^7 \text{ (with } R_s = 1 \text{ k}\Omega \text{ and } R_L = 1 \text{ k}\Omega)$$

$$\text{input impedance} > 10 \text{ k}\Omega$$

$$\text{output impedance} < 100 \text{ }\Omega$$

For simplicity, the biasing circuitry need not be shown.

SOLUTION

Because of the large power gain required, it is fairly apparent that this design will be impossible to achieve with a single transistor. However, it may be conceivable to construct this amplifier by using a two-transistor circuit design. Let's determine which circuit configuration will be the best to use to meet the stated design criteria.

Since Z_{in} is to be greater than 10 kΩ, a common-collector or a common-emitter design with an unbypassed emitter resistor will be needed. Although either of these approaches is satisfactory, we will use the common-emitter design for the input stage in this example.

On the output side of the amplifier, the requirement that Z_o is to be less than 100 Ω can really be satisfied only by using an emitter-follower circuit for the output stage. Thus, the proposed two-stage amplifier design will have the form illustrated in Figure 4.5-4a. Since we want the amplifier input impedance to be greater than 10 kΩ, we will select the value of R_{E1} in this circuit to be 100 Ω. Making it any larger than this is unwise since it will just reduce the circuit gain. In a similar fashion, to maintain Z_o below 100 Ω, R_{C1} will need to have a value of less than 9 kΩ. For simplicity in this design, let us set R_{C1} = 5 kΩ. For the values of R_{E1} and R_{C1} just selected, the amplifier input and output impedances will be

$$Z_{in} = h_{ie1} + (1 + h_{fe1})R_{E1} = 11.1 \text{ k}\Omega \qquad (4.5\text{-}12a)$$

and

$$Z_o = \frac{R_{C1} + h_{ie2}}{1 + h_{fe2}} = 59.4 \ \Omega \qquad (4.5\text{-}12b)$$

Clearly, both of these impedance levels are satisfactory, but does this amplifier have the proper power gain?

The average input power from the source v_s is

$$P_{in} = \frac{V_s^2}{(2)(12 \text{ k}\Omega)} \qquad (4.5\text{-}13)$$

To compute the average power delivered to the load, we note that

$$i_{b1} = \frac{v_s}{12 \text{ k}\Omega} \qquad (4.5\text{-}14a)$$

Figure 4.5-4
Example 4.5-2.

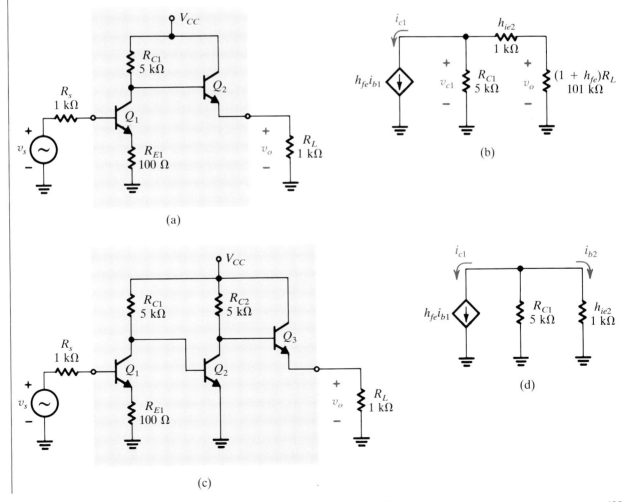

(a)

(b)

(c)

(d)

$$i_{c1} = \frac{100v_s}{12 \text{ k}\Omega} \tag{4.5-14b}$$

Following Figure 4.5-4b, it is apparent that Q_2 does not load Q_1, so that

$$v_{c1} \simeq -i_{c1}R_{C1} = -\left(\frac{100v_s}{12 \text{ k}\Omega}\right)5 \text{ k}\Omega = -41.7v_s \tag{4.5-14c}$$

and
$$v_o \simeq v_{c1} = -41.7v_s \tag{4.5-14d}$$

Using this result, the average power delivered to the load is

$$P_L = \frac{V_o^2}{2R_L} = \frac{(41.7)^2 V_s^2}{(2)(1 \text{ k}\Omega)} \tag{4.5-15}$$

Combining this expression together with that in equation (4.5-13), the overall circuit power gain is

$$G = \frac{P_L}{P_{in}} = \frac{(41.7)^2 V_s^2}{(2)(1 \text{ k}\Omega)} \frac{(2)(12 \text{ k}\Omega)}{V_s^2} = 20,866 \tag{4.5-16}$$

This power gain is substantially below the value of 10^7 that is needed for this design, so it is apparent that at least one more transistor will need to be added to this circuit as an intermediate power amplifier stage. Because of its large power gain, we will use a common-emitter circuit for this intermediate stage. The resulting three-transistor design is illustrated in Figure 4.5-4c, and for simplicity we have set both of the collector resistors (R_{C1} and R_{C2}) in this circuit equal to 5 kΩ.

Because the input and output stages of this new design are identical to those of the two-transistor circuit just examined, Z_{in}, Z_o, and P_{in} for this three-transistor amplifier are given by equations (4.5-12a), (4.5-12b), and (4.5-13), respectively. The average power delivered to the load may be calculated as follows. The base and collector currents in Q_1 are seen to be the same as those given in equations (4.5-14a) and (4.5-14b), and following the equivalent circuit given in Figure 4.5-4d,

$$i_{b2} = -\frac{5}{6}i_{C1} = -\frac{500v_s}{72 \text{ k}\Omega} \tag{4.5-17a}$$

so that
$$i_{c2} = h_{fe}i_{b2} = -\frac{5 \times 10^4 v_s}{72 \text{ k}\Omega} \tag{4.5-17b}$$

By examining Figure 4.5-4c, it should be apparent that Q_3 does not load Q_2, and as a result,

$$v_{c2} \simeq -i_{c2}R_{C2} = \frac{(5 \times 10^4)(5 \text{ k}\Omega)}{72 \text{ k}\Omega}v_s = 3472v_s \tag{4.5-18}$$

In addition, because Q_3 is an emitter follower in which R_L is large,

$$v_o \simeq v_{c2} = 3472v_s \tag{4.5-19a}$$

so that
$$P_L = \frac{(3472)^2 V_s^2}{(2)(1 \text{ k}\Omega)} \tag{4.5-19b}$$

and
$$G = \frac{P_L}{P_{in}} = \frac{(3472)^2 V_s^2}{(2)(1 \text{ k}\Omega)} \frac{(12 \text{ k}\Omega)(2)}{V_s^2} = 1.45 \times 10^8 \tag{4.5-19c}$$

This power gain is significantly larger than the minimum design value of 10 million that is needed, and therefore, the amplifier shown in Figure 4.5-4c meets all the criteria specified.

Exercises

4.5-1 In Figure 4.5-1, R_B is removed, R_E is set equal to 10 Ω, and a transformer is inserted between R_s and the base of the transistor with the lower side of the primary and the secondary grounded.

a. Find the turns ratio of the transformer that will maximize the power delivered to the load.

b. Using the answer in part (a), find the average ac power delivered to the load.

Answer (a) $1.58:1$, (b) $18.6 \ \mu\text{W}$

4.5-2 Repeat Exercise 4.5-1 if a $1:1$ turns ratio is used, effectively removing the transformer from the circuit. *Answer* $15.3 \ \mu\text{W}$

4.5-3 Find the power gain of the circuit in Figure 4.5-3a if the collector resistors in Q_1 and Q_2 are each 1 kΩ. *Answer* 2.84×10^{10}

4.5-4 For the transistors in Figure 4.5-4a, $1/h_{oe} = 300$ kΩ.

a. If R_{C1} is replaced by a transformer and R_{E1} is bypassed to ground, what turns ratio will maximize the power delivered to the load?

b. For the conditions in part (a), what is the circuit power gain?

Answer (a) $\sqrt{3}:1$, (b) 1.13×10^8

4.5-5 The two-transistor Darlington amplifier circuit shown in Figure E4.5-5 is offered as a possible solution to the design problem presented in Example 4.5-2. Find Z_{in}, Z_o, and G for this circuit if $h_{fe1} = h_{fe2} = 100$ and $h_{ie1} = h_{ie2} = 1$ kΩ. *Answer* $Z_o = 10 \ \Omega$, $Z_{in} = 10$ MΩ, $G = 10^4$

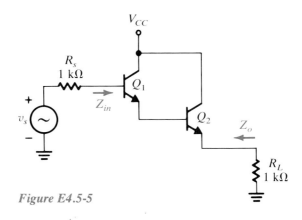

Figure E4.5-5

PROBLEMS

Unless otherwise stated consider that all transistors are silicon, with $h_{FE} = h_{fe} = 100$, $h_{ie} = (1 + h_{fe})r_e = 1$ kΩ, $h_{re} = 0$, and $1/h_{oe} = \infty$. In addition, all capacitors may be assumed to be ac shorts.

4.2-1 For the circuit in Figure 4.3-5a, $V_{CC} = 10$ V, $R_B = 930$ kΩ, and $v_{in} = 2 \sin \omega t$ mV. Sketch v_{BE}, i_B, i_C, and v_{CE}.

4.2-2 In Figure 4.2-4a, $V_{CC} = 15$ V, $R_1 = 11.8$ kΩ, $R_2 = 3.2$ kΩ, $R_E = 2.5$ kΩ, and $v_{in} = 5.7 \sin \omega t$ mV. Repeat Problem 4.2-1 for this circuit.

4.2-3 In Figure 4.2-4a, $V_{CC} = 30$ V, $R_1 = 96.5$ kΩ, $R_2 = 3.5$ kΩ, $R_E = 0$, $R_C = 1.5$ kΩ, and $v_{in} = 30 \sin \omega t$ mV. Repeat Problem 4.2-1 for this circuit.

4.2-4 To demonstrate that the transistor models given in Figure 4.2-1b and e are equivalent, substitute each of them into the circuit in Figure 4.4-1a and show that the ac loop equations obtained (with $V_{BB} = 0$ and $V_{CC} = 0$) are identical in each case.

4.2-5 Repeat the analysis of Example 4.2-1 if C_2 is removed from the circuit.

4.2-6 A bipolar junction transistor has a sinusoidal base–emitter voltage with a peak amplitude of 5 mV. Using equation (4.2-2), estimate the percent second harmonic distortion in i_E. Compare this result with that which you would have gotten by estimating this distortion using the table in Figure 4.2-2b.

4.2-7 (a) For Problem 4.2-3, estimate the amount of second harmonic distortion in i_C using the table in Figure 4.2-2b.

(b) Using the results in part (a), sketch v_{CE} and compare this result with that obtained in Problem 4.2-3, where the transistor was assumed to be perfectly linear. [*Note:* Following equation (4.2-2), the collector current will be of the form $i_C = I_C + I_1 \sin \omega t - I_2 \cos 2\omega t + \cdots$.]

4.3-1 In Figure E4.3-1 connect a 10-kΩ resistor in the branch of the circuit with the I_1 label on it, and find the h-parameters of the resulting circuit.

4.3-2 Repeat Exercise 4.3-1 if a controlled current source $5I_1$ (arrow down) is connected in parallel with the 20-kΩ resistor in Figure E4.3-1.

4.3-3 (a) Find the input impedance seen by v_{in} in Figure 4.3-5a considering that $h_{ie} = 2$ kΩ, $h_{fe} = 100$, $h_{re} = 0$, and $1/h_{oe} = \infty$.

(b) Repeat part (a) if $h_{re} = 10^{-3}$.

4.3-4 The circuit shown in Figure P4.3-4 is used to measure h_{re} and h_{oe} for the transistor in the circuit. The resistor R_B is used to establish the transistor Q point and v_t is a 1-V peak amplitude ac voltage source.

(a) What are h_{re} and $1/h_{oe}$ if the peak ac voltages V_1 and V_x are found to be 50 μV and 1 mV, respectively. Indicate any approximations that you are making.

(b) What errors, if any, are introduced in these measurements if $h_{ie} = 10$ kΩ and $h_{fe} = 100$?

4.3-5 In Figure 4.3-5a, $R_C = 1$ kΩ, $R_B = 193$ kΩ, and $v_{in} = 0.1 \sin \omega t$ V. The transistor in this circuit has the V–I characteristics given in Figure 4.3-4a and b.

(a) Find the Q point of the transistor.

(b) Using the results in part (a), determine the approximate values of the transistor hybrid parameters.

(c) Using the results in part (b), sketch i_B, v_{BE}, i_C, and v_{CE}.

(d) Repeat part (c) using graphical techniques.

4.3-6 For the graphical transistor characteristics given in Figure 4.3-4a and c, assume that the transistor is biased at $i_B = 50$ μA, $V_{BE} = 0.2$ V, $I_C = 5$ mA, and $V_{CE} = 4$ V. Determine the approximate values of h_{ie}, h_{re}, $1/h_{oe}$, and h_{fe} for this transistor.

4.3-7 (a) For the circuit shown in Figure 4.4-1, $R_B = 9$ kΩ, $R_E = 0$, and $h_{re} = 10^{-4}$. Determine the range of transistor internal gains v_{ce}/v_{be} for which the voltage source term $h_{re}v_{ce}$ may be neglected without making more than a 10% error.

(b) If $R_C = 10$ kΩ, find the overall circuit gain, neglecting h_{re}.

(c) Repeat part (b), but this time include the h_{re} term.

4.4-1 In Figure 4.2-4a, $R_1 = 100$ kΩ, $R_2 = 1$ kΩ, $R_C = 10$ kΩ, $R_E = 0$, and a resistor $R_{in} = 2$ kΩ is connected in series with v_{in}. Find the circuit small-signal ac gain v_c/v_{in}.

4.4-2 For the circuit in Figure 4.2-4a, $V_{CC} = 15$ V, $R_C = 3$ kΩ, $R_E = 2$ kΩ, and R_1 and R_2 are chosen so that $I_C = 2$ mA.

(a) What is the peak amplitude of v_{ce} for which there is no output distortion?

(b) Sketch the approximate output waveshape $v_{CE}(t)$ for $v_{in}(t) = 10 \sin \omega t$ mV.

4.4-3 (a) For the circuit shown in Figure P4.4-3, find the dc Q-point values.

(b) Draw a good ac small-signal equivalent circuit considering C_1, C_2, and C_3 to be shorts.

(c) From an analysis of the equivalent circuit, develop and carefully label the total waveshapes for v_{in}, v_{BE}, v_C, v_L, v_E, and v_{CE}.

4.4-4 In Figure 4.2-4a, $h_{fe} = 50$, $V_{CC} = 9$ V, $R_C = 3$ kΩ, $R_E = 2$ kΩ, and R_1, R_2, and v_{in} are chosen so that $i_B = (20 + 10 \sin \omega t)$ μA.

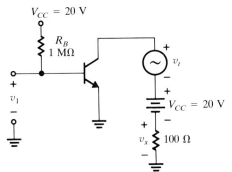

Figure P4.3-4

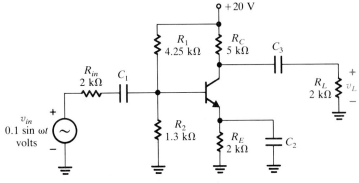

Figure P4.4-3

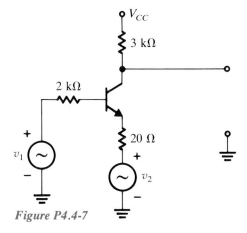

Figure P4.4-7

(a) Sketch and accurately label the waveforms for i_C [the total (dc + ac) collector current] and v_{CE} (the total collector-to-emitter voltage).

(b) i_b, the ac part of the base current, is now 20 μA(p-p). How large can it be made before clipping of the v_{CE} waveform occurs?

4.4-5 For the circuit in Figure 4.3-8a, $R_C = 500$ Ω, $R_B = 183$ kΩ, and a resistor $R_E = 100$ Ω is inserted between the emitter and ground. Carefully sketch and label v_E, i_C, v_C, and v_{CE} if $v_{in} = 0.5 \sin \omega t$ volts.

4.4-6 In Figure 4.3-5a, $v_{in} = 1 \sin \omega t$ V, $V_{CC} = 10$ V, $R_C = 2$ kΩ, and a resistor $R_E = 1$ kΩ is inserted between the emitter and ground. Sketch v_{CE} if R_B is adjusted so that $I_B = 0.02$ mA.

4.4-7 For the circuit shown in Figure P4.4-7, find v_o in terms of v_1 and v_2.

4.4-8 For the circuit in Figure 4.4-5a, R_L and C_2 are removed, $V_{CC} = 25$ V, $R_1 = 19.3$ kΩ, $R_2 = 5.7$ kΩ, $R_C = 5$ kΩ, $R_{E1} = 1$ kΩ, and $R_{E2} = 1.5$ kΩ. Given that $v_{in} = 2 \sin \omega t$ volts, sketch the total signals i_B, i_C, v_E, v_C, and v_{CE}.

4.4-9 For the circuit in Figure 4.4-7a, $R_L = 2$ kΩ, $R_C = 2$ kΩ, $R_{E1} = 1$ kΩ, $R_{E2} = 0$, and $R_2 = \infty$. Find the circuit current gain $A_i = i_L/i_{in}$.

4.4-10 (a) In Figure 4.4-7a, $V_{CC} = 20$ V, R_L and C_2 are removed, $R_C = 5$ kΩ, $R_B = 1.8$ MΩ, and $v_{in} = 0.1 \sin \omega t$ V. Find the dc Q-point values of I_B, I_C, V_C, and V_E.

(b) Find the ac voltage gains v_c/v_{in} and v_e/v_{in}.

(c) Sketch the total output voltages v_C and v_E.

(d) What is the maximum v_{in} before distortion in one of the output signals occurs? What happens to the transistor at this point?

4.4-11 For the transistor amplifier shown in Figure 4.4-1, $V_{CC} = 10$ V, $R_B = 0$, $R_E = 200$ Ω, $v_{in} = 0.2 \sin \omega t$ V, and R_C consists of the series connection of two 1-kΩ resistors. The junction point of these two resistors is ac bypassed to ground, and V_{BB} is adjusted so that $i_C = 2$ mA. Sketch the total (dc + ac) waveform for v_C.

4.5-1 In Figure 4.4-7a, $R_{C1} = 3$ kΩ, $R_{C2} = 2$ kΩ, and $R_{in} = 1$ kΩ.

(a) Find the ac Thévenin equivalent circuit to the left of the base of Q_2. (The answer for v_T will be in terms of v_{in}.)

(b) Connect the Thévenin equivalent circuit obtained in part (a) onto the second stage and find v_{c2}/v_{in}.

4.5-2 Figure P4.5-2 shows a differential amplifier commonly found in today's linear integrated circuits. Assuming that R_E is very large, prove that

(a) $Z_{in} = 2h_{ie}$

(b) $Z_o = R_c$

(c) $\dfrac{v_o}{v_s} = \dfrac{h_{fe} R_c}{R_s + 2h_{ie}}$

4.5-3 For the circuit in Figure 4.4-7a, $R_{in} = 1$ kΩ, $R_{B1} = 200$ kΩ, $R_{C1} = 2$ kΩ, $R_{B2} = 100$ kΩ, $R_{C2} = 1$ kΩ, a 10-Ω resistor is inserted between the emitter of Q_2 and ground, and a 0.5-kΩ resistor is capacitively coupled from the collector of Q_2 to ground.

(a) Find the voltage gain of the first stage, $A_{v1} = v_{c1}/v_{in}$, with Q_2 disconnected.

(b) Repeat part (a) and calculate A_{v1} with Q_2 in place.

(c) Find the voltage gain of the second stage, $A_{v2} = v_{c2}/v_{c1}$, with R_L disconnected.

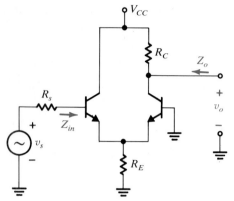

Figure P4.5-2

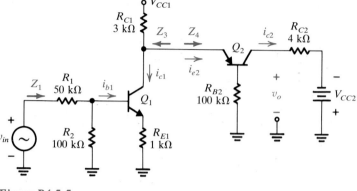

Figure P4.5-5

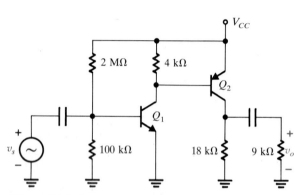

Figure P4.5-6

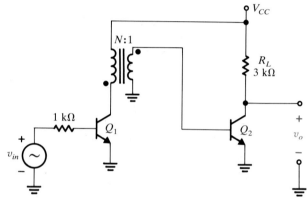

Figure P4.5-10

(d) Repeat part (c) and find A_{v2} with R_L in place.

(e) What is the overall voltage gain of the amplifier, $A_v = v_{c2}/v_{in}$? How is it related to A_{v1} and A_{v2}?

4.5-4 In Figure 4.4-8a, $R_1 = 500$ kΩ, $R_2 = R_3 = 100$ kΩ, $R_{C2} = 2$ kΩ, and 1-kΩ resistor is connected in series with v_{in}. Draw a good ac model for this circuit and find ac voltage gain v_{o2}/v_{in}.

4.5-5 Find each of the quantities indicated in Figure P4.5-5 in the following order: z_1, i_{b1}, i_{c1}, z_3, z_4, i_{e2}, i_{c2}, v_o, and the overall circuit voltage gain. Where necessary, express your answers in terms of v_{in}.

4.5-6 (a) For the circuit shown in Figure P4.5-6, draw the ac small-signal model.

(b) Find the ac gain v_o/v_s.

4.5-7 (a) For the circuit illustrated in Figure P4.5-2, develop an expression for the improvement obtained in ac power delivered to the load when transformer coupling, instead of direct coupling, is used. In particular, show that this improvement factor is equal to

$$\frac{1}{4} \frac{[(1/h_{oe}) + R_L]^2}{(1/h_{oe})R_L}$$

(b) Using the result in part (a), verify that the improvement data given in Figure 4.5-2c are correct.

4.5-8 Find the power gain of the circuit given in Figure P4.5-6 considering the 9-kΩ resistor to be the load.

4.5-9 In Figure 4.5-1, R_E is set equal to zero, and a transformer is inserted between R_s and the coupling capacitor with the lower side of the primary and the secondary grounded.

(a) Determine the turns ratio required to deliver maximum power to the transistor amplifier.

(b) Find the average power delivered to the load.

(c) Calculate the ac power gain of the circuit.

4.5-10 (a) For the circuit shown in Figure P4.5-10, $1/h_{oe1} = 1/h_{oe2} = 100$ kΩ. Select the turns ratio that will maximize the overall circuit gain.

(b) For the value of N selected in part (a), what are v_{c1}/v_{in}, v_{b2}/v_{in}, and the overall voltage gain v_o/v_{in}?

4.5-11 For the circuit in Problem 4.5-10, what is the overall circuit power gain P_L/P_{in} for transformer turns ratios of 2, 10, and 50 : 1?

CHAPTER 5
The Field-Effect Transistor

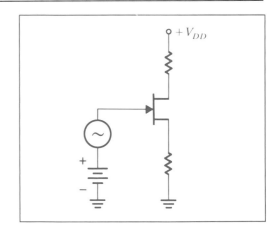

5.1 INTRODUCTION

In a bipolar junction transistor, the flow of current to the collector is controlled by the voltage applied across the base–emitter junction of the transistor. Because the current flow in a BJT involves the motion of both holes and electrons (see Figure 3.2-3), this device is referred to as a bipolar transistor, or as a transistor in which both polarities or both types of current carriers are present.

Besides the BJT, there is a second distinctly different type of transistor that is known as an FET, or field-effect transistor. In an FET the current flows through only one type of semiconductor, and this current is due solely either to the flow of holes or to the flow of electrons, depending on whether the transistor is made from p-type or n-type material, respectively. As a result, the FET is often said to be a unipolar (single polarity) type of transistor.

The operation of the FET is quite different from that of the BJT, and although a voltage applied to the input of both devices is used to control the flow of current at the output, almost no current flows into the input terminals of the FET. For this reason, the field-effect transistor is often modeled by a voltage-controlled current source since its properties are quite similar to those of an ideal transconductance amplifier (Figure 3.1-8b).

A simplified picture of one type of FET is illustrated in Figure 5.1-1. It is known as a junction field-effect transistor (JFET), and it is a three-terminal device whose gate, source, and drain terminals have characteristics that are similar to those of the base, emitter, and collector, respectively, in a BJT. The particular device illustrated in part (a) is called an n-channel JFET because the main conduction path between the drain and the source is made of n-type semiconductor material. The circuit symbol for this n-channel JFET is given in

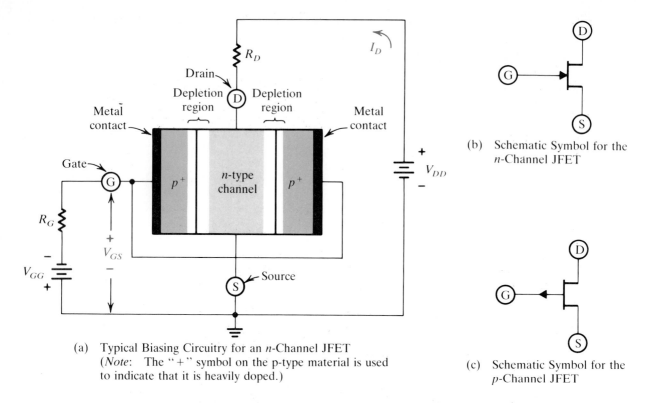

(b) Schematic Symbol for the n-Channel JFET

(a) Typical Biasing Circuitry for an n-Channel JFET
(*Note*: The "+" symbol on the p-type material is used
to indicate that it is heavily doped.)

(c) Schematic Symbol for the p-Channel JFET

Figure 5.1-1

part (b). Notice that the arrow on the gate lead points in the direction of easy current flow in the *pn* junction diode formed by the gate and the channel of the transistor.

The structure of a *p*-channel JFET is identical to that for the *n*-channel device illustrated in Figure 5.1-1a except that the roles of the *p* and *n* materials are reversed. The circuit symbol for this type of transistor is given in Figure 5.1-1c.

For the circuit in Figure 5.1-1a, the gate junction is reversed biased by the battery V_{GG}, so that the input impedance at the gate terminal of the FET is very high. The *n*-type semiconductor channel between the drain and source terminals acts like a resistance whose ohmic value depends on the voltage, V_{GS}, applied between the gate and the source. As this voltage becomes more negative, the width of the depletion region on either side of the channel grows, narrowing the channel, and increasing the effective resistance between the drain and the source. Thus, a voltage (or field) applied to the input of an FET alters the width of the channel and controls the flow of current between the drain and the source. Because of the very high input impedance of the FET, large power gains are possible in circuits using these transistors.

FETs differ from bipolar junction transistors in several important respects:

1. In an FET the current that flows from the drain to the source is transported predominantly by only one type of charge carrier, while in a BJT the current is carried by both electrons and holes. For this reason the field-effect transistor is called a unipolar device, while the BJT is often referred to as a bipolar transistor.

2. The JFET has a very high gate input resistance since its gate junction is normally reverse biased, while the BJT input impedance is typically quite low because its base–emitter junction is forward biased.

3. FETs are easier than BJTs to fabricate. In addition, they occupy less space on an IC and dissipate very little power, so that they have a much higher packing density. As a result, field-effect transistors are used almost exclusively for the design of high-density digital integrated circuits such as memories and microprocessors.

4. FETs have a wider dynamic range or range of allowed input signals than BJTs, and as a result, they produce less distortion for a given signal level than bipolar junction transistors.

5. The thermal stability of the FET is superior to that of the BJT, making it an excellent candidate for power amplifier applications, where the "thermal runaway" of BJT amplifiers has always been a major design problem (see Chapter 9).

5.2   THEORY OF OPERATION OF THE JFET

The operation of the junction field-effect transistor can be understood fairly easily because of its similarity to the behavior of the *pn* junction diode discussed in Section 1.11. In particular, the equations governing the width of the channel in an FET may be obtained directly from a consideration of the width of the depletion region in a semiconductor diode.

Under zero-bias conditions in a *pn* junction diode, that is, with $V_D = 0$, a depletion region exists near the junction in which there are no free charges present (Figure 1.11-3b). The width of this depletion region is given by equation (1.11-12) for an abrupt junction, and it is repeated here for convenience.

$$w = \sqrt{\frac{2\epsilon V_{jo}}{q}\left(\frac{1}{N_D} + \frac{1}{N_A}\right)} \tag{5.2-1}$$

When an external voltage, V_D, is applied to the diode, the width of the depletion region changes. In this case, w is still given by an expression having the same form as equation (5.2-1), but the term V_{jo} is replaced by $(V_{jo} - V_D)$, so that

$$w = \sqrt{\frac{2\epsilon (V_{jo} - V_D)}{q}\left(\frac{1}{N_D} + \frac{1}{N_A}\right)} \tag{5.2-2}$$

Following this equation, it is apparent that the width of the depletion region decreases when the diode is forward biased and gets larger when it is reverse biased. Figure 1.11-4b illustrates how the reverse-bias voltage increases the width of the depletion region by pulling the holes in the *p*-type material and the electrons in the *n*-type material away from the junction.

These results for the *pn* junction diode may be applied directly to the JFET to find out how the width of the channel depends on V_{GS} and V_{DS}, and also to determine the relationship between the drain current I_D and these voltages. To begin this analysis, consider the *n*-channel JFET illustrated in Figure 5.2-1a, and assume that this transistor is doped so that $N_A \gg N_D$. For this case, following equations (1.11-11a) and (1.11-11b), the depletion region associated with

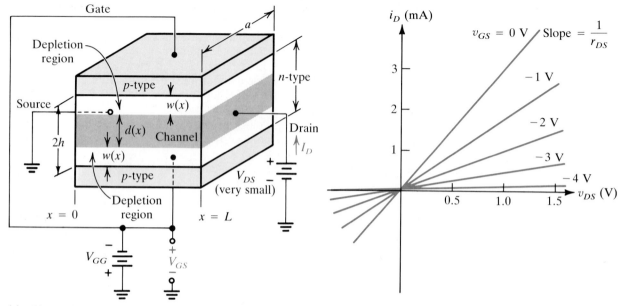

(a) Shape of the Channel in an *n*-Channel JFET for V_{GS} Negative and V_{DS} Very Small (*Note*: For simplicity in this figure all metal contacts to the semiconductor materials have been omitted.)

(b) JFET Volt-Ampere Characteristics for Small Values of V_{DS}

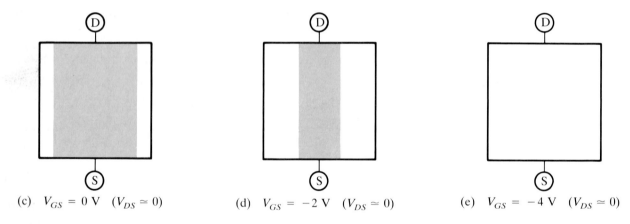

(c) $V_{GS} = 0$ V $(V_{DS} \approx 0)$

(d) $V_{GS} = -2$ V $(V_{DS} \approx 0)$

(e) $V_{GS} = -4$ V $(V_{DS} \approx 0)$

Figure 5.2-1
Characteristics of the JFET for small drain-to-source voltages.

the gate junction will expand primarily into the channel with

$$w(x) \simeq w_n(x) = \sqrt{\frac{2\epsilon[V_{jo} - V_D(x)]}{qN_D}} \tag{5.2-3}$$

Here we have written the width of the depletion region as a function of its position x in the channel, to indicate that its value is not necessarily constant since V_D can vary at different points along the gate junction. In general, we may express the junction voltage $V_D(x)$ for this circuit as

$$V_D(x) = V_{GS} - V(x) \tag{5.2-4}$$

where $V(x)$ is the potential at the point x within the channel, measured with respect to ground. Substituting this expression into equation (5.2-3), the width of the depletion region at any point x in the channel may be written as

$$w(x) = \sqrt{\frac{2\epsilon[V_{jo} - V_{GS} + V(x)]}{qN_D}} \qquad (5.2\text{-}5a)$$

Following Figure 5.2-1a, the corresponding width of the channel is

$$d(x) = 2h - 2\sqrt{\frac{2\epsilon[V_{jo} - V_{GS} + V(x)]}{qN_D}} \qquad (5.2\text{-}5b)$$

When V_{DS} is small, as it is for the biasing arrangement shown in this figure, the voltage, $V(x)$, is nearly zero at all points within the channel. As a result, the width of the channel is constant throughout its length and is approximately equal to

$$d(x) \simeq d = 2h - 2\sqrt{\frac{-2\epsilon V_{GS}}{qN_D}} \qquad (5.2\text{-}5c)$$

In this expression we have neglected the term V_{jo} since it is generally quite small in comparison to V_{GS}.

Because the width of the channel is relatively independent of the drain-to-source voltage when V_{DS} is small, we may represent the current flow through the JFET as

$$I_D \simeq \frac{1}{r_{DS}} V_{DS} \qquad (5.2\text{-}6a)$$

where r_{DS}, the effective resistance between the drain and source terminals, is

$$r_{DS} = \frac{\rho L}{A} = \frac{\rho L}{ad} = \frac{\rho L}{a(2h - 2\sqrt{-2\epsilon V_{GS}/qN_D})} \qquad (5.2\text{-}6b)$$

From this result, we can see that r_{DS} increases as V_{GS} is made more negative. This is to be expected, of course, since the channel gets narrower as the reverse-bias voltage across the gate junction increases.

For later use it is convenient to define $r_{DS(ON)}$ as the drain-to-source resistance of the JFET with V_{GS} equal to zero. Following equation (5.2-6b), we can see that

$$r_{DS(ON)} = \frac{\rho L}{2ah} \qquad (5.2\text{-}7)$$

As V_{GS} is made more and more negative, the channel gets narrower and narrower, and eventually it will close off (or pinch off) completely. At this point the drain current is zero, independent of V_{DS}, and r_{DS} is infinite. Following equation (5.2-5c), this occurs when $d = 0$ or alternatively, when the gate-to-source voltage is

$$V_{GS} = V_p = \frac{-qN_D h^2}{2\epsilon} \qquad (5.2\text{-}8)$$

This gate-to-source voltage, V_p, is known as the pinch-off voltage of the JFET.

Figure 5.2-1b illustrates the variation of I_D with changes in V_{GS} for small values of V_{DS}. In this region, to the extent that equation (5.2-6a) is valid, the characteristic curves for the transistor are straight lines with slopes equal to $1/r_{DS}$. These results emphasize the fact that a JFET operating in this region may be viewed as a voltage-controlled resistor in which the voltage applied between the gate and the source controls the resistance of the transistor between the drain and the source terminals. Notice from Figure 5.2-1b that equation (5.2-6a) is also valid for negative drain-to-source voltages as long as V_{DS} remains small.

In examining Figure 5.2-1b, it should also be clear that $V_p = -4$ V for this transistor since this is the value of V_{GS} that makes $I_D = 0$ or r_{DS} infinite. For small-signal n-channel JFETs, typical values of V_p vary from about -1 to -10 V.

Figure 5.2-1c through e shows the changes that take place in the shape of the channel as V_{GS} is varied while V_{DS} remains small. The unshaded parts of these illustrations represent the depletion region, and the shaded parts, the portions of the channel still containing free charge carriers.

For small values of V_{GS}, the effective channel cross-sectional area is large, and hence the drain-to-source resistance of the FET is low. As V_{GS} is made more negative, the channel gets narrower and the drain-to-source resistance increases. When the gate-to-source voltage is made negative enough, the depletion regions from each side of the channel meet at the center, as shown in Figure 5.2-1e, completely depleting the channel of free carriers. When this occurs, the drain current is zero independent of V_{DS}, and the effective drain-to-source resistance is infinite.

EXAMPLE 5.2-1

a. Using the characteristics of the n-channel JFET given in Figure 5.2-1, determine the drain-to-source resistance, r_{DS}, for $V_{GS} = 0, -1, -2,$ and -3 V.

b. The circuit given in Figure 5.2-2a is known as an amplitude modulator. The control voltage v_c is used to alter the amount of the input signal that appears at the output terminals by changing the resistance of the JFET. To illustrate how this is accomplished, sketch v_{GS} and v_o if v_c is a 2-V(p-p) square wave, and v_{in} a 2-V(p-p) sine wave, each having the form shown in Figure 5.2-2b.

SOLUTION

The drain-to-source resistance of the FET for various values of V_{GS} is determined directly from the slope of the characteristic curves given in Figure 5.2-1b. Specifically, for the case where $V_{GS} = 0$, $r_{DS} = (1 \text{ V})/(3 \text{ mA}) = 330 \ \Omega$. In a similar fashion, the values of r_{DS} corresponding to $V_{GS} = -1, -2,$ and -3 V are 570 Ω, 1.1 kΩ, and 2.5 kΩ, respectively.

The operation of the modulator in Figure 5.2-2a is best understood by observing that $v_{GS} = v_c - V_{GG} = v_c - 2$. This voltage is sketched in Figure 5.2-2b, and as shown, it is a square wave whose amplitude varies between -1 and -3 V. When v_{GS} is at -1 V, the FET in Figure 5.2-2a can be replaced by an equivalent resistance of 570 Ω. This model for the transistor is valid as long as v_{DS} is less than about 1 V (see Figure 5.2-1). For this value of v_{GS}, the overall equivalent circuit of the modulator has the form shown in Figure 5.2-2c, and the resulting output signal is $(0.57/2.57)v_{in} = 0.22v_{in}$. When v_{GS} drops to

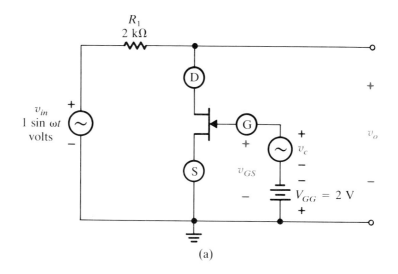

(a)

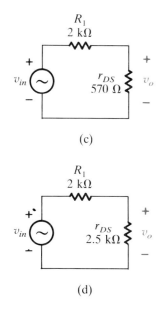

(c)

(d)

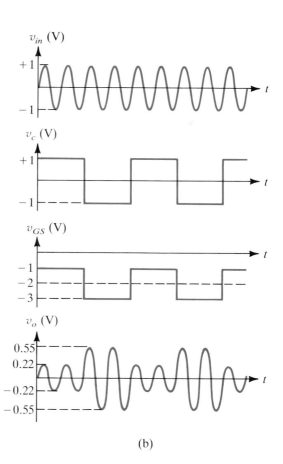

(b)

Figure 5.2-2
Modulation circuit for
Example 5.2-1.

−3 V, r_{DS} increases to 2.5 kΩ, the circuit in Figure 5.2-2d applies, and $v_o = (2.5/4.5)v_{in} = 0.55v_{in}$. The overall output that is obtained from this circuit is sketched in Figure 5.2-2b.

This type of circuit is sometimes used in electric guitar amplifiers to produce a tremolo effect. Tremelo is a slowly varying amplitude modulation of the output signal from the instrument, and for this application the control voltage in Figure 5.2-2a is usually a slowly varying sine wave in the frequency range 1 to 10 Hz. The "rate" control on the guitar amplifier changes the frequency of this sine wave, and the "depth" control changes the amplitude of the control voltage and hence the amount of modulation.

Thus far we have examined the performance of the field-effect transistor when V_{GS} is changed while V_{DS} remains small. Consider next what happens in a JFET when the V_{GS} is held at a fixed voltage and V_{DS} is varied. Let's begin by examining the specific circuit illustrated in Figure 5.2-3, in which $V_{GS} = 0$ and V_{DS} is initially equal to 1 V. Because the gate contact is grounded, the voltages at points A, B, and C along the gate are all zero. Furthermore, the potential $V(x)$ at point D inside the channel is also zero because the source is grounded, while that at point F is equal to 1 V with respect to ground. To determine the potential $V(x)$ at the point E midway down the length of the channel, we note that the partial channel resistance R_{FE} will be nearly equal to R_{ED} if the channel width is approximately uniform. In this instance, the voltage at point E will be (1 V)$(R_{ED}/(R_{ED} + R_{FE})$or about 0.5 V with respect to ground. Using these data, we may determine the potentials V_1, V_2, and V_3 along the gate junction as follows:

$$V_1 = V_A - V_D = 0 - 0 = 0.0 \text{ V} \tag{5.2-9a}$$

$$V_2 = V_B - V_E = 0 - 0.5 = -0.5 \text{ V} \tag{5.2-9b}$$

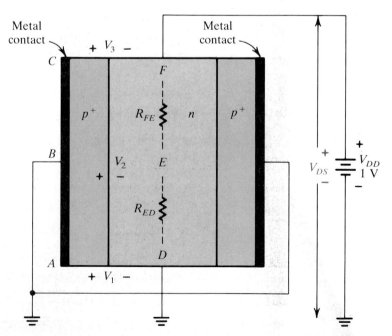

Figure 5.2-3
Potential distribution
inside an n-channel FET
with $V_{DS} = 1$ V, $V_{GS} = 0$ V.

Chapter 5 / The Field-Effect Transistor

and
$$V_3 = V_C - V_F = 0 - 1.0 = -1.0 \text{ V} \qquad (5.2\text{-}9c)$$

From these results we can see that the gate diode has zero volts across it near the source end of the transistor and that this bias voltage increases toward $-V_{DS}$ as we approach the drain terminal. Therefore, the width of the depletion region is nonuniform along the length of the channel, being narrowest near the source and expanding as we move toward the drain. The approximate shape of the depletion region profile for this case is illustrated in Figure 5.2-4b.

The sequence of sketches in Figure 5.2-4 illustrates how the shape of the depletion region changes as V_{DS} varies from 1 to 10 V with V_{GS} held fixed at 0 V. For small values of V_{DS} (less than about 1 V), the channel is almost completely open, and the drain-to-source resistance, r_{DS}, is essentially constant at $r_{DS(ON)}$ independent of V_{DS}, so that I_D initially increases linearly with V_{DS} in accordance with equation (5.2-6a). This portion of the transistor characteristic is called the ohmic region (Figure 5.2-5). As V_{DS} increases further, the channel begins to narrow more and more, so that doubling V_{DS} no longer doubles I_D because the channel resistance is increasing (Figure 5.2-4b). As a result, the I_D versus V_{DS} characteristic begins to curve downward, as shown in Figure 5.2-5.

When V_{DS} is equal to 4 V, the potential across the gate junction at the top of the channel, V_3 in Figure 5.2-4b, is equal to -4 V, so that the channel just closes off at this point. At first it might appear that this will make the drain current go to zero. This does not happen, however.

As shown in Figure 5.2-4b, the potential, $V(x)$, at the top of the triangular resistance region, labeled $R_{\Delta 1}$, must be 4 V because this is the voltage that is needed to just cause the depletion regions from each side to meet at the center of the channel. Therefore, there is a nonzero drain current $(4 \text{ V})/R_{\Delta 1}$ flowing down through the channel. This current also flows across the pinched-off por-

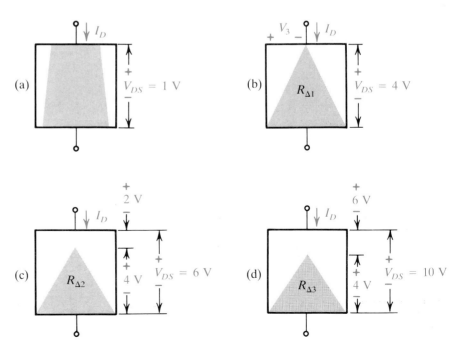

Figure 5.2-4
Variation of the depletion-region profile with changes in V_{DS} (for $V_{GS} = 0$).

tion of the channel in much the same way that the current injected from the emitter into the base of a BJT flows across the reverse-biased depletion region in the base–collector junction of a bipolar transistor.

As the drain-to-source voltage increases beyond 4 V, the length of the pinched-off portion of the channel increases slightly. However, the voltage at the top of the triangular area always remains at about 4 V since this is the potential that is needed to just cause the two sides of the depletion regions to come together. As a result, as V_{DS} increases, the drain current is given approximately by (4 V)/R_Δ, where R_Δ represents the resistance of the undepleted triangular portion of the channel. If we neglect the slight elongation of the pinch-off region as V_{DS} increases, then

$$R_{\Delta 1} \simeq R_{\Delta 2} \simeq R_{\Delta 3} \qquad (5.2\text{-}10)$$

As a result, once V_{DS} is greater than $-V_p$, the drain current remains almost constant and is independent of the drain-to-source voltage. This portion of the transistor's V–I characteristics is known as the constant-current region, and the current that flows in the transistor in this region when $V_{GS} = 0$ is called I_{DSS}. Typical values of I_{DSS} for small-signal JFETs vary from about 1 to 50 mA.

For actual transistors, I_D is not completely independent of V_{DS} in the constant-current region. As V_{DS} increases, the height of the triangularly shaped region decreases somewhat, as does its resistance, so that

$$R_{\Delta 3} < R_{\Delta 2} < R_{\Delta 1} \qquad (5.2\text{-}11)$$

Consequently, as the drain-to-source voltage increases, the drain current also goes up slightly. As a first-order approximation, however, it is generally reasonable to neglect the upward tilt of these curves.

If a negative potential is applied to the gate lead and the I_D versus V_{DS} characteristic is again sketched, the resulting curve is similar to that for the case

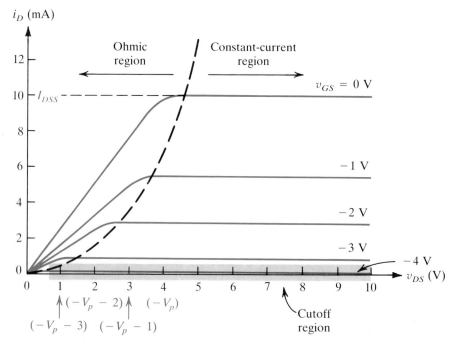

Figure 5.2-5
Volt-ampere characteristics
of an n-channel JFET.

Chapter 5 / The Field-Effect Transistor

when V_{GS} is equal to zero, except that the transition from the ohmic to the constant-current region occurs at a lower value of V_{DS} and the final plateau level for I_D is lower than for the case where V_{GS} was zero. To understand the reasons for this behavior, consider what would happen if point B in Figure 5.2-3 were connected to a -1-V battery instead of being grounded. For this case we would have $V_{GS} = -1$ V, $V_C = -1$ V, $V_F = V_{DS}$, and $V_3 = V_C - V_{DS} = (-1 - V_{DS})$. As before, the channel begins to pinch off when V_3 equals -4 V, and therefore for this case, with V_{GS} equal to -1 V, the channel will pinch off at a drain-to-source voltage of 3 V. In a similar fashion, for $V_{GS} = -2$ V and -3 V, the transition from the ohmic to the constant-current region occurs at V_{DS} values of about 2 V and 1 V, respectively. The dashed curve in Figure 5.2-5 illustrates the general shape of the boundary between the ohmic and constant-current regions.

To understand why the drain current plateaus get lower as V_{GS} is made more negative, we need to observe that the triangularly shaped areas, shown in Figure 5.2-4 for the case where $V_{GS} = 0$, will have a similar shape when V_{GS} is negative, except that their width will be narrower (e.g., see Figure 5.2-1d). As a result, the resistance of the triangular regions at pinch-off for V_{GS} less than zero will be greater than for the case where V_{GS} equals zero, and this will cause I_D to plateau out at a lower current level.

5.2-1 Development of the Generalized Volt-Ampere Equations for the JFET

The general relationship between I_D, V_{GS}, and V_{DS} may be obtained by carefully examining the n-channel JFET circuit illustrated in Figure 5.2-6. Because the drain current I_D through any cross section of the transistor is the same, it therefore follows that

$$\Delta V(x) = \text{voltage drop across an element of width } \Delta x$$

$$= I_D \Delta R(x) \tag{5.2-12a}$$

where $V(x)$ is the potential at any point x in the channel and

$$\Delta R(x) = \text{resistance of the element} = \frac{\rho \Delta x}{A} = \frac{\rho \Delta x}{ad(x)} \tag{5.2-12b}$$

Combining equations (5.2-12a) and (5.2-12b) and substituting in equation (5.2-5b) for $d(x)$, we have

$$a\left[2h - 2\sqrt{\frac{2\epsilon[V(x) - V_{GS}]}{qN_D}}\right]\Delta V(x) = \rho I_D \Delta x \tag{5.2-13}$$

Taking the limit of this expression as Δx approaches zero and integrating it along the length of the channel from $x = 0$ [where $V(x) = 0$] to $x = L$ [where $V(x) = V_{DS}$], we obtain

$$\int_0^{V_{DS}} a\left[2h - 2\sqrt{\frac{2\epsilon(V - V_{GS})}{qN_D}}\right]dV = \int_0^L \rho I_D dx \tag{5.2-14}$$

Evaluating these integrals and solving the resulting equation for I_D, we find, af-

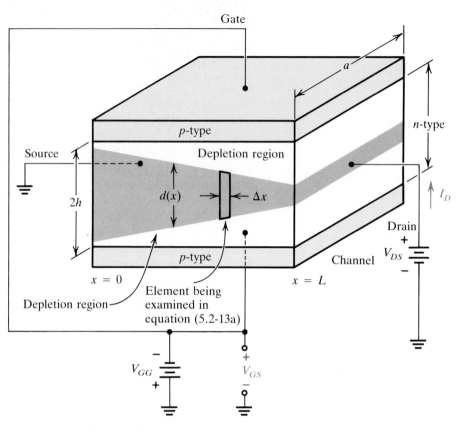

Figure 5.2-6
Shape of the channel in an n-channel JFET with v_{GS} negative and v_{DS} large and positive. (Note: For simplicity in this figure all metal contacts to the semiconductor materials have been omitted.)

ter considerable mathematical manipulation, that

$$I_D = \frac{-V_p}{r_{DS(ON)}}\left[\frac{V_{DS}}{-V_p} - \frac{2}{3}\left(\frac{V_{GS} - V_{DS}}{V_p}\right)^{3/2} + \frac{2}{3}\left(\frac{V_{GS}}{V_p}\right)^{3/2}\right] \quad (5.2\text{-}15)$$

This result is valid as long as the channel does not pinch off, or in other words, as long as

$$V_{DS} < (V_{GS} - V_p) \quad (5.2\text{-}16)$$

When the drain-to-source voltage exceeds this value, the transistor enters the constant-current region, where I_D now remains fixed as V_{DS} varies. The positions on the transistor volt-ampere characteristics where the transition from the ohmic to the constant-current region occurs are those points where $V_{DS} = V_{GS} - V_p$. Since the value of I_D at this drain-to-source voltage remains constant for all V_{DS} exceeding this value, by substituting $(V_{GS} - V_p)$ for V_{DS} in equation (5.2-15), we find that the expression for I_D in the constant-current region may be written as

$$I_D = \frac{-V_p}{3r_{DS(ON)}}\left[1 - \frac{3V_{GS}}{V_p} + 2\left(\frac{V_{GS}}{V_p}\right)^{3/2}\right] \quad (5.2\text{-}17)$$

Using the definition that I_{DSS} is the current that flows in the JFET when $V_{GS} = 0$, from equations (5.2-17),(5.2-8), and (5.2-7) we find that I_{DSS} is related to the transistor physical parameters by

$$I_{DSS} = \frac{-V_p}{3r_{DS(ON)}} = \frac{\mu q^2 N_D^2 h^3 a}{3\epsilon L} \qquad (5.2\text{-}18)$$

Furthermore, by substituting this result into equation (5.2-17), the expression for the drain current in the constant-current region may be rewritten as

$$I_D = I_{DSS}\left[1 - \frac{3V_{GS}}{V_p} + 2\left(\frac{V_{GS}}{V_p}\right)^{3/2}\right] \qquad (5.2\text{-}19)$$

Although this result is correct mathematically, as a practical matter it is difficult to work with. Fortunately, this equation is nearly equal to the quadratic expression

$$I_D = I_{DSS}\left[1 - 2\left(\frac{V_{GS}}{V_p}\right) + \left(\frac{V_{GS}}{V_p}\right)^2\right] = I_{DSS}\left(1 - \frac{V_{GS}}{V_p}\right)^2 \qquad (5.2\text{-}20)$$

This equation, which also relates I_D to V_{GS}, turns out to be much more useful than the one given in equation (5.2-19) because it is easier to sketch and easier to solve analytically. Interestingly enough, although these two expressions appear to be considerably different from one another, the actual difference between them is always less than 7% of I_{DSS}, so that the error involved in using equation (5.2-20) in place of (5.2-19) is relatively minor. A second reason for using equation (5.2-20) to represent the volt-ampere characteristics of the JFET is that the equation governing the behavior of the MOSFET is virtually identical to this expression.* Therefore, if we use equation (5.2-20) to represent the JFET when it is operating in the constant-current region, all the analysis and design techniques that we develop for use with this transistor will be directly applicable to the MOSFET.

5.2-2 DC Models for the JFET

In analyzing the performance of a JFET circuit, only its output characteristic curves are needed to determine I_D, V_{GS}, and V_{DS}. This is because the input circuit of the transistor is almost always operated with the gate junction reverse biased so that its gate current is nearly zero. Therefore, to a good approximation, the input circuit of a JFET may be represented by an open circuit. In modeling the output portion of the field-effect transistor, the form that the equivalent circuit takes depends on the region of the V–I characteristics where the device is operating.

As we saw in Example 5.2-1, when the transistor is biased in the ohmic region, it can be represented by a resistor r_{DS} whose value is determined from the slope of the particular V_{GS} characteristic curve on which the FET is operating (Figure 5.2-7b). This model is valid for all values of V_{DS} less than $(V_{GS} - V_p)$.

To place the Q point of the transistor in the cutoff region (see Figure 5.2-

*The MOSFET, or metal-oxide field-effect transistor, is a second, very important type of FET, and is discussed in Section 5.5.

State of the Transistor	Status of the Gate Junction	Reasonable Approximations	Conditions to Test for (to Ensure Are in This State)	Approximate n-Channel Model in That State
(a) Active (constant current)	Reverse biased	$I_G = 0$ $I_D = I_S$ $= I_{DSS}\left(1 - \dfrac{V_{GS}}{V_p}\right)^2$	$V_p < V_{GS} < 0$ $V_{DS} > (V_{GS} - V_p)$	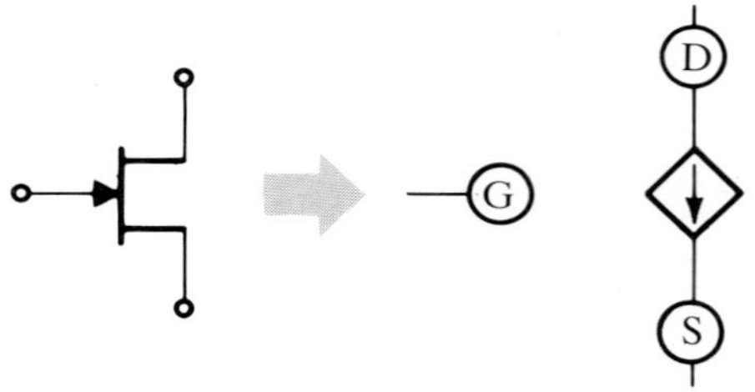
(b) Ohmic	Reverse biased	$I_G = 0$ $I_D = I_S$ r_{DS} depends on V_{GS} [eq. (5.2-6b)]	$V_p < V_{GS} < 0$ $V_{DS} < (V_{GS} - V_p)$	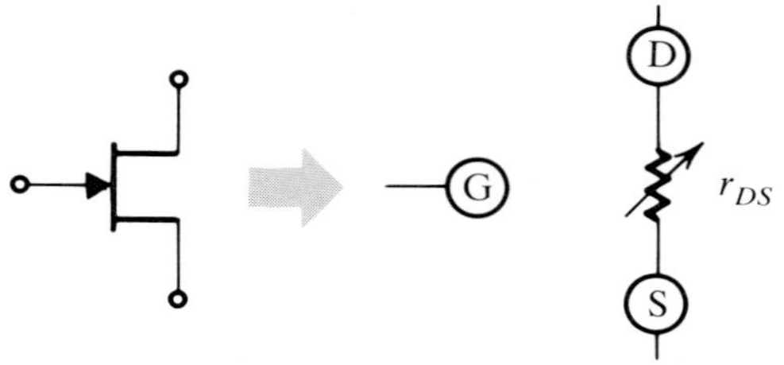
(c) Cutoff	Reverse biased	$I_G = 0$ $I_D = 0$ $I_S = 0$	$V_{GS} < V_p$	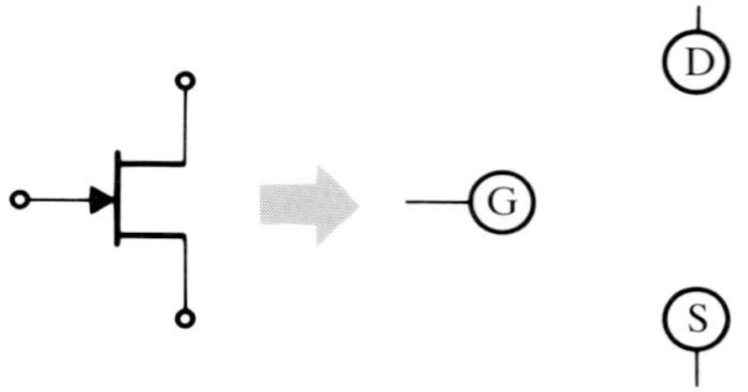

Figure 5.2-7
Three important states for the n-channel JFET transistor.

5), it is only necessary to make V_{GS} less than the pinch-off voltage. When this is done, the drain current in the FET is zero, and the device may be represented by an open circuit for all values of V_{DS} (Figure 5.2-7c).

The last region of interest on the JFET characteristics is the so-called constant-current or active region. For analog circuit design, this is the most important region of transistor operation since it is here where the FET best functions as a linear amplifier. In the active region, the transistor input still looks like an open circuit, while the output is best represented by a voltage-controlled current source whose value depends on the gate-to-source voltage (Figure 5.2-7a). In this model the relationship between the applied voltage V_{GS} and the current that flows in the controlled source is given approximately by equation (5.2-20).

EXAMPLE 5.2-2

a. Write the dc load-line equations for the JFET circuit given in Figure 5.2-8.

b. Use the results in part (a) together with the transistor graphical data given in Figure 5.2-8c to find the Q point for the circuit. Specifically, determine V_{GS}, I_G, I_D, V_{DS}, and I_S.

c. Develop the dynamic load-line equations for this circuit and use them together with the graphical data to sketch v_{GS}, i_D, and v_{DS} given that v_{in} is a 1-V sine wave.

d. Estimate the ac power delivered to the load (R_D) and repeat this calculation if V_{DD} is lowered to 3 V to place the Q point in the ohmic region.

SOLUTION

To develop the dc load-line equations, we replace the capacitor by an open circuit to obtain the dc equivalent circuit given in Figure 5.2-8b. Applying Kirchhoff's voltage law to the input and output loops of this circuit, we have

$$-V_{GG} = I_G R_G + V_{GS} \qquad (5.2\text{-}21a)$$

and

$$V_{DD} = I_D R_D + V_{DS} \qquad (5.2\text{-}21b)$$

Because the dc gate-to-source voltage is negative, the gate diode is reverse biased, and the gate current, I_G, is zero. As a result, following equation (5.2-21a), the quiescent gate-to-source voltage is -2 V.

To obtain the Q-point values of the drain current and the drain-to-source voltage, we need to plot equation (5.2-21b) on the transistor output characteristics in Figure 5.2-8c. As with the bipolar transistor, because the load-line equation is a straight line, only two points on that line are needed to be able to sketch it. The two simplest points to use the v_{DS} and i_D intercepts, and following equation (5.2-21b), these are seen to be located at $v_{DS} = V_{DD} = 10$ V and $i_D = V_{DD}/R_D = 12.5$ mA, respectively. The intersection of this load line with the $V_{GS} = -2$ V curve places the Q point for this circuit at $I_D = 3.6$ mA and $V_{DS} = 7.1$ V.

To develop the dynamic load-line equations, we begin by noting that the dc voltage across the input capacitor is $v_C = -2$ V, so that it may be replaced by a -2 V battery in Figure 5.2-8a. Applying Kirchhoff's voltage law to the input and output loops of this circuit, the dynamic load-line equations may be written as

$$v_{GS} = v_{in} - 2 \qquad (5.2\text{-}22a)$$

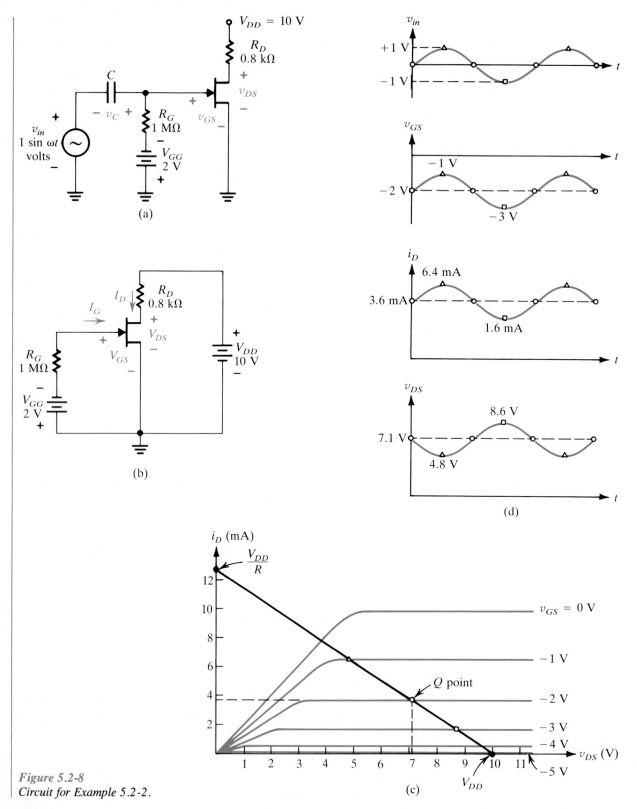

Figure 5.2-8
Circuit for Example 5.2-2.

and
$$V_{DD} = i_D R_D + v_{DS} \qquad (5.2\text{-}22b)$$

Using equation (5.2-22a), we may immediately sketch v_{GS} as illustrated in Figure 5.2-8d. The solutions for i_D and v_{DS} are obtained by plotting the output dynamic load-line equation on the transistor output characteristics. Comparing equations (5.2-22b) and (5.2-21b), we can see that the dynamic and the dc load lines for this circuit are identical. To sketch i_D and v_{DS} for this circuit, we note that as v_{in} varies between $+1$ and -1 V, v_{GS} will vary between -1 and -3 V, and the operating point will move back and forth between the points indicated by the triangle and the square, respectively, on the load line in Figure 5.2-8c. The values of i_D and v_{DS} may be read off directly from these points to complete the sketches given in Figure 5.2-8d.

The power delivered to R_D is determined as follows. For the case where the transistor operates in the constant-current region, the ac load voltage is approximately $(8.6 - 4.8) = 3.8$ V(p-p), so that the ac power delivered to the load is $(3.8/2)^2/1600 = 2.25$ mW. When V_{DD} is reduced to 3 V the load line moves down into the ohmic region with the i_D and v_{DS} intercepts at 3.75 mA and 3 V, respectively. With the load line in this position in Figure 5.2-8c, v_{ds} is about 0.5 V (p-p) so that the power delivered to the load is 39 μW. Because the ac power delivered to the load for constant-current circuit operation is 50 times larger than that achieved by operating in the ohmic region, it is clear that FETs should be biased in the constant-current region for linear circuit applications. FET operation in the ohmic region is reserved for modulator circuits, such as the one illustrated in Example 5.2-1, and for digital IC applications.

Exercises

5.2-1 Find $r_{DS(ON)}$ for an n-channel JFET where $h = 1$ μm, $a = 100$ μm, $L = 20$ μm, and $N_D = 10^{16}$ cm^{-3}. *Answer* 480 Ω

5.2-2 Find V_p for the transistor specified in Exercise 5.2-1 given that the permittivity of silicon is 10^{-10} F/m. *Answer* -8 V

5.2-3 N_D for Exercise 5.2-1 is adjusted so that $I_{DSS} = 2$ mA. Find the required doping level, N_D. *Answer* 6×10^{15} cm^{-3}

5.2-4 Find the approximate power gain of the circuit in Figure 5.2-8a if V_{GG} is changed to 3 V. *Answer* 2000

5.3   FET DC Q-POINT ANALYSIS

In this section, we develop three different methods for determining the Q point of FET circuits. Specifically, these techniques are:

1. Graphical analysis using the FET i_D versus v_{DS} volt-ampere characteristics
2. Algebraic solution of the problem using the relationship between i_D and v_{GS} given in equation (5.2-20)
3. Graphical solution using the i_D versus v_{GS} graph of equation (5.2-20)

We have already seen how to use standard graphical analysis methods with FETs in the examples of Section 5.2 (especially Example 5.2-2), and as was the case for the bipolar transistor, this is probably the best approach to use initially when learning about field-effect transistors since it works equally well regardless of the state of the transistor. On the other hand, if it is known that the transistor is operating in the constant-current region, methods 2 or 3 are generally much easier to employ, especially if the source of the FET is not connected directly to ground. Furthermore, of all these, method 3 is probably the simplest to use, and as we will see in Section 5.4, is also the one that provides us with the most information regarding the stability of the circuit's Q point. For completeness, however, we will discuss each of these methods in some detail so that you can see the merits of each technique for yourself and can use the solution method that works best for you.

The graphical Q-point analysis technique for an FET will be illustrated by examining the simple circuit given in Figure 5.3-1a. In this circuit, there are three unknowns, V_{GS}, I_D, and V_{DS}. The gate diode is assumed to be reverse biased, so that I_G is zero. To determine the Q point of this circuit, three equations will be needed to evaluate the three unknowns. Two of these equations are obtained by applying Kirchhoff's voltage law to the input and output loops

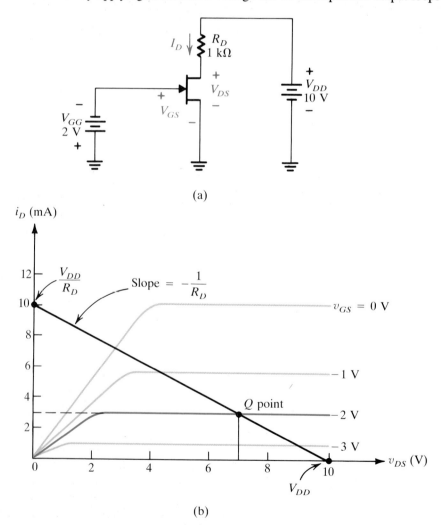

Figure 5.3-1
Graphical analysis of
a fixed-bias JFET circuit.

of the circuit, resulting in the equations

$$V_{GS} = -V_{GG} \tag{5.3-1a}$$

and

$$V_{DD} = I_D R_D + V_{DS} \tag{5.3-1b}$$

The third equation relating these three unknowns is obtained from the transistor volt-ampere characteristics given in Figure 5.3-1b. The simultaneous solution of these three equations is carried out graphically rather than algebraically since the FET volt-ampere characteristic curves are not available in a closed mathematical form. Equation (5.3-1a) indicates that the circuit Q point is located on the $V_{GS} = -2$ V line in Figure 5.3-1b. This curve has been drawn in darker than the other V_{GS} curves to emphasize the fact that the Q point must be somewhere along this curve. Notice that equation (5.3-1a) does not appear as a straight line when graphed on the I_D versus V_{DS} characteristics, despite the fact that this equation does seem to be of the form $y = mx + b$. This is because the variable V_{GS} in equation (5.3-1a) is neither the ordinate nor the abscissa variable on the graph in Figure 5.3-1b.

The exact location of the Q point on the characteristics is found by sketching equation (5.3-1b) on the same set of axes as (5.2-1a) and noting where these curves intersect. This is the simultaneous solution of all three equations: equations (5.3-1a) and (5.3-1b) and the transistor volt-ampere equations. Notice that when equation (5.3-1b) is sketched on the transistor characteristics, it is drawn as a straight line because the equation variables I_D and V_{DS} are the same as the ordinate and abscissa variables, respectively, on the transistor curves. From Figure 5.3-1b, we find that the circuit Q point is located at $V_{GS} = -2$ V, $I_D = 3.0$ mA, and $V_{DS} = 7.0$ V.

In analyzing this particular circuit, the graphical approach appears to work quite nicely. However, this type of circuit is not very popular because, in addition to the main power supply voltage V_{DD}, a second power supply, V_{GG}, is needed to reverse bias the gate diode. This same effect can be achieved by adding a resistor R_S in series with the source lead (Figure 5.3-2a). The addition of this resistor greatly reduces the construction cost because only a single power supply is now needed, but as we shall see shortly, it also significantly complicates the graphical analysis of this circuit.

Applying Kirchhoff's voltage law to the input and output loops as before, the circuit's load-line equations can be written as

$$V_{GS} = V_G - V_S = 0 - I_D R_S = -I_D R_S \tag{5.3-2a}$$

and

$$V_{DD} = I_D R_D + V_{DS} + I_D R_S = I_D (R_D + R_S) + V_{DS} \tag{5.3-2b}$$

Equation (5.3-2b) may readily be sketched on the transistor volt-ampere characteristics as a straight line with a v_{DS} intercept of V_{DD} and an i_D intercept equal to $V_{DD}/(R_D + R_S)$. The graph of equation (5.3-2a) is much more difficult to draw because it is not a straight line since V_{GS} is not an axis variable. To sketch this equation, we will first need to tabulate several points on this curve (Figure 5.3-2c) and then transfer these points onto the transistor characteristic curves. The resulting sketch for equation (5.3-2a) is shown by the dashed curve in Figure 5.3-2b. Notice that the point ($V_{GS} = 3$ V, $I_D = 3$ mA) is located far to the right, out near plus infinity (assuming a slight upward tilt of the $V_{GS} = -3$ V curve on the original transistor V–I characteristics). Furthermore, the sketch

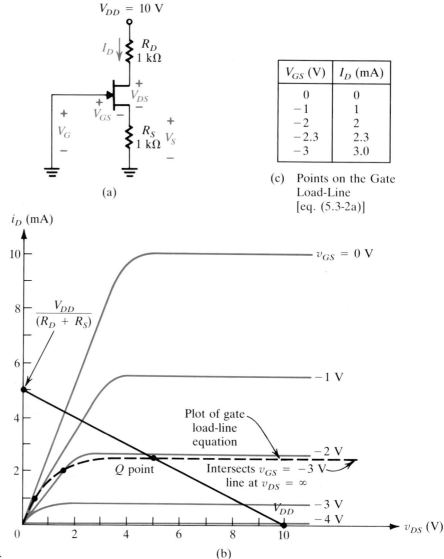

(c) Points on the Gate Load-Line [eq. (5.3-2a)]

V_{GS} (V)	I_D (mA)
0	0
−1	1
−2	2
−2.3	2.3
−3	3.0

Figure 5.3.2
Graphical analysis of a circuit containing a source resistor, R_S.

for equation (5.3-2a) reaches a plateau at a drain current level of about 2.3 mA. The intersection of the two load lines determines the circuit Q point as I_D = 2.3 mA, and V_{DS} = 5.2 V. The corresponding value of V_{GS} is found from equation (5.3-2a) to be −2.3 V. To say the least, this technique is very involved and should really be used only when first learning about FETs to get a physical understanding of the relationship between the various voltages and currents in the circuit. This method is also useful if the state of the transistor is unknown.

If, on the other hand, the transistor in the circuit being examined is known to be in the active (or constant-current) region, then the ordinary FET volt-ampere characteristics, such as those shown in Figure 5.3-2b, are not necessary. Instead, equation (5.2-20) can be used directly and the circuit can be analyzed either algebraically or graphically using this expression. Let's repeat the analysis of the circuit in Figure 5.3-2a, but this time employing an algebraic solution of the problem. As before, equations (5.3-2a) and (5.3-2b) are still valid, and if we combine (5.3-2a) with equation (5.2-20), we may solve this pair of equations for the two unknowns V_{GS} and I_D. These results can then be used to solve equation (5.3-2b) for V_{DS}. Substituting equation (5.3-2a) into (5.2-20), we obtain

$$I_D = I_{DSS}\left(1 - \frac{V_{GS}}{V_p}\right)^2 = I_{DSS}\left(1 + \frac{I_D R_S}{V_p}\right)^2 \tag{5.3-3}$$

For this transistor, $V_p = -4$ V, $I_{DSS} = 10$ mA, and from the circuit $R_S = 1$ kΩ. Inserting these values into this expression, after some algebraic manipulation, we may rewrite it as

$$I_D^2 - 9.6I_D + 16 = 0 \tag{5.3-4}$$

In this equation, because the value substituted for R_S is in kilohms, the answer for I_D will come out in milliamperes. The solution of equation (5.3-4) is obtained by using the standard quadratic formula, and this results in two possible solutions: $I_D = 7.45$ mA and $I_D = 2.14$ mA. By substituting these values for I_D into equation (5.3-2a), the corresponding solutions for V_{GS} are found to be $V_{GS} = -7.45$ V and $V_{GS} = -2.14$ V. Clearly, the $I_D = 7.45$ mA, $V_{GS} = -7.45$ V solution is extraneous because when V_{GS} is less than the pinch-off voltage, I_D is zero and not 7.45 mA. Therefore, the actual Q point for this circuit is $V_{GS} = -2.14$ V, $I_D = 2.14$ mA, and $V_{DS} = 5.7$ V. The solution for V_{DS} was obtained by substituting $I_D = 2.14$ mA into equation (5.3-2b).

A similar solution of this problem is also possible by sketching equations (5.2-20) and (5.3-2a) on the same set of axes to simultaneously solve them graphically for the circuit Q point. This method is illustrated in Figure 5.3-3. Notice that the gate load-line equation passes through the origin for this example and has a slope equal to $-1/R_S$. In addition, you should observe that equation (5.2-20) has a parabolic shape that is tangent to the v_{GS} axis at $v_{GS} = V_p$ and intersects the i_D axis at the point I_{DSS}. Although the parabola can be sketched for all values of v_{GS}, equation (5.3-3) is really a valid description of the FET behavior only for $(V_p < v_{GS} < 0)$.

When v_{GS} is greater than zero, the gate diode conducts and the source and drain currents are no longer equal. In fact, unless some external means is used to limit the gate current, forward biasing the gate diode can actually destroy a JFET. In addition, when v_{GS} falls below the pinch-off voltage, the drain current is cut off and does not increase as equation (5.2-20) would indicate but is actually zero for all gate-to-source voltages below V_p.

The graphical analysis shown in Figure 5.3-3 illustrates that there are two solutions to this problem. These solutions correspond to the two solutions found previously when we solved this problem by means of the quadratic equation formula. Of course, with this graphical picture of the problem in front of us, it is immediately apparent that the leftmost intersection of the two curves (i.e., larger values of I_D and V_{GS}) represents the extraneous solution since this part of the parabolic curve does not correctly represent the behavior of the transistor. The correct graphical solution to this problem occurs at the other intersection point of the gate load-line equation and the parabola at $V_{GS} = -2.2$ V and $I_D = 2.2$ mA. The corresponding drain-to-source voltage is obtained by substituting this answer for I_D into equation (5.3-2b), which yields $V_{DS} = 5.6$ V.

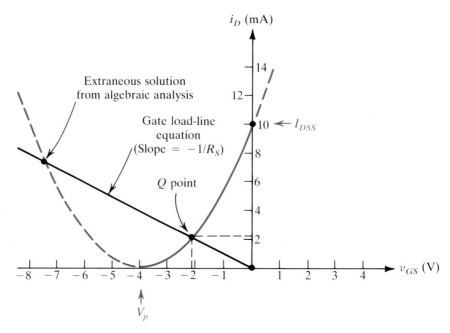

Figure 5.3-3
Graphical analysis of an FET circuit using the equation $I_D = I_{DSS}[1 - (V_{GS}/V_p)]^2$ together with the gate load-line equation.

In the figure: "Extraneous solution from algebraic analysis", "Gate load-line equation (Slope = $-1/R_S$)", "Q point", "i_D (mA)", "$10 \leftarrow I_{DSS}$", "v_{GS} (V)", "V_p".

Of the three Q-point analysis methods that we have investigated, we shall see that the last is the quickest and most useful for finding the Q-point of field-effect transistor circuits. In addition, it will prove to be an excellent way to analyze the Q-point stability of specific FET circuits due to changes in the transistor's volt-ampere characteristics.

EXAMPLE 5.3-1

The transistor shown in the circuit in Figure 5.3-4a has the I_D versus V_{GS} curve given in part (b) of the figure. Determine the Q point of the transistor.

SOLUTION

To find the circuit Q point we begin by developing an expression for V_{GS} in terms of I_D and the other relevant circuit parameters. In general, the gate-to-source voltage may be written as

$$V_{GS} = V_G - V_S \qquad (5.3\text{-}5a)$$

where for this circuit, $$V_S = I_D R_S \qquad (5.3\text{-}5b)$$

The expression for the gate voltage is more difficult to obtain. However, because R_1 and R_2 are much greater than the drain resistor R_D, a significant simplification of this problem is possible. Assuming that the R_1–R_2 feedback network does not load the drain circuit, we may write the drain voltage, V_D, as

$$V_D = V_{DD} - I_D R_D \qquad (5.3\text{-}6)$$

Using this result, the voltage V_G now follows directly as

$$V_G = \frac{R_2}{R_1 + R_2}(V_{DD} - I_D R_D) = \frac{1}{5}(15 - 2I_D) = (3 - 0.4I_D) \quad (5.3\text{-}7)$$

Combining equations (5.3-5a), (5.3-5b), and (5.3-7), the overall expression for the gate load-line equation is

$$V_{GS} = (3 - 0.4I_D) - I_D = 3 - 1.4I_D \qquad (5.3\text{-}8)$$

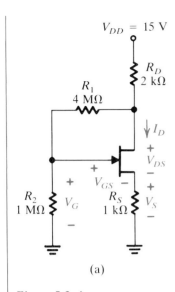

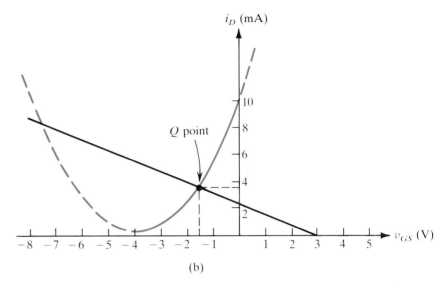

(a)

(b)

Figure 5.3-4
Example 5.3-1.

If we plot this equation on the I_D versus V_{GS} transistor characteristics (Figure 5.3-4b), we find the circuit Q point to be located at $V_G = -1.55$ V, $I_D = 3.6$ mA, and $V_{DS} = 4.2$ V, where the solution for V_{DS} was determined by direct substitution into the equation for the output loop.

Exercises

5.3-1 For the circuit in Figure 5.3-2a, a battery V_{GG} is inserted in the gate circuit between the gate lead and ground. Using the i_D versus v_{GS} characteristic in Figure 5.3-3, find the value of V_{GG} needed to make $I_D = 5$ mA. **Answer** 3.2 V

5.3-2 For the JFET circuit in Figure 5.3-4, R_1 is made adjustable and is connected to V_{DD} instead of to the drain lead of the transistor. If $I_{DSS} = 10$ mA and $V_p = 4$ V, find the value of R_1 for which $I_D = 3$ mA. **Answer** 11.6 MΩ

5.3-3 For the JFET circuit in Figure 5.3-4, V_{DD} is adjusted until $I_D = 5$ mA. If $I_{DSS} = 10$ mA and $V_p = -4$ V, find the required value of V_{DD}. **Answer** 29.15 V

5.4 DC Q-POINT STABILITY OF FET CIRCUITS

In the design of field-effect transistor circuits, we are generally interested in placing the Q point at a specific location on the transistor characteristics. For example, if maximum signal swing in the output is desired, it is necessary that the Q point be located near the center of the dynamic load line, while for minimum power consumption it needs to be positioned closer to transistor cutoff (Figure 5.4-1). Regardless of where the Q point is ultimately placed on the load line, it is usually important that its position be relatively independent of the specific characteristics of the transistor used.

As was the case for bipolar transistors, the characteristics of FETs are specified in terms of their typical and their maximum and minimum values. For example, the typical values of I_{DSS} and V_p for a particular n-channel FET might be 20 mA and -4 V, respectively, while the variation of these parame-

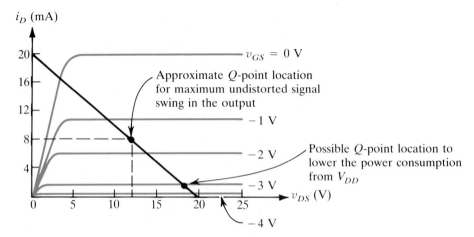

Figure 5.4-1

Positioning of the Q point on the load line.

ters from device to device over a specified temperature range could be

$$10 \text{ mA} \le I_{DSS} \le 30 \text{ mA} \tag{5.4-1a}$$

and

$$-6 \text{ V} \le V_p \le -2 \text{ V} \tag{5.4-1b}$$

Furthermore, these maximum or minimum values need not occur together. For example, a particular transistor might have an $I_{DSS} = 30$ mA and a $V_p = -6$ V or any other combination of I_{DSS} and V_p within the limits indicated in the equations above. As a result, when designing a circuit containing this type of transistor, the i_D versus v_{GS} characteristic curve for the specific transistor being used could actually be located anywhere between the maximum and minimum curves shown in Figure 5.4-2.

The basic problem created by this variation of I_{DSS} and V_p can be illustrated by examining the transistor characteristic curves shown in Figure 5.4-1. For the purpose of this discussion, let's assume that we wish to design the circuit in Figure 5.4-3a so that its Q point is located at the center of the load line. Using the nominal transistor volt-ampere characteristics given in Figure 5.4-1, we

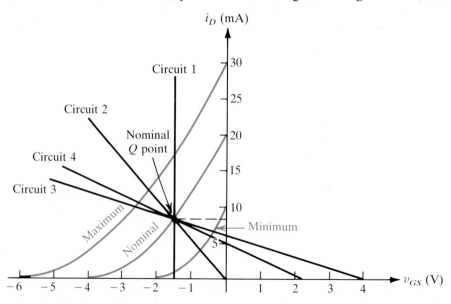

Figure 5.4-2

Load-line position for each of the circuits given in Figure 5.4-3.

Chapter 5 / The Field-Effect Transistor

can see that this requires a value for V_{GG} of about -1.5 V. When the transistor parameter variations are considered (see Figure 5.4-2), we find that there can be a considerable shift in the Q-point values of I_D or V_{DS} for this circuit. However, during all these parameter variations, V_{GS} remains constant at -1.5 V. From these results it should be clear that if we want to maintain the Q point at a particular location on the load line, then, as with a similar situation in the BJT, it is the quiescent drain current, I_D, and not V_{GS}, which must remain constant.

To develop an understanding of how best to design field-effect transistor circuits that have stable Q points, we will examine the series of FET circuits illustrated in Figure 5.4-3. Each of these circuits has approximately the same nominal Q-point position when we use values of $I_{DSS} = 20$ mA and $V_p = -4$ V to calculate I_D and V_{GS}. However, the behavior of these circuits to changes in the transistor's parameters are dramatically different.

Following the procedures developed in Section 5.3, the gate load-line equations for circuits 1 through 4, respectively, may be written as

Circuit 1: $V_{GS} = -V_{GG} = -1.5$ V   (5.4-2a)

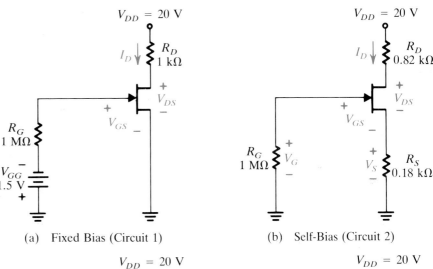

(a) Fixed Bias (Circuit 1)

(b) Self-Bias (Circuit 2)

(c) Input Divider with
     Self-Bias (Circuit 3)

(d) Input Feedback Divider
     with Self-Bias (Circuit 4)

Figure 5.4-3
Different FET biasing schemes.

5.4 / DC Q-Point Stability of FET Circuits

Circuit 2: $\quad V_{GS} = -I_D R_S = -0.18 I_D$ $\hspace{3cm}$ (5.4-2b)

Circuit 3: $\quad V_{GS} = \dfrac{R_2}{R_1 + R_2} V_{DD} - I_D R_S = 4 - 0.66 I_D$ $\hspace{1.5cm}$ (5.4-2c)

Circuit 4: $\quad V_{GS} = \dfrac{R_2}{R_1 + R_2}\left(V_{DD} - I_D R_D\right) - I_D R_S$

$$= 2.22 - 0.45 I_D \hspace{5cm} \text{(5.4-2d)}$$

Each of these load-line equations is sketched on the transistor characteristics in Figure 5.4-2, and the resulting Q-point shift data are shown in Table 5.4-1. Clearly, from the data in this table, circuit 1, the fixed-bias circuit, has the poorest Q-point stability, while circuits 3 and 4 exhibit the smallest Q-point shift with changes in the transistor parameters. As was the case for the fixed-bias BJT biasing scheme, the FET fixed-bias circuit has a very poor Q-point stability because it maintains the wrong parameter constant when the transistor characteristics change. Specifically, in circuit 1, V_{GS} is always held constant at -1.5 V regardless of the transistor parameter values, while I_D can change all over the place. The desired situation is exactly opposite from this.

If I_D is to remain constant as we change from the nominal to the maximum or the minimum transistor characteristic curves, then clearly in Figure 5.4-2 the load line of the circuit needs to be as horizontal as possible. Because circuit 1 has a vertical load line, it is unsatisfactory. In addition, circuit 2 has limited stability since its load line must pass through the origin. Circuits 3 and 4, on the other hand, have two degrees of freedom, so that by properly choosing the components, both the slope and the v_{GS} intercept can be placed at (almost) any desired location.

The excellent Q-point stability offered by circuits 3 and 4 is not obtained without cost, however. Both circuits use rather large values of source resistors, so that the ac gain of these circuits is significantly reduced. The best way to understand why increases in R_S decrease the circuit gain is to note that the ac output voltage v_d at the drain is proportional to the ac gate-to-source voltage v_{gs}. When R_S is increased, this increases the voltage drop v_s from source to ground

---

5.4-1 Q-Point Shift Data When V_p and I_{DSS} Vary for Each of the Circuits Given in Figure 5.4-3

Circuit	Nominal I_D (mA)	Maximum I_D (mA)	Minimum I_D (mA)	Percent Shift of Maximum from Nominal
1. Fixed bias	8.22	17.2	1	109
2. Self-bias	8.22	12.0	4.0	46
3. Self-bias with supply divider at gate	8.22	10.0	6.6	22
4. Self-bias with feedback divider at gate	8.22	10.5	6.0	28

and reduces v_{gs} because $v_{gs} = v_g - v_s$. As a result, increasing the source resistor R_S lowers the amplifier voltage gain between the gate and the drain. This subject is covered more fully in Section 5.6.

Circuit 4 has a further loss in gain, owing to the presence of the drain-to-gate feedback resistor R_1 (Section 7.1). Of course, the ac gain of circuits 2, 3, and 4 can be increased substantially by bypassing R_S to ground with a suitable capacitor. But even if this is done, although the ac gain will be increased, the signal swing available from circuit 1 will never be matched by circuits 2, 3, or 4 unless V_{DD} is increased.

EXAMPLE 5.4-1

The transistor in the circuit shown in Figure 5.4-4a has the characteristics given in part (b) of the figure and is to be designed for a nominal quiescent current of 2 mA. Select R_1 and R_S so that the circuit's ac gain is as large as possible while maintaining the Q-point shift from its nominal position at less than 15%.

SOLUTION

If the circuit gain is to be maximized, the value chosen for the source resistor, R_S, must be as small as possible while still satisfying the Q-point stability criteria. Furthermore, if the nominal quiescent current is to be 2 mA, the maximum and minimum allowed drain currents are $2(1 + 0.15) = 2.3$ mA and $2(1 - 0.15) = 1.7$ mA, respectively. Placing these points on the appropriate transistor characteristics, we now draw in the steepest possible straight line passing through the nominal Q point that does not exceed the allowed maximum and minimum drain currents. This line represents the load line for this circuit, which satisfies the stability criteria and has the smallest allowed value of R_S (or the biggest slope) possible.

The load-line equation for the circuit in Figure 5.4-4a is

Figure 5.4-4
Example 5.4-1.

$$V_{GS} = V_G - V_S = \frac{R_2}{R_1 + R_2} V_{DD} - I_D R_S \qquad (5.4\text{-}3)$$

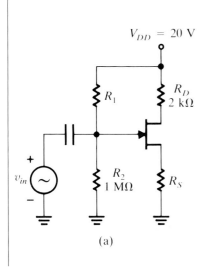

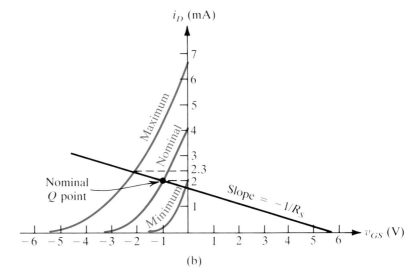

(a)

(b)

The slope of this line is $-1/R_S$, so that from the sketch

$$R_S = \frac{5.8 \text{ V} - (-1 \text{ V})}{(2 \text{ mA}) - (0 \text{ mA})} = 3.4 \text{ k}\Omega \qquad (5.4\text{-}4)$$

Furthermore, the V_{GS} intercept of this line may be used to calculate the value of R_1 as follows:

$$\frac{R_2}{R_1 + R_2} V_{DD} = \frac{1}{R_1 + 1}(20) = 5.8 \qquad (5.4\text{-}5a)$$

or
$$R_1 = \frac{20}{5.8} - 1 = 2.5 \text{ M}\Omega \qquad (5.4\text{-}5b)$$

Exercises

5.4-1 For the circuit in Figure 5.4-3a, $I_{DSS} = 20$ mA and $V_p = -4$ V.
a. Find the quiescent drain current.
b. Determine the percent change in I_D if I_{DSS} increases by 50%.
Answer (a) 7 mA, (b) 50%

5.4-2 Repeat Exercise 5.4-1 for the circuit in Figure 5.4-3b. *Answer* (a) 8 mA, (b) 18%

5.4-3 Repeat Exercise 5.4-1 for the circuit given in Figure 5.4-3d if $|V_p|$ also increases by 50%. *Answer* (a) 7.8 mA, (b) 28%

5.4-4 For the circuit given in Figure 5.4-4, $R_1 = 4$ MΩ and $R_S = 3$ kΩ. Find the worst-case percent shift in I_D from its nominal value. *Answer* 24%

5.5 METAL-OXIDE-SEMICONDUCTOR FIELD-EFFECT TRANSISTOR

In this section we discuss a second type of FET known as a MOSFET, or metal-oxide-semiconductor field-effect transistor. Like the JFET, the MOSFET is available in both n- and p-channel varieties, and each of these two categories may be further subdivided into what are known as enhancement-mode and depletion-mode devices. The depletion-type MOSFET, like the ordinary JFET (which is also a depletion-mode transistor), is one that is said to be normally ON since current can flow between the drain and the source, even when V_{GS} equals zero. To turn this type of transistor OFF, a nonzero gate-to-source voltage is required. The depletion-mode transistor gets its name from the fact that a gate voltage needs to be applied in order to "deplete" the channel of free carriers and turn the transistor OFF. For an enhancement-mode MOSFET, the situation is exactly the opposite. It is a device that is normally OFF with $V_{GS} = 0$ and is turned ON by the application of a nonzero voltage between the gate and the source.

A simplified picture of an n-channel enhancement-style MOSFET is illustrated in Figure 5.5-1a. With no gate-to-source voltage applied, the channel region between the drain and the source is filled with holes and behaves like ordinary p-type material. As a result, for $V_{GS} = 0$, the MOSFET equivalent circuit between the drain and the source looks like a pair of back-to-back diodes (Figure 5.5-1b). Since one of these diodes is always reverse biased re-

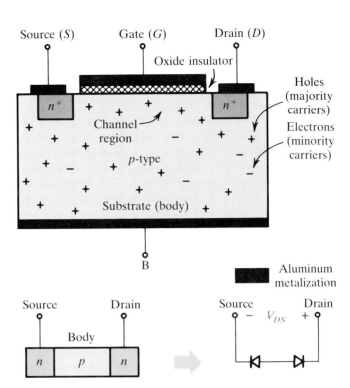

Figure 5.5-1

n-Channel enhancement MOSFET.

gardless of the polarity of V_{DS}, the drain current is always nearly zero. For this reason, an enhancement-type MOSFET is said to be a normally OFF device.

Unlike the JFET, there is no diode between the gate and the channel of a MOSFET. Instead, because the gate is separated from the channel by a thin oxide insulating layer, we may represent the input of a MOSFET as a capacitance connected between the gate and the channel. As a result, the dc input impedance of a MOSFET is theoretically infinite. As a practical matter, however, due to surface leakage effects, it is typically about 10^{10} to 10^{14} Ω.

When a positive voltage is applied to the gate of a MOSFET, the gate capacitance charges (Figure 5.5-2a), and this positive charge tends to push the holes under the gate electrode out of the channel region and down toward the bottom of the substrate while at the same time attracting the minority-carrier electrons into the channel region. As the gate voltage is increased further, more holes are driven out of the channel area, and more electrons are attracted in until for some value of V_{GS}, known as the threshold voltage (V_T), there are more free electrons than free holes in the channel region. At this point the channel region now behaves as though it is n-type material, and we say that the applied positive gate voltage has created or induced an n-channel between the drain and the source (Figure 5.5-2b). As a result, we now have a device whose equivalent circuit looks very much like that of a JFET operating in the ohmic region (Figure 5.5-2c). By making V_{GS} still more positive, additional electrons are attracted into the channel region, both making it wider and increasing the free electron density in the channel. Both of these factors tend to lower the drain-to-source resistance, r_{DS}, and to increase the drain current that flows for a given value of V_{DS}.

The volt-ampere characteristics for a typical n-channel enhancement

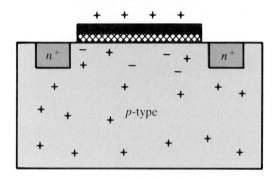

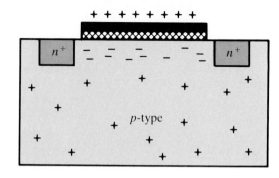

(a) Charge Distribution in Body for $0 < V_{GS} < V_T$

(b) Charge Distribution in Body for $V_{GS} > V_T$

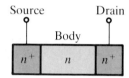

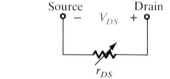

Figure 5.5-2

(c) Equivalent Circuit for $V_{GS} > V_T$ in Ohmic Region

MOSFET are given in Figure 5.5-3, and from these data we can see that the threshold voltage, V_T, for this particular transistor is about 3 V. The behavior of this MOSFET in the ohmic region follows the description just given. However, its operation in the constant-current region, while similar to that of the JFET, needs a little more explanation. As shown in Figure 5.5-4, nonzero values of V_{DS} cause the enhanced channel to develop a nonuniform shape, as was the case for the JFET examined in Section 5.2 (Figure 5.2-4). This occurs because with $V_{DS} = 5$ V, for example (Figure 5.5-4a), there is a nonuniform voltage distribution between the source and the drain across the oxide layer. Near the source the voltage V_1 is about 10 V, so that a strong n-type channel is induced in this region. If the voltage is assumed for simplicity to divide nearly evenly across the channel resistors R_A and R_B, the V_A will be about $\frac{1}{2}V_{DS}$ or 2.5 V. As a result, the voltage in the middle of the channel, V_2, is approximately $(10 - 2.5)$, or 7.5 V, while that near the drain end of the transistor, V_3, is $(10 - 5)$, or 5 V. Since all these voltages are well beyond the threshold voltage for this transistor, no severe constriction occurs in the channel. Notice also that a depletion region exists between the substrate and the channel. Because the substrate is usually maintained at the same potential as the source, the width of this depletion region is nonuniform, being widest near the drain, where the reverse-bias voltage between the channel and the substrate is the largest, and narrowest at the source end of the transistor, where this voltage is the smallest.

When the value of V_{DS} increases to 8 V, for example, the voltage drop V_3 at the drain end of the transistor is now $(10 - 8)$, or 2 V. Because this voltage is below V_T, the channel at this end of the transistor is pinched off. As a result, the effective resistance of the right-hand side of the channel increases greatly and there is a relatively large voltage drop (1 V in the figure) across this short length of the channel. The remaining portion of V_{DS}, 7 V in Figure 5.5-4b, ap-

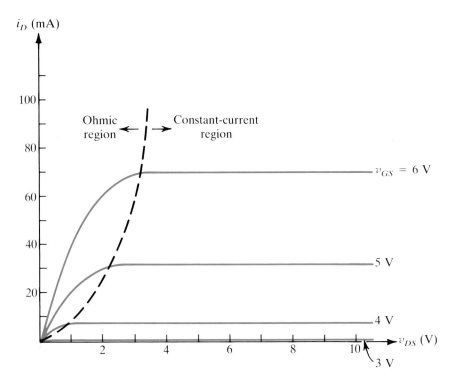

Figure 5.5-3
Volt-ampere characteristics for an n-channel enhancement MOSFET.

pears across the main part of the channel. Further increases in V_{DS}, as with the JFET transistor, do not greatly alter the shape or the resistance R_Δ of the main portion of the channel (Figure 5.5-4c). As a result, because the voltage drop across R_Δ stays at 7 V as long as $V_{GS} = 10$ V, the current flow through the channel will remain relatively constant (at 7 V/R_Δ) once the drain-to-source voltage increases beyond

$$V_{DS} = V_{GS} - V_T \qquad (5.5\text{-}1)$$

As with the JFET, this portion of the MOSFET volt-ampere characteristics (see Figure 5.5-2) is known as the constant-current region since increases in V_{DS} beyond the value specified in equation (5.5-1) produce only small increases in drain current.

To describe the behavior of the MOSFET quantitatively, consider the transistor cross section shown in Figure 5.5-5, and for later use let us define the gate capacitance per unit area, C_{ox}, as

$$C_{ox} = \frac{\epsilon}{t_{ox}} \qquad (5.5\text{-}2)$$

where ϵ is the permittivity of the oxide layer and t_{ox} is its thickness. When a small positive voltage is applied to the gate of the transistor, this capacitance charges, and the positive charges on the gate produce a field that repels the free holes under the gate, exposing the ionized acceptor atoms and producing a depletion region near the surface.

As the gate voltage continues to increase positively, electrons, which are the minority carriers in the *p*-type material, are attracted in from the substrate and a thin channel of *n*-type material is formed at the surface. This channel is also called an inversion layer because it represents a conversion of the original

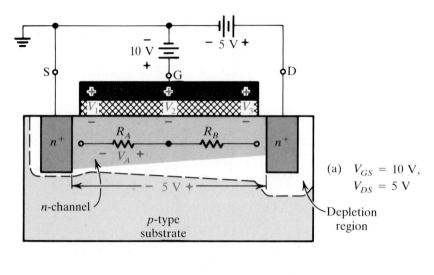

(a) $V_{GS} = 10$ V,
$V_{DS} = 5$ V

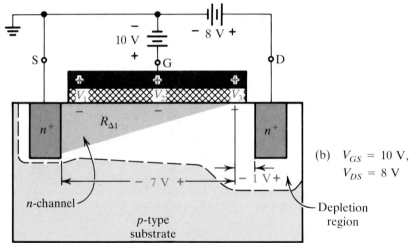

(b) $V_{GS} = 10$ V,
$V_{DS} = 8$ V

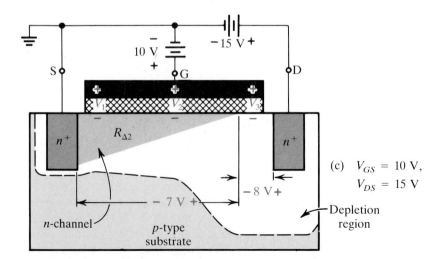

(c) $V_{GS} = 10$ V,
$V_{DS} = 15$ V

Figure 5.5-4
Channel profile for V_{GS} fixed
and V_{DS} varying.

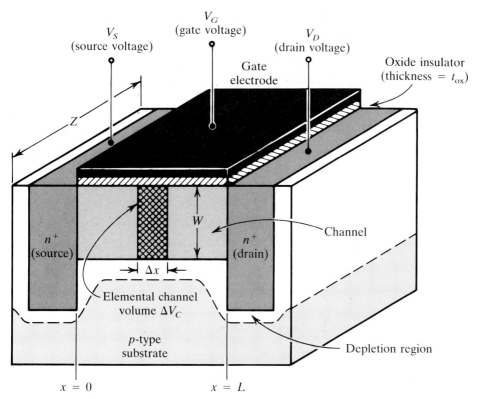

V_S
(source voltage)

V_G
(gate voltage)

V_D
(drain voltage)

Gate
electrode

Oxide insulator
(thickness $= t_{ox}$)

Z

W

n^+
(source)

n^+
(drain)

Channel

Δx

Elemental channel
volume ΔV_C

Depletion region

p-type
substrate

$x = 0$

$x = L$

Figure 5.5-5
*Cross-sectional view of
a MOSFET transistor.*

p-type material into n-type. The voltage needed to produce this inversion is called the threshold voltage. Once the channel is formed, further increases in gate voltage now tend simply to increase the carrier concentration in this region as C_{ox} charges, and because $Q = CV$ for a capacitor, to a first-order approximation the total negative charge/unit area contained in the inversion layer at any point x in the channel may be written as

$$Q(x) = C_{ox}[V_{GS} - V_T - V(x)] \tag{5.5-3}$$

where, as for the JFET, $V(x)$ is equal to the voltage in the channel at any point x.

The resistance of the elemental volume ΔV_C shown in Figure 5.5-5 may be written as

$$\Delta R(x) = \frac{\Delta V(x)}{I_D} = \frac{\rho \Delta x}{WZ} = \frac{\Delta x}{WZ\mu_n q_n} = \frac{\Delta x}{Z\mu_n Q(x)} \tag{5.5-4}$$

Substituting equation (5.5-3) into this expression and separating the variables, we have

$$\mu_n C_{ox} Z[V_{GS} - V_T - V(x)]\Delta V(x) = I_D \Delta x \tag{5.5-5}$$

Taking the limit of this expression as Δx approaches zero and integrating both sides of this equation from $x = 0$ [where $V(x) = 0$] to $x = L$ [where $V(x) = V_{DS}$], we obtain

$$\mu_n C_{ox} Z \int_0^{V_{DS}} [(V_{GS} - V_T) - V]dV = \int_0^L I_D dx \tag{5.5-6}$$

Evaluating these integrals and solving for I_D, we find that

$$I_D = \frac{\mu_n C_{ox} Z}{L} \left[(V_{GS} - V_T)V_{DS} - \frac{V_{DS}^2}{2} \right] \qquad (5.5\text{-}7)$$

This result describes the behavior of the MOSFET in the ohmic region for small values of V_{DS}, and is valid as long as the channel does not pinch off, or in other words, as long as

$$V_{DS} < V_{GS} - V_T \qquad (5.5\text{-}8)$$

Once V_{DS} exceeds this value, the transistor enters the constant-current region, where I_D now remains fixed as V_{DS} varies. The positions on the transistor volt-ampere characteristics where the transition from the ohmic to the constant-current region occur are found by substituting $(V_{GS} - V_T)$ for V_{DS} in equation (5.5-7). Because these results are also valid for all V_{DS} exceeding $(V_{GS} - V_T)$, the expression

$$I_D = \frac{\mu_n C_{ox} Z}{L} \left[(V_{GS} - V_T)^2 - \frac{(V_{GS} - V_T)^2}{2} \right] = \frac{\mu_n C_{ox} Z}{2L} (V_{GS} - V_T)^2 \quad (5.5\text{-}9)$$

is also a correct representation of the transistor's behavior in the constant-current region.

To emphasize the similarity between these results and those developed in Section 5.2 for the JFET, it is useful to define

$$I_{D2T} = \frac{\mu_n C_{ox} Z V_T^2}{2L} \qquad (5.5\text{-}10a)$$

so that we may rewrite equation (5.5-9) as

$$I_D = I_{D2T} \left(\frac{V_{GS}}{V_T} - 1 \right)^2 \qquad (5.5\text{-}10b)$$

In this expression it is apparent that the constant I_{D2T} is simply equal to the current that flows in the MOSFET when V_{GS} is set equal to $2V_T$. Equation (5.5-10b) is sketched in Figure 5.5-6, and as with the JFET, this graph of i_D versus v_{GS} will prove extremely useful for determining the Q point of various MOSFET circuit configurations.

The circuit symbols for both n- and p-channel enhancement-style MOSFET transistors are given in Figures 5.5-7a and b, respectively. The dashed line between the drain and the source in these symbols is used to indicate that there is no connection between the drain and source when $V_{GS} = 0$, or in other words, that the transistor is normally OFF. The gate portion of the symbol shows that the connection between the gate and the channel of the transistor is capacitive, while the arrow on the substrate lead indicates the direction of the current flow

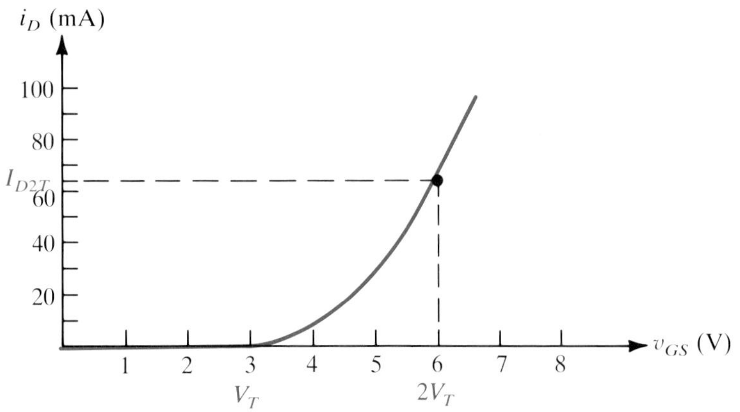

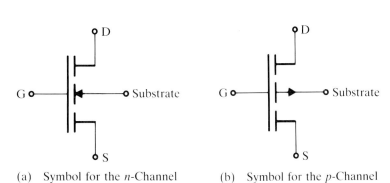

Figure 5.5-6
i_D *versus* v_{GS} *characteristic for the MOSFET in Figure 5.5-3 when it is operating in the constant-current region.*

Figure 5.5-7

(a) Symbol for the n-Channel Enhancement MOSFET

(b) Symbol for the p-Channel Enhancement MOSFET

in the parasitic *pn* junction diodes that exist between the body of the transistor and the drain and source leads.

For normal operation, in an n-channel MOSFET the substrate is connected to the most negative part of the circuit (usually, the source lead of the transistor) in order to ensure that substrate diodes remain reverse biased. This is especially important when there is more than one transistor on the same substrate, since if the substrate diodes become forward biased, the electrical isolation between the individual transistors is lost. Sometimes the substrate is connected to a separate power supply that is used to adjust the threshold voltage of the transistors, but most commonly it is connected directly to the source lead of the transistor.

In addition to the enhancement-style MOSFET, it is also possible to construct MOS transistors that have characteristics which are very similar to those of the JFET. These transistors are called depletion-type MOSFETs because they are normally ON and require the application of a specific gate-to-source voltage to deplete the channel of free carriers and turn them OFF. The conventional JFET is also an example of a depletion-style field-effect transistor.

The physical structure of an n-channel depletion MOSFET is illustrated in Figure 5.5-8. Its construction is quite similar to that of an enhancement MOSFET except that it has a built-in n-type channel in place between the drain and the source. As a result, even with no voltage applied to the gate of the transistor, a conduction path exists between the drain and the source. Therefore, this type of transistor, like the JFET, may normally be said to be ON

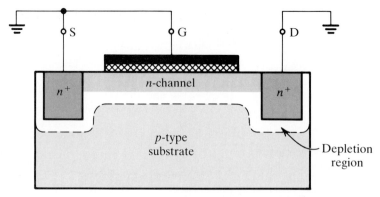

(a) Charge Distribution in a Depletion MOSFET with $V_{GS} = 0$

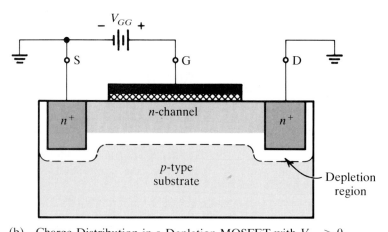

Figure 5.5-8

(b) Charge Distribution in a Depletion MOSFET with $V_{GS} > 0$

since the application of a voltage between the drain and the source will cause a drain current to flow even with $V_{GS} = 0$.

Making the gate voltage negative with respect to the source drives the free electrons out of the channel and attracts free holes from the substrate. In fact, if this voltage is made negative enough, the n-type channel will disappear and the entire region between the drain and the source will appear to be p-type material. When this occurs, the FET equivalent circuit between the drain and the source will look like a pair of back-to-back diodes (Figure 5.5-1b). For this case, no drain current can flow, and as with the JFET, we say that the transistor is pinched off.

The volt-ampere characteristics of a typical n-channel depletion MOSFET are illustrated in Figure 5.5-9. For values of V_{GS} less than or equal to zero, the characteristics are similar to those of the JFET transistor, but the curves for $V_{GS} > 0$ are very different. If V_{GS} is made greater than zero in an n-channel JFET, the gate diode forward biases and a large gate current can flow unless the external circuit limits this current. As a result, the JFET should not ordinarily be operated in a mode that forward biases the gate diode. No such problem exists for the MOSFET.

When the gate voltage on the depletion MOSFET shown in Figure 5.5-8 increases above zero, the gate capacitance charges positively (Figure 5.5-8b). This attracts additional electrons (minority carriers from the substrate) into the channel and increases the channel width and electron density in the channel region. Both of these factors decrease the channel resistance and increase the drain current that flows for a given value of V_{DS}. For this reason, the characteristic curves for positive gate-to-source voltages increase above the $V_{GS} = 0$ curve, as shown in Figure 5.5-9a.

The i_D versus v_{GS} curve for this transistor (when it is operating in the constant-current region) is shown in Figure 5.5-9b. As might be expected, its shape for values of v_{GS} less than or equal to zero is nearly identical to that of the ordinary JFET. The major difference between this curve and that for the JFET, given in Figure 5.3-3 for example, is that the MOSFET curve extends into the $v_{GS} > 0$ region because it can be operated with positive gate voltages. When a depletion-style MOSFET is operating in the constant-current region,

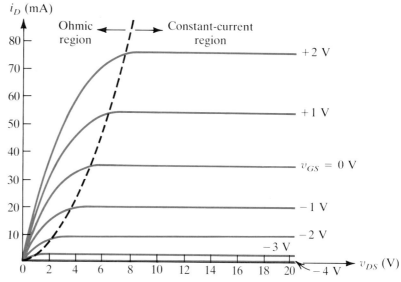

(a)   Volt-Ampere Characteristics for an n-Channel Depletion MOSFET

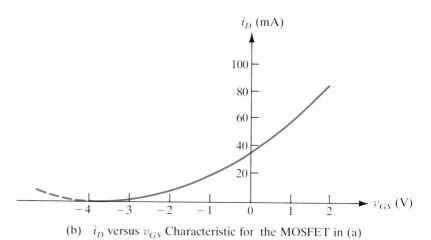

Figure 5.5-9

(b) i_D versus v_{GS} Characteristic for the MOSFET in (a)

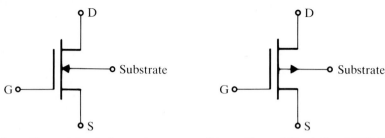

Figure 5.5-10
Schematic symbols for
a depletion MOSFET.

(a)   *n*-Channel Depletion MOSFET   (b)   *p*-Channel Depletion MOSFET

the relationship between I_D and V_{GS} also follows equation (5.2-20), which was derived for the JFET.

Figure 5.5-10 gives the schematic symbol for the depletion-style MOSFET. It is similar to that for the enhancement MOSFET except that there is a continuous line drawn between the drain and the source to indicate that a conduction path ordinarily exists with $v_{GS} = 0$.

EXAMPLE 5.5-1

The enhancement MOSFET shown in Figure 5.5-11 has a V_T equal to 2 V. In addition, the transistor behavior in the constant-current region follows equation (5.5-10b) with $I_{D2T} = 32$ mA.

a. Using the algebraic solution method, select R_D and R_S to place the transistor Q point at $V_{DS} = 4$ V and $I_D = 2$ mA.
b. Repeat part (a), but this time use the i_D versus v_{GS} curve given in Figure 5.5-11b.

SOLUTION

By direct substitution into equation (5.5-10b) we find that

Figure 5.5-11
Example 5.5-1.

$$2 = 32\left(\frac{V_{GS}}{2} - 1\right)^2 \qquad (5.4\text{-}11a)$$

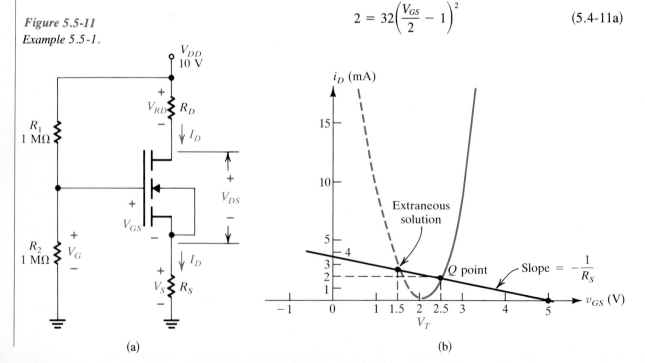

(a)

(b)

or
$$\frac{V_{GS}}{2} - 1 = \pm\frac{1}{4} \tag{5.5-11b}$$

This results in two possible solutions for V_{GS}, namely, $V_{GS} = 1.5$ V and $V_{GS} = 2.5$ V. The smaller answer, that is, $V_{GS} = 1.5$ V, represents the extraneous solution to the problem because the transistor is cut off for all values of V_{GS} below 2 V. Therefore, the required solution for V_{GS} is 2.5 V.

For this circuit, the expression for the gate-to-source voltage may be written as

$$V_{GS} = V_G - V_S = \frac{R_2}{R_1 + R_2}V_{DD} - I_D R_S \tag{5.5-12}$$

so that the required value of source resistance is

$$R_S = \frac{5 - 2.5}{2} = 1.25 \text{ k}\Omega \tag{5.5-13}$$

The drain resistance is obtained by noting that Kirchhoff's voltage law for the output loop is

$$V_{DD} = I_D R_D + V_{DS} + I_D R_S \tag{5.5-14}$$

so that
$$R_D = \frac{10 - 4 - 2.5}{2} = 1.75 \text{ k}\Omega \tag{5.5-15}$$

To begin the solution of this problem by graphical means, we must first develop the gate load-line equation for this problem. From the circuit in Figure 5.5-11a,

$$V_{GS} = V_G - V_S = \frac{R_2}{R_1 + R_2}V_{DD} - I_D R_S = 5 - I_D R_S \tag{5.5-16}$$

To sketch this equation on the transistor characteristic given in Figure 5.5-11b, we note that the V_{GS} intercept of this equation is at $V_{GS} = 5$ V, and also that the load line must also pass through the point $I_D = 2$ mA on the curve.

As illustrated, the load line intersects the transistor curve in two places. These correspond to the two solutions obtained during the algebraic analysis of this problem. Of course, the intersection of the load line with the dashed portion of the transistor curve represents the extraneous solution to the problem because the actual transistor characteristic is zero for $V_{GS} < V_T$. The actual Q point is at the intersection of the load line with the solid portion of the curve. Here too, as was the case for the JFET problems examined in Section 5.3, of the two answers obtained either algebraically or graphically, the correct one is always equal to the more positive of the two solutions. The Q point for this problem is at $V_{GS} = 2.5$ V and $I_D = 2.0$ mA, and the required value of R_S is found from the slope of the load line as

$$R_S = \frac{-(5 - 2.5)V}{(0 - 2.0) \text{ mA}} = 1.25 \text{ k}\Omega \tag{5.5-17}$$

Exercises

5.5-1 Find the gate capacitance of the MOSFET shown in Figure 5.5-5 if $L = 5 \ \mu m$, $Z = 2 \ \mu m$, and $t_{ox} = 1000$ Å. The permittivity of silicon is 10^{-10} F/m. **_Answer_** 0.01 pF

5.5-2 Using the answer in Exercise 5.5-1 and referring to equation (5.5-7), find the channel resistance of the transistor shown in Figure 5.5-5 if V_{DS} is small, $V_T = 5$ V, and $V_{GS} = 7$ V. **_Answer_** $r_{DS} = 9.6$ kΩ

5.5-3 Find I_{D2T} for the transistor in Exercise 5.5-1. **_Answer_** 0.65 mA

5.6   SMALL-SIGNAL MODELS FOR THE FET

A small-signal ac model for the field-effect transistor can be developed by several different methods. One approach involves a simple expansion of the defining equation for the FET given in (5.2-20). A second technique uses a power series expansion of the drain current about the Q point, and parallels the method that we employed when we developed the hybrid parameter model for the BJT. We use the first of these approaches to develop a simplified model for the FET and employ the second technique afterward to help us determine how to include some of the second-order effects ignored in developing the first model.

5.6-1   Simplified Small-Signal Model for the Field-Effect Transistor

When the FET is biased in the constant-current region, to a first-order approximation, as per equation (5.2-20), we may consider that the drain current is only a function of the gate-to-source voltage. To the extent that this assumption is valid, and following the same approach that we used in Section 4.3 for the bipolar junction transistor, we may express the ac portion of the drain current as

$$i_d = \left. \frac{di_D}{dv_{GS}} \right|_{Q \text{ pt}} v_{gs} \tag{5.6-1}$$

The derivative term in this expression is known as the transistor's g_m, or transconductance, and it describes how much the drain current changes for a given change in gate-to-source voltage. Following equation (5.2-20), this derivative may be evaluated as

$$\left. \frac{di_D}{dv_{GS}} \right|_{Q \text{ pt}} = \left. \frac{d}{dv_{GS}} \left[I_{DSS} \left(1 - \frac{v_{GS}}{V_p} \right)^2 \right] \right|_{Q \text{ pt}} = \left. -\frac{2I_{DSS}}{V_p} \left(1 - \frac{v_{GS}}{V_p} \right) \right|_{Q \text{ pt}} \tag{5.6-2a}$$

so that the transconductance is

$$g_m = -\frac{2I_{DSS}}{V_p} \left(1 - \frac{V_{GS}}{V_p} \right) \tag{5.6-2b}$$

The units of g_m are siemens. Clearly, from equation (5.6-2b) the transconductance of the FET is strongly dependent on the position of the Q point. This relationship can best be understood by observing that the g_m of the transistor is a measure of the slope of the i_D versus v_{GS} curve (Figure 5.6-1a). As a result, near $v_{GS} = 0$, where the slope is the steepest, the g_m is a maximum, whereas near pinch-off, where the slope is nearly horizontal, the g_m approaches zero (Figure 5.6-1b).

The transconductance expression given in equation (5.6-2b) is valid for the ordinary JFET transistor and also for the depletion-style MOSFET. For the enhancement MOSFET the results are quite similar; however, because the drain current for this type of transistor is described by equation (5.5-10b), the g_m for the enhancement MOSFET is given by

$$g_m = \frac{d}{dv_{GS}}\left[I_{D2T}\left(\frac{v_{GS}}{V_T} - 1 \right)^2 \right] = \frac{2I_{D2T}}{V_T}\left(\frac{V_{GS}}{V_T} - 1 \right) \qquad (5.6\text{-}3)$$

The variation of this transistor's g_m with Q-point position has the same shape as that illustrated in Figure 5.6-1b, except that it is shifted to the right and has a v_{GS} intercept at V_T volts.

Using these results, it is apparent that the small-signal models for all three types of FETs are essentially the same. The input portion of the transistor, to a good approximation, may be considered an open circuit. In addition, following equation (5.6-1), we can see that the drain portion of the circuit may be represented by a voltage-controlled current source because the ac drain current depends on a voltage, v_{gs}, that is not a part of the output circuit. This model is illustrated in Figure 5.6-2d.

Figure 5.6-1

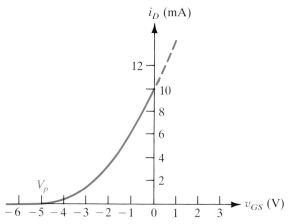

(a) Variation of Drain Current as a Function of v_{GS} for Depletion FETs ($I_{DSS} = 10$ mA, $V_p = -5$ V)

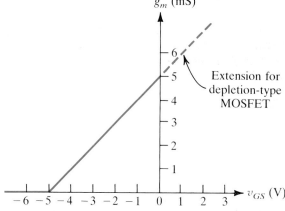

(b) Variation of Transistor Transconductance as a Function of v_{GS} for a Depletion FET

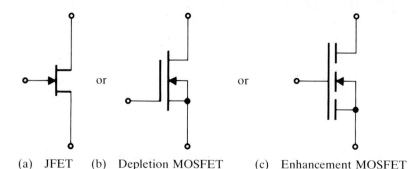

(a) JFET   (b) Depletion MOSFET   (c) Enhancement MOSFET

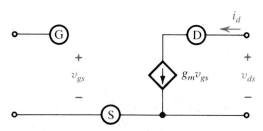

(d) Simplified Small-Signal Model: Valid for all Three Types of FETs [*Note: g_m is defined by eq. (5.6-2b) for the JFET and depletion MOSFET, and by eq. (5.6-3) for the enhancement MOSFET.*]

Figure 5.6-2

EXAMPLE 5.6-1

The JFET illustrated in Figure 5.6-3a has a pinch-off voltage of -4 V and an I_{DSS} equal to 10 mA. A dc analysis of the circuit discloses that the quiescent drain current is 5 mA.

a. Determine the g_m of the transistor at this Q point.
b. Draw the small-signal equivalent circuit.
c. Sketch i_D, v_S, v_{GS}, v_{DS}, and v_o given that the applied input signal v_{in} is 0.3 sin ωt volts.

SOLUTION

Using the fact that the quiescent drain current is 5 mA, we may immediately evaluate the dc voltages in this circuit as follows:

$$V_S = I_D R_S = 1.18 \text{ V} \tag{5.6-4a}$$

$$V_G = 0 \text{ V} \tag{5.6-4b}$$

$$V_{GS} = V_G - V_S = -1.18 \text{ V} \tag{5.6-4c}$$

$$V_D = V_{DD} - I_D R_D = 10 \text{ V} \tag{5.6-4d}$$

and $$V_{DS} = V_D - V_S = 8.82 \text{ V} \tag{5.6-4e}$$

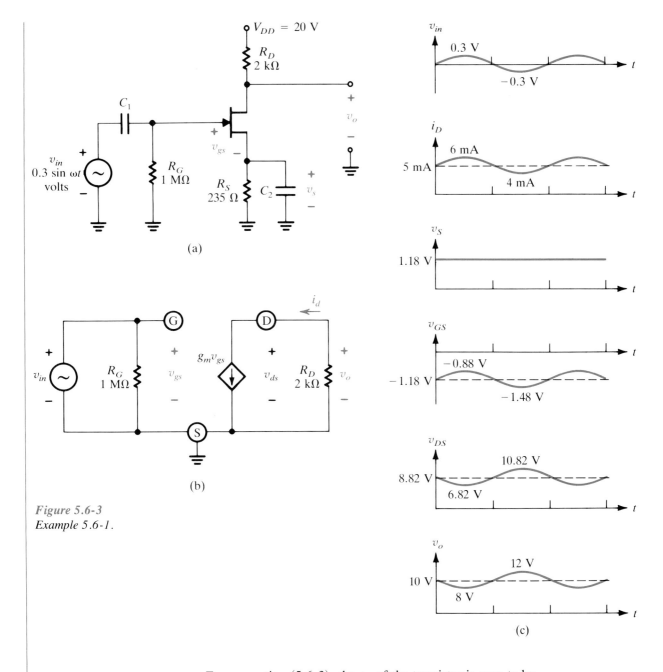

Figure 5.6-3
Example 5.6-1.

From equation (5.6-3), the g_m of the transistor is seen to be

$$g_m = \frac{(-2)(10 \text{ mA})}{-4 \text{ V}}\left(1 - \frac{-1.18}{-4}\right) = 3.54 \text{ mS} \qquad (5.6\text{-}5)$$

The small-signal model for the circuit given in Figure 5.6-3a is drawn in part (b) of the figure considering the capacitors C_1 and C_2 to be ac shorts. By analyzing this circuit, we can see that

$$v_g = v_{in} = 0.3 \sin \omega t \text{ V} \qquad (5.6\text{-}6a)$$

$$v_s = 0 \tag{5.6-6b}$$

$$v_{gs} = v_g - v_s = v_{in} = 0.3 \sin \omega t \text{ V} \tag{5.6-6c}$$

$$i_d = g_m v_{gs} = g_m v_{in} \simeq 1.0 \sin \omega t \text{ mA} \tag{5.6-6d}$$

and $$v_{ds} = v_o = -i_d R_D = -2.0 \sin \omega t \text{ V} \tag{5.6-6e}$$

Combining equations (5.6-4) and (5.6-6), the overall waveforms for the required voltages and currents may be sketched as illustrated in Figure 5.6-3c. For this circuit the amplifier voltage gain v_o/v_{in} is approximately $(-4)/(0.6) = -6.66$.

5.6-2 Limitations of the FET Small-Signal Model: Large-Signal Distortion in the Constant-Current Region

Thus far in this section, we have made use of equation (5.2-20) to develop a simplified small-signal model for the FET. This equation may also be employed to determine the range of input signals for which the small-signal model is valid. Consider, for example, the case where the applied ac input signal v_{gs} is a cosine wave of the form

$$v_{gs} = V_{gs} \cos \omega t \tag{5.6-7a}$$

In this instance, the overall gate-to-source voltage may be written as

$$v_{GS} = V_{GS} + V_{gs} \cos \omega t \tag{5.6-7b}$$

where V_{GS} represents the dc quiescent gate-to-source voltage. Substituting equation (5.6-7b) into equation (5.2-20), we obtain

$$i_D = I_{DSS}\left(1 - \frac{v_{GS}}{V_p}\right)^2 = I_{DSS}\left(1 - \frac{2v_{GS}}{V_p} + \frac{v_{GS}^2}{V_p^2}\right)$$

$$= I_{DSS}\left[1 - \frac{2(V_{GS} + V_{gs} \cos \omega t)}{V_p} + \frac{(V_{GS} + V_{gs} \cos \omega t)^2}{V_p^2}\right] \tag{5.6-8}$$

If we make use of the trigonometric identity

$$\cos^2 \omega t = \frac{1 + \cos 2\omega t}{2} \tag{5.6-9}$$

and substitute this expression into equation (5.6-8), we find, after considerable algebraic manipulation, that i_D may be written as

$$i_D = I_{DSS}\underbrace{\left[\left(1 - \frac{2V_{GS}}{V_p} + \frac{V_{GS}^2}{V_p^2}\right) + \frac{V_{gs}^2}{2V_p^2}\right]}_{\text{dc term}} - \underbrace{\frac{2I_{DSS}}{V_p}\left(1 - \frac{V_{GS}}{V_p}\right)V_{gs} \cos \omega t}_{\text{linear ac term}} + \underbrace{\frac{I_{DSS} V_{gs}^2}{2V_p^2}\cos 2\omega t}_{\text{second harmonic distortion}} \tag{5.6-10}$$

From this result, we see that the approximate percent second harmonic distortion in the output is

$$
\begin{array}{l}
\% \text{ second} \\
\text{harmonic} \\
\text{distortion}
\end{array}
=
\dfrac{\dfrac{I_{DSS}\,V_{gs}^{2}}{2V_{p}^{2}}}{\dfrac{2I_{DSS}\,V_{gs}(1 - V_{GS}/V_{p})}{V_{p}}}
=
\dfrac{\dfrac{V_{gs}}{4V_{p}}}{1 - \dfrac{V_{GS}}{V_{p}}} \times 100
\qquad (5.6\text{-}11)
$$

Let's determine the maximum allowed input signal level if the second harmonic distortion at the output is to be kept smaller than 10% of the fundamental. Following equation (5.6-11) and assuming, for example, that the FET has a $V_p = -4$ V and is biased at $I_{DSS}/2$ [where $(1 - V_{GS}/V_p)$ equals $1/\sqrt{2}$], we find that the input signal can be on the order of 1.1 V before the second harmonic distortion exceeds 10%. For a silicon bipolar junction transistor, this input level must be kept below 13 mV to stay within the same distortion level.

5.6-3 Second-Order Model for the FET

Sometimes the simplified model of the field-effect transistor we have been working with does not adequately describe its behavior. To understand why this model fails, and more important, to determine how it can be modified to make it usable for a larger class of FET circuits, we redevelop this model using the graphical characteristics of the transistor. This approach parallels the hybrid model development for the BJT presented in Section 4.3.

To illustrate this procedure, consider, for example, the n-channel JFET characteristics given in Figure 5.6-4. By examining these curves, it should be clear that i_D is a function of both v_{GS} and v_{DS}, even in the constant-current region. As a result, if v_{GS} changes by $(v_{GS} - V_{GS})$ and v_{DS} by $(v_{DS} - V_{DS})$, we may approximate the new value of i_D by the first few terms of the two-dimensional Taylor series expansion for i_D as

$$
i_D = I_D + \left.\frac{\partial i_D}{\partial v_{GS}}\right|_{\substack{Q\ \text{pt} \\ v_{DS}=\text{constant}}} (v_{GS} - V_{GS}) + \left.\frac{\partial i_D}{\partial v_{DS}}\right|_{\substack{Q\ \text{pt} \\ v_{GS}=\text{constant}}} (v_{DS} - V_{DS})
$$

$$(5.6\text{-}12)$$

Bringing I_D onto the left-hand side of the equation and defining $\Delta v_{GS} = v_{GS} - V_{GS} =$ change in gate-to-source voltage, with similar definitions for Δv_{DS} and Δi_D, we may rewrite equation (5.6-12) as

$$
\Delta i_D = \left.\frac{\partial i_D}{\partial v_{GS}}\right|_{\substack{Q\ \text{pt} \\ v_{DS}=V_{DS}}} \Delta v_{GS} + \left.\frac{\partial i_D}{\partial v_{DS}}\right|_{\substack{Q\ \text{pt} \\ v_{GS}=V_{GS}}} \Delta v_{DS}
\qquad (5.6\text{-}13)
$$

The partial derivative terms in this equation are constants. The first of these, that is, the term $(\partial i_D/\partial v_{GS})|_{Q\ \text{pt}}$, is of course the g_m of the transistor, and the second, which also has the units of 1/resistance, is denoted by $1/r_d$. Equation (5.6-13) may be rewritten in terms of these quantities as

$$
\Delta i_D = g_m\,\Delta v_{GS} + \frac{1}{r_d}\Delta v_{DS}
\qquad (5.6\text{-}14)
$$

The parameters g_m and r_d in this expression may be evaluated as follows. To

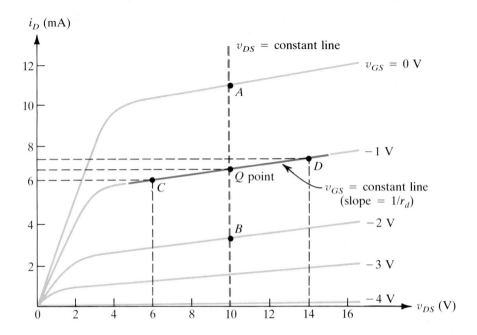

Figure 5.6-4

determine the g_m, we hold v_{DS} constant (i.e., we set $\Delta v_{DS} = 0$) and solve for the transconductance as

$$g_m = \left.\frac{\Delta i_D}{\Delta v_{GS}}\right|_{Q\,\text{pt},\,v_{DS}=\text{constant}}$$
(5.6-15a)

In the same way, r_d is evaluated by holding v_{GS} constant and solving equation (5.6-14) for $1/r_d$, which yields

$$\frac{1}{r_d} = \left.\frac{\Delta i_D}{\Delta v_{DS}}\right|_{Q\,\text{pt},\,v_{GS}=\text{constant}}$$
(5.6-15b)

Let's demonstrate how to evaluate these parameters numerically by using the i_D versus v_{DS} curves given in Figure 5.6-4 and assuming the transistor's Q point to be located at $I_D = 6.5$ mA, $V_{DS} = 10$ V, and $V_{GS} = -1$ V. In accordance with equation (5.6-15a), the transconductance is found by forming the ratio of the change in i_D to the change in v_{GS} and evaluating this ratio along the line where v_{DS} is constant. Following this definition and using points A and B in the figure, the g_m of the transistor at this Q point is $(11 - 3.3)$ mA/$[0 - (-2)]$ V, or 3.85 mS. It should be pointed out when using equations (5.6-15a) and (5.6-15b) that the greatest accuracy is obtained when the changes about the Q point are small. The 2-V change in v_{GS} is a bit large and was used only to make the changes easier for you to see on the graph. Incidentally, for comparison, the corresponding value of g_m obtained using equation (5.6-2b) is 3.75 mS.

To calculate r_d, we make use of equation (5.6-15b) and form the ratio of $\Delta v_{DS}/\Delta i_D$, evaluating it along the constant v_{GS} line in the figure that passes through the Q point. Using points C and D on this line, we find that r_d is $(14 - 6)$ V/$(7 - 6)$ mA, or 8 kΩ. It is important to mention that the tilt of the v_{GS} lines in the figure was exaggerated to simplify the calculation of r_d. As a result, the numerical value of r_d found here is smaller than that for actual FET devices. In particular, for JFETs r_d is typically on the order of 100 kΩ to 1 MΩ, while for MOSFETs it is about 10 to 100 kΩ.

Now that we understand the relationship between the graphical characteristics of the field-effect transistor and its small-signal parameters, let's return to equation (5.6-15) to see how it can be used to develop the ac model for this FET. When the input signals to the transistor are periodic, the corresponding changes in i_D, v_{GS}, and v_{DS} are also periodic signals. For this case, the expression for the drain current in equation (5.6-14) may be rewritten as

$$i_d = \underbrace{g_m v_{gs}}_{i_1} + \underbrace{\frac{1}{r_d} v_{ds}}_{i_2} \qquad (5.6\text{-}16)$$

Furthermore, the gate input current i_g is approximately zero. Using these expressions, the overall ac small-signal model for the FET, at least in block diagram form, may be drawn immediately as illustrated in Figure 5.6-5a. The input portion of this circuit should readily be apparent, but the output section may require a bit more explanation.

Equation (5.6-16) indicates that the small-signal drain current is composed of two components, i_1 and i_2. This suggests that the output section of the FET may be modeled by a circuit containing two branches as illustrated in the figure. Because the branch current i_1 depends on a voltage, v_{gs}, that is not a part of the output circuit, this branch is represented by a voltage-controlled current source $g_m v_{gs}$ (Figure 5.6-5b). On the other hand, the second branch in the circuit has a current flowing through it that is proportional to the voltage v_{ds} across it. This suggests that the branch containing the current i_2 can be modeled by a resistor whose value equals the coefficient r_d in equation (5.6-16). The complete small-signal model for the field-effect transistor is given in

Figure 5.6-5

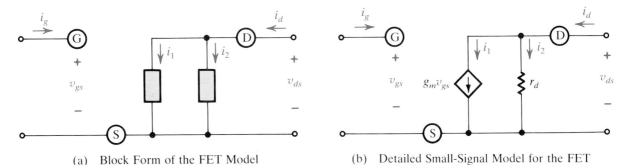

(a) Block Form of the FET Model   (b) Detailed Small-Signal Model for the FET

Figure 5.6-5b. Notice that except for the inclusion of the resistor r_d, this model is essentially the same as that developed earlier in this section (see Figure 5.6-2). It should be apparent that this resistor will be important when the external drain and load resistors are comparable to r_d. For most applications, however, r_d will usually be large enough that it can be neglected.

EXAMPLE 5.6-2

The transistor in the circuit given in Figure 5.6-6a has the volt-ampere characteristics given in part (e) of the figure. The quiescent drain current is 3 mA, and the capacitors are large enough so that they may be considered to be ac shorts.

a. Determine the dynamic load-line equations for the input and output portions of the circuit by replacing each of the circuits within the shaded regions by its Thévenin equivalent.
b. Plot the output dynamic load-line equation on the transistor characteristics, and use this result to sketch the voltages v_{DS} and v_L.
c. Find the amplifier voltage gain $A_v = v_L/v_{in}$.

SOLUTION

Using the fact that the quiescent drain current is 3 mA, the pertinent circuit dc voltages may be calculated as follows:

$$V_G = V_{C1} = \frac{R_2}{R_1 + R_2} V_{DD} = 3.5 \text{ V} \tag{5.6-17a}$$

$$V_S = V_{C2} = I_D R_S = 6 \text{ V} \tag{5.6-17b}$$

$$V_{GS} = V_G - V_S = -2.5 \text{ V} \tag{5.6-17c}$$

$$V_D = V_{DD} - I_D R_D = V_{C3} = 21 \text{ V} \tag{5.6-17d}$$

and $$V_{DS} = V_D - V_S = 15 \text{ V} \tag{5.6-17e}$$

Because the voltages across the capacitors are constant (since the capacitors are assumed to be shorts to the ac signals), they may be replaced by batteries

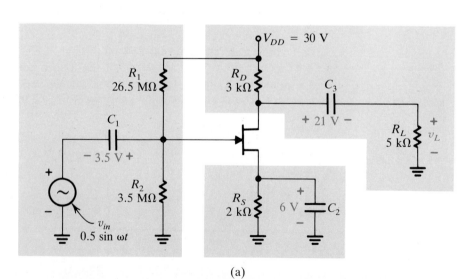

(a)

Figure 5.6-6
Example 5.6-2.

246

Chapter 5 / The Field-Effect Transistor

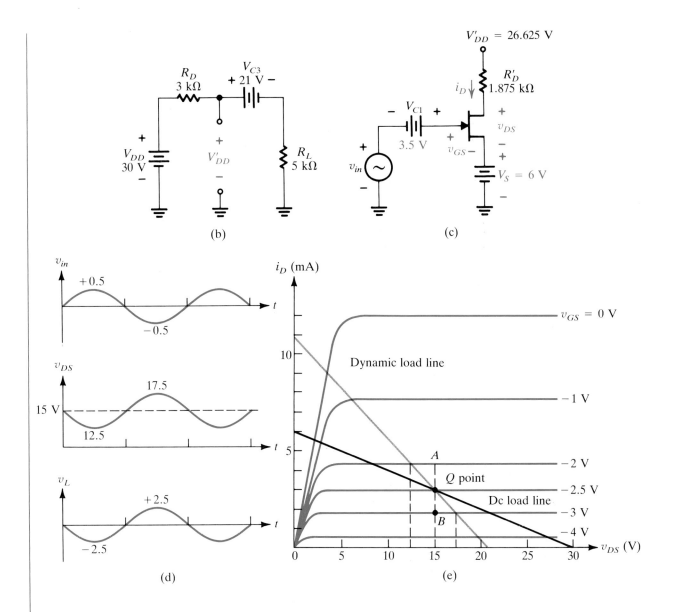

Figure 5.6-6
(Continued)

and the Thévenin equivalents of the resulting circuits may be calculated to simplify the analysis of the circuit. The Thévenin equivalents for the gate and source circuits are easily obtained (Figure 5.6-6c), but the equivalent circuit for the drain portion of the circuit needs a little more explanation.

The portion of the drain circuit within the shaded region is redrawn in Figure 5.6-6b with the capacitor C_3 replaced by a 21-V battery. Using this figure, the Thévenin voltage from drain to ground may be written by superposition as

$$V'_{DD} = V_{DD}\frac{R_L}{R_L + R_D} + V_{C3}\frac{R_D}{R_L + R_D} = 26.625 \text{ V} \qquad (5.6\text{-}18)$$

Of course, the Thévenin resistance of this circuit is just the parallel combination of R_D and R_L or 1.875 kΩ. Following Figure 5.6-6c, the input and output

5.6 / Small-Signal Models for the FET

247

dynamic load-line equations may be written directly as

$$v_{GS} = v_{in} + V_{C1} - V_S = -2.5 + v_{in} \tag{5.6-19a}$$

and $\quad\quad V'_{DD} = i_D R'_D + v_{DS} + V_S$

or $\quad\quad\quad\quad\quad 20.625 = i_D R'_D + v_{DS} \tag{5.6-19b}$

The output dynamic load-line equation is drawn on the transistor characteristics in Figure 5.6-6e.

The sketch for v_{DS} in Figure 5.6-6d is obtained by noting that as v_{in} varies between $+0.5$ and -0.5 V, the operating point of the transistor shifts back and forth along the dynamic load line between the $v_{GS} = -2$V and -3V curves. The corresponding values for v_{DS} may be read off directly from the graph. To obtain the sketch for the output voltage v_L, we note from Figure 5.6-6a that

$$v_{DS} + V_S = V_{C3} + v_L \tag{5.6-20a}$$

or that $\quad\quad\quad\quad\quad v_L = v_{DS} - 15 \tag{5.6-20b}$

As a result, the sketch for v_L is identical to that for v_{DS} except that the 15-V dc level is missing since it was blocked by the capacitor C_3. The voltage gain of the circuit is $A_v = [(-2.5) - (2.5)]/[(0.5) - (-0.5)] = -5.0$.

EXAMPLE 5.6-3

Using the appropriate small-signal model for the FET, calculate the voltage gain of the amplifier discussed in Example 5.6-2.

SOLUTION

By examining the transistor characteristics in the neighborhood of the Q point in Figure 5.6-6e, it should be clear that the small-signal model for the transistor in this circuit need not include r_d because the transistor characteristic curves near the Q point are nearly horizontal, so that r_d is very large compared with R'_D. The transconductance of the FET is obtained by determining the change in the drain current produced by a specific change in the gate-to-source voltage when the drain-to-source voltage is held constant. This calculation must be done in the neighborhood of the Q point for small changes in v_{GS}. In particular, if v_{GS} changes by ± 0.5 V (points A and B in Figure 5.6-6e), the corresponding variation in the drain current is between 4.4 and 1.8 mA. As a result, the transistor g_m at this Q point is

$$g_m = \frac{\Delta i_D}{\Delta v_{GS}}\bigg|_{\substack{Q\,\text{pt}\\ v_{DS}=\text{constant}}} = \frac{4.4 - 1.8}{(-2) - (-3)} = 2.6 \text{ mS} \tag{5.6-21}$$

The small-signal equivalent circuit for the FET amplifier shown in Figure 5.6-6a is given in Figure 5.6-7. From this figure the output voltage v_L may be

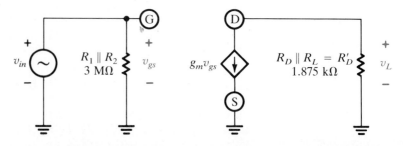

Figure 5.6-7
Example 5.6-6.

written as

$$v_L = -i_d R_D' = -g_m R_D' v_{gs} = -g_m R_D' v_{in} \qquad (5.6\text{-}22)$$

so that the voltage gain of the circuit is

$$A_v = \frac{v_L}{v_{in}} = -g_m R_D' = -4.875 \qquad (5.6\text{-}23)$$

This answer compares quite favorably with the result of -5 that was obtained in Example 5.6-2 by graphical analysis of the same problem.

Exercises

5.6-1 An experimental JFET has a drain current that is related to its gate-to-source voltage by the expression

$$i_D = K_1 v_{GS}^2 + K_2 v_{GS} + K_3$$

where $K_1 = 2$ mA/V^2, $K_2 = 10$ mA/V, and $K_3 = 20$ mA.
a. Find I_D when $V_{GS} = -2$ V.
b. Determine the transistor g_m at the Q point specified in part (a).
Answer (a) 8 mA, (b) 2 mS

5.6-2 Given the JFET with the volt-ampere characteristics shown in Figure 5.6-4, find the transistor small-signal parameters g_m and r_d if the Q point is located at $V_{DS} = 10$ V and $V_{GS} = -3$ V. *Answer* $g_m = 1.5$ mS, $r_d = 10$ kΩ

5.6-3 Find the voltage gain, v_d/v_{in}, of the circuit in Figure 5.6-6 graphically if C_3 and R_L are removed. *Answer* -8

5.7 SMALL-SIGNAL EQUIVALENT CIRCUITS FOR THE FET

The analysis of circuits containing field-effect transistors can be greatly simplified if the small-signal equivalent circuits seen looking into the gate, the source, and the drain terminals of the transistor are developed. The final circuits obtained, and the techniques employed for deriving them, will in many respects be similar to those presented for the BJT in Section 4.4. In developing these equivalent circuits we will assume that the simplified model for the FET, which neglects r_d, is valid.

The basic circuit to be analyzed is illustrated in Figure 5.7-1a, and the ac equivalent circuit with the batteries replaced by shorts is given in part (b) of the figure. Figure 5.7-1c shows the overall equivalent circuit that results when the simplified FET model is inserted in place of the transistor and we replace the current source $g_m v_{gs}$ by two equivalent current sources. At first this does not appear to have simplified the circuit very much, but its major advantage is that it clearly separates the source and drain portions of the circuit from one another.

By examining Figure 5.7-1c, it should be apparent that the equivalent circuit seen looking into the gate of the transistor is an open circuit. Therefore, the gate voltage, at least in this case, is simply equal to the applied input voltage v_{in}. To develop the source equivalent circuit, consider the source por-

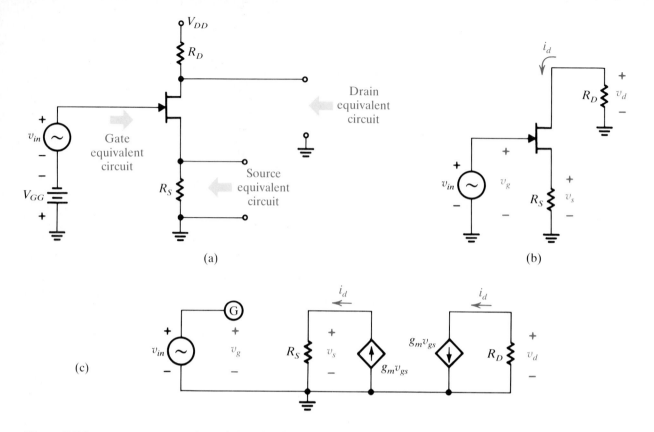

(a)

(b)

(c)

Figure 5.7-1

Developing simplified equivalent circuits for the FET.

tion of the circuit in Figure 5.7-1c and note that because $v_{gs} = v_g - v_s$, it is possible to split the current source $g_m v_{gs}$ into two equivalent sources $g_m v_g$ and $g_m v_s$ (Figure 5.7-2a). The branch of this circuit containing the current source $g_m v_s$ has the interesting property that the current flow through it is proportional to the voltage, v_s, across it. Therefore, this current source may be replaced by an equivalent resistor

$$R = \frac{v_s}{i_2} = \frac{v_s}{g_m v_s} = \frac{1}{g_m} \tag{5.7-1}$$

Taking the Thévenin equivalent of this circuit, we obtain the circuit given in part (b) of the figure. To summarize this result, the output impedance seen looking back into the source terminal of an FET is simply $1/g_m$, where g_m is equal to the transistor transconductance evaluated at the Q point. The Thévenin voltage is equal to the voltage v_g developed between the gate terminal of the transistor and ground.

Using this equivalent circuit, the voltage at the source of this transistor amplifier is seen to be

$$v_s = \frac{R_S}{R_S + 1/g_m} v_g = \frac{g_m R_S}{1 + g_m R_S} v_g \tag{5.7-2}$$

Notice that if R_S is much greater than $1/g_m$, the voltage at the source will appear to follow that at the gate. Hence this type of circuit, where the input signal is applied to the gate and the output is taken from the source, is often

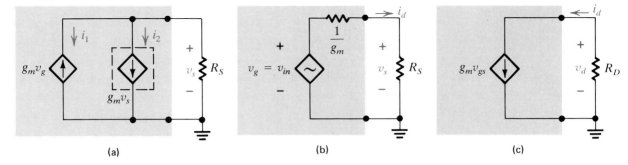

(a)   (b)   (c)

Figure 5.7-2
(a), (b) Developing the
equivalent circuit at the source;
(c) equivalent circuit at the drain.

called a source follower. Its properties are similar to those of the emitter follower discussed in Chapter 4, except that the source-follower circuit generally has a higher input impedance.

Consider next the drain equivalent circuit in Figure 5.7-2c. At first it might appear that the current source $g_m v_{gs}$ in this figure can be simplified by employing a technique similar to that used in developing the source equivalent circuit. This is not the case, however, because the current source $g_m v_{gs}$ does not depend on the voltage v_d across it, and hence it cannot be replaced by an equivalent impedance.

In summary, to analyze FET circuits where r_d is large enough to be neglected:

1. Calculate the voltage v_g appearing between the gate and ground assuming that the gate terminal equivalent circuit is an open.
2. Reflect this voltage into the source and draw the overall source equivalent circuit as illustrated in Figure 5.7-2b. Using this equivalent circuit, calculate i_d and v_s. If necessary, notice that v_{gs} may also be calculated using this figure as the voltage drop across the resistor $1/g_m$.
3. By making use of the drain equivalent circuit given in Figure 5.7-2c and the value for i_d calculated in step 2, compute v_d.
4. If necessary, v_{ds} may be obtained from the results in steps 2 and 3 by recalling that $v_{ds} = v_d - v_s$.

The application of these equivalent-circuit techniques will be illustrated in the examples that follow.

EXAMPLE 5.7-1

For the field-effect transistor amplifier shown in Figure 5.7-3a, $g_m = 1$ mS and r_d is infinite.

a. Find the amplifier gain $A_v = v_s/v_{in}$ with R_L disconnected.
b. Repeat part (a) if a load $R_L = 1$ kΩ is capacitively coupled to the source of the transistor as shown in the figure.

SOLUTION

The small-signal equivalent circuits at the gate and source terminals of the FET circuit are illustrated in Figure 5.7-3b and c, respectively. Because the gate resistor R_G is so large, the gate voltage v_g is approximately equal to the signal voltage v_{in}. The source voltage v_s may be written with the aid of Figure 5.7-3c as

$$v_s = v_{in} \frac{R_S}{R_S + 1/g_m} \tag{5.7-3}$$

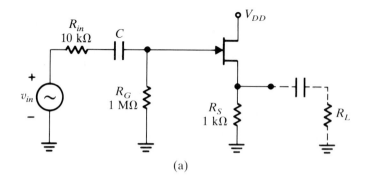

(a)

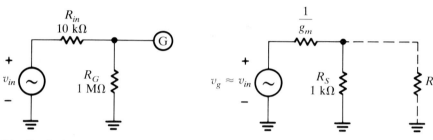

Figure 5.7-3
Example 5.7-1.

(b) Equivalent Circuit in the Gate   (c) Equivalent Circuit at the Source of the FET

As a result, the voltage gain for this circuit is

$$A_v = \frac{v_s}{v_{in}} = \frac{g_m R_S}{1 + g_m R_S} = 0.5 \qquad (5.7\text{-}4)$$

When the load resistor R_L is connected, the solution is essentially the same as that given in equation (5.7-4) if R_S is replaced by R_S', the parallel combination or R_S and R_L. For the specific resistors used in this example, the voltage gain with R_L connected is $(0.5/1.5)$, or 0.33.

EXAMPLE 5.7-2

The transconductance of the FET amplifier shown in Figure 5.7-4a is 1 mS, and the drain resistance r_d is considered to be infinite. Determine the voltage gain $A_v = v_L/v_{in}$.

SOLUTION

The gate, source, and drain equivalent circuits are given in Figure 5.7-4b, c, and d, respectively. By examining the gate circuit it should be clear that

$$v_g = v_{in} \qquad (5.7\text{-}5)$$

The drain current may be written directly from Figure 5.7-4c as

$$i_d = \frac{v_{in}}{1/g_m + R_{S1}} = \frac{v_{in}}{1.5 \text{ k}\Omega} \qquad (5.7\text{-}6)$$

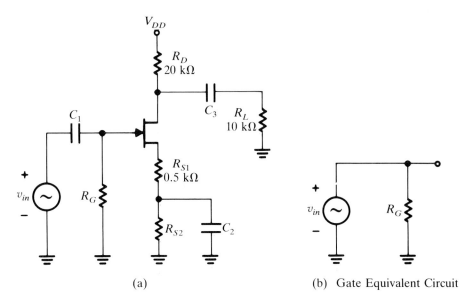

(a)

(b) Gate Equivalent Circuit

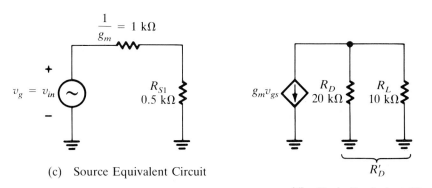

(c) Source Equivalent Circuit

(d) Drain Equivalent Circuit

Figure 5.7-4
Example 5.7-2.

Substituting this expression for i_d into the equivalent circuit in Figure 5.7-4d, the output voltage is seen to be

$$v_L = -i_d R'_D = -\frac{v_{in}}{1.5 \text{ k}\Omega}(6.66 \text{ k}\Omega) \tag{5.7-7a}$$

so that the circuit voltage gain is

$$A_v = \frac{v_L}{v_{in}} = -4.44 \tag{5.7-7b}$$

EXAMPLE 5.7-3

The two-transistor amplifier illustrated in Figure 5.7-5a combines an FET and a BJT to achieve both a high input impedance and a large voltage gain. By considering the g_m of Q_1 to be 1 mS, and h_{ie} and h_{fe} for Q_2 to be 1 kΩ and 100, respectively, determine the voltage gain v_o/v_{in} of the amplifier.

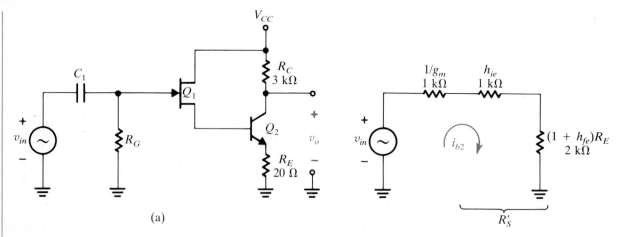

(a)

(b) Equivalent Circuit for the Source of Q_1
and the Base of Q_2

Figure 5.7-5
Example 5.7-3.

SOLUTION

The gain of this amplifier may be found rather easily by drawing the equivalent circuit seen in the source of Q_1 and the base of Q_2 (Figure 5.7-5b). Notice that R_G does not appear in this equivalent circuit. This is because the drain current in the FET depends only on the voltage appearing from gate to ground, but is independent of the impedance levels in the gate portion of the circuit.

Using the circuit in Figure 5.7-5b, the base current into Q_2 can be written down immediately as

$$i_{b2} = \frac{v_{in}}{1/g_m + h_{ie} + (1 + h_{fe})R_E} = \frac{v_{in}}{4 \text{ k}\Omega} \qquad (5.7\text{-}8)$$

Although we could also sketch the equivalent circuit for the collector portion of Q_2, at this point the analysis of the rest of this circuit should be possible by inspection. In particular,

$$i_{c2} = h_{fe}i_{b2} = \frac{100v_{in}}{4 \text{ k}\Omega} \qquad (5.7\text{-}9\text{a})$$

and

$$v_{c2} = -i_{c2}R_{C2} = -\frac{300 \text{ k}\Omega \; v_{in}}{4 \text{ k}\Omega} \qquad (5.7\text{-}9\text{b})$$

so that

$$A_v = \frac{v_o}{v_{in}} = -75 \qquad (5.7\text{-}9\text{c})$$

If you have forgotten where equations (5.7-9a) and (5.7-9b) come from, you should review Section 4.4.

Exercises

5.7-1 Determine the power gain of the circuit shown in Figure 5.7-4 if $R_G = 1 \text{ M}\Omega$ and R_L is changed to 20 kΩ. *Answer* 4400

5.7-2 Find the voltage gain of the amplifier in Figure 5.7-5 if a 2-kΩ resistor is shunted from the source of Q_1 to ground. *Answer* −55

5.7-3 Repeat Example 5.7-1 if C_2 shunts across both R_{S1} and R_{S2}. *Answer* -6.67

5.7-4 Repeat Example 5.7-3 if r_d of Q_1 is 10 kΩ. *Answer* -70

5.8 MULTISTAGE AMPLIFIER DESIGN CONSIDERATIONS

In Section 4.5 we discussed techniques for designing multistage amplifiers containing only bipolar transistors, since at that time the field-effect transistor had not yet been introduced. As we close this introductory chapter on FETs, it is therefore appropriate for us to reexamine the design of these amplifier circuits in light of the possible improvements that may be afforded by the introduction of field-effect transistors into these circuits.

Linear amplifier circuits can be constructed by making use of either BJT, JFET, or MOS transistors, and actual amplifiers are often designed using combinations of these devices. To facilitate the comparison of these three transistor types with one another, the approximate range of the small-signal parameters for each are listed in Table 5.8-1.

In general, the design procedure for nearly all multistage electronic amplifiers can be broken down into three distinct phases. Specifically, these are the design of the input, the output, and the intermediate stages. Since the fundamentals of this design procedure were carefully developed in Section 4.5, only the results of that material and its extension to include FETs are presented in this section. To begin with, the input and output stages of the amplifier are generally used to match the amplifier's impedance levels to those of the source and the load. Depending on the characteristics of the signal source and the load to be driven, this may require input and output stages having either high or low impedance levels.

For amplifier designs where a high input impedance is required, the JFET provides superior performance to the bipolar junction transistor. When very

5.8-1 Comparison of the Small-Signal Parameters of BJT, JFET, and MOSFET Transistors

Transistor Type	Range of Important Transistor Parameters	
BJT	h_{ie}	1–10 kΩ
	h_{fe}	20–500
	$g_m \sim \dfrac{1}{r_e}$	10–4000 mS
	$\dfrac{1}{h_{oe}}$	50 kΩ–1MΩ
JFET	r_{gs} (input resistance)	10–100 MΩ
	g_m	1–10 mS
	r_d	50 kΩ–1 MΩ
MOSFET	r_{gs} (input resistance)	10^{10}–10^{14} Ω
	g_m	1–50 mS
	r_d	5–100 kΩ

high input impedance in excess of 10 or 20 MΩ are needed, special JFET circuits or perhaps even MOSFET circuits may be required.

EXAMPLE 5.8-1

The JFET amplifier illustrated in Figure 5.8-1 is to be designed to have the highest possible input impedance while ensuring that the quiescent gate-to-source voltage varies by less than 10% over the temperature range from 25 to 100°C. If the gate leakage current is -1 nA at room temperature (25°C), and if it doubles for each 10°C increase in temperature, determine the highest allowed value for R_G.

SOLUTION

Because the leakage current through the reversed biased gate junction doubles for every 10°C increase, the approximate gate leakage current at 100°C may be written as

$$I_G(100) = I_G(25) \times 2^{(100-25)/10} = -0.18\ \mu\text{A} \qquad (5.8\text{-}1)$$

Using Figure 5.8-1, we can see that the quiescent gate-to-source voltage, including the effect of the gate leakage current, is

$$V_{GS} = V_G - V_S = -I_G R_G - V_{GG} \qquad (5.8\text{-}2)$$

At room temperature the voltage drop produced by I_G across R_G is negligible, so that V_{GS} is approximately -2V. At 100°C, on the other hand, this term can be significant.

For this problem the largest possible value of R_G is to be selected while still maintaining the change in V_{GS} over the specified temperature range at less than 10%. Because $V_{GS} = -2$ V at room temperature, this means that the change in V_{GS} must be less than 0.2 V when the temperature increases to 100°C. Using equation (5.8-2), this requires that R_G be less than 0.2 V/0.18 μA, or 1.1 MΩ. Thus, for JFET circuits of the type described in this example, there is an upper bound on the input impedance of the amplifier on the order of several megohms, and this value is determined by Q-point stability constraints. If input impedances in excess of 5 or 10 MΩ are needed, special bootstrap-style JFET circuits must be employed, or alternatively, MOSFET transistors, with their much smaller gate leakage currents, should be used.

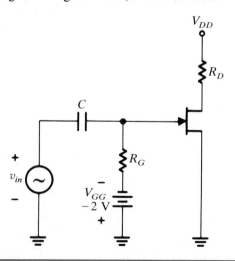

Figure 5.8-1
Example 5.8-1.

In designing the output stage of an amplifier, a low output impedance is usually desired. To achieve this design goal, some type of follower circuit is the best choice, and generally the BJT emitter follower circuit is found to have a lower output impedance than a comparable FET source follower. This is because the output impedance of an emitter follower, $(R_B + h_{ie})/(1 + h_{fe})$, is typically on the order of 10 to 50 Ω, while the output impedance of an FET, $1/g_m$, is generally about 100 to 1000 Ω. Therefore, for most amplifier designs an emitter follower is preferred for the output stage.

The circuit configuration selected for the intermediate-stage design should be the one that offers the largest possible power gain with the smallest number of components. In Section 4.5 we examined the common-emitter, common-collector, and common-base-style intermediate amplifier stages and found that the common-emitter multistage amplifier afforded the highest power gain of all three BJT circuit configurations. In the remainder of this section, we examine the power gain of FET common-source, common-gate, and common-drain multitransistor amplifier circuits and compare their power gains to that of the BJT common-emitter amplifier, to determine the best way to design the intermediate stages of an electronic amplifier.

The three three-stage FET circuits illustrated in Figures 5.8-2a, b, and c are examples of common-source, common-gate, and common-drain amplifiers, respectively. For simplicity, some of the biasing circuitry has been eliminated, and in analyzing these three circuits it will be assumed that the transistor g_m values are all 1 mS and that r_d is large enough to neglect. Furthermore, it will also be assumed that the source and load impedance for all three circuits have nominal values of 10 kΩ.

To analyze the power gain of the common-source amplifier in Figure 5.8-2a, we note that the average input power input is

$$P_{in} = \frac{V_{in}^2}{2(R_{in} + R_G)} \simeq \frac{V_{in}^2}{2R_G} \tag{5.8-3}$$

where V_{in} is the peak amplitude of the input voltage. Because the FET stages in this circuit do not load one another, it is possible to analyze the overall voltage gain of this amplifier by inspection as follows:

$$v_{g1} = \frac{R_G}{R_G + R_{in}} v_{in} \simeq v_{in} \tag{5.8-4a}$$

$$i_{d1} = g_m v_{in} \tag{5.8-4b}$$

$$v_{d1} = v_{g2} = -i_{d1}R_D = -g_m R_D v_{in} \tag{5.8-4c}$$

$$i_{d2} = g_m v_{g2} = g_m v_{d1} \tag{5.8-4d}$$

$$v_{d2} = v_{g3} = -g_m R_D v_{g2} = (g_m R_D)^2 v_{in} \tag{5.8-4e}$$

$$i_{d3} = g_m v_{g3} = g_m v_{d2} \tag{5.8-4f}$$

and

$$v_L = v_{d3} = -i_{d3}R_L = -g_m^3 R_D^2 R_L v_{in} \tag{5.8-4g}$$

Using equation (5.8-4g), the average power delivered to the load is

$$P_L = \frac{V_L^2}{2R_L} = \frac{g_m^6 R_D^4 R_L V_{in}^2}{2} \tag{5.8-5a}$$

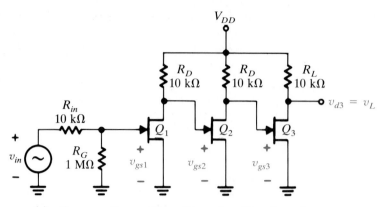

(a)   Common-Source Three-Stage Amplifier Cascade

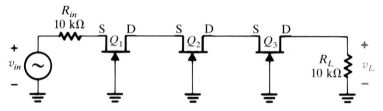

(b)   Common-Gate Three-Stage Amplifier Cascade

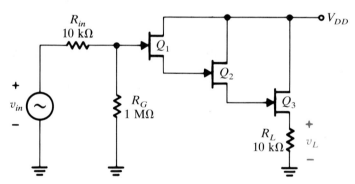

(c)   Common-Drain Three-Stage Amplifier Cascade

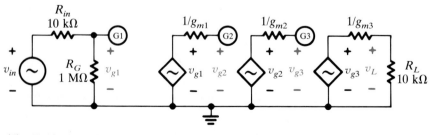

Figure 5.8-2   (d)   Equivalent Circuit for the Common Drain Three-Stage Amplifier Cascade

so that the amplifier power gain may be written as

$$G = \frac{P_L}{P_{in}} = g_m^6 R_D^4 R_L R_G \tag{5.8-5b}$$

Substituting the numerical values into this expression, we find that the power gain of this circuit is approximately 10^8.

The analysis for the common-gate amplifier in Figure 5.8-2b follows similarly. In particular, we can readily see that

$$i_{s1} = i_{d1} = i_{s2} = i_{d2} = i_{s3} = i_{d3} = \frac{-v_{in}}{R_{in} + 1/g_m} \tag{5.8-6a}$$

and

$$v_L = -i_{d3}R_L \simeq v_{in}\frac{R_L}{R_{in}} \tag{5.8-6b}$$

The average power into the amplifier and the average power delivered to the load are

$$P_{in} = \frac{V_{in}^2}{2(R_{in} + 1/g_m)} \simeq \frac{V_{in}^2}{2R_{in}} \tag{5.8-7a}$$

and

$$P_L = \frac{V_{in}^2(R_L/R_{in})^2}{2R_L} = \frac{V_{in}^2 R_L}{2R_{in}^2} \tag{5.8-7b}$$

respectively, so that the overall amplifier power gain may be written as

$$G = \frac{P_L}{P_{in}} = \frac{R_L}{R_{in}} \simeq 1 \tag{5.8-8}$$

This result clearly indicates that the common-gate amplifier should not be used as an intermediate amplifier stage since it offers virtually no power gain if the source and load impedances are comparable.

The common-drain amplifier illustrated in Figure 5.8-2c is only a little better than that of the common-gate amplifier. The overall equivalent circuit for this amplifier is given in Figure 5.8-2d, from which it follows that

$$v_{g1} = \frac{R_G}{R_G + R_{in}}v_{in} \simeq v_{in} \tag{5.8-9a}$$

$$v_{g2} = v_{g1} = v_{in} \tag{5.8-9b}$$

$$v_{g3} = v_{g2} = v_{in} \tag{5.8-9c}$$

and

$$v_L = \frac{R_L}{1/g_{m3} + R_L}v_{g3} \simeq v_{g3} = v_{in} \tag{5.8-9d}$$

Using equation (5.8-9d), the average power delivered to the load may be written as

$$P_L = \frac{V_{in}^2}{2R_L} \tag{5.8-10a}$$

and from the equivalent circuit it is also apparent that the average input power to the amplifier is approximately

$$P_{in} = \frac{V_{in}^2}{2R_G} \tag{5.8-10b}$$

Combining equations (5.8-10a) and (5.8-10b), the amplifier power gain is

$$G = \frac{P_L}{P_{in}} = \frac{R_G}{R_L} \qquad (5.8\text{-}11)$$

or approximately 100 for this example.

Of the three multitransistor FET amplifiers analyzed in this section, the common-source transistor cascade provides the highest power gain [approximately 10^8 from equation (5.8-5b)]. But when this result is compared with that for a similar three-transistor common-emitter BJT amplifier, the much higher power gain of the bipolar circuit [about 10^{12} from equation (4.5-6c)] makes it clearly preferable to the FET common-source circuit.

In summary, the design of multistage electronic amplifiers containing bipolar and field-effect transistors is best carried out by splitting the design problem into three distinct parts. Specifically, these are the design of the input, the intermediate, and the output stages. With regard to each of these design phases:

1. For amplifiers requiring a low input impedance, the common-base amplifier is the best choice for the design of the input stage. On the other hand, where a high input impedance is required, a JFET common-source amplifier is usually adequate; however, if extremely high input impedances are needed, a bootstrapped JFET amplifier, or perhaps even the use of a MOS-FET transistor, may be the best way to handle this design problem.
2. In designing the intermediate stages, the cascaded BJT common-emitter amplifier offers the highest overall power gain and is the preferred circuit configuration for this portion of the amplifier design.
3. For most amplifiers, a low output impedance is desired so that changes in the load impedance will not alter the output signal voltage. To satisfy this design requirement, an emitter-follower output stage is usually chosen because its output impedance is typically an order of magnitude smaller than that of an FET source-follower circuit.

When a constant current rather than a constant voltage is required at the amplifier output, an amplifier with a high output impedance, rather than one with a low output impedance, is needed. Common-emitter circuits and common-source JFET amplifiers have output impedances on the order of 50 to 200 kΩ, and if higher output impedances are needed, a common-base BJT amplifier can be used which can easily achieve output impedances in excess of 1 MΩ.

Exercises

5.8-1 For the amplifier in Example 5.7-3, shown in Figure 5.7-5, $R_G =$ 1 MΩ. Find the smallest load resistance permitted for which the voltage gain changes by less than 20% from its open-circuit value. *Answer* 12 kΩ

5.8-2 Repeat Exercise 5.8-1 if an emitter-follower transistor Q_3 is connected onto the amplifier output. Assume that Q_3 has the same small-signal parameters as Q_2. *Answer* 160 Ω

5.8-3 For the amplifier in Example 5.7-3, shown in Figure 5.7-5, $R_G =$ 1 MΩ. The voltage gain of this amplifier is -75 [equation (5.7-9c)].

a. If the input voltage v_{in} has a source resistance R_{in} in series with it, and if R_{in} can vary anywhere between 0 and 500 kΩ, find the maximum gain error that R_{in} can cause (compared with the circuit gain when $R_{in} = 0$).

b. Repeat part (a) if R_G is removed from ground and connected to the source of the FET instead.

Answer (a) 33%, (b) 12%

5.8-4 Find the power gain of the common-source amplifier in Figure 5.8-2a if $r_d = 10$ kΩ for all three transistors. *Answer* 1.56×10^6

PROBLEMS

5.2-1 Carry out the integration in equation (5.2-14) to demonstrate the validity of equation (5.2-15).

5.2-2 Show that $I_D \simeq V_{DS}/r_{DS}$ for small V_{DS}, where r_{DS} is given by equation (5.2-6b). [*Hint:* Expand the term $[(V_{GS} - V_{DS})/V_P]^{3/2}$ in equation (5.2-6b), noting that $(1 + x)^b \simeq (1 + bx)$ for small x.]

5.2-3 By using equations (5.2-15) and (5.2-19), derive an expression for the volt-ampere characteristics of an n-channel JFET whose gate and source are connected together. As part of this analysis, find the range of v_{DS} values for which the transistor equivalent circuit is a current source.

5.2-4 The transistor circuit shown in Figure 5.2-8a has the $V–I$ characteristics given in part (c) of the same figure, and a load resistor $R_L = 0.8$ kΩ is capacitively coupled to the drain.

(a) Sketch v_G, v_L, and i_L.
(b) Find the circuit voltage, current, and power gain.

5.2-5 Starting from equation (5.2-6b), prove that the drain-to-source resistance in the ohmic region for small values of v_{DS} is given by the expression

$$r_{DS} = \frac{r_{DS(ON)}}{1 - \sqrt{V_{GS}/V_P}}$$

5.2-6 For the circuit shown in Figure 5.2-2, the transistor has the $V–I$ characteristics given in Figure 5.2-1b.

(a) Find v_o for $v_C = 2$ V.
(b) Find v_o for $v_C = -10$ V.

5.2-7 The FET in Figure P5.2-7 has the $V–I$ characteristics in Figure 5.2-5. Given that $v_2(t) = 1.5 \sin (2\pi \times 5t)$ volts:

(a) What is v_o if $v_1 = 0$?
(b) What is v_o if $v_1 = -5$V?
(c) Sketch v_o if v_1 varies as shown in Figure P5.2-7. (*Hint:* Look at the curves and note that v_{DS} is always less than 2 V.)

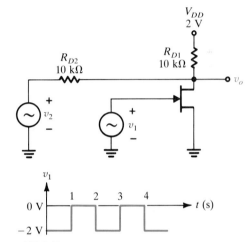

Figure P5.2-7

For all problems in this section, unless otherwise stated, $V_p = -5$ V and $I_{DSS} = 10$ mA.

5.3-1 In Figure 5.4-3c, $R_D = 0$, R_S is unknown, and R_1 is set equal to 2 MΩ and connected in series with a switch S. When S is open, $I_D = 2$ mA. Find I_D, V_G, V_S, V_{GS}, and V_{DS} when the switch is closed.

5.3-2 For the circuit in Figure 5.4-3b, $V_{DD} = 15$ V, and R_D and R_S are adjustable. Select R_D and R_S to place the Q point at $V_{DS} = 7$ V and $I_D = 5$ mA.

5.3-3 In Figure 5.4-3c, $R_1 = 9$ MΩ, $R_D = 1$ kΩ, and $R_S = 500$ Ω. Find the drain current using graphical techniques by first sketching the i_D versus V_{GS} characteristic curve.

5.3-4 In Figure 5.4-3b, $V_{DD} = 10$ V, $R_D = 4$ kΩ, and $R_S = 2.5$ kΩ. Find the drain current I_D.

5.3-5 For the circuit in Figure 5.4-3c, $V_{DD} = 15$ V, $R_D = 1$ kΩ, and R_S is made adjustable. Find the value of R_S that places the Q point at $I_D = 5$ mA and $V_{GS} = -1.5$ V.

5.3-6 For the circuit in Figure 5.4-3c, $R_1 = 4.6 \text{ M}\Omega$, $R_D = 1.5 \text{ k}\Omega$, and $R_S = 1 \text{ k}\Omega$. Algebraically determine the circuit Q point.

5.3-7 For Problem 5.3-6, considering that $|V_{GS}| \ll V_S$ and V_G (so that $V_S \approx V_G$), approximately determine I_D and V_{DS}.

5.3-8 In Figure 5.4-3d, R_D and R_1 are adjustable, $R_S = 1 \text{ k}\Omega$, and a 3-V battery (negative terminal up) is connected in series with R_2. Select R_1 and R_D so that $I_D = 1 \text{ mA}$ and $V_{DS} = 10 \text{ V}$.

5.3-9 For the circuit in Figure 5.4-3b, $V_{DD} = 30 \text{ V}$, $R_D = 5 \text{ k}\Omega$, $R_S = 1 \text{ k}\Omega$, and a resistor $R_{S2} = 2 \text{ k}\Omega$ is inserted between R_S and ground. In addition, the lower terminal of R_G is removed from ground and connected at the junction of R_S and R_{S2}. Find the Q point. (*Hint:* $I_G = 0$, as is the voltage drop across R_G.)

5.3-10 In Figure 5.4-3b, R_D is adjustable, $R_S = 0.5 \text{ k}\Omega$, and a 3-mA current source (arrow up) is connected in parallel with R_S. Select R_D to place the transistor Q point at $V_{DS} = 10 \text{ V}$.

5.3-11 In the circuit shown in Figure P5.3-11, Q_1 has $I_{DSS} = 4 \text{ mA}$ and $V_p = -1.4 \text{ V}$ and Q_2 is silicon with $h_{FE} = 100$. Because of the presence of the BJT, $V_{S1} = V_{BE2} = 0.7 \text{ V}$. Starting from this point, find, V_{GS1}, I_{D1}, I_{R1}, I_{B2}, I_{C2}, and V_{CE2}.

5.3-12 Find the Q points of Q_1 and Q_2 for the circuit shown in Figure P5.3-12, graphically, given that $I_{DSS} = 10 \text{ mA}$ and $V_p = 5 \text{ V}$ for Q_1 and Q_2.

5.3-13 Given that $I_{DSS} = 2 \text{ mA}$ and $V_p = -4 \text{ V}$ and that due to the circuit symmetry $I_{D1} = I_{D2}$, $V_{DS1} = V_{DS2}$, and $V_{GS1} = V_{SG2}$, find each of these dc quantities for the circuit shown in Figure P5.3-13. In other words, find the dc drain current and the dc drain-to-source and gate-to-source voltage on each transistor.

5.3-14 Choose R_S, R_3, and R_D to place the transistor Q points in Figure P5.3-14 at $I_{D1} = I_{D2} = 5 \text{ mA}$, $V_{GS1} = V_{GS2} = -1\text{V}$ and $V_{DS1} = V_{DS2}$ 10 V.

5.4-1 In Figure 5.4-3a, $V_{DD} = 15 \text{ V}$, $R_D = 1.5 \text{ k}\Omega$, and $V_{GG} = 2 \text{ V}$. In addition, I_{DSS} can vary 50% from its normal 10 mA value and V_p remains fixed at -5 V. Find the corresponding percent shift in V_{DS}.

5.4-2 Repeat Problem 5.4-1 for the circuit shown in Figure 5.4-3b, where $V_{DD} = 15 \text{ V}$, $R_D = 1.5 \text{ k}\Omega$, and $R_S = 0.55 \text{ k}\Omega$.

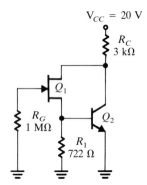

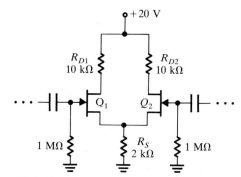

Figure P5.3-13

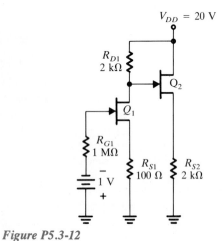

Figure P5.3-11

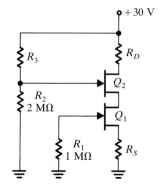

Figure P5.3-12

Figure P5.3-14

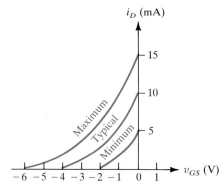

i_D (mA)

15

10

Maximum
Typical
Minimum

5

-6 -5 -4 -3 -2 -1 0 1 v_{GS} (V)

Figure P5.4-3

5.4-3 **(a)** For the circuit shown in Figure 5.4-3c, R_1 and R_S are adjustable. $V_{DD} = 10$ V, and $R_D = 0.6$ kΩ. Select R_S for maximum possible ac gain while keeping the "typical" drain current at 5 mA and the worst-case Q-point shift from this typical value to less than ±40%. (Use the transistor curves in Figure P5.4-3.)
(b) What value of R_1 is required to achieve this Q point?
(c) What is V_{DS}?

5.4-4 **(a)** In Figure 5.4-3c, R_D and R_S are adjustable, $V_{DD} = 25$ V, and $R_1 = 7.33$ MΩ. Select R_S and R_D for a nominal Q point at $I_D = 5$ mA, $V_{DS} = 10$ V. The transistor characteristics are given in Figure P5.4-3.
(b) What will be the maximum percent shift in the Q point for different transistors?

5.4-5 For the circuit in Figure 5.4-3b, $V_{DD} = 25$ V, $R_D = 5$ kΩ, and R_S is adjustable. In addition, a resistor R_{S2} is inserted between R_S and ground, and R_G is removed from ground and connected at the junction of R_S and R_{S2}.
(a) Select R_S and R_{S2} so that a nominal transistor will have its Q point at $I_D = 2$ mA and $V_{DS} = 10$ V. The transistor characteristics are given in Figure P5.4-3.
(b) Determine the worst case Q-point shift.

5.4-6 In Figure P5.3-12, R_{S1} and R_{S2} are changed to 200 Ω and 4 kΩ, respectively.
(a) Find the Q points of Q_1 and Q_2 for this circuit using the nominal transistor values given in Problem 5.4-3.
(b) Find the maximum percent production spread expected in V_{DS2}. (You may assume that the transistors take on their maximum and minimum values together.)

5.4-7 In Figure 5.4-3d, $V_{DD} = 30$ V, and R_D, R_S, and R_1 are adjustable. Select these components to place the nominal Q point at $I_D = 5$ mA and $V_{DS} = 10$ V, and also so that the maximum production spread in I_D is about ±1 mA from the nominal value. The transistor characteristics are given in Figure P5.4-3.

5.5-1 Using equation (5.5-7), derive an expression for the drain-to-source resistance of a MOSFET for small values of V_{DS}.

5.5-2 Derive an expression for the parabolic equation relating i_D and v_{DS} that defines the boundary between the ohmic and the constant-current regions in a MOSFET.

5.5-3 The integrated circuit shown in Figure P5.5-3 contains two MOSFETs. Explain how connecting the substrate to the most negative potential in the circuit keeps the transistors electrically isolated from one another.

5.5-4 Explain how the drain current is able to flow through the pinched-off region of the channel in Figure 5.5-4c.

5.5-5 **(a)** By using equation (5.5-9), derive an expression for the volt-ampere characteristics of an n-channel enhancement-style MOSFET whose gate and drain are connected together.
(b) Discuss how this circuit can be used to measure the threshold voltage of the transistor.

5.5-6 The device shown in Figure 5.5-11 is an enhancement-style MOSFET with a threshold voltage of 3 V. In this circuit $V_{DD} = 7$ V, $R_D = 0$, and $R_S = 1.2$ kΩ. When R_2 is removed, $I_D = 2$ mA. Find I_D with R_2 in place.

5.5-7 For the circuit shown in Figure P5.5-8, $I_{D2T} = 9$ mA, $V_T = 3$ V, $R_D = 2$ kΩ, $R_S = 0$, $R_1 = ∞$, and R_2 is variable.
(a) Find the circuit Q point if $R_2 = 10$ MΩ and the gate leakage current is 100 nA.
(b) Repeat part (a) if $R_2 = 50$ MΩ.

5.5-8 Select R_1 and R_D in Figure P5.5-8 to place the transistor Q point at $V_{GS} = 2$ V, $I_D = 4$ mA, and $V_{DS} = 8$ V.

5.5-9 The transistor in Figure 5.5-11a is considered to be a depletion-style device with $I_{DSS} = 4$ mA and $V_p = -3$ V. In addition, $V_{DD} = 12$ V, $R_D = 500$ Ω, and $R_S = 2$ kΩ.
(a) Find the Q point of the MOSFET.
(b) To what value must R_2 be changed to make $I_D = 1$ mA?

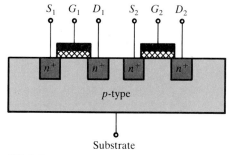

S_1 G_1 D_1 S_2 G_2 D_2

n^+ n^+ n^+ n^+

p-type

Substrate

Figure P5.5-3

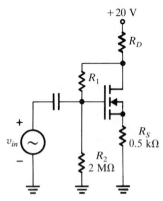

+20 V

R_D

R_1

R_S
0.5 kΩ

R_2
2 MΩ

v_{in}

Figure P5.5-8

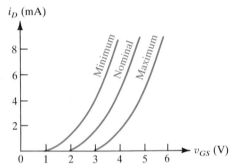

i_D (mA)

Minimum   Nominal   Maximum

v_{GS} (V)

Figure P5.5-12

5.5-10 The circuit shown in Figure P5.5-8 contains an enhancement-type transistor whose volt-ampere equation is given by

$$I_D = I_x \left(\frac{V_{GS}}{V_x}\right)^3$$

where I_x and V_x are constants having values of 2 mA and 3 V, respectively. In addition, $R_1 = 0.67$ MΩ and R_S is adjustable. Select R_D and R_S to place the Q point at $V_{DS} = 10$ V and $I_D = 16$ mA.

5.5-11 The circuit shown in Figure P5.5-8 has an enhancement MOSFET transistor with $V_T = 3$ V and $I_{D2T} = 9$ mA. In addition, $V_{DD} = 15$ V, $R_D = 2.5$ kΩ, $R_2 = 1$ MΩ, and R_S is adjustable. Select R_1 and R_S to place the Q point at $I_D = 2$ mA and $V_{DS} = 7$ V.

5.5-12 For the circuit shown in Figure P5.5-8, $V_{DD} = 15$ V, $R_2 = 1$ MΩ, and $R_D = 1$ kΩ. Using the characteristic given in Figure P5.5-12, complete the design of the circuit so that the enhancement-type FET transistor has a nominal drain current $I_D = 5$ mA. In addition, the design is to be such that the use of a transistor having production spread characteristics anywhere within the range given on the graph will result in less than a ±20% shift in its Q point from the nominal value.

Note: In this section all coupling and bypass capacitors are to be considered ac shorts.

5.6-1 Using the graphical data given in Figure 5.6-4 estimate g_m and r_d at the Q point $I_D = 1$ mA and $V_{DS} = 5$ V.

5.6-2 For the circuit in Figure 5.6-6, $V_{DD} = 15$ V, R_L and C_3 are removed, $R_D = 1.5$ kΩ, $R_S = 0.5$ kΩ, $R_1 = 13.5$ MΩ, $R_2 = 1.5$ MΩ, $I_{DSS} = 6$ mA, and $V_p = -3$ V.
(a) Find the dc Q point for the circuit given.
(b) For $v_{in} = 0.5 \sin \omega t$ volts, sketch v_G, v_{GS}, v_S, i_D, and v_{DS}.

5.6-3 In Figure 5.6-6, $V_{DD} = 15$ V, $R_L = R_D = 3$ kΩ, $R_S = 1$ kΩ, R_1 is removed, $I_{DSS} = 6$ mA, and $V_P = -3$ V. Repeat Problem 5.6-2 for this circuit. In part (b), also sketch v_L.

5.6-4 For the circuit in Figure 5.6-3, $v_{in} = 1 \sin \omega t$ V, $R_D = 3$ kΩ, $R_S = 0.67$ kΩ, and C_2 is removed. Carefully sketch and label v_{GS}, i_D, v_S, v_D, and v_{DS}. The transistor characteristics are given in Figure 5.6-1a.

5.6-5 For the circuit described in Problem 5.6-4:
(a) Using the i_D versus v_{GS} graph given in Figure 5.6-1a, determine the Q-point information.
(b) Evaluate the transistor g_m at this Q point and using the small-signal model, compute all the ac quantities needed to complete part (c).
(c) Carefully sketch and label v_{in}, v_{GS}, i_D, v_S, v_D, and v_{DS}.

5.6-6 For the circuit in Figure 5.6-6a, $v_{in} = 1 \sin \omega t$ V, $V_{DD} = 20$ V, $R_L = 3$ kΩ, $R_S = 500$ Ω, and R_1 and C_2 are removed. Carefully sketch and label v_{GS}, i_D, v_S, v_D, and v_{DS}. The transistor characteristics are given in Figure 5.6-1a.

5.6-7 For the circuit in Problem 5.6-6:
(a) Graphically determine the circuit Q point.
(b) Evaluate the transistor g_m at this Q point, and using the small-signal model, compute all the ac parameters needed to complete part (c).
(c) Carefully sketch and label v_{in}, v_{GS}, i_D, v_S, v_D, and v_{DS}.

5.6-8 In Figure 5.6-3, $R_S = 0.66$ kΩ, and R_D is replaced by the series connection of $R_{D1} = 1.33$ kΩ and $R_{D2} = 1$ kΩ (with R_{D1} on top). In addition, a bypass capacitor is connected from the junction of these two resistors to ground. Given that $V_{GS} = -2$ V and that $v_{in} = 1 \sin \omega t$ volts, sketch and label the waveforms v_G, v_S, v_{GS}, i_D, v_{DS}, and v_D. The transistor volt-ampere characteristics are given in Figure 5.6-1a.

5.6-9 For the circuit given in Figure 5.6-6, $V_{DD} = 25$ V, $R_L = R_D = 5$ kΩ, $R_S = 2.5$ kΩ, $R_1 = 2$ MΩ, and $R_2 = 500$ kΩ. Sketch the total output signals v_D and v_L given that $I_D = 2$ mA, $v_{in} = 2 \sin \omega t$ V, $r_d = 10$ kΩ, and $g_m = 2$ mS.

5.6-10 In Figure 5.6-6, $v_{in} = 0.5 \sin \omega t$ V, $V_{DD} = 16$ V, $R_D = 5$ kΩ, $R_S = 1$ kΩ, $R_1 = 3$ MΩ, $R_2 = 1$ MΩ, and R_L and C_3 are removed. In addition, consider that the JFET is replaced by an n-channel enhancement-type MOSFET operating at a Q point $V_{GS} = 3$ V. If the drain current is related to the other transistor variables by the equation

$$i_D = I_1\left[1 - 0.5\left(\frac{v_{GS}}{V_x}\right)^2 + 1.5\left(\frac{v_{GS}}{V_x}\right)^3\right]$$

where I_1 and V_x are 0.5 mA and 3 V, respectively:
(a) Determine g_m and r_d for the transistor.
(b) Using the results in part (a), sketch the total v_{DS} for the circuit.

5.6-11 For the circuit in Figure 5.6-6, $I_{DSS} = 4$ mA, $V_p = -3$ V, $V_{DD} = 20$ V, $R_L = 2$ kΩ, $R_S = 0$, $R_1 = \infty$, and C_1 is replaced by a 1-MΩ resistor. The input signal v_{in} is a square wave whose amplitude varies between 0 and -10 V. Sketch v_{GS}, i_L, v_{DS}, and v_L. (*Hint:* The average value of i_L must be zero.)

5.6-12 For the transistor amplifier shown in Figure 5.6-6, $I_{DSS} = 4$ mA, $V_p = -1.5$ V, $V_{DD} = 10$ V, $R_D = 1$ kΩ, $R_S = 0.5$ kΩ, $R_1 = 10$ MΩ, $R_2 = 1.5$ MΩ, and R_L and C_3 are removed. Find the largest allowed value of v_{in} for which v_{DS} is undistorted.

For all problems in this section, unless otherwise noted, $g_m = 1$ mS and $r_d = \infty$ for all FETs, and $h_{ie} = 1$ kΩ and $h_{fe} = 100$ for all BJTs. In addition, assume all capacitors to be ac shorts.

5.7-1 For the circuit shown in Figure P5.7-1, given that $I_D = 1$ mA and that $r_d = 10$ kΩ:
(a) Calculate the ac small-signal gain v_o/v_{in}.
(b) Sketch the total voltages v_o, v_D, and v_S versus time for $v_{in} = 0.2 \sin \omega t$.

5.7-2 In Figure 5.6-3, $R_S = 3$ kΩ, C_2 is removed, and R_D is replaced by the series connection of two 5-kΩ resistors. In addition, a bypass capacitor is connected from the junction of these two resistors to ground. Given that $I_D = 2$ mA, sketch the total voltages v_S, v_{GS}, and v_D if $v_{in} = 1 \cdot (\sin \omega t)$ volts. You may assume that $r_d = 10$ kΩ and $g_m = 5$ mS.

5.7-3 For the FET circuit shown in Figure 5.7-4a, $I_{DSS} = 10$ mA, $V_p = -4$ V, $V_{DD} = 10$ V, $R_{S1} = 100$ Ω, $R_{S2} = 134$ Ω, $R_D = 1$ kΩ, and C_3 and R_L are removed.
(a) Find the Q point: I_D, V_{DS}, and V_{GS}.
(b) Find v_D and v_{DS} given that $v_{in} = 1.0 \sin \omega t$ volts.

5.7-4 In the circuit shown in Figure P5.7-4, consider the transformer to be ideal and the FET to have the following characteristics: $g_m = 5$ mS and $r_d = 50$ kΩ.
(a) Draw a good ac model valid in midband.
(b) Find the voltage gain v_L/v_{in}.

5.7-5 For the circuit shown in Figure 5.7-4a, $V_{DD} = 30$ V, $R_D = 5$ kΩ, $R_{S1} = 2$ kΩ, $R_{S2} = 3$ kΩ, $R_L = 5$ kΩ, $R_G = 1$ MΩ, and a 500-kΩ resistor is connected in series with C_1. In addition, the lower side of R_G is removed from ground and connected instead to the junction point of R_{S1} and R_{S2}.
(a) Calculate the ac gain v_L/v_{in} given that $r_d = 10$ kΩ.
(b) If $I_D = 1$ mA, find the important dc voltages and sketch the total signals v_L, v_D, v_G, and v_S versus time for $v_{in} = 0.2 \sin \omega t$ volts. (*Hint:* To do the dc calculations, you may neglect the current flow in R_G.)

5.7-6 For the circuit shown in Figure P5.7-6, find v_o/v_{in}.

5.7-7 In the circuit shown in Figure P5.7-7, V_1 and V_2 are used to provide the dc bias. Find the circuit gains at each of the points v_{o1} and v_{o2}.

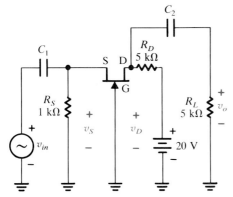

Figure P5.7-1

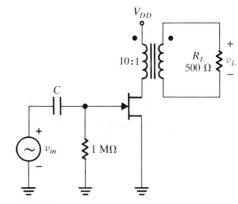

Figure P5.7-4

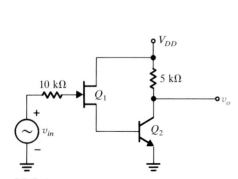

Figure P5.7-6

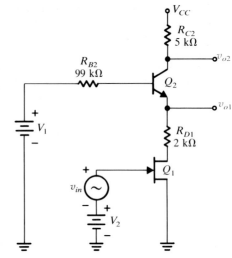

Figure P5.7-7

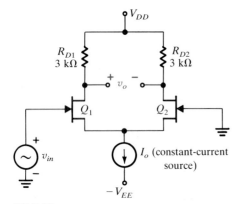

Figure P5.7-11

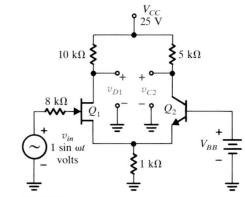

Figure P5.7-12

5.7-8 Compute the voltage gain v_o/v_{in} of the amplifier in Figure 5.7-5a if a 2-kΩ resistor is shunted from the base of Q_2 to ground and if a 5-kΩ load is connected from the collector of Q_2 to ground.

5.7-9 Find the overall voltage gain of the amplifier in Figure 5.8-2a if 500-Ω resistors are connected between the source lead of each transistor and ground. The transistor g_m's are each 1 mS.

5.7-10 For the circuit in Figure 5.8-2c, 2-kΩ resistors are connected from the source leads of Q_1 and Q_2 to ground. Find the voltage gain v_L/v_{in} if $g_m = 1$ mS for the transistors.

5.7-11 For the circuit shown in Figure P5.7-11, find the ac small-signal gain v_o/v_{in} given that $g_m = 4$ mS for Q_1 and Q_2.

5.7-12 Given that the quiescent current in each of the tran-

sistors shown in Figure P5.7-12 is 1 mA, carefully sketch and label the total (dc + ac) voltages v_{D1} and v_{C2}.

5.8-1 For the case where $R_{in} = 0$ and $R_L = \infty$, which of these circuits has the largest voltage gain (v_o/v_{in})? What is the value of this gain?

5.8-2 For the case where $R_{in} = 0$ and $R_L = 1$ kΩ, which of these circuits has the largest current gain (i_L/i_{in})? What is the value of this gain?

5.8-3 An amplifier with a voltage gain of 100 is needed. The circuit source impedance $R_{in} = 1$ MΩ and the load R_L is infinite. Which circuit should be selected, and what are the required component values?

5.8-4 An amplifier with a voltage gain of 100 is needed. The source impedance is 10 kΩ and the load impedance is 100 Ω. Which circuit should be selected, and what are the required component values?

The basic circuit to be examined in each of the problems in this section is shown in Figure P5(A). Possible amplifiers to be used in this circuit are given in Figure P5(B). For each of these circuits the drain and collector resistors can take on any values between 0 Ω and 20 kΩ. In addition, you should consider that $g_m = 2$ mS for all FETs, and $h_{ie} = 1$ kΩ and $h_{fe} = 100$ for all BJT devices.

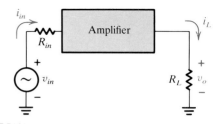

Figure P5(A)

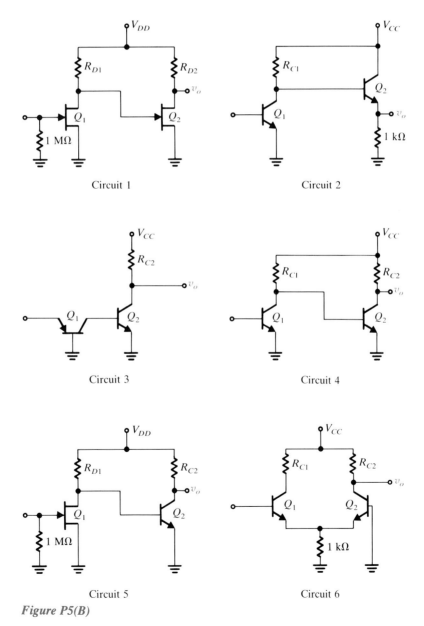

Figure P5(B)

5.8-5 An amplifier with a current gain (i_L/i_S) of 75 is needed. The source resistance varies between 50 and 500 Ω, and it is necessary that the current gain vary by less than 50% when this occurs. R_L is fixed at 1 kΩ (Figure P5.8-5). Which circuit should be selected, and what are the required components?

5.8-6 An amplifier with a voltage gain of approximately 50 is needed. The load varies between 100 Ω and 1 kΩ, and it is necessary that the circuit gain vary by less than 20% in response to these changes in R_L. The source resistance $R_{in} = 0$. Which circuit should be selected, and what are the required component values?

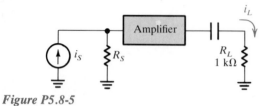

Figure P5.8-5

CHAPTER 6

Frequency Effects in Amplifier Circuits

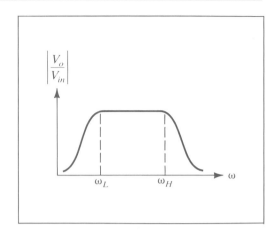

6.1 INTRODUCTION

Electronic circuits containing capacitors and inductors exhibit transfer functions that are frequency dependent because the impedance of these elements varies with the frequency of the applied signal. Sometimes these reactive elements are intentionally placed in the circuit to achieve a specific design goal, whereas in other cases, although these elements are physically present in the circuit, they were not put there intentionally by the designer.

A coupling capacitor is an example of a reactive element that is often purposefully included in a circuit to achieve dc isolation between stages. Generally, its impedance is chosen so that it is an ac short in the midband or middle range of frequencies, where signal transmission is desired. At low frequencies the impedance of this coupling capacitor rises significantly and causes the gain of the amplifier to decrease (Figure 6.1-1a).

Bypass capacitors are generally connected in parallel with the emitter or source resistors in transistor amplifiers, and are used to increase the ac gain of the amplifier. As with the coupling capacitor, the size of the bypass capacitor is chosen so that it looks like an ac short circuit in midband. At low frequencies the impedance of the bypass capacitors increase, and this reduces the amplifier gain at the low-frequency end of the spectrum, as shown in Figure 6.1-1b.

Dc (direct-coupled) amplifiers generally do not use any coupling or bypass capacitors. As a result, their frequency response does not exhibit any gain roll-off at the low-frequency end of the spectrum but is instead "flat" all the way down to dc (Figure 6.1-1c).

In addition to the coupling capacitors, bypass capacitors, and other reactive elements intentionally included in the circuit by the designer, there are numerous other "parasitic" R, L, and C elements in the circuit which affect its fre-

269

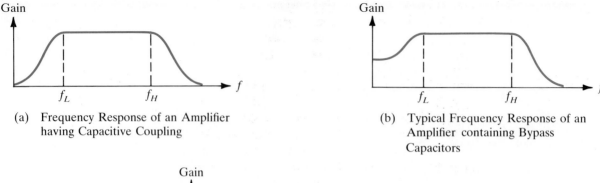

(a) Frequency Response of an Amplifier having Capacitive Coupling

(b) Typical Frequency Response of an Amplifier containing Bypass Capacitors

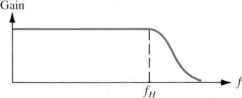

(c) Frequency Response of a DC Coupled Amplifier

Figure 6.1-1

quency response. These components, although not placed there by the designer, are actually present in the circuit. The term "parasitic," which is used to describe these components, comes from the word "parasite" and indicates that these elements are unwelcome guests in the circuit in which they are found. Most often these parasitic elements have a negligible influence on the performance of the circuit, so their presence can be ignored. However, sometimes, especially at high frequencies, their effects can be quite pronounced.

Parasitic resistive elements are associated with the finite resistance of the wiring used to interconnect the electrical components, as well as with the resistance of the component leads themselves. In general, because these resistances are only a small fraction of an ohm, they can be ignored; however, for circuits carrying large currents, such as power supplies and power amplifiers, the voltage drops produced across these elements can become important.

In addition to dc resistance, circuit wiring has inductance associated with it. However, because this inductance is only a small fraction of a microhenry, it can usually be ignored, except in circuits designed to operate at very high frequencies. For our work in this chapter we assume that the effects of both the parasitic resistances and the parasitic inductances can be neglected. However, we will not be able to ignore the parasitic circuit capacitances since their effect in electronic amplifiers, operating at even moderately high frequencies, can be significant.

In electronic circuits a parasitic capacitance exists wherever two conductors are close to one another. As a result, significant capacitance effects can occur between adjacent wires or adjacent lands on printed circuit boards. In addition, other important parasitic capacitances are associated with the transistors, ICs, and diodes in the circuit. As might be imagined, these stray capacitances are rather small (typically, on the order of 0.1 to 100 pF), and as a result, in the low- and middle-frequency ranges, their impedances are large enough that they may be considered open circuits. At high frequencies, on the other hand, their effects can be significant. As the frequency of the applied signal increases, the impedance of these elements falls, and in general the amplifier gain decreases. As a result, for all types of electronic amplifiers the gain in the high-frequency region falls off as the frequency increases (Figures 6.1-1a through c).

In addition to the direct and capacitive interstage coupling methods, it is possible to couple stages together with a transformer. As discussed in Chapter 4, transformer coupling is a good way to interconnect two stages since it offers both dc isolation and maximum power transfer. Its major drawbacks are its size, cost, and poor frequency response. At low frequencies the impedance of the transformer's primary inductance is very small and shunts most of the signal current around the transformer, significantly reducing the signal level getting through to the secondary. At high frequencies, due to flux leakage effects, the current in the primary is greatly decreased, and again the signal is blocked before it reaches the secondary. As a result, the frequency response of a transformer-coupled amplifier exhibits a roll-off at high and low frequencies that is similar to the capacitively coupled amplifier illustrated in Figure 6.1-1a.

6.2 LOW-FREQUENCY ANALYSIS TECHNIQUES FOR CIRCUITS CONTAINING COUPLING CAPACITORS

The effect of coupling capacitors on the frequency response of electronic circuits is best illustrated by examining the performance of a relatively simple amplifier such as that shown in Figure 6.2-1a. Initially, we analyze the response of this circuit by replacing the transistor by its small-signal model and then evaluate the transfer function of the resulting network. Once this formal analysis technique has been developed completely, we reexamine the problem, this time with an eye toward developing a more intuitive method for understanding the frequency response of this circuit.

The small-signal equivalent circuit for this amplifier is illustrated in Figure 6.2-1b. In developing the expressions for the transfer functions, we use capital letters with lowercase subscripts [e.g., $V_{in}(s)$] to represent the Laplace transforms of the ac variables in the problem. Following Figure 6.2-1b, we may write the transfer function for the input portion of this circuit as

$$\frac{V_g(s)}{V_{in}(s)} = \frac{R_p}{R_p + R_{in} + 1/sC} = \frac{R_p}{R_p + R_{in}} \frac{s}{s + 1/(R_p + R_{in})C} \tag{6.2-1}$$

The corresponding expression for the output voltage at the drain is given by

$$V_d(s) = -g_m R_D V_g(s) \tag{6.2-2}$$

Combining equations (6.2-1) and (6.2-2), the complete transfer function for this amplifier may be written as

$$\frac{V_d(s)}{V_{in}(s)} = \underbrace{(-g_m R_D) \frac{R_p}{R_p + R_{in}}}_{A_{\text{mid}}} \frac{s}{s + 1/(R_p + R_{in})C} \tag{6.2-3}$$

Substituting in the component values and assuming a transistor g_m of 10 mS,

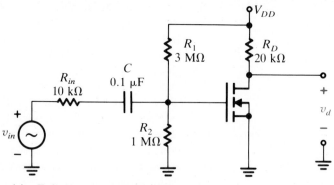

(a) Enhancement MOSFET Capacitively Coupled Amplifier

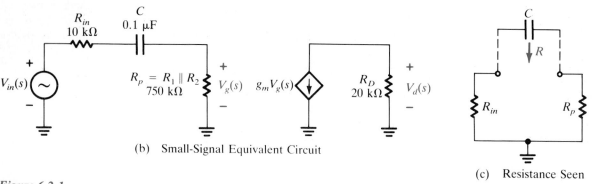

(b) Small-Signal Equivalent Circuit

(c) Resistance Seen by the Capacitor C

Figure 6.2-1

we may rewrite this expression as

$$\frac{V_d(s)}{V_{in}(s)} = \underbrace{(-200)}_{A_{mid}}\frac{s}{s + 13.3} \qquad (6.2\text{-}4)$$

The resulting gain characteristic for this amplifier is sketched in Figure 6.2-2. If you have forgotten how to make this type of sketch, you should review the material in Section 8 of Appendix I.

To understand the shape of the amplifier frequency response, consider that in the midband frequency range the capacitor can be replaced by a short circuit to calculate the value of the midband gain, A_{mid}, given on the Bode plot. Furthermore, in this frequency range $V_g(j\omega)$ is approximately equal to $V_{in}(j\omega)$, so that the output voltage $V_d(j\omega)$ may be written as

$$V_d(j\omega) = -g_m R_D V_g(j\omega) = -g_m R_D V_{in}(j\omega) \qquad (6.2\text{-}5a)$$

from which it follows that

$$A_{mid} = \frac{V_d(j\omega)}{V_{in}(j\omega)} = -g_m R_D = -200 \qquad (6.2\text{-}5b)$$

From equation (6.2-5b) the resulting midband gain magnitude in decibels is seen to be equal to $20 \log_{10} (200)$, or 46 dB.

As the frequency of the applied signal is lowered, the impedance of the ca-

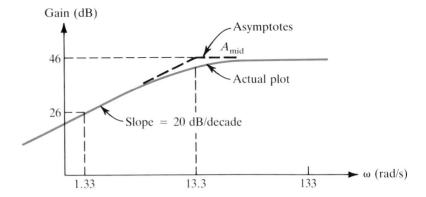

Figure 6.2-2
Low- and middle-frequency gain characteristic for the MOSFET amplifier in Figure 6.2-1.

pacitor begins to increase and eventually its effect can no longer be ignored. To find the lower 3-dB point of the amplifier, we observe that this break is associated with the pole or denominator term of the transfer function in equation (6.2-3). Furthermore, it occurs at the frequency where the magnitude of the capacitor impedance equals the impedance of the effective resistor connected across the capacitor terminals, that is, where

$$\frac{1}{\omega C} = R \tag{6.2-6a}$$

or where

$$\omega = \frac{1}{RC} \tag{6.2-6b}$$

In this expression the term R represents the resistance seen by the capacitor with all independent sources set equal to zero. For more complex networks where the value of R is not immediately obvious, it can always be found by determining the Thévenin resistance between the terminals of the capacitor. However, for this circuit these complexities are unnecessary. In particular, if we set $V_{in} = 0$ (see Figure 6.2-1c), the resistance seen by the capacitor is simply given by

$$R = R_{in} + R_p = 760 \text{ k}\Omega \tag{6.2-7}$$

Using this result together with the fact that $C = 0.1 \ \mu F$, we see that the low-frequency break for this amplifier occurs at 13.3 rad/s or at around 2 Hz.

To summarize the method for finding the overall transfer function of an amplifier containing a coupling capacitor, we observe that this capacitor introduces a term of the form $s/(s + \omega_D)$ into the transfer function and that the complete transfer function may be written as

$$A_v(s) = A_{mid} \frac{s}{s + \omega_D} \tag{6.2-8}$$

where A_{mid} represents the amplifier's midband gain. The value of A_{mid} is found by replacing the capacitor by a short circuit, and the value of the denominator break frequency in this equation is given by equation (6.2-6), where R represents the resistance seen by the capacitor. If the circuit contains multiple coupling capacitors, the transfer function will still be similar to that given by equation (6.2-8) except that each capacitor will contribute a term of the form $s/(s + \omega_D)$. This type of circuit is illustrated in the example that follows.

EXAMPLE 6.2-1

For the circuit shown in Figure 6.2-3, the capacitors C_1 and C_2 are coupling capacitors. Determine the overall transfer function (in the low- and mid-frequency regions) and sketch the gain characteristic of this amplifier. Consider that $h_{ie} = 1$ kΩ and that $h_{fe} = 100$ for the transistor.

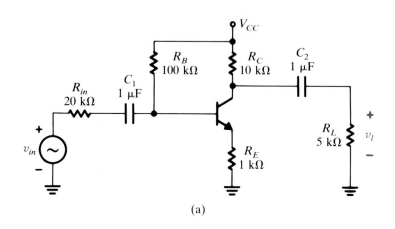

(a)

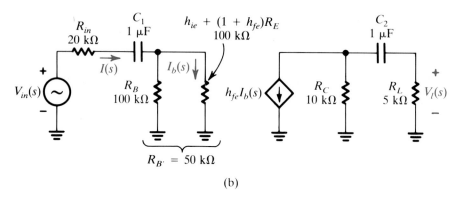

(b)

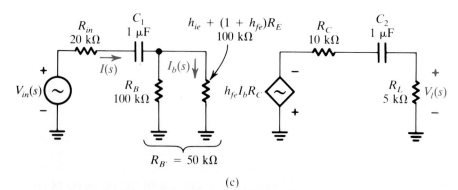

(c)

Figure 6.2-3
Example 6.2-1.

Chapter 6 / Frequency Effects in Amplifier Circuits

SOLUTION

To analyze this problem, we begin by drawing the small-signal equivalent circuit for the amplifier (Figure 6.2-3b). Notice that all the small-signal impedance transformation techniques developed in previous chapters are still valid, and thus the input impedance of the transistor can be represented by a resistor $h_{ie} + (1 + h_{fe})R_E$.

The current $I(s)$ in the input circuit is given by Ohm's law as

$$I(s) = \frac{V_{in}(s)}{R_{in} + R_B' + 1/sC} = \frac{V_{in}}{R_{in} + R_B'} \frac{s}{s + \omega_1} \qquad (6.2\text{-}9a)$$

where

$$R_B' = R_B \parallel [h_{ie} + (1 + h_{fe})R_E] \qquad (6.2\text{-}9b)$$

and

$$\omega_1 = \frac{1}{(R_{in} + R_B')C} \qquad (6.2\text{-}9c)$$

The current into the base of the transistor is found by current division to be

$$I_b(s) = I(s)\frac{R_B}{R_B + h_{ie} + (1 + h_{fe})R_E} = \frac{V_{in}(s)[R_B'/(R_B' + R_{in})]}{h_{ie} + (1 + h_{fe})R_E} \frac{s}{s + \omega_1} \qquad (6.2\text{-}10)$$

To assist in obtaining the voltage $V_l(s)$ at the output, the equivalent circuit is redrawn in Figure 6.2-3c. Using this figure, the output voltage is found by applying the voltage-divider equation as

$$V_l(s) = [-h_{fe}R_C I_b(s)]\frac{R_L}{R_L + R_C + 1/sC_2} = [-h_{fe}R_L' I_b(s)]\frac{s}{s + \omega_2} \qquad (6.2\text{-}11a)$$

where

$$R_L' = R_L \parallel R_C \qquad (6.2\text{-}11b)$$

and

$$\omega_2 = \frac{1}{(R_L + R_C)C_2} \qquad (6.2\text{-}11c)$$

Combining equations (6.2-10) and (6.2-11a), we obtain

$$\frac{V_l(s)}{V_{in}(s)} = \underbrace{\frac{(-h_{fe}R_L')[R_B'/(R_{in} + R_B')]}{h_{ie} + (1 + h_{fe})R_E}}_{A_{mid}} \frac{s}{s + \omega_1} \frac{s}{s + \omega_2} \qquad (6.2\text{-}12)$$

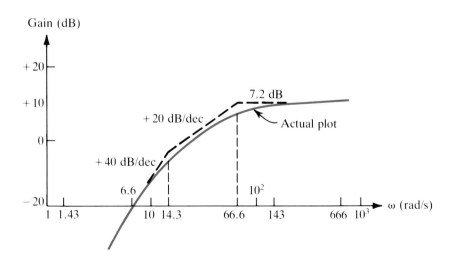

Figure 6.2-4
Gain plot for the amplifier illustrated in Figure 6.2-3.

On substituting numerical values in this problem, the overall transfer function may be written as

$$\frac{V_l(s)}{V_{in}(s)} = \underbrace{(-2.3)}_{A_{mid}} \frac{s}{s + 14.3} \frac{s}{s + 66.6} \qquad (6.2\text{-}13)$$

The gain plot for this amplifier is illustrated in Figure 6.2-4.

EXAMPLE 6.2-2

Repeat the calculation of the transfer function for the circuit in Example 6.2-1, but this time use the more intuitive analysis method.

SOLUTION

Because this transistor amplifier has two coupling capacitors, we know that the general form of the transfer function will be

$$\frac{V_l(s)}{V_{in}(s)} = A_{mid} \frac{s}{s + \omega_1} \frac{s}{s + \omega_2} \qquad (6.2\text{-}14)$$

where A_{mid} represents the midband gain of the amplifier, with C_1 and C_2 replaced by short circuits. In addition, ω_1 and ω_2 are the break frequencies associated with the capacitors C_1 and C_2 respectively. These break frequencies are determined from the equation $\omega = 1/RC$, where R represents the impedance seen by each of the capacitors.

To obtain the equivalent circuit in the midband frequency range we need only replace C_1 and C_2 in Figure 6.2-3b by shorts. For this circuit the current $I_b(s)$ may then be written as

$$I_b(s) = \frac{1}{2}I(s) = \left(\frac{1}{2}\right)\frac{V_{in}(s)}{70 \text{ k}\Omega} \qquad (6.2\text{-}15)$$

and the corresponding output voltage, $V_l(s)$, for this circuit is

$$V_l(s) = -I_c(s)(R_C \| R_L) = [-100 I_b(s)](3.3 \text{ k}\Omega) = -2.3 V_{in}(s) \qquad (6.2\text{-}16)$$

Using this result, we see that the amplifier midband gain is -2.3.

To determine the break frequencies associated with each of the capacitors, we need to evaluate the resistance seen by C_1 and C_2. This portion of the analysis is best carried out with the aid of Figure 6.2-5, in which the original equivalent circuit in Figure 6.2-3b has been redrawn with the independent source $V_{in}(s)$ set equal to zero. The resistance seen by C_1 is obviously $(R_{in} + R_B')$ or 70 kΩ, so the break associated with this capacitor occurs at

$$\omega_1 = \frac{1}{(70 \times 10^3)(10^{-6})} = 14.3 \text{ rad/s} \qquad (6.2\text{-}17)$$

The break associated with C_2 needs further explanation. Although the output circuit contains a dependent source, this source can be eliminated from the circuit because the current flowing through it does not depend on any of the voltages or currents in the output loop. Therefore, the Thévenin impedance seen by C_2 is simply $(R_C + R_L)$ or 15 kΩ, and the resulting break frequency associated with C_2 is

$$\omega_2 = \frac{1}{(15 \times 10^3)(10^{-6})} = 66.6 \text{ rad/s} \qquad (6.2\text{-}18)$$

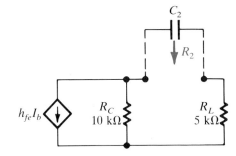

Figure 6.2-5
Example 6.2-2. Determining the break frequencies of C_1 and C_2.

It is very useful to observe that the analyses for the capacitor break frequencies could also have been carried out directly from the original circuit diagram in Figure 6.2-3a by employing the impedance concepts developed for the transistor in previous chapters. For example, the resistance seen to the left of C_1 is just equal to R_{in}, while that to the right is given by the parallel combination of R_B with the input resistance of the transistor $[h_{ie} + (1 + h_{fe})R_E]$. A similar analysis for C_2 indicates that the resistance seen to the left of C_2 is just R_C, while that to the right is R_L, and further that these two resistors are in series.

Exercises

6.2-1 Find the output signal amplitude and phase shift for the amplifier illustrated in Figure 6.2-1 if v_{in} is a 50-mV 1-Hz sine wave. **Answer** 4.26, $-115.3°$

6.2-2 The emitter resistor in Figure 6.2-3a is reduced to 500 Ω. Determine the new midband gain and denominator break terms of the amplifier. **Answer** $A_{\text{mid}} = -4.16$, $\omega_1 = 18.76$ rad/s, $\omega_2 = 66.6$ rad/s

6.2-3 Find the gain magnitude in decibels at $\omega = 20$ rad/s for the amplifier shown in Figure 6.2-3a. **Answer** -5.4 dB

6.3 LOW-FREQUENCY RESPONSE OF CIRCUITS CONTAINING BYPASS CAPACITORS

The effect produced by a bypass capacitor on the frequency response of an amplifier is somewhat different from that caused by a coupling capacitor. At low frequencies, when a coupling capacitor begins to open, the gain of the amplifier falls off toward zero because the signal path from the input to the output is broken. With a bypass capacitor, on the other hand, when the capacitor opens, depending on its location in the circuit, it will either increase or decrease the amplifier gain, but will not cause it to go to zero. As a result, an amplifier containing a bypass capacitor tends to have a frequency response that consists of a transition between two fixed-gain levels as the capacitor impedance changes from "looking like an open circuit" to "looking like a short" (see, for example, Figure 6.3-2).

The general form of the transfer function associated with this type of frequency response has break terms in both the numerator and denominator. Techniques for analyzing the gain-phase characteristics of this type of transfer function are discussed in Section 8 of Appendix I.

The circuit given in Figure 6.3-1 illustrates a single-stage JFET amplifier

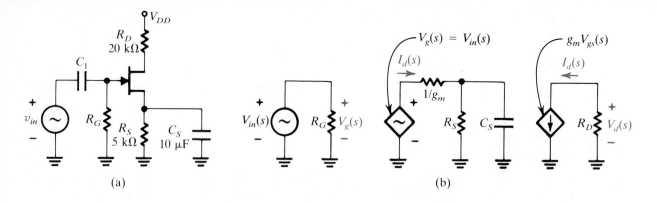

(a)

(b)

Figure 6.3-1

Frequency response of an FET circuit containing a bypass capacitor ($g_m = 1$ mS, $r_d = \infty$).

containing a source resistor R_S to provide dc bias for the circuit, and a bypass capacitor C_S to short out this resistor in order to increase the amplifier gain in the midband frequency range. In analyzing this circuit, we assume that the coupling capacitor C_1 is so large that it can be considered an ac short for all frequencies being examined. To the extent that this approximation is valid, we may draw the small-signal model for this circuit as shown in Figure 6.3-1b; this model will be valid in the low- and middle-frequency ranges. Notice that we have made use of the small-signal equivalent circuit methods developed in previous chapters to divide the overall model into three simpler subcircuits which represent, from left to right, the gate, source, and drain equivalent circuits, respectively.

From the source portion of this equivalent circuit, we can see that the drain current is

$$I_d(s) = \frac{V_{in}(s)}{(1/g_m) + [R_S(1/sC_S)/(R_S + 1/sC_S)]} = \frac{V_{in}(s)(1 + sR_S C_S)}{(1/g_m) + R_S + sR_S(1/g_m)C_S}$$

(6.3-1)

After some algebraic manipulation, this expression may be rewritten as

$$I_d(s) = g_m V_{in}(s)\frac{s + \omega_{SN}}{s + \omega_{SD}}$$

(6.3-2)

where

$$\omega_{SN} = \frac{1}{R_S C_S}$$

(6.3-3a)

and

$$\omega_{SD} = \frac{1}{(R_S \parallel 1/g_m)C_S}$$

(6.3-3b)

Using the output portion of the equivalent circuit, the drain voltage is

$$V_d(s) = -I_d(s) R_D = \underbrace{(-g_m R_D)}_{A_{\text{mid}}}\frac{s + \omega_{SN}}{s + \omega_{SD}}V_{in}(s)$$

(6.3-4)

In this expression, the term $-g_m R_D$ represents the midband gain with the capacitor replaced by a short circuit. The gain of this circuit as a function of frequency is sketched in Figure 6.3-2.

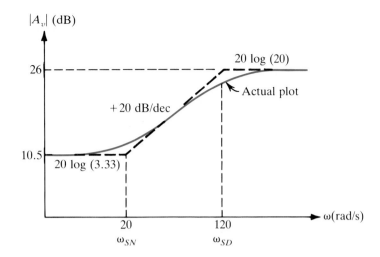

Figure 6.3-2
Gain versus frequency response for the circuit in Figure 6.3-1.

6.3-1 Techniques for Rapidly Determining the Frequency Response of Circuits Containing Bypass Capacitors

It is useful to observe that it is possible to sketch this amplifier's gain versus frequency characteristic without formally developing the circuit transfer function. To begin with, we need to recognize that the flat portions of the magnitude response represent those gain regions where the capacitor may be considered to be an open or a short circuit. Specifically, in the sketch shown in Figure 6.3-2, the flat region below 20 rad/s represents the circuit gain with the capacitor replaced by an open circuit, and the region above 120 rad/s the gain with the capacitor replaced by a short.

At very low frequencies the capacitor is open and following Figure 6.3-1b we have

$$I_d(s) = \frac{V_{in}(s)}{1/g_m + R_s} = \frac{V_{in}(s)}{6 \text{ k}\Omega} \tag{6.3-5a}$$

$$V_d(s) = -I_d(s)R_D = \left[-\frac{V_{in}(s)}{6 \text{ k}\Omega}\right](20 \text{ k}\Omega) \tag{6.3-5b}$$

and
$$A_v = \frac{V_d(s)}{V_{in}(s)} = -\frac{20}{6} = -3.33 \tag{6.3-5c}$$

For frequencies in midband, the capacitor can be replaced by a short, and in this case it follows that

$$I_d(s) = \frac{V_{in}(s)}{1/g_m} = \frac{V_{in}(s)}{1 \text{ k}\Omega} \tag{6.3-6a}$$

so that
$$V_d(s) = -I_d(s)R_D = -\frac{V_{in}(s)}{1 \text{ k}\Omega}(20 \text{ k}\Omega) \tag{6.3-6b}$$

and
$$A_v = \frac{V_d(s)}{V_{in}(s)} = -20 \tag{6.3-6c}$$

To relate the mathematical expressions for the break frequencies to the

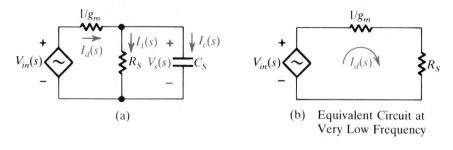

(a)

(b) Equivalent Circuit at
Very Low Frequency

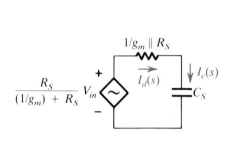

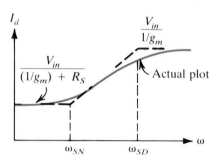

(c) Approximate Equivalent Circuit
When $Z_C \ll R_S$

(d) Variation of I_d with Frequency

Figure 6.3-3

physical performance of the circuit, we need to observe that the output voltage at the drain is directly proportional to the drain current. As a result, anything that tends to increase the drain current will increase the drain voltage and hence the gain of the circuit. From the source equivalent circuit (repeated in Figure 6.3-3a), we can see that as the impedance of C_S drops, the drain current will tend to increase.

At very low frequencies the impedance of this capacitor is so large that it has no effect on the impedance of the source circuit. In this region, where Z_C (the impedance of the capacitor) is much greater than R_S, the drain current is essentially constant, and $I_d(s)$ may be calculated using the equivalent circuit in Figure 6.3-3b. As the frequency is raised, the impedance of the capacitor begins to drop, and when its impedance becomes comparable to R_S, the overall source impedance begins to decrease and $I_d(s)$ increases. The frequency at which this occurs is when

$$\frac{1}{\omega C_S} \simeq R_S \tag{6.3-7a}$$

or when
$$\omega = \omega_{SN} = \frac{1}{R_S C_S} \tag{6.3-7b}$$

For the example being discussed, $\omega_{SN} = 20$ rad/s.

With further increases in the frequency of the applied signal, the impedance of the capacitor continues to decrease, and as its impedance becomes much less than that of R_S, it is reasonable to consider that $I_d(s)$ and $I_c(s)$ are nearly equal. When this approximation is valid, it is useful to replace the left-hand portion of the network in Figure 6.3-3a by its Thévenin equivalent (Figure 6.3-3c). It is important to note that one can always represent this network by its Thévenin equivalent, but it is only in this frequency range that the

Chapter 6 / Frequency Effects in Amplifier Circuits

current through the capacitor, $I_c(s)$, is approximately the same as the overall drain current, $I_d(s)$.

As the frequency is increased still higher, the capacitor impedance continues to fall, and eventually it becomes so low that it no longer affects the drain current. For frequencies above this value the capacitor is essentially a short circuit. The transition into this region occurs at the frequency where the impedance of the capacitor looks like the resistance of the rest of the network in Figure 6.3-3c, that is, when

$$\frac{1}{\omega C_S} \simeq R_S \parallel \frac{1}{g_m} \tag{6.3-8a}$$

or when

$$\omega = \omega_{SD} = \frac{1}{(R_S \parallel 1/g_m)C_S} \tag{6.3-8b}$$

For the circuit being examined, this break occurs at a frequency of 120 rad/s.

As an interesting aside, it is worth noting that the zero in the numerator of equation (6.3-4) may be viewed as being that value of s (in particular, $s = -1/R_S C_S$) which makes the output go to zero independent of $V_{in}(s)$. The location of this zero may be found by observing that the value of s that makes $V_d(s)$ equal to zero is the same one that makes $I_d(s)$ zero as well. From Figure 6.3-3a, $I_d(s)$ will be zero when the sum of $I_1(s)$ and $I_c(s)$ are zero. We may express these currents in terms of $V_s(s)$ as follows:

$$I_1(s) = \frac{V_s(s)}{R_S} \tag{6.3-9a}$$

and

$$I_c(s) = \frac{V_s(s)}{1/sC_S} = sC_S V_s(s) \tag{6.3-9b}$$

so that we may write

$$I_d(s) = I_1(s) + I_c(s) = C_S V_s(s)\left(s + \frac{1}{R_S C_S}\right) \tag{6.3-9c}$$

Clearly, $I_d(s)$ will be zero if $s = -1/R_S C_S$ or if the factor $s + 1/(R_S C_S)$ is zero.

In summary, the lowest break in the transfer function (the numerator break or zero for emitter or source bypass circuits) occurs when the impedance of C_S looks like the resistance R_S in parallel with it. The second break (which is a pole because it is in the denominator for this circuit) occurs when the impedance of the capacitor looks like the overall Thévenin resistance seen looking back into the network from the terminals of the capacitor. Care was taken in these derivations not to associate these breaks specifically with the numerator or the denominator, because if the bypass capacitor is in the drain or collector portion of the circuit, the roles of the numerator and denominator breaks will be interchanged; that is, the denominator break will occur before the numerator break.

From these results we can see that it is possible to sketch this amplifier's gain versus frequency response if four parameters are known. Specifically, these are the break frequencies ω_{SN} and ω_{SD}, and the amplifier gain plateaus with C_S an open and with C_S a short. However, these four parameters are not independent of one another, and as a result, if three of them are known, the fourth may easily be computed from the other three. To determine the relationship that exists between these quantities, we will reexamine the expression for the circuit transfer function [equation 6.3-4)] in two special situations. In particular, at very high frequencies (or very large values of s), the amplifier gain approaches

$$\frac{V_d(s)}{V_{in}(s)} \simeq -g_m R_D \tag{6.3-10a}$$

while for very low frequencies (where s approaches zero), C_S can be replaced by an open circuit, and we obtain

$$\frac{V_d(s)}{V_{in}(s)} \simeq -g_m R_D \frac{\omega_{SN}}{\omega_{SD}} \tag{6.3-10b}$$

Dividing equation (6.3-10a) by (6.3-10b), we see that the ratio of the circuit's gains at high and low frequencies is equal to the ratio of the break frequencies; that is,

$$\frac{A_v(\text{midband})}{A_v(\text{low})} = \frac{\omega_{SD}}{\omega_{SN}} \tag{6.3-10c}$$

6.3-2   Techniques for Approximating the Low-Frequency 3-dB Point

In the analysis and design of electronic amplifiers, it is very useful to be able to estimate ω_L, the low-frequency 3-dB point of the amplifier. To illustrate how this may be accomplished, consider, for example, the transfer function

$$H(s) = A_{\text{mid}} \frac{s + \omega_{N1}}{s + \omega_{D1}} \frac{s + \omega_{N2}}{s + \omega_{D2}} \tag{6.3-11}$$

which might result from an amplifier containing two bypass capacitors. The sketch for the frequency response of this amplifier is given in Figure 6.3-4, assuming that $\omega_{N1} < \omega_{N2} < \omega_{D1} < \omega_{D2}$. Notice that the actual location of the 3-dB break frequency will generally be somewhere to the right of the highest pole ω_{D2}. To determine its location, we rewrite $H(s)$ as

$$H(s) = A_{\text{mid}} \frac{s^2 + (\omega_{N1} + \omega_{N2}) s + \omega_{N1} \omega_{N2}}{s^2 + (\omega_{D1} + \omega_{D2})s + \omega_{D1} \omega_{D2}} \tag{6.3-12}$$

and observe for all frequencies to the right of ω_{D2} in the figure that the terms in the numerator of this expression satisfy the inequality

$$s^2 > (\omega_{N1} + \omega_{N2})s > \omega_{N1} \omega_{N2} \tag{6.3-13a}$$

while those in the denominator satisfy the inequality

$$s^2 > (\omega_{D1} + \omega_{D2})s > \omega_{D1} \omega_{D2} \tag{6.3-13b}$$

Using these results, we can approximate $H(s)$ as

$$H(s) \simeq A_{\text{mid}} \frac{s^2 + (\omega_{N1} + \omega_{N2})s}{s^2 + (\omega_{D1} + \omega_{D2})s} = A_{\text{mid}} \frac{s + (\omega_{N1} + \omega_{N2})}{s + (\omega_{D1} + \omega_{D2})} \tag{6.3-14}$$

by dropping out the small terms $\omega_{N1} \omega_{N2}$ in the numerator and $\omega_{D1} \omega_{D2}$ in the

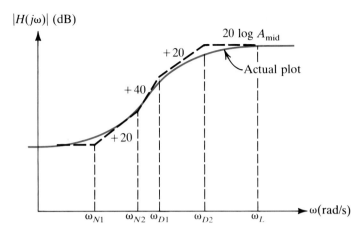

Figure 6.3-4
Magnitude of

$$H(s) = A_{mid} \frac{s + \omega_{N1}}{s + \omega_{D1}} \frac{s + \omega_{N2}}{s + \omega_{D2}}$$

versus frequency.

denominator. Using this approximation to the original transfer function, we may now find the 3-dB frequency by determining when $|H(j\omega)|$ is equal to $1/\sqrt{2}$ of its midband value or when

$$|H(j\omega)| = A_{mid} \left| \frac{j\omega + (\omega_{N1} + \omega_{N2})}{j\omega + (\omega_{D1} + \omega_{D2})} \right| = A_{mid} \frac{\sqrt{\omega^2 + (\omega_{N1} + \omega_{N2})^2}}{\sqrt{\omega^2 + (\omega_{D1} + \omega_{D2})^2}} = \frac{A_{mid}}{\sqrt{2}}$$

$$(6.3\text{-}15)$$

Solving this expression for $\omega = \omega_L$, we find that

$$\omega_L = \sqrt{(\omega_{D1} + \omega_{D2})^2 - 2(\omega_{N1} + \omega_{N2})^2} \qquad (6.3\text{-}16)$$

In actuality, this estimate for the amplifier 3-dB frequency is always a little pessimistic in that the actual location of ω_L will always be somewhat lower than this result predicts. In fact, the actual low-frequency 3-dB point can be shown to be within the range

$$\sqrt{\omega_{D,max}^2 - 2\omega_{N,max}^2} \le \omega_L \le \sqrt{(\sum_i \omega_{Di})^2 - 2(\sum_j \omega_{Nj})^2} \qquad (6.3\text{-}17)$$

where $\omega_{D,max}$ and $\omega_{N,max}$ are the largest pole and zero break terms present in the transfer function, and $\sum_i \omega_{Di}$ represents the sum of the denominator break terms and $\sum_j \omega_{Nj}$ the sum of the numerator breaks. For example, an amplifier with a transfer function

$$H(s) = (100) \frac{(s + 10)(s + 50)}{(s + 100)(s + 200)} \qquad (6.3\text{-}18a)$$

has a low-frequency 3-dB point somewhere in the range

$$\sqrt{(200)^2 - 2(50)^2} = 187 \le \omega_L \le \sqrt{(100 + 200)^2 - 2(10 + 50)^2} = 287 \qquad (6.3\text{-}18b)$$

A detailed analysis of this problem discloses that the actual 3-dB point for this amplifier is located at 229 rad/s.

The following facts about the result given in equation (6.3-17) are worth emphasizing.

1. In general, this estimate for the amplifier low-frequency 3-dB point is valid regardless of the number of poles and zeros in the transfer function, and in no way depends on whether these terms are associated with bypass or coupling capacitors in the original circuit.

2. If we can neglect the zeros of the transfer function (which is valid if they are more than a decade below the highest denominator break present), equation (6.3-17) reduces to

$$\underbrace{\omega_{D,\max}}_{\substack{\text{highest} \\ \text{individual} \\ \text{pole}}} \le \omega_L \le \underbrace{\omega_{D1} + \omega_{D2} + \omega_{D3} + \cdots}_{\text{sum of the poles}} \qquad (6.3\text{-}19)$$

This result is very useful when a rapid estimate of the amplifier's 3-dB point is needed, or when we are trying to design an amplifier having a specific half-power point.

3. If the results obtained in this discussion are extended to an amplifier containing emitter bypass and coupling capacitor interactions, it is not too difficult to show that equations (6.3-17) and (6.3-19) still apply if the denominator breaks associated with each capacitor are determined by replacing the other capacitor by a short circuit.

EXAMPLE 6.3-1

For the two-stage amplifier illustrated in Figure 6.3-5, estimate the low-frequency 3-dB point. The transistor parameters are given in the figure.

SOLUTION

This problem may be solved by inspection of the original circuit if we determine the impedances seen looking into the various points in the circuit. In particular, the important resistance values for this problem are $R_1 = 0$, $R_2 = \infty$, $R_3 = 100\ \Omega$, $R_4 = 10\ \text{k}\Omega$, and $R_5 = 1\ \text{k}\Omega$. If the origins of these resistance values are unclear to you, review Chapters 4 and 5.

Using these impedance values, the breaks associated with each of the ca-

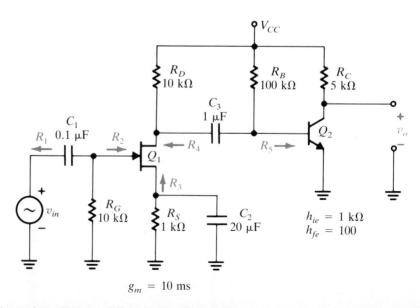

Figure 6.3-5
Example 6.3-1.

Chapter 6 / Frequency Effects in Amplifier Circuits

pacitors may be determined as follows:

$$\text{For } C_1: \quad \omega_{D1} = \frac{1}{(10 \text{ k}\Omega)(0.1 \ \mu\text{F})} = 10^3 \text{ rad/s} \qquad (6.3\text{-}20\text{a})$$

$$\text{For } C_2: \quad \omega_{N2} = \frac{1}{(10^3)(20 \ \mu\text{F})} = 50 \text{ rad/s} \qquad (6.3\text{-}20\text{b})$$

$$\omega_{D2} = \frac{1}{(10^2 \| 10^3)(20 \ \mu\text{F})} \simeq 500 \text{ rad/s} \qquad (6.3\text{-}20\text{c})$$

$$\text{and} \qquad \text{For } C_3: \quad \omega_{D3} = \frac{1}{(10^4 + 10^3)(1 \ \mu\text{F})} \simeq 100 \text{ rad/s} \qquad (6.3\text{-}20\text{d})$$

Substituting these results into equation (6.3-17), we see that the approximate range for the half-power frequency is

$$\sqrt{(10^3)^2 - 2(50)^2} = 997 \leq \omega_L \leq \sqrt{(1600)^2 - 2(50)^2} = 1598 \qquad (6.3\text{-}20\text{e})$$

$$\text{or} \qquad\qquad\qquad 159 \text{ Hz} \leq f_L \leq 254 \text{ Hz} \qquad\qquad (6.3\text{-}20\text{f})$$

For comparison, the solution of this same problem by computer-aided circuit analysis techniques yields $f_L = 194$ Hz.

EXAMPLE 6.3-2

For the amplifier circuit shown in Figure 6.3-5, C_1, C_2, and C_3 are adjustable. Complete the circuit design by selecting these components so that f_L is less than 25 Hz. Moreover, because cost minimization is a factor, and because C_2 will be the largest and hence the most expensive capacitor in the circuit, C_2 will be chosen as small as possible, breaking at or near 25 Hz. To prevent C_1 and C_3 from influencing the 3-dB point too much, their breaks are to be selected to be a decade below the desired 25-Hz 3-dB point.

SOLUTION

To begin this design problem, we select C_1 and C_3 so that their breaks occur at about 2.5 Hz. In this way, in computing the overall 3-dB point of the amplifier, these terms will have little influence on ω_L. Furthermore, because the denominator break of C_2 can now be chosen at or near the 25-Hz point, its capacitance value will be as small as is practical. Following Figure 6.3-5, the capacitors C_1 and C_3 are selected as follows:

$$\omega_1 = \frac{1}{(10 \text{ k}\Omega)C_1} \qquad\qquad (6.3\text{-}21\text{a})$$

$$\text{or} \qquad\qquad C_1 = \frac{1}{2\pi(2.5)(10 \text{ k}\Omega)} = 6.4 \ \mu\text{F} \qquad (6.3\text{-}21\text{b})$$

$$\text{and} \qquad\qquad \omega_3 = \frac{1}{(10 \text{ k}\Omega + 100 \text{ k}\Omega \| 1 \text{ k}\Omega)C_3} \qquad (6.3\text{-}21\text{c})$$

$$\text{or} \qquad\qquad C_3 = \frac{1}{2\pi(2.5)(10.99 \text{ k}\Omega)} = 5.8 \ \mu\text{F} \qquad (6.3\text{-}21\text{d})$$

If we assume that because these breaks are a decade below the amplifier 3-dB point they may be neglected, then to a good approximation, only the numerator and denominator breaks of C_2 need to be considered in computing ω_L for the amplifier. Furthermore, because the impedance seen by the numerator break of C_2 is 10 times larger than that seen by the denominator break, we can also neglect ω_{N2} in computing ω_L. Therefore, as a first-order approximation for this amplifier,

$$\omega_L \simeq \omega_{D2} = \frac{1}{[100\ \Omega\,\|\,(1\ k\Omega)]C_2} \qquad (6.3\text{-}22a)$$

from which we can estimate C_2 as

$$C_2 = \frac{1}{(2\pi)(25)(91)} = 70\ \mu F \qquad (6.3\text{-}22b)$$

Rounding up these capacitor sizes to commercially available capacitor values, we will use $C_1 = 10\ \mu F$, $C_2 = 100\ \mu F$, and $C_3 = 10\ \mu F$, and for these component values we can now recompute the breaks associated with each of the capacitors as follows:

$$\text{For } C_1: \qquad \omega_1 = \frac{1}{(10\ k\Omega)(10\ \mu F)} = 10\ \text{rad/s} \qquad (6.3\text{-}23a)$$

$$\text{For } C_2: \qquad \omega_{N2} = \frac{1}{(1\ k\Omega)(100\ \mu F)} = 10\ \text{rad/s} \qquad (6.3\text{-}23b)$$

$$\omega_{D2} = \frac{1}{(91)(100\ \mu F)} = 109.9\ \text{rad/s} \qquad (6.3\text{-}23c)$$

$$\text{and} \qquad \text{For } C_3: \qquad \omega_3 = \frac{1}{(11\ k\Omega)(10\ \mu F)} = 9.1\ \text{rad/s} \qquad (6.3\text{-}23d)$$

Substituting these results into equation (6.3-17), we may estimate the low-frequency 3-dB point of our final design as

$$\sqrt{(109.9)^2 - 2(10)^2} = 109 \le \omega_L \le \sqrt{(129)^2 - 2(10)^2} = 128.2 \quad (6.3\text{-}24a)$$

$$\text{or} \qquad\qquad 17.4\ \text{Hz} \le f_L \le 20.4\ \text{Hz} \qquad (6.3\text{-}24b)$$

Exercises

6.3-1 Determine the midband gain and the numerator and denominator breaks of the amplifier shown in Figure 6.3-1a if a 1.5-kΩ resistor is connected in series with the source lead of the transistor above the parallel combination of R_S and C_S. *Answer* $A_{\text{mid}} = 8$, $\omega_N = 20$ rad/s, $\omega_D = 60$ rad/s

6.3-2 Develop an expression for the circuit transfer function of the FET amplifier illustrated in Figure 6.3-1 if $C_1 = 1\ \mu F$, $C_S = 50\ \mu F$, and $R_G = 1\ M\Omega$. *Answer* $H(s) = -20[s(s+4)/(s+1)(s+24)]$

6.3-3 Estimate the low-frequency 3-dB point of the amplifier illustrated in Figure 6.3-5 if R_G, R_S, and R_B are changed to 50, 0.5, and 200 kΩ, respectively, and if a 1-kΩ resistor is inserted in series with the emitter lead of Q_2. **Answer** 583 rad/s $< \omega_L <$ 800.6 rad/s

6.4 HIGH-FREQUENCY SMALL-SIGNAL MODELS FOR DIODES AND TRANSISTORS

In Chapter 6 thus far we have investigated the performance of bipolar and field-effect transistor amplifiers operating in the middle- and low-frequency regions. In analyzing these circuits, we saw that we could replace the bypass and coupling capacitors by short circuits in the midband frequency range, while at low frequencies the increase in the impedance of these elements needed to be considered and would usually cause the gain of the amplifier to decrease as the frequency was lowered.

For both the low- and middle-frequency regions, the effect of any parasitic capacitive elements present in the circuit was ignored; this was permissible because the impedance of these capacitors in this frequency range was so large that they could, for all practical purposes, be assumed to be open circuits. As the frequency of the applied signal increases, the impedance of the parasitic capacitive elements eventually drops to the point where their effect on the amplifier gain must be taken into account.

In this section we develop models for both the BJT and FET that will permit us, in the sections that follow, to obtain a good quantitative understanding of the high-frequency performance of both single-stage and multistage transistor amplifiers. Because the BJT and FET are both *pn* junction devices, we begin the development of these high-frequency models by investigating the factors limiting the high-frequency performance of a simple semiconductor diode.

6.4-1 Development of the High-Frequency Model for the *pn* Junction Diode

Figure 6.4-1a illustrates the depletion region for a *pn* junction diode under zero-bias conditions. When this diode is reverse biased, the width of the de-

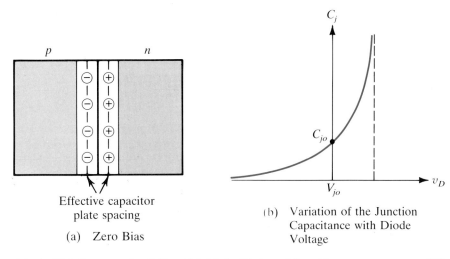

Figure 6.4-1
pn diode junction capacitance.

Effective capacitor plate spacing

(a) Zero Bias

(b) Variation of the Junction Capacitance with Diode Voltage

pletion region increases from its zero-bias value and more charge is exposed in the vicinity of the junction. To uncover this additional charge, it is necessary to remove a small number of free holes from the p-type material and free electrons from the n-type material. Associated with this charge movement there is a transient flow of current into the diode whenever the voltage across it is changed. We can model this effect by placing a parasitic capacitive element in parallel with the ordinary low-frequency model for the diode. This capacitance is known as the diode's junction or depletion region capacitance, and its value is a measure of how much additional charge must be added to or removed from the diode to alter the width of the depletion region when the voltage across the diode is changed.

To understand why the junction capacitance decreases as the reverse voltage applied to the diode is increased, consider that the "plates" of the junction capacitance are located at approximatley the center of mass of the uncovered charge in Figure 6.4-1a. As the reverse voltage across the diode is increased, the width of the depletion region is likewise increased, as is the effective spacing of the plates. As a result, because the size of a capacitor varies inversely with the spacing of its plates, the junction capacitance decreases as the reverse voltage increases.

To quantify this relationship, consider that the charge densities in the p and n regions are N_A and N_D, respectively. The total uncovered positive bound charge stored on the right-hand side of the depletion region is equal to

$$Q_J = qAN_D\, w_n \qquad (6.4\text{-}1)$$

where A is the diode cross-sectional area and w_n the width of the depletion region in the n-type material. Because the diode has overall charge neutrality, the total uncovered negative charge on the left-hand side of the junction is also equal to Q_J. Substituting equation (1.11-11a) for w_n into equation (6.4-1), we obtain

$$Q_J = qAN_D\sqrt{\frac{2\epsilon(V_{jo} - v_D)}{qN_D^2(1/N_D + 1/N_A)}} = A\sqrt{2q\epsilon(V_{jo} - v_D)\frac{N_A N_D}{N_A + N_D}} \qquad (6.4\text{-}2)$$

In this expression we have replaced the built-in voltage V_{jo} by the term $(V_{jo} - v_D)$ to account for the change in the width of the depletion region which occurs when the diode voltage is nonzero. Because the charge stored in the depletion region varies with the applied voltage, we may model this relationship by a capacitance, and following equation (6.4-2), it is also apparent that this capacitance is nonlinear. However, when the ac voltage applied to the diode is small, we may represent it approximately as a linear capacitance, C_j, where

$$C_j = -\frac{dQ_J}{dv_D}\bigg|_{Q\text{ pt}} = A\sqrt{\frac{q\epsilon}{2(V_{jo} - V_D)}\frac{N_A N_D}{N_A + N_D}} \qquad (6.4\text{-}3)$$

Here the minus sign is required because a positive change in v_D causes a positive charging current i_D that ultimately results in a reduction of Q_J.

Sometimes it is convenient to rewrite equation (6.4-3) as

$$C_j = C_{jo}\frac{1}{\sqrt{1 - V_D/V_{jo}}} \qquad (6.4\text{-}4)$$

Chapter 6 / Frequency Effects in Amplifier Circuits

to illustrate more explicitly the variation of C_j with diode voltage. Here C_{jo} represents the junction capacitance under zero bias conditions. The variation of this junction capacitance with v_D is illustrated in Figure 6.4-1b.

In modeling the diode at high frequencies, there is a second capacitive-like effect which also needs to be taken into account. When a diode is forward biased, the height of the voltage barrier at the junction is reduced. This allows many extra holes to cross the junction from the p into the n material, and many extra electrons to cross over from the n into the p material (Figure 1.11-5c). This increases the density of minority carriers at both edges of the depletion region, and the resulting steady-state diffusion of these charges toward the external leads of the diode constitutes the large forward bias current that flows in the device. The expressions for the minority-carrier charge concentrations, $p_n(x)$ and $n_p(x)$, are given by equations (1.11-16a) and (1.11-16b), respectively. Because the shape of these distributions is a function of the diode voltage, the total stored minority-carrier charge varies with v_D. This change in stored charge and transient current flow into the diode that is associated with changes in v_D constitutes a capacitance C_d, known as the diode's diffusion capacitance.

To develop a quantitative expression for C_d, let's begin by determining the total excess positive minority-carrier charge Q_B stored in the n-type material. Because $p_n(x) - p_{no}$ represents the excess positive charge concentration, Q_B may be written as

$$Q_B = qA \int_{w_n}^{\infty} [p_n(x) - p_{no}]dx \qquad (6.4\text{-}5)$$

and substituting equation (1.11-30a) into this expression, we have

$$Q_B = qAp_{no}(e^{v_D/V_T} - 1) \int_0^{\infty} e^{-(x-w_n)/L_p}dx = qAL_p p_{no}(e^{v_D/V_T} - 1) \quad (6.4\text{-}6)$$

Notice that Q_B is quite large for a forward-biased diode, whereas it is virtually zero when the diode is reverse biased.

For later use, it is convenient to relate Q_B to the current flowing in the diode. Following equation (1.11-20), we may write the diffusion current for holes at the edge of the depletion region as

$$i_{\text{diffusion,holes}}(w_n) = -qAD_p \frac{dp_n(x)}{dx}\bigg|_{x=w_n} = \frac{qAD_p}{L_p}(e^{v_D/V_T} - 1) \quad (6.4\text{-}7)$$

and combining this expression with equation (6.4-6), we may express Q_B in terms of the hole component of the diode current as

$$Q_B = \frac{L_p^2}{D_p}\left[\frac{qAD_p p_{no}}{L_p}(e^{v_D/V_T} - 1)\right] = \frac{L_p^2}{D_p} i_{\text{diffusion,holes}}(w_n)$$

$$= i_{\text{diffusion,holes}}(w_n)\tau_p \qquad (6.4\text{-}8)$$

where τ_p is the recombination time or average lifetime of holes in the n-type material. A similar expression exists for the excess electron charge in the p-type material, but for simplicity here, let us assume that we have a p^+n diode where $N_A \gg N_D$ so that almost all of the stored minority-carrier charge is associated with the excess holes in the n-type material.

Because the free hole concentration in the p-type material is so much greater than the free electron concentration in the n-type material, following equation (1.11-19), it is also apparent that the diode current flow across the junction is almost entirely due to the diffusion of holes, so that

$$i_D = i_{\text{diffusion,holes}}(w_n) + i_{\text{diffusion,electrons}}(-w_p) \simeq i_{\text{diffusion,holes}}(w_n) \quad (6.4\text{-}9)$$

As a result, equation (6.4-8) may be rewritten approximately as

$$Q_B = i_D \tau_p \quad (6.4\text{-}10a)$$

where, of course,

$$i_D = I_s(e^{v_D/V_T} - 1) \simeq \begin{cases} I_s e^{v_D/V_T} & \text{for } v_D > 0 \\ 0 & \text{for } v_D < 0 \end{cases} \quad (6.4\text{-}10b)$$

When the voltage on the diode changes, so does the stored minority-carrier charge, and the ratio of ΔQ_B to Δv_D defines the capacitance C_d as

$$C_d = \left.\frac{dQ_B}{dv_D}\right|_{Q \text{ pt}} = \left.\frac{\tau_p I_s}{V_T} e^{v_D/V_T}\right|_{Q \text{ pt}} = \frac{\tau_p}{V_T} I_s(e^{v_D/V_T}) \quad (6.4\text{-}11)$$

Clearly, when the diode is reverse biased, the diffusion capacitance is nearly zero because the stored minority-carrier charge is also very small. However, when the diode is forward biased, C_d can be extremely large. To illustrate how C_d depends on the diode quiescent current, notice that i_D is approximately equal to $I_s e^{v_D/V_T}$, so that the expression for the diffusion capacitance in equation (6.4-11) can be rewritten in terms of this current as

$$C_d = \frac{\tau_p I_D}{V_T} \quad (6.4\text{-}12a)$$

Notice further that the diode small-signal resistance

$$r_D = \frac{V_T}{I_D} \quad (6.4\text{-}12b)$$

so that we may also express C_d as

$$C_d = \frac{\tau_p}{r_D} \quad (6.4\text{-}12c)$$

where τ_p is the lifetime or average recombination for holes in the n-type material.

In summary, the total current flow in a pn junction diode may be written as

$$i_D = I_s(e^{v_D/V_T} - 1) + \frac{dQ_J}{dt} + \frac{dQ_B}{dt} \quad (6.4\text{-}13)$$

where the first term in this equation represents the ordinary expression for the diode current at low frequencies, the second term the additional current that needs to be added to alter the charge distribution at the depletion region, and the third term the additional current required to change the shape of the minority-carrier charge distributions. All three of these terms are nonlinear func-

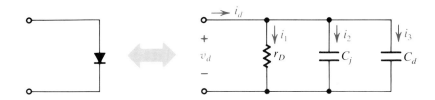

Figure 6.4-2
Small-signal high-frequency model for the pn junction diode.

tions of the voltage applied to the diode, but for the case where the time-varying part of v_D is small, equation (6.4-13) may be approximated as

$$i_D = \left(I_D + \frac{v_d}{r_D}\right) + C_j\frac{dv_d}{dt} + C_d\frac{dv_d}{dt} \qquad (6.4\text{-}14\text{a})$$

so that we have

$$i_d = \underbrace{\frac{v_d}{r_D}}_{i_1} + \underbrace{C_j\frac{dv_d}{ct}}_{i_2} + \underbrace{C_d\frac{dv_d}{dt}}_{i_3} \qquad (6.4\text{-}14\text{b})$$

This result suggests that at high frequencies the small-signal model for the semiconductor diode has the form illustrated in Figure 6.4-2. Clearly, following equation (6.4-12a), C_d can be neglected for reverse-biased diodes because i_D is nearly zero; in addition, r_D for a reverse-biased diode is nearly an open circuit. As a result, a *pn* junction diode operating at high frequencies under reverse-bias conditions can be modeled approximately by a single capacitance equal to the diode's junction capacitance. Typical values for C_j are on the order of 0.1 to 20 pF, depending on the physical size of the diode and the value of the applied reverse voltage.

For forward-bias conditions, C_d is very large and C_j can generally be neglected. Furthermore, because r_D is inversely proportional to I_D while C_d varies directly with I_D, the product $r_D C_d$ (which has units of time) is independent of the circuit's Q point. This product, the circuit time constant, is, in accordance with equation (6.4-12c), equal to the average minority-carrier recombination time, τ_p, for the diode. The recombination time is a figure of merit for a particular diode and is a measure of how well it will perform at high frequencies. Typical values for τ_p vary from about 10 ns to 1 μs. Corresponding diffusion capacitance values (for diode quiescent currents on the order of 1 mA) vary from about 400 pF to about 0.04 μF. In general, for forward-bias conditions, C_d is so much greater than C_j that the junction capacitance can be ignored. Of course, for reverse-biased conditions, the opposite situation is true.

6.4-2 High-Frequency Model Development for the Field-Effect Transistor

The small-signal high-frequency models for the FET follow almost directly from the semiconductor diode equivalent circuits just discussed. Figure 6.4-3 illustrates the physical structure of the JFET- and MOSFET-style field-effect transistors. In general, junction field-effect transistors are most often manufactured as discrete devices, while MOS transistors are usually fabricated as integrated circuits.

We begin our model development for the field-effect transistor by examin-

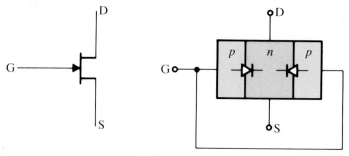

(a) Discrete JFET

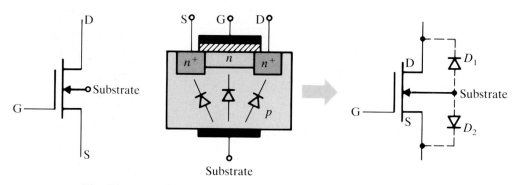

(b) Discrete and Integrated Design for a MOSFET transistor

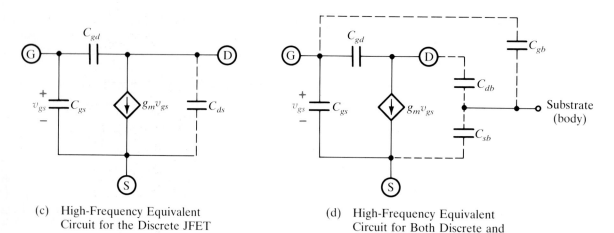

(c) High-Frequency Equivalent
Circuit for the Discrete JFET

(d) High-Frequency Equivalent
Circuit for Both Discrete and
IC-Style MOSFETs

Figure 6.4-3
High-frequency equivalent cir-
cuits for the FET.

ing the high-frequency characteristics of the junction-style FET. For normal operation the parasitic *pn* junction diodes along the gate of the JFET are reverse biased (see Figure 6.4-3a). As a result, these diodes can be represented by an array of junction capacitances distributed along the length of the channel. For simplicity, the overall effect of this distributed capacitance is usually modeled by two equivalent lumped element capacitances, C_{gs} and C_{gd} (see Figure 6.4-3c).

The capacitance C_{gs} represents the effective capacitance between the gate and the source terminals of the transistor, and C_{gd}, the capacitance between the

6.4-1 Typical JFET and MOSFET Equivalent-Circuit Parameters

Transistor Type	Transconductance, g_m (mS)	C_{gs} (pF)	C_{gd} (pF)	Body Capacitances (C_{db}, C_{gb}, C_{sb}) (pF)
JFET	0.5–40	2–20	0.3–20	—
MOSFET (enhancement or depletion)	0.5–40	1–10	0.02–5	1–5

gate and the drain. Because the magnitude of the gate-to-drain voltage is generally much greater than that from gate to source, C_{gd} is usually smaller than C_{gs}. The high-frequency equivalent circuit for the JFET is given in Figure 6.4-3c, and typical values for these circuit parameters are listed in Table 6.4-1. Sometimes, an additional parasitic capacitance, C_{ds}, is added to the model to account for stray interelectrode and wiring capacitance between the drain and source leads. Usually, the value of this capacitance for JFET transistors is so small that it need not be included in the circuit; for MOSFET devices, however, it can be much more important.

The high-frequency model for the MOSFET transistor, while quite similar to that for the JFET, has several subtle differences worth mentioning. To begin with, the manufacturing technique used to construct an MOS transistor (Figure 6.4-3b) requires that we consider the influence of the substrate on the high-frequency performance of this device. As illustrated in this figure, the main effect of the substrate is to create two parasitic diodes that are effectively connected between the substrate and the drain and the source leads of the transistor. For the n-channel device illustrated in the figure, the substrate lead on the transistor (or alternatively, the overall substrate of the chip if this transistor is part of a larger integrated circuit) would usually be connected to the most negative potential in the circuit in order to reverse bias these parasitic diodes. If the diodes are reverse biased, they can be replaced by the junction capacitances illustrated in Figure 6.4-3d. Furthermore, if, as is often the case for a single transistor device, the substrate lead is simply connected to the source terminal of the transistor, then C_{sb} is shorted out and C_{db} is effectively connected between the drain and the source.

In addition to its substrate capacitances, the MOSFET transistor has a distributed capacitance formed by the gate and the channel of the transistor. As with the JFET, for simplicity, this distributed capacitive effect is represented by two lumped capacitive elements (C_{gs} and C_{gd}) in the model. However, unlike the JFET, these capacitances have fixed values that are independent of V_{GS} and V_{GD} because they are purely electrostatic capacitances. The complete high-frequency model for the MOSFET is illustrated in Figure 6.4-3d, and typical values for the model parameters are given in Table 6.4-1.

6.4-3 BJT High-Frequency Model

The high-frequency model for the bipolar junction transistor may be developed by applying similar considerations to those employed for the FET. Figure 4.3-

3d illustrates the standard h-parameter model for the BJT; this is probably the most useful low-frequency model for analyzing bipolar transistor circuits. However, the hybrid-parameter model has considerable difficulty in properly describing the performance of the BJT at high frequencies because it assumes that the collector current is proportional to the base current, independent of the frequency of the applied signal.

From our discussion of the theory of operation of the bipolar junction transistor in Section 3.2, you should recall that the BJT operates as follows. The voltage applied to the base–emitter junction of the transistor forward biases the base–emitter diode and determines the level of current emitted by the emitter. This current diffuses across the base and is collected by the collector. Thus, the small-signal ac collector current that flows in the transistor depends on the ac base-to-emitter voltage and not the ac base current. At low frequencies there is no problem in using a model where $i_c = h_{fe}i_b$, because i_b and v_{be} are proportional. However, at high frequencies this is no longer the case, and a considerable error will result if the h-parameter model is utilized to predict the transistor's behavior. In fact, at high frequencies we actually observe that the base current flowing into the transistor increases with frequency, while the collector current actually decreases.

To develop a model for the BJT that will properly explain these phenomena, let us reexamine the hybrid model shown in Figure 4.2-3b. This equivalent circuit illustrates in greater detail the physical origins of the small-signal input resistance, h_{ie}. Basically, h_{ie} has two resistive components, $r_{bb'}$ and $r_{b'e}$. The resistance $r_{bb'}$ represents the ohmic resistance between the external base lead on the transistor and the center of the BJT device, where the actual active portion of the transistor is located. The resistance r_e models the ac small-signal resistance of the base–emitter diode, and $r_{b'e}$ is the equivalent resistance when r_e is reflected into the base of the transistor. In modifying this model to account for the observed high-frequency performance of the BJT, we begin by replacing the controlled current source, $h_{fe}i_b$, by one that depends on $v_{b'e}$, the effective voltage across the internal base–emitter junction of the transistor. Using the equivalent circuit in Figure 4.2-3b, we can see that

$$h_{fe}i_b = h_{fe}\frac{v_{b'e}}{r_{b'e}} = \frac{h_{fe}}{(1 + h_{fe})r_e}v_{b'e} = g_m v_{b'e} \qquad (6.4\text{-}15)$$

The quantity

$$g_m = \frac{h_{fe}}{(1 + h_{fe})r_e} \qquad (6.4\text{-}16a)$$

Figure 6.4-4

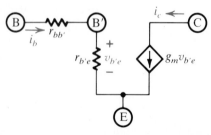

(a)   Low-Frequency Hybrid-π Model

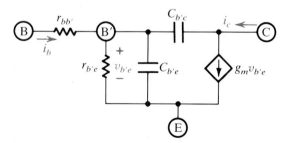

(b)   High-Frequency Hybrid-π Model for the BJT

is known as the transconductance of the transistor, and because $h_{fe}/(1 + h_{fe})$ is nearly equal to 1, we may approximate this expression as

$$g_m \simeq \frac{1}{r_e} = \frac{I_E}{V_T} \qquad (6.4\text{-}16b)$$

Using this result, we can see that typical values for g_m for small-signal transistors with quiescent currents on the order of 1 to 10 mA vary from about 40 to 400 mS, depending on the value of the quiescent emitter current. Comparing these transconductance values with those for field-effect transistors (Table 6.4-1), we can see why BJT amplifiers generally have higher voltage gains than those of comparable FET circuits. If we replace the current-controlled current source $h_{fe}i_b$ by the voltage-controlled current source $g_m v_{b'e}$, we obtain the equivalent circuit illustrated in Figure 6.4-4a. At low frequencies bipolar transistor circuits can be analyzed using either this equivalent circuit or the original hybrid model presented in Figure 4.3-3d. In general, however, the h-parameter model is easier to use at low frequencies.

Let us now investigate the additional modifications that must be made to this transistor equivalent circuit to account properly for the observed high-frequency behavior of the BJT. For normal active-region operation you should recall that the base–emitter junction of the bipolar transistor is forward biased, while the base–collector junction is reverse biased. As a result, we can model the high-frequency performance of the BJT approximately by adding two parasitic capacitances, $C_{b'e}$ and $C_{b'c}$, to the low-frequency model given in Figure 6.4-4a. This high-frequency model, known as the hybrid pi (π), is given in part (b) of the figure.

The capacitance $C_{b'c}$ represents the depletion region capacitance associated with the reverse-biased base–collector junction, and because the base–emitter junction is forward biased, the parasitic capacitance $C_{b'e}$ is the sum of two capacitive components:

$$C_{b'e} = C_{je} + C_{de} \qquad (6.4\text{-}17)$$

The capacitive term C_{je} in this expression represents the junction capacitance of the base–emitter diode, and the term C_{de} is equal to the diffusion capacitance associated with this junction. In this equation, C_{je} depends on v_{BE} much in the same way that C_j depends on v_D for a semiconductor diode [see equations (6.4-3) and (6.4-4)]. However, the equation describing the behavior of C_{de} is a bit more involved than the expression for the diffusion capacitance of a pn junction diode and may be evaluated as follows.

As with the diode analysis carried out earlier in this section, the diffusion capacitance is used to account for the additional current that flows into the transistor to alter the minority-carrier charge distributions that exist on either side of the depletion region at the emitter junction. In the BJT, because the emitter is doped much more heavily than the base, for a pnp transistor most of the transient current flow into the base when v_{BE} changes is associated with the changes that occur in the minority-carrier charge stored on the base side of the junction.

Following Figure 6.4-5, we can see that the excess minority-carrier charge stored in the base is

$$Q_B = \tfrac{1}{2}(qAW)p_b(0) = \tfrac{1}{2}qAWp_{bo}(e^{v_{EB}/V_T} - 1) \qquad (6.4\text{-}18)$$

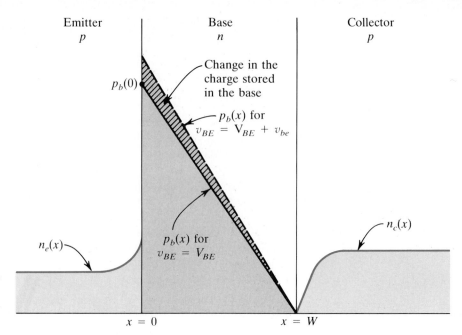

Emitter	Base	Collector
p	n	p

Change in the charge stored in the base

$p_b(0)$

$p_b(x)$ for $v_{BE} = V_{BE} + v_{be}$

$p_b(x)$ for $v_{BE} = V_{BE}$

$n_e(x)$

$n_c(x)$

$x = 0$ $x = W$

Figure 6.4-5
Minority-carrier charge distributions in a transistor that is biased for operation in the active region.

As a result, the diffusion capacitance associated with this stored charge is

$$C_{de} = \frac{dQ_B}{dv_{EB}}\bigg|_{Q_{pt}} = \frac{1}{2}\frac{qAWp_{bo}}{V_T}e^{V_{EB}/V_T} \tag{6.4-19}$$

But following equation (3.2-4a) and neglecting I_{En} because of the asymmetrical doping, we may write

$$I_E \simeq I_{Ep} = \frac{qAD_p p_{bo}}{W}e^{V_{EB}/V_T} \tag{6.4-20}$$

Combining this result with equation (6.4-19), the expression for C_{de} may be rewritten as

$$C_{de} = \frac{W^2}{2D_p}\frac{I_E}{V_T} \tag{6.4-21}$$

Because this capacitance varies directly with I_E, C_{de} is generally much greater than C_{je} when the transistor is operating in the active region. Equation (6.4-21) also indicates why high-frequency transistors are generally *npn* devices; the diffusion constant for electrons is nearly three times that for holes, so that the effective values of C_{de} are much smaller for *npn* transistors.

Manufacturers often specify the component values for the hybrid-π model indirectly. For example, one transistor parameter that is frequently given is the high-frequency 3-dB point of the common-emitter short-circuit current gain. This frequency,

$$f_\beta = \frac{1}{2\pi r_{b'e}(C_{b'e} + C_{b'c})} \tag{6.4-22a}$$

is known as the beta cutoff frequency of the transistor and is a figure of merit

6.4-2 Typical Values for the Hybrid-π Parameters for BJT Small-Signal and Power Transistors

Parameter	Small-Signal Transistor	Power Transistor
g_m	40–400 mS	400–4000 mS
$C_{b'c}$	0.2–10 pF	5–500 pF
$C_{b'e}$	0.5–200 pF	50–1000 pF
f_T	100 MHz–1 GHz	5–150 MHz
f_β	1–100 MHz	50 kHz–15 MHz
$r_{b'e}$	0.3–15 kΩ	25–500 Ω
$r_{bb'}$	10–500 Ω	0.1–10 Ω

for the high-frequency performance of the device. Typically, $C_{b'e}$ is much greater than $C_{b'c}$, so that

$$f_\beta \simeq \frac{1}{2\pi r_{b'e} C_{b'e}} \tag{6.4-22b}$$

Because $r_{b'e}$ varies inversely with I_E while $C_{b'e}$ varies directly with the quiescent emitter current, it follows that f_β is nearly independent of the transistor's operating point. Sometimes, the beta cutoff frequency is given directly by the manufacturer, but even more frequently, the transistor's f_T is given instead. This frequency, which is known as the common-emitter unity-current-gain crossover frequency, is defined as

$$f_T = \frac{1}{2\pi r_e(C_{b'e} + C_{b'c})} \tag{6.4-23a}$$

and using the fact that $r_{b'e} = (1 + h_{fe})r_e$, we may rewrite f_T in terms of f_β as

$$f_T = \frac{1 + h_{fe}}{2\pi(1 + h_{fe})r_e(C_{b'e} + C_{b'c})} = \frac{1 + h_{fe}}{2\pi r_{b'e}(C_{b'e} + C_{b'c})} \tag{6.4-23b}$$
$$= (1 + h_{fe})f_\beta$$

Typical values of f_T, f_β, and the transistor hybrid-π parameters are listed in Table 6.4-2.

Exercises

6.4-1 Find the junction capacitance of a *pn* junction diode at bias voltages of zero, -5, and $+0.4$ V given that $N_A = N_D = 10^{17}$ cm^{-3} and $A = 10^{-3}$ cm^2. *Answer* 98.7 pF, 37 pF, 140 pF

6.4-2 For a p^+n diode with $\tau_p = 100$ ns, find C_d at diode currents of 0.0, 0.1, and 10 mA. *Answer* 0, 384 pF, 0.038 μF

6.4-3 A BJT device is biased at a quiescent emitter current of 5 mA. Find $C_{b'e}$ if $f_T = 500$ MHz and $C_{b'c} = 10$ pF at this Q point. *Answer* 51.2 pF

6.5 INTRODUCTION TO HIGH-FREQUENCY ANALYSIS: THE MILLER THEOREM

In analyzing the performance of transistor circuits operating at high frequencies, we need to consider carefully the effects of the parasitic capacitors on the circuit's behavior. Those parasitic elements connected between either the input or output terminals of the transistor and ground may easily be accounted for; however, the capacitive element between the input and output terminals of the transistor is much more difficult to handle. In this section we present a general technique for approximating the response of circuits containing feedback elements. This analysis tool is known as the Miller theorem or the Miller effect and is named after its discoverer.

Basically, the Miller theorem is used to simplify the analysis of circuits of the form illustrated in Figure 6.5-1a. Here A represents the gain of the amplifier with the feedback element Z_f in place; generally, the value of A is known only approximately. By applying the Miller theorem to this problem, we are able to transform the original circuit of Figure 6.5-1a into that shown in Figure 6.5-1b. At first glance this transformation might not appear to be all that useful. However, as we shall see shortly, it can greatly simplify the analysis of feedback problems because this transformation changes the network from one containing two interacting circuit loops to one on which the two loops are considered to be independent of one another.

If the two circuits in Figure 6.5-1 are to be equivalent, the two node currents I_1 and I_2 in the original network must be equal to the currents I_1' and I_2' in the transformed network. In the original circuit,

$$I_1 = \frac{V_{in} - V_o}{Z_f} = \frac{V_{in} - AV_{in}}{Z_f} = \frac{V_{in}}{Z_f/(1 - A)} \tag{6.5-1a}$$

and

$$I_2 = \frac{V_o - V_{in}}{Z_f} = \frac{V_o - V_{in}/A}{Z_f} = \frac{V_o}{Z_f[-A/(1 - A)]} \tag{6.5-1b}$$

Similarly, in the transformed circuit we may express the branch currents I_1' and I_2' in terms of V_{in} and V_o as follows:

$$I_1' = \frac{V_{in}}{Z_1} \tag{6.5-2a}$$

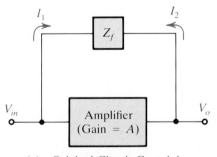

(a) Original Circuit Containing the Feedback Element

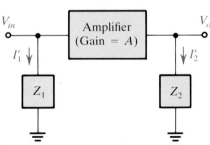

(b) "Equivalent" Amplifier with the Feedback Element Split into Two Impedances Z_1 and Z_2

Figure 6.5-1

and
$$I'_2 = \frac{V_o}{Z_2} \qquad (6.5\text{-}2b)$$

As a result, for $I_1 = I'_1$ and $I_2 = I'_2$, Z_1 and Z_2 must be chosen so that

$$Z_1 = \frac{Z_f}{1 - A} \qquad (6.5\text{-}3a)$$

and
$$Z_2 = Z_f \frac{-A}{1 - A} \qquad (6.5\text{-}3b)$$

From these results we can see that when Z_f is resistive, Z_1 and Z_2 will also be resistive, with Z_1 being much smaller than Z_f and Z_2 being approximately equal to Z_f if the amplifier gain A is reasonably large.

If, on the other hand, the feedback element Z_f is capacitive, then because

$$Z_f = \frac{1}{j\omega C_f} \qquad (6.5\text{-}4)$$

Z_1 and Z_2 will also appear to be capacitors. Because $Z_1 = 1/[j\omega(1 - A)C_f]$ the equivalent input capacitor C_1 will be

$$C_1 = C_f(1 - A) \qquad (6.5\text{-}5a)$$

Similarly, the equivalent capacitor C_2 at the output can be shown to be

$$C_2 = C_f \frac{1 - A}{-A} \qquad (6.5\text{-}5b)$$

As before, if the amplifier gain is large, Z_2 and Z_f, and hence C_2 and C_f, will be nearly equal.

In developing the Miller theorem we assumed that A (the gain between the Z_f terminals with the feedback element connected) was known. If this is the case, and if Z_1 and Z_2 are chosen in accordance with equations (6.5-3a) and (6.5-3b), the two networks illustrated in Figure 6.5-1 will be identical. However, only rarely is A explicitly known because this requires a complete detailed analysis of the circuit being examined. Instead, A is usually determined approximately by finding the circuit gain between the Z_f terminals with the feedback element removed. As we shall see in the example that follows, this approximation provides good results when the feedback element does not load the output of the circuit. When the loading of Z_f on the output cannot be ignored, a detailed network analysis of the problem is required.

EXAMPLE 6.5-1

For the BJT feedback amplifier illustrated in Figure 6.5-2, assume that the coupling capacitor is an ac short and that the transistor small-signal parameters are $h_{ie} = 1 \text{ k}\Omega$ and $h_{fe} = 100$. Using the Miller theorem, determine the overall circuit gain v_{ce}/v_{in}.

SOLUTION

Because of the presence of the feedback element, this problem is one in which the Miller theorem can be extremely useful. To apply this theorem, we first need to find the gain A so that we may split Z_f into the two equivalent elements Z_1 and Z_2. Notice that A represents the circuit gain between the two points

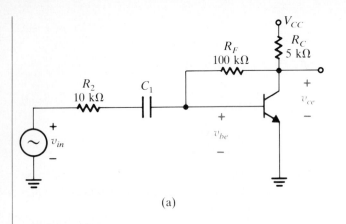

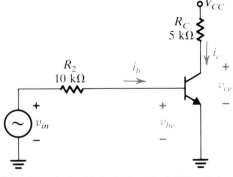

(a)

(b)   Circuit for Calculating the Miller Gain A

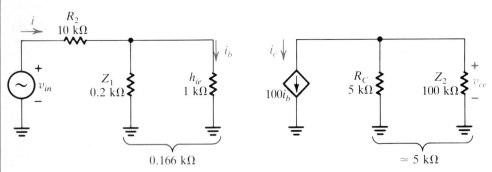

(c)   Equivalent Circuit for Calculating the Overall Amplifier Gain v_{ce}/v_{in}

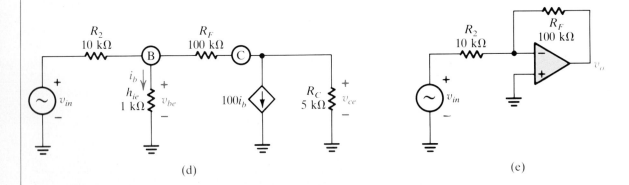

(d)

(e)

Figure 6.5-2
Example 6.5-1.

where the feedback element is connected, not the overall amplifier gain. As a result, for this specific circuit A is equal to the voltage gain from the base of the transistor to the collector, with the resistor R_F connected in the circuit. As an approximation we will assume that this gain is nearly the same when R_F is removed.

The overall amplifier circuit with R_F removed is illustrated in Figure 6.5-

2b. Using this circuit, we can readily calculate the gain A as follows:

$$v_{ce} = -i_c R_C = -h_{fe} i_b R_C = -\left(\frac{100 \, v_{be}}{1 \, \text{k}\Omega}\right) R_C = \frac{(-500 \, \text{k}\Omega) R v_{be}}{1 \, \text{k}\Omega} \qquad (6.5\text{-}6a)$$

and

$$A = \frac{v_{ce}}{v_{be}} = -500 \qquad (6.5\text{-}6b)$$

Substituting this value for A into equations (6.5-3a) and (6.5-3b), we obtain

$$Z_1 = \frac{100 \, \text{k}\Omega}{1 - (-500)} \simeq 0.2 \text{k}\Omega \qquad (6.5\text{-}7a)$$

and

$$Z_2 = (100 \, \text{k}\Omega)\left(\frac{500}{501}\right) \simeq 100 \, \text{k}\Omega \qquad (6.5\text{-}7b)$$

The equivalent circuit for calculating the overall amplifier gain is presented in Figure 6.5-2c. For this circuit we see that

$$i_b = \left(\frac{0.2}{1.2}\right) i = \left(\frac{0.2}{1.2}\right)\left(\frac{v_{in}}{10.167 \, \text{k}\Omega}\right) = \frac{v_{in}}{61 \, \text{k}\Omega} \qquad (6.5\text{-}8a)$$

and

$$v_{ce} = -i_c R_C = -h_{fe} i_b R_C = \frac{(-500 \, \text{k}\Omega) v_{in}}{61 \, \text{k}\Omega} = -8.2 v_{in} \qquad (6.5\text{-}8b)$$

In equation (6.5-8b) we neglected the loading of Z_2 on the output, and in effect, this was the same approximation that we made when we computed the Miller gain A with R_F removed from the circuit. Clearly, in this case because R_F is much greater than R_C, the loading of R_F on the output is negligible, and we therefore expect that our Miller theorem analysis using the value for A given in equation (6.5-6b) will yield good results. To illustrate this fact, we have also solved this problem exactly, with no approximations, using nodal analysis of the equivalent circuit given in Figure 6.5-2d. The solutions for v_{be} and v_{ce} are $v_{ce} = -8.11 v_{in}$ and $v_{be} = -0.017 v_{in}$. As expected, these results compare favorably with those computed approximately assuming that $A = -500$. Furthermore, if we form the ratio of v_{ce}/v_{be} using these results, we find that the exact solution for the Miller gain A is actually -477 when loading effects are included.

Although not related directly to the problem currently at hand, it is interesting to note that the approximate gain of this amplifier could have been found very rapidly if we had considered this transistor circuit to be an inverting style operational amplifier (see Figure 6.5-2e). As we shall see in Chapter 7, the gain of such an amplifier is $-(R_F/R_2)$, or -10 for this example. The discrepancy between this theoretical gain of -10 and the actual circuit gain of -8.11 is due to the fact that an ideal op amp has an infinite input impedance, whereas that of the amplifier circuit in this example is only equal to the h_{ie} of the transistor.

Exercise

6.5-1 Repeat Example 6.5-1 if a 20-Ω resistor is connected in series with the emitter of the transistor. *Answer* -5.98

6.6 HIGH-FREQUENCY RESPONSE OF SINGLE-STAGE TRANSISTOR AMPLIFIERS

The high-frequency performance of an electronic amplifier is difficult to understand because the parasitic capacitive elements couple many of the circuit nodes together. As a result, a complete analysis of such a circuit often requires the solution of a rather involved network problem. Fortunately, a detailed analysis of this problem is seldom necessary since we are usually only interested in obtaining an estimate of the circuit's high-frequency 3-dB point. In achieving this end, many excellent (and several not so excellent) approximation techniques are available to simplify this task. These methods are studied extensively in this section.

6.6-1 Common-Source and Common-Emitter Amplifiers at High Frequencies

We begin our study of transistor circuits operating in the high-frequency region by examining the common-source FET amplifier because this circuit is relatively easy to understand and will permit us to develop several powerful high-frequency analysis tools. A typical JFET common-source amplifier is illustrated in Figure 6.6-1a, and its high-frequency equivalent circuit is given in part (b) of the figure. The high-frequency model used for the FET in this circuit is that developed in Section 6.4 and illustrated specifically in Figure 6.4-3.

The approximate transfer function for this circuit can readily be developed if we apply the Miller theorem to the analysis of this problem. Using equations (6.5-5a) and (6.5-5b), we may split the feedback capacitor C_{gd} into two equivalent capacitors,

$$C_{m1} = C_{gd}(1 - A) \tag{6.6-1a}$$

and

$$C_{m2} = C_{gd}\frac{1 - A}{-A} \tag{6.6-1b}$$

(see Figure 6.6-1c). In these equations A represents the voltage gain, V_{ds}/V_{gs}, across the terminals of C_{gd} measured with the capacitor in place. As we shall show shortly, to a good approximation in the frequency range of interest, this gain is nearly the same as that measured in midband. To the extent that this approximation is valid, we may open circuit the parasitic capacitors (see Figure 6.6-1c) and calculate A as

$$A = \frac{V_{ds}}{V_{gs}} = -g_m R_D = -100 \tag{6.6-2}$$

so that

$$C_{m1} = C_{gd}(1 + g_m R_D) = 101 \text{ pF} \tag{6.6-3a}$$

and

$$C_{m2} = C_{gd}\frac{1 + g_m R_D}{g_m R_D} \simeq 1 \text{ pF} \tag{6.6-3b}$$

Using these capacitance values, it is now a relatively simple matter to determine the overall transfer function for the equivalent circuit given in Figure 6.6-1c. In particular, for this circuit

$$V_{gs} = \frac{(1/sC_1)V_{in}}{R_{in} + 1/sC_1} = \frac{1}{1 + sR_{in}C_1}V_{in} = \frac{1}{1 + s/\omega_1}V_{in} \tag{6.6-4a}$$

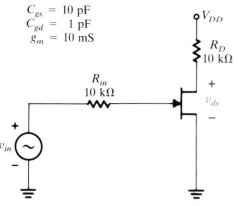

$C_{gs} = 10$ pF
$C_{gd} = 1$ pF
$g_m = 10$ mS

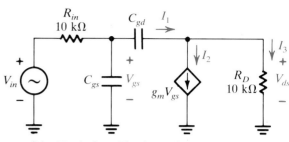

(a)   Common-Source JFET Amplifier

(b)   Equivalent Circuit at High Frequencies

$C_1 = C_{gs} + C_{m1}$
$\quad = C_{gs} + C_{gd}(1 - A)$
$\quad = 111$ pF

$C_2 = C_{m2}$
$\quad = C_{gd}\left(\dfrac{1 - A}{-A}\right)$
$\quad \simeq C_{gd} = 1$ pF

(c)   Approximate High-Frequency Equivalent Circuit Using Miller Theorem to Split C_{gd}

Figure 6.6-1
Common-source JFET amplifier at high frequencies.

where

$$\omega_1 = \frac{1}{R_{in}C_1} = 0.9 \times 10^6 \text{ rad/s} \tag{6.6-4b}$$

and

$$V_{ds} = -g_m V_{gs}\frac{R_D(1/sC_2)}{R_D + 1/sC_2} = -g_m R_D V_{gs}\frac{1}{1 + s/\omega_2} \tag{6.6-4c}$$

where

$$\omega_2 = \frac{1}{R_D C_2} = 10^8 \text{ rad/s} \tag{6.6-4d}$$

Combining equations (6.6-4a) and (6.6-4c), we may write down the overall circuit transfer function as

$$H(s) = \frac{V_{dS}}{V_{in}} = \underbrace{(-g_m R_D)}_{A_{\text{mid}}} \underbrace{\frac{1}{1 + s/\omega_1}}_{\text{break due to } C_1} \underbrace{\frac{1}{1 + s/\omega_2}}_{\text{break due to } C_2} \tag{6.6-5}$$

and substituting the data from Figure 6.6-1a into this expression, we obtain

$$H(s) = (-100)\frac{1}{1 + s/0.9 \times 10^6}\frac{1}{1 + s/10^8} \tag{6.6-6}$$

The magnitude of this transfer function is sketched as a function of frequency in Figure 6.6-2.

Let's examine the transfer function in equation (6.6-5) a bit more closely. To begin with, as indicated in this expression, the coefficient $(-g_m R_D)$ represents the midband gain of the amplifier with the parasitic elements replaced by open circuits. Furthermore, associated with each of the parasitic capacitive el-

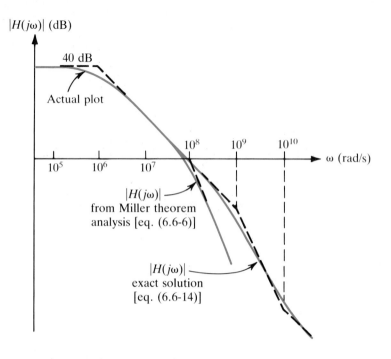

Figure 6.6-2
Common-source amplifier gain versus frequency.

ements C_1 and C_2, there is a multiplicative term of the form $1/(1 + s/\omega_x)$ in which ω_x represents the break frequency of the capacitor. For frequencies well below ω_x, the capacitor is essentially an open circuit and the term $1/(1 + s/\omega_x)$ is nearly unity. For frequencies above ω_x, the capacitor begins to behave like a short circuit, and the magnitude of the term $1/(1 + s/\omega_x)$ decreases, as does the overall gain of the amplifier. Again, as with the low-frequency analyses carried out previously, the denominator break term associated with each of the capacitors is equal to $1/RC$, where R is the Thévenin resistance seen by the capacitor with all independent sources set equal to zero, and all dependent sources "carefully" taken into account.

EXAMPLE 6.6-1

Without using conventional network analysis methods, determine the transfer function for the circuit given in Figure 6.6-1c.

SOLUTION

To write the transfer function for this circuit, we observe that it contains two capacitors, each of which reduces the amplifier gain to zero when it shorts out. As a result, the overall circuit transfer function will be of the form

$$H(s) = A_{\text{mid}} \frac{1}{1 + s/\omega_1} \frac{1}{1 + s/\omega_2} \tag{6.6-7}$$

In this equation the coefficient A_{mid} represents the amplifier midband gain V_{ds}/V_{in}. To determine A_{mid}, we note that in this frequency range the parasitic capacitive elements have such high impedances that they can be replaced by open circuits. Furthermore, although unimportant for this particular problem, it is nonetheless important to point out that had there been any coupling or bypass capacitors in the circuit, they would have been replaced by shorts in order to carry out this calculation. Using the circuit given in Figure 6.6-1c, with C_1

and C_2 open, we see that

$$V_{ds} = -g_m R_D V_{gs} = -g_m R_D V_{in} \qquad (6.6\text{-}8a)$$

or

$$A_{mid} = \frac{V_{ds}}{V_{in}} = -g_m R_D = -100 \qquad (6.6\text{-}8b)$$

To determine the break frequency ω_1 associated with C_1, we observe that with the independent source V_{in} set equal to zero, the resistance seen by C_1 is just R_{in}, so that

$$\omega_1 = \frac{1}{R_{in} C_1} = 0.9 \times 10^6 \text{ rad/s} \qquad (6.6\text{-}9)$$

The calculation for the break frequency associated with C_2 might at first appear complex because of the presence of the controlled source $g_m V_{gs}$ in parallel with C_2. This source, however, does not present any problem because the controlling parameter (the voltage V_{gs}) is not in the same part of the circuit. Therefore, this source too may simply be set equal to zero (by opening it) to calculate the resistance seen by C_2. When this is done, we find that

$$\omega_2 = \frac{1}{R_D C_2} = 10^8 \text{ rad/s} \qquad (6.6\text{-}10)$$

Combining equations (6.6-7) through (6.6-10), we can see that the transfer function obtained is the same as that found previously by network analysis techniques [see equation (6.6-6)].

When we derived the expression for the gain of the common-source amplifier, we used the Miller theorem and calculated the Miller gain A in the midband frequency range. We then used this same value of A at much higher frequencies to determine the equivalent Miller capacitances and ultimately to calculate the break frequencies associated with each of these capacitors. To demonstrate why this approximation is valid, we derive the exact expression for A. In the equivalent circuit shown in Figure 6.6-1b, the Miller gain A represents the voltage gain across the terminals of C_{gd} calculated with the capacitor in place in the circuit. From this circuit, we can see that

$$I_1 = I_2 + I_3 \qquad (6.6\text{-}11a)$$

or

$$sC_{gd}(V_{gs} - V_{ds}) = g_m V_{gs} + \frac{V_{ds}}{R_D} \qquad (6.6\text{-}11b)$$

Gathering terms, we obtain

$$A = \frac{V_{ds}}{V_{gs}} = (-g_m R_D)\frac{1 + s/\omega_N}{1 + s/\omega_D} = -100\left(\frac{1 + s/\omega_N}{1 + s/\omega_D}\right) \qquad (6.6\text{-}12a)$$

where

$$\omega_N = -g_m/C_{gd} = 10^{10} \text{ rad/s} \qquad (6.6\text{-}12b)$$

and

$$\omega_D = \frac{1}{R_D C_{gd}} = 10^8 \text{ rad/s} \qquad (6.6\text{-}12c)$$

From this result we can see that in the frequency range of interest (in particular, for frequencies at and below the high-frequency breakpoint at 1 Mrad/s), the Miller gain is constant at -100. This gain value is, of course, the same as that calculated by using the midband model.

For comparison purposes, it is instructive to contrast the Miller theorem results with those found by carrying out a detailed network analysis of the same problem. Using the equivalent circuit given in Figure 6.6-1b, this can be accomplished by writing down and solving the two node equations for V_{gs} and V_{ds}. When this is done, the following result is obtained:

$$H(s) = (-g_m R_D) \frac{1 - sC_{gd}/g_m}{1 + s[C_{gd}R_D + R_{in}(C_{gd} + C_{gs}) + g_m R_{in} R_D C_{gd}] + s^2 R_D R_{in} C_{gd} C_{gs}}$$

$$(6.6\text{-}13)$$

If we substitute in the numbers associated with this problem and factor the quadratic equation in the denominator, we find that the exact transfer function is

$$H(s) = (-100) \frac{1 - s/10^{10}}{(1 + s/0.89 \times 10^6)(1 + s/1.12 \times 10^9)} \qquad (6.6\text{-}14)$$

At first glance, this result appears to be substantially different from that found by the application of the Miller theorem to the same problem [equation (6.6-6)]. However, by comparing the two sketches in Figure 6.6-2a, we can see that in the frequency range of interest, both analysis methods produce nearly identical results. Therefore, because of its simplicity, the Miller theorem should be used in the solution of electronics problems whenever possible.

To illustrate several other very important concepts, this problem will be solved again by a third analysis technique. To begin the solution by this approach, we again examine the equivalent circuit in Figure 6.6-1b. In this circuit you should observe that an interaction exists between capacitors C_{gs} and C_{gd}. This interaction is apparent in equation (6.6-13) because the quadratic in the denominator cannot be separated into factors containing only C_{gs} in one term and C_{gd} in the other. This phenomenon is very similar to the interaction that exists at low frequencies in BJT circuits containing both bypass and coupling capacitors. Furthermore, and although it is by no means intuitively obvious, it can be shown that the transfer function given in equation (6.6-13) can be rewritten in the form

$$H(s) = A_{\text{mid}} \frac{1 - s/\omega_N}{1 + s/(\omega_{1o} \| \omega_{2o}) + s^2/[(\omega_{1s} + \omega_{2s})(\omega_{1o} \| \omega_{2o})]} \qquad (6.6\text{-}15)$$

In this equation, A_{mid} refers to the circuit's midband gain with both parasitic capacitors considered as open circuits, and

ω_{1o} = break associated with C_{gs} with C_{gd} open circuited

ω_{1s} = break associated with C_{gs} with C_{gd} short circuited

ω_{2o} = break associated with C_{gd} with C_{gs} open circuited

ω_{2s} = break associated with C_{gd} with C_{gs} short circuited

Although algebraically complex, the interested reader is invited to demonstrate the equivalence of equations (6.6-13) and (6.6-15) (Problem 6.6-8).

Of particular interest in this result is the fact that ω_N, ω_{1s}, and ω_{2s} are generally much greater than ω_{1o} and ω_{2o} for most circuits (see, e.g., Example 6.6-2). As a result, for frequencies at or below $\omega_{1o} \parallel \omega_{2o}$, the s term in the numerator and the s^2 term in the denominator of equation (6.6-15) are negligibly small, and therefore this equation is approximately equal to

$$H(s) \simeq A_{\text{mid}} \frac{1}{1 + s/(\omega_{1o} \parallel \omega_{2o})} \tag{6.6-16}$$

This result demonstrates that the high-frequency 3-dB point of this amplifier, ω_H, is approximately equal to

$$\omega_H = \omega_{1o} \parallel \omega_{2o} \tag{6.6-17}$$

where ω_{1o} and ω_{2o} represent the breaks associated with each of the capacitors, with the other parasitics in the circuit considered as opens. As with the low-frequency 3-dB point approximations developed in Section 6.3, equation (6.6-17) for ω_H is somewhat pessimistic in that the amplifier's actual 3-dB point will always be higher than this equation predicts. In particular, the actual 3-dB point is located somewhere in the range

$$\omega_{1o} \parallel \omega_{2o} \leq \omega_H \leq \text{smaller of } \omega_{1o} \text{ and } \omega_{2o} \tag{6.6-18}$$

Although this result was developed for the specific JFET circuit in Figure 6.6-1, as we demonstrate in Section 6.7, it is approximately valid for nearly any electronic circuit. In particular, for a circuit containing n such parasitic capacitors, the approximate range for ω_H is

$$(\omega_{1o} \parallel \omega_{2o} \parallel \cdots \parallel \omega_{no}) \leq \omega_H \leq \text{smaller of } \omega_{1o}, \omega_{2o}, \cdots, \omega_{no} \tag{6.6-19}$$

where $\omega_{xo} = 1/R_x C_x$ with R_x equal to the Thévenin resistance seen by C_x with all the other parasitic capacitors open circuited. It is very important to note in determining the high-frequency 3-dB point for the FET circuit in Figure 6.6-1a that either of the equivalent circuits in Figure 6.6-1b or c can be used. However, in general, it is highly preferable to work with the circuit in Figure 6.6-1c, in which the Miller theorem has already been applied.

EXAMPLE 6.6-2

Determine the approximate value for the high-frequency 3-dB point for the amplifier in Figure 6.6-1a by using the equivalent circuit given in Figure 6.6-1c, which results after the Miller theorem has been applied to the original equivalent circuit in Figure 6.6-1b.

SOLUTION

If the equivalent circuit in Figure 6.6-1c is used, the calculation of ω_H is very simple because the Thévenin resistance seen by each of the equivalent capacitors C_1 and C_2 in the figure can be determined by inspection of the circuit. In fact, it is worth mentioning that whenever an equivalent parasitic capacitor is connected between a specific circuit node and ground, the break frequency associated with that capacitor is especially easy to find because the Thévenin resistance seen by the capacitor is just the ordinary resistance seen looking into the specific terminal in question (with the parasitics considered to be open circuits).

Specifically for this case, we can see by inspection that

$$\omega_1 = \frac{1}{R_{in}C_1} = 0.9 \times 10^6 \text{ rad/s} \qquad (6.6\text{-}20a)$$

and

$$\omega_2 = \frac{1}{R_D C_2} = 10^8 \text{ rad/s} \qquad (6.6\text{-}20b)$$

Combining these two equations, the overall amplifier 3-dB point is

$$(0.9 \times 10^6 \parallel 10^8) = 0.89 \times 10^6 \text{ rad/s} \le \omega_H \le 0.9 \times 10^6 \text{ rad/s} \qquad (6.6\text{-}21)$$

Computer-aided analysis of the same problem demonstrates that the actual amplifier 3-dB point is 143 kHz or 0.898 Mrad/s, so that we see that this approximation technique does provide accurate results. Notice that ω_H for this amplifier is dominated by the input break because of the Miller magnification of the capacitor C_{gd} when it is reflected into the input. Notice also that the larger the amplifier gain, the bigger the Miller capacitor, and hence the smaller the amplifier bandwidth, illustrating the nearly universal trade-off that exists between the gain of an amplifier and its bandwidth.

EXAMPLE 6.6-3

For the common-emitter BJT amplifier circuit given in Figure 6.6-3, determine the approximate high-frequency 3-dB point. Analyze this problem by applying the Miller theorem to the feedback capacitor $C_{b'c}$. The pertinent transistor parameters are $g_m = 100$ mS, $r_{b'e} = 1$ kΩ, $r_{bb'} = 100$ Ω, $C_{b'e} = 100$ pF, and $C_{b'c} = 1$ pF.

SOLUTION

The equivalent circuit for this amplifier in the middle- and high-frequency regions is found by substituting in the hybrid-π model for the BJT (see Figure 6.4-4b). For the midband frequency range the parasitic capacitors $C_{b'c}$ and $C_{b'e}$ can be ignored, while at high frequencies they must be included. The overall equivalent circuit valid in the middle- and high-frequency regions is shown in Figure 6.6-3b. The midband model is needed to compute the Miller gain A across the terminals of the capacitor $C_{b'c}$ and using Figure 6.6-3b, with $C_{b'e}$ and $C_{b'c}$ considered to be open circuits, we have

$$V_c = V_l = -g_m R_C' V_{b'e} \qquad (6.6\text{-}22)$$

so that

$$A = \frac{V_c}{V_{b'e}} = -g_m R_C' = -500 \qquad (6.6\text{-}23)$$

Now that the Miller gain has been determined, we may split the feedback capacitor in Figure 6.6-3b into two capacitors C_{m1} and C_{m2}, with C_{m1} connected between the input node b' and ground, and C_{m2} between the collector node and ground. The resulting circuit is drawn in Figure 6.6-3c. Specifically for this circuit,

$$C_1 = C_{b'e} + C_{m1} = C_{b'e} + C_{b'c}(1 - A) = 601 \text{ pF} \qquad (6.6\text{-}24a)$$

and

$$C_2 = C_{m2} \simeq C_{b'c} = 1 \text{ pF} \qquad (6.6\text{-}24b)$$

Because each of the capacitors in Figure 6.6-3c is connected between a circuit node and ground, the break frequencies for each are especially easy to determine. In particular,

$$\omega_1 = \frac{1}{[(R_{in} + r_{bb'}) \parallel r_{b'e}]C_1} = 3.33 \times 10^6 \text{ rad/s} \qquad (6.6\text{-}25)$$

and

$$\omega_2 = \frac{1}{R'_C C_2} = 2 \times 10^8 \text{ rad/s} \qquad (6.6\text{-}26)$$

so that following equation (6.6-19) we find that

$$3.27 \times 10^6 \text{ rad/s} \le \omega_H \le 3.33 \times 10^6 \text{ rad/s} \qquad (6.6\text{-}27)$$

Figure 6.6-3
Example 6.6-3.

Computer-aided circuit analysis of the same problem demonstrates that the circuit 3-dB point is actually located at 521 kHz or 3.27 × Mrad/s.

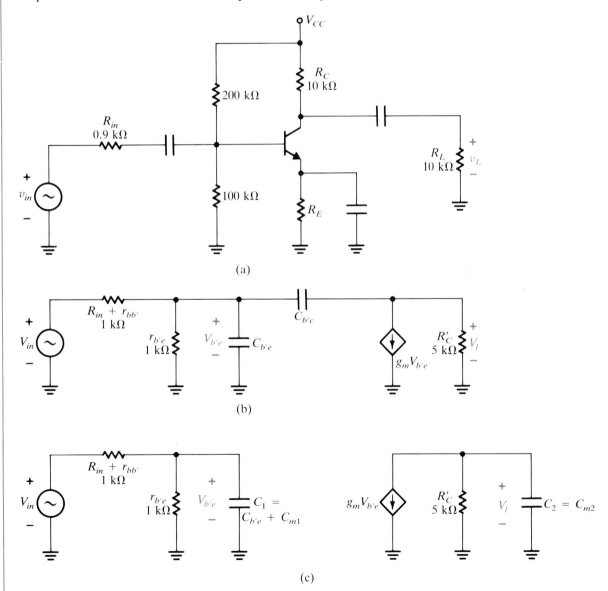

(a)

(b)

(c)

The common-emitter and common-source amplifiers provide good voltage gains, but their high-frequency response is limited because of the Miller effect. As we attempt to increase the voltage gain of these amplifiers by increasing the drain or the collector resistance, the increased Miller capacitance reflected into the amplifier input produces a corresponding decrease in the circuit bandwidth. Because of the inverse relationship between the gain and the bandwidth of an amplifier, the product of these two quantities tends to be somewhat constant for a given amplifier design. This product is called the amplifier's gain–bandwidth product and is a figure of merit for a particular amplifier design. It is defined as the product of the amplifier's midband gain and its high-frequency 3-dB point, or

$$\text{gain–bandwidth product (GBW)} = A_{\text{mid}}\omega_H \qquad (6.6\text{-}28)$$

For BJT amplifiers A_{mid} usually refers to the midband current gain of the circuit, whereas for FETs and op amps A_{mid} most often refers to the circuit's midband voltage gain.

6.6-2 Common-Gate and Common-Base Amplifiers at High Frequencies

As we have seen in Section 6.6-1, the common-emitter and common-source amplifiers have fundamental limitations on their gain–bandwidth products. In particular, as one attempts to increase the voltage gain of these circuits, the Miller effect will reduce the circuit bandwidth, tending to maintain the gain–bandwidth product constant. In principle, the common-gate and common-base transistor amplifiers possess no such fundamental limitations because, as we shall see shortly, there is no feedback capacitor from the output to the input, and hence there is no Miller effect to worry about. As a result, to the extent that our circuit models are valid, it is possible independently to vary the gain and the bandwidth of the amplifier.

To illustrate these points, consider the common-gate amplifier circuit given in Figure 6.6-4a, and let us derive a complete expression for the frequency response of this circuit. For later use, the location of the parasitic capacitors has been shown in dashed lines in the original figure. The overall high-frequency model for this circuit is presented in Figure 6.6-4b in which the controlled sourcer $g_m V_{gs}$ going from drain to source has already been split into two equivalent inverted sources, each equal to $g_m V_s$ since $V_g = 0$ in this circuit. In this figure, it is important to observe that there is no feedback capacitor between the input and output terminals of this circuit, and hence as mentioned previously, there is no Miller effect to worry about. Furthermore, because the input current source $g_m V_s$ has a voltage V_s across it, we can replace it by an equivalent resistance equal to $1/g_m$. The resulting equivalent circuit is drawn in Figure 6.6-4c, and this is the one that we will analyze.

At the input side of the figure, the voltage V_s may be calculated from the voltage-divider equation as

$$V_s = \frac{Z_1}{Z_1 + R_{in}}V_{in} = \frac{1/g_m}{R_{in} + 1/g_m}\frac{1}{1 + s/\omega_1}V_{in} \qquad (6.6\text{-}29a)$$

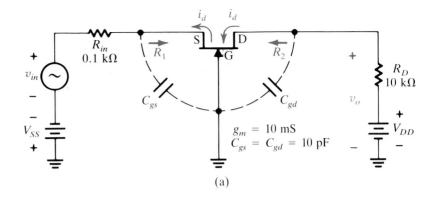

(a)

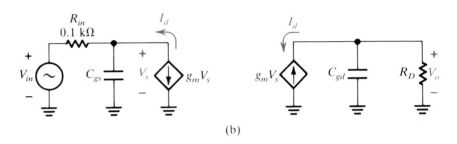

(b)

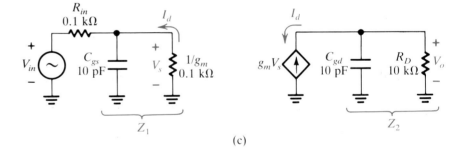

Figure 6.6-4
Common-gate amplifier at
high frequencies.

(c)

where
$$\omega_1 = \frac{1}{(R_{in} \parallel 1/g_m)C_{gs}} = 2 \times 10^9 \text{ rad/s} \qquad (6.6\text{-}29\text{b})$$

From this result, the drain current may be immediately written as

$$I_d = -\frac{V_s}{1/g_m} = -\frac{g_m}{1 + g_m R_{in}} \frac{1}{1 + s/\omega_1} V_{in} \qquad (6.6\text{-}30)$$

At the output side of this circuit, we have

$$V_o = -I_d Z_2 = -I_d \frac{R_D 1/sC_{gd}}{R_D + 1/sC_{gd}} = -I_d R_D \frac{1}{1 + s/\omega_2} \qquad (6.6\text{-}31\text{a})$$

where
$$\omega_2 = \frac{1}{R_D C_{gd}} \qquad (6.6\text{-}31\text{b})$$

Combining equations (6.6-30) and (6.6-31a), we may write the overall circuit transfer function as

$$\frac{V_o}{V_{in}} = \underbrace{\frac{g_m R_D}{1 + g_m R_{in}}}_{A_{\text{mid}}} \underbrace{\frac{1}{1 + s/\omega_1}}_{\substack{\text{break due} \\ \text{to } C_{gs}}} \underbrace{\frac{1}{1 + s/\omega_2}}_{\substack{\text{break due} \\ \text{to } C_{gd}}} \tag{6.6-32}$$

Substituting in the numbers associated with this problem, we find that

$$\frac{V_o}{V_{in}} = \underbrace{50}_{A_{\text{mid}}} \underbrace{\frac{1}{1 + s/2 \times 10^9}}_{\text{input break}} \underbrace{\frac{1}{1 + s/10^7}}_{\text{output break}} \tag{6.6-33}$$

The magnitude of this transfer function is sketched in Figure 6.6-5 as a function of frequency.

As with the case of the common-emitter and common-source amplifiers discussed previously, the high-frequency response of the common-gate (and correspondingly the common-base) transistor amplifiers may also be analyzed by inspection of the equivalent circuit. By examining equation (6.6-32), it should be clear that the circuit transfer function can be obtained by multiplying the amplifier midband gain by terms of the form $1/(1 + s/\omega_x)$ for each of the parasitic capacitors. Furthermore, in this equation the break frequency ω_x is given by $1/R_x C_x$, where R_x is the net Thévenin resistance seen by the capacitor C_x. Of even greater significance for this particular problem (because each of the capacitors is connected between a single circuit node point and ground) is the fact that the resistance R_x seen by both C_{gs} and C_{gd} can be determined by the same impedance analysis techniques developed in previous chapters.

To illustrate these points, we recompute the circuit transfer function directly from the original circuit in Figure 6.6-4a. To begin with, the input

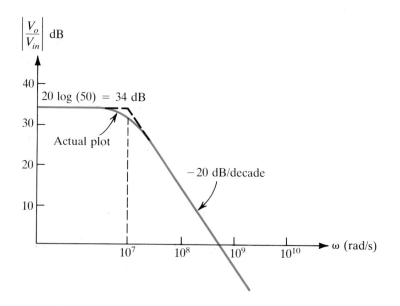

Figure 6.6-5
Frequency response of the
common-gate amplifier in
Figure 6.6-4.

Chapter 6 / Frequency Effects in Amplifier Circuits

impedance R_1 seen looking into the source of this circuit is $1/g_m = 100\ \Omega$, so that in midband the equivalent circuit in Figure 6.6-4c applies with C_{gs} and C_{gd} replaced by open circuits. As a result, in the midband frequency range

$$I_d = \frac{-V_{in}}{R_{in} + 1/g_m} \tag{6.6-34a}$$

and

$$V_o = -I_d R_D = \frac{g_m R_D}{1 + g_m R_{in}} V_{in} \tag{6.6-34b}$$

so that the amplifier midband gain is

$$A_{\text{mid}} = \frac{g_m R_D}{1 + g_m R_{in}} = 50 \tag{6.6-35}$$

The input and output break frequencies can be determined by inspection of the high-frequency equivalent circuit presented in Figure 6.6-4c. Specifically,

$$\omega_1 = \frac{1}{(R_{in} \parallel 1/g_m) C_{gs}} = 2 \times 10^9 \text{ rad/s} \tag{6.6-36a}$$

and because $R_2 = \infty$, in Figure 6.6-4a,

$$\omega_2 = \frac{1}{R_D C_{gd}} = 10^7 \text{ rad/s} \tag{6.6-36b}$$

Combining the results obtained in equation (6.6-35), (6.6-36a), and (6.6-36b), we can see that the final transfer function will be the same as that given in equation (6.6-33).

EXAMPLE 6.6-4

For the BJT common-base amplifier illustrated in Figure 6.6-6a, develop a complete expression for the transfer function V_l/V_{in} by inspection of the circuit. The transistor parameters for this problem are $g_m = 100$ mS, $r_{bb'} = 0$, $r_{b'e} = 1\ k\Omega$, $C_{b'e} = 100$ pF, and $C_{b'c} = 1$ pF. You may consider that C_E and C_L in the figure are ac shorts in the midband and high-frequency regions.

SOLUTION

Because the form of the FET circuit in Figure 6.6-4a is basically the same as that of the BJT common-base amplifier shown in Figure 6.6-6a it is reasonable to expect, following equation (6.6-32), that the general form of the frequency response for this circuit will be

$$\frac{V_l}{V_{in}} = A_{\text{mid}} \underbrace{\left(\frac{1}{1 + s/\omega_1} \right)}_{\substack{\text{break due} \\ \text{to } C_{b'e}}} \underbrace{\left(\frac{1}{1 + s/\omega_2} \right)}_{\substack{\text{break due} \\ \text{to } C_{b'c}}} \tag{6.6-37}$$

Thus, in order to determine the transfer function for this circuit, only three parameters, A_{mid}, ω_1, and ω_2, need to be evaluated.

To find the midband gain it may be easier if we evaluate it using the hybrid rather than the hybrid-π parameters. For the hybrid-π parameters given, we find that the corresponding h-parameters are

$$h_{ie} = r_{bb'} + r_{b'e} = 1\ k\Omega \tag{6.6-38a}$$

and

$$h_{fe} = g_m r_{b'e} = 100 \tag{6.6-38b}$$

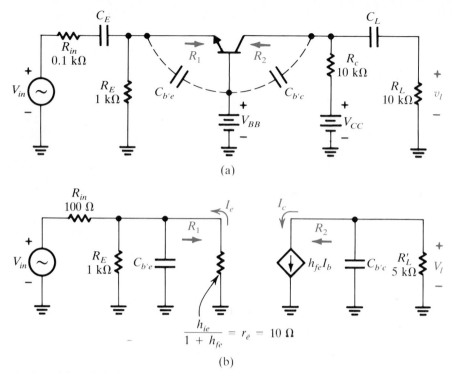

Figure 6.6-6

Example 6.6-4.

(a)

(b)

Again, although it is not strictly necessary for solving this problem, we have drawn the equivalent circuit for this amplifier that is valid in the middle- and high-frequency regions (Figure 6.6-6b). Notice that the input resistance of this amplifier is simply $h_{ie}/(1 + h_{fe})$, or about 10 Ω. To evaluate the midband gain from this figure, we note that

$$I_e \simeq -\frac{V_{in}}{R_{in} + r_e} = -\frac{V_{in}}{0.11 \text{ k}\Omega} \qquad (6.6\text{-}39\text{a})$$

$$I_c \simeq I_e = -\frac{V_{in}}{0.11 \text{ k}\Omega} \qquad (6.6\text{-}39\text{b})$$

and

$$V_l = -I_c R'_L = -\left(-\frac{V_{in}}{0.11 \text{ k}\Omega}\right)(5 \text{ k}\Omega) = 45.5 V_{in} \qquad (6.6\text{-}39\text{c})$$

from which we find that the midband gain is 45.5.

To evaluate the input and output breaks associated with the capacitors $C_{b'e}$ and $C_{b'c}$, we need to find the Thévenin resistances seen by each of these capacitors. Because each of these capacitors is connected between a single circuit node point and ground, this task is especially easy since all the transistor impedance techniques developed previously are applicable. Thus the break associated with $C_{b'e}$ is

$$\omega_1 = \frac{1}{(R_{in} \| R_E \| h_{ie}/(1 + h_{fe})C_{b'e}} \simeq 10^9 \text{ rad/s} \qquad (6.6\text{-}40\text{a})$$

where $h_{ie}/(1 + h_{fe})$ is the resistance R_1 seen looking into the emitter of the transistor in Figure 6.6-12b (or alternatively, in part (a) of the figure). Similarly, because the resistance R_2 is equal to infinity, the break associated with

Chapter 6 / Frequency Effects in Amplifier Circuits

$C_{b'c}$ can be written down directly as

$$\omega_2 = \frac{1}{R_L' C_{b'c}} = 2 \times 10^8 \text{ rad/s} \qquad (6.6\text{-}40b)$$

From the results determined in equations (6.6-39c), (6.6-40a), and (6.6-40b), we can see that

$$H(s) = 45.5 \left(\frac{1}{1 + s/10^9} \right) \left(\frac{1}{1 + s/2 \times 10^8} \right) \qquad (6.6\text{-}41)$$

6.6-3   Common-Drain and Common-Collector Amplifiers at High Frequencies

The common drain (or source follower) and common collector (or emitter follower) are used for impedance-matching purposes, where high input and low output impedances are needed. At high frequencies these impedance-matching properties are useful to extend the amplifier's bandwidth. For example, consider the amplifier shown in Figure 6.6-7a. Because of the amplifier input capacitance, there is a high-frequency break at $1/R_{in}C_{in}$ associated with C_{in}. If R_{in} is a large resistance, this break may significantly affect the amplifier's overall high-frequency 3-dB point. By adding an emitter follower between the source resistance and the capacitor, it may be possible to improve the circuit bandwidth (Figure 6.6-7b).

If we assume that $(1 + h_{fe})R_E \gg R_{in}$, the addition of the emitter follower will have little effect on the circuit's midband gain. Furthermore, the capacitor C_{in} now sees an impedance $R_{in}' = [(R_{in} + h_{ie})/(1 + h_{fe})] \| R_E$ rather than R_{in}, and typically this resistance, R_{in}', will be much smaller than R_{in}, so that the break associated with C_{in} will move up considerably.

In a similar fashion, an emitter or source follower can be added to the output of an amplifier to improve its ability to drive capacitive loads. When an amplifier attempts to drive a capacitive load (Figure 6.6-7c), the break associated with the load capacitance is $1/R_o C_L$. If the load capacitance is large, this break can seriously degrade the frequency response of the amplifier. The addition of an emitter follower at the output can significantly improve this circuit's performance by lowering the amplifier output impedance seen by the capacitor (Figure 6.6-7d).

The improvements in high-frequency performance afforded by the addition of follower circuits to the input and output of an amplifier have ignored the effect that the follower's frequency response has on the overall frequency response of the circuit. To take this into account, we need to develop a technique for analyzing the frequency response of emitter-follower and source-follower circuits.

In analyzing voltage-follower circuits, we will see that there is a parasitic feedback capacitance connected between the output and the input of the amplifier, and based on our experience with common-source and common-emitter amplifiers, we may be tempted to handle these follower circuits by applying the Miller theorem to split this feedback capacitor in two. Unfortunately, for reasons that will be apparent from the derivations to follow, this method will fail to give correct results, and a somewhat more elaborate technique will be

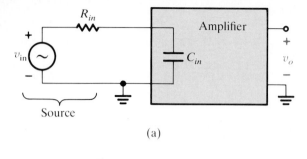

(a)

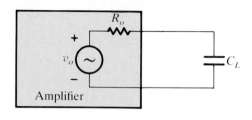

(c)   Amplifier Driving a Capacitive Load

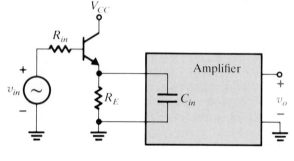

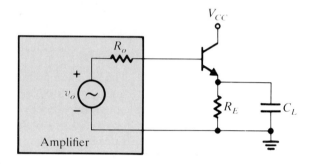

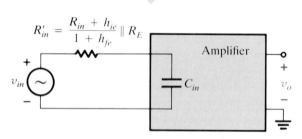

(b)   Improving the Frequency Response by Adding
an Emitter Follower at the Input

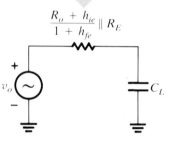

(d)   Improving the Frequency Response
by Adding an Emitter Follower
on the Output

Figure 6.6-7
*Use of source and emitter fol-
lowers to extend amplifier
bandwidth at the input and
output of an amplifier.*

needed to account for the breaks associated with the gate-to-source capacitor
in FETs and the base-to-emitter capacitor in BJTs.

Let us begin our investigation of follower circuits by initially examining the
FET source follower, and after that the BJT emitter follower. Once the high-
frequency performance of these circuits is understood, we will be able to extend these results to the analysis of common-emitter and common-source amplifiers containing unbypassed emitter and source resistors, as well as to
common-gate and common-base circuits containing resistors in their gate and
base leads. Since it is the gate-to-source feedback capacitor, C_{gs}, which creates
most of the difficulty in analyzing the high-frequency performance of a source
follower, we will initially neglect the other parasitic capacitor, C_{gd}, to reduce
the mathematical complexity because its effect can easily be included later.

Consider the source-follower circuit illustrated in Figure 6.6-8a and the
equivalent circuit at high frequencies given in part (b) of the figure. In analyzing this circuit we will initially determine the transfer function by employing a
conventional network analysis approach, and then, after careful examination of

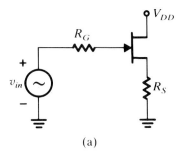

(a)

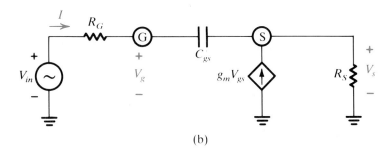

(b)

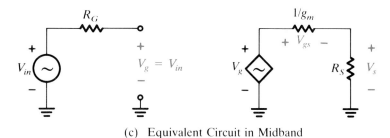

(c) Equivalent Circuit in Midband

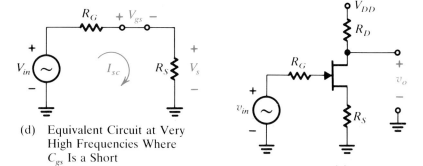

(d) Equivalent Circuit at Very High Frequencies Where C_{gs} Is a Short

(e)

Figure 6.6-8
Source follower at high frequencies ($C_{gd} = 0$).

these results, we will develop techniques for finding the circuit transfer function by inspection.

To solve this problem by network analysis, we first notice that the current I can be written in terms of V_{in} and V_s as

$$I = \frac{V_{in} - V_s}{R_G + 1/sC_{gs}} \qquad (6.6\text{-}42)$$

The gate-to-source voltage V_{gs} is given by

$$V_{gs} = I\frac{1}{sC_{gs}} \tag{6.6-43}$$

and substituting (6.6-42) into (6.6-43), we obtain

$$V_{gs} = \frac{V_{in} - V_s}{R_G + 1/sC_{gs}} \frac{1}{sC_{gs}} = \frac{V_{in} - V_s}{1 + sR_GC_{gs}} \tag{6.6-44}$$

To find the output voltage V_s, we note that

$$V_s = (I + g_mV_{gs})R_S \tag{6.6-45}$$

Combining equations (6.6-42) through (6.6-45), after some manipulation we may write down the overall circuit transfer function as

$$\frac{V_s}{V_{in}} = \frac{g_mR_S}{1 + g_mR_S} \frac{1 + sC_{gs}/g_m}{1 + s(R_G + R_S)C_{gs}/(1 + g_mR_S)} = A_{\text{mid}}\frac{1 + s/\omega_2}{1 + s/\omega_1} \tag{6.6-46a}$$

where

$$\omega_1 = \frac{1}{[(R_G + R_S)/(1 + g_mR_S)]C_{gs}} \tag{6.6-46b}$$

and

$$\omega_2 = \frac{1}{(1/g_m)C_{gs}} \tag{6.6-46c}$$

The magnitude of this transfer function is sketched as a function of frequency in Figure 6.6-9. Considerable physical interpretation of the results in this sketch is possible if we examine it together with the equivalent circuit in Figure 6.6-8b. At low frequencies the parasitic capacitor is an open, and the equivalent circuit reduces to the simplified midband source equivalent circuit given in Figure 6.6-8c. Using this figure, the low-frequency gain is therefore

$$\frac{V_s}{V_{in}} = A_{\text{mid}} = \frac{R_S}{1/g_m + R_S} = \frac{g_mR_S}{1 + g_mR_S} \tag{6.6-47}$$

At high frequencies the capacitor C_{gs} in Figure 6.6-8b becomes a short, and V_{gs}, and hence the current source g_mV_{gs}, are reduced to zero. For this case the equivalent circuit in Figure 6.6-8d applies and V_s is given by

$$\frac{V_s}{V_{in}} = \frac{R_S}{R_S + R_G} \tag{6.6-48}$$

In this sketch it is not too difficult to show that the ratio of the low- and high-frequency gains [in equations (6.6-47) and (6.6-48)] is equal to the ratio of the break frequencies ω_2 to ω_1 [in equations (6.6-46c) and (6.6-46b)]. As a result, depending on the relative sizes of R_G and $1/g_m$, ω_1 can be greater than, equal to, or less than ω_2. This means that for particular combinations of R_G and $1/g_m$, there may be no high-frequency 3-dB point, and in fact, the follower gain could actually increase with frequency (to the extent that the circuit model in Figure 6.6-8b is valid).

To apply these results, for example, to a common-source amplifier with an unbypassed source resistor (Figure 6.6-8e), we need to remember that the output voltage V_o is proportional to I_d, which, in turn, is proportional to V_{gs}, not to V_s. Thus, to understand the high-frequency behavior of this type of amplifier,

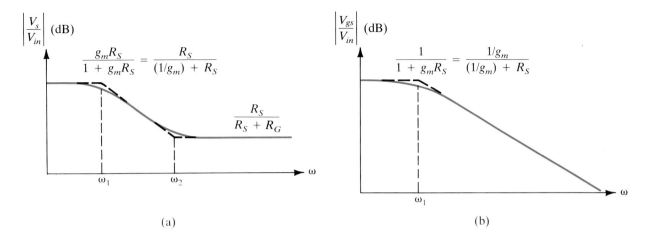

(a)   (b)

Figure 6.6-9
Source-follower high-frequency response.

we need to examine more closely how V_{gs} varies with frequency. Combining equations (6.6-44) and (6.6-46a), we have

$$\frac{V_{gs}}{V_{in}} = \frac{1}{1 + sR_G C_{gs}} - \frac{g_m R_S(1 + sC_{gs}/g_m)}{(1 + g_m R_S)\left[1 + \dfrac{s(R_G + R_S)C_{gs}}{(1 + g_m R_S)}\right]}$$

$$= \underbrace{\frac{1}{1 + g_m R_S}}_{A_{\text{mid}}} \frac{1}{1 + s/\omega_1}$$

(6.6-49)

where ω_1 is defined as in equation (6.6-46b). The frequency response of this transfer function is sketched in Figure 6.6-9b. From Figure 6.6-8c, we can see that the term $1/(1 + g_m R_S)$ represents the midband voltage gain from V_{in} to V_{gs}. The origin of the break frequency term, ω_1, is more difficult to understand.

At first, following our previous experience, we might be tempted to apply the Miller theorem to this problem in order to split the capacitor C_{gs} into two capacitors, each connected from a single node point to ground. Clearly, this approach would be desirable because then the breaks associated with each of these capacitors could easily be found. Unfortunately, the Miller theorem cannot be applied accurately here because the Miller gain A is not a constant in the frequency range where C_{gs} begins to break. This statement can be justified by developing the expression for the Miller gain for this circuit. From Figure 6.6-8b,

$$sC_{gs}(V_g - V_s) + g_m(V_g - V_s) = \frac{V_s}{R_S}$$

(6.6-50)

or

$$A = \frac{V_s}{V_g} = \frac{g_m R_S}{1 + g_m R_S} \frac{1 + s/\omega_2}{1 + s/\omega_3}$$

(6.6-51a)

where ω_2 is defined as in equation (6.6-46c), and

$$\omega_3 = \frac{1}{(1/g_m \parallel R_S)C_{gs}}$$

(6.6-51b)

Because the relative sizes of ω_2 and ω_1 in equation (6.6-46b) and (6.6-46c) depend on R_G and g_m, the Miller gain cannot generally be considered to be constant in the frequency range of interest. Therefore we cannot use the Miller theorem to simplify this problem.

Instead, we will determine the break frequency associated with C_{gs} by evaluating the expression $1/R_T C_{gs}$, where R_T is the Thévenin resistance seen by C_{gs}. If we derive a general expression for this resistance, we can use this result whenever the C_{gs} break frequency needs to be computed. In addition, we will see that the expression for the Thévenin resistance seen by $C_{b'e}$ in the equivalent BJT circuit is nearly the same as R_T for this FET problem.

The Thévenin resistance seen by the capacitor C_{gs} can readily be found by employing the series of equivalent circuits shown in Figure 6.6-10, from which it is apparent that

Figure 6.6-10
*Thévenin resistance "seen" by
C_{gs}.*

$$R_T = \frac{R_G + R_S}{g_m R_S} \,\Big\|\, (R_G + R_S) = \frac{R_G + R_S}{1 + g_m R_S} \qquad (6.6\text{-}52)$$

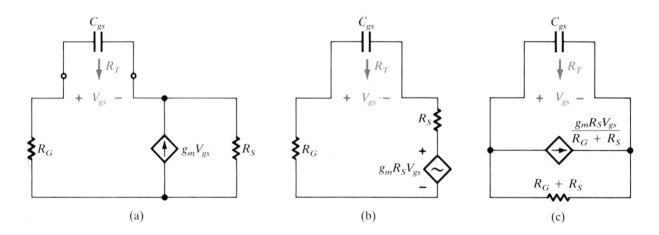

(a)   (b)   (c)

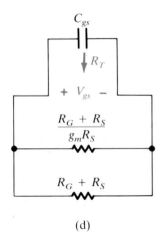

(d)

This expression can also be derived directly from Figure 6.6-8b, using the fact that $R_T = V_{oc}/I_{sc}$. Regardless of the technique employed, once R_T is known, the pole associated with the capacitor C_{gs} can now be written down directly as

$$\omega_1 = \frac{1}{R_T C_{gs}} = \frac{1}{[(R_G + R_S)/(1 + g_m R_S)]C_{gs}} \qquad (6.6\text{-}53)$$

This result is the same as that calculated by network analysis techniques in equation (6.6-46b).

Combining this method for finding ω_1 with that for computing A_{mid}, we are now in a position to determine the transfer function of voltage-follower circuits by inspection [see equation (6.6-49)]. However, to make this technique really

Chapter 6 / Frequency Effects in Amplifier Circuits

useful, we need to include the effects of the other parasitic elements, which we initially neglected.

The equivalent circuit in Figure 6.6-11b for the source follower given in Figure 6.6-11a includes both parasitic capacitances, C_{gs} and C_{gd}. By carrying out a rigorous network analysis of this problem, we can demonstrate that the transfer function for this circuit has the same general form as that associated with similar circuits analyzed earlier in this section [see equation (6.6-15)]. In particular,

$$H(s) = A_{mid} \frac{1 + s/\omega_N}{1 + s/(\omega_{1o} \parallel \omega_{2o}) + s^2/[(\omega_{1s} + \omega_{2s})(\omega_{1o} \parallel \omega_{2o})]} \qquad (6.6\text{-}54)$$

Again, because ω_N, ω_{1s}, and ω_{2s} are almost always much greater than ω_{1o} and ω_{2o}, for frequencies at and below the 3-dB point, this expression may be approximated as

$$H(s) \simeq A_{mid} \frac{1}{1 + s/\omega_{1o} \parallel \omega_{2o}} = A_{mid} \frac{1}{1 + s/\omega_H} \qquad (6.6\text{-}55)$$

where A_{mid} is the midband gain calculated with the parasitic capacitors replaced by open circuits, and

$$\omega_H = \omega_{1o} \parallel \omega_{2o} \qquad (6.6\text{-}56)$$

with ω_{1o} = break associated with C_{gd} with C_{gs} an open, and ω_{2o} = break associated with C_{gs} with C_{gd} an open. From Figure 6.6-11b,

$$\omega_{1o} = \frac{1}{R_G C_{gd}} = 10^7 \text{ rad/s} \qquad (6.6\text{-}57a)$$

and

$$\omega_{2o} = \frac{1}{R_T C_{gs}} = \frac{1}{[(R_G + R_S)/(1 + g_m R_S)] C_{gs}} \simeq 10^8 \text{ rad/s} \qquad (6.6\text{-}57b)$$

Substituting these results into equation (6.6-56), we may approximate the 3-dB point for this amplifier as

$$10^7 \parallel 10^8 = 9.1 \times 10^6 \text{ rad/s} \le \omega_H \le \omega_{1o} = 10.0 \times 10^6 \text{ rad/s} \qquad (6.6\text{-}58)$$

Computer analysis of this same problem yields $f_H = 1.45$ MHz or $\omega_H = 9.11$ Mrad/s.

Figure 6.6-11
Source follower at high frequencies (both C_{gs} and C_{gd} included).

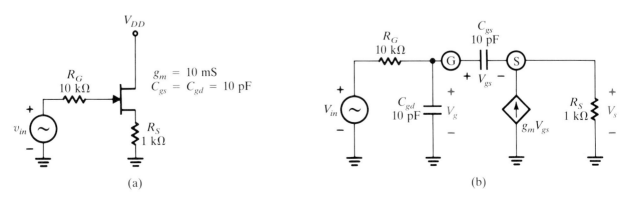

(a)

(b)

EXAMPLE 6.6-5

For the emitter-follower circuit illustrated in Figure 6.6-12a, develop an expression for the denominator break frequency associated with the parasitic capacitor $C_{b'e}$ assuming that $r_{bb'} = 0$.

SOLUTION

The equivalent circuit for this emitter follower is found by substituting in the hybrid-π model for the transistor. The resulting equivalent circuit is given in Figure 6.6-12b. As with the source-follower circuit just examined, the Miller theorem cannot be used to find the break frequency due to $C_{b'e}$. Instead, and again directly paralleling the previous development for the source follower, we will find the Thévenin resistance seen by this capacitor and then use $1/R_T C_{b'e}$ to determine the break.

By setting the independent source V_{in} equal to zero and opening the other parasitic capacitors (in this case, just $C_{b'c}$), we obtain the equivalent circuit shown in Figure 6.6-12c, and by means of the sequence of Thévenin and Norton transformations shown in parts (d) through (f), we can readily show that

Figure 6.6-12
Example 6.6-5. Emitter follower at high frequencies.

$$R_T = \frac{R_B + R_E}{1 + g_m R_E} \parallel r_{b'e} \qquad (6.6\text{-}59)$$

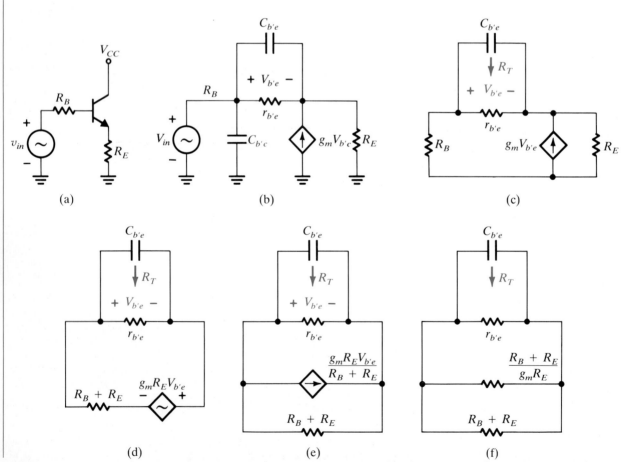

(a) (b) (c)

(d) (e) (f)

This result is not all that surprising, since the equivalent circuit for the emitter follower in Figure 6.6-12b is the same as that for the source follower in Figure 6.6-10a except for the additional resistance $r_{b'e}$ in parallel with the rest of the circuit [compare equations (6.6-52) and (6.6-59)].

Using the Thévenin resistance from equation (6.6-59), the break frequency associated with $C_{b'e}$ is

$$\omega = \frac{1}{R_T C_{b'e}} = \frac{1}{\{[(R_B + R_E)/(1 + g_m R_E)] \parallel r_{b'e}\}C_{b'e}} \qquad (6.6\text{-}60)$$

At this point in our investigation of the high-frequency performance of electronic circuits, we have examined the three most important circuit configurations—the common-source, the common-gate, and common-drain FET amplifiers—as well as their BJT counterparts. The techniques acquired in analyzing these circuits are usually all that is necessary to enable us to obtain an accurate estimate of the high-frequency 3-dB point of nearly any single- or multitransistor circuit configuration. These techniques are summarized below.

1. Whenever a parasitic capacitor is connected between a single circuit node point and ground (see, e.g., C_{gs} in Figure 6.6-13a), the high-frequency break associated with it is $1/R_T C$, where R_T is the Thévenin resistance connected across the capacitor terminals. For this case, the R_T can be found by applying the transistor impedance techniques developed in earlier chapters.

2. When R_D (or R_C for a BJT circuit) is nonzero, then as illustrated in Figure 6.6-13b, split C_{gd} (or $C_{b'c}$) into two capacitors by using the Miller theorem. This results in the creation of two capacitors, where each is now connected between a single node point and ground. The break frequencies for each capacitor is then found as explained in technique 1. Note that if these capacitors are in parallel with other capacitors (e.g., C_{gs} and C_{m1} in Figure 6.6-13b), they are simply added together to compute their overall break.

3. When R_S (or R_E) is nonzero in a common-source amplifier (see Figure 6.6-13c), the Miller capacitor is handled as explained in technique 2 (see Figure 6.6-13d). The break associated with the capacitance C_{gs} (or $C_{b'e}$ for BJTs) is found from $1/R_T C$, where

$$R_T = \frac{R_G + R_S}{1 + g_m R_S}$$

for FETs and

$$R_T = \frac{R_B + R_E}{1 + g_m R_E} \parallel r_{b'e}$$

for BJTs. The resulting circuit has three breaks to calculate, one each for C_{m1}, C_{m2}, and C_{gs}.

4. When R_G (or R_B) is nonzero in a common-gate (or common-base) amplifier, (Figure 6.6-13e), the effect of C_{gs} (or $C_{b'e}$) on the high-frequency response is accounted for as described in (3) above. The effect of the feedback capacitor C_{gd} (or $C_{b'c}$) on the amplifier's frequency response is best handled by observing that V_g is generally much smaller than V_d so that C_{gd} may be split in two as shown in Figure 6.6-13f. Here the controlled voltage source V_d is taken to be its midband value and by considering that $R_G \ll 1/\omega C_{gd}$, after considerable manipulation, the source equivalent circuit can be shown to take on the form given in Figure 6.6-13g. The break associated with the equivalent inductor in this figure is evaluated by inspection.

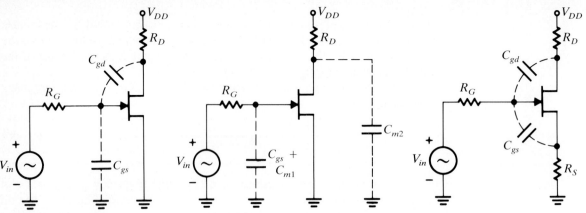

(a) Common-Source Amplifier

(b) Common-Source Amplifier after the Miller Theorem Is Applied

(c) Common-Source Style Amplifier with Nonzero R_S

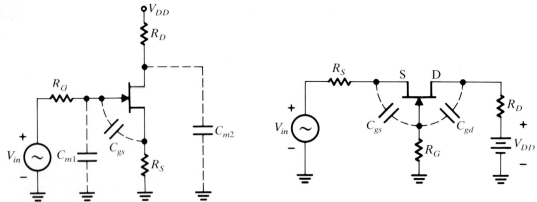

(d) Common-Source Style Amplifier with C_{gd} Split Via Miller Theorem

(e) Common-Gate Amplifier with Nonzero R_G

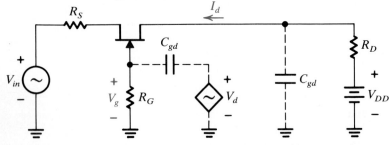

(f) Including the Effect of C_{gd} When R_G is Nonzero (C_{gs} Omitted for Simplicity)

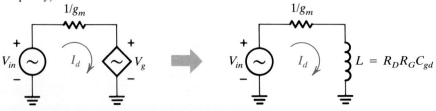

(g) Source Equivalent Circuit When R_G is Nonzero (L is an Equivalent Inductor that Represents the Effect of C_{gd} on the Input Circuit)

Figure 6.6-13

EXAMPLE 6.6-6

For the FET amplifier given in Figure 6.6-14a, estimate the midband gain and the high-frequency 3-dB point. The transistor parameters are $g_m = 10$ mS, $C_{gs} = 10$ pF, and $C_{gd} = 1$ pF. C_L is a 2-pF load capacitance that the amplifier is assumed to be driving.

SOLUTION

This problem can be analyzed either by drawing a complete equivalent circuit valid at high frequencies or simply by indicating the location of the parasitic elements on the original figure and using this circuit to solve the problem by inspection. Because of its utility in handling multitransistor problems, we will solve this problem by employing the latter approach. The two parasitic capacitors are shown by dashed lines on the original figure. To handle C_{gd}, we need to apply the Miller theorem.

The Miller gain is found with the aid of the source equivalent circuit in Figure 6.6-14b, from which

$$I_d = \frac{V_g}{0.3 \text{ k}\Omega} \tag{6.6-61a}$$

$$V_d = -I_d R_D = -\frac{(10 \text{ k}\Omega)V_g}{0.3 \text{ k}\Omega} \tag{6.6-61b}$$

Figure 6.6-14
Example 6.6-6.

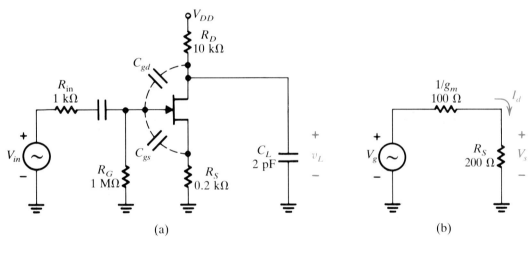

(a)   (b)

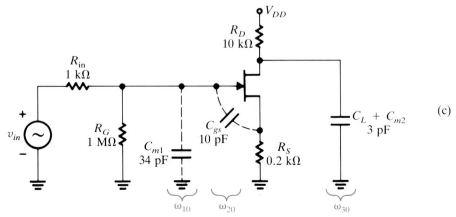

(c)

and
$$A = \frac{V_d}{V_g} = -33.3 \qquad (6.6\text{-}61c)$$

For this circuit the midband gain is equal to the Miller gain because $R_{in} \ll R_G$, so that $V_g = V_{in}$. Applying the Miller theorem, we may split C_{gd} into two equivalent capacitors ($C_{m1} = 34$ pF and $C_{m2} = 1$ pF), and on the output side of the circuit, C_{m2} is simply added to C_L as shown in Figure 6.6-14c. Using this figure, the break frequencies associated with each of these capacitors may be calculated as follows:

$$\omega_{1o} = \frac{1}{(1 \text{ k}\Omega \parallel 1 \text{ M }\Omega)(34 \text{ pF})} = 2.94 \times 10^7 \text{ rad/s} \qquad (6.6\text{-}62a)$$

$$\omega_{2o} = \frac{1}{(1.2 \text{ k}\Omega/3)(10 \text{ pF})} = 2.5 \times 10^8 \text{ rad/s} \qquad (6.6\text{-}62b)$$

and

$$\omega_{3o} = \frac{1}{(10 \text{ k}\Omega)(3 \text{ pF})} = 3.3 \times 10^7 \text{ rad/s} \qquad (6.6\text{-}62c)$$

Combining these results and using equation (6.6-19), the high-frequency 3-dB point may be estimated as

$$14.7 \times 10^6 \text{ rad/s} \leq \omega_H \leq 29.4 \times 10^6 \text{ rad/s} \qquad (6.6\text{-}63)$$

Computer-aided analysis of this problem yields $A_{\text{mid}} = -33.3$ and $\omega_H = 15.4$ Mrad/s ($f_H = 2.44$ MHz).

Exercises

6.6-1 Estimate the high-frequency 3-dB point of the amplifier shown in Figure 6.6-1 if it drives a load consisting of a 5-kΩ resistor in parallel with a 5-pF capacitor. *Answer* 2.2 Mrad/s $< \omega_H <$ 2.3 Mrad/s

6.6-2 In Example 6.6-3, R_E is a 500-Ω resistor. Estimate the high-frequency 3-dB point of this circuit if the emitter bypass capacitor across R_E is removed. *Answer* 56 Mrad/s $< \omega_H <$ 100 Mrad/s

6.6-3 For the common-gate amplifier in Figure 6.6-4, R_D is considered to be adjustable. Find the largest allowed value of R_D and the corresponding value of the gain for which the circuit bandwidth is at least 25 MHz. *Answer* 588 Ω, 2.94

6.6-4 The amplifier illustrated in Figure 6.6-11 drives a 500-Ω load in parallel with a 400-pF capacitor at its source terminal. Determine the break frequencies associated with C_{gd}, C_{gs}, and C_L for this circuit. *Answer* 10 Mrad/s, 42 Mrad/s, 32.5 Mrad/s

---

6.7 MULTISTAGE AMPLIFIERS AT HIGH FREQUENCIES

In the design of electronic amplifiers, the gain and bandwidth requirements are frequently so large that it is impossible to accomplish the design task with a single-stage circuit. At first glance it might appear that any desired gain could be obtained from a single-stage amplifier simply by increasing the collector or drain resistance. This, of course, assumes that the transistor's output impedance is infinite. Furthermore, it ignores the effect that this increased resistance has on the high-frequency 3-dB point of the amplifier. Because the high-frequency break terms are generally of the form $1/RC$, as the output resistance increases, the bandwidth of the amplifier goes down correspondingly. Thus single-stage amplifiers have upper bounds on the gains they can achieve, and similar constraints on their bandwidths. However, by cascading several low-gain amplifier stages, it is possible to extend a circuit's gain–bandwidth product beyond that attainable with a single-stage amplifier.

In this section, we examine the factors governing the design and analysis of multistage amplifiers, with particular attention being given to techniques for evaluating the high-frequency 3-dB point of such circuits. We will see that as several single-stage amplifiers are cascaded, the overall circuit bandwidth will decrease. But because the overall gain increases faster than the rate at which the bandwidth decreases, it is still possible to improve the overall circuit gain–bandwidth product by using a multistage design.

To develop an expression for the high-frequency 3-dB point of a multistage amplifier, we need to review a few facts about the bandwidth of single-stage amplifiers. In Section 6.6 we showed that the transfer function of most single-stage electronic amplifiers could be approximately written in the form

$$H(s) = A_o \frac{1}{1 + s/\omega_H} \qquad (6.7\text{-}1a)$$

where

$$\omega_H = \omega_{10} \| \omega_{20} \| \cdots \qquad (6.7\text{-}1b)$$

Consider next the case where we have a cascade of several stages. For such an amplifier, using a transfer function of the form given in equation (6.7-1a) for each stage, we may write the approximate overall transfer function as

$$H(s) = A_{\text{mid}} \left(\frac{1}{1 + s/\omega_{H1}} \right) \left(\frac{1}{1 + s/\omega_{H2}} \right) \cdots \left(\frac{1}{1 + s/\omega_{Hn}} \right) \qquad (6.7\text{-}2)$$

where A_{mid} is the midband gain of the amplifier, including all loading effects. Furthermore, ω_{H1}, ω_{H2}, and so on, represent the high-frequency 3-dB breakpoints of each stage in the amplifier, with the resistive loading of the stages on one another included. Because ω_H, the overall amplifier 3-dB frequency, will be located below the smallest single-stage 3-dB point, we may approximate $H(s)$ in the frequency region at and below ω_H as

$$H(s) \simeq A_{\text{mid}} \frac{1}{1 + s\left(\dfrac{1}{\omega_{H1}} + \dfrac{1}{\omega_{H2}} + \cdots + \dfrac{1}{\omega_{Hn}} \right)} = A_{\text{mid}} \frac{1}{1 + s/(\omega_{H1} \| \omega_{H2} \| \cdots \| \omega_{Hn})} \qquad (6.7\text{-}3)$$

Using this result and paralleling the development of equation (6.6-19), we can therefore see that the overall high-frequency 3-dB point of this amplifier is given by the expression

$$\omega_{H1} \| \omega_{H2} \| \cdots \| \omega_{Hn} \le \omega_H \le \text{smaller of } \omega_{H1}, \omega_{H2}, \ldots, \omega_{Hn} \qquad (6.7\text{-}4)$$

Alternatively, this result can also be expressed as

$$\omega_{1o} \| \omega_{2o} \| \cdots \| \omega_{mo} \le \omega_H \le \text{smaller of } \omega_{1o}, \omega_{2o}, \ldots, \omega_{mo} \qquad (6.7\text{-}5a)$$

where
$$\omega_{io} = \frac{1}{R_i C_i} \qquad (6.7\text{-}5b)$$

for $i = 1, 2, 3, \ldots, m$. Here m equals the number of independent parasitic capacitors in the overall amplifier. Furthermore, w_{io} represents the break frequency associated with the parasitic capacitor C_i, and R_i is the Thévenin resistance seen by C_i with all other capacitors considered to be open. Of course, the utility of equation (6.7-5a) depends on the extent to which we may neglect the effect of ω_N, ω_{1s}, and so on, in each amplifier stage [see equation (6.6-15)]. Fortunately, as we shall see in the examples that follow in this section, this approximation is quite good for most electronic circuits.

EXAMPLE 6.7-1

Consider an amplifier consisting of a cascade of individual noninteracting stages of the form

$$H_1(s) = A_1 \frac{1}{1 + s/\omega_1} \qquad (6.7\text{-}6)$$

Thus, each stage has a midband gain A_1 and a high-frequency 3-dB breakpoint at $\omega = \omega_1$. If n of these stages are cascaded (see Figure 6.7-1a) because they are assumed to be noninteracting, the overall transfer function will be

$$H(s) = [H_1(s)]^n = A_1^n \frac{1}{(1 + s/\omega_1)^n} \qquad (6.7\text{-}7)$$

Determine this amplifier's high-frequency 3-dB point exactly, and also by using equation (6.7-5a).

SOLUTION

The exact value of this amplifier's high-frequency 3-dB point is found by determining the value of ω that satisfies the equation

$$|H(j\omega)| = A_1^n \frac{1}{[1 + (\omega/\omega_1)^2]^{n/2}} = A_1^n \frac{1}{\sqrt{2}} \qquad (6.7\text{-}8a)$$

or
$$[1 + (\omega/\omega_1)^2]^{n/2} = 2^{1/2} \qquad (6.7\text{-}8b)$$

Solving equation (6.7-8b) for $\omega = \omega_H$, we find that

$$\omega_H = \omega_1 \sqrt{2^{1/n} - 1} \qquad (6.7\text{-}9)$$

This solution is tabulated in Figure 6.7-1b for different values of n.

The approximate solution for ω_H is found by applying equation (6.7-5a), noting that $H(s)$ for this amplifier cascade has n poles, all located at $s = -\omega_1$.

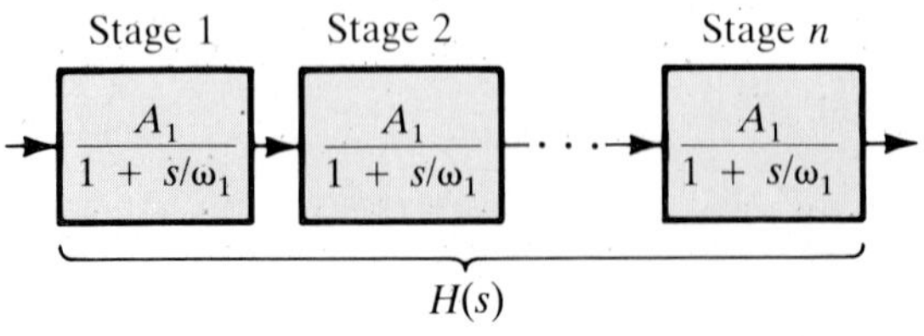

Figure 6.7-1
Example 6.7-1. Comparison of the exact and approximate bandwidth solutions for an amplifier cascade of n noninteracting stages.

(a)

Number of Stages, n	Exact ω_H [Using Eq. (6.7-15)]	Approximate ω_H [Using Eq. (6.7-5a)]
1	ω_1	$\omega_1 \leqslant \omega_H \leqslant \omega_1$
2	$0.64\omega_1$	$0.5\omega_1 \leqslant \omega_H \leqslant \omega_1$
3	$0.51\omega_1$	$0.33\omega_1 \leqslant \omega_H \leqslant \omega_1$
4	$0.43\omega_1$	$0.25\omega_1 \leqslant \omega_H \leqslant \omega_1$

(b)

As a result, this amplifier's high-frequency 3-dB point is somewhere inside the region

$$\frac{\omega_1}{n} \leq \omega \leq \omega_1 \tag{6.7-10}$$

These results are tabulated in Figure 6.7-1b, and comparing them with the exact solutions, we can see that equation (6.7-5a) appears to provide a good estimate for ω_H.

EXAMPLE 6.7-2

Figure 6.7-2
Example 6.7-2.

An amplifier having an overall gain of 50 and a high-frequency 3-dB point of at least 1 MHz is to be constructed by forming a cascade of identical amplifier stages. Each stage is considered to have a gain that can be varied by adjusting a resistor inside the amplifier. This resistance is also assumed to determine the amplifier's high-frequency breakpoint. As a result, as the resistor value is increased, the gain/stage goes up, but the bandwidth decreases. The single-stage gain–bandwidth product is 6.28×10^7 rad/s. This means, for example, that when the gain/stage is 1, its bandwidth is 10 MHz, while when the gain is 100, the bandwidth is reduced to 10 MHz/100 or 100 kHz. If the stages are assumed to be noninteractive, determine the minimum number of stages needed to satisfy the design requirement.

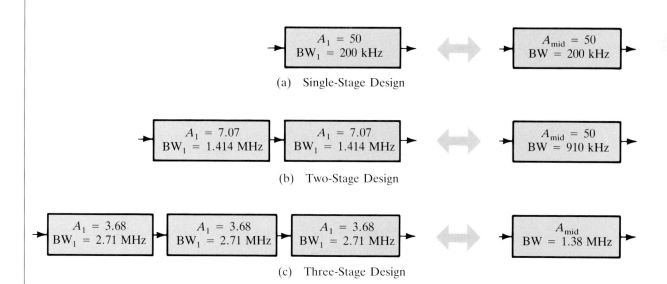

(a) Single-Stage Design

(b) Two-Stage Design

(c) Three-Stage Design

Number of Stages, n	Gain/Stage	Single-Stage Bandwidth	Overall Gain	Overall Bandwidth
1	50	200 kHz	50	200 kHz
2	7.07	1.41 MHz	50	910 kHz
3	3.68	2.71 MHz	50	1.38 MHz

(d)

We will solve this problem by trial and error; that is, first we attempt the design using only a single stage; if that fails, with two stages; and so on. The major advantage of this technique is that it vividly illustrates the relationship of the gain–bandwidth characteristics of each stage to those of the overall amplifier. Obviously, if n turns out to be a large number, this method will prove to be a tedious way to solve this problem.

For the case where $n = 1$, this single stage will need to have a midband gain of 50. Using the fact that the gain–bandwidth product of each stage is 10 MHz, the corresponding bandwidth of this single stage will be 10 MHz/50, or 200 kHz. This bandwidth is insufficient, so this design cannot be accomplished by using a single stage (Figure 6.7-2a).

Consider next the possibility of cascading two stages in order to meet the design goal (Figure 6.7-2b). For this case, the midband gain of each stage needs to be $(50)^{0.5}$, or 7.07, and the corresponding bandwidth of each stage is 10 MHz/7.07, or 1.41 MHz. The overall amplifier bandwidth is found by applying equation (6.7-9). The resulting value for f_H is 0.91 MHz, and so, as with single-stage design, a cascade of two of these amplifier stages still cannot satisfy the bandwidth requirement.

When three stages are used, the individual gain per stage needed is $(50)^{0.33}$, or 3.68. The corresponding single-stage bandwidth in this case is 2.71 MHz, and the overall amplifier bandwidth is $2.71(2^{1/3} - 1)$, or 1.38 MHz. Thus, the use of three stages, with each having the form illustrated in Figure 6.7-2c, does satisfy the bandwidth criteria of this problem. The results for the one-, two-, and three-stage designs are summarized in Figure 6.7-2d.

In the remainder of this section, we concentrate on developing techniques for the analysis and design of multitransistor circuits operating at high frequencies. In examining these circuits, we will use exactly the same methods as those employed in Section 6.6 to find ω_H for single-stage amplifiers.

To illustrate the application of this technique to multitransistor problems, we examine the performance of several different video amplifier circuits. In each case we compare the solutions for ω_H obtained using equation (6.7-5a) with the values found with the aid of a standard computer circuit analysis program.

Each amplifier to be examined has a midband voltage gain of approximately 100, is driven by a source with an impedance of 50 Ω, and is connected to a 10-pF capacitive load. Two two-transistor amplifier configurations are analyzed in Examples 6.7-3 and 6.7-4, and the data obtained from the analysis of three other two-transistor circuits is also included in order to determine which, if any, of these circuits is clearly better than the others.

EXAMPLE 6.7-3

The amplifier illustrated in Figure 6.7-3a is a cascade of two common-emitter stages. The transistor parameters are $r_{bb'} = 0$, $r_{b'e} = 1$ kΩ, $C_{b'e} = 10$ pF, $C_{b'c} = 1$ pF, and $g_m = 100$ mS. To simplify the analysis, some of the biasing components have been eliminated. For the collector resistors shown, the gain of each stage is approximately -10, and the overall amplifier gain in midband is nearly 100. Determine the approximate value of ω_H for this amplifier.

To analyze the high-frequency performance of this circuit, we begin by splitting the feedback capacitors, $C_{b'c1}$ and $C_{b'c2}$, by using the Miller theorem. The Miller gain for each stage is approximately -10, so that Miller capacitances of 11 and 1 pF are reflected into the input and output of each stage, respectively. The resulting equivalent circuit is shown in Figure 6.7-3b. The breaks associated with each of the capacitors are computed as follows:

$$\omega_{1o} = \frac{1}{(r_{b'e} \parallel R_{in})C_1} = \frac{1}{(1\ \text{k}\Omega \parallel 50\ \Omega)(21\ \text{pF})} = 9.5 \times 10^8 \text{ rad/s} \qquad (6.7\text{-}11\text{a})$$

$$\omega_{2o} = \frac{1}{(R_{C1} \parallel r_{b'e2})C_2} = \frac{1}{(100\ \Omega \parallel 1\ \text{k}\Omega)(22\ \text{pF})} = 5 \times 10^8 \text{ rad/s} \qquad (6.7\text{-}11\text{b})$$

and

$$\omega_{3o} = \frac{1}{R_{C2}C_3} = \frac{1}{(100)(11\ \text{pF})} = 9.1 \times 10^8 \text{ rad/s} \qquad (6.7\text{-}11\text{c})$$

Figure 6.7-3
Example 6.7-3.

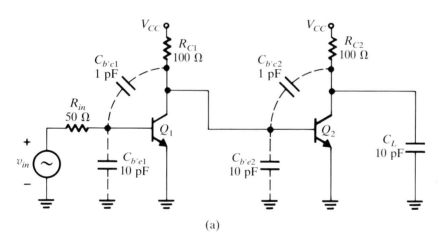

(a)

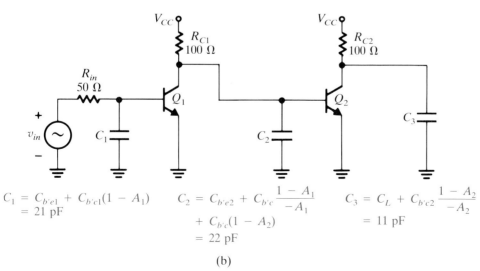

$$C_1 = C_{b'e1} + C_{b'c1}(1 - A_1)$$
$$= 21\ \text{pF}$$

$$C_2 = C_{b'e2} + C_{b'c}\frac{1-A_1}{-A_1}$$
$$+ C_{b'c}(1-A_2)$$
$$= 22\ \text{pF}$$

$$C_3 = C_L + C_{b'c2}\frac{1-A_2}{-A_2}$$
$$= 11\ \text{pF}$$

(b)

from which the high-frequency 3-dB point is

$$2.4 \times 10^8 \text{ rad/s} \leq \omega_H \leq 5 \times 10^8 \text{ rad/s} \tag{6.7-12a}$$

or $$38.3 \text{ MHz} \leq f_H \leq 79.6 \text{ MHz} \tag{6.7-12b}$$

Computer-aided analysis of the same problem results in $f_H = 49$ MHz.

EXAMPLE 6.7-4

The circuit illustrated in Figure 6.7-4 is a differential amplifier. It is very popular for use with integrated circuits because of its good Q-point stability and excellent common-mode rejection. I_o is a dc current source, and the transistor parameters are the same as those for Example 6.7-3. Using these data the circuit midband gain can be shown to be about 97.6. Find the approximate high-frequency 3-dB point for this amplifier.

SOLUTION

In analyzing this problem, we note that Q_2 is operating as a common-base amplifier, so that $C_{b'c2}$ and $C_{b'e2}$ are both connected to ground as shown in the simplified equivalent circuit in Figure 6.7-4b. Furthermore, because R_{C1} is zero, $C_{b'c1}$ is also connected to ground, and no Miller effect multiplication of this capacitance occurs. The pertinent equivalent capacitors in Figure 6.7-4b are labeled C_1 through C_4, and the breaks associated with each may be written by inspection of this circuit as follows:

$$\omega_{1o} = \frac{1}{(R_{in} \| 2h_{ie})C_1} = \frac{1}{(50 \ \Omega \| 2 \text{ k}\Omega)(1 \text{ pF})} = 2 \times 10^{10} \text{ rad/s} \tag{6.7-13a}$$

$$\omega_{2o} = \frac{1}{\{[(R_B + R_E)/(1 + g_mR_E)] \| r_{b'e}\}C_2}$$

$$= \frac{1}{\{[(50 + 10)\Omega/(1 + 1)] \| 1 \text{ k}\Omega\}(10 \text{ pF})} = 3.3 \times 10^9 \text{ rad/s} \tag{6.7-13b}$$

$$\omega_{3o} = \frac{1}{(R_1 \| R_2)C_3} = \frac{1}{(10 \| 10)(10 \text{ pF})} = 2 \times 10^{10} \text{ rad/s} \tag{6.7-13c}$$

and

$$\omega_{4o} = \frac{1}{R_{C2}C_4} = \frac{1}{(2 \text{ k}\Omega)(11 \text{ pF})} = 4.55 \times 10^7 \text{ rad/s} \tag{6.7-13d}$$

From these results we may calculate the high-frequency 3-dB point as

$$4.46 \times 10^7 \text{ rad/s} \leq \omega_H \leq 4.55 \times 10^7 \text{ rad/s} \tag{6.7-14a}$$

or $$7.1 \text{ MHz} \leq f_H \leq 7.25 \text{ MHz} \tag{6.7-14b}$$

Computer-aided circuit analysis yields $f_H = 7.23$ MHz.

Before moving on to the next example, it is worthwhile to spend a few moments examining this circuit to see what else we can learn about its behavior. To begin with, it is interesting to note that R_{C1} is zero. Besides saving the cost of a resistor, this design technique has a second, more important advantage. R_{C1} has no effect on the amplifier gain, but if present, it would create a rather large Miller capacitance at the amplifier input. In fact, its inclusion in this cir-

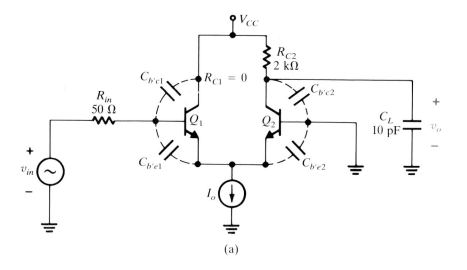

(a)

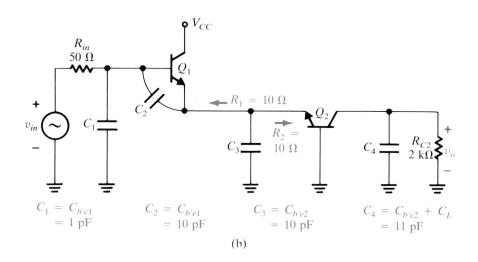

$C_1 = C_{b'c1}$
$= 1$ pF

$C_2 = C_{b'e1}$
$= 10$ pF

$C_3 = C_{b'e2}$
$= 10$ pF

$C_4 = C_{b'c2} + C_L$
$= 11$ pF

(b)

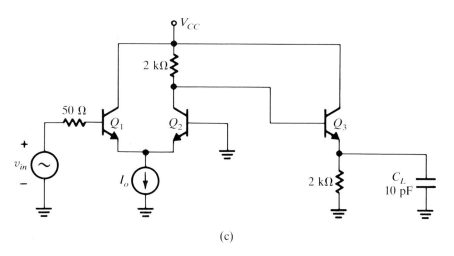

(c)

Figure 6.7-4
Example 6.7-4.

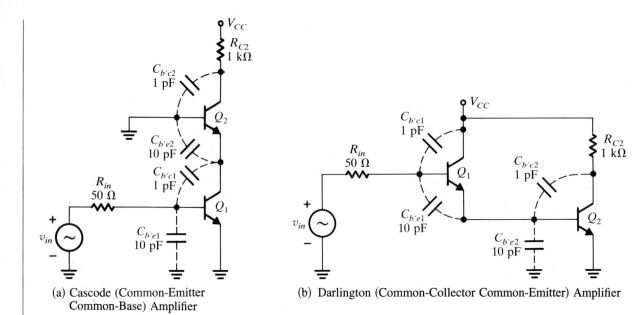

(a) Cascode (Common-Emitter
Common-Base) Amplifier

(b) Darlington (Common-Collector Common-Emitter) Amplifier

Figure 6.7-5
Additional two-transistor
circuit configurations
examined (with $C_L = 10$ pF
connected onto the collector
of Q_2).

cuit would drop ω_{1o} to 200 Mrad/s and would reduce the amplifier bandwidth to about 6 MHz.

Let's look next at possible circuit modifications we could make to improve this amplifier's bandwidth. If we examine the four capacitor breaks, it is clear that the dominant break is ω_{4o}. The overall frequency response of this amplifier can be increased significantly if the resistance seen by C_4 is reduced by adding an emitter follower at the output. In fact, when this is done (see Figure 6.7-4c), the bandwidth of the amplifier is increased to about 35 MHz. The reader should verify this result by recalculating ω_{4o} and including the additional break terms created by the transistor Q_3 to recompute ω_H for this circuit.

6.7-1 Summary of the Results from the Analysis of Several Different Two-Transistor Amplifiers

Amplifier Type	A_{mid}	BW (MHz)	GBW (rad/s)
Common-emitter cascade	86.5	49.6	26.6×10^9
Differential	97.6	7.23	4.4×10^9
Cascode	94.3	14.4	8.5×10^9
Darlington	98.9	12.0	7.5×10^9
Common-emitter common-collector cascade	97.2	22.1	13.5×10^9

The results obtained by analyzing Examples 6.7-3 and 6.7-4 and the two additional circuits shown in Figure 6.7-5 are summarized in Table 6.7-1. Also included for comparison in this table are the data from the analysis of the two transistor common-emitter common-collector circuit configuration obtained by reversing the positions of stages Q_1 and Q_2 in Figure 6.7-5b. For the particular combination of source impedance and load capacitance used with these examples, the cascaded common-emitter configuration provides the best gain–bandwidth product. However, before you generalize this result, we should point out that there is no single circuit that is "best" in all situations. As you gain design experience, you will come to realize that the best amplifier to use in solving a particular design problem depends on the specifications of the problem, that is, on values of the load and source impedances, the gain required, and the bandwidth needed, as well as many other more subtle factors to be introduced later.

Exercises

6.7-1 Five identical stages, each with a gain–bandwidth product of 50 MHz, are used to design an amplifier with an overall gain of 4000. Find the gain and the bandwidth of each stage, and the bandwidth of the overall amplifier. *Answer* 5.25, 9.52 MHz, 3.67 MHz

6.7-2 Estimate the approximate high-frequency breakpoint of the amplifier in Example 6.7-4 if C_L is removed and a 1-kΩ resistor is connected between the base of Q_2 and ground. *Answer* 10 Mrad/s

6.7-3 For the cascode amplifier given in Figure 6.7-5a, estimate the maximum size of the capacitive load that can be driven at the junction of the emitter of Q_2 and the collector of Q_1 if the pole associated with this load is to have less than a 10% effect on the amplifier high-frequency 3-db point. *Answer* 100 pF

6.7-4 In Figure 6.7-5b, the BJT Q_1 is replaced with an FET with $g_m = 1$ mS and $C_{gs} = C_{gd} = 5$ pF. Determine ω_H for the amplifier. *Answer* 14.4 Mrad/s $< \omega_H < 18$ Mrad/s

---

6.8 RELATIONSHIP BETWEEN THE TRANSIENT RESPONSE AND THE FREQUENCY RESPONSE OF AN ELECTRONIC AMPLIFIER

A knowledge of the frequency response of an amplifier provides a great deal of information about how the circuit will respond to nonsinusoidal input signals, and conversely, an examination of the circuit's response to a nonsinusoidal signal such as a square wave can give us many clues about the frequency response characteristics of the amplifier. In this section, we examine both the transient and sinusoidal steady-state frequency response of several selected circuits in order to establish the relationship that exists between these two seemingly different circuit properties. A review of both the standard solution method for differential equations and the Laplace transform method of solution are contained in Sections 5 and 6, respectively, of Appendix I.

The circuit shown in Figure 6.8-1 represents an amplifier having a midband gain A, a single high-frequency break term at $\omega_H = 1/RC$, and an overall transfer function

$$H(s) = A \frac{1}{1 + s/\omega_H} \tag{6.8-1}$$

The frequency response of this amplifier is sketched in Figure 6.8-1b.

Let us consider the response of this amplifier to an input signal v_{in}, which is a step function of amplitude V_B (Figure 6.8-2a). Using Laplace transform techniques, we see that

$$V_o(s) = H(s)V_{in}(s) = AV_B \frac{\omega_H}{s(s + \omega_H)} \tag{6.8-2}$$

and by applying partial-fraction expansion, we have

$$V_o(s) = AV_B \left(\frac{1}{s} - \frac{1}{s + \omega_H} \right) \tag{6.8-3}$$

It is convenient to rewrite this equation in terms of the circuit time constant

$$\tau_H = \frac{1}{\omega_H} = RC \tag{6.8-4}$$

and using this quantity, $V_o(s)$ is

$$V_o(s) = AV_B \left(\frac{1}{s} - \frac{1}{s + 1/\tau_H} \right) \tag{6.8-5}$$

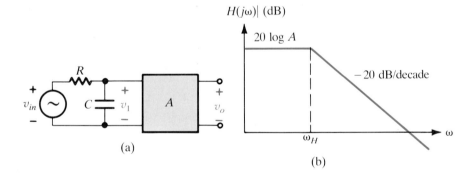

Figure 6.8-1
Amplifier with a single high-frequency break at $\omega = \omega_H$.

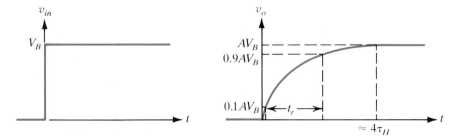

Figure 6.8-2
Step response of the amplifier in Figure 6.8-1.

Chapter 6 / Frequency Effects in Amplifier Circuits

Taking the inverse Laplace transform of this expression, we have

$$v_o(t) = AV_B \left(1 - e^{-t/\tau_H}\right) u(t) \tag{6.8-6}$$

This result is sketched in Figure 6.8-2b. Notice that although $v_o(t)$ mathematically reaches its final or steady-state value in about four time constants, we cannot say just where on the curve this actually occurs because v_o approaches the final value AV_B only asymptotically. Thus, it is difficult to determine τ_H directly by applying the four-time-constant rule. Instead, because it is much easier to measure, it is convenient to define the circuit's rise time, t_r, as the time required for the output to rise from 10% to 90% of its final value, and then to obtain τ_H from t_r. As discussed in Section 5 of Appendix I, for a simple RC network, such as the one we are examining currently, the circuit's rise time is related to its time constant by the expression

$$t_r = 2.2\,\tau_H \tag{6.8-7}$$

Therefore, once the rise time is known, it is a simple matter to compute τ_H from this equation.

Let's extend the results just obtained to examine the response of this amplifier to application of a square-wave signal of the form illustrated in Figure 6.8-3a. To determine the response to this type of input, we can think of the amplifier as being alternately connected to step inputs of $+V_B$ volts and $-V_B$ volts, with this sequence repeated forever. As a result, the response of the amplifier to this square wave should be similar to that obtained previously when the input to the amplifier was just a single-step function. The actual shape of

Figure 6.8-3

Square-wave response of the amplifier in Figure 6.8-1.

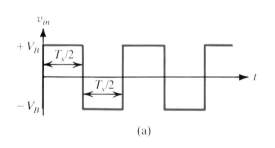

(a)

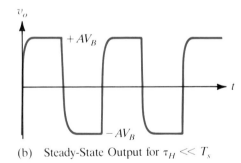

(b)   Steady-State Output for $\tau_H \ll T_s$

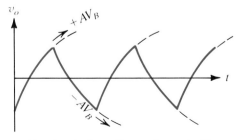

(c)   Steady-State Output for $\tau_H \geq T_s$

the steady-state response will, of course, depend on the relationship between the circuit time constant and the period, T_s, of the square wave.

If $\tau_H \ll T_s$, then, for example, during the time interval when v_{in} is positive, the capacitor will charge completely to $+V_B$ volts. Similarly, when v_{in} becomes negative, the capacitor will charge to $-V_B$ volts. As a result, the amplifier output (see Figure 6.8-3b) will look quite similar to the input square wave, except that it will be bigger and will have rounded edges.

When τ_H is much greater than T_s, the response is considerably different. For this case the capacitor does not have sufficient time to charge fully during each half-cycle of the square wave, and as a result, during each half-period the output waveshape looks like a small portion of an exponential charging curve. For this case the input square wave and the voltage at the output of the amplifier bear little resemblance to one another (Figure 6.8-3c).

Clearly, then, for high-fidelity signal amplification (i.e., for amplification where the output, except for being amplified, is a faithful reproduction of the input), it will be necessary that

$$\tau_H \ll T_s \tag{6.8-8}$$

This result may also be understood from an analysis in the frequency domain. Based on the elementary Fourier series concepts introduced earlier, you should recall that any periodic signal (such as the square wave in this problem) can be represented by a sum of sine waves at the fundamental frequency $\omega_s = 2\pi/T_s$, the second harmonic $2\omega_s$, the third harmonic $3\omega_s$, and so on, with the higher-order harmonic terms having less and less influence on the overall shape of the waveform. If this signal is to be amplified with high fidelity, it will be necessary to preserve the relative amplitude and relative phase shift of all the important frequency components of the signal. To achieve this, the amplifier gain and phase characteristics will need to be flat in the region of the fundamental and at least the first few of its harmonics. Equation (6.8-8) is just another way of stating this requirement, since if we invert both sides of this expression, we have

$$\frac{1}{\tau_H} \gg \frac{1}{T_s} \tag{6.8-9a}$$

or
$$\omega_H \gg \omega_s \tag{6.8-9b}$$

In other words, for high-fidelity signal amplification, the high-frequency 3-dB point of the amplifier must be greater than all the significant frequency components of the applied input signal.

To understand the effect that the low-frequency performance of an amplifier has on its transient response, consider the circuit illustrated in Figure 6.8-4. The transfer function for this amplifier is

$$H(s) = \frac{V_o}{V_{in}} = A\frac{s}{s + \omega_L} \tag{6.8-10}$$

where A is the midband gain and ω_L the low-frequency 3-dB point. The frequency response of this circuit is shown in Figure 6.8-4b.

To determine the transient response of this amplifier to a step input, we

Figure 6.8-4

*Amplifier with a single
low-frequency break at
$\omega = \omega_L$.*

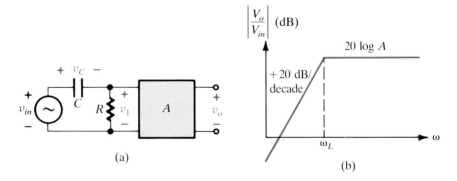

(a)

(b)

may write

$$V_o(s) = \left(A\frac{s}{s + \omega_L}\right)\frac{V_B}{s} = AV_B\frac{1}{s + \omega_L} \tag{6.8-11}$$

and taking the inverse Laplace transform of this expression, we have

$$v_o(t) = AV_B e^{-t/\tau_L} \tag{6.8-12a}$$

where

$$\tau_L = \frac{1}{\omega_L} = RC \tag{6.8-12b}$$

is the circuit time constant. This output is sketched in Figure 6.8-5.

We may understand the shape of this response by recalling that the voltage across a capacitor cannot change instantaneously. As a result, when the input voltage suddenly changes at $t = 0$, the voltage across the capacitor initially stays at zero and the voltage v_R jumps to $+V_B$ volts. Of course, corresponding to this, the amplifier output jumps to $+AV_B$ volts. As time goes on, the capacitor begins to charge and the voltage v_R (which equals $v_{in} - v_C$) begins to fall exponentially toward zero. The total decay time for this response is approximately four or five time constants.

When the input signal is a square wave, the waveshape present at the output may again be understood by considering the input to be a sequence of alternating steps (Figure 6.8-6). If the circuit time constant τ_L is much greater than the square-wave period T_s, the capacitor will hardly have any time to charge of discharge as v_{in} changes. As a result, v_C will remain zero, and when, for example, v_{in} is at $+V_B$ volts, v_R will also equal $+V_B$, and the output will remain relatively constant at $+AV_B$ volts. A similar situation will exist when v_{in} is negative. The net result in this case is that the capacitor voltage will remain very close to zero; that is, it will look like an ac short, and the output voltage, except for being amplified, will be identical to the input signal (Figure 6.8-6a).

When τ_L is only slightly bigger than T_s, the charging of the capacitor is too large to be ignored, and for this case, the output of the amplifier appears to "droop" or "tilt" as shown in Figure 6.8-6b. This apparent tilt in the output waveshape is actually the beginning of an exponential decay in the voltage across the resistor, but before it goes on for too long, the input signal changes. We will calculate the approximate size of this droop for the case where the voltage buildup on the capacitor is still relatively small.

When v_{in} is at V_B volts, the input RC network of the amplifier may be represented by the simplified circuit shown in Figure 6.8-7a, and if v_C is small, we

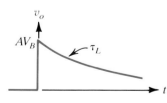

Figure 6.8-5

*Step response of the amplifier
in Figure 6.8-4.*

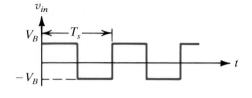

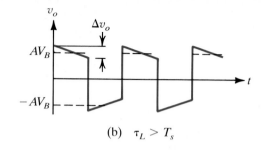

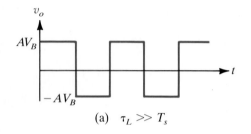

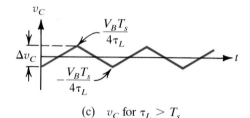

(a) $\tau_L \gg T_s$

(b) $\tau_L > T_s$

(c) v_C for $\tau_L > T_s$

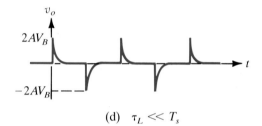

(d) $\tau_L \ll T_s$

Figure 6.8-6

Square-wave response of the amplifier in Figure 6.8-4.

may approximate the current in this circuit as

$$i \simeq \frac{V_B}{R} \tag{6.8-13}$$

Using the capacitor equation $i = C \, dv/dt$, the voltage across the capacitor is seen to be

$$v_C = \frac{1}{C} \int i \, dt \simeq \frac{V_B t}{C} + \text{const.} = \frac{V_B t}{\tau_L} + \text{const.} \tag{6.8-14}$$

Thus to the extent that the charging current remains constant, the voltage across the capacitor increases in a nearly linear fashion with time (Figure 6.8-6c). Therefore, in one half-cycle of the square wave, the total change in capacitor voltage is

$$\Delta v_C = \frac{i \, \Delta t}{C} = \frac{V_B (T_s/2)}{\tau_L} = \frac{V_B T_s}{2\tau_L} \tag{6.8-15}$$

Similarly, when the input signal goes negative (Figure 6.8-7b), the voltage across the capacitor will decrease linearly with time, and during this interval will change by an amount

$$\Delta v_C = -\frac{V_B T_s}{2\tau_L} \tag{6.8-16}$$

Therefore, the application of a square wave to a simple RC network when τ_L is

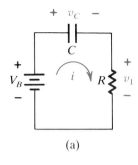

(a)

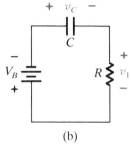

(b)

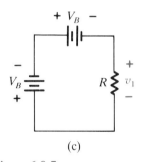

(c)

Figure 6.8-7
Equivalent circuits for the amplifier in Figure 6.8-4.

only slightly greater than T_s will produce a small triangle wave of voltage across the capacitor (Figure 6.8-6c). Because $v_R = v_{in} - v_C$, subtracting the two curves from one another, we obtain a drooping square-wave response at v_R and hence at the output (Figure 6.8-6b). Using equation (6.8-15), we may express the change in the output level as

$$\Delta v_o = A \, \Delta v_1 = A \, \Delta v_c = \frac{AV_B T_s}{2\tau_L} \tag{6.8-17a}$$

from which we can define the percent droop in the output to be

$$\% \text{ droop} = \frac{\Delta v_o}{AV_B} = \frac{T_s}{2\tau_L} \times 100\% \tag{6.8-17b}$$

Clearly, this quantity is a measure of the high-fidelity performance of the amplifier, and using this result, we can see that we will need

$$\tau_L \gg T_s \tag{6.8-18}$$

if the output waveshape is to be a faithful reproduction of the input.

This criterion for high-fidelity signal amplification can also be examined in the frequency domain. If the signal is to be amplified faithfully, once again, as with the high-frequency case examined previously, this requires that the amplifier gain and phase response be flat in the frequency region of the fundamental and all its significant harmonics. For the low-frequency case we are currently examining (see Figure 6.8-4b), this will require that ω_L be much smaller than ω_s. This inequality is just a restatement of equation (6.8-18). We can demonstrate this by inverting equation (6.8-18) which yields

$$\frac{1}{\tau_L} \ll \frac{1}{T_s} \tag{6.8-19a}$$

or

$$\omega_L \ll \omega_s \tag{6.8-19b}$$

For completeness, let us also examine what happens when the amplifier low-frequency breakpoint is significantly above ω_s, or in other words, where τ_L is much less than T_s. For this case, when v_{in} is positive (Figure 6.8-7a), the capacitor will rapidly charge to $+V_B$ volts and the voltage across the resistor, and hence the output voltage, will decay exponentially to zero. When the input changes to $-V_B$ volts, because the capacitor voltage cannot change instantaneously, V_R will jump down to $-2V_B$ volts (Figure 6.8-7c) and will then again rapidly decay to zero. Thus, the amplifier output in the steady state will consist of a sequence of alternating exponentially shaped spikes whose amplitudes are $+2AV_B$ and $-2AV_B$ volts (Figure 6.8-6d). Clearly, this output is a distorted version of the input. This occurs because ω_L is greater than ω_s, so that significant changes occur at the output in the Fourier series components of the original signal.

Thus far, we have developed expressions for the rise time and waveform droop associated with circuits in which we had only a single time constant to consider. The procedure for handling circuits containing multiple time constants, or multiple transfer function breaks, exactly parallels that for approximating the high- and low-frequency 3-dB points of such circuits. To illustrate this technique, as an example, consider the complex transfer function

$$H(s) = A_{mid} \underbrace{\left(\frac{s}{s + \omega_1} \frac{s}{s + \omega_2}\right)}_{\text{low-frequency breaks}} \underbrace{\left(\frac{1}{1 + s/\omega_3} \frac{1}{1 + s/\omega_4}\right)}_{\text{high-frequency breaks}} \qquad (6.8\text{-}20)$$

whose corresponding Bode magnitude plot is given in Figure 6.8-8a. For the middle range of frequencies, that is, for $\omega_L < \omega < \omega_H$, the transfer function $H(s)$ may be approximated as

$$H(s) \simeq A_{mid} \frac{s}{s + \omega_L} \frac{1}{1 + s/\omega_H} \qquad (6.8\text{-}21)$$

where, following equations (6.3-19) and (6.7-5a), we may describe ω_L and ω_H by the expressions

$$\omega_2 \leq \omega_L \leq \omega_1 + \omega_2 \qquad (6.8\text{-}22a)$$

and

$$\omega_3 \| \omega_4 \leq \omega_H \leq \omega_3 \qquad (6.8\text{-}22b)$$

To understand how ω_H and ω_L are related to the transient response of this function, let us examine the response of $H(s)$ to a step of amplitude V_B. Using

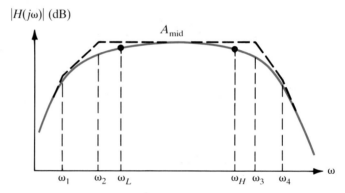

(a) Magnitude Plot for $H(s) = A_{mid}\left(\dfrac{s}{s + \omega_1}\right)\left(\dfrac{s}{s + \omega_2}\right)\left(\dfrac{1}{1 + s/\omega_3}\right)\left(\dfrac{1}{1 + s/\omega_4}\right)$

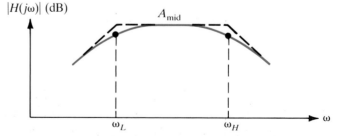

(b) Approximate Magnitude Plot for $H(s)$, Valid for $\omega_L \leq \omega \leq \omega_H$

Figure 6.8-8

Laplace transforms, the output voltage $V_o(s)$ corresponding to this input is

$$V_o(s) = A_{mid} V_B \frac{1}{s + \omega_L} \frac{\omega_H}{s + \omega_H} = A_{mid} V_B \left(\frac{C_1}{s + \omega_L} + \frac{C_2}{s + \omega_H} \right) \qquad (6.8\text{-}23)$$

where
$$C_1 = \frac{\omega_H}{\omega_H - \omega_L} \qquad (6.8\text{-}24a)$$

and
$$C_2 = \frac{\omega_H}{\omega_L - \omega_H} \qquad (6.8\text{-}24b)$$

Because ω_H is generally much greater than ω_L, the coefficients C_1 and C_2 are approximately $+1$ and -1, respectively, so that we may write the inverse Laplace transform of equation (6.8-23) as

$$v_o(t) = A_{mid} V_B (e^{-t/\tau_L} - e^{-t/\tau_H}) \qquad (6.8\text{-}25)$$

The time constants τ_L and τ_H in this expression may be related to the break frequency terms in the original transfer function through equations (6.8-22a) and (6.8-22b). Inverting equation (6.8-22a), we have

$$\frac{1}{\omega_2} \geq \frac{1}{\omega_L} \geq \frac{1}{\omega_1 + \omega_2} \qquad (6.8\text{-}26a)$$

But $\tau_1 = 1/\omega_1$, $\tau_2 = 1/\omega_2$, and $\tau_L = 1/\omega_L$, so that

$$\tau_2 \geq \tau_L \geq \tau_1 \| \tau_2 \qquad (6.8\text{-}26b)$$

Similarly, if we invert equation (6.8-22b), we obtain

$$\frac{1}{\omega_3} + \frac{1}{\omega_4} \geq \frac{1}{\omega_H} \geq \frac{1}{\omega_3} \qquad (6.8\text{-}27a)$$

or
$$\tau_3 + \tau_4 \geq \tau_H \geq \tau_3 \qquad (6.8\text{-}27b)$$

In equation (6.8-25) the time constant τ_L is much greater than τ_H (since $\omega_L \ll \omega_H$), and as a result, if we sketch the two component parts of the expression, the first exponential term in this equation takes much longer to decay than the second (Figure 6.8-9a). Furthermore, if we examine the overall

Figure 6.8-9

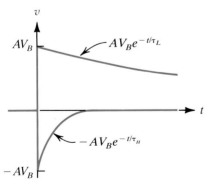

(a) Component Parts of the Step Response of the
Transfer Function $H(s) = A_{mid} \dfrac{s}{(s + \omega_L)} \dfrac{1}{(1 + s/\omega_H)}$

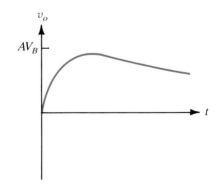

(b) Overall Step Response of the Transfer Function
$H(s) = A_{mid} \dfrac{s}{(s + \omega_L)} \dfrac{1}{(1 + s/\omega_H)}$

sketch for $v_o(t)$ in Figure 6.8-9b, we can see that the rise-time portion of the step response is determined almost entirely by τ_H, the high-frequency 3-dB point, while the later droop in the step response is associated with τ_L, or alternatively, the low-frequency 3-dB point.

By a similar analysis, we could extend these results to the case where the input to the transfer function is a square wave, and as with the step response, we would find that τ_H essentially determines the rise and fall times of the output square wave while τ_L controls its droop. In other words, the nonzero rise and fall times on the edges of the output square wave are associated primarily with the time required to charge and discharge the parasitic capacitors. During these rise and fall times, which are usually very short, the bypass and coupling capacitors (which are relatively large) have no time to charge or discharge, and as a result during these time intervals these capacitors may be treated as short circuits.

The converse is true in the interval over which we compute the square-wave droop. During this time the voltages in the circuit are only slowly changing functions of time, as are the $C \, dv/dt$ currents flowing into the circuit's parasitic capacitances because both C and dv/dt are small. Therefore, in carrying out the square-wave-droop calculations, the parasitic capacitors may be considered open circuits.

In summary, the measurement of a circuit's rise time gives us information about the parasitic capacitance effects and the high-frequency 3-dB point of the circuit [equation (6.8-7)]. During the calculation of τ_H, the bypass and coupling capacitors are treated as short circuits. On the other hand, measurement of an amplifier's square-wave tilt or droop gives us information about the coupling and bypass capacitors. In particular, it tells us the approximate value of the low-frequency 3-dB break. During the calculations for τ_L, the parasitic capacitances are considered open circuits.

| EXAMPLE 6.8-1 | For the BJT amplifier illustrated in Figure 6.8-10, the transistor parameters are $r_{bb'} = 0$, $r_{b'e} = 1 \text{ k}\Omega$, $g_m = 100 \text{ mS}$, $C_{b'e} = 10 \text{ pF}$, and $C_{b'c} = 1 \text{ pF}$. |

a. If the input signal is a 1-kHz 80 mV (p-p) square wave, sketch the output voltage for $C_3 = 100$, 1.0, and 0.001 μF.

b. How does the output change if the frequency of the applied signal is raised to 1 MHz?

SOLUTION

To sketch the overall output voltage $v_{CE}(t)$, we need the dc Q-point values for this amplifier. By inspection of the circuit, these may be found to be $I_B = 46.5 \ \mu\text{A}$, $I_C = 4.65 \text{ mA}$, and $V_{CE} = 5.35 \text{ V}$.

At a frequency of 1 kHz, the circuit rise and fall times are so fast that we would hardly see them unless the oscilloscope trace were greatly expanded. However, for later use we will calculate their values even though these data will not be used explicitly for the sketches associated with part (a) of this example. The equivalent circuit shown in Figure 6.8-10b, in which the coupling capacitor C_3 has been replaced by a short, is valid in the middle- and high-frequency regions. Notice that the Miller theorem has already been applied to split $C_{b'c}$ using the fact that the Miller gain for this circuit is -100. The high-

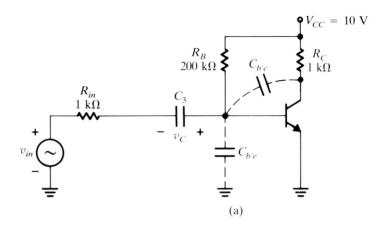

(a)

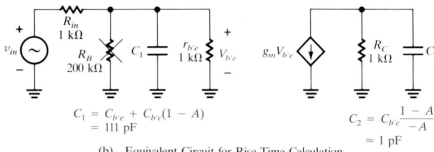

$$C_1 = C_{b'e} + C_{b'c}(1 - A)$$
$$= 111 \text{ pF}$$

$$C_2 = C_{b'c}\frac{1 - A}{-A}$$
$$\approx 1 \text{ pF}$$

(b)　Equivalent Circuit for Rise-Time Calculation

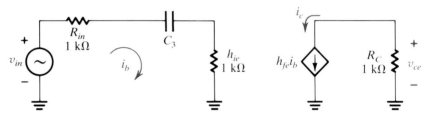

(c)　Equivalent Circuit for Droop Calculation

Figure 6.8-10
Example 6.8-1.

frequency time constants associated with the capacitors C_1 and C_2 are

$$\tau_1 = (r_{b'e} \parallel R_{in})C_1 = 55 \text{ ns} \qquad (6.8\text{-}28a)$$

and $\qquad\qquad \tau_2 = R_C C_2 = 1 \text{ ns} \qquad\qquad\qquad\qquad (6.8\text{-}28b)$

so that $\qquad\qquad \tau_1 = 55 \text{ ns} \le \tau_H \le \tau_1 + \tau_2 = 56 \text{ ns} \qquad (6.8\text{-}29)$

The circuit rise time is approximately

$$t_r = 2.2\tau_H = 120 \text{ ns} \qquad (6.8\text{-}30)$$

Taking into account that the period of a 1-kHz square wave is 1 ms or 1000 μs it is apparent that a 120-ns rounding on the edges of this square wave would scarcely be noticeable.

Consider next the effect of the coupling capacitor on the output waveshape.

The equivalent circuit shown in Figure 6.8-10c is valid in the middle- and low-frequency regions. The time constant associated with the coupling capacitor C_3 is

$$\tau_L = (R_{in} + h_{ie})C_3 \qquad (6.8\text{-}31)$$

and is equal to 200 ms, 2.0 ms, and 2 μs for C_3 = 100, 1.0, and 0.001 μF, respectively.

When C_3 = 100 μF and the time constant is 200 ms, for all practical purposes the capacitor does not charge or discharge at all during each cycle of the input. Therefore, the voltage across C_3 remains constant, and it may be considered an ac short circuit. As a result, in this instance,

$$i_b = \frac{v_{in}}{R_B + h_{ie}} = \frac{v_{in}}{2 \text{ k}\Omega} \qquad (6.8\text{-}32a)$$

$$i_c = h_{fe}i_b = \frac{100v_{in}}{2 \text{ k}\Omega} \qquad (6.8\text{-}32b)$$

and

$$v_{ce} = -i_c R_C = -50 \, v_{in} \qquad (6.8\text{-}32c)$$

The waveforms for v_{in} and v_{CE} for this case are shown in Figure 6.8-11a.

When C_3 = 1.0 μF, the low-frequency time constant is only 0.2 ms. Because it is comparable to the square-wave period, the charging and discharging of C_3 can no longer be ignored, and the base current will tend to drop off exponentially toward zero. By using equation (6.8-17b), we may calculate the approximate droop in i_b as

$$\% \text{ droop} = \frac{T_s}{2\tau_L} \times 100\% = 25\% \qquad (6.8\text{-}33)$$

Since the peak ac base current with C_3 a short is 40 mV/2 kΩ, or 20 μA, this corresponds to change in i_b of (20) $\times$ (0.25) = 5 μA(p-p). A similar droop

Figure 6.8-11

Example 6.8-1. Transistor output voltage at f = 1 kHz for different values of C_3.

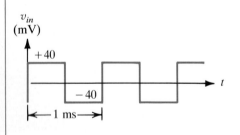

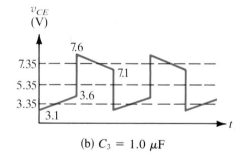

(b) C_3 = 1.0 μF

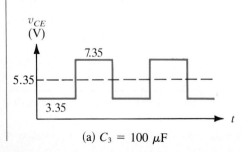

(a) C_3 = 100 μF

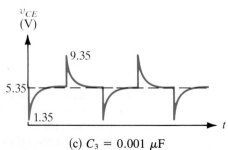

(c) C_3 = 0.001 μF

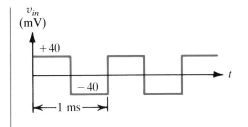

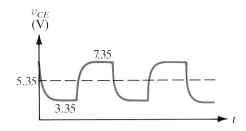

Figure 6.8-12
Example 6.8-1. Transistor output voltage at f = 1 MHz.

will occur in the waveshapes of the collector current and the collector-to-emitter voltage (Figure 6.8-11b).

In the last case, where $C_3 = 0.001\mu F$, the corresponding time constant is so short (2 μs) that the coupling capacitor will charge completely during each half-cycle of the square wave. As a result, the base current will consist of a series of alternating spikes, as will the collector-current and collector-to-emitter voltage. The waveshape for v_{CE} is illustrated in Figure 6.8-11c. We should note that the base current spikes have an amplitude of 40 and not 20 μA, as we might have expected. This occurs because with $v_{in} = -40$ mV, for example, the capacitor will charge fully to $v_C = 40$ mV before the end of that half-cycle. As a result, when the input voltage then changes from -40 mV to $+40$ mV, the initial ac base current will be $(v_{in} + v_C)/(R_{in} + h_{ie}) = 80$ mV/2 kΩ, or 40 μA.

When the frequency of the applied signal is increased to 1 MHz, the droop produced in the output by the charging and discharging of C_3 can be neglected for all three values of this capacitor because even with $C_3 = 0.001$ μF, the droop is less than 1.5%. This statement is equivalent to saying that C_3 may be neglected and considered an ac short at this frequency. However, the parasitic capacitors have a significant effect on the output waveshape at this frequency and must be taken into account. As we calculated in equation (6.8-31), the rise time of this circuit is approximately 120 ns, and since the period of the square wave is only 1 μs (or 1000 ns), this rise time represents a significant portion of the signal period. The resulting waveshape for v_{CE}, including this rise-time effect, is sketched in Figure 6.8-12.

Exercises

6.8-1 For the amplifier in Figure 6.8-10, C_3 is equal to 10 μF.

a. Estimate the percent droop at the output if a 500-Hz 50-mV peak amplitude square wave is applied.

b. Repeat part (a) if the input signal amplitude is increased to 200 mV. *Answer* (a) 5%, (b) 0%

6.8-2 The amplifier discussed in Example 6.8-1 is used to drive a 100-pF load. Estimate the output waveshape rise time if the input signal is a 50-mV-peak-amplitude 10-kHz square wave. *Answer* 222 ns $< t_r <$ 342 ns

6.8-3 Repeat Exercise 6.8-2 if an emitter follower containing a 1-kΩ emitter resistor going to ground is inserted between the BJT amplifier and the load. The small-signal characteristics of the follower may be assumed to be the same as those of the amplifier transistor. *Answer* 122 ns $< t_r <$ 129 ns

For all problems in this chapter, unless otherwise noted, $h_{ie} = 1$ kΩ and $h_{fe} = 100$ for all BJTs, and $g_m = 1$ mS for all FETs. Additionally, in Sections 6.6 and 6.7, unless otherwise noted, $r_{bb'} = 0$, $r_{b'e} = 1$ kΩ, and $g_m = 100$ mS for all BJTs.

6.2-1 For the circuit shown in Figure P6.2-1, use network analysis techniques and whatever equivalent circuits seem appropriate to derive an expression for $H(s) = V_o/V_{in}$. The final expression should be in the same form as in equation (6.2-8).

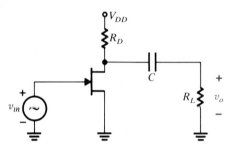

Figure P6.2-1

6.2-2 For the circuit shown in Figure 6.2-3, $R_{in} = R_E = 0$, and C_1 is an ac short. Sketch the Bode gain and phase plots for V_l/V_{in}.

(a) Considering that $C = 0.1$ μF, $R_D = 50$ kΩ, $R_L = 20$ kΩ, and $r_d = 50$ kΩ for the FET in the circuit shown in Figure P6.2-1, develop an expression for V_o/V_{in}.

(b) Sketch the gain–phase characteristics of this transfer function.

6.2-4 For the amplifier circuit in Figure 6.2-3, consider that $v_{in} = 1 \sin 2\pi ft$ volts and determine $v_o(t)$ for $f = 10$, 100, and 1 kHz.

6.2-5 For the circuit shown in Figure P6.2-5, sketch the Bode magnitude plot $|V_o/V_{in}|$ at low and midband frequencies.

6.2-6 For the circuit shown in Figure 6.2-3, $R_{in} = 0$, $R_B = 200$ kΩ, $R_E = 2$ kΩ, and $R_C = 2$ kΩ. Develop an expression for the transfer function (V_l/V_{in}).

6.2-7 Sketch the low-frequency and midrange magnitude response $|V_o/V_{in}|$ of the amplifier circuit shown in Figure given that C_1 is an ac short, R_S and C_2 are removed and the source of Q_1 is grounded, $C_3 = 100$ μF, $R_D = 9$ kΩ, and $R_C = 2$ kΩ.

6.2-8 Develop an expression for the transfer function V_l/V_{in} for the circuit shown in Figure P6.2-8.

$g_{m1} = 5$ mS
$h_{ie2} = 1$ kΩ
$h_{fe2} = 20$

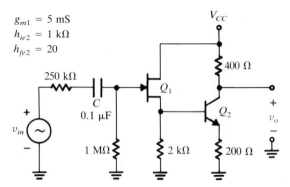

Figure P6.2-5

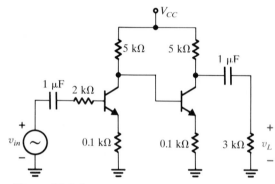

Figure P6.2-8

6.2-9 For the circuit shown in Figure 6.2-3, R_B is removed and replaced by a 0.01-mA dc current source (arrow pointing down). Derive an expression for V_1/V_{in} using network analysis techniques.

6.3-1 Sketch and carefully label the Bode magnitude plot $|V_o/V_{in}|$ for the circuit shown in Figure P6.3-1.

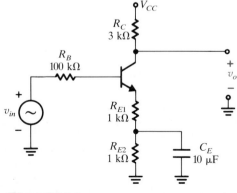

Figure P6.3-1

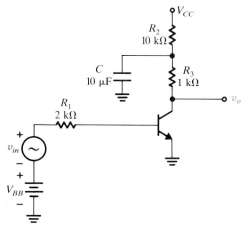

Figure P6.3-2

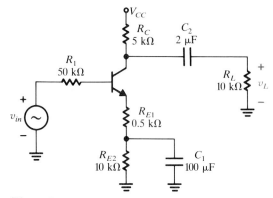

Figure P6.3-4

6.3-2 Carefully sketch and label the gain and phase plots in the low- and midband-frequency range for the circuit shown in Figure P6.3-2.

6.3-3 For the circuit shown in Figure P6.3-3:
(a) Find the midband gain V_o/V_{in}.
(b) Select C_2 to place the LF 3-dB point for V_o/V_{in} at 20 Hz assuming that C_1 is a short.
(c) Select C_1 so that the assumption in part (b) is valid.
(d) For the values of C_1 and C_2 selected, sketch $|V_o/V_{in}|$ in the low- and midband-frequency ranges.

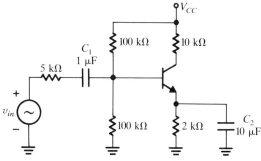

Figure P6.3-5

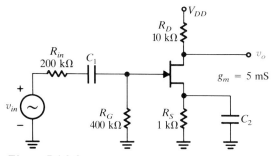

Figure P6.3-3

6.3-4 Accurately sketch the Bode magnitude plot $|V_l/V_{in}|$ for the low- and midband-frequency region in the circuit shown in Figure P6.3-4.

6.3-5 Show that the break for C_2 in the amplifier shown in Figure P6.3-5 occurs well after the break for C_1.

6.3-6 For the circuit shown in Figure P6.3-5, let $C_1 = 0.1$ μF and show that the C_2 breaks for that amplifier occur well before the break of C_1.

6.3-7 Consider the transfer function $H(s) = A_{mid}(s + \omega_N)/(s + \omega_D)$ where $\omega_N = 0.1\omega_D$. Compute the exact value of ω_L for this transfer function and then the approximate value of ω_L neglecting ω_N. Prove that the error made in ne-

glecting the zero is less than 1% when it is a decade or more below the pole location.

6.3-8 (a) Calculate the exact value of the low-frequency 3-dB point of the transfer function

$$H(s) = \frac{s(s + 10)}{(s + 5)(s + 20)}$$

(b) Repeat this calculation using equation (6.3-28).

6.3-9 An amplifier is to be built by cascading n noninteracting identical stages, each having a transfer function of the form $H_1(s) = A_1 s/(s + \omega_1)$.
(a) Derive an exact expression for the low-frequency 3-dB point of the overall amplifier.
(b) Compare this result with that obtained by estimating the amplifier's 3-dB frequency by using equation (6.3-19) for the case where $n = 1, 2, 3,$ and 4.

6.3-10 For the circuit shown in Figure 6.2-3, R_E is shunted with a capacitor C_E. Determine the value of C_E that will place the amplifier's 3-dB point at about 100 Hz. (You may assume that C_E shorts at a higher frequency than C_1.)

6.3-11 Given that $|I_o/I_s| = -7.9$ in midband for the circuit shown in Figure P6.3-11:
(a) Sketch the Bode plot of I_o/I_s versus ω. Show gain magnitudes in decibels and break frequencies in rad/s.

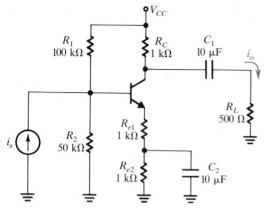

Figure P6.3-11

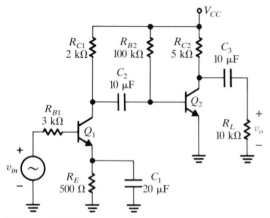

Figure P6.3-12

(b) Write the transfer function $H(s) = I_o/I_s$ that is valid at middle and low frequencies.

6.3-12 For the circuit shown in Figure P6.3-12:
(a) Determine the break frequencies associated with the capacitors C_1, C_2, and C_3.
(b) Sketch the Bode magnitude plot.

6.3-13 (a) Sketch the Bode gain plot for the transfer function $H(s) = (s + 100)/(s + 125)$.
(b) Use equation (6.3-17) to evaluate the approximate low-frequency 3-dB point of $H(s)$ and explain the results obtained.

6.4-1 For the circuit shown in Figure P6.4-1, V_1 is a dc voltage, D is a silicon diode, and v_{in} is a 100 mV (p-p) square wave. When $V_1 = +10$ V the square wave at v_o has a rise time of 100 ns, and when $V_1 = -10$ V, the output square-wave rise time increases to 500 ns.
(a) Estimate C_j and C_d for the diode.
(b) Sketch the approximate waveshapes at v_o for each

value of V_1 if the input signal frequency is 100 kHz.

6.4-2 The transistor in the circuit shown in Figure P6.4-2, has an $h_{fe} = 100$ and a short-circuit current gain of unity at 10 MHz. Given that $C_{b'c} = 5$ pF, find $C_{b'e}$.

6.4-3 For the circuit shown in Figure P6.4-2, consider that $i_{in} = 10 \sin \omega t$ μA, $C_{b'e} = 17.8$ pF, $C_{b'c} = 0$, and $R_C = R_L = 10$ kΩ. Derive expressions for $i_C(t)$ and $v_L(t)$ if $\omega = 10^8$ rad/s.

6.4-4 Given that $h_{ie} = 1.5$ kΩ, $h_{fe} = 100$, $C_{ob} = 5$ pF, and $f_\beta = 2$ MHz, determine $r_{bb'}$, $r_{b'e}$, g_m, $C_{b'e}$, and $C_{b'c}$ if the transistor is biased at a quiescent emitter current of 4 mA.

6.4-5 Manufacturers often express the parasitic capacitances of FETs in terms of C_{iss}, C_{oss}, and C_{rss}, where C_{iss} represents the common-source input capacitance with the output shorted, C_{oss} the common-source output capacitance with the input shorted, and C_{rss} the common-source reverse coupling capacitance from output to input with the input grounded. Derive expressions for C_{gs}, C_{gd}, and C_{ds} in terms of the transistor parameters C_{iss}, C_{oss}, and C_{rss}.

6.5-1 (a) Repeat Example 6.5-1 using network analysis methods.
(b) Comment on the accuracy of the Miller theorem approach in the solution of this problem.

6.5-2 (a) Change R_F to 10 kΩ and repeat Problem 6.5-1.
(b) Carry out the analysis in part (a) using the Miller theorem approximation. Why does the Miller theorem give such poor results in this case?

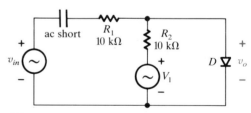

Figure P6.4-1

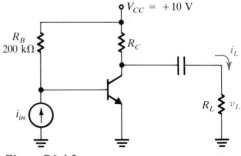

Figure P6.4-2

6.6-1 For the circuit shown in Figure 6.6-3a, $r_{bb'} = 100\ \Omega$, $r_{b'e} = 1\ k\Omega$, $C_{b'e} = 500$ pF, $C_{b'c} = 10$ pF, and $g_m = 100$ mS.
(a) Find the midband gain.
(b) Find the magnitude of the gain at 300 kHz.

6.6-2 For the circuit shown in Figure 6.6-13e, $R_S = 2\ k\Omega$, $R_D = 5\ k\Omega$, $R_G = 0$, $g_m = 1$ mS, and $C_{gd} = C_{gs} = 10$ pF. Carefully sketch and label $|V_d/V_{in}|$ versus frequency in the mid- and high-frequency regions.

6.6-3 **(a)** For the circuit shown in Figure P6.6-3, select R_L to locate the upper 3-dB break frequency for the voltage gain at $\omega = 2 \times 10^6$ rad/s.
(b) For the R_L chosen in part (a), what is the low-frequency voltage gain of the circuit?

6.6-4 For the circuit shown in Figure 6.6-6a, by considering C_E and C_L to be shorts in the frequency range of interest, sketch the Bode gain and phase plots in the mid- and high-frequency regions. You may assume that $r_{bb'} = 0$, $r_{b'e} = 1\ k\Omega$, $h_{fe} = 100$, $C_{b'c} = 5$ pF, and $C_{b'e} = 100$ pF.

6.6-5 For the amplifier in Figure 6.6-8, $R_G = 10\ k\Omega$, $R_S = 0.5\ k\Omega$, $g_m = 2$ mS, $C_{gs} = C_{gd} = 5$ pF, $C_{ds} = 0.5$ pF, and a load capacitance $C_L = 20$ pF is connected in parallel with R_S.
(a) Find the midband gain.
(b) Find the important high-frequency breaks.
(c) Sketch the mid- and high-frequency-range Bode magnitude plot $|V_s/V_{in}|$.

6.6-6 For the FET circuit shown in Figure 6.6-1, $R_{in} = 1\ k\Omega$, $R_D = 5\ k\Omega$, $C_{gs} = C_{gd} = 10$ pF, $C_{ds} = 0.2$ pF, and $g_m = 20$ mS.
(a) Determine the midband gain V_{ds}/V_{in}.
(b) Find the high-frequency breaks.
(c) Sketch the Bode magnitude plot in the mid- and high-frequency range.

6.6-7 For the circuit in Figure P6.6-3, R_S is changed to 10 $k\Omega$. As R_L is increased, the circuit midband gain also increases. Neglecting the output break (at the collector),

find the largest allowed value of R_L if f_H is to be greater than 1 MHz. The transistor parameters are $r_{bb'} = 0$, $r_{b'e} = 1\ k\Omega$, $g_m = 100$ mS, $C_{b'e} = 10$ pF, and $C_{b'c} = 1$ pF.

6.6-8 Demonstrate the equivalence of equations (6.6-15) and (6.6-13) by substituting the appropriate expressions for ω_{1o}, ω_{2o}, and so on, into equation (6.6-15) to show that it reduces to equation (6.6-13).

6.6-9 Prove the validity of equation (6.6-52) for the circuit in Figure 6.6-10a by using the relationship $R_T = V_{oc}/I_{sc}$.

6.6-10 Prove the validity of the equivalent circuit given in Figure 6.6-13g.

6.7-1 For the Darlington amplifier shown in Figure 6.7-5b, C_L is removed, $R_{in} = 0$, $R_{C2} = 5\ k\Omega$, and a 1 $k\Omega$ resistor is inserted between the emitter of Q_2 and ground. In addition, for both of the transistors, $r_{bb'} = 0$, $r_{b'e} = 1\ k\Omega$, $g_m = 100$ mS, $C_{b'e} = 10$ pF, and $C_{b'c} = 1$ pF. Find the break frequencies associated with each of the parasitics, and estimate the overall high-frequency 3-dB point of the amplifier.

6.7-2 For the amplifier in Figure 6.7-3, $R_{in} = 2\ k\Omega$, $R_{C1} = 10\ k\Omega$, $R_{C2} = 3\ k\Omega$, and C_L is removed. In addition, for both of the transistors, $r_{bb'} = 0$, $r_{b'e} = 1\ k\Omega$, $C_{b'e} = 50$ pF, $C_{b'c} = 1$ pF, and $g_m = 50$ mS.
(a) Draw a good ac model.
(b) Apply the Miller theorem and redraw the new model.
(c) Carefully sketch and label $|V_{c2}/V_{in}|$ versus ω in the mid- and high-frequency regions.

6.7-3 Find the approximate high-frequency 3-dB point for the cascode amplifier shown in Figure P6.7-3, considering that C_1 and C_2 are ac shorts. The transistor parameters are $g_m = 10$ mS and $C_{gs} = C_{gd} = 1$ pF for Q_1, and $r_{bb'} = 0$,

$r_{bb'} = 200\ \Omega$
$r_{b'e} = 1\ k\Omega$
$g_m = 99$ mS
$C_{b'c} = 5$ pF
$C_{b'e} = 500$ pF

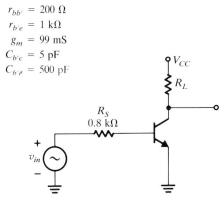

Figure P6.6-3

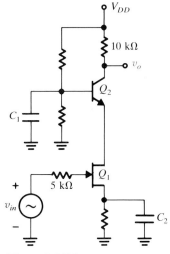

Figure P6.7-3

$r_{b'e} = 1\ k\Omega$, $g_m = 100\ mS$, $C_{b'e} = 10\ pF$, and $C_{b'c} = 1\ pF$ for Q_2. Note that this circuit may be redrawn as a common-source common-base amplifier cascade.

6.7-4 For the amplifier in Figure 6.7-3, $R_{in} = 0$, $R_{C1} = 3\ k\Omega$, $R_{C2} = 5\ k\Omega$, C_L is removed, and a 1-kΩ resistor is connected between the emitter of Q_2 and ground. In addition, $r_{bb'} = 0$, $r_{b'e} = 1\ k\Omega$, $g_m = 100\ mS$, $C_{b'e} = 10\ pF$, and $C_{b'c} = 1\ pF$. Find the break frequencies associated with each of the parasitics, and estimate the overall high-frequency 3-dB point of the amplifier.

6.7-5 An amplifier having an overall gain of 2500 is to be designed using as many stages of the form shown in Figure P6.7-5 as needed. The transistor parameters are $g_m = 10\ mS$, $C_{gs} = C_{gd} = 1\ pF$, and $C_{ds} = 0$. The 3-dB point at the high-frequency end is to be $\geq 400\ kHz$. Find the minimum number of stages and the required value of R_D needed to achieve this gain. What is the resulting 3-dB point of the amplifier? (*Hint:* Try it for $n = 1$, $n = 2$, etc.)

6.7-6 Estimate A_{mid} and ω_H for the cascode amplifier shown in Figure 6.7-5a. Consider that a load capacitance $C_L = 10\ pF$ is connected onto the output at the collector of Q_2 and that the transistor parameters are $r_{bb'} = 0$, $r_{b'e} = 1\ k\Omega$, $C_{b'e} = 10\ pF$, $C_{b'c} = 1\ pF$, and $g_m = 100\ mS$.

6.7-7 Repeat Problem 6.7-6 if an emitter follower is connected between the collector of Q_2 and C_L.

6.7-8 Repeat Problem 6.7-6 for the Darlington transistor amplifier in Figure 6.7-5b.

6.8-1 For the circuit shown in Figure P6.8-1, v_{in} is a 1-kHz, 0- to 5-V square wave. Carefully sketch and label i_B, i_C, and v_{CE}.

6.8-2 Assume that the transistor shown in Figure P6.8-2 is ideal with $I_{DSS} = 2\ mA$ and $V_p = -2\ V$. The switch has been open for a long time and is closed at $t = 0$. Sketch v_o

versus time. Include all important amplitudes and time intervals.

6.8-3 In the circuit shown in Figure P6.8-3, v_{in} is a 2-V(p-p), 1-kHz square wave. Assuming the $h_{fe} = 100$, sketch v_1, i_B, and v_o. (*Hint:* Consider the base–emitter junction an ideal diode.)

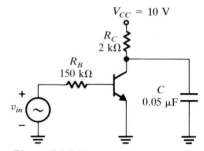

Figure P6.8-1

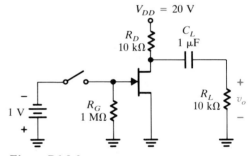

Figure P6.8-2

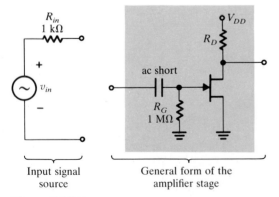

Input signal source   General form of the amplifier stage

Figure P6.7-5

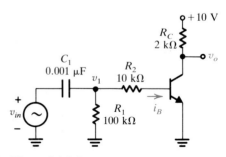

Figure P6.8-3

6.8-4 The FET in the circuit shown in Figure P6.8-4 has an $I_{DSS} = 35$ mA and a $V_p = -3$ V. Given that v_{in} is a 0- to -10-V, 150-Hz square wave, carefully sketch v_{GS}, i_L, v_{DS}, and v_L, considering that v_{DS} is approximately zero when the FET is in the ohmic region.

6.8-5 For the circuit shown in Figure P6.8-5, the switch is closed at $t = 0$. Sketch v_o versus time. (To analyze the circuit neglect the $V_{BE} = 0.7$ V drop and consider only that $h_{ie} = 1$ kΩ and $h_{fe} = 100$.)

6.8-6 The switch in the circuit shown in Figure P6.8-6a has been in position 1 for a long time. At $t = 0$ it is thrown to position 2. Sketch an accurate plot of $v_o(t)$ versus time. (*Hint:* The approximate equivalent circuit for the base–emitter junction in the active region may be assumed to be as shown in Figure P6.8-6b.)

6.8-7 For the circuit shown in Figure P6.8-7, v_{in} is a 1 kHz 100-mV(p-p) square wave. The transistor has the following parameters: $I_{DSS} = 10$ mA and $V_p = -3$ V.
(a) Determine the circuit Q point.
(b) Sketch the total signals v_{GS}, i_D, and v_{DS}. (*Hint:* Recall that $g_m = \partial i_D / \partial v_{GS}|_{Q\text{ pt.}}$.)

6.8-8 For the circuit shown in Figure P6.8-8, the switch has been in position 1 for a long time. At $t = 0$ it is thrown to position 2. What is $v_o(t)$ for $t > 0$? (*Hint:* Consider the base–emitter junction to be an ideal silicon diode.)

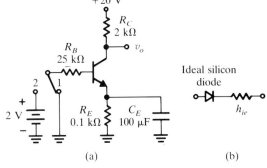

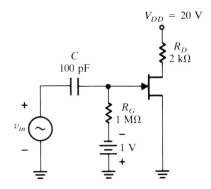

(a) (b)

Figure P6.8-6

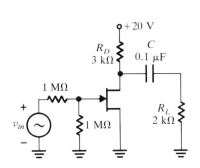

Figure P6.8-4

Figure P6.8-7

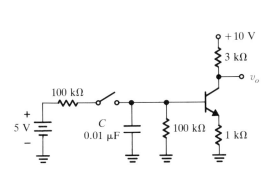

Figure P6.8-5

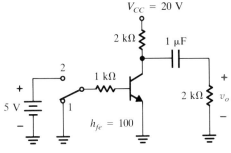

Figure P6.8-8

6.8-9 For the circuit shown in Figure P6.8-9, $v_{in} =$ 2 sin ωt volts with $\omega = 10^6$ rad/s and the transistor is silicon with $h_{fe} = h_{FE} = 100$. Find the dc voltage at the output with respect to ground. Assume the base–emitter junction to be an ideal diode. (*Hint:* R_2 and C form a low-pass filter.)

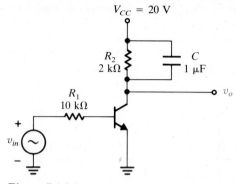

Figure P6.8-9

CHAPTER 7

Introduction to Feedback and the Operational Amplifier

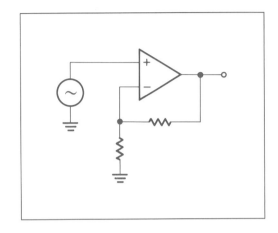

7.1 BASIC PRINCIPLES OF FEEDBACK AMPLIFIERS

A system is said to have feedback when all or part of the output of the system is fed back to the input. One of the earliest practical applications of feedback principles relates to the invention of the mechanical governor by James Watt in about 1769. It was used to control the speed of a steam engine automatically. A simplified diagram of this system is illustrated in Figure 7.1-1. Initially, the engine is set to run at a specific speed by adjusting the position of the weights on the governor. Once this set point is determined, the governor monitors the output, in this case the engine speed, and automatically feeds back information to the steam inlet valve to correct for any factors that might tend to alter this speed.

If, for example, the boiler pressure goes up, increasing the speed of the engine, the weights on the governor will move away from the shaft, closing down the steam inlet valve, reducing the steam pressure in the cylinder, and returning the engine speed to its original set point. The governor also helps to prevent speed variation when the load on the engine changes.

In a system of this type without feedback, a large load connected to the steam engine would tend to slow it down. However, because of the feedback present in this system, when loading occurs and the engine speed is reduced, the weights on the governor move in, causing the steam inlet valve to open further. This increases the steam pressure in the cylinder to compensate for the increased load and helps to bring the engine back up to speed. Thus the addition of feedback to this steam engine helps to make the speed of the engine relatively independent of the steam pressure in the boiler and the mechanical load on the engine. The addition of feedback to electronic amplifiers can provide similar improvements in their performance.

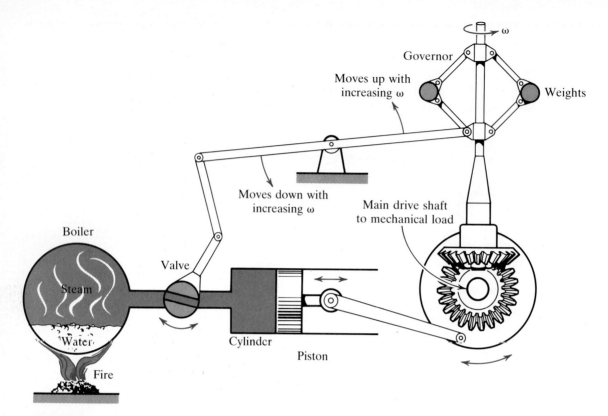

Figure 7.1-1

Addition of a governor to a steam engine to form a mechanical feedback speed control system.

The basic building blocks of a feedback-style electronic amplifier are illustrated in Figure 7.1-2a and that of a conventional amplifier without feedback in Figure 7.1-2b. In this book we use the term "open-loop amplifier" to denote an amplifier that does not have any feedback, and "closed-loop amplifier" to indicate one that does.

In Figure 7.1-2a, the overall amplifier gain is given by the ratio x_o/x_i. The block labeled A_{OL} is the amplification portion of this circuit, and because $x_o/x_i = A_{OL}$ when the feedback is removed, A_{OL} is often called the open-loop gain of the amplifier.

The feedback branch in the amplifier is usually a passive linear network and in fact most often consists of a simple resistive voltage divider. For this network the feedback factor β is defined as the ratio of x_f/x_o and is a measure of how much of the output is fed back to the input. An amplifier for which $\beta = 0$ contains no feedback, while for one in which $\beta = 1$ the entire output is fed back to the input. The product βA_{OL} defines the amplifier's loop gain and is a

Figure 7.1-2

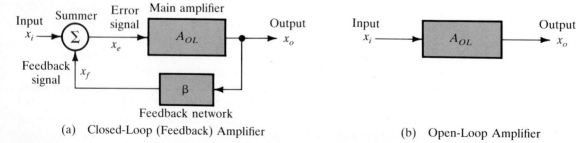

(a) Closed-Loop (Feedback) Amplifier   (b) Open-Loop Amplifier

measure of how much a signal would be amplified in making one complete pass around the loop consisting of the amplifier and the feedback network.

The circular element near the input labeled "summer" is used to combine the input and feedback signals. In a positive-feedback amplifier, these signals are added, and in a negative-feedback amplifier, they are subtracted. The output from the summer is called the error signal, x_e.

Now that we have an idea of the basic elements that make up a feedback amplifier, it is important to get a good understanding of why we would want to bother adding feedback to a circuit in the first place. Feedback offers many important advantages:

1. It makes the closed-loop gain less sensitive to variations that occur in the open-loop gain of the amplifier.
2. It "apparently" reduces the noise level at the output that is associated with noise generated within A_{OL}.
3. Feedback reduces distortion in the output due to the nonlinearity of the components within A_{OL}.
4. Feedback extends the amplifier's bandwidth and improves the transient response.
5. Feedback allows for the independent adjustment of the overall amplifier input and output impedances to nearly any desired levels.

By examining this list it is easy to see why feedback amplifiers are so popular today, but these circuit improvements are not without cost. When feedback is added to an amplifier, the closed-loop amplifier gain is much smaller than the open-loop gain. As we shall see shortly, this gain trade-off is an absolute necessity if the closed-loop gain is to be unaffected by the possibly large changes that might occur in the open-loop gain of the amplifier.

Another significant problem associated with feedback amplifiers is that they can become unstable. Prior to this chapter, all the electronic amplifiers we examined were always stable because the poles associated with their transfer functions all had negative real parts. As a result, when these amplifiers are first turned on, the transient response that results contains only decaying exponentials, and the output eventually stabilizes at the selected Q-point value. When feedback is added to this type of amplifier, it can become unstable because, as we shall see in Chapter 8, the poles move. If all the poles continue to have negative real parts, the amplifier will remain stable, but should one or more of these develop a positive real part, this will give rise to a transient response that contains growing exponentials or growing sine waves. This can result in an amplifier that oscillates or one that simply "hangs up" or saturates at one of the power supply rails. This topic is discussed in great detail in Chapter 8. For now, however, let's just be aware that this is something that we need to be careful about when we design feedback amplifiers.

At the most elementary level, there are two basic types of feedback amplifiers: one in which the feedback signal is in phase with the applied input, and one in which the feedback signal is 180° out of phase. As mentioned previously, these are called positive- and negative-feedback amplifiers, respectively. A specific example of each is illustrated in Figure 7.1-3. In this figure the input and output variables are indicated by x_i and x_o, rather than as specific voltages

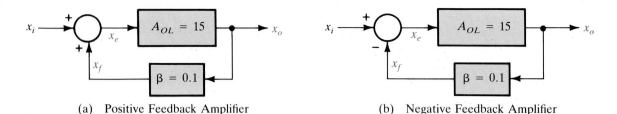

(a) Positive Feedback Amplifier   (b) Negative Feedback Amplifier

Figure 7.1-3

or currents, in order to make the results more generally applicable to all types of electronic amplifiers.

Clearly, for both of these amplifiers the open-loop gain is 15. Their gain with the feedback connected is a little more difficult to determine. For the positive-feedback amplifier illustrated in Figure 7.1-3a, the output voltage is

$$x_o = A_{OL}x_e \qquad (7.1\text{-}1a)$$

But
$$x_e = x_i + x_f = x_i + \beta x_o \qquad (7.1\text{-}1b)$$

and combining these two equations, we have

$$x_o = A_{OL}(x_i + \beta x_o) \qquad (7.1\text{-}2a)$$

or
$$A_{CL} = \frac{x_o}{x_i} = \frac{A_{OL}}{1 - \beta A_{OL}} \qquad (7.1\text{-}2b)$$

In this example because the product βA_{OL} is close to 1, we find that the closed-loop gain of this amplifier is actually bigger than the open-loop gain, with

$$A_{CL} = \frac{15}{1 - (0.1)15} = \frac{15}{-0.5} = -30 \qquad (7.1\text{-}3)$$

On the basis of this result, positive feedback appears to be very attractive since by using it we can actually increase the amplifier gain beyond its open-loop value. In fact, as the product βA_{OL} approaches 1, the closed-loop gain given by equation (7.1-2b) goes to infinity. However, positive-feedback circuits are almost never used in the design of practical amplifiers.

First, as we shall see shortly, although the closed-loop gain can be increased above the open-loop value by adding positive feedback, this gain is virtually useless because it is extremely sensitive to variations in A_{OL}. Furthermore, the addition of positive feedback to an amplifier almost guarantees that the resulting circuit will oscillate. In fact, in Chapter 8 we show that the addition of positive feedback to an amplifier is precisely the way to design an oscillator. Thus positive-feedback designs are seldom, if ever, used when stable amplifier circuits are needed.

Let's examine the negative-feedback amplifier in Figure 7.1-3b. Following the same approach as for the positive-feedback circuit just examined, we have

$$x_o = A_{OL}x_e \qquad (7.1\text{-}4a)$$

and
$$x_e = x_i - x_f = x_i - \beta x_o \qquad (7.1\text{-}4b)$$

Combining these results, we find that

$$x_o = A_{OL}(x_i - \beta x_o) \qquad (7.1\text{-}5)$$

from which we obtain the closed-loop gain for this negative-feedback amplifier as

$$A_{CL} = \frac{x_o}{x_i} = \frac{A_{OL}}{1 + \beta A_{OL}} \qquad (7.1\text{-}6a)$$

Using the data given in Figure 7.1-3b, we have

$$A_{CL} = \frac{15}{1 + (0.1)15} = \frac{15}{2.5} = 6 \qquad (7.1\text{-}6b)$$

A careful examination of equation (7.1-6a) reveals several interesting characteristics of negative-feedback amplifiers. To begin with, because the βA_{OL} term in the denominator is positive, the closed-loop gain of a negative-feedback amplifier is always smaller than the open-loop gain. Furthermore, if the product βA_{OL} is much greater than 1, the closed-loop gain of the amplifier may be approximated as

$$A_{CL} \simeq \frac{A_{OL}}{\beta A_{OL}} = \frac{1}{\beta} \qquad (7.1\text{-}7)$$

This result supports one of our earlier contentions concerning negative-feedback amplifiers, namely that by proper design, A_{CL} can be made nearly independent of A_{OL}. To express this relationship mathematically, we define the sensitivity of A_{CL} to changes in A_{OL} as

$$S_{A_{OL}}^{A_{CL}} = \frac{\Delta A_{CL}/A_{CL}}{\Delta A_{OL}/A_{OL}} = \frac{\Delta A_{CL}}{\Delta A_{OL}} \frac{A_{OL}}{A_{CL}} \qquad (7.1\text{-}8)$$

If the changes in A_{OL} are small, the term $\Delta A_{CL}/\Delta A_{OL}$ can be represented by dA_{CL}/dA_{OL}. When this approximation is valid, we may take the derivative of equation (7.1-6a) with respect to A_{OL} and substitute this into equation (7.1-8) to yield

$$S_{A_{OL}}^{A_{CL}} = \frac{dA_{CL}}{dA_{OL}} \frac{A_{OL}}{A_{CL}} = \frac{1}{1 + \beta A_{OL}} \qquad (7.1\text{-}9)$$

This result indicates that it is possible to reduce the gain sensitivity of an amplifier to changes in open-loop gain by making the quantity $1 + \beta A_{OL}$ large. However, by examining equation (7.1-6a), we can see that this will reduce A_{CL} by the same factor. Thus in feedback amplifier design, we will need to trade closed-loop gain for improved gain stability.

7.1-1 Reduction of Nonlinear Distortion by Using Negative Feedback

Besides improving its gain stability, the application of feedback to an amplifier also reduces the effect of nonlinear distortion generated within A_{OL}. In the previous chapters we always assumed that the signals present were small enough that linear models could be used for the transistors. In the output stages of an amplifier, for example, this is often not the case, and the nonuniform spacing of the transistor's characteristics as well as possible Q-point excursions into or near transistor saturation and cutoff can give rise to significant amounts of non-linear distortion. Feedback helps to minimize the effect of amplifier nonlinearity on the output signal.

To illustrate this point, let's examine the effect of feedback on an amplifier whose open-loop input–output transfer characteristic is given by the mathematical expression

$$x_o = \underbrace{100x_i}_{A_0} + 5x_i^2 \qquad (7.1\text{-}10)$$

This equation is sketched in Figure 7.1-4a. Clearly, the gain of this amplifier is not a constant but depends on the amplitude of the input signal. For small signals, where the second term on the right-hand side of this expression may be neglected, the amplifier is essentially linear and has a gain of about 100. However, for values of x_i greater than about 10, there is a significant deviation from linearity, and beyond this point an almost-square-law relationship exists between input and output signal amplitudes.

To understand how this nonlinearity affects the amplifier's output, let's consider what happens when the input is a sine wave

$$x_i = 4 \sin \omega t \qquad (7.1\text{-}11)$$

To determine this amplifier's output under open-loop conditions, we simply substitute equation (7.1-11) into (7.1-10) to yield

$$x_o = 400 \sin \omega t + 80 \sin^2 \omega t \qquad (7.1\text{-}12a)$$

Using the trigonometric identity

$$\sin^2 \omega t = \frac{1 - \cos 2\omega t}{2} \qquad (7.1\text{-}12b)$$

we may rewrite equation (7.1-12a) as

$$x_o = 400 \sin \omega t + 40 - 40 \cos 2\omega t \qquad (7.1\text{-}12c)$$

Figure 7.1-4
Distortion reduction using negative feedback.

The first term on the right-hand side of this result represents the ordinary linear amplification of the input signal, the second term represents the dc Q-point

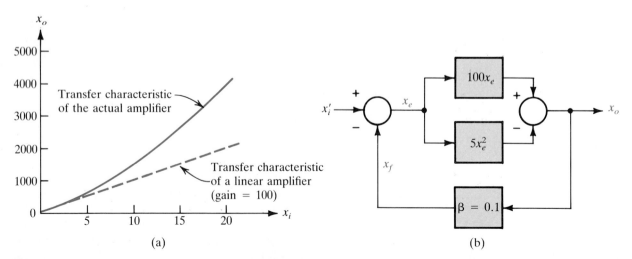

(a)   (b)

shift caused by the nonlinearity, and the last term is called the second harmonic distortion because it is a cosine wave at a frequency that is twice that of the input. For this particular example, the amount of second harmonic distortion in the output is

$$\% \text{ second harmonic distortion} = \frac{40}{400} \times 100 = 10\% \qquad (7.1\text{-}13)$$

Feedback can significantly reduce this distortion level. To simplify the analysis of this amplifier when feedback is added, it is convenient to split the amplifier gain block into two parts, the first a linear term with gain equal to 100, and the second the nonlinear portion of the amplification $5x_e^2$. This amplifier with feedback added is illustrated in Figure 7.1-4b.

To reduce the effect of the nonlinearity, it will be necessary to make A_{CL} relatively independent of A_{OL}. Following equation (7.1-9), this requires that the loop gain magnitude be large. If we consider that A_{OL} is about 100 (at least for small signals), a feedback factor of about 0.1 will result in a loop gain magnitude of about 10, and this should significantly reduce the effect of the gain nonlinearity in A_{OL}.

For small signals the gain nonlinearity in A_{OL} is negligible, and for this case it follows that

$$x_o = A_{CL}x_i' = \frac{A_{OL}}{1 + \beta A_{OL}}x_i' \simeq \frac{100}{1 + (0.1)100}x_i' = 9.09x_i' \qquad (7.1\text{-}14)$$

Because the amplifier gain with the feedback added is smaller than it was under open-loop conditions, we will add a preamplifier (or preamp) with a gain of $100/9.09 = 11$ to the circuit in order to bring the overall linear portion of the gain back up to 100. This modified design is given in Figure 7.1-5a.

To estimate the nonlinear distortion at the output in this circuit following equation (7.1-12c), we will approximate x_o by the expression

$$x_o = C_1 \sin \omega t + C_2 \cos 2\omega t \qquad (7.1\text{-}15)$$

where C_1 and C_2 are two coefficients that need to be evaluated. Following the figure, we can also see that

$$x_f = \beta(C_1 \sin \omega t + C_2 \cos 2\omega t) \qquad (7.1\text{-}16a)$$

and
$$x_e = (44 - \beta C_1) \sin \omega t - \beta C_2 \cos 2\omega t \qquad (7.1\text{-}16b)$$

By direct substitution of this expression for x_e into equation (7.1-10), we have

$$x_o = A_0 x_e + 5x_e^2 = A_0(44 - \beta C_1) \sin \omega t - A_0 \beta C_2 \cos 2\omega t$$
$$+ 5[(44 - \beta C_1)^2 \sin^2 \omega t - \underbrace{2\beta C_2(44 - \beta C_1) \sin \omega t \cos 2\omega t}_{1}$$
$$+ \underbrace{(\beta C_2)^2 \cos^2 2\omega t]}_{2} \qquad (7.1\text{-}17)$$

In this equation, terms 1 and 2 give rise to third and fourth harmonic distortion, respectively, and because their amplitudes are quite small, to a first-order approximation they may be neglected. When this is done and when equation (7.1-12b) is inserted in place of the $\sin^2 \omega t$ term, the expression for x_o may be

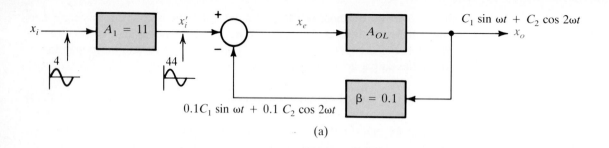

$$0.1C_1 \sin \omega t + 0.1 C_2 \cos 2\omega t$$

(a)

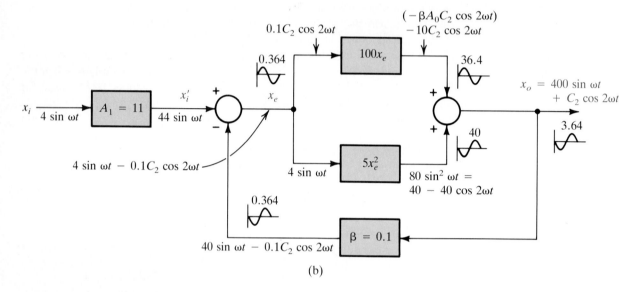

(b)

Figure 7.1-5

Reduction of second harmonic distortion by using negative feedback.

simplified to

$$x_o \simeq A_0(44 - \beta C_1) \sin \omega t - A_0\beta C_2 \cos 2\omega t - 2.5(44 - \beta C_1)^2 \cos 2\omega t$$

(7.1-18)

Equating the coefficients in this expression with the corresponding terms in equation (7.1-15), it follows that

$$C_1 = A_0(44 - \beta C_1) \qquad (7.1\text{-}19a)$$

or

$$C_1 = \frac{44A_0}{1 + \beta A_0} = 400 \qquad (7.1\text{-}19b)$$

and

$$C_2 = -2.5(44 - \beta C_1)^2 - \beta A_0 C_2 \qquad (7.1\text{-}19c)$$

or

$$C_2 = \frac{(-2.5)(16)}{1 + \beta A_0} = \frac{-40}{1 + \beta A_0} = -3.64 \qquad (7.1\text{-}19d)$$

Substituting these results back into equation (7.1-15), we may now approximate the overall feedback amplifier output as

$$x_o \simeq 400 \sin \omega t - 3.64 \cos 2\omega t \qquad (7.1\text{-}20)$$

neglecting any dc and higher-order (smaller-amplitude) harmonic terms that may also be present. In this expression the total second harmonic distortion is

Chapter 7 / Introduction to Feedback and the Operational Amplifier

only 3.64/400 or about 0.9%, compared to the nearly 10% harmonic distortion for the same amplifier running open-loop. If we compare equations (7.1-12c) and (7.1-19d), we can see that improvement obtained by the addition of feedback to the circuit is equal to the factor $1 + \beta A_{OL}$, which is approximately equal to the loop gain magnitude of the amplifier.

To obtain a better physical understanding of how the feedback acts to reduce the nonlinear distortion, let us examine the picture of this amplifier given in Figure 7.1-5b. Here because A_{CL} is approximately 9.09 for this feedback circuit, the linear part of x_o is $(11)(9.09)(4 \sin \omega t)$ or about $400 \sin \omega t$. As a result, the overall output signal may be approximated as

$$x_o = 400 \sin \omega t + C_2 \cos 2\omega t \qquad (7.1\text{-}21)$$

By tracing this signal along with the applied input signal around the loop, we may now readily evaluate C_2 by examining only those terms that ultimately contribute to the second harmonic distortion in the output. In particular, for this circuit it follows immediately that

$$x_f = \beta x_o = 400\beta \sin \omega t + \beta C_2 \cos 2\omega t = 40 \sin \omega t + 0.1 C_2 \cos 2\omega t$$
$$(7.1\text{-}22a)$$
and

$$x_e = x_i' - x_f = 4 \sin \omega t - \beta C_2 \cos 2\omega t = 4 \sin \omega t - 0.1 C_2 \cos 2\omega t$$
$$(7.1\text{-}22b)$$

In tracing the amplification of this signal by A_{OL}, both terms of this expression should be fed into both the linear and nonlinear parts of the amplifier. However, because we are interested only in terms contributing to the second harmonic distortion in the output, it will only be necessary to feed the sin ωt and cos $2\omega t$ terms into the nonlinear and the linear parts of the amplifier, respectively. The resulting second harmonic outputs from each of these portions of the amplifier are illustrated in the figure, and it is these two terms added together which produce the overall second harmonic distortion at x_o, so that we may write

$$C_2 = -\beta A_0 C_2 - 40 \qquad (7.1\text{-}23a)$$

or

$$C_2 = \frac{-40}{1 + \beta A_0} = \frac{-40}{11} = -3.64 \qquad (7.1\text{-}23b)$$

Of course, this is exactly the same result as that obtained earlier in equation (7.1-19d).

In Figure 7.1-5b, we have also indicated the approximate signals found at all important points in the circuit. Starting at the amplifier output and using equation (7.1-20), we see that we have a sine wave with an amplitude of 400 and a second harmonic distortion term of amplitude 3.64 (in describing these feedback effects the dc term in this equation will be ignored; its amplitude throughout the system is the same as that of the second harmonic). Because $\beta = 0.1$, one-tenth of this signal will be fed back to the input, and the fundamental and second harmonic signal levels at x_f will be 40 and 0.364, respectively. The error signal, x_e, which is formed by subtracting x_f from the input, x_i, will have signal amplitudes of 4 and 0.364. The linear amplification of the second harmonic term and the nonlinear amplification of the fundamental produce two second harmonic components at the amplifier output which are nearly equal in amplitude but are 180° out of phase (36.36 and −40, respectively). When these two signals are combined, the net second harmonic distortion produced in the output has only an amplitude of −3.6, not −40, which is what it was under open-loop conditions. Thus the addition of the feedback results in the creation of a rather small second harmonic term in the output which, when fed back to the input and linearly amplified, almost completely cancels the original distortion term produced by nonlinearly amplifying the fundamental.

It is important for us to notice that not only does the feedback signal cancel out $A_{OL}/(1 + \beta A_{OL})$ or 10/11 of the distortion signal, but it also removes 10/11 of the input signal. To compensate for this, we need to increase the input signal amplitude 11-fold or by

$$1 + \beta A_{OL} \tag{7.1-24}$$

7.1-2 "Apparent" Reduction of the Output Noise from an Amplifier by Using Negative Feedback

Feedback, in addition to reducing nonlinear distortion in the amplifier output, can also significantly lower the noise level present at this point as well. Electronic noise arises from two principal sources, those generated within the amplifier, and those that enter the amplifier from the outside world along with the input signal. Feedback can help to reduce the effect of noise generated inside the amplifier but is useless when the noise is essentially part of the input signal. The principal sources of noise internal to the amplifier are resistive or "thermal noise," power supply noise, and transistor and IC noise.

As we shall see shortly, the ability of feedback to reduce output noise levels depends markedly on where in the amplifier chain the noise is introduced. Interestingly enough, the mechanism by which feedback acts to reduce noise levels in the output is quite similar to the way in which it reduces nonlinear distortion. The major difference between the two is that the sources of signal distortion, being proportional to the signal level, are greatest in the stages nearest the output, whereas noise sources tend to be distributed throughout the amplifier.

Feedback is best at handling noise problems that occur near the output. As the noise source gets nearer the amplifier input, the improvement that feedback affords goes to zero.

EXAMPLE 7.1-2

The gain of each of the four amplifiers illustrated in Figure 7.1-6 is approximately 100. In Figure 7.1-6a and c the source v_{n2} represents the noise associated with the amplifier A_2 as well as any power supply noise present at this point. The amplifier A_1 in these two circuits is noiseless. For simplicity the noise sources (v_{n1} and v_{n2}) are considered to be triangle waves, and the input signal to all four amplifiers, v_{in}, is assumed to be a 1-V sine wave. In the amplifiers shown in Figure 7.1-6b and d, A_2 is assumed to be noiseless, and the source v_{n1} represents the noise associated with the amplifier A_1 or that caused by the power supply at this input node. For each of these four circuits determine the signal-to-noise ratio present at the amplifier output.

SOLUTION

By examining the two open-loop amplifiers shown in Figure 7.1-6a and b, it can readily be seen that the output voltages for each of these circuits may be

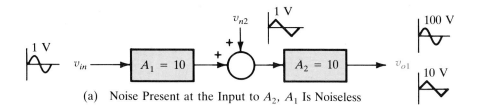

(a) Noise Present at the Input to A_2, A_1 Is Noiseless

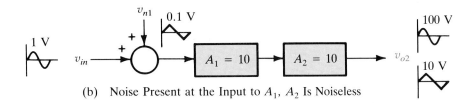

(b) Noise Present at the Input to A_1, A_2 Is Noiseless

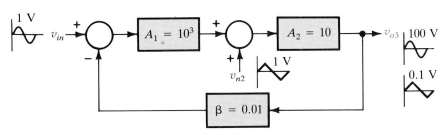

(c) Same Amplifier as in (a) with Feedback Added and
the Gain of A_1 Increased

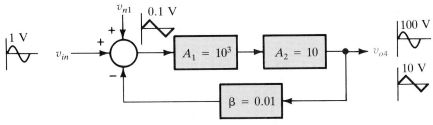

(d) Same Amplifier as in (b) with Feedback Added and
the Gain of A_1 Increased

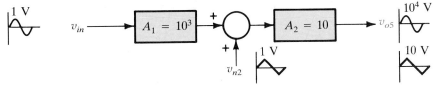

(e) Same Amplifier as in (c) with the Feedback Loop Broken

Figure 7.1-6
Example 7.1-1.

written as

$$v_{o1} = A_1 A_2 v_{in} + A_2 v_{n2} \qquad (7.1\text{-}25a)$$

and

$$v_{o2} = A_1 A_2 v_{in} + A_1 A_2 v_{n1} \qquad (7.1\text{-}25b)$$

As a result, the output signal levels for both circuits are 100-V sine waves while the noise levels are 10-V triangle waves. Therefore, the signal-to-noise ratios for both of these circuits is $(100 \text{ V})/(10 \text{ V}) = 10$ or 20 dB. The two feedback amplifiers illustrated in Figure 7.1-6c and d represent attempts to improve the signal-to-noise ratios of each of the open-loop amplifiers in Figure 7.1-6a and b by adding feedback.

Let us first consider the feedback amplifier shown in Figure 7.1-6c, in which all the noise occurs in the second stage. For this case,

$$V_{o3} = A_2 v_{i2} = A_2(v_{o1} + v_{n2}) = A_2(A_1 v_e + v_{n2}) \qquad (7.1\text{-}26)$$

But the error voltage v_e is equal to

$$v_e = v_{in} - \beta v_{o3} \qquad (7.1\text{-}27)$$

so that

$$v_{o3} = A_1 A_2(v_{in} - \beta v_{o3}) + A_2 v_{n2} \qquad (7.1\text{-}28a)$$

or

$$v_{o3} = \frac{A_1 A_2 v_{in}}{1 + \beta A_1 A_2} + \frac{A_2 v_{n2}}{1 + \beta A_1 A_2} \qquad (7.1\text{-}28b)$$

For $\beta A_1 A_2 \gg 1$, equation (7.1-28b) reduces to

$$v_{o3} \simeq \frac{1}{\beta} v_{in} + \frac{1}{\beta A_1} v_{n2} = 100 v_{in} + 0.1 v_{n2} \qquad (7.1\text{-}29)$$

so that for this circuit the output signal and noise levels are 100 and 0.1 V, respectively. The corresponding signal-to-noise ratio is 1000 or 60 dB.

For the amplifier in Figure 7.1-6d the noise source is located at the input, and a circuit analysis similar to that for the previous case yields

$$v_{o4} = \frac{A_1 A_2 v_{in}}{1 + \beta A_1 A_2} + \frac{A_1 A_2 v_{n1}}{1 + \beta A_1 A_2} \qquad (7.1\text{-}30)$$

which for $\beta A_1 A_2 \gg 1$ is approximately equal to

$$v_{o4} \simeq \frac{1}{\beta} v_{1n} + \frac{1}{\beta} v_{n1} = 100 v_{in} + 100 v_{n1} \qquad (7.1\text{-}31)$$

The signal-to-noise ratio for this case is 10 (or 20 dB). Thus in an amplifier where the noise occurs at the input, the addition of feedback does not offer any improvement in the signal-to-noise ratio (compare the amplifiers in Figure 7.1-6b and d).

However, for amplifiers where the noise occurs after the input stage, feedback does help, although the improvement may not be quite as good as it seems. In making the transition from the open-loop amplifier in Figure 7.1-6a to the closed-loop circuit in Figure 7.1-6c, not only did we add a feedback network, but we increased the gain of A_1 from 10 to 1000. In actuality, it was this gain change, not the addition of the feedback, which improved the signal-to-noise ratio.

To see why this is so, let's compare the amplifier circuit in Figure 7.1-6c with that given in Figure 7.1-6e, in which the feedback loop has been removed. For both circuits the signal-to-noise ratio is 1000. The feedback has no effect on the signal-to-noise ratio, and in fact, all that it does in this circuit is to lower the overall amplifier gain so that the output stage is not overloaded by the available input signal. Of course, this overload problem could also have been solved without using feedback, simply by reducing the input signal in the open-loop amplifier in Figure 7.1-6e until the output sine wave dropped back to 100 V, but this would have lowered the signal-to-noise ratio by the same factor. In this way feedback does help to improve the signal-to-noise ratio by allowing you to drive the amplifier with large input signals without overloading the output. However, this improvement may not be as large as expected.

For the amplifier in Figure 7.1-6a, we assumed that the preamp A_1 was noiseless, and in an actual amplifier this would certainly not be the case. Furthermore, in the feedback amplifier version of this circuit in Figure 7.1-6c because the gain of A_1 was raised substantially we would also expect to see a significant increase in the noise associated with A_1. As a result, it is possible that the noise reduction achieved in the second stage by the feedback could be offset by the additional noise produced by the input stage, because these noise signals are amplified by the stages that follow. In fact, if this preamp stage were noisy enough, it might actually cause the "improved" amplifier in Figure 7.1-6c to be noisier than the original circuit.

7.1-3 Effect of Feedback on the Input and Output Impedance of an Amplifier

Besides being useful for reducing an amplifier's closed-loop gain dependence on A_{OL}, and for minimizing noise and nonlinearity effects in the output, feedback can also be used to modify an amplifier's input and output impedances to tailor them to specific values. For certain amplifier design problems this attribute can be extremely important.

Depending on the character of the source, the source impedance may not be well defined, or if known may be nonlinear or perhaps a function of frequency. If signal amplification independent of the source impedance is desired and if, for example, the signal to be amplified is a voltage, then an amplifier with a high input impedance is needed. Conversely, if the signal to be amplified is in the form of a current, then an amplifier with a low input impedance is required for amplification independent of the source impedance level.

Sometimes the source impedance is a well-defined quantity. Examples of this include antennas, transmission lines, and certain kinds of electromechanical transducers. In amplifying the signals from these types of sources it might be desirable to adjust the amplifier's input impedance to match that of the source in order to achieve maximum power transfer or to reduce signal reflections at the input of the amplifier. Thus it is often extremely useful to be able to tailor the circuit's input impedance.

The ability to adjust the output impedance of an amplifier is similarly useful. Sometimes we want the voltage appearing across the load to be a constant independent of Z_L, the load impedance. This requires that Z_L be much greater than the amplifier output impedance, Z_{out}, or that the equivalent output circuit of the amplifier approximate an ideal voltage source. Conversely, there are situations where we desire that the load current, not the load voltage, be independent of Z_L. For this situation we would want Z_{out} to be much greater than the load impedance so that the equivalent output circuit of the amplifier approximates an ideal current source.

By properly choosing the circuit configuration of the feedback network, the amplifier's input and output impedances can be adjusted to nearly any desired values. At the amplifier output we may sample either the output current or the output voltage, depending on which one of these quantities we want to hold constant when the load or amplifier parameters vary. Voltage sampling tends to lower the amplifier output impedance (and make it look more like an ideal voltage source) and current sampling increases Z_{out} (to make the amplifier behave more like an ideal current source).

The way in which the feedback path is connected to the input determines the effect that the feedback will have on the amplifier's input impedance. If the feedback effectively looks like a voltage connected in series with the input signal, we have "voltage-series mixing (or summing)," and as we will demonstrate shortly, the amplifier's input impedance will increase. If, on the other hand, the feedback path connects back to the same node as the input signal and effectively shunts across the amplifier input, we have what is called "current-shunt mixing," and this can be shown to lower the amplifier's input impedance.

Thus by properly selecting the mixing and sampling techniques to be used, we can independently determine the effect that the feedback has on the input and output impedances of the amplifier, respectively (Table 7.1-1a). Because there are two distinct sampling methods and two types of mixing, four different feedback amplifier configurations are possible; these are listed in Table 7.1-1b. To demonstrate quantitatively how the input and output impedances of a feedback amplifier depend on the type and the amount of feedback employed, we will examine the behavior of the first entry on this table. The analysis of the other three feedback amplifier configurations will be left as an exercise for the interested reader.

The circuit illustrated in Figure 7.1-7 is a voltage sampling–voltage series feedback amplifier. By definition the input impedance of this amplifier is

$$Z_{in} = \frac{v_{in}}{i_{in}} \tag{7.1-32}$$

and in the absence of any feedback, with v_f equal to zero, we can readily see that $Z_{in} = Z_1$. When v_f is nonzero, this impedance level changes, and by applying Kirchhoff's law to the input loop, we have

$$i_{in} = \frac{v_{in} - \beta v_o}{Z_1} \tag{7.1-33}$$

If for simplicity we neglect any loading on the output, v_o is approximately equal to $A_{OL}v_1$ so that equation (7.1-33) may be rewritten as

$$i_{in}Z_1 = v_{in} - \beta A_{OL}v_1 = v_{in} - \beta A_{OL}(i_{in}Z_1) \tag{7.1-34}$$

or

$$i_{in} = \frac{v_{in}}{Z_1(1 + \beta A_{OL})} \tag{7.1-35}$$

Substituting this result into equation (7.1-32), we find that the input impedance

7.1-1 Effect of Negative Feedback on an Amplifier's Input and Output Impedances

(a)

Feedback Type		Principal Effects
Sampling method	Current sampling	Raises output impedance; makes amplifier look more like a current source to keep load current constant
	Voltage sampling	Lowers output impedance; amplifier looks more like an ideal voltage source to keep voltage constant in spite of load changes
Mixing method	Voltage-series summing	Raises input impedance: good if source is a voltage source—do not load it; gain independent of source impedance
	Current-shunt summing	Lowers input impedance: good if source is a current source—makes gain independent of source impedance

(b)

Amplifier Configuration	Z_{in}	Z_{out}
Voltage sampling–voltage series mixing	Increases	Decreases
Voltage sampling–current shunt mixing	Decreases	Decreases
Current sampling–voltage series mixing	Increases	Increases
Current sampling–current shunt mixing	Decreases	Increases

of this amplifier with the feedback in place is

$$Z_{in} = \frac{v_{in}}{i_{in}} = Z_1(1 + \beta A_{OL}) \qquad (7.1\text{-}36)$$

Since this amplifier's input impedance without feedback is just Z_1, the addition of voltage series feedback to this circuit has increased its input impedance by the factor $(1 + \beta A_{OL})$.

The circuit's output impedance may be found by applying Thévenin's theorem and finding the ratio of the circuit's open-circuit voltage to its short-circuit current. Under open-circuit conditions,

$$v_{oc} = v_o\big|_{Z_L=\infty} = A_{OL}v_1 \qquad (7.1\text{-}37)$$

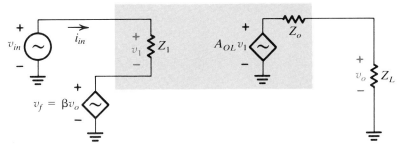

Figure 7.1-7
Voltage sampling–voltage mixing amplifier.

and from equation (7.1-35) we have that

$$v_1 = i_{in} Z_1 = \frac{v_{in}}{1 + \beta A_{OL}} \tag{7.1-38}$$

Combining equations (7.1-37) and (7.1-38), the open-circuit voltage is

$$v_{oc} = \frac{A_{OL} v_{in}}{1 + \beta A_{OL}} \tag{7.1-39}$$

To evaluate the short-circuit current we note that v_o and v_f are both equal to zero under short circuit conditions so that

$$v_1 = v_{in} \tag{7.1-40a}$$

and

$$i_{sc} = i_L \Big|_{Z_L=0} = \frac{A_{OL} v_1}{Z_o} = \frac{A_{OL} v_{in}}{Z_o} \tag{7.1.40b}$$

Combining equations (7.1-39) and (7.1-40b), we have

$$Z_{out} = \frac{v_{oc}}{i_{sc}} = \frac{Z_o}{1 + \beta A_{OL}} \tag{7.1-41}$$

It is important to note that the amplifier's output impedance without any feedback (i.e., with $\beta = 0$) is just Z_o. One could show this rigorously by forming the ratio of v_{oc}/i_{sc} for the case where $\beta = 0$, but it is also possible to determine this by inspection of Figure 7.1-7. The output impedance of an amplifier is the impedance seen looking back into the output terminals when the independent sources are set equal to zero, and the dependent sources are "carefully considered." For this circuit, because the controlled source $A_{OL} v_1$ depends on a voltage that exists elsewhere in the circuit, and because v_1 in no way depends on the output loop (for $\beta = 0$), this source $A_{OL} v_1$ may simply be set equal to zero. As a result, the amplifier output impedance with no feedback is just equal to Z_o. Thus, for this particular feedback amplifier configuration, the output impedance has decreased by $(1 + \beta A_{OL})$, the same factor by which the input impedance increased.

7.1-4 Analysis of Feedback Amplifier Circuits Containing Discrete Transistors

The following general analysis technique can be used for attacking nearly any type of feedback problem.

1. Define the amplifier input current or input voltage of interest at the feedback point.
   (a) For current shunt feedback where the feedback element and the input signal are effectively connected to the same point (and are out of phase from one another):
       (i) Use v_b as the input variable for a circuit where the input transistor is a BJT device.
       (ii) Use v_g as the input variable for a circuit where the input transistor is an FET.
   (b) For voltage series feedback [where the feedback and input signal are connected to different points (and are in phase with one another)]:

(i) Use i_b (or i_e) for a circuit with a BJT input transistor.

(ii) Use i_d for a circuit with an FET input transistor.

2. (a) Develop the input equivalent circuit at the feedback point with the feedback element removed, and label the input parameter of interest selected in step 1.

(b) Develop the Thévenin or Norton equivalent of the output circuit where the feedback element is connected, and express the output-controlled source in terms of the input parameter selected in step 1.

(c) Connect the input and output equivalent circuits together with the feedback element to form the overall equivalent circuit to be analyzed.

3. (a) Simplify the circuit obtained in step 2 by finding the Norton or Thévenin equivalent of the circuit to the right of the input parameter of interest. Use the Norton equivalent when the input parameter is a voltage and the Thévenin equivalent when it is a current.

(b) Use source elimination techniques to replace the controlled source by an equivalent resistor, and then use the resulting simplified circuit to find the input parameter of interest.

(c) In the circuit that results in step 3b, Z_{in} is often obvious by inspection of the circuit. If the value of Z_{in} is not apparent, find Z_{in} by forming the ratio of v_{in}/i_{in}.

4. Find the output parameter of interest (typically by using superposition).

5. Determine the circuit output impedance by using one of the two methods indicated below.

(a) Find Z_{out} by forming the ratio of v_{oc}/i_{sc}.

(b) Apply a voltage v_o at the output terminals and measure the current i_o that flows from this source with the independent input sources set equal to zero. In this case $Z_{out} = v_o/i_o$. As a practical matter, in this method Z_{out} is actually obtained by inspection of the circuit rather than from v_o/i_o since the application of v_o to the circuit output terminals is just a technique for eliminating the controlled source in the output circuit.

As illustrated in Table 7.1-1b, there are a total of four different types of feedback amplifier circuit configurations. The example that follows will demonstrate how to analyze one of these feedback circuits using the technique just outlined.

| EXAMPLE 7.1-3 | The amplifier illustrated in Figure 7.1-8a is an example of a voltage sampling–current summing feedback circuit. The feedback resistor R_f samples the output voltage v_o, and feeds back a current to the base terminal of Q_1 which is proportional to this voltage. Because of the type of feedback connected to this amplifier, we expect that the input and output impedances will both decrease significantly from their open-loop values (Table 7.1-1b). |

a. Given that $h_{ie} = 1$ kΩ and $h_{fe} = 100$ for all three transistors, calculate the circuit input and output impedances with and without the feedback connected.

b. Determine A_{OL} and A_{CL} for this amplifier.

SOLUTION

Before getting into the actual solution of this problem, it will be useful to demonstrate that this circuit has negative rather than positive feedback. To un-

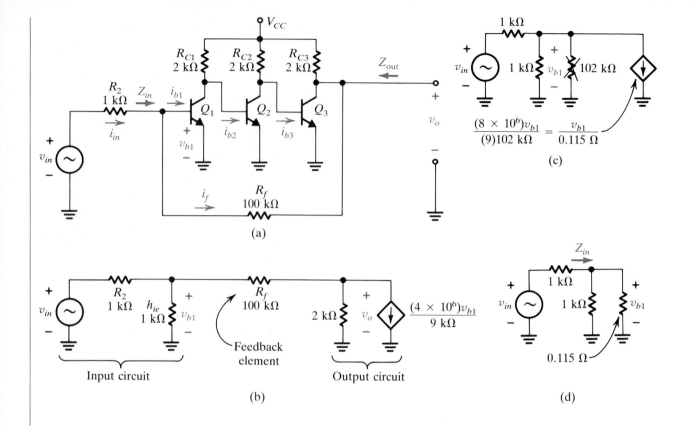

Figure 7.1-8
Example 7.1-2.

derstand this discussion, you need to recall that negative-feedback amplifiers tend to reduce the amplifier gain. They do this by feeding back a signal that is out of phase from the input signal.

With no feedback (i.e., with R_f removed) by using a voltage divider we can see that $v_{b1} = \frac{1}{2} v_{in}$. This signal is then amplified by the three-stage common-emitter transistor circuit consisting of Q_1, Q_2, and Q_3, and thus the signal appearing at the output is a large inverted version of v_{in}. When feedback is added, by using superposition, we can see that the voltage at the v_{be1} node decreases from its open-loop value because the contribution from v_o adds in out of phase. The fact that v_{be1} and hence v_o are both reduced when the feedback is added demonstrates that this circuit has negative feedback.

The input and output impedances of this amplifier in the absence of feedback (i.e., with R_f removed) can be seen from an inspection of Figure 7.1-8a to be $Z_{in} = h_{ie} = 1 \text{ k}\Omega$ and $Z_{out} = R_{C3} = 2 \text{ k}\Omega$. The open-loop-circuit voltage gain is calculated as follows:

$$i_{c1} = h_{fe} i_{b1} = h_{fe}\left(\frac{v_{in}}{2 \text{ k}\Omega}\right) = \frac{100 v_{in}}{2 \text{ k}\Omega} \qquad (7.1\text{-}42a)$$

$$i_{c2} = h_{fe} i_{b2} = h_{fe}\left(-\frac{2}{3} i_{c1}\right) = -\frac{2 \times 10^4}{3} i_{b1} \qquad (7.1\text{-}42b)$$

and

$$i_{c3} = h_{fe} i_{b3} = h_{fe}\left(-\frac{2}{3} i_{c2}\right) = \frac{4 \times 10^6}{9} i_{b1} \qquad (7.1\text{-}42c)$$

Using this value for i_{c3}, the output voltage v_o may be written as

$$v_o = -i_{c3}R_{c3} = -4.44 \times 10^5 v_{in} \tag{7.1-43a}$$

so that

$$A_{OL} = \frac{v_o}{v_{in}} = -4.44 \times 10^5 \tag{7.1-43b}$$

When the feedback resistor R_f is added to the circuit, the internal character of the amplifier is unchanged so that equations (7.1-42b) and (7.1-42c) are still valid. However, equations (7.1-42a), (7.1-43a), and (7.1-43b) need to be modified. To see how this may be accomplished, following the feedback circuit analysis scheme outlined earlier, we first observe that because this amplifier has current shunt feedback, the voltage v_{b1} should be chosen as the input variable of interest. The overall feedback circuit to be analyzed is given in Figure 7.1-10b. Notice following equation (7.1-42c) that the output-controlled source $h_{fe}i_{b3}$ has been reexpressed in terms of the variable v_{b1} as

$$i_{c3} = h_{fe}i_{b3} = \frac{4 \times 10^6}{9}\left(\frac{v_{b1}}{1\ \text{k}\Omega}\right) = \frac{4 \times 10^6 v_{b1}}{9\ \text{k}\Omega} \tag{7.1-44}$$

To express v_{b1} in terms of v_{in}, we could apply network theory directly at this point. However, much more insight into this circuit's principles of operation can be obtained by finding the Norton equivalent of the circuit to the right of the variable v_{b1}. The short-circuit current at the terminals of this network is seen to be

$$i_{sc} = \frac{-(4 \times 10^6)v_{b1}}{9\ \text{k}\Omega}\left(\frac{2\ \text{k}\Omega}{102\ \text{k}\Omega}\right) = \frac{-4 \times 10^6 v_{b1}}{459\ \text{k}\Omega} = -\frac{v_{b1}}{0.115\ \Omega} \tag{7.1-45a}$$

Furthermore, because the controlling parameter v_{b1} associated with the controlled source $(4 \times 10^6\ v_{b1})/9\ \text{k}\Omega$ is external to the network, the Thévenin impedance of this network may be written by inspection as

$$R_T = 100\ \text{k}\Omega + 2\ \text{k}\Omega = 102\ \text{k}\Omega \tag{7.1-45b}$$

The resulting circuit is given in Figure 7.1-8c. In this circuit the controlled current source may now be replaced by an equivalent resistance

$$R_{b1} = \frac{v_{b1}}{v_{b1}/0.115\ \Omega} = 0.115\ \Omega \tag{7.1-46}$$

because the current flow through this source is proportional to the voltage v_{b1} across it. The resulting equivalent input circuit is shown in Figure 7.1-8d, and following this figure, it is immediately apparent that

$$Z_{in} \simeq 0.115\ \Omega \tag{7.1-47a}$$

and

$$v_{b1} = \left(\frac{0.115}{1000.115}\right)v_{in} = 0.0001147 v_{in} \tag{7.1-47b}$$

The circuit output voltage is best obtained by taking the Thévenin equivalents of the input and output portions of the original equivalent circuit given in Figure 7.1-8b. In the resulting circuit given in Figure 7.1-9a, the output-controlled source has been expressed in terms of v_{in} by using equation (7.1-47b). Following this figure and applying the superposition principle, we have

$$v_o = (-101.98v_{in})\left(\frac{100.5}{102.5}\right) + (0.5v_{in})\left(\frac{2}{102.5}\right) = -99.98v_{in} \qquad (7.1\text{-}48\text{a})$$

so that

$$A_{CL} = \frac{v_o}{v_{in}} = -99.98 \qquad (7.1\text{-}48\text{b})$$

Because v_{oc} is already known from equation (7.1-48a), the circuit output impedance can be readily found by obtaining i_{sc}. For the circuit in Figure 7.1-9b,

$$v_{b1} \simeq \frac{v_{in}}{2} \qquad (7.1\text{-}49\text{a})$$

$$i_f \simeq \frac{v_{b1}}{100 \text{ k}\Omega} = \frac{v_{in}}{200 \text{ k}\Omega} \qquad (7.1\text{-}49\text{b})$$

Figure 7.1-9
Example 7.1-2 (continued).

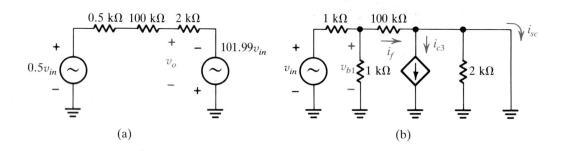

(a) (b)

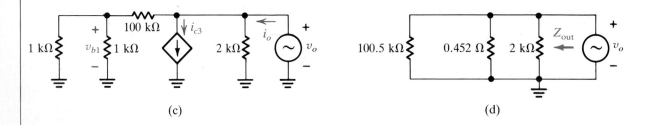

(c) (d)

Parameter	Without Feedback (R_f removed)	With Feedback (R_f connected)
Gain	$A_{OL} = -4.44 \times 10^5$	$A_{CL} = -99.98$
Input Z	$h_{ie} = 1 \text{ k}\Omega$	$0.115 \ \Omega$
Output Z	$R_{C3} = 2 \text{ k}\Omega$	$0.45 \ \Omega$

(e)

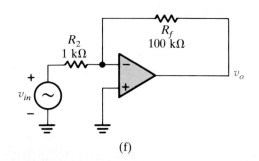

(f)

and the controlled source is

$$i_{c3} = \frac{4 \times 10^6 v_{b1}}{9 \text{ k}\Omega} = \frac{4 \times 10^6 v_{in}}{18 \text{ k}\Omega} = \frac{v_{in}}{0.0045 \ \Omega} \qquad (7.1\text{-}49\text{c})$$

Since i_{c3} is so much greater than i_f, it therefore follows that

$$i_{sc} \simeq -i_{c3} = \frac{-v_{in}}{0.0045 \ \Omega} \qquad (7.1\text{-}50\text{a})$$

and
$$Z_{out} = \frac{v_{oc}}{i_{sc}} = -\frac{99.98 v_{in}}{-v_{in}/0.0045 \ \Omega} = 0.45 \ \Omega \qquad (7.1\text{-}50\text{b})$$

For completeness, we will also find the output impedance by setting the independent source v_{in} to zero and applying a voltage v_o, as illustrated in Figure 7.1-9c. Following this figure,

$$v_{b1} = \left(\frac{0.5 \text{ k}\Omega}{100.5 \text{ k}\Omega}\right) v_o = 0.00498 v_o \qquad (7.1\text{-}51\text{a})$$

so that the controlled source i_{c3} may be written as

$$i_{c3} = \frac{4 \times 10^6 v_{b1}}{9 \text{ k}\Omega} = \left(\frac{4 \times 10^6}{9 \text{ k}\Omega}\right)(0.00498 v_o) = \frac{v_o}{0.452 \ \Omega} \qquad (7.1\text{-}51\text{b})$$

Because the current flow through this source is proportional to the voltage v_o across it, it may be replaced by an equivalent $0.452\text{-}\Omega$ resistor (Figure 7.1-9d). Notice that Z_{out} can now be obtained by inspection of this circuit as

$$Z_{out} = (0.452 \ \Omega) \parallel 2 \text{ k}\Omega \parallel 100.5 \text{ k}\Omega \simeq 0.45 \text{ k}\Omega \qquad (7.1\text{-}52)$$

and that i_o never really needs to be calculated to find Z_{out} since the output impedance can be obtained by inspection of the circuit once the controlled source has been eliminated. The results obtained from the analysis of this problem are summarized in Figure 7.1-9e.

Let us now reexamine this feedback amplifier in order to see how the closed-loop gain could have been calculated by using a more intuitive approach. To begin this analysis we need to recognize that the input and output impedances will be very low because of the type of feedback used with this circuit.

Due to the low input impedance, the voltage divider formed by R_2 and Z_{in} will cause the voltage at v_{b1} to be "virtually" at ground potential. In fact, in the jargon of the operational amplifier, this point in the circuit where the current summing occurs is frequently called the circuit's virtual ground point. To the extent that v_{b1} is virtually zero,

$$i_{in} = \frac{v_{in} - v_{b1}}{R_2} \simeq \frac{v_{in}}{R_2} \qquad (7.1\text{-}53)$$

Furthermore, because v_o is so much larger than v_{b1}, we also find that

$$i_f = \frac{v_{b1} - v_o}{R_f} \simeq -\frac{v_o}{R_F} \qquad (7.1\text{-}54)$$

A second consequence of the fact that $v_o \gg v_{b1}$ is that the feedback current is also much greater than Q_1's base current. As a result, the input and feedback currents are nearly equal, so that

$$i_f = -\frac{v_o}{R_f} \simeq i_{in} = \frac{v_{in}}{R_2} \qquad (7.1\text{-}55a)$$

or

$$A_{CL} = \frac{v_o}{v_{in}} = -\frac{R_f}{R_2} = -100 \qquad (7.1\text{-}55b)$$

In this example, because of the large loop gain, all the approximations just discussed are quite good, and as a result the approximate solution given in equation (7.1-55b) compares very favorably with the exact solution found earlier in equation (7.1-48b).

Before leaving this example, it is appropriate to mention that there are two alternative methods for attacking this same problem. With the feedback amplifier redrawn as shown in Figure 7.1-9f, we can see that this circuit may be viewed, to a good approximation, as an inverting-style operational amplifier. In fact, as we will show later in this chapter, equation (7.1-55b) exactly·describes the gain of this type of amplifier circuit.

The last analysis method for determining the closed-loop gain of this amplifier involves the simplification of this problem by splitting R_f using the Miller theorem. The input Miller resistance is $R_{m1} = R_f/(1 - A)$, where A is the gain from the base of Q_1 to the output with R_f removed. Using equation (7.1-42c) and the fact that

$$i_{b1} = \frac{v_{b1}}{1 \text{ k}\Omega} \qquad (7.1\text{-}56a)$$

and

$$v_o = -i_{c3} R_{c3} \qquad (7.1\text{-}56b)$$

we find that

$$A = -\frac{8 \times 10^6}{9} \qquad (7.1\text{-}57a)$$

and

$$R_{m1} = 0.112 \ \Omega \qquad (7.1\text{-}57b)$$

Thus the application of the Miller theorem to this problem leads to an input circuit that is identical to that given in Figure 7.1-8d and therefore ultimately to the same solution as that developed previously.

Exercises

7.1-1 For the feedback amplifier shown in Figure 7.1-5a, $A_{OL} = 100x_e + 5x_e^3$. Find the percent third harmonic distortion at x_o if $x_i = 4 \sin \omega t$. Note that $\sin^3 \theta = \frac{3}{4} \sin \theta - \frac{1}{4} \sin 3\theta$. **Answer** 0.114%

7.1-2 Find Z_{out} for the circuit given in Figure 7.1-8 if R_f is changed to 20 kΩ. **Answer** 0.09 Ω

7.1-3 Find Z_{out} for the circuit in Figure 7.1-8 if a 2-kΩ resistor is shunted from the base of Q_1 to ground. **Answer** 0.565 Ω

7.2 THE IDEAL OPERATIONAL AMPLIFIER

The term "operational amplifier" (op amp) dates back to the 1950s, and was coined to describe a special class of direct-coupled (dc) high-gain amplifiers that was originally developed for use with analog computers. By connecting negative feedback to these amplifiers, they could be used to perform a variety of mathematical operations, including the addition, subtraction, integration, and differentiation of electronic signals. In fact, the name "operational amplifier" specifically refers to the many mathematical *operations* that this type of *amplifier* is capable of carrying out.

Originally, op amps were designed using vacuum tubes, and as such they were bulky, power hungry, and costly. By way of contrast, today's modern operational amplifiers, which are most often fabricated as linear integrated circuits, are about the size of a thumbnail, consume only a few milliwatts of power, and are very inexpensive. In addition, their electrical characteristics are vastly superior to those of their vacuum-tube ancestors.

The major appeal of the operational amplifier to the system designer is that it is a low-cost, nearly ideal general-purpose amplification element that can perform a variety of circuit functions simply by modifying the input and feedback networks that are connected to the amplifier. In particular, the op amp may be used for linear signal-processing applications such as addition, subtraction, and filtering, as well as for a large number of nonlinear signal processing tasks which include level detection, rectification, and amplitude limiting to name just a few.

In this section we discuss the basic properties of the operational amplifier; in addition, we will present several elementary applications of this device. In the sections that follow we will learn more about this very popular circuit element.

Specifically, in Sections 7.3 through 7.5, we describe some of the more sophisticated linear and nonlinear signal-processing applications of this device, and in Section 7.6 we present a rather detailed picture of the circuit building blocks that make up the modern operational amplifier. The chapter concludes with a discussion of some of the more important second-order effects associated with the op amp, and describes how they alter the previously assumed ideal performance of this circuit element.

But that's getting ahead of ourselves; let's go back now, begin at the beginning, and first make sure that we fully understand what we are talking about when we use the words "ideal operational amplifier." To begin with, as with the feedback amplifiers discussed in Section 7.1, there is more than one type of operational amplifier available, and depending on the application, one or the other may be a better choice. Op amps can be obtained which have either low or high input impedances, and also either low or high output impedances. At the time of this writing, and for as long as this author can remember, the high input impedance–low output impedance style op amp has been the most popular type. Therefore, for our purposes in this book, when we use the term "ideal operational amplifier," we mean a high-gain, differential input voltage amplifier having an infinite input impedance and zero output impedance.

The schematic symbol for this type of operational amplifier is given in Figure 7.2-1a. The two leads labeled minus and plus are used to represent the inverting and noninverting input terminals of the amplifier, respectively. The third lead on the right-hand side of this circuit symbol is the output terminal of the amplifier. A signal applied to the noninverting input appears at the output

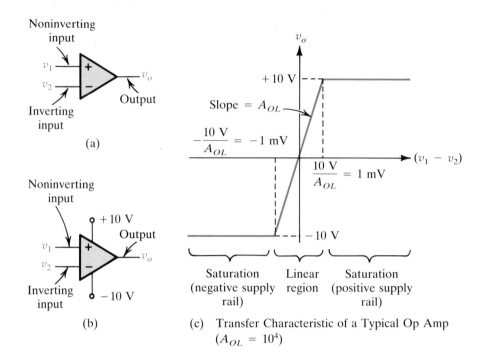

Figure 7.2-1

(a)

(b)

(c) Transfer Characteristic of a Typical Op Amp ($A_{OL} = 10^4$)

of the amplifier in phase with the original signal, while that applied to the inverting input will arrive at the output 180° out of phase.

Actual op amp devices have more than three terminals. At least two additional leads are needed for the connection of the power supplies. While op amps can operate off a single supply, it is more common to use two symmetrical power supplies in order to allow the output voltage to be able to swing equally above and below ground. In general, the signal range available at the output of the op amp is limited by the size of the power supplies. For our ideal operational amplifier, we will assume that its output can swing all the way to the power supply rails without distortion, and that if the output attempts to go beyond these levels, it simply saturates at the supply voltage (Figure 7.2-1c). For most op amp applications the simple three-terminal symbol given in Figure 7.2-1a is all that is needed to represent the amplifier, but for circuit designs where the output signal level hits the supply rails, this symbol may be expanded as shown in Figure 7.2-1b to include the power supply information. Some op amps have additional leads for dc balance adjustment, and for frequency-compensation purposes. These terms are discussed more fully in the sections that follow, and where appropriate, these additional device terminals will be indicated on the circuit symbol for the op amp.

The input–output equation for the ideal operational amplifier, when it is operating in its linear region, is given by the expression

$$v_o = A_{OL}(v_1 - v_2) \tag{7.2-1}$$

where A_{OL} is the amplifier's open-loop gain. If we assume that A_{OL} is 10^4, for example, and that the amplifier is powered by ±10-V power supplies, the transfer characteristic for this device will be similar to that presented in Figure 7.2-1c.

As indicated in this figure, because A_{OL} is so large, there is only a narrow range of input signal levels for which the amplifier will remain in the linear (or active) region. In particular, to keep the magnitude of v_o below 10 V, from equation (7.2-1) we require that

$$\frac{-10\text{ V}}{10^4} = -1\text{ mV} \le (v_1 - v_2) \le \frac{10\text{ V}}{10^4} = 1\text{ mV} \tag{7.2-2}$$

This result indicates that v_1 and v_2 must be within 1 mV of one another for linear operation of the op amp.

When additional circuit elements are added to the op amp to form a feedback amplifier, these fundamental relationships do not change. Equation (7.2-1) still applies, and v_1 and v_2 must still be very close to one another if linear signal amplification is to be achieved. In fact, in the derivations that follow, we will see that the approximation

$$v_1 = v_2 \tag{7.2-3}$$

can significantly reduce the time required to analyze the performance of the op amp circuits that we will be examining.

Let's start by developing an expression for the transfer function for the noninverting amplifier illustrated in Figure 7.2-2. Notice that the amplifier has negative feedback, and that the feedback network is a resistive voltage divider.

We begin the analysis of this circuit by deriving a complete expression for v_o/v_{in}, making no approximations other than that v_o is related to v_1 and v_2 by equation (7.2-1). By examining this circuit, it should be clear that

$$v_1 = v_{in} \tag{7.2-4a}$$

and also that

$$v_2 = \frac{R_2}{R_2 + R_f} v_o \tag{7.2-4b}$$

because this voltage is obtained by feeding the output back through the resistive network consisting of R_2 and R_f. This simple voltage-divider relationship between v_o and v_2 applies because the input terminals of this ideal op amp are assumed to be open circuits.

To obtain the overall circuit transfer function, we substitute equations (7.2-4a) and (7.2-4b) into equation (7.2-1), which yields

$$v_o = A_{OL}\left(v_{in} - \frac{R_2}{R_2 + R_f} v_o\right) \tag{7.2-5a}$$

or

$$A_{CL} = \frac{v_o}{v_{in}} = \frac{A_{OL}}{1 + A_{OL}[R_2/(R_2 + R_f)]} \tag{7.2-5b}$$

In this equation the quantity A_{CL} is known as the amplifier's closed-loop gain, or the gain with the feedback in place, in contrast to A_{OL}, the amplifier gain without feedback. The solution as it stands in equation (7.2-5b) is not all that

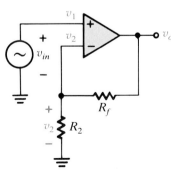

Figure 7.2-2

Noninverting op amp.

useful because it implies that A_{CL} depends strongly on the value of A_{OL}. However, if

$$A_{OL} \frac{R_2}{R_2 + R_f} \gg 1 \qquad (7.2\text{-}6a)$$

then the closed-loop gain may be approximated as

$$A_{CL} \simeq \frac{A_{OL}}{A_{OL}[R_2/(R_2 + R_f)]} = 1 + \frac{R_f}{R_2} \qquad (7.2\text{-}6b)$$

This result is extremely important because it indicates that if equation (7.2-6a) is satisfied, the amplifier closed-loop gain, to a good approximation, is determined only by the ratio of the resistors used in the feedback loop, and is essentially independent of the properties of the op amp.

Let's take a closer look at the physical meaning of the inequality given in equation (7.2-6a). Dividing both sides of this expression by $R_2/(R_2 + R_f)$, we have

$$A_{OL} \gg \frac{1}{R_2/(R_2 + R_f)} = 1 + \frac{R_f}{R_2} \qquad (7.2\text{-}7a)$$

But from equation (7.2-6b), the quantity on the right-hand side of this expression is simply equal to A_{CL}, so that for equation (7.2-6b) to be valid, we only require that

$$A_{OL} \gg A_{CL} \qquad (7.2\text{-}7b)$$

In other words, in designing a noninverting op amp circuit, the closed-loop gain of the amplifier will be independent of the actual properties of the op amp as long as the value of A_{CL} chosen is much smaller than the amplifier's open-loop gain.

EXAMPLE 7.2-1

For the noninverting amplifier illustrated in Figure 7.2-2, assume that $A_{OL} = 10^4$ and that the amplifier uses $\pm$ 10-V power supplies.

a. Calculate the exact and approximate circuit gains for R_f/R_2 equal to 100, 10^3, and 10^4.
b. If $v_{in} = 1 \sin \omega t$ volts, sketch the approximate waveshapes for v_o and v_2 for the two cases $R_f/R_2 = 4$ and $R_f/R_2 = 19$.

SOLUTION

As long as inequality (7.2-7b) is satisfied, the closed-loop gain is simply $1 + R_f/R_2$. However, the numbers in this problem were specifically chosen to illustrate that this is not always the case. For situtations where A_{OL} and A_{CL} are comparable, equation (7.2-5b) needs to be used.

The approximate and exact closed-loop gain solutions for the values of R_f/R_2 specified in part (a) of this example are tabulated in Figure 7.2-3a, and the results obtained indicate that one needs to be careful to make sure that A_{OL} is always much bigger than A_{CL} if the closed-loop gains of the op amp are to be independent of A_{OL}. Since this requires only that the designer keep A_{CL} reason-

R_f/R_2	Exact solution for A_{CL} using equation (7.2-5b)	Approximate solution for A_{CL} using $(1 + R_f/R_2)$	Percent error
10^2	99.99	101	1
10^3	909.91	1001	10
10^4	5000.25	10,001	100

(a)

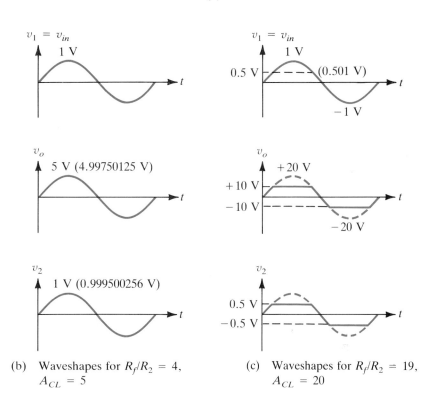

(b) Waveshapes for $R_f/R_2 = 4$, $A_{CL} = 5$

(c) Waveshapes for $R_f/R_2 = 19$, $A_{CL} = 20$

Figure 7.2-3
Example 7.2-1.

ably small, in practice this inequality is not that difficult to maintain. In particular, as illustrated by the data in the table, as long as the desired A_{CL} is 10 times smaller than A_{OL}, the actual closed-loop gain obtained will differ by less than 10% from the value predicted by equation (7.2-6b). If 1% accuracy is required, A_{OL} will need to be at least 100 times larger than the desired value of A_{CL}.

In part (b) of this example, we may use the approximate formula given in equation (7.2-6b) to compute the closed-loop gain. This equation will provide accurate results for this example because even in the second case where $R_f/R_2 = 19$ and $A_{CL} = 20$, A_{OL} is still 500 times larger than A_{CL}. Therefore, for each of the two cases to be examined, we will use equation (7.2-6b) to sketch v_o, and then equation (7.2-4b) to find v_2. In making these sketches, we need to remember that equation (7.2-6b) is valid only when the amplifier is operating in the linear region, that is, when the theoretically predicted amplitude at the op amp output terminals stays between the supply rails at ± 10 V.

For the case where $R_f/R_2 = 4$, the amplifier closed-loop gain is approximately 5 so that the waveshape for v_o is the same as that for v_{in} except that it is five times larger. The signal appearing at the inverting input terminal v_2 is obtained by applying equation (7.2-4b), from which we find that $v_2 = [R_2/(R_2 + R_f)]v_o = \frac{1}{5}v_o$. However, because $v_o = 5v_{in}$, as illustrated in Figure 7.2-3b, we also find that v_2 has almost the same waveshape as v_{in}. This result is to be expected since if the amplifier is to operate in the linear region, then, in accordance with equation (7.2-3), v_2 and v_1 must be nearly equal.

Although v_2 and v_1 must be approximately equivalent if the amplifier is to remain in the linear region, it should also be clear that v_2 cannot be identically equal to v_1 or else, in accordance with equation (7.2-1), v_o would be zero.

To determine the actual voltages that exist in this circuit when no approximations are made, we will begin by finding v_o by using equation (7.2-5b), which yields

$$v_o = \frac{10^4 v_{in}}{1 + 10^4(\frac{1}{5})} = 4.99750125 \sin \omega t \text{ volts} \qquad (7.2\text{-}8a)$$

Substituting this result into equation (7.2-4b), we find that v_2 is equal to

$$v_2 = \frac{1}{5}v_o = 0.99950025 \sin \omega t \text{ volts} \qquad (7.2\text{-}8b)$$

Thus the voltage at v_2 is slightly smaller than v_1, and is in fact just small enough so that when the difference $(v_1 - v_2)$ is multiplied by the amplifier's open-loop gain, the output voltage given in equation (7.2-8a) is obtained. To illustrate this point, we have

$$v_1 - v_2 = 1 \sin \omega t - 0.99950025 \sin \omega t$$

$$= 0.00049975 \sin \omega t \text{ volts} \qquad (7.2\text{-}9a)$$

and $\qquad\qquad v_o = A_{OL}(v_1 - v_2) = 4.9975 \sin \omega t \text{ volts} \qquad (7.2\text{-}9b)$

These exact solutions for v_o and v_2 are indicated on the sketches in Figure 7.2-3b by the terms in parentheses, and clearly the approximate solutions obtained earlier (and shown to the left of these terms in the figure) are accurate enough for most applications. These exact solutions are really needed only when we are initially trying to get an understanding of how the negative feedback operates in an operational amplifier.

When the closed-loop gain of the amplifier is increased to 20, something interesting happens because the theoretical output obtained from the amplifier exceeds the power supply rails for some values of v_{in}. For small values of v_{in}, there is no difficulty and v_o is simply equal to $20v_{in}$. However, when the input signal gets too large, the output saturates, and as a result, the upper and lower portions of the output waveform get clipped off. This effect is illustrated in Figure 7.2-3c. The dashed portion of the output waveform illustrates the waveshape that would have been obtained if there were no amplifier saturation effects. But as illustrated in Figure 7.2-1c, whenever $(v_1 - v_2)$ exceeds 1 mV, the output flattens off at either $+10$ V or -10 V, depending on the polarity of $(v_1 - v_2)$.

The input signal level at which clipping first occurs in the output may easily be determined by observing that the approximate amplifier gain is 20. Therefore, the output will just saturate when the magnitude of the input signal exceeds 10 V/20 or 0.5 V. Furthermore, because v_2 is always equal to one-twentieth of the output, it has the same waveshape as v_o and is also clipped off when the input signal amplitude exceeds 0.5 V.

Before leaving this example, it is useful to take a closer look at the value of v_{in}, where output saturation first occurs, again not because we need to know this number to a high degree of accuracy, but because a careful examination of this quantity will help us to obtain a better understanding of how this feedback amplifier actually works. To begin with, using equation (7.2-5b), we find that the actual closed-loop gain of this amplifier is not 20 but 19.96007984, and as a result output saturation actually begins when the input signal amplitude is 10 V/19.96007984 = 0.501 V, not 0.5 V as we had approximately determined. This result by itself is not that startling, but it does have some interesting implications.

According to the sketch in Figure 7.2-1c, we expect that the amplifier will just saturate at $+10$ V when $(v_1 - v_2)$ is $+1$ mV. To see that this is exactly the situation for the circuit we are currently examining, we need to observe that with the output saturated at $+10$ V, $v_2 = (\frac{1}{20})(+10)$ or $+0.5$ V. Thus with v_1 equal to 0.501 V, $(v_1 - v_2)$ is exactly equal to the required value of 1 mV.

Earlier in this section we mentioned that the approximation that $v_2 = v_1$ would prove to be the single most useful equation for rapidly analyzing linear op amp circuits. To illustrate this point, let us recompute the gain expression for the noninverting op amp given in equation (7.2-6b) by using this approximation.

If v_1 and v_2 in Figure 7.2-2 are approximately equal, then because

$$v_1 = v_{in} \tag{7.2-10a}$$

and

$$v_2 = \frac{R_2}{R_2 + R_f} v_o \tag{7.2-10b}$$

we may write

$$\frac{R_2}{R_2 + R_f} v_o = v_{in} \tag{7.2-10c}$$

or

$$v_o = \frac{R_2 + R_f}{R_2} v_{in} = \left(1 + \frac{R_f}{R_2}\right) v_{in} \tag{7.2-10d}$$

This result, which is the same as equation (7.2-6b), is valid as long as $v_1 = v_2$, or equivalently, as long as the op amp is operating in the linear region and $A_{OL} \gg A_{CL}$.

EXAMPLE 7.2-2

The operational amplifier circuit illustrated in Figure 7.2-4 is known as a voltage follower, because to a good approximation, the voltage at the output terminal follows the input. Its behavior is similar to the emitter- and source-follower circuits discussed in earlier chapters in that it has a high input impedance and a low output impedance, and as such is used primarily for buffering applications. If $A_{OL} = 10^4$, derive an expression for the exact and approximate voltage gains of this circuit.

SOLUTION

To derive an expression for the exact gain of this amplifier, we note that

$$v_2 = v_o \tag{7.2-11a}$$

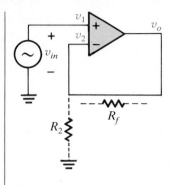

Figure 7.2-4
Example 7.2-2. Voltage follower.

because the output is connected to the inverting input terminal, and substituting this result into equation (7.2-1), we obtain

$$v_o = A_{OL}(v_1 - v_2) = A_{OL}(v_{in} - v_o) \qquad (7.2\text{-}11b)$$

so that

$$v_o = \frac{A_{OL}}{1 + A_{OL}} v_{in} = 0.99990001 v_{in} \qquad (7.2\text{-}11c)$$

We may develop the approximate expression for the gain of this amplifier by assuming that it is operating in the active or linear region, from which we have that $v_1 = v_2$. Combining this fact together with equation (7.2-11a), we therefore find that

$$v_o = v_{in} \qquad (7.2\text{-}12)$$

Thus for this circuit because the voltage v_2 follows v_1, and also because the output is directly connected to the inverting input terminal, we find, to a very high degree of accuracy, that v_o follows the applied input voltage v_{in}.

Notice that this same result could also have been obtained from the formula given in equation (7.2-10d) by substituting in the values $R_f = 0$ and $R_2 = $ infinity for these resistors. However, it is never a good idea to get too dependent on formulas, because there are many circuit configurations for which specific formulas such as that just discussed cannot be directly applied.

Besides the noninverting operational amplifier circuit we have been discussing, there is a second equally important op amp circuit configuration which we will now examine. It is illustrated in Figure 7.2-5 and is known as an inverting amplifier because the output is 180° out of phase from the input. As with the noninverting circuit examined previously, we will first calculate the gain of this amplifier by using an exact analysis method, and then afterward, for this amplifier, too, we will see that we can rapidly compute its gain by using the fact that v_1 and v_2 are approximately equal.

Referring to Figure 7.2-5a, we can see that for this circuit v_1 is zero because the noninverting input terminal is grounded. The voltage at the inverting input terminal may be determined with the aid of Figure 7.2-5b. Applying superposition to this circuit, the voltage v_2 is seen to be

$$v_2 = \frac{R_f}{R_2 + R_f} v_{in} + \frac{R_2}{R_2 + R_f} v_o \qquad (7.2\text{-}13)$$

and substituting this result into equation (7.2-1), we obtain

$$v_o = A_{OL}(v_1 - v_2) = A_{OL}\left[0 - \left(\frac{R_f}{R_2 + R_f} v_{in} + \frac{R_2}{R_2 + R_f} v_o \right) \right] \qquad (7.2\text{-}14a)$$

or

$$\frac{v_o}{v_{in}} = A_{CL} = \frac{-A_{OL}[R_f/(R_2 + R_f)]}{1 + A_{OL}[R_2/(R_2 + R_f)]} \qquad (7.2\text{-}14b)$$

As with the noninverting amplifier discussed previously, if

$$A_{OL} \frac{R_2}{R_2 + R_f} \gg 1 \qquad (7.2\text{-}15a)$$

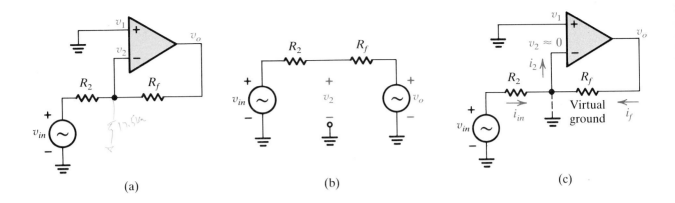

(a)

(b)

(c)

Figure 7.2-5
Inverting op amp.

or alternatively, if

$$A_{OL} \gg 1 + \frac{R_f}{R_2} \qquad (7.2\text{-}15\text{b})$$

then we may approximate the closed-loop gain of this amplifier from equation (7.2-14b) as

$$\frac{v_o}{v_{in}} = A_{CL} \simeq \frac{-A_{OL}[R_f/(R_2 + R_f)]}{A_{OL}[R_2/(R_2 + R_f)]} = -\frac{R_f}{R_2} \qquad (7.2\text{-}16)$$

The minus sign in front of this gain expression indicates that there is a 180° phase difference between v_o and v_{in}. For this result to be valid, we require that the inequality in equation (7.2-15b) be satisfied, which is the same as requiring that the open-loop gain be much greater than the magnitude of the closed-loop gain.

The result given in equation (7.2-16) may also be derived from the fact that v_1 and v_2 are nearly equal. Furthermore, a great deal of other valuable information may also be gained by using this approach.

Because v_1 is at ground potential, then since v_1 and v_2 are nearly equal, we also require that the v_2 terminal be "virtually" grounded. To be sure, this does not mean that v_2 is identically equal to zero, but instead, that in comparison to v_{in} and v_o, it is very small. As a result, because v_2 is a virtual ground, the current i_{in} flowing through R_2 in Figure 7.2-5c is approximately given by

$$i_{in} = \frac{v_{in}}{R_2} \qquad (7.2\text{-}17\text{a})$$

while that in the feedback resistor is

$$i_f = \frac{v_o}{R_f} \qquad (7.2\text{-}17\text{b})$$

The current flow i_2 into the op amp is zero because its input impedance is assumed to be infinitely large. As a result, by applying Kirchhoff's current law to the v_2 node, we have

$$i_{in} = \frac{v_{in}}{R_2} = -i_f = -\frac{v_o}{R_f} \qquad (7.2\text{-}18\text{a})$$

or

$$\frac{v_o}{v_{in}} = A_{CL} = -\frac{R_f}{R_2} \qquad \text{(7.2-18b)}$$

EXAMPLE 7.2-3

For the amplifier illustrated in Figure 7.2-5a, $A_{OL} = 10^4$, $R_f = 5$ kΩ, $R_2 = 1$ kΩ, and $v_{in} = 1 \sin \omega t$ V. Determine the exact and the approximate expressions for v_o and v_2.

SOLUTION

To the extent that we may consider the inverting input terminal to be a virtual ground, we have that

$$v_2 \simeq 0 \qquad \text{(7.2-19a)}$$

and

$$v_o = -\frac{R_f}{R_2} v_{in} = -5v_{in} = -5 \sin \omega t \text{ volts} \qquad \text{(7.2-19b)}$$

The exact solution for the output voltage v_o is found from equation (7.2-14b) to be

$$v_o = \frac{(-10^4)(\frac{5}{6})}{1 + (10^4)(\frac{1}{6})} v_{in} = -4.997001799 \sin \omega t \text{ V} \qquad \text{(7.2-20)}$$

This solution is useful because it helps us to understand how the op amp works, not because this result is significantly different from that in equation (7.2-22b). Combining this expression with equation (7.2-16), we may evaluate v_2 as

$$v_2 = \tfrac{1}{6} v_o + \tfrac{5}{6} v_{in} = -0.832833633 \sin \omega t + 0.833333333 \sin \omega t$$

$$= 0.0004997 \sin \omega t \text{ volts} \qquad \text{(7.2-21)}$$

Thus the voltage at the inverting input terminal is not identically zero. Instead, it is a small positive-going sine wave that in comparison with v_{in} or v_o is "virtually" zero but is nonetheless present and measurable at the v_2 terminal. In fact, it is precisely this voltage which, when multiplied by A_{OL}, produces the output voltage v_o. To illustrate this point, for this amplifier we have

$$v_o = A_{OL}(v_1 - v_2) = A_{OL}(0 - v_2) = -4.997 \sin \omega t \text{ volts} \qquad \text{(7.2-22)}$$

which is nearly identical to the result found in equation (7.2-20).

Exercises

7.2-1 Consider that the op amp in Figure 7.2-2 is a Norton-style amplifier. As such, its output equivalent may be modeled as a current source i_o, where $i_o = g_{OL}(v_1 - v_2)$. Given that $R_f = 10$ kΩ and $R_2 = 1$ kΩ, find the circuit voltage gain, v_o/v_{in}, for $g_{OL} = 10$ mS and 10^4 mS. *Answer* 10, 11

7.2-2 **a.** For the circuit shown in Figure 7.2-5a, $R_f = 5$ kΩ, $R_2 = 1$ kΩ, $A_{OL} = 10^4$, and a 1 kΩ resistor is connected from v_2 to ground. Find the gain v_o/v_{in}.

b. Repeat part (a) if the resistor is reduced to 1 Ω.
Answer −4.99, −3.33

7.2-3 For the circuit in Figure 7.2-5a, $R_2 = 2$ kΩ, $R_f = 50$ kΩ, and the op amp is a two-stage amplifier with a differential input stage having a gain $A_1 = 20$ and an output stage with a gain $A_2 = 10$. Find the voltage at the output of each of the stages in terms of v_{in}. *Answer* $-2.21v_{in}$, $-22.1v_{in}$

7.3   LINEAR APPLICATIONS OF THE OPERATIONAL AMPLIFIER

The low cost and design flexibility of the operational amplifier make it one of the most efficient ways to handle both linear and nonlinear signal-processing tasks in the frequency range from dc to several megahertz. At the time of this writing, microprocessors and specialized digital signal-processing chips are taking over many of the low-frequency signal-processing applications that were formally the bailiwick of the op amp. However, a good number of the analog signal processing tasks performed by the operational amplifier are just too complex and/or too time consuming to be handled by a digital signal processor. As a result, today's signal-processing systems tend to be a mixture of analog and digital hardware where the signal amplification and complex signal processing are handled by op amp circuits, and control of these subsystems, as well as some of the more-low-speed aspects of the signal processing, are taken care of by the microcomputer.

In this section and the two that follow, we will discuss some of the more important application areas of the op amp. The presentation is divided into two major parts: linear and nonlinear. For the purpose of this discussion we define a nonlinear application as one in which we employ nonlinear circuit elements (such as diodes) in the feedback or signal input paths, or as one where the output signal hits the power supply rails. Throughout this discussion we assume that the op amps being examined are ideal, and in analyzing these circuits we make extensive use of the analysis techniques developed in Section 7.2. The linear applications are presented in this section and in Section 7.4, and the nonlinear application areas of the operational amplifier are described in Section 7.5.

7.3-1   The Difference Amplifier

In the design of instrumentation systems, it is often convenient to have an amplifier whose output is proportional to the difference between two input signals. This circuit configuration is especially useful when each of the input signals is contaminated by a common-mode noise, or noise source, which is the same or common to both input signals.

A typical difference amplifier is illustrated in Figure 7.3-1a. It has two inputs and one output, and ideally, the output from this amplifier depends only on the difference between the inputs. The input signals to this amplifier have been broken down into their differential and common-mode components. The two voltages $v_{d1} = v_d/2$ and $v_{d2} = -v_d/2$ represent the signals whose difference, v_d, we are interested in amplifying, and the two voltage sources labeled v_{cm} represented a common-mode noise signal which is assumed to be present

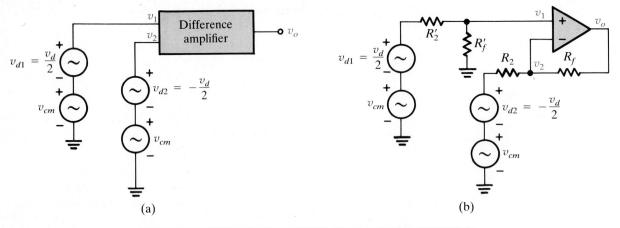

Resistor tolerance (%)	Worst case CMRR [for $R_f/R_2 = 100$]	Worst case CMRR (dB) [for $R_f/R_2 = 100$]
10	250	48
1	2,500	68
0.1	25,000	88

(c)

Figure 7.3-1
Op amp difference amplifier.

in equal amounts at both terminals of the amplifier. Typical sources of common-mode noise include ac line interference, atmospheric disturbances, power supply ripple, and radio-frequency (RF) radiation.

The function of a difference amplifier is to amplify the differential portion of the input signal while rejecting the common-mode component. The ability of a specific amplifier to accomplish this task is indicated by a figure of merit known as its common-mode rejection ratio (CMRR). Mathematically, the common-mode rejection ratio of an amplifier is defined as

$$\text{CMRR} = \frac{A_d}{A_{cm}} \qquad (7.3\text{-}1a)$$

where

$$A_d = \frac{v_o}{v_{d1} - v_{d2}} = \frac{v_o}{v_d} \qquad (7.3\text{-}1b)$$

with v_{cm} set equal to zero is the amplifier's differential gain, and

$$A_{cm} = \frac{v_o}{v_{cm}} \qquad (7.3\text{-}1c)$$

represents the common-mode gain measured with $v_{d1} = v_{d2} = 0$. Because the common-mode rejection ratio of an amplifier is frequently a large number, it is often convenient to express this quantity in decibels as

$$\text{CMRR} = 20 \log_{10} \frac{A_d}{A_{cm}} \qquad (7.3\text{-}2)$$

The circuit illustrated in Figure 7.3-1b represents a simple form of a difference amplifier. The differential gain of this op amp circuit may be found by setting $v_{cm} = 0$ in this figure. When this is done we have

$$v_1 = \frac{R_f'}{R_2' + R_f'} \frac{v_d}{2}$$

(7.3-3a)

and

$$v_2 = \frac{R_f}{R_2 + R_f} \left(\frac{-v_d}{2} \right) + \frac{R_2}{R_2 + R_f} v_o$$

(7.3-3b)

Using the fact that $v_1 = v_2$, we obtain

$$v_o = \frac{R_2 + R_f}{R_2} \frac{R_f'}{R_2' + R_f'} \frac{v_d}{2} + \frac{R_f}{R_2} \frac{v_d}{2} \simeq \frac{R_f}{R_2} v_d$$

(7.3-4a)

assuming that $R_f = R_f'$ and $R_2 = R_2'$. As a result, the differential gain of the circuit is

$$A_d = \frac{v_o}{v_d} = \frac{R_f}{R_2}$$

(7.3.4b)

By using the same analysis technique for the computation of this circuit's common-mode gain, but this time letting the two sources $v_d/2$ equal zero in Figure 7.3-1b, we can see that

$$v_o = \left(\frac{R_2 + R_f}{R_2} \frac{R_f'}{R_2' + R_f'} - \frac{R_f}{R_2} \right) v_{cm}$$

$$= \left[\left(1 + \frac{R_f}{R_2} \right) \frac{R_f'/R_2'}{1 + R_f'/R_2'} - \frac{R_f}{R_2} \right] v_{cm}$$

(7.3-5)

from which it is apparent that $A_{cm} = 0$ if the resistors are perfectly matched. As a result, if $R_f'/R_2' = R_f/R_2$, the CMRR of this amplifier is equal to infinity.

However, as a practical matter, it is extremely unlikely that the common-mode rejection of a circuit of this type will ever be anywhere near infinity because in an actual circuit it is virtually impossible to get two resistors that are identical to one another. The table in part (e) of Figure 7.3-1 illustrates the worst case CMRRs associated with different resistor tolerances. These data were computed by allowing the resistors in the amplifier in part (a) of the figure to take on their extremum values, computing A_d, A_{cm}, and the CMRR in each case. The approximate closed form mathematical expression for this amplifier's worst case CMRR can be shown to be

$$\text{CMRR} = \frac{1 + R_f/R_2}{4T}$$

(7.3-6)

where T is the resistor tolerance. This result is left as an exercise for the interested reader (see Problem 7.3-10).

These results appear to indicate that the difference amplifier that we have just discussed can be used to obtain nearly any desired common-mode rejection ratio if the resistors are carefully matched. Unfortunately, in practice, high CMRRs are difficult to achieve with this specific circuit since any signal-source output impedance will imbalance it. One way to get around this problem is to add voltage followers at the input of the difference amplifier to remove the effect of the source impedance on the circuit's CMRR. But even when this is done, the overall common-mode rejection ratio of this circuit is still limited by the matching of the differential amplifier resistors as indicated by the table in Figure 7.3-1c.

A significant improvement beyond the CMRR levels given in this table is possible by making some additional modifications to the circuit given in Figure 7.3-1b. These changes are discussed in the example that follows.

EXAMPLE 7.3-1

The circuit illustrated in Figure 7.3-2 is known as an instrumentation amplifier, and except for the addition of the buffers IC1 and IC2 and the resistors R_1 and R_3, this circuit is identical to the difference amplifier just discussed. Generally, all of the components shown within the dashed lines in this figure are contained in a single IC package, and only the resistor R_3 is added by the user to set the gain of the amplifier. Because all resistors within the IC are identical, very close matching between these components is possible and a very high common-mode rejection ratio can be achieved.

a. Show that the differential gain of this instrumentation amplifier is $(R_f/R_2)(1 + 2R_1/R_3)$.

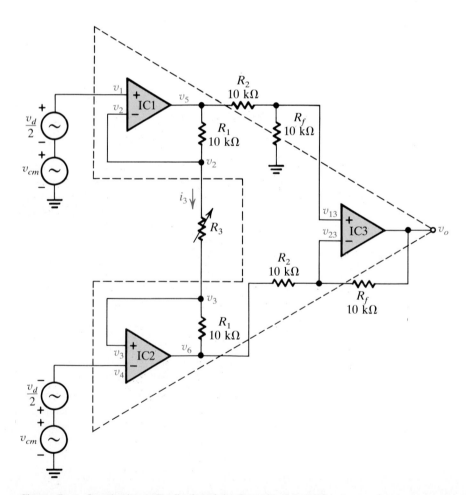

Figure 7.3-2
Instrumentation amplifier.

b. Demonstrate that the CMRR of this amplifier for a given set of resistor tolerances exceeds that of a comparable difference amplifier by the differential gain $1 + 2R_1/R_3$ of the first stage.

c. If the resistors have 0.1% tolerances and if R_3 is set at 100 Ω, determine the worst-case common-mode rejection ratio of this circuit.

SOLUTION

If we assume that all of the op amps in this instrumentation amplifier are operating in their active regions, we have

$$v_2 = v_1 = \frac{v_d}{2} + v_{cm} \tag{7.3-7a}$$

and

$$v_3 = v_4 = -\frac{v_d}{2} + v_{cm} \tag{7.3-7b}$$

so that

$$i_3 = \frac{v_2 - v_3}{R_3} = \frac{[(v_d/2) + v_{cm}] - [(-v_d/2) + v_{cm}]}{R_3} = \frac{v_d}{R_3} \tag{7.3-7c}$$

Because the input impedances of the op amps are considered to be infinite, all the current i_3 flows though the resistors labeled R_1, so that we may write

$$v_5 = v_2 + i_3 R_1 = \left(\frac{v_d}{2} + v_{cm}\right) + \frac{R_1}{R_3} v_d \tag{7.3-8a}$$

and

$$v_6 = v_3 - i_3 R_1 = \left(-\frac{v_d}{2} + v_{cm}\right) - \frac{R_1}{R_3} v_d \tag{7.3-8b}$$

Therefore, in passing through the first stage of this instrumentation amplifier, the common-mode signal has stayed at the same amplitude while the differential portion of the signal at the output of the first stage, which is equal to

$$v_5 - v_6 = \left(1 + \frac{2R_1}{R_3}\right) v_d \tag{7.3-9}$$

has increased in amplitude by the factor $1 + 2R_1/R_3$.

Using this result, together with equation (7.3-4c) for the gain of the second stage, the overall circuit differential gain of this amplifier may be written as

$$A_d = \frac{v_o}{v_d} = \underbrace{\left(1 + \frac{2R_1}{R_3}\right)}_{A_{d1}} \underbrace{\frac{R_f}{R_2}}_{A_{d2}} \tag{7.3-10}$$

Because the differential gain has increased by $(1 + 2R_1/R_3)$ while the common mode gain has remained the same, the CMRR of this instrumentation amplifier is boosted by the same factor. Thus, following equation (7.3-6), if the resistors have 0.1% tolerances the worst case CMRR of this circuit is $(1 + 200) \times (2/.004)$, or approximately 100 dB.

The analysis of this example assumed that the operational amplifiers themselves had infinite common-mode rejection ratios. Problem 7.3-11 considers the effect that op amps with finite CMRRs have on the performance of these types of difference amplifier circuits.

7.3-2 Voltage-to-Current Converter

The Thévenin equivalent at the output of the op amp circuits that we have been discussing is basically a constant-voltage source that is proportional to the applied input voltage. As such, when a load is connected to the output of this type of amplifier (Figure 7.3-3a), the current I_L that flows will depend not just on V_{in} and the circuit gain, but also on the load impedance Z_L. If the load is resistive, then as Z_L varies, the amplitude of the current will change but the waveshape will remain the same. However, when the load is complex, both the waveshape and the amplitude of the current will depend on the character of the load.

Sometimes it is desirable to produce a load current that is proportional to an applied input voltage, but is independent of the character of the load. One application area where this type of performance is useful relates to the production of linear current sweeps in magnetic deflection circuits. In a cathode ray tube (CRT) that employs magnetic deflection, a coil of wire, known as a deflection coil or a deflection yoke, is placed on the neck of the CRT as illustrated in Figure 7.3-3b. Electrons emitted by the heated cathode at the rear of the CRT are drawn to the screen by applying a high voltage, V^{++}, to the front of the tube. As the electrons strike the screen, their energy excites the phosphorescent material coating the screen and causes the emission of light.

The current flow direction indicated in the horizontal deflection coils in Figure 7.3-3b produces an upward-going magnetic field which (via $\bar{F} = q\bar{v} \times \bar{B}$) causes the electrons to be deflected to the right if i is greater than zero, and to the left if i is less than zero. To a good approximation the amount of beam deflection produced is proportional to the current that flows in the coil, and not to the voltage applied to the coil.

Figure 7.3-3

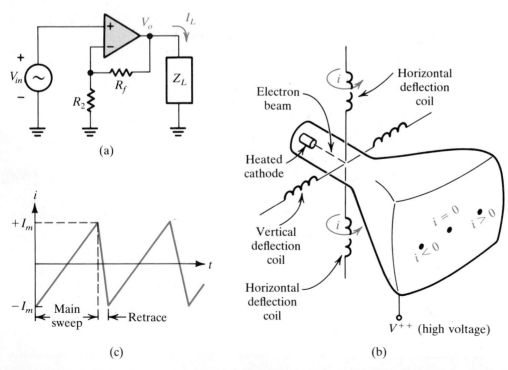

(a)

(c)

(b)

In a TV set the current waveshapes that are applied to the horizontal and vertical deflection coils of the picture tube have the basic shape illustrated in Figure 7.3-3c. This type of waveshape, applied to the horizontal deflection coils, would cause the beam to start out on the left-hand side of the screen (when i is negative) and then to sweep slowly and linearly from the left side over to the right. This constitutes the linear sweep portion of the beam deflection. Once the beam reaches the right-hand edge of the screen, the current in the coil rapidly goes from $+I_m$ to $-I_m$, and as a result, the electron beam rapidly returns or retraces back to its starting point on the left side of the screen. A similar deflection signal (at a much lower frequency) is also applied to the vertical deflection coils.

There are many different types of oscillator circuits that can produce voltage sweeps having the same basic shape as the current waveform given in Figure 7.3-3c. By connecting the output of this particular type of oscillator to a voltage-to-current converter, a complete magnetic deflection circuit may be constructed.

A simple voltage-to-current converter is shown in Figure 7.3-4a. It is basically a noninverting-style op amp circuit, and because the voltage at the node v_2 is approximately equal to that at v_1, we have

$$i_2 = \frac{v_{in}}{R_2} \tag{7.3-11}$$

Furthermore, from Kirchhoff's current law, we may also write

$$i_L = i_2 \tag{7.3-12a}$$

so that

$$i_L = \frac{v_{in}}{R_2} \tag{7.3-12b}$$

Thus to the extent that the op amp remains in the active region, the current flow into the load depends only on v_{in} and R_2 and is independent of the value of Z_L. The reason for this behavior may be understood by redrawing this circuit, as illustrated in Figure 7.3-4b.

When drawn in this way it is fairly clear that this amplifier is a current sampling–voltage series feedback circuit in which the resistor R_2 develops a voltage v_2 across it that is proportional to the current flowing in the load, and this

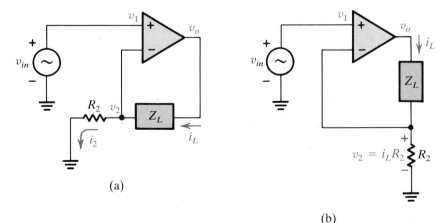

Figure 7.3-4
Op amp voltage-to-current converter.

voltage is directly fed back to the amplifier input. The net effect of the current sampling is that the current in the load is maintained constant independent of the value of Z_L.

7.3-3 Differentiators and Integrators

In the areas of instrumentation and measurement, it is often useful to have circuits that are capable of producing outputs proportional to the derivative or the integral of an input signal. For example, in respiratory physiology transducers measuring the gas flow entering and leaving a patient's lungs are often used. To obtain information about the volume of gas breathed in and out during each respiratory cycle, we can take the flow signal, which is proportional to the derivative of the volume with respect to time, and integrate this signal to obtain the desired volume information.

Circuits of the type just described can also be used to obtain information about the movement of a vehicle. For example, during the launching of a spacecraft, accelerometers on board the vehicle measure its acceleration in various directions. Integrating these signals once provides information about the velocity of the spacecraft, and a second integration enables us to track the exact position of the ship as a function of time.

When working with analog signals it is sometimes convenient to know the specific points in time where the signal reaches its maximum and minimum values. Because these two points occur where the slope of the waveshape is zero, they may be found by differentiating the signal electronically and then passing this derivative function through a comparator or zero-crossing detector (Figure 7.3-5a). Each time the comparator switches, the original waveshape has just passed through a point of zero slope, and the direction of the transition at the comparator output determines whether this transition corresponds to a maximum or a minimum value of the original signal (Figure 7.3-5b).

As we have seen from even these few examples, circuits that perform the functions of integration and differentiation can be extremely useful in basic signal-processing applications. Unfortunately, however, the practical implementation of circuits performing these functions each have their own drawbacks. As engineers it is important for us to have a clear understanding of these limitations.

Differentiators tend to enhance or magnify the presence of high-frequency noise because their "gain" is proportional to the frequency of the incoming signal. Therefore, in the practical implementation of differentiator circuits, it is a good idea to provide some filtering to reduce the gain of these circuits at high frequencies. Furthermore, differentiators tend to be unstable, and as a result, they are very prone to oscillate. This subject is discussed in great length in Chapter 8 and at that time suggestions are offered on ways to prevent oscillations in these circuits.

Integrator circuits are more stable than differentiators, and as such are more popular in practical circuit designs. However, they, too, have their problems. Integrators treat op amp offset voltages and currents (Section 7.7) as dc, or constant, input signals. Of course, the time integral of a constant is equal to the constant multiplied by t. Thus if we watch the output of an op amp integrator when it is first turned on, we will see that it linearly increases toward plus or minus infinity, continuing until the amplifier saturates at one of the power supply rails. Depending on the size of the offsets in the op amp and on the size of the integrator time constant, the time required for the output to reach the supply rails can be anywhere from a few microseconds to several seconds, after which time the circuit is useless as an integrator. Therefore, as a practical matter, it is necessary to reset the integrator periodically to prevent these offset effects from creating large errors in the output.

The circuit shown in Figure 7.3-6a is an op amp differentiator. It is basically an inverting amplifier arrangement, and as such, the voltage at the inverting input terminal is a virtual ground. Therefore, using Laplace transform notation, we may write

$$I_{in} = \frac{V_{in}}{1/sC} = sCV_{in} \tag{7.3-13a}$$

and

$$I_f = \frac{V_o}{R} \tag{7.3-13b}$$

But

$$I_{in} + I_f = 0 \tag{7.3-13c}$$

so that

$$\frac{V_o}{R} = -sCV_{in} \tag{7.3-13d}$$

or

$$V_o = -sRCV_{in} \tag{7.3-13e}$$

If we recall that multiplication by s in the "Laplace domain" is equivalent to differentiation with respect to t in the "time domain," equation (7.3-13e) may alternately be written as

$$v_o(t) = -RC\frac{dv_{in}}{dt} \tag{7.3-14}$$

$\frac{dv}{dr} = m = \phi \quad \text{Max Point}$

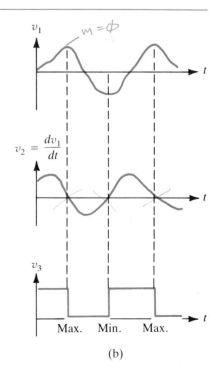

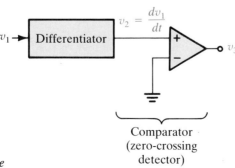

Figure 7.3-5
Use of a differentiator to determine the location of the maximum and minimum points on a signal.

Comparator
(zero-crossing
detector)

(a)

(b)

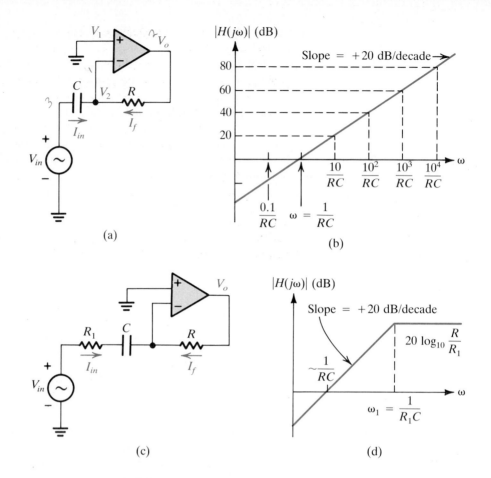

Figure 7.3-6
Op amp differentiator.

(a)

(b)

(c)

(d)

Thus the output of the op amp in Figure 7.3-6a is proportional to the derivative of the applied input signal.

As we mentioned earlier, the process of differentiation degrades the signal-to-noise ratio when the noise is some type of high-frequency interference that has gotten mixed in with the signal we are interested in differentiating. To illustrate why this occurs, we can see from equation (7.3-13e) that the transfer function of this circuit may be written as

$$H(j\omega) = \frac{V_o}{V_{in}}\bigg|_{s=j\omega} = -j\omega RC \qquad (7.3\text{-}15)$$

∪ N D ∈ R S T N ʌ N D   B O D ∈   P L ʌ T S .

The magnitude of this transfer function versus frequency is sketched in Figure 7.3-6b, and from it we can see that the effective gain of this particular differentiator is unity (or 0 dB) at $\omega = 1/RC$, and that it increases by 20 dB for each decade increase in frequency. Thus a noise signal whose frequency is 10^3 times higher than the signal we are trying to differentiate would be "amplified" 1000 times more than the main signal, and this would reduce the signal-to-noise ratio at the output by the same factor.

One way to avoid this degradation in signal-to-noise level at the differentiator output is to reduce the gain of this circuit at high frequencies. This may be accomplished by adding a small resistor in series with the capacitor as illustrated in Figure 7.3-6c. In the high-frequency region where the impedance of the capacitor is much less than R_1, the approximate gain of this circuit tapers off at $-R/R_1$ and no longer increases uniformly with frequency.

The addition of the resistor R_1 in series with the capacitor goes a long way toward preventing high-frequency noise from destroying the signal-to-noise ratio at the output of a differentiator. An additional improvement in the signal-to-noise ratio of this differentiator may be obtained by attenuating the gain of the amplifier at high frequencies rather than just allowing it to flatten out as in the circuit just discussed. This gain reduction may be accomplished by adding in a second capacitor in parallel with the feedback resistor. At high frequencies this capacitor shorts out the resistor, and the effective gain of the amplifier $(-Z_f/Z_{in})$ goes to zero.

The circuit shown in Figure 7.3-7 is an elementary integrator. It is similar to the differentiator analyzed previously except that the position of the R and C elements are interchanged. For this circuit

Figure 7.3-7
Op amp integrator.

$$I_{in} = \frac{V_{in}}{R} \qquad (7.3\text{-}16a)$$

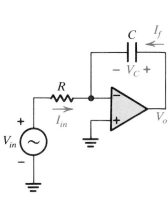

(a) Basic Integrator

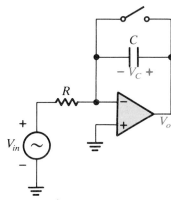

(b) Integrator with Switch
     Reset Capability

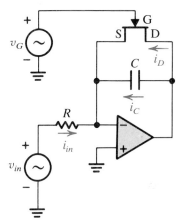

(c) Integrator with Electronic
     Reset Capability

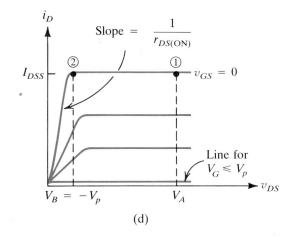

(d)

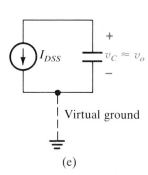

(e)

and
$$I_f = \frac{V_o}{1/sC} = sCV_o \qquad (7.3\text{-}16\text{b})$$

so that applying Kirchhoff's law, we have
$$I_f = -I_{in} \qquad (7.3\text{-}16\text{c})$$

or
$$V_o = -\frac{1}{sRC}V_{in} \qquad (7.3\text{-}16\text{d})$$

Taking the inverse Laplace transform of this expression, and recalling that division by s in the Laplace domain is equivalent to integration with respect to t in the time domain, we obtain

$$v_o(t) = -\frac{1}{RC}\int_0^t v_{in}(\tau)\,d\tau + \text{constant} \qquad (7.3\text{-}17)$$

where the constant represents the initial voltage on the capacitor, v_C, at $t = 0-$.

For most applications we want to ensure that the voltage on the capacitor is zero at $t = 0-$. This initial condition may be achieved by connecting a switch in parallel with the feedback capacitor as shown in Figure 7.3-7b. When this switch is closed, the capacitor discharges immediately making $v_C = 0$, and because the effective gain of this circuit $(-R_f/R)$ is zero as well, the output voltage will stay at zero volts regardless of the values taken on by v_{in}. When the switch is opened at $t = 0+$, integration begins and the output voltage at any time t is found by applying equation (7.3-17) with the constant set equal to zero.

The circuit given in Figure 7.3-7c is a practical implementation of an integrator having electronic reset capability. In this circuit the JFET acts as an electronically controlled switch. To understand the operation of this reset circuit, we must first realize that the source of the FET is essentially at ground potential because of the virtual ground character of the inverting input terminal of the op amp. As a result, if the control voltage v_G on the gate of the FET is maintained well below pinch-off, the FET will be cut off (the switch will be open) and the circuit will behave like a conventional integrator.

Let us suppose, after operating for some time, that the integrator output voltage has reached a level of $+V_A$ volts, and that we now want to reset the integrator output to zero volts. To do this we raise the control voltage on the gate of the FET to zero. This makes $v_{GS} = 0$ and places the Q point of the FET on the $v_{GS} = 0$ line at the point where $v_{DS} = V_A$. This is indicated by point 1 on the FET volt-ampere characteristics in Figure 7.3-7d. When operating in this region, because v_{GS} is being held at zero volts, the FET may be modeled by a constant-current source I_{DSS}. Once the FET turns on, the capacitor current is given by

$$i_C = -i_D - i_{in} = -I_{DSS} - i_{in} \approx -I_{DSS} \qquad (7.3\text{-}18)$$

if I_{DSS} is considered to be much greater than i_{in}. As a result, the capacitor now begins to discharge in accordance with the approximate equivalent circuit shown in Figure 7.3-7e. If we consider that point 2 in Figure 7.3-7d is nearly at zero volts, the approximate time required for the FET operating point to traverse from point 1 to point 2, and consequently, the time required to reset this integrator may be calculated from the expression

$$i_C = C\frac{dv_o}{dt} \tag{7.3-19a}$$

or

$$-I_{DSS} = C\frac{\Delta v_o}{\Delta t} \tag{7.3-19b}$$

Solving this equation for Δt, we find that the integrator reset time is approximately given by

$$\Delta t = -\frac{C\Delta v_o}{I_{DSS}} = \frac{(-C)(-V_A)}{I_{DSS}} = \frac{CV_A}{I_{DSS}} \tag{7.3-19c}$$

where V_A is the initial voltage on the capacitor just prior to the application of the reset (or control) pulse to the FET.

If the integrator uses an ideal mechanical switch as in Figure 7.3-7b, it can be reset in nearly zero time, whereas when a FET is used in place of the switch, a finite time is required to discharge the capacitor. By representing the FET by a constant-current source during the entire discharge, the output voltage decreases linearly with the time until v_{DS} and hence v_o finally reaches zero. The resulting calculation for the discharge time given in equation (7.3-19c) is accurate enough for most purposes. If a closer approximation to the actual discharge time is needed, the constant-current source discharge model can be used until v_{DS} drops to V_B in Figure 7.3-7d, and then the transistor may be replaced a fixed resistor $r_{DS(ON)}$ (equal to the reciprocal of the slope of the i_D versus v_{DS} curve in the ohmic region) to complete the discharge to zero.

7.3-4 Sample-and-Hold Amplifiers

A sample-and-hold (S/H) amplifier is essentially an analog memory circuit in which a voltage is temporarily stored on a capacitor. Storage times can vary anywhere from nanoseconds to minutes depending on the application. Sample-and-hold circuits are used with analog-to-digital (A/D) converters to hold the input signal constant during the conversion period, with digital-to-analog (D/A) converters for "deglitching" purposes, and for signal reconstruction in analog signal-processing and data transmission systems.

The basic form of a sample-and-hold circuit is illustrated in Figure 7.3-8a. When a sample of the input signal is desired, the switch is closed for a short period and the capacitor rapidly charges to the same voltage as the input. When the sample command is removed, the switch opens and the capacitor "holds" the signal level sampled. Figure 7.3-8b illustrates the output obtained from this type of circuit when the input signal is a sine wave and the switch control voltage is a periodic train of pulses. In this case the output signal is a staircase approximation to the original input sine wave, and the accuracy of this approximation increases with the frequency of sampling.

The idealized sample-and-hold circuit illustrated in Figure 7.3-8a has several practical problems associated with it. If the source, v_{in}, has a high output impedance, it will be difficult for the capacitor to charge completely during the

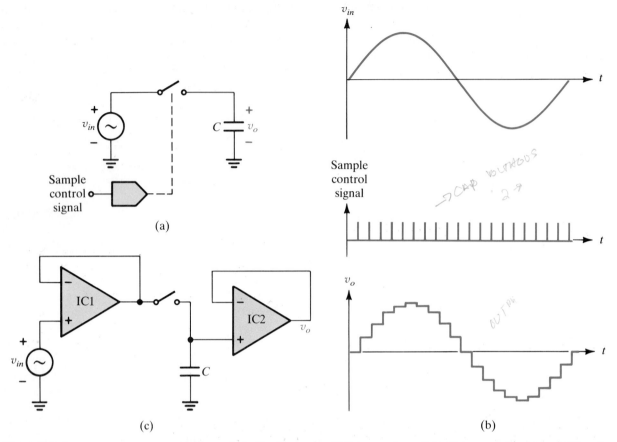

Figure 7.3-8

sample period. Furthermore, if the sample-and-hold is connected to additional circuitry, the input impedance of this circuitry will cause the voltage on the capacitor to leak off. Both of these problems can be solved by buffering the input and the output of the sample-and-hold as illustrated in Figure 7.3-8c. IC1 ensures that the source impedance has little influence on the sample time required by the circuit, and IC2 prevents output loading from discharging the hold capacitor.

To illustrate how sample-and-hold circuits can be used to advantage in digitally based analog signal-processing systems, consider the simple system illustrated in Figure 7.3-9a which consists of an A/D converter connected to a D/A converter. This circuit functions as follows. The system clock generates a periodic train of pulses, and on each pulse an A/D conversion is performed. At the end of each conversion cycle the A/D converter produces a binary code proportional to the amplitude of the original analog input signal, and the D/A converter transforms this code back into an analog voltage. This process is repeated with each tick of the system clock and results in the generation of a staircase approximation to the original waveform (Figure 7.3-9d). The accuracy of this reconstruction depends on the number of bits (or step levels) in the A/D converter, the time required to complete each A/D conversion, and the frequency content of the original signal.

For an A/D converter to function properly, it is necessary that the input to it remain relatively constant during the conversion process, and this requirement imposes a serious constraint on the maximum allowable frequency which may be applied to the system. Consider, for example, that the system illustrated in Figure 7.3-9a uses a 10-bit A/D converter with a 10-μs conversion time, and let us determine the highest frequency that may be applied to this system without introducing any errors other than those already inherent in the quantization process. A 10-bit converter quantizes an incoming signal to an accuracy of $1/2^{10}$, or about 1 part in 1000. If this converter is to maintain its full 10-bit accuracy, it is essential that the input signal amplitude change by less than 1 part in 1000 during the 10-μs conversion period. This requirement severely restricts the maximum allowed signal frequency that can be applied to the converter.

To illustrate this, let us assume that the input to the A/D converter is a sine wave,

$$v_{in} = V_1 \sin \omega t \qquad (7.3\text{-}20a)$$

For this case the rate of change of v_{in} with respect to time is

$$\frac{\Delta v_{in}}{\Delta t} \simeq \frac{dv_{in}}{dt} = \omega V_1 \cos \omega t \qquad (7.3\text{-}20b)$$

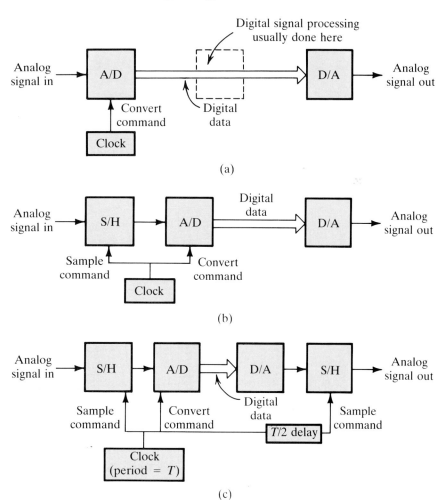

Figure 7.3-9

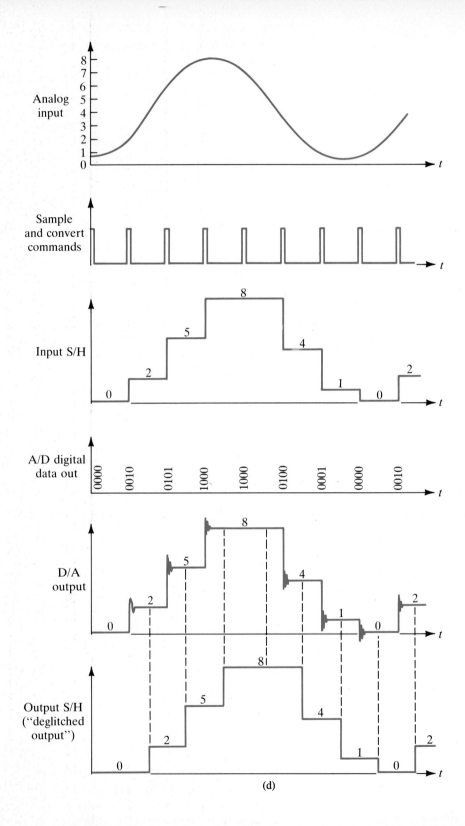

Figure 7.3-9 (Continued)

(d)

and the maximum of this function (which occurs at the zero crossings of the original sine wave) is

$$\frac{\Delta v_{in}}{\Delta t} = \omega V_1 = 2\pi f V_1 \qquad (7.3\text{-}21)$$

If v_{in} is to change by less than 1 part in 1000 during the 10-μs conversion time of the A/D converter, then following equation (7.3-21), the frequency of the input signal must be less than

$$f = \frac{\Delta v_{in}/V_1}{2\pi \, \Delta t} = \frac{0.001}{2\pi \times 10^{-5} \, s} = 16 \, Hz \qquad (7.3\text{-}22)$$

The addition of a sample-and-hold circuit in front of the A/D converter (Figure 7.3-9b) can substantially improve the system's high-frequency performance because now the input to the converter is constant all during the conversion process. As a result, the high-frequency limit is no longer determined by equation (7.3-22), but now depends instead on the constraints imposed by the sampling theorem.

According to the sampling theorem, if a signal is sampled at a speed greater than twice the highest-frequency component present in the signal, then in theory it is possible to reconstruct the original signal from these samples. For the case of a sine wave, this requires that the sampling speed by greater than or equal to twice its frequency. In other words, we must sample the sine wave at least two times during each cycle. Because the converter requires 10 μs to carry out a single A/D conversion, the maximum rate at which we can sample is once every 10 μs or 100,000 times per second, and therefore the maximum allowable input sine-wave frequency is 50,000 cycles per second, or 50 kHz. In practice, frequencies of only about one-fourth to one-half of this rate can actually be handled. However, this still represents nearly a 1000-fold improvement in high-frequency response. No wonder then that many of the higher-performance A/D converters include a sample-and-hold circuit right in the front end of the device.

Sample-and-hold circuits are also useful at the output side of an analog data processing system in order to remove "glitches" or transients that occur in the output of the D/A converter. These glitches at the edges of each of the sample transitions are caused by time delays that exist between the bits of the digital data, and also by unequal delays in the switches of the D/A converter. Most of these glitches can be removed by adding an additional sample-and-hold at the output of the D/A converter with the sample command timed to occur right in the middle of each of the original samples. In this way when these samples are taken, the glitches will have died out completely. A block diagram of this system is shown in Figure 7.3-9c, and its output waveform is illustrated at the bottom of Figure 7.3-9d.

7.3-5 Analog Signal Switching

In the switching of analog signals, the use of mechanical switches or relays is often unsatisfactory because of the switching speeds required, and in these cases some type of solid-state switch is needed. Although the BJT transistor makes a good power switch, it has several drawbacks for low-level analog

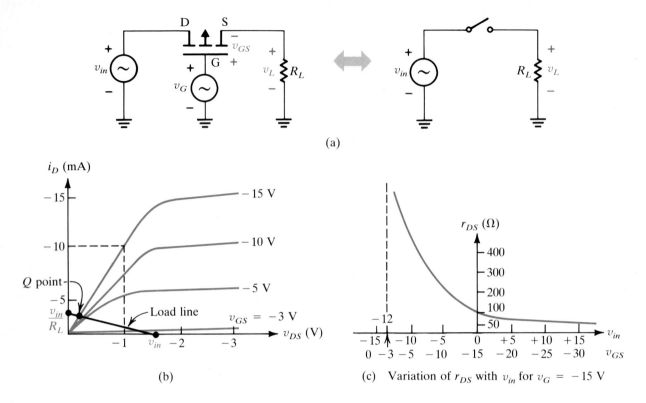

(a)

(b)

(c)   Variation of r_{DS} with v_{in} for $v_G = -15$ V

Figure 7.3-10
MOSFET analog transmission gate.

switching applications. Bipolar junction transistors function as switches by operating between saturation and cutoff, and when saturated, because their saturation voltage is nonzero, they introduce a dc offset into the output. In addition, because they have little isolation between input and output, the switch controlling voltage (usually applied to the base) feeds through into the output. FET devices have neither of these disadvantages.

The p-channel MOSFET circuit given in Figure 7.3-10a illustrates the basic form of an FET analog switch. In this circuit, v_G represents the switch control voltage, and v_{in}, the signal voltage. We assume that a control voltage of $+15$ V is used to open the switch, and -15 V to close the switch and pass the input signal through to the output. The transistor characteristics are given in Figure 7.3-10b and indicate that it is an enhancement-style MOSFET device with a threshold voltage of -3 V.

When the switch is ON, if R_L is assumed to be very large in comparison to the channel resistance, the FET is operating in the ohmic region as illustrated in Figure 7.3-10b. Therefore, the transistor may be represented by a resistance, r_{DS}, which depends on v_{GS} and which is equal to the reciprocal of the slope of the characteristic curve in the region of the Q point. For $v_{GS} = -15$ V, r_{DS} is approximately (1 V)/(10 mA) or 100 Ω, while for $v_{GS} = -5$ V it is about 200 Ω, and when v_{GS} is -3 V, it equals infinity. This resistance variation with changes in v_{GS} is important to understand because v_{GS} depends on v_{in}, and as a result, r_{DS} is a function of the input signal level (Figure 7.3-14c).

By connecting an n-channel FET in parallel with this p-channel device, we can reduce the variation of the switch resistance with signal level. The circuit that results is illustrated in Figure 7.3-11a; it is known as a CMOS or comple-

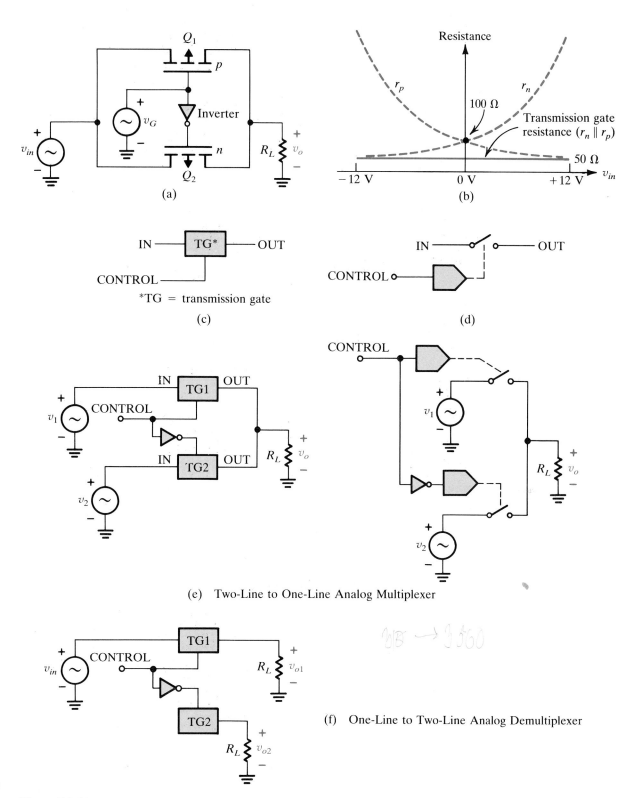

(a)

(b)

(c)

*TG = transmission gate

(d)

(e) Two-Line to One-Line Analog Multiplexer

(f) One-Line to Two-Line Analog Demultiplexer

Figure 7.3-11

405

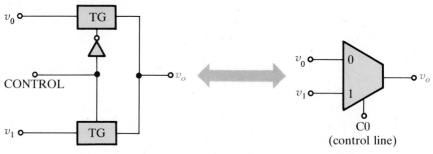

(a)   Two-Line to One-Line Analog Signal Multiplexer

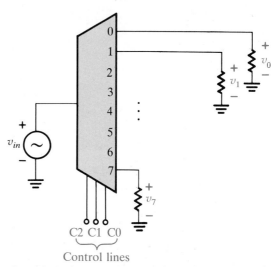

(b)   One-Line to Eight-Line Analog Signal Demultiplexer

Figure 7.3-12

Schematic notation for analog signal multiplexers and demultiplexers.

mentary MOS analog transmission gate. The inverter in this circuit ensures that with $v_G = -15$ V, both transistors are ON, while with $v_G = +15$ V, both are OFF. The resistance variation of each of the transistors as well as that of the overall transmission gate is illustrated in Figure 7.3-11b. To a good approximation the gate resistance of this switch is nearly constant at about 50 Ω independent of the signal level. The ON resistance of commercially available analog transmission gates varies from about 50 to 750 Ω, and two of the more popular schematic symbols for these devices are given in Figure 7.3-11c and d.

Transmission gates may be combined together to form analog signal multiplexers and demultiplexers. An analog signal multiplexer is a circuit that allows any one of several input signals to be connected to the output. In this way it functions like a multiposition switch. Generally, the switch position is determined by one or more digital control signals. A circuit simulating an electronically controlled SPDT (single-pole double-throw) switch is illustrated in Figure 7.3-11e.

If the FETs making up the analog transmission gates have symmetrical drains and sources, the roles of the input and output leads can be interchanged, and the same SPDT switch in Figure 7.3-11e can also be used as a signal demultiplexer. This circuit configuration is illustrated in Figure 7.3-11f. In a signal demultiplexer a single input signal can be distributed to one of several outputs. Analog signal multiplexers and demultiplexers are available that can handle up to 16 inputs or outputs.

A more compact notation for analog signal multiplexers and demultiplexers is illustrated in Figure 7.3-12. The schematic symbol for the SPDT multiplexer gate just discussed is given in Figure 7.3-12a. The control line, C0, is a digital control signal. When C0 is at a logic zero, v_0 is connected to v_o, and with C0 at a logic 1, v_1 is connected to v_o.

The schematic symbol for a one-line to eight-line signal demultiplexer is shown in Figure 7.3-16b. It has three digital control lines, and the binary code connected to these lines determines the switch position of the demultiplexer. For example, with C2, C1, and C0 all set to a logic 1, the voltage v_{in} will be connected to output 7 $[(111)_2]$.

7.3-6 Digital-to-Analog and Analog-to-Digital Converters

A digital-to-analog (D/A) converter is a circuit designed to take a digital or binary signal and convert it into a proportional analog voltage. For example, for an 8-bit D/A converter, if the binary code 11111111 produces an analog output voltage of +2.55 V, while a binary code of 00000000 produces an output of zero volts, a binary code of 10000000 should produce an analog output of +1.28 V.

Figure 7.3-13a illustrates the construction of an R-2R style D/A converter. In this circuit the position of the switches is determined by the binary digital logic signals b_0 through b_{n-1} applied to the converter. The coefficient b_0 is the least significant bit (LSB) in the binary number being converted, and b_{n-1} represents the most significant bit (MSB). In analyzing the operation of this circuit, the most important fact to observe is that the impedance to the left and to the right of any node in the circuit is always equal to $2R$. As a result, the equivalent circuit at node $(n - 1)$, for example, [considering only the current contribution from bit number $(n - 1)$] is that given in Figure 7.3-13b, and the current in this branch is

$$i_{n-1} = \frac{V_{BB}}{3R} b_{n-1} \tag{7.3-23}$$

This current splits evenly to the left- and right-hand sides of the node so that the current component of i_f due to bit b_{n-1} is $\left(\frac{V_{BB}}{3R} \frac{b_{n-1}}{2^1} \right)$. In a similar fashion the current contribution from bit b_{n-2} is $\left(\frac{V_{BB}}{3R} \frac{b_{n-2}}{2^2} \right)$, while that from the least significant bit b_0, which undergoes n such $2:1$ current divisions, is $\left(\frac{V_{BB}}{3R} \frac{b_0}{2^n} \right)$.

Applying superposition to combine the effects from all the digital bit val-

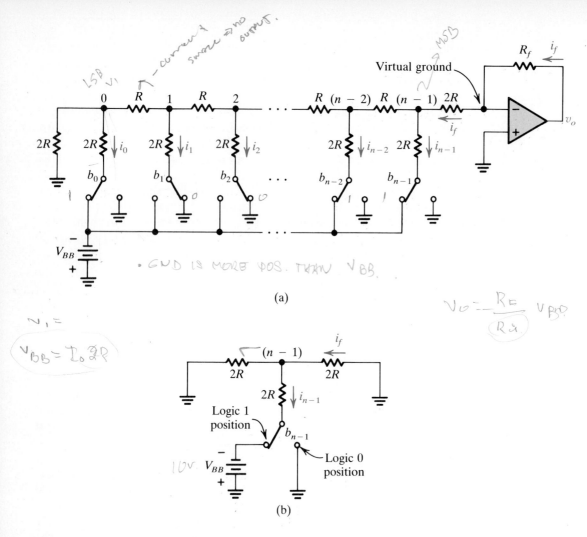

(a)

(b)

Figure 7.3-13
*R-2R resistor ladder D/A
converter.*

ues, we may write the overall output from the op amp as

$$v_o = i_f R_f = \frac{R_f}{3R} V_{BB} \left(\frac{b_{n-1}}{2^1} + \frac{b_{n-2}}{2^2} + \cdots + \frac{b_1}{2^{n-1}} + \frac{b_0}{2^n} \right) \qquad (7.3\text{-}24a)$$

which after bringing all terms under a common denominator yields

$$v_o = \frac{R_f}{3R} V_{BB} \frac{\text{decimal equivalent of } n\text{-bit binary code}}{2^n} \qquad (7.3\text{-}24b)$$

An analog-to-digital converter performs the opposite function from a D/A converter. In particular, an n-bit analog-to-digital converter (Figure 7.3-14a) is a circuit that has a single analog input line and n binary or digital output lines. It generates a binary code at the output proportional to the input analog voltage. Many different types of analog-to-digital converters have been developed over the years, but for the sake of brevity, we consider only two of the most popular types: the ramp or staircase-style A/D converter and the successive-approximation-style A/D converter.

The block diagram of a ramp-type A/D converter is depicted in Figure 7.3-14b. In analyzing this circuit's operation we will assume, prior to the start of

Chapter 7 / Introduction to Feedback and the Operational Amplifier

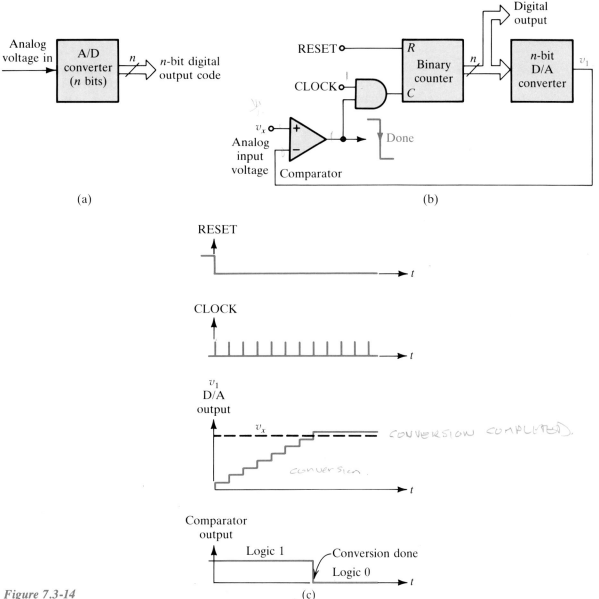

RESET

CLOCK

v_1
D/A
output

v_x

CONVERSION COMPLETED.

conversion.

Comparator
output

Logic 1

Conversion done

Logic 0

(a)

(b)

(c)

Figure 7.3-14

the A/D conversion process, that the binary counter is held in the reset mode, so that v_1 is initially zero volts. As a result, v_x, the voltage to be converted (which is assumed to be positive) is greater than v_1, and the comparator output is a logic 1. This enables the AND gate and allows the clock signal to enter the counter. The reset input to the counter is assumed to override the clock so that counting does not occur as long as the reset signal is present.

At $t = 0$ (Figure 7.3-14c) the reset signal is removed from the counter, and this allows the A/D conversion process to begin. On each tick of the clock, the counter increments and the output from the D/A converter increases. Eventually, when the count gets high enough, the analog voltage at the output of the D/A converter will exceed the input voltage v_x. When this occurs, the comparator output will switch from a logic 1 to a logic 0, inhibiting further counting and indicating that the conversion process is completed. At this point the number stored in the counter is the binary or digital equivalent of the analog input voltage.

This type of A/D conversion process is very inefficient because it can take up to 2^n clock cycles to carry out an n-bit A/D conversion. There is a much more rapid way to accomplish this conversion process, and to understand how this may be done, it is helpful to think of an A/D converter as being an automated number-guessing circuit. Viewed in this way, the job of an 8-bit A/D converter, for example, is to figure out which of the 256 possible binary output

Figure 7.3-15
Successive-approximation A/D conversion.

Question number	Question	Subtle question	Answer	Partial code produced $b_7\ b_6\ b_5\ b_4\ b_3\ b_2\ b_1\ b_0$
1	Is it ≥ 128	Is $b_7 = 1$?	Yes	1 _ _ _ _ _ _ _
2	Is it $\geq 128 + 64 = 192$	Is $b_6 = 1$?	No	1 0 _ _ _ _ _ _
3	Is it $\geq 128 + 32 = 160$	Is $b_5 = 1$?	No	1 0 0 _ _ _ _ _
4	Is it $\geq 128 + 16 = 144$	Is $b_4 = 1$?	Yes	1 0 0 1 _ _ _ _
5	Is it $\geq 144 + 8 = 152$	Is $b_3 = 1$?	Yes	1 0 0 1 1 _ _ _
6	Is it $\geq 152 + 4 = 156$	Is $b_2 = 1$?	Yes	1 0 0 1 1 1 _ _
7	Is it $\geq 156 + 2 = 158$	Is $b_1 = 1$?	No	1 0 0 1 1 1 0 _
8	Is it $\geq 156 + 1 = 157$	Is $b_0 = 1$?	Yes	1 0 0 1 1 1 0 1

(a)

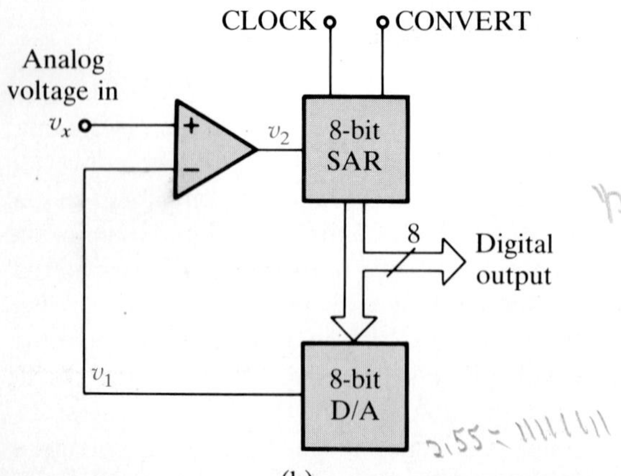

(b)

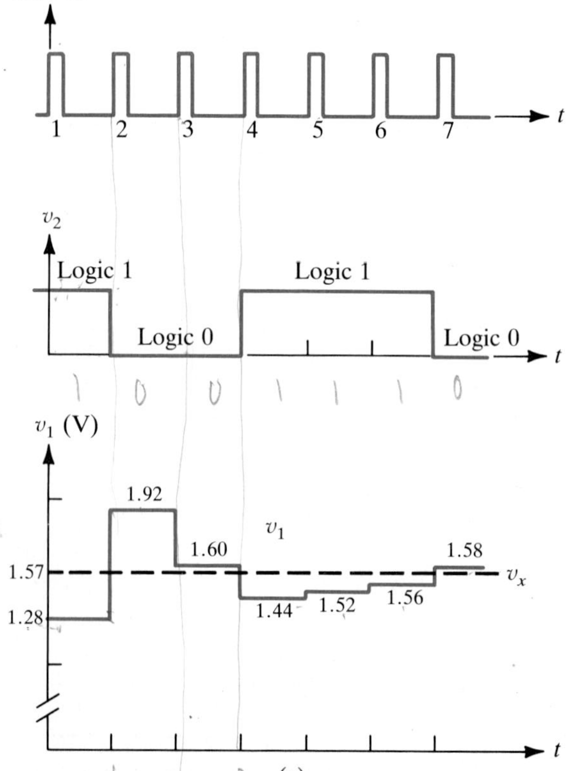

(c)

numbers is the equivalent of the analog input voltage. The ramp-style A/D converter in effect determines this unknown number by asking the following series of questions in sequence: Is the number zero? Is the number one? Is the number two? and so on. It keeps on going until it gets a match at the comparator output. For example, if the final number found were $(157)_{10}$, then 157 clock cycles or guesses would be required to carry out this conversion. Clearly, this is not the most efficient way to proceed.

To find an unknown number between 0 and 255 with the smallest number of guesses, the best approach is first to guess a number halfway up, and determine whether the unknown number is above or below this point. Depending on the answer, we divide in half the upper or lower range that remains, and then ask whether the number is above or below this point. This series of questions is continued until the unknown number is found. As illustrated in the table in Figure 7.3-15a, asking the question "Is the unknown number greater than 128?", for example, is equivalent to the more subtle question "Is the MSB equal to 1?". The other questions have similar meanings.

This technique is known as successive approximation and it requires only n clock cycles to accomplish an n-bit A/D conversion. As a result, an 8-bit successive-approximation A/D converter, for example, is 32 times faster than an equivalent ramp-style A/D converter, while the hardware required for both is comparable (Figure 7.3-15b). The major difference between the two circuits is that the binary counter in the ramp-style A/D circuit has been replaced by a successive-approximation register (SAR).

This circuit operates as follows. On the first clock pulse the MSB (b_7) of the SAR is set equal to 1. If the comparator output stays high (as it does for the case under consideration in the figure), this bit is kept at a 1; if the comparator output goes low, the MSB is reset to zero by the SAR hardware. On the next clock pulse b_6 is set equal to 1, and as before, if the comparator output remains high, this bit is also kept at a logic 1. This procedure is repeated six more times by the SAR hardware, and after a total of eight clock cycles, this 8-bit A/D conversion is completed. The important circuit waveforms during the first few clock cycles are shown in Figure 7.3-15c, assuming that the analog input voltage is 1.57 V and that the D/A output voltage is 2.55 V when a count of $(255)_{10}$ is in the SAR.

Exercises

7.3-1 Find the CMRR of the subtractor circuit given in Figure 7.3-1b if $R_f = 100$ kΩ, $R_f' = 101$ kΩ, and $R_2 = R_2' = 1$ kΩ. (Hint: Do not use formulas. Instead, just calculate A_d and A_{cm} directly from the circuit.) **Answer** 80.2 dB

7.3-2 Derive an expression for the output from the D/A converter in Figure 7.3-13 if the resistor $2R$ at the input to the op amp is replaced by a short circuit. **Answer** $v_o = (R_f/R)V_{BB}$ [decimal equivalent of the n-bit binary code]$/2^n$

7.3-3 For the circuit shown in Figure 7.3-7a, $R = 500$ kΩ, $C = 1$ μF, and $v_{in} = 2 \sin 4\pi t$ V. Find v_o at $t = 2$ s. **Answer** 0.318 V

7.4 ACTIVE FILTERS

A filter may be defined as a circuit that is designed to pass a certain band of frequencies while attenuating all others. The range of frequencies allowed through the filter is called its passband, and the range attenuated is known as its stopband. There are four major filter categories: low-pass, high-pass, band-pass, and band-elimination (or band-reject) filters. The magnitude response versus frequency for each of these filter styles is sketched in Figure 7.4-1. The phase response for each of these filters is assumed to be zero.

The filter frequency response curves illustrated in Figure 7.4-1 are all ideal, and no physically realizable filter can have precisely these characteristics, but by careful design it is possible to approach many of their attributes. Consider, for example, the design of a low-pass filter having characteristics similar to those illustrated in Figure 7.4-1a. Specifically, this filter has a flat gain response and zero phase shift in the passband, zero gain in the stopband, and an instantaneous transition from the passband to the stopband at the cutoff frequency, f_c. If this filter is constructed with lumped-circuit elements, its transfer function will be of the form

$$H(s) = \frac{(a_0)A_o}{s^n + a_{n-1}s^{n-1} + \cdots + a_3s^3 + a_2s^2 + a_1s^1 + a_0} \qquad (7.4\text{-}1)$$

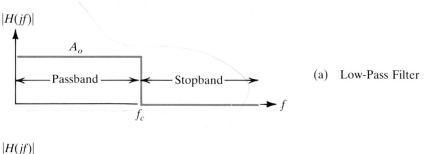

(a)   Low-Pass Filter

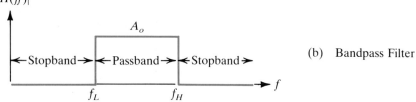

(b)   Bandpass Filter

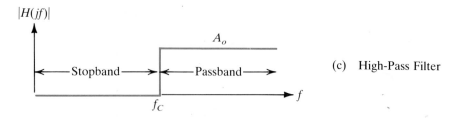

(c)   High-Pass Filter

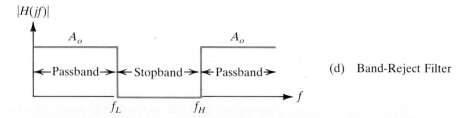

(d)   Band-Reject Filter

Figure 7.4-1
Basic filter categories.

where A_o is the low-frequency gain of the filter. The values selected for the denominator polynomial coefficients a_0, a_1, a_2, and so on, determine the shape of the filter's gain–phase characteristics and also the transient response of the filter. Unfortunately, no single set of polynomial coefficients in this equation can simultaneously optimize all the desired filter parameters, and therefore it is necessary at the outset of any filter design problem to determine which of these characteristics is most important for the problem at hand.

7.4-1 Basic Filter Types

The Butterworth low-pass filter design is one in which the coefficients of the transfer function are selected to produce a flat gain response in the passband. The magnitude and phase characteristics of this filter are sketched in Figure 7.4-2a as a function of n, the order (or number of poles) of the filter. As illustrated in the figure, as n increases, the response in the passband becomes flatter, and in addition the gain outside the passband falls off more rapidly above the cutoff frequency (at $-20n$ dB/decade). The coefficients of the denominator polynomial of equation (7.4-1) for the Butterworth design are given in Figure 7.4-2b for different values of n, for the case where $\omega_c = 2\pi f_c = 1$ rad/s. We will see later in this section that these coefficient values can easily be adjusted to change the filter cutoff frequency to any desired value.

Figure 7.4-2c is an s-plane plot of the position of the poles for an $n = 3$ Butterworth low-pass filter. Regardless of the order of the filter, all the poles lie on a semicircle of radius ω_c, with the angles between the poles equal to π/n, where n is the order of the filter. In addition, the angles between the $j\omega$ axis and the poles on the semicircle closest to the $j\omega$ axis are equal to $\pm\pi/2n$. For example, for the $n = 3$ Butterworth filter illustrated in Figure 7.4-2c, the angle between the poles is $\pi/3$ or $60°$, while the angles between s_1 and the $j\omega$ axis and s_3 and the $j\omega$ axis are each $\pi/6$ or $30°$. From this figure the specific values of these poles are seen to be

$$s_1 = \omega_c\left(-\frac{1}{2} + j\frac{\sqrt{3}}{2}\right) \tag{7.4-2a}$$

$$s_2 = -\omega_c \tag{7.4-2b}$$

and
$$s_3 = \omega_c\left(-\frac{1}{2} - j\frac{\sqrt{3}}{2}\right) \tag{7.4-2c}$$

The corresponding filter transfer function may be written as

$$H(s) = \frac{A_o\omega_c^3}{(s - s_1)(s - s_2)(s - s_3)} = \frac{A_o\omega_c^3}{(s + \omega_c)(s^2 + \omega_c s + \omega_c^2)} \tag{7.4-3}$$

where A_o is the low-frequency gain and ω_c is its cutoff frequency. Note that the denominator polynomial in this transfer function matches that given for the $n = 3$ case in Figure 7.4-2b if ω_c in this equation is set equal to 1.

For some applications a sharp transition from the passband to the stopband at f_c is more important than gain flatness in the passband, and for these situations the Chebychev filter provides a better solution than the Butterworth. The gain–phase characteristics of the Chebychev filter are given in Figure 7.4-3a.

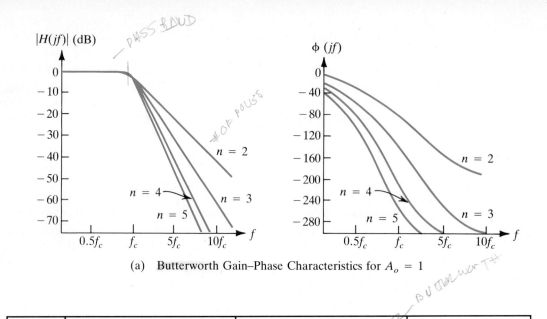

(a)　Butterworth Gain–Phase Characteristics for $A_o = 1$

Number of poles, n	$H(s)^*$					$H(s)^{**}$					Denominator roots
	a_4	a_3	a_2	a_1	a_0	a_{11}	a_{21}	a_{20}	a_{31}	a_{30}	
2	—	—	—	1.414	1.000	—	1.414	1.000	—	—	$-0.707 \pm j0.707$
3	—	—	2.000	2.000	1.000	1.000	1.000	1.000	—	—	$-0.500 \pm j0.866, -1$
4	—	2.613	3.141	2.613	1.000	—	1.848	1.000	0.765	1.000	$-0.383 \pm j0.924$ $-0.924 \pm j0.383$
5	3.236	5.236	5.236	3.236	1.000	1.000	0.618	1.000	1.618	1.000	$-0.309 \pm j0.951$ $-0.809 \pm j0.588$ -1.000

$$^*H(s) = \frac{(a_0)A_o}{a_n s^n + a_{n-1}s^{n-1} + \ldots a_1 s + a_0}$$

$$^{**}H(s) = \frac{(a_0)A_o}{(s + a_{11})(s^2 + a_{21}s + a_{20})(s^2 + a_{31}s + a_{30})}$$

(b)

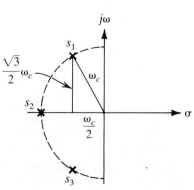

Figure 7.4-2
Data for the Butterworth low-pass filter.

(c)　Position of the Poles for an $n = 3$ Butterworth Low-Pass Filter

As illustrated in this figure, for an $n = 4$ filter, for example, the attenuation an octave above the cutoff frequency is about 38 dB, compared with only 25 dB for a similar Butterworth design (Figure 7.4-2a). This added attenuation is not obtained without cost, however. The Chebychev filter has a ripple in the passband, and the more rapid the attenuation outside the passband, the larger the ripple becomes. Therefore, in selecting a Chebychev-style filter, careful consideration must be given to the maximum amount of ripple that can be tolerated in the passband.

The location of the poles of a Chebychev filter in the s-plane is more complex than for the Butterworth filter (Figure 7.4-3c). Rather than being located on a circle, they are instead positioned on an ellipse whose major and minor axes are located at

$$\text{major axis} = \omega_c \cosh \xi \qquad (7.4\text{-}4a)$$

and

$$\text{minor axis} = \omega_c \sinh \xi \qquad (7.4\text{-}4b)$$

where

$$\xi = \frac{1}{n} \sinh^{-1} \frac{1}{\epsilon} \qquad (7.4\text{-}5)$$

and ϵ is a filter parameter that determines the shape of the transfer function. For large values of ϵ, the poles are closer to the $j\omega$ axis and the ripple in the passband is large (Figure 7.4-3d), but the filter attenuation outside the passband is also increased. The specific location of the poles in the s-plane for a particular filter order is given by the expressions

$$\sigma_m = -\sinh \xi \sin \frac{(2m - 1)\pi}{2n} \qquad (7.4\text{-}6a)$$

and

$$\beta_m = \cosh \xi \cos \frac{(2m - 1)\pi}{2n} \qquad (7.4\text{-}6b)$$

where $m = 1, 2, 3, \ldots, n$ and σ_m and β_m are the real and imaginary parts of the poles' positions, respectively.

In examining the sketch for $H(j\omega)$ given in Figure 7.4-3d for a fifth-order Chebychev filter, several interesting points can be made. First, the peak-to-peak amplitude of the ripple in the passband is

$$\text{ripple (dB)} = 20 \log \frac{1}{\sqrt{1 + \epsilon^2}} = -10 \log (1 + \epsilon^2) \qquad (7.4\text{-}7)$$

The table given in Figure 7.4-3b is for Chebychev filters with 0.5-dB (5%) and 1-dB (11%) ripple. The corresponding values of ϵ are 0.35 and 0.51, respectively.

In the sketch for $|H(j\omega)|$ the total number of peaks and valleys in the passband (five in this figure) is equal to the order of the filter. To see why this is so, consider that this fifth-order filter may be viewed as a cascade of two second-order filters (H_{15} associated with the pole pair s_1–s_5 and H_{24} with the pole pair s_2–s_4) and a single-pole filter H_3 associated with the pole at s_3. Because the poles of the Chebychev filter are closer to the $j\omega$ axis than for a comparable Butterworth filter, considerable peaking occurs in each of the second-order transfer function terms. This peaking causes the ripple in the magnitude response of $H(j\omega)$, with each bump corresponding to the resonant point in the

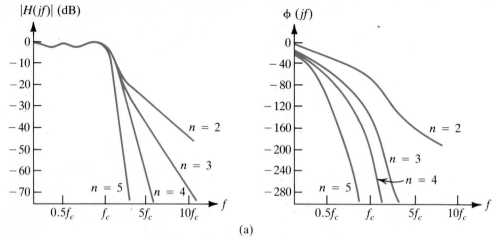

(a)

0.5-dB ripple

Number of poles, n	$H(s)^*$					$H(s)^{**}$					Denominator roots
	a_4	a_3	a_2	a_1	a_0	a_{11}	a_{21}	a_{20}	a_{31}	a_{30}	
2	—	—	—	1.427	1.516	—	1.427	1.516	—	—	$-0.713 \pm j1.004$
3	—	—	1.253	1.535	0.716	0.626	0.626	1.142	—	—	$-0.313 \pm j1.022$ -0.626
4	—	1.197	1.717	1.025	0.379	—	0.847	0.356	0.351	1.064	$-0.175 \pm j1.016$ $-0.423 \pm j0.421$
5	1.172	1.937	1.310	0.753	0.179	0.362	0.586	0.477	0.224	1.036	$-0.112 \pm j1.012$ $-0.293 \pm j0.625$ -0.290

$$*H(s) = \frac{(a_0)A_o}{a_n s^n + a_{n-1}s^{n-1} + \ldots a_1 s + a_0}$$

$$**H(s) = \frac{(a_0)A_o}{(s + a_{11})(s^2 + a_{21}s + a_{20})(s^2 + a_{31}s + a_{30})}$$

1.0-dB ripple

Number of poles, n	$H(s)^*$					$H(s)^{**}$					Denominator roots
	a_4	a_3	a_2	a_1	a_0	a_{11}	a_{21}	a_{20}	a_{31}	a_{30}	
2	—	—	—	1.098	1.103	—	1.098	1.103	—	—	$-0.549 \pm j0.895$
3	—	—	0.988	1.238	0.491	0.494	0.494	0.994	—	—	$-0.247 \pm j0.966$ -0.494
4	—	0.953	1.454	0.743	0.276	—	0.674	0.279	0.279	0.987	$-0.140 \pm j0.983$ $-0.337 \pm j0.407$
5	0.937	1.689	0.974	0.580	0.123	0.290	0.469	0.429	0.179	0.988	$-0.089 \pm j0.990$ $-0.234 \pm j0.612$ -0.290

$$*H(s) = \frac{(a_0)A_o}{a_n s^n + a_{n-1}s^{n-1} + \ldots a_1 s + a_0}$$

$$**H(s) = \frac{(a_0)A_o}{(s + a_{11})(s^2 + a_{21}s + a_{20})(s^2 + a_{31}s + a_{30})}$$

(b)

Figure 7.4-3
Data for the Chebychev low-pass filter.

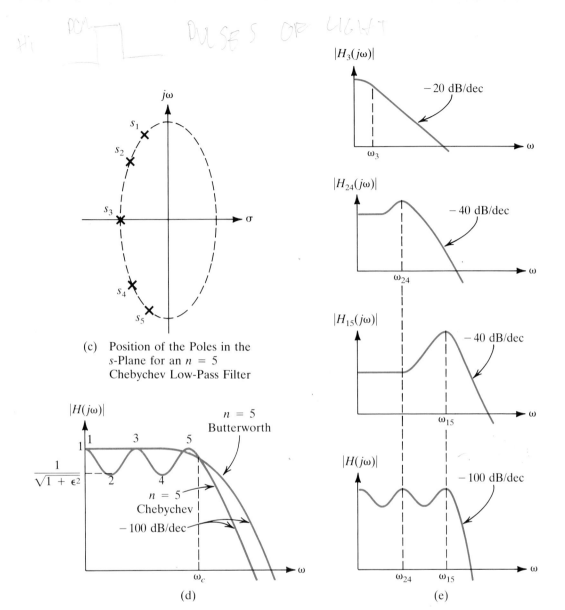

(c) Position of the Poles in the
s-Plane for an $n = 5$
Chebychev Low-Pass Filter

(d)

(e)

Figure 7.4-3 (Continued)

corresponding second-order pole pair response (Figure 7.4-3e). Because H_{15} has such a pronounced peak near ω_c, it can almost completely compensate for the gain loss due to H_3 and H_{24}, whose breaks started much earlier than ω_c. Once outside the cutoff frequency, the magnitude of H_{15} falls off rapidly, and this, along with the continued decrease in the magnitude of H_{24} and H_3, accounts for the rapid attenuation of this fifth-order Chebychev filter in comparison to that of a comparable Butterworth filter. Notice that the final slope of both filters far above ω_c are -100 dB/decade because both are fifth-order filters (Figure 7.4-3d).

For some applications, for example in the processing of pulse and video information, the preservation of the input signal waveshape may be more important than gain flatness in the passband or rapid signal attenuation outside the passband. In such cases, the phase response of the filter is more important than the gain response, and if the phase cannot be made identically zero or $-180°$,

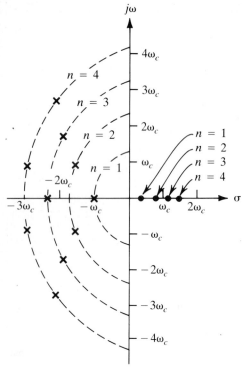

(a) Pole Positions in the s-plane for the Bessel Low-Pass Filter

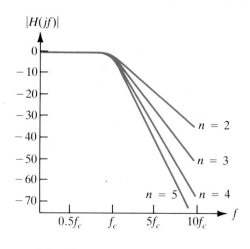

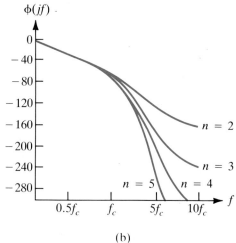

(b)

Number of poles, n	$H(s)^*$					$H(s)^{**}$					Denominator roots
	a_4	a_3	a_2	a_1	a_0	a_{11}	a_{21}	a_{20}	a_{31}	a_{30}	
2	—	—	—	3.000	3.000	—	3.000	3.000	—	—	$-1.500 \pm j0.866$
3	—	—	6.000	15.000	15.000	2.322	3.678	6.459	—	—	$-1.839 \pm j1.754$ -2.322
4	—	10.000	45.000	105.000	105.000	—	4.208	11.488	5.792	9.140	$-2.896 \pm j0.867$ $-2.104 \pm j2.657$
5	15.000	105.000	420.000	945.000	945.000	3.647	4.649	18.156	6.704	14.272	$-3.352 \pm j1.743$ $-2.325 \pm j3.571$ -3.637

$$^*H(s) = \frac{(a_0)A_o}{a_n s^n + a_{n-1}s^{n-1} + \ldots a_1 s + a_0}$$

$$^{**}H(s) = \frac{(a_0)A_o}{(s + a_{11})(s^2 + a_{21}s + a_{20})(s^2 + a_{31}s + a_{30})}$$

Figure 7.4-4
Data for the Bessel (linear-phase) low-pass filter.

it should, if possible, be made to increase (or decrease) linearly with frequency.

Butterworth and Chebychev filters have rather nonlinear phase characteristics, and therefore their transient responses tend to be poor. As a result, when filtering is needed, but when preservation of the signal's waveshape is important, a third style of filter, called the linear-phase or Bessel filter, is preferred. The s-plane plots for first-, second-, third-, and fourth-order linear-phase filters are given in Figure 7.4-4a. For each of these filters the poles are located on a circle whose center is on the positive real axis as indicated. Unfortunately, no "simple" formula can be given for the pole positions of these filters.

The gain–phase characteristics of the Bessel filter are sketched in Figure 7.4-4b, and by examining the phase characteristics of this filter and comparing them to those of the other filters described previously, it should be apparent that the phase response of this filter is much flatter in the passband than are those of the Butterworth or Chebychev designs.

7.4-2 Design of Specific Active Filter Circuits

Traditionally filters were designed using passive R, L, and C components, and in many cases these designs were excellent and are still in use. However, in certain application areas, most notably those in the audio- and subaudio-frequency regions, active filters have all but completely replaced the older passive design techniques. Table 7.4-1 compares the pertinent features of each approach.

In a passive filter, whenever complex poles are needed, it is necessary to include both inductors and capacitors in the filter design. The use of inductors in passive filters causes several problems. Of the three passive elements—the inductor, resistor, and capacitor—the inductor is the most troublesome component with which to work. Inductors, especially those used at the audio and sub-audio frequencies, are heavy, bulky, and costly. In addition, owing to their winding resistance and hysterisis losses, the performance of an actual inductor can differ significantly from that of the ideal component. The fact that active filters can be designed using only resistors and capacitors is probably the greatest advantage of the active filter over its passive component counterpart. In addition, their ease of design, tuning, and troubleshooting, as well as their ability to operate with nearly any source and load impedance, are important attributes.

In this book we design our active filter circuits by using a building-block approach in which a filter of any order is constructed by cascading a series of first- and second-order circuit blocks. For example, a sixth-order filter can be made by cascading three second-order blocks while a seventh-order filter can be constructed by cascading three second-order stages and one first-order stage. In theory, a seventh-order filter could be constructed by cascading seven first-order filters. However, this is impractical because it requires seven rather than four amplifier stages, and also because it is impossible to realize complex poles with this type of circuit design. First-order building blocks can only be used for the synthesis of transfer functions whose poles lie on the negative real axis.

The two types of first-order designs that we will need for the realization of the active filters in this book are given in Figure 7.4-5a and b. We will see

7.4-1 Comparison of Passive and Active Filter Design Features

Parameter	Passive Design	Active Design
Size and weight	Heavy and bulky due to inductors	Inductors can be eliminated through use of feedback
Source and load impedance	Frequency response critically dependent on R_L and R_s	Buffers can make frequency response independent of R_L and R_s
Overall gain	Gain is generally < 1	Active devices can be used to achieve any gain desired
Complex transfer functions	Conceptually difficult to realize because of loading effects; also difficult to tune	Can use multistage circuit, which because of interstage isolation is easy to design, tune, and troubleshoot
Applicability in different frequency ranges • Subaudio (< 20 Hz) • Audio (20 Hz–20 kHz) • 20 kHz–1 MHz • >1 MHz	 Poor Good Excellent Excellent	 Excellent Excellent Good Poor
Signal levels	Microvolts to several hundred volts	Usually limited to 10 or 20 V maximum; at low end need to consider op amp noise
Stability	Extremely stable	Pole locations can drift with time and temperature; can cause circuit to oscillate; op amp Q-point shifts can also cause dc offsets in output

later that these circuits can be used as the first-order building blocks for the design of low-pass and high-pass filters, respectively. Bandpass and band-reject filters do not require first-order building blocks since their transfer functions are always even polynomials that contain only complex-conjugate pole pairs.

For illustrative purposes, we examine the single-pole low-pass filter circuit presented in Figure 7.4-5a; the analysis of the second circuit, namely the high-pass first-order building block given in part (b) of the figure, follows similarly and is discussed quite fully in Example 7.4-4. For the first-order low-pass single-pole building block in Figure 7.4-6a, it should be apparent that

$$\frac{V_o}{V_{in}} = -\frac{Z_f}{Z_2} = \frac{-R_f(1/sC)}{(R_f + 1/sC)/R_2} = \left(-\frac{R_f}{R_2}\right)\frac{1}{1 + s/\omega_1} \tag{7.4-8}$$

where $\omega_1 = 1/R_f C$. The s-plane plot for this transfer function is given in Fig-

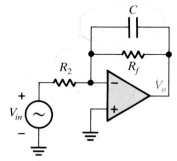

(a) Single Pole on the Negative Real Axis, First-Order Low-Pass Filter Building Block

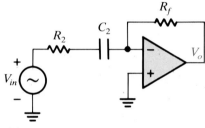

(b) Single Zero at Origin and Single Pole on the Negative Real Axis, First-Order Building Block Used for the Design of High-Pass Filters

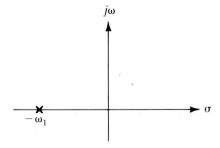

(c) Pole–Zero Plot for the First-Order Low-Pass Building Block Given in (a)

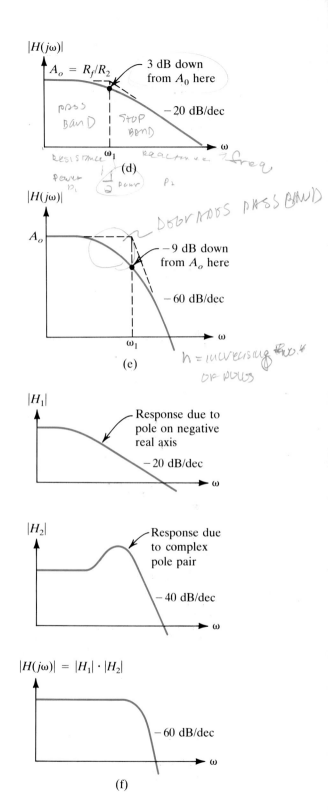

Figure 7.4-5

ure 7.4-5c, and, as required, it has a single pole on the negative real axis at $s = -\omega_1$. The magnitude of this transfer function versus frequency is sketched in part (d) of the figure. For values of $\omega \ll \omega_1$, the capacitor is effectively an open circuit and the gain is therefore equal to $-R_f/R_2$. At $\omega = \omega_1$ the reactance of C equals R_f, and for frequencies above this point, the capacitor begins to short out and the gain therefore falls off at -20 dB/decade.

It is important to understand why filter designs generally are not implemented using circuits whose transfer functions have only real poles. Consider, for example, the single-pole low-pass filter circuit that we just examined. By comparing its frequency response with that of the ideal filter given in Figure 7.4-1a, two significant differences are immediately apparent. First, the gain response of this single-pole filter is not very flat in the passband, and second, the gain attenuation outside the passband occurs only at a rate of -20 dB/decade. Of course, the rate of signal attenuation outside the passband can be improved by increasing the number of poles in the filter. Figure 7.4-5e illustrates what happens when a three-pole filter (with a triple pole at $s = -\omega_1$) is used in place of the single-pole filter we have just been discussing. In this instance, while the attenuation outside the passband has increased to -60 dB/decade, the gain flatness in the passband has been degraded even further. In fact, as long as all the poles are real, there is no way to achieve a very flat gain response. However, by making two of the poles complex (with poles close to the $j\omega$ axis) the bump in their gain response can be used to compensate for the attenuation introduced by the real pole so that the combination of the two stages produces a relatively flat gain response having an extended bandwidth (Figure 7.4-5f). Thus in the design of circuits having specifically tailored gain and phase characteristics, it will almost always be necessary to work with transfer functions that have complex poles. These components of the overall filter transfer function may be implemented by using the second-order building blocks that we will now discuss.

Many different types of second-order active filter designs are available for the realization of quadratic-style transfer functions, and while we could spend considerable time examining different filter architectures, we will, instead, work with only one specific second-order filter design that is known as a voltage controlled–voltage source active filter. This circuit uses very few components and leads to transfer functions whose coefficients are only moderately sensitive to component variation effects.

The basic form of a second-order low-pass voltage controlled-voltage source active filter is illustrated in Figure 7.4-6a, and the typical frequency response of this circuit is sketched in part (b) of the figure. The dashed lines in the sketch represent the asymptotic approximations to the gain characteristic, and the solid lines indicate what the actual response might look like depending on the position of the poles.

The transfer function for this circuit may be readily computed if we recall that V_1 and V_2 must be equal, since we may then write that

$$V_1 = \frac{V_o}{A_{CL}} = V_2 \qquad (7.4\text{-}9a)$$

where

$$A_{CL} = \frac{V_o}{V_1} = 1 + \frac{R_A}{R_B} \qquad (7.4\text{-}9b)$$

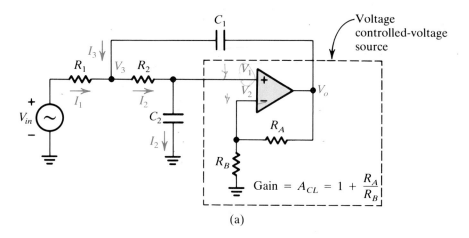

(a)

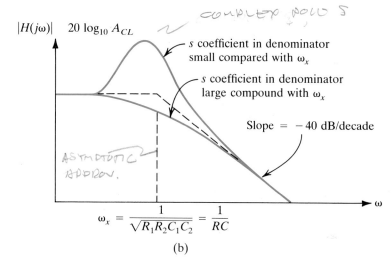

Figure 7.4-6
Second-order low-pass active filter using a voltage-controlled voltage source.

(b)

Using this result, the node voltage V_3 may be related to V_o by

$$\frac{V_o}{A_{CL}} = V_1 = \frac{1/sC_2}{R_2 + 1/sC_2}V_3 = \frac{1}{1 + sR_2C_2}V_3 \qquad (7.4\text{-}10a)$$

or

$$V_3 = \frac{1 + sR_2C_2}{A_{CL}}V_o \qquad (7.4\text{-}10b)$$

Writing the node equation for the node V_3, we obtain

$$\frac{V_{in} - V_3}{R_1} + (V_o - V_3)sC_1 = sC_2V_1 \qquad (7.4\text{-}11a)$$

and if we substitute in equations (7.4-9a) and (7.4-10b) into this expression, after some algebraic manipulation, we find that the circuit transfer function may be written as

$$\frac{V_o}{V_{in}} = \frac{(1/R_1R_2C_1C_2)A_{CL}}{s^2 + [1/R_1C_1 + 1/R_2C_1 + (1 - A_{CL})/R_2C_2]s + 1/R_1R_2C_1C_2} \qquad (7.4\text{-}11b)$$

This equation has the same form as the low-pass filter transfer function given

7.4 / Active Filters

423

in equation (7.4-1). In particular, it is a second-order low-pass filter with a low-frequency gain given by $A_{CL} = 1 + R_A/R_B$.

In working with this active filter it will be convenient for educational simplicity to let

$$R_1 = R_2 = R \qquad (7.4\text{-}12a)$$

$$C_1 = C_2 = C \qquad (7.4\text{-}12b)$$

and to define

$$\omega_x = \frac{1}{RC} \qquad (7.4\text{-}12c)$$

Making these substitutions into equations (7.4-11b), the transfer function for this second-order active low-pass filter simplifies to

$$H(s) = \frac{V_o}{V_{in}} = \frac{A_{CL}\,\omega_x^2}{s^2 + (3 - A_{CL})\omega_x s + \omega_x^2} \qquad (7.4\text{-}13)$$

If we were to use this transfer function for the design of a specific second-order low-pass filter, we would find after examining the table in Figure 7.4-2b, for example, that three coefficients (A_o, a_0, and a_1) need to be specified. However, in equation (7.4-13) there are only two parameters to be selected, and this means that if we choose A_{CL} and ω_x to match the coefficients a_0 and a_1, the filter gain may not be what we want it to be. But this problem is not all that bad since we can always add an additional stage of amplification to adjust the gain to any desired value.

Once A_{CL} and ω_x are determined, specific values now need to be chosen for R, C, R_A, and R_B. Because there are only two constraints and four variables in the problem, some freedom still exists in the selection of these components. Since the number of different capacitor sizes available is much more limited than for resistors, it is probably a good idea to select specific values for C and then solve for the required value of R from the ω_x equation. Care should be taken to ensure that the value selected for R is large enough that it will not cause circuit loading, yet small enough that it does not create dc offset problems. Generally, values of R in the range from about 1 kΩ to 1 MΩ will work satisfactorily. Similar resistance values should be used for R_A and R_B to achieve the desired A_{CL}.

In designing a particular Butterworth, Chebychev, or Bessel low-pass filter, we will need to refer to the tables in Figures 7.4-2b, 7.4-3b, and 7.4-4c, respectively. The coefficients in these tables were calculated for a low-pass filter with an assumed cutoff frequency of 1 rad/s. The coefficient values required to place the cutoff frequency at any desired location may be determined by replacing all the values of s in the filter transfer function by s/ω_c, where ω_c represents the desired filter cutoff frequency.

EXAMPLE 7.4-1

Design a second-order Butterworth low-pass filter with a cutoff frequency equal to 1 kHz and a low-frequency gain of 10. In selecting the components, let us assume that 0.01-μF capacitors are available. Compute the required values for the resistors and check that their sizes are practical.

SOLUTION

The general form of the transfer function for this filter will be

$$H_L(s) = \frac{a_0}{s^2 + a_{21}s + a_{20}} \qquad (7.4\text{-}14)$$

and from Figure 7.4-2b, we find that

$$a_{21} = 1.414 \tag{7.4-15a}$$

and

$$a_{20} = 1 \tag{7.4-15b}$$

so that

$$H_L(s) = \frac{10}{s^2 + 1.414s + 1} \tag{7.4-15c}$$

for a second-order Butterworth low-pass filter with a cutoff frequency of 1 rad/s, and a low-frequency gain of 10. To transform this filter design into one having a cutoff frequency at 1 kHz, we simply replace s in equation (7.4-15c) by s/ω_c to yield

$$H(s) = \frac{10}{s^2/\omega_c^2 + 1.414\ s/\omega_c + 1} = \frac{10\omega_c^2}{s^2 + 1.414\omega_c s + \omega_c^2} \tag{7.4-16}$$

where $\omega_c = 6.28 \times 10^3$ rad/s. Equating the denominators of equations (7.4-13) and (7.4-16), we have

$$\omega_x = \omega_c = 6.28 \times 10^3 \tag{7.4-17a}$$

and

$$3 - A_{CL} = 1.414 \tag{7.4-17b}$$

From equation (7.4-17b) the required closed-loop gain is $(3 - 1.414)$ or 1.586, and letting $R_B = 10$ kΩ, it therefore follows that $R_A = 5.86$ kΩ. In a similar fashion, from equation (7.4-17a) with $C = 0.01$ μF, the required value of R is

$$R = \frac{1}{\omega_x c} = \frac{1}{(6.28 \times 10^3)(10^{-8})} = 15.9 \text{ k}\Omega \tag{7.4-18}$$

The values selected for R, R_A, and R_B are all within the limits specified earlier. The only problem with the design so far is that the low-frequency gain of this second-order building block is 1.586 and not 10 as required. However, this problem can be easily corrected by adding an additional stage of amplification with a gain of $10/1.586 = 6.3$. The overall final design for this filter is given in Figure 7.4-7. Notice in this type of multistage filter design

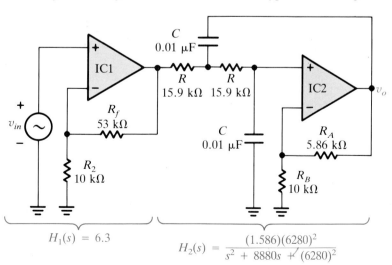

Figure 7.4-7

Example 7.4-1. Design of an $n = 2$ Butterworth low-pass filter ($f_c = 1$ kHz, $A_o = 10$).

$$H_1(s) = 6.3$$

$$H_2(s) = \frac{(1.586)(6280)^2}{s^2 + 8880s + (6280)^2}$$

that the overall circuit transfer function $H(s) = V_o(s)/V_{in}(s)$ is simply equal to the product of the individual transfer functions of the stages $[H_1(s) \times H_2(s)]$, because each op amp stage has zero output impedance, so that no loading occurs.

To assist in the design of high-pass, bandpass, and band-elimination filters, we could present tables similar to those given for the low-pass filter in Figures 7.4-2b, 7.4-3b, and 7.4-4c. However, all this information can be derived from the low-pass tables if the transformations shown in Table 7.4-2 are employed.

Because a high-pass filter is basically the opposite of a low-pass filter, it is not too surprising that s just needs to be replaced by $1/s$ to accomplish this transformation. In this way things that used to happen for small values of ω will now occur at high values of ω, and vice versa. The factor ω_c in the equation just shifts the cutoff frequency from unity to the desired value.

To convert a low-pass filter design to a bandpass design, the center frequency of the filter needs to be shifted from $\omega = 0$ to $\omega = \omega_0$. The replacement of values of s in the original low-pass transfer function by the term $(\omega_0/\mathrm{BW})(s/\omega_0 + \omega_0/s)$ provides the needed shift of the center frequency to ω_0, as well as the required scale changes. The quantity $(\omega_H - \omega_L)$ defines the bandwidth (BW) of the bandpass filter, and ω_0, the center frequency of the filter, is located at the geometric mean of ω_H and ω_L. The term

$$Q = \frac{\omega_0}{\mathrm{BW}} = \frac{\omega_0}{\omega_H - \omega_L} \qquad (7.4\text{-}19)$$

is called the Q or quality factor of the bandpass filter, and is a measure of how rapidly the gain of the filter falls off as the signal frequency moves away from ω_0.

The band-reject filter, because it has frequency characteristics which are the inverse of the bandpass filter, has a low-pass to band-reject transformation factor that is the reciprocal of that used for the low-pass to bandpass transformation. It is important to observe that both the bandpass and band-elimination

7.4-2 Transfer Function Substitutions to Transform a Low-Pass Filter to an Equivalent High-Pass, Bandpass, or Band-Reject Filter

Type of Transformation	Change Required in S	Comments
Low-pass to high-pass	$s \rightarrow \dfrac{\omega_c}{s}$	ω_c = desired cutoff frequency; inverse of low-pass filter
Low-pass to bandpass	$s \rightarrow \dfrac{\omega_0}{\mathrm{BW}}\left(\dfrac{s}{\omega_0} + \dfrac{\omega_0}{s}\right)$ $\rightarrow Q\left(\dfrac{s}{\omega_0} + \dfrac{\omega_0}{s}\right)$	ω_H = high-frequency 3-dB point ω_L = low-frequency 3-dB point $\mathrm{BW} = \omega_H - \omega_L$ $\omega_0 = \sqrt{\omega_H \omega_L}$ = center frequency Shifts filter center frequency from $\omega = 0$ to $\omega = \omega_0$
Low-pass to band-reject	$s \rightarrow \dfrac{1}{Q(s/\omega_0 + \omega_0/s)}$	Inverse of bandpass filter

transformations double the number of poles from that in the equivalent low-pass transfer function. Furthermore, all these poles occur in complex conjugate pairs.

EXAMPLE 7.4-2

a. Develop an expression for the transfer function of a third-order Chebychev high-pass filter having a ripple of 0.5 dB in the passband and a cutoff frequency at 1 rad/s. The high-frequency gain is to be unity.

b. How does this transfer function change if the characteristics of the filter are the same as those given in part (a) except that the cutoff frequency is at 500 Hz and the high-frequency gain is 200?

SOLUTION

Using the table in Figure 7.4-2b, we find that the denominator coefficients for an $n = 3$ 0.5-dB ripple Chebychev low-pass filter are $a_0 = 0.716$, $a_{11} = a_{21} = 0.626$, and $a_{20} = 1.142$. As a result, the transfer function for this low-pass filter may be written as

$$H_L(s) = \frac{0.716}{(s + 0.626)(s^2 + 0.626s + 1.142)} \tag{7.4-20}$$

The filter transfer function is written in this factored form rather than as an expanded polynomial, because in synthesizing this function we will be using a cascade of active filter circuits where each will be able to generate either a single-pole or a quadratic-style transfer function.

To transform equation (7.4-20) into an equivalent high-pass function having the same cutoff frequency at $\omega_c = 1$ rad/s, we simply replace all the s terms by $1/s$. When this is done in the original low-pass expression, we obtain

$$H(s) = \frac{0.716}{[(1/s) + 0.626][(1/s^2) + 0.626(1/s) + 1.142]} \tag{7.4-21a}$$

Multiplying above and below in this expression by s^3 and factoring out the constant 0.626 from the first term in the denominator, and 1.142 from the second, we have

$$H(s) = \frac{s^3}{(s + 1.596)(s^2 + 0.548s + 0.875)} = \frac{s}{s + 1.596} \frac{s^2}{s^2 + 0.548s + 0.875} \tag{7.4-21b}$$

This completes part (a) of the example. Incidentally, readers should verify for themselves that the high-frequency gain of this transfer function (or gain for large values of s) is 1.

To shift the cutoff frequency of this filter up to 500 Hz, we need to replace s by $s/\omega_c = s/(3.14 \times 10^3)$ in equation (7.4-21b). In addition, because the high-frequency gain of the transfer function in equation (7.4-21b) is equal to 1, we will also need to multiply this function by 200 to achieve the desired high-frequency gain. Carrying out both of these operations simultaneously, and multiplying above and below by $(3.14 \times 10^3)^3$, we obtain

$$H(s) = (200)\frac{s}{s + 5 \times 10^3} \frac{s^2}{s^2 + 1.723 \times 10^3 s + 8.639 \times 10^6} \tag{7.4-22}$$

EXAMPLE 7.4-3

a. Starting with an $n = 2$ Butterworth low-pass filter, determine the transfer function for a Butterworth-style bandpass filter of order 2, centered at 10^4 rad/s. The desired filter Q is 10.

b. Sketch the frequency response of this filter, and also determine filter bandwidth as well as the upper and lower 3-dB points.

SOLUTION

From Figure 7.4-2b we find that the coefficients for the $n = 2$ Butterworth low-pass filter are $a_{21} = 1.414$ and $a_{20} = 1.000$. Therefore, the transfer function for this low-pass filter may be written as

$$H_L(s) = \frac{1}{s^2 + 1.414s + 1.000} \tag{7.4-23}$$

assuming that the filter gain is to be 1. In accordance with Table 7.4-2, the required transformation to convert this low-pass filter into an equivalent bandpass filter is

$$s \longrightarrow \frac{\omega_0}{BW}\left(\frac{s}{\omega_0} + \frac{\omega_0}{s}\right) = Q\left(\frac{s}{\omega_0} + \frac{\omega_0}{s}\right) \tag{7.4-24a}$$

Because the bandpass filter characteristics are to be $\omega_0 = 10^4$ rad/s and $Q = 10$, the required substitution for s in equation (7.4-23) is

$$s \longrightarrow 10\left(\frac{s}{10^4} + \frac{10^4}{s}\right) = \left(\frac{s}{10^3} + \frac{10^5}{s}\right) \tag{7.4-24b}$$

Substituting this expression for s into equation (7.4-23), we obtain

$$H(s) = \frac{1}{(s/10^3 + 10^5/s)^2 + 1.414(s/10^3 + 10^5/s) + 1.000} \tag{7.4-25a}$$

which after considerable algebraic manipulation may be written as

$$H(s) = \frac{10^6 s^2}{s^4 + 1.414 \times 10^3 s^3 + 2.010 \times 10^8 s^2 + 1.414 \times 10^{11} s + 10^{16}} \tag{7.4-25b}$$

Comparing this result with the original low-pass transfer function in equation (7.4-23), we can see that the number of poles has doubled, and in addition, there are two new zeros at the origin.

In the original low-pass transfer function, $H_L(s)$, at high frequencies (i.e., frequencies very far away from $\omega = 0$) is approximately equal to $1/s^2$, so that the gain falls off at -40 dB/decade (Figure 7.4-8a). The frequency response of this bandpass filter is similar except that it falls off at this rate for frequencies very much greater than or very much smaller than ω_0 (Figure 7.4-8b). When ω is much greater than ω_0, the s^4 term dominates in the denominator of $H(s)$, and we may approximate $H(s)$ as $10^6 s^2/s^4$ or $10^6/s^2$. The magnitude of this expression falls of at -40 dB/decade with increasing frequency. In a similar fashion for frequencies well below ω_0, the denominator of $H(s)$ is approximately equal to 10^{16}, so that we may approximate $H(s)$ as $10^6 s^2/10^{16}$ or $10^{-10} s^2$. In this region the magnitude of the filter response increases by 40 dB/decade for frequencies well below ω_0.

To determine the upper and lower 3-dB points for this filter, we begin by

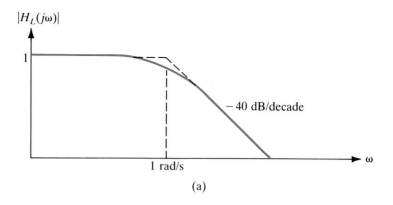

(a)

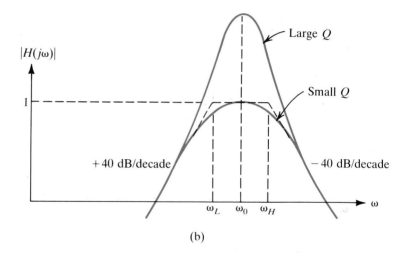

(b)

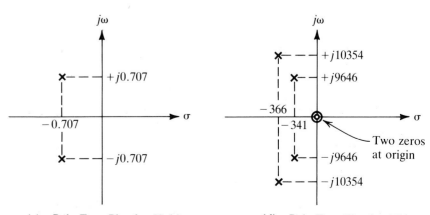

(c) Pole-Zero Plot for $H_L(s)$

(d) Pole-Zero Plot for $H(s)$

Figure 7.4-8
Example 7.4-3.

computing its bandwidth from the relationship

$$Q = \frac{\omega_0}{BW} \qquad (7.4\text{-}26a)$$

or
$$BW = \omega_H - \omega_L = \frac{\omega_0}{Q} = \frac{10^4}{10} = 10^3 \text{ rad/s} \qquad (7.4\text{-}26b)$$

Combining this result with the fact that ω_0 is the geometric mean of the half-power points or

$$\omega_0 = \sqrt{\omega_H \omega_L} \qquad (7.4\text{-}27)$$

and eliminating ω_H from both equations, we obtain

$$\omega_L^2 + 10^3 \omega_L - 10^8 = 0 \qquad (7.4\text{-}28a)$$

The solution to this quadratic equation, dropping the extraneous negative root, is

$$\omega_L = 9512 \text{ rad/s} \qquad (7.4\text{-}28b)$$

and substituting this result into equation (7.4-27), we also find that

$$\omega_H = 10{,}512 \text{ rad/s} \qquad (7.4\text{-}28c)$$

In examining these results, we see that the bandwidth is precisely equal to $(10{,}512 - 9512)$ or 1000 rad/s, as required. Another interesting fact with regard to this transfer function is that the center frequency ω_0 is approximately located at the arithmetic mean of (or halfway in between) the upper and lower 3-dB points. This occurs because the filter Q is large, and for high-Q circuits, this fact provides us with a very rapid way to estimate the values of ω_H and ω_L in bandpass circuits.

The final transfer function given in equation (7.4-25b) is not very useful for synthesis purposes because we need the transfer function in factored quadratic form, that is, as the product of two quadratic polynomials, if we are to be able to implement $H(s)$ by a cascade of two quadratic-style active bandpass filter circuits. To obtain this form of the bandpass transfer function, we need to begin by rewriting $H_L(s)$ in factored form from the table in Figure 7.4-2b as

$$H_L(s) = \frac{1}{(s + 0.7071 + j0.7071)(s + 0.7071 - j0.7071)} \qquad (7.4\text{-}29)$$

The location of the poles of this transfer function are shown in the pole–zero plot given in Figure 7.4-8c. Applying the low-pass to bandpass transformation in equation (7.4-24b), the transfer function for this bandpass filter may be written as

$$H(s) = \frac{1}{[(s/10^3 + 10^5/s) + 0.7071 + j0.7071][(s/10^3 + 10^5/s) + 0.7071 - j0.7071]} \qquad (7.4\text{-}30a)$$

or

$$H(s) = \frac{s^2}{\underbrace{[s^2 + 707.1(1 + j)s + 10^8]}_{Q_1(s)}\underbrace{[s^2 + 707.1(1 - j)s + 10^8]}_{Q_2(s)}} \qquad (7.4\text{-}30b)$$

Using the quadratic equation, we can show that the roots of the denominator polynomial $Q_1(s)$ are

$$s_1 = -341 + j9646 \qquad (7.4\text{-}31a)$$

and
$$s_2 = -366 - j10{,}354 \qquad (7.4\text{-}31b)$$

Using the same approach for $Q_2(s)$, the roots of this polynomial are found to be located at

$$s_3 = -341 - j9646 \qquad (7.4\text{-}31c)$$

and
$$s_4 = -366 + j10{,}354 \qquad (7.4\text{-}31d)$$

Notice that s_1 and s_3 and s_2 and s_4 are complex conjugates of one another.
Using these results, we may rewrite $H(s)$ as

$$H(s) = \frac{10^6 s^2}{(s + 341 + j9646)(s + 341 - j9646)(s + 366 + j10{,}354)(s + 366 - j10{,}354)} \qquad (7.4\text{-}32)$$

and by multiplying out the first and second terms and the third and fourth terms in the denominator, we obtain

$$H(s) = \frac{10^6 s^2}{(s^2 + 6825 + 9.316 \times 10^7)(s^2 + 732s + 1.073 \times 10^8)} \qquad (7.4\text{-}33)$$

The pole–zero plot for this transfer function is given in Figure 7.4-8d. In this form it is relatively easy to synthesize this transfer function by splitting it into the two quadratic-style bandpass transfer functions

$$H(s) = (10^6) \frac{s}{s^2 + 682s + 9.316 \times 10^7} \frac{s}{s^2 + 732s + 1.073 \times 10^8} \qquad (7.4\text{-}34)$$

Specific circuits for accomplishing the synthesis of this type of complex transfer function will be presented shortly.

The op amp circuit configuration for a high-pass voltage controlled–voltage source active filter is given in Figure 7.4-9a. By comparing it with the circuit in Figure 7.4-6, we can see that it is essentially the same circuit except that the position of the R's and C's have been interchanged. The transfer function for this circuit may be developed by exactly the same method that we employed for finding $H(s)$ for the low-pass filter in Figure 7.4-6, and when this is done we obtain

$$H(s) = \frac{V_o}{V_{in}} = \frac{A_{CL}s^2}{s^2 + [1/R_2 C_2 + 1/R_2 C_1 + (1 - A_{CL})/R_1 C_1]s + 1/R_1 R_2 C_1 C_2} \qquad (7.4\text{-}35)$$

If as with the second-order low-pass filter circuit examined earlier, we let $R_1 = R_2 = R$, $C_1 = C_2 = C$, and $\omega_x = 1/RC$, then the transfer function for the circuit given in Figure 7.4-9 simplifies to

$$H(s) = \frac{V_o}{V_{in}} = \frac{A_{CL}s^2}{s^2 + (3 - A_{CL})\omega_x s + \omega_x^2} \qquad (7.4\text{-}36)$$

The frequency response of this transfer function is sketched in Figure 7.4-9b. At high frequencies, that is, for large values of s, the denominator may be

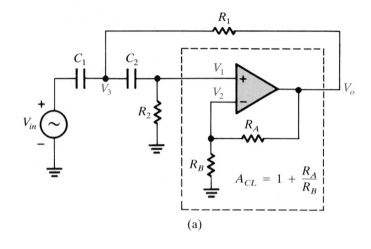

(a)

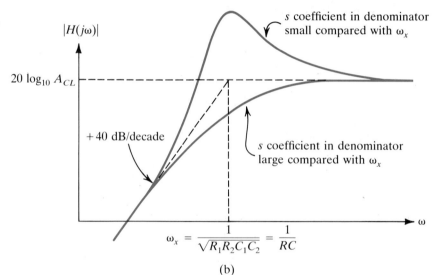

Figure 7.4-9
Voltage controlled–voltage source high-pass active filter.

(b)

approximated by s^2, and the gain therefore approaches A_{CL}, the op amp gain. By examining the circuit in Figure 7.4-9a, we can see that this result is not all that surprising since at high frequencies if we replace C_1 and C_2 by short circuits we just have an ordinary noninverting op amp circuit with $v_o = A_{CL}v_{in}$. At low frequencies the s^2 and s terms in the denominator may be neglected in comparison to the last term (ω_x^2), and therefore we find that $H(s)$ in this region is proportional to s^2, so that the gain below ω_x falls off at 40 dB for each decade decrease in frequency below ω_x. The specific shape of the frequency plot (indicated by the solid lines in Figure 7.4-9) depends on the relative size of ω_x and the coefficient in front of the s term in the denominator.

The circuit illustrated in Figure 7.4-10a is a bandpass-style voltage controlled–voltage source active filter. By using the same techniques as for the high-pass analysis just carried out, it is not too difficult to show that the trans-

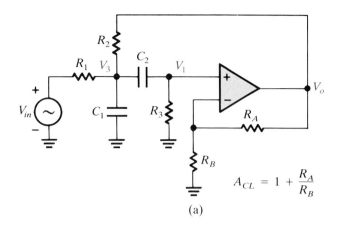

(a)

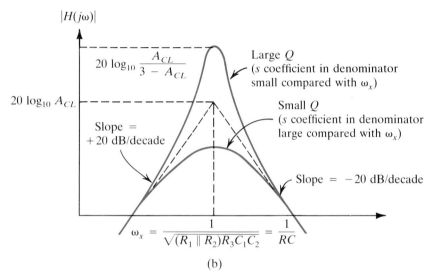

Figure 7.4-10
Active bandpass filter imple-
mented using a voltage con-
trolled–voltage source design.

(b)

fer function for this circuit is

$$H(s) = \frac{V_o}{V_{in}}$$

$$= \frac{A_{CL}(s/R_1 C_1)}{s^2 + [1/R_1 C_1 + (1/R_3)(1/C_1 + 1/C_2) + (1 - A_{CL})/R_2 C_1]s + 1/(R_1 \parallel R_2)R_3 C_1 C_2} \quad (7.4\text{-}37)$$

By making the substitutions $R_1 = R_2 = R$, $R_3 = 2R$, $C_1 = C_2 = C$, and $\omega_x = 1/RC$, this transfer function may be simplified considerably to the expression

$$H(s) = \frac{V_o}{V_{in}} = \frac{A_{CL}\omega_x s}{s^2 + (3 - A_{CL})\omega_x s + \omega_x^2} \quad (7.4\text{-}38)$$

The frequency response of this transfer function is sketched in Figure 7.4-10b. The dashed lines are the asymptotes of $H(s)$, and the solid lines indicate what the actual gain looks like and how it depends on the relative size of the coefficients in the denominator of the transfer function.

EXAMPLE 7.4-4

In Example 7.4-2 we developed the expression

$$H(s) = (200)\frac{s}{s + 5 \times 10^3}\frac{s^2}{s^2 + 1.723 \times 10^3 s + 8.639 \times 10^6} \quad (7.4\text{-}39)$$

as the transfer function of a third-order 0.5-dB Chebychev filter with a gain of 200 and a cutoff frequency of 500 Hz. Design an op amp active filter to realize this transfer function.

SOLUTION

One possible circuit that could be used to implement this particular transfer function is shown in Figure 7.4-11. The circuit containing IC2 is a standard second-order high-pass active filter, and this portion of the circuit is used to generate the second term on the right-hand side of equation (7.4-39). The single-pole filter circuit containing IC1 handles the first term on the right-hand side of this equation. Because it has fewer degrees of freedom, let's begin this design problem by determining the transfer function and the component values for the second stage of this circuit.

By matching coefficients in the denominator of the second term on the right-hand side of equation (7.4-39) with the denominator of the standard equation for the second-order high-pass filter given in equation (7.4-36), we obtain

$$\omega_x = \frac{1}{RC} = \sqrt{8.639 \times 10^6} = 2939 \text{ rad/s} \quad (7.4\text{-}40a)$$

and

$$(3 - A_{CL})\omega_x = 1.723 \times 10^3 \quad (7.4\text{-}40b)$$

If we let $C = 0.01\ \mu\text{F}$, then following equation (7.4-40a),

$$R = \frac{1}{\omega_x C} = \frac{1}{(2939)(10^{-8})} = 34 \text{ k}\Omega \quad (7.4\text{-}41)$$

Furthermore, from equation (7.4-40b) we also find that the required closed-loop gain of the second stage is

Figure 7.4-11
Example 7.4-4.

$$3 - A_{CL} = \frac{1.723 \times 10^3}{2939} = 0.586 \quad (7.4\text{-}42a)$$

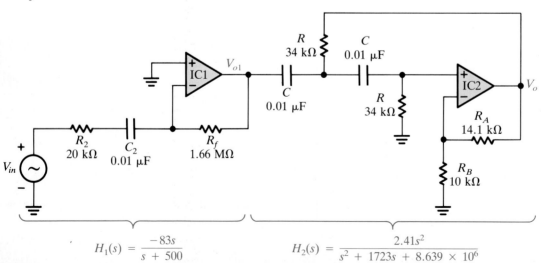

$$H_1(s) = \frac{-83s}{s + 500} \qquad H_2(s) = \frac{2.41s^2}{s^2 + 1723s + 8.639 \times 10^6}$$

or
$$A_{CL} = 3 - 0.586 = 2.41 \qquad (7.4\text{-}42b)$$

This gain can readily be implemented with values of R_A and R_B equal to 14.1 and 10 kΩ, respectively. The final transfer function of the second stage is

$$H_2(s) = \frac{V_o}{V_{o1}} = \frac{2.41s^2}{s^2 + 1.723 \times 10^3 s + 8.639 \times 10^6} \qquad (7.4\text{-}43)$$

Because IC1 is a inverting style op amp, we may write its gain in terms of the impedances R_f and Z_2 as

$$\frac{V_{o1}}{V_{in}} = -\frac{R_f}{Z_2} \qquad (7.4\text{-}44a)$$

but

$$Z_2 = R_2 + \frac{1}{sC_2} \qquad (7.4\text{-}44b)$$

so that we have

$$\frac{V_{o1}}{V_{in}} = -\frac{R_f}{R_2 + 1/sC_2} = -\frac{sR_fC_2}{sR_2C_2 + 1} = \left(-\frac{R_f}{R_2}\right)\frac{s}{s + \omega_2} \qquad (7.4\text{-}44c)$$

where $\omega_2 = 1/R_2C_2$. Thus, except for the minus sign, we can see that this circuit can be used to produce the first part of the transfer function on the right-hand side of equation (7.4-39). In particular, letting $C_2 = 0.01$ μF, the corresponding required value of R_2 is seen to be

$$R_2 = \frac{1}{\omega_2 C_2} = \frac{1}{(5 \times 10^3)(10^{-8})} = 20 \text{ k}\Omega \qquad (7.4\text{-}45)$$

Furthermore, because a voltage gain of the second stage at high frequencies is equal to 2.41, the required gain of the first stage must be

$$A_1 = -\frac{R_f}{R_2} = -\frac{200}{2.41} = -83 \qquad (7.4\text{-}46a)$$

so that

$$R_f = 83R_2 = 1.66 \text{ M}\Omega \qquad (7.4\text{-}46b)$$

Using these results, the transfer function of the first stage may be written as

$$H_1(s) = \frac{-83s}{s + 5 \times 10^3} \qquad (7.4\text{-}47)$$

Because the stages do not load one another, the overall filter transfer function is simply equal to the product of the transfer functions of the individual stages, so that

$$\frac{V_o}{V_{in}} = H(s) = H_1(s)H_2(s) = \frac{-83s}{s + 5 \times 10^3} \frac{2.41s^2}{s^2 + 1723s + 8.639 \times 10^6}$$

$$= (-200)\frac{s}{s + 5000} \frac{s^2}{s^2 + 1723s + 8.639 \times 10^6} \qquad (7.4\text{-}48)$$

We will assume that the inversion produced by the first stage is not a problem.

7.4-3 Switched-Capacitor Active Filters

Ever since the advent of active filters, there has been a continuing effort to place the entire filter design on a single integrated-circuit chip. However, there have been two major stumbling blocks to the practical implementation of this concept. First, active filter circuits in the audio and subaudio frequency regions (where these filters are most commonly used) require large RC products. Unfortunately, in a purely integrated-circuit design approach, big RC time constants are hard to come by since there are rather severe limitations on the allowable sizes of IC resistors and capacitors. Typically, the largest reproducible resistor sizes that can be achieved are about 100 kΩ, while the biggest practical capacitors are about 50 pF. As a result, the lowest active filter corner frequencies or break frequencies that can be achieved with purely integrated-circuit components are on the order of $1/(10^5)(5 \times 10^{-11}) = 2 \times 10^5$ rad/s, or about 32 kHz. Clearly, filters built using RC products that are in this range will be unsuitable for active filter designs in the audio- and subaudio-frequency ranges.

A second problem associated with the use of integrated-circuit components concerns their tolerances. The absolute accuracy of integrated-circuit resistors and capacitors is about $\pm 25\%$. As a result, active filter designs based solely on the use of these integrated-circuit components tend to have widely varying characteristics (on the order of $\pm 50\%$), and for most applications this spread in filter performance is unacceptable.

An interesting variation on the conventional active filter design approach involves a technique known as switched-capacitor filtering, and as we will demonstrate shortly, this method overcomes both of the major shortcomings associated with ordinary integrated active filter circuits. The theory of operation of the switched-capacitor filter centers on the fact that a capacitor switched in and out of a circuit at a high rate of speed has electrical characteristics that are in many ways similar to those of an ordinary resistor.

To demonstrate this equivalence, consider the circuit illustrated in Figure 7.4-12a, in which the two switches S_1 and S_2 are opened and closed 180° out of phase from one another at a frequency f_{CLK} that is much greater than the signal frequency. In the practical implementation of this circuit shown in part (b) of the figure the switches are actually seen to be MOS transistors that are driven by two nonoverlapping clock signals ϕ_1 and ϕ_2.

The current waveshapes corresponding to the operation of these two switches are indicated in Figure 7.4-12c, and because these MOS switches have finite resistances when they are ON, the currents charging and discharging C_s are seen to be exponentials rather than impulses. The area ΔQ_s underneath each of curves i_1 and i_2 represents the charge added to and then removed from C_s by the operation of switches S_1 and S_2, respectively. Thus the basic functioning of this circuit may be explained as follows. Each time that S_1 closes, C_s is charged an amount

$$\Delta Q_s = C_s v_{in} \qquad (7.4\text{-}49)$$

and then later in the cycle, when S_2 closes, it is discharged by the same amount. Because this sequence repeats every T_{CLK} seconds, there is an average current flow into this circuit equal to

$$(i_1)_{AVG} = \frac{\Delta Q_s}{\Delta t} = \frac{C_s v_{in}}{T_{CLK}} = (f_{CLK} C_s) v_{in} \qquad (7.4\text{-}50)$$

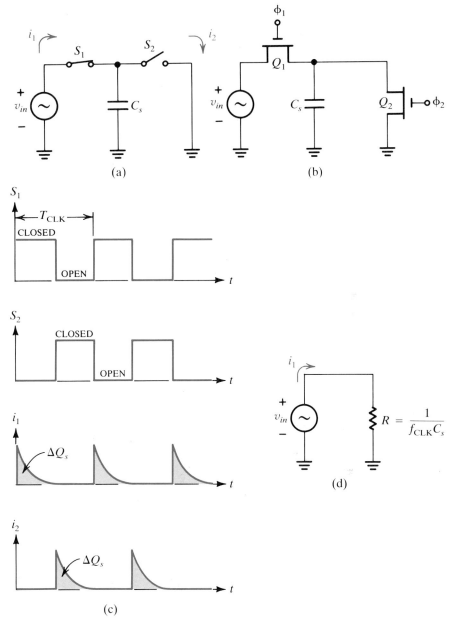

Figure 7.4-12
Switched-capacitor as a circuit element.

Since the current flow is proportional to the voltage applied to this circuit, but is independent of its frequency, this switched capacitor element "on the average" may be replaced by an equivalent resistor

$$R = \frac{v_{in}}{(i_1)_{\text{AVG}}} = \frac{v_{in}}{(f_{\text{CLK}}\, C_s)v_{in}} = \frac{1}{f_{\text{CLK}}\, C_s} \qquad (7.4\text{-}51)$$

as illustrated in Figure 7.4-12d.

7.4 / Active Filters

437

To make use of the switched-capacitor phenomenon, it is necessary to use it in a circuit that inherently responds to the average rather than to the instantaneous current flowing through it. Fortunately, the electronic integrator in Figure 7.4-13a is just such a circuit, and in part (b) of this figure we illustrate how an integrator may be implemented using a switched capacitor in place of the input resistor R. Here, each time S_1 closes, a charge equal to that given in equation (7.4-49) is placed on C_s. Later in the cycle, when S_2 closes, virtually all of this charge is transferred to the feedback capacitor C_f. To understand why this occurs, consider the input equivalent circuit shown in Figure 7.4-13c, in which C_f is assumed to be initially uncharged. In this circuit the Miller theorem has been applied to demonstrate why C_s rapidly discharges once S_1 opens and S_2

Figure 7.4-13

Switched-capacitor building blocks.

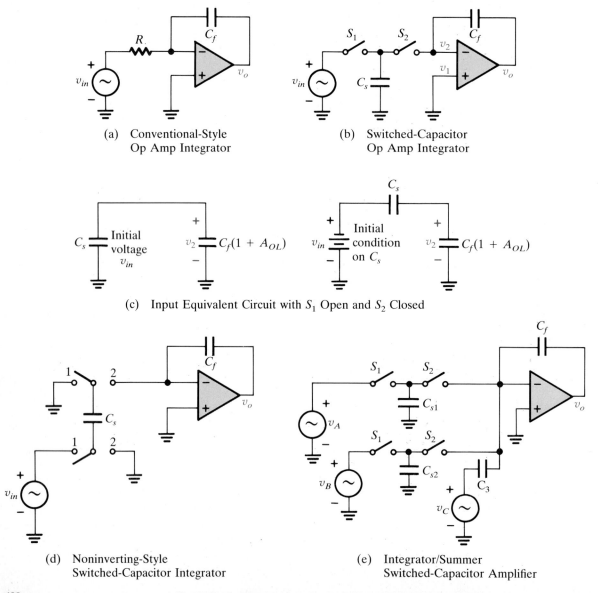

(a) Conventional-Style
Op Amp Integrator

(b) Switched-Capacitor
Op Amp Integrator

(c) Input Equivalent Circuit with S_1 Open and S_2 Closed

(d) Noninverting-Style
Switched-Capacitor Integrator

(e) Integrator/Summer
Switched-Capacitor Amplifier

closes. Since the circuit is a capacitive divider, v_2 is seen to be equal to

$$v_2 = v_{in}\frac{C_s}{C_s + C_f(1 + A_{OL})} \simeq v_{in}\frac{C_s}{C_f A_{OL}} \qquad (7.4\text{-}52)$$

Notice that because A_{OL} is large, v_2 is now "virtually" at ground potential, and as a result, C_s is almost completely discharged. Furthermore, since

$$v_o = A_{OL}(v_1 - v_2) = A_{OL}\left(0 - \frac{C_s v_{in}}{C_f A_{OL}}\right) = -\frac{C_s v_{in}}{C_f} \qquad (7.4\text{-}53)$$

it also follows that

$$\Delta Q_f = v_o C_f \simeq -\frac{C_s v_{in}}{C_f}C_f = -\Delta Q_s \qquad (7.4\text{-}54)$$

In other words, essentially all the charge that was originally on C_s has now been transferred to C_f.

In spite of all of these rather interesting facts concerning the charge transfer that occurs between C_s and C_f, it is still a bit difficult to understand how this circuit functions like an integrator. To bring this point into better focus, consider that every Δt (or T_{CLK}) seconds, a charge ΔQ_s [given by equation (7.4-49)] is dumped from C_s onto C_f. This produces a corresponding change in the output voltage equal to

$$\Delta v_o = -\frac{\Delta Q_s}{C_f} = -\frac{C_s v_{in}}{C_f} \qquad (7.4\text{-}55a)$$

so that we have

$$\frac{\Delta v_o}{\Delta t} = -\frac{C_s v_{in}}{C_f \Delta t} = -\frac{C_s v_{in}}{C_f T_{\text{CLK}}} = -\frac{C_s f_{\text{CLK}}}{C_f}v_{in} \qquad (7.4\text{-}55b)$$

In the limit as Δt approaches zero (or alternatively, as f_{CLK} becomes very large in comparison to the input signal frequency), equation (7.4-55b) reduces to

$$\frac{dv_o}{dt} = -\frac{C_s f_{\text{CLK}}}{C_f}v_{in} \qquad (7.4\text{-}56a)$$

and solving this expression for v_o, we have

$$v_o(t) = -\frac{C_s f_{\text{CLK}}}{C_f}\int_0^t v_{in}(\tau)\,d\tau \qquad (7.4\text{-}56b)$$

It is interesting to point out that this same result can also be obtained, although somewhat mechanistically, by substituting equation (7.4-51) for the equivalent resistance R into the standard expression for an op amp integrator given in equation (7.3-33).

Let's examine equation (7.4-56b) to see why this switched-capacitor integrator is so useful for the design of active filters. First, the integrator time constant for this circuit is given by the expression $(C_f/C_s)T_{\text{CLK}}$. As a result, this time constant can be made as large as needed without using large capacitor values, simply by increasing the clock period. This means that active filters having virtually any desired corner frequencies can readily be constructed since this circuit's operation does not depend on the actual size of these capacitors, but is instead, only a function of their ratios. This last fact also points out another very significant attribute of switched-capacitor active filters, namely, that the accuracy of the filter (which is really a function of the accuracy of the circuit time constants) does not depend on the tolerance ratings of the capacitors, but rather on the way in which these capacitor values track one another. Because the relative sizes of the capacitors in an IC can be held to very tight tolerances (on the order of 0.1%), it will be possible to design highly accurate switched-capacitor active filter circuits.

Another interesting point in this regard is that the circuit time constant can be adjusted by varying the speed of the clock that drives the capacitor switches. As a result, by changing the clock frequency, we can simultaneously alter every time constant in the filter by the same amount. This will allow for the corner frequencies of the filter to be varied without altering the basic shape of the filter. This type of adjustment is extremely difficult to implement in conventional active filter designs because many circuit components need to be changed at the same time.

An interesting variation on the basic inverting-style integrator circuit just discussed is shown in Figure 7.4-13d. Here, after C_s is charged to v_{in} by placing the switches in position (1), it is then discharged by throwing both of these switches into position (2). Because this effectively flips C_s upside down, a charge equal to ΔQ_s rather than $-\Delta Q_s$ is placed on C_f; in other words, this circuit is a noninverting-style integrator with

$$v_o = \frac{1}{RC_f} \int_0^t v_{in}(\tau) \, d\tau = \frac{f_{\text{CLK}} C_s}{C_f} \int_0^t v_{in}(\tau) \, d\tau \qquad (7.4\text{-}57)$$

Figure 7.4-13e illustrates one additional circuit that will prove to be very useful for the design of switched-capacitor active filter circuits. In this circuit, because v_2 is a virtual ground,

$$I_f = -(I_1 + I_2 + I_3) = -\frac{V_A}{R_1} - \frac{V_B}{R_2} - \frac{V_C}{1/sC_3} \qquad (7.4\text{-}58a)$$

so that

$$V_o = I_f \frac{1}{sC_f} = -\frac{V_A}{sR_1 C_f} - \frac{V_B}{sR_2 C_f} - V_C \frac{C_3}{C_f} \qquad (7.4\text{-}58b)$$

Substituting equation (7.4-51) for R_1 and R_2 and taking the inverse Laplace transform of equation (7.4-58b), we obtain

$$v_o(t) = -\frac{f_{\text{CLK}} C_{s1}}{C_f} \int_0^t v_A(\tau) \, d\tau - \frac{f_{\text{CLK}} C_{s2}}{C_f} \int_0^t v_B(\tau) \, d\tau - \frac{C_3}{C_f} v_C \qquad (7.4\text{-}59)$$

Thus this circuit functions like an integrator/summing amplifier where the signals applied to the switched-capacitor inputs are integrated, while that applied to the fixed capacitor C_3 is simply amplified by $-C_3/C_f$.

Now that we have an understanding of the building blocks available for the design of switched-capacitor active filters, let's take a look at how they can be applied to implement specific filter circuits. The approach that we will be using is quite similar to that employed for the solution of differential equations on analog computers. To illustrate this technique, let's consider the design of a

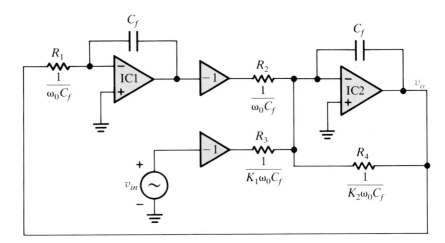

Figure 7.4-14
Initial stages in the development of a switched-capacitor second-order bandpass filter.

filter having a transfer function of the form

$$\frac{V_o}{V_{in}} = H(s) = \frac{K_1\omega_0 s}{s^2 + K_2\omega_0 s + \omega_0^2} \tag{7.4-60}$$

which is a second-order bandpass filter centered at ω_0, and having a gain equal to (K_1/K_2) at this frequency.

To construct this filter using a switched-capacitor design approach, we begin by cross-multiplying equation (7.4-60), which yields

$$(s^2 + K_2\omega_0 s + \omega_0^2)V_o = K_1\omega_0 s V_{in} \tag{7.4-61a}$$

Dividing both sides of this expression by s^2, we obtain

$$\left(1 + K_2\frac{\omega_0}{s} + \frac{\omega_0^2}{s^2}\right)V_o = K_1\frac{\omega_0}{s}V_{in} \tag{7.4-61b}$$

In general, this and any other second-order transfer function can be implemented by a cascade of two integrator/summing amplifiers. To determine the specific form of the circuit required, we begin by solving equation (7.4-61b) for V_o, which yields

$$V_o = K_1\frac{\omega_0}{s}V_{in} - K_2\frac{\omega_0}{s}V_o - \frac{\omega_0}{s}\frac{\omega_0}{s}V_o$$

$$= \left(\frac{K_1\omega_0}{s}V_{in} - K_2\frac{\omega_0}{s}V_o\right) - \frac{\omega_0}{s}\frac{\omega_0}{s}V_o \tag{7.4-62}$$

$$\qquad\qquad \uparrow \qquad\qquad \uparrow \qquad\qquad \uparrow$$

inputs to second integrator IC2   input to first integrator IC1

Defining the output voltage of IC2 in Figure 7.4-14 as v_o, we should be able to see that the circuit shown in this figure correctly implements equation (7.4-62). The translation of this type of analog active filter into an actual switched-capacitor design now only requires that we replace each of the integrator resistors by the appropriate circuit in Figure 7.4-13 (see Problem 7.4-13).

7.4-1 Using as many circuits of the type illustrated in Figure 7.4-5b as needed, design a circuit to implement the transfer function

$$H(s) = \frac{500s^2}{(s + 100)(s + 1000)}$$

All capacitors are to be 0.1 μF. **Answer** Two stages of the form given in Figure 7.4-5a are needed. Stage 1: $R_2 = 10$ kΩ, $R_f = 500$ kΩ; stage 2: $R_2 = 100$ kΩ, $R_f = 1$ MΩ.

7.4-2 Derive an expression for the transfer functon of an $n = 4$ (1 dB) Chebychev high-pass filter with a cutoff frequency of 500 rad/s and a high-frequency gain of 40. **Answer** $H(s) = (40s^4)/(s^2 + 1208s + 9 \times 10^5)(s^2 + 141s + 2.5 \times 10^5)$

7.4-3 **a.** Find the transfer function for an $n = 1$ unity-gain Butterworth band-reject filter with a center frequency ω_0 and a $Q = 1$.

b. Show that this filter can also be implemented as the sum of a low-pass and a high-pass filter.

Answer $H(s) = \dfrac{s^2 + \omega_0^2}{s^2 + s\omega_0 + \omega_0^2} = \underbrace{\dfrac{s^2}{s^2 + s\omega_0 + \omega_0^2}}_{\text{HPF}} + \underbrace{\dfrac{\omega_0^2}{s^2 + s\omega_0 + \omega_0^2}}_{\text{LPF}}$

7.4-4 For the second-order active bandpass filter whose transfer function is given in equation (7.4-38), find the sensitivity of the gain at the center frequency to changes in A_{CL} for values of A_{CL} equal to 1 and 2.9. The basic expression for the sensitivity of one circuit parameter with respect to changes in another is given in equation (7.1-9). **Answer** 1.5, 30

7.4-5 The inverter at the output of IC1 in Figure 7.4-14 is replaced by a short. Determine V_o/V_{in} for the resulting circuit. **Answer** $K_1\omega_0 s/(s^2 + K_2\omega_0 s - \omega_0^2)$

7.5 NONLINEAR APPLICATIONS OF OPERATIONAL AMPLIFIERS

When an amplifier is operating in its linear region, the input–output transfer characteristic is a straight line, and as a result, when the input signal is doubled, the output signal is also doubled. Most op amp circuits operate in the linear region. However, for some applications it is advantageous to have an amplifier that exhibits nonlinear behavior. This can be accomplished with an op amp circuit by driving the amplifier output into saturation at one of the power supply rails, or by adding nonlinear elements such as diodes to the input or feedback paths of the amplifier circuit. In this section we examine the application areas where nonlinear amplifier performance is desirable, and we also investigate techniques for designing op amp circuits that exhibit specific types of nonlinear behavior.

7.5-1   The Comparator

A comparator, as its name implies, is a circuit in which an applied input voltage is compared to a reference voltage, and depending on whether the input is greater than or less than the reference, the output is set to one of two different voltage levels.

Comparators are used in many different types of electronic circuits; specific examples include A/D converters, waveform generators, and voltage level detectors, to name just a few. Nearly every integrated-circuit manufacturer produces a line of ICs specifically designed to function as "comparators," but for many applications, an op amp circuit such as that shown in Figure 7.5-1 will serve just as well.

In trying to understand how this circuit operates, it is useful to think of it as an amplifier that follows the input–output equation

$$v_o = A_{OL}(v_1 - v_2) = A_{OL}(v_{in} - V_{REF}) \tag{7.5-1}$$

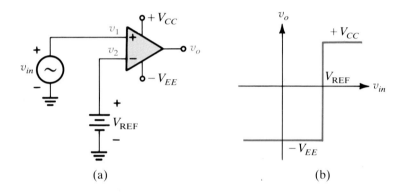

(a)   (b)

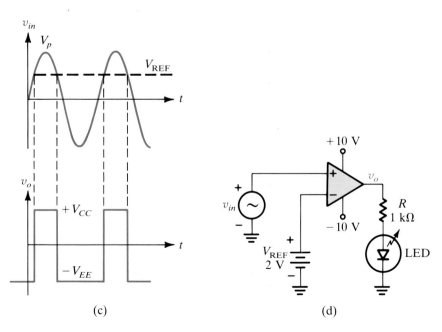

(c)   (d)

Figure 7.5-1
Using the op amp as a comparator.

as long as v_o does not exceed the power supply rails. Because A_{OL} is very large, whenever v_{in} is greater than V_{REF}, the output will attempt to go to a large positive voltage, and this will cause the amplifier to saturate at $+V_{CC}$. Conversely, whenever v_{in} is less than V_{REF}, the output will saturate at $-V_{EE}$. The transfer characteristic for this circuit is shown in Figure 7.5-1b, and the output waveshape obtained when v_{in} is a sine wave is illustrated in part (c) of the figure.

This type of circuit is useful as an alarm or threshold detector. For example, by connecting a light-emitting diode or LED to the output of the op amp illustrated in Figure 7.5-1d, whenever the input voltage, v_{in}, rises above $+2$ V, the op amp output will jump up to $+10$ V. This will forward bias the light-emitting diode with about 10 mA of current, and as a result the LED will light, signaling an alarm condition.

Notice that by minor modifications to this circuit design, the alarm can be activated for nearly any desired input signal condition. For example, by reversing the polarity of V_{REF} in this figure, the light-emitting diode will go ON whenever v_{in} exceeds -2 V, while reversing the op amp input connections in Figure 7.5-1d will cause the LED to light whenever v_{in} drops below $+2$ V.

Open-loop comparators of the type we have been discussing can exhibit rather strange behavior in the presence of input noise. Specifically, the outputs of these types of circuits tend to chatter back and forth as the input voltage approaches V_{REF}. Figure 7.5-2 will help us to understand why this occurs. Here, the input signal is assumed to be a slowly varying ramp voltage which is contaminated by a high-frequency sinusoidal noise source.

At $t = 0$ v_{in} is negative and v_o starts out at $-V_{EE}$, and at $t = t_1$, because of the noise, the op amp switches prematurely to $+V_{CC}$. However, at $t = t_2$, v_{in} again drops below V_{REF}, so that v_o temporarily switches back down to $-V_{EE}$, and then at $t = t_3$ it goes back up to $+V_{CC}$. Thereafter, it remains at this level. By examining the circuit in Figure 7.5-2a, and the waveforms in part (b) of the figure, it should be clear that this type of chattering is unavoidable with this circuit and that it can become much more severe if the amplitude of the noise gets larger.

LAB #3

The significance of the chattering depends on the application of the circuit. If the output of the op amp comparator is just being used to drive an LED, this output chatter will not be too serious because the flickering of the LED probably will not be noticed. But if, for example, this output is driving a digital counting circuit, the effect of the chattering will be disastrous because it will cause erroneous counts. Fortunately, this type of chatter can be almost completely eliminated by the use of positive feedback.

7.5-2 The Schmitt Trigger: A Positive-Feedback Comparator

neg. / pos feedback

In a negative-feedback amplifier, the feedback tends to make the two inputs v_1 and v_2 nearly equal to one another, whereas in a positive-feedback amplifier just the opposite effect occurs. When an amplifier having positive feedback is in its active region, the feedback acts to drive the two input voltages apart, and this causes the amplifier to saturate. As a result, a Schmitt trigger or positive-feedback comparator is always saturated at either the $+V_{CC}$ or $-V_{EE}$ power supply rail.

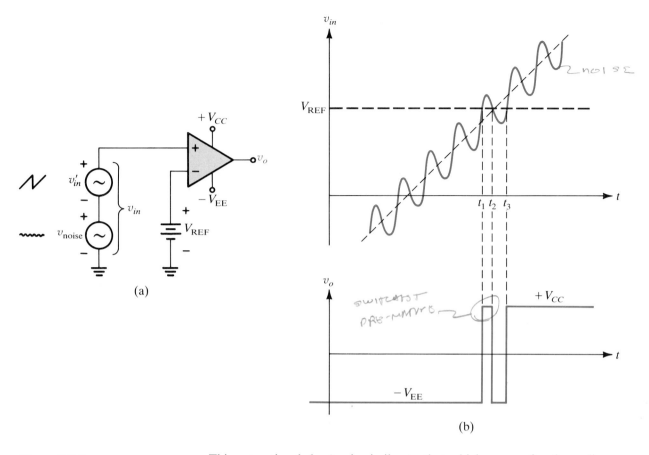

(a)

(b)

Figure 7.5-2

Output chatter in a comparator without positive feedback.

This saturation behavior is similar to that which occurs for the ordinary comparator. However, unlike the open-loop comparator, the Schmitt trigger has two different switching levels, depending on its output state. In fact, it is precisely this characteristic which gives this circuit its noise immunity. To illustrate this behavior, let's consider how a Schmitt trigger comparator prevents multiple output transitions from occurring when the input signal is noisy.

The transfer characteristics of a conventional comparator and a Schmitt trigger comparator are given in Figure 7.5-3a and b, respectively. The arrows on the Schmitt trigger transfer characteristic indicate the path taken by the output as v_{in} varies. When the circuit is initially in the $-V_{EE}$ output state and v_{in} increases beyond V_{REF1}, the output switches to $+V_{CC}$ along path (a), while when the circuit output is at $+V_{CC}$ volts, the switching back to $-V_{EE}$ occurs when the input voltage drops below V_{REF2} and traverses along path (b). The fact that the current behavior of the Schmitt trigger depends on its past history is called hysteresis.

Let's see how the Schmitt trigger responds to the noisy ramp signal shown in Figure 7.5-3c. If we assume that the comparator is initially in the $-V_{EE}$ state, then as the input voltage rises, at $t = t_1$ it will cross V_{REF1} and this will cause the output to make a transition to $+V_{CC}$ along path (a) on the transfer characteristic. Once the comparator is in the $+V_{CC}$ state, the switching threshold of the Schmitt trigger changes. It is now located at V_{REF2}. However, for $t > t_1$, at least for the time interval shown in the figure, the input never

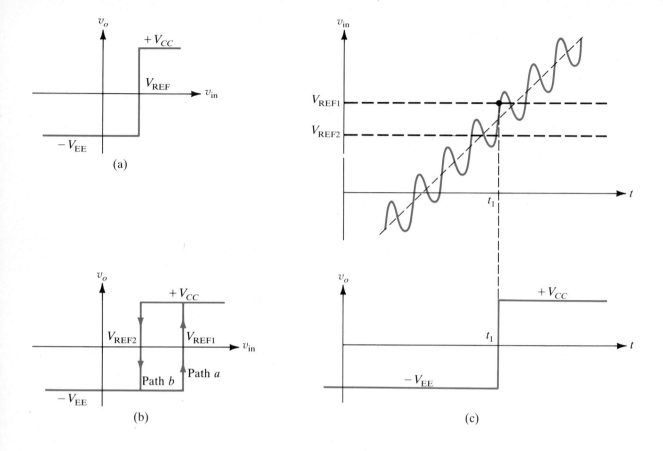

(a)

(b)

(c)

Figure 7.5-3

Elimination of output chatter by the use of a positive-feedback (Schmitt trigger) comparator.

again drops below V_{REF2}, so that the output stays at $+V_{CC}$, and no chattering occurs. From an examination of Figure 7.5-3c, we can see that the chattering will be eliminated as long as the hysteresis in the transfer characteristic, that is, the difference between V_{REF1} and V_{REF2}, is greater than the peak-to-peak amplitude of the noise voltage.

The circuit shown in Figure 7.5-4 illustrates one specific way to construct a Schmitt trigger. To determine the shape of its transfer characteristic, let's initially assume that v_{in} is a large negative voltage, much more negative than v_1 (say, -5 V). For this situation, because v_2 is less than v_1, the op amp output is saturated at $+10$ V, and therefore

$$v_1 = +V_{CC}\frac{R_2}{R_f + R_2} = +1 \text{ V} \qquad (7.5\text{-}2)$$

We are now at point 1 on the transfer characteristic in Figure 7.5-4b and on the first line of the table in Figure 7.5-4c. As v_{in} increases, nothing will happen until v_{in} is nearly equal to v_1 and the op amp enters the active region.

Let's follow the sequence of events that transpires when $v_{in} = 0.9999$ V. At this point, if v_1 still equals $+1$ V, then $(v_1 - v_2)$ drops to 0.0001 V, and v_o also drops to $10^4 \times (0.0001) = +1$ V. But if v_o drops, this will cause v_1 to drop proportionally, making $(v_1 - v_2)$ still larger, which causes v_o to drop still further, and so on. This sequence continues until the op amp saturates at the

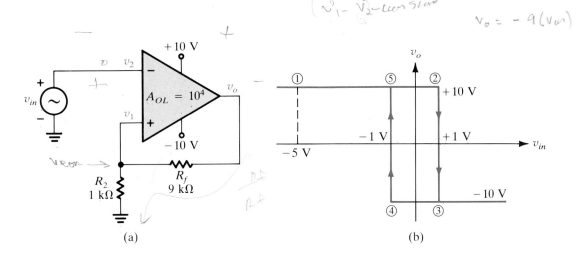

$$\left(v_1 - v_2 \leftarrow \text{term start}\right)$$

$$V_o = -9(V_{in})$$

(a)

(b)

$v_{in} = v_2$ (V)	v_1 (V)	$(v_1 - v_2)$ (V)	$A_{OL}(v_1 - v_2)$ (V)	v_o (V)	Op amp state
−5	+1	+6	60,000	+10	Saturated
−4	+1	+5	50,000	+10	Saturated
0	+1	+1	10,000	+10	Saturated
+0.9	+1	+0.1	1000	+10	Saturated
+0.9999	+1	+0.0001	1.0	+1.0	Active
+0.9999	+0.1	−0.8999	−8,999	−10	Saturated
+0.9999	−1.0	−1.9999	−19,999	−10	Saturated (stable now at −10 V supply rail)

(c)

Figure 7.5-4

Schmitt trigger—a comparator with positive feedback.

−10-V supply rail, and at this point the output is again stable. The switching sequence is illustrated by the entries on lines 5, 6, and 7 in Figure 7.5-4c.

The operation of this Schmitt trigger may be summarized as follows. When v_o is initially in the +10-V state, v_1 is given by equation (7.5-2) and the amplifier remains saturated at the positive supply rail as long as v_{in} is less than +1 V. When v_{in} gets very close to +1 V (at point 2 in Figure 7.5-4b), the op amp enters the active region, and the output rapidly switches to −10 V from point 2 to point 3 on the transfer characteristic.

Once the output is in the −10-V state, the voltage v_1 now equals −1 V, and the output will remain fixed at −10 V until v_{in} approaches −1 V at point 4 on the transfer characteristic. At this point, following a similar sequence of events to those listed in Figure 7.5-4c, the output again rapidly switches from −10 to +10 V (point 5 in Figure 7.5-4b).

EXAMPLE 7.5-1

The circuit shown in Figure 7.5-5 is another type of Schmitt trigger where the input signal is fed into the noninverting input terminal of the op amp. Determine the circuit's transfer (or hysteresis) characteristic, and sketch the output obtained when $v_{in} = 5 \sin \omega t$ volts.

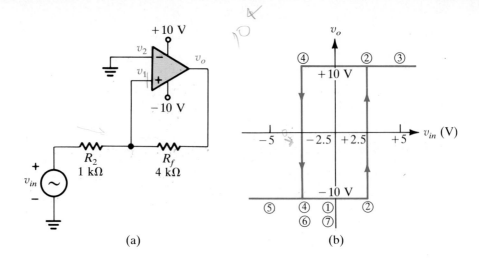

(a)

(b)

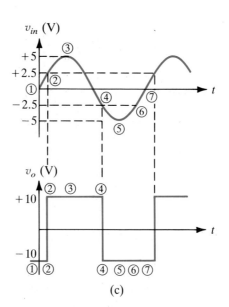

(c)

Figure 7.5-5
Example 7.5-1.

SOLUTION

For this circuit $v_2 = 0$, and the voltage v_1 may be expressed as

$$v_1 = \frac{R_f}{R_2 + R_f} v_{in} + \frac{R_2}{R_2 + R_f} v_o = \tfrac{4}{5} v_{in} + \tfrac{1}{5} v_o \tag{7.5-3}$$

If we make v_{in} very negative, eventually we will force v_o to the negative supply rail. Let's start the analysis of this circuit at this point, that is, with v_{in} very negative and v_o at -10 V. At this point, following equation (7.5-3),

$$v_1 = \tfrac{4}{5} v_{in} - 2.0 \text{ V} \tag{7.5-4}$$

Nothing will happen to the output of the op amp until it enters the active region, and this will occur when v_1 approaches zero or when v_{in} nears $\tfrac{5}{4}(2.0 \text{ V}) = +2.5$ V. At this point, due to the positive feedback, the op amp output will switch to the opposite state (the $+10$-V supply rail).

With the output at $+10$ V, in accordance with equation (7.5-3a), the expression for v_1 is now

$$v_1 = \tfrac{4}{5} v_{in} + 2.0 \text{ V} \tag{7.5-5}$$

and the output switches in this case when v_{in} approaches -2.5 V.

The hysteresis characteristic for this Schmitt trigger is given in Figure 7.5-5b, and the output response when $v_{in} = 5 \sin \omega t$ volts is sketched in part (c) of this figure assuming that the op amp was initially in the -10-V state at $t = 0$. This numbers on these sketches correspond to the position of the circuit operating point on the transfer characteristic at each instant of time.

7.5-3 Precision Rectifiers and Peak Detectors

The circuit illustrated in Figure 7.5-6a is an ideal half-wave rectifier in which the positive half of the applied input voltage appears in the output while the negative half is clipped off. In an actual circuit of this sort, the output waveform obtained can be very far from this ideal, especially when the input signal is small. Consider, for example, the silicon diode circuit shown in Figure 7.5-6b, in which the diode is modeled as a 0.7-V battery in series with an ideal diode. Here, because v_{in} is only a 1-V sine wave, the circuit will not produce very good half-wave rectification of the input signal. The efficiency of this half-wave rectifier can be improved significantly by placing the diode in the feedback loop of an op amp, as illustrated in Figure 7.5-6c.

The operation of this circuit may be explained as follows. When v_{in} is positive, if we consider that the op amp is in its active region, then because v_2 and v_1 will nearly be equal, there will be a positive diode current

$$i_R = i_D \simeq \frac{v_{in}}{R} \tag{7.5-6}$$

Thus the diode will be forward biased as long as v_2 is greater than zero. In the circuit in Figure 7.5-6b the silicon diode is forward biased when v_{in} exceeds 0.7 V. Let's determine the required turn-on voltage for the precision rectifier circuit in Figure 7.5-6c.

If we assume that the diode is ON and that the op amp is in the active region, then

$$v_{o1} = v_2 + 0.7 \text{ V} = A_{OL}(v_{in} - v_2) \tag{7.5-7}$$

or

$$v_2 = \frac{A_{OL}}{1 + A_{OL}} v_{in} - \frac{0.7 \text{ V}}{1 + A_{OL}} \tag{7.5-8}$$

Using this result, we can see that the diode is ON, or $i_D > 0$, whenever v_{in} is greater than $(0.7 \text{ V})/A_{OL}$. Thus this rectifier circuit reduces the diode turn-on voltage by the factor A_{OL}. For a typical A_{OL} of 10^4, this means that the turn-on voltage for this precision rectifier will only be 70 μV instead of 700 mV.

When v_{in} drops below this voltage, the diode goes OFF, i_R becomes zero, and as a result with v_1 negative and v_2 equal to zero, v_{o1} goes to the -10-V power supply rail. The sketches in Figure 7.5-6d illustrate the typical voltages found at v_1, v_2, and v_{o1} for the case where v_{in} is a sine wave. As is apparent

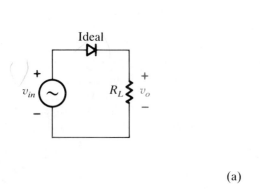

Ideal

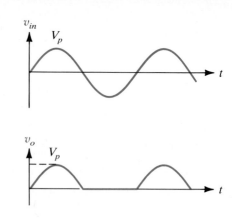

(a)

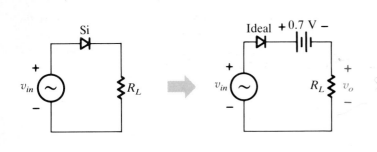

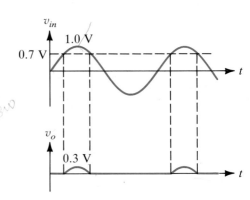

(b)

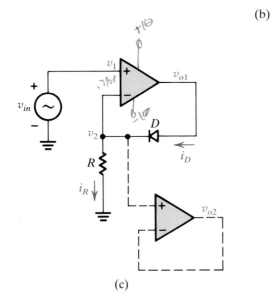

(c)

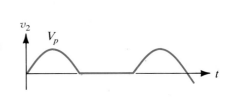

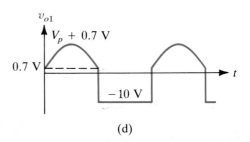

(d)

Figure 7.5-6
Precision half-wave rectifier.

from an examination of these sketches, the half-wave rectified output from this circuit is obtained across the resistor R. To avoid loading the circuit at this point, a second buffer op amp is frequently added. It is shown by the dashed lines in Figure 7.5-6c.

By making a simple modification in the half-wave rectifier circuit in Figure 7.5-6c, we can convert it into a precision peak detector. All that we need to do is replace the resistor R in this figure with a capacitor. This modified circuit is illustrated in Figure 7.5-7a, and its operation may be explained as follows. Let's assume that C is initially uncharged. At $t = 0$ when v_{in} begins to increase positively, because v_1 will become greater than v_2, the output of the op amp will rise toward $+V_{CC}$ volts. This will cause the diode D_1 to conduct and will charge the capacitor until v_2 matches v_1. Once v_2 equals v_{in}, the charging will stop and v_o will stabilize at $(v_{in} + 0.7 \text{ V})$. If we assume that the capacitor charging occurs nearly instantaneously, then as v_{in} increases, the voltage v_2 will simply follow it, as illustrated in Figure 7.5-7b.

When the input signal begins to decrease, the capacitor tries to discharge through the diode, but because the current cannot flow through the diode in the reverse direction, D_1 turns OFF and the voltage across the capacitor stays at the peak value of v_{in}. As v_{in} decreases, v_1 becomes much smaller than v_2, and therefore the output of the op amp saturates at the negative supply rail.

Figure 7.5-7
Precision peak detector.

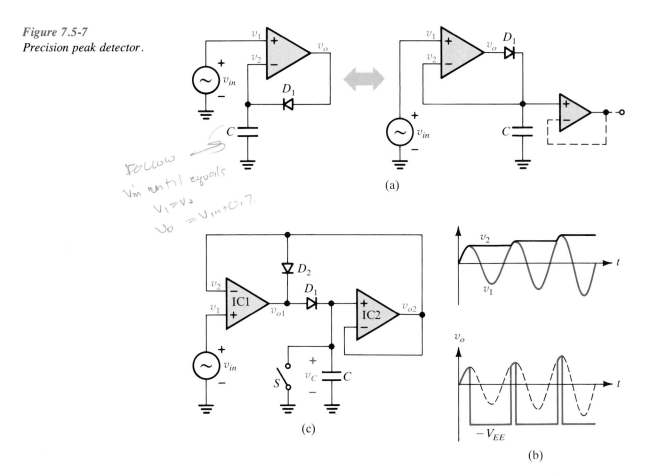

(a)

(c)

(b)

As soon as the input signal amplitude increases beyond the peak voltage currently being held on C, the op amp voltage switches positively to $(v_2 + 0.7 \text{ V})$ and the capacitor voltage again begins to track the input to the next peak. In this way the peak detector circuit captures the largest positive voltage seen at the input terminal.

For this circuit to have practical utility, some means must be provided to discharge the capacitor to initiate a new measurement cycle. This is accomplished with the switch S shown in Figure 7.5-7c. Also shown in this figure is a second diode D_2. This diode prevents the output from the first op amp from saturating when v_{in} drops. When v_{in} decreases and v_{o1} tries to go to the negative supply rail, D_2 conducts and v_{o1} is held at $(v_C - 0.7 \text{ V})$. This improves the operating speed of the circuit because it can require a significant amount of time to bring an op amp back into the active region from saturation. It is also worthwhile to point out that in this version of the peak detector, the feedback voltage into IC1 is taken from the output of the buffer IC2. This is done to prevent D_2 from discharging the capacitor when v_{o1} tries to go negative. When high-precision peak detection is required, IC2 may need to be an FET or Darlington input op amp to minimize the discharge of C by the input bias current of IC2.

7.5-4 Amplitude Limiting Circuits

As we had discussed in Section 1.6, a limiter is a circuit that uses nonlinear circuit elements to prevent the output amplitude from exceeding certain predefined values. Simple amplitude limiting circuits may be constructed by connecting the nonlinear element in the feedback loop of the op amp.

One example of this type of circuit is illustrated in Figure 7.5-8a; it is a nonprecision half-wave rectifier. For this circuit when v_{in} is positive, v_o goes negative and the voltage drop across R_f acts to reverse bias D_1 so that the diode is OFF and $v_o = -(R_f/R_2)v_{in}$. When v_{in} goes negative, and in particular when the voltage drop across R_f exceeds 0.7 V, the diode turns ON and v_o is limited to $+0.7$ V. The output for $v_{in} = 5 \sin \omega t$ volts is sketched in Figure 7.5-8a to the right of the circuit.

When amplitude limiting at a voltage level other than zero is needed, a zener diode may be added in series with D_1 (Figure 7.5-8b). As before, when v_{in} is greater than zero, the current in through R_2 cannot flow through D_1, so that it is OFF and v_o is simply equal to $-2v_{in}$. When v_{in} goes negative, the current flow through R_2 reverses, and this produces a voltage drop across R_f tends to turn D_1 ON, but this does not occur initially because the output voltage is not big enough to cause D_2 to zener. When the output voltage attempts to go beyond 5.7 V, D_1 turns ON, D_2 zeners, and the output limits at 5.7 V. The waveshapes shown to the right of the circuit in Figure 7.5-8b illustrate the output obtained with $v_{in} = 5 \sin \omega t$ volts.

For applications where symmetrical limiting is required, two zeners may be connected back to back as illustrated in Figure 7.5-8c. In this circuit when v_o attempts to go beyond 5.7 V, D_1 turns ON and D_2 zeners, while for the opposite case, when v_o tries to go below -5.7 V, D_2 turns ON and D_1 zeners. Thus, as illustrated in the figure, for this circuit the output is symmetrically clipped at ± 5.7 V. Because the circuit gain is -2, the input signal amplitude at the onset of the clipping is $(\pm 5.7/2)$ or ± 2.85 V.

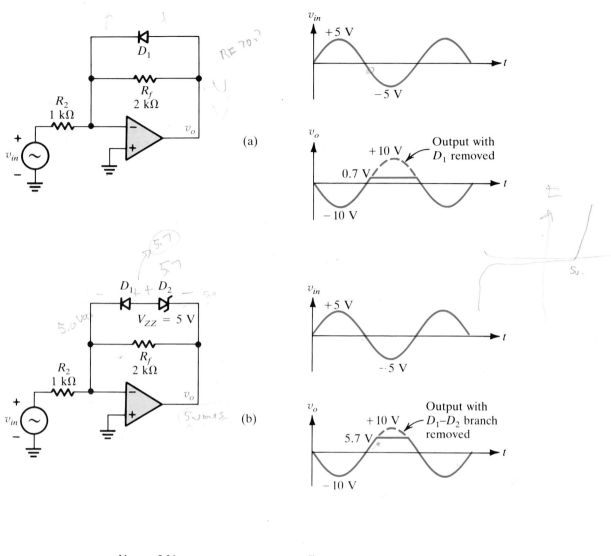

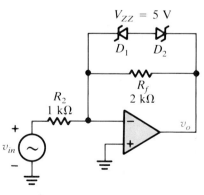

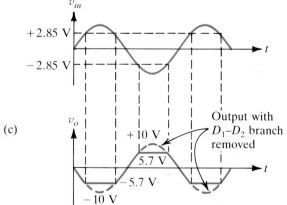

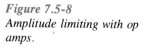

Figure 7.5-8
Amplitude limiting with op
amps.

453

All the limiting circuits we have examined thus far have the same kind of precision as the diode limiting circuits that we discussed in Section 1.6. The only advantages offered by the op amps in these circuits are that they allow us to amplify the input signal and that they eliminate loading effects on the output. However, by using circuits similar to those employed for making precision rectifiers, it is possible to construct precision limiting circuits where the effect of the turn-on voltage of the diode can be all but eliminated.

7.5-5 Nonlinear Function Generation

Operational amplifiers may be used in conjunction with diodes to generate good piecewise-linear approximations to nearly any desired nonlinear transfer function. The basic concept of nonlinear function generation is illustrated in Figure 7.5-9a. If the switch S is open for all values of v_{in}, the input–output transfer characteristic of this amplifier is linear and has the form shown in Figure 7.5-9b as long as v_o does not exceed the power supply voltages. Notice in making this sketch that we are using $(-v_{in})$ as the x variable, so that the straight lines that we sketch will have positive slopes.

Consider next what the transfer characteristic will look like if S is closed when v_o reaches a specific output voltage V_{o1}. For output voltages where $v_o < V_{o1}$, the gain characteristic will be the same as before, and for voltages beyond this point, the gain will drop down to $-(R_1 \parallel R_3)/R_4$. This overall transfer characteristic in this instance is illustrated in Figure 7.5-9c.

For the opposite case where the switch is initially closed and then opens at $v_o = V_{o1}$, the gain starts out at a low value and then increases when the switch opens. This transfer characteristic is shown in Figure 7.5-9d.

Circuits for implementing each of these switching conditions automatically are illustrated in Figure 7.5-9e and f. In analyzing these two circuits, it is helpful to consider that the diodes are ideal. These results can easily be modified later once we understand the operation of these circuits to include the effect of the diode voltage drops.

Starting with the circuit illustrated in Figure 7.5-9e, let's begin by calculating the voltage at the point v_x assuming that the diode is OFF. Applying superposition and neglecting the current flow through the diode, we obtain

$$v_x = \frac{R_3}{R_2 + R_3}(-V_{CC}) + \frac{R_2}{R_2 + R_3}v_o \tag{7.5-9}$$

If we consider that the diode is ideal, the switch S will just turn ON when $v_x = 0$ or when

$$v_o = \frac{R_3}{R_2}V_{CC} \tag{7.5-10}$$

Because of the direction of the diode, the diode turns ON (or the switch closes) when v_o rises above this voltage and is OFF when v_o drops below this value. Furthermore, when the diode is OFF, the amplifier gain is

$$\frac{v_o}{v_{in}} = -\frac{R_1}{R_4} \tag{7.5-11}$$

Once the diode goes ON, we can effectively replace the diode by a short cir-

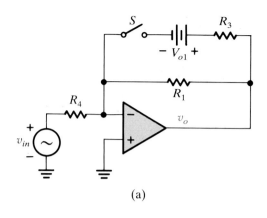

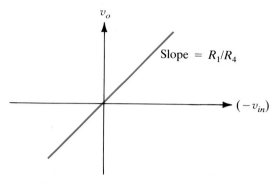

(a)

(b)   Switch Open for All Voltages

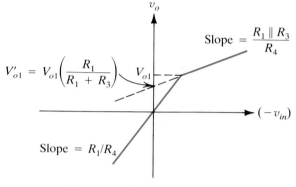

(c)   Switch Closes for $v_o > +V_{o1}$

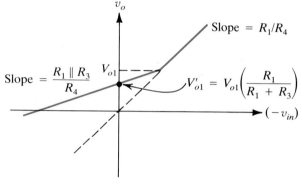

(d)   Switch Opens for $v_o > +V_{o1}$

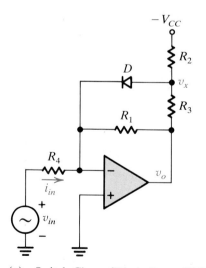

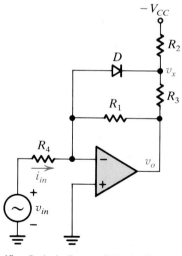

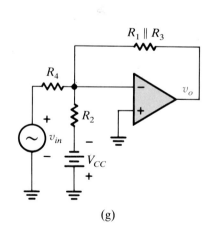

(e)   Switch Closes (Diode Turns ON)
for $v_o > V_{o1}$ $\left(V_{o1} = +\dfrac{R_3}{R_2} V_{CC}\right)$

(f)   Switch Opens (Diode Goes OFF)
for $v_o > V_{o1}$ $\left(V_{o1} = +\dfrac{R_3}{R_2} V_{CC}\right)$

(g)

Figure 7.5-9
Nonlinear function generation.

455

cuit, and using the equivalent circuit in Figure 7.5-9g, in this instance we may write v_o as

$$v_o = \underbrace{\frac{R_1 \| R_3}{R_4}}_{\text{slope}}(-v_{in}) + \underbrace{\frac{R_1 \| R_3}{R_2}V_{CC}}_{y \text{ intercept } (V'_{o1})} \qquad (7.5\text{-}12\text{a})$$

from which it follows that the small-signal gain magnitude (or slope of the transfer characteristic in this region) is

$$|\text{gain}| = \left|\frac{\Delta v_o}{\Delta v_{in}}\right| = \frac{R_1 \| R_3}{R_4} \qquad (7.5\text{-}12\text{b})$$

By applying the same analysis procedure to the circuit in Figure 7.5-9f, we find that the breakpoint is also given by equation (7.5-10). The principal difference between these two circuits is that for this one, when v_o is above this value, the diode is OFF, and it goes ON when v_o drops below $(R_3/R_2)V_{CC}$.

It is useful to observe that multiple-switch circuits of either type may be combined in the same op amp to achieve nearly any desired nonlinear transfer characteristic. The example that follows illustrates how this is accomplished.

EXAMPLE 7.5-2

By using the nonlinear function generation technique just discussed, design a circuit that has the transfer characteristic given in Figure 7.5-10a. You may assume that any diodes used in the design are ideal.

SOLUTION

The slopes on the transfer characteristic represent the small-signal amplifier gain magnitudes in each of the three linear regions of amplifier operation. In particular, the gains in each of these regions are

$$\text{gain} = \begin{cases} -10 & \text{for } v_o < +1 \text{ V} & \text{(region 1)} \\ -20 & \text{for } +1 \text{ V} < v_o < +3 \text{ V} & \text{(region 2)} \\ -3.33 & \text{for } +3 \text{ V} < v_o & \text{(region 3)} \end{cases} \qquad (7.5\text{-}13)$$

The basic circuit for implementing this design is illustrated in Figure 7.5-10b. In this circuit it is assumed that S_A is initially closed and that it opens when v_o reaches $+1$ V, while S_B is initially open and closes when v_o rises above $+3$ V.

To select the component values, let's begin by arbitrarily choosing $R_4 = 1 \text{ k}\Omega$ and noting that the equivalent circuits in regions 1, 2, and 3 are those illustrated in Figure 7.5-10c, d, and e, respectively. The design of these types of circuits should always begin in the region of greatest slope or highest gain since this is where all diodes are OFF and the circuit is the simplest.

Following this procedure, we will start the design in region 2, where the small-signal gain is equal to -20. In this region Figure 7.5-10d applies, so that we have

$$\frac{\Delta v_o}{\Delta(-v_{in})} = \frac{R_1}{R_4} = 20 \qquad (7.5\text{-}14\text{a})$$

or
$$R_1 = 20R_4 = 20 \text{ k}\Omega \qquad (7.5\text{-}14\text{b})$$

Moving next to region 1, and using Figure 7.5-11c, we find that

$$R_1 \| R_{3A} = 10R_4 = 10 \text{ k}\Omega \qquad (7.5\text{-}15\text{a})$$

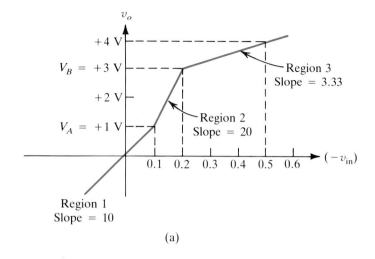

(a)

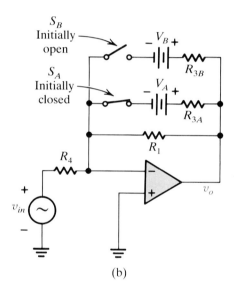

(b)

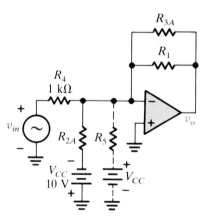

(c)   Region 1, Gain = −10

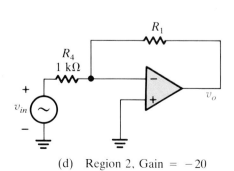

(d)   Region 2, Gain = −20

(e)   Region 3, Gain = −3.33

Figure 7.5-10
Example 7.5-2.

457

But $R_1 = 20 \text{ k}\Omega$, so that

$$\frac{1}{R_1} + \frac{1}{R_{3A}} = \frac{1}{20} + \frac{1}{R_{3A}} = \frac{1}{10} \tag{7.5-15b}$$

which yields $R_{3A} = 20 \text{ k}\Omega$. The gain for region 3 is -3.33, and following Figure 7.5-10e, it therefore follows that

$$R_1 \parallel R_{3B} = 3.33 R_4 = 3.3 \text{ k}\Omega \tag{7.5-16a}$$

or

$$\frac{1}{R_1} + \frac{1}{R_{3B}} = \frac{1}{20} + \frac{1}{R_{3B}} = \frac{1}{3.33} \tag{7.5-16b}$$

Solving this expression for R_{3B}, we have $R_{3B} = 4.0 \text{ k}\Omega$.

The overall circuit for implementing this nonlinear transfer function is obtained by combining the circuits in Figure 7.5-10 as shown in Figure 7.5-11. In this circuit R_{2A} must be selected so that D_A turns OFF when v_o exceeds $+1$ V, while R_{2B} needs to be chosen so that D_B turns ON when v_o rises above $+3$ V.

Let's start with R_{2A}. Using equation (7.5-10), we obtain

$$R_{2A} = R_{3A} \frac{V_{CC}}{v_o} = (20 \text{ k}\Omega) \frac{10 \text{ V}}{1 \text{ V}} = 200 \text{ k}\Omega \tag{7.5-17}$$

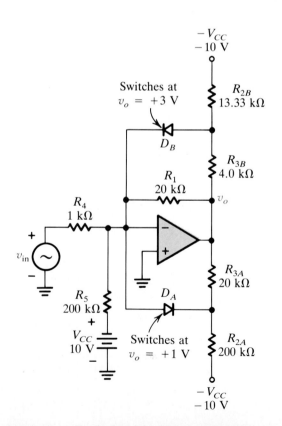

Figure 7.5-11

Final circuit design for Example 7.5-2.

In a similar fashion, the required value for R_{2B} is seen to be

$$R_{2B} = R_{3B}\frac{V_{CC}}{v_o} = (4.0 \text{ k}\Omega)\frac{10 \text{ V}}{3 \text{ V}} = 13.33 \text{ k}\Omega \qquad (7.5\text{-}18)$$

Unfortunately, the transfer characteristic for the circuit as it currently stands does not pass through the origin since when $v_{in} = 0$, following Figure 7.5-10c, v_o is equal to $[V_{CC}(R_1 \parallel R_{3A})/R_{2A}]$. By adding the input circuit shown in the dashed lines in the figure, where $R_5 = R_{2A} = 200 \text{ k}\Omega$, the transfer characteristic can be altered to pass through the origin without changing the slopes of the straight-line segments. These additional circuit elements have been included in the final version of this circuit given in Figure 7.5-11.

Exercises

7.5-1 A 2-V battery (positive terminal up) is inserted between R_2 and ground in Figure 7.5-4. Determine the values of v_{in} at which switching occurs. **Answer** 2.8 V, 0.8 V

7.5-2 a. For the circuit in Figure 7.5-9a, R_f is removed and R_2 is changed to 100 kΩ. Using the diode equation, derive an exact expression for v_o for $v_{in} < 0$ if the diode reverse leakage current is 10 nA.
b. Evaluate v_o for $v_{in} = 0.1$ V, 1 V, and 10 V.
Answer (a) $v_o = V_T \ln [(v_{in}/1 \text{ mV}) + 1]$, (b) 120 mV, 180 mV, 239 mV

7.5-3 Considering the diode in Exercise 7.5-3 to be ideal, find v_o given that $v_{in} = 2 \sin \omega t$ V and that the supply voltages are ± 10 V. **Answer** v_o is a square wave going between 0 and -10 V

7.5-4 For the circuit in Figure 7.5-9e, $R_3 = R_1 = R_4 = 10 \text{ k}\Omega$, $R_2 = 40 \text{ k}\Omega$, and $V_{CC} = 10$ V. Determine v_o when $v_{in} = -10$ V. **Answer** 6.25 V

7.6 INTERNAL STRUCTURE OF OPERATIONAL AMPLIFIERS

In Section 7.2, we described the ideal operational amplifier as a circuit having a differential input, an infinite input impedance, zero output impedance, and an infinite open-loop voltage gain. In this section, we investigate the internal structure of several actual operational amplifier circuits and we compare these devices to the ideal ones discussed earlier. In addition, we try to present some of the guidelines to be considered when designing these types of circuits. As illustrated in Figure 7.6-1, a typical op amp consists of four stages.

The input stage is generally some type of differential amplifier. This circuit is used because of its high common-mode rejection ratio, good high-frequency response, and ease of dc coupling. Most often we find that this stage is designed using bipolar transistors, but occasionally, when extremely high input impedances are required, FETs are used. BJTs are preferred over FET devices because of their higher gain.

The second stage, which is usually another differential amplifier, is used to boost the amplifier's open-loop voltage gain. As a result of the dc coupling between the stages, the dc level at the output of the second stage is generally well above ground potential. The block labeled "level shifter" in Figure 7.6-1, in

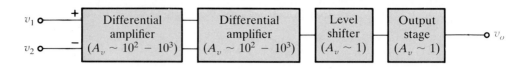

Figure 7.6-1

Block diagram of a typical operational amplifier.

between the second stage and the output stage, is used to adjust this dc level so that with both inputs grounded the dc voltage at the output of the op amp is nominally equal to zero.

The output stage acts as a buffer and gives the op amp its ability to drive low-impedance loads. Most commonly, it is some type of emitter-follower circuit, and although this stage could be designed with a single transistor, it usually consists of a pair of complementary-symmetry transistors operating in a "push-pull" configuration. Much more will be said about this particular type of design in Chapter 9 when we discuss power amplifiers, but for now it is sufficient to think of this circuit as being a pair of emitter followers, one operating on the positive half of the signal, and the other on the negative half.

From even this brief introduction to the internal organization of the operational amplifier, it is apparent that a good understanding of the inner workings of the op amp will require that we become familiar with a rather broad spectrum of electronic circuits. In this section, we specifically discuss the operation of differential amplifiers, current sources, active loads, level shifters, and emitter-follower driver circuits. Although our study will by no means be exhaustive, it should give you a good working knowledge of the internal structure of the operational amplifier.

7.6-1 Theory of Operation of the Differential Amplifier

A differential amplifier is a circuit containing two inputs, and either one or two outputs that are proportional to the difference between these input signals. If the amplifier has only one output terminal, it is said to be a single-ended or unbalanced differential amplifier. Some differential amplifiers have two outputs that are 180° out of phase from one another. These are called balanced-output amplifiers and are used to construct multistage differential amplifier circuits.

The input–output relationship of an ordinary amplifier, $v_{out} = Av_{in}$, is characterized by a single gain coefficient A, while that for an single-ended differential amplifier has two such gain coefficients. If v_1 and v_2 are the inputs and v_o is the output, one possible way to express the relationship between these quantities is

$$v_o = A_1 v_1 + A_2 v_2 \quad \text{(superposition)} \tag{7.6-1}$$

where A_1 represents the gain from the v_1 terminal to the output with v_2 set equal to zero, and A_2 the gain from v_2 to the output with v_1 equal to zero. Although equation (7.6-1) is certainly correct, it is not the best way to represent the input–output relationship for this type of amplifier.

A differential amplifier is designed to amplify the difference between the two input voltages v_1 and v_2 while rejecting any signal elements which they have in common. As a result, it is convenient to break up the input signals into their common-mode and differential components as shown in Figure 7.6-2a.

$v_1 - v_2$

$$\frac{v_d}{2} + \left(-1\frac{v_d}{2}\right) = \frac{2 v_d}{2} = v_d$$

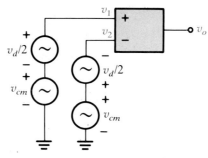

(a) Separating the Inputs into Their
     Differential and Common-Mode Components

Figure 7.6-2

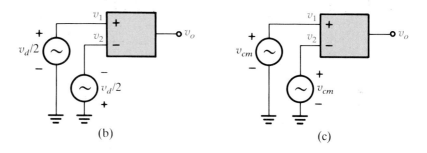

(b)   (c)

These are indicated by v_{cm} and v_d, respectively. Here since

$$v_1 = v_{cm} + \frac{v_d}{2} \tag{7.6-2a}$$

$$v_2 = v_{cm} - \frac{v_d}{2} \tag{7.6-2b}$$

we may express v_{cm} and v_d in terms of the overall input signals to the amplifier as

$$v_d = v_1 - v_2 \tag{7.6-2c}$$

and $$v_{cm} = \frac{v_1 + v_2}{2} \tag{7.6-2d}$$

Thus the component of the input signals known as v_{cm} represents the portions of v_1 and v_2 that are common to both inputs, while v_d is that part which is different.

Substituting (7.6-2a) and (7.6-2b) into (7.6-1), the output voltage may be written in terms of the common-mode and differential components of the input signal as

$$v_o = A_1\left(v_{cm} + \frac{v_d}{2}\right) + A_2\left(v_{cm} - \frac{v_d}{2}\right) = \underbrace{\frac{A_1 - A_2}{2}}_{A_d} v_d + \underbrace{(A_1 + A_2)}_{A_{cm}} v_{cm} \tag{7.6-3}$$

This expression may be rewritten in terms of the amplifier's differential and

7.6 / Internal Structure of Operational Amplifiers

461

common-mode gain coefficients as

$$v_o = A_d v_d + A_{cm} v_{cm} \tag{7.6-4a}$$

where
$$A_d = \left. \frac{v_o}{v_d} \right|_{v_{cm}=0} \tag{7.6-4b}$$

is the differential gain, and

$$A_{cm} = \left. \frac{v_o}{v_{cm}} \right|_{v_d=0} \tag{7.6-4c}$$

is the common-mode gain. Following these definitions, it is apparent that the differential gain, $A_d = v_o/v_d$, is measured by connecting one voltage source $v_d/2$ to the noninverting input terminal v_1 and another source $-v_d/2$ to the inverting input terminal v_2, as illustrated in Figure 7.6-2b. On the other hand, the common-mode gain, $A_{cm} = v_o/v_{cm}$, is measured by connecting the same voltage source v_{cm}, to both inputs, as shown in Figure 7.6-2c.

In designing a differential amplifier, we usually want the differential gain to be as large as possible, while ideally we would like the common-mode gain to be zero. This is because the differential portion of the input signal generally represents the quantity that we are interested in amplifying, while the common-mode part is most often some undesired noise signal that has inadvertently gotten mixed in with the signal that we are trying to amplify. Examples of common-mode noise sources include power supply ripple, ac line noise picked up from lights and motors, RF (or radio-frequency) interference radiated from nearby electronic equipment. If these noise signals enter the amplifier as purely common-mode signals, a circuit such as a differential amplifier, which has a small common-mode gain, can be very effective in reducing their presence at the output.

The common-mode rejection ratio (CMRR) of a differential amplifier measures its ability to reject common-mode noise while amplifying the differential portion of the input signal. It is defined as the ratio of the amplifier's differential gain to its common-mode gain, or

$$\text{CMRR} = \left| \frac{A_d}{A_{cm}} \right| \quad \begin{matrix} = \text{Large} \\ = \text{small} \end{matrix} \tag{7.6-5}$$

The CMRR of an amplifier is also frequently expressed in decibels, so that, for example, in describing an amplifier whose differential gain is 10^4 times larger than its common-mode gain, we could also say that it has a common-mode rejection ratio of $20 \log_{10}(10^4)$ or 80 dB.

To illustrate the concepts of common-mode gain, differential gain, and common-mode rejection ratio, let us consider the case of a specific differential amplifier whose input–output relationship is

$$v_o = A(v_1 - v_2) \tag{7.6-6}$$

Using this expression, we can then see that the amplifier's response to a purely differential input such as that shown in Figure 7.6-2d will be

$$v_o = A\left[\frac{v_d}{2} - \left(\frac{-v_d}{2} \right) \right] = A v_d \tag{7.6-7a}$$

Thus the differential gain of this amplifier is simply equal to the coefficient A in equation (7.6-6).

To find this amplifier's common-mode gain, we must apply the same signal to both inputs as illustrated in Figure 7.6-2d. Because $v_1 = v_2 = v_{cm}$ for this case, using equation (7.6-6) we have

$$v_o = A(v_{cm} - v_{cm}) = 0 \qquad (7.6\text{-}7b)$$

so that the common-mode gain for this amplifier is zero.

Equations (7.6-7a) and (7.6-7b) describe the performance of what might be called an ideal subtractor or ideal difference amplifier. It has a finite differential gain A and zero common-mode gain. As a result, its common-mode rejection ratio, or its ability to reject common-mode noise, is infinite.

7.6-2 Practical Differential Amplifier Circuits

Figure 7.6-3 illustrates a physically realizable version of a differential amplifier. Let's examine the factors governing its ability to amplify differential and common-mode signals. Several very useful network simplifications are possible because of the symmetry of the circuit. We start by examining the differential gain of the circuit shown in Figure 7.6-3b. Because of the antisymmetric nature of the two applied input signals, we expect that the ac portion of all the currents and voltages in Q_1 and Q_2 will be 180° out of phase from one another. As a result, we have

$$i_{b1} = -i_{b2} \qquad (7.6\text{-}8a)$$

$$i_{c1} = h_{fe} i_{b1} = -h_{fe} i_{b2} = -i_{c2} \qquad (7.6\text{-}8b)$$

$$i_{e1} = -i_{e2} \qquad (7.6\text{-}8c)$$

and

$$v_{o1} = -v_{o2} \qquad (7.6\text{-}8d)$$

Equation (7.6-8c) is interesting because it affords us an excellent opportunity to simplify this circuit. Because $i_{e1} = -i_{e2}$, the ac current through the emitter resistor R_E is zero, so that the ac voltage v_e from the transistor emitters to ground is also zero. As a result, we can consider this point in the circuit to be a virtual ground and split the circuit in half as shown in Figure 7.6-3c in order to carry out the differential gain portion of this analysis. Using this figure, we can see that

$$i_{b1} = \frac{v_d/2}{h_{ie}} = \frac{v_d}{2h_{ie}} = -i_{b2} \qquad (7.6\text{-}9a)$$

$$i_{c1} = h_{fe} i_{b1} = \frac{h_{fe} v_d}{2h_{ie}} \qquad (7.6\text{-}9b)$$

and

$$v_{o1} = -i_{c1} R_{C1} = -\frac{h_{fe} R_C v_d}{2h_{ie}} \qquad (7.6\text{-}9c)$$

A similar analysis for the output voltage, v_{o2}, at the collector of Q_2 yields

$$v_{o2} = -i_{c2} R_{C2} = \frac{h_{fe} R_C v_d}{2h_{ie}} \qquad (7.6\text{-}9d)$$

so that the differential gain for this amplifier may be written as

$$A_d = \frac{v_{o2}}{v_d} = \frac{h_{fe} R_C}{2h_{ie}} \qquad (7.6\text{-}10)$$

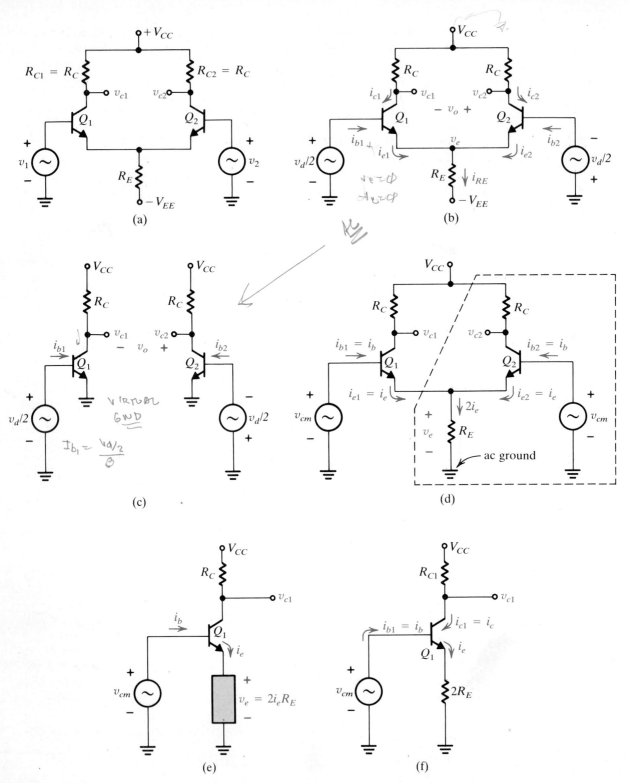

Figure 7.6-3
BJT-style differential amplifier.

In equations (7.6-9c) and (7.6-9d), we can see that the two output voltages v_{o1} and v_{o2} are 180° out of phase, but are equal in magnitude preserving the circuit's common-mode rejection ability in a cascade arrangement. Of course, this occurred because we assumed that Q_1 and Q_2 and R_{C1} and R_{C2} were identical. In an actual integrated circuit, this would certainly not be the case, and typically in such a device 5 or 10% component matching is considered good.

For later use we also define the differential input impedance of this amplifier as the effective impedance between the two base terminals when a differential voltage is applied across them. For the circuit given in Figure 7.6-3b, the voltage between the base terminals is equal to v_d and the corresponding current flow into the base of Q_1 (and out of the base of Q_2) is given by equation (7.6-9a), so that we may write down the differential input impedance for this circuit as

$$Z_{id} = \frac{v_d}{i_{b1}} = 2h_{ie} \tag{7.6-11}$$

It is useful to notice that the differential gain of this circuit may also be expressed in terms of this input impedance as

$$A_d = \frac{h_{fe} R_C}{Z_{id}} \tag{7.6-12}$$

COMMON-MODE

The common-mode gain of this circuit may be determined by examining Figure 7.6-3d and noting that because of the circuit symmetry,

$$i_{b1} = i_{b2} = i_b \tag{7.6-13a}$$

$$i_{c1} = i_{c2} = i_c \tag{7.6-13b}$$

$$v_{o1} = v_{o2} \tag{7.6-13c}$$

and

$$i_{e1} = i_{e2} = i_e \tag{7.6-13d}$$

As a result, there is a current flow equal to $2i_e$ in the emitter resistor R_E and a resulting voltage drop across it of

$$v_e = 2i_e R_E = i_e(2R_E) \tag{7.6-14}$$

To simplify the analysis of this problem, let us take the portion of the circuit shown within the dashed lines in Figure 7.6-3d, and place it inside the box in Figure 7.6-3e. The circuit within this box has the following terminal characteristics. When a current i_e flows into it from Q_1, the voltage v_e developed across it is given by equation (7.6-14). As a result, as we have seen many times before, because the voltage drop across the box is proportional to the current flow through it, we may replace the circuitry within the box by an equivalent resistor equal, in this case, to $2R_E$. The final equivalent circuit for this differential amplifier when connected to common-mode signal excitation is given in Figure 7.6-3f, and the pertinent currents and voltages in this circuit are seen by inspection of this figure to be

$$i_{b1} = i_{b2} = \frac{v_{cm}}{h_{ie} + 2(1 + h_{fe})R_E} \tag{7.6-15a}$$

$$i_{c1} = i_{c2} = \frac{h_{fe}v_{cm}}{h_{ie} + 2(1 + h_{fe})R_E} \tag{7.6-15b}$$

and $$v_{o1} = -i_{c1}R_{c1} = \frac{-h_{fe}R_C v_{cm}}{h_{ie} + 2(1 + h_{fe})R_E} = v_{o2} \tag{7.6-15c}$$

so that the common-mode gain of this differential amplifier is

$$A_{cm} = \frac{v_{o2}}{v_{cm}} = \frac{-h_{fe}R_C}{h_{ie} + 2(1 + h_{fe})R_E} \tag{7.6-16}$$

Combining equations (7.6-10) and (7.6-16), the common-mode rejection ratio of this circuit may also be written as

$$CMRR = \left|\frac{A_d}{A_{cm}}\right| = \frac{h_{ie} + 2(1 + h_{fe})R_E}{2h_{ie}} \tag{7.6-17}$$

As with the differential case examined earlier, we may define this circuit's common-mode input impedance as v_{cm}/i_{cm}, where i_{cm} is the current that flows into the amplifier when a voltage v_{cm} is applied to each of the input terminals. The common-mode input current for this BJT differential amplifier is given by equation (7.6-15a), so that the effective input impedance connected between each of the base terminals and ground is

$$Z_{icm} = \frac{v_{cm}}{i_b} = h_{ie} + 2(1 + h_{fe})R_E \tag{7.6-18}$$

Combining this result with equation (7.6-16), this amplifier's common-mode gain may be written in terms of its common-mode input impedance as

$$A_{cm} = \frac{-h_{fe}R_c}{Z_{icm}} \tag{7.6-19}$$

Using this result for A_{cm} together with a similar expression for A_d given in equation (7.6-12), we may express the amplifier's common-mode rejection ratio in terms of its differential and common-mode input impedances as

$$CMRR = \left|\frac{A_d}{A_{cm}}\right| = \left|\frac{h_{fe}R_c/Z_{id}}{-h_{fe}R_c/Z_{icm}}\right| = \frac{Z_{icm}}{Z_{id}} \tag{7.6-20}$$

By examining this result, or better still that given in equation (7.6-17), it should be clear that the CMRR of this circuit can be maximized by making R_E as large as possible. Let's review the purpose of R_E in this circuit and see what can be done to increase its size. R_E's function in this differential amplifier is to provide dc bias for the two transistors Q_1 and Q_2. The simple example illustrated in Figure 7.6-4a will help to remind you how this is accomplished. In analyzing this circuit, assume that the transistors are silicon with h_{FE} values equal to 100.

Due to the negative potential of the power supply, $-V_{EE}$, the base–emitter junctions of Q_1 and Q_2 are both forward biased, and therefore the voltage at the top of R_E is -0.7 V with respect to ground. As a result, the voltage across R_E will be $[-0.7 - (-10.7)]$ or $+10$ V. The corresponding current through R_E is $(10 \text{ V})/(1 \text{ k}\Omega)$ or 10 mA, and because of the circuit symmetry, this cur-

R_E function

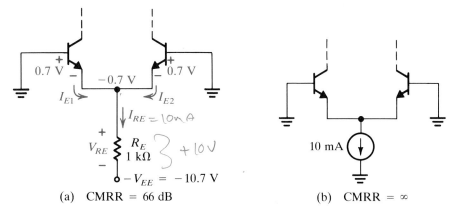

Figure 7.6-4
Increasing the CMRR of a differential amplifier.

(a)   CMRR = 66 dB   (b)   CMRR = ∞

rent will divide evenly between both transistors so that $I_{E1} = I_{E2} = 5$ mA. At this Q point the transistor emitter resistances, r_e, are $(26 \text{ mV})/(5 \text{ mA})$ or about $5.2\ \Omega$, so that $h_{ie} = (1 + h_{fe})r_e = 520\ \Omega$. Using equation (7.6-17), the corresponding circuit CMRR is $2000.52/1.04$ or about 66 dB.

To improve this circuit's CMRR, while still maintaining the same dc Q point, we will need to increase the value of R_E while simultaneously making the supply voltage more negative. As R_E gets larger and larger, the Norton equivalent circuit for the biasing network will approach the ideal 10-mA current source indicated in Figure 7.6-4b. This current source provides the dc current needed to bias the transistors but looks like an open circuit to the ac signal components, and therefore leads to an infinite common-mode rejection ratio. Techniques for designing these current sources are discussed immediately after Example 7.6-1.

EXAMPLE 7.6-1

The circuit shown in Figure 7.6-5 is a differential amplifier that employs JFETs instead of BJTs. This type of differential amplifier is used when extremely high input impedances and very low input bias currents are required. The transistor parameters are $V_p = -5$ V and $I_{DSS} = 10$ mA.

a. Determine the transistor Q points.
b. Find the circuit's differential gain, v_o/v_d, with the load R_L connected as shown.

SOLUTION

The dc performance of this circuit may be examined by setting v_1 and v_2 equal to zero to obtain the circuit shown in Figure 7.6-5b. Owing to the dc symmetry of this circuit, we can see by inspection that

$$I_{D1} = I_{D2} = \frac{I_o}{2} = 0.5 \text{ mA} \qquad (7.6\text{-}21)$$

and that
$$V_{D1} = V_{D2} = V_{DD} - I_D R_D = 20 - 5 = 15 \text{ V} \qquad (7.6\text{-}22)$$

Because $V_{D1} = V_{D2}$, we find that $V_o = (V_{D2} - V_{D1}) = 0$, so that there is no dc current in the load.

To determine V_{GS1} and V_{GS2}, we should recall that I_D and V_{GS} in an FET are related by $I_D = I_{DSS}\left(1 - \dfrac{V_{GS}}{V_p}\right)^2$, and combining this expression with equation

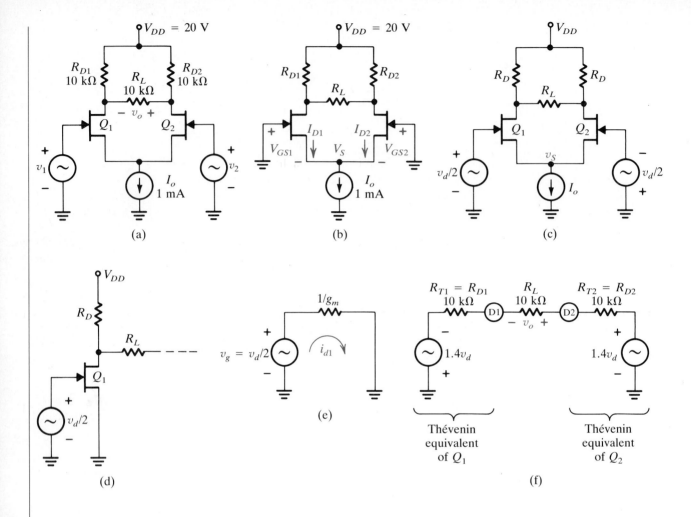

(a)

(b)

(c)

(d)

(e)

$R_{T1} = R_{D1}$ 10 kΩ R_L 10 kΩ $R_{T2} = R_{D2}$ 10 kΩ

1.4v_d 1.4v_d

Thévenin
equivalent
of Q_1

Thévenin
equivalent
of Q_2

(f)

Figure 7.6-5
Example 7.6-1.

(7.6-21) and the transistor data, we have

$$0.5 \text{ mA} = (1 \text{ mA})\left(1 - \frac{V_{GS}}{-5}\right)^2 \qquad (7.6\text{-}23)$$

from which we find that $V_{GS1} = V_{GS2} = -1.46$ V. Because the dc gate voltages are both zero, this means that V_{S1} and V_{S2} equal 1.46 V and correspondingly that V_{DS1} and V_{DS2} are both 15 − 1.46 or 13.54 V.

The differential gain of the circuit in Figure 7.6-5c is most easily found by recalling that the antisymmetry of this circuit causes v_s to be a virtual ground. As a result, we may split this circuit in two and examine the response of Q_1, for example, by grounding its source. The circuit that results from this network simplification is shown in Figure 7.6-5d, and the equivalent circuit seen in the source of this transistor is given in part (e) of the figure. Using this circuit, the drain current for Q_1 is

$$i_{d1} = \frac{g_m v_d}{2} \qquad (7.6\text{-}24a)$$

and similarly for Q_2, we may write

$$i_{d2} = -\frac{g_m v_d}{2} \tag{7.6-24b}$$

The transistor g_m values are determined from the Q-point information using equation (5.5-2b) as

$$g_m = -\frac{2I_{DSS}}{V_p}\left(1 - \frac{V_{GS}}{V_p}\right) = 0.28 \text{ mS} \tag{7.6-25}$$

The voltage across R_L will be found by first determining the Thévenin equivalent circuits looking to the left and right of the load in Figure 7.6-5c. The load is then connected to these circuits in order to find v_o. Because i_{d1} and i_{d2} are both known, the open-circuit voltages at the drains of Q_1 and Q_2 are seen by inspection to be

$$v_{d1} = -i_{d1}R_{D1} = -\frac{g_m R_{D1} v_d}{2} = -1.4 v_d \tag{7.6-26a}$$

and

$$v_{d2} = -i_{d2}R_{D2} = \frac{g_m R_{D2} v_d}{2} = 1.4 v_d \tag{7.6-26b}$$

The Thévenin impedances are just equal to the drain resistances. The overall equivalent circut for the output portion of this differential amplifier is presented in Figure 7.6-5f, from which we have that

$$v_o = \tfrac{1}{3}(2.8 v_d) \tag{7.6-27a}$$

so that the differential gain is

$$A_d = \frac{v_o}{v_d} = 0.93 \tag{7.6-27b}$$

7.6-3 Use of the Transistor as a Current Source

We have seen that the common-mode input resistance and common-mode rejection ratio of a differential amplifier can be improved significantly if the emitter biasing resistor is replaced by a dc current source. Let us now examine several of the more popular techniques for constructing these sources. As you will recall, a constant-current source is essentially a two-terminal device in which the current flow through it remains the same regardless of the voltage across it.

Constant current source

Excellent current sources can be made using BJT or FET transistors if they are operated in the flat horizontal portion of their characteristics. For example, if the bipolar junction transistor whose characteristics are given in Figure 7.6-6a is biased so that its base current is 0.02 mA, its collector current will remain nearly constant at 2 mA as long as the voltage, v_{CE}, across it stays between about 0.2 V and 50 V (the transistor breakdown voltage). The circuit presented in Figure 7.6-6b illustrates one possible circuit implementation of this current source design. With $(1 + h_{FE})R_E \gg (R_1 \| R_2)$, the collector current is essentially independent of the transistor current gain and is found as follows:

$$V_B = V_{CC}\frac{R_2}{R_1 + R_2} = (10)\frac{4.7 \text{ k}\Omega}{10 \text{ k}\Omega} = 4.7 \text{ V} \tag{7.6-28a}$$

$$V_E = V_B - V_{BE} = 4.7 - 0.7 = 4.0 \text{ V} \qquad (7.6\text{-}28\text{b})$$

$$I_E = \frac{V_E}{R_E} = \frac{4 \text{ V}}{2 \text{ k}\Omega} = 2.0 \text{ mA} \qquad (7.6\text{-}28\text{c})$$

and $\qquad I_C \simeq I_E = 2.0 \text{ mA} \qquad (7.6\text{-}28\text{d})$

Thus between point A and ground, we effectively have a 2-mA current source as long as the voltage V_A is greater than $(4 + 0.2) = 4.2$ V but less than $(4 + 50) = 54$ V.

Current sources can also be constructed using FETs simply by taking the gate and source leads of the transistor and shorting them together. To understand how this configuration behaves as a constant-current source, consider the JFET circuit illustrated in Figure 7.6-6c, along with the transistor characteris-

Figure 7.6-6

Current-source design using transistors.

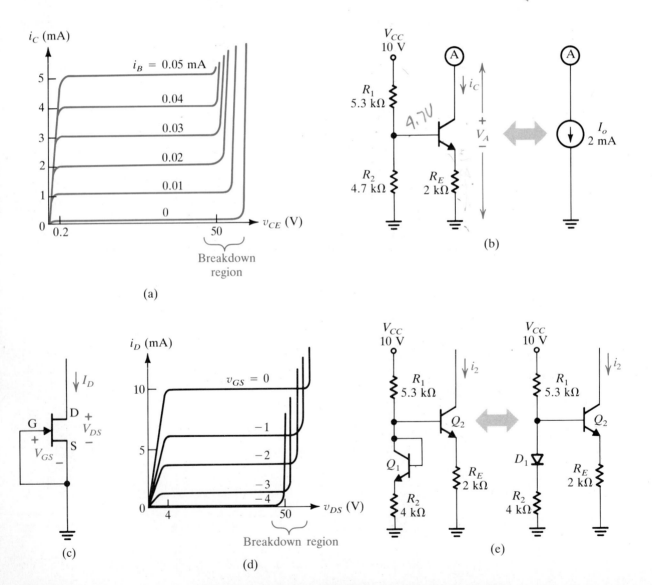

(a)

(b)

(c)

(d)

(e)

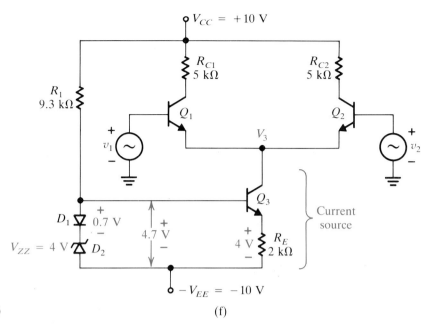

Figure 7.6-6 (Continued)

(f)

tics given in part (d) of the figure. Clearly, for this circuit V_{GS} is equal to zero, and therefore this transistor's volt-ampere behavior is described by the $V_{GS} = 0$ curve in Figure 7.6-6d. By carefully examining this curve, we can see that as long as V_{DS} is greater than $-V_p = 4$ V but less than the breakdown voltage at 50 V, then I_D is approximately 10 mA. Therefore, for all drain-to-source voltages in the range just specified, this device can be characterized as a 10-mA constant-current source. The fact that the lines are not truly horizontal requires the inclusion of a resistor in parallel with the current source in order to complete the equivalent circuit.

Substantial improvements in the performance of the BJT current source can be achieved by making minor modifications to the original circuit in Figure 7.6-6b. To begin with, the circuit's temperature stability can be greatly increased by adding a diode, or a transistor connected as a diode (see Figure 7.6-6e), in series with R_2 to cancel the Q-point shift produced by the variation of V_{BE2} with temperature. If current source independence from changes in V_{CC} is also needed, a zener diode can be inserted in place of R_2, as illustrated in Figure 7.6-6f. When this is done, the voltage across the D_1–D_2 combination will remain fixed at 4.7 V over a wide range of supply voltages.

Although conceptually easy to understand, the current source designs discussed thus far have several disadvantages. From an IC designer's viewpoint, they have too many components, and more important, too many resistors. Resistors, it turns out, take up much more space on an IC than do transistors. Therefore, it is desirable to minimize the number of resistive elements in the circuit.

A second disadvantage of this design is that the large voltage drop across R_E significantly reduces the allowed common-mode input voltage range over which a differential amplifier circuit such as that in Figure 7.6-6f can be operated. Specifically, for this circuit we find that to keep the current source in its active region, V_3 in the figure must be greater than

$$V_3 = V_{CE3} + V_{RE} - V_{EE} = 0.2 + 4 - 10 = -5.8 \text{ V} \qquad (7.6\text{-}29\text{a})$$

which means that the common-mode input voltage to this circuit cannot go below

$$V_{B1} = V_{B2} = 0.7 + V_3 = -5.1 \text{ V} \qquad (7.6\text{-}29\text{b})$$

Both of these disadvantages are overcome to a certain extent with the circuit illustrated in Figure 7.6-7a. It is known as a current mirror because, as we shall see shortly, the current in the collector of Q_2 mirrors that in the collector of Q_1.

For this circuit, although the base and collector of Q_1 are shorted together, the transistor still operates in the active region since $V_{BE} = 0.7$ V and $V_{CB} = 0.0$ V, so that V_{CE} is about 0.7 V. Furthermore, from equation (4.2-1) we should also recall that for a transistor operating in the active region, the collector current is approximately related to the base–emitter voltage by the expression

$$I_C \simeq I_{EBO}e^{V_{BE}/V_T} \qquad (7.6\text{-}30)$$

As a result, if we assume that the transistors are identical, then, since they

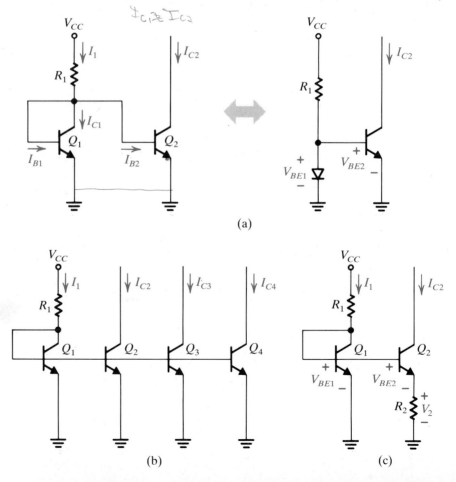

Figure 7.6-7

"Current mirror" dc current source.

(a)

(b)   (c)

have the same base–emitter voltage, the collector currents in Q_1 and Q_2 will be equal.

The current I_1 in the resistor R_1 is approximately equal to

$$I_1 \simeq \frac{V_{CC} - 0.7}{R_1} \tag{7.6-31a}$$

and is related to the transistor currents by the expression

$$I_1 = I_{B1} + I_{B2} + I_{C1} = I_{B1} + I_{B2} + I_{C2} \tag{7.6-31b}$$

Using the fact that $I_{C2} = h_{FE}I_{B2} = h_{FE}I_{B1}$, we may solve equation (7.6-31b) for I_{C2} to yield

$$I_{C2} = \frac{I_1}{1 + 2/h_{FE}} \tag{7.6-31c}$$

If the transistor h_{FE} is large enough, $I_{C2} = I_1$, and the collector current in Q_2 is said to "mirror" the current I_1 flowing through R_1.

This circuit is very popular for op amp designs because it has a minimal number of components and allows the common-mode input voltage to approach closely the negative supply rail with little degradation in the performance of the current source. Furthermore, this circuit can be readily extended to produce multiple tracking current sources, as illustrated in Figure 7.6-7b. For this circuit, to the extent that the transistors are identical, and also to the extent that h_{FE} is large, we have

$$I_{C4} = I_{C3} = I_{C2} \simeq I_1 = \frac{V_{CC} - 0.7}{R_1} \tag{7.6-32}$$

By altering the relative sizes of the transistor base–emitter geometries in Figure 7.6-7b, it is possible to produce current sources that are scaled versions of the reference current I_1. Specifically, if we recall that the I_{EBO} in equation (7.6-30) is proportional to the base–emitter cross-sectional area, then if Q_2, for example, has twice the area of Q_1 while Q_3 and Q_4 have only one-third the area, the collector currents in Q_2, Q_3, and Q_4 will be $2I_1$, $\frac{1}{3}I_1$, and $\frac{1}{3}I_1$, respectively. Because it is difficult to produce transistors having very different geometries on the same IC chip, this technique is really useful only for producing current sources whose values are all within a factor of 5 of one another.

When current ranges in excess of 5 : 1 are needed, the basic circuit shown in Figure 7.6-7c, known as a logarithmic current source, is frequently employed. This circuit is also used when low-value current sources are required. Such designs are difficult to achieve with the circuit given in Figure 7.6-7a because the size of the resistor R_1 becomes so large that the IC area needed to fabricate it is prohibitive. As a practical matter, integrated-circuit resistors in excess of 50 kΩ are considered undesirable.

If the transistors in Figure 7.6-7c are identical, we have that

$$I_1 \simeq I_{EBO}e^{V_{BE1}/V_T} \tag{7.6-33a}$$

and

$$I_{C2} \simeq I_{EBO}e^{V_{BE2}/V_T} \tag{7.6-33b}$$

so that the ratio of these currents is

$$\frac{I_{C2}}{I_1} = e^{(V_{BE2} - V_{BE1})/V_T} \tag{7.6-33c}$$

Applying Kirchhoff's voltage law to this circuit, we find that

$$V_{BE1} - V_{BE2} = V_2 \simeq I_{C2}R_2 \qquad (7.6\text{-}34)$$

and substituting this result into equation (7.6-33c), we obtain

$$\frac{I_{C2}}{I_1} = e^{-I_{C2}R_2/V_T} \qquad (7.6\text{-}35)$$

This equation is nonlinear and we cannot solve it directly for I_{C2} in terms of I_1. Fortunately, this nonlinearity will not present any difficulty in the design of circuits of this type because I_1 and I_{C2} will both be known, so that R_1 and R_2 can easily be determined from equations (7.6-31a) and (7.6-35). However, in analyzing a circuit of this type, the collector current in Q_2 will need to be found by using a trial-and-error technique to solve equation (7.6-35) for I_{C2}.

7.6-4 Construction of Differential Amplifiers Having Active Loads

Besides having a large common-mode rejection ratio, we would like to design differential amplifiers having the largest possible differential gain. This can be achieved either by raising the transistor g_m or alternatively by increasing the values of the collector resistors. To increase the transconductances (or g_m values) of the transistors, the quiescent currents must be raised correspondingly. Because low power consumption is usually also desired, this is not generally considered to be the best way to proceed. The second alternative, increasing the size of the collector resistors, also has several drawbacks.

To begin with, it is difficult to incorporate very large resistors into an IC because they simply take up too much room. Second, as the collector resistor value increases, so does the dc voltage drop across it, and if R_C is made too large, the transistors will saturate. At first glance it appears that this saturation problem may be avoided simply by reducing the quiescent current to lower the IR drop across these resistors. However, as equation (7.6-10) indicates, when I_o is reduced, h_{ie} increases, and this decreases the differential gain.

Fortunately, it is possible to design collector loads for differential amplifier circuits that have both low dc voltage drops across them and large effective ac impedances. These circuits are called "active loads" and are essentially current sources connected in the collectors of the differential amplifier. One possible active load differential amplifier circuit is illustrated in Figure 7.6-8a. Here we have simply taken a conventional differential amplifier and replaced each of the collector resistors by an independent current source. Although this circuit works fine if all the transistors are identical, it undergoes unacceptably large common-mode Q-point shifts in the output for rather small imbalances between the upper and lower current sources. This problem makes this particular circuit virtually useless for practical applications.

However, by replacing the two current sources Q_3 and Q_4 by a current mirror pair, the high amplifier gains associated with active loads can still be achieved, but without the common-mode dc output shift problem. A typical current mirror differential amplifier is illustrated in Figure 7.6-8b. In this circuit Q_3 and Q_4 form a current mirror active load, and because I_{C1} and I_{C3} are

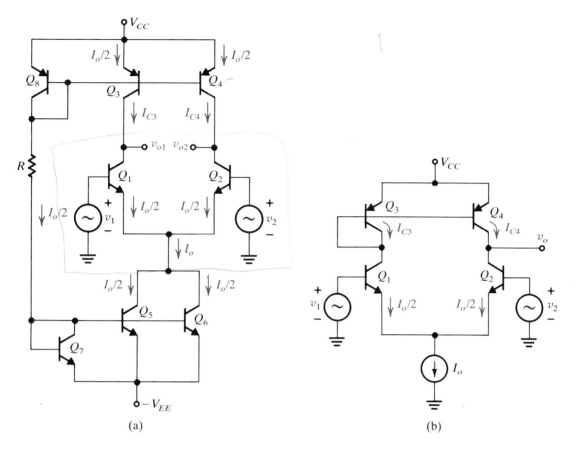

Figure 7.6-8

Differential amplifier with active loads.

equal, the current mirror will make I_{C4} nearly the same as I_{C1}. Furthermore, because Q_3 looks like a diode load to Q_1, V_{C1} stays at about $V_{CC} - 0.7$ V, and it is thus impossible for this dc Q-point voltage to change very much. V_{C2}, on the other hand, has all the same dc shift problems as the amplifier in Figure 7.6-8a, but because this shift is no longer common-mode on both sides of the amplifier, it can be compensated for by adding a small dc offset voltage to the input of either Q_1 or Q_2. The principal disadvantage of this circuit is that it does not have a differential output.

To analyze the differential gain of this amplifier, let's begin by examining the equivalent emitter circuit for Q_1 and Q_2 shown in Figure 7.6-9a. In this portion of the amplifier, because of the circuit symmetry, the effect of $1/h_{oe}$ on this equivalent circuit is negligible, and as a result, we still find that

$$v_e \simeq 0 \qquad (7.6\text{-}36)$$

Therefore, the emitters of Q_1 and Q_2 may be considered to be virtual grounds, and thus we may split the circuit in two as illustrated in Figure 7.6-9b. Because $v_{eb3} = i_{b3}h_{ie3}$, we may replace the current source $h_{fe}i_{b3}$ by a resistor h_{ie3}/h_{fe} and combine this together with h_{ie3} to form an equivalent resistor $h_{ie3}/(1 + h_{fe}) = r_{e3}$. This simplified equivalent circuit is given in Figure 7.6-9c, and in analyzing this portion of the circuit, the effect of the transistor output impedances may be neglected because they are so much larger than r_{e3}.

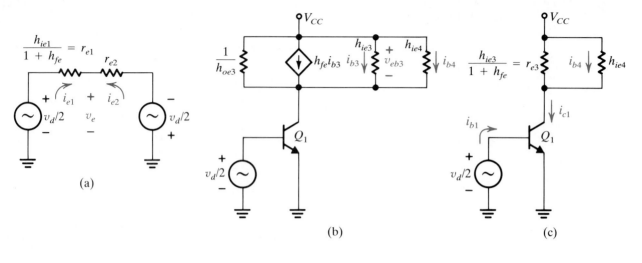

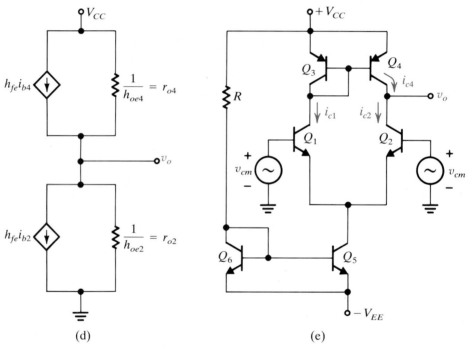

Figure 7.6-9
Differential amplifier with a current mirror active load.

Therefore, from this figure we find that

$$i_{b1} = \frac{v_d}{2h_{ie}} = -i_{b2} \tag{7.6-37a}$$

$$i_{c1} = h_{fe}i_{b1} \tag{7.6-37b}$$

and

$$i_{b4} = \frac{r_{e3}}{r_{e3} + h_{ie4}}i_{c1} = \frac{1}{2 + h_{fe}}i_{c1} \tag{7.6-37c}$$

To calculate the output voltage v_o, we must examine the equivalent circuit given in Figure 7.6-9d. Notice that in this portion of the circuit the $1/h_{oe}$

impedance terms must be included, and by applying superposition we find that

$$v_o = h_{fe}(i_{b4} - i_{b2})(r_{o2} \parallel r_{o4}) \qquad (7.6\text{-}38)$$

By combining equations (7.6-37a), (7.6-37b), (7.6-37c), and (7.6-38), we may write v_o in terms of v_d as

$$v_o = h_{fe}\left[\frac{h_{fe}(v_d/2h_{ie})}{2 + h_{fe}} - \left(\frac{-v_d}{2h_{ie}}\right)\right](r_{o2} \parallel r_{o4}) = g_m(r_{o2} \parallel r_{o4})v_d \qquad (7.6\text{-}39a)$$

so that this amplifier's differential gain is simply equal to

$$A_d = g_m(r_{o2} \parallel r_{o4}) \qquad (7.6\text{-}39b)$$

To estimate the typical gain that can be expected from this type of amplifier, assume that the transistors Q_1 and Q_2 are each biased at 0.1 mA, and that the output impedances are nominally 1 MΩ. For the quiescent current given, the transistor g_m values are approximately $I_E/V_T = (0.1 \text{ mA})/(26 \text{ mV})$ or 3800 μS. Substituting these data into equation (7.6-39b), we find that the circuit's differential gain is about 2000. This is substantially larger than that which could be obtained using ordinary resistors as the collector loads.

The common-mode gain for this circuit may be calculated by examining Figure 7.6-9e. For this case, because the excitation is symmetric,

$$i_{b1} = i_{b2} = \frac{v_{cm}}{Z_{icm}} \qquad (7.6\text{-}40a)$$

and because v_{c1} and v_{c2} will both be small, we can consider that $1/h_{oe1}$ and $1/h_{oe2}$ essentially fold down in parallel with $1/h_{oe5}$, so that to a good approximation,

$$Z_{icm} = (1 + h_{fe})(r_{o1} \parallel r_{o2} \parallel r_{o5}) \qquad (7.6\text{-}40b)$$

By combining equations (7.6-37c), (7.6-38), and (7.6-40a), we may write down the expression for the common-mode output voltage as

$$v_o = h_{fe}(i_{b4} - i_{b2})(r_{o2} \parallel r_{o4}) = -\frac{2v_{cm}}{Z_{icm}}(r_{o2} \parallel r_{o4}) \qquad (7.6\text{-}41a)$$

so that the common-mode gain for this amplifier is

$$A_{cm} = \frac{v_o}{v_{cm}} = -\frac{2(r_{o2} \parallel r_{o4})}{Z_{icm}} \qquad (7.6\text{-}41b)$$

Combining this result with equation (7.6-39b), we obtain

$$\text{CMRR} = \left|\frac{A_d}{A_{cm}}\right| = 2g_m Z_{icm} \qquad (7.6\text{-}41c)$$

Using the amplifier data presented previously, we have that

$$Z_{icm} = (101)[(1 \text{ M}\Omega) \parallel (1 \text{ M}\Omega) \parallel (500 \text{ k}\Omega)] \simeq 25 \text{ M}\Omega \qquad (7.6\text{-}42a)$$

$$A_{cm} = -\frac{2[(1 \text{ M}\Omega) \parallel (1 \text{ M}\Omega)]}{25 \text{ M}\Omega} = -0.04 \qquad (7.6\text{-}42b)$$

and
$$\text{CMRR} = \frac{2000}{0.04} = 5 \times 10^4 \qquad (7.6\text{-}42c)$$

or 94 dB. Thus a current mirror active load differential amplifier has a very high differential gain and an extremely small common-mode gain, so that its common-mode rejection ratio is excellent.

7.6-5 Putting It All Together: Typical Op Amp Internal Circuit Configurations

At the beginning of this section we discussed the characteristics of the typical op amp. We noted that it was basically a high-gain amplifier with a differential input and a single-ended output which had a high input impedance, a low output impedance, and a large common-mode rejection ratio. In addition, we observed that the internal stages were generally dc coupled and designed so that with both inputs connected to ground, ideally at least, the dc component of the output voltage was zero. Furthermore, although not specifically discussed in this chapter, we noted that it had a wide frequency response that was flat all the way down to dc.

When we consider the possible circuit designs available which can simultaneously satisfy all these design requirements, a configuration similar to that originally presented in Figure 7.6-1 usually results. Differential amplifiers are popular for the input stages of operational amplifiers because of their high common-mode rejection, ease of dc coupling, and low distortion levels. It is in these early stages where most of the voltage amplification occurs, and when very high stage gains are needed, current mirror active loads are often used. However, because these types of differential amplifiers have only a single-ended output, when they are used for the input stage, the stage that follows must be a conventional nondifferential amplifier. As an alternative, to maximize both CMRR and open-loop gain, sometimes a two-stage differential design is employed where the first stage uses resistive loading and the second stage a current mirror active load.

The level shifter circuit that follows the main gain stages is often a cascode arrangement where the upper transistor operates as an emitter follower and the lower transistor acts as a current source. A typical level shifter circuit is illustrated in Figure 7.6-10, and to demonstrate how it operates, we have applied an input to this circuit, which consists of both an ac signal v_{in} that we want to pass through this circuit, and a dc level V_{dc} that we would like to remove. To see how this is accomplished, let's assume that we want the dc component of the output from this circuit to be zero. For this to occur, because V_{E1} is at a dc potential of $(V_{dc} - 0.7)$ or 5.0 V for this example, we will need to produce a 5-V dc drop across the resistor R_1.

The lower transistor Q_2 is essentially a current source, and because $V_{E2} = -0.7$ V, we find that

$$I_{E2} \simeq I_{C2} = \frac{V_{E2} - (-V_{EE})}{R_2} = \frac{10 \text{ V}}{10 \text{ k}\Omega} = 1.0 \text{ mA} \qquad (7.6\text{-}43)$$

This is the same current that flows through R_1, so that to make the dc compo-

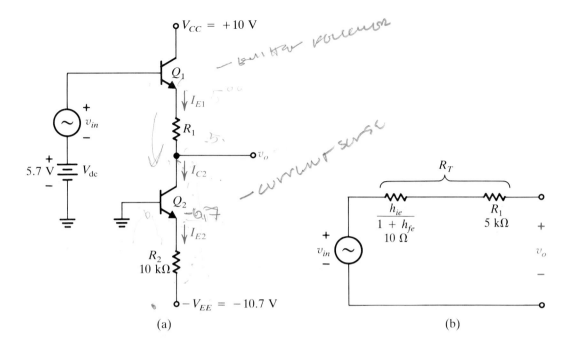

Figure 7.6-10
Cascode-style dc level shifter
circuit.

nent of v_o equal to zero, we need to choose R_1 as

$$R_1 = \frac{V_{E1} - V_o}{I_{C2}} = \frac{(5 - 0.0)\ \text{V}}{1\ \text{mA}} = 5\ \text{k}\Omega \qquad (7.6\text{-}44)$$

To illustrate why an additional stage generally follows this level shifter circuit, let us examine the ac Thévenin equivalent of this amplifier. In doing this, we consider that the transistor current gains are both 100 and that $h_{ie1} = 1\ \text{k}\Omega$. Using these data and assuming Q_2 to be an ideal dc current source, we may draw the small-signal equivalent output circuit as shown in Figure 7.6-10b. Because of the high output impedance of this circuit, the output from this stage cannot serve as the output from our operational amplifier without some type of additional buffering. In most op amp circuits this buffering, or impedance transformation, is accomplished with an emitter-follower stage. Sometimes it is a simple single-transistor emitter follower, but more commonly it is a two-transistor "push-pull" configuration containing an npn–pnp complementary-symmetry pair where one transistor handles the positive side of the signal, and the other transistor, the negative side. Push-pull arrangements are used because of their low current consumption from the power supplies and their good drive capability for a given transistor size.

In the examples that follow we present two typical operational amplifier circuits. The first to be examined in Example 7.6-2 (Figure 7.6-11) is an older-style op amp design in which the first two stages are resistively loaded differential amplifiers. The input stage uses FETs, so that the input impedance of this op amp is extremely high, and this stage is biased by a zener-regulated current source Q_3. Stage two is a BJT-style differential amplifier with a resistive biasing network in its emitters. A dc level shifter circuit identical to that just discussed follows the second differential amplifier stage, and this is connected to a single-transistor emitter follower which functions as the amplifier's output stage.

The second circuit, presented in Example 7.6-3 and illustrated in Figure 7.6-13 (p. 486), represents a more modern op amp design. The input stage is a BJT differential amplifier which contains a current mirror active load, and the second stage is a two-transistor Darlington style amplifier which is used to provide both additional gain and dc level shifting for the output stage.

The output stage consists of a pair of push-pull emitter follower transistors Q_{10} and Q_{11}, which give the amplifier its low output impedance, and transistors Q_{12} and Q_{13}, which help to make the amplifier "short-circuit proof" by providing current limiting in the output. Under ordinary operating conditions, when the load current is low, the voltage drop across the 30-Ω resistors in the output stage is so small that the base–emitter junctions of Q_{12} and Q_{13} are OFF. As a result, unless the current into the load gets too high, these transistors are essentially open circuits.

To illustrate how the current limiting occurs, consider, for example, what happens when the output voltage is positive, Q_{10} is ON, and the load current reaches about $+20$ mA. Under these circumstances the voltage drop across R_4 is 600 mV, and at this point Q_{12} just turns ON and begins to take the current being supplied by Q_9 away from the base of Q_{10}. This prevents the load current from going much beyond 20 mA and protects the driver transistors from being destroyed by accidental short circuits (or low-impedance connections) at the output.

The diodes D_1 and D_2 in the output stage are used to reduce "crossover distortion" by biasing each of the output transistor base–emitter junctions so that they are each just on the verge of turning ON. This avoids the creation of a dead zone in the output, v_o, for base input voltages at Q_{10} and Q_{11} between $+0.7$ and -0.7 V. Much more will be said about crossover distortion and the design of push-pull circuits in Chapter 9 when we discuss power amplifier construction techniques.

EXAMPLE 7.6-2

The operational amplifier illustrated in Figure 7.6-11 employs silicon transistors with the following characteristics: Q_1 and Q_2 are FETs with $I_{DSS} = 5$ mA and $V_p = -2$ V, and all the BJT transistors are identical with current gains of 100. Find all the pertinent dc quiescent currents and voltages in this circuit, and determine the small-signal open-loop voltage gain and output impedance of this amplifier.

SOLUTION

Because it will frequently be useful to refer to voltage levels measured with respect to the negative supply rail $(-V_{EE})$, we will identify such voltages by an additional subscript N. With this notation, V_{EN3}, for example, refers to the dc voltage between the emitter of Q_3 and the negative supply rail, and is equal to the voltage across the resistor R_2.

In carrying out the dc analysis of this circuit, the best place to start is at the input stage, and even more specifically, at the current source Q_3. The resistor R_1 forward biases the diode D_1 and causes the diode D_2 to zener so that

$$V_{BN3} = V_{D1} + V_{Z2} = 0.7 + 4.0 = 4.7 \text{ V} \qquad (7.6\text{-}45a)$$

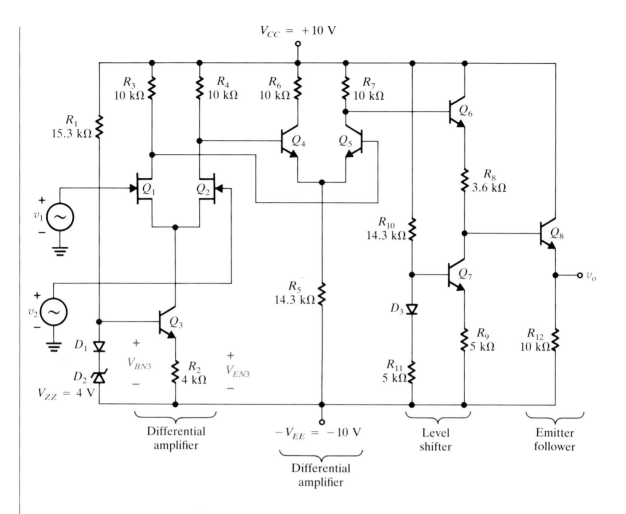

Figure 7.6-11
Example 7.6-2.

As a result, the bias current into the zener is approximately

$$I_{R1} = \frac{V_{CC} - 4.7 - (-V_{EE})}{R_1} = 1.0 \text{ mA} \qquad (7.6\text{-}45\text{b})$$

and the voltage at the emitter of Q_3 is

$$V_{EN3} = V_{BN3} - 0.7 = 4.0 \text{ V} \qquad (7.6\text{-}45\text{c})$$

Because V_{EN3} is the same as V_{R2}, the emitter current in Q_3 is seen to be

$$I_{E3} = \frac{V_{EN3}}{R_2} = \frac{4.0 \text{ V}}{4 \text{ k}\Omega} = 1.0 \text{ mA} \qquad (7.6\text{-}46)$$

Since Q_1 and Q_2 are assumed to be identical, the collector current from Q_3 splits evenly between them and we have

$$I_{D1} = I_{D2} = \frac{I_{C3}}{2} = 0.5 \text{ mA} \qquad (7.6\text{-}47\text{a})$$

Using the equation $I_D = I_{DSS} (1 - V_{GS}/V_p)^2$, we can show that the corresponding gate-to-source voltages needed to produce these drain currents are

$$V_{GS1} = V_{GS2} = -1.36 \text{ V} \qquad (7.6\text{-}47\text{b})$$

and since the dc gate voltages of Q_1 and Q_2 are both zero, $V_{S1} = V_{S2} = 1.36$ V. Furthermore, the collector-to-emitter voltage on Q_3 will be $[V_{S1} - V_{EN3} - (-V_{EE})]$ or 7.36 V.

If for the moment we neglect loading of the second stage on the first, then

$$V_{D1} = V_{D2} = V_{DD} - I_D R_D = 10 - 5 = 5.0 \text{ V} \qquad (7.6\text{-}48\text{a})$$

and

$$V_{E4} = V_{E5} = 5 - 0.7 = 4.3 \text{ V} \qquad (7.6\text{-}48\text{b})$$

at the emitters of Q_4 and Q_5. The current flow through R_5 is therefore

$$I_{R5} = \frac{V_{E4} - (-V_{EE})}{R_5} = \frac{14.3 \text{ V}}{14.3 \text{ k}\Omega} = 1.0 \text{ mA} \qquad (7.6\text{-}48\text{c})$$

and because the transistors Q_4 and Q_5 are identical, we have

$$I_{C4} = I_{C5} = \frac{I_{R5}}{2} = 0.5 \text{ mA} \qquad (7.6\text{-}48\text{d})$$

At this point it is useful to go back and check the validity of our original assumption, specifically that the second stage creates negligible dc loading on the first stage. Using equation (7.6-48d), we find that the quiescent base currents in Q_4 and Q_5 are each 0.005 mA, and because these currents are so much less than the drain currents in Q_1 and Q_2, we can see that we make only a small error by neglecting them.

We will also examine this loading problem from a second viewpoint in order to develop an equivalent circuit which will permit us to determine the circuit Q point even in the situation where the second stage does load the first. The dc voltages at the output of the first stage represent common-mode inputs into the second stage. Therefore, the loading effect of Q_5 on Q_1, for example, may be determined by doubling the value of R_5 and splitting the circuit in half as illustrated in Figure 7.6-12a. The equivalent circuit given in part (b) of this figure is then obtained by taking the Thévenin equivalent at the drain of Q_1, and connecting it to the base equivalent circuit of Q_5. Using this equivalent circuit, we may calculate the base current into Q_5, including second-stage loading effects, as

$$I_{B5} = \frac{5 - 0.7 - (-10)}{2.87 \text{ M}\Omega} = 4.98 \text{ } \mu\text{A} = 0.00498 \text{ mA} \qquad (7.6\text{-}49)$$

Because this result is essentially the same as that obtained by neglecting the loading, it is apparent that the second stage places a negligible dc load on the first.

Examining the output portion of the second differential amplifier, it is similarly apparent that Q_6 does not load Q_5 since the dc input impedance of Q_6 is at least $(1 + h_{FE})R_8$, or 360 kΩ. Therefore, the voltage at the collector of Q_5 is

$$V_{C5} = V_{CC} - I_{C5} R_7 = 5.0 \text{ V} \qquad (7.6\text{-}50\text{a})$$

so that

$$V_{E6} = V_{C5} - 0.7 = 4.3 \text{ V} \qquad (7.6\text{-}50\text{b})$$

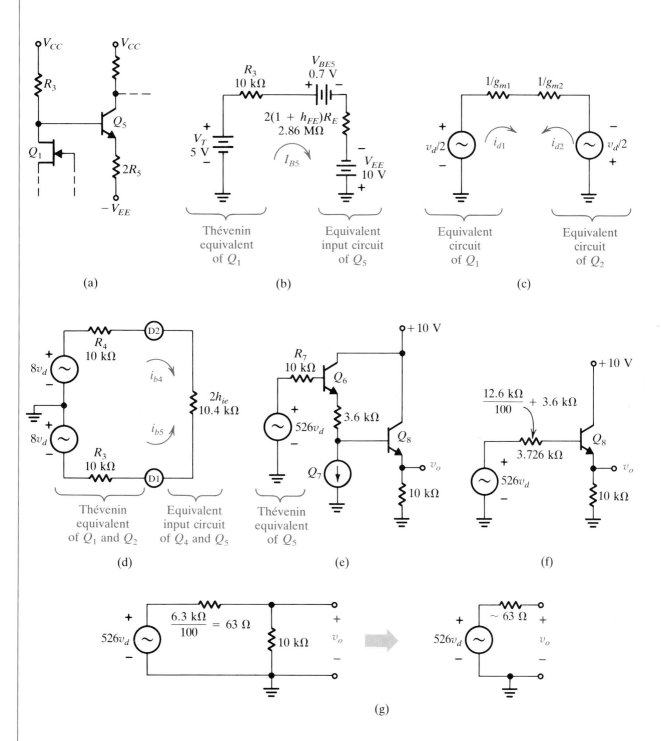

Figure 7.6-12
Example 7.6-2 (Continued).

483

The lower transistor, Q_7, in the level shifter is a current source. Because the dc input impedance of Q_7 is $(1 + h_{FE})R_9$ or 500 kΩ, we may neglect the loading of Q_7 on the R_{10}–R_{11} voltage divider, and as a result the current through R_{11} is seen to be

$$I_{R11} = \frac{V_{CC} - V_{D3} - (-V_{EE})}{R_{10} + R_{11}} = \frac{19.3 \text{ V}}{19.3 \text{ k}\Omega} = 1.0 \text{ mA} \qquad (7.6\text{-}51a)$$

so that

$$V_{BN7} = V_{D3} + I_{R11}R_{11} = 5.7 \text{ V} \qquad (7.6\text{-}51b)$$

$$V_{EN7} = V_{BN7} - 0.7 = 5.0 \text{ V} \qquad (7.6\text{-}51c)$$

and

$$I_{E7} = \frac{V_{EN7}}{R_9} = 1.0 \text{ mA} \qquad (7.6\text{-}51d)$$

Because the dc input resistance of Q_8 is about 1 MΩ, we may also neglect base current of Q_8 in comparison to I_{C7}, so that to a good approximation,

$$I_{E6} = I_{C7} = I_{E7} = 1.0 \text{ mA} \qquad (7.6\text{-}51e)$$

This result confirms our earlier assumption that we could neglect the loading of Q_6 on Q_5 since from equation (7.6-51e) we see that I_{B6} is only about 0.01 mA.

Combining equations (7.6-50b) and (7.6-51e), we obtain

$$V_{B8} = V_{E6} - V_{R8} = 4.3 - 3.6 = 0.7 \text{ V} \qquad (7.6\text{-}52a)$$

so that the output voltage is

$$V_o = V_{B8} - 0.7 = 0.0 \text{ V} \qquad (7.6\text{-}52b)$$

and

$$I_{E8} = \frac{V_o - (-V_{EE})}{R_{12}} = \frac{10 \text{ V}}{10 \text{ k}\Omega} = 1.0 \text{ mA} \qquad (7.6\text{-}52c)$$

Using equation (7.6-52c), it follows that the base current in Q_8 is about 0.01 mA, and because the collector current in Q_7 is 1 mA, we can see, in this case, too, that our assumption in neglecting the loading of Q_8 on the level shifter was also justified.

It is always important to check the validity of any assumptions that we make in analyzing electronic circuits, and to recognize that where these approximations fail, the problem can always be solved by taking loading into account. In each case, this can always be done simply by connecting the Thévenin equivalent of the previous stage to the equivalent input circuit of the next stage.

The ac analysis for this circuit proceeds in a similar fashion to the dc analysis and is simplified to a certain extent by the fact that the dc current sources Q_3 and Q_7 may be replaced by open circuits. Using the results from the dc analysis, we can approximate the transistor input impedances h_{ie4} and h_{ie5} as 5.2 kΩ and h_{ie6} and h_{ie8} as 2.6 kΩ.

Starting with the source equivalent circuit for Q_1 and Q_2 given in Figure 7.6-12c, we obtain

$$i_{d1} = -i_{d2} = \frac{v_d}{2/g_m} = \frac{g_m v_d}{2} \qquad (7.6\text{-}53a)$$

from which we can see that the open-circuit voltage at each of the drains is

$$v_{d1} = -i_{di}R_3 = -\frac{g_m R_3}{2}v_d = -8v_d \qquad (7.6\text{-}53b)$$

and

$$v_{d2} = -i_{d2}R_4 = \frac{g_m R_4}{2}v_d = +8v_d \qquad (7.6\text{-}53c)$$

using $g_m = 1.6$ mS from equation (5.6-2b). Because v_{d1} and v_{d2} are 180° out of phase, these signals act as differential inputs to the next stage and we may compute the base currents flowing into Q_4 and Q_5 by connecting the Thévenin equivalents of Q_1 and Q_2 from the first stage to the differential input impedance of the second stage (Figure 7.6-12d). Following this figure, we have

$$i_{b4} = -i_{b5} = \frac{16v_d}{30.4\text{ k}\Omega} \qquad (7.6\text{-}54a)$$

and

$$i_{c5} = h_{fe}i_{b5} = \frac{-1600v_d}{30.4\text{ k}\Omega} \qquad (7.6\text{-}54b)$$

so that the open-circuit voltage at the collector of Q_5 is

$$v_{c5} = -i_{c5}R_7 = \frac{16,000\text{ k}\Omega\ v_d}{30.4\text{ k}\Omega} = 526v_d \qquad (7.6\text{-}54c)$$

To find the output voltage of the overall op amp, we may now connect the Thévenin equivalent of the collector circuit of Q_5 to the remainder of the amplifier as shown in Figure 7.6-12e. This circuit can easily be simplified by reflecting the base circuit of Q_6 into its emitter as illustrated in Figure 7.6-12f, and then finally by reflecting the base portion of that circuit into the emitter of Q_8. The overall Thévenin equivalent of this amplifier is given in Figure 7.6-12g, and from this circuit we can see that the open-loop gain of this op amp is 526, and its output impedance is 63 Ω. Furthermore, because the input stage uses FETs, the differential input impedance of this amplifier is essentially equal to infinity.

EXAMPLE 7.6-3

For the operational amplifier illustrated in Figure 7.6-13, the transistors may all be considered to have current gains equal to 100. Determine all the pertinent Q-point information for this circuit, as well as its small-signal open-loop gain, output impedance, and differential input impedance.

SOLUTION

As with Example 7.6-2, we begin the dc analysis of this circuit by calculating the values of the current sources. Considering that $V_{BE6} = 0.7$ V, we obtain

$$I_{R1} = \frac{19.3\text{ V}}{19.3\text{ k}\Omega} = 1.0\text{ mA} \qquad (7.6\text{-}55)$$

and combining this with equation (7.6-33c), we can show (by using a trial-and-error solution) that I_{C7} and I_{C14} are each equal to 0.1 mA.

Because of the symmetry of the differential amplifier in the input stage, we find that

$$I_{C1} = I_{C2} = \frac{I_{C7}}{2} \qquad (7.6\text{-}56a)$$

$$I_{C3} + I_{B5} = I_{C1} \tag{7.6-56b}$$

and
$$I_{C4} + I_{B8} = I_{C2} \tag{7.6-56c}$$

Furthermore, because of the current mirror behavior of Q_3 and Q_4, we also have that

$$I_{C3} = I_{C4} \tag{7.6-57}$$

so that from equations (7.6-56a) and (7.6-56b), we obtain

$$I_{B5} = I_{B8} \tag{7.6-58a}$$

and
$$I_{E5} = I_{E8} \tag{7.6-58b}$$

Figure 7.6-13
Example 7.6-3.

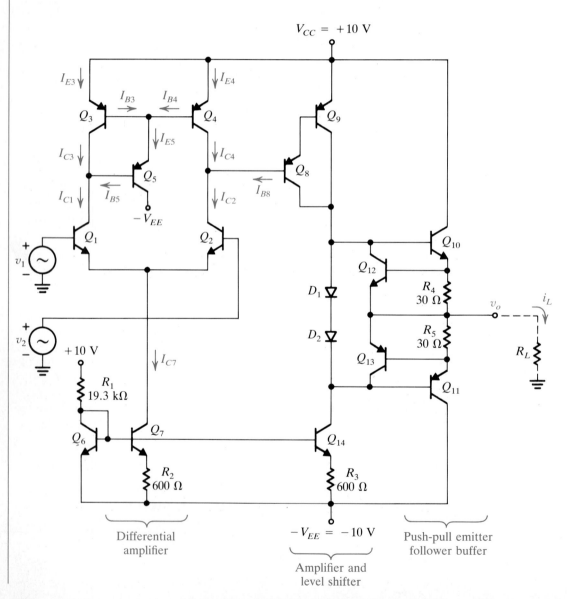

486   Chapter 7 / Introduction to Feedback and the Operational Amplifier

But from this circuit,
$$I_{E5} = I_{B3} + I_{B4} \qquad (7.6\text{-}59\text{a})$$

from which

$$I_{C9} = h_{FE} I_{B9} = h_{FE} I_{E8} = h_{FE}(I_{B3} + I_{B4}) = I_{C3} + I_{C4} \qquad (7.6\text{-}59\text{b})$$

In equations (7.6-56b) and (7.6-56c), because the h_{FE} values are large, we may, to a good approximation, neglect the base currents to yield $I_{C3} = I_{C1}$ and $I_{C4} = I_{C2}$. Substituting these results into equation (7.6-59b), we obtain

$$I_{C9} = I_{C3} + I_{C4} = I_{C1} + I_{C2} = I_{C7} = 0.1 \text{ mA} \qquad (7.6\text{-}60)$$

Thus in this circuit, not only does Q_7 bias the differential amplifier, but it simultaneously sets the quiescent current in Q_9.

The diodes D_1 and D_2 in the output stage are selected so that with 100 μA of current through them the voltage across each is on the order of 0.6 V. This ensures that Q_{10} and Q_{11}, although still OFF, are just on the verge of turning ON. At the output because I_{E10} and I_{E11} are both nearly zero, the current into the load is also zero, so that we have

$$V_o = I_L R_L = (I_{E10} - I_{E11})R_L = 0 \qquad (7.6\text{-}61\text{a})$$

$$V_{B10} = V_o + 0.6 = +0.6 \text{ V} \qquad (7.6\text{-}61\text{b})$$

and
$$V_{B11} = V_o - 0.6 = -0.6 \text{ V} \qquad (7.6\text{-}61\text{c})$$

This result assumes that the currents in Q_9 and Q_{14} are identical. If these currents do not match, then, as with the current mirror circuit discussed earlier, a substantial dc offset will be produced at the collector of Q_9 and at the output of the op amp. However, because of the very high gain of the op amp, this offset can be reduced to zero by the addition of a small voltage at either of the amplifier's input terminals.

To carry out the small-signal differential gain analysis of this circuit, we consider that the transistor output impedances are inversely proportional to their quiescent currents and that $1/h_{oe}$ is 100 kΩ at 1 mA. As a result, we approximately have $1/h_{oe4} = 1/h_{oe2} = 2$ MΩ, and $1/h_{oe7} = 1/h_{oe9} = 1/h_{oe12} = 1$ MΩ. In addition, the transistor input impedances of Q_8 and Q_9 are 2.6 MΩ and 26 kΩ, respectively, and the g_m of the transistors Q_1 and Q_2 in the input stage are each $(50 \ \mu\text{A})/(26 \text{ mV})$ or 1420 μS.

If we apply a differential signal to the input stage, because of the circuit symmetry, the emitters of Q_1 and Q_2 are at a virtual ground potential, and the circuit may be split in two as illustrated in Figure 7.6-14a. From this figure,

$$i_{b1} = \frac{v_d}{2h_{ie}} \qquad (7.6\text{-}62\text{a})$$

and
$$i_{b2} = -\frac{v_d}{2h_{ie}} \qquad (7.6\text{-}62\text{b})$$

Clearly, for this circuit the differential input impedance v_d/i_{b1} is simply $2h_{ie}$ or 104 kΩ.

Because the impedance seen looking into the base of Q_8 is equal to

$$r_{i8} = h_{ie8} + (1 + h_{fe})h_{ie9} = 5.2 \text{ m}\Omega \qquad (7.6\text{-}63)$$

the equivalent circuit shown in Figure 7.6-14b is valid, and by using a current

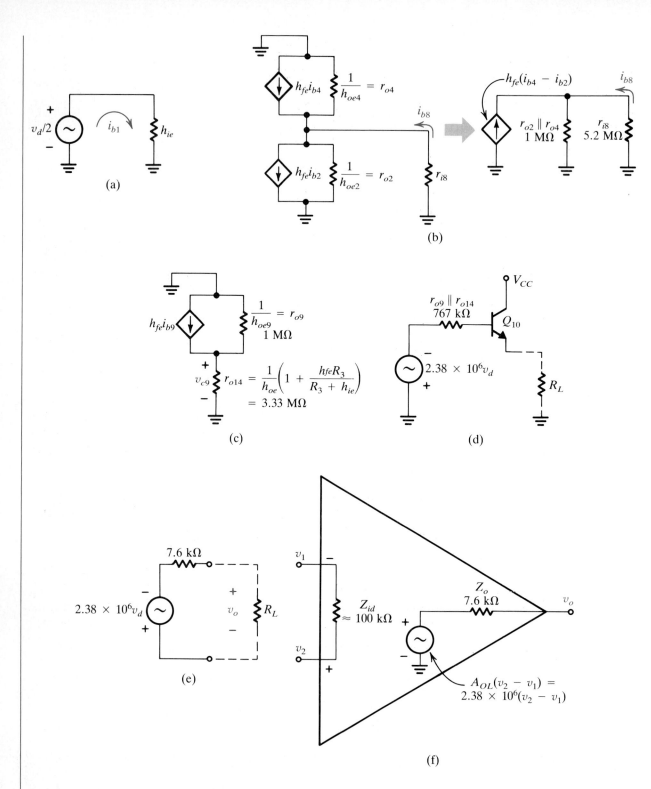

Figure 7.6-14
Example 7.6-3 (Continued).

divider we find that

$$i_{b8} = \left(-\frac{1}{6.2}\right)h_{fe}(i_{b4} - i_{b2}) \tag{7.6-64}$$

But if we recall that Q_3 and Q_4 form a current mirror load, then we find that $i_{b4} = i_{b1}$, and combining this result together with equations (7.6-62a), (7.6-62b), and (7.6-64), we obtain

$$i_{b8} = \left(-\frac{1}{6.2}\right)h_{fe}(i_{b1} - i_{b2}) = \frac{-2h_{fe}v_d}{(6.2)(2h_{ie})} = -\frac{g_m v_d}{6.2} \tag{7.6-65a}$$

and

$$i_{b9} = i_{c8} = h_{fe}i_{b8} = -\frac{100 g_m v_d}{6.2} \tag{7.6-65b}$$

The open-circuit voltage at the collector of Q_9 may be found by using the equivalent circuit in Figure 7.6-14c. Notice that because of the presence of R_3, the output impedance of Q_{14} is not simply $1/h_{oe}$ (Problem 7.6-11). As a result, the open-circuit voltage is

$$v_{c9} = h_{fe}i_{b9}(r_{o9} \parallel r_{o14}) = 2.38 \times 10^6 v_d \tag{7.6-66}$$

The output voltage and the circuit output impedance may be determined by connecting the Thévenin equivalent of the Q_9 collector circuit in Figure 7.6-14c to the output stage of the op amp. For simplicity we examine this equivalent circuit with the input signal to the Q_{10}–Q_{11} pair taken to be positive, so that we can assume that Q_{10} is ON and Q_{11} is OFF. For this case neglecting the small diode resistances of D_1 and D_2, the equivalent circuit in Figure 7.6-14d applies. If we assume that h_{ie10} has a nominal value of about 1 kΩ, the equivalent circuit at the output of the op amp is approximately that given in Figure 7.6-14e, and we can see that A_{OL} is 2.38 million and the output impedance is about 7.6 kΩ.

As a practical matter, this output impedance is rather large for an op amp, and if the output transistors Q_{10} and Q_{11} were replaced by Darlington transistor pairs, this impedance could easily be reduced by a factor of about 100 times, to 75 Ω or less.

Exercises

7.6-1 Find the CMRR at the collector of Q_2 for the circuit in Figure 7.6-3a if $R_E = 10$ kΩ, $h_{ie1} = 1$ kΩ, $h_{ie2} = 2$ kΩ, and $h_{fe1} = h_{fe2} = 100$. **Answer** 19.7 dB

7.6-2 For the current mirror in Figure 7.6-7c, $V_{CC} = 20$ V, $R_1 = 40$ kΩ, and $R_2 = 500$ Ω. Determine I_{C2}. **Answer** 0.075 mA

7.6-3 In Figure 7.6-11, set $v_1 = v_{in}$ and $v_2 = 0$, and find v_{d1} in terms of v_{in}. **Answer** $-2.74v_{in}$

7.6-4 In Figure 7.6-11, R_9 is decreased to 2 kΩ. Select R_8 so that the nominal value of V_o is still zero. **Answer** 1.44 kΩ

7.7 SECOND-ORDER EFFECTS IN OPERATIONAL AMPLIFIERS

In Sections 7.2 and 7.3, we examined the behavior of the ideal operational amplifier, and in doing so, we assumed that it had an infinite input impedance, zero output impedance, infinite bandwidth, and no dc offsets. The purpose of this section is to take a closer look at how the nonideal characteristics of an actual op amp affect its performance in different circuit configurations, and what we as designers can do to minimize the degradation in performance that occurs due to these second-order effects.

7.7-1 Input and Output Impedance

Figure 7.7-1 illustrates a second-order model for the operational amplifier which includes the amplifier's input and output impedances. In this model Z_{id} represents the differential input impedance, Z_{icm} the common-mode input impedance, and Z_o the amplifier output impedance. Typical values of these parameters are 10^5, 10^7, and 50 Ω, respectively for a BJT op amp. FET input op amps, on the other hand, have differential and common-mode input impedances approaching 10^{12} Ω. Expressions for Z_{id} and Z_{icm} were derived in equations (7.6-11) and (7.6-18).

Feedback significantly alters the input and output impedances of an op amp, and in this section we examine the impedance levels of both inverting and noninverting amplifiers. In carrying out the input impedance calculations, the amplifier output impedance will be taken as zero because its inclusion greatly increases the complexity of the analysis while providing only a third-order correction to the results. Similarly, when we compute the output impedance of the operational amplifier with feedback connected, we assume that its differential and common-mode input impedances are infinitely large.

Figure 7.7-1
Second-order model for the operational amplifier including input and output impedances.

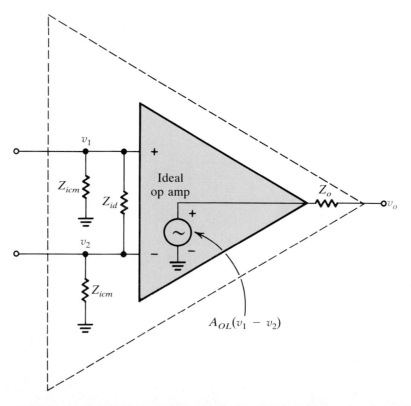

Chapter 7   /   Introduction to Feedback and the Operational Amplifier

The expression for the input impedance of an inverting amplifier may be derived with the aid of Figure 7.7-2a, and because the v_1 input terminal of this amplifier is grounded, we may simplify the circuit to that given in part (b) of the figure. The amplifier input impedance is defined as

$$Z_{in} = \frac{v_{in}}{i_{in}} \bigg|_{v_1 = 0} \tag{7.7-1a}$$

where i_{in} for this circuit is given by the expression

$$i_{in} = \frac{v_{in} - v_2}{R_2} \tag{7.7-1b}$$

However, the inverting input terminal of the op amp is a virtual ground, so that $v_2 = 0$, and therefore

$$Z_{in} = \frac{v_{in}}{v_{in}/R_2} = R_2 \tag{7.7-2}$$

Thus because the inverting input terminal of the op amp is effectively grounded, the input impedance of an inverting amplifier is simply equal to the input resistance R_2, and the common-mode and differential input impedances have practically no effect on its value. The input impedance of the noninverting amplifier, however, depends strongly on these parameters.

To develop a quantitative expression for this relationship, let's compute the input impedance of the noninverting amplifier with the aid of Figure 7.7-3a. In analyzing this circuit, it is generally true that

$$Z_{icm} \gg R_2 \tag{7.7-3}$$

so that we may drop Z_{icm} from the lower portion of this circuit as illustrated.

To find the input impedance of this circuit, we need to develop an expression for i_{in} in terms of v_{in}. By applying Kirchhoff's current law to the v_1 node, we have

$$i_{in} = i_{cm} + i_d = \frac{v_1}{Z_{icm}} + i_d = \frac{v_{in}}{Z_{icm}} + i_d \tag{7.7-4}$$

Figure 7.7-2
Input impedance of an inverting amplifier.

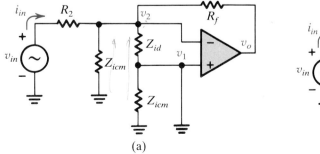

(a)

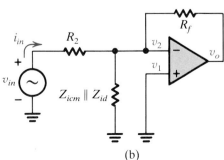

(b)

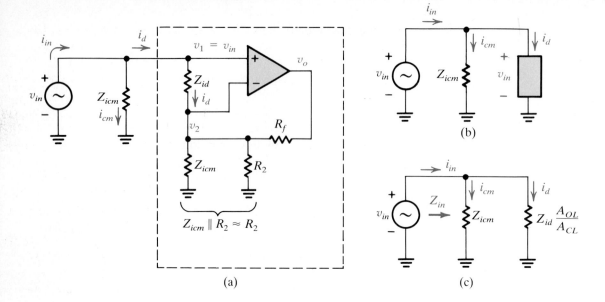

(a)   (c)

Figure 7.7-3

Input impedance of a noninverting amplifier.

so that to complete this analysis all that is needed is an expression for i_d in terms of v_{in}. To obtain this, we apply Kirchhoff's current law to the v_2 node, which yields

$$\frac{v_1 - v_2}{Z_{id}} + \frac{v_o - v_2}{R_f} = \frac{v_2}{R_2} \tag{7.7-5}$$

But

$$v_o = A_{OL}(v_1 - v_2) \tag{7.7-6a}$$

and

$$v_1 = v_{in} \tag{7.7-6b}$$

so that

$$\frac{v_{in} - v_2}{Z_{id}} + \frac{A_{OL}(v_{in} - v_2) - v_2}{R_f} = \frac{v_2}{R_2} \tag{7.7-6c}$$

This expression may readily be solved for v_2 as

$$v_2 = \frac{v_{in}(1/Z_{id} + A_{OL}/R_f)}{[(1/Z_{id}) + (1 + A_{OL})/R_f + 1/R_2]} \tag{7.7-6d}$$

From Figure 7.7-3a, the differential portion of the input current may now be written as

$$i_d = \frac{v_1 - v_2}{Z_{id}} = \frac{v_{in}}{Z_{id}}\left[1 - \frac{1/Z_{id} + A_{OL}/R_f}{(1/Z_{id}) + (1 + A_{OL})/R_f + 1/R_2}\right]$$

$$= \frac{v_{in}}{Z_{id}}\left[\frac{1/R_2 + 1/R_f}{(1/Z_{id}) + (1 + A_{OL})/R_f + 1/R_2}\right] \tag{7.7-7}$$

Approximating the denominator of this expression as A_{OL}/R_f, the equation reduces to

$$i_d = \frac{v_{in}}{Z_{id}} \frac{1 + R_f/R_2}{A_{OL}} \tag{7.7-8}$$

If we recall that

$$A_{CL} = 1 + \frac{R_f}{R_2} \tag{7.7-9}$$

is the noninverting closed-loop gain of this amplifier, we may rewrite i_d in terms of this quantity as

$$i_d = \frac{v_{in} \, A_{CL}}{Z_{id} \, A_{OL}} = \frac{v_{in}}{Z_{id}(A_{OL}/A_{CL})} \tag{7.7-10}$$

Using this result, the input impedance of this noninverting amplifier may be found by representing the input portion of the amplifier in Figure 7.7-3a by the simplified two-branch circuit given in part (b) of the figure. Here the original portion of the circuit within the dashed lines has been replaced by the box shown in Figure 7.7-3b. The voltage across this box is equal to v_{in}, and the current i_d flowing through it is given by equation (7.7-10). Because i_d is proportional to v_{in}, the circuit within the box may be represented by an equivalent resistor,

$$R = \frac{v_{in}}{i_d} = \frac{v_{in}}{v_{in}/Z_{id}(A_{OL}/A_{CL})} = Z_{id}\frac{A_{OL}}{A_{CL}} \tag{7.7-11}$$

This result leads to the equivalent circuit given in Figure 7.7-3c, and from this circuit it is clear that the input impedance of this noninverting amplifier is

$$Z_{in} = Z_{icm} \, \| \, \left(Z_{id}\frac{A_{OL}}{A_{CL}} \right) \tag{7.7-12}$$

In examining this result it is not immediately obvious why Z_{id} is apparently magnified by the factor A_{OL}/A_{CL}. The example that follows will help to explain physically why this occurs.

EXAMPLE 7.7-1

The noninverting amplifier illustrated in Figure 7.7-4a has a differential input impedance of 50 kΩ, a common-mode input impedance of infinity, and an open-loop gain of 10^4. By going back to first principles, and not simply by using equation (7.7-12), explain why the input impedance of this amplifier is much greater than 50 kΩ. Compare the solution obtained by using this approach with the results you get by a direct application of equation (7.7-12).

SOLUTION

The reason for the high input impedance of this amplifier is that i_{in} is very small for this circuit. If the op amp were running open-loop with the inverting input terminal grounded, then, as illustrated in Figure 7.7-4b, i_{in} would simply be given by the expression

$$i_{in} = i_d = \frac{v_1 - v_2}{Z_{id}} = \frac{v_{in} - 0}{50 \text{ k}\Omega} \tag{7.7-13}$$

and applying equation (7.7-1a), we would find that

$$Z_{in} = \frac{v_{in}}{i_{in}} = \frac{v_{in}}{v_{in}/50 \text{ k}\Omega} = 50 \text{ k}\Omega \tag{7.7-14}$$

The addition of feedback changes this situation considerably. If A_{OL} were

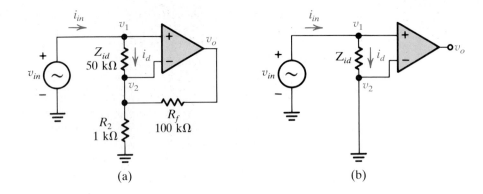

(a)   (b)

Figure 7.7-4
Example 7.7-1.

infinitely large, then v_2 in Figure 7.7-4a would exactly equal v_1, and as a result, we would find that $i_{in} = 0$ and $Z_{in} = v_{in}/0 = \infty$.

When A_{OL} is finite, v_2 is very close to v_1 but not identically equal to it, and as a result, i_{in} is very small and Z_{in} very large. In particular, for this specific circuit we find from equation (7.7-6d) that

$$v_2 = \frac{v_{in}(1/50 \text{ k}\Omega + 10^4/100 \text{ k}\Omega)}{(1/50 \text{ k}\Omega) + (1 + 10^4)/100 \text{ k}\Omega + 1/1 \text{ k}\Omega} = 0.990002969 v_{in} \qquad (7.7\text{-}15)$$

Using this result, we may write down the input current as

$$i_{in} = \frac{v_1 - v_2}{Z_{id}} = \frac{v_{in} - 0.990002969 v_{in}}{50 \text{ k}\Omega} = \frac{v_{in}}{50 \text{ k}\Omega/0.00999703} = \frac{v_{in}}{5.0015 \text{ M}\Omega} \qquad (7.7\text{-}16a)$$

so that $\qquad Z_{in} = \dfrac{v_{in}}{i_{in}} = \dfrac{v_{in}}{v_{in}/5.0015 \text{ M}\Omega} = 5.0015 \text{ M}\Omega \qquad (7.7\text{-}16b)$

For comparison, a direct application of equation (7.7-12) yields

$$Z_{in} \cong Z_{icm} \parallel \left(Z_{id}\frac{A_{OL}}{A_{CL}}\right) = (50 \text{ k}\Omega)\left(\frac{10^4}{101}\right) = 4.95 \text{ M}\Omega \qquad (7.7\text{-}17)$$

The output impedance of an operational amplifier with feedback may be found by at least two different approaches. The first employs a direct application of Thévenin's theorem with

$$Z_{out} = \frac{v_{oc}}{i_{sc}} \qquad (7.7\text{-}18)$$

and the second involves an examination of the impedance seen looking back into the output terminals of the amplifier with all the independent sources set equal to zero. Because we are already well acquainted with the first of these two techniques from numerous earlier examples, we employ the second method to find the output impedance of this op amp feedback circuit. This approach is also more useful than the straight application of Thévenin's theorem because it will clearly indicate why the output impedance of an op amp is the same regardless of whether the circuit is operating as an inverting or a noninverting amplifier (see Figure 7.7-5a and b).

The basic circuit to be examined is given in Figure 7.7-5c, and to measure Z_{out} we apply a voltage, v_x, to the output terminal of the op amp and measure the current i_x that flows with v_{in}, the independent source, set equal to zero. The ratio v_x/i_x then defines the circuit's output impedance $Z_{out} = R_T$ (Figure 7.7-5d).

A simplified version of the original op amp circuit is given in Figure 7.7-5e. For this circuit the current i_x has two components i_f and i_o, and the overall circuit output impedance is

$$Z_{out} = \frac{v_x}{i_x}\bigg|_{v_{in}=0} = \frac{v_x}{i_f + i_o} \qquad (7.7\text{-}19)$$

If i_o and i_f can be expressed in terms of v_x alone, this equation can be used to evaluate the output impedance of the amplifier.

The current i_f is seen by inspection of Figure 7.7-5c to be

$$i_f = \frac{v_x}{R_f + R_2} \qquad (7.7\text{-}20)$$

To determine the current flow i_o back into the op amp output terminal, we need to develop an expression for the controlled source $A_{OL}(v_1 - v_2)$ in terms of the voltage v_x. The voltage v_2 fed back to the inverting input terminal is

$$v_2 = v_x \frac{R_2}{R_2 + R_f} \qquad (7.7\text{-}21)$$

while $v_1 = 0$ because the noninverting input terminal is grounded. Therefore, following Figure 7.7-5c, the current i_o may be written as

$$i_o = \frac{v_x - A_{OL}(v_1 - v_2)}{Z_o} = v_x \frac{1 + A_{OL}[R_2/(R_2 + R_f)]}{Z_o} \qquad (7.7\text{-}22)$$

and substituting this result along with equation (7.7-20) into equation (7.7-19), we have

$$Z_{out} = \frac{1}{\dfrac{1}{R_2 + R_f} + \dfrac{1 + A_{OL}[R_2/(R_2 + R_f)]}{Z_o}} = (R_f + R_2) \,\|\, \frac{Z_o}{1 + A_{OL}[R_2/(R_2 + R_f)]} \qquad (7.7\text{-}23)$$

Typically, the second term on the right-hand side of this expression is much smaller than the first, so that we may approximate the output impedance of this amplifier as

$$Z_{out} \simeq \frac{Z_o}{1 + A_{OL}[R_2/(R_2 + R_f)]} \simeq Z_o \frac{A_{CL}}{A_{OL}} \qquad (7.7\text{-}24)$$

where A_{CL} is given by equation (7.7-9). Using this equation, we can see that typical values for the output impedance of an op amp with feedback will generally be a small fraction of an ohm.

7.7 / Second-Order Effects in Operational Amplifiers

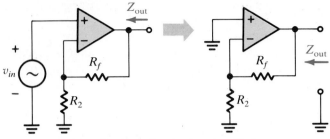

(a) Circuit to Be Analyzed to Find Z_{out} for a Noninverting Amplifier

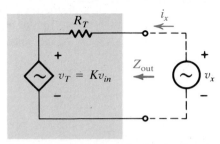

(d) Basic Technique for Measuring the Amplifier's Output Impedance ($Z_{out} = v_x/i_x|_{v_{in}=0}$)

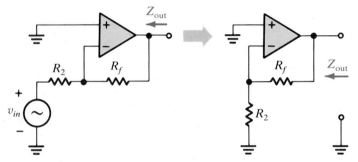

(b) Circuit to Be Analyzed to Find Z_{out} for an Inverting Amplifier

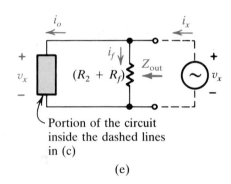

Portion of the circuit inside the dashed lines in (c)

(e)

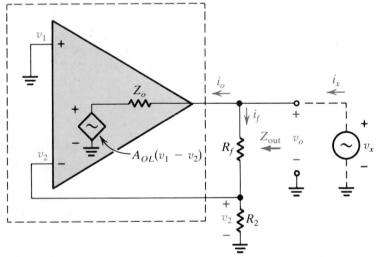

(c) Circuit to Be Analyzed Regardless of Whether Amplifier Is Inverting or Noninverting Style

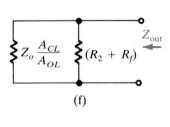

(f)

Figure 7.7-5
Output impedance of an op amp with negative feedback.

The result obtained in equation (7.7-23) may also be understood from an equivalent-circuit approach by replacing the circuit within the shaded region in Figure 7.7-5c by an equivalent circuit element. Following equation (7.7-22) because i_o, the current flow into the box in Figure 7.7-5e, is proportional to v_x, the voltage across the box, we may replace the circuit within the dashed region in Figure 7.7-5c by an equivalent resistor

$$R = \frac{v_x}{i_o} = \frac{v_x}{(v_x/Z_o)\{1 + A_{OL}[R_2/(R_2 + R_f)]\}} \simeq Z_o \frac{A_{CL}}{A_{OL}} \qquad (7.7\text{-}25)$$

The resulting equivalent circuit is given in Figure 7.7-5f, and from this figure it is clear that the overall output impedance of this circuit is the same as that given by equation (7.7-23).

7.7-2 DC Offset Effects

For the ideal op amp illustrated in Figure 7.7-6a, the input–output transfer characteristic is a straight line,

$$v_o = A_{OL}(v_1 - v_2) \qquad (7.7\text{-}26)$$

which passes through the origin (Figure 7.7-6b). This implies that the output voltage from this amplifier would be identically zero if both input terminals were grounded (Figure 7.7-6c). Although most commercially available op amps are designed to produce an output voltage of zero when the inputs are grounded, the variability of the components within these amplifiers cause a dc offset to exist in the output. As a result, the transfer characteristic, rather than passing through the origin, actually falls somewhere within the dashed lines indicated on the graph in Figure 7.7-6b. The corresponding input–output equation for such an op amp may be written as

$$v_o = A_{OL}(v_1 - v_2 \pm V_{IO}) \qquad (7.7\text{-}27)$$

where the quantity V_{IO} in this equation is called the input offset voltage of the amplifer. It represents the voltage that must be added in series with one of the inputs in Figure 7.7-6c in order to make the output zero. Typical values for V_{IO}

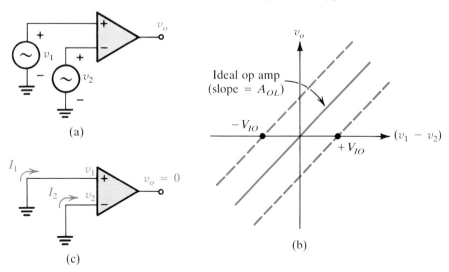

(a)

(c)

(b)

in commercially available op amps are on the order of 1 to 20 mV. The $\pm$ sign in front of the V_{IO} term in equation (7.7-27) indicates that this quantity can be either plus or minus, depending on the particular imbalances that exist in the amplifier. In addition, the value of V_{IO} specified by the manufacturer on the data sheet is usually the maximum or worst-case value that this parameter can be expected to have.

The input offset voltage of an op amp is important because, as we shall see shortly, it gives rise to dc shifts in the output. The finite values of the input bias currents in an op amp also create similar dc offsets.

The quantities labeled I_1 and I_2 in Figure 7.7-6c are called the input bias currents of the op amp, and these represent the dc currents that must flow into the input terminals of the amplifier in order to provide the proper bias for the input stage. The actual values of I_1 and I_2 are usually specified indirectly on the manufacturer's data sheets by listing the input bias current, I_{IB}, and the input offset current, I_{IO}.

The input bias current is defined as the average value of I_1 and I_2, or

$$I_{IB} = \frac{I_1 + I_2}{2} \qquad (7.7\text{-}28a)$$

and can range from the maximum value specified by the manufacturer to a minimum value of nearly zero. The input offset current is defined as the magnitude of the maximum difference between the two bias currents, or

$$I_{IO} = |I_1 - I_2| \qquad (7.7\text{-}28b)$$

Most op amp input stages are symmetrical and we would therefore expect that I_1 and I_2 would be equal, but again, because of component variabilities, these currents can be matched to an accuracy of only 10 or 20%. For example, in a typical general-purpose op amp such as the LM741C, the maximum value for the input bias current at 25°C is 500 nA, and the upper limit on the input offset current at the same temperature is 200 nA. The corresponding bias and offset currents for a JFET input op amp, such as the LFT155, are 50 and 10 pA, respectively.

Because it will be useful later, it is convenient to express I_1 and I_2 in terms of I_{IB} and I_{IO}. This may be done with the aid of equations (7.7-28a) and (7.7-28b), and solving these equations for the input bias currents I_1 and I_2, we obtain

$$I_1 = I_{IB} \pm \frac{I_{IO}}{2} \qquad (7.7\text{-}29a)$$

and

$$I_2 = I_{IB} \pm \frac{I_{IO}}{2} \qquad (7.7\text{-}29b)$$

These results represent the worst-case variations in I_1 and I_2 that can occur. It is also important to note that the plus and minus signs in I_1 and I_2 are not necessarily in phase, so that it is possible, for example, that I_1 may take on its worst-case maximum value at the same time that I_2 takes on its minimum. Using the 741 data above, if the average bias current were 500 nA, then I_1 and I_2 would each typically be somewhere within the range 400 to 600 nA. Of course, because the value of I_{IB} specified is the worst-case maximum at 25°C, the actual values obtained for I_1 and I_2 can turn out to be much smaller.

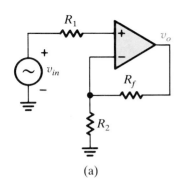

(a)

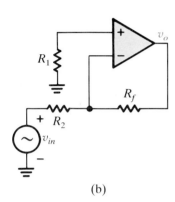

(b)

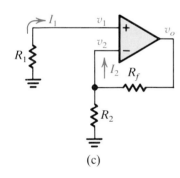

(c)

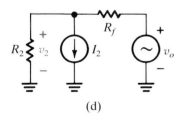

(d)

Figure 7.7-7

To develop a quantitative expression for the dc output offsets produced by the nonzero values of V_{IO}, I_1, and I_2, we will apply the superposition principle to the inverting and noninverting amplifiers given in Figure 7.7-7a and b. When v_{in} is set equal to zero in each of these amplifiers, in order to study the dc offset effects, the equivalent circuit for both of these amplifiers reduces to the circuit given in Figure 7.7-7c. As a result, the dc offsets produced in both of these circuits will be the same and may be calculated as follows.

The bias current I_1 flowing into the noninverting input terminal v_1 in Figure 7.7-7c produces a voltage

$$v_1 = -I_1 R_1 \tag{7.7-30a}$$

at this point, while the voltage v_2 at the inverting input terminal arises from two sources, I_2 and v_o (see Figure 7.7-7d). Applying superposition to this equivalent circuit, we may calculate v_2 as

$$v_2 = -I_2(R_2 \| R_f) + \frac{R_2}{R_2 + R_f} v_o \tag{7.7-30b}$$

Substituting this result, along with that in equation (7.7-30a), into equation (7.7-27), we may express the output dc offset produced by the bias currents and input offset voltage as

$$v_o = A_{OL}\left[(-I_1 R_1) - \left(-I_2(R_2 \| R_f) + \frac{R_2}{R_2 + R_f} v_o\right) \pm V_{IO}\right] \tag{7.7-31a}$$

or $\quad v_o = \dfrac{A_{OL}}{1 + A_{OL}[R_2/(R_2 + R_f)]}[I_2(R_2 \| R_f) - I_1 R_1 \pm V_{IO}] \tag{7.7-31b}$

However, because the term

$$A_{OL}\frac{R_2}{R_2 + R_f} = \frac{A_{OL}}{A_{CL}} \gg 1 \tag{7.7-32}$$

equation (7.7-31b) reduces to

$$v_o \simeq A_{CL}[I_2(R_2 \| R_f) - I_1 R_1 \pm V_{OS}] \tag{7.7-33}$$

where A_{CL} represents the closed-loop-gain expression $(1 + R_f/R_2)$, and R_1 and $R_2 \| R_f$ represent the impedances seen looking out of the noninverting and inverting input terminals of the op amp, respectively. To understand this result, it is useful to notice that V_{IO}, $I_1 R_1$, and $I_2(R_2 \| R_f)$ are all effectively input voltages produced by the offsets of the amplifier, and like any other input signals, they appear in the output amplified by the closed-loop gain of the op amp.

To minimize the dc offset in the output, it is apparent that we should keep the input impedance levels, R_1 and $R_2 \| R_f$, as small as practical so that the voltage drops produced at the input terminals by the bias currents will also be small. Furthermore, by matching these impedances, we can cause the first two terms in parentheses on the right-hand side of equation (7.7-33) to cancel, at least to the extent that I_1 and I_2 are equal. Of course, a complete cancellation is not possible because I_1 and I_2 are not identical, but, a substantial reduction in the output offset is still possible by matching these impedances because I_1 and

I_2 are very close to one another. In particular, by substituting in equations (7.7-29a) and (7.7-29b) into equation (7.7-33), we obtain

$$v_o = A_{CL}\left[\left(I_{IB} \pm \frac{I_{IO}}{2}\right)(R_2 \| R_f) - \left(I_{IB} \pm \frac{I_{IO}}{2}\right)(R_2 \| R_f) \pm V_{IO}\right] \qquad (7.7\text{-}34)$$

when $R_1 = R_2 \| R_f$. Initially, we might conclude that the first two terms inside the brackets do, in fact, cancel completely. However, because the offset current polarities are not necessarily in phase, we should add them together in order to carry out a worst-case analysis. When this is done, we obtain

$$v_o = A_{CL}[\pm I_{IO}(R_2 \| R_f) \pm V_{IO}] \qquad (7.7\text{-}35)$$

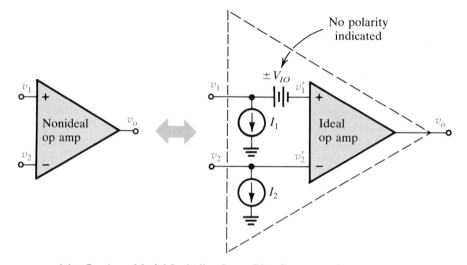

(a) Op Amp Model Including Input Bias Current and
Input Offset Voltage Effects

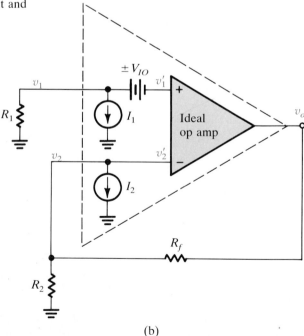

(b)

Figure 7.7-8
Dc offset analysis using a
simple op amp model.

This result assumes that the dc impedance levels seen from the inverting and noninverting input terminals are equal. For this situation, because I_{IO} is typically $\frac{1}{5}$ to $\frac{1}{10}$ of I_{IB}, we can obtain a nearly 10-fold reduction in the output offset voltage produced by the bias currents.

When op amp offset effects were neglected, equation (7.7-26) applied, and we were able to find the output voltage rapidly for nearly any op amp circuit configuration by setting $v_1 = v_2$ and solving for v_o. Because of the effect of the input offset voltage in equation (7.7-27), in a practical op amp v_1 and v_2 are not necessarily equal, and this makes it difficult to determine v_o rapidly without memorizing a large number of different equations. However, by replacing the nonideal op amp on the left side of Figure 7.7-8a with the circuit in part (b) of the figure, the offset effects can be included and v_o can still be found rapidly by using this circuit and setting v_1' and v_2' equal to one another.

We illustrate this technique by rederiving the dc offset equation (7.7-33). To begin this analysis, let us substitute the op amp dc offset model from Figure 7.7-8a into the circuit in Figure 7.7-7c. The resulting equivalent circuit is drawn in Figure 7.7-8b, and by examining this circuit, we can see that

$$v_1' = -I_1 R_1 \pm V_{IO} \tag{7.7-36a}$$

and
$$v_2' = -I_2(R_2 \parallel R_f) + \frac{R_2}{R_2 + R_f} v_o \tag{7.7-36b}$$

Because these voltages are at the input terminals of an ideal op amp,

$$v_1' = v_2' \tag{7.7-37}$$

so that we have

$$v_o = \frac{R_2 + R_f}{R_2}[I_2(R_2 \parallel R_f) - I_1 R_1 \pm V_{IO}] = A_{CL}[I_2(R_2 \parallel R_f) - I_1 R_1 \pm V_{IO}] \tag{7.7-38}$$

This is the same result that we derived in equation (7.7-33) after some rather lengthy calculations.

EXAMPLE 7.7-2

For the op amp circuit given in Figure 7.7-9a, $I_{IB} = 1$ μA, $I_{IO} = 100$ nA, and $V_{IO} = 1$ mV.

a. Derive an expression for the range of possible dc offset voltages in the output.
b. Suggest possible circuit modifications to reduce this offset.

SOLUTION

This circuit may be analyzed with the aid of the equivalent circuit in Figure 7.7-9b, from which we see that

$$v_1' = \pm V_{IO} \tag{7.7-39a}$$

and
$$v_2' = -I_2(R_2 \parallel R_f) + \tfrac{1}{2} v_o \tag{7.7-39b}$$

Equating v_1' and v_2', we obtain

$$v_o = 2\left[\left(I_{IB} \pm \frac{I_{IO}}{2}\right)100 \text{ k}\Omega \pm V_{IO}\right] \tag{7.7-40a}$$

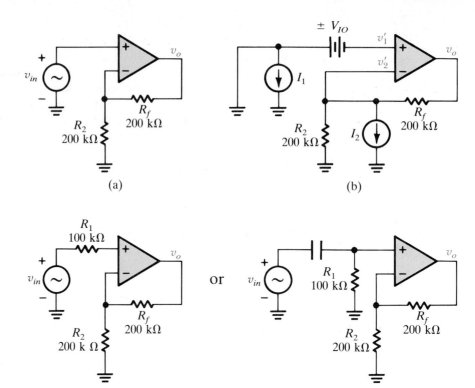

(a)   (b)

(c)   Acceptable Ways to Match the dc Impedances Seen from Both Op Amp Input Terminals

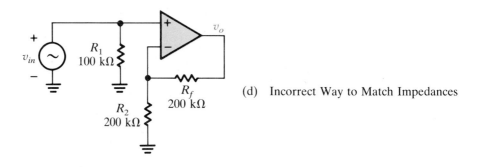

(d)   Incorrect Way to Match Impedances

Figure 7.7-9
Example 7.7-2.

and if I_{IB} takes on its maximum and minimum values, we have

$$v_o = 2[(1\ \mu A \pm 0.05\ \mu A)100\ k\Omega \pm 1\ mV] = 2[\underbrace{100\ mV}_{\substack{\text{due to}\\ I_{IB}}} \pm \underbrace{5\ mV}_{\substack{\text{due to}\\ I_{IO}}} \pm \underbrace{1\ mV}_{\substack{\text{due to}\\ V_{IO}}}] \qquad (7.7\text{-}40b)$$

and

$$V_o = 2[\underbrace{(0)100\ k\Omega}_{\substack{\text{due to } I_{IB}\\ \text{and } I_{IO}}} \pm \underbrace{1\ mV}_{\text{due to } V_{IO}}] \qquad (7.7\text{-}40c)$$

respectively. The corresponding range of possible output voltages are $+0.212$ V and -2 mV. In computing these worst-case results, the two $+$

terms involving I_{IO} and V_{IO} were assumed to add, although in an actual circuit they could just as easily have canceled each other.

By examining equation (7.7-40), we can see that most of the output offset voltage is caused by I_{IB}. The effect of this term can be eliminated by adding a 100-kΩ dc impedance path from the noninverting input lead to ground. Figure 7.7-9c illustrates two different ways to accomplish this. The circuit shown in Figure 7.7-9d will not work because when v_{in} is set equal to zero, the dc resistance path seen between the noninverting input terminal and ground is still zero ohms.

When one of the circuits in Figure 7.7-9c is employed, the voltage at v_1' changes to

$$v_1' = -I_1 R_1 \pm V_{IO} \tag{7.7-41}$$

and setting this voltage equal to v_2' in equation (7.7-39b), we have

$$v_o = 2[(-\cancel{(100\text{-k}\Omega)}\vec{I}_{IB} \pm 5\text{ mV}) + (\cancel{(100\text{-k}\Omega)}\vec{I}_{IB} \pm 5\text{ mV}) \pm 1\text{ mV}]$$

$$= 2[\underbrace{\pm 10\text{ mV}}_{\text{due to }I_{IO}}\ \underbrace{\pm 1\text{ mV}}_{\text{due to }V_{IO}}] \tag{7.7-42}$$

Thus the addition of the 100-kΩ resistor R_1 has reduced the range of dc offsets in the output to between $+22$ and -22 mV.

A further reduction in the dc offset voltage in the output is still possible. In equation (7.7-42) most of the output voltage is created by the input offset current term, and the size of this term may be reduced by lowering the circuit dc resistance levels. However, as R_f and R_2 are reduced, the loading on the output increases.

If we consider that the maximum output signal swing is 20 V while the maximum allowed output current is 10 mA, the sum of the resistors $(R_f + R_2)$ must be at least (20 V)/(10 mA), or 2 kΩ, if loading effects are to be avoided. To keep the amplifier gain the same as in the original circuit, we will choose $R_f = R_2 = 1$ kΩ, and this selection of course also requires that we make $R_1 = R_f \parallel R_2 = 500\ \Omega$ in order to minimize the offset effects. Using these parameters, the output offset voltage is now

$$v_o = 2[(-\cancel{(0.5\text{ k}\Omega)}\vec{I}_{IB} \pm 0.025\text{ mV}) + (\cancel{(0.5\text{ k}\Omega)}\vec{I}_{IB} \pm 0.025\text{ mV}) \pm 1\text{ mV}]$$

$$= 2[\underbrace{\pm 0.05\text{ mV}}_{\text{due to }I_{IO}}\ \underbrace{\pm 1.0\text{ mV}}_{\text{due to }V_{IO}}] \tag{7.7-43}$$

Thus the reduction of the resistance levels by 200 times has resulted in a 10-fold decrease in the output offset voltage, to a range between $+2.1$ and -2.1 mV. A reduction of the offset voltage below these levels will either require a better op amp, or the addition of a trimpot circuit to allow for the adjustment of the offset voltage to zero (at least at 25°C).

For some precision amplifier applications even a few millivolts of dc voltage appearing at the output terminals of the op amp may be unsatisfactory, and for these situations some type of offset adjustment circuit will be needed. Most op amps have dc offset terminals where a simple trimpot or perhaps a trimpot and a resistor are all that need be added to provide complete offset adjustment capability. However, for some op amps, especially those where there are several devices inside the same package, these terminals may not be provided. In this situation it is a relatively simple matter to construct a universal dc offset adjustment circuit with a potentiometer and a few resistors.

In designing a circuit of this type, basically what is needed is a low-impedance voltage source that can be varied over a range of plus and minus 10 or 20 mV. By connecting this source to either the inverting or noninverting input of the op amp (as illustrated in Figure 7.7-10a and b), we should be able to compensate for any dc shifts produced by the op amp's input offset voltage and input bias currents. In both of these circuits V_{BB}, just like any other input signal, is amplified by the gain of the circuit.

Figure 7.7-10c shows one possible circuit arrangement for obtaining the required variable dc voltage. However, it is unsatisfactory for several reasons. As the potentiometer, R, is adjusted, the Thévenin impedance of the circuit within the shaded region will vary, and because this impedance is in series with R_2, it will change the gain of the amplifier. To minimize this effect, R will need to be kept very small, and this will cause the trimpot to consume excessive power from the supplies.

A second problem arises from the fact that while the value of V_{BB} needed to balance the op amp circuit is only a few millivolts, V_T from the equivalent circuit within the shaded region in Figure 7.7-10c will vary between plus and minus 10 V as R is adjusted through its full range of values. As a result, circuit balance will occur somewhere in the center of the rotation of the potentiometer, and this adjustment will be extremely critical. Turning the trimpot a little too far to the left or the right will significantly alter V_T and will probably saturate the op amp.

The dc offset adjustment circuit in Figure 7.7-10d overcomes both of the problems associated with the previous circuit. Basically, it consists of a 10-kΩ potentiometer followed by a very large voltage divider. If we neglect the loading of the R_A–R_B network on the trimpot for the moment, then as R is adjusted, the voltage V_x will vary linearly between $+10$ and -10 V. Because the trimpot is rather large, only 2 mA of current is consumed from the supplies by the offset adjustment circuit.

If we look back at equation (7.7-43), we can see that a variable offset adjustment voltage at the V_y terminal of between, say, plus and minus 10 mV will more than compensate for the offset effects introduced by the op amp. This voltage range can be obtained by making the R_A–R_B network a 1000 : 1 voltage divider. In this way, as the trimpot is adjusted and V_x varies from $+10$ to -10 V, V_y will change correspondingly from $+10$ mV to -10 mV. Letting $R_A = 100$ kΩ and $R_B = 100$ Ω provides the required attenuation and ensures that the R_A–R_B network does not load the trimpot. This is important to allow V_y to vary linearly with trimpot rotation. Were the potentiometer severely loaded, then V_y would stay close to zero during most of the rotation of the trimpot and would jump up to ±10 mV right near the end of the rotation.

Let's examine the Thévenin equivalent of the offset adjustment circuit in Figure 7.7-10d illustrated in part (e) of the figure. To begin with, as required, the Thévenin voltage varies betwen $+10$ and -10 mV depending on the position of the trimpot. The Thévenin resistance is $R_B \parallel [R_A + (R_C \parallel R_D)]$, but be-

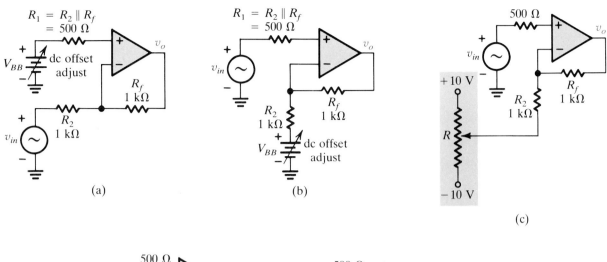

(a)

(b)

(c)

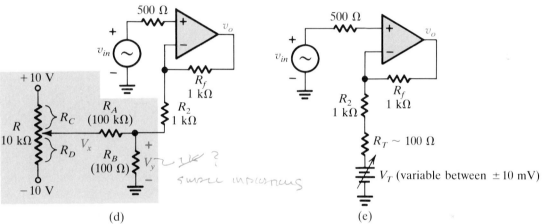

(d)

(e)

Figure 7.7-10
Development of a universal dc offset adjustment circuit.

cause R_B is so much less than R_A, this resistance is essentially equal to R_B and remains nearly constant, independent of the position of the trimpot.

The fact that $R_T = 100\ \Omega$ changes the gain of our amplifier somewhat. If this gain change is important, R_2 can simply be reduced to 900 Ω to compensate for the effect of R_T on the circuit gain.

7.7-3 Small-Signal Closed-Loop Frequency Response

The internal structure of a typical operational amplifier consists of a cascade of two or three direct-coupled stages, and as such it has a small-signal open-loop frequency response which is flat from dc up to its high-frequency breakpoint. Figure 7.7-11a illustrates the typical frequency characteristics of this type of op amp. As we can see from the figure, this amplifier's low-frequency gain is about 80 dB or 10^4, and its high-frequency 3-dB point is located at approximately 1 MHz.

Some operational amplifiers have 3-dB points which are well below that of the amplifier illustrated in Figure 7.7-11a. Generally, this is due to the inclu-

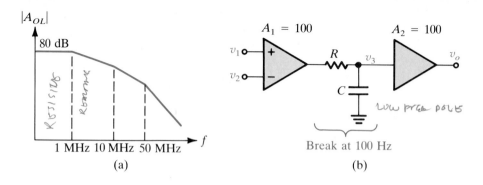

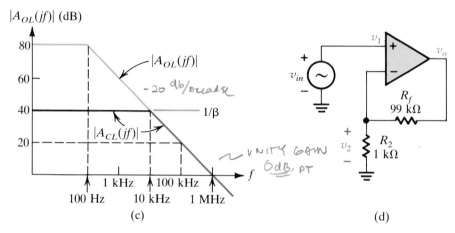

Figure 7.7-11

Closed-loop small-signal frequency response of an op amp.

sion of additional components in the op amp which are used to tailor its frequency response and prevent it from oscillating. These components are known as compensation elements. Figure 7.7-11b illustrates one such compensation circuit, and the capacitor C in this circuit is known as a compensation capacitor. Basically, this capacitor introduces a very-low-frequency pole which kills off the gain at high frequencies. The technique by which this prevents circuit oscillation is discussed in great detail in Chapter 8.

The "compensated amplifier" illustrated in Figure 7.7-11b has the open-loop frequency response given in part (c) of the figure. In sketching this response, we have assumed that the poles associated with the gain blocks A_1 and A_2 are so large that they can be neglected. As illustrated in the figure, the low-frequency gain of this amplifier is 10^4 or 80 dB, and its high-frequency 3-dB point occurs at only 100 Hz. Above this frequency the amplifier gain falls off at -20 dB/decade and crosses the unity-gain or 0-db point at 1 MHz. Let's examine what happens to this amplifier when feedback is added.

Consider, for example, the noninverting amplifier illustrated in Figure 7.7-11d, and let's try to determine the shape of its frequency response. In doing this it is useful to express the amplifier gain in terms of β, the feedback factor. As originally defined in Figure 7.1-2, β represents the amount of the output that is fed back to the input, and for the op amp circuit given in Figure 7.7-11d, this feedback factor will be given by

$$\beta = \frac{v_2}{v_o} = \frac{R_2}{R_2 + R_f} \tag{7.7-44}$$

For this specific circuit $\beta = 1/100$; in other words, 1/100 of the output voltage is fed back to the input.

To develop an expression for the closed-loop frequency response of this amplifier, if we neglect any dc offset effects, we may write

$$v_o = [A_{OL}(jf)](v_1 - v_2) = [A_{OL}(jf)](v_{in} - \beta v_o) \qquad (7.7\text{-}45a)$$

so that the amplifier closed-loop gain is

$$A_{CL}(jf) = \frac{v_o}{v_{in}} = \frac{A_{OL}(jf)}{1 + \beta A_{OL}(jf)} \qquad (7.7\text{-}45b)$$

The term jf in parentheses is used temporarily here just to remind us that both A_{OL} and A_{CL} are functions of frequency. To determine the shape of the closed-loop frequency response represented by equation (7.7-45b), it is useful to sketch $1/\beta$ on the same set of axes as $|A_{OL}(jf)|$. Because the β network is resistive, $1/\beta$ is a constant independent of frequency, and for this specific circuit it is equal to 100 or 40 dB. This is sketched as a horizontal line at the 40-dB level in Figure 7.7-11c. Notice that $|A_{OL}(jf)|$ intersects the $1/\beta$ line at $f = 10$ kHz.

To the left of the intersection point,

$$|A_{OL}(jf)| \gg \frac{1}{\beta} \qquad (7.7\text{-}46a)$$

or

$$\beta |A_{OL}(jf)| \gg 1 \qquad (7.7\text{-}46b)$$

Therefore, in this frequency region, $A_{CL}(jf)$ in equation (7.7-45b) is approximately equal to

$$A_{CL}(jf) \simeq \frac{A_{OL}(jf)}{\beta A_{OL}(jf)} = \frac{1}{\beta} = 1 + \frac{R_f}{R_2} \qquad (7.7\text{-}47)$$

To the right of the intersection at $f = 10$ kHz, the $1/\beta$ line has a higher amplitude than $A_{OL}(jf)$, so that we may write

$$|A_{OL}(jf)| \ll \frac{1}{\beta} \qquad (7.7\text{-}48a)$$

or

$$\beta |A_{OL}(jf)| \ll 1 \qquad (7.7\text{-}48b)$$

As a result, in this frequency region the closed-loop frequency response given by equation (7.7-45b) may be approximated as

$$A_{CL}(jf) \simeq \frac{A_{OL}(jf)}{1 + 0} = A_{OL}(jf) \qquad (7.7\text{-}49)$$

The overall closed-loop frequency response of this amplifier is indicated by the darkened lines in Figure 7.7-11c. As expected, the closed-loop gain at low frequencies is 100 or 40 dB. This gain is flat out to 10 kHz and then falls off at -20 dB/decade, so that the closed-loop high-frequency 3-dB point of this amplifier is now located at 10 kHz, in contrast to the open-loop 3-dB point, which was at only 100 Hz. Therefore, by adding the feedback, we have significantly increased the amplifier bandwidth. But this increase has not been obtained without cost since we had to trade away gain to get this improvement. In fact, because of the -20-dB/decade slope in $|A_{OL}(jf)|$, this is a 1 : 1 trade-off. The low-frequency closed-loop gain is 100 times smaller than the open-loop gain, but the closed-loop bandwidth has been increased by the same factor.

The technique just developed for determining the closed-loop frequency response of this op amp is valid regardless of the form of either A_{OL} or of β. Let's summarize this method:

1. Sketch $1/\beta$ on the same set of axes as $|A_{OL}(jf)|$, noting that $1/\beta$ can also be a function of frequency, depending on the character of the feedback network.
2. In the region (or regions) where $A_{OL}(jf) \gg 1/\beta$, $A_{CL}(jf) = 1/\beta$.
3. In the region (or regions) where $|A_{OL}(jf)| \ll 1/\beta$, $A_{CL}(jf) = A_{OL}(jf)$.

EXAMPLE 7.7-3

The op amp circuit illustrated in Figure 7.7-12a is assumed to have the open-loop gain characteristics given in Figure 7.7-11c.

a. Determine the high-frequency 3-dB point of this amplifier.
b. Sketch the steady-state output obtained from this circuit if a 500-kHz 10-mV(p-p) square wave is applied to the input.

SOLUTION

For this particular amplifier $\beta = 1$ because the entire output is fed back to the input. As a result, the $1/\beta$ line is drawn at the 0-dB level in Figure 7.7-11c, and it intersects the A_{OL} characteristic at 1 MHz. Therefore, the closed-loop gain of this amplifier is constant and equal to 1 all the way from dc out to 1 MHz. Beyond this point it falls off at -20 dB/decade. Clearly, the high-frequency 3-dB point for this amplifier is at 1 MHz. Furthermore, because the gain falls off at -20 dB/decade, above this frequency the amplifier may be represented by a single-pole transfer function with an equivalent time constant of $1/(2\pi \times 1 \text{ MHz})$, or 160 ns.

The square-wave response of this voltage follower can best be understood if we first examine its step response. When a step of voltage is applied to this

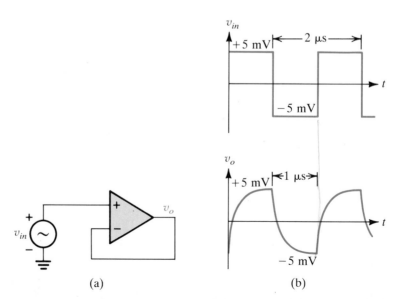

Figure 7.7-12
Example 7.7-3.

(a) (b)

Chapter 7 / Introduction to Feedback and the Operational Amplifier

amplifier, because of its single-pole transfer function, the response will be a rising exponential having a time constant of 160 ns and settling down to the same final value as the amplitude of the input step. Furthermore, if we view the applied square wave as a sequence of alternating steps, we can see that the response of the voltage follower to this input will also be a square wave of the same amplitude, but having exponentially rounded edges. The rise time of the output square wave will be approximately 2.2τ or 350 ns. This output is sketched in Figure 7.7-12b.

7.7-4 Large-Signal Frequency Response and Slew-Rate Effects in Op Amps

In an operational amplifier, the frequency response characteristics just developed are valid only when the output signal levels are small. As the amplitude and/or the frequency of the output signal increases, the waveform will begin to distort, making the amplifier virtually useless for high-fidelity applications. Owing to this phenomenon, when large-amplitude output signals are to be handled by an op amp, the frequency range of these signals must be restricted to values well below the small-signal high-frequency 3-dB point of the amplifier if output waveshape distortion is to be avoided.

The origin of this waveform distortion may be understood by using the simple model given in Figure 7.7-11b to represent our op amp. As mentioned earlier, in this circuit the capacitor C is a compensation capacitor; but for the purpose of this discussion, it could just as easily be a lumped representation of all the parasitic capacitances associated with the transistors inside this amplifier.

To illustrate how the introduction of this capacitor can give rise to signal distortion effects, we will examine the performance of this op amp when it is configured as shown in Figure 7.7-13a for a closed-loop gain of 10. We assume that the amplifier has symmetrical 15-V power supplies, and that stages A_1 and A_2 saturate when their output levels exceed ±15 V.

In assessing the behavior of this amplifier, we investigate its response to two different input signals: the first a low-frequency sine wave, and the second a step function.

The purpose of determining the response of this amplifier to a low-frequency sine wave is to review how the feedback operates when the frequency is low enough that the capacitor C may essentially be considered to be an open circuit. In particular, for this case we have that

$$v_o = \frac{A_{OL}}{1 + \beta A_{OL}} v_{in} \tag{7.7-50a}$$

where

$$\beta = \frac{R_2}{R_1 + R_2} = \frac{1}{10} \tag{7.7-50b}$$

and

$$A_{OL} = A_1 A_2 = 10^4 \tag{7.7-50c}$$

Substituting equations (7.7-50b) and (7.7-50c) into equation (7.7-50a), we obtain

$$v_o = \frac{10^4}{1 + 10^3} v_{in} = 9.99 v_{in} \tag{7.7-51}$$

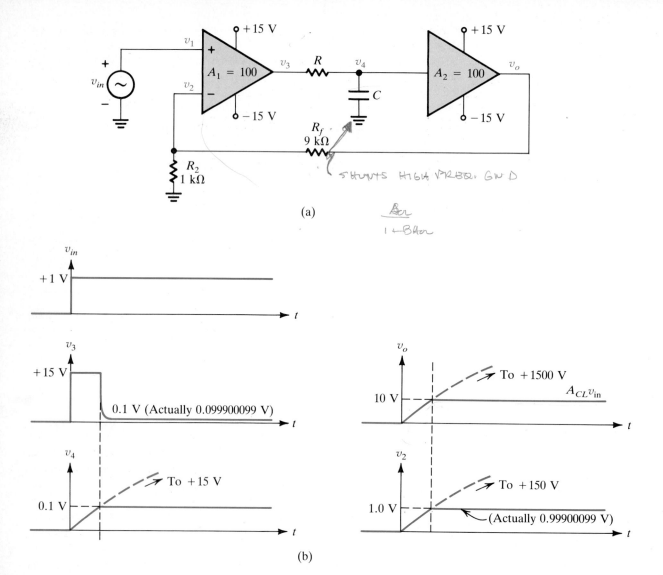

(a)

(b)

Figure 7.7-13

from which we also have that

$$v_2 = \tfrac{1}{10}v_o = 0.999v_{in} \tag{7.7-52a}$$

and

$$v_1 - v_2 = v_{in} - 0.999v_{in} = 0.001v_{in} \tag{7.7-52b}$$

The important point here, and the real reason for reviewing this low-frequency sine-wave response, is that the effect of the feedback is to make v_2 nearly the same as v_1. This keeps $v_1 - v_2$ small and prevents the amplifier from overloading.

Let's consider next what happens when a 1-V step is applied to this amplifier. Because of the rapid change in input signal amplitude that occurs at $t = 0$, the op amp response in this case is significantly different from the situation that existed when the input signal was a low-frequency sine wave. The sequence of waveshapes in Figure 7.7-13b will help to explain what happens. To begin with, when v_{in} jumps up to 1 V at $t = 0+$, it is impossible for v_2 to follow it immediately because of the finite time required to charge the capacitor C. As a result, at $t = 0+$, v_2 is still nearly equal to zero, and therefore the voltage v_3 at the output of the first stage attempts to go to

$$v_3 = A_1(v_1 - v_2) = 100(1 \text{ V} - 0 \text{ V}) = +100 \text{ V} \qquad (7.7\text{-}53)$$

However, because it is limited to the power supply voltage, instead, it just saturates at $+15$ V. With v_3 stuck at the positive supply rail, the capacitor C begins to charge exponentially toward $+15$ V with a time constant $\tau = RC$, and because $v_o = 100v_4$ (as long as A_2 stays in its linear region), the output of the op amp begins to charge exponentially toward $+1500$ V. Furthermore, because $v_2 = 0.1v_o$, the voltage at the inverting input terminal also rises exponentially with the same time constant, but to a different final voltage ($+150$ V).

This exponential charging continues until v_2 gets close to $+1$ V, and at this point the first stage again becomes active and the voltages stabilize at their final values. Once this occurs, to a good approximation,

$$v_o = A_{CL}v_{in} \simeq +10 \text{ V} \qquad [9.9900099 \text{ V}] \qquad (7.7\text{-}54a)$$

$$v_2 = \beta v_o \simeq +1 \text{ V} \qquad [0.99900099 \text{ V}] \qquad (7.7\text{-}54b)$$

and $\quad v_3 = v_4 = A_1(v_1 - v_2) = \dfrac{v_o}{A_2} \simeq 0.1 \text{ V} \qquad [0.0999001 \text{ V}] \qquad (7.7\text{-}54c)$

In these expressions the unbracketed terms on the right-hand side of each of the equations represent the approximate steady-state values of the voltages, and the terms in brackets the solutions making no approximations.

In summary, when a large-amplitude step is applied to the input of an op amp, because the feedback voltage cannot immediately follow it, the amplifier will overload internally in front of the capacitor. As a result, the capacitor will begin to charge exponentially, as will the output voltage and feedback voltage of the op amp. When the feedback voltage amplitude matches that of the applied input voltage, the amplifier will again enter the active region, and the waveshapes will flatten out at their appropriate steady-state levels. Because the output charging curve is only a very small portion of the entire exponential curve, it will appear to be nearly a straight line. The slope of this line, called the slew rate of the op amp, is a measure of the maximum rate at which the voltage in the output can change before internal amplifier overload occurs. The units for slew rate are volts/time, and typical slew rate values for general-purpose op amps vary from about 0.2 to 10 V/μs. However, high-speed op amps are available with slew rates in excess of 75 V/μs.

As evidenced by the example just discussed, the slew rate of an op amp depends on the amplifier time constants as well as the overload mechanisms that exist in the particular op amp circuit being investigated. Because the overload mechanisms in the positive and negative directions may not be the same, it is possible that the slew rates for an op amp may be different for slewing in the positive and negative directions.

EXAMPLE 7.7-4

An op amp of the type illustrated in Figure 7.7-11b has the small-signal frequency response given in part (c) of the figure. This op amp is used to con-

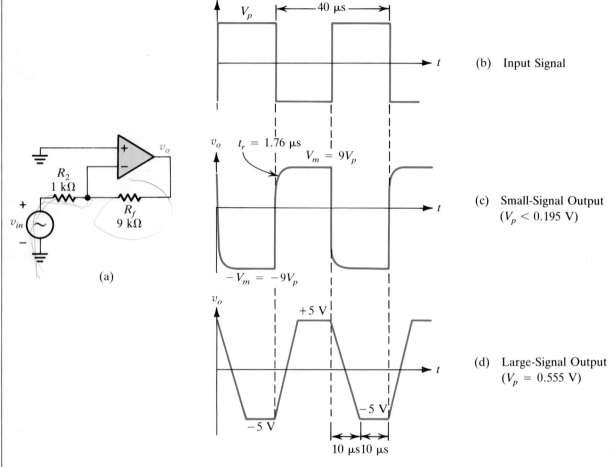

(b) Input Signal

(c) Small-Signal Output
($V_p < 0.195$ V)

(d) Large-Signal Output
($V_p = 0.555$ V)

Figure 7.7-14
Example 7.7-4.

struct the inverting amplifier circuit shown in Figure 7.7-14a. The slew rate of the op amp is equal to 1 V/μs.

a. Determine the output square-wave response to a 25 kHz input signal when the input amplitude is small enough that slew-rate limiting is not a problem.
b. Derive an expression for the largest output amplitude that can be tolerated before slew-rate limiting sets in.
c. Sketch the output obtained in part (b) under slew-rate-limited conditions.

SOLUTION

When the input amplitude is small enough that slew-rate limiting does not occur, the transient response of the circuit is determined solely by the small-signal frequency response characteristics of the amplifier. For the amplifier illustrated in Figure 7.7-14a,

$$\beta = \frac{R_2}{R_1 + R_2} = \frac{1}{10} \tag{7.7-55a}$$

and

$$A_{CL}(jf) = \frac{-[R_f/(R_2 + R_f)]A_{OL}(jf)}{1 + [R_2/(R_2 + R_f)]A_{OL}(jf)} = -\frac{[R_f/(R_2 + R_f)]A_{OL}(jf)}{1 + \beta A_{OL}(jf)} \tag{7.7-55b}$$

From equation (7.7-55b), we have

$$A_{CL}(jf) = -\frac{R_f}{R_2} \qquad\qquad \text{for } \beta|A_{OL}(jf)| \gg 1 \qquad (7.7\text{-}56a)$$

and

$$A_{CL}(jf) = -\frac{R_f}{R_2 + R_f}A_{OL}(jf) \qquad \text{for } \beta|A_{OL}(jf)| \ll 1 \qquad (7.7\text{-}56b)$$

The transition from equation (7.7-56a) to (7.7-56b) occurs at the frequency where $1/\beta = |A_{OL}(jf)|$. Because $1/\beta = 20$ dB for this amplifier, the high-frequency 3-dB break occurs at 100 kHz, and beyond this point the gain falls off at -20 dB/decade, as with the previous examples.

This type of amplifier has a single-pole transfer function whose effective time constant is

$$\tau = \frac{1}{\omega} = \frac{1}{2\pi f} = \frac{1}{2\pi \times 10^5} = 1.59 \ \mu s \qquad (7.7\text{-}57)$$

For a single-time-constant circuit of this type, the square-wave response will also be a square wave with exponentially rounded corners. As illustrated in Figure 7.7-14b, if v_{in} is a square wave with an amplitude V_p, the output will be an inverted square wave with an amplitude nine times as large. The rise time of the output exponential is

$$t_r = 2.2\tau \qquad (7.7\text{-}58a)$$

so that the approximate slope of the rising edge of the output waveshape is given by the expression

$$\text{slope} = \frac{\Delta v}{\Delta t} = \frac{2V_m}{t_r} = \frac{2V_m}{2.2\tau} \qquad (7.7\text{-}58b)$$

When the slope of any point on the output waveform exceeds the slew rate of the op amp, slew-rate limiting will occur, and the output will not rise as rapidly as would be predicted by the small-signal linear analysis. For the amplifier we are examining in this example, to avoid slew-rate limiting, we need to ensure that the slope of the exponential edges are always less than the amplifier slew rate, or from equation (7.7-58b) that

$$\frac{2V_m}{2.2\tau} \le \text{slew rate (SR)} \qquad (7.7\text{-}59)$$

Solving this equation for V_m, we have

$$V_m \le \frac{(\text{SR})(2.2\tau)}{2} = (\text{SR})(1.1\tau) \qquad (7.7\text{-}60)$$

This is the maximum-output square-wave amplitude allowed if slew-rate limiting is to be avoided. Dividing this number by the closed-loop amplifier gain gives us the maximum-input square-wave amplitude allowed. For the particular amplifier circuit under consideration in Figure 7.7-14a, these amplitude levels are

$$V_m \le \frac{1 \ V}{\mu s}(1.1)(1.6 \ \mu s) = 1.76 \ V \qquad (7.7\text{-}61a)$$

and
$$V_p \leq \frac{V_m}{9} = 0.195 \text{ V} \qquad (7.7\text{-}61b)$$

at the output and input, respectively. When the signal levels exceed these values, the output waveshape can be severely distorted.

The waveform sketched in Figure 7.7-14d illustrates the output obtained when a 0.555-V 25-kHz square wave is applied to the amplifier input. The peak output amplitude for this case is 9×0.555 or 5 V, and because it exceeds 1.76 V, slew rate limiting will occur. Were there no slew-rate-limit problem, the output waveform would have simply been a square wave with exponentially rounded edges having an approximate slope of $(10 \text{ V})/1.76 \ \mu s$ or 5.62 V/μs. But this exceeds the maximum rate at which the output voltage can change. Therefore, at each transition of the input square wave, the op amp will overload and the edges of the output waveform will slew to their final values with a slope of 1 V/μs. Because the steady-state amplitude values are ± 5 V, these transitions will each require $[5 - (-5)] \text{ V}/(1 \text{ V}/\mu s)$ or 10 μs.

Square waves are not the only type of signals that produce slew-rate-limiting problems. Any signal which requires that the output change at a rate exceeding its slewing capabilities will wind up causing distortion in the output waveform. Even a sine wave can exhibit slew-rate distortion if the amplitude and/or the frequency are large enough.

In particular, when the output is a sine wave,

$$v_o(t) = V_m \sin 2\pi f t \qquad (7.7\text{-}62)$$

the slope of the output waveshape may be obtained by differentiating this expression, which yields

$$\frac{dv_o}{dt} = 2\pi f V_m \cos 2\pi f t \qquad (7.7\text{-}63)$$

If slew-rate limiting is to be avoided, this expression must be less than the amplifier slew rate for all values of t. The maximum of this equation occurs when the cosine function is 1, so that if

$$2\pi f V_m < \text{slew rate (SR)} \qquad (7.7\text{-}64)$$

there will be no slew-rate distortion in the output. This imposes a severe restriction on the allowed maximum frequency that we can apply to the op amp; it is much more restrictive than the 3-dB high-frequency limit obtained from the small-signal analysis.

At low frequencies the output signal amplitude can extend all the way to the power supply rails before signal distortion occurs. But when the frequency of the signal is raised, following equation (7.7-64), the signal amplitude needs to be lowered if distortion effects are to be avoided. Figure 7.7-15 illustrates a typical plot of the maximum allowable output signal amplitude versus frequency for no signal distortion. The frequency f_1 in this figure represents the highest input signal frequency for which the output amplitude can go all the way to the supply rails with no slew-rate distortion. Its value may be calculated

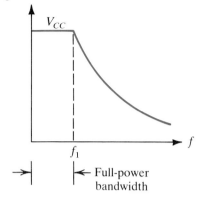

Maximum allowed
output voltage with
no signal distortion

Figure 7.7-15
Definition of the full-power
bandwidth of an amplifier.

from equation (7.7-64) as

$$f_1 = \frac{SR}{2\pi V_{CC}} \tag{7.7-65}$$

This frequency is called the full-power bandwidth of the op amp because for all signals having frequencies in the range from dc to f_1, maximum output amplitude (and hence maximum power) may be delivered to the load. Typical values for op amp full-power bandwidths are on the order of 10 to 500 kHz, depending on the amplifier slew rate.

EXAMPLE 7.7-5

The amplifier illustrated in Figure 7.7-13a has a 1-V 100-kHz sine wave applied to the input. If the amplifier slew rate is 1 V/μs, carefully sketch and label the waveshapes at points v_2, v_3, and v_o in this circuit.

SOLUTION

When a low-frequency sine wave is applied to an op amp, the output signal from the amplifier is also a sine wave. As the frequency of the applied signal is raised, gradually the output waveform begins to change from a sine wave into a triangle wave.

Were there no slew-rate distortion in this amplifier, the output would be a 10-V 100-kHz sine wave. But by applying equation (7.7-63), we can see that

$$\frac{dv_o}{dt} = (6.28 \times 10^5/s)(10 \text{ V}) \cos 2\pi ft = 6.28 \cos 2\pi ft \quad \frac{\text{V}}{\mu\text{s}} \tag{7.7-66}$$

and clearly for most of the sine-wave cycle, this quantity exceeds the slew rate of the amplifier. Figure 7.7-16 illustrates the pertinent steady-state voltages that exist in this circuit.

In sketching these waveshapes, we have assumed that the output slews over the entire cycle, and this assumption will be substantiated shortly. In Figure 7.7-16a we have sketched both v_1 and v_2 on the same set of axes to illustrate the fact that v_2 is always running to try to catch up with v_1. As a result, the output of the first stage, v_3, is always saturated at either $+15$ or -15 V, and v_o slews positively or negatively accordingly. Because the amplifier slew rate is 1 V/μs, in 5 μs the amplifier output voltage can slew 5 V, and in the steady

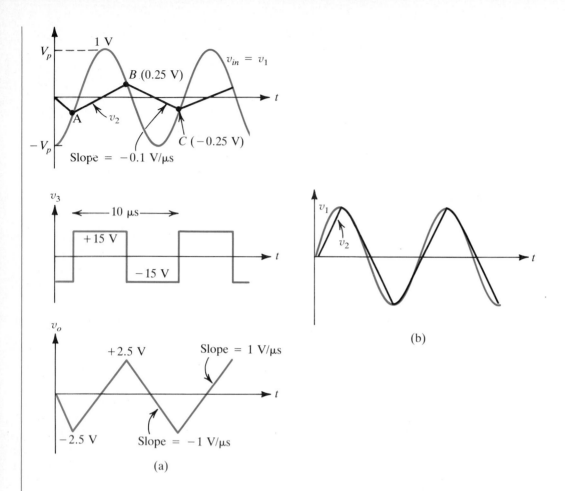

(a)

(b)

Figure 7.7-16
Example 7.7-5.

state, because of the symmetry, this leads to a 2.5-V peak amplitude triangle wave in the output. Notice that the slope of this triangle wave is ± 1 V/μs.

Using a voltage divider, the signal v_2 is also a triangle wave, but with an amplitude $\frac{1}{10}$ that of v_o. The slope of the v_2 triangle wave is 0.1 V/μs. In analyzing this circuit it is critically important to determine the slope of the input signal at the terminal v_1. By differentiating v_{in}, we have that

$$\frac{dv_1}{dt} = \frac{dv_{in}}{dt} = (2\pi f)(1 \text{ V}) \cos 2\pi ft = 0.628 \cos 2\pi ft \quad \frac{\text{V}}{\mu\text{s}} \quad (7.7\text{-}67)$$

and for all values of $v_{in} \leq 0.25$ V this slope always exceeds 0.6 V/μs. Therefore, the slope of v_1 is always greater than that of v_2. This means that v_2 can never catch up with the input signal.

Let's go through one cycle of the sine wave to illustrate what happens, starting at point A on the input signal and assuming that v_3 has just switched to +15 V. Because the slope of the v_1 waveform exceeds that of v_2, v_1 will stay well above v_2 during this part of the cycle, v_3 will remain saturated in the +15 V direction, and v_o will slew positively at 1 V/μs. Eventually, v_1 will start to come back down and will cross v_2 at point B in the figure. Because v_1 is now smaller than v_2, v_3 will switch to -15 V, and the amplifier output will begin to slew negatively. This continues until v_1 again crosses over v_2 at point C, and then the cycle repeats.

It is interesting to observe that had the frequency of the sine wave been lower, the output amplitude v_o and also v_2 would have been larger, and had the intersection occurred at a point on v_{in} where its slope was less than 0.1 V/μs, v_2, and hence v_o, would have been able to follow v_{in} over that portion of the cycle. For this situation the output would have had partially rounded sinusoidal tops and straight-line transitions in between, as illustrated in Figure 7.7-16b.

Exercises

7.7-1 An op amp has the following open-loop characteristics: $A_{OL} = 10^4$ and $Z_o = 100 \ \Omega$. It is connected as a noninverting amplifier for a theoretical closed-loop gain of 100. Find the output voltage if a 1-mV sine wave is applied to the input and a 10-Ω resistive load is connected to the output. *Answer* 90.9 sin ωt mV

7.7-2 Consider that an op amp with a resistive feedback network has an open-loop gain $A_{OL}(s) = A_o/(1 + s/\omega_1)$ and an open-loop output impedance R_o. Show that the effective closed-loop output impedance can be represented as the series combination of a resistance $R_{out} = R_o(A_{CL}/A_o)$ and an inductor $L = (R_o A_{CL}/A_o \omega_1)$.

7.7-3 For the active filter shown in Figure 7.4-7 $I_{IB} = 500$ nA, $I_{IO} = 100$ nA, and $V_{IO} = 5$ mV for each of the ICs. Find the worst-case variations in the dc output voltages from each of these op amps. *Answer* For IC1, 60.65 to -31.5 mV; for IC2, 105 to -83 mV

7.7-4 A 100-kHz 4-V(p-p) triangle wave is applied to the input of a noninverting op amp circuit with a closed-loop gain of 5. The slew rate of the op amp is 1 V/μs. For this circuit the output will also be a triangle wave. Find its amplitude and phase shift with respect to the input. *Answer* 2 V(p-p), output lags input by 45° (1.25 μs)

PROBLEMS

Unless otherwise noted, $g_m = 1$ mS and $r_d = \infty$ for all FETs and $h_{ie} = 1$ kΩ and $h_{fe} = 100$ for all BJTs.

7.1-1 For circuits 1 and 2 in Figure P7(A), v_{in} is a 1-V sine wave and v_N, a noise signal arising within the amplifier, is a 2-V(p-p) sawtooth waveshape. Determine the signal-to-noise ratio at the output of each of these circuits.

7.1-2 Repeat Problem 7.1-1 for circuits 3 and 4 in Figure P7(A).

7.1-3 Let $v_N = 0$ for circuits 1 and 2 in Figure P7(A), and for each of these circuits consider the gain block A_2 to be nonlinear and of the form

$$v_{o2} = A_2 v_{in2} + 0.05 A_2 v_{in2}^2$$

where the A_2 coefficient has the value given in the original circuit. By employing an analysis technique similar to that used in Figure 7.1-5b, estimate the percent second harmonic distortion in the output of each circuit.

7.1-4 Repeat Problem 7.1-3 for circuits 3 and 4 in Figure P7(A).

7.1-5 Find Z_{in} for the JFET circuit shown in Figure P7.1-5.

7.1-6 For the circuit shown in Figure P.7.1-6, find v_g in terms of v_{in}.

7.1-7 For the circuit shown in Figure P7.1-6 find Z_{in}, the circuit input impedance.

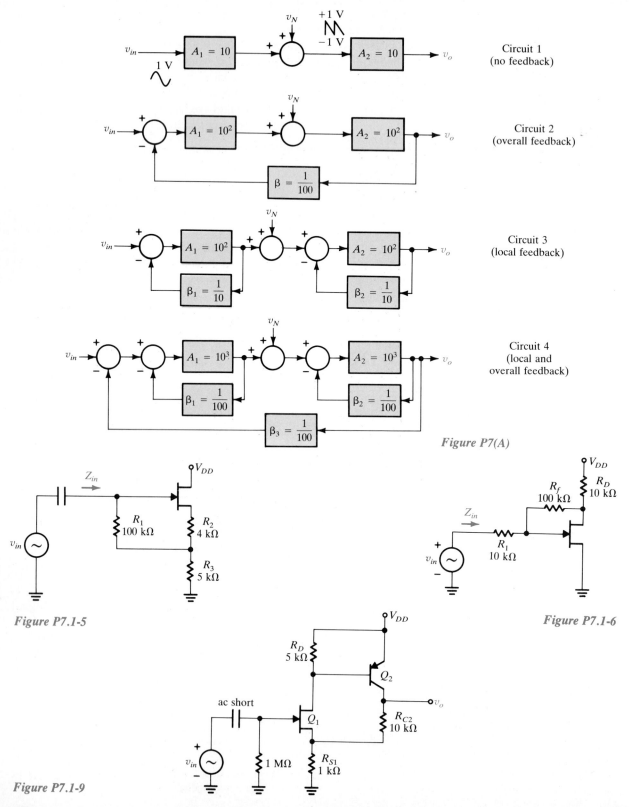

Figure P7(A)

Circuit 1
(no feedback)

Circuit 2
(overall feedback)

Circuit 3
(local feedback)

Circuit 4
(local and
overall feedback)

Figure P7.1-5

Figure P7.1-6

Figure P7.1-9

7.1-8 For the amplifier examined in Figure 7.1-5:

(a) Show by using the appropriate trigonometric identities that terms (1) and (2) in equation (7.1-17) give rise to third and fourth harmonic distortion components.

(b) Demonstrate that the amplitudes of these two terms are much smaller than those of the fundamental and the second harmonic components.

7.1-9 For the circuit shown in Figure P7.1-9, find the gain v_o/v_{in}.

7.1-10 Design an amplifier having a nominal closed-loop gain of 100 using the smallest possible open-loop amplifier gain. In this design, a 20% variation in A_{OL} is to produce less than a 1% change in A_{CL}.

7.1-11 A power amplifier, running open loop, is able to deliver 10 W of power to a 10-Ω loudspeaker with an applied input signal of 100 mV(rms). The nonlinear distortion in the amplifier output at these signal levels is 10%. Modify this circuit by adding whatever feedback and preamplifiers are necessary to reduce the nonlinear distortion to 0.1% while still delivering 10 W of power to the load with a 100-mV input signal.

7.1-12 For the circuit shown in Figure 7.1-8, $R_{C1} = 2$ kΩ, $R_{C2} = 3$ kΩ, and $R_{C3} = R_f = 5$ kΩ.

(a) Find the circuit voltage gain and output impedance.

(b) Express v_{c1}, v_{c2}, and v_{c3} in terms of v_{in}.

7.1-13 For the circuit shown in Figure P7.1.13:

(a) Find the circuit voltage gain v_o/v_{in} with R_f removed.

(b) What is the approximate gain with R_f in place?

(c) Derive an exact expression for the gain with R_f in place.

7.1-14 For the amplifier shown in Figure P7.1-14, determine the output impedance with:

(a) $R_f = \infty$

(b) $R_f = 200$ kΩ

Make all reasonable approximations.

7.1-15 (a) Compute the closed-loop gain of the circuit in Figure 7.1-8, given that $R_{C1} = R_{C2} = R_{C3} = 10$ kΩ, $R_f = 5$ kΩ, and $v_{in} = 1 \sin \omega t$ V.

(b) In this amplifier most of the nonlinear distortion will occur in Q_3 because its signal levels are much greater than those in Q_1 and Q_2. For the purpose of this example, assume that this nonlinearity can be expressed by the equation

$$i_{c3} = 100i_{b3} + 10^6 i_{b3}^2$$

(*Note*: The units of the coefficient 10^6 are 1/amperes.) By employing an analysis technique similar to that used in Figure 7.1-5b, estimate the percent second harmonic distortion in the output of this amplifier.

7.2-1 Carefully sketch and label v_o for each circuit described below given that $A_{OL} = 10^4$ and the power supplies are ±10 V.

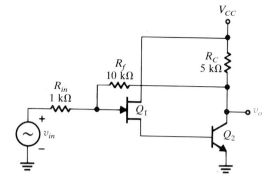

Figure P7.1-13

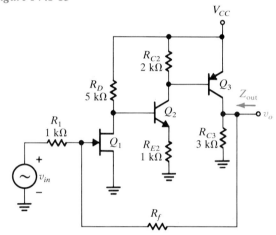

Figure P7.1-14

(a) The op amp runs open-loop with v_1 grounded and v_2 connected to a source $1 \sin \omega t$ μV.

(b) The op amp runs open-loop with v_1 connected to a source $1 \sin \omega t$ μV and v_2 connected to a 1-μV battery (positive terminal up).

(c) The op amp circuit is a noninverting amplifier with $v_{in} = 5 \sin \omega t$ V, $R_f = 4$ kΩ, and $R_2 = 2$ kΩ.

(d) The op amp is connected as shown in Figure 7.2-5a with $v_{in} = 5 \sin \omega t$ V, $R_f = 10$ kΩ, and $R_2 = 5$ kΩ. In addition, a 2-V battery (positive terminal up) is inserted between the v_1 terminal and ground.

7.2-2 The op amp in the circuit shown in Figure P7.2-2 is ideal. Find the circuit gain v_o/v_{in}.

7.2-3 Given the signal sources v_A and v_B, design an op amp circuit to produce the overall output.

$$v_o = 7v_A - 3v_B$$

7.2-4 Find v_o for each circuit shown in Figure P7.2-4 if the power supplies are ±15 V and $A_{OL} = 10^5$.

7.2-5 An op amp is connected as an inverting amplifier with $R_f = 25$ kΩ and $R_2 = 1$ kΩ. In addition, $A_{OL} = 25$ and the power supplies are ±10 V. Carefully sketch and label v_o if $v_{in} = 0.5 \sin \omega t$ V.

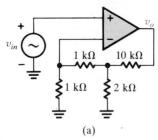

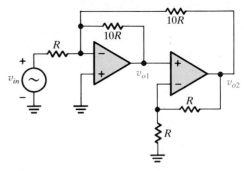

Figure P7.2-2

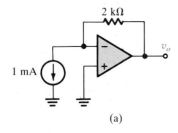

(a)

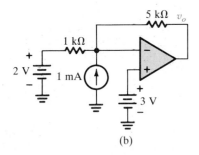

(b)

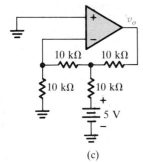

(c)

Figure P7.2-4

Figure P7.2-7

1-kΩ resistor in series with a 5-V battery (positive terminal up) is connected from v_o to ground. Find v_o and the current flow out of the op amp output terminal.

7.2-9 You have a velocity transducer whose equivalent circuit consists of a current source i connected in parallel with a resistance R. The value of R is somewhere between 10 and 100 kΩ, but is not well defined, and the current source produces an output of 2 μA/mph of velocity. Design an op amp circuit to interface this transducer to a 1-mA (full-scale) dc ammeter. The meter is to indicate full-scale deflection at a velocity of 100 mph. (You may assume that the meter has no dc resistance.)

7.2-10 The circuit shown in Figure P7.2-10 permits the measurement of the open-loop gain of an op amp. Prove that $A_{OL} \simeq 101 v_o / v_3$.

7.3-1 In Figure 7.3-1, $R_2 = R_2' = 1$ kΩ, $R_f = 100$ kΩ, and $R_f' = 105$ kΩ. By computing A_d and A_{cm} for this amplifier, determine its CMRR.

7.3-2 v_{in} is a 1-kHz 5-V(p-p) triangle wave. Sketch the steady-state output voltage for each circuit described below.
(a) The circuit is an op amp integrator with $R = 1$ kΩ, and $C = 1$ μF.
(b) The circuit is an op amp differentiator with $R = 1$ kΩ, and $C = 1$ μF.
(c) The circuit is that given in Figure 7.3-4b with $R_2 = 5$ kΩ, and Z_L a 1-H inductor.

7.3-3 For the circuit shown in Figure 7.3-7a, $R = 10$ kΩ and $C = 0.1$ μF. In addition, a resistor $R_2 = 10$ kΩ is connected in series with C. Carefully sketch and label v_o in

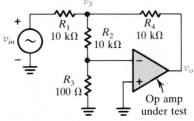

Figure P7.2-10

7.2-6 For the circuit in Figure 7.2-5a, $R_f = 50$ kΩ, and $R_2 = 10$ kΩ. In addition, the v_1 input terminal is removed from ground and connected to v_{in} through a 10-kΩ resistor. Find the gain v_o / v_{in}.

7.2-7 For the circuit shown in Figure P7.2-7, find the circuit gains v_{o1} / v_{in} and v_{o2} / v_{in}.

7.2-8 An op amp circuit is formed by grounding the v_1 input terminal and by connecting v_o to v_2. In addition, a

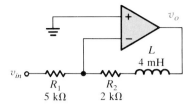

Figure P7.3-4

the interval from $t = 0$ to $t = 8$ ms if the initial charge on C is zero and v_{in} is a 500 Hz 10-V(p-p) square wave.

7.3-4 The input signal v_{in} applied to the circuit shown in Figure P7.3-4 is a 0- to 10-V sawtooth having a rise time of 8 μs and a fall time of 2 μs. Accurately sketch the resulting output voltage v_o.

7.3-5 **(a)** Derive an expression for $H(s) = V_o/V_{in}$ for the circuit in Figure 7.3-6c.
(b) Sketch the magnitude response of $H(s)$ and show that is has the basic shape given in Figure 7.3-6d.
(c) Discuss why this circuit functions like a differentiator for $\omega \ll 1/R_1C$.

7.3-6 Using the circuit shown in Figure 7.3-7c, design a sawtooth generator. You may assume that $v_{in} = -10$ V and that v_o is to be a 0 to 10-V positive-going ramp with a sweep time of 1 ms and a retrace time that is approximately 1% of the main sweep time.
(a) If the FET parameters are $I_{DSS} = 10$ mA and $V_p = -2$ V, select appropriate values for R and C.
(b) Sketch the output voltage obtained from this circuit and the required FET control voltage, v_G.

7.3-7 The circuit shown in Figure 7.3-4 is to be designed to produce a linear current sweep in the inductive load Z_L. v_{in} is a 10-V (p-p) sawtooth having a rise time of 10 ms and a retrace time of 1 ms, and L is a 500-mH inductor.
(a) Select R_2 to produce a peak positive and negative deflection current of plus and minus 10 mA, respectively.
(b) By sketching the voltage across the inductor and then v_o, determine the minimum power supply voltages needed for this circuit to work properly.

7.3-8 For the circuit shown in Figure P7.3-8 v_A and v_B are each a 0.5-Hz 5-V (p-p) triangle wave. Carefully sketch and label v_o assuming the op amp to be ideal.

7.3-9 Derive an expression for v_o in the D/A converter in Figure 7.3-13a if the resistor $2R$ going into the inverting input terminal of the op amp is replaced by a resistor R.

7.3-10 Starting with equations (7.3-4b) and (7.3-5), and using the approximations

$$\frac{1}{1-x} \simeq 1 + x, (1+x)^2 \simeq 1 + 2x, \text{ and}$$

$$(1-x)^2 \simeq 1 - 2x \text{ for small } x,$$

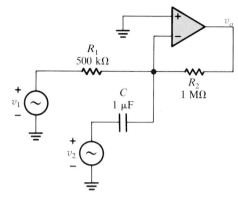

Figure P7.3-8

demonstrate that the CMRR of the circuit in Figure 7.3-1b is given by equation (7.3-6) where T is the resistor tolerance.

7.3-11 Determine the CMRR of the circuit given in Figure 7.3-2 if the resistors are perfectly matched and if the ICs each have CMRRs of 70 dB.

7.4-1 A passive second-order low-pass filter can be made by connecting the input voltage source to a series RLC circuit and taking the output voltage across the capacitor. Assume that $L = 1$ mH and $C = 0.001$ μF for this circuit, and determine the value of R needed to have a linear-phase low-pass response.

7.4-2 Design a third-order Butterworth-style high-pass filter with a gain of 40 dB in the passband and a cutoff frequency of 1 kHz. All capacitors are to be 0.1 μF.

7.4-3 Design a two-pole Butterworth low-pass filter with a 30-dB gain and a cutoff frequency of 75 kHz.

7.4-4 Prove that $H(s)$ is given by equation (7.4-35) for the high-pass active filter circuit in Figure 7.4-9a.

7.4-5 The circuit shown in Figure P7.4-5 is called a state-variable active filter. It has the interesting property that the transfer functions V_{o1}/V_{in}, V_{o2}/V_{in}, and V_{o3}/V_{in} are second-order high-pass, bandpass, and low-pass filters, respectively. Derive each of these transfer functions letting $\omega_0 = 1/RC$. (*Hint:* $V_1 \simeq V_2$.)

7.4-6 Design an active filter circuit to implement the transfer function

$$H(s) = \frac{100s^3}{(s+10)(s^2+10s+100)}$$

All capacitors are to be 1 μF.

7.4-7 Find the transfer function $V_o(s)/V_{in}(s)$ for the circuit shown in Figure P7.4-7.

7.4-8 Develop the transfer function for an $n = 2$ (four-pole) Chebychev (0.5-dB ripple) BPF having a center frequency at 1 kHz and a bandwidth of 100 Hz.

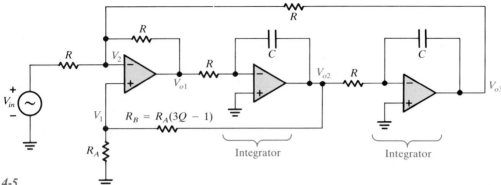

Figure P7.4-5

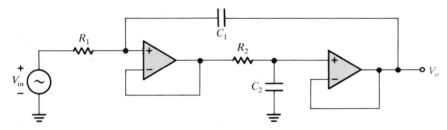

Figure P7.4-7

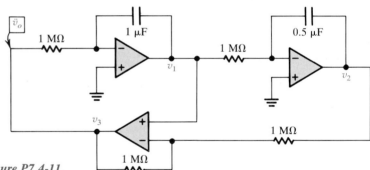

Figure P7.4-11

7.4-9 Prove that $H(s)$ for the active bandpass filter shown in Figure 7.4-10 is given by equation (7.4-37).

7.4-10 Design a fifth-order Butterworth low-pass filter with a low-frequency gain of 1 and a cutoff frequency of 1 kHz. All capacitors used are to be 0.005 μF.

7.4-11 The circuit shown in Figure P7.4-11 produces the analog computer solution of a particular differential equation.

(a) If the voltage at the upper left-hand node is defined as

$$\ddot{v}_o = \frac{d^2 v_o}{dt^2} = \text{second derivative of } v_o$$

express v_1, v_2, and v_3 in terms of v_o or its derivatives.

(b) By setting v_3 equal to $\ddot{v}_o$, determine the differential equation being solved by means of this circuit.

7.4-12 Following the technique outlined in Problem 7.4-11, design a circuit that will permit the analog computer solution of the differential equation

$$\frac{dv_o}{dt} + 10v_o = 2v_{in}$$

All capacitors used in this circuit are to be 1.0 μF.

7.4-13 Following the circuit layout given in Figure 7.4-14, design a switched-capacitor filter circuit to realize the transfer function

$$H(s) = \frac{20s}{s^2 + 5s + 100}$$

In carrying out this design, choose a clock frequency that is about 100 times higher than any frequency in the passband of the filter, and assume that all op amp feedback capacitors are equal to 20 pF.

7.4-14 (a) Design a switched-capacitor filter to implement the low-pass transfer function

$$H(s) = \frac{V_o}{V_{in}} = \frac{5 \times 10^7}{(s + 500)(s^2 + 100s + 2.5 \times 10^5)}$$

The feedback capacitors are to be 20 pF and the clock frequency is 500 kHz.

(b) What is the minimum switching frequency that is reasonable for this design? Explain.

7.5-1 In Figure 7.2-5a, $v_{in} = 2 \sin \omega t$ V, $R_f = 4$ kΩ, and $R_2 = 1$ kΩ. In addition, an ideal 5-V zener diode is shunted across R_f (anode toward v_o). Carefully sketch v_o versus time.

7.5-2 For the circuit in Figure 7.2-5a, $v_{in} = 5 \sin \omega t$ V, $R_f = 20$ kΩ, and $R_2 = 10$ kΩ. In addition, a diode (cathode toward v_1) is inserted between v_1 and ground, and a 10-kΩ resistor is connected from the v_1 terminal to v_{in}. Sketch v_o considering the diode to be ideal.

7.5-3 Sketch and carefully label v_o for each of the circuits described below, given that the power supplies are ± 10 V, and $v_{in} = 1 \sin \omega t$ volts.
(a) The circuit has the form given in Figure 7.2-5a with $R_f = 5$ kΩ and $R_2 = 1$ kΩ. In addition, an ideal diode is connected in series with R_2 (cathode on the left).
(b) The circuit is the same as in part (a) except that the diode is connected in series with R_f.

7.5-4 The circuit shown in Figure P7.5-4 is a precision full-wave rectifier. Explain how this circuit works, and in doing so, sketch v_x, v_{o1}, and v_{o2} given that $v_{in} = 2 \sin \omega t$ V.

7.5-5 In the integrator circuit in Figure 7.3-7a, v_{in} is a 5-Hz 10-V (p-p) square wave, $C = 1$ μF, and R is replaced by a network consisting of a resistor $R_1 = 200$ kΩ in series

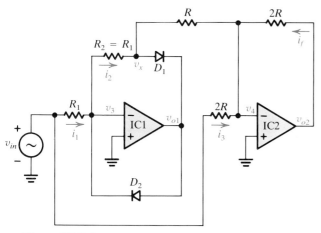

Figure P7.5-4

with a diode D_1 (cathode to the right). A second resistor–diode branch containing a resistor $R_2 = 100$ kΩ in series with a diode D_2 (cathode to the left) is connected in parallel with the R_1–D_1 branch. Sketch v_o during the first second. Consider the diodes to be ideal. The capacitor is initially uncharged.

7.5-6 In Figure 7.5-9e, $v_{in} = 7 \sin \omega t$ V, $V_{CC} = 10$ V, $R_4 = 5$ kΩ, $R_1 = R_3 = 10$ kΩ, and $R_2 = 20$ kΩ. Carefully sketch and label v_{in} and v_o if the diode is ideal and the power supplies are ± 20 V.

7.5-7 Analyze the switching of the op amp circuit shown in Figure 7.5-5a, assuming that v_o starts out at -10 V. Start with $v_{in} = -5$ V and let it increase positively. Make up a table similar to that in Figure 7.5-4 to explain what happens. Assume ± 10-V supplies.

7.5-8 For the circuit shown in Figure 7.5-5a, $R_f = 2$ kΩ, $R_2 = 1$ kΩ, and a 1.5-V battery (positive terminal up) is inserted between the v_2 op amp terminal and ground. Sketch v_o if $v_{in} = 8 \sin \omega t$ volts.

7.5-9 In Figure 7.5-4a, $R_f = 2$ kΩ, $R_2 = 1$ kΩ, and a 3-V battery (positive terminal to the right) is connected in series with R_f. Plot the hysteresis characteristic.

7.5-10 For the circuit in Figure 7.5-5a, a 2-kΩ resistor in series with a 5-V battery (positive terminal up) is connected from the v_1 op amp node to ground. Sketch the hysteresis characteristic.

7.5-11 When the diode D in Figure 7.5-9e is OFF, v_o is related to v_{in} by equation (7.5-11). Explain why v_o does not depend on R_2, R_3, and V_{CC} in this region.

7.5-12 Design an op amp circuit whose output is related to the applied input by the equation

$$v_o = \begin{cases} 3(-v_{in}) - 10 & \text{for } v_o < 2 \text{ V} \\ (-v_{in}) - 2 & \text{for } v_o > 2 \text{ V} \end{cases}$$

The circuit input resistance is to be 10 kΩ and ± 15-V supply voltages are available.

7.5-13 Design an op amp circuit whose output is related to the applied input by the equation

$$v_o = \begin{cases} 4(-v_{in}) + 4 & \text{for } v_o < -4 \text{ V} \\ 2(-v_{in}) & \text{for } -4 \text{ V} < v_0 < 4 \text{ V} \\ 4(-v_{in}) - 4 & \text{for } v_o > 4 \text{ V} \end{cases}$$

Assume that the only batteries available are the supply voltages at ± 10 V. The op amp input resistor is to be 10 kΩ.

7.5-14 In Figure 7.5-7, sketch the output at v_{o1} when the circuit in part (c) of the figure is used. Assume that D_1 and D_2 are silicon and that the peak positive input signal amplitudes in each cycle are 1, 2, and 3 V.

7.5-15 Design a triangle-to-sine wave converter using the diode nonlinear function generation technique. The input signal v_{in} applied to the converter is to be a 10-V (p-p) triangle wave, and the resulting circuit output v_o is to approximate a 20-V (p-p) sine wave. The circuit should employ two diode breaks on the positive half of the cycle and two breaks on the negative half of the cycle. The circuit input resistance is to be 10 kΩ, and the power supplies are ±15 V.

Unless otherwise noted, $g_m = 1$ mS for all FETs and $h_{ie} = 1$ kΩ and $h_{fe} = 100$ for all BJTs.

7.6-1 Find the common-mode gain (v_{c2}/v_{cm}) of the amplifier in Figure 7.6-3a if $R_C = 5$ kΩ, $R_E = h_{ie1} = 1$ kΩ, and $h_{ie2} = 1.5$ kΩ.

7.6-2 A differential amplifier of the type shown in Figure 7.6-2a has an $A_d = 100$ and an $A_{cm} = 10$. Find v_o if $v_1 = 2$ V and $v_2 = 1.8$ V.

7.6-3 Find the ac base current, i_{b3}, in terms of v_{in} for the circuit shown in Figure P7.6-3.

7.6-4 For the circuit examined in Problem 7.6-3, find the dc voltage at the point labeled V_x. Q_3 and Q_4 are silicon with $h_{FE} = 100$.

7.6-5 For the circuit shown in Figure P7.6-5, let $v_{in} = 0$ and determine I_C and V_{CE} for Q_1 to Q_5.

7.6-6 Find the small-signal ac gain v_o/v_{in} for the circuit given in Problem 7.6-5.

7.6-7 The circuit illustrated in Figure 7.6-7c has a supply voltage $V_{CC} = 10$ V and is to be designed to produce a current of 100 μA in the collector of Q_2 and a collector current of 1 mA in Q_1. Determine the values for the two resistors R_1 and R_2, assuming that the transistor current gains are large.

7.6-8 Find the dc voltage V_o for the circuit shown in Figure P7.6-8.

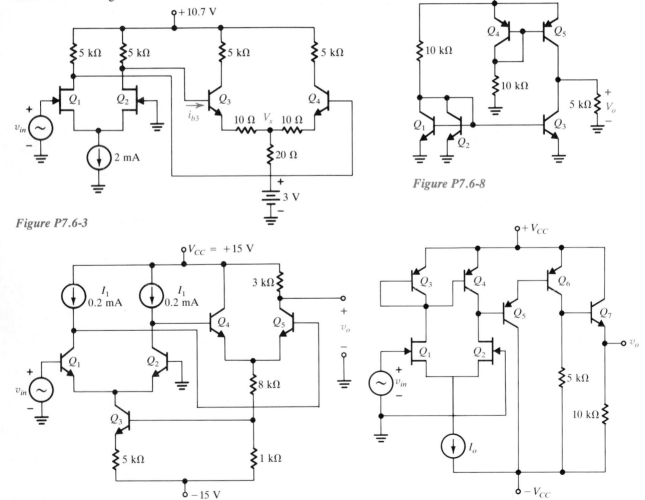

Figure P7.6-3

Figure P7.6-8

Figure P7.6-5

Figure P7.6-10

7.6-9 (a) A differential amplifier with $A_{d1} = 100$ and $A_{cm1} = 1$ drives a nondifferential amplifier with a gain $A_2 = 200$. A 1-mV sine wave is applied differentially to this two-stage amplifier in the presence of a 10-mV triangular-wave noise source. Sketch the differential and common-mode signals at the output of each amplifier stage and determine the overall circuit CMRR.

(b) Repeat part (a) if the first amplifier has a balanced output and the second stage is a differential amplifier with $A_{d2} = 100$ and $A_{cm2} = 1$.

7.6-10 Determine the small-signal voltage gain v_o/v_{in} for the circuit shown in Figure P7.6-10.

7.6-11 The collector output impedance of a grounded emitter BJT amplifier is equal to $1/h_{oe}$. Show that this impedance increases to $(1/h_{oe})[1 + h_{fe}R_E/(R_E + R_B + h_{ie})]$ when an emitter resistor is added, where R_E and R_B are the emitter and base resistances respectively.

7.7-1 In Figure 7.7-10d the op amp is ideal, $v_{in} = 500 \sin \omega t$ mV, and $R_A = R_B = 10$ kΩ. Sketch v_o when the potentiometer is in the top, middle, and bottom positions.

7.7-2 For the op amp in the circuit shown in Figure P7.7-2, $V_{IO} = 15$ mV, $I_{IB} = 1$ μA, and $I_{IO} = 0.2$ μA.
(a) Find the worst-case offsets at V_{o1} and V_{o2}.
(b) Add compensation circuit(s) to adjust both V_{o1} and V_{o2} to zero when $v_{in} = 0$.

7.7-3 (a) The circuit shown in Figure P7.7-3, is used to measure V_{IO}, I_{IB}, and I_{IO}. Discuss how V_o depends on the positions of S_1 and S_2 and on V_{IO}, I_1 and I_2.
(b) Using the data shown in Figure P7.7-3b, determine V_{IO}, I_{IB}, and I_{IO} for the amplifier under test.

7.7-4 (a) Find the gain of both circuits shown in Figure P7.7-4 with the switches in positions 1, 2, and 3.
(b) Explain the advantage of circuit (2) from a dc offset viewpoint assuming that $I_{IB} = 1$ μA, $I_{IO} = 0.2$ μA, and $V_{IO} = 1$ mV.

7.7-5 Consider the Schmitt trigger shown in Figure 7.5-5a, in which $R_f = 1$ MΩ and $R_2 = 10$ kΩ. For this circuit ideally v_o switches from -10 and $+10$ V when v_{in} exceeds $+0.1$ V. If $I_{IB} = 1$ μA, $I_{IO} = 0$, and $V_{IO} = 20$ mV, determine the range of values at which this switching could actually occur.

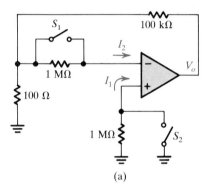

(a)

S_1	S_2	V_o (V)
Closed	Closed	5
Closed	Open	2
Open	Closed	7

Figure P7.7-3

(b)

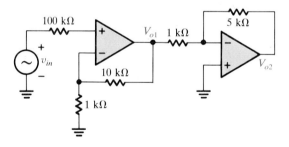

Figure P7.7-2

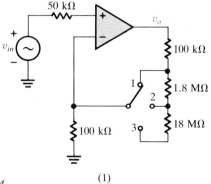

(1)

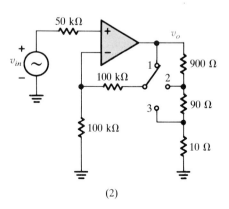

(2)

Figure P7.7-4

7.7-6 For the integrator shown in Figure 7.3-7b, $R = 1\ M\Omega$, $C = 1\ \mu F$, $I_{IB} = 10\ nA$, $I_{IO} = 0$, and $V_{IO} = 25\ mV$.

(a) Starting from first principles, derive an expression for v_o in terms of I_{IB} and V_{IO} (consider $v_{in} = 0$ to see the effect of I_{IB} and V_{IO}). The switch is opened at $t = 0$.

(b) Choosing either the plus or the minus value of V_{IO}, sketch and label $v_o(t)$ versus time in the interval from $t = 0$ to $t = 5$ seconds.

7.7-7 A noninverting op amp circuit with $R_f = 9\ k\Omega$ and $R_2 = 1\ k\Omega$ has the open-loop transfer function

$$A_{OL}(j\omega) = \frac{10^4}{(1 + j\omega/10)(1 + j\omega/100)}$$

What is the approximate bandwidth of the circuit?

7.7-8 For the small-signal output waveshape given in Figure 7.7-14c, the approximate slope of $v_o(t)$ was shown to be given by equation (7.7-58b). Prove that the maximum slope of the v_o waveshape is actually equal to $2V_m/\tau$.

7.7-9 For the differentiator circuit in Figure 7.3-6a, $v_{in} = 1 \sin 2\pi f t$ V, $R = 1\ k\Omega$, $C = 1\ \mu F$, the op amp power supplies are ± 10 V, and the slew rate is 1 V/μs. What is the highest signal frequency for which the output will faithfully look like the derivative of the input?

7.7-10 The op amps in the circuit shown in Figure P7.7-10 are operated from ± 15-V supplies and have slew rates of 1 V/μs; otherwise, they are ideal. Carefully sketch and label v_{in}, v_{o1}, and v_{o2}.

7.7-11 For the circuits described below, the open-loop op amp characteristics are $Z_{id} = 10\ k\Omega$, $Z_{icm} = 1\ M\Omega$, $Z_o = 100\ \Omega$, and $A_{OL} = 10^4$. Sketch $v_o(t)$ for $v_{in}(t) = 10^{-3} \sin \omega t$ V for each circuit.

(a) The circuit is an noninverting op amp with $R_f = 100\ k\Omega$ and $R_2 = 1\ k\Omega$. In addition, a 1-kΩ resistor is connected from v_o to ground and a 1-MΩ resistor is connected in series with the input voltage v_{in}.

(b) The circuit is an inverting op amp with $R_f = 100\ k\Omega$ and $R_2 = 1\ k\Omega$, and a 10-Ω load connected from v_o to ground.

7.7-12 Consider an inverting op amp circuit with $v_{in} = 3 \sin \omega t$ mV, $R_f = 99\ k\Omega$, $R_2 = 1\ k\Omega$, and a load R_L connected from v_o to ground. The open-loop op amp characteristics are $A_{OL} = 10^4$, $Z_o = 1\ k\Omega$ (without feedback), $Z_{id} = 50\ k\Omega$, $Z_{icm} = 1\ M\Omega$, and $i_{o,max} = 10$ mA. What is the smallest R_L that can be connected to this op amp without producing distortion?

7.7-13 Each of the op amps in the circuit shown in Figure P7.7-13 has the following open-loop characteristics $A_{OL} = 10^5$, $Z_{icm} = 1\ M\Omega$, $Z_{id} = 10\ k\Omega$, and $Z_o = 100\ \Omega$. Carefully sketch and label v_{o1} and v_{o2}.

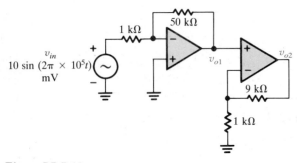

Figure P7.7-10

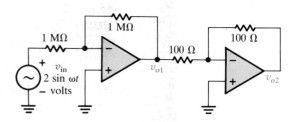

Figure P7.7-13

CHAPTER 8

Amplifier Stability and Oscillators

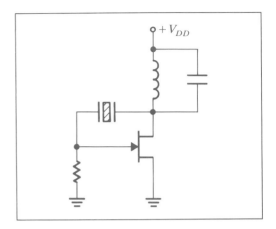

8.1 INTRODUCTION

In Chapter 7, we found that feedback could be used to improve the performance of an amplifier. In particular we saw that it permitted an alteration of the input and output impedance, a stabilization of the gain, a reduction of nonlinear distortion, and an extension of the amplifier bandwidth. As with most things in life, these improvements were not obtained without cost since a significant loss of amplifier gain occurred when the feedback was added. Furthermore, as we shall see in this chapter, the addition of feedback to an amplifier can cause an even more insidious problem—oscillation of the circuit.

Since the amplifier frequency response changes when feedback is added, it is apparent that the location of the poles (and perhaps the zeros, too) must also change. As long as the poles of the amplifier transfer function remain in the left-hand side of the complex frequency plane, the amplifier will be stable. However, if any of the poles are on the $j\omega$ axis, or worse yet, if they migrate into the right-hand plane, the amplifier will become unstable and will produce an output containing growing exponential terms.

When we are trying to design stable amplifier systems, it is essential that its poles remain in the left-hand plane. However, to build an oscillator, that is, a circuit that produces a steady-state output without the application of any input signal, quite the opposite is true. To generate a sinusoidal output from an oscillator, we require that the circuit's poles be located on or perhaps slightly to the right of the $j\omega$ axis, while for the production of nonsinusoidal outputs, such as square waves, circuits whose poles are deep in the right-hand plane will be needed.

Thus the determination of whether a circuit behaves like an amplifier or an

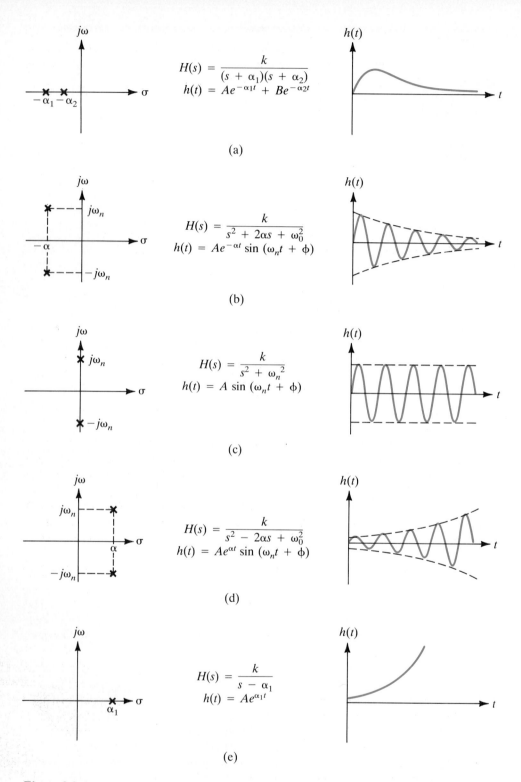

Figure 8.2-1

Transient response associated with different s-plane pole–zero locations.

oscillator depends on the location of the poles of its transfer function. The purpose of this chapter is to help us to understand how feedback affects the position of the poles and zeros of an amplifier, and more specifically, to show us what we as engineers can do to position the circuit poles where we want them. This will help to guarantee that the amplifiers we design amplify and the oscillators oscillate, and not vice versa.

8.2   FEEDBACK AND ITS EFFECT ON THE POLE–ZERO LOCATIONS

In this section, we investigate how the addition of feedback to an amplifier modifies the circuit transfer function, and more specifically, how it affects the location of poles and zeros. We do this by first considering the general feedback problem illustrated in Figure 8.2-2, and we then narrow our scope to analyze several specific circuit transfer functions, to obtain an overview of this subject.

Figure 8.2-1 illustrates the relationship that exists between the pole–zero locations of an amplifier and the transient response that results, for example, when the circuit is first turned on. In general, the form of this response, that is, whether it contains decaying exponentials, sine waves, growing sine waves, and so on, depends only on the pole locations, and is independent of the position of the zeros; however, the location of the zeros does affect the amplitude of each of the terms in the output.

For example, if all the poles of the circuit transfer function lie on the negative real axis, the transient response to a specific set of initial conditions will consist of a sum of decaying exponentials, so that the response will eventually die out. A typical circuit transient response corresponding to this situation is illustrated in Figure 8.2-1a. When the transfer function contains a pair of complex conjugate poles that lie in the left-hand plane, the inverse Laplace transform of this function will give rise to a transient response which is an exponentially decaying sinusoid, as illustrated in Figure 8.2-1b. The rate of decay of the sine wave (or rate of damping) depends on α, the negative real part of the pole positions.

When these complex poles move onto the $j\omega$ axis, the damping term is zero, and as a result, the transient response associated with this transfer function is a pure sine wave which never decays. In principle, this is where we would want to position the poles of our transfer function in order to construct a sinusoidal oscillator (Figure 8.2-1c). However, in this situation the amplitude of the oscillations would be determined by the circuit initial conditions. In a practical sinusoidal oscillator design the poles are actually positioned slightly into the right-hand plane, and for this case the transient response contains a growing sine wave. The sketch in Figure 8.2-1d indicates that the oscillation amplitude for this case will increase without bound. Although this result is correct for an ideal amplifier, in a practical circuit the output amplitude is eventually limited by the circuit nonlinearity. For example, in an op amp the output amplitude would typically increase until it hit the power supply rails.

A typical amplifier circuit containing negative feedback is illustrated in Fig-

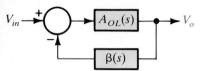

Figure 8.2-2

ure 8.2-2. The closed-loop gain for this circuit is given by the expression

$$\frac{V_o}{V_{in}} = \frac{A_{OL}(s)}{1 + \beta(s)A_{OL}(s)} \tag{8.2-1}$$

If the feedback network $\beta(s)$ has no frequency dependence, we may write that

$$A_{CL}(s) = \frac{V_o}{V_{in}} = \frac{N_A(s)/D_A(s)}{1 + \beta[N_A(s)/D_A(s)]} = \frac{N_A(s)}{D_A(s) + \beta N_A(s)} \tag{8.2-2}$$

where

$$A_{OL}(s) = \frac{N_A(s)}{D_A(s)} \tag{8.2-3}$$

represents the form of the open-loop transfer function of this amplifier. Equation (8.2-2) illustrates two important facts about the effect of feedback on the pole and zero locations of an amplifier. First, when β is a constant, the zeros of the closed-loop transfer function are in the same position as those of the open-loop amplifier. However, because the denominators of equations (8.2-1) and (8.2-2) are different, it is apparent that the poles have moved. The way in which the poles move depends on the form of the original transfer function, the type of feedback used (positive or negative), and the magnitude of feedback applied. These concepts are illustrated in the examples that follow.

8.2-1 Single-Pole Amplifier with Feedback

Let us initially consider the case of a single-pole amplifier whose open-loop transfer function contains a pole that is located on the negative real axis. The s-plane pole–zero plot for this amplifier is illustrated in Figure 8.2-3a, and its transfer function may be written as

$$A_{OL}(s) = \frac{V_o}{V_{in}} = \frac{A_o \alpha}{s + \alpha} \tag{8.2-4}$$

where A_o represents the amplifier's open-loop low-frequency gain. The Bode magnitude plot for this amplifier is presented in Figure 8.2-3b. From this sketch we can see that the amplifier's open-loop 3-dB bandwidth is equal to α.

The open-loop step response of this amplifier is readily found by noting that

$$V_o(s) = A_{OL}(s)V_{in}(s) = \frac{\alpha A_o V_B}{s(s + \alpha)} \tag{8.2-5}$$

where V_B is the amplitude of the applied input step. The inverse Laplace transform of this expression is

$$v_o(t) = A_o V_B(1 - e^{-\alpha t}) \tag{8.2-6}$$

and this result is sketched in Figure 8.2-3c.

Figure 8.2-3d illustrates the basic form of this amplifier when negative feedback is added. The closed-loop transfer function for this case is obtained directly from equation (8.2-1) as

$$A_{CL}(s) = \frac{A_{OL}(s)}{1 + \beta A_{OL}(s)} = \frac{\alpha A_o/(s + \alpha)}{1 + \beta[\alpha A_o/(s + \alpha)]} = \frac{\alpha A_o}{s + \alpha(1 + \beta A_o)} \tag{8.2-7}$$

From an examination of this result, it should be clear that the addition of the

Chapter 8 / Amplifier Stability and Oscillators

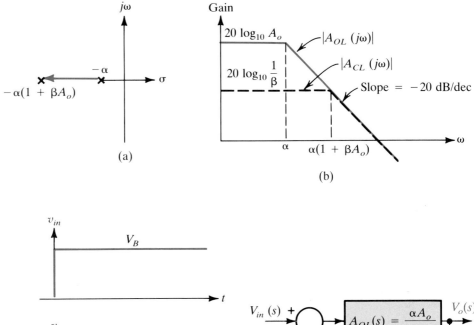

(a)

(b)

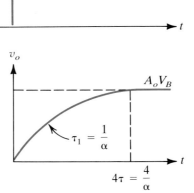

(c)   Open-Loop Step Response

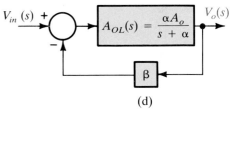

(d)

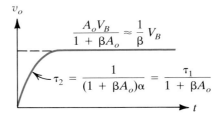

Figure 8.2-3

(e)   Closed-Loop Step Response

feedback has caused the pole of the amplifier to move from its original open-loop position at $-\alpha$ to $-\alpha(1 + \beta A_o)$. Because the pole is now farther in the left-hand side of the complex plane, the feedback has made the amplifier even more stable than it was under open-loop conditions. In addition, as illustrated by the dashed lines in Figure 8.2-3b, and as expected from our previous analyses of the effect of feedback on frequency response, its bandwidth is larger, increasing from α to $\alpha(1 + \beta A_o)$. Correspondingly, the time constant τ in the step response of the circuit has been reduced by the same factor (see Figure 8.2-3e).

8.2-2 Two-Pole Amplifier with Feedback

Before investigating the performance of a specific two-pole feedback amplifier, it will be useful first to review the general characteristics of a second-order transfer function having two poles and no zeros. The basic form of this function may be written as

$$H(s) = \frac{\omega_n^2}{s^2 + 2\delta\omega_n s + \omega_n^2} \tag{8.2-8}$$

The meaning of δ and ω_n in this expression will be clear shortly. Solving for the roots of the denominator of equation (8.2-8), we find that

$$s_1, s_2 = -\delta\omega_n \pm \omega_n\sqrt{\delta^2 - 1} \tag{8.2-9}$$

For large values of δ (i.e., for $\delta \gg 1$), the poles of the transfer function are both located on the negative real axis, while for δ less than 1, the roots become complex and are located at

$$s_1, s_2 = -\delta\omega_n \pm j\omega_n\sqrt{1 - \delta^2} \tag{8.2-10}$$

The location of these roots in the complex s-plane for this case is illustrated in Figure 8.2-4a. The distance of the poles from the origin is equal to ω_n, and the angle of the poles from the negative real axis is $\cos^{-1}\delta$. As an example, for the case where $\delta = 0.707$, we find that the poles are at a 45° angle from the real axis. The quantity δ is called the damping factor, for reasons that will be apparent shortly, and ω_n is known as the natural frequency of oscillation of the system since this is the frequency at which the system would oscillate if the damping factor were zero.

When $\delta > 1$, the system is said to be overdamped, and no oscillation occurs because the system has two negative real roots. When $\delta = 1$, the two roots coincide on the negative real axis, and the system is said to be critically damped, while for $\delta < 1$, we obtain a pair of complex conjugate roots, and in this situation the system is said to be underdamped.

The impulse response of this system for the underdamped case is obtained by taking the inverse Laplace transform of $H(s)$ in equation (8.2-8), considering the form of the roots to be those given by equation (8.2-10). In particular, for this case we have

$$h(t) = \frac{\omega_n}{\sqrt{1 - \delta^2}} e^{-\delta\omega_n t} \sin \omega_0 t \tag{8.2-11a}$$

where

$$\omega_0 = \omega_n\sqrt{1 - \delta^2} \tag{8.2-11b}$$

assuming the impulse weight to be equal to one. This response is sketched in Figure 8.2-4b for different values of the damping factor δ.

To understand the physical meaning of this result, it is useful to draw a parallel between the response of this system to the application of an impulse and the response of an automobile suspension system to a sharp sudden bump in the road. In the latter case, depending on the condition of the shock absorbers

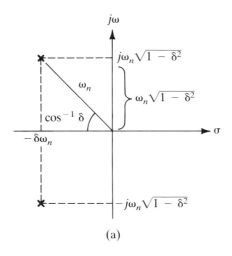

(a)

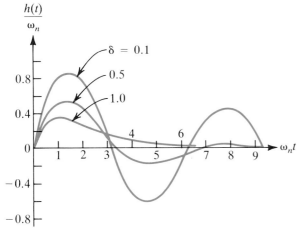

(b) Impulse Response of the Second-Order System

$$H(s) = \frac{\omega_n^2}{s^2 + 2\delta\omega_n s + \omega_n^2}$$

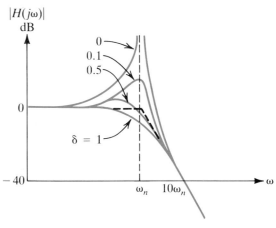

(c) Frequency Response of the Second-Order System

$$H(s) = \frac{\omega_n^2}{s^2 + 2\delta\omega_n s + \omega_n^2}$$

Figure 8.2-4
Underdamped second-order system.

(which provide the damping for this system), the car will either bounce once or twice and then stop (as for the case where δ is large), or else if the shocks are worn (and δ is small), it may oscillate up and down several times before coming to rest.

In fact, if the shock absorbers completely ceased to function and if the springs on the car had no other frictional losses, the damping factor δ would be zero, and in theory at least, if the car hit a bump, it would oscillate up and down forever at the system's natural frequency ω_n. In Figure 8.2-4a, making $\delta = 0$ is equivalent to placing the two poles on the $j\omega$ axis at $\pm j\omega_n$, and as might be expected, the inverse Laplace transform of a function having this type of pole–zero pattern is a sine wave that oscillates forever. This result can also be seen by substituting in $\delta = 0$ into equation (8.2-11a).

For later use, it is also helpful to sketch the Bode magnitude plot for this type of quadratic transfer function (Figure 8.2-4c). The asymptotic portion of the plot is relatively easy to explain if we note that $H(s)$ in equation (8.2-8) approaches 1 for values of s (or ω) much smaller than ω_n. On the other hand, for

s much greater than ω_n, the denominator of the transfer function may be approximated by the s^2 term. As a result, for $\omega \ll \omega_n$, $|H(j\omega)|$ is flat at 0 dB, and for $\omega \gg \omega_n$, it falls off at -40 dB/decade. The behavior of this function in the neighborhood of ω_n depends very strongly on the value of the damping factor. When δ is large, the system is heavily damped and the response closely follows the asymptotes, while when δ is small, a sharp resonance occurs in the neighborhood of ω_n and the Bode plot exhibits a pronounced peaking in the response near this frequency.

Let us now apply these results to the case of a second-order feedback amplifier whose open-loop amplifier poles are both assumed to be located at $s = -\alpha$. The transfer function for this amplifier under open-loop conditions is

$$A_{OL}(s) = \frac{\alpha^2 A_o}{(s + \alpha)^2} \tag{8.2-12}$$

The s-plane pole–zero plot and frequency response of $A_{OL}(j\omega)$ are given in Figure 8.2-5a and b, respectively. When feedback is added to this amplifier (Figure 8.2-5c), using equation (8.2-1), we have

$$A_{CL}(s) = \frac{\alpha^2 A_o/(s + \alpha)^2}{1 + \beta[\alpha^2 A_o/(s + \alpha)^2]} = \frac{\alpha^2 A_o}{s^2 + 2\alpha s + \alpha^2(1 + \beta A_o)} \tag{8.2-13}$$

Figure 8.2-5
Two-pole feedback amplifier.

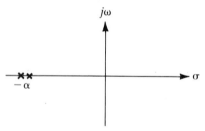

(a) s-Plane Pole Positions
for $A_{OL}(s) = \dfrac{A_o\alpha^2}{(s + \alpha)^2}$

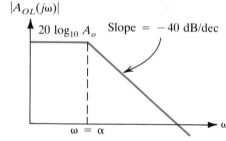

(b) Frequency Response of
$A_{OL}(s) = \dfrac{A_o\alpha^2}{(s + \alpha)^2}$

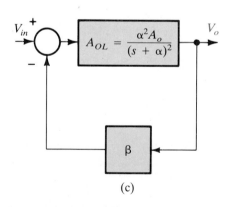

(c)

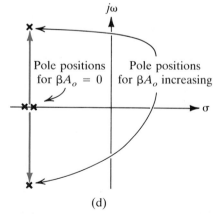

Pole positions
for $\beta A_o = 0$

Pole positions
for βA_o increasing

(d)

It is not too difficult to show that the poles of this expression will be complex for all positive nonzero values of βA_o. In fact, by comparing this expression with equation (8.2-8), we can immediately see that the natural frequency of oscillation of this system will be

$$\omega_n = \alpha\sqrt{1 + \beta A_o} \tag{8.2-14a}$$

In addition,
$$\delta\omega_n = \alpha \tag{8.2-14b}$$

or
$$\delta = \frac{1}{\sqrt{1 + \beta A_o}} \tag{8.2-14c}$$

The latter result indicates that considerable ringing (or damped oscillation) will occur for large values of βA_o. However, from equation (8.2-14b) we can see that α will be constant independent of the feedback, so that the poles of this transfer function will always remain in the left-hand side of the s-plane. As a result, the closed-loop transfer function of this second-order system will always be stable, but despite this fact, for large βA_o this amplifier will not be all that useful. When βA_o is large, δ will be small and the frequency response of the amplifier will exhibit a sharp high-gain peak in the vicinity of ω_n. Furthermore, as with the case of an automobile having poor shock absorber damping, anytime a step, impulse, or square wave is applied to the input, the output will ring (or oscillate) for a considerable length of time before settling down. Usually, this type of transient response is unacceptable.

8.2-3 Three-Pole Amplifier with Feedback

In an open-loop amplifier, whenever the excess of poles over zeros, that is, (number of poles − number of zeros), is greater than or equal to 3, the possibility exists that the amplifier will oscillate when negative feedback is added. To illustrate how this situation arises, we will examine the case of an amplifier containing three poles and no zeros. Once this analysis is completed, we will be in a position to understand just what it takes to make a circuit oscillate.

The circuit we will be examining, presented in Figure 8.2-6a, is similar to that described in Figure 8.2-5 except that this one effectively contains three instead of two RC networks. The open-loop transfer function for this amplifier is given by the expression

$$A_{OL}(s) = \frac{A_o\alpha^3}{(s + \alpha)^3} = \frac{10^4\alpha^3}{(s + \alpha)^3} \tag{8.2-15}$$

and its pole–zero plot is shown in Figure 8.2-6b.

As with the previous amplifiers that we have examined, the poles of the closed-loop transfer function will be different from those of $A_{OL}(s)$ and will be a function of the amount of feedback used in the amplifier. The location of the poles for a specific feedback situation is determined by finding the roots of the denominator of $A_{CL}(s)$. In general, this is a very arduous task, and for the particular amplifier being examined will require the solution of a cubic equation.

Fortunately, it is not usually necessary to have an exact specification of the root locations for all values of β. In general, we will be most interested in their position when the loop gain magnitude, $\beta A_{OL}(s)$, is large, because it is under these circuit conditions that the amplifier poles move into the right-hand plane

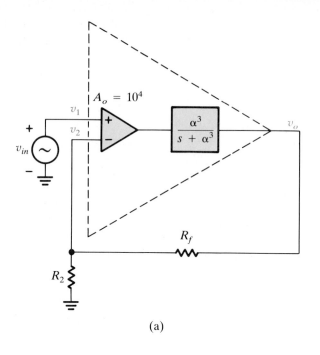

(a)

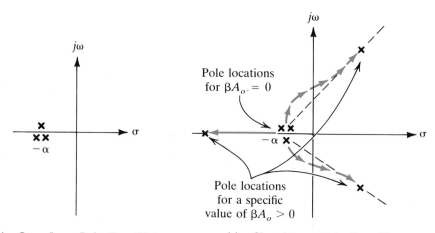

(b) Open-Loop Pole–Zero Plot   (c) Closed-Loop Pole–Zero Plot

Figure 8.2-6

and cause the circuit to oscillate. Let's return to the amplifier circuit in Figure 8.2-6 to illustrate these points.

We begin this analysis by developing an expression for the closed-loop gain of this three-pole amplifier. Substituting equation (8.2-15) into (8.2-1), we obtain

$$A_{CL}(s) = \frac{10^4 \alpha^3}{(s + \alpha)^3 + \beta A_o \alpha^3}$$
(8.2-16)

Clearly, this transfer function has no zeros and three poles. In addition, the location of these poles varies with the feedback factor, β, of the amplifier. For $\beta = 0$ all three poles of $A_{CL}(s)$ are located at $s = -\alpha$. Thus for small β the poles of $A_{CL}(s)$ start out at the open-loop pole positions. As β is increased, the location of the poles is determined by finding the roots of the equation

$$(s + \alpha)^3 + \beta A_o \alpha^3 = 0$$

or

$$s^3 + 3\alpha s^2 + 3\alpha^2 s + \alpha^3(1 + \beta A_o) = 0 \qquad (8.2\text{-}17)$$

In general, this equation is rather difficult to solve, but there are several specific solutions that are both easy to find and at the same time extremely illuminating in terms of the information they provide.

Let us consider the approximate values of the roots of this equation when the dc loop gain magnitude

$$\beta A_o \gg 1 \qquad (8.2\text{-}18)$$

For this case we find that the magnitude of the roots are very large, so that in equation (8.2-17) the cubic term will be much bigger than the quadratic and linear terms. Using these facts, for large values of loop gain (or large amounts of feedback) we may approximate equation (8.2-17) as

$$s^3 + \alpha^3 \beta A_o \simeq 0 \qquad (8.2\text{-}19)$$

Equation (8.2-19) has three solutions, and using complex arithmetic, they may be determined by solving the equation

$$s_1, s_2, s_3 = \sqrt[3]{-\alpha^3 \beta A_o}\, e^{j(2\pi n/3)} \qquad (8.2\text{-}20)$$

where n is set equal to 0, 1, and 2 to evaluate the three roots. Continuing this procedure, we have

$$s_1, s_2, s_3 = \sqrt[3]{e^{j\pi}\alpha^3 \beta A_o}\, e^{j(2\pi n/3)} = \alpha \sqrt[3]{\beta A_o}\, e^{j[(\pi + 2n\pi)/3]} \qquad (8.2\text{-}21)$$

Substituting in $n = 0$, 1, and 2, we may write each of these roots as

$$s_1 = \alpha \sqrt[3]{\beta A_o}\, e^{j\pi/3} \qquad (8.2\text{-}22a)$$

$$s_2 = \alpha \sqrt[3]{\beta A_o}\, e^{j\pi} = -\alpha \sqrt[3]{\beta A_o} \qquad (8.2\text{-}22b)$$

and

$$s_3 = \alpha \sqrt[3]{\beta A_o}\, e^{j5\pi/3} = \alpha \sqrt[3]{\beta A_o}\, e^{-j\pi/3} \qquad (8.2\text{-}22c)$$

These solutions are expressed in polar form, and the solutions indicate that each of the poles is approximately located at a distance

$$\alpha = \sqrt[3]{\beta A_o} \qquad (8.2\text{-}23a)$$

from the origin at angles of $+60$, -60, and $+180°$ from the positive real axis.

In summary, the approximate pole locations for large values of loop gain are along three asymptotes whose radial distance from the origin is given by equation (8.2-23a), and whose angular distance from the positive real axis is

$$\frac{\pi + 2n\pi}{3} \qquad (8.2\text{-}23b)$$

where $n = 0$, 1, and 2. As a practical matter, these asymptotes actually emanate from the point $s = -\alpha$, but for large values of A_o, this positional difference is negligible. The resulting plot of the path of the poles as a function of dc loop gain magnitude (βA_o) is illustrated in Figure 8.2-6c. This sketch is also frequently called the root locus of the poles. Notice that when βA_o is large enough, two of the poles (which will be complex conjugates of one another) will cross over into the right-hand side of the s-plane. Once this occurs, the system will begin to oscillate.

To determine the value of loop gain that just places the poles on the $j\omega$ axis, we note that at this value of loop gain, two of the poles will be located at

$$s_1 = j\omega_0 \tag{8.2-24a}$$

and
$$s_2 = -j\omega_0 \tag{8.2-24b}$$

where ω_0 represents the distance from the origin to the pole positions (see Figure 8.2-7). To find the value of ω_0, as well as the loop gain needed to place the poles at this position, we note that for this value of βA_o, equation (8.2-17) may be rewritten as

$$(s + j\omega_0)(s - j\omega_0)(s + \gamma) = s^3 + 3\alpha s^2 + 3\alpha^2 s + \alpha^3(1 + \beta A_o)$$

or
$$(s^2 + \omega_0^2)(s + \gamma) = s^3 + 3\alpha s^2 + 3\alpha^2 s + \alpha^3(1 + \beta A_o) \tag{8.2-25}$$

The term $(s + \gamma)$ represents the third factor of the equation, with $-\gamma$ indicating the third root along the negative real axis. Expanding the left-hand side of this expression, we have

$$s^3 + \gamma s^2 + \omega_0^2 s + \gamma \omega_0^2 = s^3 + 3\alpha s^2 + 3\alpha^2 s + \alpha^3(1 + \beta A_o) \tag{8.2-26}$$

and equating equal powers of s on the left- and right-hand sides of this equation, we obtain

$$\gamma = 3\alpha \tag{8.2-27a}$$

$$\omega_0^2 = 3\alpha^2 \tag{8.2-27b}$$

and
$$\gamma \omega_0^2 = \alpha^3(1 + \beta A_o) \tag{8.2-27c}$$

Equation (8.2-27a) indicates that the third pole has just moved to the point -3α as the other two poles cross the $j\omega$ axis. The value of ω_0 is seen from equation (8.2-27b) to be

$$\omega_0 = \sqrt{3}\alpha \tag{8.2-28}$$

Figure 8.2-7
Pole positions for amplifier in Figure 8.2-6a with $\beta A_o = 8$.

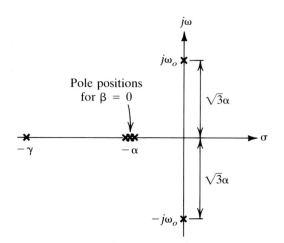

Chapter 8 / Amplifier Stability and Oscillators

Substituting equations (8.2-27a) and (8.2-27b) into (8.2-27c), we find that

$$(3\alpha)(3\alpha^2) = 9\alpha^3 = \alpha^3(1 + \beta A_o)$$

or
$$\beta A_o = 8 \qquad (8.2\text{-}29)$$

This is the value of loop gain magnitude required to just place the poles on the $j\omega$ axis. Furthermore, as we shall see shortly, with this value of βA_o the system will just begin to oscillate. Larger values of loop gain will cause the poles to move into the right-hand plane, and these values of βA_o will give rise to growing sinusoids.

For the case where $\beta A_o = 8$, two of the poles of $A_{CL}(s)$ are on the $j\omega$ axis and the closed-loop transfer function has the form

$$\frac{V_o}{V_{in}} = A_{CL}(s) = \frac{A_o \alpha^3}{(s + \gamma)(s^2 + \omega_0^2)} \qquad (8.2\text{-}30)$$

where $\gamma = 3\alpha$ and $\omega_0 = \sqrt{3}\alpha$. Partial fraction expansion of this expression yields

$$\frac{V_o}{V_{in}} = \frac{\alpha A_o}{12}\left(\underbrace{\frac{1}{s + 3\alpha}}_{e^{-3\alpha t}} - \underbrace{\frac{s}{s^2 + \omega_0^2}}_{\cos \omega_0 t} + \underbrace{\frac{\sqrt{3}\,\omega_0}{s^2 + \omega_0^2}}_{\sin \omega_0 t}\right) \qquad (8.2\text{-}31)$$

The three terms on the right-hand side of this equation represent the form of the system's "transient" response. The actual amplitude of each of these terms depends on the circuit initial conditions and the input $v_{in}(t)$, but their basic form depends only on the location of the poles. Thus, even if the input signal is zero, the same three "transient" terms will still be present, and after the transient term $e^{-3\alpha t}$ dies off, the output will be of the form

$$v_o(t) = V_o \sin (\omega_0 t + \phi) \qquad (8.2\text{-}32)$$

where V_o and ϕ will depend on the circuit's initial conditions. In other words, even with $v_{in}(t) = 0$, for almost any combination of system initial conditions, the system's steady-state output will be a sinusoid oscillating at the frequency ω_0.

To obtain a physical interpretation of the concept of system oscillation, consider the negative-feedback amplifier illustrated in Figure 8.2-8a, in which the applied input to the amplifier is assumed to be zero. To illustrate the conditions under which this system may oscillate (or produce an output with no input applied), we will break the feedback loop as shown in part (b) of the figure and inject a sine wave at the frequency ω_0 at this point. If the amplitude of this sine wave is V_B, the voltages v', v_o, and v_f' that result may be written down immediately as

$$v'(t) = v_{in} - v_f = -V_B \sin \omega_0 t \qquad (8.2\text{-}33a)$$

$$v_o(t) = -A_1 V_B \sin [\omega_0 t + \phi_A(j\omega_0)] \qquad (8.2\text{-}33b)$$

and
$$v_f'(t) = -A_1 \beta_1 V_B \sin [\omega_0 t + \phi_A(j\omega_0) + \phi_B(j\omega_0)] \qquad (8.2\text{-}33c)$$

In these expressions A_1 and β_1 and ϕ_A and ϕ_B represent the gain and phase shift at ω_0 of the amplifier and feedback network, respectively. The waveshapes il-

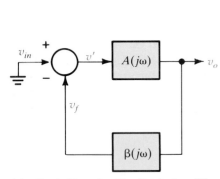

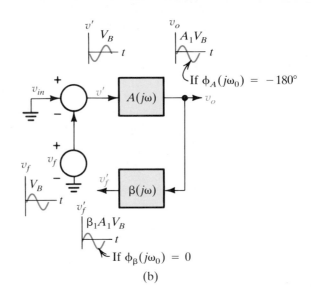

(a) Basic Negative Feedback Amplifier
with No Input Signal Applied

(b)

Figure 8.2-8
Basic form of an oscillator.

lustrated in Figure 8.2-8b were drawn assuming that $\phi_A(j\omega) = -180°$ and $\phi_B(j\omega) = 0°$ at $\omega = \omega_0$.

Examining the feedback signal $v_f'(t)$, we can see that if

$$|A(j\omega)\beta(j\omega)|_{\omega=\omega_0} = A_1\beta_1 = 1 \qquad (8.2\text{-}34a)$$

and $$\phi_A(j\omega) + \phi_B(j\omega)|_{\omega=\omega_0} = -180° \qquad (8.2\text{-}34b)$$

this signal will be exactly equal to the applied voltage $v_f(t)$. Thus once the signal is started and v_f' exists, we can remove the signal generator, reconnect v_f' to v_f, and the amplifier should be able to produce its own signals v_f, v', and v_o. This statement is just another way of saying that if equations (8.2-34a) and (8.2-34b) can be satisfied for any frequency ω_0, it is possible for the system to oscillate at that frequency.

In the example that follows we examine an op amp version of the three-pole feedback amplifier discussed previously and demonstrate that the criterion $\beta A_o = 8$ that was required to cause system oscillation by placing two of the poles on the $j\omega$ axis is precisely the same requirement as that imposed by equations (8.2-34a) and (8.2-34b).

EXAMPLE 8.2-1

The op amp circuit illustrated in Figure 8.2-9 is a three-pole feedback amplifier. In this circuit IC2, IC3, and IC4 are voltage-follower buffers and are used to isolate the RC stages from one another to simplify calculation of the overall system transfer function. In analyzing this circuit all the op amps may be assumed to be ideal.

a. Find the system transfer function V_o/V_{in}.
b. By comparing the result in part (a) with that given in equation (8.2-16), find the value of R_2 needed to place two of the system poles on the $j\omega$ axis, and thereby cause the system to oscillate.
c. For the value of R_2 determined in part (b), find the frequency of oscillation.
d. For the value of R_2 selected in part (b), demonstrate, by injecting a signal at v_f and tracing it around the loop, that the overall loop gain at the frequency ω_0 is just equal to 1.

Chapter 8 / Amplifier Stability and Oscillators

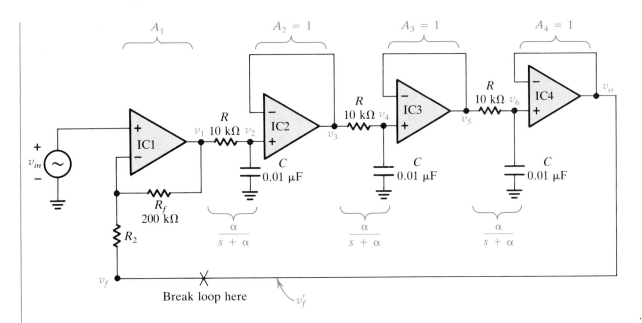

Figure 8.2-9
Example 8.2-1.

e. Determine the oscillation amplitudes at v_1, v_3, v_5, and v_o if the power supply voltages are ± 10 volts, and if R_2 is adjusted so that the poles are just slightly in the right-hand plane.

SOLUTION

Following Figure 8.2-9, we can see that

$$V_1 = \left(1 + \frac{R_f}{R_2}\right)V_{in} - \frac{R_f}{R_2}V_o \tag{8.2-35a}$$

$$V_2 = \frac{1/RC}{s + 1/RC}V_1 = \frac{\alpha}{s + \alpha}V_1 \tag{8.2-35b}$$

$$V_4 = \frac{\alpha}{s + \alpha}V_2 = \frac{\alpha^2}{(s + \alpha)^2}V_1 \tag{8.2-35c}$$

and
$$V_o = \frac{\alpha}{s + \alpha}V_5 = \frac{\alpha^3}{(s + \alpha)^3}V_1 \tag{8.2-35d}$$

Substituting equation (8.2-35a) into equation (8.2-35d), we obtain

$$V_o = \frac{\alpha^3}{(s + \alpha)^3}\left[\left(1 + \frac{R_f}{R_2}\right)V_{in} - \frac{R_f}{R_2}V_o\right] \tag{8.2-36a}$$

or
$$\frac{V_o}{V_{in}} = A_{CL}(s) = \frac{(1 + R_f/R_2)[\alpha^3/(s + \alpha)^3]}{1 + \underbrace{(R_f/R_2)[\alpha^3/(s + \alpha)^3]}_{\beta A_o}} \tag{8.2-36b}$$

Referring back to equation (8.2-16), we should recall that a value of $\beta A_o = 8$ was required to place two out of the three poles on the $j\omega$ axis. Therefore, to accomplish the same result in this amplifier, it is necessary that

$$\frac{R_f}{R_2} = 8 \tag{8.2-37a}$$

or that
$$R_2 = \frac{200 \text{ k}\Omega}{8} = 25 \text{ k}\Omega \qquad (8.2\text{-}37\text{b})$$

With R_2 set at this value, following equation (8.2-28), we find that the poles on the $j\omega$ axis are located at $\pm j\omega_0$, where

$$\omega_0 = \sqrt{3}\frac{1}{RC} \qquad (8.2\text{-}38\text{a})$$

or
$$f_0 = \frac{1}{2\pi}\omega_0 = \frac{\sqrt{3}}{2\pi RC} = 2.76 \text{ kHz} \qquad (8.2\text{-}38\text{b})$$

This, of course, is also the frequency of oscillation of this circuit with R_f/R_2 adjusted to be equal to 8.

To trace the oscillation signal around the amplifier loop, we will break the loop at the bottom of the resistor R_2 and inject a voltage v_f at this point having a frequency ω_0. In following this signal as it traverses the loop, it is useful to note that each of the RC networks has a gain

$$\frac{\alpha}{s + \alpha}\bigg|_{s=j\omega_0=j\sqrt{3}\alpha} = \frac{\alpha}{j\sqrt{3}\alpha + \alpha} = \frac{1}{1 + j\sqrt{3}} = \frac{1}{2}e^{-j\pi/3} = \frac{1}{2}\underline{/-60°} \qquad (8.2\text{-}39)$$

at the frequency ω_0. This means that each RC network reduces the signal amplitude by a factor of 2 and introduces a phase shift of $-60°$. Thus a cascade of three of these networks causes an overall signal attenuation of $(\frac{1}{2})^3$ or $\frac{1}{8}$ and a phase shift of $-180°$. Of course, this is precisely the phase shift required by equation (8.2-34b) to make v_f' come back in phase with v_f. Furthermore, the gain of 8 in IC1, followed by the attenuation of $\frac{1}{2}$ in each of the RC networks, satisfies the second requirement given in equation (8.2-34a), namely that the overall loop gain at the frequency of oscillation be equal to 1. Using this information, and assuming that

$$v_f(t) = V_B \sin \omega_0 t \qquad (8.2\text{-}40\text{a})$$

we may write the waveshapes at the other critical point in this circuit as

$$v_1(t) = -8V_B \sin \omega_0 t \qquad (8.2\text{-}40\text{b})$$

$$v_3(t) = -4V_b \sin (\omega_0 t - 60°) \qquad (8.2\text{-}40\text{c})$$

$$v_5(t) = -2V_B \sin (\omega_0 t - 120°) \qquad (8.2\text{-}40\text{d})$$

and
$$v_o(t) = v_f'(t) = -V_B \sin (\omega_0 t - 180°) = V_B \sin \omega_0 t \qquad (8.2\text{-}40\text{e})$$

In these equations, because R_2 is assumed to be precisely equal to 25 kΩ, the poles of the amplifier are located exactly on the $j\omega$ axis, and for this situation the signal amplitudes are determined by the initial conditions. For a practical oscillator design, R_2 would be adjusted to be slightly less than 25 kΩ. This would place the two complex poles in the right-hand plane, giving rise to growing sine waves whose amplitudes would ultimately be limited by the power supply rails. With R_2 adjusted in this way, v_1 (because it has the highest amplitude) would be limited to 10 V. Following equations (8.2-40c) through (8.2-40e), the other voltages v_3, v_5, and v_o would have amplitudes of 5, 2.5, and 1.25 V, respectively.

8.2-4 Feedback Systems Containing More Than Three Poles

The previous analysis demonstrated that for a negative-feedback system containing three poles and no zeros, a continued increase in loop gain would eventually cause one or more of the poles to cross over into the right-hand side of the s-plane. This, of course, produces amplifier instability and can lead to sustained system oscillation. In the remainder of this section, we demonstrate that this instability problem exists not only in three-pole systems, but for all systems where the number of poles in the closed-loop transfer function exceeds the number of zeros by three or more.

Consider, for example, an amplifier having an open-loop transfer function

$$A_{OL}(s) = A_o \frac{(s + z_1)(s + z_2) \cdots (s + z_m)}{(s + p_1)(s + p_2) \cdots (s + p_n)} \tag{8.2-41}$$

where z_1, z_2, and so on, represent the amplifier zeros, and p_1, p_2, p_3, and so on, the system poles. Here we are assuming that the number of poles exceeds the number of zeros. If the feedback network is a constant, we may write the closed-loop gain for this amplifier as

$$A_{CL}(s) = \frac{A_{OL}(s)}{1 + \beta A_{OL}(s)} = \frac{A_o(s + z_1)(s + z_2) \cdots (s + z_m)}{[(s + p_1)(s + p_2) \cdots (s + p_n)] + \beta A_o[(s + z_1) \cdots (s + z_m)]} \tag{8.2-42}$$

Clearly, the positions of the closed-loop poles for this amplifier are different from those of the open-loop case. In particular, for large values of loop gain it can be shown by finding the roots of the denominator of equation (8.2-42) that m of the poles will move toward the positions of the m zeros. The remaining $(n - m)$ poles will go out toward infinity (as for the three-pole case). The location of these poles can be found by determining the denominator roots in equation (8.2-42) for the case where both βA_o and s are large. In particular for this situation the pole locations may be determined by solving the equation

$$(s + p_1)(s + p_2) \cdots (s + p_n) + \beta A_o[(s + z_1)(s + z_2) \cdots (s + z_m)] = 0 \tag{8.2-43}$$

For s and βA_o large we may approximate this expression as

$$s^n + \beta A_o s^m = 0$$

or
$$s^{(n - m)} + \beta A_o = 0 \tag{8.2-44}$$

This equation has $(n - m)$ solutions, and using complex arithmetic, we have

$$s^{(n - m)} = -\beta A_o = e^{j\pi}\beta A_o = \beta A_o e^{j(\pi + l\pi)} \tag{8.2-45}$$

where $l = 0, 1, 2, \ldots, (n - m - 1)$. Taking the $(n - m)$th root of this expression, the solutions may be written as

$$s_1, s_2, s_3, \ldots, s_{(n-m)} = \sqrt[(n-m)]{\beta A_o}\, e^{j[(\pi + 2l\pi)/(n-m)]} \tag{8.2-46}$$

again where $l = 0, 1, 2, \ldots, (n - m - 1)$. The root loci of these excess poles are tabulated in Figure 8.2-10 for different values of $(n - m)$. These results clearly demonstrate that whenever $(n - m)$ is greater than 2, the amplifier will become unstable, or oscillate, when the loop gain is made large enough.

In analyzing these multipole amplifiers the criteria for amplifier stability is the same as that given in equations (8.2-34a) and (8.2-34b), since in deriving

Excess of poles over zeros $(n - m)$	Root locus pole angles		Comment	Root loci of the $(n - m)$ excess poles
	Coefficient	Angle		
1	$l = 0$	$\pi(-180°)$	Stable	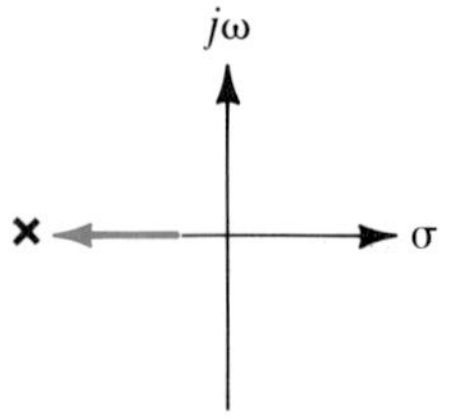
2	$l = 0$ $l = 1$	$\pi/2\ (+90°)$ $-\pi/2\ (-90°)$	Stable	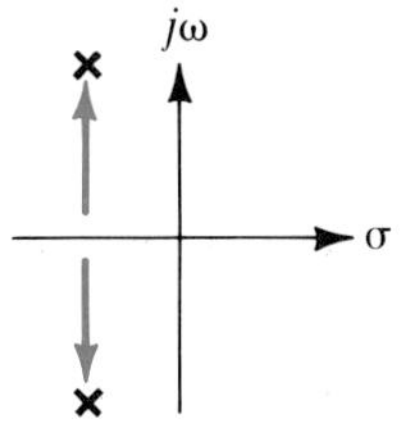
3	$l = 0$ $l = 1$ $l = 2$	$+60°$ $+180°$ $-60°$	Unstable for large loop gain	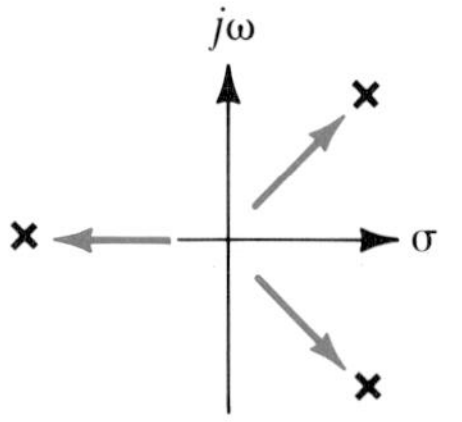
4	$l = 0$ $l = 1$ $l = 2$ $l = 3$	$+45°$ $+135°$ $+225°$ $-45°$	Unstable for large loop gain	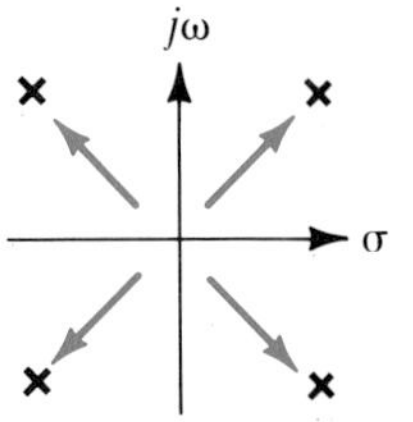
5	$l = 0$ $l = 1$ $l = 2$ $l = 3$ $l = 4$	$+36°$ $+108°$ $+180°$ $+252°\ (-108°)$ $+324°\ (-36°)$	Unstable for large loop gain	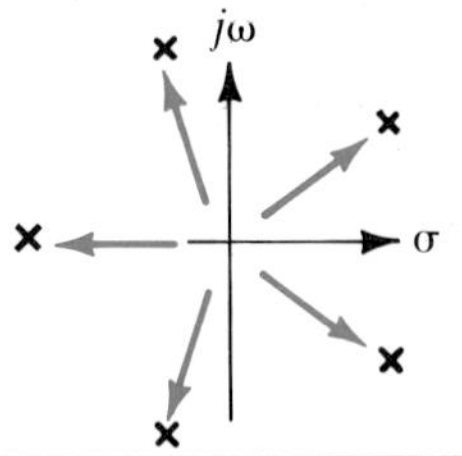

Figure 8.2-10
Approximate locations of the excess poles of a closed-loop amplifier for large values of loop gain.

these expressions, no mention was made of the number of poles or zeros in the transfer function. In fact, before closing this section it will be worthwhile to summarize what we have learned about the stability (and instability) of feedback amplifiers:

1. If all the closed-loop poles of an amplifier lie in the left-hand side of the s-plane, the amplifier will be stable (i.e., its transient response will die out).

2. If any of the closed-loop poles of the amplifier lie on the $j\omega$ axis, any nonzero initial conditions will produce a "transient response" which consists of sine and cosine terms that will continue on forever. In other words, the system will oscillate sinusoidally.

3. When the system is oscillating sinusoidally, the overall loop gain at the frequency of oscillation must just be equal to 1. In terms of $A(j\omega)$, the amplifier gain, and $\beta(j\omega)$, the feedback network gain, this requires (for a negative-feedback amplifier) that

$$|A(j\omega)\beta(j\omega)| = 1 \qquad (8.2\text{-}47a)$$

and
$$\phi_A(j\omega) + \phi_\beta(j\omega) = -180° \qquad (8.2\text{-}47b)$$

at the frequency of oscillation. When these criteria are satisfied, this means that at least two of the system poles lie on the $j\omega$ axis.

4. When the loop gain exceeds 1, the poles previously on the $j\omega$ axis will move into the right-hand side of the s-plane, and this will give rise to "transient"terms containing growing exponentials and growing sinusoids. Of course, these signals cannot grow without bound but will be limited ultimately by the amplifier nonlinearities.

Exercises

8.2-1 An op amp has an open-loop transfer function $10^7/(s + 10^3)$ and it is configured as a noninverting amplifier with $R_f = 1$ MΩ and $R_2 = 1$ kΩ. Find the small-signal step response rise time. *Answer* 200 μs.

8.2-2 An amplifier of the form illustrated in Figure 8.2-5c has the following parameter values: $A_o = 10^4$, $\beta = 10^{-3}$, and $\alpha = 10^3$ rad/s.
a. Find the frequency at which the peak in the ac gain occurs.
b. What is the gain at this frequency?
Answer (a) 3000 rad/s (b) 1.74

8.2-3 Given a feedback amplifier of the form illustrated in Figure 8.2-8 with

$$A(s) = \frac{10^4}{(s + 10)(s + 20)(s + 30)}$$

a. Find the value of β for which the circuit just oscillates.
b. Find the frequency of oscillation.
Answer (a) $\beta = 0.6$ (b) 14.14 rad/s

8.3 STABILITY IN TERMS OF BODE GAIN–PHASE PLOTS AND AMPLIFIER COMPENSATION TECHNIQUES

In Section 8.2 we saw that a feedback amplifier having poles on the $j\omega$ axis could produce sinusoidal steady-state oscillations with no input signal applied. For this to occur, we found that the denominator of the closed-loop transfer function

$$A_{CL}(s) = \frac{V_o}{V_{in}} = \frac{A_{OL}(s)}{1 + \beta(s)A_{OL}(s)} \qquad (8.3\text{-}1)$$

needed to be zero at the frequency of oscillation. This requirement is just another way of expressing the fact that the overall loop gain of the circuit must be equal to 1 at the frequency of oscillation. To see why these statements are equivalent, we note that if the transfer function has poles on the $j\omega$ axis, then

$$1 + \beta(s)A_{OL}(s)|_{s=j\omega} = 0 \tag{8.3-2}$$

or
$$\beta(j\omega)A_{OL}(j\omega) = -1 = 1e^{-j180°} \tag{8.3-3}$$

at the frequency $\omega = \omega_0$ where the poles are located. This result may be put into a more useful form by expressing $A_{OL}(j\omega)$ and $\beta(j\omega)$ in polar notation, which yields

$$[|\beta(j\omega)|e^{j\phi_\beta(j\omega)}][|A_{OL}(j\omega)|e^{j\phi_A(j\omega)}] = 1e^{-j180°} \tag{8.3-4}$$

Separating this equation into its magnitude and phase components, we have

---

$$|\beta(j\omega)||A_{OL}(j\omega)| = 1 \tag{8.3-5a}$$

and
$$\phi_\beta(j\omega) + \phi_A(j\omega) = -180° \tag{8.3-5b}$$

These results are the same as those obtained in equations (8.2-47a) and (8.2-47b).

When equations (8.3-5a) and (8.3-5b) can be satisfied at some specific frequency $\omega = \omega_0$, the feedback amplifier has poles on the imaginary axis at $\pm j\omega_0$ and will most probably oscillate at this frequency. Furthermore, if the feedback loop in this circuit if broken and a signal is injected at the frequency ω_0, this signal will, after traveling around the loop, return in phase and at the same amplitude as that of the original signal. As a result, once the oscillation is started by some circuit initial condition, the circuit will be able to oscillate on its own forever simply by removing the signal source and closing the feedback loop.

Consider next the physical implications of having loop gains that are greater than and less than 1 at the frequency $\omega = \omega_0$ where equation (8.3-5b) is satisfied. First, when

$$|A_{OL}(j\omega)||\beta(j\omega)| < 1 \tag{8.3-6}$$

at $\omega = \omega_0$, a signal injected into the loop at this frequency will return in phase with the original signal but reduced in amplitude. As a result, any initial condition producing a transient signal component at ω_0 will eventually die out as it travels around and around the loop. Therefore, for an amplifier where equations (8.3-5b) and (8.3-6) are satisfied, the system poles lie in the left-hand plane and the amplifier is stable.

If, on the other hand, the loop-gain magnitude

$$|A_{OL}(j\omega)||\beta(j\omega)| > 1 \tag{8.3-7}$$

at the frequency ω_0 where equation (8.3-5b) is satisfied, a signal injected into the loop at this frequency will return in phase but with a larger amplitude than it started with. As a result, with the loop closed, any initial condition producing a signal component at ω_0 will give rise to a growing sine wave at this frequency as the signal travels around and around the loop. Of course, this situation, that is, one for which equations (8.3-5b) and (8.3-7) are satisfied, is indicative of a feedback amplifier that has poles in the right-hand plane. This type of circuit is virtually useless for signal amplification purposes and is said to be unstable.

In summary:

1. Oscillations are possible in a feedback amplifier if the magnitude of the loop gain is greater than or equal to 1 at the frequency where the overall phase shift around the loop is 0° (or −360°).
2. For the specific case of a negative-feedback amplifier the critical phase shift through the amplifier and the feedback network is −180° rather than 0° because the circuit already has −180° of phase shift due to the signal inversion that occurs at the feedback point.
3. For a negative-feedback amplifier to be stable, that is, for its poles to be located in the left-hand side of the s-plane, the loop gain must be less than 1 by the time the phase shift through the A_{OL} and β networks reaches −180°.

Using these results, we may readily determine the stability of any particular feedback amplifier from an examination of the open-loop gain–phase characteristics of the amplifier and the gain–phase characteristics of the feedback network.

To illustrate how this analysis may be carried out, consider the simple op amp feedback circuit illustrated in Figure 8.3-1a. The open-loop gain–phase characteristics for this amplifier are given in part (b) of the figure. To assess the stability of this circuit, it is useful to rewrite equation (8.3-6) as

$$\left|A_{OL}(j\omega)\right| < \frac{1}{\left|\beta(j\omega)\right|} \tag{8.3-8}$$

and to plot $\left|1/\beta(j\omega)\right|$ together with $\left|A_{OL}(j\omega)\right|$. In addition, we will also sketch the total phase shift $\phi_T(j\omega) = [\phi_A(j\omega) + \phi_\beta(j\omega)]$ directly underneath the magnitude plots (Figure 8.3-1c). Following this procedure, for this particular circuit we have

$$\beta(j\omega) = \frac{R_2}{R_2 + R_f} \tag{8.3-9}$$

or

$$\left|\frac{1}{\beta(j\omega)}\right| = \frac{R_2 + R_f}{R_2} \tag{8.3-10a}$$

and

$$\phi_\beta(j\omega) = 0° \tag{8.3-10b}$$

Notice that because the feedback network is resistive, the gain plot for $\left|1/\beta\right|$ is a constant independent of frequency, and also that the contribution of $\phi_\beta(j\omega)$ to the overall loop phase shift, $\phi_T(j\omega)$, is zero.

In analyzing these sketches it is important to observe that the intersection of the curves for $\left|A_{OL}(j\omega)\right|$ and $\left|1/\beta(j\omega)\right|$ defines the point where

$$\left|A_{OL}(j\omega)\right| = \left|\frac{1}{\beta(j\omega)}\right| \tag{8.3-11a}$$

or where

$$\left|\beta(j\omega)A_{OL}(j\omega)\right| = 1 \tag{8.3-11b}$$

To the left of this point, $\left|A_{OL}(j\omega)\right| > \left|1/\beta(j\omega)\right|$ or $\left|A_{OL}(j\omega)\beta(j\omega)\right| > 1$, and to the right of this point $\left|A_{OL}(j\omega)\right| < \left|1/\beta(j\omega)\right|$ or $\left|\beta(j\omega)A_{OL}(j\omega)\right| < 1$. The am-

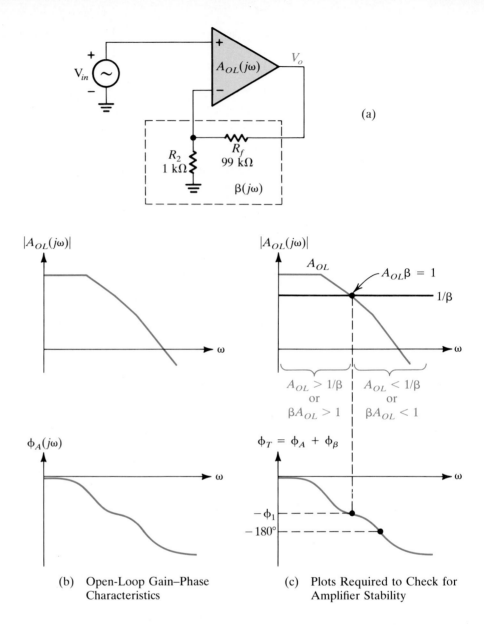

Figure 8.3-1

(a)

(b) Open-Loop Gain–Phase
Characteristics

(c) Plots Required to Check for
Amplifier Stability

plifier stability is determined by comparing the position of this intersection point to that where the total loop phase shift, $\phi_T(j\omega)$, is $-180°$.

In general, there are two different methods that can be employed to determine whether or not a particular amplifier circuit is stable. Both make use of the gain and phase information just discussed, but they analyze the data in slightly different ways. Both methods produce the same results.

To apply the first technique:

1. Using Figure 8.3-2a, determine the frequency at which the overall phase shift, $\phi_T(j\omega)$, is $-180°$.
2. Project a dashed line from this point up onto the magnitude plot and determine where it intersects $|A_{OL}(j\omega)|$ and $|1/\beta(j\omega)|$.

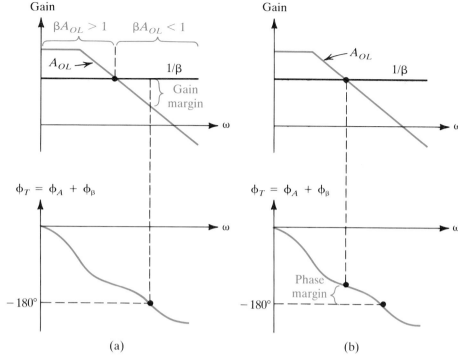

Figure 8.3-2
Gain and phase margin of a
feedback amplifier.

3. **(a)** If the dashed line passes to the left of the intersection point of $|A_{OL}|$ and $|1/\beta|$ then $|\beta(j\omega)A_{OL}(j\omega)| > 1$, where $\phi_T = -180°$, and the amplifier is unstable with poles in the right-hand plane.
   (b) If the dashed line passes to the right of the intersection, $|\beta(j\omega)A_{OL}(j\omega)| < 1$, where $\phi_T = -180°$, and the amplifier is stable with all its poles located in the left-hand plane.
4. For an amplifier that is stable, the amount by which $|1/\beta|$ exceeds $|A_{OL}|$ at the $-180°$ phase point is called the gain margin.

To apply the second technique:

1. Following Figure 8.3-2b, determine the frequency at which $|A_{OL}(j\omega)|$ and $|1/\beta(j\omega)|$ intersect.
2. Project a dashed line from this intersection point down onto the phase plot.
   (a) If at this point $|\phi_A(j\omega) + \phi_\beta(j\omega)| > 180°$, the amplifier is unstable, because to the left of this point ϕ_T will be $-180°$ where $|\beta(j\omega)A_{OL}(j\omega)| > 1$. For this condition the amplifier has poles in the right-hand plane.
   (b) If at the intersection point $|\phi_A(j\omega) + \phi_\beta(j\omega)| < 180°$, the amplifier will be stable because by the time $\phi_T(j\omega)$ reaches $-180°$ (to the right of this point), $|\beta(j\omega)A_{OL}(j\omega)|$ will be less than 1.
3. If the amplifier is stable, the amount by which the overall phase shift differs from $-180°$ at the point where $|A_{OL}| = |1/\beta|$ is called the phase margin.

In designing a particular amplifier circuit, it is clear that we must keep the amplifier poles off the $j\omega$ axis and out of the right-hand plane. However, although this is a necessary condition to guarantee that the amplifier is stable, it does not ensure that amplifier's performance will be satisfactory. If the amplifier poles, although in the left-hand plane, are "too close" to the $j\omega$ axis, several undesirable effects can occur. First, with the poles located this close to the $j\omega$ axis, small changes in the circuit's component values can cause the poles to cross over into the right-hand plane and make the circuit oscillate. Furthermore, even if the poles do not move, the fact that they are close to the $j\omega$ axis is still undesirable. In particular, this type of amplifier exhibits pronounced overshoot and ringing in its transient response, as well as a sharp peaking in its steady-state frequency response in the vicinity of ω_o (see Figure 8.2-4c). To avoid these effects, it is necessary that the amplifier poles be located a reasonable distance away from the $j\omega$ axis.

The amplifier's gain and phase margins provide us with a quantitative measure of the distance of the poles from the imaginary axis in the s-plane. When the gain and phase margins are nearly zero, this means that the $-180°$ phase point in ϕ_T almost coincides with the $|A_{OL}| = |1/\beta|$ intersection point, and this of course indicates that the amplifier poles are very close to the $j\omega$ axis. As the values of the gain and phase margins increase, the poles move farther into the left-hand side of the s-plane and the amplifier is more stable. As a general rule of thumb, it is advisable to make sure that the gain margin is at least 10 dB and that the phase margin exceeds 45°.

The use of these guidelines will help to ensure that the transient and frequency response of our amplifier designs are acceptable. For example, for the case of an amplifier that has two of its poles near the $j\omega$ axis with the others located much farther away, satisfaction of the gain and phase margins requirements just discussed will result in a frequency response that exhibits less than 3 dB of peaking, and a step response that has less that 30% overshoot.

<table>
<tr><td>EXAMPLE 8.3-1</td><td>The feedback amplifier illustrated in Figure 8.2-6 has an open-loop transfer function equal to</td></tr>
</table>

$$A_{OL}(s) = \frac{A_o \alpha^3}{(s + \alpha)^3} = \frac{10^4 \alpha^3}{(s + \alpha)^3} \qquad (8.3\text{-}12)$$

In Section 8.2 we demonstrated that sinusoidal oscillations in this circuit were possible for

$$\beta A_o = 8 \qquad (8.3\text{-}13)$$

[equation (8.2-29)]. Furthermore, for this value of dc loop gain magnitude, the frequency of oscillation was shown in equation (8.2-28) to be

$$\omega_0 = \sqrt{3}\alpha \qquad (8.3\text{-}14)$$

For the specific case where $\alpha = 10^3$ rad/s, demonstrate the correctness of equations (8.3-13) and (8.3-14) using Bode plot techniques.

SOLUTION

The open-loop gain–phase characteristics for the amplifier described by equation (8.3-12) are given in Figure 8.3-3a for the case where α is equal to 10^3 rad/s. If the feedback is adjusted to place two of the poles on the $j\omega$ axis, then for this value of β, the amplifier phase margin will just be equal to zero. With this amount of feedback in place, the $1/\beta$ line will intersect the magnitude plot of A_{OL} at the frequency where the amplifier phase shift is $-180°$.

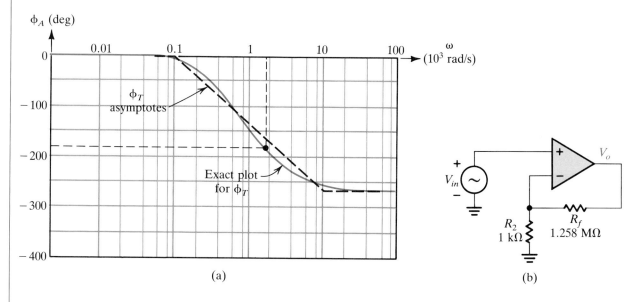

(a)

(b)

Figure 8.3-3
Example 8.3-1.

If we assume that the circuit has the form shown in Figure 8.3-3b, the $1/\beta$ line may be sketched as illustrated in Figure 8.3-3a. It is worth mentioning here that overall loop phase shift, ϕ_T, is the same as ϕ_A for this type of feedback network because ϕ_β is zero. From the magnitude plot we can see that the critical value of $1/\beta$ which places the poles on the $j\omega$ axis is 62 dB, so that

$$\frac{1}{\beta} = (10)^{62/20} = 1259 \tag{8.3-15}$$

This value of $1/\beta$ leads to a dc (or low-frequency) loop gain magnitude of

$$\beta A_o = \left(\frac{1}{1259}\right)(10^4) \simeq 8 \tag{8.3-16}$$

which in the preceding section was shown to be the exact value of βA_o required to place the closed-loop amplifier poles on the $j\omega$ axis. The frequency at which $\phi_T = -180°$ is 1.73×10^3 rad/s, and because this represents the frequency at which the loop-gain is 1, it is also precisely the frequency at which the circuit will oscillate for the given value of β. This is, of course, the same result that we would obtain if we substituted $\alpha = 10^3$ rad/s into equation (8.3-14).

EXAMPLE 8.3-2

Figure 8.3-4a through c illustrates three amplifier circuits, each having a different closed-loop gain. The open-loop transfer function for the op amp in each of these circuits is assumed to be of the form

$$A_{OL}(s) = \frac{10^4}{(1 + s/2\pi \times 10^6)(1 + s/2\pi \times 5 \times 10^6)(1 + s/2\pi \times 30 \times 10^6)} \quad (8.3\text{-}17a)$$

or

$$A_{OL}(jf) = \frac{10^4}{(1 + jf/1)(1 + jf/5)(1 + jf/30)} \quad (8.3\text{-}17b)$$

where f is the frequency in megahertz.

a. Determine the stability of each of these circuits.
b. For circuits that are stable, indicate the gain and phase margins, and comment on the amplifier's expected performance based on these results.
c. For the case of purely resistive feedback, find the allowed range of closed-loop gains for which the resulting amplifier is stable.

SOLUTION

The open-loop gain–phase characteristics are given in Figure 8.3-4d, and the $1/\beta$ lines associated with each of the amplifiers being examined are also shown on the magnitude plot in the figure. The stability of each of these circuits can be determined by examining the overall phase shift ϕ_T where the $1/\beta$ line intersects A_{OL}. These results are summarized in Figure 8.3-4e.

In particular, they demonstrate that circuits 2 and 3 will be unstable because the overall phase shift at the intersection point exceeds $-180°$. This means that at some lower frequency where the phase shift is precisely $-180°$, the loop gain will exceed 1, indicating that these circuits have poles in the right-hand plane. Interestingly enough, although these two amplifiers are unstable, circuit 1 is relatively stable. These results demonstrate that circuits

Figure 8.3-4
Example 8.3-2.

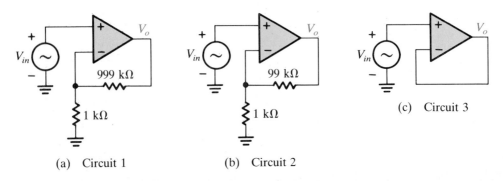

(a) Circuit 1   (b) Circuit 2   (c) Circuit 3

Figure 8.3-4 (continued)

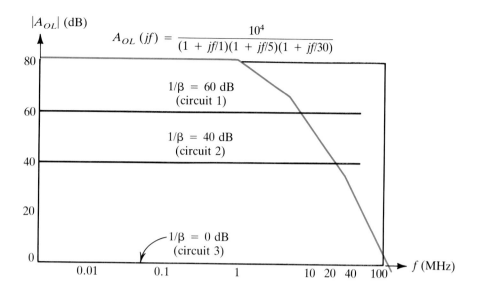

$$A_{OL}(jf) = \frac{10^4}{(1 + jf/1)(1 + jf/5)(1 + jf/30)}$$

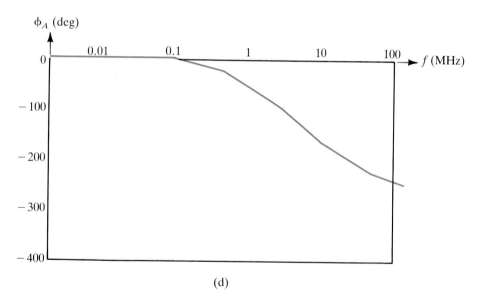

(d)

Circuit	$1/\beta$ (dB) ($\approx A_{CL}$)	$A_{OL} = 1/\beta$ **intersection** frequency (MHz)	ϕ_T **at** intersection	Comment
1	60	7.2	$-154°$	Stable
2	40	22	$-220°$	Unstable
3	0	100	$-260°$	Unstable

(e)

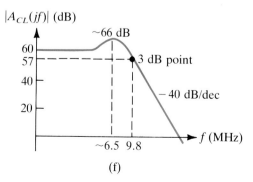

(f)

where A_{CL} (and hence $1/\beta$) is large tend to be stable, while those for which A_{CL} is small are more prone to oscillate.

Circuits with small values of closed-loop gain require large values of β. Since circuits of this type have higher loop gains, it is much more likely that $|\beta A_{OL}|$ will exceed 1 at the frequency where the overall loop phase shift is $-180°$. Therefore, circuits that have large amounts of feedback, and hence small values of A_{CL}, are the most likely candidates to oscillate. In fact, the voltage follower in Figure 8.3-4c is probably *the* amplifier that is most likely to be unstable, because in this circuit the entire output is fed back to the input. The validity of this argument is attested to by the fact that of all three circuits just examined, circuit 3 exhibits the greatest phase shift at the point where $A_{OL} = 1/\beta$.

Amplifier circuit 1, on the other hand, is seen to be stable because very little of its output is fed back to the input. As a result, by the time the overall phase shift in this amplifier reaches $-180°$, the loop gain is less than 1, corresponding to pole locations in the left-hand portion of the s-plane. However, because the phase shift at the point where $|1/\beta| = |A_{OL}|$ is $-158°$, the phase margin of this amplifier is only $22°$, which is a bit low. The corresponding gain margin (as indicated in the figure) is 6 dB, also a bit on the low side. Therefore, although this amplifier is stable, some of its poles are rather close to the $j\omega$ axis, and we should expect that its transient response will ring, and in addition that its frequency response will have a peak in the vicinity of the pole position. This fact is borne out in the closed-loop gain plot given for this amplifier in Figure 8.3-4f; the curve has a 6-dB peak at about 6.5 MHz.

It is useful to note, for the case where the amplifier is stable, that the closed-loop frequency response is approximately equal to $1/\beta(j\omega)$ below the intersection point, and to $A_{OL}(j\omega)$ above that point. As a result, following the development presented in Section 6.7 (see especially Example 6.7-1), we find for the case where the feedback is resistive and the peak in the magnitude plot is small that the amplifier 3-dB point is located approximately at

$$f_H = f_0 \sqrt{2^{1/n} - 1} \qquad (8.3\text{-}18a)$$

where

$$n = \frac{|\text{slope of } |A_{OL}| \text{ at the intersection point}|}{20 \text{ dB/decade}} \qquad (8.3\text{-}18b)$$

Applying this result to circuit 1, we would estimate the high-frequency 3-dB point of this amplifier as

$$f_H = (7 \text{ MHz})\sqrt{2^{1/2} - 1} = 4.5 \text{ MHz} \qquad (8.3\text{-}19)$$

because the slope of A_{OL} at the intersection point is approximately -40 dB/decade. The exact location of the 3-dB point for this amplifier turns out to be at about 9.8 MHz. The large discrepancy between these two results is due to the fact that the peaking in this response moves the 3-dB point out beyond the intersection frequency. For amplifiers where the gain and phase margins are within the previously defined 10-dB and $45°$ limits, peaking will be virtually nonexistent and equation (8.3-18a) will give fairly accurate results.

Let's determine the smallest closed-loop amplifier gain for which this non-inverting amplifier circuit is stable. If we assume that zero phase margin can be tolerated (which of course in reality is totally unacceptable), then the lowest

$1/\beta$ line for which the circuit is stable may be found by noting that the $-180°$ point for this amplifier occurs at a frequency of 10 MHz. The corresponding open-loop gain at this frequency is 54 dB. Thus any amplifier built with this op amp having a closed-loop gain greater than 500 (or 54 dB) will be stable. For A_{CL} values smaller than this, the amplifier poles will move into the right-hand plane and the circuit will oscillate. If a phase margin of 45° is required, the minimum allowed closed-loop gain for this circuit will be 63 dB or 1412.

Two important questions arise in regard to the general subject of feedback amplifier stability. First, since a negative-feedback amplifier circuit cannot oscillate unless there is an overall phase shift of $-180°$, why not just design the amplifier so that it never has this much phase shift? Interestingly enough, for amplifiers where the required value of A_{OL} is small, this is precisely what can be done. In a single-stage amplifier, if we assume that it has one dominant pole, at high frequencies the maximum phase shift (beyond the $-180°$ associated with the signal inversion) that can be produced in the output is about $-90°$. Such an amplifier will therefore be unconditionally stable regardless of the amount of negative feedback that is used. However, when high gains are needed (as for the case of an op amp) a multistage design is often required, and for this situation the overall phase shift at high frequencies can easily exceed $-180°$ and the circuit can oscillate.

A second important question that needs to be answered is whether or not there is anything that we as designers can do to overcome the inherent instability associated with high-gain multistage feedback amplifiers. Fortunately, the answer to this question is *yes*. The technique employed to accomplish this circuit stabilization, known as amplifier compensation, is the subject of the remainder of this section.

Several different amplifier compensation methods are available; basically, they all involve a modification of either A_{OL} or β to ensure that the magnitude of the loop gain is less than one by the time the loop phase shift reaches 0°. Of these two approaches, the one in which the open-loop gain characteristics of the amplifier are altered is most commonly used. This method is popular because it affords isolation between the compensating network and the external circuitry connected to the amplifier. A second advantage of this technique is that existing pole locations within A_{OL} can be modified rather than new ones added and old ones canceled, in order to optimize circuit performance. In general, pole movement rather than pole cancellation is easier to accomplish.

In the remainder of this section we will examine one specific compensation method, known as dominant-pole compensation, which is by far the most popular op amp compensation scheme. As we shall see, this particular compensation technique virtually guarantees that the circuit poles will remain in the left-hand plane. However, a rather high price is paid for this stability since, as we will demonstrate, the closed-loop bandwidth is extremely narrow. While many other compensation methods are available, which essentially trade increased bandwidth for reduced stability, detailed investigation of these circuits will have to be left for the problems at the end of the chapter.

In analyzing the effect of dominant-pole compensation on the performance of a particular op amp circuit, we will assume that the open-loop amplifier has the simplified block diagram form illustrated in Figure 8.3-5, where the dc or low-frequency gain $A_1A_2A_3A_4$ is equal to 10^4, and the poles f_1, f_2, and f_3 are located at 1, 5, and 30 MHz, respectively. As a result, the open-loop transfer function has the same form as that presented in Example 8.3-2 [see equation (8.3-17b)]. To reiterate this expression, we may write the transfer function for this amplifier as

$$A_{OL}(jf) = (10^4)\underbrace{\frac{1}{1 + jf/1}}_{\substack{\text{first}\\\text{stage}}}\underbrace{\frac{1}{1 + jf/5}}_{\substack{\text{second}\\\text{stage}}}\underbrace{\frac{1}{1 + jf/30}}_{\substack{\text{third}\\\text{stage}}} \qquad (8.3\text{-}20)$$

The open-loop gain–phase characteristics of this amplifier were given in Figure 8.3-4d.

In analyzing this amplifier, we showed that for it to be stable (with 45° of phase margin), the minimum allowable gain was 63 dB or 1412. Unfortunately, many circuit applications require much lower closed-loop gains, and for such an amplifier, because more of the output is fed back to the input, it will be necessary to modify A_{OL} to attenuate it at high frequencies to ensure that the loop gain falls below unity by the time the phase shift reaches $-180°$. One of the simplest ways of achieving this type of attenuation is by means of a compensation technique known as the dominant-pole method.

In the dominant-pole approach, a large capacitor is placed, for example, from the terminal labeled A in Figure 8.3-5 to ground. Because this capacitor is in parallel with C_1, it lowers the break frequency of the pole at f_1 and causes the gain of A_{OL} to begin to fall off at a much lower frequency. The idea behind this compensation scheme is to ensure that A_{OL} has fallen all the way to 0 dB before the second pole at f_2 turns on. This will guarantee that even a voltage-follower amplifier (with $|1/\beta| = 0$ dB) will have at least 45° of phase margin.

For A_{OL} to reach unity gain by the time the second pole turns on at $f_2 = 5$ MHz, it is necessary that the dominant pole be located at least 4 decades in front of f_2, or more specifically, at $f_1' = 500$ Hz, since A_{OL} starts at 80 dB and falls off at 20 dB/decade in this region (see Figure 8.3-6a). With f_1 moved to f_1' in this way, the $1/\beta$ line for a voltage follower (along 0 dB) would intersect A_{OL} at about 5 MHz. At this frequency the phase contribution from the pole at

Figure 8.3-5

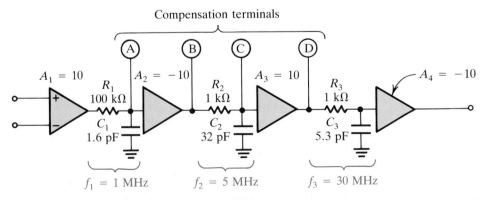

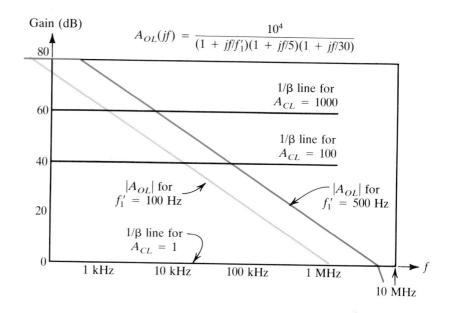

$$A_{OL}(jf) = \frac{10^4}{(1 + jf/f_1')(1 + jf/5)(1 + jf/30)}$$

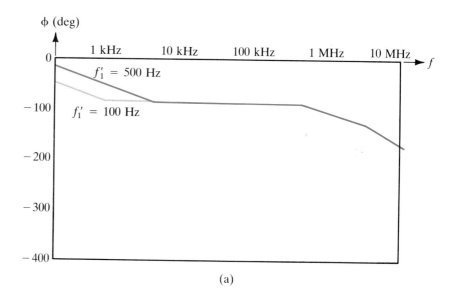

(a)

A_{CL}	Phase margin	Bandwidth	Comment
1000	90°	1 kHz	Stable
100	90°	10 kHz	Stable
10	90°	100 kHz	Stable
1	75°	1 MHz	Stable

Figure 8.3-6
Dominant-pole compensation.

(b) Amplifier Data for Dominant Pole at $f_1' = 100$ Hz

f_1' would be $-90°$, and that from f_2 would be $-45°$, leading to an apparently satisfactory overall phase shift of $-135°$. However, at a frequency of 5 MHz, there is an additional phase contribution from the pole f_3 located at 30 MHz (of about $-9.5°$), and this will cause the amplifier phase margin to fall below $45°$. Therefore, as a safety measure, we will place the dominant pole of stage 1 at 100 Hz rather than at 500 Hz.

For $f_1' = 100$ Hz, the compensation capacitor added in parallel with C_1 will need to be $1/(2\pi \times 100 \times 10^5)$ or about 0.0016 μF, and with this capacitor in place the effective open-loop amplifier transfer function will be

$$A_{OL}(jf) = \frac{10^4}{(1 + jf/0.0001)(1 + jf/5)(1 + jf/30)} \tag{8.3-21}$$

The gain–phase characteristics associated with this transfer function are also sketched in Figure 8.3-6a. In addition, the $|1/\beta|$ lines corresponding to amplifier closed-loop gains of 1000, 100, 10, and 1 are also indicated in the figure, and because each of these lines intersects A_{OL} along the -20-dB/decade portion of the curve, all these amplifiers will be stable. Furthermore, and again because the intersection occurs along the -20-dB/decade slope of A_{OL}, the closed-loop bandwidth [see equation (8.3-18a)] is equal to the value of f at the intersection point. These results are summarized in Figure 8.3-6b.

Although each of these amplifier circuits is stable with dominant-pole compensation, the bandwidth is considerably smaller than it was for the uncompensated amplifier. For example, for the case where $A_{CL} = 1000$, the uncompensated closed-loop amplifier bandwidth is approximately $7\sqrt{2^{1/2} - 1}$ or 4.5 MHz. With dominant-pole compensation, this bandwidth is reduced to only 1 kHz. Thus with this compensation scheme we have made a significant trade-off of amplifier bandwidth for amplifier stability. Of equal importance but less apparent initially is the fact that the amplifier slew rate has also been severely curtailed by the addition of this compensation capacitor.

For many instrumentation and low-frequency signal-processing applications, high slew capability and wide bandwidths are not nearly as important as the certain knowledge that the amplifier circuit you are working with "will never oscillate." Because this feature is so popular, nearly every op amp manufacturer makes one or more internally compensated operational amplifiers which have built-in dominant-pole compensation. Specific examples in this op amp category include the LM741, LM307, and LM324 devices.

In compensating the amplifier in Figure 8.3-5 for dominant-pole operation, we needed to connect a 1600-pF capacitor from point A in the circuit to ground. While this capacitor is a relatively small component when added externally to an op amp, it turns out to be a rather large capacitor to integrate directly into an IC chip. As a result, rather than connecting the capacitor between point A and ground, it it frequently connected between points A and B or A and D in the figure. In this way, the effective size of the capacitor used is magnified by the Miller gain of the circuit when it is reflected down in parallel with C_1. Specifically, for the case where it is connected between terminals A and D in the amplifier, this would allow us to use a capacitor of only 16 pF rather than 1600 pF to achieve the desired dominant-pole compensation.

EXAMPLE 8.3-3

Sometimes when an amplifier drives a capacitive load, the additional phase lag produced at the output terminal can cause the amplifier to become unstable. Demonstrate this point by reexamining the dominant-pole amplifier just discussed for the voltage-follower circuit illustrated in Figure 8.3-7, in which the amplifier is driving a 1000-pF load. In carrying out this analysis, assume that the op amp's open-loop output impedance is 100 Ω.

SOLUTION

As discussed previously (see Figure 8.3-6b), with no capacitive loading the phase margin of this voltage follower, with dominant-pole compensation, is about 75°. This leads to a good transient response and a flat frequency response out to about 1 MHz. The addition of capacitive loading creates problems because of the additional phase lag produced in the network consisting of R_o and C_L. In effect, this modifies the amplifier's open-loop gain from A_{OL} to A'_{OL}, where, as per Figure 8.3-6b, the transfer function for A'_{OL} may be written as

$$A'_{OL}(s) = [A_{OL}(s)]\frac{1/sC_L}{R_o + 1/sC_L} = A_{OL}(s)\frac{1}{1 + s/\omega_L} \tag{8.3-22}$$

where $\omega_L = 10^7$ rad/s. Substituting equation (8.3-21) into this expression, we have

$$A'_{OL}(jf) = \frac{10^4}{(1 + jf/0.0001)(1 + jf/5)(1 + jf/30)}\frac{1}{1 + jf/1.6} \tag{8.3-23}$$

Thus the capacitive loading has introduced an additional pole at a rather low frequency (1.6 MHz). The phase contribution of this pole will be significant for low-gain amplifiers such as the follower circuit that we are currently examining.

As for the case where C_L was zero, the voltage follower $|1/\beta|$ line still intersects the open-loop gain characteristic at 1 MHz (Figure 8.3-7c). However, the phase shift at this point is more severe due to the capacitive loading. The phase contributions from the poles at 100 Hz and 5 MHz are still −90° and

Figure 8.3-7
Example 8.3-3.

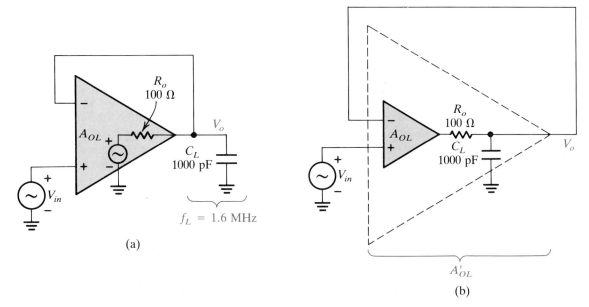

(a)

(b)

Figure 8.3-7 (*continued*).

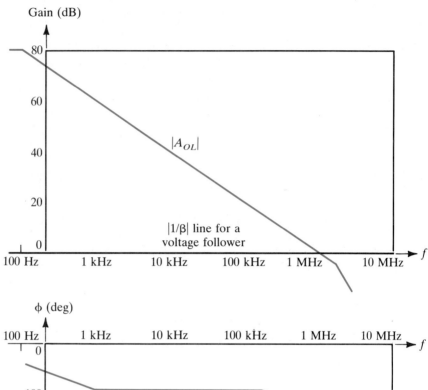

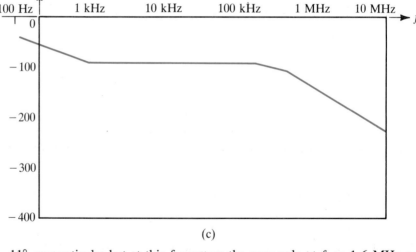

(c)

$-11°$, respectively; but at this frequency the new pole at $f_L = 1.6$ MHz contributes an additional $-32°$. As a result, the overall phase shift at 1 MHz is $-133°$ and the phase margin is $47°$. Thus, the capacitive loading on the output phase reduced the phase margin of this amplifier from $75°$ to only $47°$. Additional capacitive loading beyond this point will further degrade the transient performance of this amplifier, and if large enough could even cause it to oscillate.

Exercises

8.3-1 A noninverting amplifier with $R_f = 500$ kΩ and $R_2 = 1$ kΩ has the gain phase characteristics given in Figure 8.3-4d. Determine the circuit phase margin. **Answer** approximately $20°$

8.3-2 An amplifier having a dominant pole at 500 Hz (Figure 8.3-6a) is to be designed to have an inverting gain of 300. What are the amplifier phase margin and bandwidth? *Answer* 90°, 15 kHz

8.3-3 The op amp illustrated in Figure 8.3-5 is modified by adding a capacitor in parallel with C_1 to lower the pole at f_1 to 100 kHz. A noninverting amplifier is built using this op amp with $R_2 = 1$ kΩ and R_f adjustable. Find the smallest value of R_f for which this amplifier has gain and phase margins of at least 10 dB and 50°, respectively. *Answer* 250 kΩ

8.4   SINUSOIDAL OSCILLATORS

An oscillator is a circuit that produces an output without the application of an input signal. As we discussed in some detail in Section 8.2, this type of circuit behavior is associated with a feedback amplifier whose loop gain is adjusted to place one or more of its poles either directly on the imaginary axis or slightly in the right-hand plane. With the poles located on the $j\omega$ axis, the amplitude of oscillation is constant and is determined by the circuit's initial conditions. If, on the other hand, the poles are placed (even slightly) in the right-hand plane, the oscillation amplitude will grow until it is eventually limited by the circuit's nonlinearities. Techniques for determining the amplitude of oscillation are discussed in Section 8.5.

The purpose of this section is to examine the behavior of different types of sinusoidal oscillator circuits and to investigate the factors governing their design. In particular, we will be interested in developing techniques for determining the frequency of oscillation of these circuits as well as the amplifier gains needed to just place their poles on the $j\omega$ axis.

In Sections 8.2 and 8.3 we investigated the basic concepts involved in determining the stability (or instability) of feedback amplifier circuits, and in carrying out these investigations we found that one of several different analytical techniques could be employed. Specifically, we saw that the circuit's performance could be explained in terms of the amplifier's closed-loop transfer function, its Bode gain–phase plot, or in terms of the circuit's overall loop gain. The method to use for a particular circuit analysis or design depends on whether we are interested in the amplifier's transient behavior, its stability, or its performance as an oscillator.

If we wish to examine a feedback amplifier's transient response, then, as we saw in Section 8.2, it is probably best to use the Laplace transform technique together with a determination of the amplifier's closed-loop transfer function. If, on the other hand, we need to determine the stability of a particular feedback amplifier circuit, it should be clear from Section 8.3 that the gain–phase approach is probably the best way to proceed. Finally, for investigating the behavior of sinusoidal oscillators, the signal injection method appears to offer us the least complex way to understand the behavior of these circuits.

To carry out the signal injection approach, basically what needs to be done is to break the feedback loop at some point, inject a sinusoidal signal at the desired frequency of oscillation, ω_0, and trace its amplification as it travels around the loop. If the circuit is to be capable of oscillating at ω_0, it is necessary that this signal, after traveling completely around the loop, return in phase, and with the same amplitude as the original injected signal. If the circuit parameters are adjusted so that these criteria are satisfied, the amplifier's closed-loop transfer function will be found to have a pair of poles on the imaginary axis at $\pm j\omega_0$. If the signal returns back in phase but with a larger amplitude, these poles will be located in the right-hand plane.

Thus to build a sinusoidal oscillator, we need the loop gain to be equal to 1 only at the frequency of oscillation ω_0. A signal injected at any other frequency must, after traveling around the loop, return with an amplitude smaller than that with which it started. This, of course, requires that either the amplifier itself, or alternatively, the feedback circuit, be frequency selective so that the loop gain will be less than 1 at all frequencies other than ω_0.

The basic technique that we will employ for the analysis and design of all the oscillator circuits discussed in the remainder of this section may be summarized as follows:

1. Break the loop at some point in the circuit (Figure 8.4-1).
2. Take loading into account by connecting an impedance Z_{in} on the output side of the break (v'_{in}) equal to the impedance (Z_{in}) that formerly loaded this point in the circuit prior to breaking the loop.
3. Inject a signal, v_{in}, at the oscillation frequency ω_0, and trace the signal around the loop.
4. Select the circuit component values so that

$$v'_{in}(t) = v_{in}(t) \qquad (8.4\text{-}1)$$

at the desired frequency ω_0, and so that $v'_{in} < v_{in}$ at all other frequencies.
5. In tracing the signal around the loop, use Laplace transform notation and set $s = j\omega_0$ to evaluate equation (8.4-1) since we are interested in having a sinusoidal steady-state loop gain of 1 at the frequency ω_0. In other words, the criterion for sinusoidal circuit oscillation will be

$$\left. \frac{V'_{in}(s)}{V_{in}(s)} \right|_{s=j\omega_0} = 1 \qquad (8.4\text{-}2)$$

Figure 8.4-1

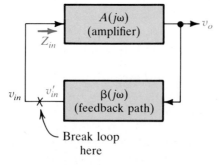

(a) Basic Form of an
Oscillator Circuit

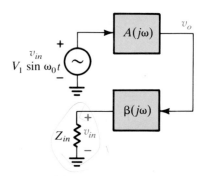

(b) General Oscillator
Analysis Technique

In the remainder of this section we examine a rather detailed series of examples in which we repeatedly apply the oscillator circuit analysis guidelines just described. Some of the oscillator circuits to be examined contain tuned circuits; others also include nonideal transformers. If you are unsure of how to handle circuits containing these types of circuit elements, it is suggested that you review these concepts by reading Sections 9 through 11 of Appendix I. Specifically, these sections of Appendix I contain material on series and parallel tuned circuits, network models for transformers, and network analysis techniques for circuits exhibiting transformer-like behavior.

We begin our examination of specific oscillators with a study of RC-style circuits. These types of oscillators are useful for low-frequency designs in the range from a fraction of a hertz to several hundred kilohertz. Above this frequency the R and C values required become so small that they are comparable to the parasitic elements present in the circuit, and in these situations, accurate control of the oscillation frequency and amplitude is no longer possible. In addition, very small R values will cause significant loading on the amplifier portion of the oscillator, making it difficult to achieve the required loop gain.

When high-frequency sinusoidal oscillators are needed in the range from several hundred kilohertz to several hundred megahertz, some type of tuned circuit oscillator is usually employed. As we shall see in Section 8.5, the accuracy of the oscillation frequency and the stability of the oscillation obtained are determined by the sharpness of tuning, or the Q, of the tuned circuit. When extremely accurate oscillation frequencies are required (better than about 1 part in 100 or 1%), crystal oscillators are usually employed. In this section we examine several different types of tuned circuit oscillators; however, we will save the crystal-controlled oscillator for Section 8.6, where we discuss oscillator frequency-stabilization techniques.

EXAMPLE 8.4-1

The RC oscillator circuit illustrated Figure 8.4-2 has been examined in Sections 8.2 and 8.3. In Example 8.2-1 we showed that under the right conditions the poles of this circuit could be placed on the $j\omega$ axis, and in Section 8.3 (Example 8.3-1) we repeated the analysis of this circuit, but from an amplifier stability viewpoint, to demonstrate that this circuit could become unstable under the right circuit conditions.

Repeat the analysis of this oscillator circuit again, but this time use the guidelines given above to determine the frequency of oscillation and the value of R_2 required to just cause this circuit to oscillate.

SOLUTION

In analyzing this circuit, we first need to choose the point where the loop is to be broken. This should be done so that the impedance, seen where V_{in} is connected, is purely resistive since this will generally lead to the simplest analysis for the loop gain. Following this guideline, we have chosen to break the loop at the output of IC4 (the spot marked by the "×" in Figure 8.4-2a), although any of the op amp inputs at V_2, V_4, or V_6 could also have been used.

Figure 8.4-2b illustrates the circuit to be examined after the loop has been broken and a voltage V_{in} connected to the input of IC1. Notice that the equivalent input impedance (R_2) seen at the V_{in} terminal has been added to the circuit output at V'_{in} in order to take loading effects into account. For this particular circuit, this loading on the output does not matter because it is connected to the

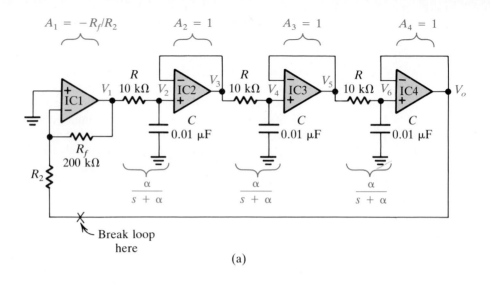

(a)

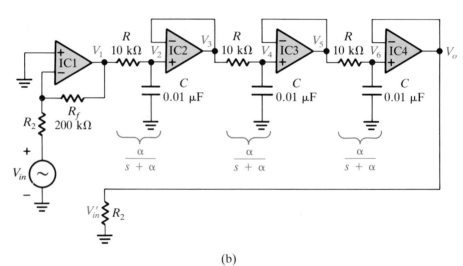

Figure 8.4-2
Example 8.4-1.

(b)

op amp IC4, which is assumed to have zero output impedance, but in other circuits we examine, the effect of loading will be extremely important.

Tracing this signal around the loop, we have

$$V_1 = -\frac{R_f}{R_2} V_{in} \tag{8.4-3a}$$

$$V_3 = \frac{\alpha}{s + \alpha} V_1 = -\frac{R_f}{R_2} \frac{\alpha}{s + \alpha} V_{in} \tag{8.4-3b}$$

$$V_5 = \frac{\alpha}{s + \alpha} V_3 = -\frac{R_f}{R_2} \left(\frac{\alpha}{s + \alpha}\right)^2 V_{in} \tag{8.4-3c}$$

and $\quad V_{in}' = V_o = \frac{\alpha}{s + \alpha} V_5 = -\frac{R_f}{R_2} \left(\frac{\alpha}{s + \alpha}\right)^3 V_{in} \tag{8.4-3d}$

For this circuit to just oscillate at the frequency ω_0, we require that

$$\left.\frac{V'_{in}}{V_{in}}\right|_{s = j\omega_0} = \left.-\frac{R_f}{R_2}\left(\frac{\alpha}{s + \alpha}\right)^3\right|_{s = j\omega_0} = 1 \qquad (8.4\text{-}4)$$

Cross-multiplying both sides by $(s + \alpha)^3$, we obtain

$$(s + \alpha)^3 = s^3 + 3\alpha s^2 + 3\alpha^2 s + \alpha^3 = -\frac{R_f}{R_2}\alpha^3 \qquad (8.4\text{-}5)$$

which on substituting $s = j\omega_0$ yields

$$-j\omega_0^3 - 3\alpha\omega_0^2 + 3j\omega_0\alpha^2 + \alpha^3\left(1 + \frac{R_f}{R_2}\right) = 0 \qquad (8.4\text{-}6a)$$

or $\qquad j(-\omega_0^3 + 3\omega_0\alpha^2) + \left[-3\alpha\omega_0^2 + \alpha^3\left(1 + \frac{R_f}{R_2}\right)\right] = 0 \qquad (8.4\text{-}6b)$

Equation (8.4-6b) is a complex quantity, and for it to be identically zero as indicated, it is necessary that both the real and the imaginary parts each be zero independently. Therefore, to satisfy equation (8.4-6b), we require that

$$\omega_0^3 - 3\omega_0\alpha^2 = 0 \qquad (8.4\text{-}7a)$$

and $\qquad 3\alpha\omega_0^2 - \alpha^3\left(1 + \frac{R_f}{R_2}\right) = 0 \qquad (8.4\text{-}7b)$

Solving equation (8.4-7a) for ω_0, we find that the frequency of oscillation of this circuit is

$$\omega_0 = \sqrt{3}\alpha \qquad (8.4\text{-}8)$$

Using the fact that $\alpha = 1/RC = 10^4$ rad/s, we find that the frequency of oscillation f_0 will be about 2.75 kHz.

Substituting equation (8.4-8) into equation (8.4-7b), we obtain

$$(3\alpha)(\sqrt{3}\alpha)^2 - \alpha^3\left(1 + \frac{R_f}{R_2}\right) = 0 \qquad (8.4\text{-}9a)$$

or $\qquad \frac{R_f}{R_2} = 8 \qquad (8.4\text{-}9b)$

which because $R_f = 200$ kΩ requires that R_2 be about 25 kΩ to just place the system poles on the $j\omega$ axis. Of course, a smaller value of R_2 will increase the loop gain beyond 1 at f_0 and will drive the poles into the right-hand plane.

Comparing these results with those presented in equations (8.2-37) and (8.2-38) from Example 8.2-1, we find that both analyses yield identical results.

EXAMPLE 8.4-2

The circuit illustrated in Figure 8.4-3 is known as a Wien bridge oscillator. It is useful as a low-frequency sinusoidal oscillator, and prior to the advent of function generators (see Section 8.6) was widely used in the design of variable-frequency signal generators.

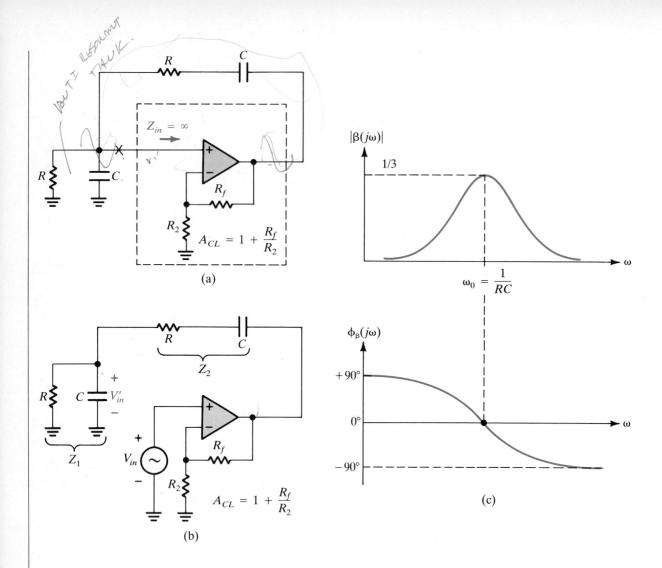

Figure 8.4-3
Example 8.4-2.

a. Determine the frequency of oscillation.
b. Find the op amp closed-loop gain needed to just make this circuit oscillate.
c. Plot the gain–phase characteristics of the feedback network, and discuss how these characteristics relate to the oscillation characteristics of the circuit.

SOLUTION

In analyzing this circuit, it is convenient to break the loop at the noninverting input to the op amp since the load seen to the right of the "×" in the figure is essentially infinite. Injecting our signal V_{in} at this point, we may redraw the circuit to be analyzed as shown in Figure 8.4-3b. Examining this circuit, it is clear that the op amp in this instance simply amplifies the signal V_{in} by

$$A_{CL} = 1 + \frac{R_f}{R_2} \qquad (8.4\text{-}10)$$

and introduces no phase shift. As a result, in tracing the signal around the loop, we have

$$V'_{in} = \frac{Z_1}{Z_1 + Z_2}V_o = A_{CL}\frac{R/(1 + sRC)}{R/(1 + sRC) + (R + 1/sC)}V_{in} \qquad (8.4\text{-}11a)$$

and multiplying above and below in this expression by $(1 + sRC) \times sC$, we obtain

$$V'_{in} = \frac{A_{CL}sRC}{(sRC)^2 + 3(sRC) + 1}V_{in} \qquad (8.4\text{-}11b)$$

For this circuit to just oscillate from equation (8.4-1), we require that $V'_{in} = V_{in}$ or

$$\left.\frac{A_{CL}sRC}{(sRC)^2 + 3(sRC) + 1}\right|_{s=j\omega_0} = 1 \qquad (8.4\text{-}12)$$

Substituting in $s = j\omega_0$, we see that

$$[-(\omega_0 RC)^2 + 1] + j(3\omega_0 RC - \omega_0 A_{CL}RC) = 0 \qquad (8.4\text{-}13)$$

For equation (8.4-13) to be satisfied, we require that the real and imaginary parts each be zero independently. Carrying out these operations leads to

$$\omega_0 = \frac{1}{RC} \qquad (8.4\text{-}14a)$$

for the frequency of oscillation, and

$$A_{CL} = 3 \qquad (8.4\text{-}14b)$$

for the op amp closed-loop gain required to just place the poles on the $j\omega$ axis.

We can obtain some additional understanding of this circuit's performance by examining the behavior of the feedback network as a function of frequency. In particular, we notice that for this circuit

$$\beta(s) = \frac{Z_1}{Z_1 + Z_2} = \frac{sRC}{(sRC)^2 + 3sRC + 1} \qquad (8.4\text{-}15a)$$

so that $\beta(j\omega)$ is given by the expression

$$\beta(j\omega) = \frac{j\omega RC}{[1 - (\omega RC)^2] + 3j\omega RC} \qquad (8.4\text{-}15b)$$

The gain–phase characteristics of $\beta(j\omega)$ are sketched in Figure 8.4-3c. At very low frequencies the Z_2 branch looks like an open circuit, so that there is no feedback. Because $\beta(j\omega) \simeq R/(1/j\omega C)$ at low frequencies, the corresponding phase shift is $+90°$. At high frequencies Z_1 looks like a short circuit, so that the feedback factor again goes to zero. Here $\beta(j\omega) \simeq (1/j\omega C)/R$, so that the phase shift is nearly $-90°$.

For the circuit to oscillate, we want V'_{in} to return in phase with the injected signal V_{in}. This requires that the phase shift through the β network be zero at the frequency of oscillation. Following Figure 8.4-3c, this can occur only at the frequency $\omega = 1/RC$. Furthermore, at this frequency the "gain" of the β network is $\frac{1}{3}$, so that the op amp circuit will need a gain of 3 if the overall circuit loop gain is to be unity.

EXAMPLE 8.4-3

Figure 8.4-4
Example 8.4-3.

The circuit shown in Figure 8.4-4 is a tuned circuit oscillator. It uses a transformer in which a capacitor is placed across the primary to form a parallel tuned circuit. Transformers are useful circuit elements in oscillators because they provide for dc isolation between the input and the output while permitting

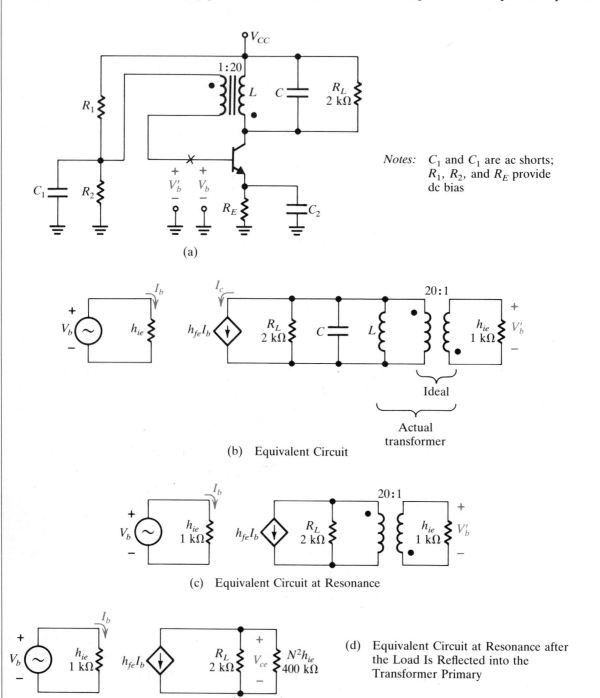

Notes: C_1 and C_1 are ac shorts; R_1, R_2, and R_E provide dc bias

(a)

(b) Equivalent Circuit

(c) Equivalent Circuit at Resonance

(d) Equivalent Circuit at Resonance after the Load Is Reflected into the Transformer Primary

the impedances to be transformed to the levels needed to maintain a high circuit Q. In addition, depending on the dot alignment between the primary and the secondary, the transformer may also be used to provide an additional 180° of phase shift. The flexibility afforded by the use of a transformer permits the selection of practical component values and allows for the design of oscillators whose frequency of oscillation is less dependent on the circuit's parasitic element values.

a. Determine the minimum value of h_{fe} for which the circuit shown in Figure 8.4-4a just oscillates. The transistor h_{ie} may be assumed to be constant at 1 kΩ.

b. Find the transformer primary inductance L and tuned circuit capacitance C for the circuit Q to be about 3 and for the oscillation frequency to be 100 kHz.

SOLUTION

If the loop is broken at the base of the transistor at the point marked "×" in Figure 8.4-4a and a signal V_b is injected at this point, we obtain the equivalent circuit illustrated in part (b) of the figure. Notice in breaking the loop that we have added an impedance h_{ie} at the transformer output in order to include the effect of the transistor loading at the point where the loop is broken. By examining this circuit, it should be clear that a nonzero signal will be fed back to the input of the oscillator at V'_b only at or near the parallel resonant frequency $\omega_0 = 1/\sqrt{LC}$. For frequencies below ω_0, the inductor looks like a short, so that V'_b is zero, and for frequencies above ω_0, the capacitor is essentially a short, so that V'_b will again be zero. At $\omega = \omega_0$ the inductor and the capacitor resonate and may be removed from the circuit. Therefore, at ω_0 the equivalent circuit takes on the form given in Figure 8.4-4c. Figure 8.4-4d illustrates the equivalent circuit after the transformer load impedance, h_{ie}, is reflected into the primary.

Because we want the Q of the tuned circuit to be 3, we find that the required transformer primary inductance is

$$L = \frac{R'_L}{\omega_0 Q} \simeq 1 \text{ mH} \tag{8.4-16a}$$

and the value of C is

$$C = \frac{Q}{\omega_0 R'_L} = 0.0024 \ \mu\text{F} \tag{8.4-16b}$$

To determine the minimum value of h_{fe} needed for this circuit to oscillate, we note following Figure 8.4-4d that

$$V_{ce} = -h_{fe} I_b R'_L = -\frac{h_{fe} V_b R'_L}{h_{ie}} \tag{8.4-17}$$

Because of the phasing of the secondary winding, the voltage returned to the base is not only reduced in amplitude, but is also shifted by 180° from the voltage at the collector, so that we may write

$$V'_b = -\frac{1}{N} V_{ce} = \frac{h_{fe} R'_L}{N h_{ie}} V_b \tag{8.4-18}$$

Therefore, for this circuit to oscillate, we require that

$$\frac{h_{fe} R_L'}{N h_{ie}} \geq 1 \qquad\qquad (8.4\text{-}19a)$$

or

$$h_{fe} \geq \frac{N h_{ie}}{R_L'} \simeq \frac{N h_{ie}}{R_L} = 10 \qquad\qquad (8.4\text{-}19b)$$

as the minimum transistor current gain needed.

EXAMPLE 8.4-4

The circuit illustrated in Figure 8.4-5a is known as a Colpitts oscillator, and it is a popular circuit for the construction of high-frequency oscillators. Rather than using a conventional transformer to feed back the output signal to the input, it makes use of a capacitive divider impedance transformer of the type discussed in Section I.11.

a. Considering that the transistor's h_{fe} is 100 and that its h_{ie} is 5 kΩ, determine the circuit's frequency of oscillation.

b. Find the minimum inductor Q for which this circuit just oscillates.

In this oscillator the tuned circuit consists of L, C_1, and C_2, and the feedback is obtained by returning part of the collector signal to the emitter through the capacitive divider, consisting of C_1 and C_2. In analyzing this circuit it is useful to break the loop at the emitter at the point labeled "×" and to inject the signal into the emitter at this point. This is probably the best place in the circuit to break the feedback loop because the impedance seen looking to the left of the point "×" is purely resistive and does not depend on the resonant characteristics of the tuned circuit. As a result, the loading effects of breaking the loop at this point can readily be taken into account.

With a voltage V_e injected into the loop at this point, the equivalent circuit for calculating the loop gain takes the form shown in Figure 8.4-5b. Notice that the effective load on the loop output is $[(h_{ie}/(1 + h_{fe})) \| R_E]$, and because R_E is so large, this is nearly the same as the transistor's emitter impedance, $h_{ie}/(1 + h_{fe})$, which is about 50 Ω. At first it might appear that this impedance would severely load the capacitive divider; however, because Z_{c2} is so small at the circuit's resonant frequency (shown below to be 10 MHz), the loading of the emitter impedance on C_2 is also small, hence the impedance transformation properties of the C_1–C_2 network developed in Appendix I may be applied to analyze this circuit.

Specifically for this case, the equivalent circuit of the collector reduces to that given in Figure 8.4-5c, in which C, the series connection of C_1 and C_2, is given by

$$C = \frac{C_1 C_2}{C_1 + C_2} \qquad\qquad (8.4\text{-}20)$$

The resistance R_L in the collector represents the parallel combination of the transformed emitter impedance and the effective parallel resistance of the inductor associated with its finite Q. As a result, we may write

$$R_L = (\omega_0 L Q_L) \, \| \left(N^2 \frac{h_{ie}}{1 + h_{fe}} \right) \qquad\qquad (8.4\text{-}21a)$$

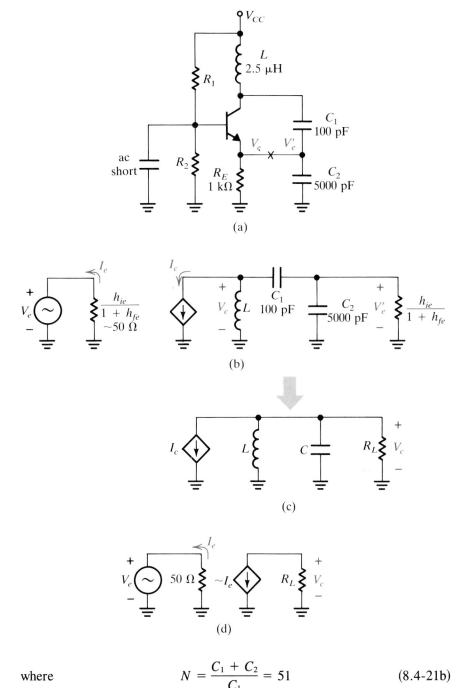

Figure 8.4-5
Example 8.4-4.

where

$$N = \frac{C_1 + C_2}{C_1} = 51 \qquad (8.4\text{-}21b)$$

represents the effective turns ratio of the C_1–C_2 divider network.

Following this equivalent circuit, we can see that a significant voltage V_e' is fed back to the emitter only at the resonant frequency

$$f_0 = \frac{1}{2\pi\sqrt{LC}} = 10 \text{ MHz} \qquad (8.4\text{-}22)$$

At all other frequencies, V_c is nearly zero, so that V'_e, which is approximately $\frac{1}{50}$ of V_c, is also zero. At resonance this circuit reduces to that shown in Figure 8.4-5d, so that we have

$$I_e = -\frac{V_e}{50\ \Omega} \tag{8.4-23a}$$

$$V_c = -I_c R_L \simeq -I_e R_L = V_e \frac{R_L}{50\ \Omega} \tag{8.4-23b}$$

and
$$V'_e = \frac{1}{N}V_c = V_e \frac{R_L}{N(50\ \Omega)} \tag{8.4-23c}$$

For this circuit to oscillate, because we need $V'_e \geq V_e$, we require that the effective collector load be at least

$$R_L \geq N(50\ \Omega) = 2.5\ \text{k}\Omega \tag{8.4-24}$$

Substituting equation (8.4-21a) into this expression, and noting that $N^2 h_{ie}/(1 + h_{fe})$ is so large that it may be neglected, we obtain

$$Q_L \omega_0 L > 2.5\ \text{k}\Omega \tag{8.4-25a}$$

or
$$Q_L > \frac{2.5\ \text{k}\Omega}{(6.28 \times 10^7)(2.5 \times 10^{-6})} \simeq 16 \tag{8.4-25b}$$

Notice that this is also the approximate Q of the overall circuit, so that the impedance transformation approximations that we have made use of in analyzing this circuit should provide highly accurate results (see Section I.11).

Exercises

8.4-1 In Figure 8.4-2, the resistor between nodes V_3 and V_4 is increased to 20 kΩ and R_2 is adjusted so that the circuit just oscillates. Find the frequency of oscillation and the required value of R_2. *Answer* 2250 Hz, 22.2 kΩ

8.4-2 Repeat Example 8.4-3b if the resistor R in Z_1, is increased to $3R$.

Answer $\omega_0 = \dfrac{1}{\sqrt{3}RC}$, $A_{CL} = 2.33$

8.4-3 Assume in Figure 8.4-4a that $R_E = 20\ \Omega$ and that C_2 is removed. Find the minimum value of h_{fe} for which the circuit just oscillates. *Answer* 30

8.4-4 For the circuit discussed in Example 8.4-4 (Figure 8.4-5), the inductor Q is 40. If the inductor is shunted with a resistive load R_L, find the smallest value of R_L for which the circuit just oscillates. *Answer* 12.25 kΩ

8.5 AMPLITUDE STABILIZATION IN OSCILLATORS

For a circuit to oscillate sinusoidally, we require that the system transfer function contain a pair of poles located either on the $j\omega$ axis or slightly in the right-hand plane. With the poles located exactly on the imaginary axis, the circuit will oscillate with an amplitude determined by the circuit's initial conditions.

If the poles are located (even slightly) in the right-hand half of the complex s-plane, then while the oscillation amplitude shortly after $t = 0$ will still be determined by the initial conditions, these oscillations will grow until their amplitude is ultimately limited by the circuit's nonlinearities.

As a practical matter, the oscillator poles must always be located somewhat in the right-hand plane. If this were not the case, slight changes in the circuit parameters due either to aging or to environmental effects could cause these poles to migrate into the left-hand part of the s-plane, and of course if this occurred, the oscillations would cease. Where the poles are relatively close to the $j\omega$ axis, the loop gain of the circuit is still fairly close to 1 and the nonlinearities observed are correspondingly small. When the poles are placed far into the right-hand plane, the small-signal loop gain is large, and as a result, the nonlinear effects observed will also be quite large.

In this section we discuss the behavior of sinusoidal oscillator circuits whose amplitude of oscillation will be adjusted by three different techniques:

1. *Manual adjustment of the loop gain:* to place the poles close to or directly on the $j\omega$ axis, to achieve nearly sinusoidal oscillations
2. *Soft amplitude limiting:* where circuitry is added that automatically adjusts the oscillator loop gain, to keep the poles on the $j\omega$ axis
3. *Hard amplitude limiting:* where the small-signal loop gain is significantly greater than 1, and where the resulting nonlinear distortion is used to reduce the effective loop gain to 1, to move the poles back onto the $j\omega$ axis

Let's begin this investigation of oscillator performance by considering that the circuits being examined have loop gains close to 1 so that their poles are located only slightly in the right-hand plane. Such an oscillator will exhibit only minute waveform distortion (to move the poles back onto the $j\omega$ axis), and in examining its waveshapes, we may to a good approximation consider that they are all perfect sinusoids.

Consider, for example, the op amp Wien bridge oscillator investigated in Example 8.4-2 (see Figure 8.4-3). In analyzing the oscillation amplitude of this circuit, let us assume that the op amp power supplies are ± 10 V, and for simplicity, let us also assume that the output clips precisely at these voltage levels.

If we were to observe the behavior of this oscillator starting at $t = 0$, what we would find is that when the circuit is first turned on, the oscillations would begin from zero and would grow exponentially in the output until limited at ± 10 V by the saturation and cutoff of the op amp's output transistors. Thus in the steady state we would find that the signal at the op amp output would be a nearly perfect sine wave oscillating $\omega_0 = 1/RC$ with an amplitude of 10 V. If we recall from Example 8.4-2 that the feedback network has zero phase shift at this frequency and a gain of $\frac{1}{3}$, then it is apparent that the steady-state waveshape at the amplifier input will be a 3.33-V sine wave that is in phase with v_o. A careful examination of the output waveshape would also disclose a small amount of waveform distortion (clipping) at the peaks. Much more will be said about this distortion when we discuss hard-limiting effects in oscillators.

In our analysis of the Wien bridge oscillator, we assumed that the positive and negative limiting levels were the same, so that symmetrical clipping occurred. For op amp–style oscillators, because the output dc level with no signal

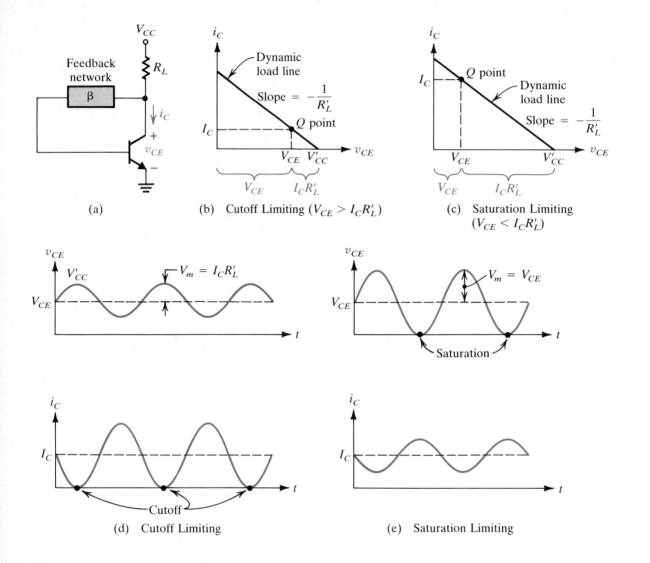

(a)

(b)   Cutoff Limiting ($V_{CE} > I_C R'_L$)

(c)   Saturation Limiting ($V_{CE} < I_C R'_L$)

(d)   Cutoff Limiting

(e)   Saturation Limiting

Figure 8.5-1

Amplitude limiting in a BJT oscillator.

applied is usually zero, symmetrical clipping will usually be the rule. However, for most other oscillators, the type of amplitude limiting that occurs will depend on the position of the Q point for the transistors in the circuit being examined. Consider, for example, the BJT oscillator circuit illustrated in Figure 8.5-1. Two distinctly different nonlinear amplitude limiting effects can occur, depending on the position of the Q point. The transistor can either saturate or cut off. In fact, if the circuit loop gain is large enough, both of these nonlinear effects can occur in the same circuit.

Following Figure 8.5-1a, we can see that this circuit's dynamic load-line output equation may be written as

$$V'_{CC} = i_C R'_L + v_{CE} \tag{8.5-1}$$

where R'_L represents the effective ac collector impedance. This load line is sketched in Figure 8.5-1b and c for two different Q point positions. If we assume that this circuit has just started to oscillate, v_{CE} will be a sine wave centered about V_{CE} whose amplitude will increase exponentially with time until limited by one or more of the circuit's nonlinearities.

For the case where

$$V_{CE} > I_C R'_L \qquad (8.5\text{-}2)$$

(see Figure 8.5-1b), the first transistor nonlinearity to occur will be cutoff, so that as illustrated in Figure 8.5-1d, the amplitude of oscillations at the output in this case will be determined by the expression

$$V_m \simeq I_C R'_L \qquad (8.5\text{-}3)$$

For the opposite situation, that is, where

$$V_{CE} < I_C R'_L \qquad (8.5\text{-}4)$$

(see Figure 8.5-1c and e), the transistor will saturate before it cuts off, and as a result, the output amplitude of oscillations will simply be equal to

$$V_m = V_{CE} \qquad (8.5\text{-}5)$$

As we shall see when we discuss the subject of hard limiting, if the signal is to be taken from the collector of the transistor, and if minimum signal distortion is desired, then limiting via cutoff rather than saturation is preferred, since all during the time that the transistor is saturated, v_{CE} will definitely be chopped off at zero volts.

A similar Q-point analysis technique may be used for the case of an FET oscillator. If in carrying out this analysis we neglect the finite ON resistance of the FET when it enters the ohmic region and assume that v_{DS} is nearly zero in this region (or alternatively, that $R'_L \gg r_{DS}$), then the theoretical development for the FET oscillator exactly parallels that for the BJT.

EXAMPLE 8.5-1

In Example 8.4-3 we examined a transformer-style oscillator circuit, in which we determined that a minimum h_{fe} of ten was required for that circuit to oscillate. This circuit is redrawn in Figure 8.5-2 with the biasing circuitry included. If the transistor $h_{ie} = 1$ kΩ and its h_{fe} is considered to be only slightly greater than 10, so that the circuit is just oscillating, sketch the waveshapes for v_C, i_C, i_B, and v_B.

SOLUTION

If we assume the signal distortion to be minimal, the dc Q-point analysis may be carried out as for any other BJT amplifier circuit. In particular, if we assume that the transformer primary and secondary windings (neglecting their ohmic resistances) are short circuits to dc, the dc circuit to be analyzed is essentially that shown in Figure 8.5-2b. Because $(1 + h_{FE})R_E \gg R_B$, it follows that

$$V_B \simeq V_{BB} = 1.2 \text{ V} \qquad (8.5\text{-}6a)$$

$$V_E = V_B - 0.7 = 0.5 \text{ V} \qquad (8.5\text{-}6b)$$

$$I_C \simeq I_E = \frac{V_E}{R_E} = 1 \text{ mA} \qquad (8.5\text{-}6c)$$

and $\qquad\qquad V_C \simeq V_{CC} = 20 \text{ V} \qquad (8.5\text{-}6d)$

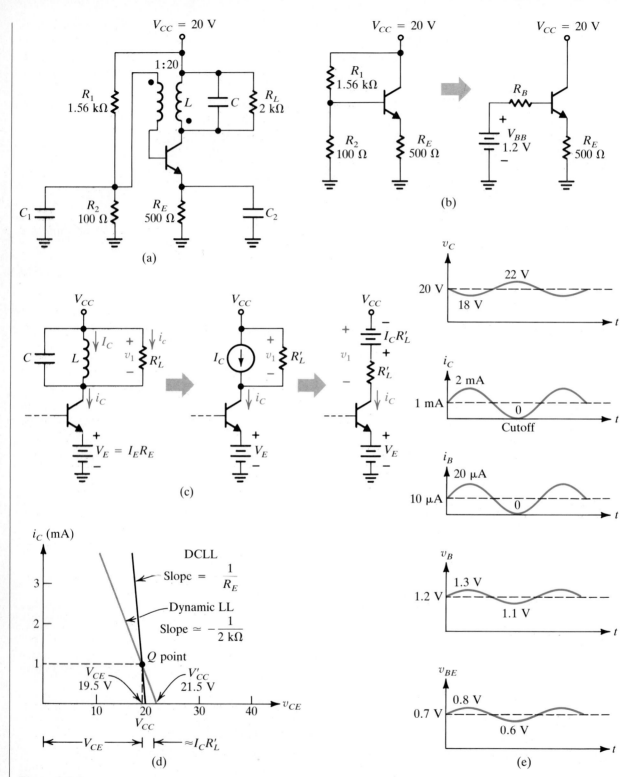

Figure 8.5-2
Example 8.5-1.

When this circuit is first turned on and begins to oscillate, the oscillations will start at the dc Q-point levels and will grow in amplitude until eventually limited by the circuit nonlinearity. To see whether this transistor saturates or cuts off first, we note that the dc load-line equation for this circuit may be written as

$$V_{CC} = I_C R_E + V_{CE} \tag{8.5-7}$$

In addition, following the left-hand portion of Figure 8.5-2c, because the transformer inductance L is assumed to be a short to dc only, the ac component of the collector current flows through R'_L. As a result, the voltage drop v_1 across the transformer primary is

$$v_1 = i_c R'_L = (i_C - I_C)R'_L \tag{8.5-8}$$

so that the dynamic load-line equation for this circuit may be written as

$$V_{CC} = v_1 + v_{CE} + V_E \simeq (i_C - I_C)R'_L + v_{CE} + I_C R_E \tag{8.5-9a}$$

or $\qquad V'_{CC} = V_{CC} + I_C R'_L - I_C R_E = i_C R'_L + v_{CE} \tag{8.5-9b}$

The same result can also be obtained by replacing the transformer inductance (and C) by a dc current source I_C. The dynamic load line then follows directly from the equivalent circuit given to the right in Figure 8.5-2c.

Both the dc and the dynamic load-line equations are sketched on the transistor characteristics in Figure 8.5-2d. In examining the dynamic load line for this circuit, notice that it is possible for the total collector-to-emitter voltage to exceed V_{CC}. This is due to the presence of the transformer, or more specifically, to the inductance of the transformer. Much more will be said about this phenomenon when we discuss transformer coupled power amplifiers in Chapter 9.

From Figure 8.5-2d it is clear that

$$V_{CE} > I_C R'_L \tag{8.5-10}$$

so that as the oscillatory sine wave grows, the first nonlinearity it encounters is when the transistor cuts off (when v_{CE} is about 21.5 V and $i_C = 0$) to the right of the Q point. This implies that the steady-state sinusoidal oscillation amplitude at the collector will be about $21.5 - 19.5 = 2$ V. Following Figure 8.5-2c, the corresponding ac collector and base currents are seen to be

$$I_c = \frac{V_{ce}}{R'_L} \simeq \frac{2 \text{ V}}{2 \text{ k}\Omega} = 1.0 \text{ mA} \tag{8.5-11a}$$

and $\qquad I_b = \frac{I_c}{h_{fe}} = 0.01 \text{ mA} = 10 \ \mu\text{A} \tag{8.5-11b}$

These results are sketched in Figure 8.5-2e, and as expected, the nonlinearity occurring at cutoff just causes the base and collector currents to go to zero. Had the nonlinearity been more severe, actual signal clipping would have occurred at these points. Furthermore, the amplitude of the ac portion of the base–emitter voltage is just obtained by dividing the ac collector signal swing by the turns ratio of the transformer.

8.5-1 Soft Limiting

In the preceding portion of this section, we assumed that the poles of the oscillator were somehow maintained on the $j\omega$ axis. In this subsection, we investigate techniques for automatically maintaining the poles at this location by a method known as soft limiting. The major advantage of this approach is that the amount of nonlinearity in the sinusoidal signals produced is very small. However, the price paid for this performance is that the steady-state signal amplitude obtained depends on the often-ill-defined characteristics of the limiting circuit and not, for example, on the more explicitly known values of V_{CC} or I_C in the circuit.

The principle involved in soft limiting is illustrated in Figure 8.5-3. Here an amplitude detection circuit is added to the basic oscillator circuit, and the output of this detector is a signal that is fed back to the oscillator in order to adjust its loop gain. As illustrated in Figure 8.5-3b, the circuit is designed so that when the oscillation amplitude is small, the loop gain is greater than 1. In this way when the circuit is first turned on (and the oscillation amplitude is zero), the poles will be in the right plane and oscillation will be guaranteed to start. As the amplitude of oscillations grows, the loop gain will automatically be reduced until in the steady-state the loop gain is precisely equal to 1, and the oscillation amplitude is V_m.

The fundamental difficulty with amplitude stabilization circuits of this type is illustrated by Figure 8.5-3b, in which we have shown (in dashed lines) the variation in the circuit characteristics that might occur, for example, when the temperature of the oscillator is changed. In this case the steady-state amplitude of the oscillator changes to V'_m when the temperature changes from T_1 to T_2. For some applications this type of long-term variation in the oscillation amplitude can be tolerated. However, for application areas where both low waveform distortion and good long-term amplitude stability are required some type

Figure 8.5-3
Soft limiting of oscillation amplitude.

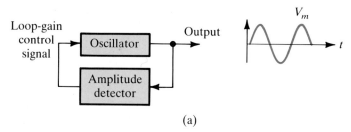

(a)

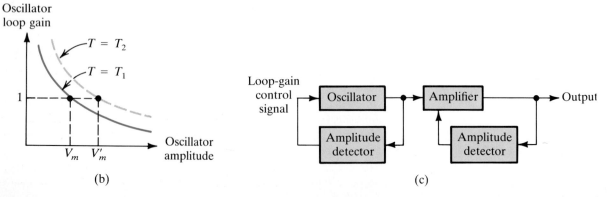

(b)

(c)

EXAMPLE 8.5-2	The circuit shown in Figure 8.5-4a is a Wien bridge oscillator in which a field-effect transistor is used to provide soft amplitude limiting. In analyzing this circuit, let us assume that v_{DS} is always small enough that the transistor operates in the ohmic region. By examining the transistor characteristics given in Figure 8.5-4b, we can see that the transistor's drain-to-source resistance varies from about 200 Ω when $v_{GS} = 0$ to nearly infinity when the transistor is pinched off at $v_{GS} = -4$ V. The resistance variation of this FET when it operates in the ohmic region is sketched in Figure 8.5-4c as a function of v_{GS}.

of AGC (automatic gain control) circuit will also be needed (Figure 8.5-3c). The AGC circuit monitors the amplitude of the output signal and automatically adjusts the amplifier gain to maintain a constant signal level at the output.

EXAMPLE 8.5-2

The circuit shown in Figure 8.5-4a is a Wien bridge oscillator in which a field-effect transistor is used to provide soft amplitude limiting. In analyzing this circuit, let us assume that v_{DS} is always small enough that the transistor operates in the ohmic region. By examining the transistor characteristics given in Figure 8.5-4b, we can see that the transistor's drain-to-source resistance varies from about 200 Ω when $v_{GS} = 0$ to nearly infinity when the transistor is pinched off at $v_{GS} = -4$ V. The resistance variation of this FET when it operates in the ohmic region is sketched in Figure 8.5-4c as a function of v_{GS}.

The portion of the circuit consisting of R_1, R_2, D, and C is a negative peak detector, where the steady-state value of v_{GS} (assuming the diode to be ideal) is a negative dc voltage proportional to the peak amplitude of the sinusoidal signal appearing at the output of the op amp. R_G allows the voltage on the capacitor to leak off slowly if the amplitude of the signal at v_3 decreases.

When the circuit is first turned on, if $v_o \simeq 0$, then the FET ohmic resistance r_{DS} will be about 400 Ω. For this situation the op amp gain will be $[1 + R_f/(R_{2A} + r_{DS})] = [1 + 2/(0.2 + 0.4)] = 4.3$, and because this oscillator only needs an amplifier gain of 3 to place the poles on the $j\omega$ axis, this will guarantee that the circuit's poles are initially in the right-hand plane and that it will start oscillating when it is first turned on.

Determine the steady-state amplitude of oscillations at the op amp output, and sketch v_o, v_1, v_2, v_3, and v_{GS} in the steady state.

SOLUTION

Following the analysis presented in Example 8.4-2, it is clear that the op amp closed-loop gain of 3 is required for this circuit to oscillate in the steady state. Because the gain of this amplifier is given by the expression

$$A_{CL} = 1 + \frac{R_f}{(R_{2A} + r_{DS})} \tag{8.5-12}$$

we find that the transistor ohmic resistance in the steady state must therefore be about 800 Ω. From Figure 8.5-4c this requires that the transistor gate-to-source voltage be -2 V. Because this voltage is equal to the negative peak of the voltage appearing at v_3, and furthermore because $v_3 \simeq \frac{1}{5} v_o$ it therefore follows that the oscillation amplitude must be about 2×5 or 10 V. If we assume that the op amp power supplies are ± 15 V, no clipping will occur in the output. The waveshape for v_o is sketched in Figure 8.5-4d.

Because the gain of the feedback network is $\frac{1}{3}$ at the frequency of oscillation, the voltage v_1 has the same form as v_o but is only one-third as big. Similarly, v_2, which is equal to $v_o[(R_{2A} + r_{DS})/(R_{2A} + r_{DS} + R_f))]$, is the same as v_1. Of course, this is not all that surprising since $v_1 \simeq v_2$ for an op amp operating in the active region. Finally, neglecting the loading of the peak detector circuit, $v_3 = [R_1/(R_1 + R_2)]v_o = v_o/5$, and v_{GS} is a dc voltage equal to the negative peak of v_3.

The capacitor in the peak detector circuit is chosen so that its time constant $(R_G C)$ is large in comparison to the period of the sine wave. This will ensure

Figure 8.5-4
Example 8.5-2.

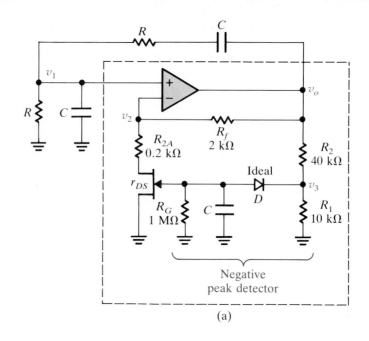

(a)

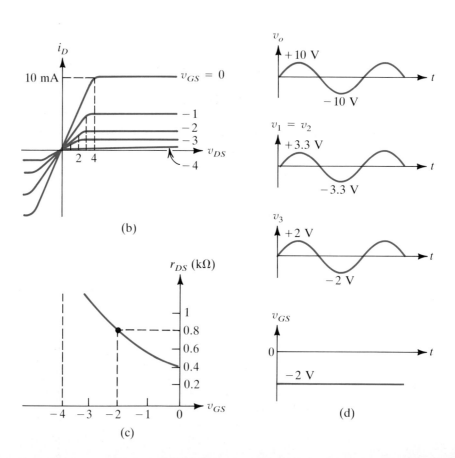

(b)

(c)

(d)

that the drain-to-source resistance of the FET is proportional to the amplitude of the sine wave appearing at the op amp output, not to its instantaneous value.

One final point worth observing is that the steady-state drain-to-source voltage across the FET is about $4v_2/5$, and hence it, too, is a sine wave having an amplitude of about 2.66 V. Noting that the FET characteristics are nearly the same for both positive or negative values of v_{DS} when this voltage is small (see Figure 8.5-4b), it is clear that the transistor is, as originally postulated, operating in the ohmic region.

8.5-2   Hard Amplitude Limiting

Sinusoidal oscillator circuits whose small-signal loop gains are significantly greater than 1 will typically exhibit significant waveform distortion at one or more points in the circuit when steady-state oscillations are finally achieved. This waveform nonlinearity is necessary to reduce the effective large signal loop gain of the oscillator to 1.

The advantage of using hard amplitude limiting instead of soft limiting is that the amplitude of oscillations is determined by the level at which the non-linearity sets in rather than being dependent on the volt-ampere characteristics of the soft-limiting device. Of course, the price paid for this rather simple oscillator design is that the waveform, at least at some point in the circuit, will be nonsinusoidal. However, as we shall see in the examples that follow, this need not result in a nonsinusoidal output from the oscillator as long as the Q of the frequency-selection circuitry is reasonably high.

The basic principle of hard amplitude limiting may be described with the aid of Figure 8.5-5. In this diagram of a typical oscillator circuit, we have artificially partitioned the amplifier's linear and nonlinear behavior into two separate boxes to illustrate how the nonlinearity, in effect, reduces the large-signal loop gain of the circuit. In analyzing this oscillator circuit, let us assume that the feedback network has a high Q so that it passes signals only at the frequency of oscillation ω_0 while rejecting all others. Furthermore, let us also assume that the small-signal loop gain of this oscillator (βA) at the fundamental frequency ω_0 is significantly greater than 1. As a result, when this circuit is first turned on, because it has poles in the right-hand plane, it will begin to oscillate at ω_0, and these oscillations will grow with time. This behavior will continue until the circuit's nonlinear effects set in. At this point the loop gain of the circuit will begin to decrease, and eventually, when the loop gain has been reduced to 1, steady-state oscillations of the form indicated in the figure will result. Let's try to understand how their signal amplitudes and the amount of signal distortion at v_B are determined.

As with all other oscillator circuits that we have examined, let's analyze this circuit's performance by breaking the loop (at the point indicated by the "$\times$"), injecting a signal v_{in} at this point, and tracing the signal around the loop. If we assume that the applied input signal is

$$v_{in}(t) = V_m \sin \omega_0 t \qquad (8.5\text{-}13a)$$

the output of the linear portion of the amplifier will be

$$v_A(t) = Av_{in}(t) = AV_m \sin \omega_0 t \qquad (8.5\text{-}13b)$$

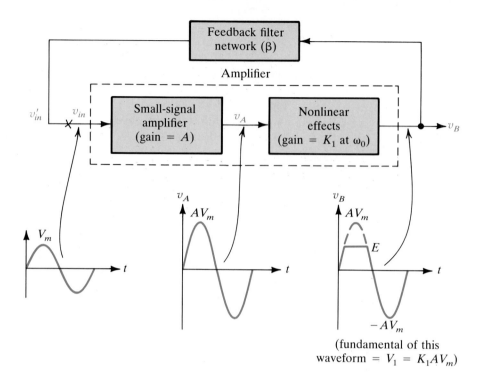

Figure 8.5-5
Basic concept of hard amplitude limiting.

(fundamental of this
waveform $= V_1 = K_1 A V_m$)

This signal is now applied to the nonlinear section of the amplifier. The gain of this circuit is considered to be 1 if the input signal amplitude is less than E, and it is assumed to clip off the positive tip of the output when its amplitude exceeds E. The dashed portion of the waveform at v_B illustrates the output that would have been obtained in the absence of any nonlinearity, and the solid portion of this sketch indicates the actual output present at this point.

Because the output $v_B(t)$ is a periodic function, we may express it in terms of its Fourier series as

$$v_B(t) = \underbrace{V_o}_{\substack{\text{dc} \\ \text{component}}} + \underbrace{V_1 \sin \omega_0 t}_{\text{fundamental}} + \underbrace{V_2 \sin 2\omega_0 t}_{\substack{\text{second} \\ \text{harmonic}}} + \underbrace{V_3 \sin 3\omega_0 t}_{\substack{\text{third} \\ \text{harmonic}}} + \dots \qquad (8.5\text{-}14)$$

In examining this series, only the fundamental component at ω_0 is of interest because all the other harmonics will be attenuated by the assumed high-Q behavior of the feedback network. As a result, the signal fed back to the input may be written approximately as

$$v_{in}' = \beta V_1 \cos \omega_0 t = (\beta A K_1) V_m \sin \omega_0 t \qquad (8.5\text{-}15)$$

where K_1 is equal to the ratio of the fundamental component of the Fourier series, V_1, to the unattenuated signal amplitude, V_m. Thus, in effect, K_1 may be thought of as the gain of the nonlinear portion of the amplifier circuit.

Following equation (8.5-15), we can see that the effective large-signal loop gain of this circuit is

$$\text{effective loop gain} = \frac{v_{in}'}{v_{in}} = \beta A K_1 \qquad (8.5\text{-}16)$$

and as with all the other oscillator circuits that we have examined, this loop gain needs to be 1 once steady-state oscillations are achieved.

When the small-signal loop gain is only slightly greater than unity, K_1 will only need to be a little less than 1 to satisfy equation (8.5-16). This means that the Fourier series fundamental component will have nearly the same amplitude as the original sine wave, or that the amount of distortion "required" to make the large-signal loop gain equal to 1 will be minimal. For circuits where βA is much greater than 1, small values of K_1 will be needed, and this will correspond to severe clipping or other types of distortion of the steady-state oscillatory waveform.

From even this brief discussion, it should be apparent that a quantitative understanding of the performance of oscillators containing hard amplitude limiting will require a careful examination of the types of waveform nonlinearity that occur in the most commonly used circuits. Fortunately, there are only two or three types of waveform distortion that will be important for our study of oscillators. These are illustrated in Figure 8.5-6, and the fundamental Fourier

Figure 8.5-6
Selected types of hard amplitude limiting.

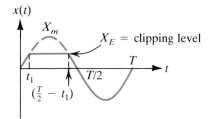

(a) Single-Ended Clipping (Sine Wave with Tip Removed, X_E Positive)

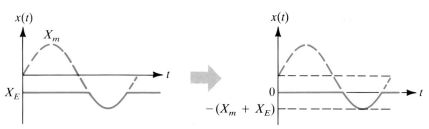

(b) Severe Single-Ended Clipping (Sine-Wave Tip, X_E Negative)

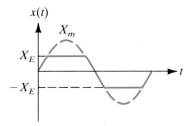

(c) Double-Ended Symmetrical Clipping (Sine Wave with Tips Removed)

8.5 / Amplitude Stabilization in Oscillators

583

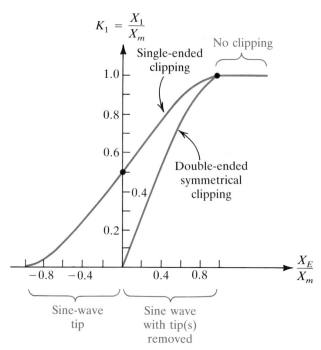

$$K_1 = \frac{X_1}{X_m}$$

Figure 8.5-7
Fundamental component of single- and double-ended symmetrically clipped sine waves as a function of the clipping level.

series components for each of these waveshapes are sketched as a function of the clipping level in Figure 8.5-7. In these graphs the generalized variable x is used rather than v for voltage because in some circuits the signal being examined may be a voltage, while in others it may be a current. The ordinate used with these graphs, X_1/X_m, represents the relative size of the fundamental component of the Fourier series, and the abscissa, X_E/X_m, the relative height of the clipping level.

To illustrate how these data were obtained, let's go through a detailed analysis for the case of single-ended clipping waveform distortion (Figure 8.5-6a). The fundamental component of this waveform is found by evaluating the integral

$$X_1 = \frac{2}{T} \int_0^T x(t)\, \sin \omega_0 t\, dt$$

$$= \frac{2}{T}\left[X_m \int_0^{t_1} \sin^2 \omega_0 t\, dt + X_E \int_{t_1}^{(T/2 - t_1)} \sin \omega_0 t\, dt + X_m \int_{(T/2 - t_1)}^{T} \sin^2 \omega_0 t\, dt \right]$$

$$(8.5\text{-}17)$$

where ω_0 = fundamental frequency of the waveform
X_1 = amplitude of the fundamental component
X_m = amplitude of the unclipped waveform
X_E = clipping level
and $t_1 = (1/\omega_0) \sin^{-1} (X_E/X_m)$

These integrals may be evaluated directly, and after considerable algebraic manipulation, we obtain

$$\frac{X_1}{X_m} = \frac{1}{\pi} \sin^{-1} \frac{X_E}{X_m} + \frac{1}{2} + \frac{1}{\pi} \frac{X_E}{X_m} \sqrt{1 - \left(\frac{X_E}{X_m}\right)^2} \qquad (8.5\text{-}18)$$

This solution is also valid for the type of single-ended clipping illustrated in Figure 8.5-6b if negative values are used for the clipping level, X_E, which yields

$$\frac{X_1}{X_m} = \frac{-1}{\pi} \sin^{-1} \frac{X_E}{X_m} + \frac{1}{2} - \frac{1}{\pi} \frac{X_E}{X_m} \sqrt{1 - \left(\frac{X_E}{X_m}\right)^2} \qquad (8.5\text{-}19)$$

The expression for the fundamental component of a symmetrically clipped sine wave may readily be obtained by observing that the waveshape given in Figure 8.5-6c is equal to the difference of the two waveforms illustrated in parts (a) and (b) of the figure. Therefore, the fundamental component of the Fourier series for a double-clipped sine wave is simply equal to the difference of equations (8.5-18) and (8.5-19), or

$$\frac{X_1}{X_m} = \frac{2}{\pi} \sin^{-1} \frac{X_E}{X_m} + \frac{2}{\pi} \frac{X_E}{X_m} \sqrt{1 - \left(\frac{X_E}{X_m}\right)^2} \qquad (8.5\text{-}20)$$

The relative amplitude, X_1/X_m, of each of these fundamental Fourier series components is sketched in Figure 8.5-7 as a function of the relative clipping level, X_E/X_m. The examples that follow will help to illustrate how these data may be employed to understand quantitatively the approximate voltages and currents that we will find when we examine oscillator circuits that contain hard amplitude limiting.

EXAMPLE 8.5-3

The circuit illustrated in Figure 8.5-8a is a Wien bridge–style oscillator in which a tuned circuit is part of the feedback loop. As a result, this circuit can oscillate only at $\omega_0 = 1/\sqrt{LC}$. Furthermore, if the Q of this parallel tuned LC network is assumed to be large, then, regardless of the form of the signal appearing at the output of the op amp, the voltage at v_x will be a sine wave. At resonance the β network looks like a voltage divider with a gain of $\frac{1}{2}$, and in addition, the op amp circuit has a closed-loop gain of $(1 + R_f/R_2) = 4$, so that the small-signal loop gain of this oscillator is much greater than 1. Therefore, this circuit initially has poles which are deep in the right-hand plane, so that the oscillation amplitude will increase exponentially until nonlinear limiting at the output of the op amp occurs.

If the limiting in the output is assumed to occur symmetrically at the positive and the negative power supply levels, determine the steady-state voltages v_o and v_x at the op amp output and input terminals, respectively.

SOLUTION

This circuit may be analyzed by breaking the loop at the point labeled "×" at the noninterverting input of the op amp, injecting a signal v_x at this point, and tracing this signal around the loop. As shown in Figure 8.5-8b, to get a physical understanding of what is happening in this circuit, it may be useful, at least initially, to represent the op amp as a linear amplifier with a gain of 4 followed by a nonlinear circuit that clips the output symmetrically whenever the input signal amplitude applied to it exceeds 10 V. The "gain" of this clipping circuit is $V_1/V_A = K_1$, where V_1 is the fundamental component of the Fourier series of v_o.

Proceeding around the loop, we have

$$V_A = AV_x \qquad (8.5\text{-}21a)$$

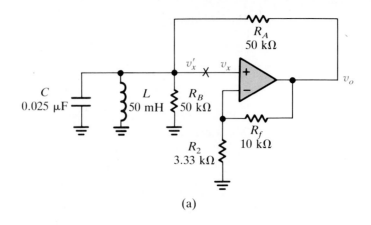

(a)

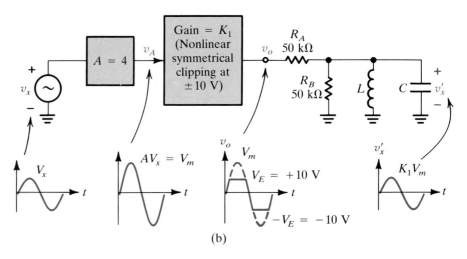

(b)

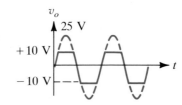

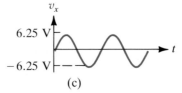

(c)

Figure 8.5-8
Example 8.5-3.

$$V_1 = K_1 V_A = K_1 A V_x \qquad (8.5\text{-}21b)$$

and
$$V'_x = \beta(j\omega_0)V_1 = K_1 A \beta(j\omega_0)V_x \qquad (8.5\text{-}21c)$$

Because of the tuned circuit, the loop gain is very small except at the frequency

$$f_0 = \frac{1}{2\pi\sqrt{LC}} = 4.5 \text{ kHz} \qquad (8.5\text{-}22)$$

and as a result, if the loop-gain requirements are satisfied, the circuit will oscillate at this frequency.

When the LC network resonates at ω_0, the gain of the β network is $\frac{1}{2}$, so that for the overall large-signal loop gain to be 1, we require that

$$K_1 = \frac{1}{\beta(j\omega_0)A} = \frac{1}{(\frac{1}{2})(4)} = \frac{1}{2} \qquad (8.5\text{-}23)$$

In other words, in this oscillator the amplitude of oscillations will increase until the fundamental component of the clipped sine wave appearing at the output of the op amp is one-half of the sine-wave amplitude which would exist there if the clipping were not present. Following Figure 8.5-7, this requires that

$$\frac{V_E}{V_m} = 0.4 \qquad (8.5\text{-}24a)$$

so that the amplitude of the unclipped sine wave will be

$$V_m = \frac{V_E}{0.4} = \frac{10 \text{ V}}{0.4} = 25 \text{ V} \qquad (8.5\text{-}24b)$$

This result permits us to sketch accurately the form of the signal appearing at the output of the op amp (Figure 8.5-8c). Furthermore, working backward from this point, we can also see that

$$V_x = \frac{V_m}{A} = 6.25 \text{ V} \qquad (8.5\text{-}25)$$

In the analysis of this oscillator circuit, we assumed that the signal appearing at the input of the oscillator was a sine wave. Because v_o is clearly nonsinusoidal, this requires that the Q of the LC tank be reasonably high in order to attenuate all the harmonics of v_o. With this circuit oscillating at 4.5 kHz, the Q of the feedback network is

$$Q = \frac{R_A \| R_B}{\omega_0 L} \simeq 35 \qquad (8.5\text{-}26)$$

By evaluating the transfer function of the β network at $s = jn\omega_0$, following equation (I.9-4), it is not too difficult to show that its gain at the nth harmonic is

$$|\beta(jn\omega_0)| = \text{gain at } n\text{th harmonic} = \frac{1}{2}\left[\frac{n/Q}{\sqrt{(n^2-1)^2 + (n/Q)^2}}\right] \qquad (8.5\text{-}27)$$

As a result, if the Fourier series of $v_o(t)$ is written as

$$v_o(t) = V_1 \sin(\omega_0 t + \phi_1) + V_2 \sin(2\omega_0 t + \phi_2) + V_3 \sin(3\omega_0 t + \phi_3) + \cdots \qquad (8.5\text{-}28a)$$

we may express the form of the signal appearing at v_x as

$$\begin{aligned} v_x(t) = {} & |\beta(j\omega_0)|V_1 \sin[\omega_0 t + \phi_1 + \phi_\beta(j\omega_0)] \\ & + |\beta(2j\omega_0)|V_2 \sin[2\omega_0 t + \phi_2 + \phi_\beta(2j\omega_0)] + \cdots \end{aligned} \qquad (8.5\text{-}28b)$$

where $|\beta(jn\omega_0)| =$ magnitude of $\beta(j\omega)$ at the nth harmonic and $\phi_\beta(jn\omega_0) =$ phase of $\beta(j\omega)$ at the nth harmonic.

Following this result and assuming that V_2 is smaller than V_1, from equation (8.5-27) we find that the second harmonic distortion of the sine wave appearing at v_x will be less than 2% of the fundamental. Furthermore, the amplitudes of the higher-order harmonic terms appearing at v_x will be even smaller than that of the second harmonic. Clearly, then, for this example the signal at v_x may, to a good approximation, be considered to be a pure sine wave.

EXAMPLE 8.5-4

Because depletion-style FETs are normally ON, oscillators containing these devices may be biased by a voltage clamping technique of the form illustrated in Figure 8.5-9a. When this oscillator is first turned on, the voltage across C_1 is initially zero, so that $v_{GS}(0+)$ will also be zero. As a result with $v_{GS} = 0$, the transistor is active, its g_m and loop gain will both be large, and the circuit will be highly likely to oscillate (Figure 8.5-9b).

As the circuit begins to oscillate and the amplitude of oscillations increases, the gate diode will conduct and the capacitor C_1 will charge up to $(V_x - 0.7)$, where V_x represents the peak voltage on the secondary of the transformer. In this way, the voltage at the gate of the transistor will be clamped negatively to zero volts (or more precisely, to $+0.7$ V). As the amplitude of oscillations increases further, the transistor will eventually cut off, and this will provide the nonlinearity necessary to reduce the large-signal loop gain to 1.

In analyzing this FET oscillator, it is useful to approximate the i_D versus v_{GS} transfer characteristic by a straight line (Figure 8.5-9b) because we can then apply the analysis techniques used previously to examine the op amp oscillator in Example 8.5-3. Admittedly, this approximation is rather crude (and leads to errors in the neighborhood of 10 or 20%), but the simplicity afforded by this approach appears to be worth it.

Following this procedure, for the circuit illustrated in Figure 8.5-9a, determine the steady-state waveshapes for v_x, v_{GS}, i_D, and v_{DS}. The transistor $I_{DSS} = 4$ mA, and $V_p = -1.5$ V.

SOLUTION

Assuming that the transistor i_D versus v_{GS} characteristic may be approximated by the straight line shown in Figure 8.5-9b, the transconductance, g_m, may then be written as

$$g_m = \frac{\Delta i_D}{\Delta v_{GS}} \simeq \frac{-I_{DSS}}{V_p} = 2.66 \text{ mS} \tag{8.5-29}$$

Injecting a signal V_x at the gate of the FET, and considering C_1 to be a short at the frequency of oscillation, we obtain the equivalent circuit given in Figure 8.5-9c. From this circuit the drain current is

$$I_d = \frac{V_x}{1/g_m} = g_m V_x \tag{8.5-30a}$$

with the fundamental component of this current given by the expression

$$I_{d1} = K_1 I_d = K_1 g_m V_x \tag{8.5-30b}$$

where K_1 is the relative size of the fundamental component of the drain cur-

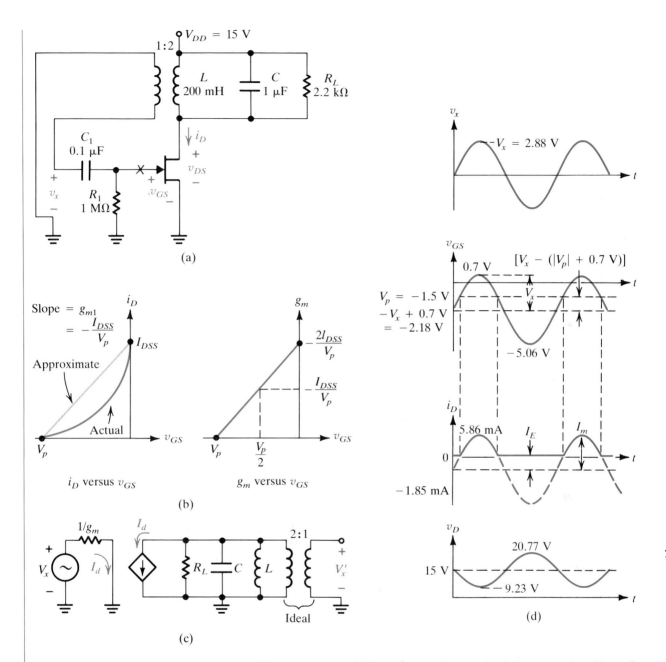

Figure 8.5-9
Example 8.5-4.

rent. For this circuit to oscillate, V_x' must be greater than or equal to V_x, and this can occur only at the frequency

$$f_0 = \frac{1}{2\pi\sqrt{LC}} = 356 \text{ Hz} \tag{8.5-31}$$

where the tuned circuit resonates. For all other frequencies, the tank circuit is a short and V_x' is nearly zero.

At the resonant frequency,

$$V_d = -I_{d1}R_L = -K_1 g_m R_L V_x \tag{8.5-32a}$$

and
$$V_x' = \frac{K_1 g_m R_L}{N} V_x \qquad (8.5\text{-}32b)$$

As a result, when the circuit is first turned on, before severe nonlinearity sets in, $K_1 = 1$, and the small-signal loop gain is $(2.66 \times 2.2)/2 = 2.93$. Therefore, because the initial loop gain is greater than 1, the amplitude of oscillations will grow until the signal is eventually limited by the FET cutting off. This nonlinearity will reduce the value of K_1 to the point where the overall large-signal loop gain is precisely equal to 1.

Following equation (8.5-32b), the required value of K_1 is

$$K_1 = \frac{I_{d1}}{I_m} = \frac{N}{g_m R_L} = \frac{1}{2.93} = 0.34 \qquad (8.5\text{-}33)$$

Using Figure 8.5-7, we find that the required clipping level corresponding to this value of K_1 is

$$\frac{I_E}{I_m} = -0.24 \qquad (8.5\text{-}34)$$

The negative sign indicates that the drain current conduction angle will be less than 180°. The basic form of the important waveshapes in this circuit are sketched in Figure 8.5-9d.

Using the straight-line approximation to the i_D versus v_{GS} characteristic, we may write

$$i_D = I_{DSS} + g_m v_{GS} \qquad (8.5\text{-}35)$$

so that the drain current corresponding to the peak gate-to-source voltage of 0.7 V will be $4 + 2.66(0.7) = 5.86$ mA, assuming that equation (8.5-35) is still valid when v_{GS} is slightly positive. By employing equation (8.5-34) and carefully examining the sketch for i_D, we can see that

$$\left| \frac{I_E}{I_m} \right| = \left| \frac{I_E}{I_E + 5.86} \right| = 0.24 \qquad (8.5\text{-}36)$$

and solving this expression for I_E, we have

$$|I_E| = 1.85 \text{ mA} \qquad (8.5\text{-}37a)$$

In addition, $\qquad I_m = |I_E| + 5.86 = 7.71 \text{ mA} \qquad (8.5\text{-}37b)$

Combining equations (8.5-33) and (8.5-37b), the fundamental component of the drain current is seen to be

$$I_{d1} = K_1 I_m = 2.62 \text{ mA} \qquad (8.5\text{-}38a)$$

so that the ac drain voltage is

$$V_d = -I_{d1} R_L = -5.77 \text{ V} \qquad (8.5\text{-}38b)$$

Furthermore, the transformer secondary voltage and the ac portion of the gate-to-source voltage are

$$V_x = -\frac{1}{N} V_d = 2.88 \text{ V} \qquad (8.5\text{-}39)$$

Using equations (8.5-37) through (8.5-39), we may now accurately sketch the waveshapes for v_x, v_{GS}, i_D, and v_{DS}. These results are given in Figure 8.5-9d.

The entire analysis of this oscillator rests on the assumption that the drain voltage will be nearly sinusoidal despite the fact that the drain current is severely distorted. Because the transistor does not saturate (see the v_{DS} sketch in Figure 8.5-9d), this approximation will be valid if the Q of the tank is reasonably large. With the circuit oscillating at 356 Hz, the Q of the tuned circuit is approximately equal to $Q = \dfrac{R_L}{\omega_0 L} = 4.9$. Following equation (8.5-27), for this value of Q the total harmonic distortion of the voltage at the drain will be less than about 15%, so that our results will be fairly accurate.

Exercises

8.5-1 For the circuit illustrated in Figure 8.5-2a, R_L is increased to 15 kΩ, C_2 is removed, and the transistor h_{fe} is assumed to be such that the circuit just oscillates. Find the ac voltages at the collector and the base of the transistor. *Answer* 15.5 V(peak) at the collector, 0.78 V(peak) at the base

8.5-2 Repeat Exercise 8.5-1 if R_L is increased to 30 kΩ. *Answer* 18.6 V(peak) at the collector, 0.93 V(peak) at the base

8.5-3 Find the dc and ac components of v_o in Figure 8.5-4 if a 1-V battery (plus on top) is connected in series, with the resistor R going from v_1 to ground. The op amp power supplies are assumed to be large enough that amplifier clipping does not occur. *Answer* dc component, 3 V; ac component, 13 V(peak)

8.5-4 The Wien bridge oscillator illustrated in Figure 8.4-3a has +10- and −20-V power supplies and an $A_{CL} = 3.25$. Find the approximate value of the steady-state voltage at the noninverting input terminal of the op amp. *Answer* 4.5 V(peak)

8.5-5 For the oscillator discussed in Exercise 8.4-1, the power supplies are ±10 V. Defining V_o as $E\ \underline{/0°}$, find E and express V_1, V_3, and V_5 in terms of V_o. *Answer* $E = 1.11$ V, $V_1 = 9E\ \underline{/-180°}$, $V_3 = 5.19E\ \underline{/-235°}$, $V_5 = 1.72E\underline{/-305°}$

8.6 FREQUENCY STABILIZATION OF OSCILLATORS

In the oscillator circuits we have discussed in previous sections of this chapter, we saw that the frequency of oscillation was equal to that frequency at which the overall phase shift of the signal traveling around the loop was zero. In carrying out these analyses, we generally considered that the frequency-determining elements were contained in the feedback loop of the oscillator, and that the amplifier circuits themselves contributed a phase shift of either 0 or −180°. This basically assumed that the amplifier portion of the oscillator was ideal and contributed a fixed phase shift independent of the frequency of oscillation of the circuit.

In this section we investigate the factors governing the frequency at which an oscillator operates in somewhat greater detail. Specifically, we will be interested in what we as engineers can do to ensure that:

1. The circuit that we design oscillates as closely as possible to the desired frequency, without any external adjustment.
2. The frequency of oscillation is relatively independent of all phase-shift effects not associated with the frequency-determining elements of the circuit.
3. The frequency of oscillation of the circuit is relatively independent of component aging, circuit loading, and circuit environmental effects.

To illustrate the basic concept of frequency stabilization, consider the tuned-circuit oscillator presented in Figure 8.6-1a. For simplicity assume that amplifiers A_1 and A_2 each have constant phase shifts of $-180°$ in the neighborhood of the resonant frequency of the parallel tuned circuit, or tank circuit as it is also called. The gain–phase characteristics of this tuned circuit are given in Figure 8.6-1b.

From an examination of Figure 8.6-1a we can see that if the circuit loop gain

$$A_1 A_2 H(s)\bigg|_{s=j\omega_0} = A_1 A_2 \frac{R_2}{R_1 + R_2} \qquad (8.6\text{-}1)$$

is equal to or slightly greater than 1, the poles will be on or close to the $j\omega$ axis and the oscillation will be nearly sinusoidal. The relative signal levels at various points in the circuit for this condition are indicated in Figure 8.6-1a. The frequency of oscillation $f_0 = 1/2\pi\sqrt{LC}$, occurs at the resonant frequency of the tank, since at this frequency the net phase shift around the loop will be zero.

The oscillation frequency of this circuit depends on the values of both L and C. As a result, because these components are typically available with tolerances on the order of $\pm 10\%$, the untrimmed value of f_0 will be accurate only to within a few percentage points of the desired oscillation frequency. Furthermore, even if one of these components is made adjustable, so that f_0 can initially be set to precisely the value desired, aging, environmental effects, and the variation of other components in the circuit will still cause f_0 to change with time.

To illustrate this effect, let's consider what will happen to the oscillator in Figure 8.6-1a if a capacitive load (shown in dashed lines in the figure) is connected at the output terminal of the amplifier A_2. For simplicity, we will assume that C_L causes an additional $-10°$ of phase shift in the amplifier A_2 at the frequency f_0, and that this phase shift is relatively constant over the passband of the tuned circuit.

Owing to this capacitive loading, the overall oscillator phase shift around the loop at f_0 is now $-370°$ (or $-10°$), not $0°$ as required for steady-state oscillation. Therefore, the circuit can no longer oscillate at the frequency f_0. Instead, for the phase shift around the loop to return to zero, the frequency of oscillation will need to drop to f_2 (see Figure 8.6-1b). At this frequency the tank will contribute a phase shift of $+10°$ to the circuit, which will just cancel the $-10°$ from A_2 and will return the net loop phase shift to zero.

The addition of capacitive loading has had two undesirable effects on our oscillator. First, it has altered the frequency of oscillation, and second, it has

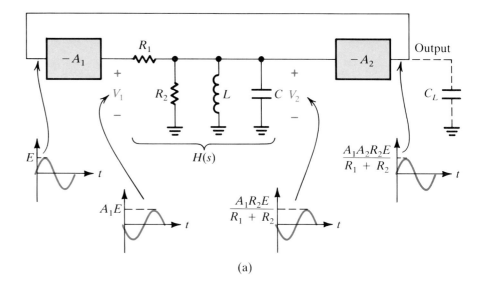

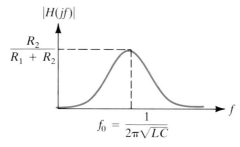

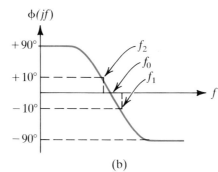

Q	$\dfrac{\Delta f}{f_0}$ (%)	$\dfrac{\Delta f}{f_0}$ (ppm)
1	8.72	87,200
3	2.90	29,000
5	1.74	17,400
10	0.872	8,720
10^2	0.0872	872
10^4	0.000872	8.72
10^5	0.0000872	0.87

(c)

Figure 8.6-1

lowered the gain of the tuned-circuit portion of our oscillator, since the tank is no longer operating at resonance. If the detuning were severe enough, it is possible that the loop-gain expression in equation (8.6-1) might no longer be greater than 1, and in this case the circuit would cease to oscillate. Before leaving this example, we should point out that although C_L is being considered an external component here, precisely the same effect will be produced by the parasitic capacitive elements in amplifiers A_1 and A_2.

In this section we investigate techniques for designing oscillators whose frequency of oscillation is nearly independent of the amplifier characteristics, circuit loading, environmental conditions, and component aging effects. To understand how to achieve these goals, it is useful again to examine the oscillator in Figure 8.6-1a. When C_L is added to this circuit, the magnitude of the frequency shift needed to compensate for the extra phase shift introduced into the loop by C_L depends strongly on the Q of the tuned circuit. For high-Q circuits the phase transition from $+90°$ to $-90°$ in $H(j\omega)$ occurs very rapidly in the region around f_0, and as a result the frequency shift needed to cancel out the phase error produced by C_L is minimal. For low-Q tuned circuits, the opposite is true.

To quantify this effect, we note that the transfer function for the tuned-circuit portion of the oscillator in Figure 8.6-1a may be written as

$$H(s) = \frac{V_2}{V_1} = \frac{R_2}{R_1 + R_2} \frac{s\omega_0/Q}{s^2 + s\omega_0/Q + \omega_0^2} \tag{8.6-2a}$$

or

$$H(j\omega) = \frac{V_2}{V_1} = \frac{R_2}{R_1 + R_2} \frac{j\omega\omega_0/Q}{(\omega_0^2 - \omega^2) + j\omega\omega_0/Q} \tag{8.6-2b}$$

where $\omega_0 = 1/\sqrt{LC}$, $Q = R/\omega_0 L$, and $R = R_1 R_2/(R_1 + R_2)$. The phase shift of $H(j\omega)$ is

$$\phi = \tan^{-1}\frac{\omega^2 - \omega_0^2}{\omega\omega_0/Q} = \tan^{-1}\frac{(\omega - \omega_0)(\omega + \omega_0)}{\omega\omega_0/Q} \tag{8.6-3a}$$

In the vicinity of $\omega = \omega_0$, the term $(\omega + \omega_0)$ is nearly equal to $2\omega_0$, and because $\tan^{-1}x \simeq x$ for small values of x, we may write

$$\phi \simeq \frac{2Q(\omega - \omega_0)}{\omega_0} = 2Q\frac{f - f_0}{f_0} \tag{8.6-3b}$$

in the vicinity of resonance, or

$$\Delta\phi \simeq 2Q\frac{\Delta f}{f_0} \tag{8.6-3c}$$

since $\phi = 0$ at f_0. This expression indicates the change in phase shift produced by a given frequency shift away from resonance. Clearly, if the Q of the tank is large, then relatively small frequency shifts away from resonance will produce large phase changes in $H(j\omega)$. Figure 8.6-1c illustrates the relative frequency changes required to produce a $10°$ phase shift in $H(j\omega)$ for different tank Q's. Both percent phase shift and ppm (parts per million) phase shift from f_0 are included. For example, if we consider that the oscillator is operating nominally at $f_0 = 1$ MHz with a tank $Q = 100$, then a $10°$ phase change in amplifier A_2, for example, will produce an 872-Hz shift away from resonance in the tank (causing a 0.0872% drop in the frequency of oscillation). This will detune the tank enough to just produce the $10°$ of phase shift needed to return the overall loop phase shift to zero. As a result, the circuit will now oscillate at 1 MHz $-$ 872 Hz or at 0.999128 MHz.

When using discrete LC components, the upper limit on attainable Q values is about 100 or 200, so that when extremely stable time bases or frequency references are needed; conventional LC tank circuits are simply inadequate. In these applications piezoelectric crystals are commonly used for electronic circuit design, and when extremely high timing accuracy is required, an atomic clock reference may actually be used. Table 8.6-1 compares the relative accuracy of each of these oscillation schemes, taking into account not only the initial accuracy of the circuit, but also its stability with time. As shown, the standard balance wheel or pendulum-style mechanism, which is a purely mechanical oscillator and is found in nearly all older-style clocks and watches, has an equivalent Q of about 100 (similar to that of an LC tuned circuit). With this mechanism used as the time base for a clock, 0.01% timing errors would be typical, leading to cumulative timing errors of about 5 to 10 s/day. Watches and clocks containing crystal oscillators can reduce these errors 100- or 1000-fold due to their very large effective circuit Q values and their excellent component stability. Today's quartz crystal watches easily attain drift errors of less than a few seconds per month.

Although used infrequently, when extreme timing accuracy is required (e.g., at the National Bureau of Standards in the United States and in similar organizations throughout the world), atomic clocks such as the cesium beam frequency standard or the hydrogen maser are employed. When incorporated into clocks, these frequency standards have timing errors of only a few milliseconds per year.

As mentioned previously, LC tuned circuit oscillators are fine when the oscillation frequency needs to be accurate only to within a few percent. For electronic applications where higher precision is required, piezoelectric crystal oscillators are most commonly employed. With these devices accuracies of 10 to 50 ppm (0.001 to 0.005%) can readily be obtained. However, before getting into the electronic details of crystal oscillators, it will be worthwhile first to review the basic principles governing their behavior.

The word "piezoelectric" comes from the Greek and literally means "pressure electricity." It is used to describe a class of materials that exhibit an interaction between their electrical and mechanical properties. Some piezoelectric materials, such as quartz, are found in nature, whereas others, including barium titanate and lead zirconium titanate (PZT), are manufactured.

If a piezoelectric crystal is compressed along a specific direction, the atoms within it are distorted and orient themselves so that a measurable voltage is produced. Conversely, if a voltage is applied to this crystal, the forces created

8.6-1 Accuracy of Different Oscillation (Time-Measurement) Schemes

Timing Source	Q	Total Drift/Day
Atomic clock	10^8	~8 μs
Temperature-compensated crystal oscillator	5×10^4	~500 μs
Crystal oscillator	5×10^4	~20 ms
Balance wheel or tank-type oscillator circuit	100	~8 s

will cause it to change its shape, and if the voltage is sinusoidal, the crystal will vibrate mechanically at the frequency of the applied electrical signal.

Because of this electromechanical interaction phenomenon, piezoelectric crystals that are cut to specific thicknesses and have conducting electrodes attached (see Figure 8.6-2a) exhibit electromechanical resonances similar to those obtained with LC tuned circuits. In fact, a crystal of this type can be shown to resonate at any frequency, f_n, where the crystal thickness is an odd number of vibratory wavelengths. Specifically, we may express the possible resonant frequencies of such a crystal as

$$f_n = \frac{nc}{t} \tag{8.6-4}$$

where c = speed of mechanical wave propagation in the crystal
t = crystal thickness
and n = 1, 3, 5, 7, etc.

Electromechanical resonance at even harmonics in piezoelectric crystals is not possible because the voltage across the crystal (which is the integral of the electric field) will always be zero if n is even.

Although the electromechanical interaction theory governing the vibration of piezoelectric crystals is beyond the scope of this book, the electrical equivalent circuit that results is still extremely useful. The basic form of this equivalent circuit is shown in Figure 8.6-2b. Here the shunt branch containing R_1–L_1–C_1 represents the crystal's resonance at the fundamental frequency;

Figure 8.6-2

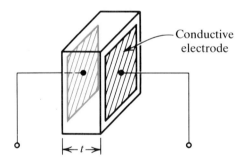

(a)   Physical Structure of a Piezoelectric Crystal Resonator

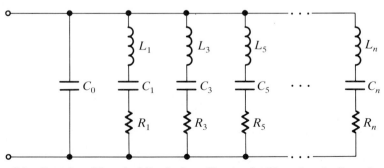

(b)   Complete Electrical Equivalent Circuit of a Piezoelectric Crystal

R_3–L_3–C_3, the third harmonic; and so on. When the crystal is operating at or near its fundamental frequency, for example, all the other resonant branches are effectively open circuits and the equivalent circuit reduces to that given in Figure 8.6-3a.

In this equivalent circuit the capacitor C_0 represents the electrostatic capacitance associated with the metallic electrodes attached to the crystal in order to make electrical contact with it. It may also include the header capacitance of the crystal, as well as any additional wiring capacitance between it and the circuit to which it is connected. The L, R, and C elements in the branch shunting C_0 are associated mainly with the mechanical properties of the crystal. The inductor L_1 is proportional to the crystal mass, C_1 is associated with the interatomic elasticity characteristics of the crystal, and R_1 takes into account the crystal losses due to the mass of the electrodes and due to collisions of the vibrating crystal surface with the air or inert-gas molecules contained inside the container housing the crystal. The magnitude of these crystal equivalent-circuit parameters depends, of course, on the physical dimensions of the crystal, and typical values for these quantities are listed in Figure 8.6-3b. The incredibly high equivalent electrical Q values of piezoelectric crystals, along with their stability with respect to time and temperature, accounts for the popularity of these crystals in electronic time-base generation circuits such as oscillators, frequency meters, and clocks.

Figure 8.6-3

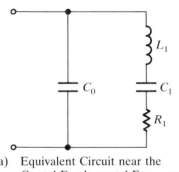

(a) Equivalent Circuit near the Crystal Fundamental Frequency

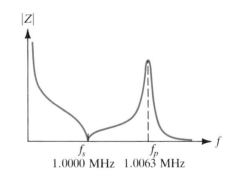

Parameter	32 kHz	1 MHz	50 MHz
L_1	500 H	500 mH	10 mH
R_1	10 kΩ	60 Ω	20 Ω
C_1	0.05 pF	0.005 pF	0.0014 pF
C_0	10 pF	4 pF	4 pF
Q	10^4	3×10^4	10^5

(b) Typical Piezoelectrical Crystal Parameters at Different Crystal Frequencies

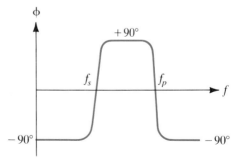

(c) Gain–Phase Characteristics for a 1 MHz Crystal with L_1 = 500 mH, C_1 = 0.005 pF, and C_0 = 4 pF

The impedance of the crystal equivalent circuit in Figure 8.6-3a may be written as

$$Z = \frac{(1/sC_0)(R_1 + sL_1 + 1/sC_1)}{(1/sC_0) + R_1 + sL_1 + 1/sC_1}$$

$$= \frac{s^2 L_1 C_1 + sR_1 C_1 + 1}{s(C_1 + C_0)\left[s^2 L_1\left(\dfrac{C_1 C_0}{C_1 + C_0}\right) + sR_1\left(\dfrac{C_1 C_0}{C_1 + C_0}\right) + 1\right]} \tag{8.6-5}$$

This transfer function has the typical gain–phase characteristic illustrated in Figure 8.6-3c for the case of 1-MHz crystal. From these curves it is apparent that the crystal exhibits a series resonance of f_s and a parallel resonance at f_p, with f_p occurring slightly above f_s. Working with the transfer function in equation (8.6-5), it is not too difficult to show that the slope of the phase plot in the vicinity of f_s and f_p follows the same form as that given in equation (8.6-3c), so that for this crystal a 10° phase change requires a frequency shift of only about $0.175/(2 \times 3 \times 10^4)$, or about 3 ppm. At 1 MHz, this corresponds to a frequency change of only 3 Hz.

A better understanding of the resonance characteristics of this crystal may be obtained if we consider that the crystal is lossless, that is, that $R_1 = 0$. For this case the equivalent circuit has the form given in Figure 8.6-4a, and the transfer function for $Z(s)$ reduces to

$$Z(s) = \frac{s^2 L_1 C_1 + 1}{s(C_1 + C_0)[s^2 L_1[C_1 C_0/(C_1 + C_0)] + 1]} \tag{8.6-6}$$

From this impedance function it is clear that the series resonance [where $Z(j\omega)$ is zero] occurs at

$$f_s = \frac{1}{2\pi\sqrt{L_1 C_1}} = 1.000000 \text{ MHz} \tag{8.6-7a}$$

where the impedances of the inductor L_1 and the capacitor C_1 cancel (Figure 8.6-4b).

Similarly, parallel resonance of this tuned circuit [where $Z(j\omega)$ is infinite] occurs at

$$f_p = \frac{1}{2\pi\sqrt{L_1[C_1 C_0/(C_1 + C_0)]}} = 1.006312 \text{ MHz} \tag{8.6-7b}$$

when the denominator of equation (8.6-6) reduces to zero. At this point the L_1–C_1 portion of the crystal equivalent circuit is slightly inductive, and this inductance resonates with the reactance of C_0 (Figure 8.6-4c).

The difference between the parallel and series resonant frequencies of the crystal may be found by subtracting equation (8.6-7b) from equation (8.6-7a), which yields

$$f_p - f_s = \frac{1}{2\pi}\left\{\frac{1}{\sqrt{L_1[C_1 C_0/(C_1 + C_0)]}} - \frac{1}{\sqrt{L_1 C_1}}\right\} \tag{8.6-8}$$

Factoring out the term $1/\sqrt{L_1 C_1}$, we have

$$f_p - f_s = \frac{1}{2\pi\sqrt{L_1 C_1}}\left(\sqrt{\frac{C_1 + C_0}{C_0}} - 1\right) = f_s\left(\sqrt{1 + \frac{C_1}{C_0}} - 1\right) \tag{8.6-9a}$$

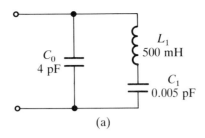

(a)

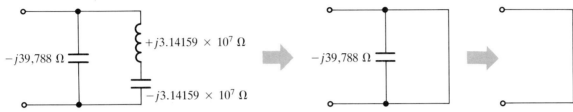

(b)   Crystal Equivalent Circuit at Series Resonant Frequency f_s = 1.0000 MHz

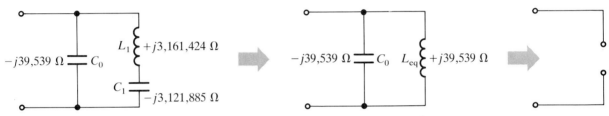

(c)   Crystal Equivalent Circuit at Parallel Resonant Frequency f_p = 1.006312 MHz

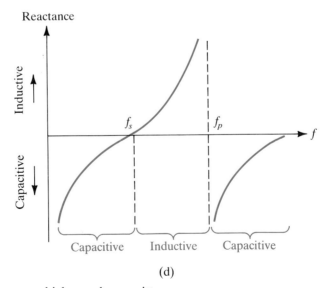

(d)

Figure 8.6-4

which may be rewritten as

$$\frac{f_p - f_s}{f_s} \simeq \frac{C_1}{2C_0} \tag{8.6-9b}$$

using the fact that $(1 + x)^{1/2} \simeq (1 + x/2)$ for small x. Using the typical component parameters given in Figure 8.6-3b, the relative difference between the series and parallel resonant frequencies for a 1-MHz crystal, for example, should be on the order of 0.06%, or about 600 ppm. Thus regardless of whether a crystal oscillates in its series or in its parallel mode, the oscillatory frequencies will be nearly the same.

The reactance of this idealized crystal is sketched as a function of frequency in Figure 8.6-4d. At low frequencies L_1 is nearly a short, so that the overall crystal reactance is associated with the parallel capacitance of C_1 and C_0. As the frequency is raised, the impedance of L_1 increases and eventually it series resonates with C_1 at f_s. With further increases in frequency, the crystal impedance takes on an inductive character because the L_1–C_1 branch has a small $+j$ impedance value and this effectively "shorts out" the C_0 branch. As the frequency is raised, the inductive reactance of the L_1–C_1 branch continues to increase until at $f = f_p$ its magnitude equals that of the C_0 branch, and parallel resonance occurs. Beyond this frequency the L_1–C_1 branch begins to open, and we are left with the capacitive reactance branch containing C_0.

Now that we have a basic understanding of the terminal characteristics of piezoelectric crystals, let's see how they can be employed in the design of electronic oscillator circuits. Figure 8.6-5 illustrates three different types of crystal oscillators. The circuit given in part (a) is essentially the same as that discussed in Example 8.4-3 except that the emitter bypass capacitor in the original circuit has been replaced by a 1-MHz piezoelectric crystal. For this circuit to have enough loop gain to oscillate, it is necessary that the emitter resistor be bypassed to ground. In Example 8.4-3 this was accomplished by bypassing R_E with a capacitor. Following this same approach, then, it is clear that this circuit can oscillate only if the crystal is operating in its series resonant mode. This can occur at any one of the frequencies 1 MHz, 3 MHz, 5 MHz, and so on; however, the tuned circuit in the collector of the transistor ensures that this circuit has loop gain only at 1 MHz and can therefore oscillate only at this fre-

Figure 8.6-5
Typical crystal oscillator circuits.

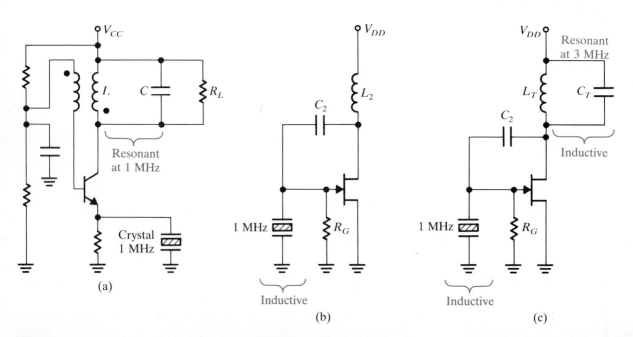

(a)

(b)

(c)

Chapter 8 / Amplifier Stability and Oscillators

quency. Notice that this circuit could also be made to oscillate at 3 MHz, for example, simply by retuning the tank to this frequency.

The circuit shown in Figure 8.6-5b is known as a Hartley-style crystal oscillator. In theory, it is possible for this circuit to oscillate at any odd harmonic of the basic crystal frequency, but we will assume that the transistor's high-frequency characteristics are such that oscillation at any frequency other than the fundamental at 1 MHz is impossible.

Clearly, for this circuit to have any loop gain, it is necessary that the crystal operate close to its parallel resonant frequency. However, this by itself will not be sufficient to guarantee that the circuit will oscillate. Because the transistor has a 180° phase shift between the gate and the drain, an additional 180° of phase shift will need to occur as the signal traverses the feedback network. If the crystal oscillates slightly below its parallel resonant frequency f_p (see Figure 8.6-4d), the crystal impedance will still be large, but it will appear inductive. For this situation the feedback network has the form of a π network, and as we shall show below, this type of network can provide the additional 180° of phase shift needed for the circuit to oscillate.

The basic form of the π feedback network used with the oscillator in Figure 8.6-5b is shown in Figure 8.6-6a. The impedance $Z(s)$ of this network can immediately be written as

$$Z(s) = \frac{sL_2(sL_3 + 1/sC_2)}{sL_2 + sL_3 + 1/sC_2} = \frac{sL_2(1 + s^2L_3C_2)}{1 + s^2(L_2 + L_3)C_2} \tag{8.6-10a}$$

or

$$Z(j\omega) = \frac{j\omega L_2(1 - \omega^2 L_3 C_2)}{1 - \omega^2(L_2 + L_3)C_2} \tag{8.6-10b}$$

At the resonant frequency of this network,

$$\omega_0 = \frac{1}{\sqrt{(L_2 + L_3)C_2}} \tag{8.6-11}$$

the impedance $Z(j\omega)$ goes to infinity and V_d is simply equal to $-I_dR_D$. The corresponding voltage V_g at this frequency is seen to be

$$\frac{V_g}{V_d} = \frac{sL_3}{sL_3 + 1/sC_2}\bigg|_{s=j\omega_0} = \frac{-\omega_0^2 L_3 C_2}{1 - \omega_0^2 L_3 C_2} \tag{8.6-12a}$$

and substituting in the expression in equation (8.6-11) for ω_0, we obtain

$$\frac{V_g}{V_d} = -\frac{L_3}{L_2} \tag{8.6-12b}$$

Notice that, as required, the π network has a 180° phase shift at its resonant frequency, and that in addition, its "gain" is determined by the ratio of the effective crystal inductance at the oscillatory frequency to that of the inductance L_2. Thus, for the oscillator circuit in Figure 8.6-4b, the overall small-signal loop gain is

$$\text{small-signal loop gain} = (g_m R_D)\left(\frac{L_3}{L_2}\right) \tag{8.6-13}$$

For this circuit to oscillate, two design criteria must be satisfied. First, L_2 and C_2 must be chosen so that $1/\sqrt{L_2 C_2}$ is greater than ω_0. This will guarantee that the crystal inductance L_3 will be able to lower the resonant frequency of the π network to precisely equal that of the crystal. Second, the value of L_3 required to accomplish this task must result in an overall small-signal loop gain [see equation (8.6-13)] which is greater than 1 if oscillations are to be self-starting.

The circuit shown in Figure 8.6-5c is essentially the same as that presented in Figure 8.6-5b except that it contains a tank circuit in the drain tuned to $3\omega_0$. As a result, this circuit will oscillate at 3 MHz, the third harmonic or third overtone of the crystal, because this is the only frequency for which the loop gain can be greater than 1. As with the circuit in Figure 8.6-5b, 180° of phase shift will be required as the signal passes through the feedback network. To accomplish this, we again want the feedback network to look like an $L–C–L$ π structure. This will again require that the crystal oscillate slightly below its parallel resonant frequency, so that it looks inductive. In addition, it is necessary that the tank circuit look inductive at the frequency of oscillation, and this requires that the resonant frequency of the tank be adjusted so that it is slightly above the resonant frequency of the crystal. In this way, at $\omega = 3\omega_0$ the reactance of L_T will be less than the reactance of C_T, and the parallel combination will appear inductive.

A π network can also be constructed using a $C–L–C$ circuit of the form illustrated in Figure 8.6-6b, and by following the same procedure as that devel-

Figure 8.6-6

L–C–L, and C–L–C π network.

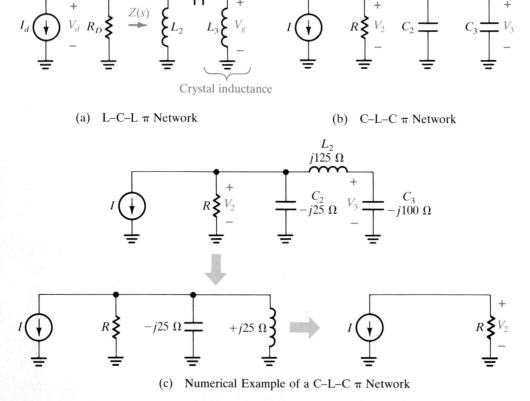

(a) L–C–L π Network

(b) C–L–C π Network

(c) Numerical Example of a C–L–C π Network

oped in equations (8.6-10) through (8.6-12), it can readily be shown that the resonant frequency for this network is

$$\omega_0 = \frac{1}{\sqrt{L_2[C_2 C_3/(C_2 + C_3)]}} \qquad (8.6\text{-}14a)$$

and that the transfer function gain at this frequency is given by the ratio of the capacitors as

$$\frac{V_3}{V_2} = -\frac{C_2}{C_3} \qquad (8.6\text{-}14b)$$

At first glance it is a little difficult to understand how these π networks provide the required 180° phase shift, and even more interestingly, how they can exhibit any gain. Let's illustrate how the C–L–C π network in Figure 8.6-6b "works" by means of a specific example. In particular, let's consider that $C_2 = 4C_3$, and assume that at the resonant frequency of the network that C_3's impedance is $-j100$ Ω. Therefore, the corresponding impedance of C_2 must be $-j25$ Ω, since this capacitor is four times larger than C_3 (see Figure 8.6-6c). If the L_2–C_3 branch is to resonate with the branch containing C_2, the impedance of L_2 at resonance must be $j125$ Ω. With this value for the reactance for L_2, the transfer function for V_3/V_2 may be written as

$$\frac{V_3}{V_2} = \frac{-j100}{j125 - j100} = \frac{-j100}{j25} = -4 \qquad (8.6\text{-}15)$$

Of course, this is the same result that we would have obtained had we applied equation (8.6-14b) directly, but this approach provides more physical insight into the actual behavior of the circuit.

EXAMPLE 8.6-1

The Pierce crystal oscillator illustrated in Figure 8.6-7a employs feedback from the drain to the gate. For this circuit to oscillate, the crystal must operate near its series resonant frequency so that the loop gain is high. In addition, because 180° of phase shift is needed in the feedback network, the crystal will need to operate slightly above its series resonant frequency where the crystal looks inductive, so that the feedback circuit forms a C–L–C π network. The FET parameters are $I_{DSS} = 8$ mA and $V_p = -2.5$ V.

a. Sketch v_{DS}, v_{GS}, and i_D assuming that the i_D versus v_{GS} characteristic has the linear form shown in Figure 8.6-7b.
b. Find the equivalent inductance of the crystal under steady-state oscillation conditions.
c. Determine the exact frequency of oscillation of this circuit if 1 MHz is the series resonant frequency of the crystal and if its equivalent-circuit parameters are $C_1 = 0.001$ pF and $C_0 = 5$ pF.

SOLUTION

Because the i_D versus v_{GS} characteristic is assumed to be linear, for all values of v_{GS} above pinch-off, the transistor g_m is approximately constant at

$$g_m = \frac{I_{DSS}}{-V_p} = 3.2 \text{ mS} \qquad (8.6\text{-}16)$$

If we break the loop at the point labeled "×" in the circuit and apply a

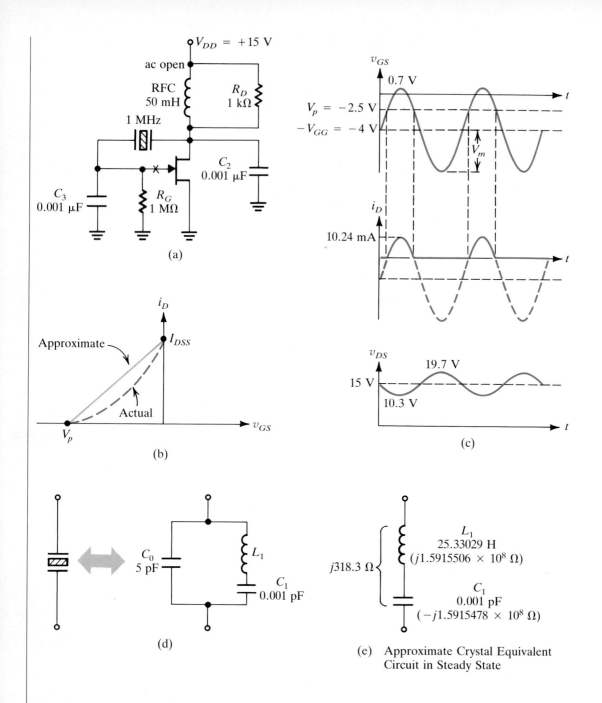

Figure 8.6-7
Example 8.6-1.

voltage V_g at this point, the corresponding drain current and drain voltage may be written as

$$I_{d1} = K_1 g_m V_g \qquad (8.6\text{-}17a)$$

and

$$V_d = -K_1 g_m R_D V_g \qquad (8.6\text{-}17b)$$

respectively. Because C_2 and C_3 are equal, following Figure 8.6-6b and equation (8.6-14b), we can see that the gain of this π network is simply equal to

Chapter 8 / Amplifier Stability and Oscillators

−1. As a result, the gate voltage V_g' is equal to

$$V_g' = -V_d = K_1 g_m R_D V_g \qquad (8.6\text{-}18)$$

For small-signal levels the factor $K_1 = 1$, so that the small-signal loop gain for this oscillator is simply equal to $g_m R_D = 3.2$. Because it is much greater than 1, this oscillator is guaranteed to self-start.

As the oscillation amplitude builds up, the gate voltage will be clamped negatively to zero volts by the gate diode (actually, to $+0.7$ V), and for those portions of the gate voltage where v_{GS} drops below -2.5 V, the transistor drain current will cut off. This will reduce the value of K_1, and in the steady state the waveshapes will have the general form indicated in Figure 8.6-7c and K_1 will equal

$$K_1 = \frac{X_1}{X_m} = 0.31 \qquad (8.6\text{-}19a)$$

In addition, from Figure 8.5-7 we obtain

$$\frac{X_E}{X_m} = -0.32 \qquad (8.6\text{-}19b)$$

By examining the sketch for v_{GS}, it is apparent that

$$V_m = V_{GG} + 0.7 \qquad (8.6\text{-}20a)$$

and using equation (8.5-19b), we may also write

$$\frac{X_E}{X_m} = \frac{V_{GG} + V_p}{V_m} = 0.32 \qquad (8.6\text{-}20b)$$

Simultaneously solving equations (8.6-20a) and (8.6-20b), we obtain

$$V_{GG} = 4 \text{ V} \qquad (8.6\text{-}21a)$$

and
$$V_m = 4.7 \text{ V} \qquad (8.6\text{-}21b)$$

Because the gain of the π network is -1, the ac portion of the drain voltage is also 4.7 V but $180°$ out of phase from the v_{GS} waveshape.

Following Figure 8.6-6b, we may express i_D in terms of v_{GS} as

$$i_D = I_{DSS} + g_m v_{GS} \qquad (8.6\text{-}22)$$

and using this equation, the drain current corresponding to the peak gate-to-source voltage of 0.7 V is $8 + (3.2)(0.7)$ or about 10.24 mA. Of course, the drain current is zero whenever v_{GS} drops below the pinch-off voltage of -2.5 V. Because $V_m = 4.7$ V, the corresponding drain current sine-wave amplitude is $I_m = g_m V_m = 15$ mA. Using this result in conjunction with equation (8.6-19a), we may compute the fundamental component of the drain current and use this to double check our solution for the drain voltage. In particular, we have

$$I_{d1} = 0.31 I_m = 4.67 \text{ mA} \qquad (8.6\text{-}23a)$$

and
$$V_d = I_{d1} R_D = 4.67 \text{ V} \qquad (8.6\text{-}23b)$$

This result compares favorably with the value for v_{DS} obtained earlier.

Let's take a closer look now at the operation of the crystal in this circuit. Neglecting any losses, its equivalent circuit has the form illustrated in Figure 8.6-7d, and because the crystal is assumed to be series resonant at precisely 1 MHz, we may calculate the equivalent crystal inductance L_1 as

$$L_1 = \frac{1}{(2\pi f_0)^2 C_1} = 25.33 \text{ H} \tag{8.6-24}$$

In the steady state, the crystal forms the inductive element of the π network in the feedback loop, and as such this network must resonate at approximately 1 MHz, since we expect that the crystal will be operating slightly above its series resonant frequency. Following equation (8.6-14a), the required equivalent inductance of the crystal in the π network is

$$L_{eq} = \frac{1}{(2\pi f_0)^2 [C_2 C_3/(C_2 + C_3)]} = 50.66 \ \mu\text{H} \tag{8.6-25a}$$

which at a frequency of about 1 MHz has an impedance equal to

$$j\omega L_{eq} = j318.3 \ \Omega \tag{8.6-25b}$$

Because the reactance of C_0 at this frequency is more than 30 kΩ, we will neglect it in finding the actual oscillatory frequency of this crystal. After making this approximation, the equivalent circuit under consideration has the form given in Figure 8.6-7e, and following this circuit, we may write

$$j\left(\omega L_1 - \frac{1}{\omega C_1}\right) = j318.3 \tag{8.6-26a}$$

or

$$\omega^2 L_1 C_1 - 318.3\omega C_1 - 1 = 0 \tag{8.6-26b}$$

Solving this quadratic, we find that the crystal is actually oscillating at 1.00000099 MHz, which, as predicted, is very close to the crystal's series resonant frequency, but slightly above it.

Exercises

8.6-1 For the circuit in Figure 8.6-7, $C_{gs} = 50$ pF.
a. Find the crystal impedance at the frequency of oscillation.
b. Determine the shift in oscillation frequency (in ppm) from the series resonant frequency stamped on the crystal.
Answer (a) $j310.888 \ \Omega$ (b) 970 ppm

8.6-2 The crystal illustrated in Figure 8.6-7d series resonates at 1 MHz.
a. Find the parallel resonant frequency.
b. Estimate the shift (in ppm of f_s and f_p) if the crystal is shunted by a 100-pF capacitor.
Answer (a) $f_p = 1.0001058$ MHz (b) f_s is the same, f_p decreases by 95 ppm

8.6-3 For the circuit in Figure 8.6-5b, the transistor g_m is such that the circuit just oscillates. If $L_2 = 100 \ \mu$H, $C_2 = 100$ pF, and the crystal properties are $C_0 = 5$ pF, $L_1 = 25.33$ H, and $C_1 = 0.001$ pF, find the exact frequency of oscillation. *Answer* $f_0 = 1.000008778$ MHz

8.7 NONSINUSOIDAL OSCILLATORS

Besides being used to produce sine waves, oscillators are employed for the generation of square, triangle, sawtooth, and many other types of special purpose nonsinusoidal waveshapes. As with the sinusoidal oscillator, these circuits are also constructed by adding positive feedback to an amplifier in order to place its poles in the right-hand plane, but there is one basic difference. In a sinusoidal oscillator the loop gain is adjusted so that the system poles lie on or near the $j\omega$ axis, whereas for nonsinusoidal oscillators the poles are usually placed on the positive real axis. As a result, the "transient response" of these circuits contains growing exponential terms which rapidly cause nonlinear amplitude limiting in the output. Figure 8.7-1 illustrates how the output of a typical two-pole oscillator will change as the system poles move farther into the right-hand

Figure 8.7-1
Effect of the pole positions on the oscillator steady-state output.

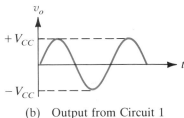

(b)   Output from Circuit 1

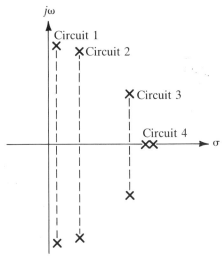

(a)   Possible Oscillator Pole Locations

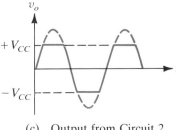

(c)   Output from Circuit 2

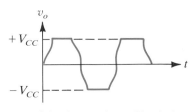

(d)   Output from Circuit 3

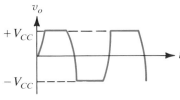

(e)   Output from Circuit 4

plane, gradually changing from purely imaginary to purely real roots. As indicated in the figure, when this transition occurs, the output will gradually change over from nearly perfect sine waves, to clipped sine waves, to clipped exponentially growing sine waves, to clipped growing exponentials.

Most nonsinusoidal oscillators contain a square-wave generator as the fundamental signal generation element, and further processing of this signal can be used to create other types of periodic signals. For example, by integrating the output from a square-wave generator, we can produce triangle waves, and by rounding off the tips of the triangle waves, signals closely approximating sine waves can be created (Problem 7.5-15). In addition, if the square-wave output described previously has a duty cycle of other than 50%, the integration of this signal will produce a sawtooth-like waveshape.

Signal generators that are capable of producing sine, square, and triangle waves are known as function generators. Until about 15 years ago, equipment of this type was extremely complex and very expensive. However, today, low-cost single-chip function generator ICs are available which contain almost all the hardware necessary for producing sine, square, and triangle waves at frequencies from near dc upward to several hundred kilohertz. The XR-2206 integrated circuit manufactured by EXAR Corporation is a good example of the state of the art in this area.

As we have already noted, at the heart of nearly all nonsinusoidal oscillators there is generally a square-wave generator circuit. Therefore, because of its importance, we begin this section by investigating techniques for generating square waves. Then we will see how these methods can be extended to design circuits capable of generating other types of nonsinusoidal waveforms.

For a circuit to be able to operate as a square-wave generator, it is necessary that it have two basic properties:

1. It must have poles in the right-hand plane (preferably on the positive real axis).
2. Once the amplitude at the output of the amplifier portion of the generator has gone into nonlinear limiting, the external circuit connected to it must eventually cause the amplifier to reenter the active region so that it can then switch to the opposite nonlinear output state.

Property 1 guarantees that the generator circuit will switch to the opposite state once it enters the active region, and property 2 ensures that the circuit will oscillate, rather than simply hanging up forever in one of the system's nonlinear states.

In our analysis of nonsinusoidal oscillators, we concentrate on op amp circuits, in which the output limiting occurs basically at the positive and negative power supply rails. However, these techniques can also be applied directly to the analysis of discrete transistor circuits and digital IC oscillators (Chapter 11).

The positive feedback circuits we examine fall into two major categories. In the first type the amplifier itself has local positive feedback and the timing network $[\beta(s)]$ is connected to the inverting input terminal of the amplifier. In the second type the local feedback is negative and the timing network $[\beta(s)]$ feeds back to the noninverting input terminal of the amplifier to provide the positive feedback needed for the circuit to oscillate.

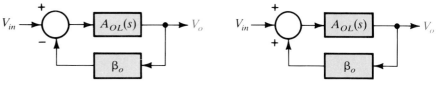

Figure 8.7-2

(a)   Negative Feedback Amplifier   (b)   Positive Feedback Amplifier

To analyze these two different types of op amp oscillators, it will be useful to redevelop the expressions for the transfer functions of amplifiers containing both negative and positive local feedback. The basic form of these circuits is illustrated in Figure 8.7-2a and b, respectively, and the transfer function for each case can readily be shown to be

$$\frac{V_o}{V_{in}} = \frac{A_{OL}(s)}{1 + \beta_o A_{OL}(s)} \tag{8.7-1a}$$

when the feedback is negative, and

$$\frac{V_o}{V_{in}} = \frac{A_{OL}(s)}{1 - \beta_o A_{OL}(s)} \tag{8.7-1b}$$

for the case of positive feedback.

The location of the poles for each of these amplifiers is found by determining the values of s for which the denominators of equations (8.7-1a) and (8.7-1b) are zero. For the negative feedback amplifier, this occurs at values of s where

$$1 + \beta_o A_{OL}(s) = 0 \tag{8.7-2a}$$

whereas for the positive feedback amplifier, the poles are located at the values of s satisfying the equation

$$1 - \beta_o A_{OL}(s) = 0 \tag{8.7-2b}$$

When the amplifier to be used is an op amp, the circuits employed to produce negative or positive local feedback generally have the form shown in Figure 8.7-3a and c, respectively. Let's determine the transfer function $V_o/V_{in} = A(s)$ for each of these circuits. In carrying out these calculations, we assume that the op amps have dominant-pole frequency compensation, so that their open-loop transfer function has the form

$$A_{OL}(s) = \frac{\alpha_o A_o}{s + \alpha_o} \tag{8.7-3}$$

where A_o is the dc open-loop gain, and α_o is the high-frequency 3-dB point.

For the negative feedback amplifier illustrated in Figure 8.7-3a, substituting equation (8.7-3) into equation (8.7-1a), we may write down this amplifier's closed-loop gain as

$$A(s) = \frac{V_o}{V_1} = \frac{A_{OL}(s)}{1 + \beta_o A_{OL}(s)} = \frac{\alpha_o A_o}{s + \alpha_o(1 + \beta_o A_o)} \tag{8.7-4a}$$

where

$$\beta_o = 1 + \frac{R_f}{R_2} \tag{8.7-4b}$$

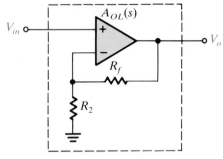

(a) Amplifier with Local Negative Feedback

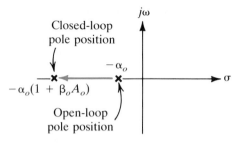

(b) Pole–Zero Diagram for an Amplifier with Local Negative Feedback

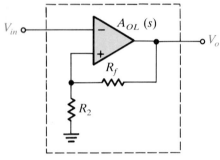

(c) Amplifier with Local Positive Feedback

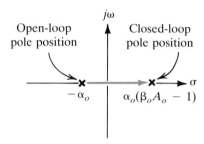

(d) Pole–Zero Diagram for an Amplifier with Local Positive Feedback

Figure 8.7-3

The pole–zero plot for this amplifier, shown in Figure 8.7-3b, indicates that the local negative feedback has taken the open-loop pole at $-\alpha_o$ and moved if farther into the left-hand plane, increasing both the bandwidth and the stability of this circuit.

Following a similar approach, but this time using equations (8.7-3) and (8.7-1b), we may write down the transfer function for the positive feedback circuit shown in Figure 8.7-3c as

$$A(s) = \frac{V_o}{V_2} = \frac{A_{OL}(s)}{1 - \beta_o A_{OL}(s)} = \frac{\alpha_o A_o}{s + \alpha_o(1 - \beta_o A_o)} \qquad (8.7\text{-}5)$$

where β_o has the same definition as in equation (8.7-4b). The pole–zero plot for this amplifier configuration is presented in Figure 8.7-3d. Notice that this amplifier (due to its built-in local positive feedback) already has a pole on the positive real axis, while the negative feedback amplifier in Figure 8.7-3a will have to wait for the addition of the positive feedback of the timing network to push this pole into the right-hand plane.

The circuits shown in Figure 8.7-4 illustrate four possible square-wave generator designs using a single resistor and a single capacitor as the timing elements. It is useful to notice that these designs represent the four circuit combinations possible using positive or negative local feedback together with an RC timing network. Interestingly enough, whereas all four of these circuits have poles on the positive real axis, only two of them are capable of generating square waves. This is because the other two circuits have outputs that hang up

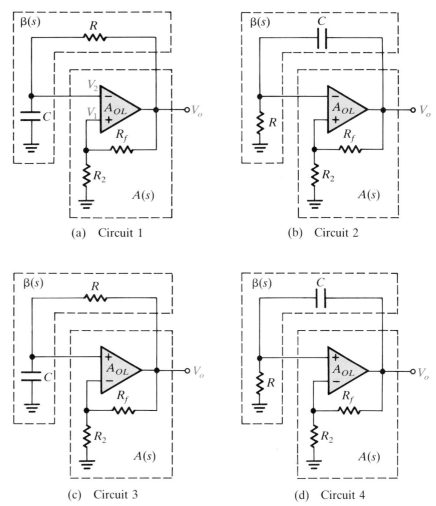

Figure 8.7-4
Possible forms of an RC feedback network square-wave generator.

(a) Circuit 1

(b) Circuit 2

(c) Circuit 3

(d) Circuit 4

at one of the power supply rails, since once the output saturates, the amplifier is unable to return to the active region.

Let's try to see which of these circuits can and which cannot oscillate, for the reason just cited. We will begin by examining circuit 1 in Figure 8.7-4a, assuming that the output has just switched to the $-V_{CC}$ voltage level at $t = 0+$. As a result, at this point the voltage v_2 at the inverting input terminal must be greater than $v_1 = -\beta_o V_{CC}$ at the noninverting input.

During the time that the output is saturated at $-V_{CC}$, the overall equivalent circuit has the form shown in Figure 8.7-5a, and by examining this circuit it should be clear that for $t > 0$, the capacitor C will begin to charge toward $-V_{CC}$. Therefore, at some point in time ($t = t_1$ in Figure 8.7-5b), v_2 will eventually equal v_1, the op amp will enter the active region, and switching to the $+V_{CC}$ output state will occur. The mechanism by which this switching to the opposite state occurs will be explained shortly.

In a similar fashion, we can show that the circuit will automatically reenter the active region from the $+V_{CC}$ state, so that switching back to $-V_{CC}$ will also be possible. As a result, this circuit is capable of generating square waves at its output.

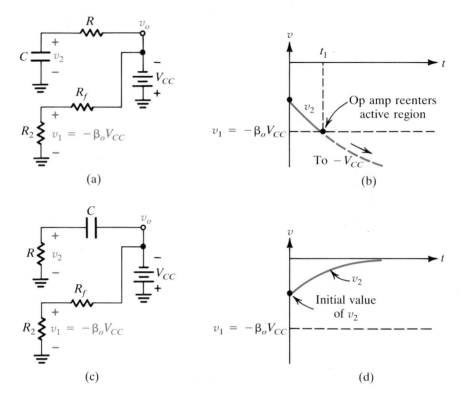

Figure 8.7-5
Analysis of circuit 1 in Figure 8.7-4a.

(a)   (b)

(c)   (d)

Consider next the circuit presented in Figure 8.7-4b, and as before, let's assume that the op amp output is again at $-V_{CC}$ at $t = 0+$. To be in this state, the voltage at the inverting input terminal v_2 must be greater than that at the noninverting terminal v_1. In addition, for v_o to be able to switch to the $+V_{CC}$ output state, v_2 must fall below $v_2 = -V_{CC}$. However, following the equivalent circuit presented in Figure 8.7-5c, we can see that as C charges, v_2 will simply decrease toward zero and v_2 will never drop below v_1. As a result, if v_o starts out in the $-V_{CC}$ state, it will remain hung up there forever.

Similar analyses for circuits 3 and 4 demonstrate that circuit 3 will not reenter the active region on its own, whereas circuit 4 will. Thus although all four of these circuits have poles on the positive real axis, only circuits 1 and 4 will be able to operate as square-wave generators. The other two will simply hang up at one of the power supply rails when turned on.

Now that we understand how circuits 1 and 4 reenter the active region, let's try to get some additional insight into how and why the output switches from one state to the other once the amplifier is active. Although this investigation will be carried out only for circuit 1, the reader is invited to carry out the parallel analysis for circuit 4.

We begin this discussion by noting that

$$\beta(s) = \frac{\dfrac{1}{sC}}{R + 1/sC} = \frac{\alpha}{s + \alpha} \qquad (8.7\text{-}6a)$$

where

$$\alpha = \frac{1}{RC} \qquad (8.7\text{-}6b)$$

In addition, we note that $A(s)$ is given by equation (8.7-5). Because the β network for this oscillator uses negative feedback, following the previous local feedback analysis [see equation (8.7-2a)], we can see that the overall poles of this circuit are located at the values of s where

$$1 - \beta(s)A(s) = 0 \tag{8.7-7a}$$

or
$$1 - \frac{\alpha}{s + \alpha} \frac{\alpha_o A_o}{s + \alpha_o(1 - \beta_o A_o)} = 0 \tag{8.7-7b}$$

For large values of dc open-loop gain, A_o, this expression may be simplified to

$$s^2 - \alpha_o \beta_o A_o s + \alpha \alpha_o A_o = 0 \tag{8.7-8}$$

and using the fact that $(1 + x)^{1/2} \approx (1 + x/2)$ for small x, the roots of equation (8.7-8) are seen to be located approximately at

$$s_1 = \alpha_o \beta_o A_o \tag{8.7-9a}$$

and
$$s_2 = \frac{\alpha}{\beta_o} = \frac{1}{\beta_o RC} \tag{8.7-9b}$$

Notice that both roots are in the right-hand half of the s-plane, with the s_1 root related to the dominant-pole capacitor in the op amp, and the s_2 root related to the timing capacitor C. Furthermore, because $s_1 \gg s_2$, the time constant $1/s_2$ associated with C will be much greater than the time constant $1/s_1$ associated with the op amp compensation capacitor. As a result, during the switching transient (which occurs during the initial period of the exponential growth of the s_1 term), the capacitor C will hardly charge or discharge at all. Thus during the switching, the circuit may be represented as shown in Figure 8.7-6a, with C replaced by a battery equal to $-\beta_o V_{CC}$. In addition, a network consisting of R_o and C_o has been connected at the output of the op amp to model its dominant-pole behavior.

In analyzing this circuit, we assume that just prior to $t = 0$, the op amp output was in the $-V_{CC}$ state, and that at $t = 0-$, the op amp entered the ac-

Figure 8.7-6

"Transient" response of circuit 1 in Figure 8.7-4a for the case where $v_o(0-) = -V_{CC}$.

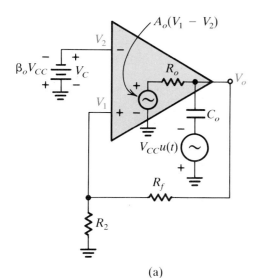

(a)

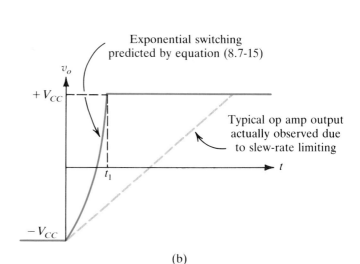

(b)

tive region. Of course, this occurs when v_2, the voltage across C, just reaches $-\beta_o V_{CC}$, the voltage at the v_1 input terminal. At this point the switching process begins, and during the switching transient, because v_C does not change much, we may replace it by a battery equal to $-\beta_o V_{CC}$.

To aid in the Laplace transform analysis of the transient response of this circuit, we have replaced the initial condition voltage on C_o by a voltage source $-V_{CC}u(t)$ connected in series with the capacitor, and using the superposition principle we may write down the output voltage $V_o(s)$ as

$$V_o(s) = A_o(V_1 - V_2)\frac{1/sC_o}{R_o + 1/sC_o} - \frac{V_{CC}}{s}\frac{R_o}{R_o + 1/sC_o}$$

$$= A_o(V_1 - V_2)\frac{\alpha_o}{s + \alpha_o} - \frac{V_{CC}}{s}\frac{s}{s + \alpha_o} \tag{8.7-10}$$

Substituting
$$V_1(s) = \beta_o V_o(s) \tag{8.7-11a}$$

and
$$V_2(s) = -\frac{\beta_o V_{CC}}{s} \tag{8.7-11b}$$

into equation (8.7-10), after some algebraic manipulation, we may rewrite the expression for $V_o(s)$ as

$$V_o(s) = \frac{-V_{CC}(s - \alpha_o\beta_o A_o)}{s[s + \alpha_o(1 - \beta_o A_o)]} \tag{8.7-12}$$

In examining this result, we should observe that as predicted earlier [see equation (8.7-9a)], this oscillator has a real pole located deep in the right-hand plane at approximately $\beta_o A_o \alpha_o$.

Using the partial fraction expansion technique, equation (8.7-12) may be rewritten as

$$V_o(s) = \frac{C_1}{s} + \frac{C_2}{s + \alpha_o(1 - \beta_o A_o)} \tag{8.7-13}$$

where the coefficients C_1 and C_2 can be readily shown to be

$$C_1 \simeq -V_{CC} \tag{8.7-14a}$$

and
$$C_2 = \frac{V_{CC}}{A_o \beta_o} \tag{8.7-14b}$$

As a result, taking the inverse Laplace transform of equation (8.7-13), we obtain

$$v_o(t) = V_{CC}\left(-1 + \frac{1}{A_o\beta_o}e^{\alpha_o\beta_o A_o t}\right) \tag{8.7-15}$$

This result is sketched in Figure 8.7-6b. Basically, $v_o(t)$ starts out at $-V_{CC}$ and grows exponentially until the output limits at the $+V_{CC}$ power supply rail. The switching time for the oscillator may be found by setting equation (8.6-15) equal to $+V_{CC}$ and solving for t. Carrying out these operations, we find that the switching time t_1 for this circuit is given by the expression

$$t_1 = \frac{\ln(2A_o\beta_o)}{\alpha_o A_o \beta_o} \tag{8.7-16}$$

which for typical op amp parameter values $A_o = 10^5$, $\beta_o = \frac{1}{2}$, and $\alpha_o = 10^2$ rad/s results in a predicted switching time of about 2.3 μs. However, in actual op amp circuits, this exponential growth pattern is rarely seen because slewing problems prevent the output from changing this fast. As a result, in most op amp square-wave generators, the switching time for the circuit is determined from the amplifier's slew rate characteristics, and the output seen during the switching is the familiar ramp associated with slew-rate limiting rather than the growing exponential shape depicted in Figure 8.7-6b. For example, with an op amp such as the 741 with a nominal slew rate of 0.5 V/μs, operating with $\pm$10-V supplies, the switching time would be about (20 V)/(0.5 V/μs) or 40 μs and would be linear rather than exponential as predicted by equation (8.7-15). This form of this response is indicated by the dashed curve in Figure 8.7-6b.

Now that we understand how and why square-wave generators switch states, it will no longer be necessary for us to go into great detail about the switching behavior of these circuits. However, before leaving this material, it may be worthwhile for us to summarize what we have learned thus far about square-wave generators:

1. If the circuit has poles in the right-hand plane, the amplifier output will grow exponentially until it self-limits at either plus or minus V_{CC} (for the case of an op amp).
2. If the amplifier reenters the active region, the circuit will switch from its current nonlinear state to the opposite state. For the simple RC oscillator circuits being examined, the switching behavior is determined primarily by the characteristics of the amplifier.
3. During the time that it takes for the circuit to switch states, the voltage across the timing capacitor does not change appreciably, so that the capacitor may be replaced by a battery during the switching transient.
4. The RC timing networks in Figure 8.7-4 may be replaced by a similar set of RL networks to construct other types of elementary square-wave generators. The analysis procedure for these oscillators is essentially the same as that carried out for the RC circuits, except that during the switching transient it is the current through the inductor, which remains constant. As a result, during the switching the inductor may be replaced by a constant-current source.

Up until now we have focused our attention on the requirements for a non-sinusoidal oscillator to switch, and on waveshapes of the switching signals. Now that we understand this portion of the oscillator's behavior, let's redirect our attention to the operation of the timing circuit and develop an understanding of just what the timing waveforms will look like when the circuit is oscillating. In carrying out this portion of the analysis, we assume that the amplifier switches to the opposite state instantaneously once it enters the active region. In addition, we assume that during the switching transient, the energy stored in the timing elements remains constant.

In the examples that follow, we apply these ideas initially to help us determine the timing waveshapes in the square-wave generator circuits we discussed previously. In Example 8.7-2 we extend these results to the analysis of other types of nonsinusoidal oscillator circuits.

EXAMPLE 8.7-1

For the square-wave generator circuit shown in Figure 8.7-4a, consider that the supply voltages are ± 10 V, that $R = 5$ kΩ and $C = 0.1$ μF, and that $R_f = R_2 = 10$ kΩ, so that $\beta_o = \frac{1}{2}$.

a. Explain how this circuit works, assuming that it switches to the opposite state once the amplifier enters the active region.
b. Sketch the voltages v_o, v_1, and v_2 as functions of time, assuming that the switching time of the op amp is zero.
c. Develop an expression for the frequency of oscillation of this circuit.

SOLUTION

To begin the analysis of this circuit's behavior, let us assume that v_o is initially in the $-V_{CC}$ state, and that v_2, although slightly larger than v_1, is just approaching $v_1 = -\beta_o V_{CC}$. Because v_2 is exponentially decreasing toward $-V_{CC}$, the op amp is on the verge of entering the active region, and if we assume that this occurs at $t = 0$, then at $t = 0+$, the output will switch to the $+V_{CC}$ state. During the switching the voltage across C remains constant at $-\beta_o V_{CC}$, and for $t > 0$, the circuit shown in Figure 8.7-7a is valid. For this circuit the voltage at v_1 is now equal to

Figure 8.7-7
Example 8.7-1.

$$v_1 = \beta_o V_{CC} = +5 \text{ V} \qquad (8.7\text{-}17a)$$

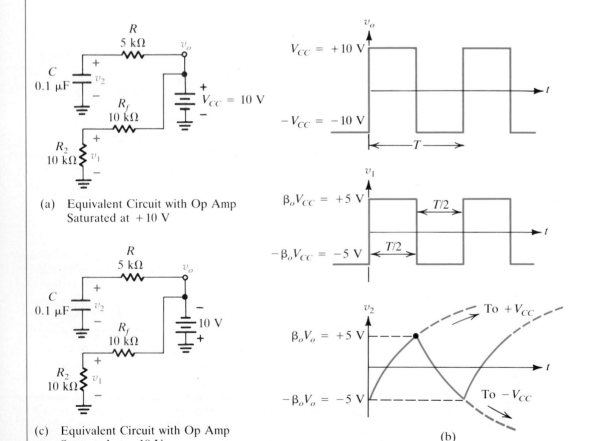

(a) Equivalent Circuit with Op Amp
      Saturated at $+10$ V

(c) Equivalent Circuit with Op Amp
      Saturated at -10 V

(b)

while v_2, the voltage across C, is given by the expression

$$v_2(t) = -5e^{-t/\tau} + 5(1 - e^{-t/\tau}) \qquad (8.7\text{-}17b)$$

where $\tau = RC$. These waveforms, along with that for v_o, are sketched in Figure 8.7-7b. As the capacitor charges, the circuit remains in $+V_{CC}$ state until the voltage v_2 reaches $+5$ V, and at this point the op amp reenters the active region and switches to the opposite output state. The time required for this $\frac{1}{2}$ cycle is found by determining how long it takes for $v_2(t)$ in equation (8.7-17b) to reach $+5$ V from its initial amplitude of -5 V. Setting $v_2(T/2) = 5$ V in this expression and solving for the half period $T/2$, we find that

$$\frac{T}{2} = \tau \ln 3 \qquad (8.7\text{-}18)$$

In a similar fashion, once the output switches to -10 V, the circuit given in Figure 8.7-7c applies and the voltage v_2 across the capacitor begins to fall exponentially toward the -10-V level, switching when it hits -5 V. Basically, the time required for v_2 to go from its initial value at $+5$ V to its switching level at -5 V is given by the same expression as in equation (8.7-18) because the circuit operation is symmetrical. As a result, the overall period of oscillation is given simply by the expression

$$T = 2\tau \ln 3 \qquad (8.7\text{-}19a)$$

while f, the oscillation frequency, is equal to

$$f = \frac{1}{T} = \frac{1}{2\tau \ln 3} \qquad (8.7\text{-}19b)$$

For the components specified in this circuit, the oscillation frequency is approximately 900 Hz.

It is interesting to observe that the operation of this circuit may also be understood from a very different viewpoint. Consider that the combination of the op amp together with R_f and R_2 in Figure 8.7-8a forms a Schmitt trigger. As a result, the op amp output will always be saturated at ± 10 V. Furthermore, us-

Figure 8.7-8

Example 8.7-1 (continued).

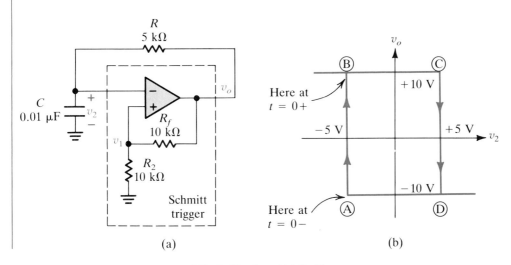

(a)

(b)

ing the component values of this example, the Schmitt trigger hysteresis characteristic can readily be seen to have the form given in Figure 8.7-8b.

To continue this analysis approach, let us assume initially for $t < 0$ that v_o is -10 V, so that the equivalent circuit given in Figure 8.7-7c applies. To be in this state, following the hysteresis curve, v_2 must be greater than -5 V; but from the equivalent circuit, it is apparent that v_2 is charging exponentially toward -10 V. Let us assume that at $t = 0-$, v_2 just reaches -5 V (point A on the hysteresis characteristic). At this time v_o will switch nearly instantaneously to $+10$ V (to point B on the hysteresis curve). Because the voltage on C cannot change instantaneously, it follows that $v_2(0+) = v_2(0-) = -5$ V.

For $t > 0$, the equivalent circuit in Figure 8.7-7a applies and therefore v_2 will now charge exponentially toward $+10$ V. Once v_2 reaches $+5$ V (point C on the hysteresis curve), v_o will switch to -10 V (at point D), and the process will repeat. The waveshapes associated with this circuit's operation were presented in Figure 8.7-7b.

<table>
<tr><td>EXAMPLE 8.7-2</td><td>The circuit shown in Figure 8.7-9 is an oscillator that generates both square and triangle waves. In this circuit IC1 functions as an integrator and IC2 as a Schmitt trigger. The power supply voltages are ± 10 V. Assuming that the circuit switching times are negligible, explain how this circuit operates, and in doing so, sketch v_{o1}, v_1, and v_{o2}.</td></tr>
<tr><td>SOLUTION</td><td>The timing analysis for this circuit closely parallels that used to analyze the previous example except that in place of the passive RC timing network, we now have an active integrator circuit. Initially, let us assume that the Schmitt trigger IC2 is in the -10-V state, and note that for this condition, v_1 must be negative, which requires that v_{o1} be less than $+5$ V (see Figure 8.7-9b). With IC2 in this state the equivalent circuit in Figure 8.7-9c is valid, and because the input to the integrator IC1 is a constant at -10 V, v_{o1}, the output from the integrator, will be a positive-going ramp (Figure 8.7-9e). Eventually, this output will reach $+5$ V, and when it does, v_1 will equal zero, IC2 will enter the active region, and its output will switch to the $+10$-V state (Figure 8.7-9b). Once this occurs, the circuit given in Figure 8.7-9d applies.</td></tr>
</table>

Just prior to the switching, $v_{o1} = +5$ V, and because IC1 is active and v_2' is a virtual ground, v_C is also approximately equal to $+5$ V. Furthermore, as discussed previously, the voltage on C does not change much during the switching, so that $v_C(0+)$ is also equal to $+5$ V. Applying the superposition principle, with $v_{o2}(0+) = +10$ V and $v_{o1}(0+) = +5$ V, we can see that $v_1(0+) = (\frac{2}{3})5 + (\frac{1}{3})10$ or 6.66 V, so that IC2 is firmly in the $+10$-V state.

Following Figure 8.7-9d, the input signal to the integrator is $+10$ V, and as a result, we can see that the voltage $v_{o1}(t)$ is given by the expression

$$v_{o1}(t) = -\frac{1}{RC} \int_0^t v_{in}'(t) \, dt + v_C(0-) = -\frac{10t}{\tau} + 5 \qquad (8.7\text{-}20)$$

Thus the output voltage from the integrator, which is also the input to the Schmitt trigger, is a negative-going ramp starting out at $+5$ V. As $v_{o1}(t)$ reaches -5 V (see Figure 8.7-9b), v_1 will again equal zero, IC2 will reenter the active region, and the Schmitt trigger will switch to the opposite state. These waveshapes are sketched in Figure 8.7-9e. The circuit operation for the

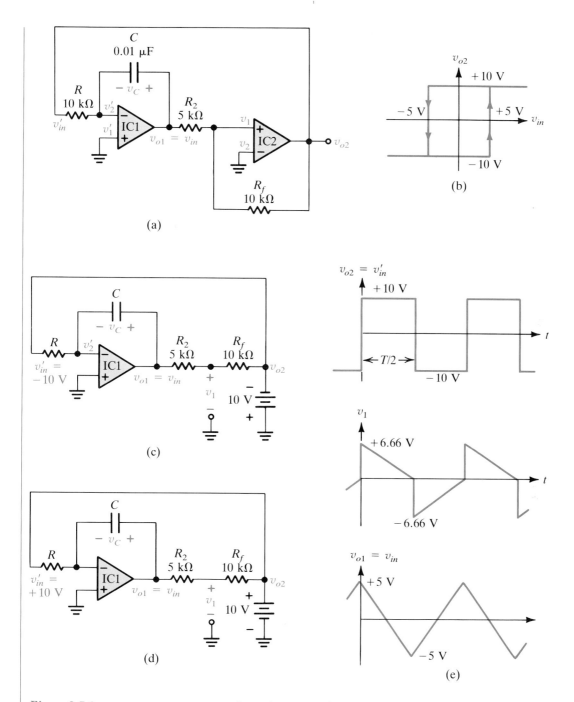

Figure 8.7-9
Example 8.7-2.

case where the output from IC2 is negative, except for the sign differences in the voltages, is identical to that just presented for the case where the output is positive, so that a detailed discussion of this portion of the cycle is unnecessary.

By carefully examining the waveshape for $v_{o1}(t)$ in Figure 8.7-9e and making use of equation (8.7-20), we can see that the half period $T/2$ in the figure may be found by setting $v_{o1}(t)$ equal to -5 V in the equation and solving for t.

Carrying out this procedure, we have

$$\frac{T}{2} = \tau \qquad (8.7\text{-}21)$$

so that the frequency of oscillation of this circuit is given by

$$f = \frac{1}{T} = \frac{1}{2\tau} = 5 \text{ kHz} \qquad (8.7\text{-}22)$$

for the component values used.

Exercises

8.7-1 For the circuit shown in Figure 8.7-8, determine the switching intervals if the supply voltages are $+5$ and -10 V. *Answer* 45.8 μs when $v_o = -10$ V, 69.3 μs when $v_o = +5$ V

8.7-2 Find the frequency of oscillation in Example 8.7-1 if a 10-kΩ resistor is connected in parallel with C. *Answer* 5.14 kHz

8.7-3 Find the frequency of oscillation in Figure 8.7-9 if R_2 is increased to 7.5 kΩ. *Answer* 6.66 kHz

---

PROBLEMS

8.2-1 An op amp is constructed by cascading two amplifier stages: A_1 a differential amplifier stage with a gain of 100 and A_2 a stage with a gain of 500. A resistor $R = 10$ kΩ is connected in series between the output of A_1 and the input of A_2, and a capacitor $C = 0.2$ μF is shunted from the input of A_2 to ground.
(a) Sketch the open-loop frequency response for this op amp with the inverting input v_2 grounded.
(b) Sketch the transient response of this amplifier to a 1-mV input step applied to the v_1 input with v_2 grounded. (Assume ±15-V supplies.)
(c) A noninverting amplifier is constructed using this op amp with $R_f = 99$ kΩ and $R_2 = 1$ kΩ. Repeat parts (a) and (b) for this circuit.

8.2-2 Repeat Problem 8.2-1 if v_{in} is a 20-mV step. (Assume ±15-V supplies.)

8.2-3 For the circuit shown in Figure P8.2-9, $R_f = 10$ kΩ and $R_2 = 90$ Ω. The open-loop transfer function for this circuit is $A_{OL}(s) = 10^3\alpha^2/(s + \alpha)^2$, where $\alpha = 1/RC = 10^3$ rad/s.
(a) Determine the closed-loop transfer function for this amplifier.
(b) Find the steady-state output obtained when a 100-Hz 1-mV peak-amplitude square wave is applied to the input.

8.2-4 Repeat Example 8.2-1 if the capacitor in the third stage is $2C = 0.02$ μF.

8.2-5 Repeat Example 8.2-1 if two additional stages, each with a transfer function $\alpha/(s + \alpha)$ are added in between $IC4$ and v_o.

8.2-6 A noninverting op amp circuit is constructed with $R_2 = 100$ Ω and R_f adjustable. If $A_{OL}(s) = 10^7/(s + 10)^3$, find the value of R_f needed to place one of the poles at $s = -20$.

8.2-7 A noninverting amplifier is constructed with $R_2 = 100$ Ω and R_f adjustable. The op amp open-loop gain is $A_{OL}(s) = 10\alpha^3/[(s + \alpha)(s + 2\alpha)^2]$.
(a) Find the value of R_f that will just place one of the closed-loop amplifier poles at $s = -3\alpha$.
(b) For the value of R_f chosen in part (a), what are the locations of the other two amplifier poles?

8.2-8 A noninverting op amp circuit is constructed with $R_2 = 100$ Ω and R_f adjustable. If $A_{OL}(s) = 10^7/(s + 10)^3$, find the value of R_f needed to place two of the poles at $s_1 = -5 + j5\sqrt{3}$ and $s_2 = -5 - j5\sqrt{3}$.

8.2-9 (a) Sketch $|A_{OL}(j\omega)|$ of the amplifier circuit shown in Figure P8.2-9 given that $1/RC = 100$ rad/s.
(b) Sketch the closed-loop frequency response of the amplifier with the feedback network in place, and estimate the amplifier bandwidth.

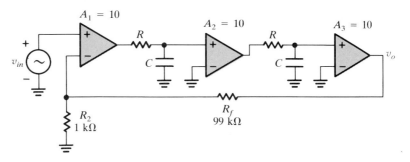

Figure P8.2-9

8.3-1 The op amp in each of the circuits described below has the gain-phase characteristic given in Figure 8.3-4d. Determine the stability of each of these circuits using the criteria: stable if $|\phi_T| = |\phi_A + \phi_B| < 135°$, marginally stable if $135° < |\phi_T| < 180°$, and unstable if $|\phi_T| > 180°$. If the circuit is stable, sketch $|A_{CL}(j\omega)|$ versus frequency

(a) The circuit is a noninverting amplifier with $R_f = 100$ kΩ and $R_2 = 1$ kΩ.

(b) The circuit is an integrator with $R = 10$ kΩ and $C = 159$ pF.

(c) The circuit is a differentiator with $R = 1$ kΩ and $C = 0.159$ μF. In addition, a capacitor $C_2 = 159$ pF is shunted across the feedback resistor.

8.3-2 Repeat Problem 8.3-1 if the pole at 1 MHz is replaced by a dominant pole at 10 kHz.

8.3-3 Repeat Problem 8.3-1 if a (phase-lag) compensation network is added to the op amp that replaces the pole at 1 MHz by a zero at 5 MHz and two poles at 100 kHz and 50 MHz.

8.3-4 Repeat Problem 8.3-1 if a (phase-lead) compensation network is connected to the op amp, which adds a zero at 5 MHz and a pole at 30 MHz to $A_{OL}(jf)$.

Problems 8.3-5 through 8.3-10 have the A_{OL} gain–phase characteristics shown in Figure P8(A).

8.3-5 An op amp integrator is constructed with $C = 0.01$ μF and R adjustable. Find the range of values of R for which it is stable.

8.3-6 A noninverting-style op amp circuit is constructed with $R_f = 1$ MΩ and R_2 replaced by a capacitor $C = 0.2$ μF. Determine the stability of this circuit, and if it is stable, estimate its gain margin.

8.3-7 An inverting-style op amp circuit is constructed with $R_2 = 10$ kΩ and R_f replaced by an inductor $L = 100$ mH. Determine the stability of this circuit. If the circuit is stable, what are its gain and phase margins?

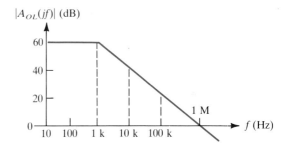

Figure P8(A)

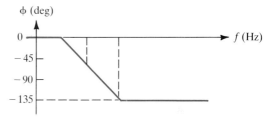

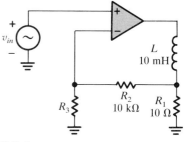

Figure P8.3-8

8.3-8 The amplifier shown in Figure P8.3-8 drives a 10-mH inductive load and uses current sampling feedback. Assuming that the R_2–R_3 network does not load R_1, find the largest value of R_3 for which this circuit is stable.

8.3-9 A noninverting op amp circuit is constructed with $R_f = 99$ kΩ and $R_2 = 1$ kΩ. The op amp open-loop output impedance is 1 kΩ and the amplifier drives a capacitive

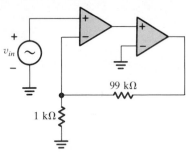

Figure P8.3-10

load $C_L = 1 \ \mu F$. Determine the amplifier phase margin with and without C_L in place.

8.3-10 Determine the overall stability of the circuit shown in Figure P8.3-10.

8.3-11 A noninverting op amp circuit is constructed with $R_f = 99 \ k\Omega$ and $R_2 = 1 \ k\Omega$. The op amp is a three stage device with the transfer functions of the first two stages given by $A_1(s) = A_2(s) = 10^5/(s + 10^4)$, and the third-stage transfer function $A_3(s) = 10^4/(s + 10^3)$. Determine the overall circuit stability.

8.4-1 An oscillator is constructed by taking a noninverting-style amplifier with $R_2 = 1 \ k\Omega$ and R_f adjustable, and feeding back the output to the noninverting input through a network

$$\beta(s) = \frac{s + 500}{s^2 + 100s + 10^5}$$

(a) Find the frequency of oscillation.

(b) Find the value of R_f for which the circuit just oscillates.

8.4-2 A Wien bridge-style oscillator similar to that shown in Figure 8.4-3b is constructed by replacing the RC feedback network with one containing resistors and inductors. In particular, the network Z_2 is now the parallel combination of a resistor R and an inductor L, and Z_1 is the series connection of the same two elements. Determine the frequency of oscillation and the value of R_f required to just place the poles on the $j\omega$ axis if $R_2 = 10 \ k\Omega$.

8.4-3 The circuit shown in Figure P8.4-3 has a closed-loop transfer function containing three poles and no zeros. Determine the value of R_f that will just place the poles on the $j\omega$ axis.

8.4-4 Determine the frequency of oscillation and the value of R_{C1} for which the circuit shown in Figure P8.4-4 just oscillates. You may assume that $h_{ie} = 1 \ k\Omega$, and $h_{fe} = 100$ for Q_1 and Q_2.

8.4-5 Find the frequency of oscillation and the minimum g_m needed for the circuit shown in Figure P8.4-5 to just oscillate.

8.4-6 Determine the minimum h_{fe} for which the circuit shown in Figure P8.4-6 just oscillates.

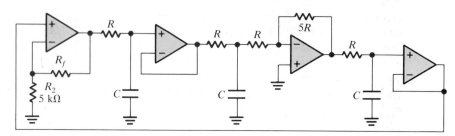

Figure P8.4-3

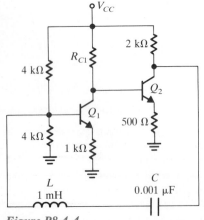

Figure P8.4-4

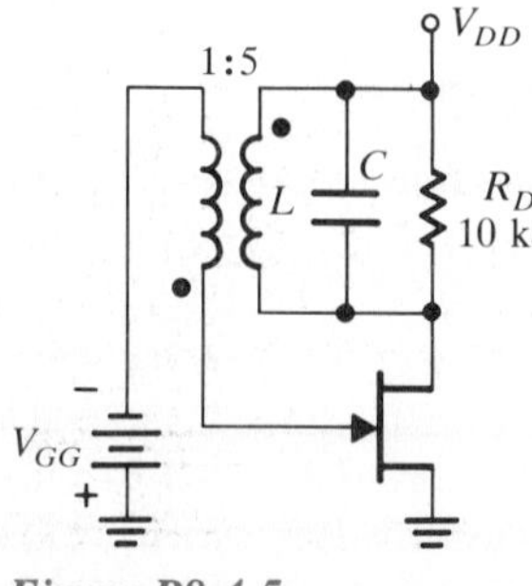

Figure P8.4-5

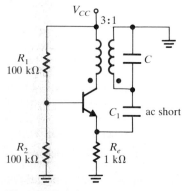

Figure P8.4-6

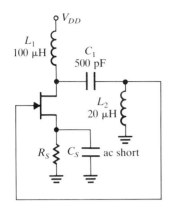

Figure P8.4-7

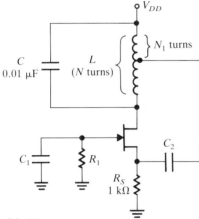

Figure P8.4-9

8.4-7 For the circuit shown in Figure P8.4-7, find the frequency of oscillation and the minimum g_m needed to produce these oscillations. L_2 has an infinite Q, and L_1 has a $Q = 100$.

8.4-8 The transistor in the circuit shown in Figure P8.4-6 has an $h_{ie} = 1$ kΩ and an $h_{fe} = 100$.
(a) The capacitor $C = 100$ pF and is connected as shown on the primary side of the transformer. Select the transformer primary inductance so that the circuit oscillates at 2 MHz.
(b) A resistor R_L is shunted across the transformer secondary. Determine the minimum value of R_L for which the circuit just oscillates.
(c) What is the inductance of the transformer secondary?

8.4-9 The circuit shown in Figure P8.4-9 is an oscillator. Find the oscillation frequency and the minimum g_m needed to produce these oscillations if $N = 100$ turns, $N_1 = 25$ turns, and $L = 1$ mH. C_1 and C_2 are ac shorts.

8.5-1 For the circuit in Figure 8.5-9a, $V_{DD} = 20$ V, $I_{DSS} = 5$ mA, $V_p = -3$ V, and R_L is adjusted until the circuit just

oscillates. Considering that C_1 is an ac short and that the i_D versus v_{GS} curve is linear as in Fig. 8.5-12b, sketch v_D and v_G.

8.5-2 In Figure 8.4-5, $V_{CC} = 20$ V, $R_1 = 14.3$ kΩ, $R_2 = 5.7$ kΩ, $C_1 = 500$ pF, and $C_2 = 4500$ pF. In addition, a resistor $R_L = 100$ Ω is connected in parallel with L. Given that the transistor h_{fe} is such that the circuit just oscillates, sketch i_C, v_E, v_C, and v_{CE} if $Z_{C2} \ll r_e$ at the frequency of oscillation.

8.5-3 For the oscillator circuit shown in Figure 8.4-4a, $V_{CC} = 10$ V, the transformer turns ratio is 2:1 instead of 20:1, $R_1 = 730$ Ω, $R_2 = 170$ Ω, $R_E = R_L = 100$ Ω, and C_1 and C_2 are ac shorts. In addition, to produce soft amplitude limiting, a light bulb is inserted between C_2 and ground. Its resistance is assumed to increase linearly with i_L (the rms current flow through it) being zero when $i_L = 0$ and 200 Ω when $i_L = 5$ mA. Using the fact that the overall circuit loop gain will be 1 when the oscillations stabilize, sketch i_C, v_E, v_B, and v_C in the steady state. In analyzing this circuit, consider that $h_{fe} = 100$ and that $h_{ie} = (1 + h_{fe})r_e$, where $r_e = 26$ mV$/I_E$.

8.5-4 For the circuit shown in Figure 8.5-4a, $R = 10$ kΩ, $C = 0.01$ μF, $R_{2A} = 0$, $R_1 = R_2 = 1$ kΩ, and D is a silicon diode. If $V_p = -8$ V and $I_{DSS} = 10$ mA, sketch v_o, v_G, v_1 and v_2.

8.5-5 Prove from first principles that the fundamental component of a double-clipped sine wave is given by equation (8.5-20).

8.5-6 For the circuit shown in Figure P8.5-6, considering that $V_T = 2$ V and that the g_m is constant at 10 mS, sketch the steady-state voltages v_D and v_G.

8.5-7 For the circuit shown in Figure 8.5-8a, $R_A = 90$ kΩ, $R_B = 10$ kΩ, $R_f = 26$ kΩ, and $R_2 = 1$ kΩ. Sketch the steady-state voltages at the op amp output and input terminals if the power supplies are ± 10 V.

8.5-8 In the Wien bridge oscillator in Figure 8.4-3a, $R_f = 30$ kΩ, $R_2 = 10$ kΩ, and the supply voltages are $+20$ V and -5 V. Sketch the approximate steady-state waveshapes at v_o and at the op amp input terminals.

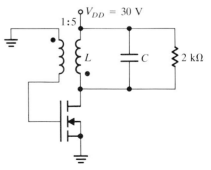

Figure P8.5-6

8.6-1 In the crystal equivalent circuit given in Figure 8.6-3a, $C_0 = 5$ pF, $L_1 = 200$ mH, $R_1 = 100$ Ω, and $C_1 = 0.002$ pF.
(a) Find its series and parallel resonant frequencies, and its equivalent impedance at each of the frequencies.
(b) Find the percent shift in the series resonant frequency if a 50-pF capacitor is connected in series with the crystal.
(c) Repeat part (b) if the capacitor is connected in parallel with the crystal.

8.6-2 For the circuit shown in Figure P8.6-2:
(a) Find the frequency of oscillation assuming zero phase shift in the amplifier.
(b) Find the frequency of oscillation if there is a $-10°$ phase shift in the amplifier at ω_0. What value of A will be needed now for this circuit to just oscillate?
(c) Repeat part (b) for $+10°$ of phase shift in the amplifier.

8.6-3 For the circuit shown in Figure 8.6-7a, $V_{DD} = 20$ V, $R_D = 4$ kΩ, $C_2 = 50$ pF, $C_3 = 100$ pF, and a 1-kΩ resistor is inserted between the transistor source lead and ground. If $V_p = -2$ V, $I_{DSS} = 2$ mA, and the transistor g_m is assumed to be constant at 1 mS, sketch v_G, v_S, and v_D considering that the circuit is just oscillating..

8.6-4 For the circuit shown in Figure 8.6-5b, $R_G = 10$ MΩ, the crystal is resonant at 5 MHz, $V_{DD} = 20$ V, $L_2 = 150$ μH, and C_2 is removed. In addition, $V_p = -3$ V, $C_{gd} = 5$ pF, and a resistor R_L is shunted across L_2 and its value adjusted so that the circuit just oscillates.

(a) What is the required crystal impedance if the circuit oscillates at 5 MHz?
(b) What are the approximate waveshapes at v_G and v_D?

8.6-5 For the circuit given in Figure 8.5-8a, $R_A = 5$ kΩ, $R_B = 10$ kΩ, $L = 2.75$ mH, $C = 0.001$ μF, and $R_2 = R_f = 10$ kΩ. In addition, a 32-kHz crystal is connected in series with R_A in the feedback loop. Determine the frequency of oscillation, and sketch v_o and the op amp input voltages if the power supplies are ±10 V.

8.6-6 In Figure 8.4-5a, $V_{CC} = 20$ V, $R_1 = 18$ kΩ, $R_2 = 2$ kΩ, $R_E = 1.3$ kΩ, $L = 1$ μH, and C_1 and C_2 are adjustable. In addition, a 1-MHz crystal is inserted into the feedback path between C_2 and R_E at the point labeled with the "$\times$." Assuming that the transistor $h_{fe} = h_{FE} = 20$ and $h_{ie} = 1$ kΩ, select C_1 and C_2 so that the circuit will just oscillate at 5 MHz. The Q of the inductor is 5.

8.7-1 For the circuit shown in Figure 8.7-4a, $R_f = 10$ kΩ, $R_2 = 5$ kΩ, $R = 10$ kΩ, and $C = 0.01$ μF.
(a) The circuit labeled $A(s)$ within the dashed lines is a Schmitt trigger. If the power supplies are $+10$ V and -5 V, sketch the transfer characteristic for v_o versus v_2.
(b) The overall circuit is an oscillator. Sketch v_o, v_1, and v_2.
(c) What is the frequency of the oscillation?

8.7-2 The circuit in Figure 8.7-4a is modified by replacing the feedback resistor R with an inductor $L = 1$ mH, and the capacitor C by a resistor $R_1 = 10$ kΩ.
(a) Sketch v_o, v_1, v_2, and i_L (the feedback current through L) if $R_f = R_2 = 5$ kΩ and if the power supplies are ±10 V.
(b) What is the oscillation frequency?

8.7-3 Repeat Problem 8.7-2 if a 5-kΩ resistor is connected in series with L.

8.7-4 For the circuit shown in Figure P8.7-4:
(a) Sketch and carefully label the steady-state voltages v_{o1}, v_3, v_4, and v_{o2}. The power supplies are ±10 V.
(b) What is the oscillation frequency?

8.7-5 For the circuit in Figure 8.7-9a, $C = 1$ μF, $R_f = 40$ kΩ, and $R_2 = 10$ kΩ. Assuming that the op amp outputs saturate at ±10 V, accurately sketch the steady-state voltage at v_{o1} and v_{o2}. Find the frequency of oscillation.

8.7-6 In the function generator circuit in Figure 8.7-9a $C = 1$ μF, and R is replaced by a network consisting of a resistor $R_1 = 100$ kΩ in series with a diode D_1 (cathode to the right). A second resistor–diode branch containing a resistor $R_2 = 1$ MΩ in series with a diode D_2 (cathode to the left) is connected in parallel with the R_1–D_1 branch. Carefully sketch and label v_{o1} and v_{o2} in the steady state considering that the diodes are ideal.

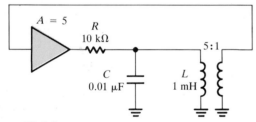

Figure P8.6-2

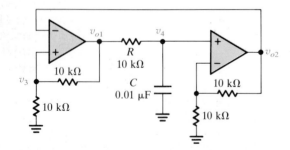

Figure P8.7-4

CHAPTER 9

Power Electronics

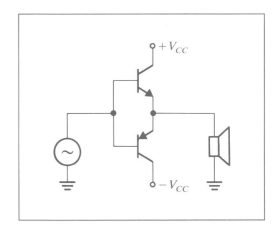

9.1 INTRODUCTION

Power electronics is a very broad and pervasive topic. It includes the design of power amplifiers which are able to handle signals from dc through radio frequencies, linear and switching regulated power supplies, and circuits designed specifically for the switching and control of ac and dc electrical power. For the purpose of this chapter we consider the problem at hand to be one of a power electronics nature when the power dissipation in the semiconductor is large enough for heat removal to become a problem. As a rule of thumb, this typically occurs when the power delivered to the load exceeds about 1 W.

The spectrum of design problems in power electronics is illustrated by the series of circuit block diagrams presented in Figures 9.1-1 and 9.1-2. In linear power amplification (Figure 9.1-1a) the basic task is to boost the power level of the input signal sufficiently to deliver the required power to the load. Because the power amplification is linear, the shape of the signal at the load is the same as that applied to the input of the amplifier.

The goal in electronic power regulation is quite different from that in power amplification. In a regulator circuit the object is to maintain a specific output parameter constant despite variations in the load, changes in the environmental conditions, and/or changes in the level of the input power connected to the regulator. In an electronic regulator it is usually the voltage or current delivered to the load that is to be held constant. Depending on the application, it is possible to construct regulator circuits for use with either ac or dc power sources.

The steam engine speed controller discussed in Section 7.1 is a good example of a regulator system. In that particular control system a mechanical governor was used to maintain the speed of a steam engine constant despite varia-

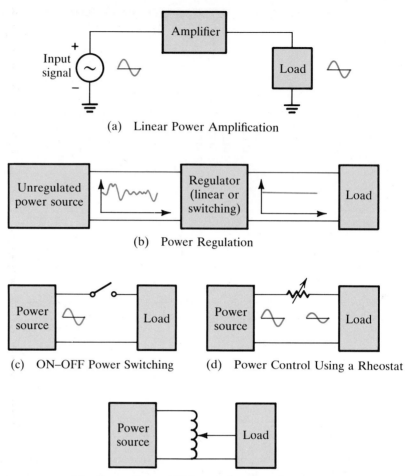

(a)   Linear Power Amplification

(b)   Power Regulation

(c)   ON–OFF Power Switching   (d)   Power Control Using a Rheostat

(e)   Power Control Using an Autotransformer

tions in the steam pressure in the boiler and changes in the load on the engine. The basic form of an electronic power regulator is illustrated in Figure 9.1-1b.

Regulators designed for the control of dc power are usually called regulated power supplies. Two types have considerable popularity in modern circuit designs; these are the so called series pass or linear regulator supply, and the more efficient (and also more complex) switching regulator dc supply. In addition to dc power regulators, it is possible to design regulator circuits for the control of ac power delivery to a load. Generally, this type of regulator employs some type of saturable core reactor or saturable transformer whose primary-to-secondary coupling can be adjusted by electronic means in order to regulate the power delivered to the secondary.

Besides being used for the amplification and regulation of electrical energy, power electronic devices also find considerable application in the areas of power switching, power control, and power conversion. Figure 9.1-1c illustrates the basic concept of ON–OFF power switching, and various types of semiconductor devices are available to handle this switching task, depending on the nature of the power source (ac or dc) and the power levels involved. Besides two-level ON–OFF power switching, it is possible to vary the power delivered to the load in a more-or-less continuous fashion between zero power and full power. Figure 9.1-1d illustrates one of the earliest methods for achieving this type of power control in which a variable resistance, known as a rheostat, is connected in series with the load. Basically, the rheostat, together with the load, forms a voltage divider. As the resistance of the rheostat is altered, so is the power delivered to the load. As should be obvious, this type of power control is very inefficient because a substantial portion of the applied input power is dissipated in the rheostat as heat.

Figure 9.1-1e shows another type of ac power control circuit in which a variable turns ratio transformer, known as an autotransformer, is used to control the power delivered to the load. Although useful in principle, in practice autotransformers have limited utility due to their large size, high cost, and rather small power-handling capabilities.

The circuits illustrated in Figures 9.1-2a and b represent two of the more modern approaches employed for the efficient control of ac electrical power. In both of these circuits, some type of electronic power switch is employed. In the first one [part (a)], the switch is closed during each ac cycle for a specific interval. In this way, the average power delivered to the load may be varied by controlling the duty cycle of the switch. This type of ac power control is useful for regulating the speed of small motors, for controlling the intensity of incandescent lights, and for regulating the temperature of electrical heaters.

The average power delivered to a heater may also be varied by using the circuit in Figure 9.1-2b, in which the power switch opens and then closes for a specific integral number of cycles of the input ac signal. Of course, in this case the duty cycle is determined by comparing the number of ON and OFF cycles. The principal advantage of this circuit over the one illustrated in Figure 9.1-2a is that it produces less RFI (or radio-frequency interference) because of the absence of any sharp edges in the output voltage waveform. As a practical matter, however, this circuit would be virtually useless in motor or lighting applications because the motor and the lamp would pulsate at the switching rate.

One other area of power electronics, that of power conversion, should also be mentioned. Typical power conversion circuits are shown in Figures 9.1-2c to e. The circuit in Figure 9.1-2c is known as an ac-to-dc converter, and of course, this type of power converter is what in Chapter 2 we simply called a dc power supply. The circuit illustrated in Figure 9.1-2d performs the opposite function. It takes dc power and converts it back into ac power. One very popular power inverter in the United States takes 12 V dc and converts it into 120 V ac at 60 Hz. The output from this type of inverter can be used to operate small ac appliances such as TV sets, drills, and power saws from the 12-V battery of an automobile or truck.

For some applications it is desirable to be able to convert from one dc voltage level to another (Figure 9.1-2e). Circuits for accomplishing this type of power conversion are known as dc-to-dc converters, and they are used when a dc voltage is needed that is different from those available from the system's main power supplies. In such a situation, rather than adding on a separate power supply to produce the required voltage, it may be more economical to make it from one of the other dc voltages which is already there by using a dc-to-dc converter.

Although many different types of semiconductor devices are available for

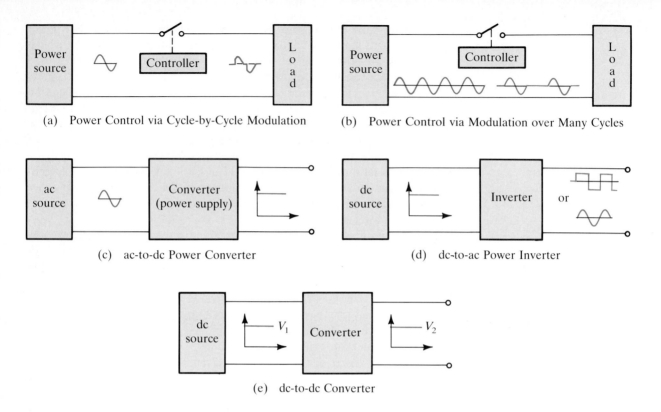

(a)   Power Control via Cycle-by-Cycle Modulation

(b)   Power Control via Modulation over Many Cycles

(c)   ac-to-dc Power Converter

(d)   dc-to-ac Power Inverter

(e)   dc-to-dc Converter

Figure 9.1-2
Spectrum of power electronics
(*continued*).

the design of power electronic systems, at the time of the writing of this book, three of these—the power BJT, the power FET, and the thyristor—are used most often. Therefore, in this chapter we restrict our discussion to these three types of power semiconductors. The major advantages and disadvantages of each of these devices are listed in Table 9.1-1.

9.1-1 Pros and Cons of BJT, FET, and Thyristor Power Devices

Device Type	Advantages	Disadvantages
BJT	1. Economical in low-voltage dc applications 2. Simple drive circuits 3. High-voltage devices available	1. Second breakdown and thermal runaway 2. Complex interfacing to CMOS and TTL logic gates
FET	1. Faster than the BJT for switching applications 2. No second breakdown 3. Huge power gain 4. Easily paralleled 5. Linear characteristics	1. Cost 2. Generally, dissipates more power than BJT 3. Lack of high-voltage devices
Thyristor	1. Can safely handle very high forward currents and reverse voltages 2. Self-commutation in ac power control applications	1. Slow 2. Hard to turn off in dc applications 3. Complex turn-on circuits

Power BJTs and power FETs are, for the most part, identical to their small-signal counterparts, so that little needs to be said about them at this point in our discussion. However, because you are probably less familiar with the thyristor, a few words are in order regarding the characteristics of this class of semiconductor devices.

There are basically three different kinds of thyristors: the SCR (or silicon-controlled rectifier), the triac (or bilateral SCR), and the GTO (or gate-turn-off) SCR. The silicon-controlled rectifier is essentially a diode whose forward turn-on characteristics can be controlled by the application of a signal to its gate. Once it has been turned ON, the gate loses control over the SCR's operation, and the device will go OFF only when the forward conduction current through it is reduced below a critical value. For ac circuit applications the SCR will turn OFF automatically when the voltage across it goes negative, but in dc circuit applications SCRs require special circuitry to assist in turning them OFF. Because of their very high maximum current and maximum voltage ratings, SCRs find extensive application in the areas of low-speed industrial ac power switching and ac power control. As indicated by the typical data presented in Table 9.1-2, of all the currently available power semiconductor devices, SCRs have the highest forward conduction current and the largest reverse blocking voltage.

The triac is similar to the SCR in that its turn-on is also controlled by the application of an external gate signal. However, the triac is bilateral, and as a result, it may be triggered into the ON state during both the positive and the negative halves of the applied ac voltage. Therefore, triacs may be used to provide full-wave ac power switching and ac power control while the SCR (unless a full-wave rectifier is connected in front of it) can be employed only in half-wave applications. Clearly, the triac is a more versatile device, and as such, we might expect that it would have made the SCR obsolete by now. However, because triacs are not yet available with current and voltage ratings comparable to those of SCRs, they can be used only in relatively low-power ac circuit designs. Typical applications for triacs include light dimmers and electric drill and appliance motor speed controllers.

Although a relatively recent entry to the thyristor family, the GTO (gate-turn-off) SCR is destined to be an important member of this family. Its intro-

9.1-2 Comparison of the Maximum Ratings of Commercially Available Power Semiconductor Devices

Device Type	Maximum Forward Current (A)	Maximum Reverse Voltage (V)	Turn-off Time
BJT	300	2000	1–5 μs
FET	60	500	50–200 ns
SCR	3000	4200	50–450 μs
Triac	40	800	1–2 μs
GTO SCR	125	1600	5–8 μs

duction removes one of the major shortcomings of SCRs in the area of ac power control: namely, the difficulty of turning the device OFF in the middle of the ac cycle once it has been turned ON. Because the GTO SCR can be turned OFF by the application of a negative trigger pulse to its gate, this device should find wide application in the areas of power control and power switching.

Although thyristors are excellent devices for the switching and (nonlinear) control of both ac and dc electrical power, they are all but useless for the linear amplification of electronic signals. In this area BJTs and, more recently, FETs are the dominant power semiconductor devices in use. At the time of this writing, high-power BJTs are generally less expensive than FETs with comparable voltage- and current-handling abilities. But device cost is not everything; FETs have huge power gains, so that the circuits needed to drive them are much simpler and much less expensive. Furthermore, the breakdown characteristics of FETs are superior to those of BJTs, and in addition, they can be easily paralleled to achieve very high output power designs.

Besides being used for linear power amplification, BJTs and FETs can also be employed in dc power-switching applications. In this area, BJTs, owing to their lower saturation voltages, are somewhat more efficient than FETs. However, because FETs are majority-carrier devices, they do not exhibit the storage-time problems of BJTs. As a result, in high-speed switching applications, field-effect power devices are usually preferred because they typically operate about 10 times faster than do comparable bipolar devices. Furthermore, the large power gain of FETs permits them to be driven directly from standard digital logic circuits such as TTL and CMOS, while comparable designs using BJT power transistors require substantial interfacing hardware.

Exercises

9.1-1 For the circuit in Figure 9.1-1c, the input is a 20-V (p-p) square wave of period T passing through zero at $t = 0$, and the load is 50 Ω.
a. Find the average power delivered to the load if the switch closes from 0 to $T/2$ and opens from $T/2$ to T during each cycle.
b. Repeat part (a) if the switch closes from 0 to $3T/4$ and opens from $3T/4$ to T during each cycle.
Answer (a) 1 W, (b) 1.5 W

9.1-2 The input signal to the circuit shown in Figure 9.1-1c is

$$v_{in} = (50 + 100 \sin \omega t) \text{ volts}$$

and the load is a 100-Ω resistor. Find the average power delivered to the load if the switch is closed when v_{in} is greater than zero, and opened when v_{in} is less than zero. *Answer* 47.7 W

9.2 DEVICE LIMITATIONS OF TRANSISTORS

In designing the transistor circuits used in earlier chapters, we ignored, for the most part, any of the device limitations of the transistors. In doing this, we basically assumed that the transistors could carry any current flow through them, could handle any voltage applied across them, and could dissipate any power (or heat) produced within them without suffering any degradation in performance or any permanent damage. Unfortunately, none of these assumptions are strictly correct, and if we are not careful about how we "push" the transistors that we are working with, it is fairly easy to damage them, sometimes permanently.

All semiconductor electronic devices have at least three fundamental device limitations: a maximum current, a maximum power, and a maximum voltage rating. In this section we discuss each of these limitations in turn as they apply to both bipolar and field-effect transistors in order to ascertain their origin and to develop an understanding of how these limitations affect the design of transistor circuits.

The power dissipation limit of a transistor is associated with the maximum internal temperature that the semiconductor can safely handle. Generally, for silicon devices this temperature must be kept below 150 to 200°C if performance degradation and premature device aging are to be avoided.

The overall power dissipated in a bipolar junction transistor, for example, may be written as the sum of the power flows into the base and collector terminals. As such, the instantaneous power dissipated in a BJT is equal to

$$P_i = v_{BE}i_B + v_{CE}i_C \qquad (9.2\text{-}1)$$

However, because the base current is much smaller than the collector current, and also because v_{BE} is generally much less than v_{CE}, the first term on the right-hand side of equation (9.2-1) is usually ignored and the instantaneous power dissipated in the transistor is written approximately as

$$P_i \simeq v_{CE}i_C \qquad (9.2\text{-}2)$$

Because v_{BE} is so much less than v_{CB}, nearly all of the power dissipation in a BJT occurs at the collector junction. As a result, in the design of power transistors the collector is usually connected to the case of the device in order to optimize the flow of heat outward, away from the transistor (see Section 9.9).

To ensure that the transistor's internal temperature stays within bounds, it is necessary that the average power dissipated in it be kept below the limit specified on the data sheet. The average power dissipated in a transistor may be computed mathematically by averaging the instantaneous power over one complete cycle of the input signal. Carrying out this operation, we obtain

$$P_D = \frac{1}{T}\int_0^T v_{CE}i_C\,dt \qquad (9.2\text{-}3)$$

For the case where v_{CE} and i_C are constants, the average and instantaneous powers are equal, and equation (9.2-3) reduces to

$$P_D = \frac{1}{T}(v_{CE}i_C)t \Big|_0^T = v_{CE}i_C \qquad (9.2\text{-}4)$$

Equation (9.2-4) is known as the transistor's power dissipation hyperbola, and it is sketched in Figure 9.2-5a for a typical BJT power transistor with a $P_{D,\max}$ rating of 200 W. In general, when the transistor's quiescent point is located underneath the maximum power dissipation hyperbola, so that the product of its dc collector current and its dc collector-to-emitter voltage is less than $P_{D,\text{max}}$, the power dissipated by the transistor will be within safe limits.

Besides the maximum power limitation in a BJT, there are also limits on the maximum current and the maximum voltage that can safely be applied. As we shall see, the current limit is a rather straightforward function of the transistor's internal structure, whereas the maximum voltage rating depends on the specific form of the circuit in which it is used.

In general, the maximum current rating, $I_{C,\text{max}}$, specified by the manufacturer is related to one of three possible transistor limitations:

1. The maximum current that the bonds and wires connecting the semiconductor to the external leads can safely handle without melting or fracturing.
2. The maximum current that can flow into the transistor while it is saturated without exceeding the power dissipation limit, $P_{D,\text{max}}$.
3. The collector current at which the h_{FE} of the transistor falls below the minimum value specified by the manufacturer (Figure 9.2-1).

Sometimes the manufacturer indicates only a single number for $I_{C,\text{max}}$, while at other times two collector current limits are specified: a maximum peak current and a maximum average current. When only the single limit is given, we consider that the transistor collector current is never to exceed this value. However, when both average and peak collector currents are specified, we assume that the quiescent current is not to exceed the maximum average current, and that the instantaneous collector current, i_C, is never to go above the specified peak transistor current.

The voltage limitation in a BJT is usually associated with avalanche breakdown in the reverse-biased collector junction of the transistor, and this breakdown phenomenon is essentially the same as that found in an ordinary pn junction diode (Figure 1.3-2). However, unlike the diode, because of the amplification properties of the transistor, the breakdown voltage depends on the circuit in which the transistor is connected.

In a diode, as the reverse voltage is raised, a critical value is reached where the velocity of the free carriers crossing the junction is high enough to ionize nearby atoms and produce additional free charge carriers. This results in a significant increase in (or multiplication of) the number of free carriers cross-

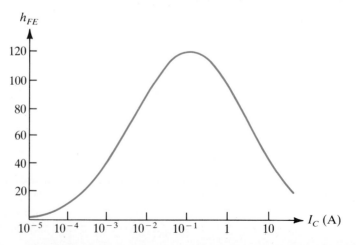

Figure 9.2-1
Variation in h_{FE} of a typical power transistor as a function of collector current.

ing the junction, and in a corresponding increase in the current flow. The factor M by which the current increases is empirically related to the reverse voltage, V, across the junction by the expression

$$M = \frac{1}{1 - (V/V_R)^n} \qquad (9.2\text{-}5)$$

where V_R is the junction reverse breakdown voltage and n is a constant from 3 to 6, depending on the form of the junction.

If we apply these results to the collector junction of a BJT, following equation (3.2-12), we may write

$$I_C = M(\alpha_F I_E + I_{CBO}) \qquad (9.2\text{-}6)$$

Using this result, we will see that it is now possible to determine the relationship between the voltage at which the transistor breaks down and the circuit in which it is connected.

Let's begin by examining the common-base circuit illustrated in Figure 9.2-2a to find the breakdown voltage of the collector–base junction of the transistor. Because I_E for this circuit is nearly constant [at $(V_{EE}-0.7)/R_E$], it is apparent from equation (9.2-6) that collector breakdown will occur, and correspondingly that I_C will rapidly increase, when M begins to grow significantly. Based on equation (9.2-5), this will occur when V_{CB} approaches V_{CBO}, the breakdown voltage of the collector junction diode.

Because the avalanche growth of I_C in equation (9.2-6) is independent of I_E (when I_E is a constant), it follows that the collector breakdown voltage for this circuit will also be nearly independent of I_E. This fact is substantiated by examining the common-base transistor characteristics illustrated in Figure 9.2-2b.

In a common-emitter circuit the emitter current that flows under breakdown conditions is no longer a constant, and this, as we shall see shortly, will make the collector breakdown voltage depend on the form of the circuit in which the transistor is connected. In developing quantitative expressions for the breakdown voltage of different types of common-emitter transistor circuits, it will be convenient to express I_C in equation (9.2-6) in terms of I_B rather than I_E. Substituting $(I_C + I_B)$ for the emitter current in this expression, we obtain

$$I_C = M[\alpha_F(I_C + I_B) + I_{CBO}] \qquad (9.2\text{-}7a)$$

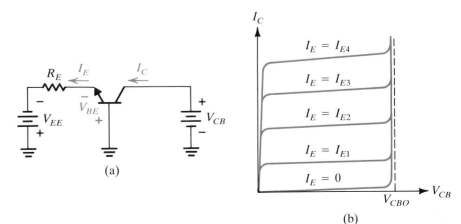

(a)

(b)

Figure 9.2-2

Collector avalanche in a common-base BJT amplifier.

or
$$I_C = \frac{\alpha_F M I_B}{1 - \alpha_F M} + \frac{M I_{CBO}}{1 - \alpha_F M} \qquad (9.2\text{-}7b)$$

This result is significant because it indicates that when I_E depends on I_C (as it does in a common-emitter amplifier), collector current multiplication, and hence collector junction breakdown, will occur when

$$1 - \alpha_F M = 0 \qquad (9.2\text{-}8a)$$

or
$$M = \frac{1}{\alpha_F} = \frac{1 + h_{FE}}{h_{FE}} \qquad (9.2\text{-}8b)$$

This result indicates that avalanche multiplication occurs for small values of M in a common-emitter amplifier, and therefore that collector junction breakdown also occurs at voltages well below those for common-base case considered previously.

When a transistor drives a resistive load, breakdown usually occurs when the transistor is near cutoff since this is when the voltage across it is the greatest. As a result, it is most useful (at least initially) to examine the breakdown behavior of common-emitter circuits when the collector current through them is small. Following Figure 9.2-1, we can see that for small values of collector current, h_{FE} increases with I_C, and using this fact together with equation (9.2-8b), it follows that the value of M needed to produce collector avalanche effects will decrease with increasing collector current, as will the collector breakdown voltage.

The collector voltage breakdown data tabulated in Figure 9.2-3a were obtained by combining equations (9.2-5) and (9.2-8b) together with the h_{FE} information presented in Figure 9.2-1. As is apparent from these results, the collector breakdown voltage for common-emitter circuits is typically about one-half that of comparable common-base circuits. The collector-to-emitter breakdown voltage at low collector currents is known as V_{CEO}.

The subscript CEO indicates that the voltage is being measured between the collector and the emitter, with the base open circuited (or $I_B = 0$). According to the data presented in the table, V_{CEO} is equal to about $0.64\, V_{CBO}$ for the particular transistor being examined.

A typical set of I_C versus V_{CE} common-emitter characteristics (including the breakdown region) is shown in Figure 9.2-3b. It is interesting to observe in this figure that once breakdown occurs, all the curves tend to coalesce to nearly the same collector–emitter voltage. This happens because, regardless of the transistor's initial base current, the final collector currents (and hence the transistor h_{FE} values) are all similar. As a result, following equations (9.2-5) and (9.2-8b) it is apparent that all the avalanche voltages will also be nearly the same. The voltage at which these curves coalesce is known as $V_{CE(\text{sus})}$, and it represents the minimum voltage necessary to sustain the transistor in the avalanche breakdown region.

In general, avalanche effects are not catastrophic unless the power or current limit of the transistor is exceeded during the breakdown. In other words, if the external circuit connected to the transistor ensures that $P_{D,\max}$ and $I_{C,\max}$ will not be exceeded, the transistor will not be permanently damaged if it should happen to enter the avalanche breakdown region.

However, in addition to collector junction avalanche, another frequently fa-

I_C	h_{FE}	M Required for Breakdown Eq. (9.2-8b)	Collector Breakdown Voltage From Eq. (9.2-5) Using $n = 4$
10 μA	5	1.200	$0.64V_{CBO}$
100 μA	15	1.066	$0.50V_{CBO}$
1 mA	40	1.025	$0.40V_{CBO}$
10 mA	70	1.014	$0.34V_{CBO}$
100 mA	120	1.008	$0.30V_{CBO}$

(a)

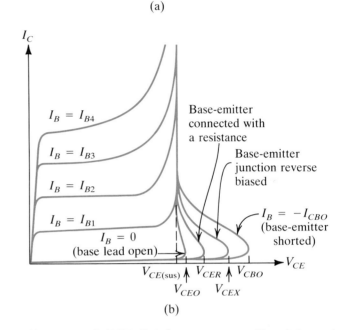

Figure 9.2-3
Collector avalanche in a common-emitter BJT amplifier.

(b)

tal breakdown effect occurs in BJTs. It is known as second breakdown, in contrast to collector avalanche, which is considered to be the primary or first voltage breakdown phenomenon in BJTs. Second breakdown is found to occur in a transistor when it is operated at high voltages and fairly high currents. It is associated with nonuniform current flow in the transistor. Where the current density is high, regions of local heating or "hot spots" are produced, and these cause the resistance of the semiconductor material in the vicinity of the heating to decrease, which in turn causes increased current flow into that region. This results in further heating, and eventually the temperature in the region can get so high that the semiconductor material may actually melt and produce a collector-to-emitter short.

The region on the i_C versus v_{CE} characteristics where the transistor may be safely operated is known as its safe operating area (SOA). It is the region bounded by the maximum collector current, the maximum collector–emitter voltage, the power dissipation hyperbola, and the transistor's second breakdown curve. A typical transistor safe operating area is illustrated in Figure 9.2-4a. This SOA region is also redrawn in part (b) of the figure using logarithmic axes for i_C and v_{CE}. Notice that the dissipation hyperbola appears as a straight line when drawn on this log-log plot, as does the second breakdown curve.

For the purpose of this chapter we assume that the transistor load line must be positioned so that it is located entirely within the SOA. In actuality, de-

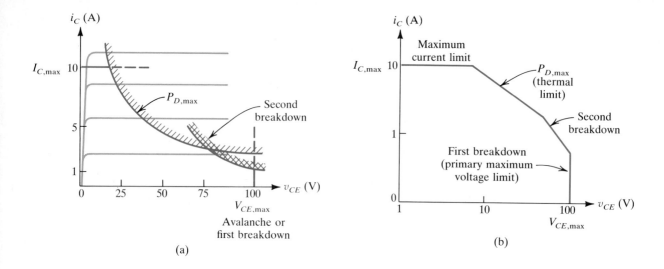

Figure 9.2-4
BJT safe operating area (SOA).

pending on the particular character of the signal, it may be possible for the operating point momentarily to exceed some of these limits without permanently damaging the transistor. But, for safety, at least in our initial investigation of power semiconductor devices, let's assume that these boundaries are rigid.

The right-hand boundary of the SOA at $V_{CE,max}$ can have values anywhere from V_{CEO} to V_{CBO}, depending on the transistor circuit configuration. However, again for safety reasons, in this chapter we always take $V_{CE,max}$ to be equal to V_{CEO}, the smaller of the two collector-to-emitter breakdown voltages. In this way, for any circuit that we design, if v_{CE} never exceeds V_{CEO}, we can rest assured that the transistor will never exhibit collector junction avalanche breakdown.

Although much newer than the power BJT, the power field-effect transistor is providing strong competition for the BJT in power control and power amplifier designs. Until a few years ago, FETs could handle power levels of only up to about 1 W. The introduction of the vertical MOS or VMOS transistor changed all that. Figure 9.2-5 illustrates the basic differences between a conventional planar-style FET and its VMOS counterpart. In a planar field-effect transistor, the channel is oriented laterally along the x-axis direction, and both the drain and the source terminals are located on the surface of the transistor. This architecture limits the maximum current that the transistor can safely carry, and causes it to exhibit a rather large ON resistance in the ohmic region.

Figure 9.2-5b illustrates the construction of an early-style V-groove power field-effect transistor. Notice that in this structure the drain terminal is located on the bottom of the transistor, while the gate and source connections are found on the top. This architecture results in a vertical rather than a lateral flow of current between the drain and the source. Because the length of the channel, L, for this structure depends on the diffusion depth and not on the accuracy of the mask alignment, much shorter effective channel lengths are possible. This results in transistors that have low ON resistances and that can carry substantial drain currents.

In recent years, the development of the self-aligned double-diffusion-transistor manufacturing technique has led to the fabrication of short-channel pla-

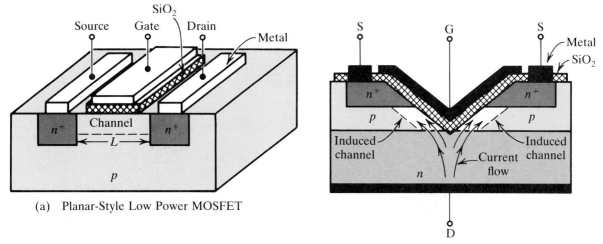

(a) Planar-Style Low Power MOSFET

(b) V-Groove Power MOSFET

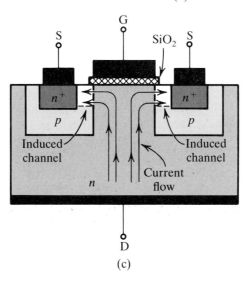

(c)

Figure 9.2-5

Comparison of the internal structure of low- and high-power MOSFETs.

nar-style FETs, and has also permitted the design of MOS power transistors of the type illustrated in Figure 9.2-5c.

The theory of operation of the MOS power transistors illustrated in Figure 9.2-5b and c is essentially the same as that for the ordinary enhancement-style planar FET shown in part (a) of the figure. For $V_{GS} = 0$ there is no direct connection between the drain and the source, and as a result, the devices are normally OFF. When a positive voltage is applied to the gate, a channel is induced in the p-type material, and current can now flow between the drain and the source. The i_D versus v_{DS} characteristics for a typical power FET are illustrated in Figure 9.2-6a. Notice that $V_{DS,max}$ for this transistor is characterized by a single breakdown voltage $V_{(BR)DSS}$, which is equal to the avalanche voltage of the drain-to-channel pn junction diode.

One rather interesting characteristic of MOS power FETs that distinguishes them from those of small-signal FETs is that the spacing of the curves for equal changes in V_{GS} is much more uniform. In other words, once you get past the

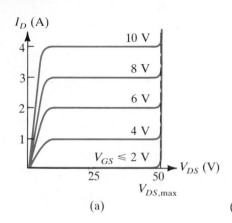

(a)

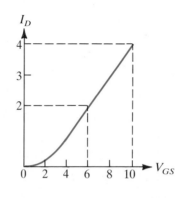

(b) I_D versus V_{GS} for Operation in
the Constant-Current Region

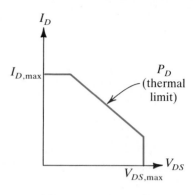

(c) SOA for a Power MOSFET

Figure 9.2-6

threshold voltage, V_T, the g_m of the power FET is nearly a constant, independent of the Q point. This fact is illustrated by the linear nature of the i_D versus v_{GS} curve in Figure 9.2-6b.

The safe operating area for a typical power FET is shown in Figure 9.2-6c. Unlike BJTs, field-effect transistors do not exhibit second breakdown. The ON resistance of an MOS transistor has a positive temperature coefficient, and as a result, if one section of the channel experiences excessive heating, the resistance of that region will increase automatically, reducing the current flow into that area. This effect tends to prevent the formation of hot spots and assures that the flow of current through the channel is relatively uniform. The other three limits that form the SOA are the same as for the BJT: the maximum current limit, the maximum power dissipation limit, and the voltage breakdown limit.

Exercises

9.2-1 A bipolar junction transistor is biased in a common-emitter circuit so that $I_B = 12 \ \mu A$ and $I_C = 1$ mA at low values of V_{CE}. If $V_{CBO} = 200$ V and $n = 5$ for this device, estimate the voltage at which the collector current will increase to about 2 mA. Assume that I_{CBO} may be neglected. **Answer** 6.3 V

9.3 SINGLE-ENDED LINEAR POWER AMPLIFICATION

In this section we study the application of BJT and FET transistors to the design of linear power amplification circuits. The fundamental difference between a power amplifier and the small-signal amplifiers that we studied in earlier chapters is that a power amplifier must be capable of delivering substantial signal power to the load. Typical loads to be driven include loudspeakers for use with audio amplifiers, piezoelectric transducers for use with sonar and ultrasonic power amplifiers, antennas for radio transmitters and radar systems, and motors for use with servomechanism-style power amplifiers.

Regardless of the type of load to be driven, an amplifier design will usually consist of several stages, and because the signals in the earlier stages will have low amplitudes, small-signal techniques will be found to work well for the "front-end" analysis of these amplifier circuits. In addition, because the quiescent currents flowing in these early stages will be small, power consumption from the dc supply and stage efficiency will generally not be that important. However, in the output stage, and possibly even in the stage driving the output stage, these factors will usually be critical.

For amplifiers delivering substantial amounts of power to a load, the circuit efficiency dramatically affects the design of both the output stage and the power supply. Basically, an amplifier's efficiency indicates how well the dc power from the supply is converted into signal power at the load. To illustrate the importance of this parameter, let's compare the performance of two different power amplifiers, one that is 10% efficient and another which is 90% efficient, and let us assume that both are to deliver 10 W of signal power to a particular load. In the circuit that is 10% efficient, only 10% of the dc power coming in from the supply makes it into the load as ac signal power. This means that 90% of the dc input power is dissipated as heat in the transistors of the output stage. Therefore, to deliver 10 W of signal power to the load with this type of amplifier, a 100-W dc power supply is required, and the transistors in the output stage need to be able to safely dissipate 90 W of heat. If the amplifier efficiency is increased to 90%, the size of the dc power supply can be reduced to $10/0.9$ or 11.1 W, and the output stage now only needs to dissipate 1.1 W. Clearly, with this second design, it will be possible to use much smaller and much less expensive transistors, as well as a significantly smaller power supply.

In the design of power amplifiers, many different circuit configurations are possible, and the one to use for a specific application depends on the often conflicting requirements of circuit efficiency, allowable signal distortion, signal power to be delivered to the load, and overall circuit cost. Power amplifier designs are most often classified into one of four broad categories: class A, class B, class C, and class D. In a class A amplifier the transistors conduct over the entire cycle of the input signal. This circuit may be operated in either a single-ended or a push-pull (two-transistor) configuration. As we shall see in Section 9.4, push-pull designs offer opportunities for a substantial reduction of output signal distortion.

A second very popular amplifier style is known as class B. For this type of amplifier the transistors are biased at or near cutoff, and each of the amplifying devices conducts for one-half of the input signal period. Class B signal amplification requires the use of a push-pull amplifier if a linear output is to be obtained. The major attribute of class B operation is that there is no power consumed from the dc supply when the input signal is zero. In addition, class B circuits are noted for their high efficiencies. Both of these performance attributes make class B power amplification especially attractive for portable equipment applications where battery operation is required.

When the transistor is active for less than 50% of the input signal period, the amplifier is said to be operating in the class C mode. This type of circuit finds little application at audio frequencies because it is difficult to reconstruct the original signal without introducing significant waveform distortion. However, this circuit is popular at radio frequencies, because of its very high efficiency.

The operation of the class D power amplifier is radically different from the class A, class B, and class C types just examined in that the amplifier itself does not operate in the linear mode at all, but instead, is switched ON and OFF at a very high frequency. Here the output is a square wave and the input signal amplitude is used to determine the duty cycle of the output waveform. In this way the effective "dc level" of the signal delivered to the load is proportional to amplitude of the original input. As a result, by passing the output from the amplifier through a low-pass filter, we can recover an amplified version of the original input signal. Class D amplifiers offer the promise of very high circuit efficiencies and are becoming more and more attractive as the switching speeds of power FETs and BJTs continue to increase.

In the remainder of this section we investigate the characteristics of single-ended class A power amplifiers in detail. Both direct-coupled and transformer designs are examined, and a careful investigation of distortion minimization techniques is carried out. In Section 9.4 these results are extended to the analysis and design of push-pull class A and class B circuits. Class C and class D power amplification is left for more advanced books in this area.

9.3-1   Class A Single-Ended Direct-Coupled Amplifier

In analyzing the amplifiers in this section, as well as those discussed in the other parts of this chapter, it is important to recognize that each of the designs examined could be accomplished by employing either BJT or FET transistors. As a result, in describing these circuits the author will attempt to give equal treatment to both. Where a significant advantage of one over the other exists, this will be pointed out, but otherwise, you should feel equally comfortable to design power amplifiers with either of these two transistor types.

A class A direct-coupled BJT design may be accomplished by placing the load in either the emitter or the collector of the transistor. It is usually inadvisable to use a common-base design for a power amplifier because its power gain is very low when the load resistance is small.

We begin by examining the performance of the common-emitter design presented in Figure 9.3-1a, and to simplify this investigation, initially, we assume that the small-signal analysis techniques developed in the earlier chapters are still valid. Although this approach will greatly ease the analysis task, we need to point out that the actual outputs obtained from these circuits are somewhat different from these theoretical calculations because the signal swings will be so large that h_{ie} and h_{fe} will actually vary considerably with the position of the operating point. With these words of caution duly noted, we begin our analysis of this problem by writing down the dynamic load-line equation for the output portion of the amplifier in Figure 9.3-1a as

$$V_{CC} = i_C R_L + v_{CE} \qquad (9.3\text{-}1)$$

This load line is sketched on the transistor's output characteristics in Figure 9.3-1b. The battery V_{BB} in the base of the transistor is used to place the Q point at the desired values of I_C and V_{CE}.

If we assume that the transistor is biased for class A operation, and furthermore that it is operating linearly, then, because the input signal is a sine wave, the base current, collector current, and collector-to-emitter voltage will also

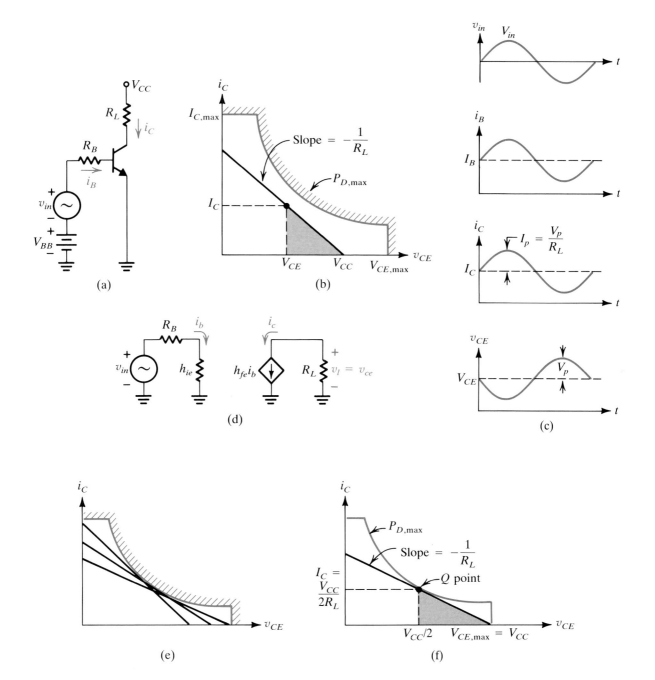

Signal Level, v_{in}	P_{IN} (W)	P_L (W)	$P_D = P_{IN} - P_L$ (W)	$P_{L,ac}$ (W)	η (%)
0	40	20	20	0	0
Maximum	40	30	10	10	25

(g)

Figure 9.3-1
Class A direct-coupled amplifier.

641

be sine waves, and will have the waveshapes illustrated in Figure 9.3-1c. Using these sketches, we may write down expressions for v_{CE} and i_C as

$$v_{CE} = V_{CE} - V_p \sin \omega t \qquad (9.3\text{-}2a)$$

and
$$i_C = I_C + I_p \sin \omega t \qquad (9.3\text{-}2b)$$

where, following Figure 9.3-1d, the ac amplitudes I_p and V_p are seen to be related to one another by the expression

$$I_p = \frac{V_p}{R_L} \qquad (9.3\text{-}3)$$

The average power delivered to the load may be obtained by integrating (or averaging) the instantaneous power dissipated in the load over one period of the input signal v_{in}. Carrying out this operation, we have

$$P_L = \frac{1}{T} \int_0^T i_c^2 R_L dt = \frac{1}{T} \int_0^T \left(I_C^2 R_L + 2I_C V_p \sin \omega t + \underbrace{\frac{V_p^2}{2R_L} \sin^2 \omega t}_{\frac{1}{2}(1 - \cos 2\omega t)} \right) dt \qquad (9.3\text{-}4)$$

$$= \underbrace{I_C^2 R_L}_{\substack{\text{dc power} \\ \text{dissipated} \\ \text{in } R_L}} + \underbrace{V_p^2/2R_L}_{\substack{\text{ac power} \\ \text{dissipated} \\ \text{in } R_L}}$$

In evaluating this integral, we have used the fact that the average value (or area under the curve) over one period of a sine wave is zero. In addition, the area under the curve of a sine wave (or of a cosine wave) of frequency 2ω is also zero when integrated over the period $T = 2\pi/\omega$. In examining equation (9.3-4), it is important to recognize that the first term in this result represents the dc power dissipated in the load, while the second term is the ac signal power delivered to the load.

In a similar fashion, we may compute the average power delivered to the circuit by the dc supply as

$$P_{IN} = \frac{1}{T} \int_0^T V_{CC} i_C dt = \frac{1}{T} \int_0^T V_{CC}(I_C + I_p \sin \omega t) dt = V_{CC} I_C \qquad (9.3\text{-}5)$$

The power dissipated in the transistor is evaluated from the expression

$$P_D = \frac{1}{T} \int_0^T v_{CE} i_C dt = \frac{1}{T} \int_0^T (V_{CE} - V_p \sin \omega t)(I_C + I_p \sin \omega t) dt$$

$$= V_{CE} I_C - \frac{I_p V_p}{2} = V_{CE} I_C - \frac{V_p^2}{2R_L} \qquad (9.3\text{-}6)$$

It is useful to observe that this result could also have been obtained (much more simply) by applying the principle of conservation of energy to this circuit. The power supplied by the battery V_{CC} basically goes into the load and into the transistor, and therefore we may write

$$P_{IN} = P_L + P_D \qquad (9.3\text{-}7a)$$

or using equations (9.3-4) and (9.3-5), we have

$$P_D = P_{IN} - P_L = V_{CC}I_C - I_C^2 R_L - \frac{I_p^2 R_L}{2} \qquad (9.3\text{-}7\text{b})$$

At first glance this result does not appear to match that obtained in equation (9.3-6). However, by substituting

$$V_{CE} = V_{CC} - I_C R_L \qquad (9.3\text{-}8)$$

into equation (9.3-6) along with equation (9.3-3) for I_p, these two expressions are seen to be identical.

As we examine the expressions for P_L, P_{IN}, and P_D in equations (9.3-4), (9.3-5), and (9.3-6), several interesting facts emerge. First, the dc input power from the supply is independent of the level of ac power in the load. As a result, it is the same under quiescent conditions as it is when delivering maximum power to the load. Second, the overall power delivered to the load increases as the signal level increases, while, by conservation of energy, the power dissipated in the transistor actually goes down. In fact, as indicated by equation (9.3-6), maximum power is actually dissipated in the transistor under quiescent conditions when $V_p = 0$.

Following equation (9.3-4), it is clear that to deliver the maximum undistorted ac power to the load, we need to have V_p as large as possible. This is achieved by placing the Q point at the center of the dynamic load line with $V_{CE} = V_{CC}/2$ and $I_C = V_{CC}/2R_L$. Substituting these results into equation (9.3-4), we have

$$P_L = \underbrace{\frac{V_{CC}^2}{4R_L}}_{\substack{\text{dc} \\ \text{power} \\ \text{into} \\ \text{load}}} + \underbrace{\frac{V_p^2}{2R_L}}_{\substack{\text{ac} \\ \text{power} \\ \text{into} \\ \text{load}}} \qquad (9.3\text{-}9)$$

With the Q point at the center of the load line, the maximum undistorted ac power at the load is obtained when $V_p = V_{CC}/2$, and for this value of V_p we have

$$P_{L,\text{ac}} = \frac{(V_{CC}/2)^2}{2R_L} = \frac{1}{2}\left(\frac{V_{CC}}{2}\right)\frac{V_{CC}}{2R_L} = \frac{1}{2}\left(\frac{V_{CC}}{2}\right)I_C \qquad (9.3\text{-}10)$$

Here the additional subscript ac is used to denote that this is the ac portion of the overall power being delivered to the load. This notation will be used only when this type of distinction is needed. Otherwise, the power delivered to the load will be denoted simply by P_L.

Equation (9.3-10), besides being equal to the ac power delivered to the load under maximum signal swing conditions, also represents the area of the shaded triangle underneath the load line in Figure 9.3-1b. Using this fact, the problem of designing a class A transistor amplifier that can deliver the maximum ac power to a load is seen to be equivalent to the problem of selecting the circuit's parameters to maximize the area inside the shaded triangle. Following Figure 9.3-1b, we can see that this area may be increased by increasing V_{CC} and decreasing R_L. Decreasing R_L raises the slope of the load line (making it more vertical), while increasing V_{CC} moves its v_{CE} intercept farther to the right. For reasons that will be apparent shortly, an upper limit on the triangle area is reached when the load line is made tangent to the maximum power dissipation hyperbola.

To see why this is the case, we first note that any load line drawn tangent to the power dissipation hyperbola is bisected at the point of tangency (Problem 9.3-1). Therefore, if the Q point is placed at the center of the load line for maximum signal swing, it will be located at the point of tangency on the dissipation hyperbola. With the circuit parameters adjusted to place the load line in this position, under quiescent conditions the power dissipated in the transistor ($I_C \cdot V_{CE}$) will simply be equal to $P_{D,\max}$. Raising the load line above this tangent position while delivering more ac power to the load will also dissipate excessive power in the transistor and will destroy it thermally. Therefore, as a general guideline, the dynamic load line should never be positioned beyond the point of tangency to the dissipation hyperbola.

As illustrated in Figure 9.3-1e, for the case where R_L and V_{CC} can both be varied, many different load-line positions are possible, all of which will deliver the same maximum power to the load. However, for our designs, when possible we will always choose the one with the smallest slope and hence the largest value of V_{CC}. This load line is preferred over all others because it results in the smallest currents flowing in the amplifier, and this will minimize the I^2R losses in the circuit. Thus, in summary, to design a class A direct-coupled amplifier that can deliver maximum power to the load:

1. Make V_{CC} as large as possible by setting it equal to $V_{CE,\max}$ (V_{CEO}) at the right-hand boundary of the SOA.
2. Move the load line up by reducing R_L until the line is tangent to the dissipation hyperbola.
3. Place the operating point at the point of tangency so that $V_{CE} = V_{CC}/2$ and $I_C = V_{CC}/2R_L$.
4. If R_L is fixed, draw a tangent line to the dissipation hyperbola with a slope $-1/R_L$. If this load line does not hit the $I_{C,\max}$ or $V_{CE,\max}$ lines, the design is completed with V_{CC} determined from the v_{CE} intercept of the load line. If the load line crosses either the $I_{C,\max}$ or $V_{CE,\max}$ limits, lower the load-line position (keeping the slope constant) until it no longer intersects them.

Several interesting observations about class A direct-coupled power amplifiers may be made if the equations for P_{IN} and P_D are expressed in a slightly different form. If we assume that the Q point is at the center of the load line, the power delivered to the load is given by equation (9.3-9), and in addition, the dc supply power and the transistor dissipation equations (9.3-5) and (9.3-6) may be rewritten as

$$P_{IN} = V_{CC}I_C = \frac{V_{CC}^2}{2RL} \qquad (9.3\text{-}11a)$$

and
$$P_D = \frac{V_{CC}^2}{2R_L} - \left(\frac{V_{CC}^2}{4R_L} - \frac{V_p^2}{2R_L}\right) = \underbrace{\frac{V_{CC}^2}{4R_L}}_{\substack{\text{dc power} \\ \text{to load}}} - \underbrace{\frac{V_p^2}{2R_L}}_{\substack{\text{ac power} \\ \text{to load}}} \qquad (9.3\text{-}11\text{b})$$

By examining these results, we can see that 50% of the dc input power is always dissipated as heat in the load. In addition, under quiescent conditions (with $V_p = 0$) the remaining 50% of the input power is dissipated as heat in the power transistor.

If we define η, the circuit efficiency, as the ratio of the ac load power to the dc input power, this quantity may be written as

$$\eta = \frac{P_{L,\text{ac}}}{P_{IN}} = \frac{V_p^2/2RL}{V_{CC}^2/2RL} = \left(\frac{V_p}{V_{CC}}\right)^2 \times 100\% \qquad (9.3\text{-}12)$$

Clearly, from this equation we can see that the circuit efficiency increases with increasing signal to the load. Furthermore, because the maximum undistorted output signal is $V_p = V_{CC}/2$, the maximum circuit efficiency is 25%.

To illustrate the significance of these results, let's consider the design requirements for a class A amplifier that is to deliver 10 W of ac signal power to a load. Because the maximum circuit efficiency is 25%, an amplifier of this type delivering 10 W of signal power to the load will require a dc input power of 40 W from the supply. For example, if V_{CC} were chosen to be 40 V, the power supply would need to be able to deliver an average current of 1 A.

Figure 9.3-1g summarizes the values of P_{IN}, P_L, and P_D obtained under zero and maximum input signal conditions. Regardless of the signal level, the dc supply delivers a constant 40 W of input power to this amplifier. With no signal applied, 50% of this power (20 W) is dissipated in the load and 50% (20 W) in the transistor. When the maximum input signal is applied, the maximum ac power is delivered to the load, and because of the assumed 25% efficiency, this increases the load power by $(\frac{1}{4})(40 \text{ W})$, or 10 W, to a total of $(20 + 10) = 30$ W. Furthermore, because the dc input power is a constant, corresponding to this 10-W increase in load power there must be a 10-W decrease in the power dissipated in the transistor. Thus, under zero-signal conditions, 20 W is dissipated in the transistor, while under full-signal conditions the power dissipation drops to only 10 W. Of course, in designing this circuit, a 20-W power transistor will be needed.

EXAMPLE 9.3-1

For the class A amplifier illustrated in Figure 9.3-2a, the transistor parameters are $h_{fe} = h_{FE} = 100$, $h_{ie} = 100 \ \Omega$, $P_{D,\text{max}} = 2.5$ W, $V_{CE,\text{max}} = 25$ V, and $I_{C,\text{max}} = 500$ mA.

a. Select V_{CC} and R_B to deliver maximum power to the load.
b. For the design values selected in part (a), what is the maximum undistorted ac power that can be delivered to R_L?

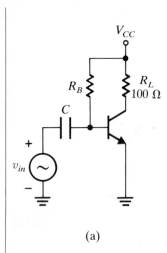

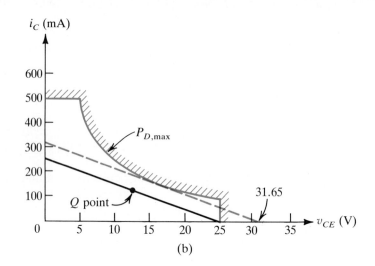

(a)

(b)

Figure 9.3-2
Example 9.3-1.

c. Find the value of v_{in} required to deliver this power to R_L, assuming the small-signal transistor models to be valid.

d. Find the circuit power gain.

The safe operating area for this transistor is sketched in Figure 9.3-2b, and if the output load line is drawn tangent to the dissipation hyperbola with a slope equal to $-1/R_L$, it intersects the $V_{CE,max}$ line and is therefore unsatisfactory since large input signals could cause v_{CE} to exceed $V_{CE,max}$ and damage the transistor (see the dashed load line in the figure). Therefore, it will be necessary to lower the position of the load line until it no longer passes through the SOA boundaries. The load line just satisfying this condition is indicated by the solid line in Figure 9.3-2b.

For this load line we have $V_{CC} = 25$ V, and if we want to place the Q point at the center of the load line, we require that

$$V_{CE} = 12.5 \text{ V} \tag{9.3-13a}$$

and

$$I_C = \frac{V_{CC} - V_{CE}}{R_L} = 125 \text{ mA} \tag{9.3-13b}$$

Because $h_{FE} = 100$ we also have

$$I_B = \frac{125 \text{ mA}}{100} = 1.25 \text{ mA} \tag{9.3-13c}$$

so that the required value of R_B is

$$R_B = \frac{24.3\text{V}}{1.25 \text{ mA}} = 19.4 \text{ k}\Omega \tag{9.3-13d}$$

Assuming that the ac collector voltage can swing equally in both directions about the Q point, the maximum undistorted v_{ce} is

$$v_{ce} = -V_p \sin \omega t = -12.5 \sin \omega t \text{ volts} \tag{9.3-14a}$$

so that
$$P_{L,\text{ac}} = \frac{V_p^2}{2R_L} = 0.78 \text{ W} \tag{9.3-14b}$$

Working backward from output, it follows that

$$i_b = \frac{i_c}{h_{fe}} = -\frac{v_{ce}/R_L}{h_{fe}} = 1.25 \sin \omega t \text{ mA} \tag{9.3-15a}$$

so that v_{in} may be written as

$$v_{in} = i_b h_{ie} = V_{in} \sin \omega t = 125 \sin \omega t \text{ mV} \tag{9.3-15b}$$

The ac input power for this circuit is equal to

$$P_{in} = \frac{V_{in}^2}{2h_{ie}} = 78 \ \mu\text{W} \tag{9.3-16}$$

so that the overall circuit power gain is

$$\text{power gain} = \frac{P_{L,\text{ac}}}{P_{in}} = 10^4 \tag{9.3-17}$$

9.3-2 Distortion Considerations in Class A Direct-Coupled Amplifiers

In spite of the superior power gain of the common-emitter amplifier, the output stages of most power amplifiers employ emitter-follower or source-follower designs. For the most part this is because of the low signal distortion of these circuits.

The distortion in a BJT power amplifier comes about from two main sources:

1. The exponential nature of the i_B versus v_{BE} characteristic (or nonlinearity of h_{ie}).
2. The nonlinear nature of the i_C versus i_B characteristic (or nonlinearity of h_{fe}).

Both of these factors cause the amplifier gain to vary with the position of the operating point, and for the case where the amplifier is handling large-signal swings, this means that the output will exhibit nonlinear distortion. We begin our analysis of the effect of these parameter variations by developing expressions for the second harmonic distortion produced in both common-emitter and common-collector circuits due to the nonlinearity of the base–emitter diode. The effect of h_{fe} nonlinearity on output signal distortion is left for the homework problems (Problem 9.3-9).

For the common-emitter amplifier shown in Figure 9.3-3a, the input load-line equation is

$$v_{BE} = V_{BE} + v_{be} = V_{BB} + v_{in} = V_{BB} + V_p \sin \omega t \tag{9.3-18}$$

so that v_{be} will be sinusoidal. For small values of V_p, the base current will be nearly sinusoidal (and equal to v_{in}/h_{ie}), but for large input signal swings, the base current (and hence i_C and v_L) will be distorted. In particular, because of the exponential character of the i_B versus v_{BE} curve, the positive-going portion

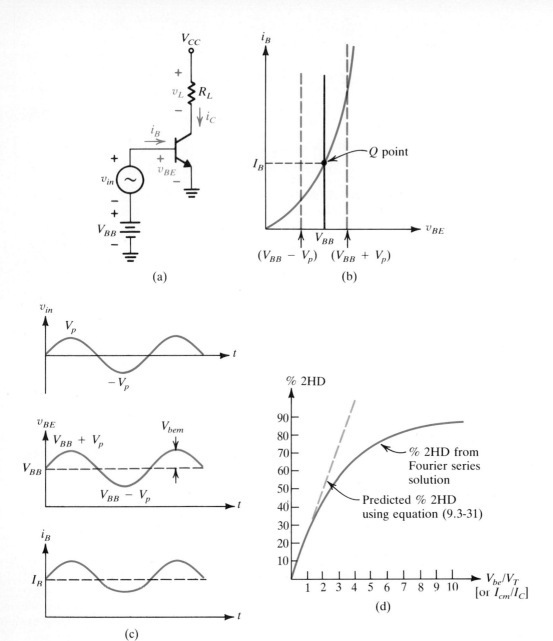

Figure 9.3-3

Second harmonic distortion in a common-emitter amplifier.

of i_B will tend to be larger than the negative part, and i_B will have the basic shape illustrated in Figure 9.3-3c.

The magnitude of the distortion may be estimated by considering that

$$i_B \simeq I_{BO}e^{v_{BE}/V_T} = I_{BO}e^{(V_{BB} + v_{in})V_T} \qquad (9.3\text{-}19)$$

When $v_{in} = 0$, $I_B = I_{BO}e^{V_{BB}/V_T}$, and therefore we may write

$$i_B = I_Be^{v_{in}/V_T} \qquad (9.3\text{-}20)$$

For small values of v_{in} we may expand the exponential as

$$i_B = I_B\left[1 + \underbrace{\frac{v_{in}}{V_T}}_{\text{linear}} + \underbrace{\frac{1}{2!}\left(\frac{v_{in}}{V_T}\right)^2}_{\substack{\text{second} \\ \text{harmonic} \\ \text{distortion}}} + \underbrace{\frac{1}{3!}\left(\frac{v_{in}}{V_T}\right)^3}_{\substack{\text{third} \\ \text{harmonic} \\ \text{distortion}}} + \cdots\right] \qquad (9.3\text{-}21)$$

Thus as the size of v_{in} increases (relative to V_T), the base current can no longer simply be represented by the dc plus the linear term; the harmonic distortion being produced needs to be considered. As v_{in} grows, the second harmonic distortion term will initially be the biggest, and for still larger values of v_{in}, the higher-order distortion terms will need to be included.

Let's estimate the percent second harmonic distortion in the base current for the case where $v_{in} = V_p \sin \omega t$. Substituting this expression into equation (9.3-21) and dropping all terms above the quadratic, we obtain

$$i_B = I_B\left[1 + \frac{V_{bem}}{V_T}\sin \omega t + \frac{1}{2}\left(\frac{V_{bem}}{V_T}\right)^2 \sin^2 \omega t\right] \qquad (9.3\text{-}22)$$

where $V_{bem} = V_p$ is the amplitude of the ac portion of the base–emitter voltage. Using the trigonometric identity

$$\sin^2 \omega t = \frac{1 - \cos 2\omega t}{2} \qquad (9.3\text{-}23)$$

we may rewrite equation (9.3-22) as

$$i_B = I_B\underbrace{\left[1 - \frac{1}{4}\left(\frac{V_{bem}}{V_T}\right)^2\right]}_{\text{dc}} + \underbrace{\left(\frac{V_{bem}}{h_{ie}}\right)\sin \omega t}_{\text{linear}} - \underbrace{\left(\frac{V_{bem}}{h_{ie}}\right)\left(\frac{V_{bem}}{4V_T}\right)\cos 2\omega t}_{\text{second harmonic}} \qquad (9.3\text{-}24)$$

If we define the second harmonic distortion (2HD) of a signal as the ratio of the second harmonic amplitude to the fundamental amplitude, we express this quantity as

$$\% \ 2\text{HD} = \frac{(V_{bem}/h_{ie})(V_{bem}/4V_T)}{(V_{bem}/h_{ie})} = \frac{V_{bem}}{4V_T} \times 100\%$$

$$= \frac{1}{4}\left(\frac{I_{cm}}{I_C}\right) \times 100\% \qquad (9.3\text{-}25)$$

where I_{cm} and I_C are the ac and dc collector current amplitudes, respectively. This result indicates that the second harmonic distortion grows linearly with I_{cm}, the amplitude of the ac current flow in the collector. Although this expression is valid for small values of I_{cm}, for large values of ac collector current the power series approximation fails and the second harmonic distortion needs to be obtained by Fourier series methods. Although this derivation is beyond the scope of this book, the results are nonetheless useful and are presented graphically in Figure 9.3-3d. Included for comparison is a sketch of the result obtained using equation (9.3-25).

Adding an emitter resistor or a base resistor in series with v_{in} in Figure 9.3-3a reduces the base current distortion because i_B is now less dependent on the nonlinear characteristics of the base–emitter junction. In particular, it can be shown (see Problem 9.3-8) that the distortion level in equation (9.3-25) is reduced to

$$\% \text{ 2HD} = \frac{1}{4}\left(\frac{I_{cm}}{I_C}\right) \times \left[\frac{h_{ie}}{R_B + h_{ie} + (1 + h_{fe})R_E}\right] \times 100\% \qquad (9.3\text{-}26)$$

by the addition of R_B and R_E to the circuit.

9.3-3 Class A Single-Ended Transformer Coupled Amplifier

The inclusion of a transformer in a class A single-ended power amplifier design offers several advantages over its direct-coupled counterpart. First, the transformer keeps the transistor's dc bias current out of the load. This is important because dc flowing in a loudspeaker, for example, statically deflects the cone away from its equilibrium position and reduces the maximum undistorted signal power that can be delivered by the loudspeaker. Furthermore, because the transformer primary winding is a short to dc, no dc power is dissipated in either the transformer or the load. By way of contrast, in a direct-coupled amplifier more than 50% of the dc power coming in from the supply is wasted in heating up the load [see equation (9.3-11b)]. Therefore, with a transformer-coupled amplifier, it is reasonable to expect a significant increase in efficiency above that for the direct-coupled case.

Another advantage of transformer coupling is that the effective load impedance at the transformer primary can be adjusted to any desired value by properly selecting the transformer turns ratio. This allows for optimal placement of the load line at a point tangent to the power dissipation hyperbola in order to achieve maximum power delivery to the load for a given-size power transistor.

The design of a typical transformer-coupled class A power amplifier is illustrated in Figure 9.3-4a. In this figure L represents the transformer's primary inductance; it is assumed to be a short to dc and an open circuit at the frequency of the applied signal. Let's begin the analysis of this circuit by developing the dc and the dynamic load-line equations using the equivalent circuits given in parts (a) and (b) of the figure, respectively.

Following Figure 9.3-4a, the dc load-line equation may be written down by inspection as

$$V_{CE} = V_{CC} \qquad (9.3\text{-}27)$$

The equivalent circuit given in part (b) of the figure for developing the dynamic load-line equation follows directly from similar material discussed in Chapter 8, and hence it will not be repeated here. If you have forgotten the basis for this equivalent circuit, you should review Example 8.5-1.

Following Figure 9.3-4b, the expression for the dynamic load-line equation is readily seen to be

$$V'_{CC} = V_{CC} + I_C R'_L = i_C R'_L + v_{CE} \qquad (9.3\text{-}28)$$

Both the dc and dynamic load lines are sketched on the transistor output characteristics in Figure 9.3-4c. In examining these sketches, several interesting facts emerge. First, the dc load line is vertical since $V_{CE} = V_{CC}$. The position of the Q point on this load line is determined from the dc analysis of the base

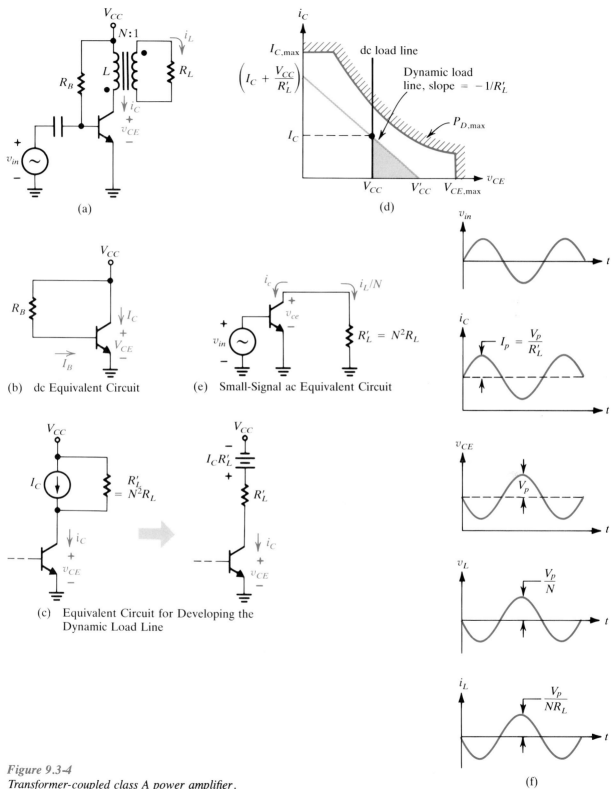

Figure 9.3-4
Transformer-coupled class A power amplifier.

circuit using the expression

$$I_C = h_{FE} I_B \tag{9.3-29}$$

Of course, the dynamic load line passes through the Q point and has a slope equal to $-1/R'_L$. Of particular interest is the fact that v_{CE} for this circuit can exceed the power supply voltage. The basic reason that this can occur is because of the inductance of the transformer. Inductors attempt to maintain the current flow through them constant, and as a result, when the collector current begins to decrease below I_C, the voltage at v_{CE} rises above its V_{CC} quiescent value in order to "push" more current into the transistor and reestablish i_C at I_C.

To analyze the performance of this circuit let us assume that v_{in} is a sine wave and that the transistor characteristics are linear so that the ac part of the collector current is also a sinusoid. Adding this current component to the dc quiescent current, we may write

$$i_C = I_C + I_p \sin \omega t \tag{9.3-30}$$

where I_p represents the ac amplitude of the collector current. In a similar fashion the collector-to-emitter voltage for this situation may also be written as

$$v_{CE} = V_{CE} - V_p \sin \omega t = V_{CC} - V_p \sin \omega t \tag{9.3-31}$$

where the ac amplitudes I_p and V_p are related by

$$I_p = \frac{V_p}{R'_L} \tag{9.3-32}$$

following the ac equivalent circuit given in Figure 9.3-4d. The pertinent waveshapes for this amplifier are sketched in Figure 9.3-4e.

The dc power supplied by V_{CC} is found from

$$P_{IN} = \frac{1}{T}\int_0^T V_{CC} i_C dt = \frac{1}{T}\int_0^T V_{CC}(I_C + I_p \sin \omega t)\, dt = V_{CC} I_C \tag{9.3-33}$$

Similarly, the average power delivered to the load is

$$P_L = \frac{1}{T}\int_0^T i_L^2 R_L\, dt = \frac{1}{T}\int_0^T \left[\left(\frac{V_p}{NR_L}\right)^2 \sin^2 \omega t \right] R_L\, dt = \frac{V_p^2}{2R'_L} \tag{9.3-34}$$

Because the transformer is lossless, this same result could have been obtained by using the equivalent circuit in Figure 9.3-4e and considering that the ac collector voltage $V_p \sin \omega t$ was applied to the effective ac load R'_L. Notice also that because the transformer is a short to dc, $V_L = 0$, and there is no dc power wasted in the load.

While the power dissipated in the transistor may be calculated from the equation (9.2-3) it is much easier to apply conservation of energy to this problem and write P_D as the difference between P_{IN} and P_L, which yields

$$P_D = P_{IN} - P_L = V_{CC} I_C - \frac{V_p^2}{2R'_L} \tag{9.3-35}$$

To deliver maximum power to the load, following equation (9.3-34) we want V_p to be as large as possible, and as for the direct-coupled case, this suggests that we place the Q point at the center of the dynamic load line. To accomplish this, we require that $V'_{CC} = 2V_{CC}$ or $I_C = V_{CC}/R'_L$.

The circuit efficiency for this transformer-coupled amplifier is found by combining equations (9.3-33) and (9.3-34), which yields

$$\eta = \frac{P_L}{P_{IN}} = \frac{V_p^2/2R'_L}{V_{CC}I_C} = \frac{V_p^2/2R'_L}{V_{CC}(V_{CC}/R'_L)} = \frac{1}{2}\left(\frac{V_p}{V_{CC}}\right)^2 \times 100\% \qquad (9.3\text{-}36)$$

With the Q point located at the center of the load line, the maximum allowed value for V_p equals V_{CC}, so that the maximum efficiency for this circuit is 50%, twice that of a direct-coupled power amplifier. This improvement in efficiency comes about because in the direct-coupled case, 50% of the dc input power is wasted in the dc heating of the load.

Paralleling the development of the analysis for the direct-coupled amplifier, it is advantageous to rewrite the expression for P_L as

$$P_L = \frac{V_p^2}{2R'_L} = \frac{1}{2}\left(\frac{V_p}{R'_L}\right)V_p = \frac{1}{2}V_p I_p \qquad (9.3\text{-}37)$$

Under maximum signal swing conditions (or maximum power delivery to the load), this expression, besides representing $P_{L,max}$, is equal to the shaded triangular area in Figure 9.3-4d. As with the direct-coupled case, it is not too difficult to show that this area, and hence the ac power delivered to the load, is maximized when the load line is drawn tangent to the power dissipation hyperbola curve.

Of all possible load lines that could be drawn tangent to the dissipation hyperbola, the one illustrated in Figure 9.3-5a is preferred because it minimizes the I^2R power losses. Hence, if possible, the load line should be placed tangent to the dissipation hyperbola with its v_{CE} intercept at $V_{CE,max}$ and its Q point at the point of tangency. With the load line drawn in this position, the following facts are apparent:

1. $V_{CC} = \frac{1}{2}V_{CE,max}$ (because the dissipation hyperbola bisects the load line at the point of tangency)
2. $I_C \times V_{CC} = P_{D,max}$, or $I_C = P_{D,max}/V_{CC}$
3. $R'_L = V_{CC}/I_C$

In addition, with the load line positioned as shown in Figure 9.3-5a, following equation (9.3-36), we can see that the maximum power that can be delivered to the load is

$$P_{L,max} = \frac{V_{CC}^2}{2R'_L} = \frac{1}{2}\left(\frac{V_{CC}}{R'_L}\right)V_{CC} = \frac{1}{2}I_C V_{CC} = \frac{P_{D,max}}{2} \qquad (9.3\text{-}38)$$

To illustrate the application of these results, consider, as for the class A direct-coupled circuit described earlier, that we wish to design an amplifier to deliver 10 W of ac power to a load. What kind of power supply and transistor will be needed if a transformer-coupled circuit is used?

The maximum circuit efficiency of a transformer-coupled amplifier is 50%,

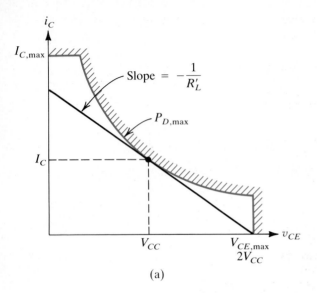

(a)

Signal Level, v_{in}	P_{IN} (W)	P_L (W)	$P_D = P_{IN} - P_L$ (W)	η (%)
0	20	0	20	0
Maximum	20	10	10	50

(b)

Figure 9.3-5

and this efficiency occurs at maximum power delivery to the load. Therefore, if $P_{L,max}$ is 10 W, the dc power supply will have to be able to deliver $2P_{L,max}$ or 20 W. The transistor power-handling requirements may be found by considering the power dissipated in the transistor under zero-signal and full-signal conditions.

The dc power input from the supply is independent of the signal level [equation (9.3-33)], and hence the same dc power (20 W in this case) is delivered to the amplifier under both zero and full-input signal conditions. When no input signal is applied to the amplifier, no ac power is delivered to the load, and because of this, under these quiescent conditions, all 20 W from the dc power supply must be dissipated in the transistor. When full ac power is delivered to the load, the transistor actually cools down, and by conservation of energy, only 10 W is dissipated in it. Therefore, in building this amplifier a transistor with a minimum power-handling capability of 20 W will be needed.

The values of P_{IN}, P_L, and P_D obtained for this transformer-coupled amplifier under both zero- and full-signal conditions are summarized in Figure 9.3-5b. In examining these results, at first glance the transformer-coupled design does not appear to be that cost-effective, considering the added expense of the output transformer. However, this may not necessarily be the case. The power supply required for the transformer-coupled design is only one-half as large as that for the direct-coupled case, and furthermore, the output transistor needed for the transformer-coupled amplifier can be smaller than for the direct-coupled circuit.

This last statement may be somewhat puzzling since, in theory at least, both amplifier designs appear to require 20-W power transistors. However, this assumes that the load lines of both circuits can be placed tangent to the dissipation hyperbola. For the transformer-coupled case, this requirement presents no problem since the turns ratio can usually be selected to achieve this. However, for the direct-coupled circuit, unless R_L is variable, it may not be possible to position the load line optimally, and as a result, a much larger power transistor may actually be needed for the direct-coupled design. Thus the additional cost of the transformer in a transformer-coupled amplifier may be more than offset by the larger power supply and larger power transistor requirements of the direct-coupled design.

EXAMPLE 9.3-2

The MOSFET in the power amplifier circuit shown in Figure 9.3-6a has the following maximum ratings: $P_{D,\max} = 25$ W, $I_{D,\max} = 3$ A, and $V_{DS,\max} = V_{(BR)DSS} = 50$ V. In addition, r_{DS} is assumed to be constant at 2 Ω throughout the ohmic region.

Figure 9.3-6
Example 9.3-2.

a. Position the load line and the Q point for maximum power delivery to the load.
b. Select R_1, R_2, V_{DD}, and N to achieve the load-line position specified in part (a).

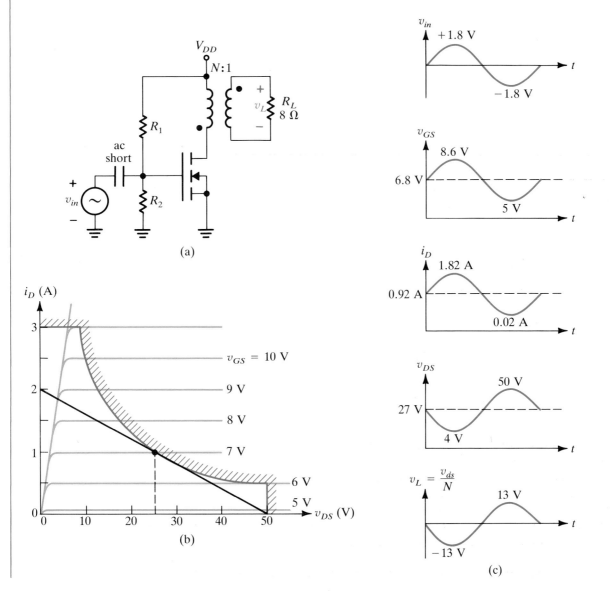

(a)

(b)

(c)

c. Determine P_{IN}, P_L, P_D, and the required value of v_{in} for the maximum power condition, and for this situation sketch v_{in}, v_{GS}, i_D, v_{DS}, and v_L assuming v_{in} to be a sine wave.

SOLUTION

If the load line is positioned on the transistor characteristics as shown in Figure 9.3-6b, the required value of V_{DD} will be $\frac{1}{2} V_{DS,\max}$ or 25 V. The corresponding values of I_D, R'_L, and transformer turns ratio for this load line and Q-point position are

$$I_D = \frac{P_{D,\max}}{V_{CC}} = \frac{25 \text{ W}}{25 \text{ V}} = 1 \text{ A} \tag{9.3-39a}$$

$$R'_L = \frac{\Delta v_{DS}}{\Delta i_D} = \frac{25 \text{ V}}{1 \text{ A}} = 25 \text{ }\Omega \tag{9.3-39b}$$

and

$$N = \sqrt{\frac{R'_L}{R_L}} = \sqrt{\frac{25}{8}} = 1.77 \tag{9.3-39c}$$

The allowed undistorted signal swing for this case will be about 21 V since the transistor will enter the ohmic region when v_{DS} is about 4 V. The maximum undistorted signal swing, and hence the ac power to the load, may be improved slightly by increasing V_{DD} to $(\frac{1}{2})(50 + 4)$ or 27 V, which for the same load-line position would require a decrease in I_D to

$$I_D = \frac{\Delta v_{DS}}{R'_L} = \frac{50 - 27}{25} = 0.92 \text{ A} \tag{9.3-40}$$

Although this new Q-point location will be slightly off the dissipation hyperbola, it will still result in more power being delivered to the load than if V_{DD} were kept at 25 V because the ac signal into the load will be larger.

Following Figure 9.3-6b, the value of V_{GS} corresponding to this second Q-point position is about 6.8 V. Arbitrarily setting $R_2 = 1$ MΩ, we find that R_1 must satisfy the equation

$$27 \left(\frac{1}{1 + R_1} \right) = 6.8 \tag{9.3-41}$$

Solving this expression for R_1, we find that it is approximately equal to 3 MΩ.

Under maximum signal swing conditions, the ac power into the load is

$$P_L = \frac{V_p^2}{2R'_L} = \frac{(27 - 4)^2}{50} = 10.6 \text{ W} \tag{9.3-42}$$

while the dc power delivered to the amplifier by V_{DD} is given by the expression

$$P_{IN} = V_{DD} I_D = (27 \text{ V})(0.92 \text{ A}) = 24.8 \text{ W} \tag{9.3-43}$$

Thus the maximum amplifier efficiency is

$$\eta_{\max} = \frac{P_{L,\max}}{P_{IN}} = \frac{10.6}{24.8} \simeq 43\% \tag{9.3-44}$$

This efficiency is slightly smaller than the maximum theoretical efficiency of

50% achievable with an ideal class A transformer-coupled amplifier because the finite ON resistance of the FET prevents the output signal from going all the way down to zero volts.

The value of v_{in} needed to deliver maximum power to the load is found by noting from Figure 9.3-6b that the required peak to peak signal swing at the gate will be about $(8.6 - 5)$ or 3.6 V. Because $v_{gs} = v_{in}$, the input sine-wave amplitude will need to be about 1.8 V. This result could have been obtained by noting that the transistor g_m is $(0.5$ A$)/(1$ V$)$ or 0.5 S, while the peak ac drain current needed is about 0.9 A, so that $V_{gs} = (0.9/0.5) = 1.8$ V. The pertinent waveshapes for this circuit corresponding to this value of v_{in} are sketched in Figure 9.3-6c.

The power dissipated in the transistor under these conditions is found from conservation of energy to be

$$P_D = P_{IN} - P_L = 24.8 - 10.6 = 14.2 \text{ W} \qquad (9.3\text{-}45)$$

Of course, this value of P_D is actually much smaller than the maximum power that the transistor needs to be able to handle. When v_{in} is reduced to zero, P_L is zero and the power dissipated in the transistor will increase to the full 24.8 W being supplied by V_{DD}. Thus a transistor with a $P_{D,\max}$ of at least 24.8 W will be needed for this design.

Exercises

9.3-1 Equation (9.3-32) was developed to predict the approximate second harmonic distortion caused by the base–emitter diode nonlinearity in BJT-style amplifiers. Estimate the % 2HD for the common-emitter amplifier in Figure 9.3-2 if a 20-Ω resistor is connected between the emitter and ground, and if $h_{FE} = h_{fe} = 100$, $R_B = 22.3$ kΩ, $v_{in} = 1 \sin \omega t$ volts, and $V_{CC} = 25$ V. **Answer** 0.16%

9.3-2 In Exercise 9.3-1, consider that v_{in} is adjustable and find the maximum power deliverable to the load if transistor distortion (other than saturation and cutoff) is neglected. **Answer** 500 mW

9.3-3 Compute P_D for the transformer-coupled power amplifier in Figure 9.3-4a by using equation (9.2-3), and show that the result obtained is the same as that given in equation (9.3-35).

9.3-4 For the circuit shown in Example 9.3-2 (Figure 9.3-6a), a 5-Ω resistor is connected in series with the source lead of the transistor.
  a. What is the optimal Q-point position for maximum power delivery to the load if the transformer turns ratio is $2:1$?
  b. Using the Q-point position determined in part (a), what is maximum power deliverable to the load?
  c. Assuming that $g_m = 0.5$ S for this transistor, find the amplitude of v_{in} needed to deliver maximum power to the load.
  Answer (a) $I_D = 0.675$ A, $V_{DS} = 25$ V, and $V_{GS} = 6.25$ V using $V_{DD} = 21.6$ V; (b) 7.3 W; (c) 1.35 V

9.4 PUSH-PULL POWER AMPLIFIERS

Single-ended power amplifiers are characterized by relatively large amounts of harmonic distortion in the signal delivered to the load as well as a high power dissipation in the amplifying transistor. The use of a push-pull amplifier design can significantly reduce both these problems.

A push-pull amplifier is basically a two-transistor circuit whose ac output currents are 180° out of phase from one another. Furthermore, the circuit is configured so that the load current is proportional to the difference between the two output currents and hence has an ac component that is twice as large as that attainable from an equivalent single-ended circuit.

The words "push-pull" used to characterize this type of electronic amplifier circuit were originally coined to describe the operation of mechanical systems having similar performance characteristics. Consider, for example, the two-man saw, a saw specifically designed to cut large logs. It looks similar to an ordinary saw except that it has two handles, one on each side. It is usually operated by two people, and to cut a log, the person on one side "pulls" the saw toward himself as the person on the other side "pushes" the saw in the same direction. This procedure is reversed to move the saw in the opposite direction. Obviously, the cutting force produced with this type of two-man saw is twice that of a standard one-man saw.

The electronic push-pull amplifier offers a similar advantage over a single-ended amplifier design in that it can deliver more power to the load than is possible with a comparable single-transistor circuit. In addition, it permits class B operation of the amplifier in which each transistor conducts for only one-half of the signal period. As we show later in this section, class B amplifiers are much more efficient than amplifiers whose operating points are always in the active region. This higher efficiency allows for the delivery of a specified load power with smaller power transistors and smaller power supplies.

Besides offering significant efficiency improvements, push-pull designs intrinsically have lower distortion levels than their single-ended counterparts. In particular, as we demonstrate, symmetrical push-pull amplifiers have no even harmonic distortion. Futhermore, and again as a natural consequence of the architecture of a push-pull amplifier, no dc current need flow in the load. This, as we have mentioned, in addition to allowing for high circuit efficiency, is especially important for loads such as loudspeakers where dc current flow can significantly reduce the ac power that the load can handle.

9.4-1 Class A Push-Pull Power Amplifier

Figure 9.4-1 illustrates the two basic types of BJT-style push-pull power amplifiers. Of course, similar circuits can also be constructed with FET devices. In examining these two designs we should observe that the circuit in part (a) is essentially a two-transistor emitter-follower or common-collector amplifier, while that in part (b) is a common-emitter push-pull design. Notice in both of these circuits that Q_1 is an *npn* transistor while Q_2 is a *pnp* device. This type of push-pull amplifier is called a complementary-symmetry design. Although a common-base push-pull amplifier can also be envisioned, as with the case of

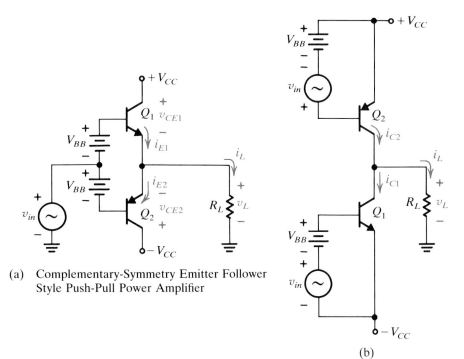

(a) Complementary-Symmetry Emitter Follower
Style Push-Pull Power Amplifier

(b)

Figure 9.4-1

single-ended designs, common-base circuits are only rarely used in push-pull
power amplifiers because of their very low power gain when driving small load
impedances.

Of the two circuits illustrated in Figure 9.4-1, the emitter-follower push-
pull design is probably the most popular. Although it has no voltage gain, its
low output impedance and its exceptionally low distortion levels make it extremely useful, for example, in audio amplifier designs. The importance of low
signal distortion for the design of high-fidelity amplifiers is obvious, but the
advantages of low circuit output impedance is somewhat less apparent. When
an amplifier has a low output impedance, its output voltage will be relatively
independent of the character of the load. This is especially important for feedback amplifier designs since a low amplifier output impedance will help to minimize the phase shifts produced by reactive loads and can thereby reduce the
likelihood of circuit instability and oscillation.

Initially, we examine the behavior of each of these push-pull circuits assuming that the small-signal transistor models developed in earlier chapters are
still valid, and then later, we extend these analyses to take into account the
nonlinear behavior of the transistors. We begin by examining the emitter-follower push-pull circuit in Figure 9.4-1a. In doing so, we assume that the transistors are symmetric and that V_{BB} is chosen so that the circuit is operating in
the class A mode with quiescent emitter currents

$$I_{E1} = I_{E2} \tag{9.4-1}$$

To the extent that this equation is valid, no dc current flows in the load, and we
have $V_L = 0$. Therefore, it also follows that

$$V_{CE1} = V_{CC} - V_L = V_{CC} \tag{9.4-2a}$$

and
$$V_{CE2} = -V_{CC} - V_L = -V_{CC} \tag{9.4-2b}$$

To carry out the ac portion of this push-pull circuit analysis, let us develop the emitter equivalent circuit seen by the load R_L. In doing this we can consider that each base is driven by a separate voltage source, v_{in}, as shown in Figure 9.4-2a. Reflecting each of these sources and their corresponding h_{ie} values into the emitter of each transistor, we obtain the equivalent circuit given in part (b) of the figure. Because points A and B in this figure are at the same potential, we may combine the two v_{in} sources together. As one last simplification, let's assume that r_{e1} and r_{e2} are identical because of the complementary symmetry of the two transistors. To the extent that this approximation is valid, we obtain the equivalent circuit given in Figure 9.4-2c, and from this circuit we have that

$$v_l = \frac{R_L}{r_e/2 + R_L} v_{in} \simeq v_{in} \tag{9.4-3}$$

if $R_L \gg r_e/2$. In addition, it is also apparent from these equivalent circuits that

$$i_{e1} = -i_{e2} = \frac{i_l}{2} \simeq \frac{v_{in}}{2R_L} \simeq i_{c1} = -i_{c2} \tag{9.4-4}$$

From Figure 9.4-1a we also have

$$v_{CE1} = V_{CC} - v_L = V_{CC} - i_l R_L = V_{CC} - 2i_{c1} R_L \tag{9.4-5}$$

But
$$i_{c1} = i_{C1} - I_{C1} \tag{9.4-6}$$

and substituting this expression into equation (9.4-5), the dynamic load-line equation for Q_1 may be written as

$$v_{CE1} = V_{CC} - 2(i_{C1} - I_{C1}) R_L \tag{9.4-7a}$$

or
$$V'_{CC} = V_{CC} + 2I_{C1} R_L = 2i_{C1} R_L + v_{CE1} \tag{9.4-7b}$$

The load line is sketched on the transistor characteristics of Q_1 in Figure 9.4-3. Following the previous discussion for the class A single-ended am-

Figure 9.4-2

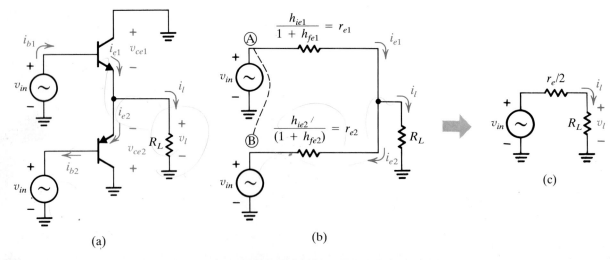

(a)   (b)   (c)

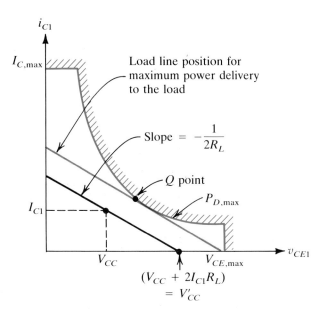

Figure 9.4-3

plifier, if R_L is variable, then to deliver maximum power to the load, V_{CC} and R_L should be adjusted to place the load line in the lower right-hand corner of the SOA at $V_{CE,\text{max}}$ and tangent to the dissipation hyperbola. Furthermore, if the Q point is placed at the point of tangency, this will lead to maximum power delivery to the load and maximum undistorted signal swing. For this case the required values of V_{CC}, I_{C1}, and R_L will be $V_{CE,\text{max}}/2$, $P_{D,\text{max}}/V_{CC}$, and $V_{CC}/2I_{C1}$, respectively. In the situation where R_L is fixed, the design is accomplished by increasing V_{CC} until the load line hits either $V_{CE,\text{max}}$ or the dissipation hyperbola. The Q point is still placed at the center of the load line in order to achieve maximum signal swing.

<table>
<tr><td>EXAMPLE 9.4-1</td><td>For the class A push-pull circuit illustrated in Figure 9.4-1a, the transistors are assumed to be complementary-symmetry devices having identical characteristics: namely, $P_{D,\text{max}} = 10$ W, $V_{CE,\text{max}} = 30$ V, and $I_{C,\text{max}} = 2$ A.</td></tr>
</table>

a. Select V_{CC}, R_L, and the transistor Q points to deliver maximum power to the load.
b. If $h_{ie} = 20\ \Omega$ and $h_{fe} = 100$ for these transistors, independent of the Q point, find the required value of v_{in} needed to deliver maximum power to the load. For this condition sketch v_{in}, i_{C1}, i_{C2}, v_L, v_{CE1}, and v_{CE2}.
c. For $v_{in} = 0$ find P_D for each transistor as well as P_L, P_{IN}, and the circuit efficiency.
d. Repeat part (c) for v_{in} at the maximum power level.

SOLUTION

The safe operating area for Q_1 is shown in Figure 9.4-4a, and a similar sketch could also be drawn for Q_2. The load line is shown positioned in the figure for maximum power delivery to the load, and corresponding to this position we have

$$V_{CC} = \tfrac{1}{2}V_{CE,\text{max}} = 15 \text{ V} \qquad (9.4\text{-}8a)$$

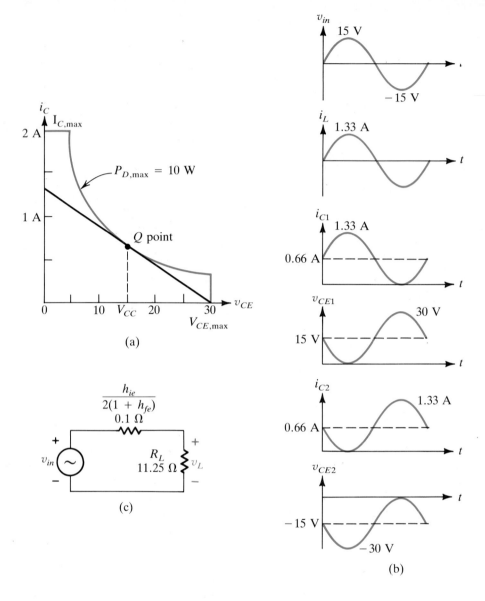

Figure 9.4-4
Example 9.4-1.

and
$$I_{C1} = I_{C2} = \frac{P_{D,\text{max}}}{V_{CC}} = \frac{10\text{ W}}{15\text{ V}} = 0.66\text{ A} \qquad (9.4\text{-}8b)$$

Because the slope of the load line is $-1/2R_L$, it follows that

$$R_L = \frac{V_{CC}}{2I_{C1}} = \frac{15\text{ V}}{1.33\text{ A}} = 11.25\ \Omega \qquad (9.4\text{-}8c)$$

Examining the position of the Q point on the load line, we can see that the maximum allowed ac signal swing in v_{ce1} is 15 V. In addition, because

$$v_{ce1} = v_{ce2} = -v_l \qquad (9.4\text{-}9)$$

when maximum sinusoidal ac power is being delivered to the load, we may therefore write

$$v_{in} \simeq v_L = 15 \sin \omega t \text{ volts} \tag{9.4-10a}$$

$$v_{CE1} = V_{CC} - v_L = (15 - 15 \sin \omega t) \text{ volts} \tag{9.4-10b}$$

$$v_{CE2} = -V_{CC} - v_L = (-15 - 15 \sin \omega t) \text{ volts} \tag{9.4-10c}$$

and
$$i_L = \frac{v_L}{R_L} = 1.33 \sin \omega t \text{ amperes} \tag{9.4-10d}$$

Because $i_{c1} = \frac{1}{2} i_L = -i_{c2}$ [equation (9.4-4)], we may also write

$$i_{C1} = I_{C1} + i_{c1} = (0.66 + 0.66 \sin \omega t) \text{ amperes} \tag{9.4-10e}$$

and
$$i_{C2} = I_{C2} + i_{c2} = (0.66 - 0.66 \sin \omega t) \text{ amperes} \tag{9.4-10f}$$

These voltages and currents are sketched in Figure 9.4-4b.

To carry out the power calculations for this example, let us write

$$v_{in} \simeq v_L = V_m \sin \omega t \tag{9.4-11a}$$

from which we also have that

$$i_L = \frac{V_m}{R_L} \sin \omega t = 2i_{c1} = -2i_{c2} \tag{9.4-11b}$$

Thus

$$P_{IN} = \frac{1}{T} \int_0^T (V_{CC} i_{C1} + V_{CC} i_{C2}) dt = 2V_{CC} I_C = 20 \text{ W} \tag{9.4-12}$$

independent of the signal level. The power delivered to the load is given by

$$P_L = \frac{1}{T} \int_0^T \frac{v_L^2}{R_L} dt = \frac{V_m^2}{R_L T} \int_0^T \frac{1 - \cos 2\omega t}{2} dt = \frac{V_m^2}{2R_L} \tag{9.4-13}$$

Using this result it is apparent that the minimum power into the load is zero when $v_{in} = 0$, and the maximum is $(15)^2/(2)(11.25) = 10$ W. Combining these results with equation (9.4-12), the corresponding efficiencies are seen to be 0 and 50%. Thus the maximum efficiency of a push-pull class A power amplifier is the same as that of a single-ended transformer-coupled design, but twice that of a single-ended direct-coupled design.

The power dissipated in the transistors, for example in Q_1, may be found by evaluating the expression

$$P_{D1} = \frac{1}{T} \int_0^T v_{CE1} i_{C1} dt = V_{CC} I_{C1} - \frac{1}{2} \left(\frac{V_m^2}{2R_L} \right) = \frac{P_{IN}}{2} - \frac{P_L}{2} = \frac{1}{2} (P_{IN} - P_L) \tag{9.4-14}$$

It is interesting to interpret the power dissipated in Q_1 by examining the final version of equation (9.4-14). When no power goes to the load (i.e., when $v_{in} = 0$), all the input power is dissipated in the two transistors, half in each of them. When ac power, P_L, is delivered to the load, then, by conservation of energy, the overall power dissipated in the transistors drops to $(P_{IN} - P_L)$, and because of the symmetry, one-half of this power goes to each transistor. Therefore, with $v_{in} = 0$, $P_{D1} = P_{D2} = \frac{1}{2} P_{IN} = 10$ W. At full power to the load, the transistor dissipation drops to $\frac{1}{2}(20 - 10)$ or $(5 \text{ W})/\text{transistor}$.

In analyzing this circuit, we assumed that v_L was nearly equal to v_{in}. With $h_{ie} = 20\ \Omega$ and $h_{fe} = 100$, the emitter equivalent circuit given in Figure 9.4-4c is valid, and thus less than a 1% error is made by assuming that v_L and v_{in} are equal.

9.4-2 Class B Push-Pull Power Amplifier

In a push-pull circuit configuration besides the class A design, it is also possible to operate the transistors in the class B mode in which they are biased at cutoff so that each transistor conducts only during one-half of the input cycle. The circuit efficiency and transistor power-handling requirements for class B circuit operation are considerably better than those for a comparable class A design. However, the distortion levels are also much greater. One way to reduce this distortion without giving up the advantages of class B is to operate the circuit in the class AB mode, where the transistors are biased close to, but not quite at, cutoff. This type of circuit has nearly the same efficiency as class B but has significantly reduced distortion levels.

We begin this discussion of class B circuit analysis techniques by examining the push-pull emitter-follower power amplifier shown in Figure 9.4-5a. Basically, this circuit is the same as the class A amplifier discussed earlier, and illustrated in Figure 9.4-1a, except that the V_{BB} biasing batteries have been removed. In analyzing this circuit, we initially neglect the base–emitter junction voltage, V_o, of Q_1 and Q_2. This approximation creates negligible error if $v_{in} \gg V_o$, and makes the analysis of the circuit much more intuitive.

The circuit operates as follows. When v_{in} is positive, Q_1 turns ON and Q_2 goes OFF, so that the circuit in Figure 9.4-5b applies. For this situation, to the extent that $R_L \gg r_{e1}$ and r_{e2}, we may write

$$v_{in} = V_m \sin \omega t \simeq v_L \tag{9.4-15}$$

for the case where the input is a sine wave. In addition, we have that

$$i_L = \frac{v_L}{R_L} = \frac{V_m}{R_L} \sin \omega t \tag{9.4-16a}$$

$$i_{C1} \simeq i_{E1} = i_L = \frac{V_m}{R_L} \sin \omega t \tag{9.4-16b}$$

and
$$i_{C2} \simeq 0 \tag{9.4-16c}$$

when $v_{in} > 0$. Similarly, when v_{in} is negative, Q_1 goes OFF and Q_2 goes ON, so that Figure 9.4-5c applies. For this circuit v_L and i_L may still be determined from equations (9.4-15) and (9.4-16a), whereas i_{C1} and i_{C2} are now given by the expressions

$$i_{C1} = 0 \tag{9.4-17a}$$

and
$$i_{C2} \simeq i_{E2} = -i_L = \frac{-V_m}{R_L} \sin \omega t \tag{9.4-17b}$$

These results are sketched in Figure 9.4-5d. Notice under quiescent conditions

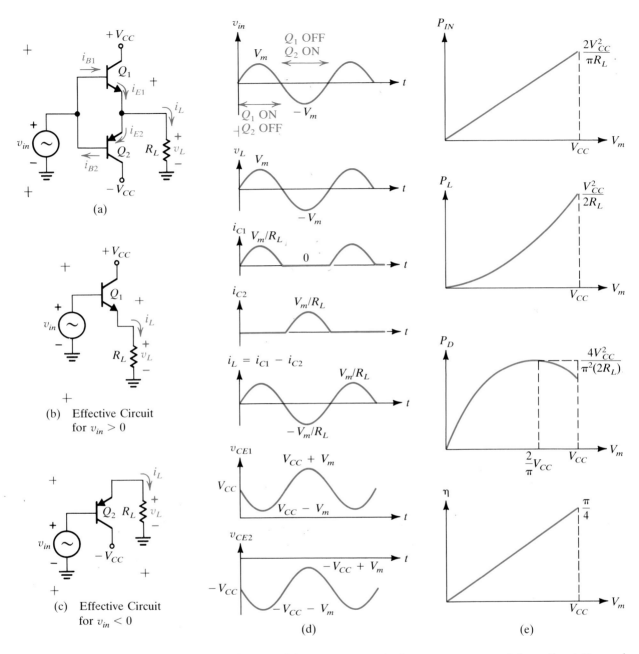

Figure 9.4-5
Class B emitter-follower push-pull power amplifier.

that both i_{C1} and i_{C2} are zero, as is the power consumed from the $+V_{CC}$ and $-V_{CC}$ supplies.

When v_{in} is nonzero, the dc input power is given by the expression

$$P_{IN} = \frac{1}{T}\int_0^T V_{CC}i_{C1}dt + \frac{1}{T}\int_0^T V_{CC}i_{C2}dt = V_{CC}\frac{V_m}{\pi R_L} + V_{CC}\frac{V_m}{\pi R_L} = \frac{2}{\pi}V_{CC}\frac{V_m}{R_L}$$

$$(9.4\text{-}18)$$

In this result it is useful to point out that the term $V_m/\pi R_L$ is just the average value of a half-wave rectified sine wave with an amplitude V_m/R_L. In a similar fashion, the power delivered to the load, the total power dissipated in both transistors, and the circuit efficiency are given by

$$P_L = \frac{1}{T}\int_0^T \frac{v_L^2}{R_L}dt = \frac{V_m^2}{2R_L} \tag{9.4-19a}$$

$$P_D = P_{IN} - P_L = \frac{2}{\pi}V_{CC}\frac{V_m}{R_L} - \frac{V_m^2}{2R_L} \tag{9.4-19b}$$

and

$$\eta = \frac{P_L}{P_{IN}} = \frac{V_m^2/2R_L}{\dfrac{2}{\pi}V_{CC}(V_m/R_L)} = \frac{\pi}{4}\frac{V_m}{V_{CC}} \tag{9.4-19c}$$

respectively.

P_{IN}, P_L, P_D, and η are sketched in Figure 9.4-5e as functions of the input signal amplitude. It is interesting to note that P_L, P_{IN}, and η all increase with increasing signal amplitude, and that a maximum theoretical circuit efficiency of 78% is achieved when $V_m = V_{CC}$. This very high circuit efficiency, together with the fact that $P_{IN} = 0$ when no signal is present, accounts for the widespread popularity of the class B push-pull power amplifier.

In both single-ended and push-pull class A circuit designs, maximum power dissipation in the transistors occurs when the input signal is zero. For class B operation, however, $P_D = 0$ under zero-signal conditions and the maximum value of P_D occurs somewhere in between $V_m = 0$ and $V_m = V_{CC}$ (see P_D curve in Figure 9.4-5e). The exact point at which this occurs may be determined by setting the derivative of equation (9.4-19b) equal to zero and solving for the value of V_m that satisfies this equation. Carrying out this operation, we have

$$\frac{dP_D}{dV_m} = \frac{2}{\pi}\frac{V_{CC}}{R_L} - \frac{V_m}{R_L} = 0 \tag{9.4-20a}$$

or

$$V_m = \frac{2}{\pi}V_{CC} \tag{9.4-20b}$$

Substituting this result into equation (9.4-19b), the corresponding total power dissipation in both transistors at this point is

$$P_{D,\max} = \frac{2V_{CC}}{\pi R_L}\left(\frac{2}{\pi}V_{CC}\right) - \left(\frac{2}{\pi}\right)^2\frac{V_{CC}^2}{2R_L} = \frac{4}{\pi^2}\frac{V_{CC}^2}{2R_L} \tag{9.4-21}$$

For reasons that will be apparent shortly, it is useful to rewrite the expressions for the maximum values of P_{IN} and P_D in terms of the maximum undistorted ac power that can be delivered to the load ($V_{CC}^2/2R_L$). Making this substitution into equations (9.4-18) and (9.4-21), we have

$$P_{IN,\max} = \frac{4}{\pi}\frac{V_{CC}^2}{2R_L} = \frac{4}{\pi}(P_{L,\max}) \tag{9.4-22a}$$

and
$$P_{D,\max} = \frac{4}{\pi^2}\frac{V_{CC}^2}{2R_L} = \frac{4}{\pi^2}(P_{L,\max}) \qquad (9.4\text{-}22b)$$

Let's apply these results to the same problem considered for each of the amplifier types examined previously by determining the power supply and transistor requirements for a class B push-pull amplifier capable of delivering 10 W of rms power to a resistive load. Following equations (9.4-22a) and (9.4-22b), if $P_{L,\max} = 10$ W, the power supplies must be able to deliver a total of $(4P_{L,\max})/\pi$ or 12.5 W to the circuit. In addition, the transistors must be capable of dissipating $4(10\text{ W})/\pi^2 \simeq 4$ W total or about (2 W)/transistor. These results are summarized in Table 9.4-1, along with the data obtained for each of the other power amplifier designs that we considered previously. Clearly, at least from an efficiency viewpoint, the class B design is superior to all other amplifier types investigated. However, because of the turn-on voltage of the transistors, which we have neglected in the analysis so far, there will also be a significant amount of output signal distortion associated with class B circuit operation.

EXAMPLE 9.4-2

a. Design a class B push-pull amplifier of the type illustrated in Figure 9.4-5a to deliver 20 W of rms power to an 8-Ω loudspeaker. Specifically, determine the required power supply voltages as well as the current that they must be capable of delivering. In addition, determine the required transistor parameters ($P_{D,\max}$, $I_{C,\max}$, and $V_{CE,\max}$).
b. Sketch the load line for Q_1 on the transistor characteristics. Is there any problem with its position?

SOLUTION

To deliver 20 W of power to an 8-Ω load, following equation (9.4-19a), we require that
$$V_m = \sqrt{2P_L R_L} = 17.8 \text{ V} \qquad (9.4\text{-}23)$$

If we neglect the transistor saturation voltage and the 0.7-V drop across v_{BE}, then V_{CC} will need to be about 18 V. Furthermore, with
$$v_{in} = V_m \sin \omega t \simeq 17.8 \sin \omega t \text{ volts} \qquad (9.4\text{-}24a)$$

9.4-1 Comparison of Different Amplifier Configurations With Each Designed to Deliver 10 W (RMS) to the Load

Amplifier Type	$P_{L,\max}$ (W)	$P_{IN,\max}$ (W)	$P_{D,\max}$/transistor (W)	$\eta_{\max}$ (%)	Comment
Class A direct-coupled	10	40	20	25	20 W dc lost in load at all times
Class A Transformer-coupled	10	20	20	50	No dc lost in load
Class A push-pull	10	20	10	50	$P_D = 20$ W total, no dc lost in load, very low distortion
Class B push-pull	10	12.5	2	78	Very high η, increased waveform distortion

the load current is

$$i_L = \frac{v_L}{R_L} \simeq \frac{v_{in}}{R_L} = 2.25 \sin \omega t \text{ amperes} \qquad (9.4\text{-}24\text{b})$$

Because i_L has the same peak value as i_{C1} and i_{C2}, $I_{C,\text{max}}$ for both transistors needs to be at least 2.25 A.

To determine the load line for this class B circuit, we note that

$$v_{CE1} = V_{CC} - v_L = V_{CC} - V_m \sin \omega t \qquad (9.4\text{-}25)$$

The collector-to-emitter voltage, as well as the collector current for Q_1 under maximum signal conditions, are both sketched in Figure 9.4-6a. According to these sketches, a transistor with a breakdown voltage of at least $2V_{CC}$ or 36 V is needed. The load line for Q_1 in this circuit is sketched on the transistor characteristics in Figure 9.4-6b. When v_{in} goes negative, $i_{C1} = 0$ and v_{CE1} increases above V_{CC} because Q_2 is ON and is making v_L negative. Because Q_1 is cut off during this interval, the load line follows along the v_{CE} axis. When v_{in} is greater than zero, Q_2 is OFF, Q_1 is ON, and

$$v_{CE1} = V_{CC} - v_L = V_{CC} - i_{C1} R_L \qquad (9.4\text{-}26)$$

This portion of the load line for Q_1 is valid in the interval from $v_{CE1} = 0$ to $v_{CE1} = V_{CC}$, and the load-line slope in this region is $-1/R_L$.

Following equation (9.4-22b), the power-handling ability of each transistor needs to be at least

$$P_{D1} = P_{D2} = \frac{2}{\pi^2} P_{L,\text{max}} \simeq (4 \text{ W})/\text{transistor} \qquad (9.4\text{-}27)$$

Figure 9.4-6
Example 9.4-2.

The power dissipation hyperbola curve for a 4-W power transistor is drawn on the transistor characteristics in Figure 9.4-6b. Notice that the load line for our

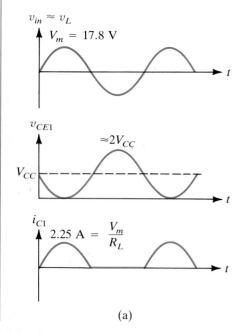

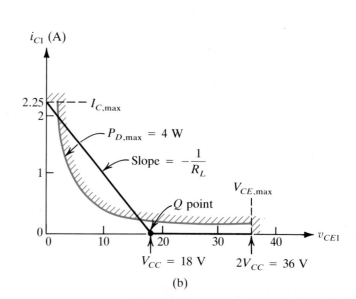

(a)

(b)

Chapter 9 / Power Electronics

class B amplifier is not tangent to this dissipation hyperbola. In fact, it cuts right through it. At first glance this might appear to indicate a serious design error on our part because for all amplifiers that we examined previously, the load line was always below, or at most just tangent to, the dissipation hyperbola. However, the position of the load line is correct as shown.

The reason that this is not a problem for the class B amplifier under consideration is that the transistors are biased at cutoff. As a result, even though the instantaneous power will exceed the 4-W limit, on the average the power dissipated in the transistor will still be less than or equal to 4 W.

9.4-3 Distortion Considerations in Push-Pull Power Amplifiers

In the remainder of this section we consider techniques for estimating the distortion levels found in push-pull amplifiers. Because of its simplicity we begin these analyses by first examining the common-emitter push-pull power amplifier, and then we will extend these results to handle other types of push-pull circuits.

Following the single-ended common-emitter distortion analysis presented in Section 9.3, for the circuit in Figure 9.4-7a, we may write

$$i_{B1} = I_{B1} e^{v_{in}/V_T} \tag{9.4-28a}$$

and

$$i_{B2} = I_{B2} e^{-v_{in}/V_T} \tag{9.4-28b}$$

If we consider that the h_{FE} and h_{fe} are the same for Q_1 and Q_2, and that their values are independent of the operating point, we also have

$$i_{C1} = h_{FE} i_{B1} = I_C e^{v_{in}/V_T} \tag{9.4-29a}$$

and

$$i_{C2} = h_{FE} i_{B2} = I_C e^{-v_{in}/V_T} \tag{9.4-29b}$$

The load current i_L in this figure is given by the expression

$$i_L = i_{C2} - i_{C1} = -I_C(e^{v_{in}/V_T} - e^{-v_{in}/V_T}) \tag{9.4-30a}$$

and therefore the load voltage, v_L, may be written as

$$v_L = i_L R_L = -I_C R_L(e^{v_{in}/V_T} - e^{-v_{in}/V_T}) \tag{9.4-30b}$$

In equation (9.4-30b), when the exponent is small, we may expand the exponential terms, which yields

$$v_L = -I_C R_L \left\{ \left[1 + \frac{v_{in}}{V_T} + \frac{1}{2!}\left(\frac{v_{in}}{V_T}\right)^2 + \frac{1}{3!}\left(\frac{v_{in}}{V_T}\right)^3 + \cdots \right] \right.$$
$$\left. - \left[1 - \frac{v_{in}}{V_T} + \frac{1}{2!}\left(\frac{v_{in}}{V_T}\right)^2 - \frac{1}{3!}\left(\frac{v_{in}}{V_T}\right)^3 \cdots \right] \right\} \tag{9.4-31}$$

and carrying out the indicated additions and subtractions, we obtain the very interesting result

$$v_L = -2I_C R_L \left[\frac{v_{in}}{V_T} + \frac{1}{3!}\left(\frac{v_{in}}{V_T}\right)^3 + \frac{1}{5!}\left(\frac{v_{in}}{V_T}\right)^5 + \cdots \right] \tag{9.4-32}$$

Because all the even terms in this power series expansion are missing, this means that a symmetrical push-pull amplifier has no even harmonic distortion. This fact significantly reduces the amplifier's overall distortion level, especially since the second harmonic term (which was the largest distortion term for single-ended amplifiers) is not even present in a well-balanced push-pull amplifier.

Notice in equation (9.4-32) that if (v_{in}/V_T) is much less than 1, we can neglect all terms above the linear one and approximate v_L as

$$v_L \simeq -2I_C R_L \frac{v_{in}}{V_T} = -2h_{fe} \frac{v_{in}}{h_{fe}\left(\dfrac{V_T}{I_C}\right)v_e} R_L = -2h_{fe}\frac{v_{in}}{h_{ie}}R_L = -2i_c R_L \qquad (9.4\text{-}33)$$

Of course, this is the same result that we would have obtained if we had carried out the small-signal linear analysis of this problem.

Since there is no second harmonic distortion in a push-pull amplifier, let's estimate the relative amplitude of the next most important distortion term (the third harmonic). In carrying out this analysis, if we consider that $v_{in} = V_m \sin \omega t = v_{be} = V_{bem} \sin \omega t$, we may rewrite equation (9.4-31) as

$$v_L = -2I_C R_L \left[\frac{V_{bem}}{V_T} \sin \omega t + \frac{1}{6}\left(\frac{V_{bem}}{V_T}\right)^3 \sin^3 \omega t + \cdots \right] \qquad (9.4\text{-}34)$$

Using the trigonometric identity

$$\sin^3 \theta = \frac{1}{4}(3\sin \theta - \sin 3\theta) \qquad (9.4\text{-}35)$$

equation (9.4-34) may be further expressed as

$$v_L \simeq -2I_C R_L \left[\frac{V_{bem}}{V_T} \sin \omega t - \frac{1}{24}\left(\frac{V_{bem}}{V_T}\right)^3 \sin 3\omega t + \cdots \right] \qquad (9.4\text{-}36)$$

Thus the cubic term in equation (9.4-32) leads to third harmonic distortion (3HD), the fifth-order term to fifth harmonic distortion, and so on, and from equation (9.4-36) the approximate level of third harmonic distortion in this amplifier [at least for small values of (V_{bem}/V_T)] is

$$\% \ 3\text{HD} = \frac{\text{third harmonic amplitude}}{\text{fundamental amplitude}} \times 100 \simeq \frac{1}{24}\left(\frac{V_{bem}}{V_T}\right)^2 \times 100\%$$
$$= \frac{1}{24}\left(\frac{I_{cm}}{I_C}\right)^2 \times 100\% \qquad (9.4\text{-}37)$$

where I_{cm} equals the amplitude of the fundamental component of the collector current and I_C its dc component.

For values of I_{cm}/I_C approaching or greater than 1, this result is no longer valid and a complete Fourier series analysis of equation (9.4-30b) is required to determine correctly the third harmonic content of v_L. Although the mathematical details of this proof are beyond the scope of this book, the results of this analysis may be useful and are presented graphically in Figure 9.4-7b. For comparison the results obtained using the small-signal power series solution [equation (9.4-37)] are also included.

Let's next consider the harmonic distortion produced in a BJT common-collector push-pull amplifier of the type illustrated in Figure 9.4-8a. Notice

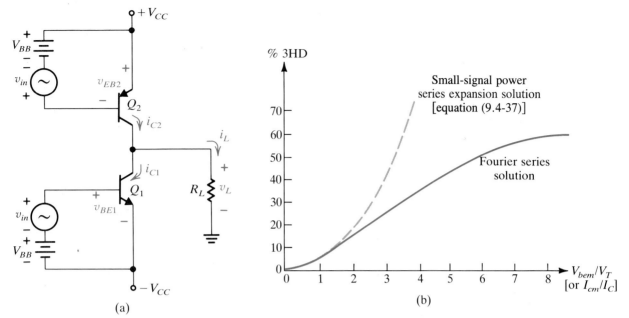

Figure 9.4-7.
Distortion in a
common-emitter push-pull
power amplifier.

that depending on the value of V_{BB}, this circuit can either be a class A, class AB, or a class B amplifier. Following our earlier analysis of single-ended amplifier distortion we should expect that the addition of the emitter resistance R_L significantly reduces the harmonic distortion because i_B and i_E are now less dependent on the nonlinear characteristics of the base–emitter junction. In fact, this is exactly what happens, and it can be shown (see Problems 9.4-10 and 9.4-11) that the distortion level in equation (9.4-37) is reduced by the factor $(h_{ie}/2)/[R_B/2 + h_{ie}/2 + (1 + h_{fe})R_L]$, where R_B represents any base resistance in series with v_{in}. As a result, the overall third harmonic distortion for a common-collector push-pull amplifier (or for that matter for any BJT-style push-pull amplifier) may be written as

$$\% \ 3HD = \frac{1}{24}\left(\frac{I_{cm}}{I_C}\right)^2 \times \frac{h_{ie}/2}{R_B/2 + h_{ie}/2 + (1 + h_{fe})R_L} \times 100\% \qquad (9.4\text{-}38)$$

for small values of I_{cm}/I_C.

For large values of I_{cm}/I_C, the distortion level in the output may be determined in a very simple way. With I_C small in comparison to I_{cm}, the circuit in Figure 9.4-8a is essentially operating class B (with $V_{BB} = 0$), and as a result, the voltage v_{BE1} will nearly look like a square wave with an amplitude of $V_o = 0.7$ V as sketched in part (b) of the figure. The output waveform distortion produced in this situation is known as crossover distortion and results from the fact that the base–emitter diodes effectively do not turn ON until v_{BE} exceeds about 0.7 V. Until that level is reached, $v_L = 0$ and a small dead zone in the output is produced.

The magnitude of the third harmonic distortion produced in this situation is especially easy to estimate. Because the base–emitter voltage, v_{BE1}, is a square wave, the Fourier series for this voltage may be written as

$$v_{BE1} = \frac{4V_o}{\pi}\left(\sin \omega t + \frac{1}{3}\sin 3\omega t + \frac{1}{5}\sin 5\omega t + \cdots\right) \qquad (9.4\text{-}39)$$

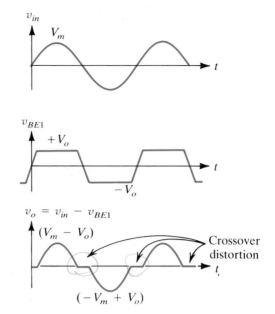

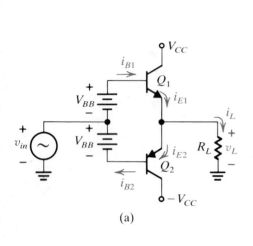

(a)

(b) Waveshapes for Class B Operation
$(V_{BB} = 0)$

Figure 9.4-8
*Distortion in an
emitter-follower push-pull
power amplifier.*

If we for the moment ignore all terms above the third harmonic, then following Figure 9.4-8a (with $V_{BB} = 0$), we may write

$$v_L = v_{in} - v_{BE1} = \left(V_m - \frac{4V_o}{\pi}\right)\sin \omega t - \frac{4V_o}{\pi}\sin 3\omega t \qquad (9.4\text{-}40)$$

from which the upper limit on the third harmonic distortion is seen to be

$$\% \text{ 3HD} = \frac{4V_o/3\pi}{V_m - 4V_o/\pi} \times 100 \qquad (9.4\text{-}41)$$

The data for third harmonic distortion levels for intermediate values of I_{cm}/I_C are found by determining the Fourier series components of equation (9.4-30b) with $v_{in} = V_m \sin \omega t$, and again, while this can easily be done by numerical integration, it is also beyond the scope of this book. However, because these data are useful, the results are presented graphically in Figure 9.4-9.

In examining these results, it is useful to observe that class B operation occurs when $V_{BB} = 0$, and for this value of V_{BB} the corresponding quiescent collector current is $(1 + h_{FE})I_{CBO}$. If I_{cm} is on the order of 1 A and $(1 + h_{FE})I_{CBO}$ about $1\mu A$, then an I_{cm}/I_C of about 10^6 defines class B operation. Following Figure 9.4-9, the distortion level of such a class B amplifier will be on the order of 5%. This value of third harmonic distortion is unacceptably high, especially for audio amplifier designs, where distortion levels of less than 1% are generally required for "high-fidelity" sound reproduction.

For class A operation, I_C must be greater than I_{cm}, or I_{cm}/I_C must be less than 1, and as illustrated in the figure, distortion levels for pure class A operation are extremely low. However, the price paid for this distortion reduction is rather high in terms of circuit efficiency.

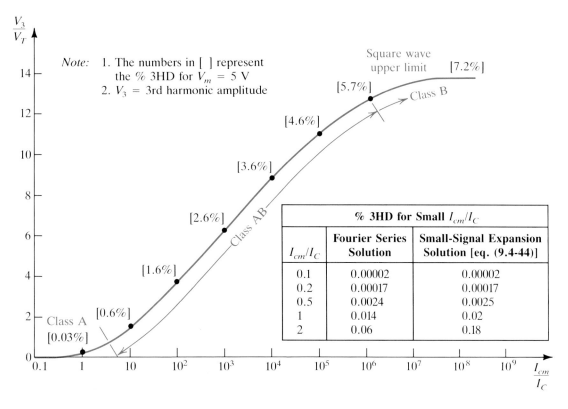

	% 3HD for Small I_{cm}/I_C	
I_{cm}/I_C	Fourier Series Solution	Small-Signal Expansion Solution [eq. (9.4-44)]
0.1	0.00002	0.00002
0.2	0.00017	0.00017
0.5	0.0024	0.0025
1	0.014	0.02
2	0.06	0.18

Figure 9.4-9
Third harmonic distortion in an emitter-follower push-pull power amplifier.

Class AB operation provides an excellent compromise between the high efficiency of class B circuits and the low distortion levels associated with class A designs. For example, by operating at a quiescent current of about $\frac{1}{10} I_{cm}$, the third harmonic distortion can be reduced to about 0.6%, a factor nearly 10 times smaller than for pure class B operation. Furthermore, operating at this dc current level does not raise the power supply or output transistor requirements much beyond the values needed for class B operation. In addition, if necessary, a further reduction in the distortion level is possible by adding negative feedback to the amplifier.

EXAMPLE 9.4-3

The circuit shown in Figure 9.4-10 is an inexpensive phonograph amplifier. It consists of an op amp driving a pair of complementary transistors operating in the class B mode.

a. Sketch v_{in}, v_A, and v_L assuming that v_{in} is a 100-mV sine wave.
b. Find the rms power delivered to the load, and the power dissipated in Q_1 and Q_2.
c. Estimate the third harmonic distortion in the loudspeaker.
d. Repeat parts (a) and (c) if R_f is removed from the point v_A and reconnected to v_L (as indicated by the dashed lines in the figure).

SOLUTION

With the feedback taken from the output of the op amp v_A is simply equal to 100 times v_{in} or 10 sin ωt volts. The waveshape for v_L is nearly identical to that for v_A except for the 0.7 V of crossover distortion associated with the

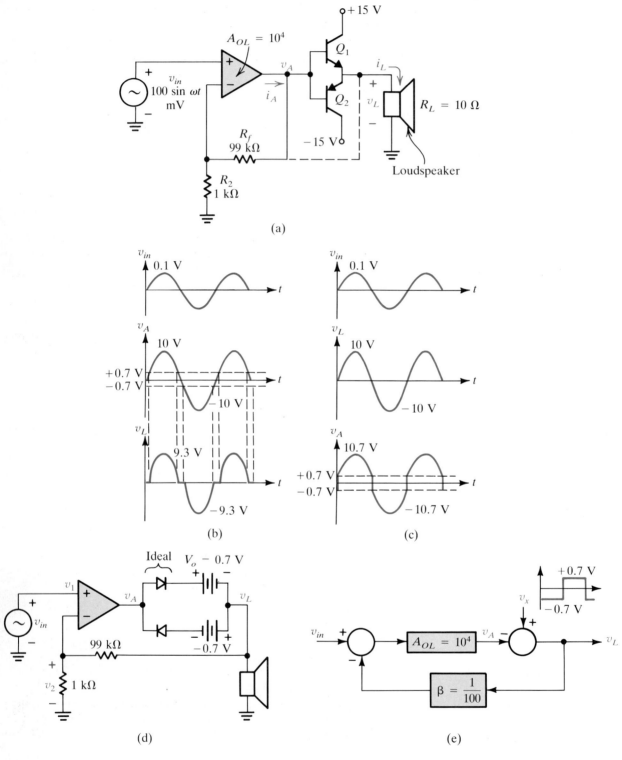

Figure 9.4-10
Example 9.4-3.

base–emitter turn-on voltages of Q_1 and Q_2. All three of these waveshapes are sketched in Figure 9.4-10b.

In analyzing the power delivered to the load and the power dissipated in the output transistors, we will neglect the crossover distortion and assume that v_A and v_L are nearly equal. To the extent that this approximation is valid, following equations (9.4-18), (9.4-19a), and (9.4-19b), we have that

$$P_L = \frac{V_m^2}{2R_L} = \frac{100}{20} = 5 \text{ W} \qquad (9.4\text{-}42a)$$

$$P_{IN} = 2V_{CC}\frac{I_{cm}}{\pi} = 9.5 \text{ W} \qquad (9.4\text{-}42b)$$

and

$$P_D = P_{IN} - P_L = 4.5 \text{ W} \qquad (9.4\text{-}42c)$$

for both transistors, or (2.25 W)/transistor.

Because the voltage across the base–emitter junctions of the transistors is nearly a 1.4-V (p-p)-amplitude square wave, following equation (9.4-39), the third harmonic component of this square wave is

$$V_3 = \frac{4}{3\pi}V_o \simeq 0.3 \text{ V} \qquad (9.4\text{-}43a)$$

and because this third harmonic signal appears directly in the output, we may express the overall third harmonic distortion at the loudspeaker as

$$\% \text{ 3HD} \simeq \frac{V_3}{V_m} \times 100 = 3\% \qquad (9.4\text{-}43b)$$

When the position of the feedback resistor is moved to the point v_L, the circuit is now effectively a giant op amp with Q_1 and Q_2 serving as its output stage. As a result, to a first-order approximation, we may write

$$v_L = \left(1 + \frac{R_f}{R_2}\right)v_{in} = 100v_{in} \qquad (9.4\text{-}44)$$

and v_L is therefore an undistorted sine wave. But how can v_L possibly be a perfect sine wave when the transistors Q_1 and Q_2 cannot turn ON until the input voltage applied to them exceeds 0.7 V?

To answer this question, let's work backward from the loudspeaker and determine the waveshape that v_A must have if v_L is to be a perfect sine wave. To begin with, when v_{in} is greater than zero, Q_1 is ON, so that $v_A = (0.7 + v_L)$, while when v_{in} is negative, Q_2 is ON and $v_A = (-0.7 + v_L)$. Thus v_A looks like v_L but it has ± 0.7 V jumps at the zero-crossover points (Figure 9.4-10c).

To explain how these jumps come about, notice that when v_{in} goes through zero, ideally $v_A = 0$, and both Q_1 and Q_2 are OFF. At this point the feedback path is broken and the op amp is running open-loop. Therefore, $v_A = A_{OL}(v_1 - v_2) = 10^4 v_{in}$. As a result, as v_{in} increases, v_A rises very rapidly toward $10^4 v_{in}$ until v_A reaches $+0.7$ V, and at this point Q_1 goes ON, closing the feedback loop so that $v_A = 100v_{in}$. Thus the apparently instantaneous jump of 0.7 V is actually a small portion of a very high amplitude (100 V) sine wave.

Although, as a first-order approximation, we can assume that the distortion has been reduced to zero, in reality some distortion will remain in the output.

Its level may be computed with the aid of Figure 9.4-10d, in which D_1 and D_2 are assumed to be ideal. For $v_{in} > 0$, D_1 turns ON, while for $v_{in} < 0$, D_2 is ON. Let's compute v_L for the case where v_{in} is greater than zero and D_1 is ON. Following part (d) of the figure, for this situation we have

$$v_L = v_A - V_o \tag{9.4-45a}$$

$$v_A = A_{OL}(v_1 - v_2) = A_{OL}(v_{in} - v_2) \tag{9.4-45b}$$

and
$$v_2 = \beta v_L = \beta(v_A - V_o) \tag{9.4-45c}$$

Combining these equations yields

$$v_A = \frac{A_{OL}}{1 + \beta A_{OL}} v_{in} + \frac{\beta A_{OL}}{1 + \beta A_{OL}} V_o \simeq 100 V_{in} + V_o \tag{9.4-46a}$$

and
$$v_L = \frac{A_{OL}}{1 + \beta A_{OL}} v_{in} - \frac{V_o}{1 + \beta A_{OL}} \simeq 100 V_{in} - 10^{-2} V_o \tag{9.4-46b}$$

For negative values of v_{in}, the signs on the V_o terms are reversed. Thus basically the effect of the base–emitter junction nonlinearities is to add in a square wave with an amplitude V_o at the op amp output, and an inverted one with an amplitude $V_o/(1 + \beta A_{OL}) = 7$ mV at v_L. With the feedback connected at v_A, the effective inverted square wave added in series at the output v_L was 700 mV. Thus the movement of the feedback point from v_A to v_L reduced the third harmonic distortion by the factor $(1 + \beta A_{OL})$ from 3% to 0.03%. This same analysis could have been carried out by using the feedback techniques developed in Chapter 7 in conjunction with the block diagram of this amplifier presented in Figure 9.4-10e.

EXAMPLE 9.4-4

Figure 9.4-11a illustrates a source-follower-style push-pull power amplifier. The transistors are assumed to have linear i_D versus v_{GS} characteristics once their threshold voltages are exceeded with $|V_T| = 3$ V and $g_m = 1$ S.

a. Sketch i_{D1}, i_{D2}, i_L, and v_L for the case where R_2 is equal to infinity.
b. For the conditions in part (a), determine the second and third harmonic distortion appearing in the output.
c. Show by properly selecting R_2 that the harmonic distortion can be made to go to zero.

SOLUTION

In analyzing this circuit we carefully investigate the performance of Q_1 in the upper half of this circuit, and then because of the assumed symmetry, we directly extend these results to cover the behavior of Q_2. When v_{in} is positive enough to turn Q_1 ON, Q_2 is OFF, and it therefore follows that

$$v_L = i_L R_L = g_m R_L(v_{G1} - v_L - V_T) = g_m R_L(v_{in} - v_L - V_T) \tag{9.4-47a}$$

or
$$v_L = \frac{g_m R_L}{1 + g_m R_L}(v_{in} - V_T) \tag{9.4-47b}$$

A similar equation may be developed for v_L when $v_{in} < -V_T$ and Q_2 is ON. As a result, the overall expressions for the quantities v_L, i_L, i_{D1}, and i_{D2} may be written as

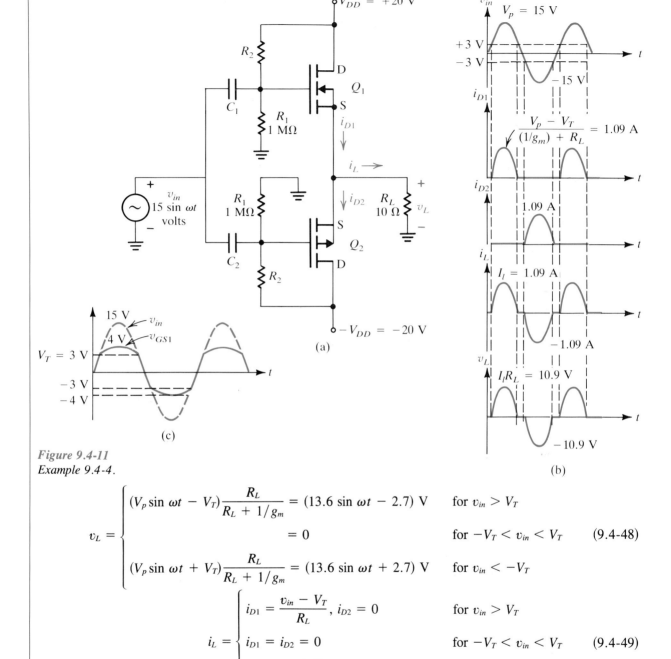

Figure 9.4-11
Example 9.4-4.

$$
v_L = \begin{cases}
(V_p \sin \omega t - V_T)\dfrac{R_L}{R_L + 1/g_m} = (13.6 \sin \omega t - 2.7)\ \text{V} & \text{for } v_{in} > V_T \\[2mm]
\hspace{3.2cm} = 0 & \text{for } -V_T < v_{in} < V_T \quad (9.4\text{-}48) \\[2mm]
(V_p \sin \omega t + V_T)\dfrac{R_L}{R_L + 1/g_m} = (13.6 \sin \omega t + 2.7)\ \text{V} & \text{for } v_{in} < -V_T
\end{cases}
$$

$$
i_L = \begin{cases}
i_{D1} = \dfrac{v_{in} - V_T}{R_L},\ i_{D2} = 0 & \text{for } v_{in} > V_T \\[2mm]
i_{D1} = i_{D2} = 0 & \text{for } -V_T < v_{in} < V_T \quad (9.4\text{-}49) \\[2mm]
i_{D2} = \dfrac{v_{in} + V_T}{R_L},\ i_{D1} = 0 & \text{for } v_{in} < -V_T
\end{cases}
$$

The waveshapes for the drain and load currents as well as that for v_L are sketched in Figure 9.4-11b.

The harmonic distortion at the v_L output may be rapidly determined by using the same procedure as that employed for analyzing the BJT push-pull emitter-follower-style power amplifier. We begin by developing an expression for

v_{GS1} (which is the same as that for v_{GS2} for $R_2 = \infty$). Because $v_{GS1} = v_{GS2} = (v_{in} - v_L)$ it therefore follows from equation (9.4.48) that

$$
v_{GS1} = v_{GS2} = \begin{cases} (1.4 \sin \omega t + 2.7) \text{ volts} & \text{for } v_{in} > V_T \\ 15 \sin \omega t \text{ volts} & \text{for } -V_T < v_{in} < V_T \\ (1.4 \sin \omega t - 2.7) \text{ volts} & \text{for } v_{in} < -V_T \end{cases} \quad (9.4\text{-}50)
$$

The waveshape for v_{GS1} is sketched in Figure 9.4-11c.

In determining the harmonic distortion in the output, we may approximate v_{GS1} as a square wave of height V_T, and because $v_L = v_{in} - v_{GS1}$, we can see that the harmonic content of v_L is basically the same as that of v_{GS1}. Since the voltage v_{GS1} is (almost) a square wave, and because a square contains only odd harmonics, the second harmonic distortion at v_L will be zero. In addition, the third harmonic distortion will be

$$
\% \ 3 \ \text{HD} = \frac{(4/3\pi)V_T}{V_m - (4/\pi)V_T} \simeq 11.4\% \quad (9.4\text{-}51)
$$

where the numerator contains the square-wave harmonic, V_m is the input signal, and the denominator contains the square-wave fundamental.

By properly selecting R_2, the level of third harmonic distortion can be reduced dramatically. In particular, if R_2 is chosen so that the dc level at v_{G1} is equal to V_T, then Q_1 will turn ON as soon as v_{in} is greater than zero. If Q_2 is similarly biased (at $-V_T$), Q_2 will turn ON as soon as $v_{in} < 0$. The value of R_2 required to accomplish this is found by solving the equation

$$
V_{G1} = V_T = 3 = 20\left(\frac{1}{1 + R_2}\right) \quad (9.4\text{-}52)
$$

from which we find that $R_2 = 5.66 \text{ M}\Omega$. With this value of R_2 in place, we now have that

$$
v_L = \frac{R_L}{1/g_m + R_L} v_{in} \quad (9.4\text{-}53)
$$

and the crossover distortion is zero.

9.4-1 The push-pull amplifier illustrated in Figure 9.4-1a is biased for class A operation and drives a 5-Ω load. If the transistors have ratings of $P_{D,\max} = 20$ W, $V_{CE,\max} = 50$ V, and $I_{C,\max} = 4$ A, select V_{CC} and determine the maximum power deliverable to the load. *Answer* 14.14 V, 20 W

9.4-2 The class B push-pull amplifier illustrated in Figure 9.4-5a has ±40-V power supplies and drives a 10-Ω load. If v_{in} is a 25-V peak amplitude sine wave, determine P_L, P_{IN}, η, and P_D in each transistor. *Answer* $P_L = 31.25$ W, $P_{IN} = 63.66$ W, 49%, $P_D = 16.2$ W

9.4-3 Estimate the % 2HD and 3HD for the circuit examined in Exercise 9.4-2. *Answer* 2 HD = 0%, 3HD = 1.23%

9.4-4 For the circuit in Figure 9.4-11a, the third harmonic distortion is about 11.4% when R_2 is infinite. Determine the value of R_2 needed to reduce this distortion level to about 5%. In carrying out this calculation, neglect the rounding at the top of the v_{GS} waveshape. *Answer* $R_2 = 14.2$ MΩ

A voltage regulator is a circuit that is designed to maintain a constant output voltage in spite of changes in the input voltage, load, or environmental conditions in which the circuit is operating. There are basically two different types of electronic voltage regulators: linear and switching. In a linear regulator the regulating device operates in the linear or active region, while in a switching regulator, as its name implies, the regulating element is switched back and forth between saturation and cutoff. The theory of operation of the linear voltage regulator is presented in this section, and the functioning of the switching regulator is the subject of Section 9.6.

Linear voltage regulators may be subdivided into series and shunt devices. In a series voltage regulator the regulating element is placed in series with the load, while with a shunt circuit the regulating device is placed in parallel with the load. For most circuit applications the series regulator is more versatile and has a higher efficiency than the shunt regulator, and hence shunt regulators are seldom used in modern circuit designs.

The zener diode circuits discussed in Sections 1.9 and 2.3 are good examples of shunt-type linear regulators, and prior to the development of inexpensive voltage-regulator ICs, they were frequently used in applications where low-cost circuits having moderate regulation capabilities were needed. This type of zener diode shunt regulator is shown in Figure 9.5-1a, and it will be used to illustrate the basic concepts of linear regulated power supply construction.

Ideally, a regulator's output voltage should be a constant, independent of variations in the input voltage and the load. As a measure of how close an actual circuit design comes to achieving these goals, we may define the power supply's load and line regulation as

$$\text{load regulation} = \frac{\text{no-load voltage} - \text{full-load voltage}}{\text{no-load voltage}} \times 100\% \qquad (9.5\text{-}1a)$$

$$\text{line regulation} = \frac{\text{change in load voltage}}{\text{change in input voltage}} \times 100\% \Bigg|_{\text{at full load}} \qquad (9.5\text{-}1b)$$

Thus a supply's load regulation indicates how much the output voltage changes when the load is varied, while its line regulation describes the effect of changes in the unregulated input voltage on the dc output voltage. This second quantity is called the supply's line regulation because most power supplies operate off the ac line, so that changes in the unregulated voltage are usually directly related to changes in the ac line voltage. Notice that the load and line regulation of an ideal power supply are both zero.

Let's determine the values of each of these parameters for the circuit illustrated in Figure 9.5-1a, which for the purpose of this discussion, will be considered to be a 10-V 1-A supply. The zener diode in this circuit is a 10-V device having an internal impedance of 2 Ω, and the unregulated dc input voltage, v_{UR}, is nominally 20 V. The load regulation of this supply may be computed by finding the Thévenin equivalent of the circuit within the shaded

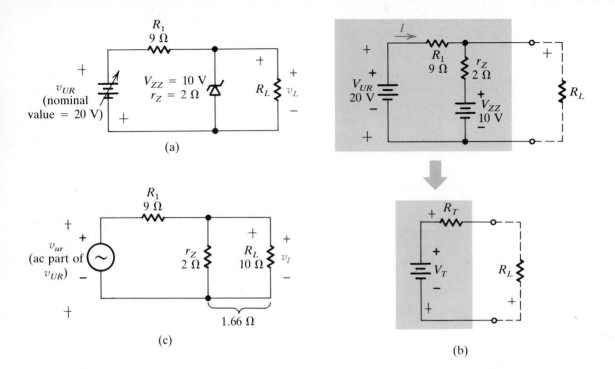

(a)

(b)

(c)

Figure 9.5-1
Zener diode shunt regulator.

region in Figure 9.5-1b, and then measuring v_L under no-load and full-load conditions. Following this figure, the Thévenin voltage and Thévenin resistance are

$$V_T = Ir_Z + V_{ZZ} = \frac{V_{UR} - V_{ZZ}}{R_1 + r_Z}r_Z + V_{ZZ} = 11.18 \text{ V} \qquad (9.5\text{-}2a)$$

and $\qquad R_T = r_Z \parallel R_1 = 1.64 \ \Omega \qquad (9.5\text{-}2b)$

Using the right-hand side of Figure 9.5-1b, the load regulation expression given in equation (9.5-1a) may be rewritten as

$$\text{load regulation} = \frac{V_T - \left(\dfrac{R_L}{R_L + R_T}\right)V_T}{V_T} \times 100\% = \frac{R_T}{R_L + R_T} \times 100\% \qquad (9.5\text{-}3)$$

Substituting $R_L = 10 \ \Omega$ and $R_T = 1.64 \ \Omega$ for the circuit being examined, the load regulation is $(1.64/11.64) \times (100)$ or 14.1%.

To compute the line regulation, we need to reexamine the circuit in Figure 9.5-1b under full-load conditions, with the unregulated input voltage considered to be a variable. For this situation we may use the equivalent circuit in part (c) of the figure to compute the change in v_L produced by a given change in v_{UR}, and carrying out this procedure, it follows that

$$\text{line regulation} = \frac{v_l}{v_{UR}} \times 100\% = \frac{r_Z \parallel R_L}{R_1 + r_Z \parallel R_L} \times 100\% \qquad (9.5\text{-}4)$$

or 15.6%.

This shunt regulator circuit is at best a marginal voltage regulator. Its load and line regulation are quite large, and in addition, for the supply circuit

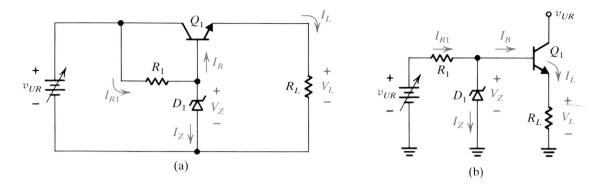

(a)   (b)

Figure 9.5-2
Series-pass linear voltage regulator.

shown, a rather high-wattage zener diode is required. This is because the entire input current flows through the zener when the load is disconnected. As a result, a zener with a minimum power-handling capability of $(1.1\ \text{A}) \times (10\ \text{V})$ or 11 W will be needed in this circuit.

The power-handling requirements of the zener can be reduced considerably by using a series-pass voltage regulator circuit of the type illustrated in Figure 9.5-2a, in which the transistor Q_1 functions as the series-pass element. By redrawing this regulator as shown in part (b) of the figure, it is apparent that this circuit is the same as the original zener shunt regulator shown in Figure 9.5-1a except that an emitter-follower buffer has been connected between the load and the zener. One of the principal effects of introducing the transistor into this circuit is that the effective load current drawn from the zener portion of the circuit is reduced by the factor $(1 + h_{FE})$, thereby reducing the power requirements of the zener by about the same amount.

EXAMPLE 9.5-1

For the circuit illustrated in Figure 9.5-2, the transistor $h_{FE} = 100$, and the unregulated input voltage is a sawtooth that ripples between 21 and 19 V.

a. Design a 10-V 1-A regulated power supply having the form given in Figure 9.5-2 assuming that I_{ZK} for the zener is 10 mA.
b. Calculate the power supply efficiency, $\eta = P_L/P_{IN}$, under full-load conditions.

SOLUTION

If V_L is to be 10 V, the zener diode voltage, because it is one base–emitter drop above V_L, will need to be 10.7 V. Because $I_{ZK} = 10$ mA, the zener current must be at least equal to this value, regardless of the load resistance connected. Minimum zener current flow occurs when the load current is a maximum, and under these conditions we have

$$I_Z = I_{R1} - I_B = I_{R1} - \frac{I_L}{1 + h_{FE}} = I_{R1} - 10\ \text{mA} \qquad (9.5\text{-}5)$$

Thus for I_Z to be equal to 10 mA here, we need the minimum value of I_{R1} to be 20 mA. Using the minimum value for v_{UR}, we see that the required value for R_1 is therefore about

$$R_1 = \frac{v_{UR,\text{min}} - 10.7}{20\ \text{mA}} = \frac{19 - 10.7}{20\ \text{m}} = 415\ \Omega \qquad (9.5\text{-}6)$$

Maximum power dissipation in the zener occurs when the load is disconnected and $I_B = 0$. Under these conditions, $I_{R1} = I_Z$, and the average power dissipated in the zener will be

$$P_D \simeq I_Z \times V_{ZZ} = 20 \text{ mA} \times 10.7 \text{ V} = 215 \text{ mW} \tag{9.5-7}$$

For completeness it is also useful to calculate the power-handling requirements of the series-pass transistor. Since the collector-to-emitter voltage (neglecting the ripple) is constant at $(20 - 10)$ or 10 V, the transistor dissipation will be largest when the collector current is at its maximum. Substituting in the data for this circuit, we therefore have

$$P_D = I_C \times V_{CE} = 1 \text{ A} \times 10 \text{ V} = 10 \text{ W} \tag{9.5-8}$$

Thus we see that the transistor has taken up the burden from the zener in this circuit as the dissipative element. Because high-power transistors are more readily available and less expensive than power zeners, this type of circuit is a more practical design than a simple zener diode shunt regulator.

The overall efficiency of this circuit under full-load conditions may be found by noting that the current consumed from v_{UR} is nearly equal to I_C. Therefore, using this approximation, the input power to this circuit is

$$P_{IN} \simeq I_C V_{UR} = 20 \text{ W} \tag{9.5-9}$$

For these same load conditions, the power delivered to the load is

$$P_L = I_L V_L = 10 \text{ W} \tag{9.5-10}$$

so that the overall circuit efficiency is

$$\eta = \frac{P_L}{P_{IN}} \times 100\% = \frac{10 \text{ W}}{20 \text{ W}} \times 100\% = 50\% \tag{9.5-11}$$

The load and line regulation for this circuit can be shown to be 10% and 2.4%, respectively (Problem 9.5-5).

The addition of feedback to the simple series-pass regulator can go a long way toward improving its load and line regulation, especially if the open-loop gain of the feedback amplifier is large. Besides upgrading the supply's regulation abilities, the addition of feedback also allows for a rather simple adjustment of the power supply's output voltage.

The circuit shown in Figure 9.5-3 is one form of a series-pass feedback-style voltage regulator circuit. In this circuit the Darlington *npn* transistor Q_1 serves as the series-pass element, D_1 as a voltage reference, against which the feedback voltage βv_L is compared, and Q_2 as the feedback amplifier. The resistor R_1 provides the base current for Q_1, and R_2 biases D_1 at the proper point in the zener region.

The functioning of this circuit may be understood by considering what happens when the load resistance is suddenly decreased. Under normal operating conditions V_{B2} adjusts itself so that the transistor Q_2 is in the active region with I_{C2} equal to the difference between I_{R1} and I_{B1}. When the load resistance decreases, V_L will tend to drop proportionally. However, when V_L decreases, this will reduce the feedback voltage to Q_2 and will lower V_{B2}. As a result, I_{C2} will

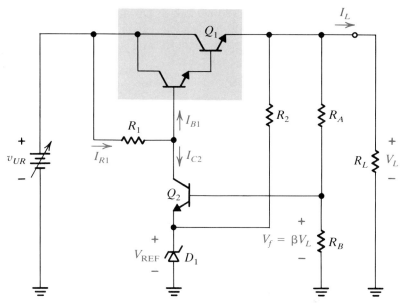

Figure 9.5-3

Elemental form of a series-pass feedback-style voltage regulator.

decrease, and this will increase I_{B1} and hence I_L. This increase in I_L will compensate for the reduction in R_L and will restore V_L nearly to its original value. Thus the addition of negative feedback to this regulator causes the output voltage variation due to load changes to be much smaller than it would be without the feedback in place. In other words, the feedback has lowered the effective output impedance of the power supply. This type of negative feedback voltage regulator circuit is examined in Example 9.5-2, which follows.

EXAMPLE 9.5-2

A 10-V 1-A regulated power supply of the type illustrated in Figure 9.5-3 is to be constructed. The zener diode parameters are $V_{ZZ} = 4.3$ V, $I_{ZK} = 10$ mA, and $r_Z = 10$ Ω, while the transistor characteristics are $h_{FE1} = 1000$, $h_{FE2} = 100$, $h_{ie1} = 100$ Ω, and $h_{ie2} = 1$ kΩ. Furthermore, the nominal value of the unregulated input voltage is about 20 V. Choose the component values to complete this design.

SOLUTION

In designing this circuit, let us observe that I_{B1} varies from 1 mA under full-load conditions to about 0 mA when $R_L = \infty$. Thus if I_{R1} is selected to be about 2 mA, this will provide ample base current for Q_1 while leaving 1 or 2 mA for the collector of Q_2. Because $V_L = 10$ V, the voltage at the base of Q_1, being two diode drops higher, will be equal to about 11.4 V, so that

$$R_1 = \frac{V_{UR} - V_{B1}}{2 \text{ mA}} = \frac{8.6 \text{ V}}{2 \text{ mA}} = 4.3 \text{ k}\Omega \qquad (9.5\text{-}12)$$

Under maximum load conditions $I_{B1} = 1$ mA and $I_{C2} = 1$ mA, so that I_{B2} will be about 0.01 mA. If we choose $(R_A + R_B) = 1$ kΩ, 10 mA will flow down through this branch of the circuit. Because I_{B2} is so much less than 10 mA, the loading of Q_2 on this circuit may be neglected and V_{B2} can therefore be approximately written as

$$V_{B2} = V_f = \beta V_L = \frac{R_B}{R_A + R_B} V_L \qquad (9.5\text{-}13)$$

The notation β has been used here to represent the fraction of V_L fed back to V_{B2} because of its similarity to the feedback factor β used with the op amp circuits examined in Chapter 7.

Because

$$V_{E2} = V_{REF} = 4.3 \text{ V} \qquad (9.5\text{-}14a)$$

it also follows that $\qquad V_{B2} \simeq V_{E2} + 0.7 = 5.0 \text{ V} \qquad (9.5\text{-}14b)$

As a result, if we combine equations (9.5-13) and (9.5-14b), we obtain

$$\frac{R_B}{R_A + R_B} V_L = 5.0 \text{ V} \qquad (9.5\text{-}15a)$$

or $\qquad\qquad V_L = \frac{R_B + R_A}{R_B}(5.0 \text{ V}) \qquad (9.5\text{-}15b)$

Having previously set $(R_A + R_B) = 1 \text{ k}\Omega$, it therefore follows that R_A and R_B must each equal 500 Ω if V_L is to be 10 V. A detailed analysis of this circuit demonstrates that V_L is actually 10.075 V. In addition, it can also be shown that the load and line regulation for this circuit are about 0.5 and 1%, respectively (Problem 9.5-7).

By increasing the amplifier gain in the feedback loop in Example 9.5-2, the performance characteristics of the regulator may be improved substantially. As the loop gain grows, not only will the load and line regulation decrease, but the dynamic performance of the amplifier will also change, and can be optimized so that the supply will be able to respond to load and line changes more rapidly.

However, a word of caution should be added here. As with any feedback amplifier, the addition of too much loop gain can move the poles into the right-hand plane and cause the circuit to oscillate. The criteria for assessing the stability of a feedback-style dc power supply is exactly the same as for a conventional amplifier, so all the guidelines and compensation techniques developed in Chapter 8 still apply. Along these lines it is worthwhile noting that most IC voltage regulator chips are internally compensated, so that, at least for conventional types of loads, they are unconditionally stable.

Figure 9.5-4a illustrates one form of a high-loop-gain op-amp-style regulated power supply. The operation of this circuit is easy to understand, especially when we redraw it as shown in part (b) of the figure. Here we can see more clearly that this circuit is very similar to the power amplifiers that we examined in the earlier sections of this chapter, where the op amp together with the power transistor combined to form a high-power operational amplifier. The principal difference between the amplifiers considered previously and the one illustrated in Figure 9.5-4 is that input to this power supply circuit is a constant dc voltage. As a result, only a single power transistor is needed at the output since the output voltage always has the same polarity.

The actual value of the output voltage obtained from this type of power supply circuit is especially easy to determine. If the op-amp gain is high, V_1 and

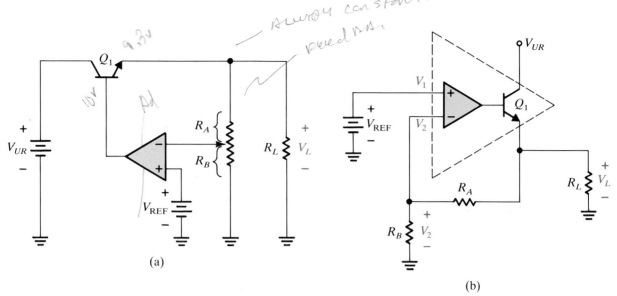

Figure 9.5-4

V_2 will nearly be equal, so that

$$V_{REF} \simeq \frac{R_B}{R_A + R_B} V_L \tag{9.5-16a}$$

or

$$V_L = \underbrace{\left(1 + \frac{R_A}{R_B}\right)}_{A_{CL}} V_{REF} \tag{9.5-16b}$$

Exercises

9.5-1 In Example 9.5-1, the BJT is replaced by an enhancement-style MOS-FET whose $V_T = 2$ V and $I_{D2T} = 2$ A.
a. Find V_L for $R_L = 10 \, \Omega$.
b. Find the peak-to-peak ripple voltage at the output.
Answer (a) 6.83 V, (b) 43 mV(p-p)

9.5-2 Find the approximate and the exact value of V_L for the power supply shown in Figure 9.5-3 if R_B is changed to 300 Ω. *Answer* 13.33 V, 11.49 V

9.5-3 Find the exact value of V_L under full-load conditions in Figure 9.5-4 if $A_{OL} = h_{FE} = 100$, $V_{REF} = 5$ V, and $R_A = R_B = 10$ kΩ. *Answer* 9.79 V

9.6 SWITCHING REGULATOR POWER SUPPLIES

Linear voltage regulator power supplies have long been the most popular method for producing high-quality dc power in electronic systems. However, in recent years a number of factors have combined to make the switching regulator a strong competitor to the linear design.

As indicated in Table 9.6-1, linear voltage regulators have rather low circuit efficiencies. This is because the regulation is achieved by placing a pass transistor in series with the load and varying the voltage drop across this transistor. As a result, the pass transistor operates in the active region and dissi-

9.6-1 Comparison of the Typical Characteristics of Linear and Switching Regulators

Parameter	Linear (Series-Pass) Power Supply	Switching Regulator Power Supply
Efficiency (%)	30–50	60–85
Power out/volume of the supply (W/cm^3)	0.03	0.12
Load and line regulation (%)	~0.1	~0.2
Output ripple (mV)	2	50
Noise (mV)	—	100 (switching)
Input voltage range (%)	±10	±20
Holdup time (ms)	2	25
Transient response time (μs)	50	500

pates considerable power, causing the overall efficiency of the regulator to be low.

Switching regulators operate on an entirely different principle, and rather than running the regulator transistor in the active mode, it is switched ON and OFF between saturation and cutoff. As a result, the power dissipation in this transistor is significantly lower than in the linear regulator circuit, and the efficiency of the switching supply is correspondingly greater.

In power supply circuits high efficiency means more than just a saving of electrical power. Inefficient circuits require lots of input power to deliver a specific amount of power to the load, and for the case of a dc power supply operating off the ac line, this means that oversized power transformers, input rectifiers, and input filters will be needed. In addition, low-efficiency supplies generate considerable amounts of heat, and thus large heat sinks and possibly even fans may be needed to properly cool the circuit's power transistors (see Section 9.9). Switching supplies offer several advantages in this area. First, because the switching transistor is either cut off or saturated, its power dissipation will be minimal, so that smaller power transistors and heat sinks can be used. Second, switching supplies can be run directly off the power line, so that large, heavy power transformers are unnecessary. Thus switching supplies are not only more efficient, but are also much smaller and lighter than their linear counterparts.

Furthermore, even when a transformer is needed to provide electrical isolation, the switching regulator design still affords a significant advantage. By placing the transformer after the switching transistor, the transformer only needs to pass the switched dc voltage, and because the frequency of this switched signal is typically much greater than 20 kHz, the transformer can be made physically smaller and lighter than a comparable 50- or 60-Hz unit.

Modern power supplies also need to be able to handle large variations in line voltage without losing regulation since fluctuations of this sort are common in the ac power provided by utility companies. These changes in line voltage can cause significant problems in linear voltage regulator power supplies. This is because the unregulated input voltage to the series-pass transistor of a linear supply is usually maintained quite close to the load voltage to minimize the power dissipated in the pass transistor. With V_{UR} close to V_L, very little decrease in input voltage can occur before the supply drops out of regulation. Thus, while linear voltage regulators have excellent line regulation, they cannot tolerate much variation in the line voltage. Switching supplies have a much greater tolerance of line voltage fluctuations. Because their efficiency does not depend on the input–output voltage differential, V_{UR} can be made quite large and the supply can usually function quite well even in the face of ±20% fluctuations in line voltage.

Another parameter that is indirectly related to the power supply's line regulation is known as its holdup time. It is a measure of the amount of time that the dc output voltage will remain in regulation after the ac input power is interrupted. A long holdup time is important in power supplies that are used with digital computer systems since it will allow time for the saving of important information on the system disk before the main semiconductor memory goes down when the ac power fails. Linear supplies have relatively small holdup times because V_{UR} can fall only a small amount before the supply drops out of regulation. Switching supplies, on the other hand, typically have long holdup times because V_{UR}, the voltage on the input filter capacitor, being fairly large can discharge for some time before the circuit drops out of regulation.

Although switching regulators have many advantages over linear regulated supplies, they also have their drawbacks. Switching supplies are more complex than their linear counterparts, although the development of controller ICs for switching regulators is making this less of a problem. In addition, switching supplies generate considerable electrical noise at the switching frequency. This fact tends to make linear supplies the choice for powering sensitive low-level linear amplifier circuits. Switching supplies, on the other hand, are ideal where large amounts of dc power are needed, but load requirements are not too severe. As such, switching supplies are finding considerable application in digital systems.

Switching regulated supplies are also more prone to oscillate than are comparable linear voltage regulators because of the extra phase shift produced in the *LC* filter networks used with switchers. To stabilize these circuits, frequency compensation is needed in the feedback loop. All of this filtering, both in the forward and the feedback loops, makes the response time of switching supplies rather sluggish. As a result, the transient response of these regulators to line and load fluctuations is considerably slower than that of a linear supply.

The relative merits of switching and linear regulated power supplies are listed in Table 9.6-1. The essence of these data may be summarized as follows: Linear regulators have a long history of providing high-quality dc power; their principal drawbacks are low efficiency and large size and weight. Switching supplies, on the other hand, have much higher efficiencies and are lighter and smaller. The main disadvantages of switchers are circuit complexity and switching noise problems. Because the advantages afforded by switching designs significantly outweigh the drawbacks, switching power supplies are playing an increasingly important role in the generation of dc electrical power.

In the remainder of this section we examine the basic principles of switching power supplies; more specifically, we investigate the operation of three of the most popular switching regulator circuits:

1. Step-down or buck-style converter
2. Step-up or boost-style converter
3. Transformer-style flyback converter

The basic design of each of these three switching supplies is illustrated in Figures 9.6-1a, b, and c, respectively.

9.6-1 Step-Down or Buck-Style Switching Regulator

Let's begin our investigation by examining the step-down circuit shown in Figure 9.6-1a. In analyzing this circuit we will assume that it is operating in the steady state, that the diode is ideal, and that the capacitor is large enough so that the ripple in the output voltage at the switching frequency may be neglected. Before getting into a detailed mathematical analysis of this circuit's operation, it might be a good idea to develop a qualitative understanding of its behavior.

Following Figure 9.6-2a, we can see that when the switch closes, $v_A = V_{UR}$, so that the diode is reversed biased and the inductor current begins to increase linearly. When the switch is opened at $t = T_1$, the current flow through the inductor cannot change instantaneously, and as a result, v_A starts toward $-\infty$, but when it reaches zero (really -0.7 V), the diode turns ON and

Figure 9.6-1
Basic types of switching power supplies.

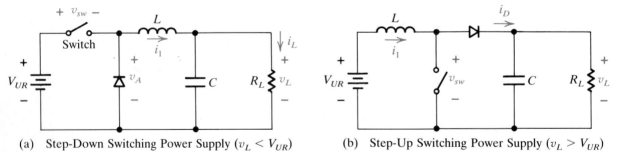

(a) Step-Down Switching Power Supply ($v_L < V_{UR}$)   (b) Step-Up Switching Power Supply ($v_L > V_{UR}$)

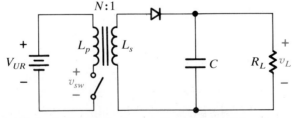

(c) Flyback-Style Transformer-Coupled Switching Power
Supply (Step-Up or Step-Down, v_L Positive or Negative)

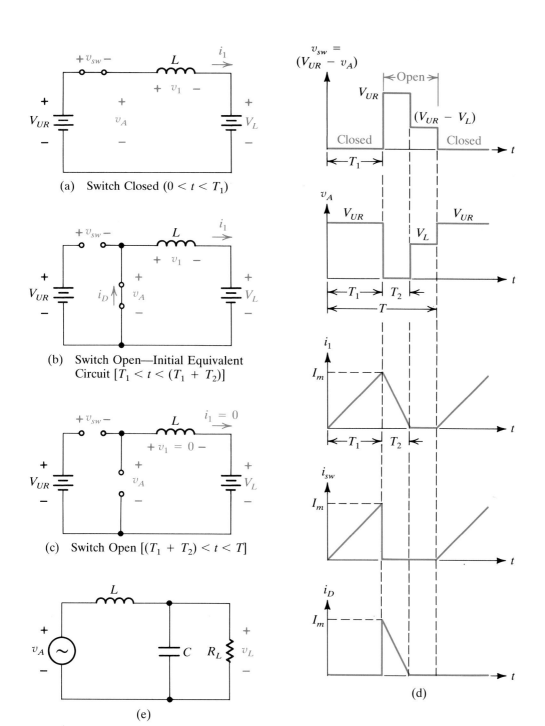

(a) Switch Closed $(0 < t < T_1)$

(b) Switch Open—Initial Equivalent Circuit $[T_1 < t < (T_1 + T_2)]$

(c) Switch Open $[(T_1 + T_2) < t < T]$

(e)

(d)

Figure 9.6-2

Step-down switching supply (see Figure 9.6-1a for the original circuit).

i_1 continues to flow. For $t > T_1$ but less than $(T_1 + T_2)$, the circuit in Figure 9.6-2b applies, and i_1 thus begins to decrease linearly.

In analyzing this circuit, we assume that i_1 goes completely to zero before the switch closes again. This is known as the discontinuous mode of circuit operation, in contrast to the continuous mode, which will be examined in the example that follows. The relative merits of each of these modes of operation will be discussed at the end of Example 9.6-1. However, for the purpose of this part of our discussion the fact that i_1 goes to zero before the switch again closes simply means that the initial current in the inductor at $t = 0+$ is zero.

Returning now to our qualitative analysis, once i_1 reaches zero the diode will go OFF and the circuit in Figure 9.6-2c will apply so that i_1 will remain zero thereafter, as will the inductor voltage. At this point with $v_1 = 0$, v_A will jump up to V_L and will remain there until the switch again closes and the cycle repeats.

The relationship between the switch duty cycle and the resulting dc output voltage from this supply may readily be determined with the aid of Figure 9.6-2e, in which we have explicitly represented the voltage into the L-C-R_L network by a voltage source v_A having the same waveshape as that given in part (d) of the figure. Viewed in this way it is apparent that this passive network behaves like a low-pass filter, and if we assume that C is so large that the filter's cutoff frequency is well below the switching frequency, only the dc component of v_A will appear in the output. Using the v_A waveshape indicated, the dc average of this waveform is

$$V_L = \frac{1}{T}\int_0^T v_A \, dt = \frac{1}{T}[V_{UR}T_1 + 0 \cdot T_2 + V_L(T - T_1 - T_2)] = V_{UR}\frac{T_1}{T_1 + T_2}$$

$$(9.6\text{-}1)$$

Let's reexamine the behavior of this circuit in somewhat greater detail in order to develop an expression for numerically calculating V_L [since T_2 is really an unknown in equation (9.6-1)]. Following Figure 9.6-2a, we have that

$$v_1 = V_{UR} - V_L = L\frac{di_1}{dt} \qquad (9.6\text{-}2a)$$

or

$$i_1(t) = \frac{1}{L}\int v_1 \, dt = \frac{(V_{UR} - V_L)t}{L} + i_1(0-) \qquad (9.6\text{-}2b)$$

Because the discontinuous current mode is assumed, $i_1(0-) = 0$, so that

$$i_1(t) = \frac{(V_{UR} - V_L)t}{L} \qquad (9.6\text{-}3)$$

Thus, when the switch is closed, the current grows linearly with time in the inductor. At $t = T_1$ it reaches its maximum,

$$I_m = \frac{(V_{UR} - V_L)T_1}{L} \qquad (9.6\text{-}4)$$

and after the switch opens, this current continues to flow through the inductor, but this time through the diode and not through the switch.

For simplicity, let us now redefine $t = 0$ as the point just after this switch opens, and following Figure 9.6-2b, the expression for the current flow through the inductor may now be written as

$$v_1 = -V_L = L\frac{di_1}{dt} \qquad (9.6\text{-}5a)$$

or

$$i_1 = \frac{-V_L t}{L} + i_1(0-) \qquad (9.6\text{-}5b)$$

Because $i_1(0-) = I_m$, we may express $i_1(t)$ as

$$i_1 = \frac{-V_L t}{L} + I_m \qquad (9.6-6)$$

Thus in this region the inductor current decreases linearly toward zero. The time, T_2, for this current to reach zero is found by setting equation (9.6-6) equal to zero and solving for t. Carrying out this procedure, we have

$$T_2 = \frac{I_m L}{V_L} \qquad (9.6-7)$$

The relationship between I_m, T_2, V_L, and the circuit parameters may be found by applying conservation of energy to the circuit in Figure 9.6-1a. Because no power is dissipated in the switch, the diode, the inductor, or the capacitor, the average power, P_{IN}, supplied by the battery must be equal to the average power, P_L, delivered to the load. The average input power is

$$P_{IN} = \frac{1}{T} \int_0^T V_{UR} i_{sw} \, dt = \frac{V_{UR}}{T} \left(\frac{1}{2} I_m T_1 \right) \qquad (9.6-8a)$$

which on substituting equation (9.6-4) for I_m yields

$$P_{IN} = \frac{V_{UR}(V_{UR} - V_L)T_1^2}{2TL} \qquad (9.6-8b)$$

To the extent that the output voltage remains constant, the average power delivered to the load is simply given by the expression

$$P_L = \frac{V_L^2}{R_L} \qquad (9.6-9)$$

Setting this result equal to equation (9.6-8b), we obtain the quadratic equation

$$V_L^2 + V_L \frac{V_{UR} T_1^2}{2\tau T} - \frac{V_{UR} T_1^2}{2\tau T} = 0 \qquad (9.6-10a)$$

which has the solution

$$V_L = V_{UR} \frac{T_1^2}{2\tau T} \left(\sqrt{1 + \frac{4\tau T}{T_1^2}} - 1 \right) \qquad (9.6-10b)$$

where $\tau = L/R_L$. Substituting this result back into equation (9.6-4), we can find I_m, and then with this value of I_m, T_2 can be computed with the aid of equation (9.6-7).

9.6-2   Step-Up or Boost-Style Switching Regulator

Besides being used as step-down voltage regulators, switching power supplies can also be employed to boost the level of the output dc voltage above that of the unregulated input voltage. A typical circuit for achieving this is shown in Figure 9.6-3a. During the time that the switch is closed, the current i_1 in the

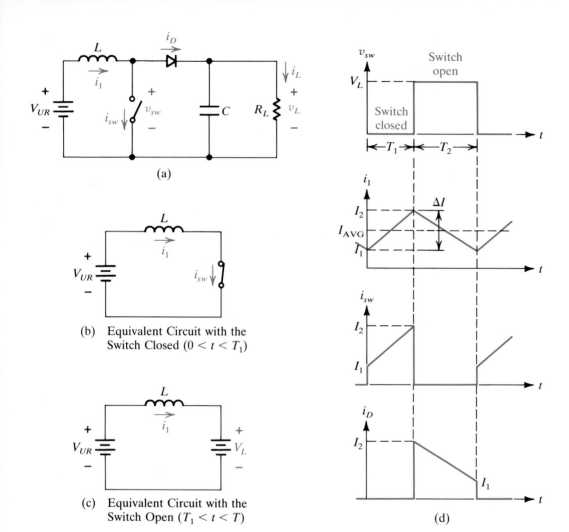

(a)

(b) Equivalent Circuit with the Switch Closed ($0 < t < T_1$)

(c) Equivalent Circuit with the Switch Open ($T_1 < t < T$)

(d)

Figure 9.6-3

Boost-style or step-up switching power supply.

inductor increases linearly, and when the switch opens, because i_1 cannot change instantaneously, v_{sw} rises rapidly toward infinity until the diode turns ON. After this, i_1 discharges linearly through the output circuit.

As with the step-down supply circuit just examined, the boost-style switching supply can also be made to operate in either the continuous or discontinuous current mode, depending on the selection of the circuit parameters. In our investigation of this circuit's behavior we assume that the inductor current flow is continuous, and then, as mentioned previously, at the end of Example 9.6-1 we take a look at the relative merits of each of these modes of circuit operation.

When the switch closes at $t = 0$, the equivalent circuit of Figure 9.6-3b applies and we may write the current in the inductor as

$$i_1(t) = \frac{V_{UR}t}{L} + I_1 \qquad (9.6\text{-}11)$$

where I_1 is equal to the initial current in the inductor at $t = 0-$ and represents the leftover current that remained in L at the end of the previous cycle since we are assuming continuous-mode operation of this circuit. If we define I_2 as the current level reached just prior to the opening of the switch at $t = T_1$, we have that

$$I_2 = \frac{V_{UR} T_1}{L} + I_1 \tag{9.6-12a}$$

or

$$\Delta I = I_2 - I_1 = \frac{V_{UR} T_1}{L} \tag{9.6-12b}$$

When the switch opens at $t = T_1$, v_{sw} rises, turning the diode ON, and in this region the circuit shown in Figure 9.6-3c is valid. If we redefine the $t = 0$ point as the time at which the switch opens, then, assuming that V_L is greater than V_{UR}, the current in the inductor will begin to decrease in accordance with the expression

$$i_1(t) = -\frac{V_L - V_{UR}}{L} t + i_1(0-) = -\frac{V_L - V_{UR}}{L} t + I_2 \tag{9.6-13}$$

After decreasing for $T_2 = (T - T_1)$ seconds, the switch again closes and the cycle repeats. The current flowing in the inductor at this point is equal to I_1, so that

$$I_1 = -\frac{V_L - V_{UR}}{L} (T - T_1) + I_2 \tag{9.6-14a}$$

or

$$\Delta I = I_2 - I_1 = \frac{V_L - V_{UR}}{L} (T - T_1) \tag{9.6-14b}$$

In the steady state the change in current that occurs during the time that the switch is closed must equal the change that occurs during the time that the switch is opened. Therefore, setting equations (9.6-12b) and (9.6-14b) equal to one another, we obtain

$$\frac{V_{UR} T_1}{L} = \frac{V_L - V_{UR}}{L} (T - T_1) \tag{9.6-15a}$$

or

$$V_L = V_{UR} \frac{T}{T - T_1} \tag{9.6-15b}$$

Thus if C is large enough so that V_L is essentially constant, V_L is determined by setting the duty cycle in accordance with equation (9.6-15b) and is independent of R_L.

The average current flow in the inductor, I_{AVG} may be determined by setting the average input power from the unregulated power supply equal to the average power delivered to the load. Carrying out this procedure, we have

$$V_{UR} I_{AVG} = \frac{V_L^2}{R_L} \tag{9.6-16}$$

and substituting in equation (9.6-15b) for V_L and solving for I_{AVG} yields

$$I_{AVG} = \frac{V_{UR}}{R_L}\left(\frac{T}{T - T_1}\right)^2 \qquad (9.6\text{-}17)$$

The pertinent waveshapes for this circuit are sketched in Figure 9.6-3d.

EXAMPLE 9.6-1

For the step-up switching power supply illustrated in Figure 9.6-4a the transistor $h_{FE} = 100$. The input signal shown in part (b) of the figure is applied to the base of the transistor.

a. Find the steady-state output voltage.
b. Sketch the steady-state current flow in the inductor and the collector-to-emitter voltage across the transistor.
c. Select the capacitor C so that the peak-to-peak output ripple is less than 25 mV.

SOLUTION

If we assume that this circuit is operating in the continuous current mode, all the results just developed for the step-up switching supply apply, and using equations (9.6-15b) and (9.6-17), the output voltage and average current flow in the inductor may be written as

$$V_L = 10\left(\frac{50\ \mu s}{25\ \mu s}\right) = 20\ \text{V} \qquad (9.6\text{-}18a)$$

and

$$I_{AVG} = \frac{10\ \text{V}}{100\ \Omega}\left(\frac{50}{25}\right)^2 = 0.4\ \text{A} \qquad (9.6\text{-}18b)$$

respectively.

Because the average value of a triangle wave is located midway between its peaks, applying equation (9.6-12b), we find that the peak-to-peak change in the inductor current is

$$\Delta I = \frac{(10\ \text{V})\ (25\ \mu s)}{2.5\ \text{mH}} = 0.1\ \text{A} \qquad (9.6\text{-}19a)$$

and therefore that

$$I_2 = I_{AVG} + \frac{\Delta I}{2} = 0.45\ \text{A} \qquad (9.6\text{-}19b)$$

and

$$I_1 = I_{AVG} - \frac{\Delta I}{2} = 0.35\ \text{A} \qquad (9.6\text{-}19c)$$

Using these results, the waveshapes for i_1 may be sketched as illustrated in Figure 9.6-4b.

The required value of C may be rapidly determined if we note that the portion of the power supply consisting of the diode, the capacitor, and the load R_L is basically a half-wave rectifier circuit. It is effectively driven by a constant-current source of about 0.4 A during the time that the transistor is cut off and is connected to ground reverse biasing the diode during the time that the transistor is saturated. The ripple voltage during the time that the diode is OFF is approximately equal to

$$V = \frac{I\ \Delta t}{C} = \frac{(V_L/R_L)(T - T_1)}{C} \qquad (9.6\text{-}20a)$$

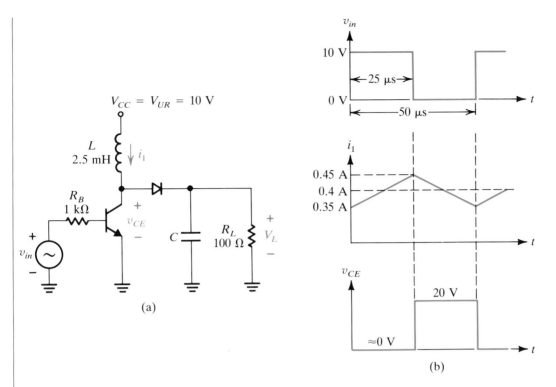

Figure 9.6-4
Example 9.6-1.

so that the minimum value of C for a ripple voltage of less than 25 mV(p-p) must be

$$C = \frac{(V_L/R_L)(T - T_1)}{\Delta V} = \frac{(20 \text{ V}/100 \text{ }\Omega)(25 \text{ }\mu s)}{25 \text{ mV}} = 200 \text{ }\mu F \qquad (9.6\text{-}20b)$$

Before examining the last switching regulator circuit on our list it may first be a good idea to take a closer look at the relative merits of the continuous and discontinuous current modes. The principal advantage of the continuous-current-flow mode is that the load voltage depends only on V_{UR} and the switch duty cycle and is independent of the load. Furthermore, for a specific amount of power being delivered to the load, the peak current flow in the transistor is smaller than in the discontinuous case. One last advantage of this operating mode is that the output ripple for a given-size filter capacitor will also be smaller than for the case where the current flow is discontinuous.

But operation in the discontinuous-current-flow mode has advantages, too. Because larger current flows are required in discontinuous designs, a smaller inductor can be used in these circuits. In addition, due to the fact that the switch voltage settles to a low final value after i_1 runs down to zero, a smaller value of V_{CEO} is needed than in the continuous mode. Finally, in the discontinuous mode the inductor transfers all its energy to the load each cycle, and this energy level is independent of the value of R_L. This attribute makes the discontinuous-mode-switching regulator especially useful in special circuit applications such as battery chargers.

9.6 / Switching Regulator Power Supplies

695

The last switching supply circuit that we examine in this section is known as a flyback-style power supply. This particular type of circuit has been used extensively in television receivers to produce the high voltage needed to accelerate the electron beam in the TV picture tube. With circuits of this basic type it is possible to produce stepped-up voltages in excess of 20,000 V from input voltages on the order of several hundred volts. This circuit is also used to derive the horizontal sweep signal for the deflection coil on the neck of the picture tube. The name flyback is derived from the fact that the energy stored in the inductive part of this circuit is delivered to the load during the retrace (or flyback) portion of the horizontal sweep signal.

Figure 9.6-5a illustrates the basic structure of a flyback-style transformer-coupled switching regulator power supply. During the time that the switch is closed, energy builds up and is stored in the primary inductance of the transformer. The phasing of the transformer is chosen so that the voltage produced in the secondary at this time reverse biases the diode.

When the switch opens, the primary current goes to zero, the secondary voltage reverses polarity (on its way toward plus infinity), and the diode turns ON so that current flows in the secondary.

The operation of the flyback-style converter is characterized by an increase in primary current during the time that the switch is closed. When the switch opens, the energy stored in the primary is transferred to the secondary, and a decrease in secondary current with time occurs. Depending on the component values selected, as with the circuits examined earlier, the flyback switching supply can operate in either the continuous or discontinuous current modes.

The operation of the flyback-style converter may be understood by examining Figure 9.6-5a together with the series of equivalent circuits presented parts (b) and (c) of the figure. We assume that the circuit is operating in the discontinuous current mode, so that $i_p(0-) = 0$. At $t = 0$, when the switch closes, v_s goes negative, the diode turns OFF, and the equivalent circuit in Figure 9.6-5b applies. As a result, we find that the current builds up linearly to a final value

$$I_{mp} = \frac{V_{UR} T_1}{L_p} \tag{9.6-21}$$

where L_p is the transformer's primary inductance and T_1 the time that the switch is closed. During this interval the transformer's primary and secondary voltages are

$$v_p = -V_{UR} \tag{9.6-22a}$$

and
$$v_s = \frac{V_p}{N} = \frac{-V_{UR}}{N} \tag{9.6-22b}$$

At the end of the runup time there is an energy

$$E = \tfrac{1}{2} L_p I_{mp}^2 \tag{9.6-23}$$

stored in the magnetic field of the transformer.

At $t = T_1+$, the switch opens abruptly and the primary current goes to zero. In an effort to maintain the same flux storage in the magnetic field, both the primary and secondary voltages rise sharply. When the secondary voltage,

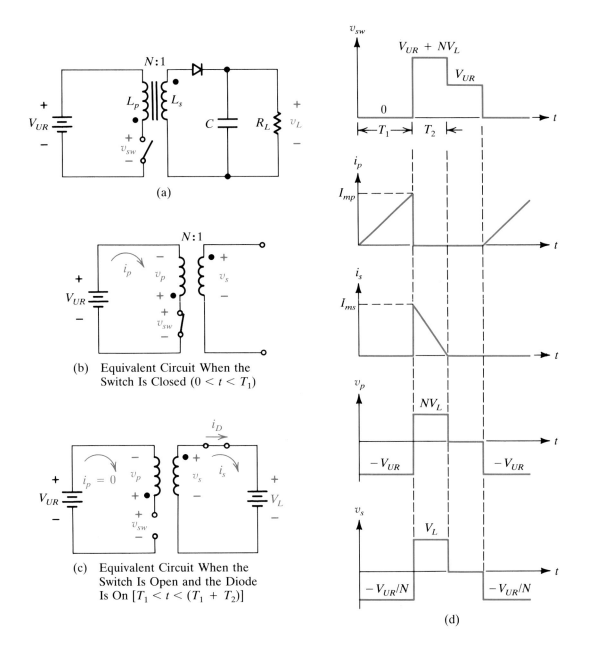

(a)

(b) Equivalent Circuit When the Switch Is Closed ($0 < t < T_1$)

(c) Equivalent Circuit When the Switch Is Open and the Diode Is On [$T_1 < t < (T_1 + T_2)$]

(d)

Figure 9.6-5
Flyback-style switching power supply.

v_s, exceeds V_L, the diode turns ON, holding v_s constant at V_L, and v_p correspondingly at NV_L (Figure 9.6-5c). Thus at this point the switch voltage is now equal to

$$v_{sw} = V_{UR} + V_p = V_{UR} + NV_L \qquad (9.6\text{-}24)$$

The current that flows in the secondary at $t = T_1+$ is found by applying conservation of energy, which requires that the energy stored in the magnetic field of the transformer cannot change instantaneously, so that

$$\tfrac{1}{2}L_p I_{mp}^2 = \tfrac{1}{2}L_s I_{ms}^2 \qquad (9.6\text{-}25a)$$

which using the fact that $L_p/L_s = N^2$ yields

$$I_{ms} = NI_{mp} \qquad (9.6\text{-}25\text{b})$$

for the initial current flow in the secondary.

Again, for simplicity, redefining $t = 0$ as the point at which the switch opens, the secondary current may be written as

$$i_s(t) = -\frac{V_L t}{L_s} + I_{ms} \qquad (9.6\text{-}26)$$

so that the run downtime is

$$T_2 = \frac{I_{ms} L_s}{V_L} \qquad (9.6\text{-}27)$$

Once i_p and i_s are both zero, v_p and v_s are also zero, so that the switch voltage then drops down to V_{UR} and remains at that value until the switch again closes at the beginning of the next cycle. The waveshapes for i_p, i_s, v_p, v_s, and the voltage across the switch are sketched in Figure 9.6-5d.

The steady-state output voltage V_L is found by applying conservation of energy to the input and output sides of this circuit. The average power delivered to the power supply by V_{UR} is equal to

$$P_{IN} = \frac{1}{T} \int_0^T i_p V_{UR} dt = \frac{V_{UR}}{T}\left(\frac{1}{2} I_{mp} T_1\right) \qquad (9.6\text{-}28\text{a})$$

while the power delivered to the load is

$$P_L = \frac{V_L^2}{R_L} \qquad (9.6\text{-}28\text{b})$$

Because the rest of the circuit is assumed to be lossless, P_{IN} and P_L must be equal, and combining these results with equation (9.6-21), we obtain

---

$$V_L = \frac{V_{UR} T_1}{\sqrt{2\tau T}} \qquad (9.6\text{-}29)$$

---

where $\tau = L_p/R_L$.

EXAMPLE 9.6-2

For the MOS flyback-style power supply circuit illustrated in Figure 9.6-6a, a pulse-width modulator feedback circuit has been added to stabilize the output voltage against shifts in V_L due to load and line variations. The MOSFET is assumed to be OFF when $v_4 = 0$ V and to be fully ON (a short circuit) when $v_4 = +10$ V. The gain coefficient for the pulse-width modulator is $K = 4$ μs/V, and its frequency is fixed at 20 kHz.

a. Derive an expression for V_L in terms of the specified circuit parameters.
b. Sketch the steady-state waveforms for v_4, i_{D1}, i_s, and v_{D1}.

SOLUTION

Following Figure 9.6-6a, we have

$$V_3 = A(V_1 - V_2) = A(V_{\text{REF}} - V_L) \qquad (9.6\text{-}30)$$

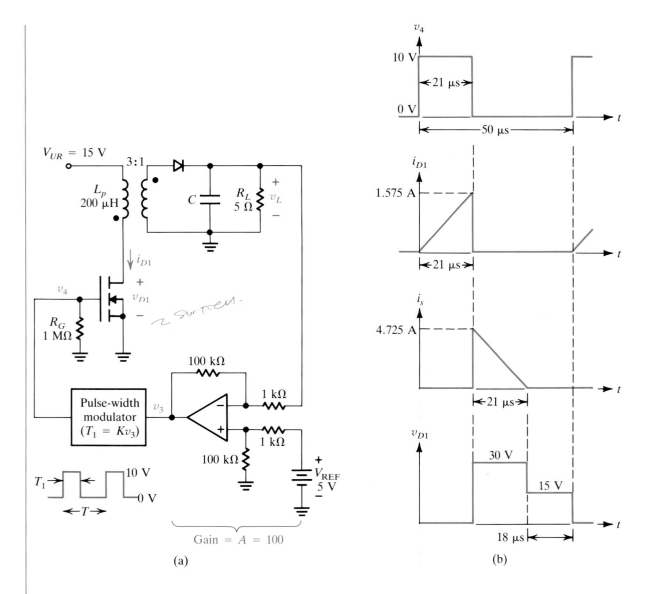

Figure 9.6-6
Example 9.6-2.

where $A = R_f/R_2 = 100$. This voltage is applied to the pulse-width modulator and produces a 20-kHz 10-V amplitude waveform v_4 at the output whose pulse width is

$$T_1 = KV_3 = KA(V_{\text{REF}} - V_L) \tag{9.6-31}$$

Substituting equation (9.6-29) into this expression, we obtain

$$V_L = \frac{V_{UR}KA(V_{\text{REF}} - V_L)}{\sqrt{2\tau T}} \tag{9.6-32a}$$

which after some algebraic manipulation yields

$$V_L = \frac{(V_{UR}KA/\sqrt{2\tau T})V_{\text{REF}}}{1 + V_{UR}KA/\sqrt{2\tau T}} \tag{9.6-32b}$$

Examining this result, it is apparent that $V_L \simeq V_{REF}$ if $V_{UR} KA/\sqrt{2\tau T} \gg 1$. Thus if the circuit loop gain is large, the pulse width, T_1, will adjust itself so that the output voltage is nearly equal to V_{REF}.

For the circuit under consideration, $V_{UR} KA/\sqrt{2\tau T} = 94.87$, so that, from equation (9.6-32b),

$$V_L = \frac{(94.87)5}{95.87} = 4.948 \text{ V} \tag{9.6-33}$$

The approximate waveshapes for the important voltages and currents in this circuit may be determined as follows. For $R_L = 5 \ \Omega$, in accordance with equation (9.6-29), the steady-state pulse width at v_4 is equal to

$$T_1 = \sqrt{2\tau T} \frac{V_L}{V_{UR}} = (63.2 \ \mu s)\left(\frac{1}{3}\right) \simeq 21 \ \mu s \tag{9.6-34}$$

During this runup time the current in the primary increases linearly to a final value of

$$I_{mp} = \frac{V_{UR} T_1}{L} = 1.575 \text{ A} \tag{9.6-35}$$

The rundown of the secondary current begins at NI_{mp} or 4.725 A. Following equation (9.6-32), the rundown time for this waveshape is

$$T_2 = \frac{I_{ms} L_s}{V_L} = \frac{1}{N} \frac{I_{mp} L_p}{V_L} = 21 \ \mu s \tag{9.6-36}$$

During the rundown $v_s = 5$ V, so that v_p and v_{D1} are 15 and 30 V, respectively. Once the secondary current goes to zero, v_s and v_p both become zero, and v_{D1} drops down to $V_{UR} = 15$ V. The waveshapes for this circuit are sketched in Figure 9.6-6b.

Exercises

9.6-1 For the circuit in Figure 9.6-1a the switching frequency is 15 kHz, $T_1 = 10 \ \mu s$, $V_{UR} = 20$ V, and $L = 45 \ \mu H$. Find V_L if $R_L = 1 \ \Omega$. (*Note:* The circuit will be in the continuous-current mode for this value of load resistance.) ***Answer*** 3.0 V

9.6-2 In Figure 9.6-3a, $V_{UR} = 10$ V, $L = 1$ mH, and $R_L = 100 \ \Omega$.
a. Find P_{IN} and V_L if the circuit operates at 10 kHz in the discontinuous mode with $T_1 = 20 \ \mu s$.
b. What is the time required for the current stored in the inductor to discharge to zero under the conditions in part (a)?
c. Find the frequency at which the circuit is just on the edge of continuous-mode operation.
Answer (a) .2 W, 14.1 V, (b) 48.7 μs, (c) 14.6 kHz

9.6-3 In Exercise 9.6-2, R_L is changed to 5 kΩ. What is the switching frequency at which the circuit crosses over from discontinuous- to continuous-mode operation? *Answer* 51.95 kHz

9.6-4 For the flyback circuit in Figure 9.6-5a, the transformer has a 1 : 5 primary-to-secondary turns ratio, $V_{UR} = 20$ V, $L_p = 1$ mH, $L_s = 25$ mH, and $R_L = 10$ kΩ. If the switch operates at 30 kHz with a 50-50 duty cycle, estimate V_L. *Answer* 182 V

9.7 POWER SUPPLY OVERLOAD PROTECTION

Because of their high voltage and high current levels, power supply circuits and the loads that they drive can be damaged easily unless some type of guard circuitry is designed into the supply. In this section we discuss three different types of power supply protection schemes:

1. *Current limiting:* to protect the pass transistor from excessive current flow
2. *Thermal limiting:* to turn off the power supply when its temperature exceeds a predetermined level
3. *Overvoltage shutdown:* to protect the load from damage should the power supply fail and cause the output voltage to rise to an unsafe level

Each of these three protection methods is important in its own right in power supply design, and whereas their goals are similar—to protect the supply and its surrounding circuitry from destruction—the techniques by which these goals are achieved are somewhat different in each case.

Power supply current limiters are very similar to the current-limiting circuits used in power amplifiers, and basically, their job is to prevent the load current from exceeding some predetermined level. Because the power dissipated in the pass transistor depends not only on the current flow through it, but also on the voltage drop across it, more elaborate current-limiting schemes, known as current foldback circuits, have also been developed that automatically reduce the limit current as the voltage across the pass transistor increases. For a given-size pass transistor, power supplies having foldback-current limiting can handle load currents much greater than those employing conventional constant-current short-circuit protection.

A power supply that is working fine at room temperature may fail completely when the ambient temperature surrounding the circuit rises, even though the maximum load current from the supply was never exceeded. Typically, this occurs when the junction temperature of the power control transistor goes beyond its safe limit, and is due to the fact that this temperature depends not only on the power dissipated in the transistor but also on the ambient temperature surrounding it (see Section 9.9). To prevent the thermal destruction of a power supply, we can design sensing circuits that will shut the supply down when the temperature of the power control transistor exceeds a predetermined level. For IC designs, because all the components are located on the same substrate, the measurement of the power transistor's temperature by the sensing circuit is a relatively easy matter. However, for larger designs where the power control transistor is a separate device, the thermal sensing circuits need to be mounted on the same heat sink as the power transistor in order to be in intimate thermal contact.

Often, the electronic circuitry connected to a power supply can be severely damaged if the output voltage from the supply rises above its nominal value by more than a few percent. A voltage increase of this type can occur for any one of several possible reasons. The power supply itself can fail, so that the output voltage rises, for example, to the level of the unregulated input voltage. This type of failure is common and can occur if the pass transistor shorts. Fast line transients can also cause the output voltage to increase when the input voltage changes so rapidly that the feedback control circuits do not have time to respond. Finally, and not all that uncommonly, if the supply is adjustable, someone may accidentally set the output voltage too high.

Unfortunately, even if the power supply's output voltage remains above its nominal level for only a few milliseconds, significant irreversible damage can still be done to the electronic hardware connected to the supply. To protect sensitive loads from this sort of damage, overvoltage protection circuits are used. These circuits monitor the load voltage, and when an overvoltage condition is detected, they shut the supply down, typically within a few microseconds. This protection is usually accomplished by placing a virtual short circuit across the power supply's input or output terminals, reducing the voltage to zero until the circuit's fuses have time to blow.

9.7-1   Current Limiting

There are basically two different popular current-limiting techniques. The first of these, which we call constant-current limiting, is the simplest and is the same as that used for short-circuit protection in audio power amplifiers. The second technique is known as foldback-current limiting, and while it is somewhat more complex than constant-current limiting, it does offer significant advantages. The volt-ampere output characteristics of two power supplies, one containing constant-current limiting, and the second foldback-current limiting, are illustrated in Figure 9.7-1a and b, respectively.

In the power supply with constant-current short-circuit protection, the output voltage remains constant until the current limit (I_{SC}) is reached. For this particular supply, current limiting begins when the load connected across its output terminals is (15 V)/(1 A), or 15 Ω. When the load resistance drops below 15 Ω, the supply enters the constant-current region, and the output voltage begins to fall. For example, if $R_L = 7.5$ Ω, v_L will be (1 A) × (7.5 Ω) or 7.5 V, while when $R_L = 0$, the output voltage will also drop to zero.

For the supply with foldback-current limiting, once the current limit is exceeded, both the output voltage and the output current drop simultaneously. This type of foldback behavior helps to minimize the power dissipated in the pass transistor once the current-limiting region is entered. To see how this is accomplished, we note that the power dissipated in the pass transistor in Figure 9.7-1c is equal to

$$P_D = (V_{UR} - V_L)I_C \simeq (V_{UR} - V_L)I_L \tag{9.7-1}$$

Under normal operating conditions (i.e., for $R_L > 15$ Ω for the supplies in Figure 9.7-1) the supply output voltage will be 15 V, and if the unregulated input

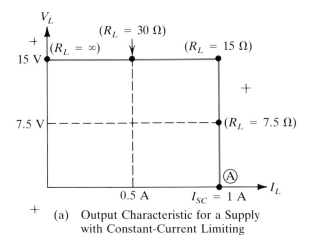

(a) Output Characteristic for a Supply with Constant-Current Limiting

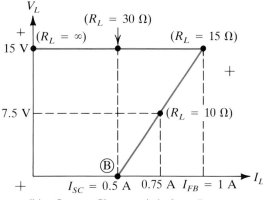

(b) Output Characteristic for a Power Supply with Foldback Current Limiting

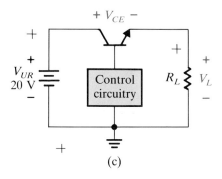

(c)

Figure 9.7-1

voltage, V_{UR}, is 20 V, the voltage drop across the pass transistor will be only (20 V − 15 V) or 5 V. Therefore, in this region the power dissipated in the pass transistor will be relatively small regardless of the current-limiting scheme employed. In particular, as R_L varies between infinity and 15 Ω, the load current will change from 0 to 1 A, and the dissipation will vary between 0 and 5 W.

To compare the performance of these two power supplies in their current-limit regions, let's examine what happens to each of them when their outputs are shorted. When V_L is made equal to zero, the Q point of the constant-current supply will shift to point A in the figure, while that of the foldback supply will be located at point B. The power dissipated in the pass transistor in this instance is significantly different for each of these circuits. For the supply employing constant-current short-circuit protection $P_D = [(20 − 0)$ V] $\times$ $(1 \text{ A}) = 20$ W, while for the foldback circuit the power dissipated is only $[(20 − 0)$ V] $\times$ (0.5 A) or 10 W. Thus power supplies with foldback-style current limiting instead of constant-current limiting can use smaller pass transistors, and since the pass transistor along with its heat sink are some of the most expensive components in the supply, the use of foldback-current limiting can be very cost-effective.

The circuit in Figure 9.7-2 is a series-pass power supply in which Q_1 is the pass transistor and Q_2 functions as the current-sensing element. The output volt-ampere characteristics for this supply exhibit constant-current limiting. In

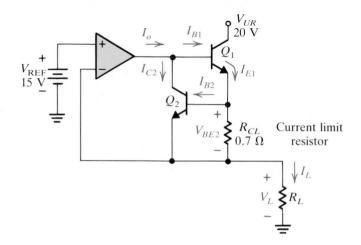

Figure 9.7-2

Regulated power supply with constant-current short-circuit protection.

particular, for load currents below 1 A, V_{BE2} is less than 700 mV, Q_2 is OFF, and the supply output is essentially constant at 15 V. When the load resistance drops below 15 Ω, Q_2 enters the active region, and this limits the current available from Q_1 to a constant value of about 1 A. As a result, the volt-ampere characteristics of this power supply are nearly identical to those presented in Figure 9.7-1a. A detailed analysis of this power supply is carried out in Problem 9.7-3.

EXAMPLE 9.7-1

The 10-V switching power supply illustrated in Figure 9.7-3 is similar to those discussed earlier in Section 9.6 except that a current-limiting circuit has been added to the power supply in this example. The transistor Q_1 has an $h_{FE} = 100$, a constant $h_{ie} = 1$ kΩ, and a base–emitter turn-on voltage $V_o = 0.6$ V. If the switching frequency is constant at 15 kHz and if the circuit operates in the discontinuous current mode, sketch the output volt-ampere characteristic for this power supply.

SOLUTION

With the op amp operating in the active region, the input voltages V_1 and V_2 must be nearly equal, so that

$$\frac{100 V_{REF}}{101} = \frac{V_L}{2}\left(\frac{100}{101}\right) + I_{C1}\left(\frac{100 \text{ k}\Omega}{101}\right) + V_3\left(\frac{1}{101}\right) \qquad (9.7\text{-}2a)$$

or

$$V_3 = 100\left(V_{REF} - \frac{V_L}{2}\right) - (100 \text{ k}\Omega)I_{C1} = A_{CL}\left[\left(V_{REF} - \frac{V_L}{2}\right) - (1 \text{ k}\Omega)I_{C1}\right] \qquad (9.7\text{-}2b)$$

where A_{CL} represents the closed-loop gain of the op amp. For load currents of less than 1 A, the emitter–base voltage on Q_1 will not be large enough to turn Q_1 ON, and in this current range I_{C1} in equation (9.7-2b) will be zero. Once the load current exceeds 1 A, Q_1 will enter the active region, where the base and collector currents in this transistor may be written as

$$I_{B1} = \frac{V_{EB1} - 0.6}{1 \text{ k}\Omega} = \frac{I_L R_{CL} - 0.6}{1 \text{ k}\Omega} \qquad (9.7\text{-}3a)$$

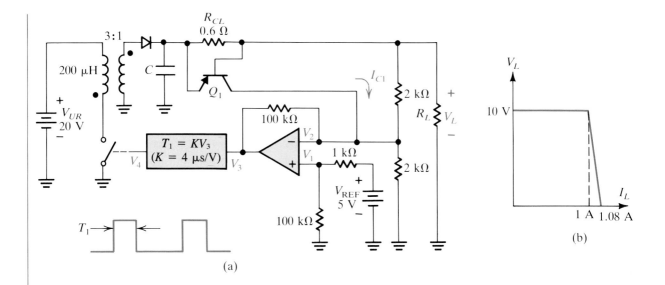

(a)

(b)

Figure 9.7-3
Example 9.7-1.

and

$$I_{C1} = h_{FE}I_{B1} = \frac{I_L R_{CL} - 0.6}{10\ \Omega} \tag{9.7-3b}$$

The pulse width T_1 at the output of the pulse-width modulator is given by $T_1 = KV_3$, and combining this expression with equations (9.7-2b) and (9.7-3b), we obtain

$$T_1 = KA_{CL}\left[\left(V_{REF} - \frac{V_L}{2}\right) - 100(I_L R_{CL} - 0.6)\right] \tag{9.7-4}$$

Substituting this result into equation (9.6-29), the voltage at the load may be written as

$$V_L = \frac{V_{UR}T_1}{\sqrt{2\tau T}} = \frac{V_{UR}}{\sqrt{2\tau T}}KA_{CL}\left[\left(V_{REF} - \frac{V_L}{2}\right) - 100(I_L R_{CL} - 0.6)\right] \tag{9.7-5a}$$

or

$$V_L = \frac{(V_{UR}KA_{CL}/\sqrt{2\tau T})[V_{REF} - 100(I_L R_{CL} - 0.6)]}{1 + V_{UR}KA_{CL}/2\sqrt{2\tau T}} \tag{9.7-5b}$$

For

$$\frac{V_{UR}KA_{CL}}{\sqrt{2\tau T}} \gg 1 \tag{9.7-6}$$

equation (9.7-5b) is approximately equal to

$$V_L \simeq 2[V_{REF} - 100(I_L R_{CL} - 0.6)] \tag{9.7-7}$$

This expression is valid only for currents greater than 1 A. For load currents that are smaller than this, Q_1 is cut off and the second term inside the brackets is zero. Using these facts together with the data for this problem, the equations for the output volt-ampere characteristic of this power supply may be written as

$$V_L = \begin{cases} 10\ V & \text{for } I_L < 1\ A \tag{9.7-8a} \\ 130 - 120 I_L & \text{for } I_L > 1\ A \tag{9.7-8b} \end{cases}$$

These results are sketched in Figure 9.7-3b. The short-circuit current for this power supply is found by substituting $V_L = 0$ into equation (9.7-8b) and solving for I. Carrying out this procedure, we obtain $I_{SC} = (130 \text{ V})/(120 \text{ }\Omega) = 1.08$ A. Thus, for all practical purposes, the current limiting in this switching supply is essentially of the constant-current type.

The major problem associated with constant-current-style current limiting is that the power dissipation in the pass transistor under short-circuit conditions is very large. Consider, for example, the supply illustrated in Figure 9.7-2, whose short-circuit current is about 1 A. When this power supply is operating in its constant-voltage region, the power dissipation in the pass transistor is always less than 5 W. However, under short-circuit conditions, this dissipation level increases to more than 20 W. Thus, to be able to handle sustained short circuits, Q_1 needs to be able to dissipate nearly four times the power that it would have to if short circuits never occurred. This is principally due to the fact that the current flow through the pass transistor stays about the same when the output is shorted, while the voltage drop across it increases by about four times (from 5 V to 20 V).

In a foldback-current-limiting circuit, the power dissipation under short-circuit conditions is reduced considerably by automatically lowering the load current as the output voltage drops. A typical output volt-ampere characteristic for a power supply containing foldback-current limiting is illustrated in Figure 9.7-4a. Here I_{FB} represents the current at which the foldback limiting begins, and I_{SC} the short-circuit current that flows when R_L is reduced to zero. One circuit for accomplishing this type of foldback limiting is illustrated in Figure 9.7-4b, and except for the addition of the resistors R_1 and R_2 at the input of Q_2, it is identical to the constant-current limiting circuit just examined.

Figure 9.7-4

Regulated power supply containing foldback-current limiting.

Let's derive an expression for the current-limiting behavior of this circuit in order to demonstrate that it exhibits linear current foldback. If V_L is the voltage across the load resistor, the voltage at the emitter of Q_1 is approximately equal

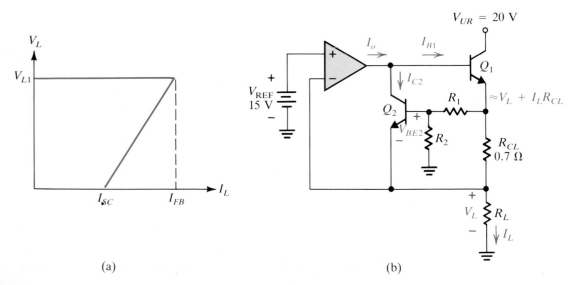

(a)   (b)

$$V_{E1} = V_L + I_{E1} R_{CL} \simeq V_L + I_L R_{CL} \qquad (9.7\text{-}9)$$

neglecting I_{E2}, which is assumed to be much less than I_{E1}. This expression also neglects the loading of the R_1–R_2–Q_2 network on Q_1. To the extent that these approximations are valid, V_{BE2} may be written as

$$V_{BE2} = V_{B2} - V_{E2} = \frac{R_1}{R_1 + R_2}(V_L + I_L R_{CL}) - V_L = I_L \frac{R_{CL} R_1}{R_1 + R_2} - V_L \frac{R_2}{R_1 + R_2} \qquad (9.7\text{-}10)$$

If we recall that the current limiting occurs in this type of circuit when V_{BE2} reaches about 0.7 V, we can see that the limiting in this case depends on the values of both I_L and V_L.

The value of I_{FB} may be found by substituting in $V_{BE2} = 0.7$ V and $V_L = V_{REF}$ into equation (9.7-10) and solving for I_L. Because the second term in this expression effectively cancels out part of the first, a rather large value can be obtained for the foldback current. The load resistance corresponding to power supply operation at this Q point is V_{REF}/I_{FB}. As the load on the supply increases beyond this point, V_L begins to fall and under short-circuit conditions is zero. Solving equation (9.7-10) with $V_L = 0$ yields I_{SC}, and typically this current will be much less than I_{FB}.

The foldback nature of this supply's current-limiting circuitry may be explicitly illustrated by setting $V_{BE2} = 0.7$ V in equation (9.7-10) and solving this expression for V_L, which yields

$$V_L = \frac{R_{CL} R_1}{R_2} I_L - (0.7 \text{ V}) \frac{R_1 + R_2}{R_2} \qquad (9.7\text{-}11)$$

This, of course, is just the straight-line equation for the foldback portion of the volt-ampere characteristic of the power supply, in which the coefficient in front of the I_L term represents the slope of the foldback line, and the second term the V_L intercept.

9.7-2 Thermal Limiting

Thermal limiting circuits in a power supply are generally used to monitor the junction temperature of the pass transistor, and to shut the supply down when this temperature exceeds a predetermined limit. These high temperatures can arise from the presence of long-term short circuits at the output, from overloads occurring at high ambient temperatures, or from inadequate heat sinking of the pass transistor. In designing a thermal protection circuit, it is important that the sensing device be placed in intimate thermal contact with the pass transistor. Most thermal cutout circuits are designed to trip at a temperature in the neighborhood of 175°C, since temperatures in excess of this value can result in permanent damage to silicon transistors.

Figure 9.7-5 illustrates the design of a typical voltage regulator circuit that has both current and thermal shutdown capabilities. In this circuit Q_4 is the pass transistor, Q_3 provides the current limiting, and Q_1–Q_2 and the zener diode D_1 form the thermal protection portion of the circuit. Because of the biasing arrangement, Q_1 always operates in the active region. By choosing R_2

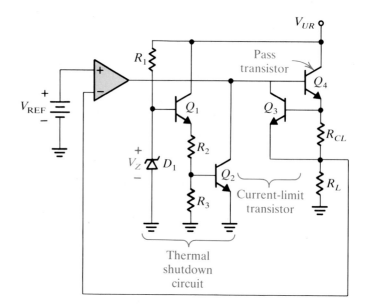

Thermal protection for a voltage-regulated power supply.

and R_3 properly, the circuit can be designed so that Q_2 is cut off at room temperature, and active when its temperature reaches a specific predetermined value, typically in the neighborhood of 175°C. Once Q_2 enters the active region, it shunts current away from Q_4 and shuts the supply down.

In selecting values for R_2 and R_3, the following information is important. The base–emitter voltage of a silicon transistor has a temperature coefficient of about $(-2 \text{ mV})/\text{°C}$. Therefore, at 25°C, if we assume that this transistor is cut off when V_{BE} is less than 600 mV and active when V_{BE} reaches 700 mV, at 175°C these voltages will shift to 300 and 400 mV, respectively. In a similar fashion, if we assume that the temperature coefficient of the zener diode is $(+3 \text{ mV})/\text{°C}$ and that its zener voltage is 6.5 V at room temperature (25°C), at 175°C it will increase to about 6.95 V.

Using this information, we now want to select R_2 and R_3 so that Q_2 is cut off at 25°C and fully ON at 175°C. At room temperature the zener voltage is 6.5 V, and $V_{BE1} = 0.7$ V. Therefore, $V_{E1} = (6.5 - 0.7)$ or 5.8 V. If we select a bias current of about 10 mA for Q_1 and neglect the base current into Q_2, then

$$R_2 + R_3 = \frac{5.8 \text{ V}}{10 \text{ mA}} = 580 \text{ } \Omega \tag{9.7-12}$$

Furthermore, if Q_2 is to be cut off at this temperature, it is necessary V_{BE2} be less than 600 mV or that

$$\frac{R_3}{R_2 + R_3}(5.8 \text{ V}) \leq 600 \text{ mV} \tag{9.7-13}$$

At 175°C we have $V_Z = 6.95$ V and $V_{BE1} = 400$ mV, so that $V_{E1} = 6.55$ V, and if we again neglect I_{B2} in comparison to I_{E1}, then for Q_2 to be strongly active at 175°C, we require that

$$\frac{R_3}{R_2 + R_3}(6.55 \text{ V}) = 400 \text{ mV} \tag{9.7-14}$$

or that $R_3 = 35.4 \ \Omega$ and $R_2 = 544.6 \ \Omega$. Substituting these values for R_2 and R_3 back into equation (9.7-13), we find that V_{BE2} will be equal about 350 mV, and therefore, as required, Q_2 will be cut off at room temperature.

9.7-3 Overvoltage Protection

In some situations the output voltage from a power supply can rise well above its intended value. This can be the result of a component failure in the supply, or due to the presence of rapidly occurring line or load transients, or even due to something as simple as operator adjustment of the output voltage level. Unfortunately, regardless of the reason for its occurrence, the creation of an overvoltage situation can be disastrous, especially for the load connected to the supply. Some digital logic IC systems, for example, can only tolerate a 10% increase in supply voltage beyond its nominal value, before the functional performance of the ICs are permanently degraded. Therefore, depending on the application where the power supply is to be used, some type of overvoltage protection may be required.

The basic technique employed in an overvoltage protection circuit is to increase the load on the supply when a higher-than-normal output voltage is detected, in order to lower this voltage back to a safe level. One of the most popular methods for achieving this type of protection is to connect an SCR directly across the power supply input or output terminals (Figure 9.7-6). As we describe much more fully in Section 9.8, an SCR is basically a diode that is normally OFF but can be made to conduct by the application of a suitable gate current. Once triggered into the ON state, the diode continues to conduct even when the gate signal is removed.

Figure 9.7-6 illustrates a typical SCR overvoltage protection circuit. When an overvoltage condition is sensed (and in particular when $V_1 > V_2$ in the figure), the comparator output goes high, and this turns the SCR ON, producing a virtual short circuit across the power supply output terminals. This so-called "crowbar effect" can take place in a matter of a few microseconds and is therefore very effective in protecting the load from prolonged exposure to excessively high voltages.

Figure 9.7-6

SCR "crowbar" overvoltage protection for a power supply.

Once the SCR turns ON, large currents can flow in both the regulator and the input diodes, especially if the overvoltage condition occurred because of a

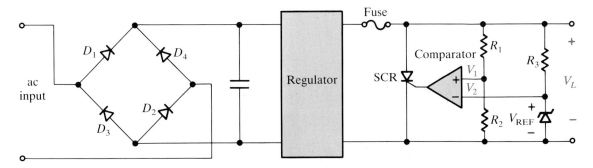

shorted pass transistor. This situation can damage the SCR and the input diodes unless some type of secondary protection is provided for the supply.

The simplest type of secondary protection is to place a fuse on either the input or the output side of the regulator. But fuses are slow and their operation may not be fast enough to save the rectifier diodes in the event that an input to output short occurs in the regulator. To protect against this type of hardware failure, a second and somewhat better approach is to replace two of the full-wave rectifier diodes (D_3 and D_4, for example) with SCRs. The gates of these SCRs can then be connected through an inverter to the same comparator output as the crowbar SCR. In this way, when an overvoltage situation is detected, the crowbar SCR is turned ON, protecting the load, and simultaneously the rectifier circuit is shut OFF to prevent any additional damage to the rectifiers, the regulator, or the crowbar SCR.

Exercises

9.7-1 The series-pass voltage regulator shown in Figure 9.7-1c has the foldback characteristic given in part (b) of the same figure.
a. Find P_D for the pass transistor when $R_L = 20\ \Omega$ given that $V_{UR} = 25$ V.
b. Repeat part (a) for $R_L = 12\ \Omega$.
Answer (a) 7.5 W, (b) 12.5 W

9.7-2 For the foldback circuit in Figure 9.7-4a, $R_{CL} = 2\ \Omega$ and $R_1 = R_2 = 500\ \Omega$. Find the maximum power dissipated in Q_1 as R_L is varied.
Answer 57.2 W

9.7-3 For the circuit in Figure 9.7-4, $R_2 = 300\ \Omega$. Select R_1 and R_{CL} so that $I_{FB} = 1$ A and $I_{SC} = 0.25$ A. *Answer* 49 Ω, 3.26 Ω

9.7-4 In Figure 9.7-5, find the temperature at which thermal shutdown occurs if $R_2 = 400\ \Omega$ and $R_3 = 25\ \Omega$. *Answer* 181°C

9.8   AC POWER CONTROL TECHNIQUES

The task at hand in the area of ac power control is the regulation of the ac electrical power delivered to the load. In achieving this goal, it is not usually necessary for the output waveshape to look like the original ac input voltage that was applied, since we are typically interested only in controlling the average power delivered to the load. As we discussed in Section 9.1, the most efficient way to achieve this type of ac power control is to use switching rather than linear circuit techniques (see Figure 9.1-1c through e and 9.1-2a and b).

The simplest type of power switching is illustrated in Figure 9.1-1c. In this type of circuit when the switch is in the open position, no power is delivered by the source since the load current is zero, and when the switch is closed, full power is delivered to the load. Because the voltage drop across the switch is zero, no power is dissipated in the switch. Therefore, ideally at least, the system is 100% efficient. This is the major advantage of using power switching rather than linear power control techniques.

Simple ON–OFF switching only allows zero or full power to be delivered to the load. However, by electronically controlling the opening and closing of the switch, intermediate power levels can also be obtained by varying the switch duty cycle, either on a cycle-by-cycle basis, as illustrated in Figure 9.1-2a, or over a large number of cycles as shown in Figure 9.1-2b.

Power switching of the type just discussed can be accomplished with BJTs or FETs, but there is another entire class of semiconductor devices known as "thyristors" whose performance in ac power control applications is far superior to that attainable with transistors. Basically, a thyristor is a voltage- or current-controlled switch which has two stable states: the ON (or low-impedance) state and the OFF (or high-impedance) state. The device is normally in the OFF state, and once triggered into the ON state, it remains there even when the trigger signal is removed. To turn off a thyristor, the current flow through it must be interrupted momentarily.

Before getting into the details of the theory of operation of thyristors, it will first be useful to make sure that we understand why they and not transistors are the primary devices that are used for the control of ac electrical power. First and foremost is the fact that transistors are primarily dc devices that can be made to handle ac voltages only with some difficulty. On the other hand, thyristors are particularly well suited to ac circuit applications.

Transistors (at least at low frequencies) are also much less efficient than comparably sized thyristors. One reason for this is that for a transistor switch to remain in the ON state, the switching signal must also be kept on. This uses up extra power (at least for BJT devices) and lowers circuit efficiency. With thyristors, on the other hand, the input signal may be removed once the device is initially triggered into the ON state. In addition, the input currents required to trigger thyristors are several orders of magnitude smaller than those needed for BJT devices handling similar amounts of power.

Besides being more efficient, thyristors are also much less expensive than BJT or FET devices having comparable forward current and reverse blocking voltage capabilities. In summary, thyristors are the best way to proceed when controlled amounts of low-frequency ac power need to be delivered to a load. In the remainder of this section we examine the theory of operation of several of the most popular types of thyristor devices, and we also illustrate how they can be applied to the design of ac power control circuits.

9.8-1   Theory of Operation of Thyristor Devices

The operation of most types of thyristors can be understood if we first develop a detailed working knowledge of the *pnpn* four-layer diode illustrated in Figure 9.8-1a. This diode structure is best analyzed by splitting it in half and redrawing it as a two-transistor positive-feedback amplifier (Figure 9.8-1b). When viewed in this way, we can see that the possibility exists for this device to become unstable if its loop gain is made large enough.

When a positive voltage is applied between the anode and cathode terminals of this diode, junctions J_1 and J_3 become forward biased, whereas J_2 (the collector junction for Q_1 and Q_2) is reverse biased. Therefore, for $V_D > 0$, both transistors in the equivalent circuit are in their active regions, and following equation (3.2-12), the collector currents flowing in each of them may be written as

$$I_{C1} = \alpha_1 I_{E1} + I_{CBO1} \qquad (9.8\text{-}1a)$$

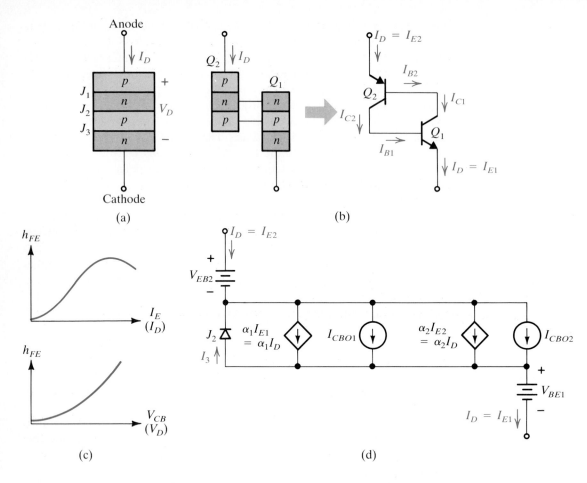

Figure 9.8-1
pnpn four-layer diode.

and
$$I_{C2} = \alpha_2 I_{E2} + I_{CBO2} \tag{9.8-1b}$$

Adding these two equations, and noting that

$$I_D = I_{C1} + I_{C2} = I_{E1} = I_{E2} \tag{9.8-2}$$

the diode current, I_D, can then be expressed in terms of the reverse leakage currents I_{CBO1} and I_{CBO2} as

$$I_D = \frac{I_{CBO1} + I_{CBO2}}{1 - (\alpha_1 + \alpha_2)} \tag{9.8-3}$$

This result may also be written in terms of the transistor h_{FE} values as

$$I_D = \frac{(1 + h_{FE1})(1 + h_{FE2})(I_{CBO1} + I_{CBO2})}{1 - h_{FE1}h_{FE2}} \tag{9.8-4}$$

by substituting $h_{FE}/(1 + h_{FE})$ for α.

To understand how I_D varies with V_D, the diode voltage, it is useful to recall that current gain of a transistor is actually a function of both the current flowing through it and the voltage across it. The graphs in Figure 9.8-1c illustrate the typical variation of h_{FE} with transistor emitter current and collector-to-base voltage. Using these data and relating them back to the four-layer diode under investigation, we can see that when V_D (or alternatively, V_{CB}) is small, the transistor h_{FE} values will be low and I_D will nearly be zero, so that the device may be considered to be in the OFF state (Figure 9.8-2a). As V_D is increased, the h_{FE} values will also increase, and when the voltage reaches the point where

$$\alpha_1 + \alpha_2 = 1 \qquad (9.8\text{-}5a)$$

or alternatively, where the product

$$h_{FE1}h_{FE2} = 1 \qquad (9.8\text{-}5b)$$

then, in accordance with equation (9.8-4), the terminal current should grow without bound.

Besides being indicative of an infinite device current, it can also be shown that the amplifier loop gain is just equal to 1 when equations (9.8-5a) or (9.8-5b) are satisfied (see Problem 9.8-1). As a result, when $(\alpha_1 + \alpha_2)$ or $h_{FE1}h_{FE2}$ is greater than 1, the circuit loop gain is also greater than 1, and the amplifier has poles in the right-hand plane. When this occurs, the device becomes unstable, and as with the Schmitt trigger investigated in Chapter 8, each time it enters the active region it will switch from one amplifier nonlinear state to the other.

For example, if the device is initially in the OFF state, then when the voltage across it increases to the point where $h_{FE1}h_{FE2}$ becomes greater than 1, the diode will switch into the ON state. In a similar fashion, when the diode is ON and V_D is reduced to the point where the transistors again become active, switching to the OFF state will occur.

It is interesting to note in the ON state that Q_1 and Q_2 will both be saturated and that all three of the junctions J_1, J_2, and J_3 in Figure 9.8-1a will be forward biased. To demonstrate this fact, following Figure 9.8-1d, we note that

Figure 9.8-2

$$I_D = \alpha_1 I_D + I_{CBO1} + \alpha_2 I_D + I_{CBO2} - I_3 \qquad (9.8\text{-}6a)$$

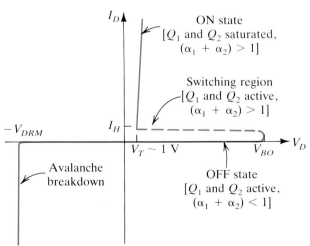

(a) Volt-Ampere Characteristics of the Four-Layer Diode and the Silicon Unilateral Switch

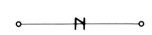

(b) Schematic Symbol for the Four-Layer Diode and for the Silicon Unilateral Switch

or
$$I_D = \frac{(I_{CBO1} + I_{CBO2}) - I_3}{1 - (\alpha_1 + \alpha_2)} \tag{9.8-6b}$$

Thus once $(\alpha_1 + \alpha_2)$ is greater than 1, the only way that I_D can be greater than zero is if I_3 is positive and greater $(I_{CBO1} + I_{CBO2})$ or if the junction J_2 is forward biased. Of course, since J_1 and J_3 are already forward biased, once J_2 also becomes forward biased, Q_1 and Q_2 will both be saturated.

The volt-ampere characteristics of a typical four-layer diode, or Shockley diode as it is also called, are illustrated in Figure 9.8-2a, and its schematic symbol is given in part (b) of the figure. When V_D is small and positive, $(\alpha_1 + \alpha_2)$, and hence I_D, are both nearly zero. As V_D increases, so do the transistor alphas, and when their sum approaches 1, the poles of the device enter the right-hand plane. This causes a rapid switching to the ON state in which Q_1 and Q_2 both saturate. The critical voltage at which the switching occurs is called V_{BO}, the diode breakover voltage.

Referring back to Figure 9.1-1a, we can see that

$$V_D = V_{EB2} + V_{CE1} = V_{CE2} + V_{BE1} \tag{9.8-7}$$

so that with both transistors saturated the voltage, V_D, across the diode will be about 1 V. Furthermore, once this four-layer diode turns ON, the current that flows through it will usually be sufficient to ensure that equation (9.8-5) is satisfied even though V_D has dropped to a low value. As a result, once ON, the diode will remain in the ON state until the current flow through it is reduced to a value that is low enough to cause Q_1 and Q_2 to reenter the active region. This current level is indicated by I_H on the volt-ampere characteristics, and is known as the diode's holding current. When I_D falls below I_H, the transistors are again active with $(\alpha_1 + \alpha_2) > 1$, and because of the positive feedback, the diode rapidly switches into the OFF state.

When V_D is made negative, J_2 is forward biased and J_1 and J_3 are reverse biased. In this situation almost no current flows through the device until V_D becomes negative enough to cause avalanche breakdown in the two reverse-biased junctions. The volt-ampere behavior in this region therefore parallels that of an ordinary diode.

There are several other devices which have electrical characteristics that are very similar to those of the four-layer diode. The silicon unilateral switch (or SUS), although having a different internal structure, has terminal characteristics that are virtually identical to the four-layer diode. In fact, the electronic symbol for this device is the same as that given in Figure 9.8-2b for this diode. The major difference between the two is that the SUS has a much lower breakover voltage, typically in the range 5 to 10 V, in comparison with V_{BO} values of several hundred volts for four-layer diodes.

By connecting two SUS diodes in parallel, as illustrated in Figure 9.8-3a, we can form a device known as a silicon bilateral switch (SBS) that has symmetrical volt-ampere characteristics (Figure 9.8-3b). A third device, known as a diac, although having a somewhat different internal structure, has terminal characteristics that are similar to those of the SBS. Basically, it may be thought of a *pnpn* four-layer diode connected in parallel with an *npnp* four-layer diode. Its schematic symbol and volt-ampere characteristics are illustrated in Figure 9.8-3c and d, respectively. Typical breakover voltages for SBS devices are on the order of 6 to 10 V while those for diacs are around 20 to 40 V. In addition,

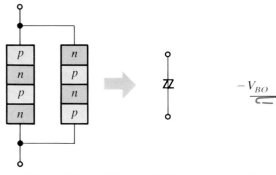

(a)   Silicon Bilateral Switch (SBS)

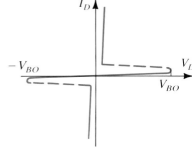

(b)   Volt-Ampere Characteristics of the SBS

(c)   Diac

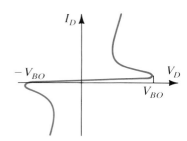

(d)   Volt-Ampere Characteristics of the Diac

Figure 9.8-3

once the SBS fires, the voltage drop across it falls to about 1 V, while that across the diac stays much higher, typically being about 10 or 20 V.

By adding a third terminal to the four-layer diode, we obtain a very interesting device known as a silicon-controlled rectifier (SCR). The physical structure, symbol, and volt-ampere characteristics for the SCR are given in Figure 9.8-4a, b, and c, respectively. The operation of the SCR is essentially the same as the four-layer diode except that the breakover voltage can be reduced by the application of a current to the third terminal, or gate, as it is also called. Following the same development as for the four-layer diode, and using the two-transistor equivalent circuit in Figure 9.8-4a, it is not too difficult to show that the terminal current, I_D, may be written in terms of gate current, I_G, and the transistor leakage currents as

$$
\begin{aligned}
I_D &= \frac{\alpha_1 I_G + (I_{CBO1} + I_{CBO2})}{1 - (\alpha_1 + \alpha_2)} \\
&= \frac{(1 + h_{FE2})[h_{FE1} I_G + (1 + h_{FE1})(I_{CBO1} + I_{CBO2})]}{1 - h_{FE1} h_{FE2}}
\end{aligned}
\tag{9.8-8}
$$

Following these results, we can therefore see that the injection of a gate current into the SCR can significantly increase I_D, and hence I_{C1} and I_{C2}. These increases in the transistor collector currents result in corresponding increases in h_{FE1} and h_{FE2}, and cause the switching of the device into the ON state to occur at a lower value of V_D. In fact, if the gate current is made large enough (see curve labeled I_{G3} in Figure 9.8-4c), the SCR volt-ampere behavior is nearly the

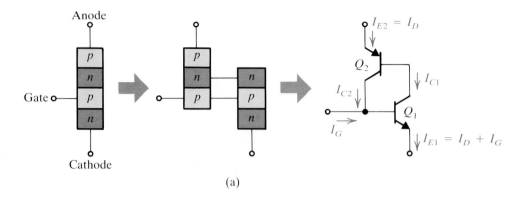

(a)

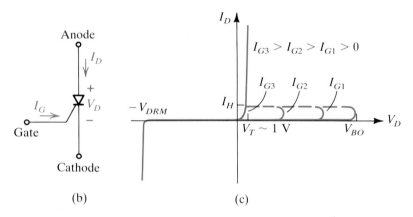

(b)   (c)

Figure 9.8-4
SCR (silicon-controlled rectifier).

same as that for a regular diode, and the SCR essentially turns on as soon as V_D goes positive.

As with the four-layer diode, an SCR in the ON state has a voltage drop across it of about 1 V, and once turned on (either by making V_D greater than the breakover voltage, or by injecting enough gate current to cause it to break over at a lower value of V_D), the device remains in the ON state until the current flow through it drops below the holding current, I_H.

Silicon-controlled rectifiers are basically half-wave devices in that they can be turned on only when the voltage across them is positive. As such, they can only be used to switch between 0 and 50% of the available ac load power unless a full-wave rectifier is connected in front of the SCR. By effectively taking two SCRs and connecting them together back to back, as illustrated in Figure 9.8-5a, a device known as a triac can be fabricated which is capable of directly switching the ac voltage applied to the load on both the positive and negative halves of the ac input signal. As such, by using either single-cycle or multiple-cycle modulation techniques, a triac can be employed to regulate the average ac power delivered to a load over the full range between 0 and 100%.

The volt-ampere characteristics of a typical triac are illustrated in Figure 9.8-5b, and to a first-order approximation we may consider that the device possesses identical characteristics for both positive and negative current flows. Although the triac can be triggered with both positive and negative gate currents regardless of the polarity of the voltage across it, for simplicity we use positive gate currents in quadrant 1 and negative gate currents in quadrant 3.

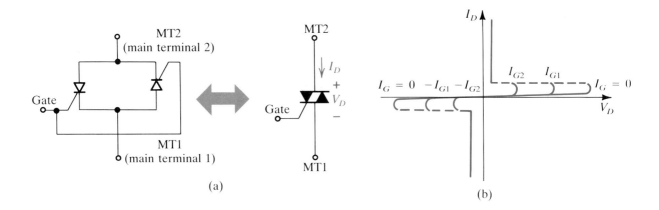

(a)

(b)

Figure 9.8-5
Triac.

Furthermore, we also assume that the magnitude of the gate current needed to trigger the triac in each of these quadrants is identical.

9.8-2 Applications of the SCR

Now that we have an understanding of the most important types of power switching devices, let's see how they can be used for the control of ac power. We will begin by investigating the use of SCRs in half-wave and full-wave ac power control circuits, and we will then extend these results to develop an understanding of the behavior of triacs in full-wave ac power control applications.

Initially, let us consider the simple SCR circuit illustrated in Figure 9.8-6a. To analyze the performance of this circuit, we notice from Figure 9.8-4c that the gate current required to trigger an SCR depends to a certain extent on the anode-to-cathode voltage across the diode. This relationship is illustrated explicitly in Figure 9.8-6b for a typical low-power silicon-controlled rectifier. In this figure I_{GT} stands for the gate threshold current, or the current that will just cause the SCR to fire at a particular value of V_D. Because I_{GT} varies only slowly with V_D, we assume for simplicity that this trigger current is essentially a constant for a given SCR, and that it is independent of the voltage across the device. In particular, for the SCR in this example, we consider that $I_{GT} = 20 \ \mu$A. Furthermore, and again for simplicity, we also assume that $v_G \approx 0.7$ V when J_1, the base–emitter junction of Q_1 in Figure 9.8-4, is forward biased. Using these data, let's attempt to sketch v_L, v_D, i_D, i_G, and v_G for the circuit in question.

When v_{in} is very small, the SCR is OFF, and therefore, because R_G is so much larger than R_L, v_L will be approximately zero and v_D will be nearly equal to v_{in}. As v_{in} increases, the gate current into the SCR will grow sinusoidally in accordance with the equation

$$i_D = \frac{v_D - v_G}{R_G} = \frac{v_{in} - 0.7}{5 \ \text{M}\Omega} \approx \frac{v_{in}}{5 \ \text{M}\Omega} \qquad (9.8\text{-}9)$$

When v_{in} reaches 100 V, the gate current will equal 20 μA and the SCR will fire. This will occur when

$$170 \sin \omega t_1 = 100 \qquad (9.8\text{-}10a)$$

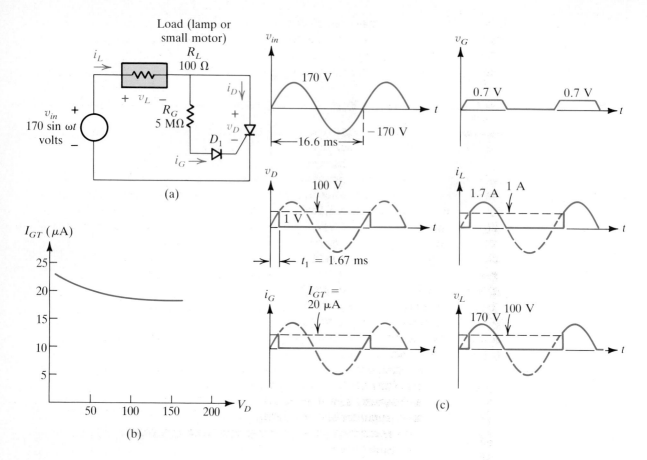

Figure 9.8-6
Simple SCR ac power control circuit.

or at

$$t_1 = \frac{1}{377} \sin^{-1} \frac{100}{170} = 1.67 \text{ ms} \qquad (9.8\text{-}10b)$$

After the SCR turns ON, the voltage across it will drop to about 1 V. The corresponding values of v_G, i_G, v_L, and i_L at this point are

$$v_G \simeq 0.7 \text{ V} \qquad (9.8\text{-}11a)$$

$$i_G = \frac{v_D - 0.7}{5 \text{ M}\Omega} = \frac{1 - 0.7}{5 \text{ M}\Omega} \simeq 0 \qquad (9.8\text{-}11b)$$

$$v_L = v_{in} - v_D \simeq v_{in} = 100 \text{ V} \qquad (9.8\text{-}11c)$$

and

$$i_L = \frac{100 \text{ V}}{100 \text{ }\Omega} = 1 \text{ A} \qquad (9.8\text{-}11d)$$

Once the SCR fires, it will remain in the ON state as long as i_D is greater than the holding current, which is assumed to be 100 μA for this SCR. The value of v_{in} for which the SCR turns off may be determined from the expression

$$v_{in} \simeq I_H R_L + V_T = (100 \text{ }\mu\text{A})(100 \text{ }\Omega) + 1 \text{ V} \simeq 1 \text{ V} \qquad (9.8\text{-}12)$$

To a good approximation the SCR therefore turns off as v_{in} passes through zero. All during the time that v_{in} is less than zero, the SCR will remain in the

OFF state, and as a result, i_D, i_L, and v_L will all be approximately zero. The diode D_1 is used to ensure that the gate junction of the SCR does not go into avalanche breakdown and guarantees when v_{in} is negative that both i_G and v_G will be zero. When v_{in} again goes positive, the cycle repeats.

EXAMPLE 9.8-1

a. For the SCR circuit in Figure 9.8-6a, discuss the range of possible conduction angles if R_G is variable.

b. Develop an expression for the average power delivered to the load as a function of the SCR conduction angle.

SOLUTION

When R_G is made very small, the SCR will turn ON almost as soon as v_{in} passes through zero going in the positive direction. The conduction angle corresponding to this situation will be nearly 180°. As the value of R_G is made larger, triggering of the SCR will occur later and later in the cycle. The value of R_G for which turn-on occurs at the positive peak of v_{in} is $(170 \text{ V})/(20 \text{ } \mu\text{A})$, or about 8.5 MΩ, and the conduction angle corresponding to SCR turn-on at this point is 90°. If R_G increases beyond 8.5 MΩ, the conduction angle drops abruptly to zero since for values of R_G beyond this point the gate current never reaches I_{GT}, and therefore the SCR never turns ON. Thus the range of possible conduction angles for this circuit is between 90 and 180°.

If θ_1 defines the angle at which the SCR turns ON, then, following Figure 9.8-7a, the average power delivered to the load may be written as

$$P_L = \frac{1}{2\pi}\int_{\theta_1}^{2\pi} \frac{v_L^2}{R_L}\,d\theta = \frac{1}{2\pi}\int_{\theta_1}^{2\pi} \frac{V_m^2}{R_L}\sin^2\theta\,d\theta = \frac{V_m^2}{2R_L}\left(\frac{\pi - \theta_1}{2\pi} + \frac{\sin 2\theta_1}{4\pi}\right) \qquad (9.8\text{-}13)$$

Figure 9.8-7
Example 9.8-1.

This result is sketched in Figure 9.8-7b as a function of the conduction angle, $(\pi - \theta_1)$. For later use the results for full-wave ac power control are also given.

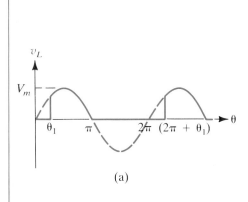

(a)

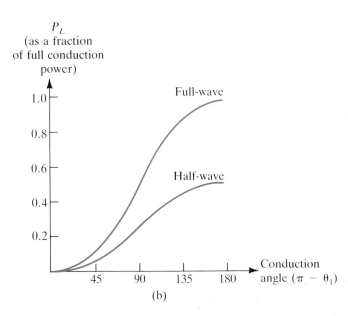

(b)

EXAMPLE 9.8-2

The silicon-controlled switch in the SCR phase control circuit given in Figure 9.8-8a has a breakover voltage of 10 V, and a negligible voltage drop across it once it fires (Figure 9.8-2a). The holding current for both the SCR and the SUS are each 20 μA, and the gate input resistance for SCR and the ON resistance of the SUS may both be considered to be about 100 Ω.

a. Explain how the circuit works and in doing so, sketch v_G, v_C, i_G, v_D, and v_L for the case where $R_1 = 50$ kΩ.

b. In a half-wave SCR ac power control circuit, the available power to the load can theoretically be adjusted to any value between zero and 50% of the maximum power available from the source. To do this, the conduction angle needs to be varied between 0 and 180°, respectively. However, it is not really necessary to have such a wide range of conduction angle adjustment. Following the sketch in Figure 9.8-7b, a conduction angle of 30°

Figure 9.8-8
Example 9.8-2.

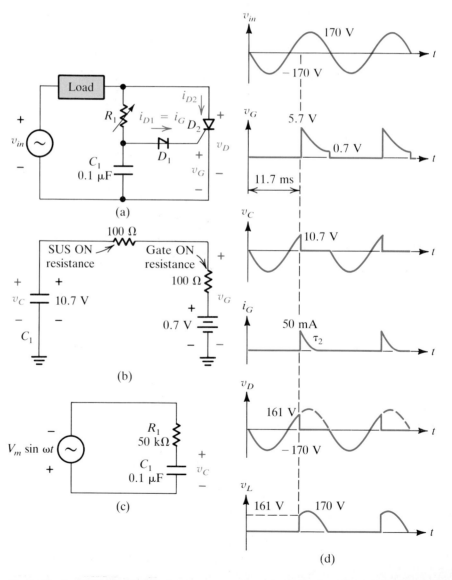

(a)

(b)

(c)

(d)

corresponds to a 1.5% power delivery to the load, while a 150° conduction angle permits 48.5% of the available power to go to the load. Therefore, for most applications, being able to vary the conduction angle between these limits provides for satisfactory control over the half-wave power that can be delivered to the load. Determine the range of R_1 values that will permit the power into the load to be adjusted to any value between 1.5 and 48.5% of full power.

SOLUTION

Let's begin the analysis of this circuit at the point in time where v_{in} goes negatively through zero so that the SCR has just turned off. With the SCR OFF, $v_D = v_{in}$, and as this voltage goes negative and then positive, the capacitor voltage, v_C, will follow v_{in} but will be delayed in time. When the voltage across the capacitor goes positive, nothing will happen to the SCR until v_C reaches about $(10 + 0.7)$ or 10.7 V. At this point the SUS will turn ON, and because the equivalent circuit in Figure 9.8-8b now applies, the capacitor will rapidly discharge through the ON resistances of the silicon-controlled switch and the SCR. The peak gate current when the diode first turns ON will be about $(10 \text{ V})/(200 \ \Omega) = 50$ mA and it will decay exponentially to zero in about 4τ or 80 μs. Thus, as soon as the SUS turns ON, the large spike of gate current produced will also turn ON the SCR regardless of its I_{GT} value, as long as it is less than 50 mA. This is the main advantage of adding the SUS to the circuit.

As soon as the gate current drops below 20 μA, D_1 will turn OFF and both i_{D1} and i_G will go to zero. However, because the SCR will stay ON until v_{in} goes to approximately zero (and i_{D2} drops below the holding current), v_G will remain at about 0.7 V all during this interval.

To determine the turn-on time for the SCR, we have to find $v_C(t)$ by solving the differential equation of the circuit in Figure 9.8-8c. The steady-state solution for this circuit is

$$v_{C,ss}(t) = \frac{-V_m}{\sqrt{1 + (\omega R_1 C_1)^2}} \sin (\omega t - \phi) \qquad (9.8\text{-}14a)$$

where

$$\phi = \tan^{-1} (\omega R_1 C_1) \qquad (9.8\text{-}14b)$$

and substituting in the circuit component values, we have

$$v_{C,ss}(t) = -79.8 \sin (\omega t - 62°) \text{ volts} \qquad (9.8\text{-}15)$$

The complete solution for the capacitor voltage, including the transient response, may be written as

$$v_C(t) = -79.8 \sin (\omega t - 62°) + Ae^{-t/R_1 C_1} \qquad (9.8\text{-}16a)$$

Because $v_C(0-) \simeq 0$, this expression may be rewritten as

$$v_C(t) = -79.8 \sin (\omega t - 62°) - 70.4e^{-t/5 \text{ ms}} \qquad (9.8\text{-}16b)$$

The waveshape for v_C is sketched in Figure 9.8-8d along with the other required waveforms. The delay time, t_1, for the turn-on of the SCR and the diode is found by setting equation (9.8-16b) equal to 10.7 V and solving it numerically, which yields $t_1 = 11.7$ ms. The corresponding amplitude of v_{in} at this point is 161 V.

To deliver 1.5% and 48.5% of the available power to the load, conduction angles of 30 and 150°, respectively, are needed. For a 60-Hz input voltage, this corresponds to turn-on times of 9.67 ms and 15.23 ms. To determine the values of R_1 that will produce these turn-on times, it is necessary to solve the following equation:

$$\frac{-170 \sin (\omega t - \phi)}{\sqrt{1 + (\omega R_1 C_1)^2}} - 170 \sin \phi \, e^{-t/R_1 C_1} = 10.7 \qquad (9.8\text{-}17)$$

Here ϕ is given by equation (9.8-14b), $\omega = 377$ rad/s and $C_1 = 0.1 \ \mu\text{F}$. Solving this expression numerically with $t = 9.67$ ms and 15.23 ms, we find that values of R_1 between 12 and 95 kΩ will permit us to adjust the conduction angle to any value between 30 and 150°.

9.8-3 Applications of the Triac in AC Power Control

The ac power control circuits discussed thus far are interesting from an educational viewpoint, but are not all that practical because they do not allow us to adjust the ac power delivered to the load over the full range between 0 and 100%. To achieve this type of power control with SCR devices it will be necessary either to use two SCRs with one operating on each half of the input sine wave, or alternatively, to full-wave rectify the ac voltage and then apply it to a single SCR.

Full-wave SCR circuits are useful in that they do provide for variation of the conduction angle over the full range possible. However, the extra hardware needed to construct them makes these circuits impractical except in very high

Figure 9.8-9

Full-wave ac power control using a triac.

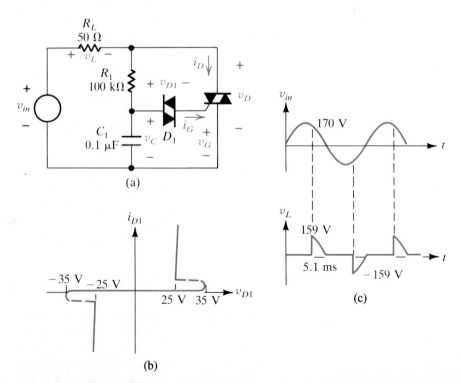

Chapter 9 / Power Electronics

power applications. For low and intermediate power levels, the triac is a much better choice when full-wave ac power control is needed. The triac is inherently a full-wave device, and by combining it with a properly designed trigger circuit, it can be used to provide ac power control over the full range of conduction angles between 0 and 360°. Figure 9.8-9a illustrates a typical triac-style ac power control circuit in which a diac (Figure 9.8-9b) is used as part of the trigger circuit (much as the SUS was used with the SCR) to guarantee that the triac will fire at the desired point in the ac cycle regardless of its I_{GT} value. The waveshape for the voltage across the load is sketched in Figure 9.8-9c and the detailed operation of this circuit is examined in Problem 9.8-7.

9.8-4   RF Interference from Thyristors

Thyristors make excellent ac power switches. In fact, in some ways their switching characteristics are too good. Because SCRs and triacs are able to switch from OFF to ON in 1 or 2 μs, the load voltages and currents that result can have very steep edges. Unfortunately, the Fourier series for this type of waveshape contains many high-frequency harmonic terms, and these high-frequency components can cause significant electrical interference in communication equipment. Television and FM radio receivers because of their high operating frequencies tend to be relatively immune from this type of interference, but AM radio receivers operating in the range from 500 kHz to 1.6 MHz can be severely affected.

The noise emanating from a thyristor circuit may be divided into two categories: airborne and lineborne noise. Airborne noise is noise that is actually radiated into space by the thyristor circuit and picked up by the antenna of the radio receiver. Generally, this type of interference is the least offensive and can easily be controlled by placing the thyristor circuit in a metal box.

Lineborne noise is a bit more difficult to eradicate since it travels over the ac power lines and can cause significant interference problems long distances away from the thyristor noise source. This type of noise is best controlled by placing a LC filter, such as the one illustrated in Figure 9.8-10a, in series with the thyristor. Basically, this filter acts to increase the rise time of the load current, and thereby to reduce the high-frequency components of the Fourier series of the waveform. The inductor in this circuit acts to prevent the current in the thyristor from rising too quickly. The capacitor, which is initially charged to the voltage v_{in} just before the thyristor switch closes, tends to remain at this voltage after the switch is closed, and this also helps to prevent i_L from rising too fast.

If the input voltage v_{in} is considered to be constant during the time that the switch transient occurs, the load current may be written as

$$I_L(s) = \frac{V_{in}}{R_L} \frac{1/LC}{s[s^2 + s(1/R_L C) + 1/LC]} \tag{9.8-18}$$

where V_{in} is the amplitude of the input voltage at the time the switch is closed. For the load resistance values normally used with this type of circuit, the roots of the quadratic in the denominator tend to be real and result in a circuit time constant that is nominally on the order of the $R_L \times C$. For example, if this cir-

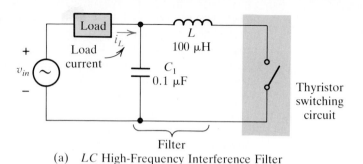

Filter

(a)   *LC* High-Frequency Interference Filter

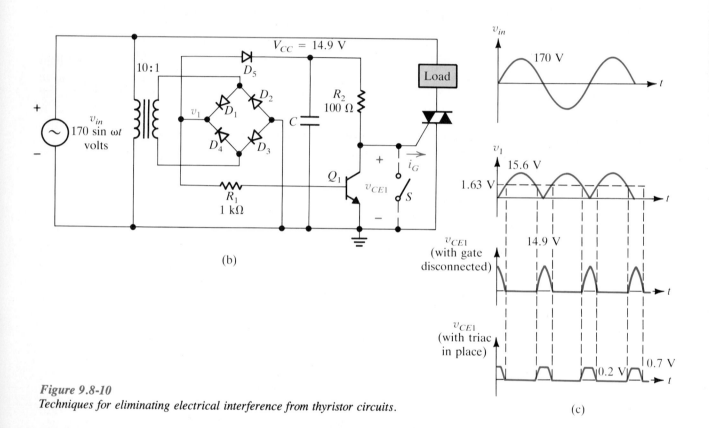

(b)

Figure 9.8-10
Techniques for eliminating electrical interference from thyristor circuits.

(c)

cuit were used with a 100-W lamp (whose resistance is $120^2/100$ or 144 Ω), the corresponding load current rise time would be about

$$T = 2.2R_L C \simeq 32 \ \mu s \tag{9.8-19}$$

which is 10 or 20 times longer than without the filter.

Another way to avoid the creation of high-frequency interference in thyristor circuits is to turn the thyristor on at the zero crossings of the input voltage. The circuit in Figure 9.8-10b illustrates one way to accomplish this. The combination of the bridge rectifiers D_1–D_4 and the capacitor C form a full-wave rectifier circuit that produces a dc voltage of about $(17 - 2.1) = 14.9$ V across the capacitor. The diode D_5 decouples the diode bridge from the capacitor when v_1 is less than V_{CC} and allows the voltage at v_1 to be a full-wave rectified sine wave, as illustrated in part (c) of the figure.

Besides producing the dc voltage at V_{CC}, v_1 is also used to switch the transistor Q_1 ON and OFF in synchronism with the line voltage. When v_1 is less than 0.7 V the transistor is cut off, and with the triac disconnected, v_{CE1} rises to V_{CC}. As v_{in} rises above 0.7 V, the transistor turns on, and when it exceeds

$$v_1 = I_{B,\min} R_1 + 0.7 = \frac{I_{C,\text{sat}}}{h_{FE1}} R_1 + 0.7 = 1.63 \text{ V} \qquad (9.8\text{-}20)$$

it saturates, assuming an $h_{FE1} = 100$.

The waveshape for v_{CE1}, with and without the triac connected, is illustrated in Figure 9.8-10c. When Q_1 is saturated, v_{CE1} is about 0.2 V, the triac gate current will be zero, and the triac will remain OFF. When the transistor cuts off at the zero crossings of v_{in}, positive gate current can flow into the triac through R_2, and if I_{GT} for this device is less than $(15.6 - 0.7)/100 = 149$ mA, the triac will fire. It is important to recognize that this positive gate current can be used to trigger the triac during both the positive and negative halves of the ac line voltage cycle.

The switch shown in dashed lines at the gate of the triac in Figure 9.8-10b can represent either a transistor switch or an actual mechanical switch that could be used to control the turn-on of the triac in a simple ON–OFF switching application. With the switch thrown into the closed position, the triac would complete its current power cycle and would remain off thereafter since its gate voltage would be identically zero. When the switch is opened, the triac would begin to conduct on the next zero crossing of the ac line voltage and would be retriggered thereafter at each zero crossing as long as the switch remained in the open position.

When proportional ac power control is needed in order to be able to deliver intermediate power levels to the load, this can also be accomplished with the zero-voltage switching technique just described. However, it does present one major problem. To achieve this type of power control, the voltage applied to the load needs to be switched ON and then OFF for an integral number of cycles. This relatively slow ON–OFF cycling makes zero-voltage switching unsuitable for the proportional control of incandescant lamps since the low frequency of the ON–OFF pulsing creates a noticeable flicker in the lamp.

However, this technique is excellent for heater control systems, traffic light controllers, or emergency light flashers which can be switched ON and then OFF at a relatively slow rate. For these applications the switch S in Figure 9.8-10b is replaced by a transistor, and the state of this transistor is used to control the ON–OFF cycling of the triac.

Exercises

9.8-1 The SCR in the circuit in Figure 9.8-6 has a breakover voltage of 100 V and an I_{GT} and I_H of 20 μA. Find the average power delivered to the load if $R_G = \infty$. *Answer* 68.7 W

9.8-2 Repeat Exercise 9.8-1 if $R_G = 100$ Ω, making any reasonable approximations. *Answer* 144.5 W

9.8-3 Estimate the conduction angle of the circuit in Figure 9.8-8 if $R_1 = 50$ kΩ. (*Hint:* Write a small Basic program.) *Answer* 113°

9.9 THERMAL CONSIDERATIONS IN POWER SEMICONDUCTOR DESIGNS

The power dissipated in a semiconductor device causes its internal temperature to rise above the ambient temperature that surrounds it, and if this internal temperature gets too high, permanent damage to the semiconductor may result. The relationship between the heat produced within a semiconductor, and its temperature may readily be understood by developing a simple electrical analog of this thermal problem.

The duality that exists between certain electrical and thermal quantities is illustrated in Figure 9.9-1. Temperature has the same role in a "thermal circuit" that voltage has in an electrical circuit, and thermal heat flow is directly analogous to electrical current flow. Therefore, just as we can define the resistance of a branch of an electrical circuit as the ratio of the voltage drop across that branch to the current flow through the branch, we can similarly define the resistance of a branch in a thermal circuit as

$$R_{\Theta 12} = \text{thermal resistance between points 1 and 2} = \frac{T_1 - T_2}{\dot{Q}} \qquad (9.9\text{-}1)$$

where T_1 and T_2 are the temperatures at the endpoints of the branch and $\dot{Q}$ is the heat flow through the branch measured in joules/second or watts. Because temperature is usually specified in degress Celsius, the most common units of thermal resistance are °C/watt.

In developing the equivalent circuit for a semiconductor device that is dissipating electrical power as heat, the power being dissipated may be considered to be a source of heat flow, and hence it can be represented by a current source. The thermal system illustrated in Figure 9.9-2a presents the basic form of the problem at hand. The power source on the left supplies electrical power to the semiconductor device, and only a part of that input power is delivered to the load. The remainder (P_D) is dissipated in the semiconductor device as heat. This heating of the semiconductor raises its internal or junction temperature

Figure 9.9-1
Comparison between electrical and thermal systems.

Electrical System	Meaning	Thermal System
Voltage (V)	Difference in this quantity at two points causes flow	Temperature (°C)
Current = rate of flow of charge (A)	Flow	$\dot{Q}$ = heat flow (J/s) = power (W)
Resistance (Ω) $R = \dfrac{V_2 - V_1}{I}$ $V_2 \qquad V_1$ $\rightarrow I$ 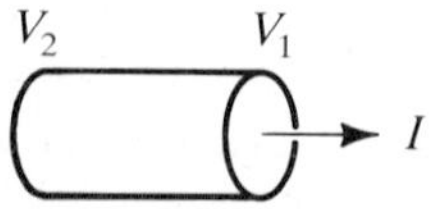	System opposition to flow	Resistance (°C/W) $R_\theta = \dfrac{T_2 - T_1}{\dot{Q}}$ 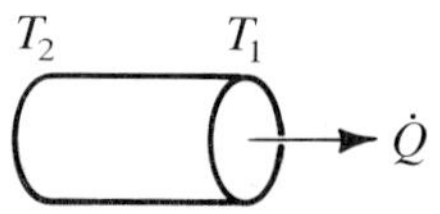 $T_2 \qquad T_1$ $\rightarrow \dot{Q}$

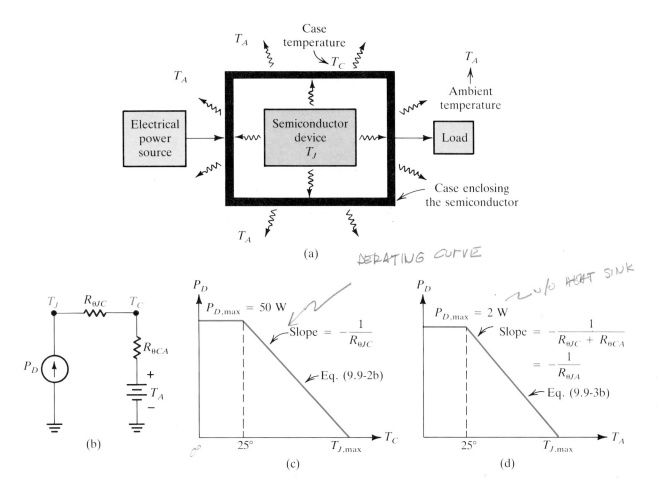

DERATING CURVE

$P_{D,max} = 50$ W

Slope $= -\dfrac{1}{R_{\theta JC}}$

Eq. (9.9-2b)

$25°$ $T_{J,max}$ T_C

(c)

~ 2 W/o HEAT SINK

$P_{D,max} = 2$ W

Slope $= -\dfrac{1}{R_{\theta JC} + R_{\theta CA}}$

$= -\dfrac{1}{R_{\theta JA}}$

Eq. (9.9-3b)

$25°$ $T_{J,max}$ T_A

(d)

Figure 9.9-2

(T_J), and as the heat flows out through the surrounding case structure into the environment, the case temperature (T_C) also rises. If we define $R_{\theta JC}$ as the junction-to-case thermal resistance and $R_{\theta CA}$ as the case-to-ambient resistance, then in the steady state this thermal problem may be analyzed by using the electrical equivalent circuit given in Figure 9.9-2b. Because the ambient temperature, T_A, is assumed to be a constant independent of the power being dissipated in the semiconductor device, it is represented by a battery or constant-temperature source.

In describing the thermal characteristics of a particular semiconductor device, the manufacturer may give specific values for $R_{\theta JC}$, $R_{\theta CA}$, and $T_{J,max}$, the maximum allowed junction temperature. As an alternative, this information may also be presented graphically in one of the forms illustrated in Figure 9.9-2c and d. These graphs are known as the device's derating curves, and basically, they describe how the maximum power the device can dissipate varies with its case or ambient temperature.

If the device is meant to be used with a heat sink, the curve in part (c) of the figure is generally given by the manufacturer, and from these data we can see that the maximum power that this device can dissipate appears to be about 50 W. However, as a practical matter, achieving this power dissipation level without damage to the semiconductor will be nearly impossible since the case temperature will have to be maintained at 25°C or less, and if T_A is also at 25°C, this will require an infinitely large heat sink. Thus because its case temperature will almost definitely be above 25°C, the power-handling capability of this device will need to be derated to the value specified for that case temperature by the graph.

When P_D is given as a function of ambient temperature, as in Figure 9.9-2d, the semiconductor is intended to be operated without a heat sink. This specific device would be listed on the data sheets as having a 2-W power dissipation capability. However, it can dissipate 2 W only when T_A is less than 25°C. When the ambient temperature exceeds this value, P_D must be derated to the level specified by the curve.

These graphs basically provide the same information as that given by the electrical equivalent circuit in part (b) of the figure. For example, in the equivalent circuit, the junction temperature is related to the power being dissipated by the equation

$$T_J = P_D R_{\theta JC} + T_C \tag{9.9-2a}$$

and solving this expression for P_D, we obtain

$$P_D = -\underbrace{\frac{1}{R_{\theta JC}} T_C}_{\text{slope}} + \underbrace{\frac{T_J}{R_{\theta JC}}}_{P_D \text{ intercept}} \tag{9.9-2b}$$

This equation is the diagonal portion of the derating curve in Figure 9.9-2c, and therefore the slope of this line is just $-1/R_{\theta JC}$, and its T_C intercept is equal to $T_{J,\max}$ (since when $T_C = T_{J,\max}$, P_D must equal zero or else T_J will exceed $T_{J,\max}$). For BJT devices $T_{J,\max}$ is on the order of 175°C, for MOSFETs in the neighborhood of 150°C, and for SCRs and triacs it is typically 100 to 125°C.

The model in Figure 9.9-2b may also be related to the derating curve in part (d) of the figure by observing that the junction temperature in the equivalent circuit can also be written in terms of T_A, the ambient temperature, as

$$T_J = P_D(R_{\theta JC} + R_{\theta CA}) + T_A \tag{9.9-3a}$$

or

$$P_D = \frac{-1}{R_{\theta JC} + R_{\theta CA}} T_A + \frac{T_J}{R_{\theta JC} + R_{\theta CA}} \tag{9.9-3b}$$

This last expression is the equation for the diagonal portion of the derating curve in Figure 9.9-2d, and as such the slope of this derating curve is simply $-1/(R_{\theta JC} + R_{\theta CA})$ and its T_A intercept is equal to $T_{J,\max}$.

The flat horizontal portion of both derating curves indicates that an absolute limit exists for the maximum power that the semiconductor can dissipate, even when the case or ambient temperature is lowered to the point where the maximum junction temperature is not being exeeded. This limit is associated with thermal stresses that develop within the device due to the large temperature differentials that exist across it. This upper bound on P_D must be kept in mind when using thermal equivalent circuits to solve problems since it is not explicitly taken into account in these models.

As we will see in the examples that follow, semiconductor thermal problems may be solved by using either the equivalent circuit approach, or by graphical means using the derating curves. As we have mentioned, the derating curve in which T_C is the abscissa will prove most useful for handling thermal problems where the device is attached to a heat sink, while that where T_A is the abscissa will work best for situations where the device radiates its heat directly into the surrounding environment.

EXAMPLE 9.9-1

A BJT power transistor has the derating curve given in Figure 9.9-3a, and in addition it is known that $T_{J,max} = 175°C$, $R_{\Theta JC} = 2°C/W$, and $R_{\Theta CA} = 13°C/W$.

   a. Determine the maximum power that the transistor can safely dissipate at an ambient temperature of 50°C using both graphical and equivalent-circuit techniques.

   b. Using the equivalent-circuit approach, find the case temperature for the conditions specified in part (a).

   c. Repeat part (a) at $T_A = 0°C$.

SOLUTION

For the case where $T_A = 50°C$, the maximum power dissipation permitted in the power transistor without exceeding $T_{J,max}$ may be read off directly from the derating curve as about 8.3 W. This same result can also be obtained from the model in Figure 9.9-3b by noting that the node labeled T_J can have a maximum temperature of 175°C. If the ambient temperature is 50°C, the temperature drop across the two thermal resistances will be $(175 - 50) = 125°C$, so that the maximum allowed heat flow or power dissipation in this circuit may be calculated using Ohm's law to be

$$P_D = \frac{T_J - T_A}{R_{\Theta JC} + R_{\Theta CA}} = \frac{125°C}{15°C/W} = 8.33 \text{ W} \tag{9.9-4}$$

The case temperature at an ambient temperature of 50°C and a power dissipation level of 8.33 W can readily be determined with the aid of the equivalent circuit in Figure 9.9-3b. In particular, we have that

$$T_C = P_D R_{\Theta CA} + T_A = 158°C \tag{9.9-5}$$

If the ambient temperature is reduced to 0°C, the derating curve indicates that the maximum allowed power dissipation in the transistor is now 10 W. However, a reapplication of the modeling technique carried out in part (a) sug-

Figure 9.9-3
Example 9.9-1.

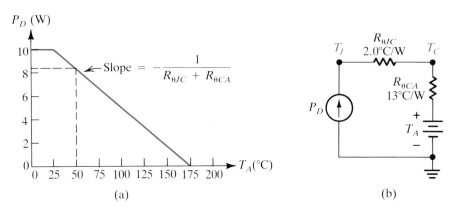

(a)

(b)

gests that the power dissipation can be raised to $(175 - 0)/15$ or 11.7 W. Although it is true that this power level will not exceed the $175°$ T_J limit when $T_A = 0$, it does nonetheless violate the upper bound of 10 W on P_D specified by the derating curve.

When the semiconductor material inside the case of a transistor, integrated circuit, rectifier, or thyristor heats up, this heat may be removed in one of three basic ways: conduction, radiation, or convection. The flow of heat through a material by conduction depends on the physical dimensions of the material and its thermal conductivity K. The equation for the thermal resistance of a conductor is

$$R_\Theta = \frac{L}{KA} \tag{9.9-6}$$

where A is the cross-sectional area and L is its length. It should be immediately apparent that this equation is very similar to that for the electrical resistance of a material. Metals have high thermal conductivities, and therefore they tend to have low thermal resistances, whereas materials such as air have low conductivities and are therefore very poor thermal conductors.

Heat, besides traveling by thermal conduction, can also leave a body by radiation, and the amount of thermal energy radiated depends on its temperature, its emissivity, and its surface area. The emissivity is a measure of how good a radiator the surface is in comparison with that of a blackbody, and the emissivity factor depends on the texture and color of the surface of the radiator. Therefore, heat sinks and semiconductors that are designed to be good radiators most often have unpolished rough surfaces that are black or some other dark color.

Whenever a body is in contact with a fluid (such as air) heat may also be removed by a third process known as convection. In natural convection the fluid near the surface is heated by the body, and this reduces its density so that the heated fluid rises and carries the heat away from the surface with it. The quantity of heat removed by convection depends on the temperature difference between the body and the fluid, and also on the surface area in contact with the fluid. When the situation warrants it, the heat removal by convection can be increased significantly by using forced convection to move the fluid past the surface of the body more rapidly.

In analyzing the heat flow out of a transistor or other semiconductor device, it is useful to observe that most of the flow between the semiconductor material and the case of the device occurs by thermal conduction, whereas the heat transfer from the case to the environment is mainly due to convection and radiation.

Small transistors have sufficiently large surface areas in comparison to the power being dissipated in them that they can usually operate without heat sinks. However, most larger power transistors do not have sufficient surface area to properly carry away the power dissipated in them without overheating. For these devices a heat sink is usually required. Basically, a heat sink is nothing more than a large piece of thermally conducting material that is attached to the semiconductor device to increase its effective surface area and

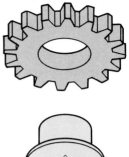

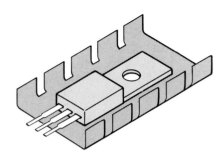

(b)   Heat Sink for a T0-220
      Case Style
      ($R_{\theta CA}$ = 17°C/W)

(a)   Push-On Fin-Type Heat Sink
      for a T0-5 Case
      ($R_{\theta CA}$ = 48°C/W)

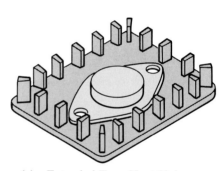

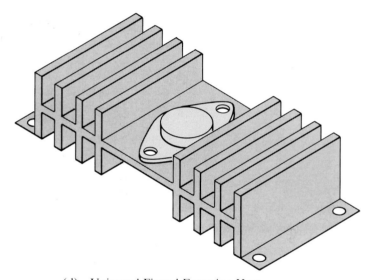

(c)   Extruded-Type Heat Sink
      for a T0-3 Case Style
      ($R_{\theta CA}$ = 7.5°C/W)

(d)   Universal Finned Extrusion Heat
      Sink for All Larger Case Styles
      (T0-3, T0-66, etc.)
      ($R_{\theta CA}$ = 1.8°C/W)

Figure 9.9-4
Examples of typical heat sink designs.

help it to better get rid of the heat that it produces by radiating and convecting it away to its surroundings. Figure 9.9-4 illustrates several different types of heat sinks. In general, they are all made from metals such as copper, aluminum, or magnesium that are excellent thermal conductors. They are almost always unpolished and anodized or painted with a dark color such as black in order to increase their radiation efficiency. In addition, most of them have fin structures to maximize their surface area for a given heat sink volume.

The fact that heat sinks are usually made of metal creates an electrical insulation problem since the collectors of most BJTs, for example, are connected directly to the case of the device, and this collector is often at a high electrical potential with respect to ground. This can cause a serious shock hazard when the heat sink is attached to the transistor, and in addition, it also means that the heat sink cannot be mounted directly on the system chassis since the chassis is usually held at ground potential. To get around both of these difficulties, it is common practice to electrically insulate the semiconductor device from the heat sink with a thin insulating washer. However, because all the heat flow from the semiconductor must pass through this washer, it is necessary that the material from which it is made, while being an electrical insulator, must still be a good thermal conductor. A material known as heat sink grease is also usually applied to both sides of the insulating washer to maximize the thermal contact between the surfaces and thereby ensure the optimal heat flow between the semiconductor and the heat sink.

The thermal equivalent circuit for a heat sink assembly containing an insulated washer is drawn in Figure 9.9-5a, and comparing it with the original free air thermal circuit given in Figure 9.9-2b, we can see that both circuits are essentially the same except that the single thermal resistance $R_{\theta CA}$ in Figure 9.9-2 has been replaced by two resistances, $R_{\theta CS}$ and $R_{\theta SA}$. In this new equivalent circuit $R_{\theta CS}$ represents the thermal resistance between the case and the heat sink. Figure 9.9-5b gives typical values for this component of the thermal resistance for situations where the semiconductor is in direct contact with the heat sink as well as when an electrically insulating washer is employed.

By examining the data sheets for power semiconductor devices, we find that the typical values for $R_{\theta CA}$ when the device radiates its heat directly into free space are on the order of 60°C/W for the TO-220 case style and about 30°C/W for TO-3-style devices. Comparing these results with those obtained when a heat sink is added (see, for example, the thermal resistance values for the heat sinks in Figure 9.9-4), we see that $R_{\theta CA}$ can be reduced considerably by the use of a heat sink. This means that a semiconductor connected to a heat sink can remove the heat that it produces much more efficiently than if it radiates directly into free space.

In terms of the equivalent circuit in Figure 9.9-5a, a lower effective value for $R_{\theta CA}$ means that for a given upper limit on T_J more power can be dissipated in the device before this limit is reached. The example that follows will help to illustrate this point.

Figure 9.9-5

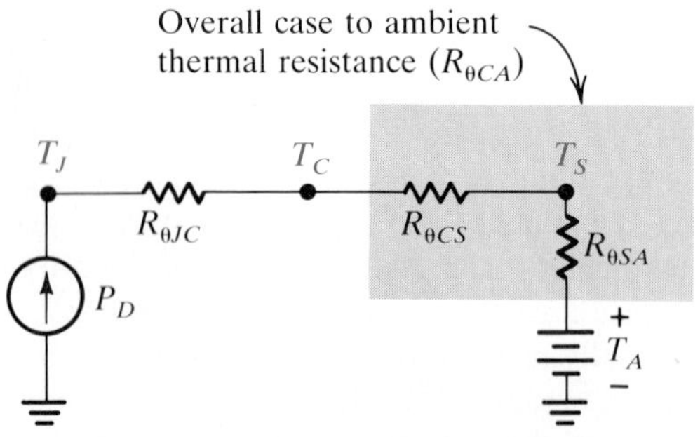

	T0-220 Case Style	T0-3 Case Style
No insulator, no grease	1.3	0.55
No insulator with thermal grease	1.0	0.15
0.003-in. mica insulator, no thermal grease	3.5	1.3
0.003-in mica insulator with thermal grease	1.75	0.35

(a) Equivalent Circuit for a Semiconductor Device Attached to a Heat Sink

(b) Effect of Thermal Grease and Insulators on the $R_{\theta CS}$ Resistance for Selected Case Styles (°C/W)

EXAMPLE 9.9-2

A TO-3-style power field-effect transistor has a junction-to-case thermal resistance of 3.125°C/W, a $T_{J,\max} = 150°C$, and a $P_{D,\max} = 40$ W when its case temperature is 25°C. The circuit is required to operate satisfactorily in ambient temperatures from -10 to 50°C.

a. If the free air thermal resistance, $R_{\Theta CA}$, is 40°C/W for this transistor, determine the maximum power that it can dissipate with no heat sink attached.

b. Using the data in Figure 9.9-5b, select a heat sink that will permit this transistor to safely dissipate 20 W of power if an insulating washer is to be used between the transistor and the heat sink.

c. If two of these transistors are mounted on the heat sink selected in part (b), how much power can each transistor now safely dissipate?

SOLUTION

When the transistor radiates its heat directly into the surrounding environment, the thermal circuit in Figure 9.9-6a applies. Here we have set $T_A = 50°C$ because this ambient temperature will cause $T_{J,\max}$ to occur at the lowest value of P_D. Following this figure, if $T_{J,\max} = 150°C$, the largest allowed power dissipation in this transistor will be

Figure 9.9-6
Example 9.9-2.

$$P_D = \frac{T_J - T_A}{R_{\Theta JC} + R_{\Theta CA}} = \frac{(150 - 50)°C}{43.15°C/W} = 2.3 \text{ W} \qquad (9.9\text{-}7)$$

For this transistor to be able to dissipate any more power than 2.3 W, it will be necessary to add a heat sink. The equivalent circuit with the heat sink and electrical insulating washer in place is illustrated in Figure 9.9-6b for the case where P_D is to be 20 W. Here to make $R_{\Theta CS}$ as small as possible, we have

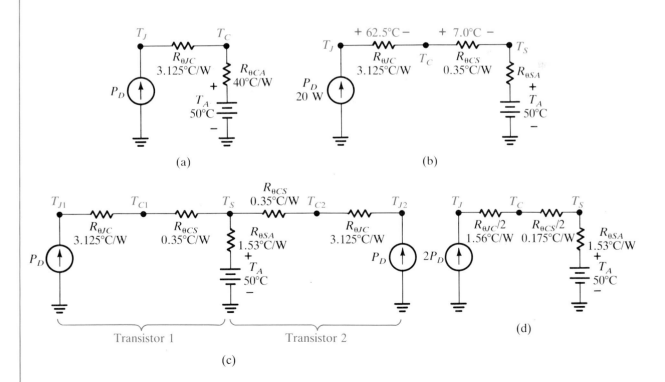

(a)   (b)

(c)

Transistor 1   Transistor 2

(d)

assumed that thermal grease has been used on the insulating washer between the case and the heat sink. In addition, as in part (a), we are using the upper limit on T_A as the worst-case condition for the ambient temperature. Following this circuit, with 20 W of power flowing out of the transistor, we can see that there is a temperature drop of 62.5°C across $R_{\Theta JC}$ and a 7.0°C drop across $R_{\Theta CS}$. If T_J is to be less than 150°C, this means that the temperature drop across the heat sink must be less than $(150 - 62.5 - 7 - 50) = 30.5$°C, so that the maximum allowed thermal resistance of the heat sink is $R_{\Theta SA} = 30.5/20 = 1.53$°C/W.

If two of these transistors are mounted on the same heat sink, and if we assume that the circuit is linear, the equivalent circuit given in Figure 9.9-6c applies. Because the circuit is symmetrical and $T_{J1} = T_{J2}$, these two points may be connected together, and the simplified equivalent thermal circuit in Figure 9.9-6d results. Thus to keep T_J below 150°C, the maximum allowed total power dissipation in both transistors is

$$2P_D = \frac{(150 - 50)°\text{C}}{1.56 + 0.175 + 1.53} = 30.62 \text{ W} \tag{9.9-8}$$

or about 15.3 W per transistor.

Exercises

9.9-1 A small 1-W power transistor has an $R_{\Theta JC} = 10$°C/W and an $R_{\Theta CA} = 20$°C/W. Find the maximum allowed power dissipation in the transistor if $T_{J,\max} = 175$°C and $T_A = 75$°C. *Answer* 1 W

9.9-2 A transistor having the derating curve shown in Figure 9.9-2c with $T_{J,\max} = 175$°C is required to dissipate 40 W of power when $T_A = 50$°C. Find the largest allowed heat sink resistance assuming that a "thermal short" exists between the transistor case and the heat sink. *Answer* 2.8°C/W

9.9-3 Two TO-3-style transistors Q_1 and Q_2 are thermally mounted on the same heat sink. The thermal resistance of the heat sink is 2°C/W. Q_1 is directly mounted on the heat sink and Q_2 uses an ungreased 0.003-in. mica insulator. In addition, transistor Q_1 dissipates 10 W and Q_2, 15 W. Find the case temperature of each transistor if $T_A = 25$°C. *Answer* $T_{C1} = 80.5$°C, $T_{C2} = 94.5$°C

PROBLEMS

For Problems 9.1-1 through 9.1-4, a switched power source v_{in} is applied to a 50-Ω load R_L.

9.1-1 Determine the average power delivered to the load if v_{in} is a 120-V(rms) sine wave of period T and the switch controller:
(a) Closes the switch continuously.
(b) Closes the switch on the negative half-cycles of v_{in}.
(c) Closes the switch from $t = T/4$ to $T/2$ and from $3T/4$ to T during each cycle.

9.1-2 The input signal v_{in} is a 120-V(rms) sine wave and the switch opens and closes for integral numbers of cycles. If the switch is closed for m cycles and opened for n cycles, determine the smallest allowed values of m and n for which the average power delivered to the load is 45% (± 2%) of the maximum power that would be delivered to R_L with the switch closed continuously.

9.1-3 Determine the average power delivered to the load if v_{in} is a 100-V peak amplitude triangle wave of period T that is zero at $t = 0$ and if:

(a) The switch is closed continuously.

(b) The switch is closed from 0 to $T/6$ and from $5T/6$ to T during each cycle.

9.1-4 Repeat Problem 9.1-3 for a 100-V peak amplitude ramp signal of period T passing through zero at $t = T/2$.

9.1-5 The circuit shown in Figure 9.1-2e is a dc-to-dc converter which is 70% efficient. $V_1 = 20$ V, $V_2 = 10$ V, and a 5-Ω load is connected across the V_2 terminals. Find P_L, the average input current, P_{IN} from V_1, and the power dissipated in the converter.

9.2-1 A power transistor has maximum power, current, and voltage ratings of 20 W, 2 A, and 100 V, respectively.

(a) Carefully sketch and label the safe operating area (SOA) for this transistor using linear scales for the voltage and current axes.

(b) Repeat part (a) if logarithmic scales are used for the axes.

9.2-2 A grounded emitter BJT has a collector resistance of 10 kΩ that is connected to a power supply V_{CC}. The base of the transistor is driven by a square-wave current source whose amplitude varies between 0 and 100 μA. Find the maximum allowed value of V_{CC} if breakdown of the collector-to-base junction is to be avoided. The transistor characteristics are $I_{CBO} = 10$ μA, $h_{FE} = 100$, $V_{CBO} = 80$ V, and $n = 2.5$ [in equation (9.2-5) for the breakdown voltage].

9.2-3 The power FET in the circuit shown in Figure P9.2-3a has the graphical characteristics shown in part (b) of the figure. In addition, $I_{D,\max} = 5$ A, $V_{DS,\max} = 40$ V, and $P_{D,\max} = 10$ W.

(a) Sketch the dc load line on the volt-ampere characteristics.

(b) Find the power dissipated in the transistor and the load for $v_{in} = 0$ V, 6 V, and 10 V.

(c) Is it possible to damage the transistor using values of v_{in} anywhere between 0 and 10 V? Explain your answer.

9.2-4 For the circuit shown in Figure P9.2-3a a capacitor $C_L = 1$ μF is connected in parallel with R_L. In addition, the FET has the same characteristics as the transistor used in Problem 9.2-3.

(a) Sketch i_D and v_{DS} versus time if v_{in} is a 50-μs wide pulse whose amplitude starts at 0 V, goes to 10 V at $t = 0$, and then back to 0 V at $t = 50$ μs.

(b) Using the results in part (a), sketch the transistor's dynamic operating path (i_D versus v_{DS}) on the volt-ampere characteristics. (*Note:* It is not a straight line.)

(c) Is it possible for the transistor to be damaged in this circuit? Explain your answer.

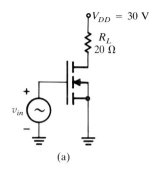

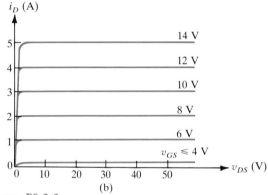

Figure P9.2-3

9.2-5 Repeat Problem 9.2-4 if the capacitor is shunted across the transistor instead of across R_L.

9.2-6 Repeat Problem 9.2-4 if a 300-μH inductor having a 10-Ω winding resistance is shunted across R_L in place of C_L. In part (a) of the solution, include a sketch of the inductor current.

9.3-1 Prove that a load line drawn tangent to the power dissipation hyperbola is bisected at the point of tangency.

9.3-2 For the circuit shown in Figure P9.3-2, the transistor has an $h_{ie} = 10$ Ω and an $h_{fe} = 100$. The input voltage $v_{in} = 10 \sin \omega t$ mV.

(a) Carefully sketch and label i_B, i_C, and v_{CE}.

(b) Determine the ac power delivered to the load.

(c) Find the power delivered by the supply.

(d) Find the power dissipated in the transistor.

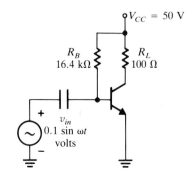

Figure P9.3-2

9.3-3 Repeat Problem 9.3-2 if R_B is reduced to 6.43 kΩ and R_L is moved from the collector to the emitter. The input voltage $v_{in} = 15 \sin \omega t$ volts.

9.3-4 For the circuit shown in Figure P9.3-2, V_{CC} is changed to 30 V and R_B and R_L are made adjustable.
(a) Given that $P_{D,max} = 15$ W, $h_{fe} = 100$, and $h_{ie} = 30 \ \Omega$, select R_L and R_B to deliver maximum power to the load, R_L.
(b) What is the value of v_{in} that is required to deliver this power to the load?

9.3-5 For the circuit shown in Figure P9.3-2, $R_L = 150 \ \Omega$, $h_{ie} = 100 \ \Omega$, $h_{fe} = 100$, and R_B is adjusted so that $I_C = 100$ mA. Determine the average power delivered to the load if v_{in} is a 200 mV(p-p) amplitude sawtooth waveshape.

9.3-6 For the circuit in Figure P9.3-2, $V_{CC} = 30$ V, $R_B = 3$ kΩ, $R_L = 10 \ \Omega$, $h_{ie} = 50 \ \Omega$, and $h_{fe} = 100$.
(a) Sketch the total base current, collector current, and collector to emitter voltage if v_{in} is a 250-mV(p-p) amplitude square wave.
(b) Find the average power dissipated in the transistor.

9.3-7 The power MOSFET shown in Figure P9.3-7 has a $V_T = 4$ V and a constant g_m equal to 1 S.
(a) Sketch v_{GS}, i_D, and v_{DS}.
(b) Find the power delivered by V_{DD}.
(c) Determine the circuit efficiency.

9.3-8 **(a)** Starting with equation (9.3-24) and letting $i_B = I_B + I_{bm} \sin \omega t + I_{b2} \cos 2\omega t$ show by equating the $\cos 2\omega t$ terms that for the purpose of making second harmonic calculations, the BJT base–emitter junction may be modeled as a resistor h_{ie} in series with a controlled voltage source $(V_{bem})^2/4V_T$.
(b) Using the model developed in part (a), verify the correctness of equation (9.3-26).
(c) Estimate the % 2HD for the amplifier in Problem 9.3-2 if a 2-Ω resistor is connected from the emitter lead to ground and v_{in} is increased to 200 $\sin \omega t$ mV.

9.3-9 Neglecting the base–emitter nonlinearity but considering that h_{FE} is nonlinear, and in particular that

$$i_C = 100 i_B + 5 i_B^2$$

where i_B is in milliamperes, repeat Problem 9.3-8c for the circuit in Figure P9.3-2.

9.3-10 Estimate the % 2HD at R_L in Problem 9.3-7 if i_D is related to v_{GS} by the equation

$$i_D = 1(v_{GS} - 4) + 0.5(v_{GS} - 4)^2 \qquad \text{for } v_{GS} \geq 4 \text{ V}$$

and is zero for $v_{GS} < 4$ V.

9.3-11 For the circuit shown in Figure 9.3-6, $V_{DD} = 30$ V, $N = 2$, $R_1 = 24$ kΩ, $R_2 = 6$ kΩ, and $R_L = 4 \ \Omega$. In addition, a bypassed 2-Ω resistor is inserted between the source lead and ground.
(a) Sketch and carefully label v_{GS}, i_D, v_S, v_D, and v_L given that $V_T = 4$ V, $g_m = 1$ S, and $v_{in} = 0.5 \sin \omega t$ volts.
(b) Find P_L, P_D, and P_{in}.

9.3-12 For the circuit shown in Figure 9.3-4a, $R_L = 10 \ \Omega$, $V_{CC} = 20$ V, and the transistor is silicon with $h_{fe} = 100$, $h_{ie} = 1$ kΩ, $1/h_{oe} = 1$ kΩ, and $P_{D,max} = 20$ W. Assuming class A operation, find the turns ratio required to deliver maximum power to the load.

9.3-13 Assuming a 5:1 turns ratio for the circuit described in Problem 9.3-12 and letting $R_B = 10$ kΩ, what is the maximum undistorted power that can be delivered to the load considering that $1/h_{oe} = \infty$?

9.4-1 The circuit shown in Figure P9.4-1 is a class B power amplifier. Consider that all transistors have h_{fe} values equal to 100, and that $h_{ie1} = 1$ kΩ and $h_{ie2} = h_{ie3} = 100 \ \Omega$.
(a) Select R_1 and R_2 to place Q_2 and Q_3 at the edge of turning on ($V_{BE2} = V_{BE3} = 0.6$ V). You may assume at this

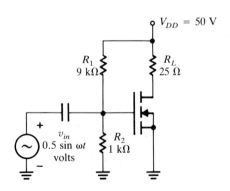

Figure P9.3-7

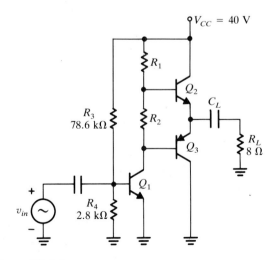

Figure P9.4-1

point that $I_{B2} = I_{B3} \simeq 0$, and that R_1 and R_2 are to be chosen so that $|V_{CE2}| = |V_{CE3}|$.

(b) Choose C_L for audio operation.

(c) Find the power delivered to the load when $v_{in} = 0.36$ sin ωt volts.

9.4-2 For Problem 9.4-1, consider that $R_1 = 1.3$ kΩ, $R_2 = 66$ Ω, and $R_3 = 98.25$ kΩ and calculate the maximum undistorted power that can be delivered to the load if v_{in} is a sinusoidal signal.

9.4-3 For the circuit shown in Figure 9.4-11, R_f is changed to 9 kΩ and connected to the load, $R_L = 4$ Ω, Q_1 and Q_2 are silicon transistors with $h_{ie} = 10$ Ω and $h_{fe} = 100$, and v_{in} is a 2-V(p-p) square wave.

(a) Sketch v_L, i_L, v_A, i_A, and v_{CE1}.

(b) Determine the circuit efficiency and the power dissipated in Q_1.

9.4-4 For the class B amplifier given in Figure 9.4-5a, neglect the 0.7-V drops across the base–emitter diodes and consider that $h_{ie} = 0$ and that $h_{fe} = 50$ for both transistors. Determine the circuit efficiency if $v_{in} = 5$ sin ωt volts, $R_L = 20$ Ω, and $V_{CC} = 10$ V.

9.4-5 Repeat Example 9.4-1 if a common-emitter push-pull circuit of the type illustrated in Figure 9.4-1b is used in place of the common-collector circuit. Include the sketches for the transistor base currents i_{B1} and i_{B2} for the case of maximum power delivery to the load. The transistors may be assumed to be identical to those used in Example 9.4-1.

9.4-6 For the circuit shown in Figure P9.4-6:

(a) Select R_1 so that the diodes remain ON for all values of v_{in} when v_L has a 25-V peak amplitude.

(b) If the diode is an ideal silicon device and if the transistors have the base–emitter characteristics shown, estimate the % 3HD when the output amplitude is 25 V peak.

(c) Select C for audio operation of the amplifier.

9.4-7 For the circuit shown in Figure P9.4-7, the characteristics of Q_3 and Q_4 are $|V_T| = 4$ V and $g_m = 1$ S.

(a) Select R_1 to eliminate the crossover distortion in the output.

(b) Find the amplitude of the sine wave at v_{in} needed to deliver 40 W(rms) to R_L.

(c) Sketch v_L, i_L, i_{D3}, i_{D4}, v_{GS3}, v_{GS4}, v_{G3}, and v_{G4} for the value of v_{in} selected in part (b).

9.4-8 For $R_1 = 350$ Ω in Problem 9.4-7, estimate the % 2HD and % 3HD for an input sine-wave amplitude of 4 V.

9.4-9 Repeat Problem 9.4-8 for $R_1 = 300$ Ω, $g_{m1} = 1$ S and $g_{m2} = 1.2$ S.

9.4-10 (a) Using the same technique as that presented in Problem 9.3-8(a), develop a third harmonic distortion model for the BJT push-pull amplifier, starting with equation (9.3-21) with $v_{in} = V_{bem}$ sin ωt, but this time

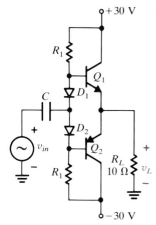

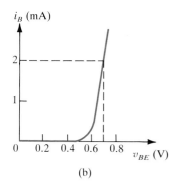

$h_{FE} = 100$ for Q_1 and Q_2

(a)

(b)

Figure P9.4-6

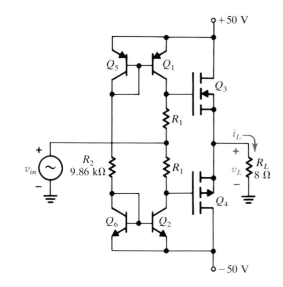

Figure P9.4-7

neglecting the quadratic term since it will cancel in the load current anyway. In particular, you are to show that this model consists of a resistor h_{ie} in series with a controlled voltage source $V_{bem}^3/24V_T^2$.

(b) In Figure 9.4-1b two resistors R_B are added in series with the base leads of Q_1 and Q_2. Using the model developed in part (a) of this problem, prove that the % 3HD from this circuit is now given by equation (9.4-38) with $R_L = 0$.

9.4-11 Using the model developed in Problem 9.4-10(a), prove that the % 3HD for the push-pull amplifier in Figure 9.4-1a is given by equation (9.4-38) with $R_B = 0$.

9.5-1 The power supply shown in Figure P9.5-1 is a shunt regulator and it is to be designed to deliver a maximum current of 1 A at 5 V to the load. The diode and transistor characteristics are $I_{ZK} = 10$ mA and $r_Z = 10\ \Omega$ for D_1, and $h_{fe} = 100$ and $h_{ie} = 20\ \Omega$ for Q_1.

(a) What zener voltage is needed?

(b) Determine the largest value of R allowed for proper circuit operation if v_{UR} varies from 15 V to 20 V at full load.

(c) Specify the transistor power rating.

(d) Calculate the output ripple voltage under full-load conditions.

9.5-2 For the power supply circuit shown in Figure 9.5-3, $V_{UR} = 25$ V, $R_1 = 400\ \Omega$, $R_A = 2$ kΩ, $R_B = 1$ kΩ, and $V_{ZZ} = 4$ V for D_1. In addition, for Q_2, $h_{fe} = 100$ and $h_{ie} = 1$ kΩ, and Q_1 is replaced by a BJT conventional device with $h_{fe} = 50$ and $h_{ie} = 20\ \Omega$.

(a) Accurately determine V_o in terms of V_{UR} and V_{REF} with R_L removed.

(b) Compare the result in part (a) with what you would expect on an intuitive basis.

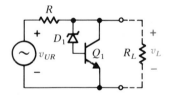

Figure P9.5-1

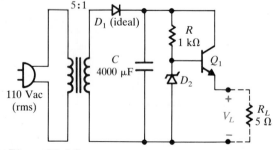

Figure P9.5-3

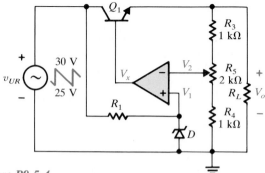

Figure P9.5-4

9.5-3 For the circuit shown in Figure P9.5-3, $h_{fe} = 100$ for Q_1, and $V_{ZZ} = 5.7$ V and $r_Z = 10\ \Omega$ for D_2. Estimate the dc and the ripple voltage across the load.

9.5-4 The circuit shown in Figure P9.5-4 is a variable-voltage, 0- to 1-A power supply. For this circuit the open-loop gain of the op amp is 10^4, the transistor Q_1 has an $h_{fe} = 100$ and an $h_{ie} = 100\ \Omega$, and $V_{ZZ} = 5$ V, $I_{ZK} = 10$ mA and $r_Z = 10\ \Omega$ for the zener diode. In addition, v_{UR} varies between 25 and 30 V.

(a) Find the required value of R_1.

(b) Find the range of voltages available at the output as R_5 is adjusted.

(c) Determine the power-handling requirements of D and Q_1.

(d) What is the circuit load regulation with R_5 set at the bottom if the open-loop output impedance of the op amp is 100 Ω?

9.5-5 For the circuit designed in part (a) of Example 9.5-1:

(a) Determine the load regulation if the zener resistance is 10 Ω and the transistor h_{ie} is constant at 100 Ω.

(b) Find the circuit's line regulation, and sketch v_L under full-load conditions.

9.5-6 (a) The circuit shown in Figure P9.5-6 is known as a dual-polarity-tracking voltage regulator. Develop expressions for V_{o1} and V_{o2} in terms of R_1 through R_4 and V_{ZZ}, the zener voltage of D_1.

(b) What are V_{o1} and V_{o2} if $V_{ZZ} = 5$ V at 25°C?

(c) What are V_{o1} and V_{o2} at 75°C if the temperature coefficient of the zener is +5 mV/°C, and those of the base–emitter junctions of Q_1 and Q_2 are −2 mV/°C?

9.5-7 For the power supply designed in Example 9.5-2:

(a) Determine the exact output voltage of this circuit.

(b) Find the load and line regulation.

9.5-8 For the circuit shown in Figure P9.5-8, D_1 is a 5-V zener diode and the transistor has an $h_{FE} = 100$. Find the range of values of R_L for which this circuit operates like a constant-current source.

9.5-9 For the regulated power supply shown in Figure P9.5-9, $h_{FE1} = 20$ and $h_{FE2} = 100$.

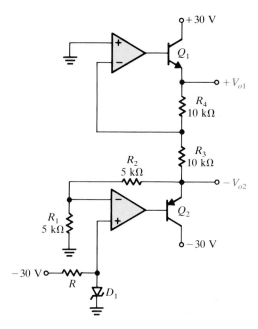

Figure P9.5-6

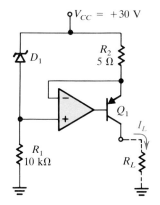

Figure P9.5-8

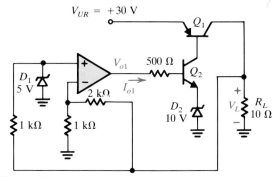

Figure P9.5-9

(a) Find V_L, V_{o1}, and I_{o1}.

(b) Determine the power dissipated in Q_1 and Q_2.

9.6-1 Using the circuit shown in Figure 9.6-1a with the switch replaced by a BJT (emitter on the right), design a continuous-mode BJT-style 5-V step-down switching regulator power supply to drive a 2-Ω load. The unregulated input voltage is 15 V and the switching frequency is to be 20 kHz.

(a) What is the required pulse width of the switching signal?

(b) Select L if the ripple current in the inductor is to be less than 10% of the dc load current.

(c) The switching oscillator drives the base of the transistor between 0 and V_p volts through a 100 Ω resistor. What is the minimum amplitude of the switching signal that will guarantee saturation of the switching transistor if $h_{FE} = 100$?

(d) Select C so that the output ripple voltage is less than 50 mV.

9.6-2 The constant drive signal of Problem 9.6-1 is replaced by the pulse-width modulator circuit shown in Figure P9.6-2, and $L = 10$ mH and has a dc resistance of 0.5 Ω. The modulator frequency is constant at 20 kHz and the pulse width T_1 is related to V_x by the expression

$$T_1 = 50\left(1 - \frac{V_x}{2}\right) \quad \mu s$$

(a) Select R_1 so that $V_L = 5$ V when $R_L = 5$ Ω.

(b) Estimate the output ripple for the value of R_1 chosen in part (a).

(c) Determine V_L when R_L changes to 6 Ω.

9.6-3 Repeat Problem 9.6-2 if the wire entering the pulse-width modulator is broken and the circuit shown in Figure P9.6-3 is inserted.

9.6-4 For the circuit in Figure 9.6-4, v_{in} is removed and replaced by a 20-kHz 0- to 15-V amplitude pulse width modulator that is connected between the v_L and v_{in} terminals. The output pulse width T_1 from the modulator is given by the expression

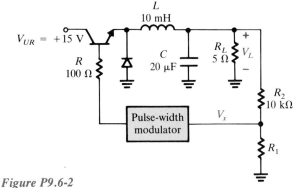

Figure P9.6-2

Figure P9.6-3

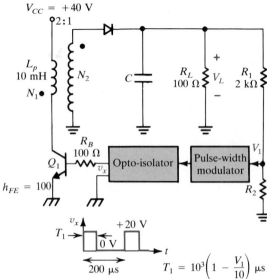

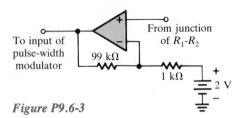

Figure P9.6-5

$$T_1 = 10^3\left(1 - \frac{v_L}{50}\right) \quad \mu s$$

where v_L is in volts. Furthermore, for this circuit $V_{CC} = 12$ V, $R_B = 500\ \Omega$, $L = 10$ mH, $h_{FE} = 100$, and $C = 50\ \mu$F.

(a) Sketch the steady-state voltages and currents i_B, i_C, v_{CE}, and v_L.

(b) Estimate the circuit efficiency given that the transistor and the diode are both silicon devices.

9.6-5 The circuit shown in Figure P9.6-5 is known as a forward converter. The opto-isolator provides electrical isolation between the secondary and the line, and the waveshape v_G has the form shown, where

$$T_1 = 10^3\left(1 - \frac{V_L}{50}\right) \quad \mu s$$

The transformer is ideal, as is the FET, which turns on fully for $v_G = 15$ V and off for $v_G = 0$.

(a) Find V_L considering that the diodes are ideal and that C_1 and C_2 are ac shorts.

(b) Sketch the steady-state current in the inductor and in the FET.

9.6-6 For the circuit shown in Figure 9.6-6, $V_{UR} = 30$ V, $N = 2$, $L_p = 5$ mH, $R_L = 1$ kΩ, and the switch operates at a 4-kHz rate and is closed for 50 μs during each cycle.

$V_{CC} = +40$ V

Figure P9.6-7

(a) Find v_L and sketch the primary current assuming that C is an ac short.

(b) An additional secondary circuit is added that is identical to the first but with $R_{L2} = 2$ kΩ and with half the number of turns on the secondary. Repeat part (a) for this circuit and find the output voltage on this secondary.

9.6-7 For the circuit shown in Figure P9.6-7:

(a) Select R_2 so that $V_L = 20$ V assuming that Q_1 saturates when $v_x = +20$ V and that the energy stored in the transformer goes to zero during each cycle.

(b) Sketch v_x and the transformer primary and secondary currents for the conditions in part (a).

(c) Repeat part (b) if V_{CC} drops to $+30$ V.

9.6-8 For the circuit discussed in Example 9.6-1:

(a) Demonstrate that the transistor is saturated all during the time that v_{in} is high, and find the required value of V_{CEO}.

(b) If the transistor saturation voltage and the diode forward-bias voltages are 0.2 V and 0.7 V, respectively, estimate the efficiency of this power supply.

9.6-9 For the MOS flyback-style power supply circuit examined in Example 9.6-2:

(a) Estimate the output ripple produced if the unregulated input voltage ripples between 14 and 15 V and if C is ineffective at the line frequency.

(b) Calculate the circuit's load regulation for load resistances from 5 Ω and infinity, and by doing so, estimate the output impedance of the supply.

9.7-1 Using a 10-V zener, design a 15-V regulated power supply having the basic form shown in Figure 9.7-2 that will limit the load current to a maximum value of 2 A. In

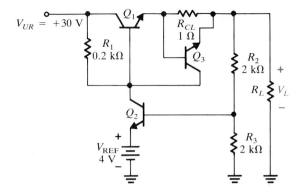

Figure P9.7-2

particular, sketch the required circuit diagram and determine the power-handling capability of the pass transistor. Assume that $V_{UR} = 30$ V and that $I_{ZK} = 5$ mA for the zener. The transistor h_{FE} values are all 50.

9.7-2 For the circuit shown in Figure P9.7-2, $V_o = 0.6$ V and $h_{ie} = 100$ Ω for Q_3.
(a) Find V_L for $R_L = 20$ Ω.
(b) Determine the approximate value of R_L at which current limiting begins.
(c) Find all circuit voltages and currents when $R_L = 0$.

9.7-3 For the regulated power supply illustrated in Figure 9.7-2, consider that the op amp current limits at 20 mA, and that Q_1 and Q_2 both have h_{FE} values equal to 100. $V_o = 0.65$ V and h_{ie} is constant at 1 kΩ for Q_2.
(a) Find the load current (and the corresponding value of R_L at which current limiting occurs.
(b) Determine the load current at which Q_2 just begins to enter the active region.
(c) Derive an expression for the power dissipated in the pass transistor in the constant voltage and constant current regions, and sketch P_D as a function of R_L.

9.7-4 For the foldback-current-limiting characteristic shown in Figure 9.7-4b, prove that the maximum power dissipation in the pass transistor in part (a) of the figure is given by the expression

$$P_{D,\,max} = \frac{(V_{UR} + R_m I_{SC})^2}{4R_m}$$

where R_m is equal to the slope of the current-limit curve in the foldback region.

9.7-5 (a) For the switching regulator power supply in Example 9.7-1, find T_1 under short-circuit conditions. Assume that the diode is ideal and also that C is large enough that the ripple may still be neglected.
(b) Sketch T_1 and P_{IN} as a function of I_L (the load current).

9.7-6 For the foldback-current-limiting circuit of Figure 9.7-4b, the unregulated input voltage is 20 V, the desired

output voltage is $V_L = 15$ V, and $P_{D,\,max}$ for Q_1 is 20 W.
(a) Use the expression given in Problem 9.7-4 to determine the largest allowed values of I_{SC} and I_{FB} for this supply.
(b) The values of I_{FB} and I_{SC} determined in part (a) lead to a foldback characteristic that is not much better than conventional constant-current limiting. By making I_{SC} one-fourth of the value computed in part (a), find the corresponding maximum allowed I_{FB} and sketch the resulting foldback characteristic.
(c) Design a circuit of the form illustrated in Figure 9.7-4b to achieve the foldback limit values specified in part (b).

9.8-1 By breaking the loop at the base of Q_1 in Figure 9.8-1b and injecting a current I at this point, show that the loop gain of this four-layer diode just exceeds 1 when $h_{FE1} h_{FE2} = 1$.

9.8-2 For the four-layer diode shown in Figure P9.8-2, $V_{BO} = 50$ V, $I_H = 200$ μA, $V_T = 1$ V, and the ON resistance of the diode is 100 Ω. Discuss how this circuit functions as an oscillator, and in doing so, sketch v_D, i_R, and i_D.

9.8-3 For the circuit shown in Figure 9.8-6, D_1 is replaced by a short, $R_G = 75$ kΩ, and the SCR is replaced by a triac.
(a) Sketch i_G, v_D, and v_L given that $I_{GT} = 1$ mA.
(b) Determine the average power delivered to the load.

9.8-4 The circuit shown in Figure P9.8-4 is an SCR time-delay power switch. If the SCR is initially off, and the switch is closed at $t = 0$, sketch v_C, v_G, v_L, and v_D.

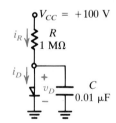

Figure P9.8-2

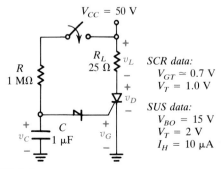

Figure P9.8-4

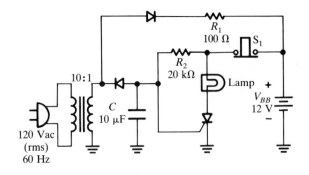

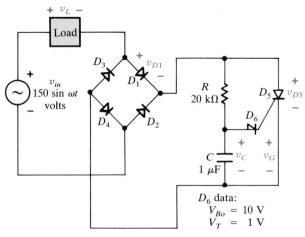

Figure P9.8-5

Figure P9.8-8

9.8-5 The circuit shown in Figure P9.8-5 is an emergency lighting system. As long as the ac power is available, the lamp remains off and when the power fails the lamp turns on and is powered by the battery.

(a) Explain why the SCR remains off as long as the ac power is present by estimating v_G under these conditions.

(b) Sketch the battery charging current for the conditions shown.

(c) Assuming that I_{GT} for the SCR is 100 μA, explain why the SCR turns on when the ac power fails.

(d) What is the function of the switch S_1?

9.8-6 For the circuit in Figure 9.8-8, replace D_1 by a short and let v_{in} be a 60-Hz 120-V ac (rms) sine wave, $R_L = 100\ \Omega$, and $R_1 = 470\ k\Omega$. Given that I_{GT} for the SCR is 20 μA, sketch and carefully label v_D, i_G, and v_G.

9.8-7 The circuit illustrated in Figure 9.8-9a employs a triac for full-wave ac power control. The triac has symmetrical trigger characteristics and the maximum gate current required to trigger it into the ON state is 50 mA. In addition, V_T for the triac is 1 V and its gate resistance is 50 Ω. The diac characteristics are illustrated in Figure 9.8-9b and the diac ON resistance is equal to 100 Ω. The triac and diac holding currents are both 20 μA.

(a) Explain how this circuit operates, and in doing so, sketch v_D, v_C, i_G, v_G and v_L. As a rough approximation, consider that the voltage, v_C, across the capacitor remains relatively constant during the time that the triac is ON.

(b) Develop an expression for the average power delivered to the load as a function of the triac conduction angle.

(c) Use the result in part (b) to compute the average power delivered to the load for the circuit in Figure 9.8-9a.

9.8-8 The bridge diodes in Figure P9.8-8 are ideal, and $I_{GT} = 0.1$ mA for the SCR. In addition, its gate resistance is 100 ohms and $V_{GT} \cong 0.7$ V when it is on. Sketch v_{in}, v_L, v_{D1}, v_{D5}, v_C, and v_G in the steady-state.

9.9-1 Mica is often used as an electrical insulator between the case of a power semiconductor and the heat sink to which it is attached. Its thermal resistivity (ρ) is about $66°C \cdot in./W$.

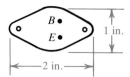

Figure P9.9-1

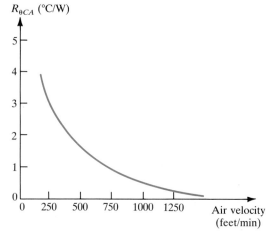

Figure P9.9-2

(a) Estimate the thermal resistance of a 3-mil-thick mica insulator designed for use with the TO-3-style transistor whose case outline is shown in Figure P9.9-1.

(b) If the transistor is mounted onto a heat sink using thermal grease with a conductivity of 0.02 W/°C · in., and if the average grease thickness is 0.3 mil on each side of the insulator, estimate the overall thermal resistance between the case of the semiconductor and the heat sink.

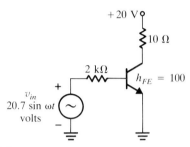

Figure P9.9-3

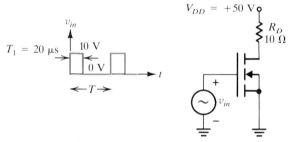

Figure P9.9-4

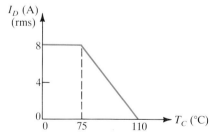

Figure P9.9-6

4.4°C/W. Determine the required air velocity that must be produced by the fan if the transistor junction temperature is to remain below 150°C at an ambient temperature of 25°C.

9.9-3 (a) Estimate the power dissipated in the transistor for the circuit shown in Figure P9.9-3.

(b) Determine the value of the largest heat sink resistance that will permit the circuit to operate properly using the thermal characteristics given in Figure 9.9-2c with $P_{D,\,max} = 10$ W and $T_{J,\,max} = 175$°C.

9.9-4 The power transistor in the circuit shown in Figure P9.9-4 has the following characteristics: $V_T = 3$ V, $g_m = 1$ S, $r_{DS(ON)} = 2\ \Omega$, $T_{J,max} = 175$°C, and $R_{\Theta JA} = 40$°C/W. Find the maximum allowed pulse repetition frequency for which the maximum junction temperature will not be exceeded if $T_A = 40$°C.

9.9-5 The thermal characteristics of SCRs are sometimes given graphically with rms current as the variable. Find the ratio of $I_D(\text{RMS})/I_D(\text{AVG})$ for a half-wave rectified sine wave.

9.9-6 A typical TO-220 case-style triac has an ON-state voltage of 2 V. The maximum allowed rms current is sketched in Figure P9.9-6 as a function of case temperature.

(a) What is $T_{J,max}$ for this device?

(b) Estimate $R_{\Theta JC}$ for this triac.

9.9-7 For the circuit shown in Figure P9.9-7, the resistor R_1 is adjusted so that the conduction angle is 90° on each half-cycle.

(a) Determine the average power dissipated in the triac assuming that the ON-state voltage is 1.5 V.

(b) If $R_{\Theta JC} = 2.2$°C/W, what kind of heat sink will be needed to lower the junction temperature to 125°C? The ambient temperature is 50°C.

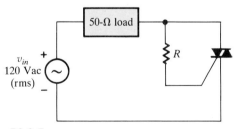

Figure P9.9-7

9.9-2 Forced-air convection (a fan) is used to cool a power transistor that is dissipating 20 W. The characteristics of the heat sink are shown in Figure P9.9-2 and $R_{\Theta JC} =$

9.9-8 For the circuit shown in Figure P9.9-3, v_{in} is set equal to 2.7 V dc and R_C is changed to 50 Ω. Assume that

$I_{CBO} = 20 \ \mu A$ at 25°C, and that it doubles for every 10°C increase in junction temperature.

(a) Sketch I_C versus T_J for junction temperatures from 0 to 200°C.

(b) Sketch P_D versus T_J.

(c) Considering that $R_{\theta JC} = 5°C/W$ and that $R_{\theta CA}$ is 80°C/W, derive a second expression [in addition to the sketch given in part (b)] that relates P_D and T_J. Graphically determine the thermal operating point of the circuit in the steady state (T_J, T_C) and estimate the transistor operating point (I_C, V_{CE}, and P_D).

(d) Repeat part (c) if the transistor is mounted on a heat sink with a thermal resistance of 15°C/W.

9.9-9 For the circuit described in Problem 9.5-4 the transistor thermal parameters are $R_{\theta JC} = 1°C/W$ and $R_{\theta CA} = 8°C/W$.

(a) Estimate the power dissipated in Q_1 if $R_L = 20 \ \Omega$ and the variable resistor R_5 is set at its midpoint.

(b) Find the transistor junction temperature for the conditions in (a) if $T_A = 25°C$.

(c) Determine the thermal resistance of the heatsink needed to lower T_J to 100°C if $T_A = 25°C$.

CHAPTER 10

Digital Electronics

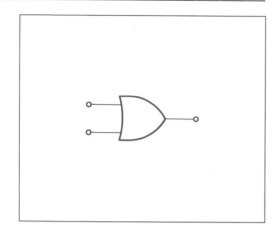

10.1 INTRODUCTION

10.1-1 Fundamentals of Digital Logic Circuit Design

In the analog circuits that we examined in previous chapters, the voltages and currents could take on an infinite number of levels between their positive and negative limits. By way of contrast, in a digital circuit these quantities usually take on only one of two possible values. The use of this binary or two-state technique for the representation of digital data is at first rather surprising since nearly all the mathematics of the modern world is based on the decimal rather than the binary number system. Why, then, don't we use digital circuits in which there are 10 distinct voltage or current levels instead of just two, to correspond to the 10 mathematical symbols 0 through 9 of the decimal number system?

The motivation for taking the binary or two-state approach in earlier electromechanical computers is easy to understand when we recognize that relays are basically binary devices; that is, devices that are either ON when current flows through them, or OFF when their current flow is zero. However, the reason for the overwhelming popularity of binary logic systems in modern electronic digital computers is less apparent.

Why, for example, don't we design digital logic circuits that have 10 rather than two discrete voltage levels? The answer to this question is that competent variabilities make it difficult to maintain the required accuracy. Consider, for example, the inverter shown in Figure 10.1-1a. In theory we could use this circuit as a decimal inverter by dividing the output logic swing into 10 regions,

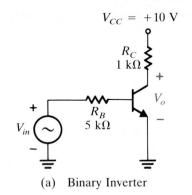

(a)   Binary Inverter

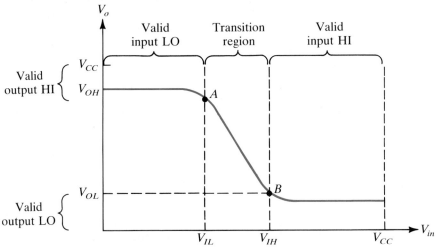

(b)   Actual Transfer Characteristic of a Typical Digital Logic Gate

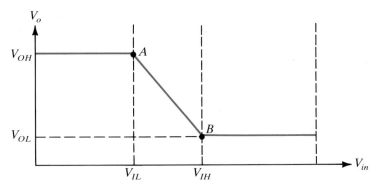

Figure 10.1-1

(c)   Transfer Characteristic Obtained for the Same Logic Gate as in (b) Using Piecewise Linear Models for the Transistors and Diodes

each 1 V wide, to represent the 10 different logic levels 0 through 9, and we could similarly associate specific values of V_{in} with each of these output levels. However, this V_{in}–V_o relationship would be a strong function of the transistor h_{FE} and the resistor tolerances, and would therefore be of little practical utility.

The reliability of this inverter circuit can be improved significantly by reducing the number of logic levels represented at the output, and maximum reliability will be achieved when only two different levels are used. For such systems the most positive output voltage from the binary logic element is called a HI and the most negative voltage a LO, and in a "positive logic" system the logic 1 state is associated with the HI and the logic 0 state with the LO. On the other hand, for a system employing "negative logic," the logic 1 state corresponds to the most negative voltage and the logic 0 state to the most positive. Most modern digital systems use positive logic, and in this book, unless otherwise stated, all the circuits that we examine will be assumed to be operating under the positive-logic convention.

In the design of binary logic gates, such as the inverter circuit currently being considered, the two binary logic levels are generally chosen to correspond to cutoff and saturation of the transistor, since operation in these two states is least affected by circuit component variations. For example, for the inverter illustrated in Figure 10.1-1a, R_B has been chosen to be 5 kΩ so that the transistor will saturate when V_{in} is equal to V_{CC} and cutoff when V_{in} is equal to zero. In this way the output logic levels will be almost independent of the circuit component values.

Besides improving circuit reliability, operating the transistor between saturation and cutoff and keeping it out of the active region significantly reduces the power dissipated in the transistor and allows it to be a much smaller device. This is especially important for integrated-circuit designs, where many transistors may be located on the same substrate.

To ensure that the transistor stays out of the active region, it is necessary to take care that intermediate voltage levels are never applied to the input of the gate. Such voltages (which are neither valid HIs nor valid LOs) can cause the transistor to enter the active region and can dissipate excessive amounts of power both in the gate transistor itself and in the other logic elements being driven by the gate.

Figure 10.1-1b illustrates the basic shape of the transfer characteristic for the inverting-style logic gate being examined. In this sketch the slope of the transfer characteristic at each point along the curve represents the circuit small-signal gain, and for later use it is convenient to identify two specific points A and B on this characteristic, which represent the points where the small signal gain is -1. The coordinates of points A and B are (V_{IL}, V_{OH}) and (V_{IH}, V_{OL}), respectively, and these voltages are formally defined as follows:

V_{IL} = maximum input voltage that is still a valid logic 0 at the input

V_{IH} = minimum input voltage that is still a valid logic 1 at the input

V_{OH} = minimum output voltage that will appear at the gate output
when the output is supposed to be a logic 1

and V_{OL} = maximum output voltage that will appear at the gate output
when the output is supposed to be a logic 0

Some of the parameters just defined are a function of the fan-out, or the number of gates being driven by the circuit.

When we analyze specific logic gates using the piecewise linear models for transistors and diodes that we developed in earlier chapters, we will find that the transfer characteristics of these gates will consist of a series of straight-line segments (Figure 10.1-2c) rather than being a smooth curve (Figure 10.1-2b). As a result, V_{IL}, V_{IH}, and so on, will generally be easy to determine once the transfer characteristic has been sketched since the points where the gain passes through -1 will occur at the intersection points of the line segments.

EXAMPLE 10.1-1

For the circuit illustrated in Figure 10.1-1a, consider that the transistor is silicon.

a. Determine the values of V_{IL}, V_{IH}, V_{OL}, and V_{OH} for the case where the transistor $h_{FE} = 100$, and the gate fan-out is zero.
b. Repeat part (a) for a gate fan-out of 10.
c. Discuss why V_{OH} must be greater than V_{IH}, and also why V_{OL} must be less than V_{IL} if the circuit is to be a useful logic element.

SOLUTION

For the $N = 0$ or zero-fan-out case illustrated in Figure 10.1-2a, the transistor is cut off when V_{in} is less than 0.7 V, and for this value of V_{in} the corresponding value of V_o is $V_o = V_{CC} = 10$ V. Therefore, in this instance we have $V_{IL} = 0.7$ V and $V_{OH} = 10$ V.

When the transistor saturates, the output voltage is approximately 0.2 V, so that $V_{OL} = 0.2$ V. The minimum input voltage required to saturate the transistor may be found by noting that

$$I_{C,\text{sat}} = \frac{10 - 0.2}{1 \text{ k}\Omega} = 9.8 \text{ mA} \tag{10.1-1a}$$

so that

$$I_{B,\text{min}} = \frac{I_{C,\text{sat}}}{h_{FE}} = 0.098 \text{ mA} \tag{10.1-1b}$$

The input voltage just producing this base current determines V_{IH}, and following Figure 10.1-2a, this quantity is given by the expression

$$V_{IH} = I_{B,\text{min}} R_B + V_{BE} = (0.098)(5 \text{ k}\Omega) + 0.7 = 1.19 \text{ V} \tag{10.1-2}$$

The circuit's transfer characteristic is given in Figure 10.1-2b for the $N = 0$ case and the pertinent circuit data are summarized in part (c) of the figure.

The equivalent circuit for the case when the fan-out is 10 ($N = 10$) is shown in Figure 10.1-2d, and by inspection of this circuit it is apparent that V_{IL} is still 0.7 V, as for the $N = 0$ case. Furthermore, because all the transistors Q_1 through Q_{10} are cut off when Q_0 saturates, to the extent that we can neglect the reverse leakage current through the base–emitter junctions of the transistors, V_{OL} and V_{IH} for gate 0 are still 0.2 V and 1.19 V, respectively. However, when the output of gate 0 is HI, V_{OH}, the output voltage with Q_0 cut off, is significantly lower than that obtained for the zero-fan-out case because of the loading of Q_1–Q_{10} on Q_0.

To compute the value of V_{OH}, let's make use of the equivalent circuit presented in Figure 10.1-2d and observe with Q_0 cut off (and $I_C = 0$) that all the base–emitter diodes conduct. Therefore, under these conditions

$$I_{RC} = NI_B = 10I_B = \frac{10 - 0.7}{1 \text{ k}\Omega + 0.5 \text{ k}\Omega} = 6.2 \text{ mA} \tag{10.1-3a}$$

and

$$V_o = V_{CC} - I_{RC} R_C = 10 - 6.2 = 3.8 \text{ V} \tag{10.1-3b}$$

Once V_{in} exceeds 0.7 V, Q_0 enters the active region and V_o begins to de-

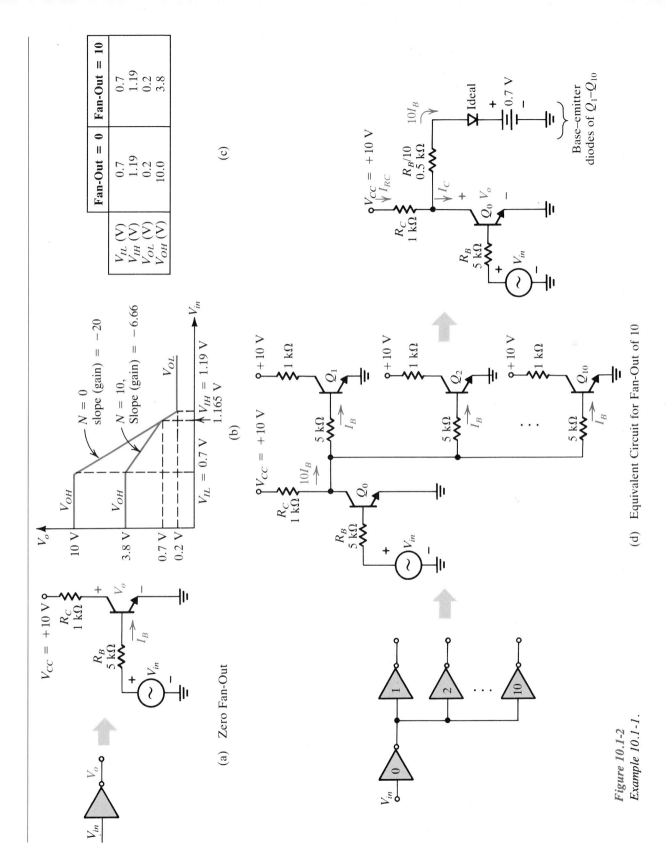

	Fan-Out = 0	**Fan-Out = 10**
V_{IL} (V)	0.7	0.7
V_{IH} (V)	1.19	1.19
V_{OL} (V)	0.2	0.2
V_{OH} (V)	10.0	3.8

(c)

(a) Zero Fan-Out

(b)

(d) Equivalent Circuit for Fan-Out of 10

Figure 10.1-2
Example 10.1-1.

749

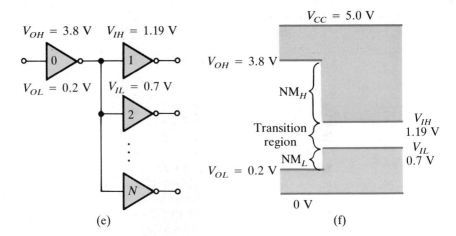

$V_{OH} = 3.8$ V $V_{IH} = 1.19$ V

$V_{OL} = 0.2$ V $V_{IL} = 0.7$ V

$V_{CC} = 5.0$ V

$V_{OH} = 3.8$ V

NM_H

Transition region

NM_L

$V_{OL} = 0.2$ V

0 V

V_{IH} 1.19 V

V_{IL} 0.7 V

Figure 10.1-2 (continued)

(e)   (f)

crease linearly. This linear decrease continues until V_o reaches 0.7 V, where the $N = 10$ line intersects the $N = 0$ line in Figure 10.1-2d. Once V_o drops below 0.7 V, the base–emitter junctions of Q_1–Q_{10} turn off, and thereafter the transfer characteristic follows the $N = 0$ curve. The input voltage at which the intersection occurs may be found as follows. With $V_o = 0.7$ V, the ideal diodes in Figure 10.1-2d are all off, and as a result,

$$I_C = I_{RC} = \frac{10 - 0.7}{1 \text{ k}\Omega} = 9.3 \text{ mA} \qquad (10.1\text{-}4)$$

Since the transistor is still in the active region, the required base current is $9.3/100$ or 0.093 mA, and the corresponding value of V_{in} is

$$V_{in} = I_B R_B + 0.7 = 1.165 \text{ V} \qquad (10.1\text{-}5)$$

When V_{in} increases beyond this point, no base current flows into Q_1–Q_{10}, and as a result, the remainder of the transfer characteristic follows the $N = 0$ curve.

The pertinent data for both the $N = 0$ and $N = 10$ cases are summarized in Figure 10.1-2c. In examining these results, it is very important to recognize that for a logic family to have compatible inputs and outputs, V_{OH} must be greater than V_{IH}, and V_{OL} must be smaller than V_{IL}. The inverter currently being examined satisfies both of these inequalities. The physical basis for each of these requirements may be understood with the aid of Figure 10.1-2e.

Recalling that V_{IH} represents the lowest input voltage that the gate will recognize as a valid input HI, V_{OH} must therefore be greater than this quantity if the output from gate 0 in the figure is to look like a HI to the input of gates 1 through N. In a similar fashion, because V_{IL} represents the largest allowed input voltage to the gate that can be recognized as a valid input LO, V_{OL} at the output of gate 0 must be less than the V_{IL} required by the gates being driven if it is to appear as a valid LO at the input to these gates. Thus, to be able to cascade the output of one logic gate into the input of another, it is necessary that

$$V_{OH} > V_{IH} \qquad (10.1\text{-}6a)$$

and $V_{OL} < V_{IL}$ (10.1-6b)

In the last example we saw that one digital logic device could be connected to another without incurring any logic-level incompatibility as long as equations (10.1-6a) and (10.1-6b) were satisfied. These results, however, are valid only when the logic circuit is noise free.

To estimate the ability of a logic circuit to perform in an electrically noisy environment, it is useful to define

$$NM_L = V_{IL} - V_{OL} \qquad (10.1\text{-}7a)$$

and

$$NM_H = V_{OH} - V_{IH} \qquad (10.1\text{-}7b)$$

as the gate's noise margins for LOs and HIs, respectively. These definitions are handy because they permit us to determine the maximum amount of noise that can safely appear at the output of a logic gate without causing the logic circuits connected onto this output to incorrectly interpret the logic signal that is present there.

As a practical matter, this type of logic noise is frequently associated with voltage differences that exist between the ground terminals or the power supply terminals of individual digital logic elements. In addition, it can arise from the radiation of nearby logic or noise signals into the gates themselves or into the wires connecting these gates together. Regardless of its origin, this type of electrical noise may generally be represented as a voltage source connected in series with the output of the driver gate.

Consider, for example, the addition of a noise voltage source v_N in series with the output of gate 0 in Figure 10.1-2e. When the output of gate 0 is LO, its output voltage will have a maximum value of 0.2 V, and because $V_{IL} = 0.7$ V for this gate, the noise margin for LOs will be about $(0.7 - 0.2)$ or 0.5 V. As a result, as long as the noise voltage, v_N, is less than this value, the input to gate 1 will always be less than 0.7 V, a valid logic 0. Similarly, if the output from gate 0 is HI, its output voltage will always be greater than 3.8 V (for a fan-out ≤ 10). As a result, because the noise margin for HIs for this circuit is $(3.8 - 1.19)$ or 2.61 V, noise signals with amplitudes of less than 2.61 V occurring at the output of gate 0 will have no effect on gate 1 when the input to it is a HI.

The noise margins for this inverter are illustrated in Figure 10.1-2f and by examining this figure it should be clear that both NM_H and NM_L can be increased by decreasing the size of the transition region. Basically, this requires an increase in the gain of the gate during the time that the inverter transistor is in the active region. The noise margin symmetry may also be improved by placing the center of the transition region at the middle (or at the arithmetic mean) of V_{OH} and V_{OL}. When this is done, if the width of the transition region can be neglected, we have $NM_H = NM_L = (V_{OH} - V_{OL})/2$.

In this chapter and Chapter 11 we examine the electronic structure of digital logic devices. Although there are thousands of different digital ICs available commercially, for the most part our investigation will focus on the behavior of the more popular elemental gates, flip-flops, and memory elements. The reason for this restriction is twofold. First, once a firm understanding of the behavior of these devices is obtained, it will be a relatively simple matter to extend this knowledge to more complex digital circuits. Second, with so many special-purpose digital ICs available, a detailed examination of even just the terminal characteristics of these ICs would easily fill a book many times the size of this one.

For review purposes the logic symbols for some of the more commonly encountered digital logic gates are illustrated in Figure 10.1-3 together with the truth tables for each of these gates. As we have already discussed, the inverter or NOT gate (Figure 10.1-3a) is a logic device having a single input and a single output. The circle drawn at the output of the symbol for the inverter is used to indicate that the output is negated or inverted. The truth table for this NOT gate is shown to the right in the figure and illustrates that an inverter functions logically by complementing or negating the applied input signal so that when the input is a LO (or logic 0), the output is a HI (or logic 1), and vice versa.

An AND gate is a digital logic device having two or more inputs and a single output. The output of an AND gate is a logic 1 only when all its inputs are

Figure 10.1-3
Elemental digital logic gates.

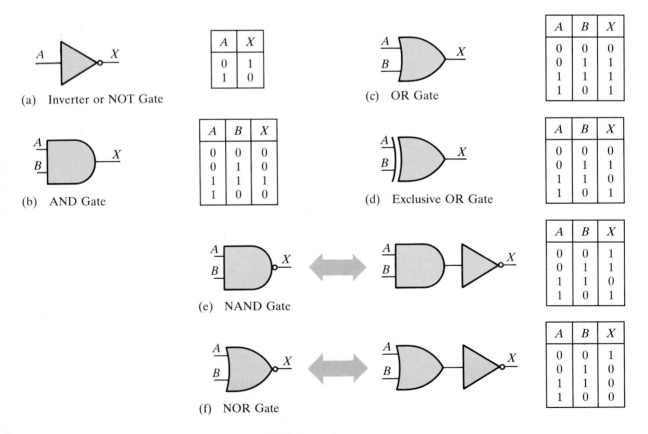

(a) Inverter or NOT Gate

A	X
0	1
1	0

(b) AND Gate

A	B	X
0	0	0
0	1	0
1	1	1
1	0	0

(c) OR Gate

A	B	X
0	0	0
0	1	1
1	1	1
1	0	1

(d) Exclusive OR Gate

A	B	X
0	0	0
0	1	1
1	1	0
1	0	1

(e) NAND Gate

A	B	X
0	0	1
0	1	1
1	1	0
1	0	1

(f) NOR Gate

A	B	X
0	0	1
0	1	0
1	1	0
1	0	0

1, and is 0 for all other input combinations. The logic symbol for a two-input AND gate is shown in Figure 10.1-3b. The truth table for this gate is also given to the right in the figure and demonstrates the AND nature of this gate since the output of this two-input gate is a logic 1 only when inputs *A and B* are both at the logic 1 level.

An OR gate is a digital logic device that also has two or more inputs and a single output. In general, the output of an OR gate is a logic 1 when one or more if its inputs are a logic 1. A typical two-input OR gate is shown in Figure 10.1-3c along with its truth table. Following this table, we can see that the output of this gate is a logic 1 if the inputs *A or B*, *or* both *A* and *B*, are at the logic 1 level. Conversely, the output from this OR gate is a logic 0 only when *A* and *B* are both simultaneously at a logic 0.

An exclusive-OR gate, or EXOR gate as it is also called (Figure 10.1-3d), is sinilar to an OR gate in that its output is a logic 1 when one of the inputs is 1, and is 0 when both inputs are 0. However, unlike the OR gate, the output from the exclusive-OR is also a logic 0 when both inputs are at a logic 1. As such, the EXOR gate output is 1 only when the inputs are different from one another. Because the output from the exclusive-OR gate is a logic 1 only when the inputs are unequal, these gates are frequently used in digital comparator applications.

The NAND and NOR gates illustrated in Figure 10.1-3e and f, respectively, are probably the two most popular logic gates employed in digital logic circuit design. This is because it is possible to synthesize any type of combinational logic circuit using only NAND gates or NOR gates alone. As shown in the figure, the NAND, or negated AND, is formed by adding an inverter to the output of an AND gate, and as indicated by its truth table, the output of this two-input NAND gate is 0 only when both of its inputs are 1. As with the NAND gate, a NOR gate may be constructed similarly by adding an inverter at the output of an OR gate. As the truth table in Figure 10.1-3f illustrates, the output from this two-input NOR gate is 1 only when all of its inputs are 0, and is 0 for all other input combinations.

The basic technique for synthesizing combinational logic circuits using only NAND gates or NOR gates is first to lay out the circuit using whatever logic gates seem to be most appropriate. Once this has been accomplished, any inverters required in the design may simply be replaced by NAND gates or NOR gates whose inputs are tied together. AND gates in the circuit may be replaced either by NAND gates connected to inverters or by NOR gates with inverters on their inputs. Similarly, any OR gates in the circuit may be replaced by NOR gates followed by inverters or by NAND gates having inverters on their inputs. These basic equivalent circuit building blocks are illustrated in Figure 10.1-4, and the proof of the equivalences shown is left as an exercise for the interested reader.

Combinational circuits are digital circuits that are composed only of logic gates and that have no feedback from the output back to the input. For such circuits the current or present output depends uniquely on the particular combination of variables present at the circuit inputs. In other words, the state of this type of circuit is determined once the input combination is specified.

By adding feedback to a combinational circuit, or by using circuit elements such as flip-flops that have local internal feedback, it is possible to construct a

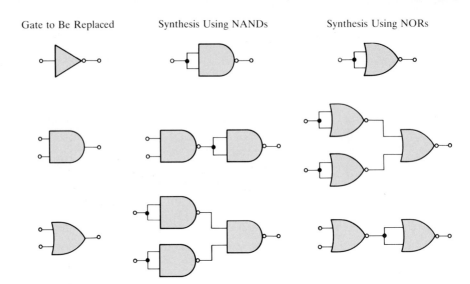

| Gate to Be Replaced | Synthesis Using NANDs | Synthesis Using NORs |

Figure 10.1-4
Synthesis of basic logic functions using only NAND gates or NOR gates.

sequential circuit or a circuit that has memory. For such a circuit the present state of the system depends not only on the current value of the system input variables, but also on the past history of the circuit.

In a sequential circuit each unique combination of output variables defines a particular state of the system, and the movement of the system from one state to another can occur in one of two different ways. In an asynchronous sequential circuit this transistion takes place as soon as the input variables change, whereas in a synchronous sequential circuit a clocking signal must be applied before a state transition can occur. The clock is a signal that is generally applied simultaneously to all flip-flops in the circuit, so that all changes in the outputs of the flip-flops occur at approximately the same time. The use of a clock helps to ensure that the state transitions which occur depend only on the previous state of the system, and on the inputs present just prior to the onset of the clocking signal. In asynchronous systems and in poorly designed synchronous systems, a "fast-changing" flip-flop may alter the input to another flip-flop and cause it to make a transition to the wrong state.

Most sequential circuits are synchronous, and the basic building block employed in these circuits is the clocked flip-flop. The schematic symbols and corresponding truth tables for several different clocked and unclocked flip-flops are given in Figure 10.1-5.

In an SR or set-reset flip-flop the values of Q and $\overline{Q}$ (the complement of Q) depend on the signal levels applied to the S and R inputs. Normally, both S and R are held at a logic 0 level, and in this mode, known as the memory mode, the value previously stored in the flip-flop will remain there forever as long as power is applied. If the S input is momentarily brought to a logic 1, the flip-flop will be set and Q will become a logic 1. When S is brought back to 0, the flip-flop will return to the memory mode and Q will stay at its logic 1 value. Similarly, if the R, or reset line, is momentarily raised to a logic 1, the flip-flop output will be reset to 0. The combination where S and R are both simultaneously made equal to 1 results in an indeterminate output from the flip-flop, and is therefore considered to be a forbidden input combination. The logic

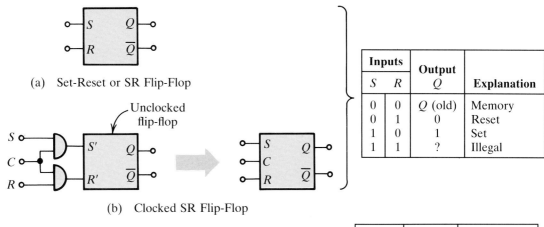

(a)   Set-Reset or SR Flip-Flop

Inputs		Output	
S	R	Q	Explanation
0	0	Q (old)	Memory
0	1	0	Reset
1	0	1	Set
1	1	?	Illegal

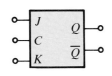

(b)   Clocked SR Flip-Flop

(c)   JK Clocked Flip-Flop

Inputs		Output	
J	K	Q	Explanation
0	0	Q (old)	Memory
0	1	0	Reset
1	0	1	Set
1	1	$\overline{Q}$	Toggle

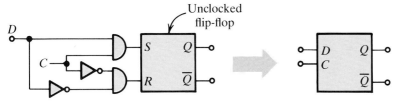

(d)   D-Style Clocked Flip-Flop

Input D	Output Q
0	0
1	1

Figure 10.1-5
Basic types of clocked and un-clocked flip-flops.

symbol for the SR flip-flop and its corresponding truth table are shown in Figure 10.1-5a.

The operation of the clocked SR flip-flop in Figure 10.1-5b is identical to that of the unclocked flip-flop just discussed except that the state transitions indicated on the truth table cannot occur until the clocking condition is satisfied. For the simple "AND gate input" clocked SR flip-flop indicated in the figure, this is seen to occur as soon as the clock goes HI.

The flip-flop illustrated in Figure 10.1-5c is known as a JK flip-flop, and its truth table is very similar to that of the clocked SR flip-flop just discussed, with the J and K inputs having roles similar to the S and R inputs, respectively. The major distinction between these two devices is that the 1–1 input combination is allowed in the JK design. In particular, when the clocking signal occurs with both J and K equal to a logic 1, the flip-flop toggles or flips to the output state opposite that it had prior to the occurrence of the clocking signal.

The last device in this category that we will examine is known as a D-style flip-flop and is illustrated in Figure 10.1-5d. Its operation is very easy to explain. Basically, when the clocking condition is satisfied, the flip-flop memorizes the signal on the D input line. Flip-flops of this type are frequently grouped together to form circuits known as latches or registers, and these are used to capture and store groups of binary digital signals simultaneously.

Exercises

10.1-1 For the circuit shown in Figure 10.1-2d, V_{in} equals zero. Find V_{CE0} and V_{CE1} for a fan-out of 20 if $h_{FE} = 20$ for all the transistors. **Answer** $V_{CE0} = V_{CE1} = 2.56$ V

10.1-2 A two-inverter cascade is formed by adding a second identical inverter onto the output of the circuit shown in Figure 10.1-1a. The transistors have h_{FE}'s equal to 20, $V_{CC} = 5$ V, and v_{in} is a 0- to 5-V periodic pulse train that is HI for one-fourth of each cycle. Find the power dissipated in each transistor and compute P_{IN} from the power supply. **Answer** $P_{D1} = 0.39$ mW, $P_{D2} = 1.1$ mW, $P_{IN} = 25.4$ mW

10.2 LARGE-SIGNAL SWITCHING CHARACTERISTICS OF DIODES AND TRANSISTORS

In designing digital circuits the system's operating speed is frequently so high that the parasitic capacitive elements in the semiconductors and the circuit wiring may create important second-order effects that need to be considered. Figure 10.2-1 illustrates how the output v_o from a typical digital logic inverter changes with the frequency of the applied input signal, v_{in} (Figure 10.2-1a). At low frequencies the output will look identical to the input except for the logical signal inversion, and as the frequency of the input signal is raised, the output signal will suffer both delay and waveshape distortion. At moderately high frequencies (Figure 10.2-1b) the output of the inverter still has the character of a binary digital logic signal even though there may be considerable gate delay and rounding of the edges of the output signal waveshape. As the frequency is raised still further (Figure 10.2-1c), the output signal's rise and fall times will eventually becomes so large in comparison to the signal period that the output waveshape will no longer reach its steady-state logic 0 and logic 1 values. As a result, at and beyond these frequencies the inverter is virtually useless as a digital logic element.

The purpose of this section is to examine the large-signal switching characteristics of diode and transistor circuits in order to determine the factors that govern their speed of response. To carry out this objective, we will basically be extending the small-signal-modeling techniques presented in Section 6.4 to permit us to handle properly the large signal swings associated with digital logic circuits. Because a firm understanding of this subject depends so heavily on your being comfortable with the various models that we will be using, it may well be worth your while at this point to go back and reread Section 6.4.

In developing large-signal models for the diode, the BJT, and the FET, certain simplifying assumptions will need to be made. Many of the circuit elements used in the models described in earlier chapters have component values that depend on the position of the operating point. As a result, because the signal swings in switching circuits are generally very large, these elements should

Figure 10.2-1

Performance of a typical digital logic gate as a function of applied input signal frequency.

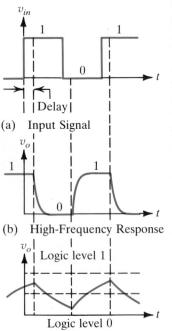

(a) Input Signal

(b) High-Frequency Response

(c) Very-High-Frequency Response

be considered to be nonlinear functions of the instantaneous position of the operating point. Unfortunately, this type of nonlinear model is very difficult to work with and provides little insight into the device's circuit behavior. Therefore, as an alternative, we neglect these nonlinearities and assume instead that the semiconductor devices are still representable by linear models whose equivalent-circuit elements are constants. This simplification will permit us to carry out a rather straightforward linear analysis of the switching performance of circuits containing diodes and transistors, which interestingly enough, provides fairly good quantitative results.

10.2-1   Switching Characteristics of the PN Junction Diode

The basic form of the large-signal model for the *pn* junction diode is presented in Figure 10.2-2a. In this model C_J denotes the diode junction capacitance, r_D its resistance, and C_D the diffusion capacitance. The charge on C_J represents the uncovered bound charge in the vicinity of the junction, while that on C_D is equal to the excess minority-carrier charge stored just outside the depletion region. Capital letters are used for the subscripts to indicate that these are large-signal equivalent-circuit components.

C_J is a nonlinear function of the voltage across the junction [equation (6.4-4)] and typically has a value on the order of 0.1 to 10 pF. C_D and r_D both depend on the diode current, and C_D varies linearly with i_D, while r_D is inversely proportional to it [equations (6.4-12a) and (6.4-12b)]. As a result, the product $r_D C_D$, known as the diode's recombination time, τ_r, is nearly independent of the current flow through it. [In this chapter we use τ_r rather than τ_p to represent the recombination time of the diode. The coefficient τ_p (the recombination time for holes), used in Section 6.4, is equal to the diode's overall recombination time only for an asymmetrically doped p^+n diode.] Typical values for r_D and C_D under diode forward-bias conditions are 0.1 to 10 Ω for r_D and 500 pF to 10 μF for C_D. When the diode is reverse biased, i_D is nearly zero, so that r_D is equal to infinity and C_D is zero.

Using this information, let's see how to simplify the diode model given in Figure 10.2-2a. When the diode is reverse biased, both r_D and C_D may be replaced by open circuits, and this leads to the model illustrated in Figure 10.2-2b. Under forward-bias conditions, C_D is much larger than C_J, and because these two capacitors are essentially connected in parallel, we may neglect C_J. When this is done, we obtain the simplified forward-bias large-signal diode

Figure 10.2-2

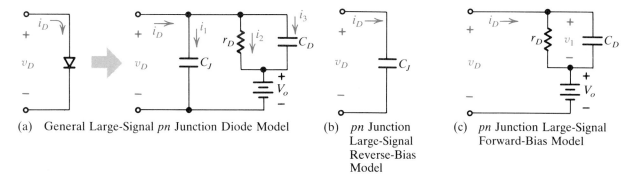

(a)   General Large-Signal *pn* Junction Diode Model

(b)   *pn* Junction Large-Signal Reverse-Bias Model

(c)   *pn* Junction Large-Signal Forward-Bias Model

model given in Figure 10.2-2c. For simplicity, in all of these final large-signal diode models we will assume that C_J, r_D, and C_D are constants independent of the diode's operating point.

In analyzing the switching performance of a diode circuit, the problem at hand is relatively easy to solve once we understand which model to use at any given point in time, and under what conditions to switch from one model to the next. Basically, a diode remains reverse biased as long as the voltage across it is less than zero (or more accurately V_o) volts. Because of the presence of the junction capacitance, this means that the diode may not be able to turn on immediately even though the applied input voltage switches to a positive value, since C_J needs to charge to a positive value first. Similarly, when a diode is forward biased, there will be excess minority-carrier charge stored on C_D, and as a result, when the input signal switches to a negative value, the diode will not be able to turn off immediately since C_D will need to be discharged completely first.

To illustrate these points, consider the elementary diode circuit shown in Figure 10.2-3a. Here we are assuming [see part (f) of the figure] that v_{in} has been equal to $-V_R$ volts for a long time prior to $t = 0$ so that the diode is initially reverse biased. As a result, at $t = 0-$ the equivalent circuit shown in Figure 10.2-3b is valid, and C_J is charged to an initial value of $-V_R$ volts.

At $t = 0$ when v_{in} switches to $+V_F$ volts, the diode does not immediately become forward biased; C_J needs to discharge first, and this takes time. Following the equivalent circuit in Figure 10.2-3c, it is apparent that the voltage across C_J will rise exponentially from $-V_R$ toward $+V_F$ with a time constant

Figure 10.2-3
Transient response of a pn junction diode.

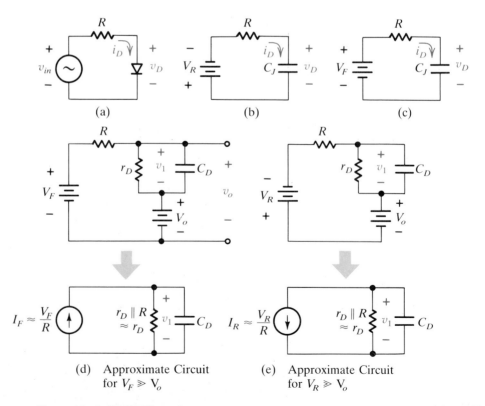

(d) Approximate Circuit
for $V_F \gg V_o$

(e) Approximate Circuit
for $V_R \gg V_o$

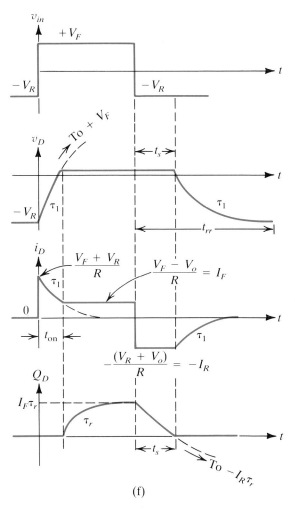

Figure 10.2-3
(continued)

(f)

$\tau_1 = RC_J$. Once v_D reaches zero (or more precisely $+V_o$ volts), the diode will turn on and the forward-bias model in part (d) of the figure will apply for the remainder of the turn-on transient.

Although the diode models are extremely helpful for describing the circuit behavior of the diode, it is also useful to think about what is actually going on within the diode when the input voltage switches to $+V_F$. The removal of negative charge from C_J gives rise to a current component i_1 in Figure 10.2-2a that is associated with the charge movement that must occur in order to reduce the size of the depletion region as the diode approaches the forward-bias condition.

Once the voltage on C_J equals V_o, the portion of the diode current associated with the buildup of the minority-carrier charge distribution outside the depletion region [$(i_2 + i_3)$ in Figure 10.2-2a] becomes much larger than the charging current of C_J. As a result, the forward-bias model for the diode is now more appropriate, and the equivalent circuit shown in Figure 10.2-2c applies. At this point the diode has just gone ON, and as C_D charges, the minority-carrier charge builds up toward its steady-state value $I_F \tau_r$ (Figure 10.2-3f).

When the input voltage reverses at $t = T$, the diode initially stays on, and therefore the equivalent circuit shown in Figure 10.2-3e is valid. To turn the diode off it is necessary to remove all the excess minority-carrier charge stored on C_D. But this cannot be done instantaneously, and as long as there is any charge left on C_D, v_D will stay approximately equal to V_o and the diode will remain forward biased. Because the input voltage is now negative and the diode is still on, a large negative current flows out of the diode associated with the removal of the minority-carrier charge from C_D. Once this charge is completely removed, the diode enters the reverse-bias region and the equivalent illustrated in Figure 10.2-3b applies. As a result, at this point C_J now charges exponentially toward $-V_R$, and when C_J is fully charged, i_D will be zero and the diode will again be off.

In Figure 10.2-3f, t_{on}, t_s, and t_{rr} are known as the diode's on-time, storage time, and reverse recovery time, respectively. For a small-signal diode having square-wave input signal excitation, typical values for these switching parameters are $t_{on} = 7$ ns, $t_s = 700$ ns, and $t_{rr} = t_s + 4t_1 \simeq 700$ ns. These results indicate that the storage time usually dominates the transient response of a diode. A detailed analysis of this type of diode circuit is carried out in Problem 10.2-1.

10.2-2 Switching Characteristics of the Bipolar Junction Transistor

As with the *pn* junction diode, the large-signal switching performance of the BJT may be readily understood once the appropriate extensions of the small-signal models have been made. The small-signal hybrid-π transistor model is shown in Figure 6.4-4 and by examining this model you should recall that

$$r_{bb'} = \text{base spreading resistance} \tag{10.2-1a}$$

$$r_{b'e} = (1 + h_{fe})r_e \tag{10.2-1b}$$

$$r_e = \frac{V_T}{I_E} \left(\text{proportional to } \frac{1}{I_E} \right) \tag{10.2-1c}$$

$$C_{b'e} = C_{je} + C_{de} \tag{10.2-1d}$$

$$C_{je} = \text{emitter junction capacitance (a nonlinear function of } v_{B'E}) \tag{10.2-1e}$$

$$C_{de} = \text{emitter diffusion capacitance (proportional to } I_E) \tag{10.2-1f}$$

$$C_{b'c} = C_{jc} = \text{collector junction capacitance} \atop \text{(a nonlinear function of } v_{CB'}) \tag{10.2-1g}$$

and

$$g_m = \frac{h_{fe}}{r_{b'e}} = \frac{h_{fe}}{(1 + h_{fe})r_e} \simeq \frac{I_E}{V_T} \text{ (proportional to } I_E) \tag{10.2-1h}$$

For simplicity, in extending this model to handle large signals, it will be convenient to consider that $r_{bb'} = 0$ and that $C_{b'c}$ is also zero. Nonzero values of $C_{b'c}$ may be handled by applying the Miller theorem (see Problem 10.2-7); however, in using this approach, we will need to assume that g_m is relatively constant even though it actually varies directly with I_E.

By setting $C_{b'c}$ and $r_{bb'}$ equal to zero, and including the voltage drop V_o across the base–emitter junction, we obtain the equivalent circuit shown in

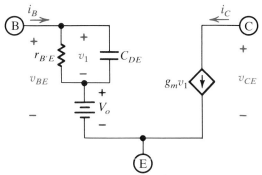

(a) Approximate BJT Active-Region Large-Signal Model

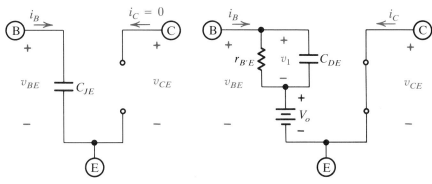

Figure 10.2-4
Approximate BJT large-signal models.

(b) BJT Large-Signal Model
in Cutoff

(c) BJT Large-Signal Model
in Saturation

Figure 10.2-4a. This is the equivalent circuit that we will use to represent the BJT when it is in the active region. Here $r_{B'E}$ is used to represent the average value of $r_{b'e}$ over all values between cutoff and saturation. In a similar fashion, C_{DE}, C_{JE}, and where needed, $r_{BB'}$ will also be used to represent the average values of C_{de}, C_{je}, and $r_{bb'}$, respectively. As with the diode, the junction capacitance C_{JE} has been neglected in comparison to the diffusion capacitance C_{DE} when the base–emitter junction is forward biased.

When the transistor is cut off, C_{JE} can no longer be neglected since C_{DE} is now nearly zero. Furthermore, in this region, with $I_E \simeq 0$, the transistor g_m is also zero. Combining these facts, in cutoff the BJT model has the approximate form given in Figure 10.2-4b.

The model that is valid when the transistor saturates is essentially the same as that when it is active except that v_{CE} is nearly zero, and as a result this equivalent circuit has the form shown in Figure 10.2-4c. This representation is somewhat inaccurate since in reality, when the transistor saturates the effective time constant τ_s is different than τ_B, but in the interest of simplicity we use the same recombination time constant to represent the transistor's performance in both the active and saturation regions.

The overall switching performance of a BJT device in a given circuit is handled in much the same way as with the *pn* junction diode, by substituting in the appropriate model for the transistor and solving for the desired voltages and currents. As with the diode, the circuit components used in these models are considered to be constants in order to simplify the analysis.

Furthermore, as long as v_{BE} is less than V_o, the transistor will be assumed to be cut off. Once v_{BE} reaches V_o, the transistor will be considered to be in the active region as long as i_C is less than $I_{C,\text{sat}}$. Once i_C attempts to exceed this value, the transistor will saturate, and the model in Figure 10.2-4c will apply.

To illustrate how to calculate the important switching times for the bipolar junction transistor, let's carefully examine the circuit illustrated in Figure 10.2-5a. In carrying out a quantitative analysis of this circuit, we will make extensive use of the technique presented in Appendix I.5 regarding the transient analysis of simple RC networks by inspection. If you have forgotten how to do this, you should reread this material, paying particular attention to equation (I.5-1). In analyzing this circuit we will assume that the models given in Figure 10.2-4 may be used to represent the transistor. In addition, V_F and V_R will be considered to be much larger than V_o, so that V_o may be neglected and switching between active and cutoff can be assumed to take place at $v_{BE} = 0$.

If we consider that the input voltage has been equal to $-V_R$ for a long time, the transistor will initially be cut off, and at $t = 0-$ the model in Figure 10.2-5b will apply with $v_{BE}(0-) = -V_R$, $i_C(0-) = 0$, and $v_{CE}(0-) = V_{CC}$. At $t = 0$ the input signal changes to $+V_F$, but because the junction capacitance cannot charge instantaneously, the transistor remains off (for a while at least), the equivalent circuit in Figure 10.2-5c applies, and at $t = 0+$ we have

$$v_{BE}(0+) = v_{BE}(0-) = -V_R \tag{10.2-2a}$$

and $$i_B(0+) = \frac{V_F - v_{BE}(0+)}{R_B} = \frac{V_F + V_R}{R_B} \tag{10.2-2b}$$

Using equation (I.5-1), we may write down $v_{BE}(t)$ and $i_B(t)$ by inspection as

$$v_{BE}(t) = \underbrace{-V_R e^{-t/\tau_1}}_{\text{initial value}} + \underbrace{V_F(1 - e^{-t/\tau_1})}_{\text{final value}} \tag{10.2-3a}$$

and $$i_B(t) = \frac{V_F + V_R}{R_B} e^{-t/\tau_1} \tag{10.2-3b}$$

These waveshapes are sketched in Figure 10.2-5d.

All during the time that v_{BE} is less than zero (actually V_o) volts, the transistor is cut off, and $i_C = 0$ and $v_{CE} = V_{CC}$. The time required for C_{JE} to charge all the way up to zero volts is called t_d, the transistor's delay time. Strictly speaking, t_d is actually defined as the time needed after the switching occurs

Figure 10.2-5
Large-signal switching perfor-
mance of a BJT.

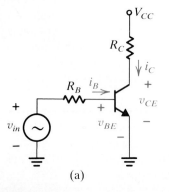

(a)

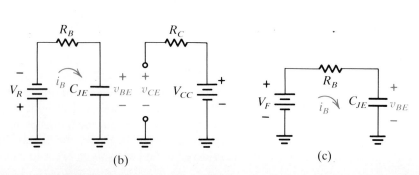

(b)   (c)

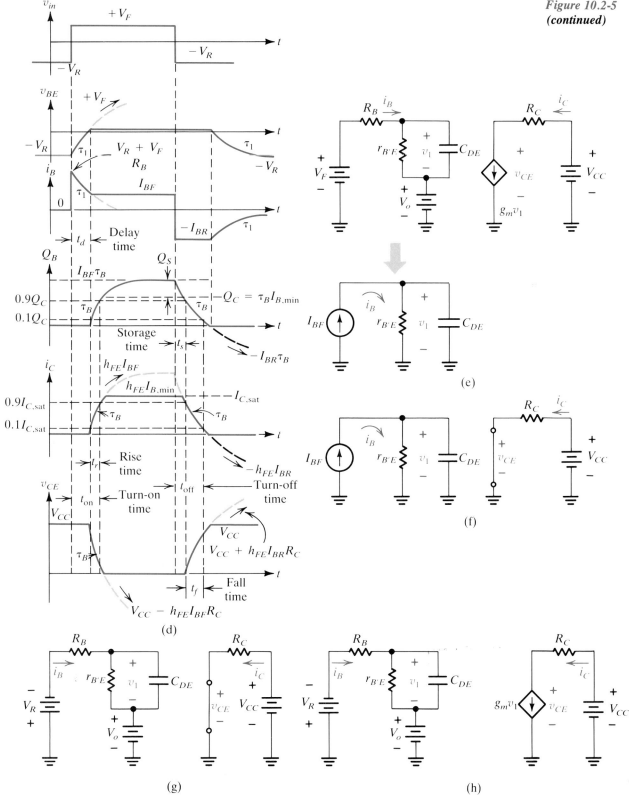

Figure 10.2-5
(continued)

763

for i_C to reach 10% of its final value; but for simplicity, we approximate it as the time required for the transistor to just turn ON. Using this approximation for t_d, this time interval may be found by setting v_{BE} equal to zero in equation (10.2-3a) and solving for t. Carrying out this procedure, we obtain

$$t_d = \tau_1 \ln \frac{V_F + V_R}{V_F} \qquad (10.2\text{-}4a)$$

where
$$\tau_1 = R_B C_{JE} \qquad (10.2\text{-}4b)$$

When v_{BE} reaches zero the active model presented in Figure 10.2-5e applies, and assuming that R_B is much greater than $r_{B'E}$,

$$i_B \simeq \frac{V_F - V_o}{R_B} \simeq \frac{V_F}{R_B} = I_{BF} \qquad (10.2\text{-}5a)$$

and
$$v_1(t) = I_{BF} r_{B'E}(1 - e^{-t/\tau_B}) \qquad (10.2\text{-}5b)$$

where
$$\tau_B = r_{B'E} C_{DE} \qquad (10.2\text{-}5c)$$

The quantity τ_B is a fundamental transistor parameter known as the minority-carrier lifetime in the base and is similar to the recombination time τ_r in diodes. By using equation (6.4-33b), it is not too difficult to show that τ_B can also be written as

$$\tau_B = \frac{1}{2\pi f_B} = \frac{1 + h_{FE}}{2\pi f_T} \qquad (10.2\text{-}5d)$$

where f_T is the common-emitter unity-gain crossover frequency.

With the transistor now operating in the active region, the collector current may be written as

$$i_C(t) = g_m v_1(t) = g_m r_{B'E} I_{BF}(1 - e^{-t/\tau_B}) \qquad (10.2\text{-}6a)$$

By using equation (10.2-1h) and substituting $h_{FE}/r_{B'E}$ for g_m, this expression reduces to the more familiar form

$$i_C(t) = h_{FE} I_{BF}(1 - e^{-t/\tau_B}) \qquad (10.2\text{-}6b)$$

In addition, the collector-to-emitter voltage may also be written as

$$v_{CE}(t) = V_{CC} - i_C R_C = V_{CC} - h_{FE} I_{BF} R_C(1 - e^{-t/\tau_B}) \qquad (10.2\text{-}7)$$

If I_{BF} is greater than $I_{B,\min}$, the transistor will saturate before the collector current reaches the final value predicted by equation (10.2-6b). When this occurs, as illustrated in Figure 10.2-5d, the collector current will flatten out at $I_{C,\text{sat}}$, and v_{CE} will also flatten off at zero.

The charge on C_{DE}, which is equal to the excess minority carrier charge stored in the base, is given by

$$Q_B(t) = C_{DE} v_1(t) = I_{BF} \tau_B(1 - e^{-t/\tau_B}) \qquad (10.2\text{-}8a)$$

and for reasons that will be apparent shortly, it is also convenient to note that this expression for the base charge can be rewritten using equation (10.2-6b) as

$$Q_B(t) = \frac{\tau_B}{h_{FE}} h_{FE} I_{BF}(1 - e^{-t/\tau_B}) = \frac{\tau_B}{h_{FE}} i_C(t) \qquad (10.2\text{-}8b)$$

As the base charge increases (if $I_{BF} > I_{B,\min}$), the transistor will eventually saturate. This will occur when the collector current level is

$$i_C = I_{C,\text{sat}} \simeq \frac{V_{CC}}{R_C} \tag{10.2-9a}$$

and

$$i_B = I_{B,\min} = \frac{I_{C,\text{sat}}}{h_{FE}} \tag{10.2-9b}$$

The corresponding minority-carrier charge stored in the base at this point is given by equation (10.2-8b) as

$$Q_c = \frac{\tau_B I_{C,\text{sat}}}{h_{FE}} = \frac{\tau_B h_{FE} I_{B,\min}}{h_{FE}} = \tau_B I_{B,\min} \tag{10.2-9c}$$

The transistor rise time, or the time for the transistor collector current to go from 10% to 90% of its final value, may be determined by using equation (10.2-6b) to find the times required for $i_C(t)$ to reach 0.1 and $0.9 I_{C,\text{sat}}$, setting their difference equal to t_r. Carrying out this procedure, we obtain

$$t_r = \tau_B \ln \frac{I_{BF} - 0.1 I_{B,\min}}{I_{BF} - 0.9 I_{B,\min}} \tag{10.2-10}$$

The sum of t_d and t_r is known as the transistor turn-on time, t_{on}.

After the transistor saturates, although i_C flattens off at $I_{C,\text{sat}}$, following the equivalent circuit in Figure 10.2-5f, we find that C_{DE} continues to charge toward $I_{BF}\tau_B$. The additional charge added beyond Q_c is called the storage charge (Q_s), and later when we try to turn the transistor off, we will see that all of this charge must be removed from the base before the transistor can come out of saturation.

When v_{in} switches to $-V_R$, the equivalent circuit shown in Figure 10.2-5g applies until all the storage charge Q_s is removed from the base. If $R_B \gg r_{B'E}$ and also if $V_R \gg V_o$, then here

$$i_B = \frac{V_o - V_R}{R_B} \simeq \frac{-V_R}{R_B} = -I_{BR} \tag{10.2-11a}$$

Thus

$$v_1(t) = I_{BF} e^{-t/\tau_B} - I_{BR} r_{B'E}(1 - e^{-t/\tau_B}) \tag{10.2-11b}$$

and

$$Q_B(t) = C_{DE} v_1 = I_{BF}\tau_B e^{-t/\tau_B} - I_{BR}\tau_B(1 - e^{-t/\tau_B}) \tag{10.2-11c}$$

redefining $t = 0$ as the point where v_{in} switches to $-V_R$. The time required to reduce $Q_B(t)$ to 90% of Q_c is called t_s, the storage time. Following equation (10.2-9c), we may write

$$Q_B(t_s) = 0.9 Q_c = 0.9\tau_B I_{B,\min} = I_{BF}\tau_B e^{-t_s/\tau_B} - I_{BR}\tau_B(1 - e^{-t_s/\tau_B}) \tag{10.2-12a}$$

and solving this expression for t_s, we have

$$t_s = \tau_B \ln \frac{I_{BR} + I_{BF}}{I_{BR} + 0.9 I_{B,\min}} \tag{10.2-12b}$$

After all the storage charge is removed, the transistor comes out of saturation, reenters the active region, and the equivalent circuit given in Figure 10.2-5h applies. Because the input circuit has not changed, $i_B(t)$, $v_1(t)$, and $Q_B(t)$ are still given by equations (10.2-11a), (10.2-11b), and (10.2-11c), respectively.

Therefore, because the transistor is now active, collector current and collector-to-emitter voltage may be immediately written as

$$i_C(t) = g_m v_1 = h_{FE} I_{B,\min} e^{-t/\tau_B} - h_{FE} I_{BR}(1 - e^{-t/\tau_B}) \qquad (10.2\text{-}13a)$$

and

$$v_{CE}(t) = V_{CC} - i_C R_C = V_{CC} - h_{FE} R_C[I_{B,\min} e^{-t/\tau_B} - I_{BR}(1 - e^{-t/\tau_B})] \qquad (10.2\text{-}13b)$$

The decrease in collector current predicted by equation (10.2-13a) continues until $Q_B(t)$ [or, alternatively, $i_C(t)$] reaches zero. At this point the base–emitter junction goes off and the transistor enters cutoff. Starting at the point where the transistor reenters the active region, and calling this point $t = 0$, the expressions for the $Q_B(t)$ and $i_C(t)$ may now be written as

$$Q_B(t) = I_{B,\min} \tau_B e^{-t/\tau_B} - I_{BR} \tau_B(1 - e^{-t/\tau_B}) \qquad (10.2\text{-}14a)$$

and
$$i_C(t) = h_{FE} I_{B,\min} e^{-t/\tau_B} - h_{FE} I_{BR}(1 - e^{-t/\tau_B}) \qquad (10.2\text{-}14b)$$

The transistor fall time, or the time for the transistor collector current to go from 90% to 10% of $I_{C,\text{sat}}$, may be determined by using equation (10.2-14b) to find the times required for $i_C(t)$ to reach 0.9 and $0.1 I_{C,\text{sat}}$, setting their difference equal to t_f. Carrying out this procedure, we obtain

$$t_f = \tau_B \ln \frac{I_{BR} + 0.9 I_{B,\min}}{I_{BR} + 0.1 I_{B,\min}} \qquad (10.2\text{-}15)$$

The sum of the storage and fall times is known as the turn-off time, t_{off}, of the transistor.

Once $Q_B(t)$ reaches zero, the base–emitter junction is reversed biased, and the circuit model in Figure 10.2-5b again applies. In this region $i_C = 0$ and $v_{CE} = V_{CC}$. In addition, because the initial voltage on the capacitor is $v_{BE}(0+) = 0$ [defining $t = 0$ as the point at which the transistor just cuts off], we may write

$$v_{BE}(t) = -V_R(1 - e^{-t/\tau_1}) \qquad (10.2\text{-}16a)$$

and
$$i_B(t) = \frac{-V_R}{R_B} e^{-t/\tau_1} \qquad (10.2\text{-}16b)$$

Of course, the recovery of these voltage and current waveshapes takes about $4\tau_1$ (Figure 10.2-5d).

10.2-3 Switching Characteristics of the Field-Effect Transistor

The basic FET switching problem is illustrated in Figure 10.2-6a. Here C_{gs} and C_{gd} are the parasitic capacitances of the FET and C_L represents the combination of the transistor drain-to-source capacitance and any stray capacitance associated with the wiring or the next stage connected onto the transistor.

For the JFET you should recall that C_{gs} and C_{gd} are nonlinear functions of the voltage between the gate and the channel, while for the MOSFET these capacitances are relatively constant. In addition, for both MOSFETs and JFETs the drain current is nonlinearly related to the gate-to-source voltage (see, e.g., Figure 10.2-6b). These nonlinearities make it difficult to analyze the transient

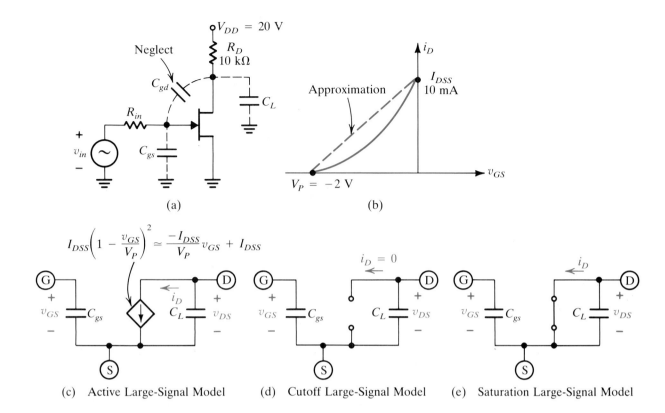

(a)

(b)

$$I_{DSS}\left(1 - \frac{v_{GS}}{V_P}\right)^2 \simeq \frac{-I_{DSS}}{V_P}v_{GS} + I_{DSS}$$

(c)   Active Large-Signal Model

(d)   Cutoff Large-Signal Model

(e)   Saturation Large-Signal Model

Figure 10.2-6
Large-signal FET models.

large-signal switching performance of FETs, and therefore to simplify this investigation, the following assumptions will be made:

1. The parasitic capacitances are constants independent of the transistor voltages.
2. The Miller feedback capacitance, C_{gd}, is negligible. This capacitance can easily be included later by applying the Miller theorem (see Problem 10.2-10).
3. The transistor g_m is a constant independent of v_{GS}. This is equivalent to saying that the i_D versus v_{GS} curve may be approximated by a straight line (see the dashed line in Figure 10.2-6b).

In addition, and again to ease the mathematical details, two more simplifying approximations will also be made:

4. The power supply voltage V_{DD} is much greater than the magnitude of the pinch-off voltage, V_p, and as a result, when the transistor enters the ohmic region, we may consider that v_{DS} is nearly zero.
5. The input and output time constants are widely separated from one another, so that the shorter of the two can be ignored.

Using these approximations, the transient analysis of the field-effect transistor is quite straightforward, and the equivalent circuits given in Figure 10.2-6c, d, and e are seen to represent the FET when it is active, cutoff, and saturated, respectively. In the example that follows we examine the FET switching circuit illustrated in Figure 10.2-6a for the case where $C_L \gg C_{gs}$. The opposite situation, namely that where $C_{gs} \gg C_L$, is left as an exercise for the homework (Problem 10.2-10).

EXAMPLE 10.2-1

Figure 10.2-7
Example 10.2-1.

The transistor in the switching circuit in Figure 10.2-6a has the i_D versus v_{GS} characteristics shown in part (b) of the figure. In addition, $C_L = 15$ pF and C_{gs} is so small that it may be neglected. If v_{in} is a 500-ns-wide periodic pulse train switching between 0 and -10 V, sketch v_{GS}, i_D, and v_{DS}.

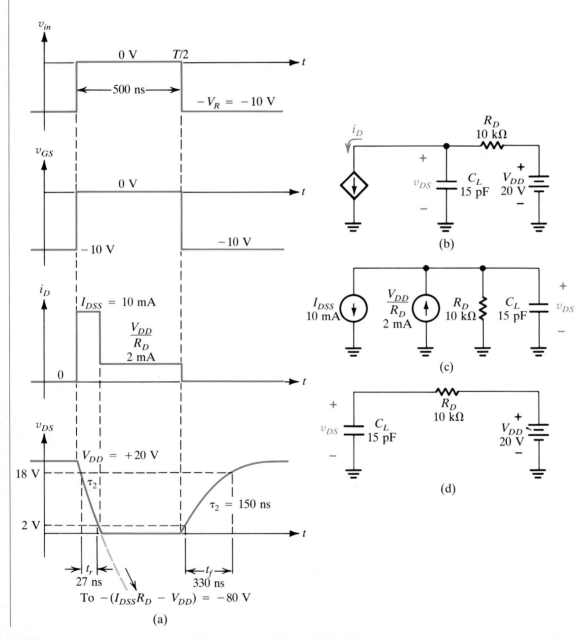

(a)

(b)

(c)

(d)

For this case, with $C_{gs} = 0$, v_{GS} follows v_{in} (see Figure 10.2-7a). The basic form of the output equivalent circuit is illustrated in part (b) of the figure. If we assume that the period of the switching signal v_{in} allows sufficient time for the circuit waveshapes to reach steady state, then at $t = 0-$ with $v_{in} = -V_R$ just before it switches to zero, $v_{DS}(0-) = V_{DD}$.

When v_{in} switches to zero, i_D jumps up to I_{DSS} and the output equivalent circuit is given by Figure 10.2-7c (after a Thévenin-to-Norton transformation has been made). By inspection of this circuit, the drain-to-source voltage is seen to be

$$v_{DS}(t) = V_{DD}e^{-t/\tau_2} - (I_{DSS}R_D - V_{DD})(1 - e^{-t/\tau_2}) \qquad (10.2\text{-}17)$$

with $\tau_2 = R_D C_L$, so that the circuit rise time (or turn-on time) is given by

$$t_r = \tau_2 \ln \frac{I_{DSS}R_D - 0.1V_{DD}}{I_{DSS}R_D - 0.9V_{DD}} \simeq 27 \text{ ns} \qquad (10.2\text{-}18)$$

Once v_{DS} reaches zero, the transistor enters the ohmic region, and to the extent that we can neglect the transistor ON resistance, i_D immediately drops to $V_{DD}/R_D = 2$ mA.

At $t = T$ when the input switches from zero to $-V_R$ volts, the transistor cuts off and the drain current immediately goes to zero. The output equivalent circuit during this interval is that given in Figure 10.2-7d. If we redefine $t = 0$ at this switching point, and note that $v_{DS}(0-) = 0$ for this circuit, it immediately follows that

$$v_{DS}(t) = V_{DD}(1 - e^{-t/\tau_2}) \qquad (10.2\text{-}19)$$

Thus the waveshape for $v_{DS}(t)$ in this region is a full exponential and the fall time is therefore about $2.2\tau_2$, or 330 ns, while the turn-off time (or time for v_{DS} to reach 90% of its final value) is equal to $2.3\tau_2$ or about 345 ns for this particular circuit.

Exercises

10.2-1 For the circuit in Figure 10.2-3a, v_{in} is a 2-V(p-p) square wave, $C_J = 10$ pF, and $\tau_r = 1$ μs. Determine the diode turn-on time if $R = 1$ kΩ. (*Note:* Because the signal amplitude is small, the 0.7-V diode drop cannot be neglected.) ***Answer*** 19 ns

10.2-2 Determine the diode storage time for the circuit described in Exercise 10.2-1. ***Answer*** 163 ns

10.2-3 The parameters for the BJT in the circuit in Figure 10.2-5a are $h_{FE} = 20$, $C_{JE} = 20$ pF, and $\tau_B = 100$ ns. If $R_B = 10$ kΩ, $R_C = 2$ kΩ, and $V_{CC} = 10$ V, find the time required for v_{CE} to reach 0.2 V when v_{in} changes from 0 to 5 V. ***Answer*** 114 ns

10.3 INTRODUCTION TO DIGITAL INTEGRATED CIRCUITS AND DIGITAL LOGIC FAMILIES

In the 1940s the first electromechanical computing equipment was introduced. Its development was spurred on by the advent of World War II, and during the war these machines were used for automating gunnery trajectory calculations. By the end of this war (in the mid-1940s) the first all-electronic vacuum-tube-style digital computers had also been developed. By today's standards these computer systems were dinosaur-like in that they were huge, slow machines that consumed tremendous amounts of electrical power. For example, a vacuum-tube computer containing, say, 1000 bytes of memory could easily require 20 kW of electricity to run it, and would probably have been large enough to fill an entire room.

The invention of the transistor in the late 1940s dramatically reduced both the size and the power consumption of digital electronic equipment. Because the physical size of a transistor is about 1/100 that of a vacuum tube and also because it consumes about 100 times less power, using transistors, our 1-kilobyte computer could now be "shrunk down" to fit inside a large cabinet.

The next major development in electronics occurred in the early 1960s when the first integrated circuits were introduced. With these devices it was now possible to put entire digital logic functions onto a single chip. This freed the designer from having to worry about the details of the circuit design, and instead, allowed him to conceptionalize the problem at the "chip level." By doing so, complex digital system designs could now be accomplished by simply interconnecting these digital building blocks.

At first, only one or perhaps several gates could be placed inside a single IC package, but later as manufacturing techniques improved, it became possible to fabricate flip-flops, registers, and other complex digital functions in a single integrated-circuit package. ICs containing up to 10 elemental digital logic functions (or gates) are usually classified as SSI or small-scale integrated circuits while those having between 10 and 100 gates are known as MSIs (medium-scale ICs). Digital integrated circuits containing from 100 to 1000 logic gates are termed LSIs (large-scale integrated circuits), and those whose gate densities exceed the 1000-gate level are called VLSIs (very large scale integrated circuits).

Let's examine our original 1-kilobyte computer design in the light of these later developments in digital IC technology. By designing our computer with SSI and MSI components, its physical size could probably be reduced sufficiently so that it could now be made to fit into a desktop cabinet. Furthermore, if this same hardware was constructed using LSI and VLSI devices, this computer could probably be built as a hand-held unit. Besides having drastically reduced the size of our computer, the use of LSI technology has also permitted a dramatic reduction in its price. Using a conservative estimate of $5 per vacuum tube, our original 1940-style computer would probably have cost between $50,000 and $100,000 to build. By way of contrast, because transistors in VLSI circuits cost about $0.02 each, our hand-held IC version of this computer could probably be built for less than $25, including batteries.

10.3-1 Overview of the History of Digital Logic Families

When the first digital logic ICs were manufactured, depending on their internal structure, some of these digital logic gates and flip-flops could be directly interconnected, whereas others required special interfacing circuits. This was true not only for ICs made by different companies but also for different devices made by the same manufacturer. As time went on, manufacturers recognized the importance of being able to connect one digital device to another without the need for additional interfacing hardware, and as a result, they began to develop "logic families" or groups of gates, flip-flops, registers, and so on, having compatible logic levels, power requirements, and speeds.

In this section we take a look at the historical development of digital logic circuits, focusing our attention on DCTL (direct-coupled transistor logic), RTL (resistor-transistor logic), and DTL (diode-transistor logic) devices. These three logic forms were the first discrete transistorized digital logic families, and later both RTL and DTL were also available in integrated-circuit form. Although all three of these logic families are basically obsolete today, it is still a good idea to take a quick look at them in order to obtain a historical perspective of the development of digital electronics.

10.3-2 Direct-Coupled Transistor Logic

The use of transistors in place of vacuum tubes in digital electronic circuits presented new opportunities for the design of very simple logic circuits. Unlike vacuum tubes, transistors have compatible input and output voltage levels so that the output of one circuit can, in theory at least, be directly connected to the input of the next without the need for any interfacing hardware.

DCTL, or direct-coupled transistor logic, was a natural outgrowth of this minimum-component design philosophy. A typical BJT-style DCTL gate is illustrated in Figure 10.3-1a. In this gate V_1, V_2, and V_3 are the logic inputs, and V_o is the gate output. If we consider that this gate operates with the positive-

Figure 10.3-1
DCTL NOR gate.

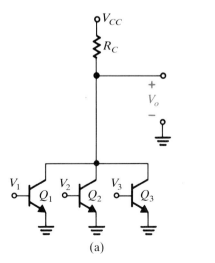

V_1	V_2	V_3	V_o
0	0	0	1
0	0	1	0
0	1	1	0
0	1	0	0
1	1	0	0
1	1	1	0
1	0	1	0
1	0	0	0

(a) (b)

logic convention, the logic function implemented by it may be understood as follows.

When V_1, V_2, and V_3 are all LO (i.e., < 0.7 V), Q_1, Q_2, and Q_3 will be cut off, and V_o will be equal to V_{CC}. On the other hand, if any of the inputs goes HI, the transistor corresponding to that input will saturate and this will bring V_o down to a logic 0 level at about 0.2 V. The output V_o will be similarly affected when two or more of the inputs goes HI. These results are summarized in the truth table given in Figure 10.3-1b, and by examining the entries in this table it should be apparent that this digital logic circuit is a three-input NOR gate.

Direct-coupled transistor logic works fine in theory but has serious practical design problems. The basic difficulty with DCTL is that when one logic gate drives two or more additional gates, "current hogging" can occur, in which all the available current from the driving gate is taken by one of the gates being driven. This occurs because the base–emitter turn-on voltages of the transistors being driven can differ substantially from one another. Typically, this problem arises in discrete logic circuit designs due to the fact that supposedly identical transistors may have slightly different doping levels and device geometries. However, this problem can also occur in IC designs.

Consider, for example, the simple digital logic circuit examined earlier in Figure 10.1-2d. Here even though all these inverters may be integrated-circuit devices, they could easily be located in different IC packages that were manufactured at different times, in different places, and even, in fact, by different manufacturers. Hence, for these inverters, the turn-on voltages for Q_1 through Q_{10} might be substantially different from one another.

Let's examine what happens in this circuit if, for example, the 5-kΩ base resistors are replaced by shorts as they would be for the case of DCTL, and if V_{o1} is 0.65 V for Q_1 while Q_2 through Q_{10} all have base–emitter turn-on voltages of 0.7 V. For this situation, when Q_0 is cut off (goes to a logic 1) Q_1 through Q_{10} are supposed to saturate, causing their outputs to go to a logic 0. However, because of the difference in their turn-on voltages only Q_1 will saturate and its base current will be $(10 - 0.65)/1$ k$\Omega = 9.35$ mA. Q_2 through Q_{10}, on the other hand, will remain cut off since their base–emitter voltages are also 0.65 V, which is below their turn-on thresholds. This is an extreme case of current hogging, in which all of the available current flows into the transistor having the smallest base–emitter turn-on voltage. By adding series base resistors, as in the original circuit in Figure 10.1-2d, a more even distribution of the available drive current is produced even when the base–emitter turn-on voltages of the transistors differ substantially from one another.

10.3-3 Resistor-Transistor Logic

The addition of base resistors in DCTL to reduce current hogging essentially converts this logic family into RTL (resistor-transistor logic) (Figure 10.3-2a). RTL was the first widely used digital logic family and was very popular in the mid-1960s and early 1970s. It was initially available in discrete form on printed-circuit cards and was later produced in integrated-circuit form by several semiconductor manufacturers. The major disadvantages of RTL are its low noise immunity and its low speed.

EXAMPLE 10.3-1

The digital logic circuit shown in Figure 10.3-2b is a cascade of three RTL inverters of the type illustrated in part (a) of the figure. The transistors in all these gates are silicon devices with h_{FE}'s = 20. Find the noise margins of gate G_1 in this circuit.

SOLUTION

For the circuit in Figure 10.3-2c, it is fairly apparent that V_{IL} and V_{OL} for G_1 are 0.7 V and 0.2 V, respectively. The value of V_{OH} from gate 1 is found by noting that Q_1 is cut off under these conditions, so that

$$V_{OH} = V_{o1} = V_{CC} - IR_{C1} = 3 - \frac{3 - 0.7}{1.5 \text{ k}\Omega}(1 \text{ k}\Omega) = 1.47 \text{ V} \qquad (10.3\text{-}1)$$

To compute the minimum value of V_{o1} that will still cause Q_2 to saturate, we note that $I_{B,min}$ for Q_2 is equal to

$$I_{B,min} = \frac{I_{C,sat}}{h_{FE}} = \frac{2.8 \text{ mA}}{20} = 0.14 \text{ mA} \qquad (10.3\text{-}2)$$

The corresponding minimum value for V_{o1} needed to guarantee that at least that much base current flows in Q_2 is

$$V_{IH} = V_{BE} + I_{B,min} R_{B2} = 0.77 \text{ V} \qquad (10.3\text{-}3)$$

Figure 10.3-2
Examples 10.3-1 and 10.3-2.

These input and output logic gate threshold voltages are summarized in

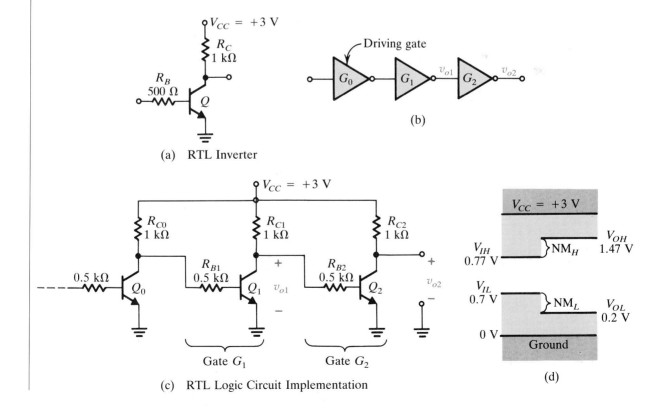

(a) RTL Inverter

(b)

(c) RTL Logic Circuit Implementation

(d)

Figure 10.3-2d, and using these data it follows immediately that

$$\text{NM}_L = V_{IL} - V_{OL} = 0.7 - 0.2 = 0.5 \text{ V} \qquad (10.3\text{-}4a)$$

and
$$\text{NM}_H = V_{OH} - V_{IH} = 1.47 - 0.77 = 0.7 \text{ V} \qquad (10.3\text{-}4b)$$

The results obtained in Example 10.3-1 for a fanout of 1 serve to illustrate one of the main disadvantages of RTL logic, its rather poor noise immunity. The situation for increased fanout is even more severe; for example, with a fanout of 5, the NM_H is reduced to only 0.14 V. A second shortcoming of RTL is its low switching speed. Example 10.3-2 will help to illustrate this point. However, before getting into this example it will first be a good idea to say a few words about how best to analyze the switching characteristics of logic circuits.

In examining the switching performance of BJT transistor circuits we basically employ the simplified linear switching models that we developed in the preceding section. However, when using these models it will be most important to recognize that the voltage v_1 across $r_{B'E}$ and C_{DE} in Figure 10.2-4a is always much smaller than V_o. This fact might tend to escape us in some of the circuits that we will be examining since when i_B is large it also appears that $i_B r_{B'E}$ can be quite large. However, when we recognize that $r_{B'E}$ is not really constant but is actually inversely proportional to i_B, it becomes apparent that v_1 is always very small in our equivalent circuit. This fact is very important in our switching analyses since it will permit us to make very simple approximations for the charging and discharging currents in the base of the transistor. Once these currents (I_{BF} and I_{BR}) are known, we can directly apply equations (10.2-4a), (10.2-10), (10.2-12b), and (10.2-15) to calculate t_d, t_r, t_s, and t_f, respectively.

EXAMPLE 10.3-2

The three-inverter circuit shown in Figure 10.3-2b is implemented using the RTL logic gates in part (c) of the figure. The transistors are all silicon devices with $h_{FE} = 20$, $C_{JE} = 2$ pF, $C_{DE} = 80$ pF, and $r_{B'E} = 500 \ \Omega$. Notice that for the values of C_{DE} and $r_{B'E}$ specified that $\tau_B = 500 \times (80 \times 10^{-12}) = 40$ ns. Determine the output voltage waveshapes at v_{o1} and v_{o2} when the input to gate 1 (or G_1) changes from a logic 1 to a logic 0.

SOLUTION

In analyzing this circuit, let us assume that Q_0 is cut off for some time prior to $t = 0$ so that for $t < 0$ a current

$$I_{BF} = \frac{(3 - 0.7) \text{ V}}{1.5 \text{ k}\Omega} = 1.5 \text{ mA} \qquad (10.3\text{-}5)$$

flows into the base of Q_1. Furthermore, because

$$I_{C,\text{sat}} = \frac{(3 - 0.2) \text{ V}}{1 \text{ k}\Omega} = 2.8 \text{ mA} \qquad (10.3\text{-}6a)$$

and
$$I_{B,\text{min}} = \frac{1.0}{20} = 0.05 \text{ mA} \qquad (10.3\text{-}6b)$$

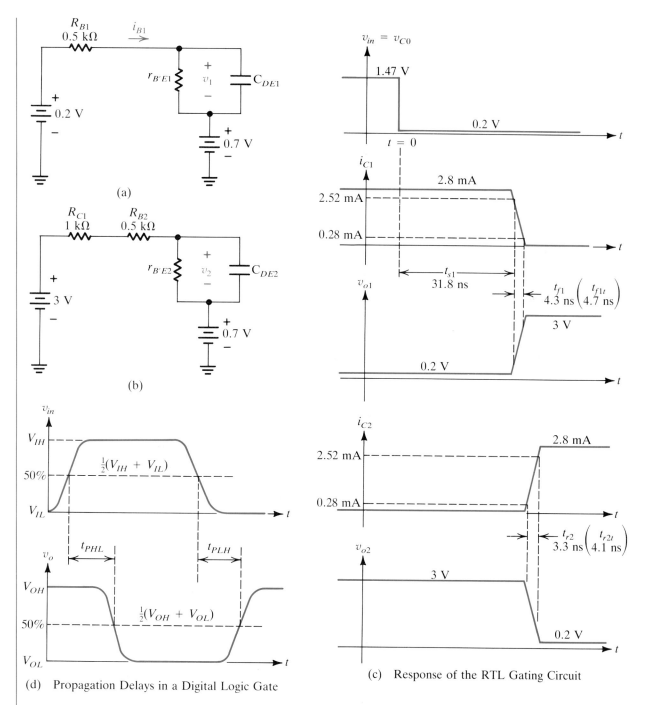

(a)

(b)

(d) Propagation Delays in a Digital Logic Gate

(c) Response of the RTL Gating Circuit

Figure 10.3-3
Example 10.3-2 (continued).

Q_1 is saturated when the input to this gate is a logic 1. In addition, with $v_{o1} = 0.2$ V, Q_2 will initially be cut off and v_{o2} will therefore equal 3 V.

When Q_0 saturates at $t = 0$ its collector voltage drops to 0.2 V, the input to G_1 becomes a logic 0, and following the equivalent circuit in Figure 10.3-3a, we have

$$I_{BR} = \frac{(0.7 - 0.5) \text{ V}}{0.5 \text{ k}\Omega} = 1 \text{ mA} \tag{10.3-7}$$

In developing this expression for I_{BR}, as per the discussion preceding this example, we have neglected the voltage drop across $r_{B'E}$ and C_{DE}.

Substituting the data from equations (10.3-5), (10.3-6b), and (10.3-7) into equations (10.2-12) and (10.2-15), the storage and fall time for Q_1 are found to be

$$t_{s1} = \tau_B \ln \frac{I_{BR} + I_{BF}}{I_{BR} + 0.9I_{B,\text{min}}} = 31.8 \text{ ns} \tag{10.3-8a}$$

and

$$t_{f1} = \tau_B \ln \frac{I_{BR} + 0.9I_{B,\text{min}}}{I_{BR} + 0.1I_{B,\text{min}}} = 4.3 \text{ ns} \tag{10.3-8b}$$

Once Q_1 comes out of saturation v_{o1} will begin to increase and the transistor Q_2 will start to turn ON. As a first-order approximation, we will neglect the delay associated with the charging of the junction capacitance. In addition, for simplicity, we will assume that the turn-on of Q_2 does not begin until Q_1 is OFF completely, although both the turn-off of Q_1 and the turn-on of Q_2 occur simultaneously. Using this approximation, for $t > [31.8 + 1.1(4.3)]$ or 36.5 ns the equivalent circuit shown in Figure 10.3-3b is valid for the base of Q_2, and following this figure it is apparent that I_{BF} for the turn-on of Q_2 is the same as that given in equation (10.3-5) for Q_1. Therefore, following equation (10.2-10), the rise time for Q_2 is seen to be

$$t_{r2} = \tau_B \ln \left[\frac{I_{BF} - 0.1 \, I_{B,\text{min}}}{I_{BF} - 0.9 \, I_{B,\text{min}}} \right] = 3.2 \text{ ns} \tag{10.3-9a}$$

Notice that because the rising portion of the i_{C2} waveshape is fairly linear, we may estimate the total time for this current to rise as

$$t_{r2t} = \left(\frac{1}{0.8} \right) t_{r2} = 1.25 t_{r2} \tag{10.3-9b}$$

or about 4.1 ns. A similar definition can be used for t_{ft}, the total fall time of the waveshape, except that the correction factor is $1/0.9$ or about 1.1 since the solution for the storage time is valid all the way down to $0.9I_{C,\text{sat}}$. The pertinent voltage and current wavehsapes are given in Figure 10.3-3c.

For later use it is convenient at this point to define the meaning of propagation delay in a digital logic circuit. For the purpose of this discussion we have illustrated two different propagation delay terms in Figure 10.3-3d. The time interval t_{PHL} is the gate propagation delay when the gate output changes from a HI to a LO, and conversely, t_{PLH} defines the propagation delay that occurs when the output goes from a LO to a HI. Both of these time intervals are measured with respect to the 50% points on the input and output waveshapes. To facilitate the comparison of different types of logic gates, it is also useful to define the average gate propagation delay as $t_P = (t_{PHL} + t_{PLH})/2$.

For the specific circuit being examined in this example, t_{PLH} is equal to $(31.8 + 4.3/2)$ or about 34 ns using the G_1 data. Similarly, the value of t_{PHL} for G_2 is 4.6 ns, so that the average propagation delay is about 19.3 ns for this RTL logic gate.

Sometimes we are interested in being able to quickly determine the logic function performed by a particular digital circuit design. This type of functional analysis of digital logic circuits can be rapidly carried out if we make use of two or three very useful and very simple digital circuit analysis concepts.

1. When a signal enters a common-emitter (or common-source) style amplifier, the output signal, when taken from the collector (or drain), is the logical negation of the input signal. For example, if the input to the RTL gate in Figure 10.3-4a is the variable A, the output at the collector can be immediately labeled as $\bar{A}$.

2. Assuming a positive logic convention, if the collector outputs from two logic gates are connected together, a "wired AND" function is effectively performed at the output. Thus the RTL circuit illustrated in Figure 10.3-4b is logically equivalent to the gating circuit given in part (c) of the figure. The reason for the presence of this "wired AND" gate at the junction point in the output may be understood as follows: If either Q_1 or Q_2 is ON in Figure 10.3-4b, the output F will be zero. The only way for F to be a logic 1 is for Q_1 AND Q_2 to both be OFF, or for $\bar{A}$ and $\bar{B}$ simultaneously to be 1.

   Once this "wired ANDing" concept is understood, the overall logic function performed by a given gate structure may be rapidly developed simply by labeling the logic signals present at all important points in the circuit. This technique may be illustrated by again using the RTL gating circuit in Figure 10.3-4b. The signal entering Q_1 is called A, so the output at the collector of Q_1 is labeled $\bar{A}$. In a similar fashion the input and output at Q_2 are B and $\bar{B}$, respectively. Because Q_1 and Q_2 are joined at their collectors, a wired AND gate is formed, and therefore the output variable F is simply given by the expression $F = \bar{A} \cdot \bar{B}$.

3. Some digital logic circuits use emitter-follower outputs to achieve a low output impedance (Figure 10.3-4d). In this gate the emitter-follower transistor acts only as a buffer and does not provide any logical inversion of the signal. If the outputs from two of these types of gates are connected together, a "wired OR" gate is formed as illustrated in Figure 10.3-4e.

   To understand how this "wired OR" gate works, we first need to observe that the two base–emitter diodes on Q_3 and Q_4 together with their emitter resistors effectively form an OR gate at the output. In particular, when both $\bar{A}$ and $\bar{B}$ are low at the collectors of Q_1 and Q_2, the base–emitter diodes on both Q_3 and Q_4 will be reverse biased, both transistors will be cut off, and F will be a logic 0. On the other hand, if either or both $\bar{A}$ and $\bar{B}$ are HI, the corresponding base–emitter diodes will be ON, and a logic 1 will be produced in the output. Thus the wiring together of the emitters of Q_3 and Q_4 produces a wired OR gate at the output.

4. The two Boolean algebra equivalences

$$x + y = (\overline{\bar{x} \cdot \bar{y}}) \qquad (10.3\text{-}10a)$$

$$x \cdot y = (\overline{\bar{x} + \bar{y}}) \qquad (10.3\text{-}10b)$$

are also very useful for the analysis of combinational digital networks. Both of these identities are relatively easy to prove (see truth tables in Figure

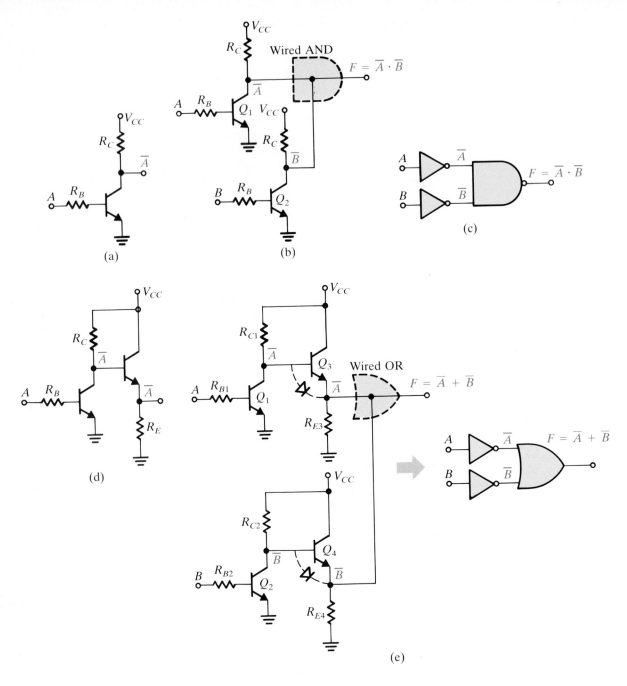

Figure 10.3-4

Functional analysis of digital logic circuits.

10.3-4f and g), and basically they state that an OR gate may be replaced by an AND gate having inverted inputs and outputs [equation (10.3-10a)], and also that an AND gate can be replaced by an OR gate if its inputs and outputs are also inverted [equation (10.3-10b)]. Of course, these two equations, known as DeMorgan's theorems, are just algebraic statements of the gate equivalences originally presented in Figure 10.1-3. These theorems are given again here because they are frequently useful for the simplification of specific logic circuit functions. For example, the output from the logic circuit illustrated in Figure 10.3-4b is expressed as

$$F = (\bar{A}) \cdot (\bar{B}) \tag{10.3-11a}$$

However, by applying equation (10.3-10b), we may rewrite this result as

$$F = [\overline{(\bar{\bar{A}}) + (\bar{\bar{B}})}] = [\overline{A + B}] \tag{10.3-11b}$$

from which it is clear that this RTL circuit is simply a NOR gate. In a similar fashion, by applying equation (10.3-10a), the output from the logic circuit in Figure 10.3-4e can be rewritten as

$$F = \bar{A} + \bar{B} = [\overline{(\bar{\bar{A}}) + (\bar{\bar{B}})}] = [\overline{A \cdot B}] \tag{10.3-12}$$

in order to demonstrate that this circuit functions as a NAND gate.

10.3-5 Diode-Transistor Logic

In DCTL and RTL the transistors are used to perform both the logic and buffering functions within the gate. A third logic type, diode-transistor logic (DTL), uses diodes to carry out the ANDing and ORing operations, and transistors to provide the buffering and signal inversion functions. A typical DTL gating circuit is shown in Figure 10.3-5a. Basically, it consists of a diode AND gate followed by a BJT inverter, and as such, the overall logic function implemented by this circuit is that of a NAND gate.

By examining this circuit, it should be readily apparent that the transistor Q_1 functions as a digital logic inverter. However, the AND gate operation of the diode portion of this circuit may require a bit more explanation. To assist us in this description, let's remove the diode network from the DTL circuit and examine its operation separately as shown in Figure 10.3-5b. In this circuit, if V_1, V_2, or both V_1 and V_2 are zero, one or both of these diodes will conduct making the voltage at V_x nearly zero. When both V_1 and V_2 are at a logic 1 level ($+V_{CC}$), D_1 and D_2 will be reverse biased, and V_x will rise to $+V_{CC}$. Thus V_x will be a logic 1 only when V_1 AND V_2 are 1. Therefore, this diode circuit functions as a digital logic AND gate.

The operation of the DTL gating circuit in Figure 10.3-5a is quite similar

Figure 10.3-5

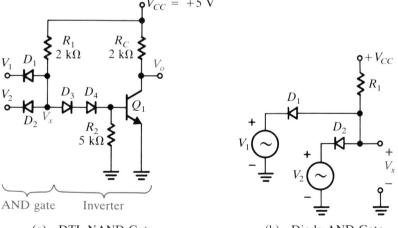

(a)   DTL NAND Gate   (b)   Diode AND Gate

to that of the diode AND gate just discussed. When V_1, V_2, or both V_1 and V_2 are at a logic 0, D_1 and/or D_2 will be ON, and as a result, the voltage at the node V_x in this circuit will be about 0.9 V, assuming that $V_{OL} = +0.2$ V from the gate driving this logic gate. Due to the presence of the diodes D_3 and D_4, this voltage will be insufficient to turn on the base–emitter junction of Q_1, and therefore Q_1 will be cut off and V_o will be at $V_{CC} = +5$ V.

When both V_1 and V_2 are at a logic 1, D_1 and D_2 will be reversed biased, and current will now be able to flow down through R_1 into D_3 and D_4, forward biasing the base–emitter junction of Q_1 and saturating this transistor. Thus in this circuit the output will be a logic 0 when both V_1 and V_2 are at a logic 1 level, and will be at a logic 1 level for all other input combinations. Clearly, based on this description, this circuit functions as a two-input NAND gate.

In this DTL circuit it is interesting to observe that the gate would still function even with D_4 and R_2 removed from the circuit. Two diodes rather than one are connected in series with the base of Q_1 in order to improve the noise immunity with respect to logic 0 signals connected at the input. With D_4 removed from the gate in Figure 10.3-5a, NM$_L$ drops from 1.2 V to only 0.5 V. The resistor R_2 has little to do with the static operation of this gate. It is there to improve the circuit switching speed by providing a discharge path for the stored charge in the base of the transistor when V_x goes LO. Sometimes this resistor is returned to a negative power supply in order to further speed up the removal of charge from the base when Q_1 needs to be turned off.

EXAMPLE 10.3-3

RTL has a much simpler gate structure than DTL, yet in spite of this, DTL rapidly replaced RTL in most digital logic applications as soon as it was introduced. This is because the noise immunity and switching speed of DTL are significantly better than those of RTL.

a. Develop the transfer characteristic for a DTL inverter gate for a circuit fan-out of 1 (see Figure 10.3-6a). Assume that the gate has the same design as that given in Figure 10.3-5a, except that it has only one input. Further assume that all semiconductors are silicon and that $h_{FE} = 20$ for the transistor.
b. Determine the gate noise immunity characteristics for the conditions in part (a).
c. Estimate t_{PLH}, t_{PHL}, and t_P for these DTL inverter gates. As with Example 10.3-1, consider that $\tau_B = 40$ ns, and neglect the charging times associated with the junction capacitances of the transistors and diodes. In addition, consider that the storage times of D_2, D_3, D_5, and D_6 are much greater than those of the transistor, so that during the removal of the stored base charge from the transistor, these diodes will remain ON and can be treated as 0.7-V batteries. Furthermore, let $r_{BB'} = 25$ Ω for the transistor and assume that the diode ON resistances and transistor saturation resistances are also 25 Ω.

SOLUTION

When the input to the gate G_1 is a logic 1, D_1 will be reverse biased and D_2, D_3, and the base–emitter junction of Q_1 will conduct so that V_{x1} in Figure 10.3-6a will be 2.1 V and

$$I_{R1} = \frac{(5 - 2.1)\ \text{V}}{2\ \text{k}\Omega} = 1.45\ \text{mA} \tag{10.3-13a}$$

Because

$$I_{R2} = \frac{0.7\ \text{V}}{5\ \text{k}\Omega} = 0.14\ \text{mA} \tag{10.3-13b}$$

Figure 10.3-6
Example 10.3-4.

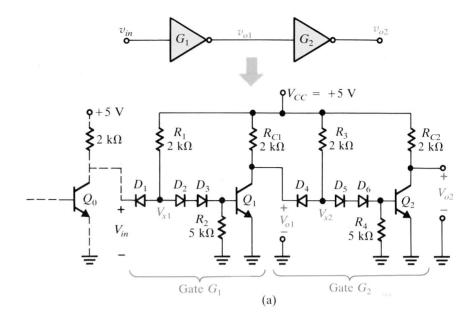

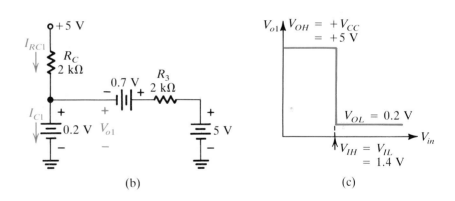

(b)

(c)

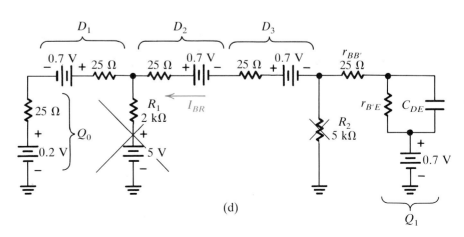

(d)

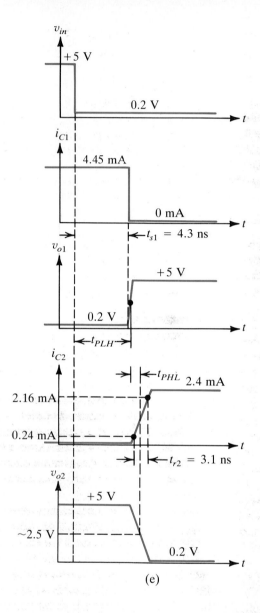

Figure 10.3-6
(continued)

(e)

the base current into Q_1 is

$$I_{B1} = I_{R1} - I_{R2} = 1.31 \text{ mA} \qquad (10.3\text{-}13c)$$

If we assume that this current is sufficient to saturate Q_1, then V_{o1} will be about 0.2 V, and this will turn ON the input diode D_4 in gate G_2 so that the equivalent circuit given in Figure 10.3-6b will be valid. Following this figure, the collector current in Q_1 is seen to be

$$I_{C1} = \frac{5 - 0.2}{2 \text{ k}\Omega} + \frac{5 - 0.7 - 0.2}{2 \text{ k}\Omega} = 2.4 + 2.05 = 4.45 \text{ mA} \qquad (10.3\text{-}14a)$$

so that

$$I_{B1,\min} = \frac{4.45}{20} = 0.223 \text{ mA} \qquad (10.3\text{-}14b)$$

Comparing equations (10.3-13c) and (10.3-14b), it is apparent that Q_1 will be saturated when the input to it is a logic 1. Notice under these conditions that

$$V_{OL} = 0.2 \text{ V} \tag{10.3-15a}$$

and

$$V_{IH} = V_x - V_{D1} = 2.1 - 0.7 = 1.4 \text{ V} \tag{10.3-15b}$$

since D_1 will be reverse biased for all input voltages exceeding this value.

When V_{in} goes LO, D_1 forward biases and this lowers V_{x1} to about $(0.2 \text{ V} + 0.7 \text{ V}) = 0.9 \text{ V}$, which reverse biases D_2, D_3, and the base–emitter junction of the inverter transistor Q_1 in gate G_1. With Q_1 cut off the gate output voltage V_{o1} will rise, reverse biasing input diode D_4 in gate G_2. This will remove the load from the output of G_1 and will allow V_{o1} to rise all the way up to $+V_{CC}$. As a result, we have

$$V_{OH} = V_{CC} = +5.0 \text{ V} \tag{10.3-16}$$

The corresponding maximum allowed input voltage to this gate that will guarantee that Q_1 stays off may be determined as follows. For Q_1 to be OFF, it is necessary that V_{x1} remain below 2.1 V in that gate. This requires that D_1 conduct and also that V_{in} be below $(2.1 - 0.7)$ volts, so that

$$V_{IL} = 2.1 - 0.7 = 1.4 \text{ V} \tag{10.3-17}$$

The transfer characteristic for this DTL inverter is sketched in Figure 10.3-10c. Notice that the slope of this curve is infinite during the time that the transistor passes through its active region. This result is not all that surprising if we recall that this slope is equal to the gate small-signal voltage gain during the time that Q_1 is active. Because we have assumed that the diodes and transistor base–emitter junction may be represented by batteries, we have also tacitly assumed that $r_{b'e}$, $r_{bb'}$, and the diode small-signal resistances are all zero. To the extent that these approximations are valid, the small-signal voltage gain of this gate is minus infinity, as the slope of the gate transfer characteristic indicates.

The transient response of this circuit may be found by using the same analysis technique that we employed to study the RTL circuit examined earlier in this section. With V_{in} a logic 1, D_1 in gate G_1 is reversed biased and this turns D_2, D_3, and Q_1 ON, so that

$$I_{BF} = \frac{5 - 2.1}{2 \text{ k}\Omega} - \frac{0.7}{5 \text{ k}\Omega} = 1.31 \text{ mA} \tag{10.3-18}$$

for Q_1. Because $I_{B1,min} = 0.223$ mA for Q_1, this transistor will be saturated, and with $V_{o1} = 0.2$ V, Q_2 will be cut off.

When the transistor in the driving gate saturates, as illustrated in Figure 10.3-6d, V_{in} will drop to about 0.2 V, and this will begin to turn OFF Q_1. If we assume that the turn-off time of the diodes is much longer than that associated with the transistors, then during the transistor turn-off transient, D_2 and D_3 will look like 0.7-V batteries. In calculating I_{BR} for this circuit, the branches containing R_1 and R_2 have only a small effect, and deleting them, the reverse base current may be written as

$$I_{BR} = \frac{(2.1 - 0.9) \text{ V}}{125 \text{ }\Omega} = 9.6 \text{ mA} \tag{10.3-19}$$

Substituting these results for I_{BF} and I_{BR} into equation (10.2-14b), the storage time for Q_1 is seen to be

$$t_{s1} = \tau_B \ln \frac{I_{BR} + I_{BF}}{I_{BR} + 0.9I_{B,\min}} = 4.3 \text{ ns} \qquad (10.3\text{-}20)$$

The fall time for the collector current is much smaller (0.7 ns) and will be assumed to be nearly zero.

Once Q_1 cuts off, the output voltage will rapidly rise to +5 V, and to the extent that we can neglect the junction capacitances across the diodes and transistors, the base–emitter junction of Q_2 will turn on instantaneously and I_{BF} will have the same value as that given in equation (10.3-18). Because $I_{C,\text{sat}} = [(5 - 0.2) \text{ V}]/(2 \text{ k}\Omega) = 2.4 \text{ mA}$ for this transistor, $I_{B,\min} = 2.4/20$ or 0.12 mA and the collector current rise time will be

$$t_{r2} = \tau_B \ln \frac{I_{BF} - 0.1I_{B,\min}}{I_{BF} - 0.9I_{B,\min}} = 40 \ln \frac{1.31 - 0.01}{1.31 - 0.11} = 3.1 \text{ ns} \qquad (10.3\text{-}21)$$

These results are sketched in Figure 10.3-6e, and following this figure, the circuit propagation delays are seen to be

$$t_{PLH} = 4.3 \text{ ns} \qquad (10.3\text{-}22a)$$

$$t_{PHL} = \frac{1.25t_{r2}}{2} \simeq 1.9 \text{ ns} \qquad (10.3\text{-}22b)$$

and
$$t_P = \frac{1}{2}(t_{PLH} + t_{PHL}) = 3.1 \text{ ns} \qquad (10.3\text{-}22c)$$

In analyzing this DTL circuit, we found that R_2 had little effect on the gate switching characteristics. However, for DTL circuits where the diode and transistor switching times are comparable, the resistor R_2 will have a pronounced effect on the circuit's transient performance.

Exercises

10.3-1 Demonstrate the current hogging effect in DCTL by letting $R_B = 0$ in Figure 10.1-2d and considering that $V_o = 0.6$ V for Q_1 in G_1 and 0.7 V for the transistors in G_2 through G_{10}. In particular, determine I_{B1} through I_{B10} when the output of G_0 is a logic 1. *Answer* $I_{B1} = 9.4$ mA, I_{B2} through I_{B10} equal zero.

10.3-2 For the DTL NAND gate in Figure 10.3-5, find I_{R1}, I_{R2}, and I_{B1} when $V_1 = +5$ V and $V_2 = 1.0$ V. *Answer* $I_{R1} = 1.65$ mA, $I_{R2} = 0.06$ mA, $I_{B1} = 0$

10.3-3 a. Determine the turn-off time for the DTL circuit in Figure 10.3-5a if τ_r for the diodes is much smaller than τ_B for the transistor.
b. Repeat the analysis in part (a) if R_2 is returned to a -5-V supply instead of to ground.
Answer (a) 94 ns, (b) 31 ns

10.4 TRANSISTOR-TRANSISTOR LOGIC

Diode transistor logic was very popular in the 1960s and was available in both discrete and integrated forms. In fact, this logic style represented one of the first attempts at the commercial manufacturing of integrated circuits.

As we saw in Section 10.3, DTL logic has a relatively good noise margin and fan-out and a fairly high switching speed. However, this high-speed performance requires that the recombination time of the input diodes (D_2 and D_3 in Figure 10.3-6a) be much greater than that of the inverter transistor. In discrete circuit designs the diodes could be chosen to achieve this, but in an IC, all the semiconductors on the substrate have nearly identical recombination times. This fact significantly increases the gate propagation delays in IC-style DTL.

To see why this is the case, let's reexamine the switching performance of the DTL inverter shown in Figure 10.3-6a, assuming that D_2 and D_3 turn off immediately as soon as the input to the gate goes LO. In this instance, I_{BR} is $(0.7\text{ V})/(5\text{ k}\Omega)$, or 0.14 mA, instead of the 9.6 mA that we obtained when the diodes stayed ON during the transistor turn-off transient [equation (10.3-19)]. Correspondingly, the storage time increases to more than 50 ns, compared with the 4.3 ns computed in equation (10.3-20). As a practical matter, commercially available DTL has an average propagation delay that is typically on the order of 15 to 20 ns.

Transistor-transistor logic (TTL) was an attempt to improve on the speed, noise immunity, fan-out, and capacitive load drive ability of DTL. The development of the circuit design of a typical TTL logic gate will be carried out in three stages to better illustrate its evolution from DTL. Figure 10.4-1a shows the layout of a conventional DTL logic gate, and part (b) of the figure contains a primitive TTL version of the same gate in which the two diodes D_1 and D_2 of the DTL gate have been replaced by the transistor Q_1.

From an IC layout viewpoint, the TTL circuit represents an improvement over its DTL counterpart because the transistor Q_1 takes up less area than the diodes that it replaces. The use of a transistor in place of the diodes also offers a significant speed advantage. In the DTL circuit, when the input goes LO, once D_2 turns OFF, the base charge on Q_2 must now leak off through R_2 before the transistor itself can turn OFF. For the TTL version of this gate, the circuit operation is quite different. When the input goes LO, Q_1 enters the active region, and this transistor draws a considerable amount of current out of the base of Q_2, rapidly turning it OFF. This is explained in much greater detail later in this section.

Multiple inputs are handled in DTL by connecting additional diodes to form a diode AND gate structure (Figure 10.4-1c). The discrete component TTL version of this circuit is shown in Figure 10.4-1d. Here the D_1–D_3 and D_2–D_3 diode combinations are replaced by the transistors Q_1 and Q_2, respectively. Because Q_1 and Q_2 have their base and collector leads in common, when this circuit is fabricated in IC form, both of these transistors may be replaced by a single multiemitter transistor. These multiemitter devices offer a significant real estate saving in the IC layout.

In summary, functionally speaking, the TTL multiemitter transistor is logically equivalent to a diode AND gate. The major difference between the two is that the transistor action in Q_1 in the TTL gate permits the rapid removal of the base charge from the transistor being driven by this device.

Two other improvements are also incorporated into TTL in order to enhance its performance. First, an additional transistor (Q_2 in Figure 10.4-1f) is usually added between the input and output transistors. This transistor buffer significantly increases the base drive current into Q_3, thereby increasing the

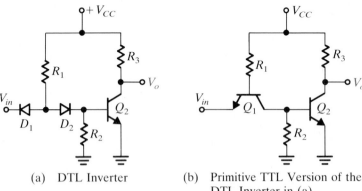

(a) DTL Inverter

(b) Primitive TTL Version of the DTL Inverter in (a)

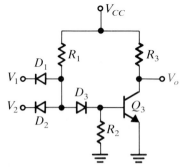

(c) Multiple-Input DTL NAND Gate

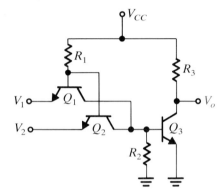

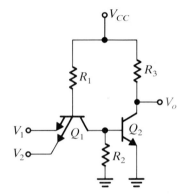

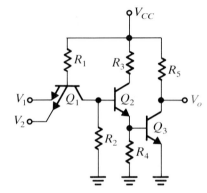

(d) Discrete Transistor Multiple Input TTL NAND Gate

(e) TTL Logic Gate with a Multiemitter Input Transistor

(f) TTL Logic Gate with a Multiemitter Input Transistor and a Buffer Transistor Q_2

Figure 10.4-1

maximum fan-out of the circuit. It also adds an additional diode drop to the gate threshold voltage, which improves the circuit noise immunity with respect to logic 0 signals applied at the input.

Besides having fan-out problems, DTL also has significant difficulty driving capacitive loads. As we will see later in the examples in this section, a resistive pull-up output circuit such as that found in DTL or in the primitive TTL circuits discussed thus far does fine when the output transistor turns ON and the output is forced to a LO. However, when this transistor cuts off and the output attempts to go high, it can take a long time for the collector resistor to charge the load capacitance.

Figure 10.4-2a illustrates the design of an actual TTL logic gate. The specific circuit shown is a two-input NAND gate. To improve its capacitive drive ability, an "active pull-up" output circuit has replaced the resistive pull-up circuit just discussed. Sometimes this active pull-up circuit is also called a totem-pole output because the transistors are stacked one on top of the other. In this circuit Q_2 still functions as a buffer, but it also acts as a phase inverter since v_{C2} is 180° out of phase from the base signal, while v_{E2} is in phase with the base voltage. In this way, when the base voltage of Q_2 goes high, Q_3 will turn ON and Q_4 will go OFF. Conversely, when v_{B2} goes low, Q_3 will cut off and Q_4 will turn ON. Because either Q_3 or Q_4 is always ON, the circuit output

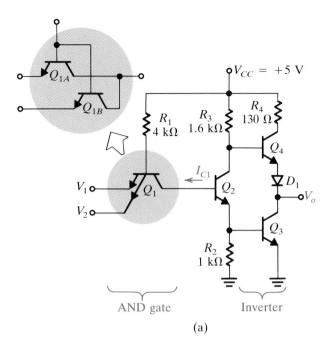

Parameter	Worst-Case Parameter Value
I_{OH}	-400 μA (min.)
I_{OL}	16 mA (min.)
I_{IH}	40 μA (max.)
I_{IL}	-1.6 mA (max.)
V_{OH}	2.4 V (min.) (driving 10 TTL loads)
V_{OL}	0.4 V (max.) (driving 10 TTL loads)
V_{IH}	2.0 V (min.)
V_{IL}	0.8 V (max.)

(b) Typical Manufacturer's Specifications for a 7400-Style TTL Logic Gate

(a)

Figure 10.4-2

Typical 7400 family TTL logic gate (a two-input NAND gate).

impedance is low regardless of the output logic state, and as a result, the output time constant $Z_{out} C_L$ will always be small. Besides lowering the gate rise and fall times associated with capacitive loading, the use of an active pull-up output circuit also significantly reduces the internal gate propagation delay.

10.4-1 Basic Operation of a TTL Logic Gate

Let's briefly examine the gating circuit in Figure 10.4-2a in order to get an elemental understanding of how it operates. Very simply, when either or both of the inputs to the gate is a logic 0, Q_1 saturates because I_{C1} is nearly zero so that I_{B1} is much greater than $I_{B1,min}$. As a result, in this instance V_{B2} will be about 0.2 V and Q_2 will be cut off. With Q_2 OFF, Q_3 is also cut off, Q_4 is ON, and V_o is therefore a logic 1.

When both V_1 and V_2 are HI, the emitter junctions of Q_{1A} and Q_{1B} are both reverse biased, and the base–collector diodes of these transistors conduct, turning Q_2 and Q_3 ON and making V_o LO. Thus V_o is a logic LO only when both inputs are HI; in other words, this circuit behaves like a digital logic NAND gate.

Let's reexamine the operation of this gate, but this time with an eye toward developing a good quantitative understanding of how this circuit operates. When V_1 and V_2 are both HI, the base–collector junction of Q_1 is ON as well as the base–emitter junctions of Q_2 and Q_3, so that

$$V_{B3} = 0.7 \text{ V} \tag{10.4-1a}$$

$$V_{B2} = 0.7 + V_{B3} = 1.4 \text{ V} \tag{10.4-1b}$$

and

$$V_{B1} = 0.7 + V_{B2} = 2.1 \text{ V} \tag{10.4-1c}$$

Thus
$$I_{R1} = \frac{(5 - 2.1)\ \text{V}}{4\ \text{k}\Omega} = 0.725\ \text{mA} \tag{10.4-2a}$$

and
$$I_{B2} \simeq I_{R1} = 0.725\ \text{mA} \tag{10.4-2b}$$

neglecting any current flowing into the gate input terminals. If Q_2 is saturated, then
$$V_{C2} = V_{CE2} + V_{E2} = 0.2 + 0.7 = 0.9\ \text{V} \tag{10.4-3}$$

and with V_{C2} at this value, Q_4 is cut off, so that
$$I_{R3} = I_{C2} = \frac{(5 - 0.9)\ \text{V}}{1.6\ \text{k}\Omega} = 2.56\ \text{mA} \tag{10.4-4}$$

If we assume that the h_{FE} of the gate transistors is 20, then
$$I_{B2,\text{min}} = \frac{2.56}{20} = 0.128\ \text{mA} \tag{10.4-5}$$

which confirms the fact that Q_2 is saturated.

To prove that Q_3 is also saturated regardless of the load being driven, let us observe that TTL logic gates have a maximum fan-out capability of 10. In other words, the logic gate we are examining must be able to drive anywhere from 0 to 10 TTL logic loads. When the output is supposed to be a logic LO, we want Q_3 to remain saturated regardless of the load being driven. Clearly, when no other gates are connected, Q_3 will be saturated because I_{C3}, and hence $I_{B3,\text{min}}$, will both be zero. The case where the output is driving 10 other logic gates requires a more detailed investigation.

As illustrated in Figure 10.4-4b, if the output transistor Q_3 is saturated, a current
$$I_{IL} = -\frac{(5 - 0.9)\ \text{V}}{4\ \text{k}\Omega} = -1.03\ \text{mA} \tag{10.4-6}$$

will flow into the input of each of the logic gates connected onto Q_3. (*Note:* The minus sign indicates that this current actually flows out of the input lead of the gate.) As a result, when driving 10 such logic loads,
$$I_{C3,\text{sat}} = -10(I_{IL}) = 10.3\ \text{mA} \tag{10.4-7}$$

and a base current of at least $(10.3\ \text{mA})/20$ or $0.515\ \text{mA}$ will be needed if Q_3 is to remain saturated. From Figure 10.4-2a,
$$I_{R2} = \frac{0.7\ \text{V}}{1\ \text{k}\Omega} = 0.7\ \text{mA} \tag{10.4-8a}$$

and
$$I_{E2} = I_{C2} + I_{B2} = 2.56 + 0.725 = 3.285\ \text{mA} \tag{10.4-8b}$$

so that
$$I_{B3} = I_{E2} - I_{R2} = 2.585\ \text{mA} \tag{10.4-8c}$$

Therefore, Q_3 is definitely saturated, and V_o remains at about 0.2 V even when this gate is driving 10 additional TTL logic loads.

When Q_3 is saturated and $V_o = 0.2$ V, it is interesting to observe that $V_{E4} = 0.2 + 0.7 = 0.9$ V, and because $V_{B4} = V_{C2} = 0.9$ V from equation (10.4-3), V_{BE4} is zero and Q_4 is cut off. If D_1 were not present in the circuit,

V_{BE4} would be 0.7 V and both Q_3 and Q_4 would be on simultaneously. Thus the function of D_1 in this circuit is to ensure that Q_4 is cut off when V_o is LO.

The actual gate output voltage, V_o, that is measured when the output is supposed to be a logic LO is known as V_{OL}, and its worst-case maximum value is tabulated in Figure 10.4-2b along with several other important TTL input–output parameters.

In our original qualitative examination of this TTL gate, we assumed when the inputs were HI that the base–emitter junctions of Q_1 were both reverse biased and that no current flowed into the gate terminals. In reality, when V_1 and V_2 are both HI, Q_1 behaves like an inverted transistor because the base–collector junction is forward biased, while the base–emitter junction is reverse biased. As a result, the transistor is operating in the inverse active region, and although emitter junctions are reverse biased, a "collector" current

$$I_{C1A} = I_{C1B} = h_{FC}I_{B1A} = h_{FC}I_{B1B} \qquad (10.4-9)$$

flows into the gate terminals, where h_{FC} equals the transistor reverse current gain. Because of the asymmetrical doping used in the input transistor, this gain is very small and is typically on the order of 0.02. As a result, since I_{B1} is about 0.725 mA in this instance, the gate input currents are not identically zero as the simple theory would predict, but are typically on the order of $(0.02)(0.725 \text{ mA})$ or about 15 μA. On data sheets this current is referred to as I_{IH} (the gate input current when the input to the gate is HI), and as the table in Figure 10.4-2b indicates, the maximum worst-case value of this current is about 40 μA.

For the case where V_1, V_2, or both V_1 and V_2 are LO (0.2 V), Q_1 will saturate in this instance since I_{B1} is given by equation (10.4-6) while I_{C1} and hence $I_{B1,min}$ are nearly zero. As a result,

$$V_{C1} = V_{B2} = \underbrace{0.2}_{\substack{V_{CE1,sat}}} + \underbrace{0.2}_{\substack{V_{OL} \text{ of} \\ \text{driving} \\ \text{gate}}} = 0.4 \text{ V} \qquad (10.4-10)$$

The low value of this voltage keeps both Q_2 and Q_3 OFF since at least 0.7 V is needed to turn ON Q_2 alone, while nearly 1.4 V is required to turn both Q_2 and Q_3 ON.

With Q_2 cut off, V_{C2} rises, and depending on the value of the load connected onto the output of the gate, Q_4 will either be active or saturated. When the load resistance R_L is large (Figure 10.4-3a), Q_4 will be active and the equivalent circuit for the gate output in this situation will therefore have the form shown in Figure 10.4-3b. As the load on the output increases, eventually Q_4 will saturate as v_{CE4} decreases. The output current from the gate which just causes Q_4 to saturate may be determined by noting that

$$V_{E4} = 4.3 - (0.08 \text{ k}\Omega)I_L \qquad (10.4\text{-}11a)$$

and
$$V_{C4} = 5 - (0.13 \text{ k}\Omega)I_C \simeq 5 - (0.13 \text{ k}\Omega)I_L \qquad (10.4\text{-}11b)$$

Therefore,
$$V_{CE4} = V_{C4} - V_{E4} = 0.7 - (0.05 \text{ k}\Omega)I_L \qquad (10.4\text{-}11c)$$

Of course, Q_4 saturates when V_{CE4} drops to 0.2 V, and setting equation (10.4-

11c) equal to this value, we find that this occurs at a load current of about

$$I_L = \frac{(0.7 - 0.2) \text{ V}}{0.05 \text{ k}\Omega} = 10 \text{ mA} \tag{10.4-12}$$

For later use, the gate output voltage at which the saturation of Q_4 occurs is found directly from Figure 10.4-3b to be

$$V_o = 3.6 - (0.08 \text{ k}\Omega)I_L = 3.6 - (0.08 \text{ k}\Omega)(10 \text{ mA}) = 2.8 \text{ V} \tag{10.4-13}$$

In addition, the value of R_L just causing Q_4 to saturate is seen to be (2.8 V)/ (10 mA), or 280 Ω. Once Q_4 saturates, the equivalent circuit for the gate output takes on the form illustrated in Figure 10.4-3c.

As the load on the output continues to increase, V_o will fall. The smallest value of V_o for which the gate output is still considered to be a valid logic 1 is defined as V_{OH}, and for a TTL logic gate, $V_{OH} = 2.4$ V. Following Figure

Figure 10.4-3

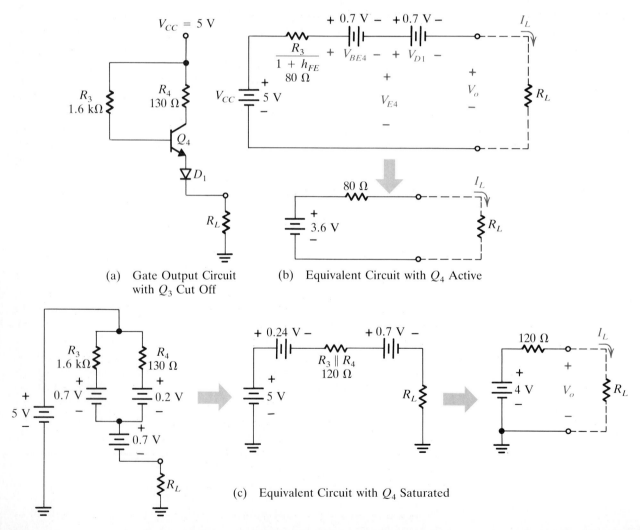

(a) Gate Output Circuit with Q_3 Cut Off

(b) Equivalent Circuit with Q_4 Active

(c) Equivalent Circuit with Q_4 Saturated

10.4-3c, the maximum allowed load current corresponding to this voltage level is seen to be

$$I_L = \frac{(4 - 2.4)\ \text{V}}{0.12\ \text{k}\Omega} = 13.3\ \text{mA} \qquad (10.4\text{-}14)$$

On the data sheets this quantity is usually specified as $-I_{OH}$, where I_{OH} is the maximum guaranteed current that the gate can sink and still have the output remain at a valid logic 1 voltage level (Figure 10.4-2b).

EXAMPLE 10.4-1

For the TTL logic gate shown in Figure 10.4-2a:

a. Determine the gate noise margins using the fact that $V_{OH} = 2.4$ V and $V_{OL} = 0.2$ V.
b. If the transistor h_{FE}'s are all assumed to be 20, compute the maximum allowed gate fan-out when the output is a logic 1 and also when it is a logic 0.

SOLUTION

To determine the gate noise margins we must first compute the values of V_{IH} and V_{IL}. V_{IH} is defined as the lowest input voltage that still acts like a valid input HI to the gate. When the input signal is a HI, the emitter of Q_1 is reverse biased, and when this occurs, the base–collector junction of Q_1, along with the base–emitter junctions of Q_2 and Q_3 are all forward biased. As a result, $V_{B1} = 2.1$ V, and for the emitter junction of Q_1 to remain reverse biased requires that V_{E1} be greater than or equal to $(2.1 - 0.7)$ or 1.4 V.

V_{IL} is defined as the largest allowed input voltage that still looks like a valid input LO to the logic gate. When the gate input is a LO, Q_1 saturates and $V_{B2} = (V_{IL} + 0.2$ V$)$. To keep Q_2 and Q_3 from turning ON, V_{B2} needs to be less than 1.4 V, so that $V_{IL} = 1.2$ V.

The gate noise margins may now be determined directly by substituting the results just obtained into equations (10.1-7a) and (10.1-7b), which yields

$$\text{NM}_L = V_{IL} - V_{OL} = 1.2 - 0.2 = 1.0\ \text{V} \qquad (10.4\text{-}15a)$$

and

$$\text{NM}_H = V_{OH} - V_{IH} = 2.4 - 1.4 = 1.0\ \text{V} \qquad (10.4\text{-}15b)$$

For informational purposes, the noise margins of commercially available TTL logic gates are guaranteed to be at least 0.4 V. These noise margins are more conservative than our calculations indicate because they include the effects of temperature and component variations.

In analyzing the gate fan-out, we are basically interested in computing the maximum number of gates that may be connected onto the output without causing V_o to deviate beyond its V_{OH} and V_{OL} limits. When the gate output is a logic 1, the theoretical fan-out is infinite if we neglect the current that flows into the emitter of the input transistor. When this current is included, the overall equivalent circuit has the form shown in Figure 10.4-4a. Notice that we have used the output equivalent circuit given in Figure 10.4-3c because at maximum fan-out $V_o = V_{OH} = 2.4$ V, and with V_o less than 2.8 V, Q_4 is saturated. Following this figure, we may write

$$V_o = 4 - (N I_{IH})(0.12\ \text{k}\Omega) \qquad (10.4\text{-}16)$$

Using a worst-case value of 40 μA for I_{IH} and setting $V_o = 2.4$ V, we find that

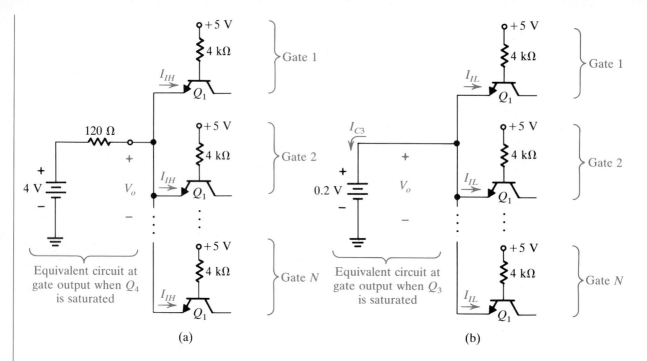

(a)   (b)

Figure 10.4-4
Example 10.4-1.

$N = 333$ gates. Thus for all practical purposes the fan-out when the gate output is a logic 1 is infinite.

In determining the gate fan-out when the output is a logic 0, we note that V_{OL} will stay at 0.2 V as long as Q_3 remains saturated. Following Figure 10.4-4b,

$$I_{IL} = -\frac{(5 - 0.7 - 0.2) \text{ V}}{4 \text{ k}\Omega} \approx -1 \text{ mA} \qquad (10.4\text{-}17a)$$

Therefore, $I_{C3,\text{sat}} = -NI_{IL} = N \text{ mA}$ \qquad (10.4-17b)

and $I_{B3,\text{min}} = \dfrac{N \text{ mA}}{20}$ \qquad (10.4-17c)

However, from equation (10.4-8c), we have $I_{B3} = 2.585$ mA, so that for Q_3 to remain saturated, we require that

$$2.585 > \frac{N}{20} \qquad (10.4\text{-}18)$$

or $N < 51$.

Thus to the extent that the gate characteristics are accurately represented by the circuit given in Figure 10.4-2a, a maximum of up to 51 logic gates may be connected onto the output of this TTL gate without creating any loading problems. If temperature effects, noise margin problems, and component variations were included in this analysis, the maximum allowed gate fan-out would be reduced significantly. As mentioned previously, commercially available 7400 TTL logic gates are guaranteed to work satisfactorily as long as the fan-out is less than or equal to 10.

Most of the time digital logic gates are used strictly as standard digital logic devices, and for these applications it is sufficient to know, for example in the case of an inverter, that the application of a HI to the input produces a LO at the output, and vice versa. However, there are other applications where a detailed knowledge of the gate transfer characteristic is useful. Typical examples of these include the use of logic gates as oscillators, pulse-forming circuits, and even as linear amplifiers.

The basic form of the TTL logic gate to be examined is given in Figure 10.4-5a, and the transfer characteristic of this inverter is shown in part (b) of the figure. For conventional digital logic applications, the gate operating point is located either in region I or in region IV. In region I, where V_{in} is a valid logic 0, Q_1 saturates. When this occurs, as long as V_{in} is less than 0.5 V, V_{B2} will be less than 0.7 V and both Q_2 and Q_3 will be cut off. The gate equivalent circuit that is valid in region I is shown in Figure 10.4-5c, and the state of transistor Q_4 in this case depends on the load connected on the gate output. Ordinarily, this load is small, and as a result, Q_4 will usually be active and V_o will typically be about $(5 - 0.7 - 0.7)$ or 3.6 V in this region.

Once V_{in} exceeds 0.5 V, the base voltage on Q_2 will exceed 0.7 V and this transistor will become active (region II on the transfer characteristic). In this region, where Q_2 and Q_4 are active, for simplicity we will analyze the slope of the transfer characteristic (or the gain of this circuit) by small-signal analysis techniques. Although this approach is not strictly valid, since the voltage changes are actually quite large, it will be sufficiently accurate for our purposes. Along these lines we will assume that $h_{ie} = 1$ kΩ and $h_{fe} = 20$ for these transistors when they are in the active region, and also that the small-signal resistance of D_1 is 50 Ω.

The equivalent circuit in region II is shown in Figure 10.4-5d, and in order to calculate the small-signal gain of this circuit we will represent v_{in} as having a dc part V_{BB} and an ac part, v_i. Using this approach, we can immediately write that

$$v_{e2} \simeq v_i \tag{10.4-19a}$$

and
$$i_{e2} = \frac{v_{e2}}{R_2} = \frac{v_i}{R_2} \simeq i_{c2} \tag{10.4-19b}$$

Neglecting the loading of Q_4 on the Q_2 circuit, we also have

$$v_{c2} = -i_{c2}R_3 = -v_i\frac{R_3}{R_2} \tag{10.4-19c}$$

and because Q_4 is an emitter follower, v_o is also nearly the same as v_{c2}, so that

$$\frac{v_o}{v_i} = -\frac{R_3}{R_2} = -1.6 \tag{10.4-19d}$$

Equation (10.4-19d) represents the small-signal gain of this TTL logic gate when it is operating in region II, and this, of course, is equal to the slope of the transfer characteristic in this region in Figure 10.4-5b.

As V_{in} increases beyond 0.5 V in region II, V_o decreases because the effective equivalent circuit consists of an inverting amplifier, Q_2, followed by an

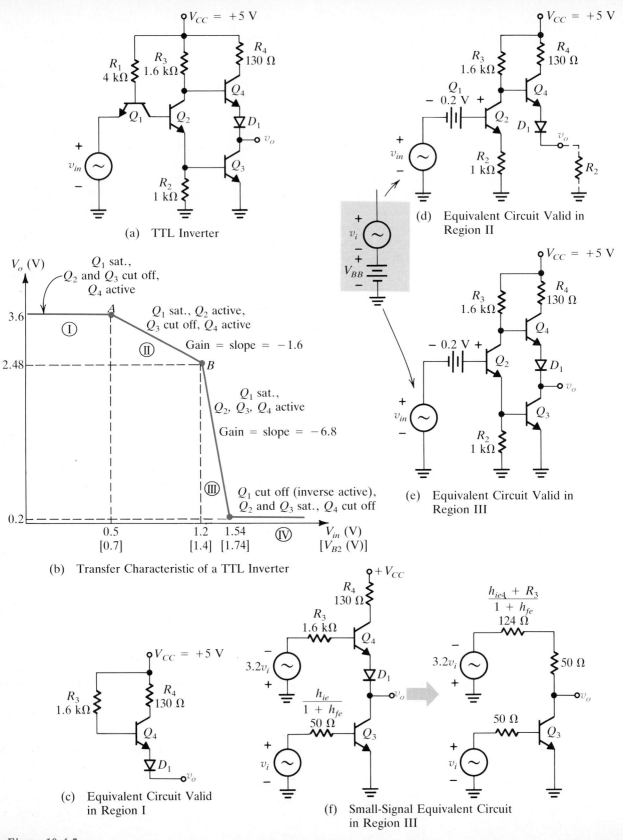

(a) TTL Inverter

(b) Transfer Characteristic of a TTL Inverter

(c) Equivalent Circuit Valid in Region I

(d) Equivalent Circuit Valid in Region II

(e) Equivalent Circuit Valid in Region III

(f) Small-Signal Equivalent Circuit in Region III

Figure 10.4-5

emitter follower, Q_4. This gate equivalent circuit is valid until we reach point B on the transfer characteristic. At this point the voltage drop across R_2 increases sufficiently to just turn on transistor Q_3, and here $V_{B3} = 0.7$ V, $V_{B2} = (0.7 + 0.7)$ or 1.4 V, and therefore V_{in} is just $(1.4 - 0.2)$ or about 1.2 V. The corresponding value of V_o at point B is found by using the fact that the slope in region II is -1.6. As a result, when V_{in} changes from 0.5 V to 1.2 V (a change of 0.7 V), the corresponding change in V_o will be (-1.6×0.7) or -1.12 V. Thus at point B $v_o = (3.6 - 1.12)$ or 2.48 V.

When Q_3 enters the active region the complexity of the circuit increases somewhat and the equivalent circuit takes the form indicated in Figure 10.4-5e. Insofar as the small-signal analysis of this circuit is concerned, we may replace the transistor Q_2 by its collector and emitter equivalent circuits as indicated in part (f) of the figure. Notice that here the value of the ac open-circuit voltage at the collector of Q_2 has been doubled to $-3.2v_i$ because R_2 is now in parallel with h_{ie3}, which halves the value of the effective emitter resistor connected to Q_2, thereby doubling i_{c2}. Furthermore, because Q_4 is an emitter follower, we may reflect the collector equivalent circuit of Q_2 into the emitter of Q_4, resulting in the final equivalent circuit given in Figure 10.4-5f. Examining the lower portion of this circuit, it is apparent that

$$i_{c3} = h_{fe}i_{b3} \simeq \frac{h_{fe}v_i}{1 \text{ k}\Omega} \qquad (10.4\text{-}20)$$

and by applying superposition to this problem, the expression for v_o may be written immediately as

$$v_o = -\left(\frac{R_3 + h_{ie}}{1 + h_{fe}} + r_D\right)i_{c3} - 3.2v_i = -6.8v_i \qquad (10.4\text{-}21)$$

Hence the slope of the transfer characteristic in region III is -6.8. As V_{in} increases further, region III continues until Q_3 saturates; that is, until V_o drops to about 0.2 V. The value of V_{in} at which this occurs is found by noting that the change in V_{in} corresponding to the change in V_o of $(2.48 - 0.2)$ V is $2.46/6.8$ or about 0.34 V. Thus the value of V_{in} at which Q_3 saturates (and at which region IV begins) is $(1.2 + 0.34)$ or about 1.54 V. Increasing V_{in} beyond this point produces no further changes in V_o.

10.4-3   Open-Collector TTL Logic Gates

Besides the totem-pole-style output circuit, some TTL logic gates are available with open collector outputs. Figure 10.4-6a illustrates this type of output circuit structure for the case of a TTL noninverting buffer gate. Open collector devices of this sort are often used as interfaces for driving loads connected to voltages that are greater than V_{CC}. For example, the output transistor in the 7407 gate shown in the figure is guaranteed to operate satisfactorily with load voltages up to 30 V.

A second advantage of open collector outputs over the more conventional totem-pole TTL output is that these gates may be "wire ANDed" together. Figure 10.4-6b illustrates a typical "wire AND" application in which five of these noninverting buffer gates are being used to drive 15 conventional TTL logic gates. For this circuit V_o will be a logic 1 (or $+5$ V) only when V_1 through V_5

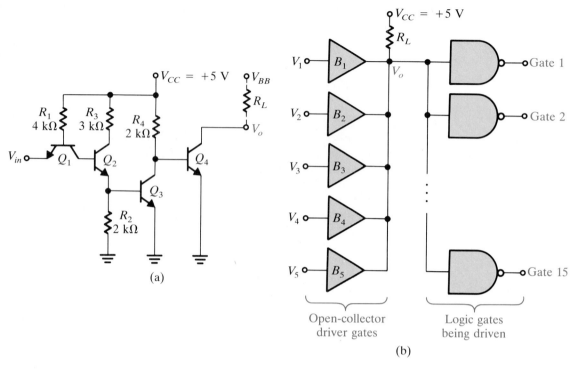

Figure 10.4-6
An open-collector TTL logic gate.

are all at the logic 1 voltage level. If this same type of "wire AND" connection were used with conventional "totem-pole"-style TTL logic gates, serious damage to the gates could result if one gate output attempted to drive V_o HI while one of the other gates tried to make it LO (see Problem 10.4-12).

10.4-4 Tri-State TTL Logic Gates

Besides having output circuits with active and passive pull-up, TTL ICs are also available in a third form known as three-state or tri-state logic. Three-state logic combines the advantages of active and passive pull-up circuits together, in that it has the low output impedance drive capability of an active pull-up circuit while allowing for "wire ANDing" of the gate outputs.

Basically, tri-state logic differs from ordinary logic in that the gate output can take on any one of three different output states: HI, LO, and disconnected. A typical tri-state inverter and its equivalent circuit are shown in Figure 10.4-7a, and basically the device operates as follows. When the enable signal (labeled EN in the figure) is a logic 1, the switch S is closed, and the device functions like an ordinary TTL inverter. In this instance the output can be either HI or LO, depending on the level of the input signal. On the other hand, when EN = 0 the switch is open and the device enters its high-impedance state. In this state the output lead is effectively disconnected from the internal logic gate.

Tri-state logic is very popular in computer system designs since multiple devices can almost be arbitrarily connected onto the same output line as long as only one tri-state device is activated or enabled at any given time. Because 8-, 16-, and 32-bit bus architectures are often used in computers, many differ-

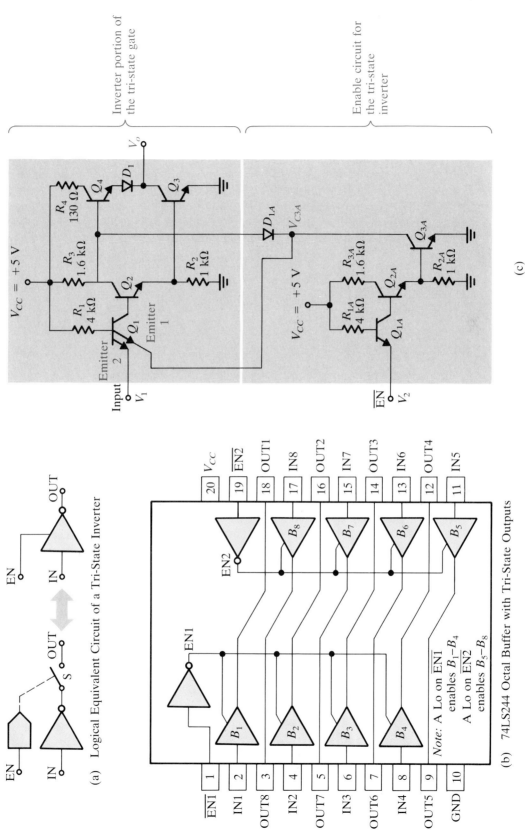

(a) Logical Equivalent Circuit of a Tri-State Inverter

(b) 74LS244 Octal Buffer with Tri-State Outputs

Note: A Lo on $\overline{EN1}$ enables $B_1 - B_4$

A Lo on $\overline{EN2}$ enables $B_5 - B_8$

(c)

Figure 10.4-7

ent types of "octal" tri-state logic devices are available in which eight three-state logic gates are contained in the same package with all eight devices activated by one or two enable signals (Figure 10.4-7b).

The internal structure of a typical three-state TTL logic gate is shown in Figure 10.4-7c. As the figure illustrates, the upper portion of this device is an ordinary TTL inverter to which some additional hardware has been added to permit tri-state operation. When V_2 (the $\overline{EN}$ line) is a logic 0, Q_{1A} saturates, cutting off Q_{2A} and Q_{3A}. As a result, no current can flow through emitter 1 on Q_1 or through D_{1A}. With both of these "diodes" reverse biased, these leads are effectively open circuits and the gate functions normally (in this case as an inverter). When V_2 (or $\overline{EN}$) is a logic 1, Q_{1A} is cut off, causing the base–collector junction of Q_{1A} to conduct; this saturates Q_{2A} and Q_{3A} so that $V_{C3A} = 0.2$ V. The LO input to emitter 1 of Q_1 saturates this transistor and cuts off Q_2 and Q_3. In addition, with $V_{C3A} = 0.2$ V, D_{1A} is ON, so that $V_{C2} = V_{B4} = (0.2 + 0.7) = 0.9$ V. This voltage is insufficient to turn ON the Q_4–D_1 combination, so that Q_4 is also cut off. Thus when $\overline{EN} = 1$, both Q_3 and Q_4 are OFF and the gate output is in the high-impedance state.

10.4-5 Switching Characteristics of a TTL Logic Gate

The switching analysis of a typical TTL logic gate is extremely involved, but interestingly enough, its performance characteristics can be explained fairly well by applying the techniques developed in Section 10.2. Admittedly, some of the fine details of this circuit's operation will be missed when this method is used, but the qualitative analysis that follows should provide you with a good feeling for how the switching transients in this circuit proceed.

In investigating the internal switching characteristics of TTL logic gates, we will, for simplicity, neglect the transistor junction capacitances and focus our attention instead on the time required to charge and discharge the emitter diffusion capacitance, C_{DE}. Consider for example the TTL inverter illustrated in Figure 10.4-8a. Notice that the output of this gate has been connected to a capacitor C_L, a 400-Ω resistor R_L, and four diodes D_2–D_5 in order to simulate the effect of a 10-gate load on the output. For the present, C_L will be assumed to be zero, although its actual value is about 15 pF.

In calculating the response of this circuit to the input voltage shown in Figure 10.4-8b, the transistor h_{FE}'s will be assumed to be equal to 20. Furthermore, because most standard TTL ICs use gold-doped transistors (a technique by which the recombination time may be significantly reduced), we assume that τ_B for these transistors is about 20 ns, instead of the 40-ns value that was used with the logic families examined earlier.

When v_{in} is LO, the transistor Q_1 is saturated, and when v_{in} changes to a HI, the transistor switches to inverse active. Because this transistor never really cuts off, the base charge of Q_1 only needs to be redistributed, and does not have to be reduced to zero. To a first-order approximation, we may therefore assume that this transistor switches back and forth between saturation and inverse active instantaneously. As a result, we find at $t = 0+$ that $I_{B2F} = (5 - 2.1)/4$ k$\Omega = 0.73$ mA, so that i_{C2} immediately begins to charge exponentially toward $h_{FE}I_{B2F}$ or 14.6 mA with a time constant τ_B. Furthermore, Q_2 saturates when i_{C2} equals 2.56 mA.

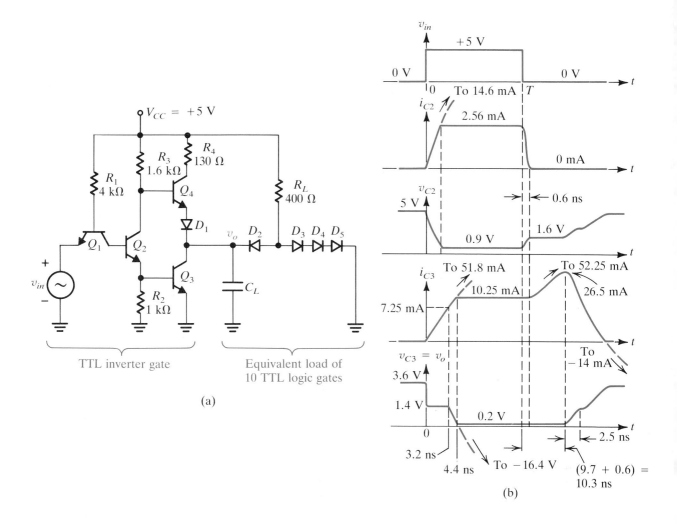

Figure 10.4-8

Switching characteristics of a TTL logic gate.

As i_{C2} begins to increase at $t = 0+$, v_{C2} falls and Q_4 cuts off almost immediately since the charge stored in the base of Q_4 is negligible. Because i_{E2} equals ($i_{B2} + i_{C2}$) the emitter current in Q_2 will initially be about 0.7 mA and will grow exponentially, leveling off at 3.29 mA in the steady state. Since i_{E2} starts out at 0.7 mA, v_{BE3} will initially be about 0.7 V, and Q_3 will turn on almost immediately. Therefore, at $t = 0+$ i_{B3} will begin growing exponentially, leveling off at a final value of $(3.29 - 0.7) = 2.59$ mA. In a similar fashion, i_{C3} will also grow exponentially, rising toward an effective value of 51.8 mA, but saturating when it reaches $(5 - 0.9)$ V/400 Ω or 10.25 mA.

Since Q_4 turns OFF at $t = 0+$ and since D_2 in Figure 10.4-8a does not turn ON until v_{C3} falls below $(2.1 - 0.7) = 1.4$ V, the effective load on Q_3 will initially be infinite. As a result, v_{C3} will rapidly drop to 1.4 V, and will remain at this level until i_{C3} increases to 7.25 mA at $t = 3.2$ ns. Once i_{C3} reaches this current level D_3, D_4, and D_5 turn OFF, and beyond this point the effective load on Q_3 is now the series connection of R_L and D_2. For this circuit, with i_{C3} rising toward 51.8 mA, v_{C3} can now be shown to discharge exponentially toward -16.4 V, saturating when it reaches 0.2 V at $t = 4.4$ ns.

For TTL circuits, because $(1/2)(V_{OH} + V_{OL}) \simeq 1.5$ V, t_{PHL} is frequently defined as the time interval between the point where the input rises about 1.5 V and the point where the output falls below this same value. Using this definition it is apparent that t_{PHL} for this logic gate is about 0 ns, due to the fact that we have neglected the junction capacitances and the effect of capacitive loading on the output. The actual value of t_{PHL} for commercial TTL logic gates is typically on the order of 7 ns.

When v_{in} switches LO at $t = T$, Q_1 cannot saturate until Q_2 turns OFF, and as a result, Q_1 is initially active so that $I_{B2R} = h_{FE}I_{B1} = 21.5$ mA. Because I_{B2R} is so large, Q_2 cuts off quite rapidly (in about 0.6 ns). Once Q_2 is off, v_{C2} begins to rise and because Q_3 is still saturated, this will cause Q_4 to enter the active region with $I_{B4F} = 2.1$ mA and $v_{B4} = 1.6$ V. At this time the collector current in Q_3 is the sum of the exponentially increasing emitter current from Q_4 and the current from the load R_L (10.25 mA). This growth in i_{C3} continues until it exceeds $h_{FE}i_{B3}$, whereupon Q_3 comes out of saturation. This occurs about 9.7 ns from the time that Q_2 cuts off, and at this point i_{C3} is about 26.5 mA. Once Q_3 enters the active region i_{C3} begins to decrease exponentially from 26.5 mA toward $-h_{FE}I_{B3R} = -14$ mA. Corresponding to this, v_{C3} begins to grow exponentially with the same time constant. The time required for v_{C3} to reach 1.5 V is about 2.5 ns, so that t_{PLH} is $(0.6 + 9.7 + 2.5) = 12.8$ ns. For commercially available TTL logic gates, t_{PLH} is typically about 11 ns.

The output current pulse down through the totem-pole transistors Q_3 and Q_4 during turnoff is a problem in TTL circuits. This current spike causes glitches (or momentary drops) in the power supply voltage and since many gates can switch at the same time in a digital circuit, this effect can be quite serious. Power supply glitching effects can be reduced considerably by the use of by-pass capacitors.

EXAMPLE 10.4-2

Besides having an inherently high internal switching speed, TTL logic also has an excellent capacitive load drive ability. Demonstrate this fact by comparing the output rise and fall times produced when a DTL and then a TTL logic gate are each used to drive a 150-pF capacitive load. In carrying out this analysis, assume for simplicity that the internal switching characteristics of each gate are ideal, and also that the transistor h_{FE}'s are all equal to 20.

SOLUTION

The DTL and TTL circuits to be examined are shown in Figure 10.4-9a and b. Prior to $t = 0$, Q_1 in the DTL gate and Q_3 in the TTL gate are both cut off so that the output voltages from each are +5 V and +3.6 V, respectively. When v_{in} on the DTL circuit rises to +5 V at $t = 0+$, D_1 turns off, D_2 and D_3 conduct, and the base current into Q_1 may be written as

$$i_{B1} = I_{R1} - I_{R2} = \frac{5 - 2.1}{2 \text{ k}\Omega} - \frac{0.7}{5 \text{ k}\Omega} = 1.31 \text{ mA} \qquad (10.4\text{-}22)$$

In the absence of C_L, this current would be more than sufficient to saturate Q_1. However, with C_L connected, v_o stays above 0.2 V for some time, and as a result with $i_{C1} > 0$ and v_{CE1} also much greater than 0.2 V, the transistor is solidly

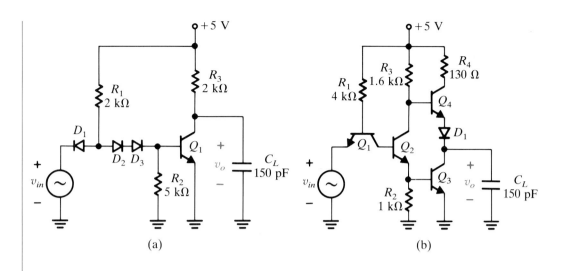

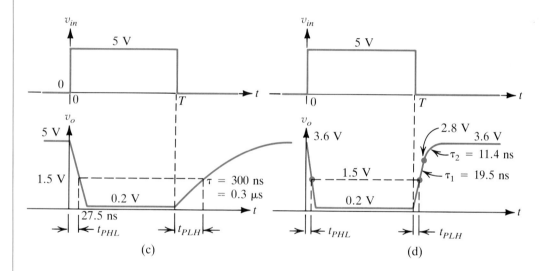

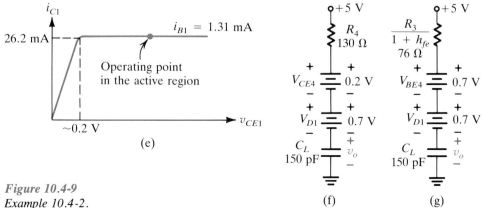

Figure 10.4-9
Example 10.4-2.

in the active region (see Figure 10.4-9e). Because Q_1 is active,

$$i_{C1} = h_{FE} i_{B1} = 26.2 \text{ mA} \tag{10.4-23}$$

Most of this current flows into the capacitor and to a first-order approximation because $i = C(dv/dt)$, the change in voltage across C_L with time may be written as

$$\Delta v = \frac{i \Delta t}{C} = -\frac{i_{C1} \Delta t}{C_L} \tag{10.4-24a}$$

Furthermore, because i_{C1} is constant, the voltage across C_L will decrease linearly with time as illustrated in Figure 10.4-9c. The time required for Q_1 to saturate is found by setting $\Delta v = (5 - 0.2) = 4.8$ V, which yields

$$\Delta t = \frac{C_L \Delta v}{i_{C1}} = \frac{(150 \text{ pF})(4.8 \text{ V})}{26.2 \text{ mA}} = 27.5 \text{ ns} \tag{10.4-24b}$$

When v_{in} drops to zero volts at $t = T$, Q_1 cuts off immediately and C_L simply charges exponentially toward +5 V. Because the time constant in this instance is quite large ($R_3 C_L = 300$ ns), the turn-off transient takes a very long time to reach the steady state. As with TTL, if we use the 1.5-V output level for defining the propagation delays in this DTL gate, too, it is not very difficult to show that

$$
\left.
\begin{array}{ll}
t_{PHL} = 20.0 \text{ ns} & \quad (10.4\text{-}25a) \\
t_{PLH} = 107.0 \text{ ns} & \quad (10.4\text{-}25b)
\end{array}
\right\} \text{ for DTL}
$$

and

Thus, when driving capacitive loads, DTL's most serious problem is the long time required for the gate output to reach its steady-state value when the output transistor turns OFF. TTL overcomes this problem by using the active pull-up transistor Q_4 in place of a resistor.

The corresponding analysis for the TTL logic gate in Figure 10.4-9b proceeds as follows. Initially, Q_3 is cut off, and when v_{in} switches to +5 V at $t = 0+$, Q_1 cuts off, Q_2 saturates (turning Q_4 off), and Q_3 enters the active region with

$$i_{B3} = 2.58 \text{ mA} \tag{10.4-26a}$$

and

$$i_{C3} = h_{FE} i_{B3} = 51.6 \text{ mA} \tag{10.4-26b}$$

As with the DTL circuit just examined, there is an almost linear rundown in v_{CE3}, and following equation (10.4-24b), the time required for Q_3 to saturate is seen to be

$$\Delta t = \frac{C_L \Delta v}{i_{C3}} = \frac{(150 \text{ pF})(3.4 \text{ V})}{51.6 \text{ mA}} = 9.9 \text{ ns} \tag{10.4-27}$$

(Figure 10.4-9d).

The situation that occurs when v_{in} switches from a HI to a LO at $t = T$ is quite different from that for the DTL inverter just examined due to the active pull-up circuit in the output of the logic gate. When v_{in} drops to zero volts, Q_1 saturates and Q_2 and Q_3 cut off immediately. Furthermore, because v_o is less than 2.8 V, Q_4 saturates so that the output equivalent circuit has the approximate form shown in Figure 10.4-9f. Following this figure, it is therefore apparent that v_o will begin to rise exponentially toward 4.1 V with a time constant of $R_4 C_L = 19.5$ ns. This will continue until v_o reaches 2.8 V and at this point, Q_4 will reenter the active region. For the remainder of the transient in v_o, the equivalent circuit has the form given in Figure 10.14-9g so that the output will continue to rise exponentially, but now the final charging voltage will be 3.6 V and the circuit time constant $\tau_2 = (76 \ \Omega)(150 \ \text{pF}) = 11.4$ ns.

The overall output waveshape for this TTL logic gate is sketched in Figure 10.4-9d, and it is not too difficult to show that the propagation delays indicated in this figure are

$$t_{PHL} = 3.2 \ \text{ns} \left. \right\} \quad \text{for TTL} \qquad (10.4\text{-}28\text{a})$$

and

$$t_{PLH} = 8.4 \ \text{ns} \left. \right. \qquad (10.4\text{-}28\text{b})$$

Comparing these results with those in equations (10.4-25a) and (10.4-25b), it is clear that the capacitive drive capability of TTL logic gates is far superior to those of a conventional DTL circuit due primarily to the use of the active pull-up circuit in the output stage. It is also worth noting that much smaller propagation delays could have been obtained with the DTL gate by reducing R_3 to, say, 130 Ω. However, this would substantially increase the gate power consumption and is therefore not really an acceptable solution to the problem. Active pull-up circuits have the double advantage of both low quiescent power consumption and low output drive impedance. These attributes cannot be matched by a simple passive pull-up circuit.

10.4-6   Important Subfamilies of Standard TTL Logic

Despite the fact that TTL has been around for nearly 20 years, it remains a very viable logic form, and at the time of the writing of this manuscript (1987), it is still the most widely used digital IC logic family. One of the principal reasons for the continued popularity of TTL has been its ability to grow and change as new developments in electronics have taken place. The TTL logic family is composed of many subfamilies, and as each new subfamily has been introduced, great care has been exercised to ensure that it is compatible with the previous subfamilies developed.

Because of the part numbering used to identify ICs within the TTL family, standard TTL is often known as the 7400 or 74xx series of digital logic. In the remainder of this section we take a look at some of the more important TTL subfamilies that have been introduced over the last 20 years. Specifically, we examine Schottky TTL (74Sxx), Low-Power Schottky TTL (74LSxx), Advanced Schottky TTL (74ASxx), and Advanced Low-Power Schottky TTL (74ALSxx). In describing the part number of a particular TTL IC, the first two numbers and the letters that follow identify the subfamily to which the device belongs. The "xx" portion of the part number indicates the logic function that the device performs. For example, a 7490 IC is a standard TTL decade counter and a 74LS90 is a Low-Power Schottky version of the same device.

Other than standard TTL, all the subfamilies that we examine in the remainder of this section are one form or another of Schottky TTL. For the most part, Schottky logic gates differ from standard TTL devices in that the gate transistors have Schottky diodes connected across their base–collector junctions. As we shall see shortly, the use of these diodes prevents the transistors from saturating, and this dramatically reduces the gate propagation delays.

Figure 10.4-10a and b illustrate the basic principle of the Schottky transistor. In part (a) we show a germanium diode and a silicon diode connected in parallel, and for this specific circuit, because of its lower turn-on voltage, the germanium diode will be forward biased while the silicon device will remain OFF. In particular, the diode Q points will be $V_{D1} = V_{D2} = 0.3$ V, $I_{D1} = 5$ mA, and $I_{D2} = 0$. In a similar fashion, if a germanium diode is connected across the base–collector junction of a silicon transistor, as shown in part (b), it will prevent the base–collector junction of this transistor from becoming forward biased. This means, then, that the transistor can never saturate since the lowest voltage that v_{CE} can ever drop to is $v_{BE} - v_D = 0.7 - 0.3$ or about 0.4 V. This result has important implications for the design of BJT-style digital logic gates because in these gates the storage time of the BJTs is often responsible for a substantial portion of the overall gate propagation delay.

While discrete-style digital circuits can employ germanium diodes to prevent transistor saturation, digital integrated-circuit designs cannot mix silicon and germanium semiconductors on the same substrate. Fortunately, there is an alternative to the use of germanium diodes. Some metal–semiconductor junctions have rectifying properties that are similar to those of germanium diodes; they are known as Schottky diodes after their discoverer. In the construction of this type of diode, the metal (which is usually aluminum or platinum) acts as

Figure 10.4-10
Schottky-style BJT.

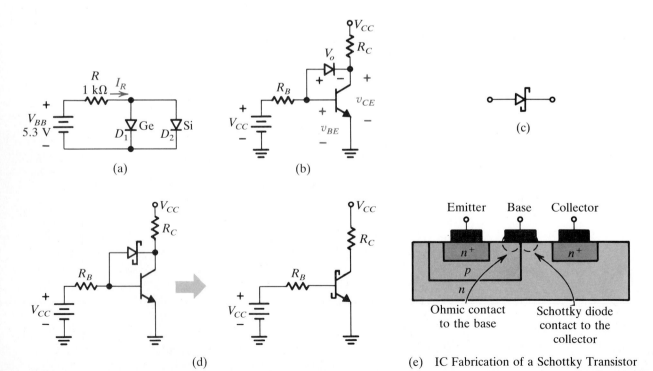

(a)

(b)

(c)

(d)

(e)   IC Fabrication of a Schottky Transistor

a p-type material, and when it is contacted onto a lightly doped piece of n-type material, a Schottky diode is formed. If the doping level of the n-type material is too high, an ohmic rather than a rectifying contact is formed. The voltage drop across the diode is typically between 0.1 and 0.5 V, depending on the type of metal used and the concentration of n-type material employed.

The Schottky diode operates by the flow of electrons in both the metal and the semiconductor, and thus the electrons are the majority carriers in both materials. Majority carriers move mainly by drift in an applied electric field, and the time required for these charges to redistribute themselves when the voltage across the diode changes is much less than that needed for the redistribution of the minority-carrier charge in a pn junction diode in a similar situation. As a result, typical Schottky diode redistribution times are on the order of 10 to 100 ps, which is several orders of magnitude faster than the corresponding recombination time of a conventional pn junction diode. Therefore, when analyzing Schottky TTL logic circuits, the switching times of the Schottky diodes in these gates may be assumed to be essentially zero.

Figure 10.4-10c illustrates the symbol for the Schottky diode, and part (d) of the figure shows a BJT transistor inverter with a Schottky diode connected across its base–collector junction. Because this diode–transistor combination occurs so frequently in Schottky logic circuits, a special symbol, shown to the right in Figure 10.4-10d, has been developed to represent the "Schottky transistor."

Figure 10.4-10e shows the construction technique employed for the manufacturing of Schottky transistors. Basically, the layout of this device is the same as that of an ordinary npn integrated-circuit transistor, and the only difference between the two is that the metallic base contact is allowed to overlap at the base–collector boundary. On the base side, this connection is simply an ordinary ohmic contact because the base is made of p-type material. However, on the collector side of the boundary, a Schottky diode is formed because the collector is composed of lightly doped n-type material.

EXAMPLE 10.4-3

In the TTL waveshapes in Figure 10.4-8b, it appears that the storage time delays are only a very small part of the overall gate propagation delays. Principally, this is because we used conservative transistor h_{FE}'s of only 20. In an actual logic gate, these current gains can be substantially larger, heavily saturating the transistors and making the storage times a significant portion of the overall gate propagation delay. By placing Schottky diodes across the transistors, these storage times can be reduced to zero.

For the simple inverter circuit given in Figure 10.4-11a, sketch the transistor collector current and collector-to-emitter voltage both with and without the Schottky diode D_1 in place. Also determine t_{PLH} for this circuit in each case. Consider that the transistor has an $h_{FE} = 20$, an $r_{B'E} = 1\ \text{k}\Omega$, and a $\tau_B = 20$ ns. The forward drop for the diode is 0.3 V.

SOLUTION

If we assume that v_{in} has been equal to $+5$ V for a long time prior to $t = 0$, then

$$v_{CE}(0-) = (0.7 - 0.3) = 0.4\ \text{V} \qquad (10.4\text{-}29a)$$

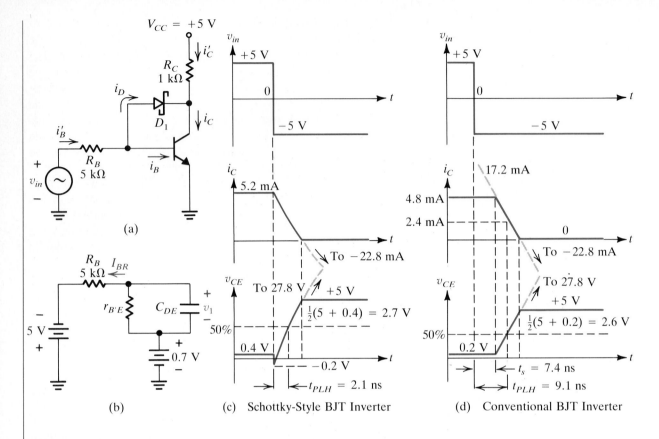

Figure 10.4-11
Example 10.4-3.

$$i_B'(0-) = \frac{5 - 0.7}{5 \text{ k}\Omega} = 0.86 \text{ mA} \qquad (10.4\text{-}29b)$$

and

$$i_C'(0-) = \frac{5 - 0.4}{1 \text{ k}\Omega} = 4.6 \text{ mA} \qquad (10.4\text{-}29c)$$

But

$$i_B = \frac{1}{h_{FE}} i_C = \frac{1}{h_{FE}} (i_C' + i_D) = \frac{1}{h_{FE}} (i_C' + i_B' - i_B) \qquad (10.4\text{-}30a)$$

or

$$i_B = \frac{(i_C' + i_B')}{(1 + h_{FE})} \qquad (10.4\text{-}30b)$$

so that

$$i_B(0-) = I_{BF} = \frac{4.6 + 0.86}{21} = 0.26 \text{ mA} \qquad (10.4\text{-}30c)$$

and

$$i_C(0-) = h_{FE} i_B(0-) = 5.2 \text{ mA} \qquad (10.4\text{-}30d)$$

In addition, it is also apparent that

$$i_D(0-) = i_B'(0-) - i_B(0-) = 0.6 \text{ mA} \qquad (10.4\text{-}30e)$$

so that the Schottky diode is initially forward biased and the transistor is in the active region.

At $t = 0$ when v_{in} drops to -5 V, the current flow through R_B reverses. When this happens i_D also tries to go negative, but instead, D_1 just rapidly turns OFF while Q_1 remains in the active region. The input equivalent circuit for this case is illustrated in Figure 10.4-11b, and as a result,

$$I_{BR} = \frac{5 + 0.7}{5 \text{ k}\Omega} = 1.14 \text{ mA} \tag{10.4-31a}$$

$$i_B(t) = I_{BF}e^{-t/\tau_B} - I_{BR}(1 - e^{-t/\tau_B}) \tag{10.4-31b}$$

$$i_C(t) = h_{FE}i_B(t) = 28e^{-t/\tau_B} - 22.8 \tag{10.4-31c}$$

and $$v_{CE}(t) = V_{CC} - i_C R_C = 27.8 - 28e^{-t/\tau_B} \tag{10.4-31d}$$

The waveshapes for $i_C(t)$ and $v_{CE}(t)$ are sketched in Figure 10.4-11c. The gate propagation delay for this Schottky inverter is found by setting equation (10.4-31d) equal to 2.7 V and solving for t, which yields

$$t_{PLH} = \tau_B \ln \frac{28}{27.8 - 2.7} = 2.1 \text{ ns} \tag{10.4-32}$$

When the diode is removed from the circuit, the analysis is essentially the same except that the initial conditions are different. In particular, in this instance,

$$i_B(0-) = I_{BF} = i_B'(0-) = 0.86 \text{ mA} \tag{10.4-33a}$$

and $$i_C(0-) = I_{C,\text{sat}} = \frac{5 - 0.2}{1 \text{ k}\Omega} = 4.8 \text{ mA} \tag{10.4-33b}$$

Because $$I_{B,\text{min}} = \frac{4.8}{20} = 0.24 \text{ mA} \tag{10.4-34}$$

the transistor is initially saturated and $v_{CE}(0-) = 0.2$ V.

At $t = 0$ when v_{in} switches to -5 V, I_{BR} has the same value as in equation (10.4-31a) and thus for $t > 0$, the expression for $i_C(t)$ may be written as

$$i_C(t) = h_{FE}I_{BF}e^{-t/\tau_B} - h_{FE}I_{BR}(1 - e^{-t/\tau_B}) = 40e^{-t/\tau_B} - 22.8 \tag{10.4-35a}$$

Of course, this equation is valid only for $i_C < 4.8$ mA, since when i_C attempts to exceed this value, the transistor saturates and i_C remains constant at 4.8 mA. The corresponding expression for v_{CE} is

$$v_{CE} = V_{CC} - i_C R_C = 27.8 - 40e^{-t/\tau_B} \tag{10.4-35b}$$

These results are sketched in Figure 10.4-11d. The value of t_{PLH} for this gate may be determined by finding the amount of time that it takes for equation (10.4-35a) to decrease to 2.4 mA. Carrying out this procedure, we obtain

$$t_{PLH} = \tau_B \ln \frac{40}{27.8 - 2.4} = 9.1 \text{ ns} \tag{10.4-36}$$

Comparing the two results in equations (10.4-32) and (10.4-36), it is apparent that the inverter with the Schottky transistor switches much faster than the conventional BJT-style inverter. This is because the base current in the conventional BJT inverter is much larger than it needs to be, and this results in a considerable storage charge in the base. When v_{in} goes negative, all this charge must be removed before the transistor can reenter the active region. For this particular circuit the time required to accomplish this is the same as the time required for the expression for i_C in equation (10.4-35a) to decrease to

4.8 mA. This time interval is equal to

$$t_s = \tau_B \ln \frac{40}{22.8 + 4.8} = 7.4 \text{ ns} \qquad (10.4\text{-}37)$$

The remaining time required for this expression to decay to 2.4 mA is very nearly equal to the propagation time of the Schottky inverter. Thus the addition of a Schottky diode to a BJT transistor basically improves the switching performance of the transistor by reducing its storage time to zero.

Figure 10.4-12a shows the design of a typical Schottky-style TTL NAND gate. This gate has the same basic layout as that of a standard 7400 logic gate; however, there are several apparent circuit differences. First, all the transistors, with the exception of Q_4, have been replaced by Schottky transistors. Q_4

Figure 10.4-12

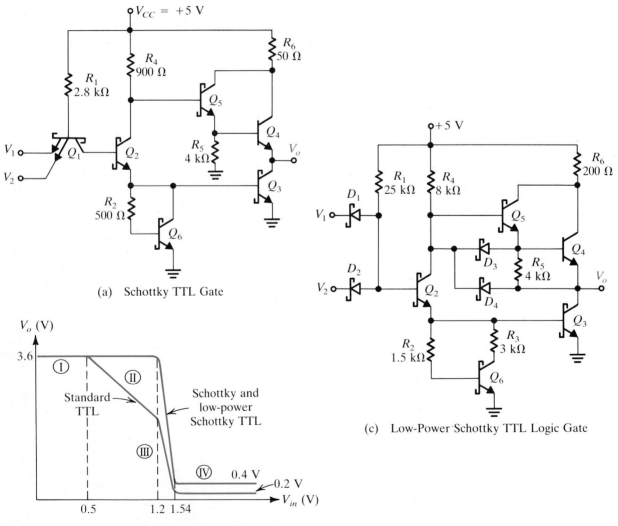

(a) Schottky TTL Gate

(b) Gate Transfer Characteristic with $V_1 = V_{in}$, $V_2 = +5$ V

(c) Low-Power Schottky TTL Logic Gate

does not need to be a Schottky device since it is impossible for this transistor to saturate with Q_5 connected across its base–collector junction.

In addition, the Q_4–D_1 combination in the standard TTL logic gate illustrated in Figure 10.4-5a has been replaced by the Darlington pair Q_4–Q_5. This circuit change lowers the gate output impedance when these transistors are in their active region and helps to reduce the influence of capacitive loading on the gate propagation delay. Notice that even though D_1 has been removed, the two base–emitter diode drops of Q_4 and Q_5 still ensure that Q_4 will be cut off when Q_2 and Q_3 saturate since V_{C2} will only be 1.1 V in this instance.

This Schottky logic gate also uses an active pull-down transistor Q_6 in place of a resistive pull-down found in standard TTL. This circuit modification offers two principal advantages. First, it helps to "square up" the gate transfer characteristic illustrated in Figure 10.4-12b. In a conventional TTL logic gate this initial drop in V_o (labeled region II in the figure) is associated with the early turning on of Q_2 before Q_3 as V_{in} increases above about 0.5 V. With an active pull-down transistor circuit used in place of the single resistor, Q_2 cannot turn on until V_{B3} exceeds 0.7 V, and as a result the gate output voltage will not begin to fall until V_{in} exceeds about $(1.4 - 0.2)$ or 1.2 V. This simultaneous turn-on of both Q_2 and Q_3 effectively eliminates region II from the transfer characteristic and results in a characteristic of the type shown in Figure 10.4-12b for this Schottky-style logic gate. The removal of region II from the transfer characteristic is important because it improves the gate noise immunity when the input to the gate is a logic 0.

Besides enhancing the gate noise performance, Q_6 also helps to turn Q_3 off more rapidly than would be possible with a comparable resistance. This is because Q_6 has a very small Thévenin impedance when it is operating in its active region, and as a result any charge on the parasitic capacitances of Q_3 can be rapidly removed by Q_6. The speed of Schottky TTL has also been increased by reducing the size of the gate resistances in order to lower the circuit time constants. These circuit changes make Schottky devices much faster than regular TTL; specifically, the propagation delays of Schottky logic gates are about one-third of those of conventional TTL devices or about 3 ns. Reducing the resistor sizes much below the values indicated in Figure 10.4-12a is impractical because of increased gate power dissipation. In fact, because the resistance values used in a Schottky gate are about one-half those found in a regular TTL, it turns out that the average power dissipation in a Schottky device is about 20 mW, twice that of a standard TTL logic gate.

For applications where the speed of regular TTL is adequate, considerable power savings can be realized if the gate resistances of the standard TTL logic gate are increased, while the gate transistors are replaced by Schottky devices. The increase in the resistor values lowers the gate power dissipation and, of course, tends to increase the gate time constants. However, the effect of these increased time constants is more than offset by the fact that the transistor storage times are reduced to nearly zero. The logic family that results from making these changes is known as Low-Power Schottky, and typically, this form of TTL logic operates at about the same speed as regular TTL but consumes about one-fifth the power. A two-input LS NAND gate is shown in Figure 10.4-12c.

Notice that besides simply increasing the resistance values, several other circuit improvements have also been added. On the input side of the gate, the multiemitter transistor, which was one of the trademarks of standard TTL, has been replaced by a DTL-style diode AND gate. This is permissible because Q_2 is now a Schottky transistor, and since it cannot saturate, an active pull-down device at the input really is not needed to ensure that it turns off quickly. In addition, the fact that the diodes are also Schottky devices actually permits this AND gate to operate faster than an equivalent multiemitter transistor. Furthermore, these Schottky diodes have higher breakdown voltages than the heavily doped base–emitter junctions of the multiemitter transistors that they are replacing. As a result, LS gates can handle input voltages of up to 7 V while conventional TTL and Schottky logic gates have maximum input voltage ratings of about 5.5 V. This higher breakdown voltage is important in high-speed logic circuits, where glitches, transmission-line reflections, and ringing (see Section 10.7) can easily cause the gate input voltage levels to exceed V_{CC}.

Because Low-Power Schottky also uses an active pull-down transistor Q_6, the transfer characteristics of an LS gate are nearly identical to those of a regular Schottky gate. In particular, the output begins to fall when V_{in} reaches about $(1.4 - 0.3)$ or 1.1 V, and flattens out at about 0.4 V when V_{in} exceeds 1.5 V.

LS circuit performance has also been improved by connecting R_5 to the collector of Q_3 instead of to ground. With R_5 connected to ground, a significant amount of power is dissipated in this resistor, especially when V_o is at a logic 1. By connecting R_5 across the base–emitter junction of Q_4 instead, the power dissipation in this resistor is reduced to nearly zero when V_o is HI. Furthermore, because the function of R_5 is to assist in turning OFF Q_4, this change has little effect on this aspect of the circuit performance.

Also included in the LS version of this gate are two Schottky diodes D_3 and D_4. These devices are known as speed-up diodes, and their primary function is to assist in pulling the output LO to reduce t_{PHL}. Specifically, D_3 is used to discharge the input capacitance of Q_4 when Q_2 turns on, and in a similar fashion, diode D_4 (which is not included in all LS logic gates) helps to discharge any load capacitance connected between V_o and ground. When Q_2 cuts off and V_{C2} rises, D_3 and D_4 have no effect since they are both reverse biased.

In the early 1980s, two additional TTL digital logic subfamilies were introduced, known as Advanced Schottky (AS) and Advanced Low-Power Schottky (ALS). Basically, AS and ALS were designed to replace Schottky and Low-Power Schottky devices, respectively, and each of these advanced TTL subfamilies offers higher speed and lower power consumption than the logic that it replaces. The propagation delay, power consumption, speed–power product, and other pertinent parameters of the more important TTL subfamilies are presented for comparison in Table 10.4-1.

Both AS and ALS take advantage of recent advances in IC fabrication technology which permit the construction of transistors having reduced parasitic capacitances. In addition, there have also been several circuit design changes incorporated into the Advanced Schottky series of logic gates. Figure 10.4-13a illustrates the design of an Advanced Schottky two-input NAND gate and Figure 10.4-13b that of an ALS version of the same gate. The AS gate architecture shown in part (a) of the figure is nearly the same as that of a conventional Schottky gate except that the multiemitter transistor at the input is replaced by a diode AND gate structure. As with the LS gate design discussed earlier, the principal reason for making this change is the higher switching speed and increased breakdown voltage offered by the use of these diodes.

Logic Subfamily	t_P (ns)	P_D (mW)	Speed–Power Product (pJ)	Maximum Input Frequency (MHz)	I_{IH} (μA)	I_{IL} (mA)[a]	I_{OH} (μA)[a]	I_{OL} (mA)
74xx	10	10	100	35	40	−1.6	−400	16
74Sxx	6.5	20	130	125	50	−2.0	−400	20
74LSxx	10	2	20	45	20	−0.36	−400	8
74ASxx	1.5	10	15	175	20	−0.5	−2000	20
74ALSxx	5	1	5	50	20	−0.1	−400	8

[a]The minus sign indicates that the current flows out of the terminal of the gate.

Figure 10.4-13

The Advanced Low-Power Schottky gate given in Figure 10.4-13b is quite similar to the LS version of this design presented in Figure 10.4-12c. The principal differences between the two are that the resistor values in the ALS gate have been increased and also that the transistors Q_1, Q_2, and Q_3 have been added to the input circuit of this gate. By doubling the values of R_1 and R_4 in the ALS gate, the power consumption is reduced to about half of that of a con-

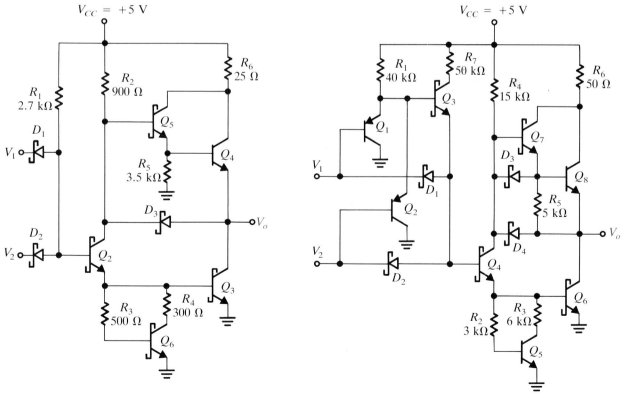

(a)　Advanced Schottky TTL Two-Input NAND Gate

(b)　Advanced Low-Power Schottky TTL Two-Input NAND Gate

ventional Low-Power Schottky device. However, despite these resistance increases, due to processing advances and architectural changes at the gate input, the ALS gate actually operates faster than its LS counterpart. For example, even though R_1 is large, the turn-on transient for Q_4 still occurs quite rapidly when V_1 and V_2 both go HI because Q_3 reduces the impedance seen by the base of Q_4 by the factor $(1 + h_{FE})$.

The addition of the transistor Q_3 also increases the input switching threshold of the gate from 1.1 V for Low-Power Schottky TTL to about $(2.1 - 0.7)$ or 1.4 V for this ALS gate. This increases the noise margin for input LOs from 0.7 V to 1.1 V, and this seemingly modest improvement in NM_L is quite important for high-speed logic gates such as ALS devices, where high-voltage glitches and interconnect reflections are common.

The *pnp* transistors Q_1 and Q_2 are essentially emitter followers, and they are used to reduce I_{IL} and also to steer the current from resistor R_1 either into the gate input terminals or into the base of Q_3, depending on the values of V_1 and V_2. Because these transistors always operate in their active regions, it is unnecessary for them to be Schottky devices. The reduction of I_{IL} to 0.1 mA for ALS devices increases the fan-out for output LOs to more than 80 gates, in comparison to a maximum allowed fan-out of only 10 gates for conventional TTL.

Exercises

10.4-1 Find I_{B1}, I_{B2}, I_{B3}, and I_{B4} for the gate in Figure 10.4-2a if a 1-kΩ load is connected onto the gate output, and if $V_1 = +5$ V and $V_2 = 1.0$ V. The transistor h_{FE}'s are 20. *Answer* $I_{B1} = 0.825$ mA, $I_{B2} = 0.025$ mA, $I_{B3} = 0$, $I_{B4} = 0.123$ mA

10.4-2 In Figure 10.4-6, find the value of V_{in} for which $V_{C2} = 4$ V. Consider that Q_1 is saturated and that the transistor h_{FE}'s are 20. *Answer* 1.156 V

10.4-3 Find the noise margins when the TTL inverter in Figure 10.4-5 is used to drive five DTL inverters of the type shown in Figure 10.3-5a. All transistor h_{FE}'s are 20. *Answer* $NM_L = 1.2$ V, $NM_H = 2.2$ V

10.4-4 Find the maximum fan-out when a TTL logic gate (Figure 10.4-5a) is used to drive Low-Power Schottky logic gates (Figure 10.4-12c). Consider the worst-case transistor h_{FE}'s to be equal to 10. *Answer* 143

10.4-5 Determine the LO-to-HI propagation delay for an LS logic gate (Figure 10.4-12c) driving a 20-pF load. Consider the switching characteristics of the gate to be ideal and the transistor h_{FE}'s to be 10. *Answer* $t_{PLH} = 2.2$ ns

10.5 MOS AND CMOS DIGITAL LOGIC DEVICES

MOS digital logic circuits are constructed using field-effect transistors. MOSFET transistors are attractive for digital ICs because of their small physical size, since typically, they occupy an area that is about one-fifth that required by a comparable BJT device. In addition, the power dissipation of an MOS logic gate is generally much smaller than that of a similar BJT gate. Despite these advantages, MOS digital logic has found little application at the individual gate level and the small-scale digital IC level, due principally to the poor drive capabilities of MOS devices. Of course, the drive ability of a gate can be improved by increasing the size of the transistors, but doing this takes away the two major advantages of MOS: its small physical size and its low power consumption.

Although MOS is not very well suited to small-scale integrated circuit designs, it has proven to be the nearly ideal logic family for constructing LSI circuits such as memories, microprocessors, and shift registers. This is because large-scale integrated circuits have relatively few outputs that need to be driven in comparison to the total number of digital logic elements that are used to construct the completed IC. Therefore, most of the circuit can be fabricated using small-area, low-power MOS transistors, and then, where needed, a few larger MOS devices can be used to interface to the outside world.

Initially, PMOS digital logic circuits were used almost exclusively for manufacturing MOS digital ICs because they were easier to fabricate than NMOS devices. However, PMOS has several disadvantages. P-channel MOS transistors are inherently slower than their NMOS counterparts because of their higher resistivity. Furthermore, PMOS requires negative power supplies, and this makes it difficult to interface PMOS logic to other popular forms of digital logic such as TTL. Therefore, as soon as technological developments permitted, there was a rapid changeover to NMOS digital ICs, and today most memories, microprocessors, and other large-scale digital integrated circuits are manufactured using NMOS technology.

Complementary MOS (CMOS) is a third type of MOS digital logic that is made up of both p-channel and n-channel transistors fabricated on the same substrate. Although it was not originally that popular when it was introduced in the early 1970s, it has already taken over many of the application areas formerly handled by TTL and conventional MOS logic. The major attributes of CMOS are its high noise immunity and its very low power consumption. However, despite these advantages, for many years CMOS remained far behind TTL and conventional MOS in popularity. The delay in the acceptance of CMOS was due principally to its high cost, large gate surface area, and relatively low speed operation. Today, however, this picture is quite different. Technological advances, while maintaining CMOS's low power consumption and high noise immunity, have permitted a significant improvement in the speed of these devices, to the point where their propagation delays are now comparable to those of TTL. Furthermore, there has also been a significant reduction in the chip area needed to implement complementary MOS logic functions, and this has made CMOS competitive with NMOS in memory, microprocessor, and other LSI applications.

10.5-1 Static and Dynamic Characteristics of MOS and CMOS Inverters

In the design of digital logic circuits, the inverter is the basic building block, and by examining the operation of this rather simple digital logic element, we can get a good understanding of the performance characteristics of a particular digital logic family. Figure 10.5-1a illustrates the basic design of an NMOS inverter. The schematic symbol used to represent the transistor in this figure is a simplified way to characterize an enhancement or depletion-style n-channel MOSFET. The arrow on the lower terminal indicates that it is the source lead

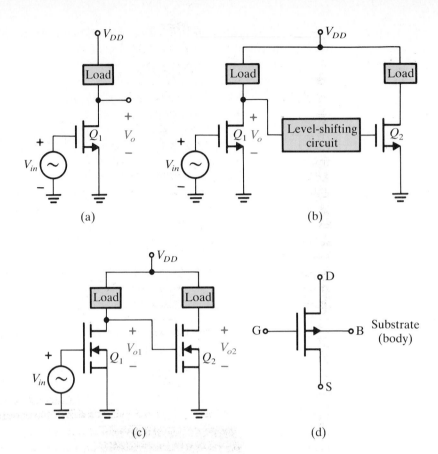

Figure 10.5-1
Basic structure of an MOS inverter.

(a)　(b)

(c)　(d)

of the transistor, and the direction of the arrow (which shows the direction of current flow through the channel) identifies this device as an n-channel transistor. The symbol for a p-channel transistor is identical except that the arrow direction is reversed. Sometimes when it is important to identify the substrate lead on the transistor, the alternative schematic symbol shown in Figure 10.5-1d is used instead.

To complete the design of the MOSFET inverter shown in Figure 10.5-1a, we need to determine whether the driver transistor Q_1 should be an enhancement or a depletion-mode device, and also, we need to specify the specific form of the load.

Let's initially examine the question of whether to use a depletion- or an enhancement-mode device for Q_1. Basically, this decision is related to the amount of interfacing circuitry needed to make the input and output voltage levels compatible. As we discussed in Section 5.5, depletion-mode NMOS devices are normally on and require negative input voltages to turn them off. As a result, it will not be possible to interconnect the output of one depletion-mode inverter directly to the input of the next since doing this will cause the second inverter to always be ON regardless of the output state of the first stage. To be able to couple two depletion-style MOSFET inverter stages together, some type of dc level shifting circuitry will be needed (Figure 10.5-1b), and the additional complexity that this creates makes the use of depletion-mode drivers impractical.

As an alternative, let's consider the possibility of using an enhancement-mode driver transistor instead. Enhancement-style NMOS transistors are normally OFF when the gate voltage applied to them is zero, and a positive gate voltage is required to turn them ON. These input voltage levels are directly compatible with the output voltages available from our digital logic inverter, and as a result, no interfacing hardware is needed to connect two of these inverters together. To illustrate this point, let's examine the two-inverter cascade shown in Figure 10.5-1c. When V_{in} is a logic 0 at zero volts, Q_1 will be cut off and V_{o1} will be a logic 1 at about $+V_{DD}$ volts. If this voltage is greater than V_T, the transistor threshold voltage, Q_2 will turn ON and its output will go to a logic 0. Conversely, when V_{in} switches to a logic 1, the exact opposite situation will prevail in which V_{o1} will be a logic 0 and V_{o2} will be a logic 1. Therefore, when enhancement transistors are used as the drivers, a complete inverter circuit consists of a driver transistor connected in series to V_{DD} by a suitable load. No additional interfacing circuitry is required.

Let's consider next possible circuit configurations that can be used for the load element in a digital logic inverter. To begin with, the load may be either passive or active. Figure 10.5-2a illustrates the passive design case, in which

Figure 10.5-2

Possible loads for use in an NMOS inverter.

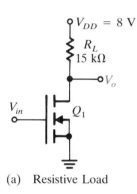

(a) Resistive Load

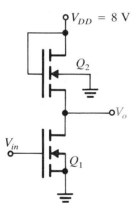

(b) Constant-Current Enhancement Load (NMOS) with Gate Tied to V_{DD}

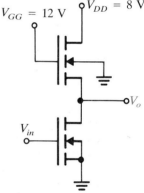

(c) Ohmic Enhancement Load (NMOS), Gate Tied to Separate V_{GG} Power Supply

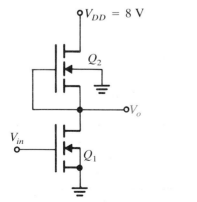

(d) NMOS Depletion Load, Gate and Source Connected

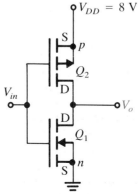

(e) CMOS Inverter, p-Channel Enhancement Load

the load is simply a resistor R_L. The remaining circuits in this figure show different types of active loads that can also be used for the design of MOS circuits. In particular, in parts (b) and (c) of the figure, the load used is an NMOS enhancement transistor, in part (d) a depletion-style NMOS device, and in part (e) a PMOS enhancement transistor. Of these different inverters two, the ohmic-style MOS load and the p-channel CMOS load, turn out to have the best switching and overall performance characteristics. As a result, the operation of these two circuits will be investigated in some detail in this section, while the analysis of the other three types of inverters will have to be left for the homework problems at the end of this chapter. Before beginning a formal analysis of these two inverters, it will be useful to review some of the pertinent characteristics of MOS transistors.

Let's begin by reexamining the volt-ampere characteristics of a typical enhancement-style NMOS transistor (Figure 10.5-3a). For a given value of V_{GS}, the transistor operates on the flat or constant-current portion of the characteristics when

$$V_{DS} > V_{GS} - V_T \tag{10.5-1a}$$

In this region I_D is related to V_{GS} by the expression

$$I_D = \begin{cases} I_{D2T}\left(\dfrac{V_{GS}}{V_T} - 1\right)^2 & \text{for } V_{GS} > V_T \\ 0 & \text{for } V_{GS} < V_T \end{cases} \tag{10.5-1b}$$

where $I_{D2T} = 1.35$ mA and $V_T = +3$ V for this particular transistor.

When V_{DS} is less than $(V_{GS} - V_T)$, the transistor enters the ohmic region, and here, following equations (5.5-10) and (5.5-13a),

$$I_D = I_{D2T}\left[\left(\frac{V_{GS}}{V_T} - 1\right)\frac{V_{DS}}{V_T} - \frac{1}{2}\left(\frac{V_{DS}}{V_T}\right)^2\right] \tag{10.5-2a}$$

Because of the complexity of this expression, for simplicity, we approximate this portion of the transistor characteristics by the linear equation

$$I_D = \frac{I_{D2T}}{V_T}\left(\frac{V_{GS}}{V_T} - 1\right)V_{DS} \tag{10.5-2b}$$

(Figure 10.5-3a). Using this expression we may represent this region of the transistor characteristics by the dashed straight lines shown in the figure. To the extent that this approximation is valid, for a given value of V_{GS}, I_D varies linearly with V_{DS} so that the transistor may be represented by a constant resistance

$$R_n = \frac{V_{DS}}{I_D} = \frac{V_T^2}{I_{D2T}(V_{GS} - V_T)} \tag{10.5-2c}$$

as long as

$$V_{DS} < V_{GS} - V_T \tag{10.5-2d}$$

Thus far we have reviewed only the static characteristics of the n-channel MOS transistor. But since we will be examining the transient performance of

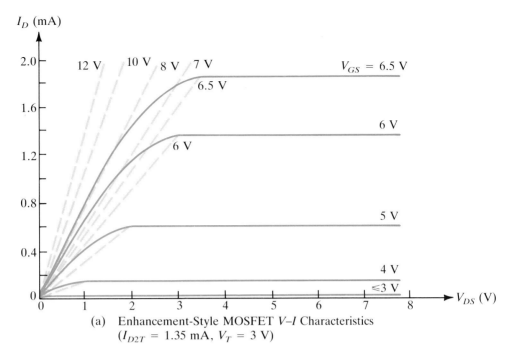

(a) Enhancement-Style MOSFET V–I Characteristics
($I_{D2T} = 1.35$ mA, $V_T = 3$ V)

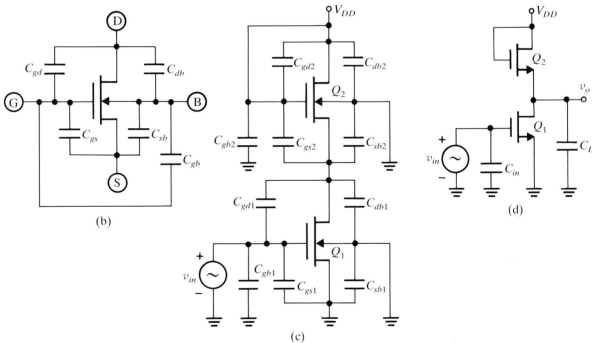

(b)

(c)

(d)

Figure 10.5-3

digital MOS logic circuits, it will also be important for us to have a transistor model that will properly account for its dynamic behavior. The essence of this model was discussed in Section 6.4, and for simplicity the equivalent circuit that we developed there has been redrawn in Figure 10.5-3b. In this circuit the elements C_{gd}, C_{gs}, and C_{gb} are all fixed electrostatic capacitances that are associated with the metal gate of the MOSFET. C_{db} and C_{sb}, on the other hand, represent the drain-to-substrate and source-to-substrate junction capacitances, and as such, their specific values are nonlinear functions of the voltages across them. However, in this chapter, for simplicity, we will also consider these capacitances to be constants.

Because of the sheer number of capacitances in the equivalent circuit shown in Figure 10.5-3b, a detailed (noncomputer-assisted) analysis of the transient performance of digital MOS circuits will be virtually impossible. As an alternative, we make several major simplifications in the digital circuits that we will be examining in this section which will permit us to carry out a rather straightforward analysis of the transient response of these circuits.

The basic technique to be used is best described by examining a specific example such as the MOSFET inverter shown in Figure 10.5-3c. To say the least, this circuit is rather complex, and while its analysis is amenable to computer solution, it is difficult to get any physical understanding of its dynamic performance, at least in its present form. To simplify this circuit, all of the parasitic capacitances will be replaced by two equivalent capacitances C_{in} and C_L connected to the inverter's input and output terminals, respectively (Figure 10.5-3d). To a first-order approximation, by inspection of the original circuit in Figure 10.5-3c, we may write each of these capacitances as

$$C_{in} = C_{gs1} + C_{gb1} + C_{gd1} \qquad (10.5\text{-}3a)$$

and
$$C_L = C_{db1} + C_{gd1} + C_{gs2} + C_{sb2} \qquad (10.5\text{-}3b)$$

In estimating these capacitance values, we have neglected the magnification of C_{gd1} due to the Miller effect because Q_1 will either be cut off or saturated during a good portion of the gate transient.

The technique that we have just used to simplify the inverter in Figure 10.5-3c to the final form given in Figure 10.5-3d may also be applied to all the other MOSFET gating circuits to be examined in this section simply by shunting the gate input and output terminals by C_{in} and C_L, respectively. Depending on the specific character of the gate, the size of these parasitic elements may change, but the basic form of the model to be used for analyzing the dynamic performance of the gate will remain the same. This completes our MOSFET review.

Let's return now to our examination of the performance of the inverters in Figure 10.5-2 beginning first with the ohmic-style active load inverter shown in part (c) of the figure. This circuit is basically the same as the constant-current inverter given in part (b) of the figure except that the gate of the transistor, rather than being tied to V_{DD}, is connected instead to a separate power supply, V_{GG}. As a result,

$$V_{GS2} = V_{G2} - V_{S2} = V_{GG} - V_o \qquad (10.5\text{-}4a)$$

and
$$V_{DS2} = V_{D2} - V_{S2} = V_{DD} - V_o \qquad (10.5\text{-}4b)$$

Solving equation (10.5-4a) for V_o and substituting this result into equation (10.5-4b), we have

$$V_{DS2} = V_{GS2} - (V_{GG} - V_{DD}) \qquad (10.5\text{-}5)$$

Comparing this result with equation (10.5-2d), we can see that as long as

$(V_{GG} - V_{DD})$ is greater than V_T (which it is for the circuit in question), Q_2 will be operating in the ohmic region. The effective load resistance of Q_2 is given by equation (10.5-2c), and the resulting load line for this circuit, sketched on the Q_1 volt-ampere characteristic shown in Figure 10.5-4a, was developed with the aid of the table presented in part (c) of the figure assuming that $I_{D2T2} = 0.135$ mA and $V_{T2} = 3$ V. In designing this circuit, I_{D2T2} has purposely been chosen to be much smaller than I_{D2T1} in order to make the effective load resistance of Q_2, and hence the inverter gain in the active region, large. The transfer characteristic for this inverter may be readily obtained once the load line is drawn by entering different values of V_{in} (or V_{GS1}) onto this load line and reading off the corresponding values of V_o (or V_{DS1}). This transfer characteristic is sketched in Figure 10.5-4b. For comparison purposes, the transfer characteristic of the constant-current inverter given in Figure 10.5-2b has also been included.

By examining the curve for the ohmic-style inverter it is apparent that this inverter's gain in the active region is comparable to that of the constant-current MOSFET inverter since both have nearly the same slope in the transition region. This fact may be substantiated by carrying out a small-signal analysis in the active region, in which the load transistor Q_2 is represented by a fixed re-

Figure 10.5-4
MOSFET inverter with an enhancement load.

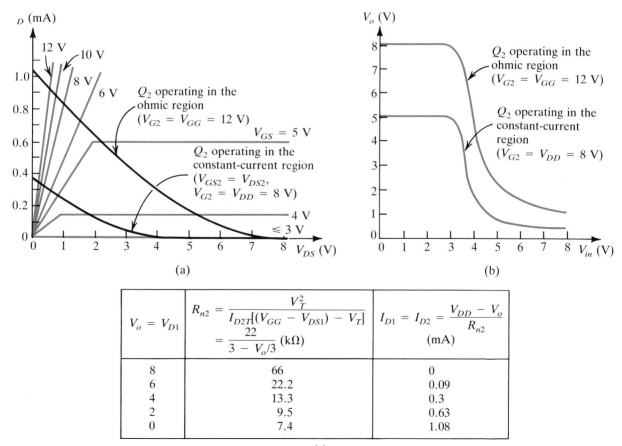

(a)

(b)

$V_o = V_{D1}$	$R_{n2} = \dfrac{V_T^2}{I_{D2T}[(V_{GG} - V_{DS1}) - V_T]}$ $= \dfrac{22}{3 - V_o/3}$ (kΩ)	$I_{D1} = I_{D2} = \dfrac{V_{DD} - V_o}{R_{n2}}$ (mA)
8	66	0
6	22.2	0.09
4	13.3	0.3
2	9.5	0.63
0	7.4	1.08

(c)

sistance having a value somewhere in between the two extremes indicated in the table in Figure 10.5-4c. Because the average gate-to-source voltage across this transistor is

$$V_{GS} = \frac{V_{GS,\text{max}} + V_{GS,\text{min}}}{2} = \frac{(V_{GG} - o) + (V_{GG} - V_{DD})}{2} = V_{GG} - \frac{V_{DD}}{2} \qquad (10.5\text{-}6a)$$

the corresponding average ohmic resistance of Q_2 may be approximated as

$$R_{n2} = \frac{V_T^2}{I_{D2T2}(V_{GG} - V_{DD}/2 - V_T)} = 13.3 \text{ k}\Omega \qquad (10.5\text{-}6b)$$

Of course, this is precisely the entry on the table in Figure 10.5-4c corresponding to the midpoint value of V_o at 4 V. Using this result, and evaluating the g_m of Q_1 at the center of the transition region using equation (5.6-3), the small-signal gain may then be written as

$$A = -g_{m1}R_{n2} = (-0.3 \text{ mS})(13.3 \text{ k}\Omega) = -3.9 \qquad (10.5\text{-}7)$$

By a similar analysis technique, the small-signal gain of the constant-current inverter can be shown to be -4, so that as expected from the shape of the transfer characteristics, the small-signal gains of these two inverters are virtually identical.

In comparing the static performance of the constant-current and ohmic-style active load MOSFET inverters, it is apparent that neither is vastly superior to the other. The constant-current inverter has reduced signal swing and noise immunity, which is overcome in the ohmic-style inverter. But the price paid for these improvements is rather high. An extra power supply is needed to achieve ohmic region operation of the load, and while the power requirements from this supply are small, multiple voltage power supply designs are generally considered to be undesirable.

EXAMPLE 10.5-1

For the ohmic-style active load inverter illustrated in Figure 10.5-2c, consider the $I_{D2T1} = 1.35$ mA, $I_{D2T2} = 0.135$ mA, and also that $V_T = 3$ V for Q_1 and Q_2, so that the gate transfer characteristic has the form previously given in Figure 10.5-4b.

a. Determine V_{OH} and V_{OL} for this logic gate.
b. Find t_{PLH} when v_{in} switches from $+8$ V to 0 V, assuming that C_{in} and C_L are each 10 pF for this gate (Figure 10.5-5a).

SOLUTION

From the transfer characteristic it is immediately apparent that V_{OH} and V_{OL} are equal to 8.0 and 1 V, respectively. Using these results, t_{PLH} may be defined as the time required for the output to rise to 50% of the signal swing or to the voltage

$$V_o = \tfrac{1}{2}(V_{OH} + V_{OL}) = \tfrac{1}{2}(8 + 1) = 4.5 \text{ V} \qquad (10.5\text{-}8)$$

Because v_{in} is assumed to be an ideal voltage source, v_{GS1} will simply be equal to v_{in}, and C_{in} will have no effect on the transient performance of this circuit. Furthermore, following equation (10.5-6b), we assume that Q_2 can be modeled approximately as a 13.3-kΩ fixed resistor. To the extent that this ap-

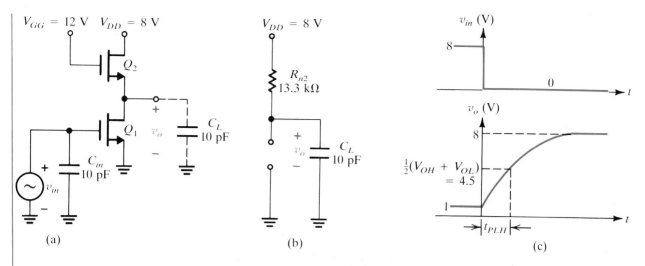

Figure 10.5-5
Example 10.5-1.

proximation is valid, the equivalent circuit in Figure 10.5-5b applies, and $v_o(t)$ may be written immediately as

$$v_o(t) = e^{-t/\tau_1} + 8(1 - e^{-t/\tau_1}) = 8 - 7e^{-t/\tau_1} \qquad (10.5\text{-}9)$$

The waveshape for $v_o(t)$ is shown in Figure 10.5-5c, and from this sketch it is apparent that

$$t_{PLH} = \tau_1 \ln \frac{7}{8 - 4.5} = 92 \text{ ns} \qquad (10.5\text{-}10)$$

Before leaving this example, it is interesting to point out that computer-aided analysis of this problem (taking into account the nonlinear resistance of Q_2) yields a solution of 66 ns for the propagation delay. A similar computer-aided analysis for the constant-current inverter using the same data for Q_1 and Q_2 yields $t_{PLH} = 157$ ns.

The next MOSFET inverter that we examine is the one illustrated in Figure 10.5-2e, which employs an enhancement style NMOS driver transistor Q_1, and an enhancement style PMOS active load transistor Q_2. This CMOS inverter circuit is repeated in Figure 10.5-6a for convenience. Because the gates of both transistors are connected to the input signal V_{in}, and also because the source of the PMOS transistor is connected to $+V_{DD}$, this inverter has some very interesting circuit properties. In particular, and as we shall see shortly, when Q_1 is ON, Q_2 is OFF, and vice versa, so that in its static mode of operation this circuit consumes virtually no power from the dc supply.

To understand how this logic gate operates, it is useful to note that Q_1 is OFF when V_{in} is less than $V_{T1} = 3$ V, and turns ON when V_{in} exceeds this value. In a similar fashion, Q_2 is OFF when $|V_{GS2}|$ is less than $|V_{T2}| = 3$ V, and ON when $|V_{GS2}|$ exceeds this voltage. One last point worth mentioning is that $V_{GS2} = (V_{G2} - V_{S2}) = (V_{in} - V_{DD})$ for this circuit.

Let's examine what happens to V_o for HI and LO values of V_{in}. When $V_{in} = 0$ V, $V_{GS1} = 0$ and $V_{GS2} = -V_{DD}$. As a result, for this value of input voltage Q_1 is OFF, Q_2 is ON, and V_o is HI at $+V_{DD}$ volts. Conversely, when

821

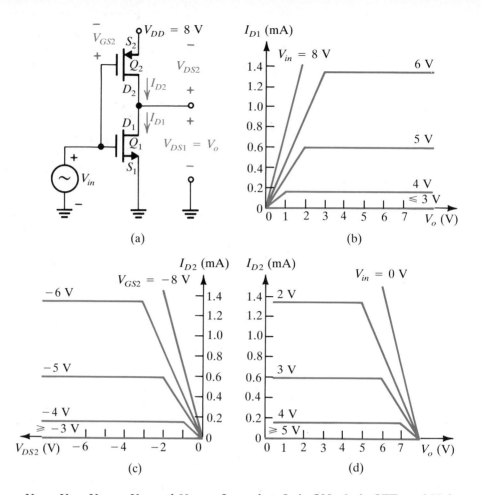

Figure 10.5-6
Static characteristics of CMOS logic gates.

$V_{in} = V_{DD}$, $V_{GS1} = V_{DD}$ and $V_{GS2} = 0$, so that Q_1 is ON, Q_2 is OFF, and V_o is LO at about zero volts. Of course, following these results, it is apparent that this circuit, too, functions like a digital logic inverter. Next we will examine the behavior of this CMOS inverter in greater detail by carefully investigating the graphical performance of this circuit.

To begin with, because $V_{GS1} = V_{in}$ and $V_{DS1} = V_o$, the volt-ampere characteristics of Q_1 may be relabeled in terms of V_{in} and V_o as illustrated in Figure 10.5-6b. For reasons that will be apparent shortly, it is also convenient to express the volt-ampere characteristics of Q_2 in terms of V_{in} and V_o rather than V_{GS2} and V_{DS2}, as is usually done. If we assume that Q_2 has identical characteristics to Q_1, except for the fact that it is a PMOS rather than an NMOS device, then its volt-ampere characteristics have the form given in Figure 10.5-6c. Because

$$V_{GS2} = V_{in} - V_{DD} \tag{10.5-11a}$$

the input voltages, V_{in}, producing specific values of V_{GS2} may be determined by rewriting this expression as

$$V_{in} = V_{GS2} + V_{DD} = V_{GS2} + 8 \tag{10.5-11b}$$

For example, a V_{GS2} value of -3 V corresponds to an input voltage of $+5$ V being applied to the gate. The volt-ampere characteristics of Q_2 are redrawn in Figure 10.5-6d with the curves relabeled to reflect their dependence on V_{in}.

In a similar fashion, the relationship between V_{DS2} and V_o is readily seen to be

$$-V_{DS2} + V_o = V_{DD} \tag{10.5-12a}$$

or

$$V_o = V_{DD} + V_{DS2} \tag{10.5-12b}$$

Thus if we want to express the Q_2 volt-ampere characteristics in terms of V_o instead of V_{DS2}, V_{DD} volts must be added to all the V_{DS2} values to convert them to their corresponding values of V_o. Basically, this means that we may relabel the x axis of the Q_2 characteristics as V_o if we shift all the curves to the right by V_{DD} volts, as illustrated in Figure 10.5-6d. Furthermore, because $I_{D1} = I_{D2}$ for this CMOS inverter, these volt-ampere characteristic curves for Q_1 and Q_2 may now be superimposed on top of one another, as shown in Figure 10.5-7a, to find the circuit operating point for different values of V_{in}.

Following this figure, for values of V_{in} less than 3 V, the operating point is

Figure 10.5-7

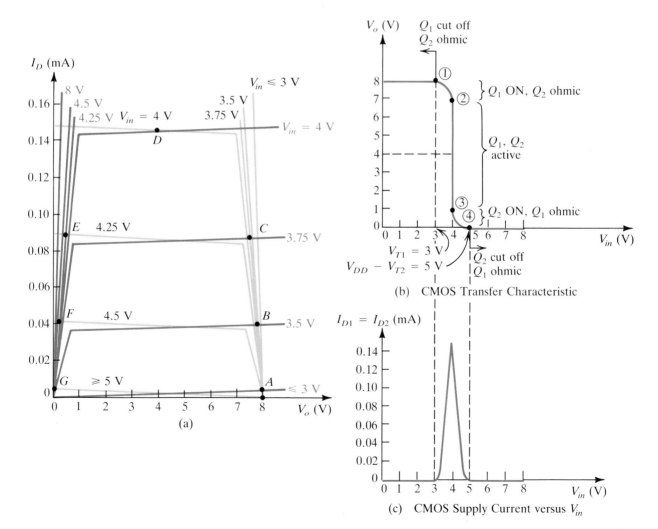

(a)

(b) CMOS Transfer Characteristic

(c) CMOS Supply Current versus V_{in}

located at point A in the lower right-hand corner of the composite characteristics with V_o approximately equal to $+8$ V and I_{D1} and I_{D2} nearly zero. Because Q_1 is cut off here and Q_2 is in the ohmic region, the equivalent circuit has the form shown in Figure 10.5-8a, where

$$R_{p2} = \frac{|V_{T2}|^2}{I_{D2T2}(-V_{GS2} + V_{T2})} = \frac{9}{(1.35)(8 - 3)} = 1.33 \text{ k}\Omega \qquad (10.5\text{-}13)$$

when $V_{in} = 0$.

When V_{in} increases beyond 3 V, transistor Q_1 enters the active region while Q_2 remains in the ohmic region until V_{in} reaches $+4$ V (see, e.g., points B and C along the right-hand side of Figure 10.5-7a). Because Q_2 is resistive, the equivalent circuit here has the form shown in Figure 10.5-8b, so that

$$A = -g_{m1} R_{p2} \qquad (10.5\text{-}14a)$$

where

$$R_{p2} = \frac{|V_{T2}|^2}{I_{D2T2}[-(V_{in} - 8) + V_{T2}]} \qquad (10.5\text{-}14b)$$

In this interval, as V_{in} increases so do g_{m1} and R_{p2}, and as a result, the slope of the transfer characteristic also increases correspondingly between points 1 and 2 in Figure 10.5-7b.

When V_{in} reaches 4 V, both transistors enter the active region, and the equivalent circuit has the form illustrated in Figure 10.5-8c. To the extent that r_{d1} and r_{d2} may be neglected, the small-signal gain in this region will be infinite. This infinite gain is indicated in Figure 10.5-7b by the vertical nature of the transfer characteristic between points 2 and 3.

Figure 10.5-8
Equivalent circuits for the CMOS inverter in Figure 10.5-6a.

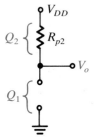

(a) Equivalent Output Circuit for $V_{in} < 3$ V

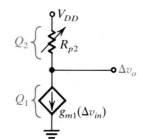

(b) Equivalent Small-Signal Output Circuit for $3 \text{ V} < V_{in} < 4 \text{ V}$

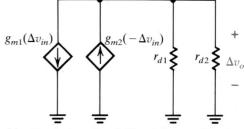

(c) Equivalent Small-Signal Output Circuit When Q_1 and Q_2 Are Both Active ($V_{in} \simeq 4$ V)

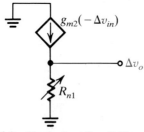

(d) Equivalent Small-Signal Output Circuit for $4 \text{ V} < V_{in} < 5 \text{ V}$

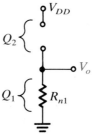

(e) Equivalent Output Circuit for $V_{in} > 5$ V

At point 2 on the characteristic, Q_2 just makes the transition from ohmic to constant-current operation since at this point $V_o = 7$ V and $V_{in} = 4$ V, V_{DS2} is just equal to -1V (Figure 10.5-7a). For values of V_o below 7 V, both transistors enter the active region and stay there until point 3 on the curve is reached at $V_o = 1$ V, where Q_1 enters the ohmic region.

Because the interval between points 2 and 3 is vertical, just as V_{in} reaches 4 V there will be a nearly instantaneous drop in the output voltage of this inverter from 7 V to about 1 V. As V_{in} increases beyond 4 V, Q_1 becomes resistive while Q_2 remains active so that the equivalent circuit in Figure 10.5-8d applies with

$$A = -g_{m2} R_{n1} \tag{10.5-15a}$$

and
$$R_{n1} = \frac{|V_{T1}|^2}{I_{D2T1}(V_{in} - V_{T1})} \tag{10.5-15b}$$

Here V_o decreases as shown between points 3 and 4, mirroring the circuit behavior between points 1 and 2 when Q_1 was active and Q_2 resistive. For values of V_{in} greater than 5 V, Q_2 cuts off because $|V_{GS2}|$ is less than 3 V here. As a result, in this region the equivalent circuit has the form shown in Figure 10.5-8e with V_o nearly zero and

$$R_{n1} = \frac{9}{(1.35)(8 - 3)} = 1.33 \text{ k}\Omega \tag{10.5-16}$$

when V_{in} reaches $+8$ V.

Both the transfer characteristic in Figure 10.5-7b and the sketch of current versus V_{in} in Figure 10.5-7c may readily be obtained from the composite volt-ampere characteristics of Q_1 and Q_2 given in Figure 10.5-7a. All that needs to be done to carry out this procedure is to read off the appropriate data for the drain current and output voltage at points A through G corresponding to different values of V_{in}.

Let's take a look at the noise margin of the CMOS inverter gate. By examining the specific form of the transfer characteristic in Figure 10.5-7b, it should be apparent that

$$V_{OH} = V_{DD} = 8 \text{ V} \tag{10.5-17a}$$

$$V_{OL} = 0 \text{ V} \tag{10.5-17b}$$

$$V_{IH} = 5 \text{ V} \tag{10.5-17c}$$

and
$$V_{IL} = 3 \text{ V} \tag{10.5-17d}$$

Using these results, it therefore also follows that

$$NM_H = V_{OH} - V_{IH} = 3 \text{ V} \tag{10.5-17e}$$

and
$$NM_L = V_{IL} - V_{OL} = 3 \text{ V} \tag{10.5-17f}$$

These noise margins are substantially higher than those of TTL logic gates, which are typically on the order of 0.4 V.

Besides having very high noise margins, as we have already noted, the power consumption of CMOS logic gates is also exceptionally low. The reason for this is apparent if we examine I_D versus V_{in} sketch in Figure 10.5-7c. Here, as long as V_{in} is a valid logic 0 or a valid logic 1, the gate current will be zero and hence the power consumed from the V_{DD} supply will also be zero. As a practical matter, the power consumption of a CMOS logic gate is never identically zero but is typically on the order of 10 to 100 nW due to the leakage currents that flow through the transistors when they are cut off.

Even though the static power consumption of a CMOS logic gate is very small, a significant amount of power can still be consumed by the gate under dynamic operating conditions. As v_{in} changes from a logic 0 to logic 1, and vice versa, current will flow briefly through Q_1 and Q_2 as v_{in} passes through the 3 to 5 V region (Figure 10.5-7c). The duration of this current spike varies with the rise and fall times of the input signal, and hence this component of the gate power consumption depends on the shape of v_{in} as well as its frequency.

EXAMPLE 10.5-2

Figure 10.5-9
Example 10.5-2.

a. Plot the power consumption of the CMOS gate just examined as a function of input signal frequency if v_{in} is considered to be a 0- to 8-V square wave with linear total rise and fall times of 100 ns.

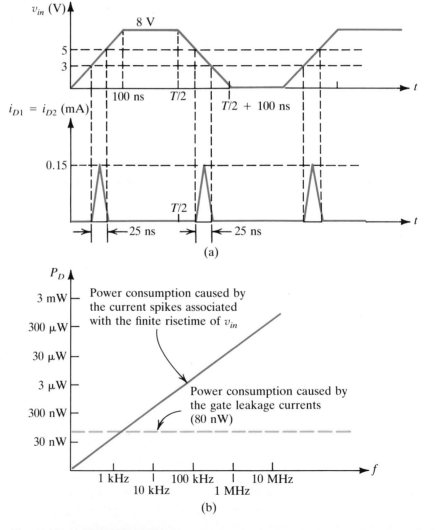

(a)

(b)

b. Determine the static power consumption of this inverter if the gate leakage current is 10 nA, and compare the result obtained with the power consumption component associated with the finite rise time of the input signal.

SOLUTION

If the rise and fall times of v_{in} are linear, as shown in Figure 10.5-9a, triangular current pulses with widths equal to $[(5 - 3)/8]$ 100 or 25 ns will be produced in the gate on both the rising and falling edges of v_{in}. As a result, because two current spikes are produced during each cycle of v_{in}, the average power consumption from the V_{DD} supply may be written as

$$P_D = \frac{1}{T}\left(2 \int_0^{T/2} V_{DD} i_{D2} \, dt\right) = \frac{2V_{DD}}{T}\underbrace{\left(\int_0^{T/2} i_{D2} \, dt\right)}_{\text{area}} \tag{10.5-18}$$

$$= (2V_{DD}f)[\tfrac{1}{2}(25 \text{ ns})(0.15 \text{ mA})] = (30 \text{ pJ})(f)$$

This expression for P_D is sketched in Figure 10.5-9b as a function of frequency. It is useful to observe that P_D represents the overall power dissipated in the gate during the intervals when Q_1 and Q_2 are simultaneously conducting, and because these two devices are assumed to have identical characteristics, one-half of the input power will be dissipated in each transistor.

From the current waveshape sketched in Figure 10.5-9a it is apparent that the current flow in from the power supply is nearly zero when v_{in} is a valid logic 1 or a valid logic 0. However, as we have already mentioned, although it is rather small under quiescent conditions, there is still a minute leakage current flow between V_{DD} and ground. If we assume that this current is about 10 nA, the power supplied by V_{DD} will be about 80 nW under static conditions. When v_{in} is a relatively low frequency signal, this portion of the gate power consumption can be an important component of the overall power delivered by V_{DD}, but at higher frequencies it can generally be neglected (see Figure 10.5-9b).

In the analysis just completed in Example 10.5-2, we neglected the effect of capacitive loading in computing the power consumed by the gate. As we shall see in Example 10.5-4, which follows shortly, the charging and discharging of this capacitance actually consumes more power from V_{DD} than the current glitch that flows between Q_1 and Q_2 during the brief instant when both transistors are active. However, before we can properly carry out this analysis, we first need to develop a good understanding of the transient performance of a CMOS inverter. Example 10.5-3 should help us to obtain this knowledge and will be useful in its own right since it will permit us to compare the performance of this CMOS logic gate with the other MOSFET inverters that we examined earlier in this section.

EXAMPLE 10.5-3

The circuit shown in Figure 10.5-10a has the same gate characteristics as the CMOS inverter that we have just been discussing. As a result, its composite volt-ampere characteristics are identical to those given in Figure 10.5-7a so that the equivalent circuits presented in Figure 10.5-8 may be applied where appropriate. Furthermore, the transistor parameters of Q_1 and Q_2 are identical with $I_{D2T1} = I_{D2T2} = 1.35$ mA and $V_{T1} = -V_{T2} = 3$ V.

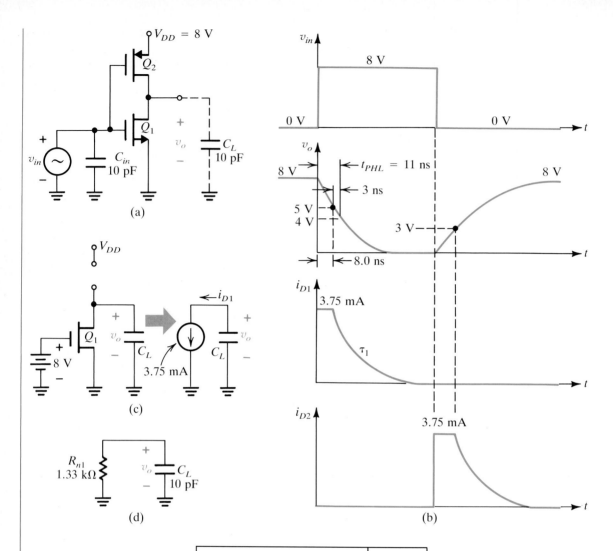

Inverter Type	t_{PLH} (ns)
Resistive load ($R_L = 15$ kΩ)	104
Constant-current enhancement load	165
Ohmic-style enhancement load	92
Depletion load	393
CMOS	11

(e)

Figure 10.5-10
Example 10.5-3. Transient response of a CMOS inverter.

Considering that C_{in} and C_L are each equal to 10 pF, carefully sketch v_o, i_{D1}, and i_{D2} if v_{in} is assumed to have the form given in Figure 10.5-10b.

SOLUTION

In analyzing this circuit, let's consider that v_{in} has been equal to zero for a long time so that $v_o(0-) = 8$ V. At $t = 0$ when v_{in} jumps to $+8$ V, Q_1 will turn ON immediately and Q_2 will instantly cut off despite the fact that the input is shunted by the capacitance C_{in}. This is because v_{in} is ideal voltage source so that the input time constant is zero.

Because $v_o = v_{DS1} = 8$ V at $t = 0+$, Q_1 will initially be in the constant-current region with

$$i_{D1} = I_{D2T1} \left(\frac{V_{GS1}}{V_T} - 1 \right)^2 = 1.35 \left(\frac{8}{3} - 1 \right)^2 = 3.75 \text{ mA} \qquad (10.5\text{-}19)$$

so that the equivalent circuit shown in Figure 10.5-10c applies. Therefore,

$$C_L \frac{dv_o}{dt} = -i_{D1} = -I_1 \qquad (10.5\text{-}20a)$$

and

$$v_o(t) = \frac{-I_1 t}{C_L} + \text{constant} = V_{CC} - \frac{I_1 t}{C_L} \qquad (10.5\text{-}20b)$$

This linear discharge of the voltage across C_L continues until it drops to $(v_{GS1} - V_{T1}) = (8 - 3) = 5$ V, and Q_1 enters the ohmic region. Following equation (10.5-20a), the time required for this to occur is

$$t_1 = \frac{C_L \, \Delta v_o}{I_1} = \frac{(10 \text{ pF})(3 \text{ V})}{3.75 \text{ mA}} = 8 \text{ ns} \qquad (10.5\text{-}21)$$

Once Q_1 enters the ohmic region, the equivalent circuit shown in part (d) of the figure is valid with

$$R_{n1} = \frac{V_{T1}^2}{I_{D2T1}(V_{GS1} - V_{T1})} = 1.33 \text{ k}\Omega \qquad (10.5\text{-}22a)$$

so that

$$v_o(t) = 5e^{-t/\tau_1} \qquad (10.5\text{-}22b)$$

where

$$\tau_1 = R_{n1} C_L = (1.33 \text{ k}\Omega)(10 \text{ pF}) = 13.3 \text{ ns} \qquad (10.5\text{-}22c)$$

In this expression $t = 0$ has been redefined as the point at which Q_1 enters the ohmic region.

For this logic gate $V_{OH} = V_{DD}$ and $V_{OL} = 0$, and therefore t_{PHL} is equal to the time needed for the output to fall from its initial value at 8 V to the voltage

$$V_o = \tfrac{1}{2} V_{DD} = 4 \text{ V} \qquad (10.5\text{-}23)$$

The time required for v_o to fall from its value at 5 V (when Q_1 just enters the ohmic region) to 4 V may be found by substituting this voltage into equation (10.5-22b) and solving for t. Carrying out this procedure, we obtain

$$t_2 = \tau_1 \ln \tfrac{5}{4} = 3 \text{ ns} \qquad (10.5\text{-}24a)$$

and combining this result with that for t_1 in equation (10.5-21), we have

$$t_{PHL} = t_1 + t_2 = 11 \text{ ns} \qquad (10.5\text{-}24b)$$

The current flow through Q_1 and Q_2 is also shown in Figure 10.5-10b for completeness.

Because this circuit is symmetrical, the same sort of transient response will occur when v_{in} switches from a HI to a LO at $t = T$ seconds. This portion of the analysis is left as an exercise for the reader (Problem 10.5-3), but the waveshapes are included in Figure 10.5-10b. Due to the symmetry, we therefore expect that

$$t_{PLH} = t_{PHL} = 11 \text{ ns} \qquad (10.5\text{-}25)$$

also. Using computer-aided circuit analysis of this same problem, we find that $t_{PHL} = 10.8$ ns. The propagation delays of all the MOSFET inverters illustrated in Figure 10.5-2, are tabulated in Figure 10.5-10e for the case where $C_L = C_{in} = 10$ pF, and on the basis of these results it is apparent that the CMOS inverter is clearly superior, at least in terms of its high-speed operation.

In the example just completed we saw that the parasitic and load capacitances caused rather large currents to flow as they charged and discharged. Because these currents have much higher amplitudes and longer decay times than the current glitches caused by the finite rise and fall times of the gate input signal, the actual power consumption of a CMOS logic gate will be much larger than that calculated in Example 10.5-2. In Example 10.5-4 we examine how P_D depends on C_L and the switching frequency, and it will demonstrate that the power dissipation of a CMOS logic gate can actually exceed that of a comparable TTL device at high switching frequencies.

EXAMPLE 10.5-4

a. Develop an expression for the power consumed by a CMOS logic gate driving a capacitive load C_L (Figure 10.5-11a). In carrying out this analysis, consider that the output equivalent circuit has the form shown in part (b) of the figure when $v_{in} = 0$, and like that given in part (c) of the same figure when $v_{in} = V_{DD}$. [This approximation neglects the fact that Q_1 and Q_2 are actually in the constant-current region for a brief portion of the transient (see Figure 10.5-10b). A complete analysis, including this constant-current excursion, is left for the homework (Problem 10.5-4).]
b. Determine the manner in which the power consumed from V_{DD} is distributed between Q_1 and Q_2.

SOLUTION

Following Figure 10.5-11a, when v_{in} goes LO, Q_1 cuts off immediately and Q_2 enters the ohmic region so that the equivalent circuit given in part (b) of the figure applies with v_o given by the expression

$$v_o(t) = V_{DD}(1 - e^{-t/\tau_1}) \qquad (10.5\text{-}26a)$$

and

$$i_{D2}(t) = \frac{V_{DD}}{R_{p2}} e^{-t/\tau_1} \qquad (10.5\text{-}26b)$$

where

$$\tau_1 = R_{p2} C_L \qquad (10.5\text{-}26c)$$

Because i_{D2} is zero during the second half of the cycle, the average power consumed from the V_{DD} supply is

$$P_D = \frac{1}{T}\left(V_{DD}\int_0^\infty i_{D2}\ dt\right) = \frac{1}{T}\left[\frac{V_{DD}^2}{R_{p2}}(-\tau_1 e^{-t/\tau_1})\Big|_0^\infty\right] = f(C_L V_{DD}^2) = f(640\text{ pJ}) \qquad (10.5\text{-}27)$$

Notice in this expression that the quantity $(C_L V_{DD}^2)$ represents the energy/cycle that flows in from V_{DD}.

During each cycle of the input waveform, C_L is periodically charged and discharged by Q_2 and Q_1, respectively, and when C_L is fully charged the stored energy in the capacitor is

$$\text{energy stored in } C_L = \tfrac{1}{2}C_L V_{DD}^2 \qquad (10.5\text{-}28)$$

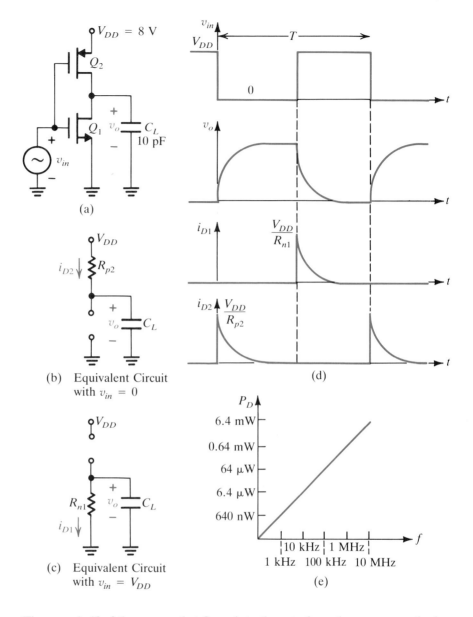

Figure 10.5-11
Example 10.5-4.

(a)

(b) Equivalent Circuit with $v_{in} = 0$

(c) Equivalent Circuit with $v_{in} = V_{DD}$

(d)

(e)

Thus one-half of the energy that flows into the gate from the power supply during each cycle is stored in C_L, while the other half is lost as heat in Q_2.

On the next half of the cycle, C_L discharges through Q_1 and all the energy stored in the capacitor $(\frac{1}{2}C_L V_{DD}^2)$ is dissipated as heat in Q_1. Therefore, one-half of the energy that flows in from the power supply is dissipated in Q_1 and the other half in Q_2, regardless of the relative sizes of R_{p2} and R_{n1}. Furthermore, in accordance with equation (10.5-27), the power flow in from the supply increases with frequency as illustrated in Figure 10.5-15e because the rate at which energy is added to the circuit and then dissipated as heat similarly increases.

Comparing equation (10.5-27) with (10.5-18), it is apparent that the gate power dissipation caused by the charging and discharging of the parasitic capacitances is typically much greater than that associated with the finite rise time of the input waveform. Therefore, unless the gate rise and fall times are very large, it is common to neglect their effect and consider that almost all of the gate power consumption is due to the charging and discharging of the load capacitance.

10.5-2 CMOS as a Logic Family

Although CMOS has been around since the early 1970s, it is only within the last few years that this logic family has made serious inroads into application areas that were formerly the domain of TTL. Initially, because of its low speed and high cost, CMOS was only used in highly specialized applications where its low power consumption and high noise immunity gave it a significant advantage over TTL. However, over the years the cost of CMOS has steadily dropped, so much so in fact that it is now considered to be the logic of choice for SSI and MSI system designs in the dc-to-low-megahertz frequency range.

In the area of LSI, however, the penetration of CMOS has been much slower. The basic structure of CMOS logic devices is much more complex than that of an equivalent NMOS or PMOS circuit element and so, initially at least, CMOS simply could not come close to the component density of conventional MOS designs. Figure 10.5-12a illustrates the basic integrated circuit architecture of one of the early styles of CMOS digital logic. Because CMOS requires that we be able to fabricate both p-channel and n-channel devices on the same substrate, CMOS IC layouts tend to be rather complex. In particular, for the IC illustrated in Figure 10.5-12a, an extra and rather large well of p-type material must be diffused into the substrate in order to house the n-channel transistor, and in addition, channel stops, or guard rings, as they are also called, are needed. These channel stops are necessary to prevent the generation of inversion layers near the surface, which could cause substantial leakage currents to flow, for example, between the p-tub of Q_1 and the drain lead of Q_2. They also help to reduce the possibility of gate latch-up due to the activation of a "parasitic SCR" located between Q_1 and Q_2.

This SCR *pnpn* device is formed by the drain of Q_2, the n-type substrate, the p-well of Q_1, and the drain of Q_1, and under the proper conditions (typically, when the gate input voltage exceeds V_{DD}) this SCR can turn ON, creating a virtual short circuit from V_{DD} to ground. This latch-up phenomenon is a common source of damage to CMOS circuits, and its occurrence is minimized by the use of guard rings around each transistor.

Because of the presence of the p-well and the guard rings, the surface area needed for this style of CMOS logic gate is significantly larger than that required for a comparable gate in almost any other digital logic family. For example, it is nearly twice the size of a similar LS gate and almost six times bigger than a comparable PMOS or NMOS digital logic gate. This vast size difference initially made it very expensive and almost impossible to manufacture large-scale integrated circuits in CMOS. The development of the self-aligned gate, and then the self-aligned silicon gate techniques for manufacturing MOS and CMOS circuits, has significantly changed this situation.

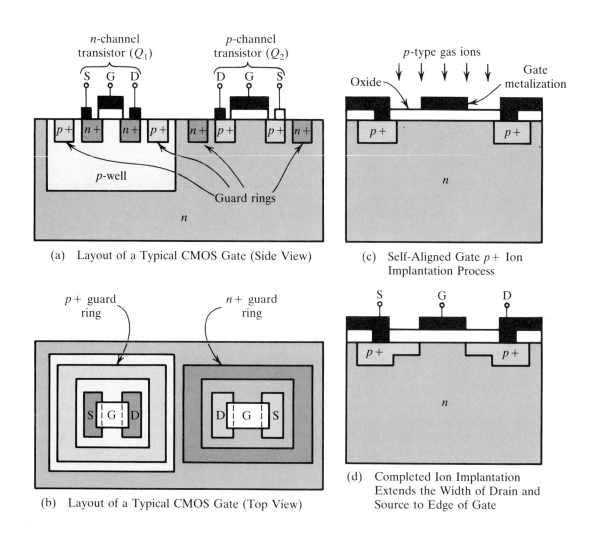

(a) Layout of a Typical CMOS Gate (Side View)

(c) Self-Aligned Gate $p+$ Ion Implantation Process

(b) Layout of a Typical CMOS Gate (Top View)

(d) Completed Ion Implantation Extends the Width of Drain and Source to Edge of Gate

Figure 10.5-12

In the fabrication of conventional metal gate MOS devices, the drain and source wells are first diffused into the substrate, and this is followed by the application of the gate oxide and then the gate metallization. The typical IC layout that results when this process is used to construct a CMOS gate has already been illustrated in Figure 10.5-12a. Here, because of mask alignment problems, it is necessary to extend the gate metallization over the drain and source diffusion wells to be sure that the gate electrode covers the entire length of the channel. This gate overlap creates significant parasitic capacitances between the gate electrode and the transistor drain and source terminals, and this slows down the operating speed of the transistors.

In addition to the gate overlap problem created by mask alignment errors, these mask positioning problems also make it difficult to fabricate short-channel transistors. This is important for two reasons. First and most obvious is the fact that the transistors, and hence the ICs produced from them, tend to be larger than might otherwise be necessary. Second, because the transistor g_m is inversely proportional to the channel length, both the gain and speed of these transistors are reduced. The use of the self-aligned gate manufacturing technique permits the fabrication of short-channel devices which have low parasitic capacitances, high gain, and occupy minimal surface area.

Figure 10.5-12c through e illustrates the basic steps used in the self-aligned gate MOSFET fabrication process. Initially, the drain and source wells are diffused into the substrate and then a thin oxide layer is deposited over the entire surface of the device. Next, the gate metallization is applied over the oxide layer, and its length is purposely made a little shorter than the distance between the drain and the source. High-voltage ion implantation is then used to diffuse in additional $p+$ material to extend the width of the drain and source wells of the transistor (Figure 10.5-12c). Because the ions are unable to penetrate the metal gate, a very well defined boundary is formed at the edge of the gate metallization, in which the edges of the drain and source are now perfectly, and automatically, aligned with the edge of the gate (Figure 10.5-12d). A final etching of the oxide and then a second contact metallization process completes the formation of the transistor (Figure 10.5-12e).

The nearly perfect alignment of the drain and source boundaries with the edges of the gate metallization significantly reduces the transistor parasitic capacitances, and this of course substantially increases the operating speed of any logic gates formed using these transistors. Furthermore, because mask misalignment effects are minimized, it is also possible to make the channel length much shorter. This, as we have already mentioned, correspondingly increases the transistor gain and speed while simultaneously reducing the surface area occupied by the transistor. An additional improvement in circuit performance results when the metal used for the gate of the MOSFET is replaced by polycrystalline silicon. This lowers the transistor threshold voltage, which also increases its gain and allows for a further increase in operating speed [see equations (10.5-20) through (10.5-25)].

The net effect of all the CMOS processing improvements outlined in the last few paragraphs is that the area needed for today's modern high-speed CMOS logic devices is about four times smaller than that of the conventional

10.5-1 Comparison of Low-Power Schottky, Advanced Low-Power Schottky, Standard CMOS, and High-Speed CMOS Logic Families

	High-Speed Silicon Gate CMOS	Buffered Metal Gate 4000 Series CMOS	74LSxx	74ALSxx
P_D (at 100 kHz; mW)	0.15	0.1	2	1
$t_P(C_L = 15$ pF; ns)	8	125	10	5
Speed–power product (pJ)	1.2	12.5	20	5
Flip-flop clock speed, (mHz)	40	5	45	50
I_{OL}(mA at $V_o = 0.4$ V)	4	0.36	8	8
I_{OH}(mA at $V_o = 2.5$ V)	-4	-0.36	-0.4	-0.4

4000 series of metal gate CMOS. In addition, it is able to operate at nearly 10 times the speed. Table 10.5-1 compares the relative performance characteristics of LS, ALS, 4000 series CMOS, and modern high-speed silicon gate CMOS logic devices.

10.5-3 Basic Architecture of CMOS Logic Gates

Nearly all complementary MOS circuits are implemented using one of four basic gating structures: the inverter, the NAND, the NOR, and the transmission gate. Typical elementary designs for each of these gate types is shown in Figure 10.5-13. The operation of the CMOS inverter has already been covered in great detail earlier in this section, so nothing further needs to be said about this logic element. The operation of the remaining three logic circuits will now be briefly described.

The functioning of the NAND gate shown in Figure 10.5-13b may be understood with the aid of the series of equivalent circuits presented in Figure 10.5-14. Here, for simplicity, we have assumed that the n-channel transistors are ON and may be represented as short circuits when the voltage applied to their gates is equal to V_{DD}, and that they are OFF and may be replaced by open circuits when the voltage applied to them is zero. The converse is assumed to be true for the p-channel devices, namely that they are ON when the voltage on their gates is zero, and OFF when their inputs are at V_{DD}. By examining the output voltages produced for the input combinations shown in Figure 10.5-14, it should be apparent that this circuit behaves like a NAND gate.

A similar series of equivalent circuits is shown in Figure 10.5-15 for the CMOS NOR gate illustrated in Figure 10.5-13c. Because the output voltage from this logic gate is high only when both V_1 and V_2 are low, we can readily see that this particular CMOS logic element is a NOR-gate structure.

Figure 10.5-13

Basic CMOS logic gate building blocks.

The operation of the CMOS transmission gate was already described briefly in Section 7.3, and there we were concerned only with it analog signal trans-

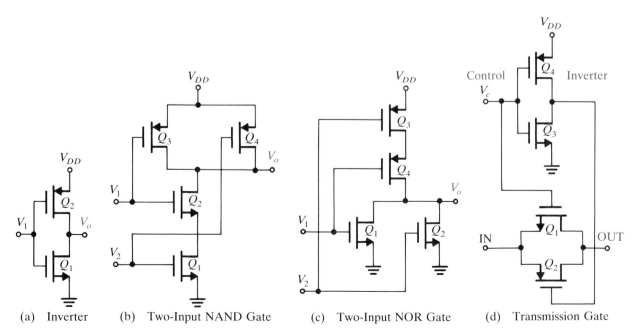

(a) Inverter   (b) Two-Input NAND Gate   (c) Two-Input NOR Gate   (d) Transmission Gate

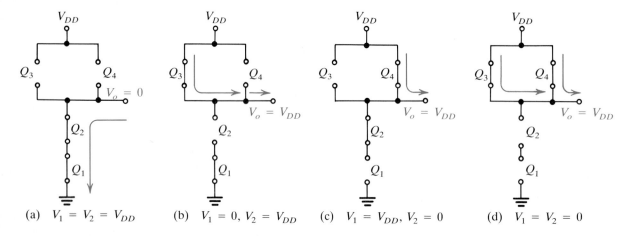

(a) $V_1 = V_2 = V_{DD}$ (b) $V_1 = 0, V_2 = V_{DD}$ (c) $V_1 = V_{DD}, V_2 = 0$ (d) $V_1 = V_2 = 0$

Figure 10.5-14
Equivalent circuits for different input combinations in the two-input NAND gate in Figure 10.5-13b.

mission properties. Because this device also finds considerable application as a digital logic element, it will again be worthwhile for us to examine its operating characteristics. The basic circuit to be analyzed is shown in Figure 10.5-16a. Here we assume that $V_{DD} = +7.5$ V, $V_{SS} = -7.5$ V, and also that the threshold voltages of Q_1 and Q_2 are 2.5 V and -2.5 V, respectively.

When the control voltage, V_c, is equal to V_{DD}, the inverter output will be $-V_{SS}$. As a result, we may write

$$V_{GS1} = V_{DD} - V_{in} \qquad (10.5\text{-}29)$$

from which it is apparent that Q_1 is ON for all

$$V_{DD} - V_{in} > V_{T1} \qquad (10.5\text{-}30a)$$

or
$$V_{in} < V_{DD} - V_{T1} = 5 \text{ V} \qquad (10.5\text{-}30b)$$

In a similar fashion, for the p-channel transistor Q_2 we also have

$$V_{GS2} = -V_{SS} - V_{in} \qquad (10.5\text{-}31a)$$

Figure 10.5-15
Equivalent circuits for different input combinations for the two-input NOR gate in Figure 10.5-13c.

so that Q_2 will also be ON for all

$$V_{in} > -V_{SS} - V_{T2} = -5 \text{ V} \qquad (10.5\text{-}31b)$$

Hence both of these transistors will be ON as long as $|V_{in}| < 5$ V.

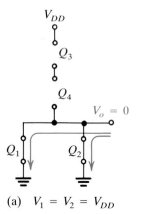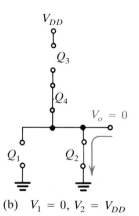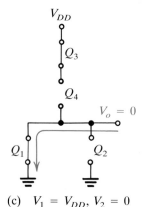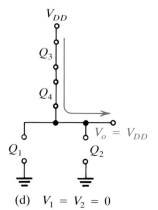

(a) $V_1 = V_2 = V_{DD}$ (b) $V_1 = 0, V_2 = V_{DD}$ (c) $V_1 = V_{DD}, V_2 = 0$ (d) $V_1 = V_2 = 0$

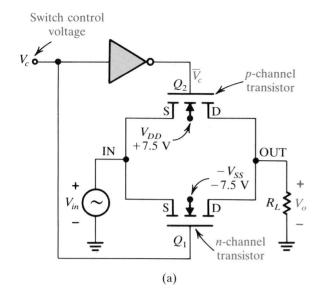

Switch control voltage

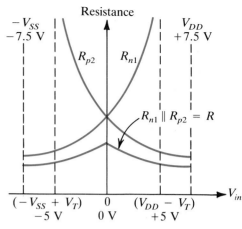

(a)

(b) Resistance of the CMOS Transmission Gate as a Function of $V_{in}(V_c = +7.5$ V)

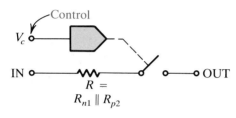

(c) Functional Equivalent Circuit of the Transmission Gate

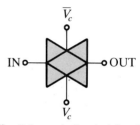

(d) Schematic Symbol for the CMOS Transmission Gate

Figure 10.5-16
CMOS transmission gate.

Because the drain and source terminals of Q_1 and Q_2 are connected in common, R_{n1} and R_{p2} are in parallel and the voltage drop across these two transistors may be written as

$$V_{DS} = \frac{R}{R + R_L} V_{in} \qquad (10.5\text{-}32\text{a})$$

where

$$R = R_{n1}\|R_{p2} \qquad (10.5\text{-}32\text{b})$$

Because R_L is generally much bigger than R, V_{DS} will usually be small, and both of the transistors will operate in their ohmic regions. In this case, R_{n1} and R_{p2} may be written from equation (10.5-2c) as

$$R_{n1} = \frac{V_{T1}^2}{I_{D2T1}(V_{DD} - V_{in} - V_{T1})} \qquad (10.5\text{-}33\text{a})$$

and

$$R_{p2} = \frac{V_{T2}^2}{I_{D2T2}(+V_{SS} + V_{in} + V_{T2})} \qquad (10.5\text{-}33\text{b})$$

The variation of R_{n1}, R_{p2}, and R is sketched as a function of V_{in} in Figure 10.5-16b. Because R_{p2} increases when R_{n1} decreases, and vice versa, the overall resistance of this transmission gate is relatively constant over the entire input voltage range from $-V_{SS}$ to $+V_{DD}$. The input voltage should not be allowed to exceed these limits since this will forward bias the substrate diodes. Typical values for R when the gate is ON are about 300 to 500 Ω.

When the control voltage is switched to $-V_{SS}$, $\overline{V}_c$ will change to V_{DD}, and because the transistor gate-to-source voltages will now be

$$V_{GS1} = -V_{DD} - V_{in} \tag{10.5-34a}$$

and
$$V_{GS2} = +V_{SS} - V_{in} \tag{10.5-34b}$$

Q_1 and Q_2 will both be cut off for all values of V_{in} less than V_{DD} and greater than $-V_{SS}$. Of course, in this instance V_o will be zero and the gate will effectively be an open circuit. Figure 10.5-16c illustrates a functional model that can be used to represent the transmission gate when carrying out a specific circuit analysis, and Figure 10.5-16d gives the schematic symbol that is most commonly used to represent this logic element.

10.5-4 Buffered CMOS Logic

The CMOS NAND and NOR gates illustrated in Figure 10.5-13b and c have two rather undesirable characteristics. The output drive capability and noise margins of both of these gates depends on the particular logic combinations applied to their inputs. To illustrate why this is the case, let's examine the behavior of the NAND gate shown in Figure 10.5-13b. When V_2, for example, is a logic 1, Q_4 is cut off and Q_2 is operating in the ohmic region. Under these conditions the simplified version of this gate shown in Figure 10.5-17 is valid, and

$$I_{D1} = I_{D3} \tag{10.5-35}$$

If we assume that V_{DS2} is small, that $V_{T1} = |V_{T3}| = V_T$, and also that $I_{D2T1} = I_{D2T3}$, it follows directly that both transistors are in the constant current region when

$$I_{D2T1}(V_{in} - V_T)^2 = I_{D2T3}(V_{DD} - V_{in} - V_T)^2 \tag{10.5-36a}$$

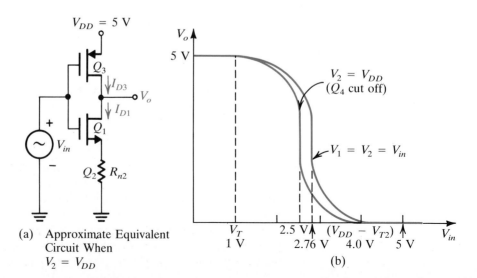

Figure 10.5-17
Variation of the transfer characteristic with input excitation pattern in A-series unbuffered CMOS.

(a) Approximate Equivalent Circuit When $V_2 = V_{DD}$

(b)

or
$$V_{in} = \frac{V_{DD}}{2} = 2.5 \text{ V} \qquad (10.5\text{-}36b)$$

for $V_{DD} = 5$ V. Thus, in this instance, the vertical portion of the transfer characteristic passes precisely through the point $V_{in} = V_{DD}/2$ (see Figure 10.5-17b).

When both inputs are connected to V_{in}, the situation changes somewhat. In this instance, all four transistors are simultaneously in the constant-current region when the gate transfer characteristic passes through the vertical high-gain region. In this interval we may therefore write

$$I_{D2} = I_{D1} = I_{D3} + I_{D4} \qquad (10.5\text{-}37a)$$

or

$$I_{D2T2}(V_{in} - V_T)^2 = I_{D2T3}(V_{DD} - V_{in} - V_T)^2 + I_{D2T4}(V_{DD} - V_{in} - V_T)^2 \qquad (10.5\text{-}37b)$$

If we consider that I_{D2T} is the same for all the transistors and also that the transistor threshold voltages are all equal to 1 V, then in this case we obtain

$$V_{in} - V_T = \sqrt{2}\,(V_{DD} - V_{in} - V_T) \qquad (10.5\text{-}38a)$$

or

$$V_{in} = \frac{(V_{DD} - V_T)\sqrt{2} + V_T}{1 + \sqrt{2}} = 2.76 \text{ V} \qquad (10.5\text{-}38b)$$

For the case of an eight-input NAND gate, this threshold would shift even farther to the right to about 3.2 V. These shifts to higher crossover voltages tend to raise the gate noise immunity for input LOs and to reduce it with respect to input HIs. The situation for a NOR gate is exactly the opposite of that just obtained for the NAND-style gate in that the transfer characteristic crossover point shifts to lower voltages when multiple inputs are simultaneously driven.

Besides causing a shift in the crossover point of the gate, the output drive ability of the gate is also a function of the way in which the inputs are excited. For example, for the two-input NAND gate in Figure 10.5-13b several different input combinations will result in the gate output being a logic 1 but not all of them will produce the same equivalent output circuit (Figure 10.5-14). When only one of the inputs is a logic zero, only the p-channel transistor corresponding to that input is in the ohmic region, and under these conditions the effective gate output impedance is R_p (where R_p is the ohmic resistance of a single p-channel FET). When both inputs are simultaneously LO, as shown in Figure 10.5-14d, both Q_3 and Q_4 will be in the ohmic region, and in this instance the gate output impedance will be reduced to $R_p/2$. This means that the gate operating in this second situation can deliver twice as much output current to a load for a given output voltage, as in the first instance when only one of the p-channel transistors is ON. In addition, the transient response of the gate in this second situation (at least for the ohmic portion of the gate transient) will also be twice as fast. Obviously, this variation of the gate output characteristics with the applied input signal pattern is even more pronounced in gates with larger numbers of inputs.

To get around the difficulties that we have just been discussing, an improved version of the original 4000A series of CMOS logic gates was developed in which two CMOS inverters were inserted between the logic portion of

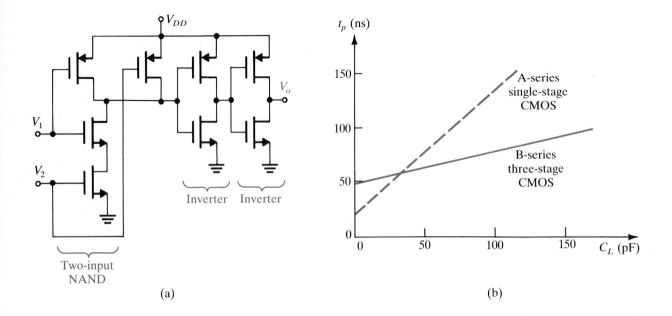

Two-input
NAND

(a)

Inverter   Inverter

(b)

Figure 10.5-18
Buffered B-series CMOS
digital logic.

the gate and the gate output terminal. This logic family is known as the B series or buffered series of CMOS digital logic. A typical two-input buffered
NAND gate is illustrated in Figure 10.5-18a. Of course, the addition of two
inverters to the logic gate has no effect on the digital logic function performed
by it, but it does significantly alter the gate behavior.

First, the gate drive characteristics are now independent of the particular
input pattern applied. In addition, because the input signal must now propagate
through three different stages, the propagation delay (at least for small capacitive loads) is somewhat greater. However, because the output transistors in
buffered CMOS gates are designed to have low ohmic resistances, the propagation delay in B series gates driving heavy capacitive loads is actually smaller
than that for a comparable unbuffered CMOS logic gate (Figure 10.5-18b).
Besides improving the gate input–output isolation, the additional gain supplied
by the two extra inverters makes the gate transfer characteristic much squarer,
and this significantly improves the noise immunity.

10.5-5 Interfacing CMOS to Other Forms of Digital Logic

When connecting one CMOS logic gate to another, the static fan-out is nearly
infinite, and the actual maximum fan-out that should be used is governed primarily by capacitive loading effects and speed considerations. Consider, for example, the situation in which one CMOS inverter is being used to drive a large
number of similar gates. Here, each gate input may be modeled approximately
as a fixed capacitance C_{in} so that the connection of N such gates onto the output
of the first inverter effectively loads it with a capacitance NC_{in}. For such an interfacing problem, the gate rise and fall times, and the propagation delays, have
the same form as the expressions developed in Example 10.5-3.

Assuming that the typical input capacitance of a CMOS logic gate is on the
order of 5 pF, the effect of fan-out on the overall gate propagation delay may

readily be determined with the aid of Figure 10.5-18b. Consider, for example, the case where a B series CMOS logic gate is being used to drive 20 similar gates. In this instance the total effective load on the first gate would be about 5 × 20 or about 100 pF, and following the graph in the figure, the resulting propagation delay in the driver gate would be about 75 ns.

When one attempts to interface CMOS and TTL logic devices together, the problem is much more complex, and careful attention needs to be paid to the voltage and current levels at both the inputs and the outputs of each of these gates. Let's consider the specific problem of the interfacing of CMOS and low-power Schottky TTL to one another assuming that the power supply voltages of both are equal to 5 V. The basic form of this problem is illustrated in Figure 10.5-19a, and for simplicity each of the gates has been shown as an inverter, but practically speaking, they could just as easily be any other type of digital logic gate. The pertinent data related to this interfacing problem are summarized in Figure 10.5-19b.

For the case where the B series CMOS logic gate is being used to drive a single LS load, no special interfacing circuitry is needed. When the output from the CMOS gate is a logic 0, V_{OL} is a maximum of 0.4 V when the gate is sinking 0.36 mA. Because V_{IL} for the Schottky gate is 0.8 V at an input current of −0.36 mA, the CMOS gate can just drive one LS gate at this logic level. When the output from the CMOS gate is a logic 1, its output voltage is guaranteed to be at least 2.5 V even when it is sourcing up to 0.36 mA. Because V_{IH} for the low-power Schottky gate is 2.0 V, and furthermore, because I_{IH} for this gate is at most only 20 μA, the CMOS gate will have no difficulty directly driving one LS TTL logic gate. Unfortunately, the converse is not true.

When a TTL gate drives a CMOS logic gate, no problem is encountered when the output of the TTL gate is a logic 0. In this instance V_{OL} of the TTL gate will be less than 0.4 V and this is substantially smaller than the CMOS V_{IL} of 1.5 V. In addition, the LS gate does not need to be able to sink any current

Figure 10.5-19

Interfacing CMOS and LSTTL together.

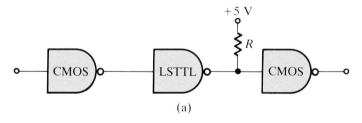

(a)

	B-Series CMOS	LS TTL
V_{OH}	5 V (I_{OH} = 0 mA) 2.5 V min. (I_{OH} = −0.36 mA)	2.7 V min. (at I_{OH} = −400 μA)
V_{OL}	0 V (I_{OL} = 0 mA) 0.4 V max. (I_{OL} = 0.36 mA)	0.4 V max. (at I_{OL} = 8 mA)
V_{IH}	3.5 V min.	2.0 V min. (at I_{IH} = 20 μA)
V_{IL}	1.5 V max.	0.8 V max. (at I_{IL} = −0.36 mA)

(b)

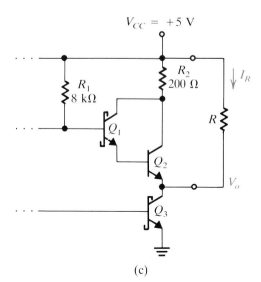

(c)

from the CMOS gate so that the allowed fan-out in this instance is virtually infinite. The situation when the output of the TTL gate goes to a logic 1 is quite different. For a low-power Schottky TTL gate, V_{OH} is only guaranteed to be above 2.7 V, but CMOS requires a voltage in excess of 3.5 V to ensure that the applied input signal will be recognized as a logic 1. Therefore, the TTL output voltage is too low for the CMOS gate. One way to raise this output voltage is to connect an external pull-up resistor from the TTL gate output to the +5-V power supply as illustrated in Figure 10.5-19a. The effect of this resistor on the performance of the TTL gate may best be understood by examining the output portion of this gating circuit that is shown in part (c) of the figure.

When Q_3 is cut off, V_o must rise to +5 V in the absence of any loading on the gate. If this were not the case, a current I_R would flow through the external pull-up resistor R. However, since Q_3 is OFF, I_R has no place to flow through and therefore, $I_R = 0$, $V_o = +5$ V, and Q_1 and Q_2 are both cut off. Thus with this pull-up resistor in place, V_{OH} for the TTL gate will be +5 V.

In selecting a value for R, the major requirement is that Q_3 remain saturated when the gate output is at a logic 0. To guarantee that this is the case, it is necessary that the current flow through the resistor R not exceed I_{OL}, or 8 mA for this LS gate. This places a lower limit on R of about $[(5 - 0.4) \text{ V}]/(8 \text{ mA})$, or 575 Ω. The upper limit on R is determined by the required rise and fall times at the interface.

Exercises

10.5-1 Find t_{PHL} for the inverter circuit in Example 10.5-1 (Figure 10.5-5). *Answer* 10.2 ns

10.5-2 A voltage source v_{in} drives a CMOS three-inverter cascade and a 20-kΩ load is connected onto the output of the last inverter. The input capacitance of each gate is 20 pF and the power supply is +5 V. Find the power consumed from the supply if the input signal is a 1-MHz 0 to 5-V square wave. Assume complete charging and discharging of the parasitic capacitors during each cycle. *Answer* 1.625 mW

10.5-3 a. Compute the transition voltage for an eight-input unbuffered NAND gate when $V_1 = V_{in}$ and $V_2 = V_3 = \cdots = V_8 = +5$ V. The gate structure is the same as that given in Figure 10.5-13b, and $|V_T|$ and I_{D2T} are the same for both the p- and n-channel devices. In particular, $|V_T| = 1$ V.
b. Repeat part (a) when $V_1 = V_2 = \cdots = V_8 = V_{in}$.
Answer (a) 2.5 V, (b) 3.21 V

10.6 EMITTER-COUPLED LOGIC

The speed of digital logic circuits containing bipolar junction transistors is limited mainly by transistor saturation effects. In Section 10.4 we saw that the addition of a Schottky diode in parallel with the base–collector junction prevented transistor saturation and allowed Schottky TTL devices to operate much faster than standard TTL logic. Another way to achieve high-speed operation in BJT digital logic circuits is to design them in such a way that the transistors are prevented from saturating.

In theory, this goal may be achieved by using a BJT circuit of the form illustrated in Figure 10.6-1a and simply keeping the base current into this transistor within acceptable limits. However, in practice this kind of circuit is extremely difficult to design. For example, if we define the logic 0 and logic 1 output voltages as shown in Figure 10.6-1b, then, assuming a transistor h_{FE} of 100, we can readily demonstrate that the input and output voltage levels of this inverter will be compatible only if R_{B1} is somewhere in between 47.2 and 49.7 kΩ. Furthermore, if we include the fact that the h_{FE} will actually vary considerably from one device to the next, it is apparent that this circuit will never really be able to perform satisfactorily.

To illustrate this point, suppose that the transistor h_{FE} can take on any values between 50 and 300, and let $R_{B1} = 48$ kΩ for computational purposes. When V_{in} is a logic 1 at 8.5 V and the transistor h_{FE} equals 50, I_C will only be 2.25 mA, and as a result, V_o will equal 7.75 V, well above its allowed V_{OL} value of 5.0 V. Conversely, when the h_{FE} equals 300 and V_{in} takes on its maximum logic 1 value of 10 V, I_C will exceed $I_{C,\text{sat}}$ and the transistor will saturate. Thus this single-transistor common-emitter logic circuit will have great difficulty maintaining the proper logic 0 output voltage levels. Principally, this is due to the variation in the transistor h_{FE}. A common-emitter circuit maintains the base current and not the collector current constant as the h_{FE} varies, and so to keep V_o within acceptable limits when the transistor is active, the h_{FE} must be tightly controlled.

Figure 10.6-1

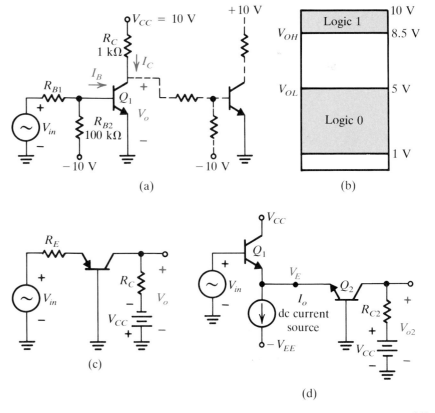

(a)   (b)

(c)

(d)

Another way to prevent transistor saturation is illustrated in Figure 10.6-1c. Here the input signal is applied to the emitter rather than to the base, and because I_C is nearly the same as I_E, it is relatively easy to keep this transistor from saturating. However, this circuit has several disadvantages that prevent it from being used in digital logic circuit designs, at least in the simplified form that we have discussed thus far. To begin with, a common-base circuit is a noninverting amplifier, and hence it cannot be used as a digital logic inverter. Second, the input and output voltage levels of this circuit are vastly different from one another.

A more practical version of this circuit is shown in Figure 10.6-1d. It overcomes both of the disadvantages of the simple common-base circuit while still achieving the desired output voltage invariance to changes in h_{FE}. In this circuit Q_1 acts like an emitter follower, and Q_2 as a common-base amplifier.

The basic operation of this circuit, assuming positive logic, may be understood as follows. When V_{in} is a logic 0 (and less than zero volts), Q_1 is cut off, and as a result, the dc current I_o flows through the emitter of Q_2 turning on this transistor. If R_{C2} is chosen properly, while the voltage V_{o2} will be a logic 0, the drop across R_{C2} will be small enough so that Q_2 does not saturate.

When V_{in} is a logic 1 (and greater than zero volts), Q_1 will turn ON, V_E will rise, and this will reverse bias the base–emitter junction of Q_2, cutting off this transistor. In this case, of course, I_{C2} will be zero so that V_{o2} will be equal to V_{CC}, a logic 1.

By redrawing this circuit as shown in Figure 10.6-2a, it is apparent that this common collector–common base circuit is just an ordinary differential amplifier. Notice that a resistor R_{C1} has been added in series with the collector of Q_1 in this figure to provide for an inverting as well as a noninverting digital logic output from this gate. We have also added batteries in series with the base leads of the transistors to shift the input threshold voltage of the gate to make it compatible with the output voltages. Let's examine the operation of this differential amplifier circuit a bit more closely.

In accordance with the discussion presented originally in Section 3.1, the collector current flowing in a transistor is related to the voltage applied to its

Figure 10.6-2

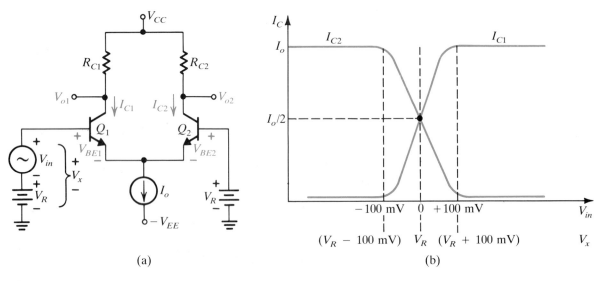

(a) (b)

base–emitter junction by the expression

$$I_C = \alpha_F I_{EBO}(e^{V_{BE}/V_T} - 1) + I_{CBO} \tag{10.6-1}$$

where I_{EBO} is the emitter current that flows when the collector is open circuited. When the base–emitter junction is forward biased, the exponential term dominates so that we may write this current as

$$I_C \simeq I_{EBO} e^{V_{BE}/V_T} \tag{10.6-2}$$

where, for simplicity, we have also set $\alpha_F = 1$. When $V_{in} = 0$, then by symmetry, $V_{BE1} = V_{BE2}$, and because

$$I_{C1} + I_{C2} = I_o \tag{10.6-3}$$

the current flow in each of the transistors will be the same and will be equal to $I_o/2$. When V_{in} is nonzero, the relationship between I_{C1}, I_{C2}, and V_{in} may be obtained by noting that

$$I_{C1} = I_{EBO} e^{V_{BE1}/V_T} \tag{10.6-4a}$$

and

$$I_{C2} = I_{EBO} e^{V_{BE2}/V_T} \tag{10.6-4b}$$

Applying Kirchhoff's voltage law to the base–emitter loop, it is also apparent that

$$V_{in} + V_R = V_{BE1} - V_{BE2} + V_R \tag{10.6-5a}$$

or

$$V_{in} = V_{BE1} - V_{BE2} \tag{10.6-5b}$$

Dividing equation (10.6-4a) by (10.6-4b), we find that

$$\frac{I_{C1}}{I_{C2}} = e^{(V_{BE1} - V_{BE2})/V_T} \tag{10.6-6a}$$

which on substituting equation (10.6-5b) yields

$$\frac{I_{C1}}{I_{C2}} = e^{V_{in}/V_T} \tag{10.6-6b}$$

Combining this result with equation (10.6-3), it is not too difficult to show that

$$I_{C1} = \frac{I_o}{1 + e^{-V_{in}/V_T}} \tag{10.6-7a}$$

and

$$I_{C2} = \frac{I_o}{1 + e^{V_{in}/V_T}} \tag{10.6-7b}$$

These collector currents are sketched as functions of V_{in} in Figure 10.6-2b.

To illustrate the basic shape of this curve, let's consider the relationship between I_{C1} and I_{C2} when $V_{in} = +100$ mV and also when it is equal to -100 mV. Following equations (10.6-7) for $V_{in} = +100$ mV, we have

$$I_{C1} = \frac{I_o}{1 + e^{-3.8}} = \frac{I_o}{1.022} \simeq I_o \tag{10.6-8a}$$

and

$$I_{C2} = \frac{I_o}{1 + e^{3.8}} = 0.021 I_o \simeq 0 \tag{10.6-8b}$$

Conversely, when $V_{in} = -100$ mV we obtain

$$I_{C1} = \frac{I_o}{1 + e^{3.8}} \simeq 0 \qquad\qquad (10.6\text{-}9\text{a})$$

and

$$I_{C2} = \frac{I_o}{1 + e^{-3.8}} \simeq I_o \qquad\qquad (10.6\text{-}9\text{b})$$

To summarize these results, when V_{in} is greater than 100 mV (or alternatively, when V_x exceeds V_R by more than 100 mV), Q_2 will be cut off and Q_1 will be on with I_{C1} equal to I_o. Conversely, when V_{in} is less than -100 mV (or when V_x is 100 mV or more below V_R), Q_1 will cut off and Q_2 will be active with $I_{C2} = I_o$. Thus, in this circuit, V_{in} acts as a control voltage to switch the current I_o back and forth between Q_1 and Q_2.

10.6-1 The Static Operation of an ECL Logic Gate

The differential amplifier circuit that we have just been examining is the basic building block that is used in the design of emitter-coupled logic (ECL). ECL was introduced by Motorola in 1962. MECL I, as it was originally called, was characterized by propagation delays on the order of 8 ns and gate power dissipations of about 40 mW per gate. An improved version of this ECL logic known as MECL II was introduced in 1966 and featured propagation delays of about 4 ns. Both of these early versions of emitter-coupled logic are obsolete today and have been replaced by MECL III and MECL 10000 (MECL 10K). MECL III operates extremely fast with typical propagation delays in the vicinity of 1 ns. Power dissipation in these gates averages about 60 mW. For applications that do not require the utmost in speed, a slightly slower version of the logic family, MECL 10K, was introduced in 1971. Gate propagation delay and power dissipation in this logic subfamily are 2 ns and 25 mW, respectively.

Figure 10.6-3 illustrates the circuit design of a typical MECL 10K logic gate. The operation of this gate may be understood as follows. When both V_A and V_B are at a logic 0 (less than V_{BB} volts), Q_{1A} and Q_{1B} will be cut off while Q_2 will be active. As a result, V_{C1} will be HI and V_{C2} will be LO, and because Q_4 and Q_5 are emitter followers, their outputs will also be HI and LO, respectively. If either or both V_A and V_B go HI, Q_{1A} and/or Q_{1B} will turn on and this will divert the current away from Q_2, cutting this transistor off. With Q_2 off, V_{o2} will be HI and correspondingly V_{o1} will be LO. In summary, then, when V_A and V_B are both zero, V_{o1} is a logic 1 and V_{o2} is a logic 0. When either or both of these inputs is 1, V_{o1} is a logic 0 and V_{o2} is a logic 1. Thus the outputs at V_{o1} and V_{o2} implement the NOR and OR logic functions, respectively.

Before getting into a detailed analysis of the electronic operation of this specific logic gate, it may first be useful to point out some of the general design characteristics of ECL. Because both the logic function and its complement are generally available in ECL logic gates, the need for inverters is reduced to a minimum. In addition, as discussed in Section 10.3 (see especially Figure 10.3-4), when emitter followers are used as the output elements, it is possible to directly connect these outputs together to form "wired OR" gates. Notice also that two separate V_{CC} power supply leads are brought out of the gate, one from the functional portion of the gate and the other from the emitter-follower drivers. This is done to minimize noise effects in the logic portion of the gate.

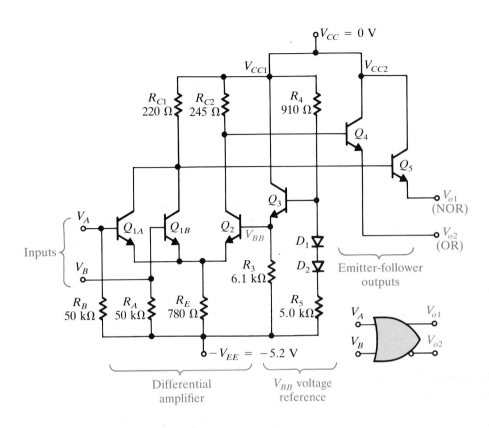

Figure 10.6-3
ECL 10K logic gate.

When ECL gates are driving heavy loads, large current transients can be produced in the collectors of the emitter followers. By routing V_{CC2} over a separate path back to the power supply from that used for V_{CC1}, any voltage spikes produced along the V_{CC2} wiring path will not find their way into the logic portion of the gate. These precautions are necessary because of the relatively small noise immunity of ECL.

To obtain a quantitative understanding of the functioning of the logic gate in Figure 10.6-3a, let us assume that the V_{CC1} and V_{CC2} are both connected to ground, while the supply voltage, $-V_{EE}$, is set equal to -5.2 V. These values of V_{CC} and $-V_{EE}$ are those typically employed with ECL logic gates. At first glance it may appear somewhat strange that ECL is operated with a negative power supply voltage; but basically, this is done to improve the ability of this circuit to reject power supply noise. This additional noise rejection is especially important in ECL circuits because the noise margins are quite small. This subject is examined in greater detail later in this section.

The analysis of this circuit's operation is best begun by initially determining the value of the differential amplifier reference voltage V_{BB}. If we neglect the loading of Q_3 on the R_4–R_5 voltage divider, then

$$I_{R5} = \frac{0 - (-5.2) - 1.4}{5.91 \text{ k}\Omega} = 0.643 \text{ mA} \qquad (10.6\text{-}10a)$$

so that $\qquad V_{B3} = 1.4 + (0.643)5 + (-5.2) = -0.585 \text{ V} \qquad (10.6\text{-}10b)$

and thus $\qquad V_{BB} = V_{E3} = -0.585 - 0.7 = -1.285 \text{ V} \qquad (10.6\text{-}10c)$

When inputs V_A and V_B are less than V_{BB} by more than 10-0 mV, Q_{1A} and Q_{1B} will both be cut off so that Q_2 will be active with

$$V_{E2} = V_{BB} - 0.7 = -1.985 \text{ V} \tag{10.6-11a}$$

and
$$I_{RE} = \frac{V_{E2} - (-V_{EE})}{R_E} = 4.12 \text{ mA} \tag{10.6-11b}$$

Because
$$I_{C1A} = I_{C1B} = 0 \tag{10.6-12a}$$

and
$$I_{C2} \simeq I_{RE} = 4.12 \text{ mA} \tag{10.6-12b}$$

it also follows directly that

$$V_{C1} = 0 \text{ V} \tag{10.6-13a}$$

$$V_{C2} = -I_{C2}R_{C2} = -1.01 \text{ V} \tag{10.6-13b}$$

$$V_{o1} = V_{C1} - 0.7 = -0.7 \text{ V} \tag{10.6-13c}$$

and
$$V_{o2} = V_{C2} - 0.7 = -1.71 \text{ V} \tag{10.6-13d}$$

When V_A and/or V_B are 100 mV or more above V_{BB}, Q_{1A} and/or Q_{1B} will be ON and Q_2 will be cut off. Because the voltages applied to this gate input will generally be the output voltages from a similar logic gate, let us consider, for example, that $V_A = -0.7$ V and $V_B = -1.71$ V. In this instance Q_{1A} will be on while both Q_{1B} and Q_2 will be cut off, so that we will have

$$V_E = -0.7 - 0.7 = -1.4 \text{ V} \tag{10.6-14a}$$

$$I_{RE} = \frac{-1.4 - (-5.2)}{780} = 4.87 \text{ mA} \tag{10.6-14b}$$

$$V_{C2} = 0 \tag{10.6-14c}$$

$$V_{C1} = -I_{C1A}R_{C1} = -1.07 \text{ V} \tag{10.6-14d}$$

$$V_{o1} = -1.77 \text{ V} \tag{10.6-14e}$$

and
$$V_{o2} = -0.7 \text{ V} \tag{10.6-14f}$$

Notice that the resistor R_{C1} is purposely made smaller than R_{C2} because I_{RE} is slightly larger when the inputs to V_A and V_B are HI than when they are LO.

By examining equations (10.6-13c), (10.6-13d), and (10.6-14f), it should be apparent that V_{OH} and V_{OL} for this logic gate are approximately equal to

$$V_{OH} = -0.7 \text{ V} \tag{10.6-15a}$$

and
$$V_{OL} = -\frac{1.77 + 1.71}{2} = -1.74 \text{ V} \tag{10.6-15b}$$

In examining these results, it is important to observe that V_{BB} is located midway between these voltage levels. Therefore, as required, a voltage $V_{OH} = -0.7$ V applied to the input of another ECL logic gate will definitely look like logic 1 to that gate input. Conversely, an input voltage of $V_{OL} = -1.74$ V will also be considered to be a valid logic 0 by this gate.

EXAMPLE 10.6-1

For the ECL logic gate illustrated in Figure 10.6-3a, determine the noise margins and find the gate fan-out for low-speed gate operation.

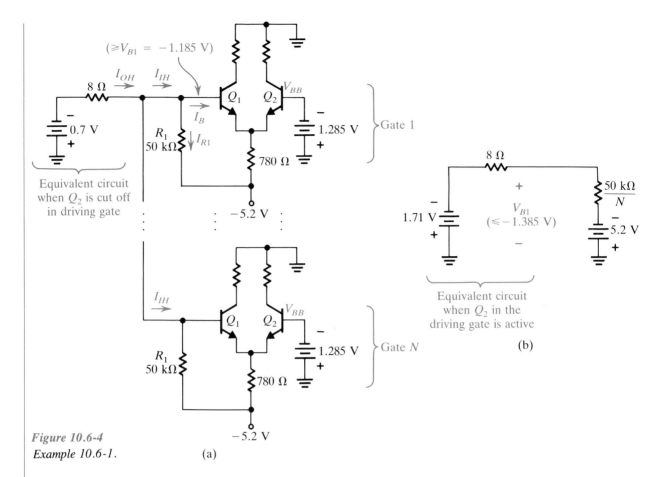

Figure 10.6-4
Example 10.6-1.

(a)

(b)

SOLUTION

By considering that $V_{BB} = -1.285$ V, it is apparent that

$$V_{IH} = V_{BB} + 100 \text{ mV} = -1.185 \text{ V} \qquad (10.6\text{-}16a)$$

and

$$V_{IL} = V_{BB} - 100 \text{ mV} = -1.385 \text{ V} \qquad (10.6\text{-}16b)$$

Combining these results with those in equations (10.6-15a) and (10.6-15b), we may write

$$NM_L = V_{IL} - V_{OL} = -1.385 - (-1.74) = 355 \text{ mV} \qquad (10.6\text{-}17a)$$

and

$$NM_H = V_{OH} - V_{IH} = -0.7 - (-1.175) = 485 \text{ mV} \qquad (10.6\text{-}17b)$$

The fan-out of an ECL gate may be found by determining the equivalent circuit at the output of the logic gate. This equivalent circuit is then connected onto the input terminals of N similar logic gates and the loading effects are examined. The output equivalent circuit shown in Figure 10.6-4a is that obtained when the V_{o2} output is HI in Figure 10.6-4, assuming that the transistor h_{FE} is 30. For Q_1 in the gates being driven to be ON, it is necessary that V_{B1} be greater than or equal to -1.185 V. With this value of V_{B1} as the limiting case, it therefore follows that

$$I_{C1} = I_{RE} = \frac{-1.885 - (-5.2)}{0.78 \text{k}\Omega} = 4.25 \text{ mA} \qquad (10.6\text{-}18a)$$

$$I_{B1} = \frac{I_{C1}}{h_{FE}} = 0.142 \text{ mA} \qquad (10.6\text{-}18b)$$

and
$$I_{R1} = \frac{-1.185 - (-5.2)}{50 \text{ k}\Omega} = 0.08 \text{ mA} \qquad (10.6\text{-}18c)$$

Thus the load presented by each logic gate is

$$I_{IH} = 0.142 + 0.08 = 0.222 \text{ mA} \qquad (10.6\text{-}19)$$

At the gate output the maximum allowed output current that will still maintain the output voltage above -1.185 V is

$$I_{OH} = \frac{0.7 - (-1.185)}{8} = 60.6 \text{ mA} \qquad (10.6\text{-}20a)$$

so that
$$N \le \frac{60.6}{0.222} = 292 \text{ gates} \qquad (10.6\text{-}20b)$$

The fan-out when the gate output voltage is LO may be similarly computed using the equivalent circuit given in Figure 10.6-4b. Here we have noted that V_{OL} is equal to -1.71 V under open-circuit conditions, and that V_{B1} can have a maximum value of -1.385 V if the gate input transistors are to remain cut off. Furthermore, with the differential amplifier input transistors cut off, the equivalent input circuit of each gate may simply be represented by a 50-kΩ resistor. Because V_{B1} will always be less than -1.385 V regardless of the value of N selected, the fan-out when the gate output is LO is infinite.

10.6-2 The Use of Negative Power Supplies with ECL Logic

In examining the basic ECL logic gate in Figure 10.6-3, we observed that it is usually operated from a negative supply voltage by grounding V_{CC} and connecting $-V_{EE}$ to -5.2 V. This is done to minimize the effects of logic and power supply noise on the gate output signals. The basic problem at hand may be illustrated with the aid of Figure 10.6-5a. Here, depending on whether we ground the system at point A or at point B, we will either be operating with a negative or a positive power supply voltage, respectively.

In this circuit two sources of noise are indicated. The first, labeled v_{n1}, is that associated with the power supply itself. This may be due to changes in the nominal 5.2-V dc level produced by the supply, or even due to ripple or other high-frequency sources of noise that the power supply generates. The second source of logic noise is that associated with transient signal currents produced by the gates in the circuit. These currents cause voltage drops along the length of the circuit wiring that can enter the logic gate under examination and can appear at the output as false logic signals. For simplicity we have represented the circuit wiring resistance and wiring inductance as lumped elements. Of course, in reality these elements are distributed along the length of the wire. To the extent that this lumped-element representation is valid, their effect may be represented by introducing a second voltage source, $v_{n2} = i_{n2}(R_{w1} + R_{w2}) + (L_{w1} + L_{w2})di_{n2}/dt$.

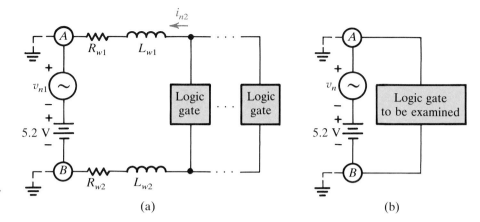

Figure 10.6-5
Effect of the ground-point position on the output logic noise produced.

(a)

(b)

In Figure 10.6-5b we have combined v_{n1} and v_{n2} as a single noise source v_n, and when a detailed analysis of the effect of this noise signal on the output is carried out for a typical ECL gate such as that shown in Figure 10.6-3, we find that the presence of noise at the gate output is significantly reduced when point A rather than point B is grounded. Principally, this is because a noise voltage in series with the emitter resistor of an emitter follower has much less of an effect on the output at the emitter than a noise voltage of similar amplitude that appears in series with the base (see Problem 10.6-5). As a result, for ECL logic circuits, it is advantageous to use negative power supplies.

10.6-3 Transient Performance of an ECL Logic Gate

In the remainder of this section we examine the factors governing the switching behavior of ECL logic gates. Basically, because the transistors in an ECL gate never saturate during normal gate operation, we will be able to carry out this analysis by using the frequency response techniques that we developed in Chapter 6. Specifically, this technique will be applied to determine the parasitic element or elements that are mainly responsible for the rise, fall, and delay times in ECL gates.

In carrying out these analyses, we assume that each of the transistors has the following equivalent circuit parameters: $r_{B'E} = 250\ \Omega$, $r_{BB'} = 0$, $h_{FE} = 30$, and $C_{B'C} = C_{B'E} = 1$ pF. In addition, C_{CS}, the collector-to-substrate capacitance, is equal to 2 pF. As a rough approximation, all parameters will be assumed to be constant independent of the position of the operating point.

The circuit to be examined is illustrated in Figure 10.6-6a, and as in Section 6.8, the time constant associated with each of these parasitic capacitances will be estimated by finding its RC product, where R is the resistance seen by the capacitance in question with all the other capacitances replaced by open circuits. To carry out this procedure, it is useful to split the base-to-collector capacitances by applying the Miller theorem. A Miller gain of about -5 will be used to compute the equivalent input and output capacitances, as this is approximately equal to the slope of the input–output transfer characteristic in the active region. Carrying out this procedure, we obtain the equivalent circuit

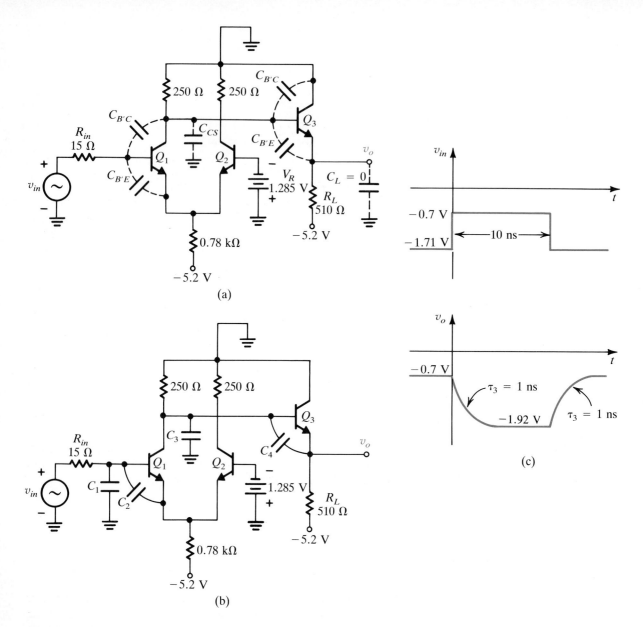

Figure 10.6-6
Transient behavior of an ECL logic gate.

shown in Figure 10.6-6b, where

$$C_1 = C_{B'C}(1 - A) = 6 \text{ pF} \tag{10.6-21a}$$

$$C_2 = C_{B'E} = 1 \text{ pF} \tag{10.6-21b}$$

$$C_3 = \underbrace{C_{B'C} + C_{CS}}_{\text{from } Q_1} + \underbrace{C_{B'C}}_{\text{from } Q_3} = 4 \text{ pF} \tag{10.6-21c}$$

and

$$C_4 = C_{B'E} = 1 \text{ pF} \tag{10.6-21d}$$

The time constants associated with each of these capacitors are

$$\tau_1 = R_{in} C_1 = (15 \text{ }\Omega)(6 \text{ pF}) = 0.03 \text{ ns} \tag{10.6-22a}$$

$$\tau_2 = C_2\left[r_{B'E}\middle\|\frac{R_{in} + r_{B'E}/(1 + h_{FE})}{1 + g_m r_{B'E}/(1 + h_{FE})}\right]$$

$$= (1 \text{ pF})(11.6 \text{ }\Omega) = 0.012 \text{ ns} \tag{10.6-22b}$$

$$\tau_3 \simeq R_{C1} C_3 = (250 \text{ }\Omega)(4 \text{ pF}) = 1 \text{ ns} \tag{10.6-22c}$$

and

$$\tau_4 = C_4\left[r_{B'E}\middle\|\frac{R_{C1} + R_L/(1 + h_{FE})}{1 + g_m R_L/(1 + h_{FE})}\right]$$

$$= (1 \text{ pF})(12.6 \text{ }\Omega) = 0.013 \text{ ns} \tag{10.6-22d}$$

By examining these results, it is apparent that the transient performance of this logic gate is dominated by the time constant associated with the collector capacitance C_3. Therefore, to a good approximation the output voltage rise and fall times will be equal to $2.2\tau_3$ or about 2.2 ns. The pertinent waveshapes for this circuit, assuming all time constants except τ_3 to be zero, are illustrated in Figure 10.6-6c.

Exercises

10.6-1 For the primitive ECL logic gate shown in Figure 10.6-2a, let $R_{C1} = R_{C2} = 1 \text{ k}\Omega$, $I_o = 2 \text{ mA}$, and $V_{CC} = V_{EE} = 5 \text{ V}$. Find V_{o1} and V_{o2} if $V_{in} = 50 \text{ mV}$. **Answer** $V_{o1} = 3.26 \text{ V}$, $V_{o2} = 4.75 \text{ V}$

10.6-2 Estimate the power consumption of the logic gate shown in Figure 10.6-3, assuming that the voltage reference portion of this circuit is also used to feed three other similar logic devices located on the same chip. **Answer** 23 mW

10.6-3 Compute the slope at the center of the transfer characteristic for v_o versus v_{in} in Figure 10.6-6a. Carry out this analysis using small-signal techniques and consider that all of the parasitic capacitances are zero and that $h_{ie} = 250 \text{ }\Omega$ and $h_{fe} = 30$. **Answer** -15

10.7 TRANSMISSION-LINE EFFECTS IN DIGITAL LOGIC CIRCUITS

In the early days of electronics, little attention was paid to the interconnections used between circuits because the signals propagated from one point to the other "nearly instantaneously" in comparison to the operating speeds of the active elements. Today, however, the situation is quite different since the rise and fall times associated with many digital logic devices are smaller than the propagation times down the wires connecting them. In these instances the wires and cables interconnecting the circuit components cannot simply be represented as short circuits; instead, they must be considered to be transmission lines.

Because of its importance in modern digital system design, we have included this section on transmission-line effects in digital logic circuits. In addition, for those people unfamiliar with transmission-line principles, we have included a review of the basic properties of transmission lines in Appendix II. In the next few paragraphs we summarize the major points covered in this appendix. However, the author encourages those who are unsure of any of these principles to carefully read through the appendix.

Transmission-line effects become important when the time required for a signal to propagate down a wire is comparable to the rise and fall times of the signal on that wire. When, on the other hand, the propagation time is short in comparison to the signal rise and fall times, transmission-line effects are suppressed and the transmission line behaves like a wire, that is, a short circuit.

The performance of a uniform transmission line can be characterized by specifying its characteristic impedance and its propagation velocity. The characteristic impedance of the line is equal to

$$Z_o = \sqrt{\frac{L_o}{C_o}} \qquad (10.7\text{-}1)$$

where L_o is the line inductance/length and C_o is the line capacitance length. The propagation velocity may be written as

$$c_p = \frac{1}{\sqrt{L_o C_o}} \qquad (10.7\text{-}2)$$

The parameter
$$T_o = \sqrt{L_o C_o} \qquad (10.7\text{-}3a)$$

defines the propagation time/length along the line, and if x_o is the length of the line, then

$$T_d = T_o x_o \qquad (10.7\text{-}3b)$$

represents the time required for a signal to travel down the entire length of the line.

Figure 10.7-1 illustrates the four most popular types of transmission lines. For the interconnection of electronic components on individual circuit boards, conventional printed circuit wiring is probably the most widely used connection method. In this process, insulating boards (typically, $\frac{1}{16}$- or $\frac{1}{8}$-in.-thick fiberglass-epoxy material) are coated with copper on one or both sides. The unwanted copper is then etched away, leaving the desired wiring pattern on the surface of the board. This type of printed circuit wiring has a transmission-line behavior similar to that of isolated point-to-point wiring. The propagation time of PC wiring is typically on the order of 1.4 to 1.8 ns/ft. Here it is important to notice that the character of these printed-circuit-style transmission lines are rather nonuniform since Z_o and c_p will depend rather strongly on the position of adjacent wiring. This problem is further complicated when double-sided and multilayer boards are used.

Microstrip is basically a highly controlled form of printed circuit wiring. Here, most of the PC lands are routed on the upper side of the board, and the lower side of the board is used to form a ground reference plane. Because of the close proximity of the ground plane to the wiring, Z_o and c_p tend to be less affected by the position of adjacent wires than with ordinary PC boards. The ground plane of the microstrip also provides a shielding effect so that noise pickup and crosstalk from signals on adjacent wires are reduced.

Type of Transmission Line	Z_o	T_o	Additional Remarks
Wire near a ground plane 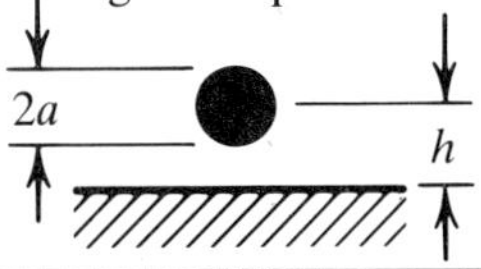 	$\dfrac{60}{\sqrt{\epsilon_r}} \ln \dfrac{2h}{a}$	$\sqrt{\epsilon_r}$ ns/ft	For a wire in air with $a = 0.025$ in., and $h = 1$ in., $Z_o \sim 263\ \Omega$, $C_o = T_o/Z_o = 3.8$ pF/ft, $T_o \sim 1$ ns/ft, $L_o = T_o Z_o = 0.263\ \mu$H/ft
Coaxial cable 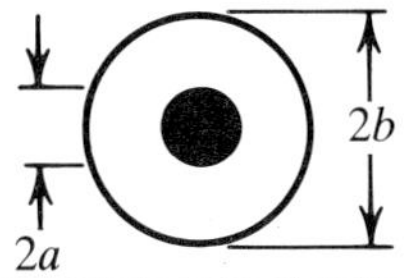 	$\dfrac{60}{\sqrt{\epsilon_r}} \ln \dfrac{b}{a}$	$\sqrt{\epsilon_r}$ ns/ft	For a polyethylene dielectric ($\epsilon_r = 2.3$) with $2a = 0.05$ in., and $2b = 0.25$ in. $Z_o = 64\ \Omega$, $C_o = 24$ pF/ft, $T_o \simeq 1.5$ ns/ft, $L_o = 0.097\ \mu$H/ft
Microstrip etched wiring 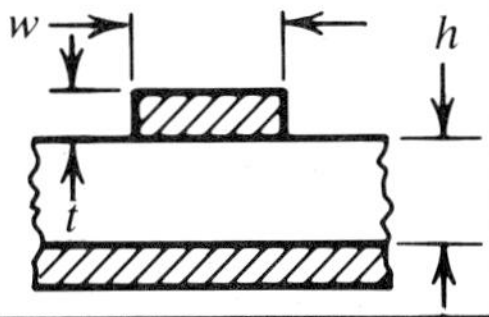 	$\dfrac{87}{\sqrt{\epsilon_r + 1.41}} \ln \dfrac{5.98h}{0.8w + t}$	$\sqrt{0.475\,\epsilon_r + 0.67}$ ns/ft	For a $\frac{1}{16}$ in. G-10 epoxy-glass PC board, with $\epsilon_r = 5$, $w = 50$ mils, and $t = 0.0015$ in., $Z_o = 75\ \Omega$, $C_o = 24$ pF/ft, $T_o = 1.77$ ns/ft, $L_o = 0.13\ \mu$H/ft
Twisted-wire pair  			For No. 24-28 gauge wire, with 30 twists/ft, $Z_o = 100\ \Omega$, $C_o = 14$ pF/ft, $T_o = 1.4$ ns/ft, $L_o = 0.14\ \mu$H/ft

Figure 10.7-1
Properties of popular transmission lines.

Because the ground plane adds extra capacitance, the characteristic impedance and the propagation velocity of microstrip boards are both smaller than for conventional printed-circuit-board wiring. Propagation times are also correspondingly longer and are typically on the order of 1.8 to 2.6 ns/ft.

When signals need to be run between circuit boards or between pieces of equipment, twisted-wire pairs or shielded cables are most commonly employed. The use of twisted-wire pairs offers the most economical solution, and these pairs can be bundled together, with 20 to 30 of them easily fitting into a cable that is $\frac{1}{2}$ in. in diameter. Coaxial cable is used when shielding and preservation of the original signal waveshape are important.

Regardless of the type of transmission line chosen for a particular application, the basic properties of all of them are really quite similar. The characteristic impedance will nearly always be 50 to 300 Ω, and the propagation delay will be about 1 or 2 ns/ft. Interestingly enough, these numbers vary only slightly with the geometry and the dimensions of the transmission line because of the logarithmic nature of the equations for Z_o. To illustrate this point, we note, for example, that a 0.05-in.-diameter circular wire located $\frac{1}{4}$ in. above a ground plane will have a characteristic impedance of about 180 Ω. If the wire is raised to a position 10 ft above the ground plane, Z_o will only increase to about 540 Ω. Thus, for all practical transmission-line dimensions, the characteristic impedance will always be rather small. This is important because it indicates that considerable current may be needed for a digital device to drive a transmission line properly.

In high-speed logic circuits it is useful to interconnect all digital devices with wiring that has known transmission-line characteristics. This allows for the use of proper termination resistors where indicated, and minimizes the generation of reflections at the source and the load.

In general, when a signal travels along a transmission line, reflections occur at both ends of the line, and the amplitude and polarity of these reflections depends on the values of the reflection coefficients at each of these points. Following Appendix II, these coefficients can be shown to be

$$\rho = \text{reflection coefficient at the load} = \frac{Z_L - Z_o}{Z_L + Z_o} \qquad (10.7\text{-}4a)$$

$$\text{and} \qquad \rho' = \text{reflection coefficient at the source} = \frac{Z_S - Z_o}{Z_S + Z_o} \qquad (10.7\text{-}4b)$$

where Z_L and Z_S represent the load and source impedances, respectively.

The size of the reflections on a transmission line may be minimized by proper impedance-matching techniques at the source and the load. For digital circuit designs the reduction of these reflections to acceptable levels is important for several reasons. First, these signal excursions may be interpreted as valid logic transitions by the digital circuit. Second, the signal voltages may go well above V_{CC} or below ground, and this can lead to gate damage.

In addition, sometimes the reflection coefficients are such that the input to a gate is held at a voltage level somewhere between a logic 0 and a logic 1. This can bias the logic circuit in the active region and cause gate damage. However, even more important, placing the gate in the active region may cause the entire system to oscillate due to the natural feedback that often exists in digital logic systems.

Now that we have a good understanding of the types of problems that can arise if we are not careful to account properly for the transmission-line nature of the digital systems that we will be designing, several other factors need to be considered before we begin wiring together our logic gates, flip-flops, shift registers, and microprocessors. First, for the logic family that we will be working with, we will want to know the maximum wire lengths that can be used before transmission-line effects need to be taken into account. Second, for situations where these effects are significant, we need to have a good understanding of the circuit designs available for use with these transmission lines.

In Appendix II we showed that transmission-line effects were minimal when the rise time of the signal traveling on the transmission line was greater than the round-trip propagation delay of the line. Considering that epoxy-glass printed circuit boards have propagation times on the order of 1.6 ns/ft, and using 20 in. as the maximum wire length on a typical PC board, we can see that all logic families with device rise times smaller than about 5 ns may exhibit transmission-line problems in single-board digital designs.

The relative speeds of several of the more important digital logic families are listed in Table 10.7-1, and by examining these data it is apparent that

10.7-1 Relative Speeds of Digital Logic Families

Logic Family	t_t (Smaller of t_r and t_f; ns)	Maximum Unterminated Line Length (Using $t_r \leq 2T_d$)[a]
74xx TTL	5	18″
74Sxx TTL	1.5	6″
74LSxx TTL	6	24″
74ALSxx TTL	5	18″
74ASxx TTL	1	4″
10K ECL	2	8″
High-speed CMOS	20	6 feet

[a] $T_0 = 1.6$ ns/ft.

transmission-line effects can be important, even in single-board designs, for nearly all the logic families indicated, with the possible exception of CMOS. Furthermore, extra-special care will need to be exercised when working with the ECL, Schottky, and advanced Schottky families since unterminated line lengths of only a few inches can cause problems with these devices.

Now that we have an idea of when it will be important to consider the transmission-line nature of the wires and cables that interconnect the digital circuits we will be working with, it is equally important to develop an understanding of the types of transmission-line circuits that are available. One of the most commonly used transmission-line circuits is shown in Figure 10.7-2a. If for the moment we assume that the gates connected to this line are ideal, that is, that they have zero output impedance and infinite input impedance, then in theory, reflections on this line can be eliminated completely if Z_L is simply made equal to Z_o. In this instance ρ is zero, and the load is said to be matched to the characteristic impedance of the line. The resistor R_L in this circuit is called a shunt or parallel line termination resistance. Reflection effects can also be eliminated by matching the source impedance to the characteristic impedance of the line using a circuit of the form illustrated in Figure 10.7-2b. The waveshapes in parts (c) and (d) of the figure illustrate the basic differences between these two approaches.

For the shunt-matched case, the voltage waveshape, except for a time delay, is the same at all points along the line. As a result, additional gates can be connected anywhere along this line with no problems. For the case of the series-matched line, the situation is quite different. Here the voltage on the line takes on an intermediate value equal to one-half of its final value. Depending on whether this voltage is recognized as a logic 0 or a logic 1 by the gate connected at that point, an additional time delay can be introduced by this staircase effect. Furthermore, this intermediate voltage level may bias the gate in the active region, and as mentioned earlier, this situation can lead to circuit oscillation or even to gate damage. Thus, in series-matched transmission-line cir-

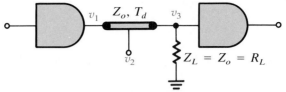

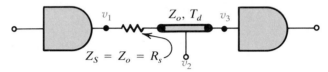

(a) Shunt-Style Impedance Matching at the Load
     (*Note*: For simplicity we represent the
     transmission lines by single wires.)

(b) Series-Style Impedance Matching at the Source

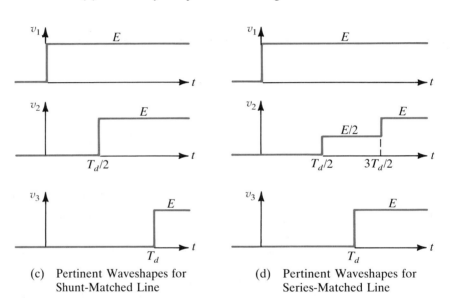

(c) Pertinent Waveshapes for
     Shunt-Matched Line

(d) Pertinent Waveshapes for
     Series-Matched Line

Figure 10.7-2

cuits it is inadvisable to connect logic gates anywhere except directly at the load end of the line (Figure 10.7-3a). The constraint makes it difficult to use series-terminated transmission lines when multiple gates need to be driven since these gates will typically be located at random positions on the PC board.

The circuits shown in Figure 10.7-3b through d illustrate several alternative methods for driving multiple receiver gates. If all the gates to be driven lie along a single relatively straight-line path, it is possible to connect multiple taps on a single shunt-matched transmission line as shown in Figure 10.7-3b. When the gates to be driven are distributed randomly throughout the board, an approach such as that shown in part (c) or (d) of the figure needs to be used. In both of these circuits, each receiver gate is driven by its own transmission line, so that these gates can be located anywhere on the printed circuit board. The only problem with these arrangements is that the dynamic input impedance

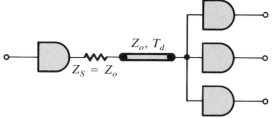

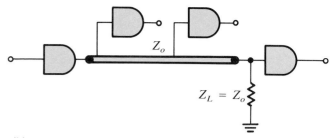

(a) Only Acceptable Way to Drive Multiple Gates on a Single Series-Matched Transmission Line

(b) Use of Multiple Taps on a Shunt-Matched Line

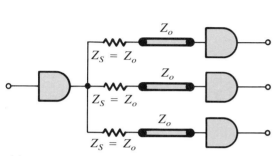

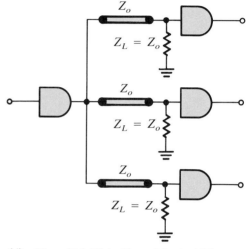

(c) Driving More Than One Gate Using Multiple Series-Matched Lines

(d) Use of Multiple Shunt-Matched Lines

Figure 10.7-3
Techniques for driving multiple gates.

seen by the driving gate will be reduced (e.g., to $Z_o/3$ in Figure 10.7-3d). As a result, loading of the driver gate can become a problem.

One obvious solution to this difficulty is to increase the characteristic impedance of each of the transmission lines (e.g., to $3Z_o$ in Figure 10.7-3d) to make the overall input impedance seen by the driving gate equal to Z_o. However, it may not be possible to fabricate transmission lines having such high impedances, and furthermore, even when their construction is feasible, they may still be unsuitable because of their increased susceptibility to crosstalk interference from adjacent lines. Thus there is no single "best" transmission-line circuit that should always be used for the interfacing of all high-speed digital logic circuits. The "best" circuit to use depends on the particular problem at hand.

10.7-1 Transmission-Line Effects in TTL Circuits

The analysis of transmission-line problems in TTL logic circuits is complicated by the fact that the gates do not operate in the active region. As a result, both their input and output volt-ampere characteristics are highly nonlinear, and graphical analysis methods will be required to obtain a good quantitative understanding of the behavior of these circuits. Figure 10.7-4 illustrates the input and output volt-ampere characteristics of a typical 7400-style TTL logic gate.

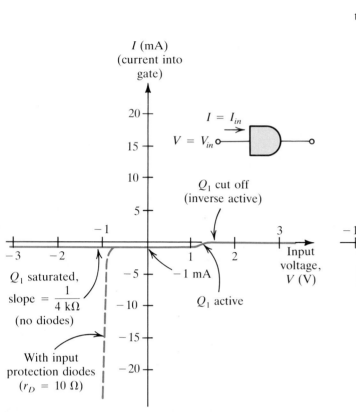

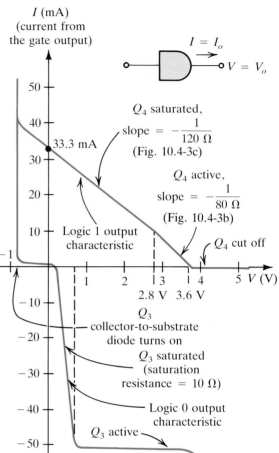

(a) Input Characteristic of a 7400 Style TTL Logic
Gate with and without Input Protection Diodes

(b) Output Volt-Ampere Characteristics of a 7400 Style
TTL Logic Gate

Figure 10.7-4

The shape of these curves may be understood with the aid of the partial gate circuits given in Figure 10.7-5.

Let's examine the input characteristic first. When $V = V_{in}$ is greater than about 1.4 V, Q_1 is cut off and the current into the gate is essentially zero. For V_{in} less than about $(1.4 - 0.2)$ or 1.2 V, Q_1 is saturated, and a current out of the gate equal to about $(5 - 0.7 - V_{in})/(4 \text{ k}\Omega)$ is produced. Thus the slope of the V–I characteristic in this region is about $1/(4 \text{ k}\Omega)$. For reasons that will be apparent shortly, it is useful to add protection diodes between the gate input terminals and ground as shown in Figure 10.7-5b. This helps to reduce ringing on gates interconnected by "long" wires. The effect of the addition of these diodes on the V–I characteristic is indicated by the dashed portion of the curve in Figure 10.7-4a. Basically, when the gate input voltage attempts to go below -0.7 V, these diodes turn on and clamp V_{in} at this level.

The output volt-ampere characteristic curves are a bit more complex because of the larger number of transistors involved and also because two different curves are obtained, one when the gate output is a logic 1, and a second when it is a logic 0. When the output is supposed to be a logic 1, Q_2 and Q_3 in Figure 10.7-5a are both cut off. For values of V_o greater than

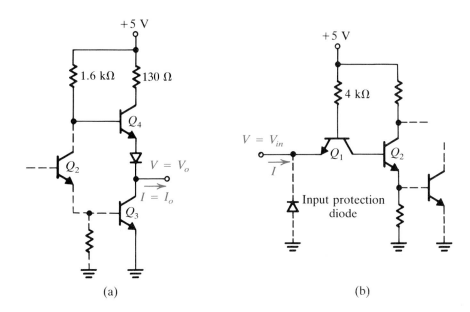

Figure 10.7-5

(a)   (b)

$(5 - 0.7 - 0.7) = 3.6$ V, Q_4 is cut off and $I = I_o = 0$. When V_o falls below 3.6 V, Q_4 enters the active region. The Thévenin resistance in this region is about 80 Ω (see Figure 10.4-3b). Once V_o falls below 2.8 V, Q_4 saturates and the Thévenin resistance increases to about 120 Ω (see Figure 10.4-3c). The saturation region continues until the output voltage falls below about -0.7 V, and at this point the parasitic collector-to-substrate diode of Q_3 turns on and clamps this voltage at about one diode drop below ground.

When the gate output is supposed to be a logic 0, transistors Q_2 and Q_3 are saturated and Q_4 is cut off. Under these conditions, as per equation (10.4-6c), the base current into Q_3 is about 2.6 mA, so that Q_3 will remain saturated as long as I_{C3} is less than 52 mA assuming an h_{FE} of 20. Therefore, as V_o becomes positive, I_o (which is equal to $-I_{C3}$) will decrease linearly with a slope equal to the transistor saturation resistance until the collector current reaches 52 mA. At this point Q_3 will enter the active region and I_o will level off at about -52 mA. When V_o goes negative, not much current flows until V_o reaches -0.7 V, and at this point, as before, the collector-to-substrate diode of Q_3 will turn on, clamping V_o at a potential about one diode drop below ground.

To understand the effect that these nonlinear volt-ampere characteristics have on the performance of TTL gating circuits, it is best to examine this problem by using the graphical analysis method presented in Appendix II. This approach is necessary because conventional transmission-line analysis methods become extremely complex when the source and/or the load are nonlinear. If you are unfamiliar with this graphical analysis procedure, it is essential that you study the material in Section II.3 before proceeding further.

To illustrate the basic problem at hand, consider, for example, the circuit shown in Figure 10.7-6a and let's determine the voltages produced at v_1 and v_2 when the output of G_0 goes from a logic 1 to a logic 0. In this circuit G_0 is driving its maximum load of 10 logic gates. Nine of these gates are wired directly to the output of the driver gate, and the tenth gate is connected through a 200-Ω transmission line.

Figure 10.7-6

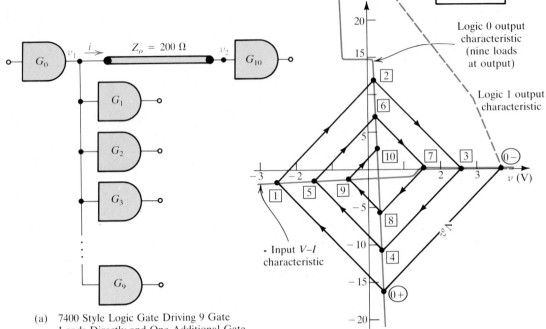

(a) 7400 Style Logic Gate Driving 9 Gate Loads Directly and One Additional Gate through a 200-Ω Transmission Line

(b) Circuit Impedance Diagram for Gates with No Input Protection Diodes

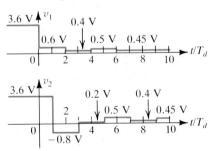

(c) Circuit Impedance Diagram for Gates with Input Protection Diodes

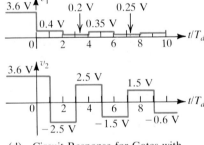

(d) Circuit Response for Gates with No Input Protection Diodes

(e) Circuit Response for Gates with Input Protection Diodes

In analyzing the performance of this circuit, let's initially assume that the gates do not have input protection diodes, in order to develop an appreciation for why they are actually included in TTL logic circuits. The voltage waveshapes for this circuit may be obtained with the aid of the gate input–output volt-ampere characteristics given in Figure 10.7-6b. These curves are also known as impedance diagrams, and we will use both of these terms interchangeably throughout the remainder of this section. Notice that the V–I characteristic curve for the case where the gate output is a logic 0 crosses the current axis at about 14.4 mA. This is because of the additional current contributed by gates G_1–G_9. When v_1 is at zero volts, Q_3 of the driving gate will almost be cut off, but each of gates G_1 through G_9 can contribute a worst-case current of about 1.6 mA, so that the maximum total current on the line at this point can be as high as (9×1.6) or 14.4 mA.

To analyze the performance of this circuit when the output of G_0 switches from a 1 to a 0, let's assume that G_0 has been a logic 1 for some time prior to $t = 0$ so that the circuit operating point is initially located at the point labeled $0-$ on the impedance diagram. The analysis of this problem for $t > 0$ is begun by drawing a line with a slope of $1/200\ \Omega$ through point $0-$. Because of the scaling used on the voltage and current axes, this line makes an angle of $45°$ with the voltage axis. The intersection of this line with the logic 0 output characteristic gives the new values of the voltage and current at the input to the line at $t = 0+$. From the graph these quantities are seen to be $v = 0.35$ V and $i = -17$ mA. The difference in the voltage coordinates at the points $0-$ and $0+$ represents the amplitude of the wave launched onto the transmission line at $t = 0+$.

After T_d seconds this wave strikes the input of gate G_{10}. The load voltage and load current produced when this occurs are found by drawing a line with a slope of $-1/200\ \Omega$ (or $-45°$) through point $0+$. The intersection of this line with the input V–I characteristic of G_{10} indicates that the values of this voltage and current are $v = -2.6$ V and $i = -1.6$ mA.

To continue the solution of this problem, a straight line with a slope of $1/200\ \Omega$ is next drawn through point 1, and its intersection with the logic 0 gate output characteristic shows that the voltage and current on the line at the output of G_0 will be 12 mA and 0.2 V shortly after $t = 2T_d$. This procedure can now be repeated again and again until sufficient steady-state accuracy is achieved.

Figure 10.7-6d illustrates the voltage waveshapes produced at the input and output sides of the transmission line when the output of gate G_0 makes a HI-to-LO transition. While the voltage at the output of the driver gate is well behaved, the input voltage at G_{10} leaves a lot to be desired. To begin with, this voltage goes well below zero and causes excessive current to flow out of the input transistor of G_{10}, which can damage the gate. Furthermore, there are two or three positive transitions of v_2 that could be interpreted by G_{10} as momentary logic 1 signals, and these could cause logic glitches (or errors) in the output from this gate.

Both of these problems can be eliminated by connecting a diode between the gate input terminal and ground as shown in Figure 10.7-5b. The major effect that this diode has on the circuit performance is that it prevents the input voltage from going significantly below ground potential. But, in addition, as we will demonstrate, it also prevents the occurrence of the positive-going glitches that appeared in the previous example. To illustrate these points, let's again examine the gating circuit shown in Figure 10.7-6a for the case where the output of G_0 goes from a logic 1 to a logic 0, but this time let's consider that the gates have diode-protected inputs. In this case the problem solution proceeds exactly as before except that the graphical results have the form given in Figure 10.7-6c. Using the data from this graph, the waveshapes versus time for v_1 and v_2 may be drawn as shown in Figure 10.7-6e.

One of the nice features of diode-protected TTL is that these logic components can be interconnected together with little concern for impedance matching. In fact, the only components needed to interface TTL logic properly are the logic elements themselves and the wires or transmission lines that connect them together. These statements are basically true as long as the transmission-line impedance is not too small. Example 10.7-1 will help to illustrate the problems that arise when a TTL logic gate drives a low-impedance 50-Ω transmission line.

EXAMPLE 10.7-1

Two TTL logic gates are connected by a 50-Ω transmission line as illustrated in Figure 10.7-7a. Sketch the voltages v_1 and v_2 for 0-to-1 and 1-to-0 logic transitions at the output of the driver gate.

Figure 10.7-7
Example 10.7-1.

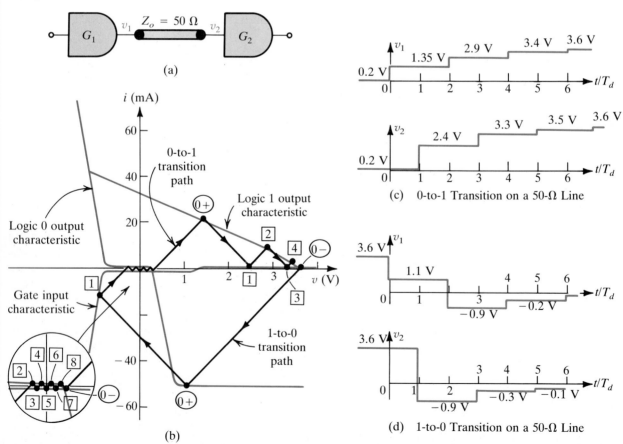

SOLUTION

This problem is analyzed graphically in Figure 10.7-7b. The results for the case of the HI-to-LO transition are indicated by the path with the circled numbers, and those for the LO-to-HI transition by the numbers with the squares around them. The voltage waveshapes of v_1 and v_2 versus time for each of these cases are shown in parts (c) and (d) of the figure. The major undesirable feature in these waveshapes is that the intermediate voltage level at v_1 might be interpreted as a logic 0 or a logic 1, depending on the gate threshold. This is a problem because it could cause an unpredictable delay in the response of any additional gates connected at the output of the driver gate. In addition, these intermediate voltage levels might bias the receiver gates in the active region, and depending on the overall feedback present in the circuit, this could cause the system to oscillate.

In this example we have seen that ordinary TTL logic gates are basically unsuitable for driving low-impedance transmission lines since the receiver gates can only be connected at the end of the line. As a result, when TTL circuits need to be connected by transmission lines, it is recommended that a characteristic impedance of at least 100 Ω be used. For applications where TTL logic needs to drive low-impedance lines, special line driver ICs are available. These circuits have low symmetrical output impedances and high drive-current capabilities.

10.7-2 Transmission-Line Effects in Emitter-Coupled Logic

Unlike transistor-transistor-logic, ECL is designed to interface easily to all types of transmission lines. Its emitter-follower driver transistors have low output impedances, so that very little voltage is lost when they drive transmission lines, and as a result, signals transmitted on them settle down to their final values almost immediately. In addition, emitter-coupled logic circuits operate in the active region, so that their input and output impedances are nearly constant at all signal levels. This makes impedance matching much easier than for nonlinear digital logic families such as TTL.

ECL can directly drive 50-Ω transmission lines using shunt-style impedance matching if the load resistance is connected to a negative power supply (Figure 10.7-8a). Some type of negative pull-down voltage is needed at R_L to ensure that the output transistor in the driver gate remains on for both states of the logic gate. Connection to a −2 V supply rather than to the −5.2 V logic supply voltage is preferred because this keeps the power dissipation in the driver gate down to a reasonable value. To illustrate this point, consider that $R_L = 50$ Ω and that the ECL driver gate has a logic 1 in its output (−0.7 V). When R_L is tied to −5.2 V, the current flow through the driver transistor is 90 mA, whereas it is only 26 mA when connected to a −2 V power supply.

At first it might appear reasonable to reduce the V_{TT} power supply to even less than −2 V to cut down further on the power dissipation in the driver gate. However, if V_{TT} is made much smaller than −2 V, the output transistor in the driver gate could be cut off when the output of this gate makes a HI-to-LO logic transition. To illustrate this point, we notice that when the output of the

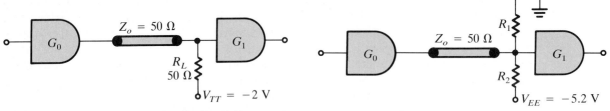

(a) Shunt Matching of an ECL Digital Logic Circuit Using a Separate Power Supply, V_{TT}

(b) Shunt Matching of an ECL Digital Logic Circuit Using the Main Logic Supply, V_{EE}

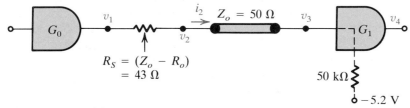

(c) Basic Circuit for Series Matching of an ECL Digital Logic Circuit

Figure 10.7-8

driver gate is a logic 1, the steady-state current out of G_0 will be

$$I_{dc} = \frac{2 - 0.7}{50} = 26 \text{ mA} \tag{10.7-5}$$

When the output of G_0 switches to a logic 0, a wave v_a will be launched onto the transmission line equal to the voltage change occurring at the gate output. Correspondingly, a current wave

$$i_a = \frac{v_a}{Z_o} = \frac{-1.0 \text{ V}}{50 \ \Omega} = -20 \text{ mA} \tag{10.7-6}$$

will also be present on the line. The total current flow out of G_0 will be equal to the sum of $(I_{dc} + i_a)$, or 6 mA. As a result, in this instance the driver transistor remains active. Had a V_{TT} of -1.5 V been used in place of a -2-V supply, then for the previous example I_{dc} would have been 16 mA, and the driver transistor would have been turned OFF when its output went to a logic 0. The importance of keeping this transistor in the active region is examined in great detail in Example 10.7-2, which follows shortly.

As an alternative to using a separate V_{TT} power supply, a similar result can also be achieved by using a resistive divider of the type illustrated in Figure 10.7-8b. In this circuit R_1 and R_2 need to be chosen so that the Thévenin equivalent of this circuit turns out to be equal to the R_L–V_{TT} combination used in the original circuit presented in part (a). For the case of a shunt-matched 50-Ω transmission line, we find that this equivalence can be achieved by setting $R_1 = 81 \ \Omega$ and $R_2 = 130 \ \Omega$. Although the circuits in parts (a) and (b) are now identical insofar as their digital logic performance is concerned, they are, nonetheless, different. The second circuit uses more parts, and in addition, it dissipates more electrical power. Therefore, for systems with a large number of gates, it soon becomes cost-effective to use a separate V_{TT} power supply.

Series termination can also be used for impedance matching in ECL logic gates. However, the classic form of this circuit shown in Figure 10.7-8c will not work very well. In theory the 50-kΩ pull-down resistor in the receiver gate G_1 will provide sufficient current flow out of G_0 to keep its driver transistor in the active region. However, as we have just discussed for the shunt-matching circuit in Figure 10.7-8a, the driver transistor will turn off on HI-to-LO logic transitions since there will be a considerable backflow of current into the driver gate when the line discharges.

On several occasions we have indicated that it is important for the output transistor in the driver gate to remain in the active region, especially for the case of a series-matched circuit such as that currently being discussed. If this does not occur, then when the driver transistor turns OFF, ρ' will be $+1$ instead of zero, and since ρ is nearly equal to 1 as well, the signal will bounce back and forth many times before a steady-state logic 0 at v_1 and v_3 is finally obtained. This has two undesirable side effects. First, the effective circuit propagation delay is greatly increased because of the large number of reflections needed to reach steady state. Second, because of the very slow rise time of v_3 at the input to G_1, this receiver gate can be biased in the active region for a considerable time, and this can result in system instability and oscillations.

EXAMPLE 10.7-2

a. Analyze the response of the ECL logic circuit shown in Figure 10.7-8c by using impedance diagrams. In particular, consider the case where G_0 makes an output transition from a logic 1 to a logic 0.

b. Sketch the approximate voltages at v_1, v_2, v_3, and v_4 considering that G_0 and G_1 are noninverting buffer gates.

SOLUTION

Before beginning the actual solution of this problem it will be necessary to develop the input and output volt-ampere characteristics of an ECL logic gate. The internal structure of a typical ECL gate is shown in Figure 10.6-3. When the gate output is a logic 1, the driver transistor is in the active region for all output voltages less than -0.7 V, and when the output voltage exceeds this value, the transistor is cut off. The behavior of the driver transistor when the gate output is a logic 0 is the same except that transistor cutoff occurs at a voltage of about -1.7 V.

These output V–I characteristics are sketched in Figure 10.7-9a, and include the effect of the resistor R_S that is connected in series with the driver gate G_0 in Figure 10.7-8c. Notice that the slope of these curves is equal to $-1/(50\ \Omega)$ when the driver transistor is ON, and zero when it is cut off. This result assumes a gate output resistance of about 7 Ω, which is typical for ECL logic gates.

The gate input characteristics has a slope of about $1/(50\ \text{k}\Omega)$ when the voltage applied to it less than about -1.4 V since the gate input transistor is cut off. The x intercept of this portion of the curve is located at -5.2 V. For input voltages greater than -1.4 V but less than -1.2 V, both transistors in the input differential amplifier are active, and for input voltages above -1.2 V, Q_1 is active and Q_2 is cut off. With Q_2 cut off the effective gate input impedance is 50 k$\Omega \| [(1 + h_{FE})0.78\ \text{k}\Omega]$ or about 16 kΩ. Therefore, the slope of the V–I characteristic in this region is $1/(16\ \text{k}\Omega)$.

To be able to sketch the shape of the gate input characteristic in the transition region, it will be necessary to determine the gate input current corre-

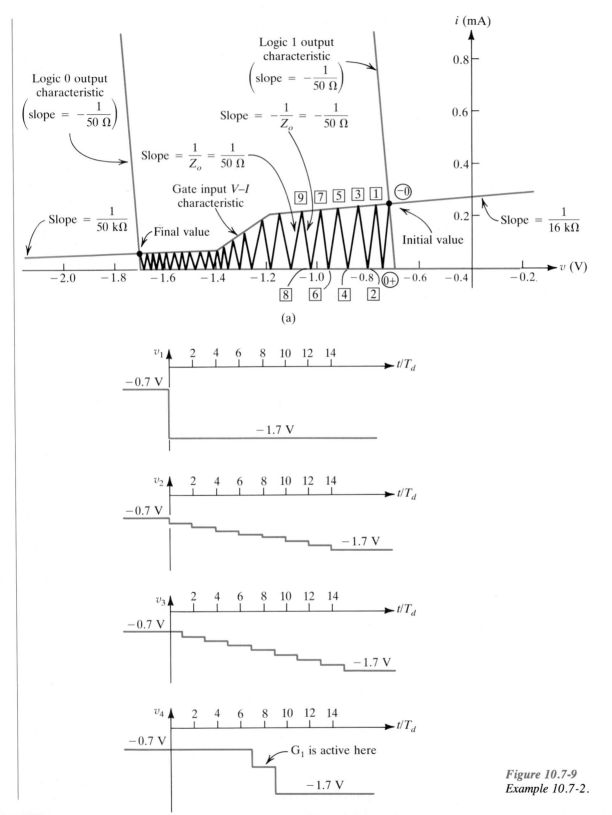

Figure 10.7-9
Example 10.7-2.

(b)

sponding to the endpoint voltages that define this region. This may be carried out as follows. With the gate input voltage equal to -1.2 V, for example, the current is the sum of that flowing into the 50-kΩ resistor plus that going into the base of Q_1. The current in the resistor is equal to $(5.2 - 1.2)/(50 \text{ k}\Omega)$ or 80 μA, and that into Q_1 is $(5.2 - 1.9)/(30 \times 0.78) = 141$ μA. Thus the total current flow into the gate when $v = 1.2$ V is 221 μA. In a similar fashion, the current flow at the other end of the transition region where $v = -1.4$ V can readily be shown to be about 76 μA. This completes our analysis of the shape of the volt-ampere characteristics of an ECL logic gate.

Returning now to the problem at hand, with G_0 initially at a logic 1, the operating point of this circuit starts out at the intersection of the gate input characteristic with the logic 1 output characteristic. As indicated in the figure, this is located at the point $v = 0.73$ V and $i = 250$ μA. When the output from G_0 changes to a logic 0 at $t = 0+$, the voltage at v_1, neglecting the 7-Ω output impedance of the gate, drops to -1.7 V immediately. The voltages elsewhere in the circuit take much longer to change because of the large number of reflections that must occur on the transmission line before steady state is finally reached. These reflections are indicated on the impedance diagram by the series of lines drawn with alternating slopes of $+1/Z_o$ and $-1/Z_o$. The intersection of these lines with the gate input characteristic gives the voltage and current at the input to G_1 in Figure 10.7-8c, while the intersections with the output characteristic determines the voltage and current at the input of the line.

Notice that because the intersections with the logic 0 output characteristic are nearly all along the voltage axis, $i_2 = 0$ and the driver transistor in G_0 is cut off for most of the transition time. In fact, this is the reason that there are such a large number of reflections on the line before steady state is finally reached. Had this transistor remained in the active region, these reflections would have stopped after the first one came back from the load, since the source impedance would then have been matched to the line impedance. Instead, what happens is that $\rho' = +1$, and with ρ also nearly equal to 1, when v_1 changes to a logic 0, many reflections will be needed before this change settles down and the final operating point on the left-hand side of Figure 10.7-9a is reached.

The approximate voltages found in this circuit are shown in Figure 10.7-9b. In making these sketches larger step sizes (and hence a smaller number of steps) were used to better illustrate the discrete nature of the steps in these waveshapes. The net effect of the large number of steps actually needed before steady state occurs is that a considerable propagation delay is created in the output of G_1. In addition, as illustrated in the v_4 waveshape, G_1 will be placed in the active region for a substantial period of time.

In Example 10.7-2 we saw that serious problems occur when a series-matched transmission-line circuit of the type illustrated in Figure 10.7-8c is used to interconnect two ECL logic gates. In particular, what happens is that the output transistor in the driver gate is cut off, and this causes many reflections to occur in response to a step change in the output of the driver gate. This leads to an exaggerated rise time in the voltage at the input of the receiver gate, which greatly increases the circuit propagation delay and can cause system oscillation. Interestingly enough, this problem can easily be overcome by adding an emitter pull-down resistor at the output of the driver gate, as shown in Figure 10.7-10a. In the example that follows we will determine the size of R_E needed to ensure that the output transistor in the driver gate stays on when the gate output goes from a logic 1 to a logic 0.

EXAMPLE 10.7-3

For the circuit illustrated in Figure 10.7-10a, determine the range of values of R_E for which the output transistor in G_0 will remain active during HI-to-LO transitions in the output. In carrying out this analysis, assume that the output impedance, R_o, of the driver gate is very small, and that R_E is very large in comparison to Z_o. You may also consider that the input impedance of the receiver gate is infinite.

SOLUTION

The basic problem involved in the driving of the transmission line without having an emitter resistor R_E in the circuit is that the logic 1 dc current flowing out of gate G_0 into gate G_1 is very small. When the output of G_0 changes to a logic 0 a large reverse current flows back into this gate due to the discharging of the transmission line. The net current flow out of the driver gate is the sum of these two currents, and if the backflow current is bigger than the initial dc quiescent current, the gate current will attempt to go negative, and this will cut off the output transistor. The addition of the emitter resistor R_E connected to a negative pull-down voltage increases the dc current flow out of the driver gate, so that the total current flow from G_0 will always be positive.

When G_0 is a logic 1 the steady-state circuit has the form given in Figure 10.7-10b, so that the dc emitter current in Q_1 may be written as

Figure 10.7-10
Example 10.7-3.

$$I_{dc} = \frac{5.2 - 0.7}{R_E + R_o} \simeq \frac{4.5 \text{ V}}{R_E} \qquad (10.7\text{-}7)$$

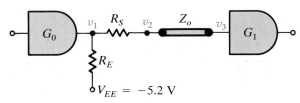

(a) Addition of an Emitter Resistor to Keep G_0 in the Active Region

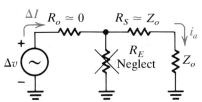

(c) Equivalent Circuit for Computing the Initial ac Current Flow Out of G_0

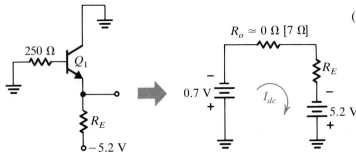

(b) Equivalent Circuit for Computing the dc Current Flow Out of G_0

When the output from G_0 changes to a logic 0, a wave is launched onto the transmission line, and the current flow associated with this wave may be determined with the aid of the equivalent circuit in Figure 10.7-10c. If we assume that R_o is negligible, and also that R_E is much greater than $2Z_o$, the approximate change in current flow from G_0 may be written as

$$\Delta I = i_a \simeq \frac{\Delta v}{2Z_o} = \frac{-1\text{ V}}{2Z_o} \tag{10.7-8}$$

The total current flow out of G_0 at $t = 0+$ is given approximately by the sum of equations (10.7-7) and (10.7-8), and therefore, for Q_1 to remain active, it is necessary that the sum of these two currents be greater than zero, or that

$$R_E \leq \frac{2Z_o}{1\text{ V}}(4.5\text{ V}) = 9Z_o \tag{10.7-9}$$

Thus, for a series-matched ECL gating circuit interconnected by a 50-Ω transmission line the largest value of R_E that will ensure that Q_1 stays on is about 450 Ω.

Thus far in this section we have indicated that ECL circuits containing transmission lines must be matched if the circuit is to work properly. In the remainder of this section we examine the amount of matching error that can actually be tolerated before problems arise.

In Section 10.6 we demonstrated that the theoretical noise immunity of an ECL logic gate was about 400 mV [see equations (10.6-17a) and (10.6-17b)]. When component variations and temperature effects are included, this parameter is found to have a worst-case value of 125 mV in commercially available ECL. Similarly, in our simplified analysis of an ECL logic gate, we found that the logic swing at the output was about 1.0 V. In commercial ECL this number is closer to 800 mV. These numbers are being included at this point because they will help us to better understand some of the engineering "rules of thumb" that have been developed in conjunction with ECL circuit designs.

When perfect impedance matching is not achieved, the voltage waveshape at the receiver gate can exhibit an undershoot, or an overshoot, or a combination, as illustrated in Figure 10.7-11. For ECL circuits to work properly, both of these quantities must be kept within bounds.

Figure 10.7-11

Definition of overshoot and undershoot on a transmission line.

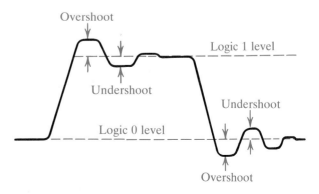

Too much overshoot at the logic 1 level can saturate the input transistor of the receiver gate, and this can significantly alter the gate input impedance. To prevent this from occurring, the overshoot should be limited to a maximum of about 35% of the logic swing.

Undershoot reduces the gate noise immunity. If the undershoot is more than 125 mV (or about 15% of the logic swing), the circuit worst-case noise immunity will be reduced to zero. To prevent this from happening, the maximum undershoot should be limited to about 15% of the logic swing.

In Figure II.2-1, when $T_R = 2T_d$ we can see that the undershoot is about 25%. Recalling that the rise time of a signal is defined as the time required for the input to rise from 10% to 90% of its final value, we can see that the rise time, t_r, of this input waveshape is about $0.8T_R$ or $1.6T_d$. If this analysis were repeated for an input with a rise time of $2.7T_d$, we can see from the geometry of the problem that the undershoot would decrease by the factor $1.6/2.7$. Therefore, the projected undershoot for the case where $t_r = 2.7T_d$ will be about $(25\%)(1.6/2.7)$ or 15%. By again referring to Figure II.2-1, we can also see that the corresponding overshoot in this instance will be $(50\%)(1.6/2.7)$ or about 30%. Thus if the input signal rise time is bigger than $2.7T_d$, the resulting overshoot and undershoot will always be within acceptable limits, and transmission-line matching components will not be needed.

EXAMPLE 10.7-4	What is the longest unterminated wire length that is permitted on an epoxy-glass PC board containing ECL logic? The rise time of the ECL may be considered to be 2 ns, and propagation delay of the wiring can be assumed to be about 0.15 ns/in.
SOLUTION	For the reflections on the line to be within acceptable limits, it is necessary to have

$$T_d = T_o x_o < \frac{t_r}{2.7} = 0.75 \text{ ns} \qquad (10.7\text{-}10)$$

or $$x_o < \frac{0.75 \text{ ns}}{T_o} = \frac{0.75 \text{ ns}}{0.15 \text{ ns/in.}} = 5 \text{ in.} \qquad (10.7\text{-}11)$$

Exercises

10.7-1 Find the characteristic impedance and the propagation delay of a 10-in. piece of stripline deposited on a $\frac{1}{16}$-in. G-10 epoxy-glass PC board. The width of the stripline is 0.1 in. and the copper thickness is 0.003 in. *Answer* 51.7 Ω, 1.45 ns

10.7-2 **a.** Determine the time required for a pulse to propagate down a 10-ft length of coaxial cable whose outside diameter is 0.25 in. and whose center conductor has a diameter of 0.05 in. The dielectric is polyethylene.
b. Find the amplitude of the first reflected wave at the end of this cable if it is terminated in a 50-Ω load. The incident wave amplitude is E.
Answer (a) 15 ns, (b) $-0.12E$

10.7-3 A 1-m transmission line with a characteristic impedance of 75 Ω and a capacitance, C_o, of 100 pF/m drives 10 CMOS logic gates, each with an input capacitance of 5 pF.

a. What is L_o for this line?

b. Find t_{PHL} for a pulse transmitted on the line if a 75-Ω load is connected at the end and if the gates are distributed uniformly along the line. Consider that the gates increase C_o by C_T/x_o, where C_T equals the total gate capacitance.

c. Repeat part (b) if all the gates are connected at the end of the line. (Connect the output-side Thévenin equivalent of the line to the lumped equivalent capacitance of the gates.)

Answer (a) 0.56 μH/m, (b) 9.2 ns, (c) 8.8 ns

10.7-4 For the circuit shown in Figure 10.7-8b, $Z_o = 200\ \Omega$.

a. Select R_1 and R_2 for no reflections on the line and also for an effective pull-down voltage, V_{TT} of -2 V.

b. Determine the overall power dissipation in R_1 and R_2. Assume that the output signal from the driving gate has a 50% duty cycle.

c. R_1 and R_2 are replaced by a single 200-Ω resistor connected to a -2-V power supply. For the conditions in part (b), what is the power dissipated in this resistor?

Answer (a) $R_1 = 325\ \Omega$, $R_2 = 520\ \Omega$, (b) 35.2 mW, (c) 3.2 mW

PROBLEMS

10.1-1 Because the gain of a logic gate is usually greater than 1, the rise and fall times of a signal can be improved by passing it through a series of logic gates.

(a) Sketch the transfer characteristic of the logic gate shown in Figure P10.1-1, assuming that $R_1 = 20$ kΩ, $R_C = 5$ kΩ, $V_{CC} = 5$ V, and $h_{FE} = 20$ for the transistor.

(b) A two-inverter cascade (G_1–G_2) is built using the inverter circuit given in part (a). Carefully sketch and label the output from each inverter if v_{in} is a signal that is zero for $t < 0$, rises linearly to 5 V in 100 μs starting at $t = 0$, remains at 5 V for 300 μs, and then falls linearly back to zero in 100 μs. For simplicity, neglect the loading of one stage on the other.

10.1-2 For the inverter shown in Figure P10.1-1, $R_1 = 10$ kΩ, $R_C = 1$ kΩ, $V_{CC} = 10$ V, and a resistor $R_3 = 5$ kΩ is inserted between R_1 and the base. In addition, a resistor R_2 in series with a battery V_{EE} (negative terminal up) is connected from the junction of R_1 and R_3 to ground. The resulting inverter circuit is to be used in a ternary (three-state) logic system in which the output levels are defined as follows:

$$\text{Logic 2:} \quad 7\text{ V} < V_o < 10\text{ V}$$

$$\text{Logic 1:} \quad 3\text{ V} < V_o < 7\text{ V}$$

$$\text{Logic 0:} \quad 0\text{ V} < V_o < 3\text{ V}$$

(a) Select R_2 and V_{EE} so that Q_1 in the inverter just saturates at $V_{in} = 7$ V, and just cuts off at $V_{in} = 3$ V. The transistor is silicon with an $h_{FE} = 50$.

(b) For the values of R_2 and V_{EE} determined in part (a), find the range of V_{in} values for which V_o is a valid logic 0, logic 1, and logic 2.

(c) Are the input and output logic levels of this inverter compatible? Explain.

10.1-3 Repeat Problem 10.1-1 using the MOSFET inverter shown in Figure P10.1-3, assuming that $R_D = 1$ kΩ, $V_{DD} = 5$ V, $V_T = 2$ V, and $g_m = 4$ mS.

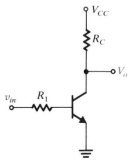

Figure P10.1-1

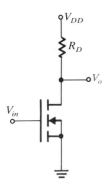

Figure P10.1-3

10.1-4 In Figure P10.1-1, $V_{CC} = 10$ V, $R_1 = 2$ kΩ, $R_C = 1$ kΩ, and a 10-kΩ resistor R_2 is shunted from base to ground. Assuming that V_{OH} and V_{OL} occur when the transistor is cut off and saturated, respectively, find NM_H and NM_L for fan-outs of 1 and 10. The transistor $h_{FE} = 100$.

10.1-5 The FET in the inverter circuit shown in Figure P10.1-3 has an ON resistance of 100 Ω. In addition, $V_{DD} = 5$ V, $R_D = 1$ kΩ, $V_T = 2$ V and $g_m = 5$ mS. Find NM_H and NM_L for a gate fan-out of 100.

10.1-6 A gating circuit is constructed using inverters of the form given in Figure P10.1-1 with $R_1 = 50$ kΩ, $R_C = 1$ kΩ, and $V_{CC} = 5$ V. In this circuit a voltage source v_{in} is connected to inverter G_0 and the output from this inverter drives gate G_1 and the two-inverter cascade G_2–G_3.
(a) Estimate the average power consumed by the logic circuit if v_{in} is a 10-kHz 0- to 5-V square wave.
(b) In part (a), determine the power dissipation in the transistors.
(c) Repeat parts (a) and (b) if v_{in} is held constant at 1.9 V.

10.1-7 A buffer is constructed by cascading two inverters of the form given in Figure P10.1-1 with $R_1 = 10$ kΩ, $R_C = 2$ kΩ, and $V_{CC} = 5$ V. In addition, a 20-kΩ resistor R_2 is shunted from the base to ground in each inverter.
(a) Determine the buffer transfer characteristic assuming that $h_{FE} = 100$ for the transistors.
(b) What is the effective small-signal gain of this buffer when it is in the active region (assume that $h_{ie} = 0$). Compare this result with the slope of the transfer characteristic in part (a).

10.1-8 In Figure P10.1-1, $R_1 = 10$ kΩ, $R_C = 2$ kΩ, $V_{CC} = 5$ V, and $h_{FE} = 50$ for the transistor. In addition, a resistor $R_2 = 50$ kΩ in series with a 5-V battery (negative terminal up) is connected from base to ground. Find the maximum fanout for this inverter.

10.1-9 In Figure P10.1-3, $R_D = 5$ kΩ, $V_{DD} = 5$ V, $V_T = 2$ V, $r_{DS(ON)} = 500$ Ω, and $g_m = 5$ mS.

(a) Plot the transfer characteristic.
(b) Determine NM_H and NM_L.

10.1-10 Design a logic circuit using only three-input NOR gates to implement the logic function $F = AB\bar{C} + \bar{B}C + (\bar{A}C) \cdot (B + D)$.

10.2-1 Find the total recovery time of the silicon diode in the circuit shown in Figure P10.2-1, if $C_J = 10$ pF, $\tau_r = 0.2$ μs, and v_{in} is a step that switches from 10 V to 0 V at $t = 0$.

10.2-2 For the circuit in Figure P10.2-1, a 3-kΩ resistor R_2 shunted by a 200-pF capacitor is connected in series with R_1. Find the diode storage time if $\tau_r = 400$ ns and v_{in} is a step that switches from 20 V to −5 V at $t = 0$.

10.2-3 Repeat the analysis of Problem 10.2-2 with the 200-pF speedup capacitor removed.

10.2-4 For the circuit shown in Figure P10.1-1, $R_1 = 20$ kΩ, $R_C = 1$ kΩ, $V_{CC} = 20$ V, and the transistor $h_{FE} = 100$. Carefully sketch and label i_B, i_C, and v_{CE} if $\tau_B = 1$ μs, considering that v_{in} is a step going from 20.7 V to 5.7 V at $t = 0$.

10.2-5 For the circuit shown in Figure P10.1-1, $R_1 = 50$ kΩ, $R_C = 5$ kΩ, $V_{CC} = 20$ V, $C_{JE} = 50$ pF, $\tau_B = 1$ μs, and $h_{FE} = 100$. In addition, a resistor $R_2 = 100$ kΩ is connected in series with a 10-V battery (negative terminal up) from base to ground. Carefully sketch and label v_T (the Thévenin voltage of the circuit to the left of the base terminal), v_{BE}, i_B, i_C, and v_{CE}. Consider that v_{in} is a step going from 0 V to 10 V at $t = 0$.

10.2-6 In Figure P10.1-1, $R_1 = 10$ kΩ, $R_C = 200$ Ω, and $V_{CC} = 10$ V. In addition, a diode is connected from base to ground (cathode on top). The diode and the transistor are both silicon devices with $\tau_r = 2$ μs and $C_J = 100$ pF for the diode, and $C_{JE} = 100$ pF, $\tau_B = 1$ μs, and $h_{FE} = 100$ for the BJT. Carefully sketch and label i_{in} (the current from v_{in}), v_B, i_{C1}, and v_{CE1}. Consider that v_{in} is a step going from −20 V to +10 V at $t = 0$.

10.2-7 Find the turn-on time of the circuit discussed in Problem 10.2-4, given that $h_{FE} = 20$, $\tau_B = 800$ ns, $f_T = 20$ MHz, $C_{B'C} = 10$ pF, $C_{JE} = 10$ pF, and v_{in} is a step going from −10 V to +15 V at $t = 0$. As an aid in carrying out this analysis:

1. Consider that $C_{B'C}$ is essentially in parallel with C_{JE} during transistor cutoff since $R_C \ll R_B$.

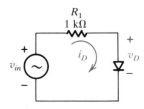

Figure P10.2-1

2. Apply the Miller theorem to $C_{B'C}$ when the transistor enters the active region.
3. Neglect the effects of C_{JE} and $C_{B'C}$ when the transistor is saturated.

10.2-8 Compute the turn-off time of the circuit discussed in Problem 10.2-7 if v_{in} is a step going from 20 V to -5 V at $t = 0$.

10.2-9 In the circuit shown in Figure P10.1-3, $R_D = 20$ kΩ, $V_{DD} = 20$ V, $V_p = -5$ V, $I_{DSS} = 10$ mA, $r_{DS(ON)} \simeq 0$, and a capacitor $C_L = 10$ pF is connected from drain to ground. Carefully sketch and label v_{GS}, i_D, and v_{DS} if the transistor is a depletion-style device and if v_{in} is a step going from -2.5 V to -5.0 V at $t = 0$.

10.2-10 **(a)** Compute the overall turn-on time of the FET in the circuit shown in Figure P10.1-3 if $R_D = 10$ kΩ, $V_{DD} = 20$ V, $C_{gd} = 0$, $C_{gs} = 5$ pF, $V_p = -5$ V, $g_m = 5$ mS, and a resistor $R_{in} = 1$ kΩ is connected in series with v_{in}. Consider that the transistor is a depletion-style device and that v_{in} is a step going from -15 V to 0 V at $t = 0$.
(b) Repeat part (a) if $C_{gd} = 5$ pF. In carrying out this analysis, neglect the effect of C_{gd} when the transistor is cut off since the current flow through it will be small with $R_D \gg R_{in}$.

10.3-1 Discuss the performance of the gating circuit in Figure 10.1-2d if the 5 kΩ base resistors are replaced by shorts, considering that $V_o = 0.65$ V for Q_1 and 0.7 V for Q_2 through Q_{10}.
(a) In particular, explain how this logic circuit works (or fails to work) when V_{in} is a logic 0 (Q_0 cutoff) and then when it is a logic 1 (Q_0 saturated).
(b) Repeat part (a) when the series base resistors are set equal to 500 Ω to reduce current-hogging effects.

10.3-2 For the RTL-style logic gate shown in Figure P10.1-1, $R_1 = 2$ kΩ, $R_C = 4$ kΩ, $V_{CC} = 3.6$ V, and $h_{FE} = 25$. Find the maximum fan-out such that a noise margin of at least 0.33 V is maintained.

10.3-3 Derive an expression for the transfer characteristic (V_o versus V_{in}) for the RTL logic gate in Figure 10.3-2a for a fanout of 10 if the transistor $h_{FE} = 20$.

10.3-4 For the inverter circuit shown in Figure P10.1-1, $R_1 = 4$ kΩ, $R_C = 1$ kΩ, $V_{CC} = 5$ V, and $h_{FE} = 100$. In addition, two silicon diodes are connected in series with R_1 and a resistor $R_2 = 10$ kΩ is shunted from the base to ground.
(a) Calculate NM_H and NM_L.
(b) Find the circuit fan-out if a noise margin of at least 0.5 V is required.

10.3-5 The DTL NAND gate shown in Figure 10.3-5 is to be used to drive a bank of four light-emitting diodes (LEDs). Each LED is connected to the $+5$-V supply (anode on top) through a 500-Ω resistor. Assume that the LED has

a 1.5-V drop across it when it is ON and that at least 3 mA of current flow is required to illuminate it properly.
(a) Determine the minimum transistor h_{FE} needed to drive this four-LED load.
(b) For the conditions in part (a) with V_1 and V_2 both HI, what is the output voltage at V_o and the power dissipated in Q_1?

10.3-6 For the logic gate shown in Figure P10.3-6, all transistors are silicon with $h_{FE} = 100$.
(a) Identify the logic function performed by this circuit.
(b) With $V_B = +5$ V, determine I_{B1} and I_{B2} if $V_A = 0.2$ V.
(c) Repeat part (b) if $V_A = 2.5$ V.

10.3-7 Repeat Example 10.3-2 for the case where the logic gates are DCTL-style devices. In this instance the circuit to be examined is basically that given in Figure 10.3-2c except that the series base resistors are zero and V_{CC} is reduced to 1.2 V. Also consider that the transistor saturation resistances $r_{CE,sat}$ and base spreading resistances $r_{BB'}$ are each 25 Ω.

10.3-8 Consider in Problem 10.3-6 that a 100-pF capacitance is connected from the output of the gate to ground.
(a) Find the fall time assuming zero fan-out, when both v_A and v_B simultaneously go HI.
(b) Find the rise time when v_A goes LO, again assuming zero fan-out.

10.3-9 **(a)** Repeat the analysis of Example 10.3-3c assuming that the diodes are ideal (except for their 0.7-V forward drops).
(b) Repeat part (a) if the 5-kΩ resistors (R_2) in the bases of Q_1 and Q_2 in Figure 10.3-6a are removed.

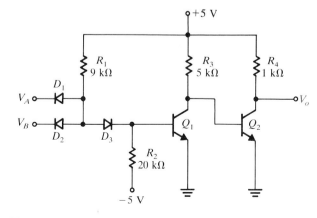

Figure P10.3-6

10.4-1 The transistor circuit shown in Figure P10.1-1 (with $R_1 = 10$ kΩ, $R_C = 1$ kΩ, and $V_{CC} = 5$ V is to drive five TTL logic gates. Given that $h_{FE} = 10$ for the transistor, find V_{in} such that each driven gate sees a LO input logic level.

10.4-2 (a) Sometimes to produce a logic 0 at the input of a logic gate it is convenient to connect it to ground through a resistor. If in Figure 10.4-5a, v_{in} is replaced by a resistor R_A, find the largest value of R_A for which the TTL gate input is still a valid logic 0 (see Figure 10.4-2b).

(b) Sometimes TTL gates directly drive resistive loads. If in Figure 10.4-5a, a resistor R_B is connected from v_o to ground, find the smallest value of R_B for which v_o is still a valid logic 1 with v_{in} a logic 0.

10.4-3 For the circuit shown in Figure 10.4-6b, determine the range of R_L values for which the circuit will perform satisfactorily if operating speed is not a consideration. Assume that the transistor h_{FE}'s are all equal to 20.

10.4-4 For the Schottky transistor circuit shown in Figure 10.4-10d, $R_B = 10$ kΩ, $R_C = 1$ kΩ, and $V_{CC} = 10$ V.

(a) Find the operating point of the circuit both with and without the Schottky diode in place. In analyzing the two forms of this circuit, consider that the diode forward drop is 0.3 V and that the transistor $h_{FE} = 20$.

(b) Repeat part (a) for the case where $R_B = 30$ kΩ.

10.4-5 For the logic gate shown in Figure P10.4-5, all transistors are silicon with $h_{FE} = 20$.

(a) Identify the logic function performed by this circuit.

(b) With $V_B = +5$ V determine I_{B2} if $V_A = 0.2$ V.

(c) Repeat part (b) if $V_A = 2.5$ V.

10.4-6 For the circuit described in Problem 10.4-5, determine the maximum gate fan-out.

10.4-7 A 200-pF load is connected to the output of the logic gate in Problem 10.4-5 and V_B is maintained at $+5$ V.

(a) Assuming that the transistors can switch instantaneously, find the time required for the output voltage to rise to 2.4 V when V_A changes from a LO to a HI.

(b) Find the time required for the output to fall from 5 to 0.2 V when V_A changes from a HI to a LO.

10.4-8 The Low-Power Schottky gate in Figure 10.4-12c drives a 100-pF load. Considering the switching characteristics of the gate to be ideal, find t_{PHL} assuming all transistor h_{FE}'s to be 20.

10.4-9 For the LS gate in Figure 10.4-12c, find the Q points of all the transistors when $V_{in} = 1.0$ V. Assume that the transistor h_{FE}'s are all equal to 20.

10.4-10 Repeat Problem 10.4-9 when $V_{in} = 1.6$ V.

10.4-11 (a) For the TTL logic gate discussed in Section 10.4-5, derive expressions for, and carefully sketch the waveshapes of, v_{C1}, v_{C2}, v_{E2}, i_{C2}, i_{C3}, and v_o when v_{in} goes from a logic 0 to a logic 1. In this analysis consider that all junction capacitances are zero, and that $\tau_B = 20$ ns and $h_{FE} = 20$ for the transistors. Make any reasonable approximations.

(b) Repeat part (a) for the case where v_{in} goes from a logic 1 to a logic 0.

10.4-12 Two totem-pole-style TTL inverters (Figure 10.4-5a) accidentally have their outputs connected together.

(a) Find the output voltage and the current flows between the gates if the input to one inverter is 0 V and the input to the other is 5 V.

(b) Estimate the power dissipated in each gate. The normal power dissipation in this type of logic gate is about 10 mW.

10.4-13 Find the noise margins of the Advanced Schottky logic gate in Figure 10.4-13a.

10.5-1 In the circuit shown in Figure 10.5-1b, Q_1 and Q_2 are depletion-mode FETs with $I_{DSS} = 2$ mA and $V_p = -5$ V. The loads are each 10-kΩ resistors and the level-shifting network consists of a resistor R_1 connected from the drain of Q_1 to the gate of Q_2 and a resistor $R_2 = 100$ kΩ connected in series with a battery V_{EE} (negative on top) connected to ground. Select the resistor R_1 and the battery V_{EE} so that the input and output logic levels are compatible, and in particular so that V_{G2} will change from 0 to -5 V when V_{D1} switches from $+10$ V to 0.

10.5-2 For the circuit discussed in Example 10.5-1, let $C_L = 0$ and connect a 1-kΩ resistor in between v_{in} and C_{in}. Sketch v_{in}, v_{G1}, i_{D1}, and i_{D2}, and find t_{PLH} for this modified circuit using the same transistor parameters as in the original example.

10.5-3 Repeat Example 10.5-3 for the case where v_{in} switches from $+8$ V to 0.

10.5-4 Repeat the analysis of Example 10.5-4 considering that the transistor is initially in the constant-current region.

10.5-5 For the circuit shown in Figure 10.5-2b, V_{DD} is changed to 20 V, and $V_T = 5$ V and $I_{D2T} = 15$ mA for Q_1 and Q_2. In addition, a 100-pF capacitor C_L is connected from V_o to ground.

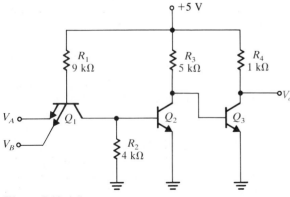

Figure P10.4-5

(a) Find $v_o(0-)$ if v_{in} has been zero for a long time prior to $t = 0$.

(b) Carefully sketch and label $v_o(t)$ for all $t > 0$ if v_{in} changes to 15 V at $t = 0$.

10.5-6 Repeat Problem 10.5-5 for the case where $v_{in} = 15$ V prior to $t = 0$ and then switches to zero volts at $t = 0$.

10.5-7 Sketch the transfer characteristic for the CMOS inverter in Figure 10.5-13a for the case where $V_{DD} = 15$ V, $I_{D2T1} = I_{D2T2} = 1$ mA, $V_{T1} = 1$ V, and $V_{T2} = -3$ V.

10.5-8 Repeat Example 10.5-1 for the constant-current inverter shown in Figure 10.5-2b. In carrying out this analysis, do not attempt to model Q_2 by an equivalent fixed resistance. This solution tends to give rather inaccurate results because the effective resistance of Q_2 goes to infinity as v_o approaches $(V_{DD} - V_T)$. As an alternative method, solve this problem mathematically by considering that equation (10.5-1b) may be used to represent Q_2 all during the charging of the load capacitance.

10.5-9 (a) Compute the transition voltage for an eight-input unbuffered NOR gate when $V_1 = V_{in}$ and $V_2 = V_3 = \cdots = V_8 = 0$ V. The gate structure is the same as that given in Figure 10.5-13c, with $V_{DD} = 5$ V, and with $|V_T| = 1$ V and $I_{DSS} = 1$ mA for both the p- and n-channel devices.

(b) Repeat part (a) when $V_1 = V_2 = \cdots = V_8 = V_{in}$.

10.5-10 Carefully sketch and label v_C, v_{o1}, and v_{o2} for the CMOS inverter circuit shown in Figure P10.5-10. Consider that the gain of each logic gate is -5 when it is in the active region and that its threshold is centered about $V_{DD}/2$.

10.6-1 For the circuit shown in Figure P10.6-1, the transistors are all silicon with $h_{FE} = 100$. Find V_o if $V_{in} = -2.0$ V.

10.6-2 Compute the fan-out for the simplified ECL logic gate shown in Figure P10.6-1. Assume that all transistors have $h_{FE} = 20$ and that a noise immunity of 0.3 V for both HIs and LOs is required.

10.6-3 Let $V_B = -1.7$ V and $V_A = V_{in}$ for the ECL logic gate shown in Figure 10.6-3.

(a) Carefully sketch the transfer characteristic (V_{o1} versus V_{in}) for values of V_{in} between -2 V and zero.

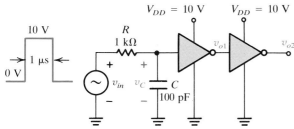

Figure P10.5-10

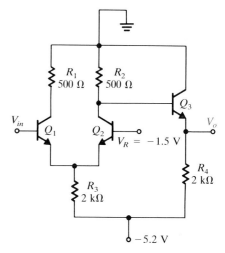

Figure P10.6-1

(b) Repeat part (a) for V_{o2} versus V_{in}.

10.6-4 Design a circuit to interface ECL to TTL logic. Consider that $V_{OH} = -0.7$ V and $V_{OL} = -1.71$ V for the ECL gate, and that $V_{IH} = 2.0$ V and $V_{IL} = 0.8$ V for the TTL gate.

10.6-5 (a) Using Figure 10.6-5b together with the circuit layout for the ECL logic gate given in Figure 10.6-3, express v_{o2} in terms of v_n for the case where point A in Figure 10.6-5b is grounded and v_A and v_B are at a logic 1 and then a logic 0.

(b) Repeat part (a) for the case where point B in Figure 10.6-5b is grounded.

10.6-6 The logic noise in ECL circuits is small not only because of the use of negative power supplies but also because there is a minimal change in power supply current when the output changes from a 0 to a 1 and then back.

(a) Verify this statement by sketching the power supply current versus time (neglecting any transients) when the input to v_A in Figure 10.6-3 is a square wave varying between -1.7 and -0.7 V, and $v_B = -1.7$ V.

(b) For comparison purposes, repeat part (a) for the case where a 0- to 5-V square wave is applied to the input of the TTL logic gate in Figure 10.4-5a.

10.6-7 For the ECL logic gate in Figure 10.6-6a, consider that all the transistor parasitic capacitances are zero and also that $h_{FE} = h_{fe} = 100$, $h_{ie} = 250$ Ω, and $C_L = 100$ pF. If v_{in} is constant at -1.7 V for $t < 0$, switches to -0.7 V for 50 ns, and then back to -1.7 V thereafter, sketch i_{C1}, v_{C1}, and v_o.

10.6-8 For the logic gate in Figure 10.6-6a, consider that $C_{B'C} = C_{B'E} = 1$ pF, $C_{CS} = 2$ pF, $r_{B'E} = 250$ Ω, $h_{FE} = 30$, and $R_{in} = 1$ kΩ. In addition, assume that v_{in} is a 20-MHz square wave that varies between -1.7 and -0.7 V. Carefully sketch and label v_{B1}, i_{B1}, i_{C1}, v_{C1}, and v_o.

10.7-1 The logic gates shown in Figure 10.7-2b are CMOS and $v_1(t) = 5u(t)$. Carefully sketch and label $v_2(t)$ and $v_3(t)$ if a capacitive load C_L is connected from the v_3 terminal to ground.

10.7-2 An ideal diode (cathode down) is connected in parallel with a load resistance Z_o across the output terminals of a transmission line of length x_o with a characteristic impedance Z_o and a propagation delay T_d. The source impedance is equal to $2Z_o$.
(a) Sketch $v(x_o/2, t)$ in the interval from $t = 0$ to $t = 4T_d$ if a narrow pulse of height E is applied.
(b) Repeat part (a) if the input is a step of height E.

10.7-3 For the BJT circuit shown in Figure P10.1-1, $R_1 = 1\ k\Omega$, $R_C = 2\ k\Omega$, $V_{CC} = 10\ V$, $h_{FE} = 100$, and v_{in} is connected to R_1 through a transmission line with $Z_o = 250\ \Omega$ and a propagation delay T_d. Considering the transistor base–emitter diode to be ideal, sketch the transmission line output voltage, the base–emitter voltage v_{BE}, and v_o if v_{in} is normally 0 V, goes to 10 V for T seconds (with $T \ll T_d$), and then back to 0 V thereafter.

10.7-4 For the circuit shown in Figure P10.7-4, G_1, G_2, and G_3 are 7400-style TTL logic gates, and the line is 20 ft long and has a $T_o = 1.5$ ns/ft.
(a) Carefully sketch and label v_1 through v_5 when v_1 goes from 0 to 5 V if the gates have no input protection diodes.
(b) Repeat part (a) if the gates are protected with input diodes.

10.7-5 One 7400-style logic gate G_1 drives another (G_2)

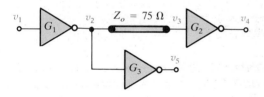

Figure P10.7-4

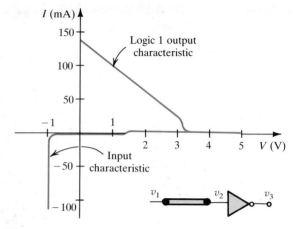

Figure P10.7-6

through a 100-Ω transmission line. The gates have no input protection diodes. Sketch the transmission line and G_2 output voltages in the interval $0 < t < 8T_d$ when the output of G_1 goes from a logic 1 to a logic 0.

10.7-6 The curves in Figure P10.7-6 represent the volt-ampere characteristics of a typical 7400-style Advanced Schottky logic gate. If two of these gates are interconnected with a 2-ft length of 30-Ω twisted pair cable, sketch v_1, v_2, and v_3 when v_1 makes a 0-to-1 logic transition.

10.7-7 Review Example 10.7-3 and discuss whether the transition from 0 to 1 in the gate output of G_0 imposes any restriction on the selection of R_E.

10.7-8 G_0 and G_1 in the circuit shown in Figure 10.7-8c are assumed to be ECL logic gates and R_S is changed to 10 Ω. Sketch v_1, v_2, v_3, i_2 and i_3 (the current into G_1) when the output of G_0 makes a 0-to-1 logic transition.

10.7-9 Two ECL logic gates are interconnected by a transmission line with a length x_o and a delay time unit length equal to T_o. Prove graphically that the undershoot at input to the driven gate will be less than 15% when $x_o < (t_r/2.7T_o)$ in.

CHAPTER 11

Digital Timing and Memory Circuits

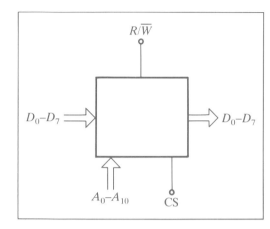

11.1 INTRODUCTION TO TIMING, PULSE SHAPING, AND SWITCH DEBOUNCING CIRCUITS

Invariably, the input applied to a digital system needs to be some type of digital logic signal that varies in one fashion or another between the logic 0 and logic 1 voltage levels. At some points in the system periodic waveforms may be required, and in this instance a digital clock circuit or digital oscillator will be needed. In Section 11.2 we discuss the design of three different types of digital clocks: astable multivibrators, crystal-controlled oscillators, and propagation delay oscillators.

For some applications it may also be desirable to alter the duration of the input signal being applied to the digital system, or to delay it by a specified amount of time. Basically, there are two different techniques for accomplishing these tasks. The first makes use of *RC* networks and standard digital logic gates connected together in an open-loop fashion, and the second approach basically uses the same circuit elements, but this time connects them together as part of a digital feedback loop. The latter technique results in a class of pulse-shaping circuits that may loosely be called monostable multivibrators and these are discussed in some detail in Section 11.3.

Sometimes the input signal to a digital system is mechanical rather than electronic in nature and involves the opening or closing of a switch. At first glance it might appear to be rather easy to convert this type of input information into an equivalent digital electronic signal. However, as we shall see when we discuss this subject more fully, the contact bounce associated with the switch operation can cause significant problems. In Section 11.4 we discuss a technique for eliminating these contact bounce problems that makes use of Schmitt triggers.

11.2 DIGITAL CLOCK CIRCUITS

Nearly all digital systems are driven by some type of clock or digital logic oscillator which generally produces a periodic rectangular output that varies between the logic 0 and logic 1 voltage levels. The purpose of the clock is to ensure that all flip-flops in the system change state at the same time, so that their new state will depend only on the current inputs to the system and the present state of the system, not on the new values assumed by the flip-flops as they change states. There are three basic ways to construct digital system clocks. The operating principles of the first two (the astable multivibrator and the crystal-controlled oscillator) are similar to the nonsinusoidal oscillators that we discussed earlier in Section 8.7, and the third (the propagation delay oscillator), although also similar, at first appears to represent an entirely new approach.

The astable multivibrator is one of a class of three different kinds of multivibrator circuits. The other two are known as bistable multivibrators (or flip-flops) and monostable multivibrators (or one-shots). The prefix used in each of these names denotes the number of stable states (or stable operating points) possessed by the circuit. Thus a bistable multivibrator has two stable operating points or stable states, a monostable has one, and an astable none.

The typical multivibrator is basically a positive feedback circuit containing two inverters, and in the bistable multivibrator the inverters are directly coupled to one another as illustrated in Figure 11.2-1a. If we break the feedback loop in this circuit by opening the switch S, the transfer characteristic relating v_o to v_{in} has the typical form illustrated in Figure 11.2-1b. Furthermore, if we neglect the loading of G_1 on the output of G_2, this transfer characteristic also

Figure 11.2-1

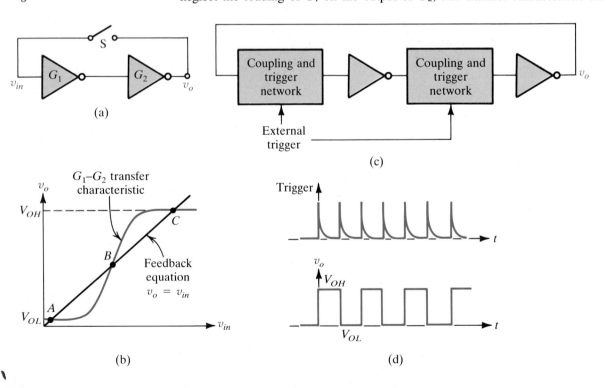

applies when S is closed. The circuit operating point(s) with the switch in the closed position may now be determined graphically by sketching the feedback circuit equation, $v_o = v_{in}$, on the same set of axes as the transfer characteristics and noting the intersection points (Figure 11.2-1b). As is apparent from this figure, the circuit has three possible operating points, A, B, and C. At point B both inverters are operating in their active regions, and based on our previous analysis of these types of circuits in Chapter 8, we know that this kind of positive feedback system will have poles in the right-hand plane. As a result, this operating point will be unstable and the output voltages will grow exponentially until one of the nonlinear regions is hit and the effective loop gain is reduced to less than 1. Thus, after the initial transient settles down, the final operating point of this circuit will be either at point A or at point C, depending on the initial conditions.

Consider next the slightly more generalized multivibrator circuit illustrated in Figure 11.2-1c. Here we have added two additional blocks to the circuit, whose function will now be clarified. Let's begin by explaining the purpose of the trigger signals, and let's relate their use to the bistable multivibrator circuit that we have just been discussing. For this type of multivibrator, as we have already mentioned, direct coupling can be used between the gates, and initially this circuit will be found in either state A or state C, depending on the initial conditions. Let's assume that it is in state A and consider what happens to this circuit when an external trigger is applied to both gate inputs.

Basically, the function of the trigger pulse is to cause the flip-flop's inverters to enter their active regions. Once this occurs, then as per our discussion of positive feedback in Section 8.7, the circuit will automatically switch to the opposite stable state at point C. The application of a second trigger pulse will cause it to switch back to state A, and the use of a periodic train of trigger pulses will cause the circuit to flip back and forth between states A and C, as illustrated in Figure 11.2-1d. Specific techniques for the triggering of flip-flops and design procedures for constructing different types of bistable circuits are discussed in Section 11.6.

By adding ac coupling between the inverter stages, the generalized multivibrator circuit being discussed can be reconfigured into either the monostable or astable modes, depending on the type of dc biasing used. To illustrate this point, consider the circuit presented in Figure 11.2-2a. In the steady state we can assume that the capacitor is an open circuit and therefore that the output of G_2 will be LO since the input to this gate is pulled up to $+V_{CC}$ through the resistor R. If the input trigger signal is normally LO, then with LOs at both of its inputs, the output of G_1 will normally be HI. Thus the steady-state operating point of this particular multivibrator will always be located at point A in Figure 11.2-1b, and because this circuit has only one stable operating point, it is termed a monostable multivibrator. If by properly triggering this circuit, it can be made to enter the active region or can be driven hard enough so that the output of G_1 goes LO and G_2 goes HI, then the circuit operating point will move to point C in Figure 11.2-1b, and it will remain there for a time determined by the properties of the RC coupling network (Figure 11.2-2b). The performance characteristics of several different types of monostable multivibrators are examined carefully in Section 11.3.

By making a slight revision in the form of the monostable multivibrator

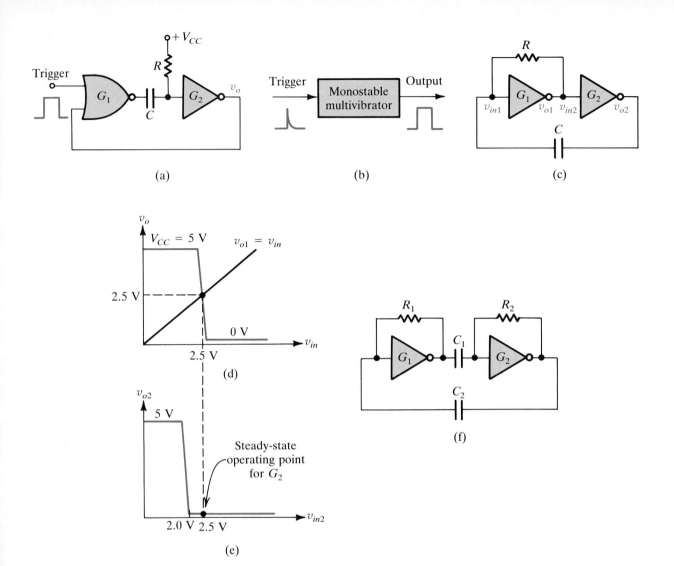

Figure 11.2-2

shown in Figure 11.2-2a, we can convert it into an astable multivibrator. This is accomplished by changing the biasing network as shown in part (c) of the figure so that the circuit is biased in the active region in the steady state. For the case of a TTL logic gate, this can be accomplished with a resistor having a value between 0 Ω and about 2.4 kΩ, while for a CMOS logic gate, almost any value of R from about 1 kΩ to several megohms will work quite nicely.

To see why a circuit of this form will, at least in theory, place G_1 and G_2 in their active regions, for simplicity, let's consider the case where the logic gates are CMOS. Figure 11.2-2d illustrates the transfer characteristic of a typical CMOS inverter. Because the input impedance of these gates is virtually infinite, almost no current flows through R and $v_{o1} = v_{in1}$ for G_1. As a result, the operating point of this logic gate is located at about $v_{o1} = v_{in1} = 2.5$ V. If G_2 is identical to G_1, then with $v_{in2} = v_{o1} = 2.5$ V, v_{o2} will also be at 2.5 V. Thus, in the absence of any oscillation, both inverters would be biased at the center of their active regions. Of course, due to the positive feedback, this circuit will have poles in the right-hand plane, and v_{o2}, for example, will rapidly switch to either 0 V or to +5 V. Once in this state, as we will show in the examples to follow, this circuit will recover on its own back into the active region, and when this happens it will rapidly switch to the opposite state. Because this process will occur over and over, in the steady state, the output voltages at both v_{o1} and v_{o2} will be square waves. Furthermore, the frequency of these signals will depend on the values of R and C in the circuit.

As a practical matter, if the transfer characteristics of the two inverters are significantly different from one another, as they might be, for example, if G_1 and G_2 were located on different ICs, the astable multivibrator shown in Figure 11.2-2c might not self-start. To understand how this could occur, consider what would happen if G_2 had the characteristic illustrated in part (e) of the figure. In this instance, with G_1 biased as shown in Figure 11.2-2d, v_{o1} would still be 2.5 V, and this would also be equal to v_{in2}. However, because the threshold voltage of G_2 is so much less than that of G_1, G_2 would saturate and v_{o2} would be about 0 V. More important, in this region the small-signal gain of G_2 would be nearly zero, so that the overall loop gain of the circuit would be less than 1. With its poles in the left-hand plane, this circuit would not oscillate and it would remain hung up at $v_{o1} = 2.5$ V and $v_{o2} = 0$ V forever. This situation can be avoided by biasing each gate individually as shown in Figure 11.2-2f. While this circuit requires more components, it is guaranteed to oscillate. In addition, it provides for greater flexibility in adjusting the output waveshape since one half of the output waveform is affected by the R_1–C_1 time constant, and the other half by the R_2–C_2 time constant.

EXAMPLE 11.2-1

The astable multivibrator originally illustrated in Figure 11.2-2c is redrawn in Figure 11.2-3a as it is usually seen in the literature. Here two-input NAND gates have been made to look like inverters by connecting their inputs together.

a. Assuming that the power supply voltage $V_{DD} = 5$ V and that the CMOS gates are ideal with threshold voltages at $V_{DD}/2$, carefully sketch and label the steady-state voltages v_1, v_2, and v_3.
b. Repeat the analysis in part (a), but this time consider that the logic gates have input protection circuits of the type illustrated in Figure 11.2-3b in which the diodes may be assumed to be ideal.

SOLUTION

To begin the analysis of this problem, consider initially that the output at G_1 has just gone LO and correspondingly that the output of G_2 has just gone HI. As a result, the equivalent circuit for $t > 0$ has the form illustrated in Figure 11.2-3c. This equivalent circuit will be valid as long as $v_1(0+)$ is greater than $V_{DD}/2$ or 2.5 V. Assuming this to be the case initially, the voltage at the input to G_1 will be given by the expression

$$v_1(t) = V_m e^{-t/\tau} \tag{11.2-1a}$$

where

$$\tau = RC \tag{11.2-1b}$$

and V_m is yet to be determined.

Once v_1 drops down to $V_{DD}/2$, v_2 takes on the same value and G_1 and G_2 enter the active region simultaneously. For later use we note that the voltage v_C

11.2 / Digital Clock Circuits

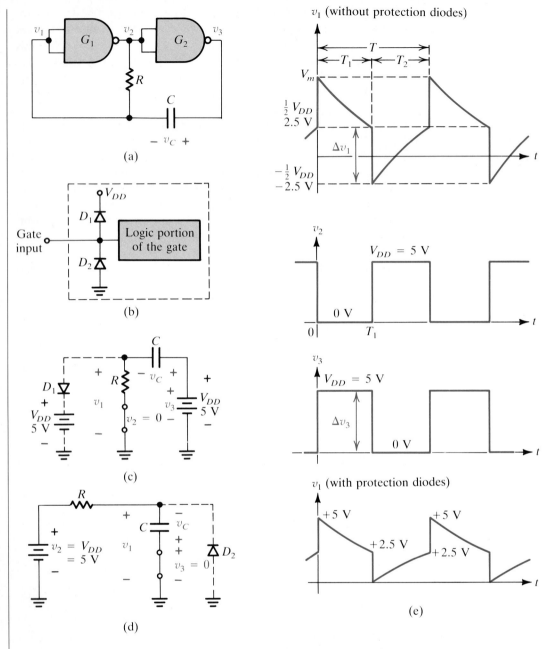

Figure 11.2-3
Example 11.2-1.

across the capacitor is also now equal to $V_{DD}/2$ or 2.5 V. At this point the circuit switches to the opposite state with the output of G_1 going to a logic 1 and that of G_2 to a logic 0. If we neglect the parasitic capacitances, this switching will take place instantaneously.

Besides understanding this switching process in terms of positive feedback, differential equations and poles in the right-hand plane, the switching can also be explained physically by inspection of the circuit as follows. Once v_1 drops to about 2.5 V, both G_1 and G_2 enter the active region. Because the small-signal gain of each of these gates is large and negative, a small decrease in v_1

produces a large increase in v_2. This voltage change produces an even greater decrease in v_3. If this change occurs rapidly enough, the capacitor C will not have any time to charge or discharge so that it will look like a short to this changing voltage. As a result, the drop in v_3 will directly couple through into v_1, causing it to decrease even further. These changes continue to propagate around the loop until eventually v_1 is a valid logic LO, v_2 is HI, and v_3 is LO. The waveshapes for v_1, v_2, and v_3 are illustrated in Figure 11.2-3e.

The size of the change in v_1 that occurs when the circuit switches at $t = T_1$ may be found by examining the equivalent circuit shown in Figure 11.2-3d. Clearly from this figure, $v_1(T_1+) = -v_C(T_1+)$, but because the voltage across the capacitor cannot change instantaneously, it is also equal to $-v_C(T_1-)$ or $-V_{DD}/2 = 2.5$ V. This result could have been obtained in a somewhat simpler fashion by noting that the change in v_1 (Δv_1 in Figure 11.2-3e) must be equal to the change in v_3 (or Δv_3 in the same figure) if the voltage across the capacitor remains constant during the switching.

For $t > T_1$ the equivalent circuit in Figure 11.2-3d still applies and the voltage v_1 therefore increases exponentially toward $+V_{DD}$ in accordance with the equation

$$v_1(t) = -\frac{V_{DD}}{2}e^{-(t-T_1)/\tau} + V_{DD}(1 - e^{-(t-T_1)/\tau}) = V_{DD} - \frac{3}{2}V_{DD}e^{-(t-T_1)/\tau} \qquad (11.2-2)$$

This charging continues until v_1 reaches 2.5 V. At this point the circuit again enters the active region, switching to the opposite state occurs, and the equivalent circuit in Figure 11.2-3c again applies. Because v_3 increases by V_{DD} volts during the switching, v_1 will increase by the same amount from its value at $V_{DD}/2$ just prior to the switching. We also note this voltage is the same as that at $t = 0+$, so that we have

$$V_m = v_1(0+) = v_1(T+) = \frac{V_{DD}}{2} + V_{DD} = \frac{3}{2}V_{DD} = 7.5 \text{ V} \qquad (11.2-3)$$

This switching sequence repeats over and over again, so that the steady-state voltages at v_2 and v_3 are square waves. Because the gate input protection diodes have a significant effect on the performance of this circuit, we will wait until their effects are included in the analysis before we calculate the relationship between the oscillation frequency and the circuit parameters.

The effect of the input protection diodes may be taken into account by including them as shown in the dashed portions of the equivalent circuits in Figure 11.2-3c and d. In each case, for simplicity, only the diode that will be conducting current is shown. Let's begin this analysis at $t = 0$ again at the point where the output of G_1 is ready to go LO and that of G_2 is just on the verge of going HI. This will occur when v_1 just reaches 2.5 V or alternatively, when $v_C = -2.5$ V since $v_3 = 5$ V here. After both gates switch, v_2 will equal zero volts, v_3 will equal $+5$ V, and the equivalent circuit in Figure 11.2-3c will apply. Following this equivalent circuit, v_1 will attempt to jump up to 7.5 V. However, as soon as v_1 tries to go above 5 V, D_1 will turn ON and this will discharge v_C to zero volts almost instantaneously. Therefore, for all practical purposes we can consider that the equivalent circuit in Figure 11.2-3c is valid for $t > 0$ but with $v_C = 0$, so that

$$v_1(t) = V_{DD}e^{-t/\tau} \qquad (11.2-4)$$

In a similar fashion, during the second half of the cycle, the equivalent circuit given in Figure 11.2-3d applies with $v_C(T_1-) = +2.5$ V. Here, too, as soon as the circuit switches, the voltage on the capacitor will be immediately discharged to zero by D_2, so that $v_1(T_1+)$ will be zero and not -2.5 V as indicated previously. Using this fact, the voltage at the input of G_1 during the time interval from $t = T_1$ to $t = T$ is readily seen to be

$$v_1(t) = V_{DD}[1 - e^{-(t-T_1)/\tau}] \tag{11.2-5}$$

Because of the waveform symmetry it is relatively apparent that $T_2 = T_1$, so that $T = 2T_1$. A simple expression for T_1 and f may be obtained directly from equation (11.2-4) by noting that T_1 is the time required for v_1 to reach 2.5 V, so that

$$2.5 = 5e^{-T_1/\tau} \tag{11.2-6a}$$

or
$$T_1 = \tau \ln 2 = 0.693\tau \tag{11.2-6b}$$

and
$$f = \frac{1}{T} = \frac{1}{2T_1} \simeq \frac{1}{1.4RC} \tag{11.2-6c}$$

In Example 11.2-1 we developed an expression for the oscillation frequency that depended on the R and C values of the circuit, but nothing was said about the allowed values of these components or about the maximum frequency of oscillation that was attainable. Of course, following equation (11.2-6c), the highest oscillation frequency will occur when the smallest possible values are used for R and C. In selecting a capacitance value for C, it should be chosen so that it is much larger than the circuit parasitic capacitances. Because the input and output capacitances of most logic gates are on the order of 5 to 30 pF, a minimum value for C of about 300 pF seems reasonable for both the CMOS and TTL versions of this circuit. The smallest resistance value to use depends on the gate output impedance, and to prevent significant loading on the gate outputs, minimum resistance values of 2 kΩ and 200 Ω are suggested for CMOS and TTL circuits, respectively. Substituting these values into equation (11.2-6c), a maximum oscillation frequency of about 1 MHz can be achieved with CMOS logic gates, while this number increases to about 10 MHz with TTL devices.

A second rather interesting type of astable multivibrator uses a digital-style Schmitt trigger and employs a circuit that is nearly identical to that given in Figure 8.7-4a. Basically, digital Schmitt triggers are quite similar to the op-amp-style Schmitt triggers that we discussed in Chapter 7. However, they do differ in several important ways. First, the hysteresis characteristic of a digital Schmitt trigger IC is fixed by internal components that are part of the integrated circuit. These levels are chosen to be within the logic range of the logic family of which the device is a part. Second, the input loading characteristics and output drive capabilities of the digital Schmitt trigger are designed to be similar to those of other logic gates within that same family.

Figure 11.2-4 illustrates the hysteresis curve of a 7414 TTL Schmitt trigger inverter. Notice that the output voltages, V_{OH} and V_{OL}, are the same as those found in an ordinary TTL logic gate, and also the threshold voltages V_{T+} and V_{T-} are both within the transition region of a standard TTL gate. Example

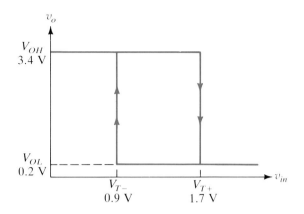

Figure 11.2-4

11.2-2 that follows provides an illustration of how a Schmitt trigger can be used as an astable multivibrator and Problem 11.2-3 at the end of the chapter illustrates its use in the monostable mode.

EXAMPLE 11.2-2

The circuit shown in Figure 11.2-5a is an astable multivibrator.

a. Sketch the voltages v_{in} and v_o.
b. Find the oscillation frequency.
c. If the threshold voltages can vary by 20%, estimate the maximum shift in oscillation frequency that can occur.

SOLUTION

In analyzing this circuit, let's assume that steady-state oscillations have already been achieved and that v_o has just switched from a logic 0 to a logic 1 as illustrated in Figure 11.2-5d. Following the hysteresis curve in Figure 11.2-4, this switching will occur when v_{in} just drops below V_{T-} or at about 0.9 V. Arbitrarily defining the time at which this occurs as $t = 0$, for $t > 0$ the equivalent

Figure 11.2-5
Example 11.2-2.

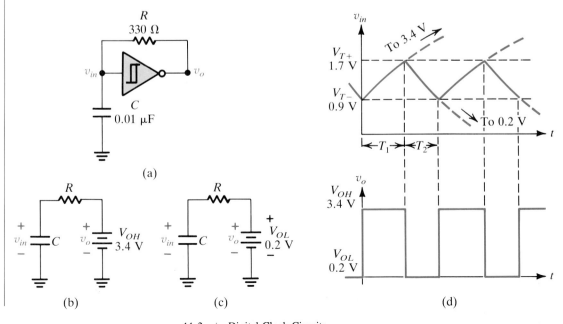

(a)

(b)

(c)

(d)

circuit given in Figure 11.2-5b applies so that the voltage across the capacitor may be expressed as

$$v_{in}(t) = V_{T-}e^{-t/\tau} + V_{OH}(1 - e^{-t/\tau}) \qquad (11.2\text{-}7)$$

This exponential growth continues until v_{in} reaches V_{T+}, and at this point v_o switches to a logic 0. The time, T_1, required for this to occur is found by setting equation (11.2-7) equal to V_{T+} and solving for t, which yields

$$T_1 = \tau \ln \frac{V_{OH} - V_{T-}}{V_{OH} - V_{T+}} \qquad (11.2\text{-}8)$$

Substituting the data for this problem into equation (11.2-8), we find that $T_1 = 1.27 \ \mu s$.

For $t > T_1$ the equivalent circuit illustrated in Figure 11.2-5c applies so that the voltage across C may now be written as

$$v_{in}(t) = V_{T+}e^{-t/\tau} + V_{OL}(1 - e^{-t/\tau}) \qquad (11.2\text{-}9)$$

where $t = 0$ has been redefined as the point where v_o switches from logic 1 to logic 0. This discharge portion of the cycle continues until v_{in} drops down to V_{T-}, and setting equation (11.2-9) equal to this voltage and solving for t, we find that this time interval may be expressed as

$$T_2 = \tau \ln \frac{V_{T+} - V_{OL}}{V_{T-} - V_{OL}} \qquad (11.2\text{-}10)$$

Substituting in the data for this problem, we find that $T_2 = 2.52 \ \mu s$, so that the overall period and frequency of oscillation are 3.79 μs and 264 kHz, respectively.

As V_{T+} and V_{T-} vary from their nominal values, the frequency of oscillation will change. The worst-case frequency shift occurs when V_{T+} decreases by 20% to 1.36 V, while V_{T-} at the same time increases by 20% to 1.08 V. Substituting these new values into equations (11.2-8) and (11.2-10), we find that the frequency of oscillation is now 748 kHz, which represents an increase of about 183% above the nominal oscillation frequency. Typically, the shift in frequency will not be nearly as large as that just calculated since V_{T+} and V_{T-} will tend to track one another. When the analysis is repeated taking this fact into account, the worst-case frequency shift is reduced to about 10%.

Example 11.2-2 clearly indicates that Schmitt-trigger-style oscillators cannot be used when a specific frequency of oscillation is required. Unfortunately, too, although we did not explicitly discuss this fact, gate-style oscillators also exhibit similar inaccuracies due to their ill-defined threshold voltages. For example, in the CMOS oscillator that we analyzed in Example 11.2-1, we assumed that the gate threshold voltages were precisely equal to $0.5V_{DD}$. However, in actual CMOS logic gates these thresholds can be anywhere between $0.33V_{DD}$ and $0.66V_{DD}$. This variation in V_T also gives rise to an uncertainty in the oscillation frequency of about 10%. Similar uncertainties exist for TTL astables; consequently, astable multivibrators are used only where the required oscillation frequency is not that critical.

For digital applications needing greater control over the frequency of oscillation, either a sinusoidal tuned circuit oscillator followed by a Schmitt trigger or a tuned circuit gate-style oscillator are recommended. With these types of oscillators, frequency accuracies on the order of 1% can be achieved. By replacing the tuned circuit with a piezoelectric crystal, the accuracy in the frequency of oscillation can be increased to about 0.02% or to 200 parts per million. When piezoelectric crystals were first introduced, they were so expensive that their use in an oscillator circuit had to be clearly justified. Today, however, their costs are comparable to those of the tuned circuits that they replace, so they are almost always used whenever accurate control over the frequency of oscillation is needed.

Crystal-controlled oscillators can be built using digital logic gates. The basic concept involved is quite similar to that discussed in Chapter 8 regarding the design of regular sinusoidal oscillators. Figure 11.2-6 illustrates two of the more popular types of crystal-controlled digital logic gate oscillators. The one shown in part (a) of the figure uses only a single inverter, and therefore for this circuit to oscillate, an additional 180° of phase shift will be needed in the feedback circuit. This phase shift is obtained by using a π network consisting of C_2, C_3, and the crystal that behaves inductively. At the frequency of oscillation, as derived in Section 8.6 [equation (8.6-14)], the gain of the π network is

$$\frac{v_{in}}{v_o} = -\frac{C_2}{C_3} \qquad (11.2\text{-}11)$$

The digital clock circuit shown in Figure 11.2-6b uses two inverters, and as such, it oscillates when the feedback network has zero phase shift at the series resonant frequency of the crystal. In this circuit the feedback resistors are chosen to place the gates in the active region. For the low-power Schottky gates used in this oscillator, resistors equal to 2.7 kΩ will place the quiescent gate input voltages at about 1.4 V. Furthermore, using the LS transfer characteristics given in Figure 10.4-12b, the overall gain of this two gate amplifier is

Figure 11.2-6
Crystal-controlled gate-style digital oscillators.

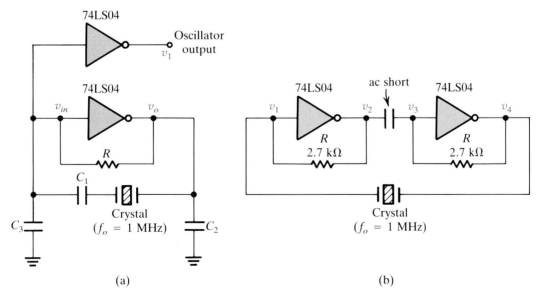

(a)

(b)

seen to be about $(-9.5)^2$ or 90. At the frequency of oscillation, f_o, the feedback factor is unity because the crystal looks like a short circuit. Therefore, the loop gain is much greater than one at the oscillation frequency. Consequently, the inverter output voltages will be square waves that are 180° out of phase from one another. In addition, because the feedback network acts like a band-pass filter at the fundamental frequency, the voltage at the input of the first inverter will ideally be a sine wave whose amplitude is approximately equal to the fundamental component of the square wave at v_4. Thus the peak amplitude of the sine wave at v_1 is about $(4/\pi)(1.6 \text{ V}) = 2.0$ V.

Figure 11.2-7 shows an oscillator whose structure is different from those previously examined in that the feedback loop contains a delay element. The loop gain of this circuit may be written as

$$\beta(s)A_{OL}(s) = -Ae^{-sT_d} \tag{11.2-12a}$$

or
$$\beta(jf)A_{OL}(jf) = -Ae^{-j2\pi fT_d} = Ae^{-j(2\pi fT_d + \pi)} \tag{11.2-12b}$$

At $f_0 = 1/2T_d$ we notice that

$$\beta(jf_0)A_{OL}(jf_0) = Ae^{-j2\pi} = A \tag{11.2-13}$$

so that a sinusoidal signal at this frequency injected at v_1 will return in phase at v_1'. As the reader will no doubt recall from Chapter 8, this result implies that this circuit can oscillate at the frequency f_0 if A is made greater than 1. However, because equation (11.2-13) is valid not just for f_0, but also for all of its odd harmonics, the oscillation, rather than simply being a clipped sine wave, will be a square wave at the frequency f_0.

To gain a better understanding of the operation of this circuit, let's examine its behavior from a slightly different viewpoint. If we assume that $v_1(t)$ is a square wave having a period T, then $v_2(t)$ may be written as

$$v_2(t) = \overline{v_1(t)} = v_1\left(t - \frac{T}{2}\right) \tag{11.2-14}$$

since inverting a square wave is equivalent to delaying it by 180° or $T/2$ seconds. Applying this signal to the input of the delay element, we also find that

$$v_1'(t) = v_2(t - T_d) = v_1\left(t - T_d - \frac{T}{2}\right) \tag{11.2-15}$$

Since $v_1'(t)$ must be equal to $v_1(t)$, it therefore follows that

$$-T_d - \frac{T}{2} = -T \tag{11.2-16a}$$

so that the oscillation period may be written as

$$T = 2T_d \tag{11.2-16b}$$

For completeness it is useful to point out that equation (11.2-15) can actually be satisfied for any square waves with periods

$$T = \frac{2T_d}{2n - 1} \tag{11.2-16c}$$

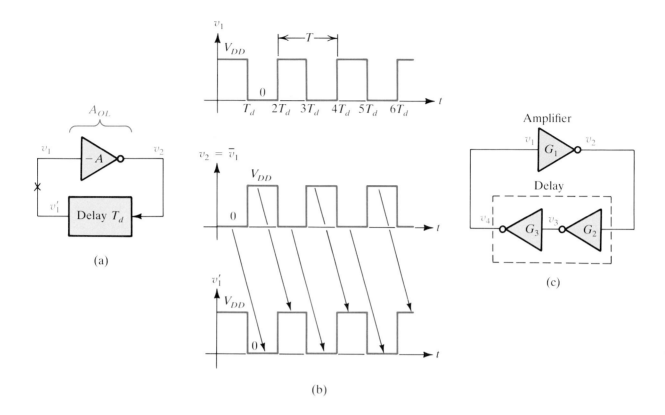

Figure 11.2-7

where $n = 1, 2, 3$, and so on. However, the circuit initial conditions will always be such that only the $n = 1$ oscillation at $T = 2T_d$ will occur.

The steady-state waveshapes for this oscillator are sketched in Figure 11.2-7b. The arrows between the v_2 and v_1' waveshapes were drawn to emphasize the fact that the v_1' waveshape is the same as that of v_2, but just delayed by T_d or $T/2$ seconds.

One of the best ways to produce small delays such as the one used in the oscillator circuit just discussed is to employ buffer-style logic gates or an even number of inverting-style logic gates. This concept is illustrated by the three-inverter oscillator circuit shown in Figure 11.2-7c, which is known as a "ring oscillator." In this circuit, G_1, for example, may be thought of as the inverting-style amplifier, and G_2 and G_3 as the noninverting delay element. Of course, in this circuit both the amplification and the delay are actually distributed among all three of the inverters. If a careful analysis of this circuit is carried out (Problem 11.2-3), it is not too difficult to show that the frequency of oscillation is $1/6t_p$, where t_p is the inverter gate propagation delay.

Exercises

11.2-1 Develop an expression for the frequency of oscillation of the circuit in Figure 11.2-3a if the gates have no input protection diodes. **Answer** $1/(2.2RC)$

11.2-2 Find the frequency of oscillation of the circuit in Figure 11.2-3a if a resistor R_1 is connected in series with the input to gate G_1. Assume that the gates have ideal input protection diodes and the $R_1 = 2R = 1$ MΩ and $C = 1000$ pF. *Answer* 1.88 kHz

11.2-3 For the circuit shown in Figure 11.2-7c, G_1 and G_2 have $t_p = 20$ ns and $t_p = 30$ ns for G_3. Find the frequency of oscillation. *Answer* 7.14 MHz

11.3 PULSE-FORMING CIRCUITS

In the design of digital systems it is sometimes necessary to delay a pulse by a specified amount of time or to produce a pulse of a particular width in response to the application of a trigger signal. These tasks can be accomplished with open-loop pulse-forming circuits using logic gates or discrete transistors, or by using closed-loop digital feedback circuits known as monostable multivibrators or one-shots.

Although they may be somewhat less complex, open-loop pulse-forming circuits are generally unsuited to digital logic system designs because they produce output waveshapes that have unacceptably long rise times and fall times. A second disadvantage of an open-loop pulse-shaping circuit is that its output may chatter as the gate input passes through the active region if there is any logic noise present.

The problems associated with open-loop pulse-shaping circuits may be corrected by adding positive feedback. First, positive feedback improves the circuit switching speed because it moves the dominant system poles into the right-hand plane. Second, it eliminates the output chatter associated with noise on the input signal since it adds hysteresis to the transfer characteristic.

Sometimes the feedback path is hidden inside the digital device, as it is in the case of Schmitt-trigger-style logic gates, while in other circuits the feedback connection from the output back to the input is visually apparent. All pulse-forming circuits containing positive feedback can be classified as monostable multivibrators or one-shots.

As with the astable multivibrator, the monostable multivibrator is basically a two-state positive-feedback circuit. However, unlike the astable, which has no stable states, the one-shot does have a single stable resting state. In order to cause the one-shot to enter its other state, a trigger pulse must be applied. Once the one-shot is triggered into this second state, it will remain there until it reenters the active region on its own, whereupon it will flip back to its resting state. The time required for this to occur is usually determined by an RC network that is part of the circuit.

The basic operation of a monostable multivibrator is illustrated in Figure 11.3-1. If we assume that this one-shot triggers on the rising edge of the input signal, its output will have the waveshape illustrated in part (b) of the figure, where the pulse width T will depend on the values selected for R and C. If a second input pulse comes in before the output pulse time-out is completed, the response of the one-shot to this signal will depend on whether or not it is re-triggerable.

When a monostable is nonretriggerable, the application of a second trigger pulse during the time-out of the circuit will be ignored. This type of circuit operation is illustrated in Figure 11.3-1c. In a retriggerable one-shot, the recep-

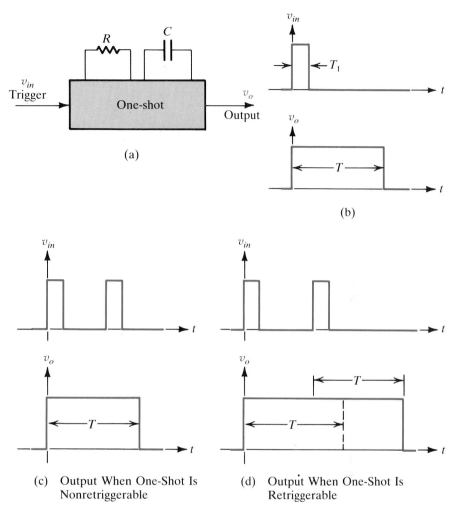

Figure 11.3-1
Fundamental characteristics of the monostable multivibrator.

(a)

(b)

(c) Output When One-Shot Is
     Nonretriggerable

(d) Output When One-Shot Is
     Retriggerable

tion of a second trigger pulse before the time-out has finished restarts the timing and effectively extends the width of the output pulse (Figure 11.3-1d).

Some monostable circuits have trigger inputs that are level sensitive, while others operate only on the rising or falling edges of the input waveform. This distinction is important because the application of a trigger pulse that is wider than the desired output pulse can cause problems in level-triggered circuits since the one-shot may be retriggered at the end of the time-out if the input pulse is still present. Most IC-style monostables are designed to be edge-triggered devices in order to avoid any output pulse dependence on width of the input pulse.

The circuit illustrated in Figure 11.3-2a is a one-shot that has been constructed from two CMOS NAND gates. When no trigger pulse is applied, the input is held at a logic 1 at V_{DD}. In the steady state the capacitor C will act like an open circuit, and because the input of G_2 is pulled down to ground through R, v_2 will be zero and v_3 will be equal to V_{DD}. Since the two inputs to G_1 will both be at V_{DD}, v_1 will be equal to zero in the resting state.

When v_{in} drops to zero, v_1 will immediately rise to $+V_{DD}$ regardless of the

11.3 / Pulse-Forming Circuits

893

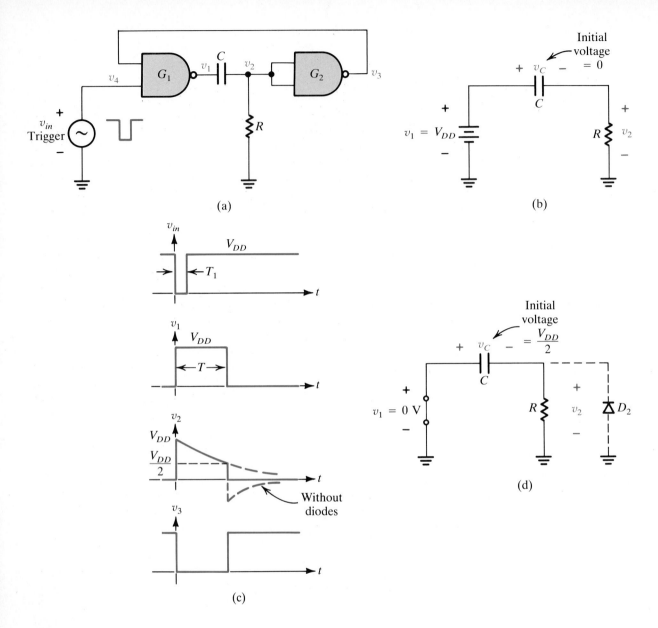

Figure 11.3-2
CMOS gate-style one-shot.

value of v_3. Because the voltage across C cannot change instantaneously, Δv_2 will be equal to Δv_1, and as a result, v_2 will jump up to $+V_{DD}$ volts at $t = 0+$. Since G_2 is wired as an inverter, v_3 will respond to this increase in v_2 by dropping to zero. This drop appears to occur instantaneously in the waveform sketches in Figure 11.3-2c, but as a practical matter it actually takes about $2t_p$ for the input trigger signal to propagate through G_1 and G_2. Once v_3 is a valid LO the trigger pulse can be removed and the circuit will remain in its quasistable state with $v_1 = V_{DD}$ and $v_3 = 0$. Thus the width of the trigger pulse, T_1, must be equal to at least two gate propagation delays for this circuit to function properly.

During the one-shot time-out period T, the equivalent circuit shown in Figure 11.3-2b applies and thus

$$v_2(t) = V_{DD}e^{-t/RC} \qquad (11.3-1)$$

since the initial voltage across C was zero. Assuming that the gate thresholds are at $V_{DD}/2$, once v_2 drops down to this value the circuit will reenter the active region and will switch back to its resting state. This switching process may be explained on a physical basis as follows. When v_2 drops below $V_{DD}/2$, G_2 enters the active region and v_3 rapidly rises. Once v_3 rises above $V_{DD}/2$, G_1 also enters the active region and v_1 begins to fall rapidly. Since v_C cannot change instantaneously, the change in v_1 further reduces v_2, causing v_3 to increase still more. The net effect of this sequence of voltage changes is that v_1 rapidly switches to zero, and v_3 to $+V_{DD}$, so that the circuit reenters the resting state.

Since $v_2(T-) = V_{DD}/2$, following the equivalent circuit in Figure 11.3-2b, it is apparent that $v_C(T-)$ is equal to $V_{DD}/2$ as well. Because the voltage across the capacitor cannot change instantaneously, $v_C(T+)$ will also be equal to $V_{DD}/2$. The equivalent circuit after the switching occurs at $t = T$ is shown in Figure 11.3-2d. If there were no input protection diodes on G_2, v_2 would initially drop below zero to $-V_{DD}/2$ at $t = T+$ and would then decay exponentially to zero. However, due to the gate input diode D_2, C will discharge to zero immediately through the diode and v_2 will be zero for all $t > T$.

This one-shot circuit that we have been discussing works fine as long as $T_1 < T$. However, when the trigger pulse is wider than the output pulse, v_{in} will still be equal to zero when v_3 goes back to V_{DD}. This prevents the output of G_1 from returning to a logic 0 and effectively lengthens the pulse width at this point in the circuit to that of the input trigger pulse. By adding a differentiator circuit to the input of the one-shot we can remove this dependence on the width of the trigger pulse (Problem 11.3-3).

Gate-style astable and monostable multivibrators have several inherent disadvantages. For both TTL and CMOS devices, the gate thresholds are not well defined, and this causes significant variations in the waveshapes produced by these circuits. Special ICs have been developed for use as digital oscillators and one-shots, and have been specifically designed so that their circuit performance is virtually independent of the power supply voltage.

One of the most popular ICs in this category is the 555 timer. It can be used as both an astable and a monostable multivibrator, and can also be configured as a a pulse-width modulator or a voltage-controlled oscillator. In its astable mode it is useful for generating frequencies from a fraction of a hertz to several hundred kilohertz, and in its monostable mode it can be used to produce output pulse widths anywhere from 10 μs to several hundred seconds.

The block diagram of the 555 is given in Figure 11.3-3a. Notice, among other things, that this circuit contains two comparators and an SR flip-flop. The comparator C_2, for example, produces a logic 0 at its output when the threshold input voltage at pin 6 is less than $\frac{2}{3}V_{CC}$. When this threshold voltage exceeds $\frac{2}{3}V_{CC}$, the comparator output is a logic 1. Comparator C_1 operates in a similar fashion except that its switching occurs when the trigger voltage at pin 2 is equal to $\frac{1}{3}V_{CC}$.

The behavior of the SR flip-flop was discussed earlier in Section 10.1, and

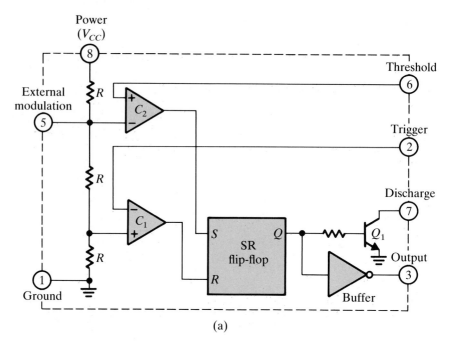

Power
(V_CC)

Threshold

External
modulation

Trigger

Discharge

Output

Ground

Buffer

(a)

Condition	Result
Trigger $< \frac{1}{3} V_{CC}$	Flip-flop is reset, $Q = 0$, Q_1 cut off, output $= V_{CC}$
Threshold $> \frac{2}{3} V_{CC}$	Flip-flop is set, $Q = 1$, Q_1 ON, output $= 0.0$ V

Figure 11.3-3
555 timer IC.

(b)

for the purpose of explaining the operation of the 555, this flip-flop's perfor-mance characteristics may be described as follows. When S and R are both at a logic 0, the flip-flop is in the memory mode and retains the last information stored in it until either the S or the R input logic levels are changed. When the S input is raised (even momentarily) to a logic 1, a 1 is stored in the Q output. In a similar fashion, raising R to a logic 1 stores a 0 in the Q output. The out-put of the flip-flop directly drives a discharge transistor Q_1 and it therefore fol-lows that Q_1 will be cut off when Q is a logic 0 and will be on when Q is a logic 1.

The most important operating characteristics of the 555 timer are summa-rized in Figure 11.3-3b. In the example that follows, we illustrate the perfor-mance of the 555 in its monostable mode.

EXAMPLE 11.3-1

The circuit shown in Figure 11.3-4a illustrates one type of monostable multivi-brator that can be built using the 555 timer IC. Explain how this circuit works, and in doing so, sketch the voltages at v_C, v_o, and the flip-flop output Q.

SOLUTION

In analyzing this circuit, we will make use of the table in Figure 11.3-3b. To begin with, in the absence of any trigger pulse the voltage on the trigger line is equal to V_{CC} so that the flip-flop cannot be reset. When the circuit is first turned

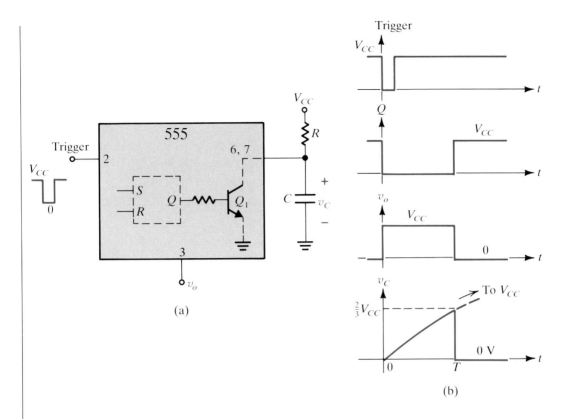

Figure 11.3-4
Example 11.3-4.

on, its initial state is arbitrary. If Q happens to be a logic 1, Q_1 will be on, v_C will remain zero, as will v_o, until a trigger pulse is applied. If the flip-flop happens to start out with Q at a logic 0, then Q_1 will initially be cut off and v_C will charge exponentially toward V_{CC}. When it reaches $\frac{2}{3}V_{CC}$, C_2 will set the flip-flop, Q_1 will turn on, and v_C and v_o will be zero. Thus, after the circuit is powered up, regardless of its initial state, it will eventually settle down into a resting state where Q is a logic 1, Q_1 is on, and v_C and v_o are both zero.

When the trigger input signal drops below $\frac{1}{3}V_{CC}$ at $t = 0$, the flip-flop is reset, turning Q_1 OFF. Once the discharge transistor cuts off, the voltage on the capacitor C begins to charge exponentially toward V_{CC} in accordance with the equation

$$v_C(t) = V_{CC}(1 - e^{-t/RC}) \tag{11.3-2}$$

This charging continues until v_C reaches $\frac{2}{3}V_{CC}$, and at that point, comparator C_2 sets the flip-flop turning on transistor Q_1. If we assume that $h_{FE}I_{B1}$ is very large compared with V_{CC}/R, the capacitor will discharge almost instantaneously and v_C, v_o, and Q will have the waveshapes shown in Figure 11.3-4b. At this point the flip-flop will remain set until another trigger pulse is applied and the cycle repeats.

The output pulse width of the one-shot is found by setting equation (11.3-2) equal to $\frac{2}{3}V_{CC}$, which yields

$$T = \tau \ln 3 = 1.1RC \tag{11.3-3}$$

11.3-1 Determine the width of the pulses produced at v_1 and v_3 by the one-shot in Figure 11.3-2a if $T_1 = 400$ μs, $R = 100$ kΩ, and $C = 1000$ pF. *Answer* 400 μs at v_1, 69 μs at v_3

11.3-2 For the circuit in Figure 11.3-2a, $V_{DD} = 5$ V, $R = 5$ kΩ, $C = 1000$ pF, and the gates are LS inverters (Figure 10.4-12) with input gate switching voltages of 1.2 V. Find the output pulse width at v_3 if a narrow trigger pulse is applied at v_{in}. *Answer* 6 μs

11.3-3 Find T in Example 11.3-1 (Figure 11.3-4) if $V_{CC} = 10$ V and an 8-V battery is connected between pin 5 and ground. *Answer* 1.61 RC

11.4 MECHANICAL SWITCH DEBOUNCING CIRCUITS

Mechanical switches are often used as input devices in digital systems. Unfortunately, they possess a property that makes them rather difficult to interface to digital logic circuits. When a switch is opened, and also when it is closed, there is considerable bouncing of the mechanical contacts before the final electrical connection is made or broken. Typically, the contacts bounce for about 20 ms or so before steady state is finally reached.

To see the effect that this can have on the operation of an electronic system, consider the simple circuit illustrated in Figure 11.4-1a, and let's examine what happens to the voltage at v_o when switch S is closed and then opened. If the switch has been open for a long time prior to $t = 0$, then v_o will be zero for $t < 0$. If the switch is thrown into the closed position at $t = 0$, the contacts will open and close several times before they finally settle down into the closed position. During this time interval, each time the contacts meet v_o will jump up to $+V_{CC}$, and when the contacts bounce open this voltage will drop to zero. Thus, associated with the mechanical bouncing of the switch contacts there will be a bouncing of v_o back and forth between 0 and V_{CC} before it finally settles down at V_{CC} volts. A similar series of pulses will also be generated when the switch is thrown into the open position at $t = T$. The resulting waveshape at v_o is indicated in Figure 11.4-1b.

If the resistance R in this circuit were a light bulb and if the switch was being used to turn this light ON and OFF, the contact bounce would have little effect on the perceived illumination of the bulb because the thermal time constant of the bulb is much longer than 20 ms. In a similar fashion, if this switch were being used to control the ac power into an audio amplifier or a TV set,

Figure 11.4-1

Effect of contact bounce in a switch.

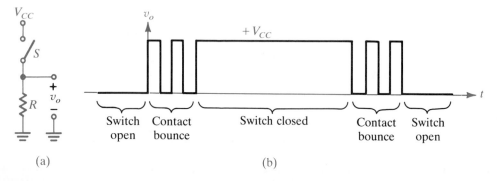

the switch contact bounce would again have little observable effect on the performance of these systems because of the long time constants of their power supplies.

If, however, the circuit in Figure 11.4-1a were being used to generate a digital logic signal at v_o, this contact bounce could have catastrophic effects on the performance of the system. Consider, for example, the case where the digital system being connected to this switch is a counter that is supposed to record the number of times that switch S is closed by an operator. If the waveshape illustrated in Figure 11.4-1b occurred each time the operator closed and then opened the switch, five counts, and not the desired count of one, would be added to the total already in the counter each time the switch was actuated.

In the example that follows we illustrate a technique for producing one, and only one pulse each time the switch S is closed and then opened.

EXAMPLE 11.4-1

The circuit shown in Figure 11.4-2 is a Schmitt-trigger-style switch debouncer. Sketch v_1 and v_2 if the logic gate has the hysteresis characteristic given in Figure 11.2-4. The maximum bounce time of the switch is 20 ms.

Figure 11.4-2
Example 11.4-1.

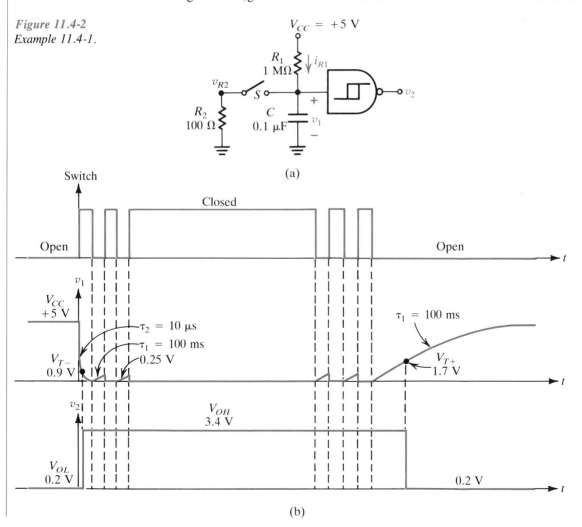

(a)

(b)

SOLUTION

If the switch has been open for a long time for $t < 0$, then at $t = 0-$, $v_1 = 5$ V and $v_2 = 0.2$ V. When the switch is closed at $t = 0$ the capacitor discharges to almost zero volts with a time constant $\tau_2 = (R_1 \parallel R_2)C \approx 10$ μs. (The actual final value of v_1 is $V_{CC}[R_2/(R_1 + R_2)]$, or about 500 μV.) Because this time constant is so small, the capacitor discharges fully during the time interval of the first contact closure.

When the switch bounces into the open position, v_1 begins to rise exponentially towards $+5$ V with a time constant $\tau_1 = R_1 C = 100$ ms. Because this time constant is so large, the voltage across C will not increase very much during the time that the switch is open. As a result, i_{R1} will be nearly constant at $V_{CC}/R_1 = 5$ μA. If we replace R_1 by a constant-current source, the voltage v_1 will increase linearly with time, and using this approximation the peak amplitude of this ramp may be calculated from the capacitor equation $i = C\, dv/dt$ as $\Delta v = I \Delta t / C = 0.25$ V, using a contact break time of 5 ms.

On the next contact closure the capacitor again discharges almost immediately to zero, as illustrated in Figure 11.4-2b. This slow ramp-style increase in v_1 occurs during each momentary contact opening, and when the switch finally opens permanently, v_1 rises exponentially towards $+5$ V in about $\frac{1}{2}$ s.

Working with the hysteresis characteristic for this Schmitt trigger, it is apparent that nothing happens to v_2 until v_1 falls below 0.9 V. At this point v_2 switches to $+3.4$ V almost instantaneously due to the built-in positive feedback of the IC. The output of the Schmitt trigger remains at 3.4 V until v_1 rises above 1.7 V, and at this voltage, v_2 switches back to its logic 0 output state of 0.2 V. It is most interesting to observe that although the input signal rise time is exceedingly slow when the switch is opened, the output fall time is nearly instantaneous because of the Schmitt triggers positive feedback.

Exercises

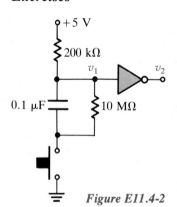

Figure E11.4-2

11.4-1 For the circuit shown in Figure 11.4-2a, the Schmitt trigger is replaced by a CMOS inverter having a gain of -50 in the active region. Estimate the fall time of the output pulse produced when the switch is closed and then opened. *Answer* 4 ms

11.4-2 The gate shown in Figure E11.4-2 is an ideal CMOS inverter.
a. What is the width of the pulse produced when the switch is pressed at $t = 0$ if the switch has been open for a long time prior to $t = 0$?
b. What pulse width will be produced if the switch is pushed for 500 ms, 1 s after it was pushed the first time?
Answer (a) 13.6 ms, (b) 4.6 ms

11.5 INTRODUCTION TO SEMICONDUCTOR MEMORIES

Memory devices were the first large-scale integrated circuits to be manufactured commercially because of their wide market appeal, and also because of their relatively simple repetitive structures. Memories may be divided into two broad categories: read-only memories (ROMs), whose memory contents cannot be altered, and read/write memories (RWMs), which can be read from and written to with equal ease.

Basically, a memory is a place for storing information, and over the years many different types of electronic memory circuits have been developed. In the first computers of the 1940s, information was stored in relays and in vacuum-tube flip-flops. However, relays were slow and vacuum tubes, while much faster, were power hungry and expensive. Furthermore, the data stored in both of these types of memories was volatile, so that the data was lost when the power to the circuit was removed.

The development of magnetic core memory in the 1950s was a significant improvement over these earlier memory structures. Basically, it relied on the magnetization of tiny doughnut-shaped pieces of material. To store a logic 1 in the memory, the doughnut was magnetized in one direction and to store a logic 0, the direction of the magnetization was simply reversed. Core memory had several features to recommend it. First, as with all magnetic storage media, it was nonvolatile, so that the data remained in it even when the power applied to the memory was turned off. Second, core memory was extremely fast, at least in its day, having typical access times on the order of a microsecond.

Like its predecessors, core memory was a true RAM or random access memory since the time required to find a particular piece of information in the memory was independent of its location. Both read-only and read/write semiconductor memories are also random access devices. However, because core is a read/write memory, the term RAM traditionally has been applied only to read/write devices. Since this usage is so widespread, we will also use the terms RAM and read/write memory interchangeably, even though a ROM is also a random access device.

By way of contrast to the random access memory structures that we have been discussing, magnetic tape and magnetic disk, while also nonvolatile storage media, are serial and serial/parallel access devices, respectively. As such the access time of information from these devices is a strong function of the position of the data in the memory, and is generally much greater than that required for RAM. Because of their long access times these last two types of memories are almost never used as the main memory in a digital system but are instead generally employed for the mass storage of large programs and large files of data. When needed, this information may be transferred into the system main memory where the actual program execution and data processing can take place.

By the 1960s the transistor was well established in computer system design, and yet in spite of this, magnetic core was still "the" random access memory of choice because of its nonvolatility, and its cost and speed advantages over transistorized memory circuits. In the early 1970s with the development of large-scale integrated circuits, this picture began to change, and today for all practical purposes core has been completely replaced by faster and much less expensive semiconductor memory. Even the nonvolatility issue, which was the core's last major selling point, is no longer that significant since modern semiconductor RAM (with its very low power consumption) can effectively be made nonvolatile by adding battery backup.

In digital electronics, information is usually stored in a binary format. Each

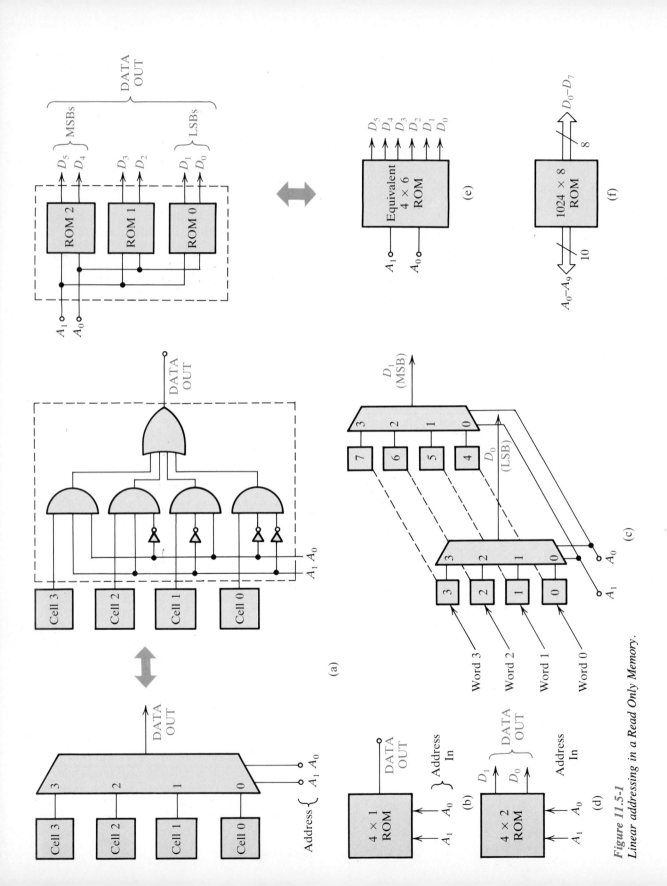

Figure 11.5-1
Linear addressing in a Read Only Memory.

elemental memory cell stores 1 bit of data, and a typical semiconductor memory IC consists of a large array of these elemental cells. To locate the position of a particular cell in the memory, it is necessary to specify the address of that cell.

For small memory arrays the linear addressing scheme illustrated in Figure 11.5-1a can be used. The memory shown in this figure is said to be made up of four words, and each word in this instance is 1 bit long. Because there is no way to write information into the cells of this memory, it is clear that this device is a ROM or read-only memory. In describing the size of this memory, it is common to say that it is a 4×1 memory chip, meaning that it contains (4 words) $\times$ (1 bit/word) for a total memory size of 4 bits. The gating circuitry shown within the dashed lines in the right-hand portion of this figure is known as the memory's address decoder, and it determines which cell's data appears on the DATA OUT line. In this elemental 4-bit memory, this decoder circuit is logically equivalent to a 4-line-to-1-line data multiplexer. Notice that two address leads are needed in this ROM to uniquely identify any one of the four cells in the memory. In general, n address leads can directly address 2^n different memory cells. A block diagram of this memory is given in Figure 11.5-1b.

Figure 11.5-1c illustrates the design of a 4×2 ROM, and part (d) of the figure gives the block diagram for this memory. Notice, as before, that only two address leads A_0 and A_1 are needed to uniquely identify each of the four different words in this memory. If a larger number of bits/word is needed, several ROMs may be connected in parallel. Figure 11.5-1e illustrates how three 4×2 ROMs can be connected together to form a memory containing four 6-bit words. Here ROM 0 supplies the two least significant bits or LSBs of the word, ROM 1 bits D_2 and D_3, and ROM 2 the two most significant bits or MSBs.

When the number of address or data bits in a memory becomes large, it is common to illustrate the wires entering or leaving the chip by an arrow with a number written next to the arrow that indicates the total number of leads associated with that bus or group of wires. This technique is illustrated by the ROM shown in Figure 11.5-1f. Because this memory chip has 10 address input leads, it contains a total of 2^{10} or 1024 different words. Furthermore, because the data bus is 8 bits wide, the total size of this ROM is 1024×8 or 8192 bits. For short, this memory would be referred to as a $1K \times 8$ ROM, where the letter K denotes a factor of 1000. Of course, this is slightly incorrect since this memory actually has 1024, not 1000 words.

For memories containing large numbers of cells, the linear addressing scheme illustrated in Figure 11.5-1a is unsatisfactory because the address decoder is difficult to construct and also because it takes up too much room on the IC. As an alternative approach, consider the coincident addressing scheme illustrated in Figure 11.5-2. In this type of memory the location of a particular cell is specified by giving its x and y coordinates, that is, by dividing the address up into its X and Y components. Here the Y address decoder is used to enable or turn ON one particular column of cells in the 4×4 array. The data from these cells then appears at the inputs to the row multiplexer, and the X portion of the address determines which of these signals will appear on the DATA OUT line. For example, if the overall YX address were 1101, the $(11)_2$ Y portion of the address would cause the data from cells 12, 13, 14, and 15 to

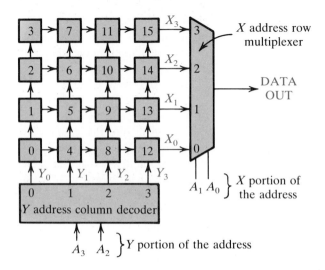

appear at the input to the row multiplexer. Because the X portion of this address is a $(01)_2$, the data from cell 13 will be selected by the multiplexer and it will appear on the data output line of the ROM. Of course, this is exactly what should happen since $(1101)_2$ is in fact the address corresponding to cell 13.

Read/write or RAM-style memories have architectures that are quite similar to those of ROMs except that provision must also be made for writing into the memory cells. Figure 11.5-3 illustrates one type of read/write memory that has been constructed by using D-style flip-flops. You should recall that a D-type flip-flop operates by memorizing the data on its D input line when the clocking signal occurs. In this figure the circle on the clock input lead of the flip-flops indicates that the data is loaded into the memory when the clocking signal goes LO.

Basically, this read/write memory design is identical to the ROM memory circuit presented in Figure 11.5-1a except that an additional address decoder, in the form of a 1-line-to-4-line data demultiplexer has been added at the input to ensure that the data is only written into the flip-flop that is addressed when a write pulse occurs. Here it is assumed that all unselected demultiplexer output lines are at a logic 1 level.

Also shown in the figure is an additional input known as CS or chip select. This signal is important for the design of multichip memories. Effectively, this memory chip is enabled or selected when CS is made equal to 1, since write pulses are then free to pass through the input OR gate, and also since the output tri-state buffer is enabled, connecting the output from the addressed memory cell to the DATA OUT lead. When CS = 0 the OR gate is disabled and it is impossible to write data into this memory chip. In addition, the output tri-state buffer is also disabled so that the DATA OUT line is effectively disconnected from the multiplexer.

Chip selects are also used in a similar fashion with ROMs except, of course, that they affect only the data output lines of the ROM. Figure 11.5-3b and c show the generalized block diagrams for read/write and read-only memory ICs, respectively. In passing it is useful to note that sometimes a chip enable rather than chip select lead is used. The principal difference between the two is that a chip enable, in addition to its chip select function, also acts to

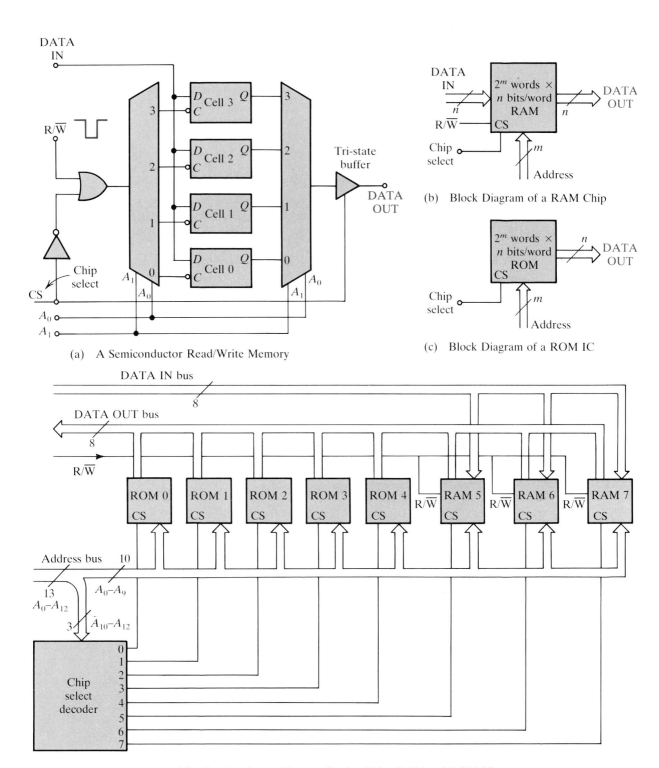

(a) A Semiconductor Read/Write Memory

(b) Block Diagram of a RAM Chip

(c) Block Diagram of a ROM IC

(d) Semiconductor Memory Design Using RAM and ROM ICs

Figure 11.5-3

"power down" selected parts of the IC when CE is made equal to zero. This can permit a significant power saving, depending on the overall size of the memory.

The use of chip-select-style memory ICs permits the system memory to be expanded to any desired number of words effectively by connecting the ICs in parallel. Consider, for example, the design of an 8-bit memory containing 3K of RAM and 5K of ROM, assuming that only 1K $\times$ 8 chips are available. Basically, as illustrated in Figure 11.5-3d, this may be accomplished by using three of the 1K RAM ICs and five of the 1K ROMs. Note that all the address and data out leads of these ICs are connected together, and that in addition the data in and the R/$\overline{\text{W}}$ lines of the RAMs are connected in parallel. To properly address 8K of memory, a total of 13 address lines (A_0 through A_{12}) are needed. A_0 through A_9 are directly connected to the corresponding address inputs on all eight of the memory chips and A_{10}, A_{11}, and A_{12} go to the input of an encoder IC that is used to activate only one of these memories at a time. For example, if $A_{10} = 0$, $A_{11} = 1$, and $A_{12} = 0$, then encoder output line 2 will go HI and ROM 2 will be selected. Thus the 8K memory space spanning hexadecimal addresses 0000 through 1FFF will be divided among the eight 1K memory chips.

To design memory systems properly, it is necessary to understand the timing relationships that exist between the data, address, and control signals that are applied to the memory. Figure 11.5-4a illustrates the sequence of events that occurs when data is read from an IC memory chip. If we assume that CS = 1, then after a specific address is applied to the memory, a time t_{AA} will elapse before the contents of the addressed memory location appear on the DATA OUT lines. This time interval is known as the memory's address access time, and typical values for t_{AA} range from about 15 ns for ECL memories to

Figure 11.5-4

Important time intervals in a memory.

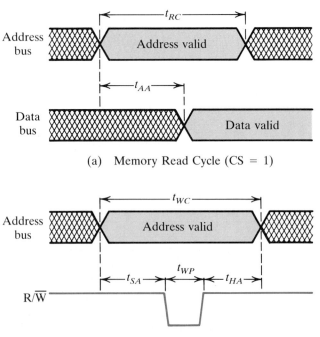

(a)   Memory Read Cycle (CS = 1)

(b)   Memory Write Cycle (CS = 1 and Input Data Stable)

100 to 250 ns for NMOS-style memories. Although not shown explicitly in the figure, we can similarly define t_{ACS} as the access time from the application of a chip select to the appearance of valid data at the memory output. Chip select access times can be considerably smaller than those associated with address selection since CS often just enables the output tri-state buffers during a memory read operation.

The quantity t_{RC} shown in the figure is known as the read cycle time. It defines the smallest time interval allowed between successive memory read operations. For semiconductor memories the read cycle time is basically the same as the memory access time. The reason for using two different names to describe the same time interval is a carryover from the days of magnetic core memory. In core memory the readout process is destructive, so that the data read from the memory need to be written back in, and this overall sequence defines the read cycle time so that t_{RC} for magnetic core is much longer than the read access time. However, for semiconductor memory these two quantities are identical since the readout process is nondestructive.

The timing sequence needed to carry out a memory write cycle is illustrated in Figure 11.5-4b, and here too we are assuming that CS = 1 during the entire write operation. In this figure the time interval t_{SA} is known as the address setup time and is equal to the minimum amount of time that the address must be present before the R/$\overline{\text{W}}$ line is allowed to go LO. This delay is needed to ensure that the proper memory cell is selected before the write operation begins. The time t_{WP} defines the minimum allowable width of the write pulse, and t_{HA} is the address hold time or the time that the address must remain valid after the write pulse ends.

The maximum rate at which data can be written into the memory is determined by its write cycle time, t_{WC}. From the figure it is apparent that $t_{WC} = (t_{SA} + t_{WP} + t_{HA})$. Besides having setup and hold times associated with changes in address bus information, similar definitions also exist for the information on the DATA IN and CS lines. In particular, t_{SD} and t_{HD} define the setup and hold times for the input data being applied to the memory, and t_{SCS} and t_{HCS} the setup and hold times with regard to the application of the chip select. In calculating the smallest allowed write cycle time, the larger of t_{SA}, t_{SD}, or t_{SCS}, and t_{HA}, t_{HD}, or t_{HCS} should be used to compute t_{WC}.

Exercises

11.5-1 The following information is provided for a particular static random access memory.

Read cycle data: $t_{RC} = 100$ ns, $t_{AA} = 100$ ns, $t_{ACS} = 70$ ns
Write cycle data: $t_{WC} = 100$ ns, $t_{WP} = 75$ ns, $t_{SD} = 5$ ns,
  $t_{HD} = 0$ ns, $t_{SA} = 0$ ns, $t_{HA} = 0$ ns

a. During a memory read cycle CS = 1. How long does it take to obtain valid data from the RAM once the address is valid?
b. During a memory read cycle the address information becomes valid at $t = 0$ and 50 ns later, CS goes HI. At what point in time (relative to $t = 0$) will the output data from the RAM be valid?
c. For a memory write cycle in this RAM with CS = 1, $t = 0$ is defined as the time when the address and the input data become valid.

i. How much time must elapse before the beginning of the write pulse?

ii. What is the required width of the write pulse?

Answer (a) 100 ns, (b) 120 ns, (c)(i) 0 ns, (ii) 75 ns

11.5-2 a. Determine the number of transistors needed to design a single-level 10-bit linear address decoder using 10-input NAND gates. Assume that each gate requires 11 transistors and do not forget that inverters (two transistors/gate) will be required to produce the complements of the address inputs.

b. Repeat part (a) for the case where a 10-bit 5×5 coincident address decoder array is used. Consider that each five-input NAND gate requires six transistors.

Answer (a) 11,284, (b) 404

11.6 STATIC RANDOM ACCESS MEMORY

There are two major types of semiconductor read/write memory: static and dynamic. In a static RAM the information is stored in flip-flops and no input signals other than dc power need to be applied to ensure that the information is retained by the memory. Dynamic memory is significantly different from static memory in that the information, rather than being stored in a flip-flop, is stored as a charge on a capacitor. Because there is always some resistance in parallel with these capacitors, the charge, and hence the information associated with it, will leak off as a function of time. As a result, in dynamic memories the information stored on the capacitor will periodically have to be refreshed or restored to its original value if its loss is to be avoided. The constant need for the presence of these refresh signals gives dynamic memory its name.

Flip-flops, or bistable multivibrators as they are also called, are used extensively in both static and dynamic memories. In static semiconductor memory the memory cells themselves are actually flip-flops, while in dynamic memories the sense amplifier that is used to refresh (or recharge) the information that is stored on the capacitors in the memory cells is also in reality a flip-flop. Because of its importance in the design of semiconductor random access memories it will be useful for us to examine the performance characteristics of a simple flip-flop.

One of the most common types of flip-flops is shown in Figure 11.6-1a. It consists of a set of cross-coupled inverters. Although noninverting amplifiers can also be used to design a flip-flop, inverters are more convenient because they use fewer components, and also because they provide for both Q and $\overline{Q}$ outputs. To analyze the performance of this circuit, let us consider that the inverters are CMOS devices with a gain $-A$, an output impedance R, and an input capacitance C. For such a flip-flop, if the power supply voltage is equal to V_{CC}, the logic 0 and logic 1 voltage levels will be very close to zero and V_{CC} volts, respectively.

Let's assume that the flip-flop initially has a logic 1 stored in it, and consider what will be needed to trigger this flip-flop to the opposite state. With $Q = 1$ capacitors C_1 and C_2 will initially be charged to 0 and V_{CC} volts, respectively. To cause this circuit to switch to the opposite state a trigger signal (not shown in the figure) will have to change the voltage on both of these capacitors. Let us represent the voltages on C_2 and C_1 just after the trigger pulse is

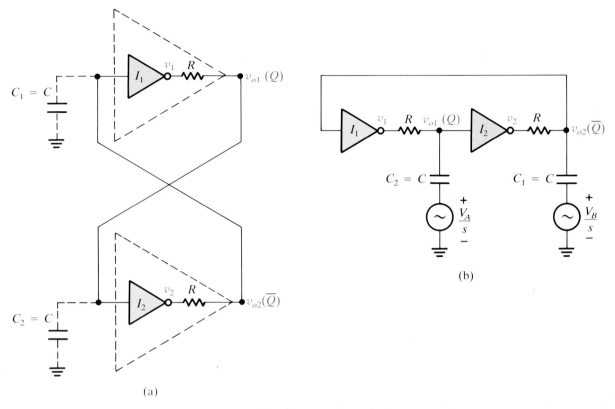

(b)

(a)

Figure 11.6-1

A two inverter-style flip-flop.

removed as V_A and V_B, respectively, and let's attempt to determine the values that these voltages must have if the circuit is to flip to the opposite state.

To set this problem up for solution by Laplace transforms, we may redraw this circuit as illustrated in Figure 11.6-1b. Here the initial conditions on C_1 and C_2 just after the trigger has been removed are represented as two voltage sources in series with the capacitors.

If we assume that the trigger signal, in addition to changing the initial conditions on the capacitors, has also placed I_1 and I_2 in their active regions, the following relationships will exist between the circuit voltages:

$$V_{o1}(s) = V_1(s)\frac{\alpha}{s + \alpha} + \frac{V_A}{s + \alpha} \tag{11.6-1a}$$

$$V_{o2}(s) = V_2(s)\frac{\alpha}{s + \alpha} + \frac{V_B}{s + \alpha} \tag{11.6-1b}$$

$$V_1(s) = A V_{o2}(s) \tag{11.6-1c}$$

and
$$V_2(s) = A V_{o1}(s) \tag{11.6-1d}$$

where
$$\alpha = \frac{1}{RC} \tag{11.6-1e}$$

Solving these equations for V_{o1}, we obtain the general result

$$V_{o1}(s) = \frac{sV_A + \alpha(V_A + AV_B)}{(s + \alpha)^2 - \alpha^2 A^2} \tag{11.6-2}$$

11.6 / Static Random Access Memory

and letting A take on a specific value, we can get a better idea of the form of the solution. In particular, for $A = -10$ we have

$$V_{o1}(s) = \frac{sV_A + \alpha(V_A - 10V_B)}{s^2 + 2\alpha s - 99\alpha^2} = \frac{sV_A + \alpha(V_A - 10V_B)}{(s - 9\alpha)(s + 11\alpha)} \qquad (11.6\text{-}3)$$

and using partial fraction expansion, this expression may be rewritten as

$$V_{o1}(s) = \frac{(V_A - V_B)/2}{s - 9\alpha} + \frac{(V_A + V_B)/2}{s + 11\alpha} \qquad (11.6\text{-}4)$$

In examining this result, it is the first term that will dominate in $v_{o1}(t)$ since the inverse Laplace transform of this term will be an exponential with a positive exponent. The inverse Laplace transform of equation (11.6-4) may be written by inspection as

$$v_{o1}(t) = \frac{V_A - V_B}{2} e^{9\alpha t} + \frac{V_A + V_B}{2} e^{-11\alpha t} \qquad (11.6\text{-}5a)$$

and because of the circuit symmetry, it is also apparent that

$$v_{o2}(t) = \frac{V_B - V_A}{2} e^{9\alpha t} + \frac{V_A + V_B}{2} e^{-11\alpha t} \qquad (11.6\text{-}5b)$$

By examining equation (11.6-5a), we can see that $v_{o1}(t)$ will grow toward plus infinity if $V_A > V_B$, stopping when it hits the $+V_{CC}$ power supply rail. On the other hand, if $V_A < V_B$, this voltage will grow exponentially toward minus infinity, stopping at zero volts. Thus if we recall that $v_{o1}(t)$ started out at V_{CC} prior to the application of the trigger pulse, it is apparent that the trigger must cause V_A to drop below V_B if switching to the opposite state is to occur.

Now that we have an understanding of how flip-flops operate, it is a good idea to examine the basic architectures of some of the more popular static random access memories. Both BJT and MOS RAM chips are available, with the bipolar devices generally being somewhat faster than their MOS counterparts. However, in spite of this speed advantage, BJT random access memories, because of their correspondingly high power consumption and low bit density per chip, find application only in specialized high-speed designs. When medium-speed large memory arrays are needed, NMOS or CMOS memories are generally used.

When MOS memories were first introduced in the early 1970s there were vast speed differences between MOS and BJT devices, but as MOS technology has continued to improve while bipolar technology has made only modest gains, the speed gap between the two has narrowed to the point where it is now on the verge of extinction. Because of the dominance of MOS technology in random access memory designs, we will focus our attention on these devices; the subject of BJT-style RAM architectures will have to be left for more specialized texts on digital memory design.

Figure 11.6-2 illustrates three different versions of MOS static memory cells. All three of these circuits will be examined simultaneously since their operating principles are similar. The flip-flop shown in part (a) is an NMOS memory cell. Transistors Q_1 and Q_2 are enhancement-style transistors, and together with the active load transistors Q_3 and Q_4, they form the basic flip-flop circuit. To achieve maximum switching speed, Q_3 and Q_4 are depletion loads

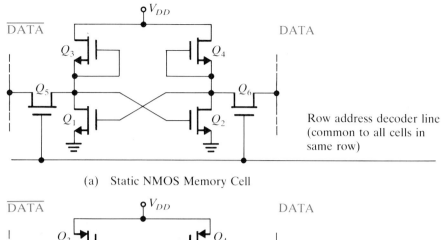

(a)   Static NMOS Memory Cell

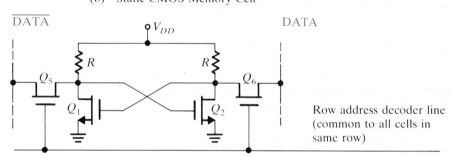

(b)   Static CMOS Memory Cell

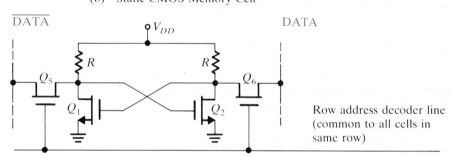

Row address decoder line
(common to all cells in
same row)

Row address decoder line
(common to all cells in
same row)

Figure 11.6-2

(c)   Static NMOS Memory Cell with Polysilicon Resistive Loads

with large values of I_{DSS}. Transistors Q_5 and Q_6 are enhancement devices, and they act as transmission gates to switch the flip-flop information out onto the data lines when they are enabled by the row address decoder.

This type of memory cell consumes considerable power because a current I_{DSS} from Q_3 or Q_4 flows down through one side or the other of the flip-flop, depending on whether it has a logic 0 or a logic 1 stored in it. For example, assuming that $V_{DD} = 5$ V and that $I_{DSS} = 100$ μA, this static flip-flop dissipates an average power of 500 μW. Although this number appears to be rather small, it is significant to observe that a 1K RAM chip made from these NMOS memory cells will dissipate about 500 mW, not including the additional power needed by the address decoders and I/O buffers.

By replacing the NMOS transistors Q_3 and Q_4 by p-channel devices this circuit may be converted to the CMOS static flip-flop shown in Figure 11.6-2b. For this memory cell the static power consumption is reduced to only a few microwatts since no direct current path exists between V_{DD} and ground. Therefore, the only current that flows in this circuit under static conditions is the transistor leakage current. If this current is assumed to be on the order of 1 μA per transistor, the static power consumption of this cell will be about 10 μW. Of course, the power consumption under dynamic operating conditions is considerably larger. In particular, if we assume parasitic capacitive loads of 1 pF between the drain and the ground of Q_1 and Q_2, then following equation (10.5-27), the additional power consumed by this flip-flop will be

$$P_D = 2fCV_{DD}^2 \tag{11.6-6}$$

At a toggle rate of 1 MHz, the power dissipated in this memory cell will therefore increase to about 50 μW. This does not mean, however, that the average power consumption of this type of CMOS memory will exhibit a similar 50-fold increase in power consumption when data is being written in at a 1-MHz rate. It must be remembered that only one cell at a time is being switched during a memory write operation. However, the actual power consumption of a CMOS memory does increase substantially with frequency; but this growth is associated with increased power dissipation in the address decoders and I/O buffer circuits, rather than just in the memory cells themselves.

The low power consumption of CMOS-style flip-flops is offset by the large area occupied by these memory cells. However, a similar power savings can also be achieved by replacing the p-channel load transistors by very high value polysilicon resistors (Figure 11.6-2c). In comparison to the flip-flops shown in Figure 11.6-2a and b, the polysilicon resistive-style memory cells are considerably smaller, yet operate at comparable speeds. This is because both sides of the flip-flop are driven during a memory write operation so that the switching speed is relatively independent of the value of the load resistors. If R is on the order of 500 kΩ, the average power consumed by this style of memory cell (assuming a 5-V supply) will be about $(5 \text{ V})^2/0.5 \text{ M}\Omega$, or 50 μW.

Figure 11.6-3 illustrates the design of a 4 $\times$ 4 static RAM that can be used with any of the MOS memory cells in Figure 11.6-2. Because these cells only contain the switching transistors for use with the row address decoders, it is necessary to add an additional set of column decoder switches at the bottom of the memory. For this circuit, when a particular row address is selected, say X_1, for example, the row select switches on all the memory cells in that row will be enabled, and in this case the data contained in cell 1, 5, 9, and 13 will come out of the memory array and appear at the inputs to the column select switches. Depending on the Y portion of the address, only one set of column select switches will be enabled. If, for example, Y_2 is a logic 1, the data from cell 9 will pass down onto the DATA and $\overline{\text{DATA}}$ lines.

If the operation to be carried out is a memory read, the R/$\overline{\text{W}}$ line will be HI, the transmission gate (or signal switch) Q_1 will be enabled, and the contents of cell 9 will appear on the DATA OUT line. Sometimes sophisticated sense amplifiers are used to process the differential data signals that come out of the addressed memory cell. This type of amplification helps to eliminate crosstalk and other kinds of common-mode noise. In addition, buffers are also generally used between Q_1 and the DATA OUT lead to restore the correct voltage levels to the data and to provide additional load drive capability. How-

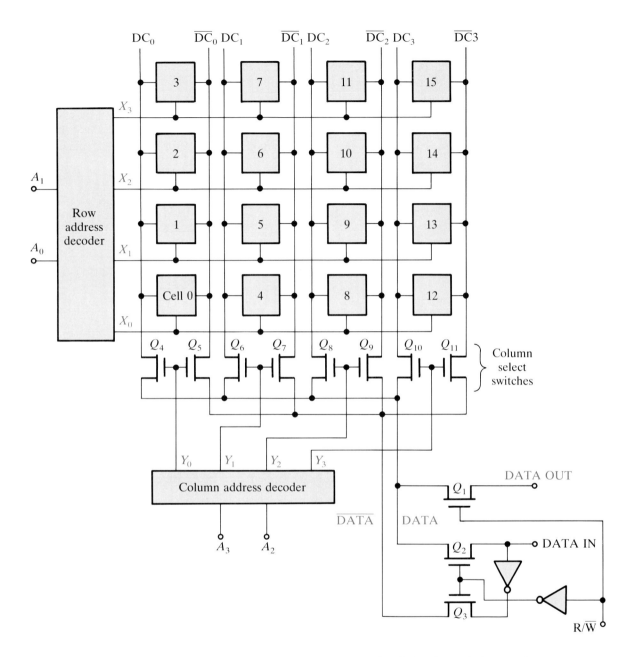

Figure 11.6-3
4 × 4 MOS static random access memory.

ever, in principle, transistor Q_1 is all that is needed to perform a proper memory read.

A similar situation exists during a memory write. When the R/$\overline{W}$ line goes LO, Q_2 and Q_3 are enabled and the information present on the DATA IN line appears on the DATA and $\overline{DATA}$ lines of the column select transistors. Because these transistors are bidirectional devices, this information will pass through the selected set of column switches (Q_8 and Q_9) and will appear at the inputs to all the memory cells in this column (cells 8, 9, 10, and 11). Because only row X_1 is enabled, this data will enter cell 9 and will set or reset this flip-flop as required. It is interesting to note that two transistor switches are used to write the data into both sides of the selected cell, despite the fact that a single switch could, in theory, be used to accomplish this task. The reason for using this two-transistor approach is discussed in the example that follows.

EXAMPLE 11.6-1

Figure 11.6-4
Example 11.6-3.

Consider the resistive load MOS flip-flop illustrated in Figure 11.6-2c and assume that $V_{DD} = 5$ V, $R = 500$ kΩ, and that Q_1 and Q_2 have parasitic gate-to-source capacitances of 1 pF. In addition, $V_T = 1$ V for all the MOSFETs, and $I_{D2T1} = I_{D2T2} = 0.005$ mA and $I_{D2T5} = I_{D2T6} = 0.05$ mA.

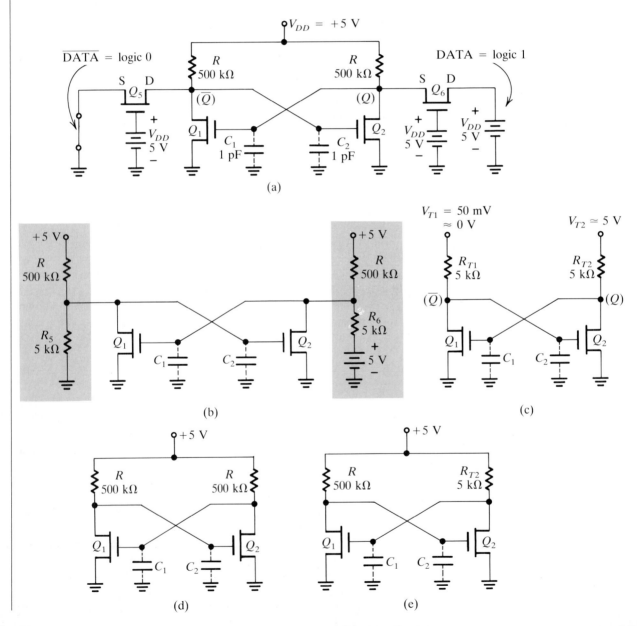

a. Assuming that the flip-flop originally has a logic 0 stored in it, explain how a logic 1 is written into this cell.

b. Repeat part (a) for the case where Q_5 is removed from the circuit.

c. For the cases discussed in parts (a) and (b), determine the minimum amount of time that valid data must be present at the cell input to guarantee that this information will be correctly written into the cell.

In discussing the performance of this flip-flop, we will assume that the cell is being driven by ideal voltage sources. The basic circuit for the case where a 1 is to be written into the cell is shown in Figure 11.6-4a. Because this cell initially has a logic 0 in it, $\overline{Q} = 1$. Thus, for $t < 0$, transistor Q_1 is OFF and Q_2 is ON, so that $v_{C1}(0-) = 0$ V and $v_{C2}(0-) = 5$ V. When the R/$\overline{W}$ line goes to a logic 0, the voltages shown in the figure are applied to the DATA and $\overline{\text{DATA}}$ lines in order to write a 1 into this cell.

Based on the values of these voltages and those at the drains of Q_1 and Q_2, the drain and source leads of the transmission gate transistors Q_5 and Q_6 have the orientation shown in the figure since the position of the drain and source in a MOSFET is not absolutely defined, but depends on the relative values of the voltages applied to the leads. For an NMOS transistor the source end of the channel is located at the side having the lower potential. Using this information it is apparent, initially at least, that $v_{GS5} = v_{GS6} = 5$ V. If we assume, for simplicity, that equation (10.5-2c) can be applied to both of these transistors, then

$$R_5 = R_6 = \frac{V_T^2}{I_{D2T}(V_{GS} - V_T)} = \frac{(1 \text{ V})^2}{(0.05 \text{ mA})(4 \text{ V})} = 5 \text{ k}\Omega \qquad (11.6\text{-}7)$$

As a practical matter, this result is only very approximate since v_{GS} will not stay at 5 V for Q_6, and also because these transistors will be in the ohmic region only when $v_{DS} < (v_{GS} - V_T)$. However, because the qualitative results will be the same, it is very useful to consider that these transistors are constant resistances. Assuming this approach to be valid, the equivalent circuit shown in Figure 11.6-4b applies, and taking the Thévenin equivalents of the portions of this circuit within the shaded regions, the circuit given in Figure 11.6-4c is obtained. Because of the low impedances of R_{T1} and R_{T2}, regardless of the initial states of Q_1 and Q_2, eventually C_1 and C_2 will charge up to 5 V and 0.05 V, respectively. Of course, this will cut off Q_2 and turn Q_1 ON, so that, as desired, a logic 1 will be stored in the flip-flop. When the R/$\overline{W}$ line goes HI, the DATA and $\overline{\text{DATA}}$ lines will float and the memory cell circuit will effectively look like that given in Figure 11.6-4d. Because the voltages on the parasitic capacitances will remain constant during this transition period, the data stored in the flip-flop will remain at the logic 1 level.

In principle, data can also be written into this memory cell using only a single transmission gate. Consider, for example, how this could be accomplished if Q_5 were removed from the flip-flop in Figure 11.6-4a. In this instance, if, as before, a zero were initially stored in the memory cell, $v_{C1}(0-)$ and $v_{C2}(0-)$ would again be 0 and 5 V, respectively. The circuit for writing a 1 into this flip-flop is identical to that given in Figure 11.6-4b except that the 5 kΩ resistor on the left is absent so that the final simplified form of this circuit is that given in Figure 11.6-4e. Here, because of the low impedance of R_{T2}, C_1 will charge to nearly 5 V regardless of the state of Q_2, so that Q_1 will turn ON.

Once Q_1 goes ON, v_{C2} drops to zero and this will cut off Q_2. This completes the switching cycle and results in a logic 1 being stored in the flip-flop.

The major difference between driving one side or both sides of the flip-flop in the writing of data is the amount of time needed to accomplish the switching. To examine the switching time, let's begin by considering the case where only a single row-select transistor is used. In this case the switching time depends on whether a zero or a 1 is to be written into the flip-flop. Because it represents the worst case, let us consider the situation where a 1-to-0 data transition is desired. If a 1 is currently stored in the flip-flop, $v_{C1}(0-) = 5$ V and following Figure 11.6-5a, $v_{C1}(t)$ may be written as

$$v_{C1}(t) = 5e^{-t/R_{T2}C1} \tag{11.6-8}$$

Figure 11.6-5
Example 11.6-3 (continued).

Because of the large drain resistance on Q_1 and the resulting high gain of this inverter, v_{D1} will remain at zero volts until v_{C1} drops below the threshold

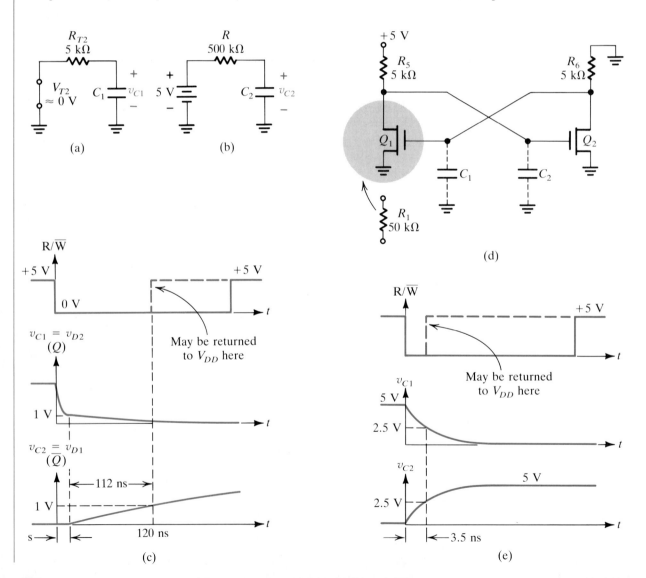

voltage at 1.0 V. The time required for this to occur is found by setting equation (11.6-8) equal to 1 V and solving for t, which yields

$$t_1 = R_{T2}C_1 \ln 5 = 8.0 \text{ ns} \qquad (11.6\text{-}9)$$

Once Q_1 cuts off, the equivalent circuit given in Figure 11.6-5b may be used to calculate the voltage across C_2 as

$$v_{C2}(t) = 5(1 - e^{-t/RC_2}) \qquad (11.6\text{-}10)$$

The time constant for this waveshape is 100 times bigger than that associated with v_{C1}. As soon as v_{C2} exceeds 1 V, Q_2 will turn on, and therefore after this point in time the R/$\overline{\text{W}}$ signal may be returned to a logic one level to terminate the write operation. The time required for Q_2 to turn on is found by setting equation (11.6-10) equal to 1 V and solving for t. Carrying out this procedure we obtain

$$t_2 = RC_2 \ln \left(\tfrac{5}{4}\right) = 112 \text{ ns} \qquad (11.6\text{-}11)$$

Thus the minimum-allowed write pulse width is dominated by the charging time of C_2, due to the very large value of the charging resistance. The waveshapes for the operation of this circuit are sketched in Figure 11.6-5c.

The circuit response time can be improved considerably by adding the second row-select transistor Q_5 to the circuit. When this is done, the basic operation of the circuit may be understood with the aid of Figure 11.6-5d. Here despite the fact that Q_1 is ON and operating in its ohmic region, it will have little effect on the circuit performance since its resistance (50 kΩ) is much larger than R_1. As a result, for $t > 0$, v_{C1} will discharge exponentially toward zero and v_{C2} will increase exponentially toward 5 V simultaneously, as illustrated in Figure 11.6-5e. The time constant for both of these exponentials is $\tau = R_5C_2 = R_6C_1 = 5$ ns. The minimum required pulse width of the R/$\overline{\text{W}}$ pulse is equal to the time needed for v_{C1} and v_{C2} to reach 2.5 V. Setting equation (11.6-8) equal to this voltage and solving for t, we find that this minimum time is $\tau \ln 2$ or about 3.5 ns from the time the row-select transistors turn on.

Exercises

11.6-1 The circuit shown in Figure E11.6-1 is an early BJT-style flip-flop. Given that the transistor $h_{FES} = 100$ and that the flip-flop has a logic 1 stored in it, so that Q_2 is cut off, compute I_{B1} and V_{B2}. **Answer** 148 μA, -0.667 V

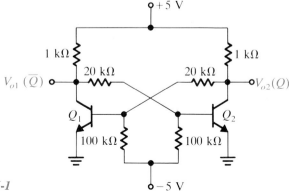

Figure E11.6-1

11.6 / Static Random Access Memory

11.6-2 Calculate the power dissipated in a 256-bit BJT-style RAM whose cells have the form shown in Figure E11.6-1. Neglect the power dissipated in the address decoders. *Answer* 6.53 W

11.6-3 How many transistors are needed to design a 10-bit tree-style decoder using two-input NAND gates containing three transistors per gate? The basic concept of a tree-style decoder is illustrated in Figure E11.6-3 for the case of a 3-bit address decoder. *Answer* 6152

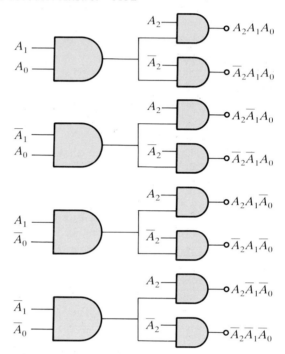

Figure E11.6-3

11.7 DYNAMIC RANDOM ACCESS MEMORY

Dynamic memory differs from the static memory circuits investigated in Section 11.6 in that the information is stored temporarily on a capacitor rather than being permanently stored in a flip-flop. As a result, the information in a dynamic memory needs to be refreshed, or reestablished, continually, to prevent it from being lost. This requires additional hardware both internal and external to the memory chip, and in addition, makes the memory unavailable for regular use while it is being refreshed.

Despite these apparently significant disadvantages, dynamic memory is exceedingly popular, its yearly sales easily exceeding those of conventional static RAM. The reason for this is not too difficult to understand if we examine the dynamic RAM cell illustrated in Figure 11.7-1. The memory cell of a modern dynamic RAM uses only one transistor and one capacitor. This design is considerably simpler than the six-transistor static RAM cells discussed earlier and illustrated in Figure 11.6-2. In addition, because the capacitor is formed vertically above the transistor as shown in Figure 11.7-1b, the IC real estate savings of dynamic over static RAMs can actually be on the order of 6:1.

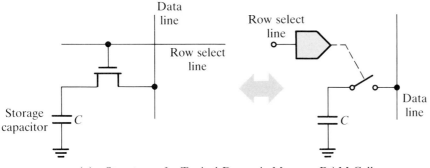

(a) Structure of a Typical Dynamic Memory RAM Cell

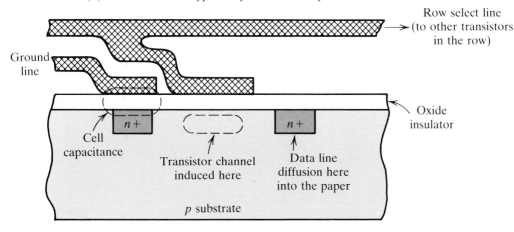

(b) IC Layout of the Memory Cell in an NMOS Dynamic RAM

Figure 11.7-1

Besides offering cell density advantages, dynamic RAMs also consume less power since no dc bias is required to maintain the data in the memory cells. As far as the cells themselves are concerned, dc power is needed only to refresh the data stored in them. This process is quite similar to the charge–discharge cycles in CMOS, so that the average power dissipation per cell is CV_{DD}^2f assuming that the cell discharges completely between refresh cycles.

Figure 11.7-2 illustrates the basic structure of a 16-bit 4×4 dynamic RAM. In many ways the circuit layout is quite similar to that of the MOS static RAMs discussed in Section 11.6. The major difference between this circuit and that shown in Figure 11.6-3 is that there are special sense/refresh amplifiers connected onto the column line of the dynamic RAM. Their function and their operation may best be understood with the aid of Figure 11.7-3. In part (a) of this figure we have drawn the overall circuit of column Y_2 from the 4×4 RAM illustrated in Figure 11.7-2. This circuit operates as follows.

There is one sense amplifier per column, and half of the cells from each column are connected onto either side of the sense amplifier. In addition to the regular memory cells, there are two additional cells, known as dummy cells, connected on either side of the sense amplifier. These cells are identical to the structure of the other cells in the memory, and they are enabled, or connected to the sense amplifier, when a row on the opposite side of the sense amplifier is selected. Thus the dummy cell on the left-hand side of the sense amplifier is enabled when cell 9 or cell 8 is selected. Conversely, the dummy cell on the

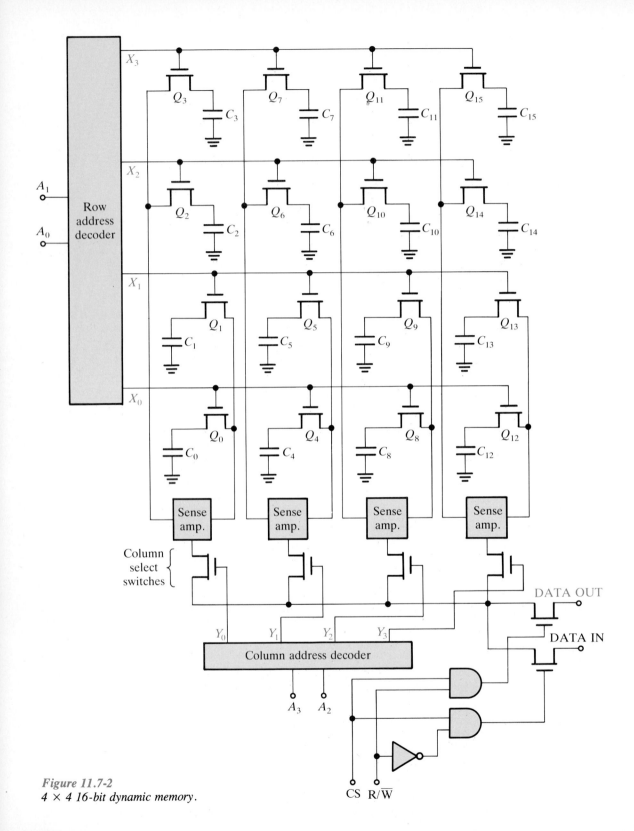

Figure 11.7-2
4 × 4 16-bit dynamic memory.

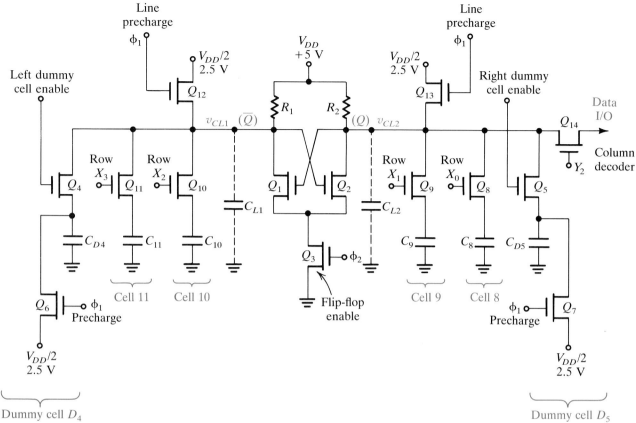

(a) Overall circuit for Column Y_2 for Dynamic RAM in Figure 11.7-2

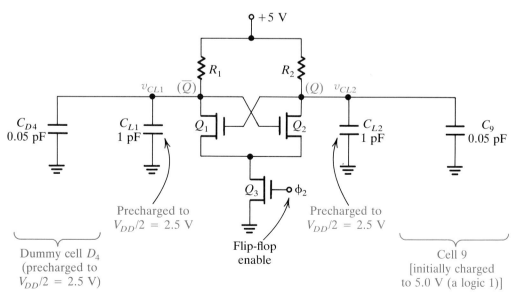

(b) Equivalent Circuit for Column Y_2 for the Dynamic Ram in Figure 11.7-2 during a Memory Read of Cell 9

Figure 11.7-3

right-hand side of the sense amplifier is enabled when cell 11 or cell 10 is selected. In this way the sense amplifier is always connected to the addressed cell on one side and to the dummy cell on the other. The memory cell capacitors and those of the dummy cells are nominally about 0.05 pF.

By examining the sense amplifier, it should be clear that it is just a flip-flop that can be turned on or off by the control transistor Q_3. The capacitances C_{L1} and C_{L2} shown at both inputs to the flip-flop are the parasitic capacitances associated with the column lines connecting the memory cells to the sense amplifiers. Typically, they have a value that is about 20 times that of a memory cell capacitor, or around 1 pF. They are included in this figure because they have a profound effect on the overall behavior of the circuit.

Let's examine the operation of this circuit by performing a memory read from cell 9. To accomplish this, address 9 is applied to the memory along with the CS signal. Besides enabling the chip, the CS pulse is used to produce the pulses ϕ_1 and ϕ_2 that control the operation of the sense amplifiers. When CS first goes HI, ϕ_1 is produced and this momentarily turns ON Q_6 and Q_7, which precharge the dummy cells on both sides of the sense amplifier to $V_{DD}/2$. In a similar fashion, the line capacitances are also charged to $V_{DD}/2$ through transistors Q_{12} and Q_{13} by ϕ_1. Once these precharging operations are completed, cell 9 and the left-side dummy cell D_4 are connected onto the column lines. The effective equivalent circuit for this case is shown in Figure 11.7-3b. The initial charges on each of the capacitances are indicated in the figure, and the final voltages on each of the equivalent parallel capacitances may be computed from conservation of charge as

$$v_{CL1}(0+)(C_{L1} + C_{D4}) = 2.5C_{L1} + 2.5C_{D4} \qquad (11.7\text{-}1a)$$

or
$$v_{CL1}(0+) = \frac{(2.5)(1) + (2.5)(0.05)}{1.05} = 2.50 \text{ V} \qquad (11.7\text{-}1b)$$

and
$$v_{CL2}(0+)(C_{L2} + C_9) = 2.5C_{L2} + 5C_9 \qquad (11.7\text{-}2a)$$

or
$$v_{CL2}(0+) = \frac{(2.5)(1) + (5)(0.05)}{1.05} = 2.62 \text{ V} \qquad (11.7\text{-}2b)$$

Thus the initial voltages on both sides of the flip-flop will be very close to one another regardless of whether cell 9 contains a zero or a 1. Of course, this difference could be increased by increasing the cell capacitances, but this would increase the size of the cell correspondingly, so that it is not very practical.

As we have demonstrated, when the cell contains a logic 1, the voltage at C_{L2} (2.62 V) will be slightly greater than the reference voltage at C_{L1} (2.50 V). A similar calculation for the case where the cell contains a logic 0 yields a voltage at C_{L2} of 2.38 V that is slightly below the reference. These voltages are the initial conditions on the flip-flop capacitances just prior to Q_3 enabling the flip-flop.

Once the flip-flop turns ON, it will immediately be in the active region since both v_{DS1} and v_{DS2} are at about $V_{DD}/2$, and therefore, per our discussion in Section 11.6 [see especially equations (11.6-5a) and (11.6-5b)], it will rapidly switch to a state determined by the capacitor initial conditions. For example, if $v_{CL1} < v_{CL2}$, Q_1 will turn ON and Q_2 will go OFF. With Q_2 OFF, v_{CL2} will rise to V_{DD}, and this will restore (or refresh) the logic 1 voltage level originally contained in cell 9. If, on the other hand, v_{CL1} is greater than v_{CL2}, Q_2 will turn ON and Q_1 will go OFF. With Q_2 ON, v_{CL2} will be about zero, and this will restore the logic 0 voltage level to cell 9. At the same time that the sense amplifier is enabled, the Y address decoder selects the output from the sense amplifier associated with column Y_2 in Figure 11.7-2. In Figure 11.7-3a this corresponds to the enabling of transistor Q_{14} on the right-hand side of the figure. For the case where cell 9 contains a logic 1 (or $+5$ V on the collector of Q_2), this would cause a logic 1 signal to appear on the data I/O line, which during a read operation would travel to the DATA OUT pin of the memory IC.

Almost the same procedure is followed during a memory write to cell 9 except that the data to be written into this cell is placed onto the data I/O line. Consider, for example, what would happen if the input data was a logic 1. In this case if Q_{14} has a low ON impedance, then when this transistor turns ON, v_{CL2} will be forced to $+5$ V regardless of the initial state of the flip-flop. This will turn Q_1 ON and cut Q_2 OFF, so that, as required, a logic 1 will be stored in memory cell 9.

EXAMPLE 11.7-1

Figure 11.7-4
Example 11.7-1.

The circuit shown in Figure 11.7-4a is a simplified version of the sense amplifier given in Figure 11.7-3b. If C_1 and C_2 are approximately 1 pF and $A = -10$ for each of the inverters, determine $v_{o1}(t)$ and $v_{o2}(t)$ if $v_{C1}(0-) = 2.50$ V and $v_{C2}(0-) = 2.62$ V. The power supply voltage $V_{DD} = 5$ V.

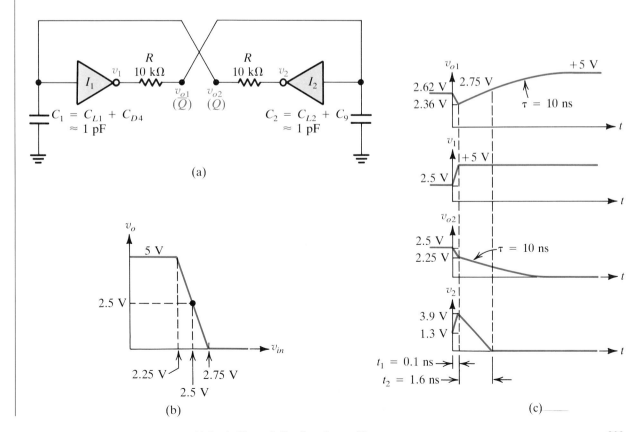

The transfer characteristics of each of the inverters in the flip-flop have the form given in Figure 11.7-4b. For $t > 0$, the inverters will be in the active region and thus the solutions for v_{o1} and v_{o2} will have the form developed in equations (11.6-5a) and (11.6-5b) with $V_A = v_{C2}(0-) = 2.62$ V and $V_B = v_{C1}(0-) = 2.50$ V. Substituting this data into the equations for v_{o1} and v_{o2}, we obtain

$$v_{o1}(t) = \frac{2.62 - 2.5}{2} e^{9t/\tau} + \frac{2.62 + 2.5}{2} e^{-11t/\tau}$$

$$= + 0.06 e^{9t/\tau} + 2.56 e^{-11t/\tau} \qquad (11.7\text{-}3a)$$

and $\qquad v_{o2}(t) = -0.06 e^{9t/\tau} + 2.56 e^{-11t/\tau} \qquad (11.7\text{-}3b)$

where $\tau = RC_1 = RC_2 = 10$ ns. Thus, as anticipated, v_{o1} will grow exponentially toward $+\infty$ while v_{o2} will grow negatively toward $-\infty$.

These positive feedback-style exponential growths will continue until one or both of the inverters leaves the active region. This occurs when the output of v_{o1} exceeds 2.75 V or when v_{o2} falls below 2.25 V. Solving both of these expressions numerically, we find that a decrease in the second term of equation (11.7-3b) causes v_{o2} to reach 2.25 V in about 0.1 ns. The corresponding value of v_{o1} at this point is 2.36 V. Interestingly enough, initially, v_{o1} falls, due to the drop in the second term before the positive exponent term takes over (Figure 11.7-4c). Once v_{o2} drops to 2.25 V, I_1 cuts off the v_1 stays at +5 V. As a result, beyond that point the positive feedback is lost and v_{o1} grows exponentially toward +5 V with a time constant $\tau = RC_2 = 10$ ns. When this voltage reaches 2.75 V, the output of I_2 will saturate at 0 V. The time required for this to occur is $t_2 = \tau \ln (2.64/2.25) = 1.6$ ns.

In examining the operation of the dynamic RAM in Figure 11.7-2, it is useful to notice that every time a CS signal is applied to the chip, all the cells contained in the row currently being addressed have their data refreshed or restored to their full logic 0 or logic 1 voltage levels. As a result, for the 4 × 4 RAM in this figure, only four CS pulses, along with the proper sequencing through the four row addresses 00, 01, 10, and 11, are needed to refresh the entire memory. Generally, this refresh sequence through all the row addresses needs to be repeated about once every millisecond to prevent the data from being lost.

Most commercially available dynamic RAMs operate slightly differently from the unit just discussed. To conserve on pins most of the popular dynamic RAMs, time-multiplex the information on the address pins, and use these pins for a double function. Initially, they are used to input the row portion of the address, and this data is latched into the row address decoder by applying the RAS or row address strobe. Shortly thereafter, the column address data must be provided, and it is used to select the proper sense amplifier. This is accomplished by applying the CAS or column address strobe. During a read or write operation the RAS–CAS sequence must occur while the chip select is a 1. However, during a refresh operation, only the row address and RAS need to be applied (CS is not required). This is important because it indicates that it is possible, simultaneously, to refresh the same row in all chips of the memory by doing a memory read on any one of them. The timing sequences for the refresh cycle, read cycle, and write cycle on a typical dynamic RAM are indicated in Figure 11.7-5. Note for this particular memory chip that the address and data enter the RAM on the falling edges of $\overline{RAS}$, $\overline{CAS}$, and R/$\overline{W}$ control signals.

With processor-based dynamic RAM systems, three methods are commonly used for refreshing the memory:

1. Processor interrupt refresh service routine (software burst mode)

Figure 11.7-5
Typical timing sequences in a dynamic RAM.

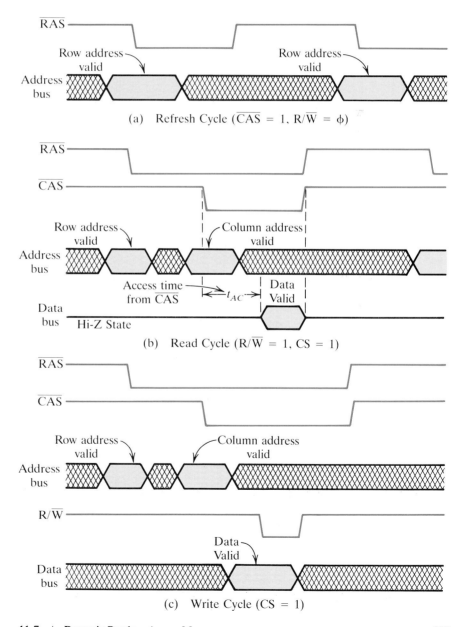

(a) Refresh Cycle ($\overline{CAS}$ = 1, R/$\overline{W}$ = ϕ)

(b) Read Cycle (R/$\overline{W}$ = 1, CS = 1)

(c) Write Cycle (CS = 1)

2. Hardware burst mode

3. Refresh between memory reads and writes during a normal processor instruction cycle

The first of these methods is illustrated in Figure 11.7-6a, and as is apparent from the figure, the hardware needed to implement this approach is minimal; a 1-kHz oscillator is all that is needed. This clock is used to interrupt the processor periodically, at the appropriate refresh interval, which in this case is assumed to be once every millisecond. When interrupted, the processor will finish the instruction on which it is currently working, and it will then enter the interrupt service routine outlined by the flowchart given in Figure 11.7-6b. Basically, this software refreshes the memory by stepping through each of the row addresses by doing a sequence of memory reads. This refresh method is said to operate in a software burst mode because once the routine begins, all the rows in the entire memory are refreshed before control is returned to the processor application software.

If we assume that the memory chips are 64K devices containing 256 rows each, the subroutine will need to go around the loop in the flowchart 256 times. If we further assume that each block in the flowchart requires one processor instruction to implement, and that the instruction execution time is 0.5 μs, this routine will take approximately (1.5 μs/pass) $\times$ 256 passes or about 384 μs to execute. This means that during 384 of every 1000 μs, about 38.4% of the time, the processor and the memory will be tied up carrying out memory refreshing. This represents very inefficient usage of both the processor and the memory, but if the time demands on the processor are low, it is a very inexpensive system to implement.

Figure 11.7-6c illustrates how this type of memory refresh may be accomplished much faster, but at the expense of additional hardware. This refresh circuit operates as follows. The 1-kHz refresh request oscillator (RFRQ) controls the cycling of the hardware. When its output goes to a logic 1 this sets flip-flop Q_1 and causes the processor to enter a HOLD state after completing the instruction on which it is currently working. When the processor enters the HOLD state, further CPU activity is suspended and an HLDA or hold acknowledge signal is generated by the processor. The HI on HLDA releases the row counter from reset and switches the multiplexers M_1 and M_2 so that the memory is now addressed from the counter, and also so that RAS is generated from the 4-MHz row counter clock. Assuming that CNTR is a 9-bit binary counter and that DONE is connected to the MSB, this bit will go HI after all 256 rows have been refreshed. This signal resets Q_1 and releases the processor from its HOLD state to resume normal operation.

If we assume that the memory has a 250-ns read cycle time so that the row counter clock can run at 4 MHz, all 256 rows can be refreshed in (256 $\times$ 250 ns) or 64 μs. Thus, with this hardware-intensive approach, the memory and the processor are unavailable for normal data processing functions for only 64/1000 or 6.4% of the time.

In the third approach to memory refreshing, advantage is taken of the fact that during certain parts of the instruction cycle on most processors, the memory will never be accessed. As a result, during these intervals it is possible to switch in an external counter and carry out the memory refresh during a normal CPU instruction cycle. The basic hardware and the timing diagram for

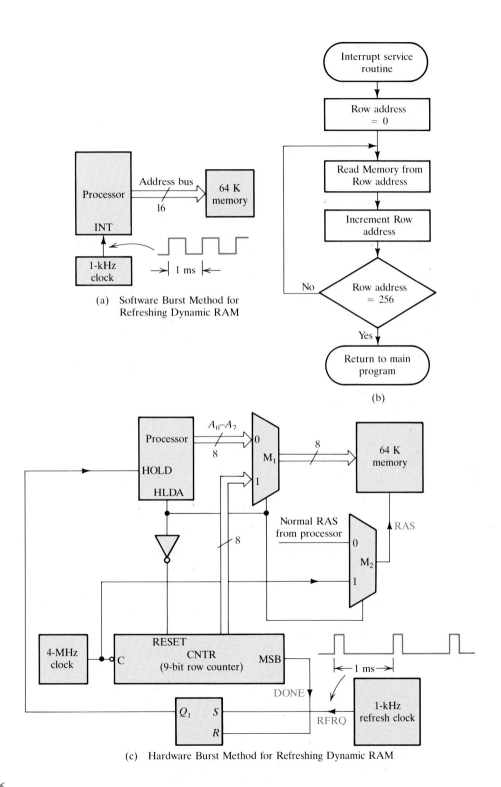

(a) Software Burst Method for Refreshing Dynamic RAM

(b)

(c) Hardware Burst Method for Refreshing Dynamic RAM

Figure 11.7-6

927

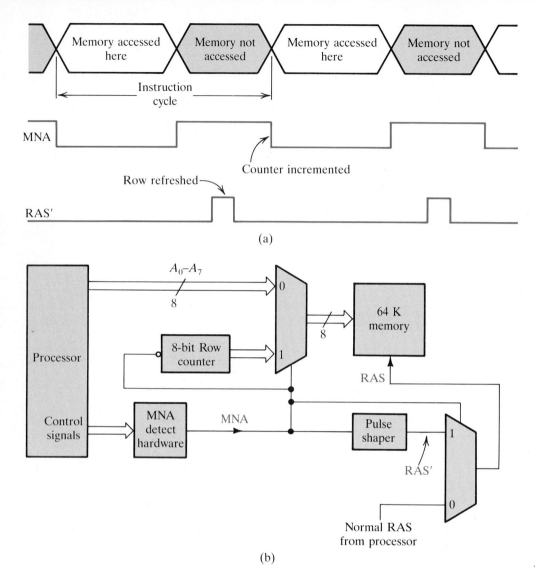

(a)

(b)

Figure 11.7-7
"Transparent" memory
refreshing during unused
portions of the instruction
cycle.

achieving this type of dynamic memory refresh are given in Figures 11.7-7a and b, respectively. Each time the processor enters the portion of the instruction cycle where its memory is never accessed, a signal, MNA, will be produced and it will be used to switch over the row addressing to the external counter. It is also used to generate the RAS signal, which will refresh the row currently being addressed by the counter and then bump the counter to the next row position. Because this type of memory refresh is carried out during time intervals where the processor will never access the memory, this refresh is "transparent" to the processor; in other words, the memory is available to the processor 100% of the time just as it would be if the memory chips were static instead of dynamic devices.

Exercises

11.7-1 The following information is provided for the refresh cycle of a particular 1 Mbit × 1 dynamic RAM chip (2048 columns × 512 rows).

$$t_{RAS} = \text{minimum RAS pulse width} = 120 \text{ ns}$$

$$t_{ASR} = \text{address setup time before RAS begins} = 0 \text{ ns}$$

$$t_{RAH} = \text{address hold time after RAS ends} = 15 \text{ ns}$$

$$t_{RP} = \text{RAS precharge time (time between RAS pulses)} = 90 \text{ ns}$$

a. Determine the smallest amount of time required to refresh this RAM if high-speed refresh hardware is available.

b. What percent of the memory's time is spent doing refresh operations if the memory must be refreshed once every 8 ms?

Answer (a) 0.10752 ms, (b) 1.34%

11.7-2 Repeat Exercise 11.7-1 if the memory is refreshed by the software burst technique. Assume that the average instruction cycle time is 500 ns, and use the flowchart in Figure 11.7-6b. *Answer* (a) 768 μs, (b) 9.6%

11.8 READ-ONLY MEMORY

A read-only memory (ROM) is a memory element that is used for the storage of nonchanging information. Like the read/write memory, it is generally a random access device in that the time required to read information from it is independent of the location of the information in the memory. One of the major attributes of ROMs is that the information stored in them is nonvolatile, so that if the power is lost, the information stored in the ROM is retained. ROMs are fabricated using both bipolar and MOS technology, with the bipolar devices being faster and the MOS offering lower power dissipation and higher packing density. ROMs have numerous applications in digital system designs. In computers they are used for high-speed code converters, microcode instruction sequencers, bootstrap loaders, and lookup tables. In addition, the development of the microprocessor has brought with it extensive use of read-only memory for program storage.

A ROM may also be used as a piece of programmable logic to perform the same functions as hardwired gates and flip-flops. In this application a single ROM can replace many SSI and MSI packages and can be used to design extremely complex sequential and combinational circuits. Besides offering substantial size, cost, and power advantages, ROM-based digital logic is also popular because it is easy to make "circuit" changes. With conventional hardwired logic designs, changes, at best, require the cutting of PC lands and the soldering in of jumper wires. Furthermore, all too often this type of board repair may be impossible, and an entirely new design may need to be made. With programmed logic the situation is quite different. To change the "circuit" the ROM is simply pulled and replaced by another one in which the required design changes have been made.

Several different kinds of read-only memories are available. Mask-programmed ROMs are the cheapest, at least in high-volume applications. These types of read-only memories have their information entered by the manufacturer during the IC fabrication stage. Figure 11.8-1 illustrates the design of a typical MOS-style mask programmed ROM. In manufacturing this type of read-only memory, the information is entered by one custom masking opera-

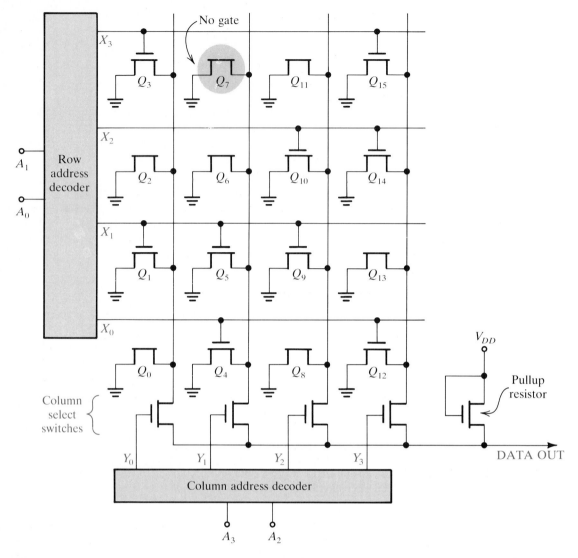

Figure 11.8-1
Mask-programmed MOS-style ROM.

tion at some point in the manufacturing process. With MOS ROMs, this is typically done during the final metallization of the ROM. Depending on whether a logic 0 or a logic 1 is to be entered at a particular memory cell, a metallized gate is either placed onto the cell transistor or is left off. For the ROM shown in Figure 11.8-1, the selection of a transistor having a metallized gate will cause the DATA OUT line to be pulled to ground, whereas a transistor without any gate will behave like an open circuit, so that DATA OUT will be at V_{DD}. Thus, in this ROM, logic 0's are entered by placing the gate metallization on the cell transistor, and logic 1's by leaving it off.

Because a special masking step is required to fabricate a conventional ROM, the manufacturer's setup charges for this procedure are expensive and can be justified only when these charges can be amortized over a large number of devices. Furthermore, the user must be absolutely certain of the validity of the data being entered before it is committed to ROM. Although this sounds like a minor point, this can be an expensive and time-consuming lesson, considering that it takes only a one bit error out of, say, 264,000 bits to render useless the ROM that you have ordered. Furthermore, if at a later date it is desired to make any changes in the ROM data, a new ROM will again be needed. Because of these drawbacks, mask-programmed ROMs should be used only in high-volume applications, where the data to be entered are well defined and are unlikely to be changed in the future.

Mask-programmed ROMs are currently available in MOS and CMOS versions with densities up to about 1 megabit and access times on the order of 150 to 200 ns. Bipolar ROMs are also available in somewhat smaller sizes having access times as low as 10 to 20 ns.

The programmable read-only memory (PROM) was one of the first alternatives developed to the mask-programmed ROM. This device is user programmable, so that a single stock item PROM IC can be used for many different tasks. Figure 11.8-2 illustrates the cell design employed in a fusible-link programmable read-only memory.* Both metals, such as NiCr (nichrome), and semiconductors, such as polycrystalline silicon, are used as fuse materials.

Regardless of the fuse material used, the programming operation is the same. The fuse to be blown is selected by applying the appropriate address to the memory, and a higher-than-normal voltage is connected to the PROM. This causes a large current to flow in the fuse material, which melts it and produces an open circuit.

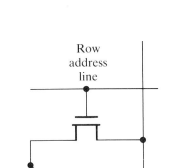

Figure 11.8-2
Fusible-link PROM.

When, for example, programmable memory cells of the type illustrated in Figure 11.8-2 are used in place of transistors Q_0 through Q_{15} in the memory in Figure 11.8-1, the selection of a particular memory cell by the row address decoder places a HI on the gate of the transistor in that cell. If the fuse in that cell is intact and if the column select transistor associated with that cell has also been enabled, this will produce a logic 0 on the DATA OUT line. If, on the other hand, the fuse associated with the cell being addressed is blown, there will be no voltage drop across the pullup resistor and the DATA OUT line will be a logic 1.

Because of their increased cell complexity, PROMs tend to be less dense, slower, and somewhat more costly than mask-programmed ROMs. However, their ease of use make them a strong competitor of conventional ROMs. One of the principal disadvantages of fuse-link PROMs is that they cannot be reused or reprogrammed. Of course, if the change that needs to be made in the PROM involves a link that should have been blown but was not, this can simply be done on the original PROM. But where multiple ROM bit changes are needed, it is inevitable that a blown link will need to be reconnected, and this simply cannot be done.

For applications where the ROM needs to be reprogrammed again and again, the erasable programmable read-only memory (EPROM) is the indicated choice. Originally used for microprocessor software development applications, these types of ROMs have become so inexpensive that they are now shipped as integral parts of finished products. EPROMs are MOS devices and as such their access speeds are significantly slower than conventional BJT-style ROMs, typically being about 200 to 450 ns.

*As a practical matter, fusible link PROMs are usually BJT devices. However, this MOS version is satisfactory for illustrative purposes.

Figure 11.8-3 illustrates the layout of one of the more popular types of EPROM memory cells. It uses the FAMOS or floating-gate avalanche injection MOS memory cell design. Other than the introduction of a polysilicon floating gate, this transistor looks just like any other enhancement-style MOS device.

When the floating gate is uncharged, the I_D versus V_{GS} characteristic for this transistor is similar to that of an ordinary MOSFET except for the fact that its threshold voltage is a little higher, due to the shielding effect of the floating gate (Figure 11.8-3b). If a high voltage (of about 25 V) is applied to the drain and the gate simultaneously, the transistor will be active and substantial current will flow from the drain to the source. Furthermore, the high electric field in the drain–substrate region causes avalanche breakdown across this junction, and this will contribute additional current in the form of high-speed electrons. Because of their high energy, some of these electrons, attracted by the high voltage on the control gate, will penetrate the thin oxide and enter the polysilicon floating gate. Once on the floating gate, these electrons will lose their energy due to collisions, and they will be trapped on the gate. Because of the excellent insulating properties of silicon dioxide, these charges can remain stored on the gate for many years.

The introduction of these negative charges onto the floating gate of the transistor tends to cancel out the effect of any positive voltage applied to the control gate, and this effectively increases the threshold voltage of the transistor, as shown in Figure 11.8-3b.

To convert the ROM in Figure 11.8-1 into an EPROM, the memory cell transistors Q_0 through Q_{15} are replaced by FAMOS devices. When no charge is stored on their floating gates, their control gate threshold voltages will be $+2$ V, and therefore a read operation will turn on the transistor of the selected cell, making DATA OUT a logic 0. When the floating gate is fully charged, the control gate threshold voltage for the transistor will be about $+10$ V, and thus when this cell is addressed by a $+5$ V row address signal, the transistor will remain OFF and DATA OUT will be a logic 1.

To erase the contents of the EPROM, the IC is exposed to ultraviolet (UV) light. To facilitate this exposure, the top of these types of EPROMs are fitted

Figure 11.8-3
Erasable programmable read-only memory (EPROM).

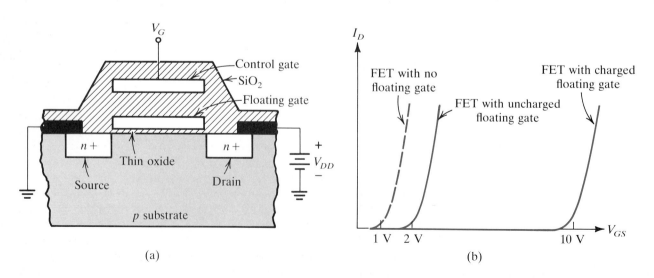

(a)   (b)

with quartz windows. The high-energy photons emitted from this light source cause electron–hole pair generation in the silicon dioxide, making it photoconductive and allowing the charge stored on the floating gate to leak off. Typically, 15 or 20 minutes of exposure is required to remove completely the charge stored on the floating gate.

512K EPROMs are currently available in both NMOS and CMOS versions, and their costs are such that it is becoming increasingly difficult to justify the use of either mask-programmed ROMs or fuse-link PROMs unless high speed is required. One drawback of EPROMs is that they must be removed from the circuit during reprogramming. This means that the system must be shut down, the EPROM removed from the socket and reprogrammed or replaced, and the system reinitialized.

To get around this difficulty, a new type of ROM has been developed known as an electrically erasable programmable read-only-memory (E^2PROM). Because this PROM can be both programmed and erased by means of electrical control signals, reprogramming of this IC can occur while the device remains in its socket in the system. Applications for this type of ROM include the updating of data stored in remotely controlled systems such as space probes and navigational aids. These ROMs are also useful in point-of-sale terminals for storing sales receipts and the latest product prices.

As Figure 11.8-4a illustrates, the design of the memory transistor of an E^2PROM is quite similar to that of an ordinary EPROM except that the floating gate is placed very close to the drain electrode. Because this spacing is so narrow (on the order of 200 Å), reversible tunneling can occur between the floating gate and the drain, and this will allow the cell to be programmed and erased electrically.

To charge the floating gate, a programming voltage of about +20 V is applied to the control gate while the drain is maintained at ground potential (Figure 11.8-4b). This large electric field between the control gate and the drain allows the tunneling of electrons across the thin oxide and onto the floating gate. Of course, as with the FAMOS EPROM transistor cell examined earlier, once the programming voltage is removed, it is reasonable to expect that the charge on the floating gate will remain there for a long time due to the excellent insulation properties of the SiO_2 surrounding the gate. Furthermore, and again as with the FAMOS EPROM transistor, the presence of this negative charge on the floating gate effectively increases the threshold voltage of this transistor so that it will always appear as an open circuit.

To remove the charge from the floating gate, the polarity of the programming voltage connected between the drain and the control gate is reversed, as illustrated in Figure 11.8-4c. In this situation tunneling again occurs, but this time from the floating gate to the drain. This depletes the charge from the floating gate—so much so, in fact, that the threshold voltage can actually become negative.

Figure 11.8-4d illustrates the design of an electrically erasable programmable read-only memory cell. To convert the 4×4 ROM in Figure 11.8-1 into an E^2PROM device, 16 such cells are required. In Figure 11.8-4d, Q_A is the cell data storage transistor and Q_B is used to carry out the cell selection. Q_B is needed because the data storage transistor Q_A will not necessarily turn off when the row decoder voltage is zero since it can have a negative

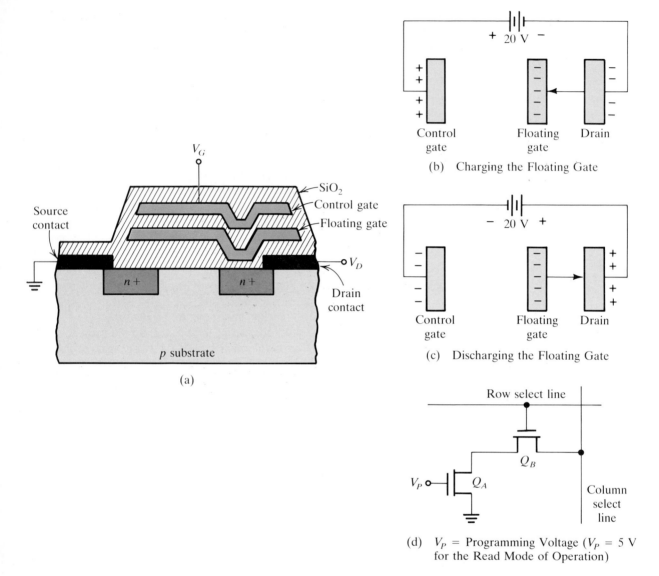

V_G

Source
contact

SiO$_2$
Control gate
Floating gate

V_D

Drain
contact

$n+$ $n+$

p substrate

(a)

+ 20 V −

Control
gate   Floating
gate   Drain

(b)   Charging the Floating Gate

− 20 V +

Control
gate   Floating
gate   Drain

(c)   Discharging the Floating Gate

Row select line

Q_B

V_P Q_A

Column
select
line

(d) V_P = Programming Voltage (V_P = 5 V
for the Read Mode of Operation)

Figure 11.8-4
Electrically erasable
programmable read-only
memory (E^{2}PROM).

threshold voltage. Q_B, on the other hand, is an ordinary enhancement-style transistor, and as such it can be switched on and off reliably with the 0-V and 5-V row address decoder output voltage levels.

Currently, 64K E^{2}PROMs with 200-ns read and write times are available, and significant size increases should occur in the next few years since these devices are near the beginning of their learning curve. In addition, although they are considerably more expensive than UV-style EPROM at the present time, it is reasonable to expect that the two will soon be competitively priced and that electrically erasable PROMs will inevitably replace UV EPROMs as the most popular type of read-only memory.

In the two examples that follow we illustrate several uses of read-only memories, in addition to their most obvious application, as a program storage element. We begin with an example that illustrates how a ROM can be employed as a function generator to produce one or more specialized waveshapes, and then in the second example demonstrate how ROMs can be used to replace random logic in the design of combinational and sequential circuits.

EXAMPLE 11.8-1

Design a ROM-based circuit that will produce the four periodic waveforms illustrated in Figure 11.8-5a.

SOLUTION

The basic concept involved in the design of this type of function generation circuit is to divide the waveform up into a sequence of equal time increments and to associate each interval with a corresponding word in the memory. For the case of a binary waveform, the number of bits required per word is equal to the number of waveforms to be produced. For the case of analog waveshapes, however, the number of bits that will be needed depends on the number of quantizing levels used. For example, if 1% amplitude accuracy were required, at least 7 bits would be needed for each waveshape sample to be stored in the ROM.

Once the word size is determined, which for this example would simply be 1 bit per waveform, or a total of 4 bits per word, the sequence of waveshape amplitudes is simply stored in sequential memory locations in the ROM. By reading this data out sequentially, using a counter, for example, the correct waveshapes will be obtained directly at the ROM output lines for the case of binary waveforms. For the case of an analog waveform that has been digitized and stored in the ROM, the binary output data from the ROM would need to be passed through a D/A converter in order to reconstruct the original analog waveshape.

Figure 11.8-5
Example 11.8-1.

Figure 11.8-5b illustrates the basic hardware needed to produce the waveforms, and part (c) of the figure gives a listing of the required ROM contents.

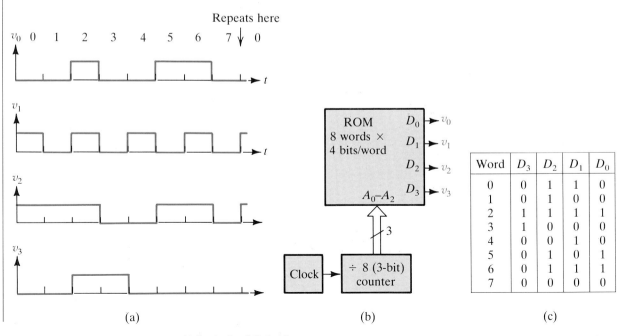

(a) (b) (c)

EXAMPLE 11.8-2

The state diagram given in Figure 11.8-6a is to be implemented using a ROM-based sequential circuit design. The variable X is the system input signal, and Z_0 and Z_1 are the desired outputs. As is apparent from an examination of the state diagram, the system has four states, S_0, S_1, S_2, and S_3, which for simplicity will be designated as 00, 01, 10, and 11, respectively. In this diagram the labels within the circles are of the form Q_1Q_0/Z_1Z_0, where Q_1Q_0 is the state number, and Z_1 and Z_0 are the outputs associated with the present state of the system. In addition, in this diagram the labels next to the arrows represent the value of the input variable causing the state transition indicated. When an arrow does not have a label associated with it, the transition occurs unconditionally on receipt of the next clock pulse. Using a ROM and as many D-type flip-flops as needed, complete the design of this circuit.

SOLUTION

The basic concept of using a ROM to implement a sequential circuit design is to couple it with a latch (or set of D-type flip-flops). The number of flip-flops needed to accomplish this task is equal to n, where 2^n is greater than or equal to the number of states in the system. Thus, for this four-state sequential circuit design, two flip-flops will be needed.

The contents of the flip-flops determines the current state of the system, and this state information is fed back to the input of the ROM to form part of its address. The remainder of the address is obtained from the system inputs.

Figure 11.8-6
Example 11.8-2.

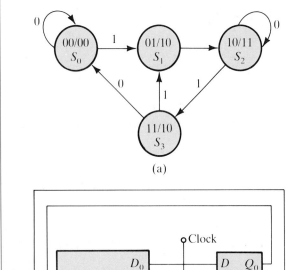

(a)

(b)

Address			Data			
A_2	A_1	A_0	D_3	D_2	D_1	D_0
(Input X)	$\begin{pmatrix} \text{Present State} \\ Q_1 \quad Q_0 \end{pmatrix}$		$\begin{pmatrix} \text{Outputs} \\ Z_1 \quad Z_0 \end{pmatrix}$		$\begin{pmatrix} \text{Next State} \\ Q_1 \quad Q_0 \end{pmatrix}$	
0	0	0	0	0	0	0
0	0	1	1	0	1	0
0	1	1	1	0	0	0
0	1	0	1	1	1	0
1	1	0	1	1	1	1
1	1	1	1	0	0	1
1	0	1	1	0	1	0
1	0	0	0	0	0	1

(c)

This design is illustrated in Figure 11.8-6b. The combination of the current state of the system and the present inputs to the system determines the next state of the system, and the system outputs are a function of the present state of the system. Both the system outputs and the next state of the system are determined by the data stored at the ROM location currently being addressed.

All of this sounds a lot more complicated than it really is, and it can probably be most easily understood if we examine parts (a), (b), and (c) of Figure 11.8-6 simultaneously, starting at the point where Q_1 and Q_0 are initially both zero, or in other words, with the system in state S_0. In this situation, if $X = 0$, we are addressing word 000 in the ROM. Furthermore, following the state diagram, with the system in state S_0 and with $X = 0$, the system should stay in that state with the system outputs at $Z_1 = Z_2 = 0$. Because D_2 and D_3 are directly connected to the output lines, both of these bits should be zero. In addition, the next state of the system will be determined by the values loaded into the flip-flops when the next clock pulse occurs. Because the system is to stay in state S_0 as long as $X = 0$, the values of D_1 and D_0 also both need to be zero. Thus the data entries 0000 at memory location 000 will keep the system in state zero with both outputs zero as long as $X = 0$.

When X changes to 1, the ROM address switches to 100, and corresponding to this value of X, two things should occur. First, since the system is still in state 00, the outputs must remain at $Z_1 = 0$ and $Z_0 = 0$. Second, on the next clock pulse the system is to move to state S_1 ($Q_1 = 0$ and $Q_0 = 1$). Both of these tasks can be accomplished by making the ROM contents at memory location 100 equal to 0001, as indicated in the ROM listing in Figure 11.8-6c.

We continue this analysis for one more entry to be sure that the technique is completely understood. With the system in state S_1, the state diagram indicates that a transition will take place to state S_2 on the next clock pulse regardless of the value of X, and also that the system output in state S_1 will be 10. With the system in state S_1, the ROM word being addressed will be 001 or 101, depending on whether X is a 0 or a 1. Since the transition to the next state and the system outputs are to be independent of the value of X here, both of these ROM words should contain a 1010. The remaining ROM data may by computed similarly.

Exercises

11.8-1 For the system in Figure 11.8-6, X is held at a logic 1 level continuously and the clock period is 5 μs. Find the duty cycle and the frequency of the output at Z_0 in the steady state. *Answer* 33%, 66.7 kHz

11.8-2 Find the ROM size needed to create a sine lookup table accurate to 0.01%. *Answer* ROM size $\simeq$(16K $\times$ 14) using symmetry considerations

---

11.9 PROGRAMMABLE LOGIC ARRAYS AND GATE ARRAYS

In some respects a programmable logic array (PLA) may be considered to be a generalized type of ROM. The reason for this will be apparent after we compare the internal structure of these two types of LSI arrays using the diagrams in Figures 11.9-1 and 11.9-2, respectively. In Figure 11.9-1a we have drawn the block diagram of a ROM containing 2^4 or 16 one-bit words. In some respects this ROM looks a bit different from the ones that we drew earlier, in that the X and Y address decoders have been combined into a single linear address decoder. This is done to simplify our discussion of the address decoding process, and should not be interpreted as a significant change in the actual structure of the ROM.

In the particular read-only memory shown in the figure, the gate metallizations have been left off Q_0, Q_2, and Q_{14}. As a result, a logic 1 output is obtained from the ROM when the X_0, X_2, or X_{14} address combinations are applied. For all other input address combinations, one of the data transistors will turn ON and the F_0 output will be a logic 0.

If we view a ROM as being a combinational circuit composed of an array of AND gates followed by an array of OR gates, the equivalent circuit of the ROM just described may be drawn as indicated in Figure 11.9-1b. Here the AND gates represent the function performed by the address decoder, and because the addresses in a ROM are fully decoded, all possible address combinations appear at the AND gate outputs. Note, for simplicity, that the inputs to the gates are drawn as single lines, and the fact that they actually represent multiple input connections is indicated by the presence of more than one dot on the input line.

The OR gate in the equivalent circuit represents the combination of the data transistors and the active load pull-up transistor at the top of the column in the original ROM. In addition, the address decoder lines that drive the open gate transistors show up in the equivalent circuit as connections between their corresponding AND gate outputs and the OR gate input. Following this figure, the Boolean expression for the output F_0 from this ROM may be written as a product of sums, where

$$F_0 = X_0 + X_2 + X_{14} = (\bar{A}_3\bar{A}_2\bar{A}_1\bar{A}_0 + \bar{A}_3\bar{A}_2A_1\bar{A}_0 + A_3A_2A_1\bar{A}_0) \qquad (11.9\text{-}1)$$

In a ROM each possible address combination is represented by a gate output in the AND gate array, and then those combinations which are the minterms of the output are simply ORed together to produce the overall expression for the output data. For the specific ROM in Figure 11.9-1c, the terms selected to appear in F_0 are indicated by placing dots on the OR gate input line. In a ROM, no Boolean algebra or Karnaugh map simplification techniques are employed to reduce the complexity of the combinational circuit produced, and as a result, a ROM equivalent circuit is basically a line-for-line realization of the original truth table. Although the greatest flexibility is achieved by using this approach (in that any Boolean output function can be realized), the circuit complexity is at a maximum.

The programmable logic array was developed as an alternative to the use of ROMs for situations where the number of inputs is large, yet the number of product terms needed in the final Boolean output expression is relatively small. Figure 11.9-2a illustrates the architecture of a typical PLA, and comparing this design with that of the ROM in Figure 11.9-1c, we can see that a PLA has the same basic structure as a ROM except that the input address decoder is replaced by a programmable AND gate array. As such, because the number of

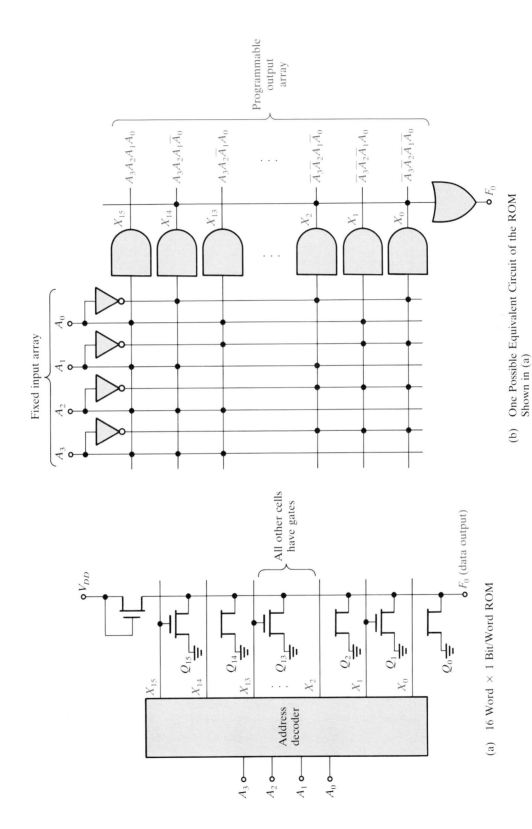

(b) One Possible Equivalent Circuit of the ROM Shown in (a)

(a) 16 Word × 1 Bit/Word ROM

Figure 11.9-1

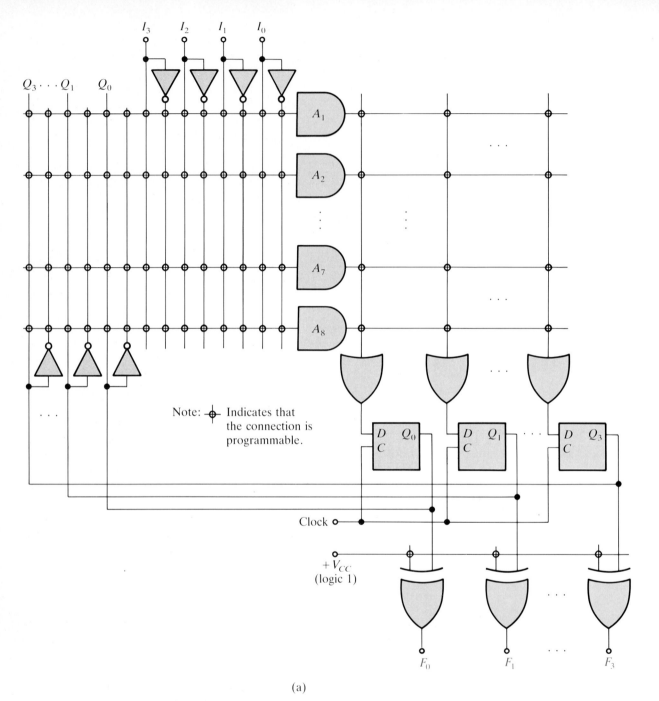

Figure 11.9-2
Basic architecture of a PLA (a four-input, four-output, eight-product term PLA).

(a)

Control IN	IN	OUT	
0	0	0	} Noninverting
0	1	1	
1	1	1	} Inverting
1	0	0	

(b)

inputs to the AND gate array is no longer fixed, the product terms at the outputs of these gates, rather than being simple minterms, can now be prime implicants or simplified Karnaugh map expressions that cover many words in an equivalent ROM. Because of this, the number of AND gates, or the number of product terms in a PLA, is usually much smaller than the 2^n gates found in the address decoder of an n-input ROM. However, despite this, a small PLA can often perform the same logic replacement function as a much larger read-only memory.

Consider, for example, the case of a 14-input, 8-output, 96-product-term PLA. This type of logic array can be used to implement, simultaneously, up to eight different Boolean logic functions containing 14 variables each as long as the total number of product terms in the simplified forms of these output expressions in less than 96. A ROM of equivalent logic replacement power would need 2^{14} eight-bit words, for a total ROM size of 131,072 bits. Although this type of ROM is not terribly expensive today, another factor to consider is that all these bits need to be programmed in order to use this device. Because the equivalent PLA has a much smaller number of product terms, its programming is also much simpler.

The PLA shown in Figure 11.9-2a besides having the programmable AND gate and OR gate arrays already discussed, also has programmable output inverters and output latches whose signals can be fed back to the input AND gate array. The latches are for storing the current state of the system, and as we shall see in Example 11.9-1, their use will permit the design of PLA-based sequential circuits that use only a single chip.

The exclusive OR gates in series with the latch outputs are programmable inverters. Because one of the inputs to the EXOR gates in programmable, following Figure 11.9-2b it is apparent that this gate will act like a noninverting buffer when the control input is zero, and will invert the latch output signal when the control line signal is a logic 1. This output inversion ability is useful because the complement of the output function can frequently be implemented with fewer product terms than the function itself. In other words, it may be easier to cover the zeros in the Karnaugh map representation of a particular Boolean logic function than to cover the 1's.

EXAMPLE 11.9-1

Design a programmable logic array circuit that will be able to generate the three waveforms shown in Figure 11.9-3a. The period of the waveforms is $7T$, where T is the shortest interval during which the outputs are constant.

SOLUTION

Basically, to produce the three waveforms indicated, we will need to design a seven-state counter-style sequential circuit and a companion combinational circuit that will produce the required outputs when the system is in each state. Because three flip-flops or latches will be needed to represent the seven possible states of the system, the basic structure of the PLA circuit to be designed will need to have the form shown in Figure 11.9-3b.

Based on the waveforms to be generated and the state numbers 0 through 6 listed on the top waveshape in part (a) of the figure, the table given in Figure 11.9-3c may readily be constructed. Note that an entry has been included for

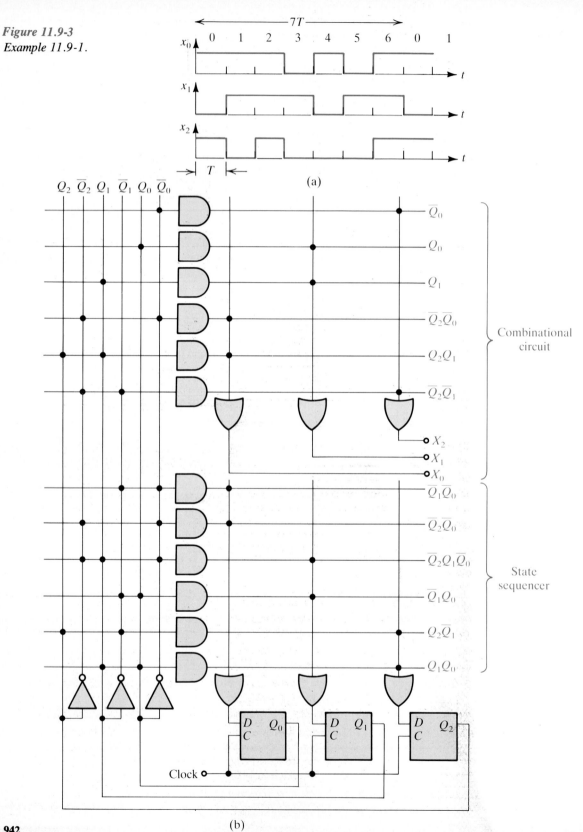

Figure 11.9-3
Example 11.9-1.

(a)

(b)

Figure 11.9-3
(continued)

Present State Q_2 Q_1 Q_0			Next State Q_2 Q_1 Q_0			Outputs in Present State X_2 X_1 X_0		
0	0	0	0	0	1	1	0	1
0	0	1	0	1	0	1	1	0
0	1	0	0	1	1	1	1	1
0	1	1	1	0	0	0	1	0
1	0	0	1	0	1	1	0	0
1	0	1	1	1	0	0	1	0
1	1	0	0	0	0	1	1	1
1	1	1	ϕ	ϕ	ϕ	ϕ	ϕ	ϕ

(c)

Q_0

Q_2 \ Q_1Q_0	00	01	11	10
0	1	0	0	1
1	1	0	ϕ	0

X_0

Q_2 \ Q_1Q_0	00	01	11	10
0	1	0	0	1
1	0	0	ϕ	1

Q_1

Q_2 \ Q_1Q_0	00	01	11	10
0	0	1	0	1
1	0	1	ϕ	0

X_1

Q_2 \ Q_1Q_0	00	01	11	10
0	0	1	1	1
1	0	1	ϕ	1

Q_2

Q_2 \ Q_1Q_0	00	01	11	10
0	0	0	1	0
1	1	1	ϕ	0

X_2

Q_2 \ Q_1Q_0	00	01	11	10
0	1	1	0	1
1	1	0	ϕ	1

(d)   (e)

state 7 even though it will never occur during normal operation of the circuit. For the present this will be listed as a don't-care entry in order to simplify the final PLA design. This will be okay as long as the final values selected for these don't-cares allow the system to move out of state 7 should it inadvertently find itself there. Of course, this situation could arise during the initial power-up since it is equally likely that the flip-flops will start out in any one of the eight possible states. Therefore, when we are done with the design, we will need to check that the state 7 entries for Q_2, Q_1, and Q_0 are not 111. This will ensure that the system will not hang up in this state if it initially finds itself there, and will also ensure that the circuit will recover to normal operation should a glitch momentarily force it into this illegal state.

As with the design of a ROM-based sequential circuit, the signals fed into the D-flip-flops are just the values for the next state that is desired. The Boolean expressions for these control signals may be simplified with the aid of the Karnaugh maps given in Figure 11.9-3d, which yield

$$Q_0 = \overline{Q_1}\overline{Q_0} + \overline{Q_2}\overline{Q_0} \tag{11.9-2a}$$

$$Q_1 = \overline{Q_1}Q_0 + \overline{Q_2}Q_1\overline{Q_0} \tag{11.9-2b}$$

and

$$Q_2 = Q_2\overline{Q_1} + Q_1Q_0 \tag{11.9-2c}$$

Based on these results, a next-state output of 100 will be obtained when the system is in state 7, so that the circuit will always recover on its own from an inadvertent transition to this illegal state.

In the design of the combinational part of this circuit, the data on the right-hand side of this table would have been used directly to form the input words in a ROM-based design. In our PLA design, this data can also be simplified by using the Karnaugh maps given in Figure 11.9-3e since this will result in fewer product terms. Note that don't cares have also been used in state 7 to aid further in this simplification process. Following these maps, the simplified forms for X_0, X_1, and X_2 may be written immediately as

$$X_0 = \overline{Q_2}\overline{Q_0} + Q_2Q_1 \tag{11.9-3a}$$

$$X_1 = Q_0 + Q_1 \tag{11.9-3b}$$

and

$$X_2 = \overline{Q_0} + \overline{Q_2}\overline{Q_1} \tag{11.9-3c}$$

Combining these results with the sequential portion of the design, we may draw the completed PLA circuit as illustrated in Figure 11.9-3b. For simplicity the sequential and combinational parts of this array have been drawn separately. However, in an actual PLA design, common product terms (such as $\overline{Q_2}\overline{Q_0}$) should be combined to reduce the circuit complexity. Not shown in this figure is the external clock circuit that is needed to sequence the PLA. The frequency of this clock determines the overall period of the waveshapes, and of course the clock period also determines the interval T.

Another type of logic array that has become increasingly popular in recent years is the gate array or uncommitted logic array. This type of array is known as a semicustom IC because elemental logic gates of the inverter, NAND, and NOR gate variety are already fabricated on the substrate, and then one or more custom metallizations is applied to the IC to interconnect the gates and form the final circuit. Because it is possible for the manufacturer to fabricate and stockpile the unconnected forms of these gate arrays, the cost and the time needed to produce these semicustom ICs are substantially less than for fully customized versions of these chips.

Gate array ICs are presently available in CMOS, ECL, and TTL versions, and because of the reduced interconnect distances, compared with standard SSI/MSI versions of the same circuit, gate array devices tend to operate much faster than their SSI/MSI counterparts. Table 11.9-1 illustrates the current state of the art in gate arrays. Because most applications do not require the speed of ECL or TTL, CMOS, because of its very low power consumption, tends to be the most popular type of gate array.

Figure 11.9-4a illustrates the layout of a typical CMOS gate array IC. It consists of columns of uncommitted p-channel and n-channel transistors grouped together in cells as shown in part (b) of the figure. As indicated in this figure, the gates of each complementary transistor pair are connected with polysilicon, and pads are brought out on both sides of the column for later connection to other parts of the chip. In addition to the polysilicon gate connections, two vertical metallizations run the length of the channel. These serve as ground and power supply connection points. Because these metallizations partially obscure the view of the polysilicon, an expanded view of one of these transistor pairs is shown in Figure 11.9-4c with the metal overlay removed.

Note that considerable space between the columns is provided for running the customized wiring that will interconnect the transistors and ultimately convert this uncommitted logic array into a semicustom digital LSI circuit. An example of how this is carried out is shown in Figure 11.9-5. Here we have il-

11.9-1 Typical Characteristics of Gate Array ICs

Parameter	Logic Family		
	CMOS	**ECL**	**TTL (ALS)**
Number of equivalent gates/chip	400–4000	300–3000	250–2000
Delay time (ns)/gate	2–3	0.5–1	2–2.5
Output buffer delay time (ns)	10–15	0.8–1.7	5–7
Input buffer delay time (ns)	3–5	0.5–1.0	1.7–2
Number of input buffers	33–188	56–180	42–80
Number of output buffers	36–168	28–80	30–68
Power dissipation	30 μW/MHz/ gate, 1.5 mW/ output buffer	1.1–5.4 mW/gate	1.1–1.4 mW/gate

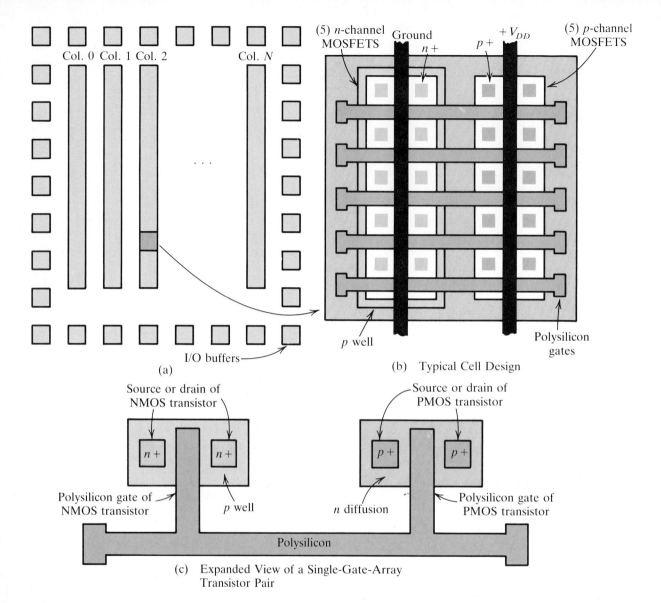

(a)

(5) *n*-channel MOSFETS   Ground $+V_{DD}$ (5) *p*-channel MOSFETS

n+   *p*+

Col. 0   Col. 1   Col. 2   Col. *N*

I/O buffers

p well   Polysilicon gates

(b)   Typical Cell Design

Source or drain of NMOS transistor

Source or drain of PMOS transistor

n+   *n*+   *p*+   *p*+

Polysilicon gate of NMOS transistor

p well   *n* diffusion

Polysilicon gate of PMOS transistor

Polysilicon

(c)   Expanded View of a Single-Gate-Array Transistor Pair

Figure 11.9-4

Architecture of a typical CMOS gate array.

lustrated the interconnections needed to convert this cell into a three-input AND gate. As a practical matter, this has been accomplished by cascading a three-input NAND gate and an inverter as shown in parts (a) and (b). The actual circuit layout is given in part (c). Note that either side of the transistors may be used as the drain or the source since it is the applied biasing that determines the location of these terminals.

Most of the layout of this circuit is fairly straightforward, with the exception of the transistor pair Q_1–Q_5. The gates of these transistors are permanently wired to V_{DD} and this cuts off Q_5 and permanently turns on Q_1. This is done to allow the drain-to-source connection of Q_1 to serve as a low-impedance path under the ground metallization to interconnect OUT_1 to IN_4.

One of the interesting aspects of gate array design is that computer-aided design (CAD) techniques are used almost exclusively to ensure a successful cir-

Figure 11.9-5
Metallization to convert a cell in a CMOS gate array into a three-input AND gate.

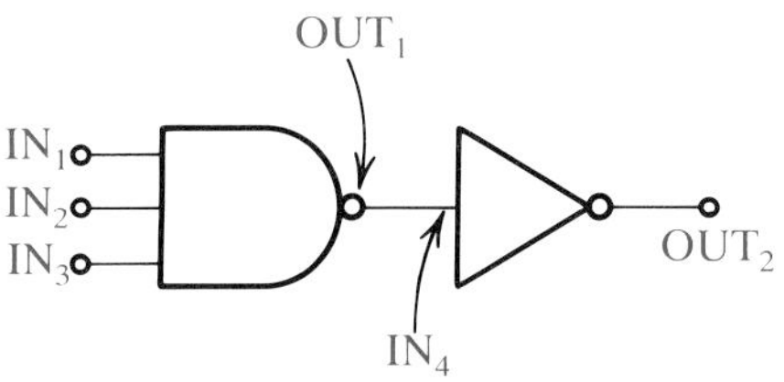

(a)

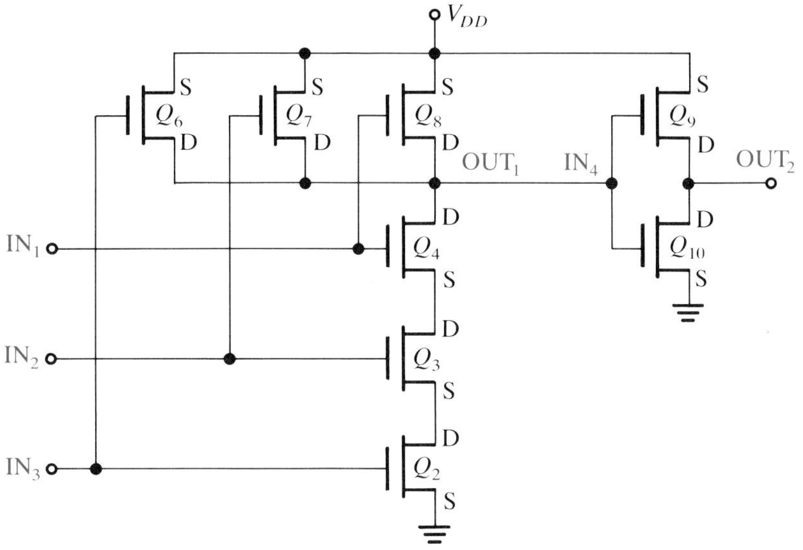

(b)

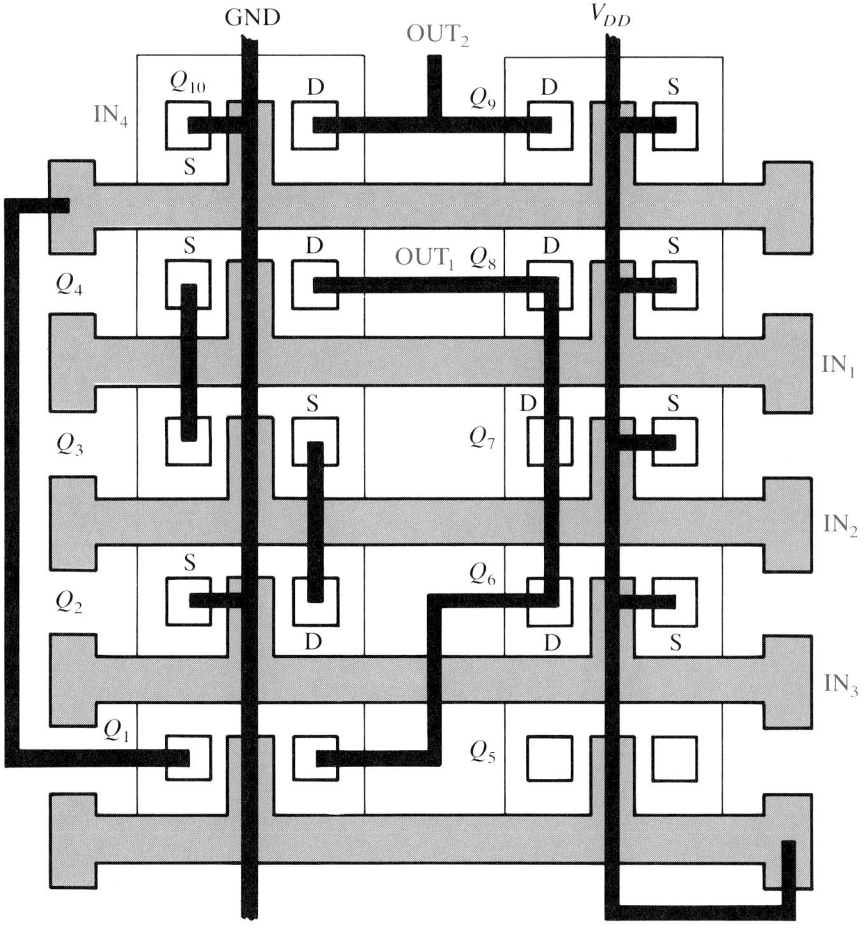

(c)

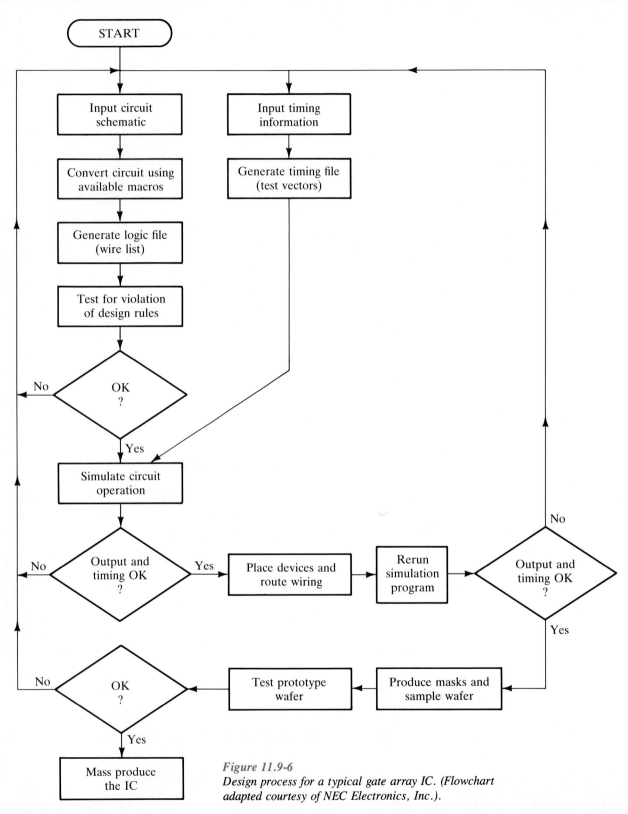

Figure 11.9-6
Design process for a typical gate array IC. (Flowchart adapted courtesy of NEC Electronics, Inc.).

cuit design on the first pass. The design sequence of a typical gate array IC is presented in the flowchart in Figure 11.9-6. In this design process automatic schematic capture programs are commonly employed that directly convert the schematic of the intended circuit into the required wire list. In developing the schematic, it is not necessary to draw the circuit at the transistor or gate level. Instead, each manufacturer has a list of allowed "macros," such as AND, NAND, OR, NOR, and EXOR gates, as well as latches, flip-flops, multiplexers, demultiplexers, adders, counters, and RAMs. If the schematic is drawn using the appropriate symbols for these devices, the proper component interconnections needed to form these macros can be implemented automatically.

Once the schematic has been converted to a stored computer file (or wire list), a logic rule checking program takes over and tests the fan-out on each device to make sure that it is within the proper limits. In addition, it computes the propagation delays to various points within the circuit.

Now that it has been determined that the circuit design does not violate any logic rules, it is time to see if the circuit actually operates as it is supposed to. This is accomplished by supplying a set of test vectors (or input signals) to the simulation program. This program then exercises the circuit using the test vectors and determines the output signals (or set of output vectors) that the circuit produces. In conventional circuit design, this part of the development process is akin to building a breadboard of the circuit. However, the simulation program is superior to the breadboard design technique in that it carries out a worst-case analysis by using the typical, maximum, and minimum gate delays associated with each part in the circuit. Prototype breadboards are almost never tested to this extent.

Once the basic performance of the chip design has been shown to be correct, the IC is ready to be laid out. This is accomplished with a placement and routing program that optimizes the positioning of the gates and routes the wiring and interconnects the devices to form the final IC. The simulation program is then run again, including the capacitive and resistive effects of the wiring. If this test is satisfactory, the metal masks are laid out automatically and a sample wafer is fabricated. Once the wafer is tested and found to be operating satisfactorily, the semicustom gate array that was designed is ready to be mass produced.

Exercise

11.9-1 For the PLA circuit in Figure 11.9-3b, a connection is made from the $Q_1 Q_0$ AND gate output to the OR gate input on flip-flop Q_0. Find the frequency of the waveshape produced at Q_0 if the clock is 10 MHz. *Answer* 1.67 MHz.

PROBLEMS

11.2-1 For the CMOS oscillator in Figure 11.2-3, $R =$ 100 kΩ, $C = 1000$ pF, and a resistor $R_1 = 50$ kΩ in series with a diode (cathode up) is connected in parallel with

R. Carefully sketch and label v_1, v_2, and v_3 assuming that the gates have input protection diodes.

11.2-2 An oscillator is made by cascading three CMOS inverters G_1–G_3 and connecting a resistor R from the output of G_3 to the input G_1 and a capacitor C from the output of G_2 to the input of G_1. Sketch the gate input and output voltages in the steady state assuming that the gates have input protection diodes.

11.2-3 For the three-inverter CMOS oscillator in Figure 11.2-7c, carefully sketch and label each of the gate outputs if the gate propagation delays are each t_p seconds.

11.2-4 A ring oscillator is formed by cascading 5 inverters G_1 through G_5 and feeding the output of G_5 back to the input of G_1. Carefully sketch and label the steady state output of each inverter if t_p is the gate propagation delay. What is the oscillation frequency?

11.2-5 Paralleling the development in Problem 11.2-4, sketch the gate output voltages if the gates are CMOS and if each has an output resistance R and an input capacitance C_{in}. Determine the circuit oscillation frequency considering the internal gate delays, other than those associated with the R's and the C's to be zero and the gate thresholds to be at $V_{DD}/2$.

11.2-6 (a) The gates in the Schmitt trigger shown in Figure P11.2-6a are CMOS. Sketch the hysteresis characteristic for this circuit.

(b) Find the frequency of oscillation for the circuit shown in Figure P11.2-6b, and sketch v_1, v_2, and v_o. The Schmitt trigger in this circuit is that given in part (a) of the figure.

11.3-1 Repeat the analysis of the one-shot in Figure 11.3-2 if G_1 and G_2 are replaced by NOR gates, R is removed from ground and connected to V_{DD}, and the polarity of the trigger is reversed so that it is normally zero and rises to V_{DD} for T_1 seconds.

11.3-2 Repeat the analysis of the one-shot in Figure 11.3-2 if $T_1 = 2T$.

11.3-3 For the one-shot in Figure 11.3-2, the gate propagation delays are each 50 ns, and R and C are 10 kΩ and 0.01 μF, respectively. In addition, a capacitor C_1 is connected in series with v_{in} and a resistor R_1 is connected from v_4 to $+V_{DD}$. Select appropriate values for R_1 and C_1 to trigger the one-shot properly and indicate the reasons for your choices. The width of the trigger pulse is 100 μs.

11.3-4 Using one or more 555 timer ICs and any other logic gates that are required, design a circuit that will delay an incoming 10-V, 100-μs pulse by 1 ms.

11.3-5 The input signal v_{in} is a 10-V pulse having a width equal to T seconds. Using as many 555 timer ICs and logic gates as necessary, design a circuit that will produce a 100 μs wide 10-V pulse at v_{o1} on the rising edge of v_{in} and a second 100 μs wide 10-V pulse at v_{o2} on the falling edge of v_{in}.

11.3-6 The input signal v_{in} is a 10-V pulse having a width equal to T seconds. Using as many 555 timer ICs and logic gates as needed, design a circuit that will produce a 10 μs wide output pulse on the falling edge of v_{in} only when 100 μs $< T <$ 200 μs.

11.3-7 For the circuit shown in Figure P11.3-7:
(a) Select R_B to produce a 50% duty cycle, assuming the diodes to be ideal.
(b) Determine the frequency of oscillation for the conditions in part (a).

11.4-1 For the circuit shown in Figure P11.4-1, carefully sketch and label v_1, v_2, v_3, and v_o if the switch is closed at $t = 0$ and is opened at $t = 1$ s. Assume a 20 ms bounce time.

11.4-2 Repeat the analysis of Example 11.4-1 (Figure 11.4-2) if the input to the Schmitt trigger is v_{R2} instead of v_1. Include a sketch of the voltage v_{R2}.

11.4-3 For the circuit shown in Figure P11.4-3, carefully sketch and label v_1, v_2, and v_3 when the switch is closed at $t = 0$. Assume that the gate is CMOS and that the switch bounce time is 20 ms.

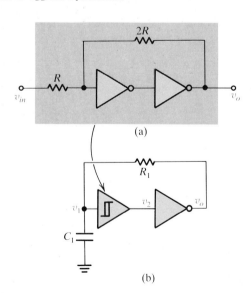

Figure P11.2-6

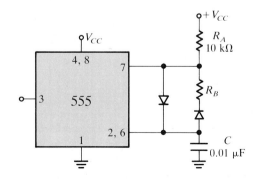

Figure P11.3-7

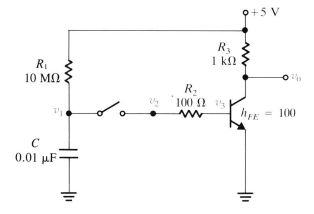

Figure P11.4-1

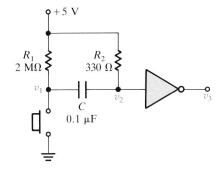

Figure P11.4-3

11.5-1 Design a 32K microcomputer memory containing 16K of ROM and 16K of RAM, considering that the computer and the RAM chips have common data input and output paths. Assume that 4K ROM and RAM chips are available, and that these memory IC's each have an output enable (OE) line which together with the CS signal controls the output data tristates. Include all address decoding. The microprocessor can directly address 64K of memory and the read and write control signals are to be labeled MEMR and MEMW, respectively.

11.5-2 A memory is designed using five 64K RAM chips and four 32K ROM ICs.
(a) If the memory is one byte wide, how many bits are there in each RAM chip?
(b) Repeat part (a) for the ROMs.
(c) What are the total number of bytes in the memory?
(d) What are the total number of bits in the memory?

11.5-3 Design a 256K × 16-bit microcomputer memory using 64K × 8 RAM chips. Consider that the computer and the RAM chips have common data input and output paths, and that the memory IC's each have an output en-

able (OE) line which together with the CS signal controls the output data tristates. The microprocessor can directly address 1 megabyte of memory and the read and write control signals are designated as MEMR and MEMW, respectively.

11.6-1 For the circuit in Figure 11.6-1, $A = -5$, $R = 2$ kΩ, and $C = 20$ pF. Furthermore, initially $v_{o1} = V_{CC}$ and $v_{o2} = 0$.
(a) Sketch v_1, v_{o1}, v_2, and v_{o2} if the trigger signal results in $V_A = 0.51V_{CC}$ and $V_B = 0.49V_{CC}$.
(b) Repeat part (a) if $V_A = 0.49V_{CC}$ and $V_B = 0.51V_{CC}$.

11.6-2 For the flip-flop in Figure 11.6-1a additional switching transistors (not shown in the figure) are connected into the circuit during the application of a trigger pulse. In effect, with Q a logic 1 they introduce a resistor $R/10$, connected from v_{o1} to ground and a second resistor, also $R/10$, going from v_{o2} to V_{CC}.
(a) How long must the trigger be applied to guarantee switching of flip-flop? In carrying out this analysis consider that $V_{CC} = 10$ V, $C_1 = C_2 = 5$ pF, $R = 10$ kΩ, and that the inverter gains are -10.
(b) Sketch v_1, v_{o1}, v_2, and v_{o2} if the trigger pulse width is twice that computed in part (a).

11.6-3 (a) Repeat part (a) of Problem 11.6-2.
(b) Sketch v_1, v_{o1}, v_2, and v_{o2} if the trigger pulse width is one-half that computed in part (a).

11.6-4 Repeat Problem 11.6-2 if the resistor $R/10$ connected from v_{o1} to ground is removed.

11.6-5 In the beginning of Section 11.6 we analyzed the switching characteristics of a set of cross-coupled inverters (Figure 11.6-1). Repeat this derivation and determine the results obtained for V_{o1} and V_{o2} when the inverters are replaced with noninverting buffers each having gains of 10.

11.6-6 (a) Design a 256-word address decoder using eight-input NAND gates. How many transistors are needed if each NAND gate and inverter require nine and two transistors, respectively?
(b) Repeat part (a) if the eight-input NAND gates are replaced by dual four-input gating circuits of the type shown in Figure P11.6-6.

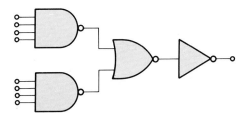

Figure P11.6-6

11.7-1 Repeat Example 11.7-1 if the inverter gains are each equal to -2.

11.7-2 The two circuits shown in Figure P11.7-2 are examples of static and dynamic address decoders, respectively. Little needs to be said about the operation of the static decoder. But for the dynamic device, the signal ϕ_1, which is derived from the RAS pulse, turns on Q_8, and this transistor then attempts to charge C_L.

(a) Considering that the decoder transistors have ON resistances of about 500 Ω, explain how the dynamic address decoder circuit operates.

(b) Find the address for which V_0 in each of these circuits is HI.

(c) Estimate the average power consumed by each of these decoders. The refresh rate is 1 kHz.

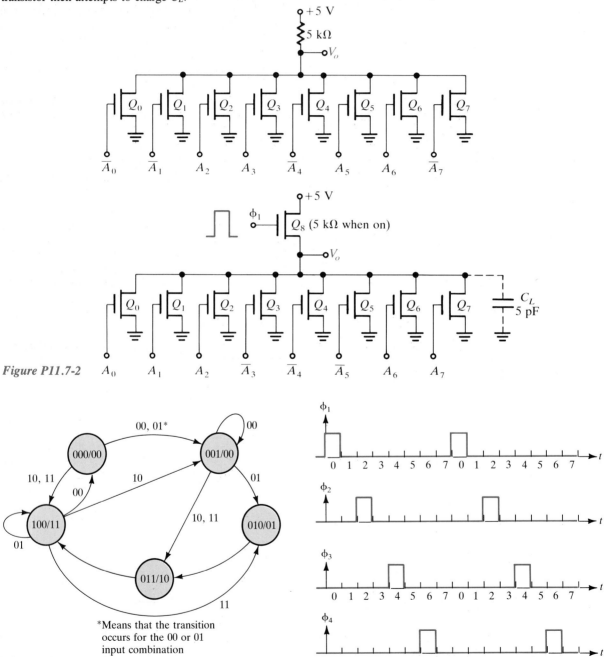

Figure P11.7-2

Figure P11.8-2

*Means that the transition occurs for the 00 or 01 input combination

Figure P11.8-3

11.7-3 Estimate the average power consumption of a 5-V 256-kilobit dynamic RAM having 0.5-pF gate storage capacitors and 10-pF row capacitors. Assume that each cell is refreshed once every millisecond and neglect the power dissipated in the address decoders.

11.8-1 **(a)** Redesign the circuit in Figure 11.8-5b to permit the following switch-selectable waveform sequences to be generated (the numbers refer to the time intervals specified in the original wave-shapes in Figure 11.8-5a).

Sequence 1: 0, 1, 2, 3, 4, 5, 6, 7, 0, 1, . . .

(the same as before)

Sequence 2: 0, 1, 2, 3, 0, 1, 2, 3, 0, . . .

Sequence 3: 0, 2, 4, 6, 0, 2, 4, 6, 0, . . .

(b) Show the connections of the selector switch that is needed.

11.8-2 Design a ROM-based sequential circuit to implement the state diagram shown in Figure P11.8-2.

11.8-3 Design a four-phase eight-state ROM-based clock generator producing the waveforms shown in Figure P11.8-3.

11.8-4 Design a decimal keyboard encoder using a ROM. The circuit inputs are to be 10 SPST switches representing

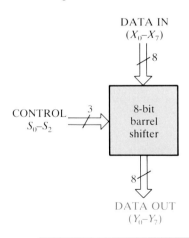

DATA IN
(X_0-X_7)

8

CONTROL
S_0-S_2

3

8-bit barrel shifter

8

DATA OUT
(Y_0-Y_7)

Control	Data Out						
S_2 S_1 S_0	Y_7	Y_6	Y_5	Y_4	Y_3	Y_2	Y_1 Y_0
0 0 0	X_7	X_6	X_5	X_4	X_3	X_2	Y_1 X_0
0 0 1	X_6	X_5	X_4	X_3	X_2	X_1	Y_0 X_7
0 1 0	X_5	X_4	X_3	X_2	X_1	X_0	X_7 X_6
0 1 1	X_4	X_3	X_2	X_1	X_0	X_7	X_6 X_5
1 0 0	X_3	X_2	X_1	X_0	X_7	X_6	X_5 X_4
1 0 1	X_2	X_1	X_0	X_7	X_6	X_5	X_4 X_3
1 1 0	X_1	X_0	X_7	X_6	X_5	X_4	X_3 X_2
1 1 1	X_0	X_7	X_6	X_5	X_4	X_3	X_2 X_1

Figure P11.9-4

the numbers 0 through 9, with a 0 volt input occurring when the switches are open and +5 V when they are closed. The circuit is to have 5 outputs (D_0 through D_4) with D_0–D_3 representing the binary code of the switch that has been pressed, and D_4 a signal called "key-pressed (KP)" that goes to a logic 1 when any key is struck.

11.8-5 Repeat Example 11.8-1 but add a start-stop switch. When the switch is in the start (or logic 1) position, the output waveshapes are to be generated continuously, and when the switch in the stop (or logic 0) position, v_0 through v_3 are to remain at the logic 0 level. In addition, when the switch is thrown into the stop position in the middle of a waveshape sequence, the sequence is to finish before the outputs go to a logic zero.

11.8-6 Design a ROM-based binary counter that counts backward from $(15)_{10}$ to 1 in accordance with the sequence

15, 13, 11, 9, . . . 5, 3, 1, 15, 13, 11, . . .

11.9-1 Repeat Problem 11.8-3 using a PLA.

11.9-2 Explain the difference between ROM and a PLA, indicating the advantages and disadvantages of each.

11.9-3 Repeat Problem 11.8-2 using a PLA.

11.9-4 A barrel shifter combinational circuit is similar to an end-around shift register. The major difference between the two is that no clocking signal is required in the combinational circuit. For the barrel shifter shown in Figure P11.9-4, the number of bits that the data are shifted is determined by the control signals S_0 to S_2 in accordance with the truth table shown in the figure. Design a PLA circuit to implement this 8-bit barrel shifter.

11.9-5 Lay out the gate array design for a cell containing a three-input OR gate following the procedure used in Figure 11.9-5 for a three-input AND gate.

APPENDIX I

A Review of Basic Network Concepts

In this appendix we will briefly review several network simplification techniques and network analysis methods that will be used over and over again in this text. If you have completed a course in network theory, you should already be familiar with each of the subjects to be covered, and in this case the short review being presented should be sufficient to jog your memory on each of these subjects. However, if this material is unfamiliar to you, then you should also consult the references listed at the end of this appendix.

I.1 VOLTAGE AND CURRENT DIVIDERS

For simple single-loop series circuits, such as the one shown in Figure I.1-1, the applied input voltage divides proportionally across each of the resistors. These proportionality constants are known as the circuit's voltage-divider coefficients with the voltage across R_2, for example, being obtained by taking the applied input voltage V_{BB} and multiplying it by $R_2/(R_1 + R_2)$, the ratio of the resistor whose voltage drop is desired divided by the sum of the resistors in the loop. In this example V_2 is therefore equal to $10 \times (2/5)$ or 4 volts, and in a similar fashion the voltage V_1 may be calculated as $10 \times (3/5)$ or 6 volts.

For parallel circuits, such as the one shown in Figure I.1-2, the applied input current divides proportionally through each of the resistive branches. These proportionality constants are known as the current-divider coefficients, and for the simple two-resistor parallel circuit shown in this figure the coefficient associated with the current flow through one of the resistors is equal to the ratio of the opposite resistance divided by the sum of the resistors. Thus, for example, the current through the branch containing R_2 may be obtained by

954

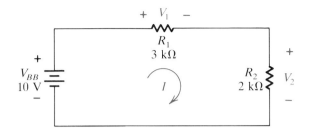

Figure I.1-1
Circuit to illustrate the voltage-divider concept.

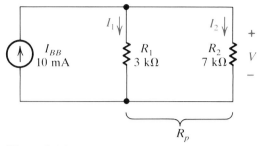

Figure I.1-2
Circuit to illustrate the current-divider concept.

multiplying the applied current I_{BB} by the current-divider coefficient $R_1/(R_1 + R_2)$, so that I_2 in this specific example is 3 mA. In a similar fashion I_1 is readily found to be 7 mA.

I.2   THE SUPERPOSITION PRINCIPLE

Voltage and current-divider concepts are very useful for handling single-loop single-source problems, but when the network contains multiple loops and/or multiple sources, the situation is more complex and may require the solution of several simultaneous network equations. Often this sort of problem can be solved more easily by applying the superposition principle in combination with the previously described voltage and current-divider techniques.

The principle of superposition may be stated as follows:

1. In a linear network containing two or more independent sources, the overall response of the network may be determined by adding the responses due to each source acting alone, that is, with all the other sources set equal to zero.
2. To set a source equal to zero, its magnitude is reduced to zero and the source is replaced by the equivalent circuit element that remains. As a result, when a voltage source is reduced to zero, it is replaced by a short circuit; when a current source is reduced to zero, it is replaced by an open circuit.
3. If the network contains dependent (or controlled) sources, they are left alone and are not set equal to zero.

EXAMPLE I.2-1

To illustrate the application of the superposition principle to a specific problem, consider the circuit given in Figure I.2-1a and find the voltage V_2.

SOLUTION

Conventional analysis of this problem would involve the solution of a two-loop circuit which at best would be tedious and time consuming to develop. Superposition provides us with a much simpler technique that has the added advantage of showing how much each of the independent voltage sources V_A and V_B contribute to the overall voltage at V_2. The portion of V_2 due to V_A is obtained by setting $V_B = 0$ (Figure I.2-1b). Combining R_2 and R_3 in parallel produces a 2.67 kohm equivalent resistance, and if we make use of the voltage-divider relationship, we can calculate the contribution of V_A to V_2 as 5.71 volts. A simi-

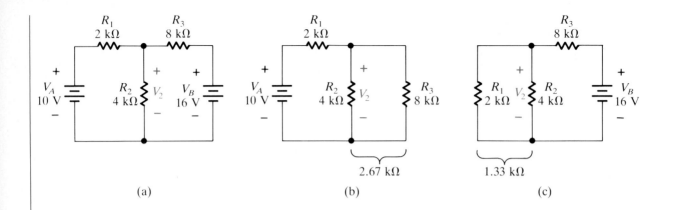

Figure I.2-1
Example I.2-1.

lar calculation for the contribution of V_B to V_2 using the equivalent circuit of Figure I.2-1c yields 2.28 volts. Combining these two results together, we obtain the overall voltage at V_2 as $(5.72 + 2.28) = 8.0$ volts. To convince yourself that superposition really does work you should also solve this problem by using a more conventional networks approach, such as loop or nodal analysis.

I.3 THÉVENIN AND NORTON EQUIVALENT CIRCUITS

Any linear two-terminal network containing voltage sources, current sources, and resistors can be replaced by a single resistor in series with a voltage source (the Thévenin equivalent) or by a current source in parallel with the same resistance (the Norton equivalent). These two equivalent-circuit forms are illustrated in Figures I.3-1a and I.3-1b, and the element values in these circuits are determined as follows:

1. The voltage source v_T in the Thévenin equivalent is equal to the open-circuit voltage (v_{OC}) at the terminals of the network, that is, the voltage v with network 2 removed.
2. The current source i_N in the Norton equivalent is equal to the short circuit current (i_{SC}) at the terminals of the network with the external circuitry (network 2) removed.
3. The resistor value R_T to be used in these equivalent circuits may be found by one of several different approaches.
   a. If the network contains only independent sources, R_T, the Thévenin resistance, is simply equal to the equivalent resistance "seen" between the terminals of the circuit when the external network is removed and the internal independent sources are set equal to zero (i.e., voltage sources shorted, and current sources opened).
      When the network contains controlled sources, this method is difficult to apply since these sources cannot be set equal to zero (unless the voltage or current controlling them is external to the network). Therefore, for networks containing controlled sources, one of the two alternate methods listed below is suggested for finding R_T.
   b. The Thévenin resistance can always be written as the ratio of the open-

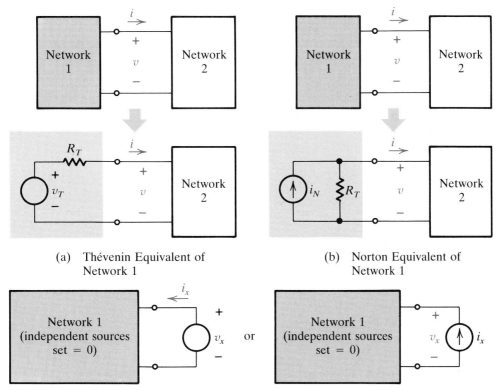

(a) Thévenin Equivalent of
    Network 1

(b) Norton Equivalent of
    Network 1

(c) Alternative Method for Finding the Thévenin
    Resistance of Network 1

Figure I.3-1
Basic concepts of Norton and
Thévenin equivalents.

circuit voltage to the short circuit current or.

$$R_T = \frac{v_{OC}}{i_{SC}} \qquad (I.3\text{-}1)$$

c. R_T can also be computed by setting the independent sources inside of the network to zero and applying an external source, v_x, to the network, measuring the current i_x which flows (Figure I.3-1c). R_T in this case is written as

$$R_T = \frac{v_x}{i_x} \qquad (I.3\text{-}2)$$

Alternatively, a current source i_x can be applied and the resulting voltage v_x measured with equation (I.3-2) again applied to compute R_T.

I.4   ROOT MEAN SQUARE

The RMS or root mean square value of a time-varying voltage is equal to the equivalent dc voltage that will produce the same heating effect, or average power dissipation, in a resistor. Mathematically, this dc equivalent heating

Waveform Description	Sketch of the Waveform	RMS
Sine wave	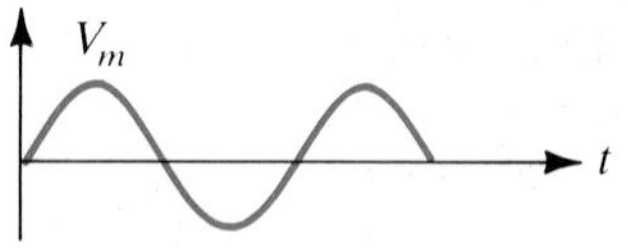	$\dfrac{V_m}{\sqrt{2}}$
Half-Wave rectified sine wave		$\dfrac{V_m}{2}$
Full-wave rectified sine wave	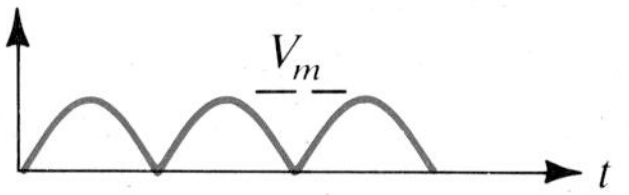	$\dfrac{V_m}{\sqrt{2}}$
Square wave	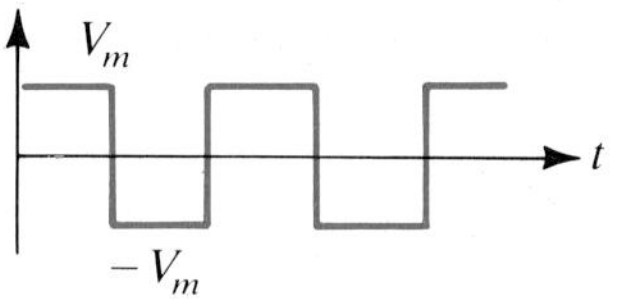	V_m
Partial sine wave	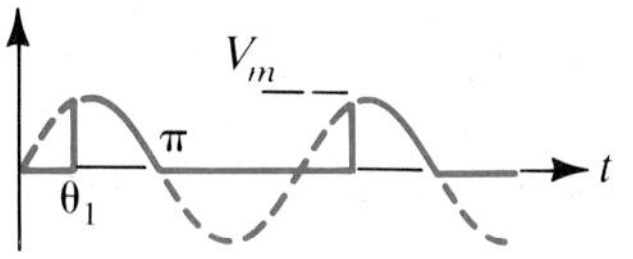	$V_m\sqrt{\dfrac{1}{8} - \dfrac{\theta_1}{4\pi} - \dfrac{1}{8\pi}\sin 2\theta_1}$

Figure I.4-1
Rms of selected waveshapes.

voltage, V_{RMS}, can readily be shown to be given by the expression

$$V_{\text{RMS}} = \underbrace{\sqrt{\underbrace{\frac{1}{T}\underbrace{\int_0^T v^2(t)dt}}}}_{\text{Root \quad Mean \quad Square}} \tag{I.4-1}$$

The RMS value for the current is defined similarly by replacing $v^2(t)$ by $i^2(t)$ in the equation.

The table in Figure I.4-1 gives the RMS value of several important waveforms. Consider for example the sine wave entry on the first line of this table. This data indicates that a sine wave with amplitude V_m produces the same heating effect in a resistor as a battery with a dc voltage equal to $V_m/\sqrt{2}$. This result can be verified mathematically by substituting $V_m \sin(2\pi t/T)$ into equation (I.4-1) and solving for V_{RMS} using the trigonometric identity $\sin^2 x = (1 - \cos 2x)/2$.

I.5 DETERMINING THE TRANSIENT RESPONSE OF SIMPLE ELECTRICAL NETWORKS BY INSPECTION

The analysis of the transient response of simple RC and RL circuits containing batteries and/or dc current sources typically involves the solution of first-order differential equations whose drivers on the right-hand side are constants. In general the transient response solution for all these types of circuits has the same form, and as a result it is possible to determine the voltage or current at specific points in these circuits by inspection without ever explicitly writing down the circuit's differential equation. To accomplish this we note that the differential equation solution, say for the voltage across a particular element in an RC or RL circuit, can always be written in the form

$$v(t) = V_I e^{-t/\tau} + V_F(1 - e^{-t/\tau}) \qquad \text{(I.5-1)}$$

where V_I and V_F represent the initial and final values of the voltage respectively and τ is the circuit time constant.

In examining this solution initially we observe that for small values of t the exponential terms are each equal to one so that $v(t)$ starts off at

$$v(0+) = V_I e^0 + V_F(1 - e^0) = V_I \qquad \text{(I.5-2)}$$

as expected. Furthermore, for large values of t the exponential terms each approach zero so that

$$v(\infty) = V_I e^{-\infty} + V_F(1 - e^{-\infty}) = V_F \qquad \text{(I.5-3)}$$

Thus, as originally postulated, V_I and V_F represent the initial and the final values of $v(t)$ respectively. Because V_I and V_F can generally be determined by inspection of the circuit, it is therefore apparent that the solution for $v(t)$ can be obtained without requiring a formal solution of the circuit differential equation.

In equation (I.5-1) the transition from the initial to the final value of the voltage takes about 4 or 5 time constants, and the specific form of the solution depends on the relationship between V_I and V_F. For these types of waveshapes the response rise time, T_R, (or fall time, T_F) is defined as the time required for the voltage to go from 10% to 90% of the difference between V_F and V_I (see, for example, Figure I.5-1b). By simple substitution of these 10% and 90% voltage levels into equation (I.5-1) it is not too difficult to demonstrate that the overall system rise (or fall) time is equal to about 2.2τ.

Figure I.5-1
Example I.5-1.

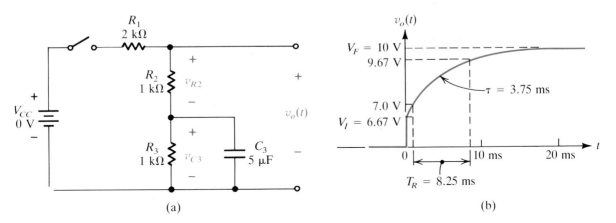

(a)   (b)

EXAMPLE I.5-1

The switch in the circuit illustrated in Figure I.5-1a has been open for a long time and is closed to $t = 0$. Using the inspection method just discussed sketch the voltage $v_o(t)$ for all $t > 0$.

SOLUTION

Because the capacitor is initially uncharged, the voltage across it prior to closing the switch at $t = 0$ is zero, and therefore the voltage $v_{C3}(0+)$ is also equal to zero. As a result, at $t = 0+$ when the switch is closed the input voltage divides across the resistors R_1 and R_2 so that $v_{R2}(0+) = (20)(1/3) = 6.67$ volts. Because v_{C3} is initially zero, it also follows that

$$V_I = v_o(0+) = v_{R2}(0+) + v_{C3}(0+) = 6.67 \text{ V} \qquad (\text{I.5-4})$$

To determine the final value for the voltage v_o we simply replace the capacitor by an open circuit in the steady-state from which we see that

$$V_F = v_o(\infty) = \left(\frac{R_2 + R_3}{R_1 + R_2 + R_3} \right) V_{CC} = 10 \text{ V} \qquad (\text{I.5-5})$$

In order to accurately sketch $v_o(t)$ the only additional information needed is the circuit time constant. This of course is equal to RC_3 where R is the resistance seen by C_3 with V_{CC} set equal to zero. For this circuit $R = R_3 \parallel (R_1 + R_2) = 750$ ohms, and the time constant is therefore $(5 \times 10^{-6}) \cdot (750) = 3.75$ ms. The output voltage is sketched as a function of time in Figure I.5-1b. For this circuit the rise time T_R is $2.2(3.75)$ or 8.25 ms and is equal to the time required for $v_o(t)$ to increase from $6.67 + .1(10 - 6.67) = 7.00$ V to $6.67 + .9(10 - 6.67) = 9.67$ V.

I.6 THE USE OF LAPLACE TRANSFORMS FOR THE ANALYSIS OF ELECTRICAL NETWORKS

Laplace transforms can be used as a differential equation solution method and can also be employed to obtain the transient and steady-state response of an electrical network without ever explicitly writing down the system differential equation. The Laplace transform method of differential equation solution proceeds as follows:

1. Take the Laplace transform of each of the terms in the original differential equation in accordance with Table I.6-1.
2. Solve the equation algebraically for the dependent variable.
3. If necessary, use partial fraction expansion to get the expression for the dependent variable in a form suitable for direct term by term lookup in the table.
4. Write down the time response for the dependent variable as the sum of the inverse Laplace transforms of each of the terms in item (3).

The basic process involved in the solution of differential equations by Laplace transform techniques may be illustrated by solving the differential

I.6-1 Laplace Transforms of Selected Time Functions

Function Name	$f(t)$	$F(s)$
Impulse	$\delta(t)$	1
Step	$u(t)$	$\dfrac{1}{s}$
Exponential	$e^{-\alpha t}u(t)$	$\dfrac{1}{s+\alpha}$
Sine wave	$(\sin \beta t)\,u(t)$	$\dfrac{\beta}{s^2+\beta^2}$
Cosine wave	$(\cos \beta t)\,u(t)$	$\dfrac{s}{s^2+\beta^2}$
Linearity	$f_1(t) + f_2(t)$	$F_1(s) + F_2(s)$
Multiplication by a constant	$af(t)$	$aF(s)$
Derivative	$\dfrac{df}{dt}$	$sF(s) - f(0^-)$
Integral	$\int F(t)\,dt$	$\dfrac{1}{s}F(s)$

equation

$$\tau \frac{dv_o}{dt} + v_o = \frac{v_{in}}{2} + \frac{\tau}{3}\frac{dv_{in}}{dt} = 10u(t) + \frac{20\tau}{3}\delta(t) \qquad \text{(I.6-1)}$$

for the circuit in Figure I.5-1a that we examined earlier. Here the circuit time constant τ is equal to $[(R_1 + R_2) \parallel R_3]C_3$ or 3.75 mS. Taking the Laplace transform for each of the terms in this equation we obtain the algebraic expression

$$s\tau V_o(s) + V_o(s) = \frac{10}{s} + \frac{20\tau}{3} \qquad \text{(I.6-2)}$$

and solving this equation for $V_o(s)$ we have

$$V_o(s) = \frac{\dfrac{10}{s} + \dfrac{20\tau}{3}}{s\tau + 1} = \frac{(30 + 20\tau s)}{3\tau s(s + 1/\tau)} \qquad \text{(I.6-3)}$$

The inverse Laplace transform of $V_o(s)$ on the left-hand side of this equation is $v_o(t)$; but, unfortunately, the inverse transform of the expression on the right-hand side of this equation cannot be evaluated immediately because it is not listed explicitly in Table I.6-1.

In order to evaluate the inverse Laplace transform of equation (I.6-3) it will be necessary to rewrite the algebraic expression on the right-hand side of this equation as a sum of the more elementary functions that are listed in the table. Fortunately, this expression may be simplified by using the partial fraction expansion method and rewriting equation (I.6-3) as

$$V_o(s) = \frac{C_1}{s} + \frac{C_2}{(s + 1/\tau)} \tag{I.6-4a}$$

where
$$C_1 = \left.\frac{(30 + 20\tau s)s}{3\tau s(s + 1/\tau)}\right|_{s=0} = 10 \tag{I.6-4b}$$

and
$$C_2 = \left.\frac{(30 + 20\tau s)(s + 1/\tau)}{3\tau s(s + 1/\tau)}\right|_{s=-1/\tau} = -3.33 \tag{I.6-4c}$$

so that
$$V_o(s) = \frac{10}{s} - \frac{3.33}{s + 1/\tau} \tag{I.6-5}$$

By looking up the inverse Laplace transform of each of the terms on the right-hand side of this expression and applying the linearity principle, the overall inverse Laplace transform of $V_o(s)$ may be written as

$$v_o(t) = 10u(t) - 3.33e^{-t/\tau}u(t) = [10 - 3.33e^{-t/\tau}] \cdot u(t) \tag{I.6-6}$$

This result, of course, is identical to that obtained earlier by using the inspection solution method [see equations (I.5-1), (I.5-4), and (I.5-5) as well as Figure I.5-1b].

Besides being useful for solving differential equations, the Laplace transform method also provides a way to determine the relationship between the input and output variables of an electrical network without ever explicitly writing down the network's differential equation.

To carry out this procedure we begin by noting that the impedance of an inductor and a capacitor may be written in Laplace transform notation as

$$Z_L = sL \tag{I.6-7a}$$

and
$$Z_C = \frac{1}{sC} \tag{I.6-7b}$$

respectively.

Using these definitions for Z_L and Z_C the transfer function (or input-output relationship), $H(s)$, for any electrical network can now be obtained algebraically by employing the same network analysis techniques that we would have used if the network were purely resistive. In particular, if we define $V_{in}(s)$ and $V_o(s)$ as the Laplace transforms of the circuit's input and output signals, then $H(s)$ is related to these quantities by the equation

$$\frac{V_o(s)}{V_{in}(s)} = H(s) \tag{I.6-8}$$

Using this result we can see that $V_o(s)$ can be readily found if $V_{in}(s)$ and $H(s)$ are known; and furthermore, once $V_o(s)$ has been determined, then $v_o(t)$ can be obtained by finding the inverse Laplace transform of $V_o(s)$.

To illustrate this approach let us repeat the analysis of the circuit given in Figure I.5-1a by using the transfer function concept. For this circuit if we represent the impedance of the capacitor by $Z_{C3} = 1/sC_3$, the Laplace transform of the output voltage may be written as

$$V_o(s) = H(s) V_{in}(s) = \left[\frac{R_2 + \dfrac{R_3 Z_{C3}}{R_3 + Z_{C3}}}{R_1 + R_2 + \dfrac{R_3 Z_{C3}}{R_3 + Z_{C3}}} \right] \cdot V_{in}(s) \qquad \text{(I.6-9a)}$$

or

$$V_o(s) = \left[\frac{R_2 + R_3 + sR_2 R_3 C_3}{(R_1 + R_2 + R_3) + s(R_1 + R_2)R_3 C_3} \right] \cdot \left(\frac{20}{s} \right) = \left[\frac{30 + 20\tau s}{3s(1 + s\tau)} \right] \qquad \text{(I.6-9b)}$$

Because this result is identical to that obtained from the Laplace transform solution of the differential equation [see equation (I.6-3)], the time response $v_o(t)$ will likewise be identical to that given by equation (I.6-6). Thus, the analysis of electrical network problems can be accomplished by utilizing transfer function concepts without having to explicitly write down the system differential equation.

I.7   THE SINUSOIDAL STEADY STATE

In the analysis and design of electronic circuits the sinusoidal steady-state frequency response has importance for several reasons. To begin with, it is easy to measure in the laboratory. In addition, once the sinusoidal response to different frequencies has been determined, then, in principal at least, the system's response to any arbitrary input signal can also be predicted by applying Fourier series concepts.

The sinusoidal steady-state frequency response of any electrical system may be determined directly by writing down and then solving the system differential equation. As a practical matter however this approach is very difficult to apply for all but the simplest circuits. As an alternative, this problem too may be solved by the application of the Laplace transform transfer function method.

When a sine wave is applied to the input of a linear electrical network, the steady-state output (or steady-state solution to the differential equation for this network) is also a sine wave at the same frequency, differing from the input only in amplitude and phase (see Figure I.7-1). Furthermore, the output amplitude V_2 and phase shift ϕ are independent of the circuit initial conditions and can be obtained directly from the transfer function as

$$V_2 = |H(j\omega)| \cdot V_1 \qquad \text{(I.7-1a)}$$

and

$$\phi = \underline{/H(j\omega)} = \theta(j\omega) \qquad \text{(I.7-1b)}$$

In both these expressions ω is the frequency of the applied input signal. Furthermore, $|H(j\omega)|$ and $\underline{/H(j\omega)}$ represent the magnitude and phase of $H(s)$ and can be readily obtained by substituting $j\omega$ for s and writing the transfer function in polar form.

Figure I.7-1
*Sinusoidal steady-state descrip-
tion of an electrical network.*

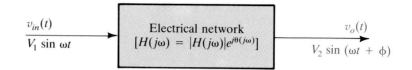

$$v_{in}(t) \qquad \boxed{\begin{array}{c}\text{Electrical network} \\ [H(j\omega) = |H(j\omega)|e^{j\theta(j\omega)}]\end{array}} \qquad v_o(t)$$

$$V_1 \sin \omega t \qquad\qquad\qquad\qquad\qquad\qquad\qquad V_2 \sin (\omega t + \phi)$$

I.8   SKETCHING THE GAIN AND PHASE RESPONSE OF ELECTRICAL NETWORKS

As we have seen from Section I.7 of this appendix, a determination of the steady-state output produced by the application of a sinusoidal signal to a linear electrical network basically involves an evaluation of the magnitude and the phase of the transfer function at the frequency of the applied input signal. While many different methods exist for obtaining these two quantities, the use of the gain–phase plots originally proposed by Bode have proven to be of the greatest value.

Usually when making these sketches we plot the magnitude in decibels and the phase directly in degrees. By using this approach and by also using a logarithmic scale along the frequency axis, the sketches for most transfer functions can be accurately represented by a series of straight line asymptotic approximations.

The basic technique for sketching the Bode gain-phase characteristics of a particular transfer function may be outlined as follows:

1. Determine the numerator and denominator break frequencies and also examine the transfer functions behavior at very high and very low frequencies.
2. To sketch the gain plot:
   a. If possible begin the sketch in a frequency region where the gain magnitude is constant and increase (or decrease) this frequency until the first break is encountered.
   b. A break term in the numerator contributes a positive (or upward) slope to the magnitude plot of +20 dB/decade that starts at the break frequency. Before that point this term contributes a slope of 0 dB/decade.
   c. A break term in the denominator contributes a negative (or downward) slope of −20 dB/decade for all frequencies above the break. Below the break frequency its contribution to the overall slope of the curve is 0 dB/decade.
   d. The overall asymptotic sketch for the transfer function magnitude is obtained by noting that the total slope of the curve at any specific frequency is simply equal to the sum of the slope contributions from each of the individual break terms.
   e. The largest deviation between the actual gain plot and its asymptotic approximation occurs at the break frequencies where, for a single break term, the two curves differ by about 3 dB. For multibreak transfer functions the gain error between the actual and the asymptotic magnitude plots is additive.
3. To sketch the phase plot:
   a. If possible begin the sketch in a frequency region where the phase is a constant and increase (or decrease) the frequency until a point is

reached that is one decade below (or above) the first break. This is the point at which the height of the phase plot will begin to change.

b. A numerator break term contributes a slope of +45 degrees/decade to the total phase in the region from one decade below to one decade above the break frequency. At frequencies less than one tenth of the break frequency the phase contribution is zero, and for frequencies more than one decade above the break the phase contribution is +90 degrees.

c. A denominator break term contributes a slope of −45 degrees/decade to the total phase in the region from one decade below to one decade above the break frequency. At frequencies less than one tenth of the break frequency the phase contribution is zero, and for frequencies more than one decade above the break the phase contribution is −90 degrees.

d. The overall asymptotic sketch for the phase plot is obtained by noting that the total slope of the curve at any specific frequency is simply equal to the sum of the slope contributions from each of the individual break terms.

e. The greatest errors between the actual and asymptotic approximations to the phase plot occur at the intersection of the asymptotes one decade above and one decade below the break frequency. At each of these points there is a 6 degree difference between the two curves. For multi-break transfer functions the overall phase differences between the actual and asymptotic phase plots are additive.

In the example which follows we will illustrate the technique for sketching the Bode gain-phase plots.

EXAMPLE I.8-1

Sketch the Bode gain and phase plots for the transfer function

$$H(s) = \frac{10^5(s + 10)}{(s + 10^3)} \qquad (\text{I.8-1})$$

SOLUTION

This transfer function has one break in the numerator at 10 rps, and one in the denominator at 10^3 rps. For very low frequencies (or for values of s well below 10 rps) the transfer function is approximately equal to $(10^5(10)/(10^3) = 10^3$ or 60 dB. Because $H(s)$ is a constant here at 60 dB, there is no phase shift in this region. For frequencies above 10 rps the numerator break turns on, and the gain plot begins to rise with a +20 dB/decade slope. Of course, the corresponding phase shift also increases (with a +45 degree/decade slope) beginning a decade below the break at 1 rps and terminating a decade above at 10^2 rps with a +90 degree phase contribution.

When the frequency increases beyond 10^3 rad/sec, the denominator break turns on, and its negative slope contribution at −20 dB/decade cancels out the numerator contribution at +20 db/decade so that the magnitude plot above the 10^3 rps point is flat at the 100 dB level. The final asymptotic value of the magnitude plot can also be obtained by noting that for high frequencies (or for large values of s) the transfer function approaches $(10^5)(s)/(s) = 10^5$ or 100 dB.

The phase contribution associated with the denominator break at 10^3 rps has a slope of −45 degrees/decade beginning at 100 rps and ending at 10^4 rps. Because the numerator slope contribution ends at 100 rps the net slope in the region from 10^2 to 10^4 rps is −45 degrees/decade. Beyond 10^4 rps the phase is

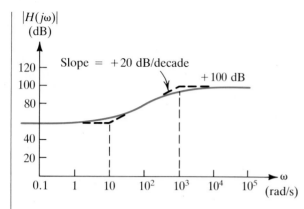

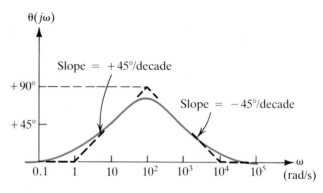

Figure I.8-1
Example I.8-1.

constant at zero degrees. The latter result is to be expected since $H(s)$ is constant at 10^5 for large values of s.

The gain and the phase plots are sketched in Figure I.8-1. In these sketches the actual curves are shown in solid color while the asymptotic approximations to these curves are indicated by the dashed black lines.

I.9 PARALLEL RESONANCE

An understanding of the performance of parallel-tuned RLC networks of the type illustrated in Figure I.9-1a is extremely useful for the design of high frequency oscillator circuits. Because of the analogy between electrical energy storage in circuits of this type and the mechanical storage of potential energy in water towers, these parallel RLC networks are also frequently called "tank circuits." The basic operation of this tank circuit may be understood by examining the variation of its impedance with frequency (Figure I.9-1c).

At very low frequencies the impedance of the inductor is so small that it may be considered to be a short circuit making $Z(j\omega)$, the impedance of the tank, approximately zero. A similar situation exists at very high frequencies where the capacitor behaves like a short. For some intermediate frequency ω_o, known as the resonant frequency of the circuit, the impedance of the capacitor and the inductor are equal so that the magnitude of the current flow through each of these elements is the same. However, because the current flow through the inductor is proportional to the integral of the voltage across it while that through the capacitor is proportional to its derivative, the two currents, i_L and i_C, will be 180 degrees out of phase. As a result, when these two equal-amplitude but out-of-phase currents are added, their sum is zero, so that in effect at resonance the parallel combination of the L and the C may be eliminated from the circuit leaving only the resistance R (Figure I.9-1b). Therefore, a parallel-tuned circuit is characterized by having a rather high impedance at its resonant frequency, and a low impedance away from resonance.

Let us examine the behavior of this circuit a bit more closely. Following Figure I.9-1a we have

$$Z(s) = \frac{1}{\dfrac{1}{R} + sC + \dfrac{1}{sL}} = R\left[\frac{s/RC}{s^2 + s\left(\dfrac{1}{RC}\right) + \dfrac{1}{LC}}\right] \tag{I.9-1}$$

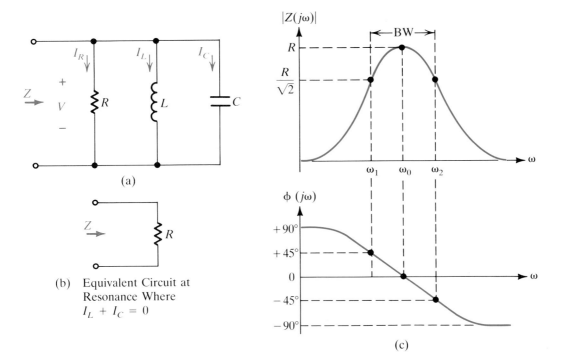

Figure I.9-1
Parallel tuned RLC circuit.

(a)

(b) Equivalent Circuit at Resonance Where $I_L + I_C = 0$

(c)

In order to better understand the behavior of this tank circuit it is useful to define

$$\omega_o = \frac{1}{\sqrt{LC}} \qquad (\text{I.9-2a})$$

as its resonance frequency, and

$$Q = \frac{R}{\omega_o L} = \frac{R}{\left(\dfrac{1}{\omega_o C}\right)} = \omega_o RC \qquad (\text{I.9-2b})$$

as its Q or "quality factor". Notice that this definition of the circuit Q actually represents the ratio of R, the tank impedance at resonance, to the impedance of either of the reactive elements at the resonant frequency.

Using equations (I.9-2a) and (I.9-2b) we may rewrite equation (I.9-1) in terms of the parameters ω_o and Q as

$$Z(s) = R\left[\frac{\dfrac{\omega_o s}{\omega_o RC}}{s^2 + s\dfrac{\omega_o}{\omega_o RC} + \omega_o^2}\right] = R\left[\frac{s\omega_o/Q}{s^2 + s\omega_o/Q + \omega_o^2}\right] \qquad (\text{I.9-3})$$

The variation of the tank impedance with frequency is obtained by substituting $s = j\omega$ into this expression which yields

$$Z(j\omega) = R\left[\frac{j\omega\omega_o/Q}{(\omega_o^2 - \omega^2) + j\omega\omega_o/Q}\right] \qquad (\text{I.9-4})$$

The magnitude and phase of $Z(j\omega)$ are sketched in Figures I.9-1c. As expected, at the resonant frequency, ω_o, the L and C branches effectively disappear leaving only the resistance R. Therefore, at ω_o the voltage across the tank and the current flow through it are in phase. For frequencies well above or well below ω_o the tank impedance goes to zero with either L or C looking like a short. The rapidity with which the circuit impedance falls to zero away from the resonant frequency is determined by the Q of the circuit.

The frequencies ω_1 and ω_2 shown in the figure are called the half-power frequencies of the tank circuit because an applied current source at the frequency ω_1 or ω_2 will only deliver $1/2$ the power to R that it would deliver at the frequency ω_o. This is due to the fact that $Z(j\omega)$ is equal to $R/\sqrt{2}$ at these frequencies so that $V(j\omega)$ is also $1/\sqrt{2}$ of its value at resonance. Besides being referred to as the network's half-power frequencies, ω_1 and ω_2 are also known as the circuit's three dB points or three dB frequencies because of these frequencies the magnitude of the tank impedance is down 3 dB from its value at ω_o.

As indicated in Figure I.9-1c the phase shift at the half-power frequencies is plus or minus 45 degrees. Furthermore, at very low frequencies the inductor is almost a short so that $Z(j\omega) \simeq Z_L = j\omega L$, and the phase shift is about $+90$ degrees. At very high frequencies, on the other hand, where the capacitor impedance is nearly zero, we find that $Z(j\omega) = 1/j\omega C$ so that the phase shift is about -90 degrees. The difference between the half-power points, ω_1 and ω_2, is known as the bandwidth (BW) of the tuned circuit.

The relationship between ω_1, ω_2, ω_o, and Q may be determined by setting the magnitude of equation (I.9-4) equal to $R/\sqrt{2}$ and solving for the resulting values of ω. When this is done, the following results are obtained

$$\omega_1 = \omega_o\sqrt{1 + \left(\frac{1}{2Q}\right)^2} - \omega_o/2Q \qquad \text{(I.9-5a)}$$

$$\omega_2 = \omega_o\sqrt{1 + \left(\frac{1}{2Q}\right)^2} + \omega_o/2Q \qquad \text{(I.9-5b)}$$

Figure I.9-2
Relationship between the shape of the impedance characteristic and the Q of the circuit.

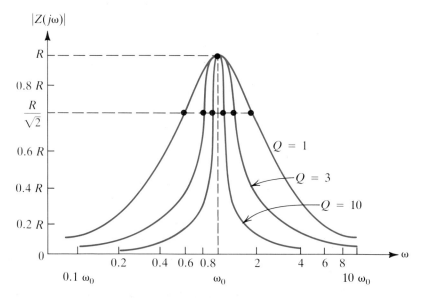

Appendix I / A Review of Basic Network Concepts

and
$$BW = \omega_2 - \omega_1 = \frac{\omega_o}{Q} \qquad \text{(I.9-5c)}$$

These results are important because they allow us to relate the Q of the circuit, which is a function of the circuit component values [see equation (I.9-2b)], to the resonant frequency and the bandwidth of the network. By examining Figure I.9-2 we can see that circuits having high Qs exhibit sharp peaking in the vicinity of ω_o and a rapid falloff in impedance away from resonance. To achieve these high Qs [see equation (I.9-2b)] it is necessary that the impedances of the capacitor and the inductor at ω_o be much less than that of the resistor R. In this way, while Z_L and Z_C appear to be an open circuit at ω_o, for frequencies slightly off resonance $Z(j\omega)$ the parallel combination of R, Z_C, and Z_L will rapidly go to zero because both Z_L and Z_C are small.

I.10 PARALLEL-TUNED CIRCUITS CONTAINING NONIDEAL INDUCTORS

In the design of practical tuned circuits we may generally consider that the capacitive elements are ideal. Inductors, on the other hand, tend to exhibit losses which cannot be neglected. This occurs because inductors are wound from wire, such as copper, and this wire exhibits resistive losses. These losses can best be modeled by placing a resistor in series with the inductor as illustrated in Figure I.10-1a. In analyzing this type of circuit it is useful to define the Q of the inductor as

$$Q_L = \frac{\omega_o L_1}{r_1} \qquad \text{(I.10-1)}$$

as a way to describe the losses associated with this element. At first, on the basis of this equation, we might expect that the inductor's Q would increase linearly with frequency. However, r_1 is not simply equal to the dc ohmic resistance of the wire; it represents the effective ac resistance of the inductor at the frequency where the Q is being measured. This resistance tends to increase with frequency because at higher frequencies, due to the skin effect, the current in the wire flows along the outer surface effectively decreasing the cross-sectional area of the wire. As a result, the Q of an inductor tends to be relatively constant over the frequency range where it is designed to operate.

Figure I.10-1
Approximating a parallel-tuned circuit containing a nonideal inductor by an equivalent RLC circuit.

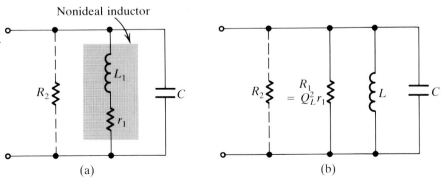

(a)   (b)

If Q_L is reasonably large, it can be shown that the two networks in Figure I.10-1a and b have nearly the same transfer function if $L = L_1$ and if

$$R_1 = Q_L^2 r_1 \qquad (I.10\text{-}2)$$

Thus, a parallel LC circuit containing a nonideal inductor may be approximated by an RLC parallel-tuned circuit in which the same inductor is used, and the equivalent parallel resistance R_1 is determined by the relationship given in equation (I.10-2). If any additional resistance R_2 is shunted across the circuit in Figure I.10-1b, it is simply connected in parallel with $Q_L^2 r_1$ in order to find the overall equivalent resistance in parallel with L and C. In determining the Q of the tank circuit, equation (I.9.2b) needs to be used with the resistance R set equal to the effective overall resistance of the tank $[R_2 \parallel (Q_L^2 r_1)]$. Furthermore, for these approximations to be valid it is necessary for the overall Q of the tank to be reasonably large. A Q of 3 or more will insure that the errors obtained in approximating the circuit in Figure I.10-1a by that given in Figure I.10-1b will be less than 10%, at least in the vicinity of $\omega = \omega_o$.

It is important to note for the case where $R_2 = \infty$ that Q of the tank may be written as

$$Q = \frac{R}{\omega_o L} = \frac{Q_L^2 r_1}{\omega_o L} = \frac{Q_L^2}{\left(\dfrac{\omega_o L}{r_1}\right)} = Q_L \qquad (I.10\text{-}3)$$

so that the maximum Q of a parallel-tuned circuit is equal to the Q of the inductor. If R_2 is finite, then the Q of the tank will be less than Q_L.

I.11   TRANSFORMERS AND TRANSFORMER-LIKE CIRCUITS

The behavior of the ideal transformer was originally discussed in Section 2.1, and for completeness the results obtained there are summarized in Figure I.11-1a. In developing the input–output equations for this transformer it was assumed that the coefficient of coupling between the primary and the secondary was unity, and that the primary and secondary inductances were infinitely large with their ratio determining the transformer turns ratio $[N = (L_1/L_2)^{1/2}]$.

The nonideal transformer may be analyzed with the aid of Figure I.11-1b where L_1 and L_2 represent the primary and secondary inductances respectively, and M the mutual inductance linking these two windings. The mutual inductance is related to L_1 and L_2 by the expression

$$M = k\sqrt{L_1 L_2} \qquad (I.11\text{-}1)$$

where k, the coefficient of coupling, is a measure of how much of the flux generated in the primary links the secondary, and vice versa. When the primary and secondary windings are very far apart $k = 0$, and when all of the flux from the primary links the secondary $k = 1$.

Applying Kirchhoff's voltage law to this transformer we may write down

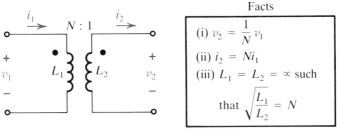

(a)　Ideal Transformer

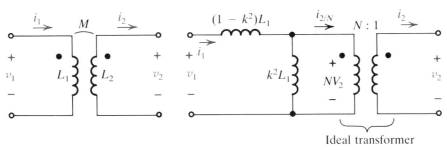

(b)　Nonideal Transformer

(c)　Model for the Nonideal Transformer

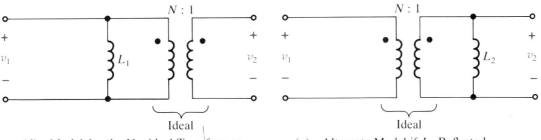

(d)　Model for the Nonideal Transformer
　　　If $k \cong 1$

(e)　Alternate Model if L_1 Reflected
　　　into the Secondary

Figure I.11-1

Equivalent circuits for the non-ideal transformer.

its circuit equations as

$$v_1 = L_1 \frac{di_1}{dt} - M \frac{di_2}{dt} \tag{I.11-2a}$$

and

$$v_2 = -L_2 \frac{di_2}{dt} + M \frac{di_1}{dt} \tag{I.11-2b}$$

It is not too difficult to show that this transformer may also be represented by the equivalent circuit shown in part (c) of Figure I.11-1, since its behavior is described by equations identical to (I.11-2a) and (I.11-2b). In this figure the turns ratio of the ideal transformer is given by the expression

$$N = k \sqrt{\frac{L_1}{L_2}} \tag{I.11-3}$$

It is useful to notice in this expression that because L_1 is proportional to N_1^2

(where N_1 = the number of turns on the primary winding) and L_2 to N_2 (where N_2 = the number of turns on the secondary winding) that N is k times the actual transformer turns ratio.

For simplicity in our analysis of oscillator circuits containing transformers we will assume that the coefficient of coupling k is nearly unity. For this situation the transformer model reduces to that given in Figure I.11-1d. Notice that by reflecting the primary inductance L_1 into the secondary the equivalent circuit for the transformer with $k = 1$ may also be represented as illustrated in Figure I.11-1e, since the equivalent reflected inductance will be $L_1/N^2 = L_1/(L_1/L_2) = L_2$. Thus the equivalent circuit of a transformer whose coefficient of coupling is nearly one may be represented by an ideal transformer shunted by an inductor L_1 if the inductance is included in parallel with the primary winding, or L_2 if it is in parallel with the secondary.

In the analysis of tuned circuits there are several other networks which, although they do not contain transformers, exhibit transformer-like properties under certain conditions. One such network is illustrated in Figure I.11-2a; it is known as a capacitive divider-tuned circuit. If R_2 does not significantly load the capacitive divider $C_1 - C_2$, then it is not too difficult to show that near its resonant frequency this circuit may be approximated by the simplified parallel-tuned circuit given in part (b) of the same figure where

$$R \simeq \left(\frac{C_1 + C_2}{C_1}\right)^2 R_2 \qquad \text{(I.11-4a)}$$

and

$$C \simeq \left(\frac{C_1 C_2}{C_1 + C_2}\right) \qquad \text{(I.11-4b)}$$

Figure I.11-2

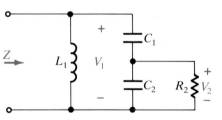

(a)   Capacitively Tapped Tuned Circuit

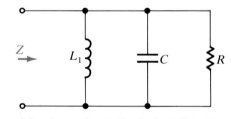

(b)   Approximate Equivalent Circuit
       Near Resonance

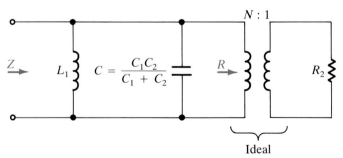

(c)   Equivalent Circuit with C_1–C_2 Divider
       Viewed as an Impedance Transformer
       $(N = (C_1 + C_2)/C_1)$

Appendix I   /   A Review of Basic Network Concepts

In equation (I.11-4a) the term $C_1/(C_1 + C_2)$ represents the voltage divider ratio V_2/V_1 which would occur in the circuit in Figure I.11-2a if the loading on the divider by R_2 were small. When considered in this way we may view the $C_1 - C_2$ divider as an impedance transformer, converting the impedance R_2 into an equivalent impedance $R = [(C_1 + C_2)/C_1]^2 R_2$. When this approximation is valid, the effective equivalent circuit for the original network presented in Figure I.11-2a may be approximated by the circuit given in part (c) of the same figure.

REFERENCES

1. W. H. Hayt and J. E. Kemmerly, *Engineering Circuit Analysis*, 4th ed., New York, N.Y.: McGraw-Hill Book Company, 1986.
2. L. Huelsman, *Basic Circuit Theory With Digital Computations*, Second Edition, Englewood Cliffs, N.J.: Prentice-Hall Inc., 1984.
3. M. E. Valkenburg, *Network Analysis*, 3rd ed., Englewood Cliffs, N.J.: Prentice-Hall Inc., 1974.

APPENDIX II
Transmission-Line Effects

II.1   INTRODUCTION

In most instances a wire, or a cable, or a land on a printed circuit board may be thought of as being a nearly ideal conductive element posessing zero resistance, zero inductance, and zero capacitance. These approximations are usually excellent in low-frequency circuits because the parasitic elements associated with the conductors are much smaller than the other circuit components. Therefore, at low frequencies most conductors can usually be considered to be short circuits.

As the frequency of the applied signal increases, eventually the parasitic inductance and capacitance of the wire can no longer be neglected. Typically, this occurs when the signal period, or the signal rise and fall times in the case of nonsinusoidal inputs, become comparable to the signal propagation time down the length of the wire. Because electrical signals travel in wires at nearly the speed of light, typical transmission-line propagation times are on the order of 3 ns/m. Therefore, for example, for signals traveling on a 1-ft length of wire, the effect of the parasitic elements associated with that wire will need to be taken into account, for sinusoidal inputs, when the applied frequency exceeds about $1/(1 \text{ ns})$ or 1000 MHz. When the input signal is nonsinusoidal, these parasitic elements will need to be included in the overall circuit model whenever the signal rise and fall times are comparable to the propagation time down the wire, which in this case is about 1 ns.

In developing a high-frequency model for a conductor, we need to recognize that it is possible for the input and output voltage to exhibit large phase differences from one another at high frequencies, due to the finite time that is required for changes in the input to reach the output. As a result, when rapidly

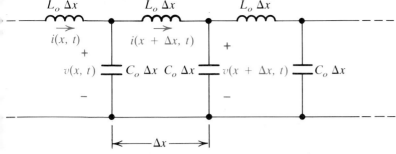

(a)   A Distributed Parameter Model for a Transmission Line.

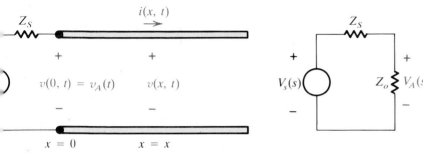

(b)   An Infinitely Long Transmission Line.

(c)   Input Equivalent Circuit for an Infinitely Long Transmission Line.

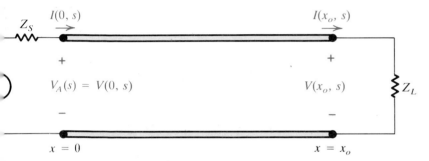

(d)   A Finite Length Transmission Line.

(II.1-1b) with respect to t, we can also readily show that

$$\frac{\partial^2 i(x, t)}{\partial x^2} = \frac{1}{c_p^2}\frac{\partial^2 i(x, t)}{\partial t^2} \qquad\text{(II.1-2c)}$$

Equations (II.1-2a) and (II.1-2c) are known as the transmission-line wave equations, and their solution will generally be in the form of waves traveling to the left and to the right along the line. The specific solution to these equations is best found by applying the Laplace transform analysis method to this problem. If we define $V(x, s)$ as the Laplace transform of the voltage along the line and write this relationship as

$$v(x, t) \longleftrightarrow V(x, s) \qquad\text{(II.1-3a)}$$

varying input signals are applied to a conducting elemen
rent along the length of the conductor will be highly noi
impossible to use lumped models to describe its behavio
distributed parameter model will be required, and in thi:
ble, or PC board land can no longer be represented by
now be considered to be a transmission line.

In this appendix we examine the basic characteristic
to develop an understanding of when a wire is a wire an
sion line. In addition, we investigate the techniques avai
transmission-line problems. These methods will include
solution of the transmission-line differential equations, 1
lattice) diagrams for the analysis of transmission-line pi
opment of a graphical technique that will facilitate the a
line circuits connected to nonlinear sources and load
gates.

To develop the transmission-line equations, cor
parameter model illustrated in Figure II.1-1a. Here we
mission line into segments of length Δx, each contain
$L_o \Delta x$ and a shunt capacitance $C_o \Delta x$, where L_o and C_o
ductance per unit length and capacitance per unit length
has been assumed to be lossless since there are no resis
the model. This approximation is reasonable for all bu:
the resistance of a typical transmission-line wire is onl'
ohm per meter.

By applying Kirchhoff's voltage and current laws t(
the limit as $\Delta x \to 0$, it is not too difficult to show that

$$\frac{\partial i(x,\ t)}{\partial x} = -C_o \frac{\partial v(x,\ t)}{\partial t}$$

and

$$\frac{\partial v(x,\ t)}{\partial x} = -L_o \frac{\partial i(x,\ t)}{\partial t}$$

These two expressions are known as the transmissi
equations. Their meaning may be understood as foll(
states that the loss in current per unit length along the l
that flows into the shunt capacitance C_o. In a sir
(II.1-1b) indicates that the loss in voltage per unit leng
to the voltage drop across the series inductance L_o.

By differentiating equation (II.1-1a) with respect
respect to x, and combining the resulting expressions,

$$\frac{\partial^2 v(x,\ t)}{\partial x^2} = \frac{1}{c_p^2} \frac{\partial^2 v(x,\ t)}{\partial t^2}$$

where $c_p = \dfrac{1}{\sqrt{L_o C_o}}$ = wave propagation velocity c

In a similar fashion, by differentiating equation (II.1-1

then if the initial conditions are zero, it follows directly that

$$\frac{\partial v(x, t)}{\partial t} \longleftrightarrow sV(x, s) \qquad \text{(II.1-3b)}$$

and

$$\frac{\partial^2 v(x, t)}{\partial t^2} \longleftrightarrow s^2 V(x, s) \qquad \text{(II.1-3c)}$$

Using these results, the Laplace transform of the differential equation (II.1-2a) may then be written as

$$\frac{\partial^2 V(x, s)}{\partial x^2} = \frac{s^2}{c_p^2} V(x, s) \qquad \text{(II.1-4)}$$

By inspection, the solution of this equation is seen to be of the form

$$V(x,s) = \underbrace{Ae^{-sx/c_p}}_{\substack{\text{wave} \\ \text{traveling} \\ \text{to the right}}} + \underbrace{Be^{sx/c_p}}_{\substack{\text{wave} \\ \text{traveling} \\ \text{to the left}}} \qquad \text{(II.1-5)}$$

where the coefficients A and B are determined by the boundary conditions. Here, as we shall see shortly, the first term in this expression represents a wave traveling to the right along the line, and the second term a wave traveling to the left.

The corresponding expression for the current along the line may be found by taking the Laplace transform of equation (II.1-1b), which yields

$$\frac{\partial V(x, s)}{\partial x} = -sL_o I(x, s) \qquad \text{(II.1-6)}$$

and substituting the solution for $V(x, s)$ into this equation, we have

$$I(x, s) = \frac{1}{Z_o} (Ae^{-sx/c_p} - Be^{sx/c_p}) \qquad \text{(II.1-7a)}$$

where

$$Z_o = \sqrt{\frac{L_o}{C_o}} = \text{characteristic impedance of the line} \qquad \text{(II.1-7b)}$$

To find the values of the coefficients A and B, we need to examine a specific problem. Consider, for example, the case of an infinitely long line illustrated in Figure II.1-1b. In this case, if we assume that the source is connected onto the line at the left at the $x = 0$ boundary, it will cause a wave to be launched from left to right along the line. However, if the line is infinitely long, there will never be a reflected wave, and hence the B coefficient will be zero. The test of whether or not this assumption is correct depends on whether the boundary conditions can be satisfied with $B = 0$ for an infinitely long line.

If B is, in fact, assumed to be zero for this case, then

$$V(x, s) = Ae^{-sx/c_p} \qquad \text{(II.1-8a)}$$

and

$$I(x, s) = \frac{A}{Z_o} e^{-sx/c_p} \qquad \text{(II.1-8b)}$$

The ratio of $V(x, s)/I(x, s)$ defines the equivalent impedance or equivalent circuit seen looking to the right of the point x. For the case of an infinitely long line,

$$\frac{V(x, s)}{I(x, s)} = \frac{Ae^{-sx/c_p}}{Ae^{-sx/c_p}/Z_o} = Z_o \qquad \text{(II.1-9)}$$

at all points on the line. This result also applies at the $x = 0$ boundary, and therefore the equivalent circuit given in Figure II.1-1c may be used to calculate the voltage at $x = 0$. If we denote this voltage as $v_A(t)$, it is apparent that

$$V_A(s) = V_s(s)\frac{Z_o}{Z_o + Z_S} \qquad \text{(II.1-10)}$$

Since equation (II.1-8a) represents the voltage at all points along the line, it follows that

$$V_A(s) = V(x, s)\Big|_{x=0} = Ae^0 = A = \frac{Z_o}{Z_o + Z_S}V_s(s) \qquad \text{(II.1-11a)}$$

and substituting this result back into equation (II.1-8a), we have

$$V(x, s) = V_A(s)e^{-sx/c_p} = \frac{Z_o}{Z_o + Z_S}V_s(s)e^{-sx/c_p} \qquad \text{(II.1-11b)}$$

Taking the inverse Laplace transform of both sides of this expression and noticing that

$$f(t - T) \longleftrightarrow F(s)e^{-sT} \qquad \text{(II.1-12)}$$

we obtain
$$v(x, t) = \frac{Z_o}{Z_o + Z_S}v_s\left(t - \frac{x}{c_p}\right) \qquad \text{(II.1-13)}$$

Thus the voltage $v(x, t)$ at any point x along the transmission line is in this instance a delayed, reduced-amplitude version of the source voltage $v_s(t)$ applied at the $x = 0$ boundary. Of course, the time delay is just x/c_p.

When the length of the transmission line is finite, some of the main wave propagating to the right along the line will usually be reflected at the load, and in this instance the wave equation can only be satisfied by considering that the B coefficient is nonzero. The general transmission-line problem now to be considered is illustrated in Figure II.1-1d. If the incident wave at $x = x_o$ is

$$V_i = Ae^{-sx/c_p} = V_A(s)e^{-sx/c_p} \qquad \text{(II.1-14a)}$$

the overall voltage at $x = x_o$ will be

$$V(x_o, s) = \underbrace{Ae^{-sx_o/c_p}}_{V_i} + \underbrace{Be^{sx_o/c_p}}_{V_r} \qquad \text{(II.1-14b)}$$

where V_i and V_r represent the incident and the reflected waves respectively. Following equation (II.1-7a), the current at this point is also seen to be

$$I(x_o, s) = \frac{1}{Z_o}(Ae^{-sx_o/c_p} - Be^{sx_o/c_p}) \qquad \text{(II.1-14c)}$$

By examining Figure II.1-1d, we can see that the ratio of the voltage to the current at $x = x_o$ must of course be equal to the load impedance Z_L, so that we may write

$$Z_L = \frac{V(x_o, s)}{I(x_o, s)} = Z_o \frac{Ae^{-sx_o/c_p} + Be^{sx_o/c_p}}{Ae^{-sx_o/c_p} - Be^{sx_o/c_p}} = Z_o \frac{1 + (Be^{sx_o/c_p}/Ae^{-sx_o/c_p})}{1 - (Be^{sx_o/c_p}/Ae^{-sx_o/c_p})} \quad \text{(II.1-15)}$$

If we define
$$\rho = \frac{V_r}{V_i} = \frac{\text{total reflected wave}}{\text{total incident wave}} = \frac{Be^{sx_o/c_p}}{Ae^{-sx_o/c_p}} \quad \text{(II.1-16)}$$

we may rewrite equation (II.1-15) as

$$\frac{Z_L}{Z_o} = \frac{1 + \rho}{1 - \rho} \quad \text{(II.1-17a)}$$

from which we have
$$\rho = \frac{Z_L - Z_o}{Z_L + Z_o} \quad \text{(II.1-17b)}$$

The quantity ρ is called the reflection coefficient at the $x = x_o$ boundary, and it is a measure of how much of the total incident wave is reflected by the load. Using this definition of ρ and following equations (II.1-14a) and (II.1-16), we may write the expression for the B coefficient as

$$B = \rho Ae^{-2sx_o/c_p} = \rho V_A(s)e^{-2sx_o/c_p} \quad \text{(II.1-18)}$$

Notice that if $Z_L = Z_o$, both ρ and $B = 0$, and there is no reflected wave. In this instance the load is said to be matched to the line, and for this case, as for the case where the line is infinite, $v(x, t)$ is given by equation (II.1-13).

However, when Z_L is not equal to Z_o, the overall solution for $v(x, t)$ [ignoring for the moment the fact that there may be additional reflections returning from the $x = 0$ boundary at later times] may be written as

$$V(x, s) = Ae^{-sx/c_p} + Be^{sx/c_p} = V_s(s)\frac{Z_o}{Z_o + Z_S}[e^{-sx/c_p} + \rho e^{-s(2x_o-x)/c_p}] \quad \text{(II.1-19a)}$$

or
$$v(x, t) = \frac{Z_o}{Z_o + Z_S}\left[\underbrace{v_s\left(t - \frac{x}{c_p}\right)}_{\text{incident wave}} + \underbrace{\rho v_s\left(t - \frac{2x_o - x}{c_p}\right)}_{\substack{\text{reflected wave at the} \\ x = x_o \text{ boundary}}}\right] \quad \text{(II.1-19b)}$$

This result indicates that the overall solution is the sum of two components. The first of these is the incident wave traveling to the right. It has the same shape as the applied input voltage but is delayed in time by x/c_p, the time that it takes for the wave to reach the point x being examined. The second part of the solution is a wave traveling to the left, and it is produced by the reflection that occurs at the $x = x_o$ boundary. Its amplitude is ρ times that of the incident wave, and the overall delay of this signal $[(2x_o - x)/c_p]$ represents the time required for this wave to propagate to the $x = x_o$ boundary (x_o/c_p seconds) and then back to the point x on the line [an additional distance $(x_o - x)$ or a time $(x_o - x)/c_p$].

In carrying out the analysis thus far we have neglected the fact that there can be additional reflections of the reflected wave at the $x = 0$ boundary. Depending on the reflection coefficient at $x = 0$, this can lead to a nearly infinite number of reflected waves going back and forth between the $x = 0$ and $x = x_o$ boundaries. Because these reflections come back at later and later times, as the order of the reflection gets higher and higher, we can ignore them initially if we simply limit the size of the time interval being examined.

In equation (II.1-9) we noted, for an infinite transmission line, that the ratio of $V(x, s)/I(x, s)$ was equal to Z_o. This fact led to the use of the simplified input equivalent circuit shown in Figure II.1-1c for calculating $v_A(t)$, and was an extremely valuable tool in our analysis of the infinite transmission-line problem. Although we did not prove its validity explicitly, this equivalent circuit was then also used to calculate the amplitude of $v_A(t)$ even for the case where the line length was finite. This approach is justified because "the line does not know that it is noninfinite" until the first reflections come back from the load. However, for $t > 2x_o/c_p$ (after the first reflections from the load reach the $x = 0$ boundary) the equivalent circuit in Figure II.1-1c and equation (II.1-9) are no longer valid.

To account for the reflection from the $x = 0$ boundary, a third term needs to be added to $V(x, s)$ in equation (II.1-19a). Because this reflected wave is delayed by $2x_o/c_p$ seconds from the input voltage $v_s(t)$, it is reasonable to assume a solution for $V(x, s)$ of the form

$$V(x, s) = V_s(s)\left(\frac{Z_o}{Z_o + Z_S}\right)\left[e^{-sx/c_p} + \rho e^{-s(2x_o-x)/c_p}\right] + \underbrace{Ce^{-s(x+2x_o)/c_p}} \qquad \text{(II.1-20)}$$

new term to account
for the reflection at
the $x = 0$ boundary

In addition, from equation (II.1-6) it also follows directly that the current on the line in this instance may be expressed as

$$I(x, s) = \frac{V_s(s)}{Z_o + Z_S}\left[e^{-sx/c_p} - \rho e^{-s(2x_o-x)/c_p}\right] + \frac{C}{Z_o}e^{-s(x+2x_o)/c_p} \qquad \text{(II.1-21)}$$

At the $x = 0$ boundary, following Figure II.1-1d, it is apparent that

$$V_s(s) = I(0, s)Z_S + V(0, s) \qquad \text{(II.1-22)}$$

Evaluating equations (II.1-20) and (II.1-21) at $x = 0$ and substituting these results into equation (II.1-22), after some algebraic manipulation, we find that

$$C = \underbrace{V_s(s)\frac{Z_o}{Z_o + Z_S}}_{V_A(s)}\rho\frac{Z_S - Z_o}{Z_S + Z_o} \qquad \text{(II.1-23)}$$

Defining

$$\rho' = \frac{Z_S - Z_o}{Z_S + Z_o} \qquad \text{(II.1-24)}$$

as the reflection coefficient at the $x = 0$ boundary, we may now write the expression for the voltage at any point along the line as

$$V(x, s) = V_A(s)e^{-sx/c_p} + \rho V_A(s)e^{-s(2x_o-x)/c_p} + \rho\rho'V_A(s)e^{-s(2x_o+x)/c_p} \qquad \text{(II.1-25)}$$

The origin of the individual terms in this solution may be understood with

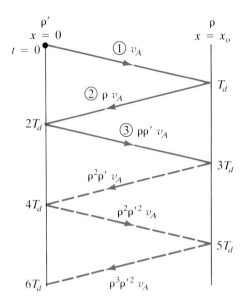

Figure II.1-2
Analysis of a finite-length transmission line using reflection diagrams.

the aid of the reflection or lattice diagram in Figure II.1-2. In this diagram the two vertical lines represent the $x = 0$ and $x = x_o$ boundaries of the transmission line, and time is measured in the vertical direction with $t = 0$ located at the top of the diagram and time increasing vertically downward. Waves traveling to the right are indicated by lines with arrows on them pointing to the right, and waves traveling to the left by lines with arrows on them pointing to the left. The quantity listed next to the arrow represents the amplitude of that component of the wave, and the time markings indicated on the vertical boundary lines represent the time that it takes for that wave component to reach the boundary. T_d is equal to the time that it takes for the wave to make one pass down the length of the line (x_o/c_p seconds).

Returning now to the specific problem being examined, the initial incident wave is launched on the line at $t = 0$ and has the form

$$v_1(x, t) = v_A\left(t - \frac{x}{c_p}\right) \tag{II.1-26}$$

where $v_A(t)$ is found from the equivalent circuit in Figure II.1-1c. At $t = T_d$ seconds the wave strikes the $x = x_o$ boundary, and a reflected wave

$$v_2(x, t) = \rho v_A\left[t - \left(\frac{2x_o}{c_p} - \frac{x}{c_p}\right)\right] \tag{II.1-27}$$

is initiated that travels to the left with time. The total voltage on the line is now equal to

$$v(x, t) = v_1(x, t) + v_2(x, t) = v_A\left(t - \frac{x}{c_p}\right) + \rho v_A\left[t - \left(\frac{2x_o}{c_p} - \frac{x}{c_p}\right)\right] \tag{II.1-28}$$

At $t = 2T_d$ the wave $v_2(x, t)$ strikes the $x = 0$ boundary, producing a second reflected wave

$$v_3(x, t) = \rho\rho'v_A\left[t - \left(\frac{x}{c_p} + \underbrace{\frac{2x_o}{c_p}}_{2\,T_d}\right)\right] \tag{II.1-29}$$

that travels in the same direction as the original incident wave. The total voltage on the line is now

$$v(x, t) = v_1(x, t) + v_2(x, t) + v_3(x, t)$$

$$= v_A\left(t - \frac{x}{c_p}\right) + \rho v_A\left(t - \frac{2x_o - x}{c_p}\right) + \rho\rho'v_A\left(t - \frac{2x_o + x}{c_p}\right) \tag{II.1-30}$$

Although we did not prove it mathematically, it should be fairly apparent that this process can be continued ad infinitum as indicated by the dashed lines shown on the reflection diagram. The amplitude of each successive reflected wave is equal to the amplitude of the wave incident on the boundary times the reflection coefficient at that boundary. The example that follows will help to illustrate these points.

EXAMPLE II.1-1

Sketch $v(x_o, t)$ for the transmission line shown in Figure II.1-3a, assuming that $v_s(t) = 10u(t)$ volts.

Figure II.1-3
Example II.1-1

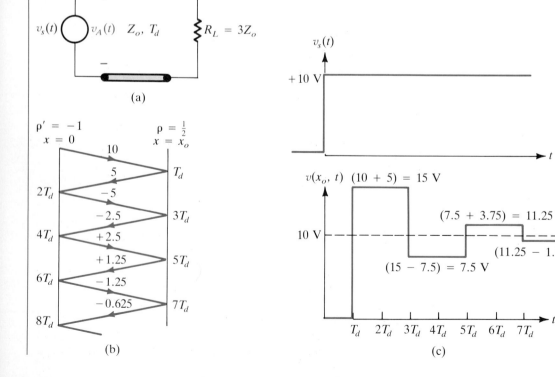

(a)

(b)

(c)

For this circuit $\rho' = -1$ and $\rho = \frac{1}{2}$, so that the reflection diagram has the form shown in Figure II.1-3b because $v_A = v_s$ for this circuit. The output waveshape for $v(x_o, t)$ is obtained from the reflection diagram with $v(x_o, t) = 0$ for $t < T_d$, $v(x_o, t) = 10 + 5 = 15$ V for $T_d < t < 3T_d$, and so on. Replacing the transmission line by a wire in the steady state, it is also apparent that the final value of the voltage across the load will be equal to $+10$ V. The overall waveshape for $v(x_o, t)$ is sketched in Figure II.1-3c.

II.2 DETERMINING WHEN TRANSMISSION-LINE EFFECTS NEED TO BE CONSIDERED

In the analysis of high-speed digital logic circuits one of the most important questions to be answered is: When is a wire a wire, and when is it a transmission line? In this portion of the appendix we demonstrate that the transmission-line nature of wires, cables, and PC board lands needs to be considered when the signal rise and fall times become comparable to the propagation time, T_d, down the line. When these rise and fall times are much larger than T_d, we will show that the reflections that occur will tend to cancel each other out, so that the transmission line, in this instance, can effectively be replaced by a short circuit. These concepts are illustrated by the example that follows.

EXAMPLE II.2-1

Repeat Example II.1-1 for the case where the input signal is considered to be a 10-V step with a total rise time of T_R seconds.

SOLUTION

As in Example II.1-1, the reflection diagram in Figure II.1-3b still applies except that the waves propagating on the line rather than being ideal 10-V steps are now steps with rise times equal to T_R. Following this diagram, and defining $v_r(t)$ as a unit step with a rise time T_R, the voltage at the output of the line may now be written as

$$v(x_o, t) = 15v_r(t - T_d) - 7.5v_r(t - 3T_d) + 3.75v_r(t - 5T_d)$$
$$- 1.875v_r(t - 7T_d) + \cdots \qquad \text{(II.2-1)}$$

The overall waveshape of the output voltage may now be obtained by adding all these individual waves together.

Figure II.2-1 illustrates how $v(x_o, t)$ depends on the rise time of the input signal. The dashed waveshapes in each of these figures shows the output that would have been obtained if the transmission line, except for the inclusion of the delay, had been replaced by an ideal wire. From these results it is apparent that transmission-line effects need to be considered when T_R is less than or equal to T_d, but are of little importance when the rise time is much longer than T_d.

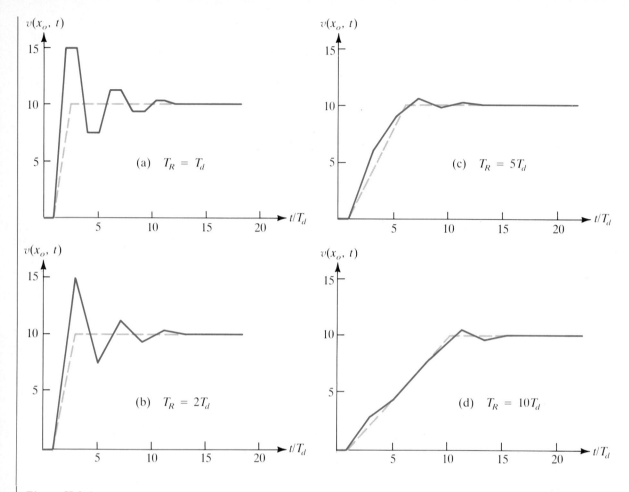

Figure II.2-1
Example II.2-1.

II.3 GRAPHICAL ANALYSIS OF TRANSMISSION-LINE PROBLEMS USING THE SOURCE AND LOAD VOLT–AMPERE CHARACTERISTICS

In many transmission-line problems the load and/or the source may be nonlinear, and in this case the simple analysis methods that we have been using up until now become extremely difficult to apply. In the remainder of this appendix we consider a relatively simple graphical analysis method that is especially useful for attacking this type of nonlinear transmission-line problem.

To graphically solve a problem of the type illustrated in Figure II.3-1, begin by sketching the source and load volt-ampere characteristics on the same set of axes as shown in Figure II.3-2, and then proceed as follows.

1. Determine the values of v and i on the line prior to $t = 0$. This provides

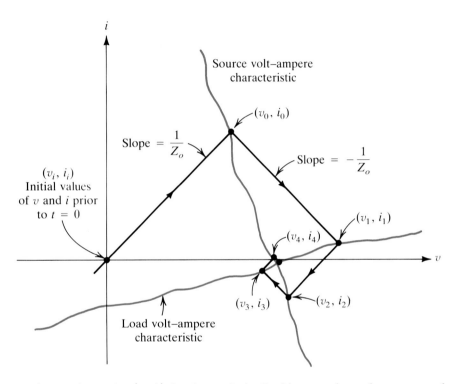

Figure II.3-1
The generalized transmission-line problem to be graphically analyzed.

Figure II.3-2
Combining the source and load volt-ampere characteristics on the same set of axes to rapidly analyze the circuit in Figure II.3-1.

the starting point (v_i, i_i) for the analysis. In this example we have assumed that v_i and i_i are both zero.

2. Draw a line with a slope of $+1/Z_o$ passing through (v_i, i_i). The intersection of this line with the source volt-ampere characteristic (evaluated at $t = 0+$) determines the initial wave amplitudes launched onto the line. These quantities are labeled v_0 and i_0 in the figure.

3. After a delay of T_d seconds, these waves reach the load boundary. Thus, neglecting any signal attenuation on the line v_0 and i_0 are the incident wave amplitudes at the load. To determine the voltage and current at the load after the wavefront reaches it, draw a line with a slope of $-1/Z_o$ through the point (v_0, i_0). The intersection of this line with the load volt-ampere characteristic determines the next value of v and i, namely, v_1 and i_1. Here the subscript 1 refers to the fact that this signal appears at the output after $(1 \times T_d) = T_d$ seconds.

4. Draw a line with a slope of $+1/Z_o$ through (v_1, i_1). Its intersection with the source V–I characteristic determines (v_2, i_2) (the voltage and current at the source end of the line after $t = 2T_d$ seconds).

5. Repeat steps 3 and 4 again and again until the desired steady-state accuracy is achieved.

This analysis technique may appear to be rather complex at first, but once the basic concept is understood, the actual procedure is relatively easy to implement. The example that follows will help to illustrate this point.

EXAMPLE II.3-1

Use the technique just discussed to analyze the nonlinear transmission-line problem shown in Figure II.3-3a.

SOLUTION

For $t < 0$, $v_s(t) = 0$ and the line is uncharged, so that the initial values of v and i are both equal to zero for all positions on the line. The volt-ampere equation for the source end of the transmission line is

$$v(0, t) = v_s(t) = 10u(t) \qquad \text{(II.3-1)}$$

To sketch the volt-ampere characteristic of the load we note that for $v_L < 10$ V, the diode is OFF, so that

$$v_L = 3Z_o i_L \qquad \text{(II.3-2)}$$

For $v_L > 10$ V, the diode is ON, and the Thévenin equivalent of the load has the form given in Figure II.3-3b, so that

$$v_L = 1.5Z_o i_L + 5 \qquad \text{(II.3-3)}$$

The source and load volt-ampere characteristics are sketched in Figure II.3-3c. The analysis of this problem begins by drawing a straight line with a slope of $1/Z_o$ through the initial condition point at the origin. (Notice that the scaling of the voltage and current axes has been chosen purposely so that lines with slopes of plus and minus $1/Z_o$ make angles of plus and minus 45° with the voltage axis. This greatly simplifies the graphical construction of the solution.)

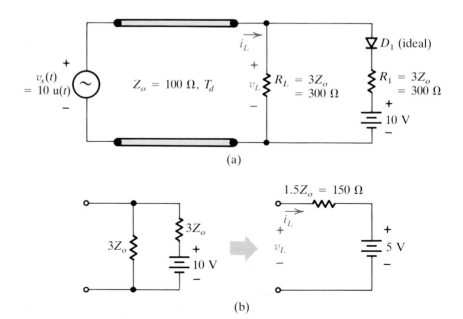

(a)

(b)

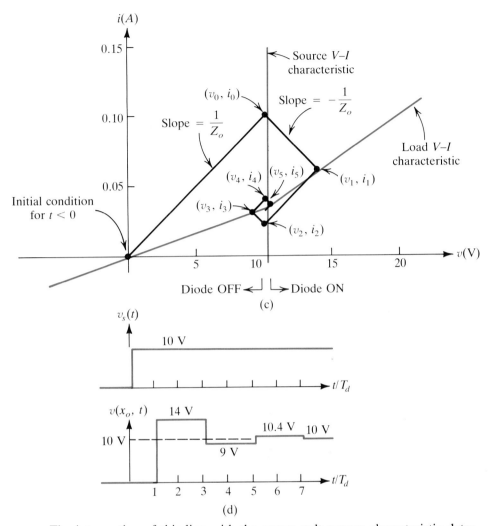

Figure II.3-3
(continued).

The intersection of this line with the source volt-ampere characteristic determines the initial voltage at the $x = 0$ boundary at $t = 0+$ after v_s jumps to $+10$ V. As is apparent, the coordinates of this point are (10 V, 0.1 A), and these two quantities represent the amplitudes of the voltage and current wave that is launched from the $x = 0$ boundary at $t = 0$.

After T_d seconds this wave strikes the load boundary. The load voltage and load current produced when this occurs are found by drawing a straight line with a slope of $-1/Z_o$ through the point (v_0, i_0). From its intersection with the load volt-ampere curve, we find that $(v_1, i_1) = (14$ V, 0.06 A).

To continue the problem solution further, a straight line with a slope of $+45°$ (or $1/Z_o$) is drawn next through the point (v_1, i_1). The intersection of this line with the source V–I characteristic determines (v_2, i_2), the voltage and current at the $x = 0$ boundary of the line shortly after $t = 2T_d$. This procedure is repeated ad infinitum until sufficient steady-state accuracy is obtained (Figure II.3-3c). The transmission-line output voltage is sketched as a function of time in Figure II.3-3d.

Answers to Selected Odd Problems

CHAPTER 1

1.3-1. (a) 360 mV (b) 180 mV (c) 18 mV   **1.3-3.** 60 mV   **1.3-5.** (a) 0.38 V (b) 10.4 V (c) −18 mV
1.4-1. (a) 40 Ω (b) 45 Ω (c) ∞   **1.4-3.** 2 mA, 4 V   **1.4-5.** 5 mW   **1.4-7.** (a) 50 mA (b) 49.3 mA
(c) 0.53 Ω, 48.8 mA   **1.4-9.** (a) 40 mA (b) 26 mA (c) 1 Ω, 21.3 mA
1.4-11. 5 mA, $V_{D1} = 0.75$ V, $V_{D2} = 0.4$ V, $V_R = 0.35$ V
1.4-13. $I_{D1} = 49.45$ mA, $I_{D2} = 0$, $V_{D1} = V_{D2} = 0.547$ V   **1.5-1.** (a) $v_D = v_{in}$ for $v_{in} > 0$, $v_D = 0$ for $v_{in} < 0$
(b) $v_D = v_{in}$ for $v_{in} > -0.7$ V, $v_D = 0.7$ V for $v_{in} < -0.7$ V
(c) same as (b) except that for $v_{in} < -0.7$ V, v_D is a ramp starting at 1.3 V for $v_{in} = -15$ V and ending at $v_D = 0.7$ V
for $v_{in} = -0.7$ V   **1.5-5.** $v_o = v_{in}$ for $v_{in} < 0$, $v_o = 0.67\, v_{in}$ for $v_{in} > 0$
1.5-7. (a) 110 mA (extending the diode curve) (b) 0 mA (c) nonsinusoidal—flattened on the bottom
1.5-9. $v_D = v_{in}$ for $v_{in} < 0.7$ V, $v_D = 0.47$ V $+ 0.33\, v_{in}$ for $v_{in} > 0.7$ V
1.6-1. i_D is a 20 mA (p-p) triangle wave sitting on a 10 mA dc level
1.6-3. v_{o1} is a 20 V (p-p) triangle wave that is clipped off above $+3$ V and below -6 V
1.7-1. $v_C = 15$ V, $v_o = (v_{in} - 15$ V)   **1.7-3.** $v_D = (6.67 - 6.67 \sin \omega t)$ V
1.7-5. $v_o = (1.5\, V_m + V_m \sin \omega t)$ V   **1.8-1.** (a) 13 Ω (b) 23 Ω   **1.8-3.** $v_D = -(1 + 0.05 \sin \omega t)$ V
1.8-5. $v_D = (0.65 + 0.011 \sin \omega t)$ V, $i_D = (23 + 6.6 \sin \omega t)$ mA   **1.9-1.** (a) −5 mA
(c) the diode is OFF and the Q point depends on the specific characteristics drawn   **1.9-3** (a) 20 V (b) 10 V
(c) 5 mA   **1.9-5.** 5.6 Ω to 20 Ω   **1.9-7.** $i_D = 10$ mA for $v_{in} = +10$ V and 0 mA for $v_{in} = -10$V
1.9-9. i_Z is a sawtooth having maximum and minimum values of 20 and 10 mA respectively
1.10-1. $6.2 \times 10^7/\text{cm}^3$ **1.10-3.** 6.92×10^{15} **1.10-5.** b   **1.10-7.** a   **1.10-9.** c
1.11-1. 0.416 cm   **1.11-3.** 0.037 Ω   **1.11-7.** (a) 820 mV (b) 578 mV   **1.11-9.** (a) 0.26 Ω (b) 5.21 Ω

CHAPTER 2

2.1-1. (a) $I_m/1.73$ (b) 0 (c) $I_m/2.45$ (d) $I_m/4$ **2.1-3.** (a) 0.857 V (b) 1.85 V (c) 3.43 mW
2.1-5. (a) 2.15 V (no load), 2.11 V (full load) (b) 2.33 %   **2.1-7.** (a) 0.65 A (b) 7.8 W (d) 0.42 Ω
2.2-1. approximately a sawtooth with a peak amplitude of 17 V, a ripple of 3.47 V, and a period of 8.67 ms

2.2-3. i_D is a pulse train with an amplitude of 2 A and a pulse width of 1 ms   **2.2-5.** (a) 0.63 A (b) 3.64 W

2.2-7. $v_o = v_s$ when $v_s = +10$ V and v_o is a sawtooth falling from 10 V to 9 V when $v_s = -10$ V   **2.2-9.** 0.4 A

2.2-11. $V_o = 20.8$ V, $V_r = 0.88$ V

2.2-13. $v_{C1} = 10$ V, $v_{D1} = (-10 - 10 \sin \omega t)$ V, $v_o = v_{C2} = 20$ V, $v_{D2} = (-10 + 10 \sin \omega t)$ V

2.3-1. (a) $9.5\ \Omega < R_1 < 14.24\ \Omega$, let $R_1 = 12\ \Omega$, $P_D = 27$ W (b) 39,600 μF (c) $> 10\ \Omega$ **2.3-3.** (a) 17 V

(b) 8 V (c) 24 W (d) $> 3.2\ \Omega$ **2.3-5.** 10 %

CHAPTER 3

3.1-3. transconductance amplifier: output circuit is a controlled voltage source $R_m i_i$ in series with R_o where $R_m = 1$ MΩ, current amplifier: output circuit is a controlled current source $A_{io} i_i$ in parallel with R_o where $A_{io} = 200$, transconductance amplifier: output circuit is a controlled current source $G_m v_i$ in parallel with R_o where $G_m = 20$ mS

3.1-5. (a) 16.7 for circuit 1, 6667 for circuit 2   **3.1-7.** (a) 3.162 (b) 4.5 W   **3.1-9.** 3×10^{11} **3.2-1.** d

3.2-3. a   **3.2-11.** $v_{CB} \approx (-10 + 10 \sin \omega t)$ V   **3.3-1.** (a) $I_B = 0$, $V_{BE} = -2$ V, $I_C = 0$, $V_{CE} = 15$ V

(b) $I_B = 0.13$ mA, $V_{BE} = 0.7$ V, $I_C = 1.3$ mA, $V_{CE} = 5.9$ V

(c) $I_B = 0.13$ mA, $V_{BE} = 0.7$ V, $I_C = 2.14$ mA, $V_{CE} = 0.2$ V

3.3-3. $R_E = 1.33$ kΩ, let $R_2 = 4.7$ kΩ then $R_1 = 13.3$ kΩ

3.3-5. $I_B = 0$, $V_{BE} = -1.43$ V, $I_C = 0$, $V_{CE} = 19.14$ V   **3.3-7.** $I_{B,min} = 0.057$ mA, $I_B = 1.06$ mA

3.3-9. $V_{CE} = 13.1$ V   **3.3-11.** $I_B = 0.05$ mA, $I_C = 5$ mA, $V_{CE} = 5$ V   **3.3-13.** $I_{C2} = 71$ mA, $V_{C2} = 8.45$ V

3.3-15. $I_B = 0.138$ mA, $V_B = 6.55$ V, $I_C = 2.79$ mA, $V_C = 6.05$ V

3.4-3. max $I_C = 1.05$ mA, min $I_C = 0.993$ mA   **3.4-5.** 1.4 mA, 0.8 mA   **3.4-7.** 1.775 V

3.4-9 (a) $I_B = 0.051$ mA, $I_C = 5.1$ mA, $V_{CE} = 7.25$ V (b) 57 %

3.4-11. (a) at 25° C: $I_B = 1\ \mu$A, $I_C = 0.1$ mA, $V_{CE} = 15$ V; at 100° C: $I_B = 1\ \mu$A, $I_C = 0.284$ mA, $V_{CE} = 5.85$ V

(b) at 100° C: $I_B = 1\mu$A, $I_C = 0.396$ mA, $V_{CE} = 0.2$ V   **3.5-1.** $V_{CE} = (11 - 5 \sin \omega t)$ V

3.5-3. (a) $V_{BB} = 2$ V, $R_B = 66.7$ kΩ (b) $V_{BE} = 0.65$ V, $I_B = 25\ \mu$A, $I_C = 2.5$ mA, $V_{CE} = 3.5$ V (c) 10 μA, 40 μA

3.5-5. (a) cutoff for $v_{in} < 1.4$ V, saturated for $v_{in} > 16.4$ V

(b) v_o is an out of phase square wave with amplitudes of 11.4 and 14.9 V

3.5-7. (a) $i_C = (1.3 + \sin \omega t)$ mA but clips above 2 mA, $v_{CE} = (3.5 - 5 \sin \omega t)$ V but clips below 0 V

(b) $V_{BB} = 1.7$ V, $v_{CE} = (5 - 5 \sin \omega t)$ V (c) -5 **3.6-1.** (a) $2 = (18$ k$\Omega)I_B + V_{BE}$

3.6-3. (a) $20 = (1$ k$\Omega)I_C + V_{CE}$ (b) $16.25 = (0.5$ k$\Omega)i_C + v_{CE}$ (c) $v_{CE} = (12.5 - 2.5 \sin \omega t)$ V

3.6-5. $v_{CE} \approx (8.85 - 7.6 \sin \omega t)$ V, slight transistor saturation occurs near $v_{CE} = 0$

3.6-7. (a) $I_C = 4$ mA, $V_{CE} = 8$ V (b) $16 = (2$ k$\Omega)i_C + v_{CE}$ (c) $v_{CE} = (8 - 4 \sin \omega t)$ V

CHAPTER 4

4.2-1. $v_{BE} = (0.7 + 0.002 \sin \omega t)$ V, $v_{CE} = (5 - \sin \omega t)$ V   **4.2-3.** $v_{CE} = (14.5 - 4.5 \sin \omega t)$ V

4.2-5. $v_{CE} = (8.16 - 0.026 \sin \omega t)$ V, gain $= -4.9$ **4.2-7.** (a) approximately 24 %

(b) $v_{CE} = (14.5 - 4.5 \sin \omega t + 1.08 \cos 2\omega t)$ V [max $v_{CE} = 17.8$ V, min $v_{CE} = 8.92$ V]

4.3-1. $h_{11} = 15$ kΩ, $h_{12} = h_{21} = 0.5$, $h_{22} = 0.1$ mS   **4.3-3.** (a) 2 kΩ (b) 1.5 kΩ **4.3-5.** (a) $V_{CE} = 8$ V

(b) $h_{ie} = 2.4$ kΩ, $h_{fe} = 100$, $h_{re} \approx 0$, $1/h_{oe} = 6.4$ kΩ **4.3-7.** (a) magnitude less than 1000 (b) -1000

(c) $-1,111$ **4.4-1.** -200 **4.4-3.** (a) $I_E = I_C = 2$ mA, $V_C = 10$ V, $V_E = 4$ V

(c) $v_E = 4$ V, $v_{CE} = (6 - 2.86 \sin \omega t)$ V, $v_L = 2.86 \sin \omega t$ V   **4.4-5.** $v_{CE} = (14 - 2.72 \sin \omega t)$ V

4.4-7. $-60(v_1 - v_2)$ **4.4-9.** -25 **4.4-11.** $(6 - 0.95 \sin \omega t)$ V   **4.5-1.** (a) $v_T = -150\ v_{in}$, $R_T = 3$ kΩ

(b) 7500   **4.5-3.** -100 (b) -50 (c) -50 (d) -16.67 (e) 833   **4.5-5.** $v_o = -1.5\ v_{in}$ **4.5-9.** (a) 2.236 : 1

(b) 75 μW (c) 15,000   **4.5-11.** 5.53×10^8, 3.75×10^9, 5.53×10^8

CHAPTER 5

5.2-3. $I_D = I_{DSS}$ for $V_{DS} > -V_p$, $I_D = [-V_p/r_{DS(ON)}][V_{DS}/-V_p) - (2/3) \times (V_{DS}/-V_p)^{3/2}]$ for $V_{DS} < -V_p$

5.2-7. $v_o = (0.7 + 0.056 \sin \omega t)$ V for $v_1 = 0$, $v_o = (1 + 0.75 \sin \omega t)$ V for $v_1 = -5$ V

5.3-1. $I_D = 5.8$ mA, $V_{GS} = -1.3$ V, $V_{DS} = 12$ V   **5.3-3.** 6.1 mA   **5.3-5.** 0.9 kΩ

5.3-7. $I_D = 3.57$ mA, $V_{DS} = 11.07$ V   **5.3-9.** $I_D = 2.5$ mA, $V_{GS} = -2.5$ V, $V_{DS} = 10$ V

5.3-11. $V_{CE2} = 8$ V   **5.3-13.** $I_D = 0.85$ mA, $V_{GS} = -3.4$ V, $V_{DS} = 8.1$ V

5.4-1. nominal $v_{DS} = 9.6$ V, shift is approximately ± 28 %   **5.4-3.** $R_S = 0.4$ kΩ, $R_1 = 11.5$ MΩ, $V_{DS} = 5$ V
5.4-5. (a) $R_S = 1.1$ kΩ, $R_{S2} = 1.4$ kΩ (b) ± 1 mA   **5.4-7.** $R_1 = 6.7$ MΩ, $R_D = 3.4$ kΩ, $R_S = 0.6$ kΩ
5.5-1. $r_{DS} = L/[\mu_n C_{ox} Z(V_{GS} - V_T)]$ **5.5-5.** $I_D = 0$ for $V_{DS} < V_T$, $I_D = I_{D2T}(V_{DS}/V_T - 1)^2$ for $V_{DS} > V_T$
5.5-7. (a) $V_{DS} = 20$ V (b) $V_{DS} = 2$ V   **5.5-9.** (a) $I_D = 3.2$ mA, $V_{GS} = -0.4$ V, $V_{DS} = 4$ V (b) 43 kΩ
5.5-11. $R_S = 1.5$ kΩ, $R_1 = 350$ kΩ **5.6-1.** $g_m = 1.25$ mS, $r_d = 12$ kΩ
5.6-3. $V_{DS} = (9 - 1.5 \sin \omega t)$ V, $v_L = -1.5 \sin \omega t$ V   **5.6-5.** (a) $I_D = 3$ mA, $V_{GS} = -2$ V, $V_{DS} = 9$ V
(b) $g_m = 2.3$ mS (c) $v_{DS} = (9 - 3.3 \sin \omega t)$ V   **5.6-7.** (a) $I_D = 2$ mA, $V_{GS} = -2.5$ V
(b) $g_m = 2$ mS, $v_{DS} = (6 - 2.2 \sin \omega t)$ V   **5.6-9.** $v_D = (15 - 8 \sin \omega t)$ V, $v_L = -8 \sin \omega t$ V
5.6-11. v_{DS} is a square wave whose voltage levels go between 16.4 V and 11.6 V, v_L is a 4.8 V (p-p) square wave
5.7-1. $v_S = (1 + 0.2 \sin \omega t)$ V, $v_D = (15 + 0.44 \sin \omega t)$ V, $v_o = 0.44 \sin \omega t$ V
5.7-3. (a) $I_D = 5$ mA, $V_{GS} = -1.17$ V, $V_{DS} = 3.83$ V (b) $v_D = (5 - 2.6 \sin \omega t)$ V   **5.7-5.** (a) -0.483
(b) $v_D = (25 - 0.483 \sin \omega t)$ V   **5.7-7.** $v_{o1} = -v_{in}$, $v_{o2} = -5 v_{in}$ **5.7-9.** -296 **5.7-11.** -12
5.8-1. circuit 4, 1.9×10^5 **5.8-3.** circuit 5 is OK, let $R_{D1} = 1$ kΩ then $R_{C2} = 4$ kΩ
5.8-5. circuit 3, $R_{C2} = 3.26$ kΩ

CHAPTER 6

6.2-1. $g_m(R_D \| R_L)\{1/[1 + 1/(R_D + R_L)C]\}$ **6.2-3.** $-11.1[s/(s + 222)]$ **6.2-5.** $-1.08[s/(s + 8)]$
6.2-7. $180[s/(s + 1)]$ **6.2-9.** $-2.78 [s/(s + 8.33)][s/(s + 66.7)]$ **6.3-1.** $-1.5[(s + 100)/(s + 150)]$
6.3-3. (a) -33.3 (b) 48 μF (c) 0.13 μF for C_1 break at 2 rps   **6.3-5.** $\omega_1 = 22$ rps, $\omega_{2N} = 50$ rps
6.3-7. approximate 3 dB point is ω_D, exact 3 dB point is 0.99 ω_D **6.3-9.** (a) $\omega_1/\sqrt{2^{1/n} - 1}$
6.3-11. (b) $-7.9 [s/(s + 66.7)][(s + 100)/(s + 175)]$
6.3-13. ω_L is imaginary, circuit gain never falls 3 dB below midband gain
6.4-1. (a) $C_d = 1.6$ nF, $C_j = 45$ pF   **6.4-3.** $i_C(t) = [4.65 + 0.707 \sin (\omega t - 45°)]$ mA
6.4-5. $C_{gd} = C_{rss}$, $C_{ds} = C_{oss} - C_{rss}$, $C_{gs} = C_{iss} - C_{rss}$ **6.5-1.** $v_{be} = 0.017 v_{in}$, $v_{ce} = -8.11 v_{in}$
6.6-1. (a) -250 (b) -48 **6.6-3.** (a) 1 kΩ (b) -44.5
6.6-5. input break $= 2 \times 10^7$ rps, C_{gs} break $= 3.8 \times 10^7$ rps, output break $= 2 \times 10^8$ rps   **6.6-7.** 1.65 kΩ
6.6-9. $V_{oc} = V_{in}/(1 + g_m R_S)$, $I_{sc} = V_{in}/(R_G + R_S)$
6.7-1. input break $= \infty$, C_{be1} break $= 10^{10}$ rps, break at base of $Q_2 = 1.7 \times 10^{10}$ rps, C_{be2} break $= 10^{10}$ rps, output
break $= 2 \times 10^8$ rps   **6.7-3.** 4.8×10^7 rps $< \omega_L < 9.5 \times 10^7$ rps   **6.7-5.** $n = 2$, 493 kHz $< f_H < 600$ kHz
6.7-7. $A_{mid} = -100$, using $R_E = 1$ kΩ : 0.48×10^9 rps $< \omega_H < 10^9$ rps
6.8-1. v_{CE} is an inverted square wave whose amplitudes go between 10 V and 4.26 V. The edges are exponentially
rounded with rise and fall times of 220 μs.
6.8-3. v_o is normally at $+10$ V. On the rising edge of v_{in}, v_o goes to 0 V for 13.8 μs and then rises back exponentially towards $+10$ V with a time constant of 10 μs.
6.8-5. v_o starts out at $+10$ V and drops exponentially towards $+5$ V with a time constant of 0.33 ms
6.8-7. (b) v_{DS} is nominally 11.1 V. On the rising edge of v_{in}, v_{DS} drops to 10.2 V and then rises exponentially back to
11.1 V with a time constant of 100 μs. In a similar fashion, on the falling edge of v_{in}, v_{DS} jumps to 12 V and then exponentially returns to 11.1 V.   **6.8-9.** 7.27 V

CHAPTER 7

7.1-1. 10 : 1 for circuit 1, 100 : 1 for circuit 2   **7.1-3.** 25 % for circuit 1, 0.025 % for circuit 2
7.1-5. 205 kΩ **7.1-7.** 20 kΩ **7.1-9.** 10.76
7.1-11. the final circuit looks like that in Figure 7.1-5a with A_{OL} the connection of the original amplifier together
with a preamp of gain 10, $\beta = 0.099$, and A_1 a preamp with a gain of 10   **7.1-13.** (a) -250 (b) -9.42
7.1-15. (a) -4.999980035 (b) 0.00003 %   **7.2-1.** (a) $-10 \sin \omega t$ mV (b) $(10 \sin \omega t - 10)$ mV
(c) 15 $\sin \omega t$ V but clips when output exceeds ± 10 V (d) $(-10 \sin \omega t + 6)$ V but clips when output exceeds $+10$ V
7.2-3. One design uses an amplifier of the type shown in Figure 7.2-5a with $R_f = 70$ kΩ, $R_2 = 23.3$ kΩ, and a resistor $R_3 = 17.5$ kΩ connected from the inverting input to ground. In this circuit, source $v_2 = v_{in}$ and source v_1 is connected between the noninverting input and ground.   **7.2-5.** $-6.127 \sin \omega t$ V   **7.2-7.** -3.33, -6.66
7.2-9. Connect the transducer directly to the inverting input with $R_f = 5$ kΩ and the noninverting input grounded.
Connect the meter in series with a 1 kΩ resistor from the op amp output to ground.   **7.3-1.** 2120
7.3-3. v_o is a 5-V (p-p) triangle wave sitting on a -2.5 V dc level   **7.3-5.** $sRC/(1 + sR_1C)$ **7.3-7.** (a) 1 kΩ
(b) $+11$ V, -20 V   **7.3-9.** $V_{BB}(R_f/2R)[(\text{decimal equivalent of the number})/2^n]$ **7.3-11.** 96 dB

7.4-1. 1.73 kΩ

7.4-3. two stage cascade: 1st stage—a noninverting amplifier with $R_f = 18.94$ kΩ, $R_2 = 1$ kΩ; 2nd stage—a second order LPF as in Figure 7.4-9a with $C_1 = C_2 = 0.001$ μF, $R_1 = R_2 = 2.12$ kΩ, $R_B = 10$ kΩ, and $R_A = 5.86$ kΩ

7.4-5. $V_{o1}/V_{in} = -(sRC)^2/[(sRC)^2 + sRC/Q + 1]$, $V_{o2}/V_{in} = sRC/[(sRC)^2 + sRC/Q + 1]$, $V_{o3}/V_{in} = -1/[(sRC)^2 + sRC/Q + 1]$ **7.4-7.** $1/[s^2 R_1 R_2 C_1 C_2 + sR_2 C_2 + 1]$

7.4-11. $d^2 v_o/dt^2 + 2\,dv_o/dt + 2\,v_o = 0$

7.4-13. The basic design is that given in Figure 7.4-14 with R_1 replaced by a 1.25 pF switched capacitor, R_4 by a 0.625 pF switched capacitor, R_2 by an inverting-style 1.25 pF switched capacitor, and R_3 by an inverting-style 2.5 pF switched capacitor. The switching frequency used is 160 Hz.

7.5-1. $-8 \sin \omega t$ V but clips when v_o goes above 0 V or below -5 V

7.5-3. (a) $-5 \sin \omega t$ V but clips when v_o goes below 0 V (b) -10 V for $v_{in} > 0$, $-5 \sin \omega t$ V for $v_{in} < 0$

7.5-5. the output is a series of 100 ms duration ramps starting at 0 V and ramping down by 2.5 V and then up by 5 V in each 200 ms interval

7.5-9. For v_{in} very positive, $v_o = -10$ V. As v_{in} decreases v_o switches to $+10$ V at $v_{in} = -4.33$ V. With $v_o = +10$ V, v_o switches to -10 V at $v_{in} = 2.33$ V.

7.5-11. with D OFF R_2, R_3 and $-V_{CC}$ are all connected to the op amp output, and since the op amp is ideal its output impedance is zero and v_o is only a function of R_1, R_4, and v_{in}

7.5-13. the circuit is similar to that given in Figure 7.5-13 with $R_4 = 10$ kΩ, $R_1 = R_5 = \infty$, $R_{3A} = R_{3B} = 40$ kΩ, $R_{2A} = R_{2B} = 100$ kΩ, and the supply connected to R_{2B} equal to $+10$ V

7.5-15. Consider that the triangle wave and sine wave amplitudes are each 20 V (p-p) and that the diode breaks occur at $v_o = \pm 5$ V and $v_o = \pm 8.67$ V. The overall circuit is similar to that given in Figure 7.5-10e with $R_4 = 10$ kΩ and $R_1 = 15$ kΩ except that it has 4 diode-R_2-R_3 subcircuits. For the break at ± 5 V: $R_{3A} = 40.95$ kΩ, $R_{2A} = 122.8$ kΩ, and $-V_{CC} = -15$ V; for the break at 8.67 V: $R_{3B} = 6.34$ kΩ, $R_{2B} = 10.95$ kΩ, and $-V_{CC} = -15$ V. The other two subcircuits for negative values of v_o are identical to those just given except that the diode directions and power supply polarities are reversed.   **7.6-1.** -1.25 **7.6-3.** $2.5\, v_{in}/(7$ kΩ$)$ **7.6-5.** $V_{CE5} = 3$ V

7.6-7. $R_1 = 9.3$ kΩ, $R_2 = 600$ Ω

7.6-9. (a) the output is the sum of a 20-V sine wave and a 2-V triangle wave (CMRR $= 100$) (b) the output is the sum of a 20-V sine wave and a 10-mV triangle wave (CMRR $= 2 \times 10^4$)

7.7-1. top: $(0.583 \sin \omega t - 0.833)$ V, middle: $0.571 \sin \omega t$ V, bottom: $(0.583 \sin \omega t + 0.833)$ V

7.7-3. $V_{IO} = 5$ mV, $I_1 = 3$ nA, $I_2 = 2$ nA   **7.7-5.** $+90$ to $+130$ mV   **7.7-7.** 1000 rps   **7.7-9.** 5 kHz

7.7-11. (a) $33.7 \sin \omega t$ mV (b) $-90.9 \sin \omega t$ mV   **7.7-13.** $v_{o1} = -2 \sin \omega t$ V, $v_{o2} = 2 \sin \omega t$ V

CHAPTER 8

8.2-1. (a) flat at low frequencies at 94 dB and falls off at -20 dB/dec beyond 500 rps (b) $v_o(t) = 50(1 - e^{-t/(2\text{ ms})})$ V, clipping at $+15$ V at $t = 713$ μs
(c) flat at low frequencies at 40 dB and falls off at -20 dB/dec beyond 250 krps, $v_o(t) = 100(1 - e^{-t/(4\text{ μs})})$ mV

8.2-3. (a) $10^3 \alpha^2/(s^2 + 2\alpha s + 10\alpha^2)$ (b) The response to a 1 mV step is $100[(1 - 1.05e^{-t/(1\text{ ms})}\sin(3000t + 72°)]$ mV. If we consider that the square wave input is a series of alternating steps, then the response is basically a 200 mV (p-p) square wave with ringing on the edges of the form just given.

8.2-5. (a) $(1 + R_F/R_2)[\alpha^5/(s + \alpha)^5]/[1 + (R_F/R_2) \times \alpha^5/(s + \alpha)^5]$ (b) 69 kΩ (c) 7280 rps
(d) $V_1 = 10\underline{/0°}$, $V_2 = 8.08\underline{/-36°}$, $V_4 = 6.52\underline{/-72°}$, $V_6 = 5.275\underline{/-108°}$, $V_8 = 4.26\underline{/-144°}$, $V_{10} = V_f = 3.44\underline{/-180°}$

8.2-7. (a) 400 Ω (b) $\alpha + j\alpha$

8.2-9. (a) flat at 60 dB at low frequencies and falls off at -40 dB/dec beyond 100 rps (b) 203 rps

8.3-1. (a) unstable (b) unstable (c) marginally stable   **8.3-3.** (a) stable (b) unstable (c) stable

8.3-5. $R > 160$ Ω for $\phi_T < 135°$ **8.3-7.** unstable   **8.3-9.** 45° without C_L, 24° with C_L

8.3-11. $\phi_T \cong -180°$, circuit is on the edge of instability   **8.4-1.** (a) 223.6 rps (b) 99 kΩ   **8.4-3.** 13 kΩ

8.4-5. $1/\sqrt{LC}$, 0.5 mS   **8.4-7.** 4×10^6 rps, 0.122 mS   **8.4-9.** 3.16×10^5 rps, 0.33 mS

8.5-1. $v_G = (-1.85 + 1.85 \sin \omega t)$ V, $v_{DS} = (20 - 3.7 \sin \omega t)$ V

8.5-3. $i_C \cong (10 + 5 \sin \omega t)$ mA, $v_C = (10 - 0.5 \sin \omega t)$ V

8.5-7. $v_o(t) = 33 \sin \omega t$ V but clips when v_o exceeds ± 10 V, $v_1(t) = 1.22 \sin \omega t$ V

8.6-1. (a) $f_s = 7.9577471$ MHz, $f_p = 7.959339$ MHz, $|Z(j\omega_s)| = 100$ Ω, $|Z(j\omega_p)| = 160$ kΩ (b) 0.00182 % (c) 0

8.6-3. $v_G = 2 \sin \omega t$ V, $v_S = (1 + \sin \omega t)$ V, $v_D = (20 - 4 \sin \omega t)$ V

8.6-5. $f_o = 96$ kHz, $v_o = 16.7 \sin \omega t$ V but clips when v_o exceeds $+10$ V, $v_1 = 8.33 \sin \omega t$ V

8.7-1. The waveshapes are similar to those in Figure 8.7-7b except that v_o goes between $+10$ V and -5 V, and v_1 and v_2 between $+3.33$ V and -1.67 V. In addition, the charging time when $v_o = +10$ V is 56 μs and that when $v_o = -5$ V is 91.6 μs so that $f_o = 6.8$ kHz

8.7-3. (a) The waveshapes are similar to those in Figure 8.7-7b except that v_2 charges towards ±6.7 V (b) 3.9 MHz

8.7-5. v_{o2} is a 100-Hz 20-V (p-p) square wave and v_{o1} is an out of phase 100-Hz 5-V (p-p) triangle wave

CHAPTER 9

9.1-1. (a) 288 W (b) 144 W (c) 144 W   **9.1-3.** (a) 66.7 W (b) 9.88 W

9.1-5. $P_L = 20$ W, $I_{IN} = 1.43$ A, $P_{IN} = 28.6$ W, $P_D = 8.6$ W

9.2-3. (a) v_{DS} intercept = 30 V, i_D intercept = 1.5 A (b) 0 W, 10 W, 0 W (c) yes, for values of v_{in} about 5 V to 6 V

9.2-5. (a) For t < 0, $v_{DS} = 30$ V. For t > 0, $v_{DS} = (60\,e^{-t/(20\,\mu s)} - 30)$ V and v_{DS} decreases exponentially towards -30 V flattening off at 0 V after 13.9 μs. For $t > 50\ \mu$s v_{DS} charges exponentially towards 30 V with a time constant of 20 μs.

(c) the instantaneous power dissipation in the transistor is as high as 90 W and this could damage the FET especially if the v_{in} pulsing were periodic   **9.3-3.** (a) $v_{CE} = (20 - 15\sin\omega t)$ V, $i_C = (300 + 150\sin\omega t)$ mA (b) 1.125 W (c) 15 W (d) 4.875 W   **9.3-5.** 2 W

9.3-7. (a) $v_{GS} = (5 + 0.5\sin\omega t)$ V, $i_D = (1 + 0.5\sin\omega t)$ A, $v_{DS} = (25 - 12.5\sin\omega t)$ V (b) 50 W (c) 6.25 %

9.3-9. 0.11 %   **9.3-11.** (a) $i_D = (0.67 + 0.5\sin\omega t)$ A, $v_L = -4\sin\omega t$ V

(b) $P_L = 2$ W, $P_{IN} = 20.1$ W, $P_D = 17.2$ W   **9.3-13.** 0.8 W   **9.4-1.** (a) $R_1 = 776\ \Omega$, $R_2 = 48\ \Omega$

(b) 4750 μF for break at about 2 Hz (c) 11.1 W

9.4-3. (a) v_L is a 20 V (p-p) square wave, v_A is a 21.9 V (p-p) square wave, v_{CE1} is an out of phase square wave with amplitudes of +5 V and +25 V (b) $\eta = 66.7$ %, 6.25 W/transistor

9.4-5. (a) $V_{CC} = 15$ V, $R_L = 11.25\ \Omega$, $I_C = 0.67$ A, $V_{CE} = 15$ V (b) $v_{in} = 134\sin\omega t$ mV

(c) $P_D = 10$ W/transistor, $P_L = 0$, $P_{IN} = 20$ W (d) $P_{IN} = 20$ W, $P_L = 10$ W, $P_D = 5$ W/transistor

9.4-7. (a) 400 Ω (b) 28.5 V

(c) $v_L = 25.3\sin\omega t$ V, $i_{D3} = 3.16\sin\omega t$ A for $v_{in} > 0$ and 0 for $v_{in} < 0$, $v_{in} = 28.5\sin\omega t$ V, $v_{G3} = (4 + 28.5\sin\omega t)$ V, $v_{GS3} = (4 + 3.2\sin\omega t)$ V   **9.4-9.** 3HD = 15.6 %, 2HD $\approx$ 1 % neglecting the crossover distortion

9.5-1. (a) 4.7 V (b) 5 Ω (c) 15 W for R = 5 Ω and $v_{UR} = 20$ V (d) 0.27 V (p-p)

9.5-3. $V_L = 5$ V, $V_l = 41$ mV (p-p)   **9.5-5.** (a) 10 %

(b) 2.4 %, v_L is a 48 mV (p-p) sawtooth riding on a 10 V dc level   **9.5-7.** (a) 10.075 V

(b) load regulation = 0.5 %, line regulation = 1 %   **9.5-9.** (a) $V_L = 15$ V, $I_{o1} = 0.75$ mA, $V_{o1} = 11.075$ V

(b) $P_{D1} = 22.5$ W, $P_{D2} = 1.45$ W   **9.6-1.** (a) 16.7 μs (b) 668 μH (c) 18.125 V (d) 40 μF   **9.6-3.** (a) 6.7 kΩ

(b) 6.6 mV (p-p) (c) approximately 5 V, V_L is independent of R_L in a 1st order analysis of this supply

9.6-5. (a) 49.63 V (b) The inductor current is a 0.36 A (p-p) triangle wave sitting on a 0.49 A dc level. The FET current is a ramp going from 0.155 A to 0.355 A during T_1 and is zero during the remainder of the cycle.

9.6-7. (a) 1.64 kΩ

(b) The waveshape of v_x is as shown in Figure P9.6-7 with $T_1 = 100\ \mu$s. The other waveshapes are similar to those given in Figure 9.6-7 with the peak in $i_{D1}\,(i_{C1}) = 0.4$ A, and the peak secondary current = 0.8 A.

(c) $V_L = 19.7$ V, the waveshapes are the same as in (b) except that $T_1 = 112.2\ \mu$s, the peak in $i_{C1} = 0.337$ A, and the peak secondary current = 0.674 A   **9.6-9.** (a) 4 mV (p-p) (b) 1 %, 0.052 Ω

9.7-1. In Figure 9.7-2 insert a 5 kΩ feedback resistor in between V_L and the inverting input terminal (V_2) of the op amp. Also connect a 10 kΩ resistor from V_2 to ground. Replace V_{REF} by the 10-V zener and connect a 2 kΩ resistor from V_{UR} to the noninverting op amp input terminal. Set $R_{CL} = 0.35\ \Omega$. $P_{D1,max} = 60$ W.

9.7-3. (a) 1.073 A, $R_L = 14\ \Omega$ (b) 0.93 A (c) $75/R_L$ W for $R_L > 14\ \Omega$, $(21.4 - 1.14\,R_L)$ W for $R_L < 14\ \Omega$

9.7-5. (a) 6.8 μs (b) for $I_L < 1$ A: $T_1 = (44.7\sqrt{I_L})\ \mu$s and $P_{IN} = (10 \times I_L)$ W

9.7-7. (a) $I_{SC} = 1$ A, $I_{FB} = 1.75$ A (b) $I_{SC} = 0.25$ A, $I_{FB} = 2.86$ A

(c) let $R_1 + R_2 = 1$ kΩ to prevent loading on Q_1, then $R_1 = 487\ \Omega$, $R_2 = 512\ \Omega$, $R_{CL} = 5.47\ \Omega$

9.8-3. (a) let $v_{in} = 170\sin\theta$ then $v_L = 0$ for $\theta < 0.46$ rad and for 3.14 rad $< \theta < 3.6$ rad, and $v_L = v_{in}$ for all other values of θ (b) 141 W   **9.8-5.** (a) $v_G \approx -16.3$ V

(b) the battery charging current is zero when the transformer secondary voltage is less than 12.7 V and equals $[(17\sin\omega t - 12.7)\ \text{V}]/(100\ \Omega)$ when the secondary voltage exceeds 12.7 V

(c) When the ac voltage fails the current flow through R_2 will be about 565 μA. If we assume that the current divides evenly between the lower diode and the gate of the SCR, then the SCR will fire.

(d) used to turn the lamp off when ac power is restored

9.8-7. (a) The firing angle is about 1.93 radians and represents the time required for v_C to reach 35.7 V. The waveshape for v_L is given in Figure 9.8-9c. (b) double the half-wave solution given in equation (9.8-13) (c) 319 W

9.9-1. (a) 0.2° C/W (b) 0.23° C/W   **9.9-3.** (a) 3.87 W (b) 23.8° C/W   **9.9-5.** $\pi/2$ **9.9-7.** (a) 1.62 W

(b) 44.1° C/W   **9.9-9.** (a) 8.75 W (b) 103.75° C (c) 7.57° C/W

10.1-1. (a) $v_o = 5$ V for $v_{in} < 0.7$ V, slope $= -5.2$ for 0.7 V $< v_{in} < 1.66$ V, and $v_o = 0.2$ V for $v_{in} > 1.66$ V (b) v_{o1} is an inverted waveshape with rise and fall times of 19.2 μs, v_{o2} is in phase with v_{in} with rise and fall times of 3.8 μs

10.1-3. (a) $v_o = 5$ V for $v_{in} < 2$ V, slope $= -4$ for 2 V $< v_{in} < 3.25$ V, and $v_o = 0$ (neglecting the ohmic region) for $v_{in} > 3.25$ V (b) v_{o1} is an inverted waveshape with rise and fall times of 25 μs, v_{o2} is in phase with v_{in} with rise and fall times of 6.25 μs

10.1-5. $NM_L = 1.55$ V, $NM_H = 2.09$ V

10.1-7. (a) $v_o = 0.2$ V for $v_{in} < 1.23$ V, slope $= 343$ for 1.23 V $< v_{in} < 1.244$ V, and $v_o = 5$ V for $v_{in} > 1.244$ V (b) 333

10.1-9. (a) $v_o = 5$ V for $v_{in} < 2$ V, slope $= -25$ for 2 V $< v_{in} < 2.18$ V, and $v_o = 0.45$ V for $v_{in} > 2.18$ V (b) $NM_L = 1.55$ V, $NM_H = 2.82$ V **10.2-1.** 233 ns **10.2-3.** 591 ns

10.2-5. v_T goes from -3.33 V initially to $+3.33$ V at $t = 0$, $R_T = 33.3$ kΩ, v_{CE} stays at $+20$ V for 1.55 μs and then begins to fall exponentially towards -19.5 V with a time constant of 1 μs stopping at 0.2 V when the transistor saturates. The total fall time is 690 ns. **10.2-7.** approximately 1.1 μs (the transistor does not saturate)

10.2-9. $i_D = 1$ mA and goes to zero at $t = 0$, $v_{DS} = 0$ at $t = 0$ and rises exponentially topwards 20 V with a time constant of 200 ns

10.3-1. (a) when Q_0 saturates, $Q_1 - Q_{10}$ cut off as required; when Q_0 cuts off, Q_1 saturates but $Q_2 - Q_{10}$ remain cut off (b) when Q_0 is cut off, $I_{B1} = 0.98$ mA, $I_{B2} - I_{B10} = 0.88$ mA; since $I_{B,\,min} = 0.098$ mA, $Q_1 - Q_{10}$ are all saturated.

10.3-3. Using $h_{FE} = 20$, $v_o = 0.81$ V for $v_{in} < 0.7$ V, slope $= -1.9$ V for 0.7 V $< v_{in} < 0.7575$ V, slope $= -40$ for 0.758 V $< v_{in} < 0.77$ V, $v_o = 0.2$ V for $v_{in} > 0.77$ V **10.3-5.** (a) 22 (b) $V_o = 0.2$ V, $P_D = 6.7$ mW

10.3-7. when Q_0 saturates at $t = 0$, Q_1 turns off in 1.75 ns and Q_2 saturates 4.25 ns after that

10.3-9. $t_{PLH} = 74$ ns, $t_{PHL} = 1.9$ ns, $t_p = 38$ ns **10.4-1.** >10.625 V **10.4-3.** 173 $\Omega < R_L < 4.33$ kΩ

10.4-5. (a) AND gate (b) 0 (c) 0.225 mA **10.4-7.** (a) 131 ns (b) 65.4 ns

10.4-9. with $V_{in} = V_1 = 1$ V and $V_2 = +5$ V, D_2 is OFF, D_1 is ON, Q_2, Q_3, and Q_6 are cut off, and Q_5 and Q_4 are active when $V_o < 3.6$ V

10.4-11. See Figure 10.4-8b for i_{C2}, v_{C2}, i_{C3} and v_{C3} waveshapes. At $t = 0+$ v_{C1} jumps almost immediately from 0.2 V to 1.4 V and v_{E2} also jumps almost immediately from 0 V to 0.7 V. When v_{in} drops to 0 V at $t = T$, v_{C1} drops to 0.2 V at about the same time that Q_2 turns OFF (or $i_{C2} = 0$), and v_{E2} drops to 0 V after Q_3 turns OFF ($i_{C3} = 0$)

10.5-1. $R_1 = 100$ kΩ, $V_{EE} = 10$ V

10.5-3. $v_o = [(0.375$ V/ns $\times t]$ and $i_{D2} = 3.5$ mA for $t < 8$ ns; $v_o = [(8 - 5e^{-(t-8)/(13.3\text{ ns})}]$ V and $i_{D2} = 6.83e^{-t/13.3\text{ ns}}$ mA for $t > 8$ ns

10.5-5. (a) 15 V (b) $v_o = (5 + 25e^{-4t/(8.25\text{ ns})})/(1 + e^{-4t/(8.25\text{ ns})})$ for $t > 9.2$ ns, $v_o = [6.77 + 4.62e^{-(t-9.2)/(6.7\text{ ns})}]/[1 + 0.139e^{-(t-9.2)/(6.7\text{ ns})}]$ for $t > 9.2$ ns

10.5-7. for V_{in} varying from 1 to 7 V in 1 V steps, the corresponding values of V_o are 15, 14, 10.5, 2.5, 2, 1.5, and 0.75 V **10.5-9.** (a) 2.5 V (b) 1.784 V **10.6-1.** -1.45 V

10.6-3. $V_{o1} = -0.7$ V and $V_{o2} = -1.71$ V for $V_{in} < -1.385$ V, V_{o1} falls linearly to -1.74 V and V_{o2} rises linearly to -0.7 V as V_{in} increases from -1.385 V to -1.185 V, $V_{o2} = -0.7$ V for all $V_{in} > -1.185$ V, V_{o1} falls from -1.74 V to -1.8 V as V_{in} increases from -1.185 V to -0.6 V, V_{o1} rises from -1.8 V to -1.2 V as V_{in} increases from -0.6 V to 0 V

10.6-5. Assume that $h_{fe} = 20$, $h_{ie} = 1$ kΩ, and that there is a 10 kΩ load on the gate output. (a) with $V_A = V_B =$ logic 1, $v_{o2} = 0.0062$ v_n; with $V_A = V_B =$ logic 0, $v_{o2} = 0.25$ v_n (b) with $V_A = V_B =$ logic 1, $v_{o2} \approx v_n$; with $V_A = V_B =$ logic 0, $v_{o2} = 0.73$ v_n

10.6-7. When v_{in} jumps to -0.7 V, v_o falls exponentially from -0.7 V towards -5.2 V with a time constant of 51 ns stopping at -1.92 V after 16.1 ns. When v_{in} drops to -1.7 V, v_o rises exponentially from -1.92 V to -0.7 V with a time constant of 0.5 ns.

10.7-1. $v_2(t) = 2.5\, u(t - T_d/2) + (2.5 - 5\, e^{-[(t-1.5T_d)/Z_o C_L]}) \times u(t - 1.5T_d)$, $v_3(t) = 5\,(1 - e^{-[(t-T_d)/Z_o C_L]}) \times u(t - T_d)$

10.7-3. for $p(t) =$ unit pulse of width T; voltage at the end of the line $= 16\, p(t - T_d) - 12\, p(t - 3T_d) + 9.6\, p(t - 5T_d) - 7.2\, p(t - 7T_d) + \ldots$, $v_{BE}(t) = 0.7\, p(t - T_d) - 12\, p(t - 3T_d) + 0.7\, p(t - 5T_d) - 7.2\, p(t - 7T_d) + \ldots$, $v_o(t) = 10 - 9.8\, p(t - T_d) - 9.8\, p(t - 5T_d) + \ldots$

10.7-5. The transmission line output is at 3.6 V for $t < T_d$, -2.4 V for $T_d < t < 3T_d$, $+0.7$ V for $3T_d < t < 5T_d$, -0.1 V for $5T_d < t < 7T_d$, $+0.2$ V for $7T_d < t < 9T_d$. The gate output at G_2 is 0.2 V for $t < T_d$ and 3.6 V for $t > T_d$.

10.7-7. no; however, making R_E too small will dissipate excessive power in R_E and in the gate output transistor

10.7-9. graphically the gate input voltage at $t = 5T_d$ will be $(2 - 4T_d/T_R)$ times the input step size, and for less than 15 % undershoot this requires $T_d < t_r/2.78$

11.2-1. The waveshapes are similar to those in Figure 11.2-3e except that $T_1 = 23.1$ μs and $T_2 = 69.3$ μs assuming ideal diodes

11.2-3. The outputs of G_1, G_2, and G_3 are each square waves of period $6T_p$. The output at G_2 is delayed $4T_p$ from that at G_1 and the output at G_3 is delayed $4T_p$ from that at G_2. **11.2-5.** $T_p = 0.69\,RC_{in}$, $f = 1/(6.9\,RC_{in})$

11.3-1. The trigger pulse produces a positive pulse at v_3 and a negative pulse at v_1, each of width $T = RC \ln 2$. When triggered v_2 goes to zero and rises towards V_{DD} with a time constant RC, switching back to $+V_{DD}$ after it reaches $V_{DD}/2$ at $t = T$.

11.3-3. need 100 ns $<< R_1 C_1 <<$ 100 μs; let $\tau_1 = 5$ μs, and $C_1 = 0.001$ μF and $R_1 = 5$ kΩ

11.3-5. The circuit consists of two identical 555 one-shots whose outputs are connected to a 2-input OR gate to produce the required output. One 555 is triggered by v_{in} and the other by its complement. Each 555 circuit has the form given in Figure 11.3-4a with $V_{DD} = 10$ V, $C = 0.001$ μF, and $R = 91$ kΩ. The triggers to each 555 are coupled through a 0.001 μF capacitor to pin 2 and pin 2 is pulled up to $+10$ V through a 10 kΩ resistor.

11.3-7. (a) 10 kΩ (b) 7.2 kHz.

11.4-1. When the switch is closed v_o goes to 0.2 V for about 7 μs and then returns exponentially to 5 V. When the switch is opened v_o stays at 5 V.

11.4-3. When the switch closes v_3 goes from 0 to 5 V for about 23 μs and then returns to zero.

11.5-1. The data bus leads ($D_0 - D_7$) and the address bus leads ($A_0 - A_{11}$) go to all RAM and ROM chips. A 3-line to 8-line decoder is used with $A_{12} - A_{14}$ as the inputs and A_{15} connected to $\overline{\text{CS}}$ on the decoder. The decoder outputs go to the CS lines on the memory chips and MEMR is connected to the memory OE lines. In addition, MEMW is inverted and connected to the R/$\overline{\text{W}}$ lines.

11.5-3. Two banks of 4 RAM chips each are used for the memory. $D_0 - D_7$ are connected to the upper bank [IC0 $-$ IC3] and $D_8 - D_{15}$ to the lower bank [IC4 $-$ IC7]. $A_0 - A_{11}$ go to all chips. MEMW is inverted and $\overline{\text{MEMW}}$ and MEMR are connected to all R/$\overline{\text{W}}$ and OE lines respectively. A 2-line to 4-line decoder is used with A_{12} and A_{13} as the inputs, and a 6-input OR gate ($A_{14} - A_{19}$) is connected to the decoder $\overline{\text{CS}}$ line. Decoder output line zero goes to IC0 and IC4, line 1 to IC1 and IC5, etc.

11.6-1. (a) $v_{o1} = 0.5\,V_{CC}\,e^{-6t/\tau} + 0.01\,V_{CC}\,e^{4t/\tau}$, $v_{o2} = 0.5\,V_{CC}\,e^{-6t/\tau} - 0.01\,V_{CC}\,e^{4t/\tau}$ where $\tau = 40$ ns until the inverters saturate at about $t = 30$ ns. Thereafter, v_{o1} rises exponentially towards 5 V and v_{o2} falls exponentially to 0 V with a time constant of 40 ns, the flip-flop switches states in about 200 ns

(b) v_{o1} and v_{o2} equations are opposite from those given in (a), the flip-flop does not switch but returns to the original state in about 200 ns

11.6-3. (a) neglecting the effect of v_1 and v_2 on v_{o1} and v_{o2} the critical trigger pulse width is about 3.5 ns

(b) $v_{o1} = 5 - 1.5^{-t/\tau}$, $v_{o2} = 1.5e^{-t/\tau}$ where $\tau = 50$ ns

11.7-1. $v_{o1} = 0.06e^{3t/\tau} + 2.56e^{-t/\tau}$, $v_{o2} = -0.06e^{3t/\tau} + 2.56e^{-t/\tau}$ with $\tau = 10$ ns until the inverters saturate at about $t = 40$ ns. Thereafter v_{o1} rises to 5 V exponentially and v_{o2} falls to 0 V exponentially with time constants of 10 ns.

11.7-3. 43 mW

11.8-1. The circuit is the same as in Figure 11.8-5 except that the 2-pole 3-position selector switch activates ROM address leads A_3 and A_4. In switch position 1: $A_3 = A_4 = 0$, in position 2: $A_3 = 1$ and $A_4 = 0$, and in position 3: $A_3 = 0$ and $A_4 = 1$. Starting at address zero and increasing sequentially the required hexadecimal ROM data is 6, 4, F, 8, 2, 5, 7, 0, 6, 4, F, 8, 6, 4, F, 8, 6, F, 2, 7, 6, F, 2, 7

11.8-3. The circuit is basically that given in Figure 11.8-5b with $\phi_1 = v_0$, $\phi_2 = v_1$, $\phi_3 = v_2$, and $\phi_4 = v_3$. The required hexadecimal ROM data stored at addresses 0 through 7 is 1, 0, 2, 0, 4, 0, 8, 0.

11.8-5. The ROM size is 16 $\times$ 7, and the counter is replaced by three D-style flip-flops Q_6, Q_5, and Q_4 whose D-input lines are connected to the D_6, D_5, and D_4 data output lines from the ROM. The Q_6, Q_5, Q_4 flip-flop output lines are connected to the A_2, A_1, and A_0 address leads of the ROM respectively, and the switch goes to A_3. The hexadecimal ROM data stored at addresses 00 through 15 is 16, 24, 3F, 48, 52, 65, 77, 70, 16, 24, 3F, 48, 52, 65, 77, 00

11.9-1. The PLA has the form given in Figure 11.9-3 with 4 output lines X_3, X_2, X_1 and X_0. The Boolean expression for the flip-flop inputs are $D_0 = \overline{Q}_0$, $D_1 = Q_1\overline{Q}_0 + \overline{Q}_1 Q_0$, and $D_2 = Q_2\overline{Q}_0 + Q_2\overline{Q}_1 + \overline{Q}_2 Q_1 Q_0$. The Boolean expressions for the output variables are $X_3 = Q_2 Q_1 \overline{Q}_0$, $X_2 = Q_2\overline{Q}_1\overline{Q}_0$, $X_1 = \overline{Q}_2 Q_1\overline{Q}_0$, and $X_0 = \overline{Q}_2\overline{Q}_1\overline{Q}_0$.

11.9-3. The PLA has the form given in Figure 11.9-3 with two output lines X_1 and X_0, three D-style flip-flops, and two external inputs Y_1 and Y_0. The Boolean expressions for the output variables are $X_1 = Q_2 + Q_1 Q_0$ and $X_0 = Q_2 + Q_1\overline{Q}_0$. The Boolean expressions for the flip-flop inputs are $D_2 = Q_1 Q_0 + Y_1\overline{Q}_2\overline{Q}_1\overline{Q}_0 + \overline{Y}_1 Y_0 Q_2$, $D_1 = Q_1 Q_0 + Y_0\overline{Q}_1 Q_0 + Y_1\overline{Q}_1 Q_0 + Y_1 Y_0 Q_2$, and $D_0 = \overline{Y}_1 Q_2\overline{Q}_0 + \overline{Y}_1\overline{Y}_0\overline{Q}_2\overline{Q}_1 + Y_1\overline{Q}_1 Q_0 + \overline{Q}_2 Q_1\overline{Q}_0 + Y_1\overline{Y}_0 Q_2$

Index